Materials Handbook

OTHER McGRAW-HILL REFERENCE BOOKS OF INTEREST

Handbooks

Baumeister and Avallone: MARKS' STANDARD HANDBOOK FOR MECHANICAL ENGINEERS
Considine: PROCESS INSTRUMENTS AND CONTROLS HANDBOOK
Crocker and King: PIPING HANDBOOK
Dean: LANGE'S HANDBOOK OF CHEMISTRY
Fink and Beaty: STANDARD HANDBOOK FOR ELECTRICAL ENGINEERS
Fink and Christiansen: ELECTRONICS ENGINEERS' HANDBOOK
Hicks: STANDARD HANDBOOK OF ENGINEERING CALCULATIONS
Juran: QUALITY CONTROL HANDBOOK
Kaufman and Seidman: HANDBOOK FOR ELECTRONICS ENGINEERING TECHNICIANS
Lange: HANDBOOK OF METALS FORMING
McLellan and Shand: GLASS ENGINEERING HANDBOOK
Maynard: INDUSTRIAL ENGINEERING HANDBOOK
Merritt: STANDARD HANDBOOK FOR CIVIL ENGINEERS
National Association of Purchasing Managers: ALJIAN'S PURCHASING HANDBOOK
Perry and Green: PERRY'S CHEMICAL ENGINEERS' HANDBOOK
Rohsenow, Hartnett, and Ganic: HANDBOOK OF HEAT TRANSFER FUNDAMENTALS
Rohsenow, Hartnett, and Ganic: HANDBOOK OF HEAT TRANSFER APPLICATIONS
Rosaler and Rice: STANDARD HANDBOOK OF PLANT ENGINEERING
Rothbart: MECHANICAL DESIGN AND SYSTEMS HANDBOOK
Seidman, Mahrous, and Hicks: HANDBOOK OF ELECTRIC POWER CALCULATIONS
Teicholz: CAD/CAM HANDBOOK
Tuma: ENGINEERING MATHEMATICS HANDBOOK

Dictionaries

McGraw-Hill: DICTIONARY OF SCIENTIFIC AND TECHNICAL TERMS

Encyclopedias

McGraw-Hill: CONCISE ENCYCLOPEDIA OF SCIENCE AND TECHNOLOGY
McGraw-Hill: ENCYCLOPEDIA OF PHYSICS
McGraw-Hill: ENCYCLOPEDIA OF ENGINEERING
McGraw-Hill: ENCYCLOPEDIA OF ENERGY

Materials Handbook

An Encyclopedia for Managers, Technical Professionals, Purchasing and Production Managers, Technicians, Supervisors, and Foremen

GEORGE S. BRADY
(Deceased)

HENRY R. CLAUSER
Materials Consultant
Former Editor, Materials Engineering

TWELFTH EDITION

McGraw-Hill Book Company
New York St. Louis San Francisco Auckland Bogotá Hamburg
London Madrid Mexico Montreal New Delhi
Panama Paris São Paulo Singapore Sydney Tokyo Toronto

The Library of Congress cataloged the first printing
of this title as follows:

Brady, George Stuart, date
Materials handbook; an encyclopedia for purchasing agents, engineers, executives, and foremen. 1st– ed.
New York, McGraw-Hill, 1929–
v. illus., maps. 18–24 cm.

1. Materials—Dictionaries. I.Title.
TA403.B75 29—1603 rev*
Library of Congress [r51u1]
ISBN 0-07-007071-7

4567890 DOC/DOC 898

ISBN 0-07-007071-7

The editors for this book were Harold B. Crawford and Vivian Koenig, the designer was Mark E. Safran, and the production supervisor was Teresa F. Leaden. It was set in ITC Baskerville by University Graphics, Inc.

Printed and bound by R. R. Donnelley & Sons Company.

Contents

Preface

The twelfth edition of the *Materials Handbook* is the cumulative development of an idea that originated with the late Colonel George S. Brady more than fifty years ago. Aware then of the steadily increasing number of materials being developed, Colonel Brady set out to prepare a one-volume encyclopedia that would intelligently describe the important characteristics of commercially available materials without involvement in details—an encyclopedia that would meet the job needs of managers and executives, purchasing and manufacturing managers, supervisors, engineers, students, and others. Attesting to both the validity of his idea and the success of its execution is the fact that over the years the *Materials Handbook* has come to be recognized as the leading reference work of its kind in the world.

The philosophy behind the *Materials Handbook,* as well as its purpose and scope, is best expressed in this excerpt from the foreword to the tenth edition:

> All materials, infinite in possible numbers, derive from only 92 natural elements. It is as basically simple as that, but the varying forms and usages are so intertwined in all the industries that a person can have no real comprehension of the characteristics and economics of any one of the materials which he procures for his own use unless he has an intelligent overall grasp of its varying forms and usages. The selection of data and of examples in the *Materials Handbook* has been made with a view toward giving the reader an intelligent overall insight. It is not

the purpose of the book to provide an exhaustive treatise on any materials, as it is assumed that the reader will consult producers of the materials for detailed specifications.

General information, with the most commonly accepted comparative figures, is given on materials in their group classifications in order to give a general picture; selected processed materials and patented and trade-named materials are then described to give a more specific understanding of commercial applications. The relative position and the length of description of proprietary materials are for purposes of illustration and bear no relation to the relative merits of the products of any one producer.

Since the first edition of the *Materials Handbook* was published, a virtual revolution in materials has been taking place, resulting in new types and grades of materials—plastics, metal alloys, rubbers, textiles, finishes, foodstuffs, chemicals, and animal products—that are being developed at an exponential rate. Each new edition of the *Materials Handbook* reflects this phenomenal proliferation in number and variety. Whereas the first edition covered only a few thousand materials, this latest edition describes some 14,000 different materials. Despite this manyfold increase in coverage, it is virtually impossible to include all commercially available materials in a one-volume work of this kind. Nevertheless, descriptions of the most important and most widely used of the thousands of materials introduced every year are added to each edition of this handbook.

After publication of the tenth edition, Colonel Brady retired and transferred to me the task of preparing this and subsequent editions of the *Materials Handbook.* In serving as co-author I have endeavored to hold to his standards of accuracy and readability. His help and counsel were invaluable to me in maintaining the original purpose and character of the handbook.

HENRY R. CLAUSER

Acknowledgments

The sources of information for *Materials Handbook* are too numerous to permit mention of all who furnished data and counsel, but acknowledgment is made of the ready and willing cooperation of producers, manufacturers, technical authorities, industrial-paper editors, and government bureaus in this country and abroad. A deep sense of appreciation is felt for the encouragement and advice given to the work by the many members of the National Association of Purchasing Management and the American Ordnance Association.

Materials Handbook

1

Materials, Their Properties and Uses

ABLATIVE AGENTS. Materials used for the outward dissipation of extremely high heats by mass removal. Their most common use is as an external heat shield to protect supersonic aerospace vehicles from an excessive buildup of heat caused by air friction at the surface. The ablative material must have a low thermal conductivity in order that the heat may remain concentrated in the thin surface layer. As the surface of the ablator melts or sublimes, it is wiped away by the frictional forces that simultaneously heat newly exposed surfaces. The heat is carried off with the material removed. The less material that is lost, the more efficient is the ablative material. The ablative material, in addition to low thermal conductivity, should have a high thermal capacity in the solid, liquid, and gaseous states; a high heat of fusion and evaporation; and a high heat of dissociation of its vapors. The ablative agent, or **ablator,** is usually a carbonaceous organic compound, such as a plastic. As the dissociation products are lost as liquid or vapor, the char is held in place by the refractory filler fibers, still giving a measure of heat resistance. The effective life of an ablative is short, calculated in seconds per millimeter of thickness for the distance traveled in the atmosphere.

Single ablative materials seldom have all of the desirable factors, and thus composites are used. Phenolic or epoxy resins are reinforced with asbestos fabric, carbonized cloth, or refractory fibers such as asbestos, fused silica, and glasses. The refractory fibers are incorporated not only for mechanical strength but have a function in the ablative process, and surface-active agents may be added to speed the rate of evaporation. **Ablative paint,** for protecting woodwork, may be organic silicones which convert to silica at temperatures above 2000°F (1093°C). **Pyromark,** of Tempil Corp., is a paint of this type.

Metals can resist temperatures higher than their melting point by **convection cooling,** or **thermal cooling,** which is heat protection by heat exchange with a coolant. Thus, tungsten can be arc-melted in a copper kettle which is cooled by circulating water. The container metal must have high thermal conductivity, and the heat must be quickly carried away and stored or dissipated. When convection cooling is difficult or not possible, cooling may be accomplished by a heat sink. **Heat-sink cooling** depends on the heat absorption capability of the structural material itself or backed up by another material of higher heat absorption. Copper, beryllium, graphite, and beryllium oxide have been used. A heat-sink material should have high thermal conductivity, high specific heat and melting point, and for aerodynamic applications, a low specific gravity.

ABRASIVES. Materials used for surfacing and finishing metals, stone, wood, glass, and other materials by abrasive action. The natural abrasives include the diamond, emery, corundum, sand, crushed garnet and quartz, tripoli, and pumice. **Artificial abrasives,** or **manufactured abrasives,** are generally superior in uniformity to natural abrasives, and are mostly silicon carbide, aluminum oxide, boron carbide, or boron nitride, marketed under trade names. Artificial diamonds are also now being produced. The massive natural abrasives, such as sandstone, are cut into grinding wheels from the natural block, but most abrasive material is used as grains or built into artificial shapes.

For industrial grinding, artificial abrasives are preferred to natural abrasives because of their greater uniformity. Grading is important because uniform grinding requires grains of the same size. The abrasive grains are used as a grinding powder, are made into wheels, blocks, or stones, or are bonded to paper or cloth. **Abrasive cloth** is made of cotton jean or drills to close tolerances of yarns and weaves, and the grains are attached with glue or resin. But the **Fabricut cloth** of the 3M Co. is an open-weave fabric with alumina or silicon-carbide grains of 100 to 400 mesh. The open weave permits easy cleaning of the cloth in an air blast. **Abrasive paper** has the grains, usually aluminum oxide or silicon carbide, glued to one side of 40- to 130-lb kraft paper. The usual grain sizes are No. 16 to No. 500.

Abrasive powder is usually graded in sizes from 8 to 240 mesh. Coarse grain is to 24 mesh; fine grain is 150 to 240. **Blasting abrasive** for blast cleaning of metal castings is usually coarse grain. **Arrowblast,** of the Norton Co., is aluminum oxide with grain sizes 16 to 80 mesh. **Grinding flour** consists of extremely fine grains separated by flotation, usually in grain sizes from 280 to 600 mesh, used for grinding glass and fine polishing. **Levigated abrasives** are fine powders for final burnishing of metals or for metallographic polishing, usually processed to make them chemically neutral. **Green rouge** is levigated chromic oxide, and **mild polish** may be levigated tin oxide; both are used for burnishing soft metals. **Polishing powder** may be aluminum oxide or metal oxide powders of ultrafine particle size down to 600 mesh. **Micria AD,** of the Monsanto Co., is alumina; **Micria ZR** is zirconia; and **Micria TIS** is titania. **Gamal,** of the Fisher Scientific Co., is a fine aluminum oxide powder, the smaller cubes being 1.5 μm, with smaller particles 0.5 μm. **Cerox,** of the Lindsay Div., is cerium oxide used to polish optical lenses and automobile windshields. It cuts fast and gives a smooth surface. **Grinding compounds** for valve grinding are usually aluminum oxide in oil.

Mild abrasives, used in silver polishes and window-cleaning compounds, such as chalk and talc, have a hardness of 1 to 2 Mohs. The milder abrasives for dental pastes and powders may be precipitated calcium carbonate, tricalcium phosphate, or combinations of sodium metaphosphate and tricalcium phosphate. Abrasives for metal polishes may also be pumice, diatomite, silica flour, tripoli, whiting, kaolin, tin oxide, or fuller's earth. This type of fine abrasive must be of very uniform grain in order to prevent scratching. **Cuttle bone,** or **cuttlefish bone,** is a calcareous powder made from the internal shell of a Mediterranean marine mollusk of the genus *Sepia,* and is used as a fine polishing material for jewelry and in tooth powders. **Ground glass** is regularly marketed as an abrasive for use in scouring compounds and in match-head compositions. **Lapping abrasives,** for finish grinding of hard materials, are diamond dust or boron carbide powder.

Aluminum oxide wheels are used for grinding materials of high tensile strength. **Silicon carbide** is harder but is not as strong as aluminum oxide. It is used for grinding metals that have dense grain structure and for stone. **Vitrified wheels** are made by molding under heat and pressure. They are used for general and precision grinding where the wheel does not exceed a speed of 6,500 surface ft/min (33 m/s). The rigidity gives high precision, and the porosity and strength of bond permit high stock removal. **Silicate wheels** have a silicate binder and are baked. The silicate bond releases the grains more easily than the vitrified, and is used for grinding edge tools to reduce burning of the tool. Synthetic resins are used for bonding where greater strength is required than is obtained with the silicate, but less openness than with the vitrified. Resinoid bonds are used up to 16,000 surface ft/min (81 m/s), and are used especially for thread grinding and cutoff wheels. Shellac binder is used for light work and for high finishing. Rubber is used for precision grinding and for centerless-feed machines.

Grading of abrasive wheels is by grit size number from No. 10 to No. 600, which is 600 mesh; by grade of wheel, or strength of the bond, which is by letter designation, increasing in hardness from A to Z; and by grain spacing or structure number. The ideal condition is with a bond strong enough to hold the grains to accomplish the desired result, and then release them before they become too dull. Essential qualities in the abrasive grain are: penetration hardness, body strength sufficient to resist fracture until the points dull and then break to present a new edge, and an attrition resistance suitable to the work. Some wheels are made with a porous honeycombed structure to give free cutting and cooler operation on some types of metal grinding. Some diamond wheels are made with aluminum powder mixed with a thermosetting resin, and the diamond abrasive mix is hot-pressed around this core wheel. Norton diamond wheels are of three types: metal bonded by powder metallurgy, resinoid bonded, and vitrified bonded.

ABRASIVE SAND. Any sand used for abrasive and grinding purposes, but the term does not include the sharp grains obtained by crushing quartz and used for sandpaper. The chief types of abrasive sand include **sandblast sand, glass-grinding sand,** and **stone-cutting sand.** Sand for stone sawing and for marble and glass grinding is usually ungraded, with no preparation other than screening, but it must have tough, uniform grains. **Chats** are sand tailings from the Missouri lead ores, used for sawing stone. **Banding sand** is used for the band grinding of tool handles, and for the grinding of plate glass, but is often replaced by artificial abrasives. Banding-sand grains are fine, 95% being retained on a 150-mesh screen. **Burnishing sand,** for metal polishing, is a fine-grained silica sand with rounded grains. It should pass a 65-mesh screen, and be retained on a 100-mesh screen.

ABS PLASTICS. The letters ABS identify the family of **acrylonitrile-butadiene-styrene.** Common trade names for these materials are **Cycolac, Kralastic,** and **Lustran.** They are opaque and distinguished by a good balance of properties, including high impact strength, rigidity, and hardness over a temperature range of −40 to 230°F (−40 to 110°C). Compared to other structural or engineering plastics, they are generally considered to fall at the lower end of the scale. Medium impact grades are hard, rigid, and tough, and are used for appearance parts that require high strength, good fatigue resistance, and surface hardness and gloss. High impact grades are formulated for similar products where additional impact strength is gained at some sacrifice in rigidity and hardness. Low- temperature impact grades have high impact strength down to −40°F (−40°C). Again, some sacrifice is made in strength, rigidity, and heat resistance. Heat-resistant, high-strength grades provide the best heat resistance—continuous use up to about 200°F (93°C), and a 264 lb/in^2 (2 MPa) heat distortion temperature of around 215°F (102°C). Impact strength is about comparable to that of medium impact grades, but strength, modulus of elasticity, and hardness are higher. At stresses above their tensile strength, ABS plastics usually yield plastically instead of rupturing, and impact failures are ductile. Because of relatively low creep, they have good long-term load-carrying ability. This low creep plus low water absorption and relatively high heat resistance provide ABS plastics with good dimensional stability. ABS plastics are readily processed by extrusion, injection molding, blow molding, calendering, and vacuum forming. Resins have been developed especially for cold forming or stamping from extruded sheet. Typical applications are helmets, refrigerator liners, luggage tote trays, housings, grills for hot air systems and pump impellers. Extruded shapes include tubing and pipe. ABS plated parts are now in wide use, replacing metal parts in the automotive and appliance field.

ACAROID RESIN. A gum resin from the base of the tufted trunk leaves of various species of *Xanthorrhoea* trees of Australia and Tasmania. It is also called **gum accroides** and **yacca gum. Yellow acaroid** from the *X. tateana* is relatively scarce, but a gum of the yellow class comes from the tree *X.*

preissii of Western Australia, and is in small hollow pieces of yellow to reddish color. It is known as **black boy resin,** the name coming from the appearance of the tree. **Red acaroid,** known also as **red gum** and **grass tree gum,** comes in small dusty pieces of reddish-brown color. This variety is from the *X. australis* and about 15 other species of the tree of southeastern Australia. The resins contain 80 to 85% resinotannol with **coumaric acid,** which is a hydroxycinnamic acid, and they also contain free cinnamic acid. They are thus closely related chemically to the balsams. Acaroid resin has the property unique among natural resins of capacity for thermosetting to a hard, insoluble, chemical-resistant film. By treatment with nitric acid it yields picric acid; by treatment with sulfuric acid it yields fast brown to black dyes. The resins are soluble in alcohols and in aniline, only slightly soluble in chlorinated compounds, and insoluble in coal-tar hydrocarbons. Acaroid has some of the physical characteristics of shellac, but is difficult to bleach. It is used for spirit varnishes and metal lacquers, in coatings, in paper sizing, in inks and sealing waxes, in binders, for blending with shellac, in production of picric acid, and in medicine.

ACETAL RESINS. Highly crystalline resins that have the repeating group $(OCH_2)_x$. The resins are **polyformaldehyde.** The natural acetal resin is translucent white and can be readily colored. There are two basic types: a homopolymer (**Delrin**) and a copolymer (**Celcon**). In general, the homopolymers are harder, more rigid, and have higher tensile flexural and fatigue strength, but lower elongation. The copolymers are more stable in long-term high-temperature service and have better resistance to hot water. Special types of acetals are glass filled, providing higher strengths and stiffness, and tetrafluoroethylene (TFE) filled, providing exceptional frictional and wear properties.

Acetals are among the strongest and stiffest of the thermoplastics. Their tensile strength ranges from 8,000 to about 13,000 lb/in^2 (55 to 89 MPa), the tensile modulus of elasticity is about 500,000 lb/in^2, (3,445 MPa), and fatigue strength at room temperature is about 5,000 lb/in^2 (34 MPa). Their excellent creep resistance and low moisture absorption (less than 0.4%) give them excellent dimensional stability. They are useful for continuous service up to about 220°F (104°C). Acetals' low friction and high abrasion resistance, though not as good as nylon's, rates them high among thermoplastics. Their impact resistance is good and remains almost constant over a wide temperature range. Acetals are attacked by some acids and bases, but have excellent resistance to all common solvents. They are processed mainly by molding or extruding. Some parts are also made by blow and rotational molding. Typical parts and products made of acetal include pump impellers, conveyor links, drive sprockets, automobile instrument clusters, spinning reel housings, gear valve components, bearings, and other machine parts. Delrin, of Du Pont, is used for mechanical and electrical parts. It has a specific gravity of 1.425, a tensile strength of 10,000 lb/in^2 (68 MPa) with elongation of 15%, dielectric strength of 500 volts per mil (19.6×10^6 volts per meter), and Rockwell hardness M94. It retains its mechanical strength close to the melting point of 347°F (175°C). Celcon, of the Celanese Corp., is a thermoplastic linear acetal resin produced from **trioxane,** which is a cyclic form of formaldehyde. The specific gravity is 1.410, flexural strength 12,000 lb/in^2 (82 MPa), Rockwell hardness M76, and dielectric strength 1,200 volts per mil (47×10^6 volts per meter). It comes in translucent white pellets for molding.

ACETIC ACID. Also known as **ethanoic acid.** A colorless, corrosive liquid of pungent odor and composition $CH_3 \cdot COOH$, having a wide variety of industrial uses as a reagent, solvent, and esterifier. A carboxylic acid, it is employed as a weak acid for etching and for soldering, in stain removers and bleaches, as a preservative, in photographic chemicals, for the manufacture of cellulose acetate, as a solvent for essential oils, resins, and gums, as a precipitant for latex, in tanning leather, and in making artificial flavors. Acetic acid is found in the juices of many fruits, and in combination in the stems or woody parts of plants. It is the active principle in **vinegar,** giving it the characteristic sour taste, acid flavor, and pungent odor. It is made commercially by the oxidation of ethyl alcohol, and also produced in the destructive distillation of wood. It is also made by the reaction of methanol and carbon monoxide. Its specific gravity is 1.049, its boiling point is 118°C, and it becomes a colorless solid below 16.6°C. The pure 99.9% solid is known as **glacial acetic acid.** Standard and laundry special grades contain 99.5% acid, with water the chief impurity. Standard strengths of water solution are 28, 56, 70, 80, 85, 90%.

Acetic anhydride, $CH_3COOCOCH_3$, a colorless liquid with boiling point 139.5°C, is a powerful acetylating agent, and is used in making cellulose acetate. It forms acetic acid when water is added. **Hydroxyacetic acid,** $HOCH_2COOH$, or **glycolic acid,** is produced by oxidizing glycol with dilute nitric acid and is intermediate in strength between acetic and formic acids. It is soluble in water, is nontoxic, and is used in foodstuffs, dyeing, tanning, electropolishing, and in resins. Its esters are solvents for resins. **Diglycolic acid,** $O(CH_2CO_2H)_2$, is a white solid melting at 148°C. It is stronger than tartaric or formic acids, and is used for making resins and plasticizers. **Thioacetic acid** has the formula of acetamide but with HS replacing the NH_2. It is a pungent liquid used for making esters for synthetic resins.

Chloroacetic acid, $CH_2ClCOOH$, is a white crystalline powder melting at 61.6°C and boiling at 189°C. It is used for producing carboxymethylcellulose, dyes, and drugs. **Sequestrene,** used as a clarifying agent and water softener in soaps and detergents, and to prevent rancidity in foods and sulfonated oils, is ethylene bisaminodiacetic acid, $(HOOCCH_2)_2$-$NCH_2CH_2N(CH_2COOH)_2$. It is a liquid, but in the form of its sodium salt is a water-soluble white powder. **Trifluoroacetic acid,** CF_3COOH, is one of the strongest organic acids. It is a colorless, corrosive liquid, boiling at 71.1°C and freezing at −15.3°C. It is used in the manufacture of plastics, dyes, pharmaceuticals, and flame-resistant compounds.

Paracetic acid, $CH_3 \cdot O \cdot COOH$, is a colorless liquid of strong odor with the same solubility as acetic acid. It has 8.6% available oxygen and is used as a bleaching agent, a polymerization catalyst, for making epoxy resins, and as a bactericide. **Acetin** is an ester of acetic acid made from glycerin and acetic acid, used as a solvent for basic dyes and tannins. It is a neutral straw-colored liquid of specific gravity 1.20 and boiling point 133 to 153°C. It is also used in low-freezing dynamites and smokeless powder. The triacetic ester, **triacetin,** is a water-white liquid of specific gravity 1.16 and flash point 271°F (133°C), soluble in aromatic hydrocarbons. It is used as a plasticizer.

Phenylacetic acid, $C_6H_5CH_2COOH$, is a white flaky solid melting at 74.5°C. The reactive methylene group makes it useful for the manufacture of fine chemicals. **Cyanoacetic acid,** $CN \cdot CH_2 \cdot COOH$, has an active methylene group and an easily oxidized cyano group, and is used for producing caffeine, while the derivative **ethyl cyanoacetate,** $NC \cdot CH_2COO \cdot CH_2 \cdot CH_3$, a liquid boiling at 207°C, is used for making many drugs. **Malonic acid,** $CH_2(COOH)_2$, is a very reactive acid sometimes used instead of acetic acid for making plastics, drugs, and perfumes. It decomposes at 160°C, yielding acetic acid and carbon dioxide. **Methyl acetic acid,** CH_3CH_2COOH, is **propionic acid** or **propanoic acid,** a by-product in the extraction of potash from kelp. Modifications of this acid are used for cross-linking plastics.

ACETONE. An important industrial solvent, used in the manufacture of lacquers, plastics, smokeless powder, for dewaxing lubricating oils, for dissolving acetylene for storage, for dyeing cotton with aniline black, and as a raw material in the manufacture of other chemicals. It is a colorless, flammable liquid with a mintlike odor and is soluble in water and in ether. The composition is $CH_3 \cdot CO \cdot CH_3$, specific gravity 0.790, boiling point 56°C, and solidification point −94°C. Acetone is made from isopropyl alcohol, or by a special fermentation of grain. The oily residue from the distillation is called **acetone oil** and is used as a denaturant of alcohol. **Diacetone,** or **diacetone alcohol,** is a colorless liquid of the composition $CH_3 \cdot CO \cdot CH_2 \cdot COH(CH_3)_2$ with a pleasant odor. It is used as a solvent for nitrocellulose and cellulose acetate, for gums and resins, in lacquers and thinners, and in ink removers. Because of its low freezing point and miscibility with castor oil it is used in hydraulic brake fluids. The specific gravity is 0.938, boiling point 166°C, and freezing point −54°C. Synthetic **methyl acetone** is a mixture of about 50% acetone, 30 methyl acetate, and 20 methanol, used in lacquers, paint removers, and for coagulating latex. **Dihydroxy-acetone,** a colorless crystalline solid produced from glycerin by sorbose bacteria reaction, is used in cosmetics, and in preparing foodstuff emulsions, plasticizers, and alkyd resins. It is soluble in water and in alcohol.

ACETYLENE. A colorless gas of the composition HC:CH, used for welding and flame cutting of metals, and for producing other chemicals. It contains 92.3% carbon, and is therefore nearly gaseous carbon. When pure, it has a sweet odor, but when it contains hydrogen sulfide as an impurity, it has a disagreeable odor. Acetylene burns brightly in the air, and was widely used for theater stage lighting before the advent of the electric light. When mixed with oxygen as **oxyacetylene** for flame cutting and welding, it gives a temperature of 3500°C. In air it is an explosive gas. The maximum explosive effect is with a mixture of 7.7% gas and 92.3% air. Acetylene has a specific gravity of 0.92. It is nontoxic, and is soluble in water, alcohol, or acetone. It liquefies under a pressure of 700 lb/in^2 (5 MPa) at 70°F (21°C). It is easily generated by the action of water on calcium carbide, but is also produced from petroleum. It is marketed compressed in cylinders, dissolved in acetone to make it nonexplosive. One volume of acetone will dissolve 25 volumes of acetylene at atmospheric pressure, or 250 volumes at 10 atm (10.3 kg/cm^2). **Prest-O- Lite** is a trade name of Union Carbide Corp. for acetylene dissolved in acetone. **Acetylene snow,** or solid acetylene, is produced by cooling acetylene below the melting point and compressing. It is insensible to shock and flame, and is thus easier to transport. A replacement for acetylene for producing plastics is **methyl acetylene propadiene,** which contains 70% methyl acetylene and 30 of the isomer **propadiene.** It has the reactions of both acetylene and its isomer. **Mapp,** of the Dow Chemical Co., for metal cutting is **methyl acetylene,** $CH:C \cdot CH_3$. It is safer to handle and gives about the same flame temperature.

ACRYLIC RESINS. Colorless, highly transparent, thermoplastic, synthetic resins made by the polymerization of acrylic derivatives, chiefly from the esters of **acrylic acid,** $CH_2:CH \cdot COOH$, and **methacrylic acid,** $CH_2:C(CH_3) \cdot COOH$, **ethyl acrylate** and **methyl acrylate. Glacial acrylic acid** is the anhydrous monomer with less than 2% moisture. It can be esterified directly with an alcohol. **Vinyl acrylic acid,** $CH_2:CHCH:CHCOOH$, with a melting point of 80°C, is made from acrolein and malonic acid. It polymerizes on heating.

The resins vary from soft, sticky semisolids to hard, brittle solids, depending upon the constitution of the monomers and upon the polymerization. They are used for adhesives, protective coatings, finishes, laminated glass, transparent structural sheet, and molded products. Acrylic resins, or **acrylate resins,** are stable and resistant to chemicals. They do not cloud or fade in light when used as laminating material in glass and are used as air-curing adhesives to seal glass to metals or wood. Water-based acrylics are used for the formulation of calks and sealants. They have better adhesion and weather resistance than butyl rubbers and dry more quickly. The sealants usually contain about 80% solids.

Most **acrylic plastics** are based on polymers of **methyl methacrylate,** which may be modified by copolymerizing or blending with other monomers. Noted for excellent optical properties, they have a light transmission of about 92%. Besides the transparent grades, they can be obtained in translucent or opaque colors as well as the natural color of water white. Moldings have a deep luster and high surface gloss, and for this reason are widely used for decorative parts. Acrylics have excellent weathering characteristics. Because they are little affected by sunlight, rain, and corrosive atmospheres, they are well suited for outdoor applications. In general, the majority of grades can be used up to about 200°F (100°C). Thermal expansion is relatively high.

Acrylics are hard and stiff. They are also a relatively strong plastic; their tensile strength ranges from 5,000 to about 11,000 lb/in^2 (34 to about 75 MPa). However, regular grades are somewhat brittle. High impact grades are produced by blending with rubber stock. The high strength is useful only for short-term loading. For long-term service, to avoid crazing or surface cracking, tensile stresses must be limited to about 1,500 lb/in^2 (10 MPa).

Acrylic plastics are available as cast sheets, rods, tubes, and blocks. They are also processed by injection or compression molding. Sheets are produced in thicknesses from ⅛ to ⅜ in (0.32 to 0.95 cm) and in sizes up to 10 by 12 ft (3 by 4 m). A special process that produces molecular orientation in the cast product is used to make crack-resistant aircraft cabin windows and fighter plane canopies. Acrylic moldings as large as 1 yd^2 (1 m^2) have been produced. A **lead-filled acrylic** sheet, produced by Kyowa Gas Chemical Industry Co., Japan, is used as a radiation shielding material. It is transparent and has better impact strength and fabricability than leaded glass. Typical moldings include knobs, handles, escutcheons, parts for vending machines, and a wide variety of lenses for light control, signal lamps, and the like.

Tough molding resins are made by copolymerizing methyl methacrylate with styrene. **Zerlon 150,** of the Dow Chemical Co., is such a molding resin with a flexural strength of 17,600 lb/in^2 (123 MPa). **Thiacril,** of the Thiokol Chemical Corp., is an **acrylate rubber** having a tensile strength of

2,500 lb/in^2 (16 MPa) and an elongation of 350%. It is used for gaskets, wire insulation, and hose.

Allyl methacrylate is a liquid of the empirical formula $C_7H_{10}O_2$, boiling at 63°C, and insoluble in water. It can be polymerized to form liquid or hard solid resins, but is used chiefly as a cross-linking agent for other resins to raise the softening point and increase hardness. **Gafite,** of the General Aniline & Film Corp., is **polymethyl alphachloroacrylate,** $(CH_2{:}CCl \cdot COOCH_3)_x$. It is a transparent and craze-resistant resin used for aircraft windows. The heat distortion point is 260°F (127°C), and it has higher tensile and flexural strength than other acrylics. **Cyclohexyl methacrylate** has optical properties similar to crown glass, and is used for cast lenses, where its softness and low softening point, 160°F (71°C), are not objectionable.

Lucite is methyl methacrylate of E. I. du Pont de Nemours & Co., Inc., marketed as molding powder and in rods, tubes, and cast and molded sheets. **Lucitone** is this material molded in dentures in pink and translucent. **Lucite HM-140** is this material compounded for high-temperature injection molding. **Acrylic syrup** is a liquid Lucite for use as a low-pressure laminating resin. It produces strong, stiff, tough laminates adaptable to translucent or bright colors. Reinforced with glass fibers, a panel with contact cure has a flexural strength of 25,000 lb/in^2 (172 MPa), elongation 1.5%, distortion point at 233°F (112°C), Rockwell hardness R121, and light transmission up to 65%. **Crystalite,** of Rohm & Haas Co., is an acrylic molding powder. **Plexiglas,** of this company, is transparent methyl methacrylate in sheets and rods. **Perspex** is a similar English acrylic resin. All of these plastics are used for aircraft windows. **Plexiglas V** is for injection molding, while **Plexiglas VM** is a molding powder to resist heat distortion to 174°F (79°C). **Vernonite,** of Rohm & Haas Co., is an acrylic denture resin. The **Acryloid resins,** of the same company, are **acrylic copolymer** solid resins, and the **Acrysol resins** are solutions for coatings. **Plexene M,** of the same company, is a **styrene-acrylic** resin for injection molding. The specific gravity of the molded resin is 1.08, the dielectric strength is 350 volts per mil (14×10^6 volts per meter), and tensile strength 15,000 lb/in^2 (103 MPa). **Rhoplex resins,** of the same company, and **Crilicon,** of the Jersey State Chemical Co., are acrylic resin emulsions for paints, textile finishes, and adhesives. **Polyco 296,** of the American Polymer Corp., is a water-soluble acrylic copolymer used for thickening natural or synthetic rubber latex for paper and textile coatings. Coatings made with acrylics have good adhesion and gloss, are resistant to oils and chemicals, and have good dielectric strength. **Carboset 511,** of the B. F. Goodrich Chemical Co., is a water solution of acrylic resin for protecting polished metal surfaces and precision parts against scratching. It is resistant to water, but can be washed off with soap and water. **Cavalon,** of Du Pont, is a **polyacrylic** resin for coatings that has high hardness and resistance to abrasion.

Volan, of the same company, is a **methacrylate-chromic oxychloride,** $CH_2:C(CH_3)C(OH)(OCrCl_2)_2$, in which methacrylic acid is joined with two $CrCl_2$ groups to form **resonant bonds.** It is a dark-green liquid with a specific gravity of 1.02, boiling at 180°F (82°C). When applied to negatively charged surfaces such as cellulose, polyamides, or silica materials, the chromium complex is strongly held while the chlorine is lost. In attaching to glass, the CrO forms a chemical bond to the silica of the glass, Cr·O·Si. With polyamides, the CrO attaches to a carbon atom, Cr·O· C. It is thus used to obtain strong bonds in plastic laminates.

ACRYLONITRILE. Also called **vinyl cyanide** and **propene nitrile.** A liquid of the composition $CH_2:CHCN$, boiling at 78°C, used in insecticides and for producing plastics and other chemicals. It is made by the addition of hydrocyanic acid to acetylene, by using propylene as the starter and reacting with ammonia, or from petroleum. **Acrylonitrile fiber,** originally developed in Germany as a textile staple fiber and as a monofilament for screens and weaving, and known as **Redon,** has good dimensional stability and high dielectric strength, and is resistant to water and to solvents. The polymerized acrylonitrile has a molecular structure that can be oriented by drawing to give fibers of high strength. **Orlon,** of Du Pont, is a polymerized acrylonitrile fiber. It is nearly as strong as nylon, and has a softer feel. It can be crimped to facilitate spinning with wool. It is used for clothing textiles and for filter fabrics. **Dynel,** of Union Carbide Corp., is an **acrylonitrile-vinyl chloride** copolymer staple fiber. It produces textiles with a warmth and feel like those of wool. It has good strength, is resilient, dyes easily, and is mothproof. **Verel,** of the Eastman Chemical Products, Inc., is a similar acrylic fiber produced from acrylonitrile and vinylidene chloride, and **Creslan,** of the American Cyanamid Co., called **Exlan** in Japan, is an acrylic fiber. **Acrilan,** of the Chemstrand Corp., is a similar textile fiber, and is an **acrylonitrile-vinyl acetate** copolymer. **Bakelite C-11** is an **acrylonitrile-styrene** copolymer for injection molding and extruding that produces rigid thermoplastic parts of higher tensile strength than those of the methacrylates, and has good dimensional stability and scratch resistance. **Saran F-120,** of the Dow Chemical Co., is a similar material.

Zefran, of the Dow Chemical Co., is an acrylonitrile polymerized with vinyl pyrrolidone or other dye-receptive monomer. The fiber has a molecular structure called a **nitrile alloy,** with a continuous polyacrylonitrile backbond with close-packed hydrophilic groups which hold the dye molecules. It resists heat to 490°F (254°C). **Fostacryl,** of the Foster Grant Co., Inc., is a crystal-clear **styrene-acrylonitrile** copolymer used for molding such articles as dinnerware and food containers. Acrylonitrile- styrene copolymers are also combined with alkyl-substituted phenolic resins to produce hard, glossy, flexible coatings. **Itaconic acid,** or **methylene suc-**

cinic acid, $CH_2:CCH(COOH)_2$, is also polymerized with acrylonitrile to produce fibers. When this acid is polymerized with styrene it produces transparent plastics of good optical properties. For jet aircraft tires an extremely wear-resistant rubber is made of **acrylonitrile-butadiene** with an organometalic catalyst that has alternating groups in the copolymer.

Acrylonitrile reacts with cellulose to form a wide range of resins from soluble ethers useful for textile finishes to tough, resistant materials useful for fibers. It can be reacted directly with cotton to improve the fiber. **Krilium,** of the Monsanto Co., and **Agrilon,** of the Borden Chemical Co., are sodium salts of acrylonitrile used as soil conditioners. They are more efficient than peat moss.

ACTIVATED CHARCOAL. A nearly chemically pure amorphous carbon made by carbonizing and treating dense material such as coconut shells, peach pits, or hardwood. When made from coal, or in the chemical industry, it is more usually called **activated carbon,** or **filter carbon.** It may be made by dry distillation, or by leaching the charcoal with steam or by treatment with zinc chloride or potassium thiocyanate. It is used as an adsorbent material for gas masks, for cigarette filters, and for purifying acids, recovering solvents, and decolorizing liquids. **Coconut charcoal,** valued for gas masks, is an activated charcoal usually made by heating coconut shells in a closed retort, crushing, and steam treating. An activated charcoal made from coconut shell will adsorb 68% of its weight of carbon tetrachloride. A requirement of activated charcoals, besides high adsorbing power, is that they possess strength to retain a porous structure to pass the air or liquid. **Activated carbon CXC4-6,** of Union Carbide Corp., produced from petroleum and used as a catalyst support, is in 3/16-in (0.48-cm) pellets of high hardness and strength. Powdered activated charcoal is usually ground to 300 mesh. For water purification it should be fine enough to wet easily, but not so light that it will float on the top. For decolorizing or deodorizing oils and chemicals it is mixed in the liquid and settles out in a few hours. A single drop of water will hold 10,000 particles of powdered charcoal. In sugar and oil refining it removes color but does not bleach like chemicals. Color removal is measured by the **molasses number,** which is the index of color removed per gram of carbon when tested on a standard molasses solution.

Hydrodarco, of the Atlas Chemical Industries, is a powdered activated carbon. **Norit,** of the American Norit Co., Inc., is a group of highly adsorptive activated carbons. **Kelpchar** is activated carbon made from seaweed. **Tec-Char,** of the Tennessee Eastman Co., is a by-product charcoal obtained in wood distillation, and in graded grains for various uses. **Nuchar** is an activated carbon of the West Virginia Pulp & Paper Co. The activated carbon of the Masonite Corp. is made by subjecting wood chips to high steam pressure and disintegrating by sudden release of the pressure. The doughy mass is briquetted and carbonized.

Activated carbon derived from coal is harder than organic carbons and does not crumble easily, permitting a higher flow of liquid to be filtered. It has a high density and high activity. **SGL carbon,** of the Pittsburgh Coke & Chemical Co., has a density of 0.46, and an iodine number of 1,000 compared with 650 for ordinary carbons. Its color-removal index is about 40% higher than that of organic carbons. **Filt-o-cite,** of the Shamokin Filter Co., is finely ground anthracite used to replace sand as a filtering agent for industrial wastes.

ADHESIVES. Materials employed for sticking, or adhering, one surface to another. Forms are liquid, paste, powder, and dry film. The commercial adhesives include pastes; glues; pyroxylin cements; rubber cements; latex cement; special cements of chlorinated rubber, synthetic rubbers, or synthetic resins; and the natural mucilages. Adhesives are characterized by degree of tack, or stickiness, by strength of bond after setting or drying, by rapidity of bonding, and by durability. The strength of bond is inherent in the character of the adhesive itself, particularly in its ability to adhere intimately to the surface to be bonded. Adhesives prepared from organic products are in general subject to disintegration on exposure. The life of an adhesive depends usually upon the stability of the ingredient that gives the holding power, although otherwise good cements of synthetic materials may disintegrate by the oxidation of fillers or materials used to increase tack. Plasticizers usually reduce adhesion. Some fillers such as mineral fibers or walnut-shell flour increase the thixotropy and the strength, while some such as starch increase the tack but also increase the tendency to disintegrate.

Adhesives can be grouped into five classifications based on chemical composition. **Natural adhesives** include vegetable- and animal-base adhesives and natural gums. They are inexpensive, easy to apply, and have a long shelf life. They develop tack quickly, but provide only low strength joints. Most are water soluble. They are supplied as liquids or as dry powders to be mixed with water. **Casein-latex adhesive** is an exception. It consists of combinations of casein with either natural or synthetic rubber latex. It is used to bond metal to wood for panel construction and to join laminated plastics and linoleum to wood and metal. Except for this type, most natural adhesives are used for bonding paper, cardboard, foil, and light wood.

Thermoplastic adhesives can be softened or melted by heating and hardened by cooling. They are based on thermoplastic resins (including asphalt and oleoresin adhesives) dissolved in solvent or emulsified in water. Most of them become brittle at subzero temperatures and may not be used under stress at temperatures much above 150°F (65°C). Being relatively soft materials, thermoplastic adhesives have poor creep strength. Although lower in strength than all but natural adhesives and suitable only for non-critical service, they are also lower in cost than most adhesives. They are also odorless and tasteless and can be made fungus resistant.

Elastomeric adhesives, based on natural and synthetic rubbers, are available as solvent dispersions, latexes, or water dispersions. They are primarily used as compounds which have been modified with resins to form some of the adhesive "alloys" discussed below. They are similar to thermoplastics in that they soften with heat but never melt completely. They generally provide high flexibility and low strength and, without resin modifiers, are used to bond paper and similar materials.

Thermosetting adhesives soften with heat only long enough for the cure to initiate. Once cured, they become relatively infusible up to their decomposition temperature. Although most such adhesives do not decompose at temperatures below 500°F (260°C), some are useful only to about 150°F (65°C). Different chemical types have different curing requirements. Some are supplied as two-part adhesives and mixed before use at room temperature; some require heat and/or pressure to bond.

As a group, these adhesives provide stronger bonds than the other three groups. Creep strength is good and peel strength is fair. Generally, bonds are brittle and have little resilience and low impact strength.

Alloy adhesives are adhesives compounded from resins of two or more different chemical families, e.g., thermosetting and thermoplastic, or thermosetting and elastomeric. In such adhesives the performance benefits of two or more types of resins can be combined. For example, thermosetting resins are plasticized by a second resin resulting in improved toughness, flexibility, and impact resistance.

Paste adhesives are usually water solutions of starches or dextrins, sometimes mixed with gums, resins, or glue to add strength, and containing antioxidants. They are the cheapest of the adhesives, but deteriorate on exposure unless made with chemically altered starches. They are widely employed for the adhesion of paper and paperboard. Much of the so-called **vegetable glue** is **tapioca paste.** It is used for the cheaper plywoods, postage stamps, envelopes, and labeling. It has a quick tack, and is valued for pastes for automatic box-making machines. **Latex pastes** of the rub-off type are used for such purposes as photographic mounting, as they do not shrink the paper as do the starch pastes. **Glues** are usually water solutions of animal gelatin, and the only difference between animal glues and edible gelatin is in the degree of purity. Hide and bone glues are marketed as dry flake, but fish glue is liquid. **Mucilages** are light vegetable glues, generally from water-soluble gums.

Rubber cements for paper bonding are simple solutions of rubber in a chemical solvent. They are like the latex pastes in that the excess can be rubbed off the paper. Stronger rubber cements are usually compounded with resins, gums, or synthetics. An infinite variety of these cements is possible, and they are all waterproof with good initial bond, but they are subject to deterioration on exposure, as the rubber is uncured. This type of cement is also made from synthetic rubbers which are self-curing. **Curing cements** are rubber compounds to be cured by heat and pressure or by chemical curing agents. When cured, they are stronger, give better adhesion to metal surfaces, and have longer life. **Latex cements** are solvent solutions of rubber latex. They provide excellent tack and give strong bonds to paper, leather, and fabric, but they are subject to rapid disintegration unless cured.

In general, natural rubber has the highest cohesive strength of the rubbers, with rapid initial tack and high bond strength. It also is odorless. **Neoprene** has the highest cohesive strength of the synthetic rubbers, but it requires tackifiers. **Gr-S rubber** (styrene-butadiene) is high in specific adhesion for quick bonding, but has low strength. Reclaimed rubber may be used in cements, but it has low initial tack and needs tackifiers.

Pyroxylin cements may be merely solutions of nitrocellulose in chemical solvents, or they may be compounded with resins, or plasticized with gums or synthetics. They dry by the evaporation of the solvent, and have little initial tack, but because of their ability to adhere to almost any type of surface they are called **household cements.** Cellulose acetate may also be used. These cements are used for bonding the soles of women's shoes. The bonding strength is about 10 lb/in^2 (0.07 MPa), or equivalent to the adhesive strength of the outer fibers of the leather to be bonded. For hot-press lamination of wood the plastic cement is sometimes marketed in the form of thin sheet.

Polyvinyl acetate-crotonic acid copolymer resin is used as a **hot- dip adhesive** for book and magazine binding. It is soluble in alkali solutions, and thus the trim is reusable. **Polyvinyl alcohol,** with fillers of clay and starch, is used for paperboard containers. **Vinyl emulsions** are much used as adhesives for laminates.

Epoxy resin cements give good adhesion to almost any material and are heat-resistant to about 400°F (204°C). An epoxy resin will give a steel-to-steel bond of 3,100 lb/in^2 (22 MPa), and an aluminum-to-aluminum bond to 3,800 lb/in^2 (26 MPa).

Some **pressure-sensitive adhesives** are mixtures of a phenolic resin and a nitrile rubber in a solvent, but adhesive tapes are made with a wide variety of rubber or resin compounds.

Furan cements, usually made with **furfural-alcohol resins,** are strong and highly resistant to chemicals. They are valued for bonding acid-resistant brick and tile.

Structural adhesives have come to mean those adhesives used to bond metals to other metals, to wood, or to rigid plastics, where bond strength is a critical requirement. They are generally of the alloy or thermosetting type. Three of the most commonly used are the modified epoxies, neoprene-phenolics and vinyl formal-phenolics. **Modified epoxy adhesives** are thermosetting and may be of either the room-temperature curing type, which cure by addition of a chemical activator, or the heat-curing type. They have high strength and resist temperatures up to nearly 500°F (260°C). **Neoprene-phenolic adhesives** are alloys characterized by excellent peel strength, but lower shear strength than modified epoxies. They are moderately priced and offer good flexibility and vibration absorption. **Vinyl formal-phenolic adhesives** are alloys whose properties fall between those of modified epoxies and the thermoset-elastomer types. They are supplied as solvent dispersions in solution or in film form.

Acrylic adhesives are solutions of rubber-base polymers in methacrylate monomers. They are two-component systems and have characteristics similar to those of epoxy and urethane adhesives. They bond rapidly at room temperature, and adhesion is not greatly affected by oily or poorly prepared surfaces. Other advantages are: low shrinkage during cure, high peel and shear strength, excellent impact resistance, and good elevated temperature properties. They can be used to bond a great variety of materials, such as wood, glass, aluminum, brass, copper, steel, most plastics, and dissimilar metals.

Ultraviolet cure adhesives, of Loctite Corp., are anaerobic structural adhesives formulated specifically for glass bonding applications. The adhesive remains liquid after application until ultraviolet light triggers the curing mechanism.

A **ceramic adhesive** developed by the Air Force for bonding stainless steel to resist heat to 1500°F (816°C) is made with a porcelain enamel frit, iron oxide, and stainless-steel powder. It is applied to both parts and fired at 1750°F (954°C), giving a shear strength of 1,500 lb/in^2 (10 MPa) in the bond. But ceramic cements that require firing are generally classed with ordinary adhesives. **Wash-away adhesives** are used for holding lenses, electronic crystal wafers, or other small parts for grinding and polishing operations. They are based on acrylic or other low-melting thermoplastic resins. They can be removed with a solvent or by heating.

Electrically conductive adhesives are made by adding metallic fillers, such as gold, silver, nickel, copper, or carbon powder. Most conductive adhesives are epoxy-based systems, because of their excellent adhesion to metallic and nonmetallic surfaces. Silicones and polyimides are also frequently the base in adhesives used in bonding conductive gaskets to housings for electromagnetic and radio-frequency interference applications.

ADIPIC ACID. Also called **butane dicarboxylic acid.** A white crystalline solid of the composition $HOOC(CH_2)_4COOH$, used as a plasticizer in synthetic resins and coatings, and in the production of nylon. It is made by the treatment of fatty acids with nitric acid, or can be made by oxidizing cyclohexanol. The melting point is 152°C. It is soluble in alcohol and slightly soluble in water. Many other dibasic acids useful for making synthetic resins are produced readily from fatty oils. **Suberic acid,** $HOOC(CH_2)_6COOH$, is made by the oxidation of castor oil. It is the same as the **octane-dioic acid** made from butadiene. **Sebacic acid,** $HOOC(CH_2)_8COOH$, is produced by heating castor oil with sodium hydroxide. **Azelaic acid,** $HOOC(CH_2)_7COOH$, is a strong dibasic acid with melting point at 106°C, made by the oxidation of oleic acid, and is used as a substitute for phthalic anhydride to react with glycerin to form alkyd resins less hard and brittle than those made with phthalic anhydride. It is also used instead of sebacic acid for producing the high-temperature lubricant **ethylhexyl sebacate.** Another substitute for this acid is pelagonic acid. **Petroselic acid,** which is an isomer of oleic acid with the double bond in a different position, is made by the hydrogenation of the ricinoleic acid of castor oil, and is then oxidized to produce adipic acid. Adipic acid can be used as a substitute for citric acid for the acidulation of beverages, but is less water-soluble. It is also used in protein foods to control the gelling action.

ADMIRALTY METAL. An alloy containing about 70% copper, 1 tin, and the remainder zinc, sometimes termed **admiralty bronze.** The tin increases the hardness and strength but decreases the ductility. The hard-drawn tube has a tensile strength of 100,000 lb/in^2 (690 MPa) and elongation 3%. The weight is 0.305 lb/in^3 (8.4 g/cm^3). It machines readily, especially when it contains lead. The alloy has been standard for condenser tubes, but for condenser tubes and other uses where there may be fouling from algae or dezincification from biotic attack, a 70–30 cartridge brass with a small amount of mercury may be used.

ADSORBENT. A material used to remove odor, taste, haze, and color from oils, foods, pharmaceuticals, or chemicals by selective adsorption of the impurities. Such materials are also called **adsorbates.** Adsorption is distinct from absorption in that it is the process of adhesion of the molecules of the substance to the surface of the adsorbent. The common adsorbents are **activated carbon,** or **activated clays, alumina, magnesium silicate,** or **silica gel.** The noncarbonaceous adsorbents are used for decolorizing vegetable, animal, or mineral oils, but activated carbon may also be used in conjunction with clays to adsorb color bodies not removed by the clay. Granular adsorbents are employed as filter beds, but powdered adsorbents are stirred into the liquid and are usually more effective. Adsorption from a gas is usually done with activated carbon. Silica gel is usually employed for removing trace quantities of water from water- insoluble liquids, while activated carbon is used for removing trace quantities of oils or chemicals from water. Adsorbents are normally recovered and are regenerated for reuse by heating, steaming, or burning off the adsorbed material.

Adsorbents are also used to separate chemicals of different molecular diameters without regard to their boiling points. A double hydrated aluminum calcium silicate marketed by the Union Carbide Corp. as a **zeolite** will pass chemicals with molecular diameters less than 5 angstroms and retain larger ones by selective adsorption. Such adsorptives are called **molecular sieves.** However, a material used in the separation of liquid mixtures whose components boil too close together for simple fractional distillation is called an **azeotrope.** It is a solvent added to the mixture to increase the relative volatility of one of the components so that it can be separated. The solvent may be alcohols, glycols, or nitrobenzene.

The adsorbents used in vacuum tubes to adsorb or combine with residual gases are called **getters. Flash getters** are pellets or strips of barium or barium alloy used to shorten the exhaust period. The getter is evaporated by induction heating during tube exhaust, and condenses on the tube walls, adsorbing the gas residues. Later, at operating temperatures of 300 to 400°F (150 to 204°C), the getter formed on the tube wall traps gases liberated during tube life. **Bulk getters** are sheets or wires of zirconium, tantalum, or columbium mounted on the hot electrode to trap gases at temperatures of 900 to 2200°F (482 to 1204°C). **Thorium** or **thorium-misch metal** may be used as getters for high temperatures by a coating sintered on the tube anode.

AEROSOLS. A dispersion of particles in air, particularly the chemical dispensing of a liquid or a finely divided powder substance by a gas propellant under pressure. The common aerosol can system was developed during World War II for dispensing insecticides. Substances commonly dispensed by the aerosol process include resins, paints, waxes, and cosmetics. Chlorofluorocarbons (fluorocarbons), hydrocarbons, and carbon dioxide are used as propellants. The use of fluorocarbons is being questioned because the possibility exists that when they reach the stratosphere they are subject to attack by ultraviolet radiation, which frees their chlorine atoms, which in turn react with the ozone and thus deplete the protective ozone layer. The use of carbon dioxide as a propellant has been limited because of problems in filling aerosol cans on a mass production basis. **Aerothene MM,** of Dow Chemical, is a **methylene chloride** that has properties needed for the efficient functioning of carbon dioxide and hydrocarbon propellants in cosmetic aerosol applications. In noncosmetic aerosol applications, such as paints and insecticides, hydrocarbons claim 45% of the market. Principal objection to their use for cosmetic aerosols has been their flammability.

AGAR-AGAR. The dried bleached gelatinous extract from various species of seaweed, *Algae,* mostly species from the Pacific and Indian Oceans. It is the only one of the seaweed products classed as a strategic material because of its use in medicine, but its use is small compared with the use of the products from other seaweeds. The word agar means seaweed. Translations of double words from the primordial languages, such as Malay, Carib, or Gaelic, must be made by taking the first word as a superlative adjective or the second word as a cognate verb. Thus, agar-agar means best-quality agar.

When dissolved in hot water, agar forms a transparent jelly, and is used for fixing bacteria for counts, as a stabilizer in toilet lotions, and in medicines. It has high thickening power, but, unlike most other seaweed extracts, it is indigestible and is not used in foodstuffs. **Kantan** is a variety of agar from the **tengusa seaweed,** *Gelidium corneum,* of Japan. **Australian agar** is from the abundant seaweed *Gracilaria confervoides.* Commercial agar is colorless, yellowish, or pink to black. It is marketed in strips, blocks, or shredded, and is obtained by boiling the dry seaweed and straining out the insoluble matter.

Most of the American production of agar, as distinct from the algins of the Atlantic, is from the giant **kelp,** *Macrocystia pyrifera* and *Gelidium cartilagineum* of the coast of California and Mexico, but it is not valued as highly for bacteriologic use as the Asiatic. The kelp grows straight up in water 60 to 100 ft (18.3 to 30.5 m) deep and then spreads out on the water another 60 to 80 ft (18.3 to 24.4 m). It is cut about 3 ft (0.9 m) below the surface, and three crops are harvested annually. The plant is 90% water. The dried kelp is washed with boiling water, cooked with soda ash, filtered, sterilized, and treated with muriatic acid to extract the agar.

AGATE. A natural mixture of crystalline and colloidal silica, but consisting mainly of the mineral chalcedony. It usually occurs in irregular banded layers of various colors derived from mineral salts, and when polished has a waxy luster. The specific gravity is about 2.6, and the mineral is sometimes harder than quartz. Agate is used for knife-edges and bearings of instruments, for pestles and mortars, for textile rollers, and for ornamental articles; and the finer specimens are employed as gemstones. The finest of the massive agates come from Uruguay and Brazil. Much agate encloses dendritic, or fernlike, patterns of manganese oxide or iron oxide, suggestive of moss. The **moss agates** of Montana and the yellow-green moss agate of California known as **amberine** are used as gemstones. Agate is a water-deposited stone, and often occurs in the form of stalactites and in petrified wood. **Agatized wood** of Wyoming and Arizona has a green fluorescence. It is cut into ornaments. Clear translucent yellow agates are called **sard,** while the clear reddish ones are **carnelian.** Both are cut as gemstones. **Sardoine** is a brownish carnelian. **Iris agate,** with rainbow colors, from Montana and Oregon, is highly prized. **Moss opal** of Nevada and California is moss agate intergrown with opal. **Blue moonstone** of California is not a true moonstone but is a blue agate of opalescent luster. Commercial agates may be artificially stained with mineral salts, dyed, or treated with acids to bring out color differences. **White agate** is a cream-colored chalcedony with a more waxy appearance than agate.

AJOWAN OIL. Also called **ptychotis oil.** A yellow essential oil distilled from **ajwan seed** of the herbaceous plant *Carum copticum,* or *Ptychotis ajowan,* of India. The seed yields 3 to 4% oil containing up to 50% **thymol** and some **cymene,** most of the thymol separating out on distillation. Thymol is known as **ajwan ka phul,** meaning flowers of ajwan, and the latter part of the name is Anglicized to thymol. Ajowan oil has a specific gravity of 0.900 to 0.930. It is used in pharmceuticals. Thymol, $(CH_3)_2CHC_6H_3(CH_3)OH$, is a white crystalline solid with a strong thyme odor, soluble in alcohol, and melting at 50°C. It is used in antiseptics and as a deodorant for leather. Thymol is also obtained from horsemint oil and from eucalyptus oil, or can be made synthetically from metacresol. It was originally distilled from the **thyme plant,** *Thymus vulgaris,* of the Mediterranean countries, the dried leaves of which are used as a condiment. **Cymene,** $(CH_3)_2CHC_6H_4CH_3$, is used as a scent in soaps, and has high solvent properties. It is also obtained from spruce turpentine. It is a liquid of specific gravity 0.861, boiling at 177°C.

ALBUMIN. The water-soluble and alcohol-soluble protein obtained from blood, eggs, or milk, and used in adhesives, textile and paper finishes, leather coatings, varnishes, as a clarifying agent for tannins, and in oil emulsions. Crude **blood albumin** is a brown amorphous lumpy material obtained by clotting slaughterhouse blood and dissolving out the albumin. The remaining dark-red material is made into **ground blood** and marketed as a fertilizer. Blood albumin is sold as clear, pale, amber, and colored powders. Blood albumin from human blood is a stable, dry, white powder. It is used in water solution for treatment of shock. The material of **egg white** is sometimes spelled **albumen.** Egg white is a complex mixture of at least eight proteins, with sugar and inorganic salts. More than half of the total is the protein **ovalbumin,** a strong coagulating agent, and another large percentage consists of **conalbumin** which forms metal complexes and unites with iron in the human system. Two of the proteins not so desirable in the human body are **ovomucoid,** which inhibits the action of the digestive enzyme trypsin, and **avidin,** which combines with and destroys the action of the growth vitamin biotin. **Egg albumin** is prepared from the dried egg white, and is marketed in yellowish amorphous lumps or powdered. The complexity of proteins is illustrated by the fact that the formula for egg albumin is $C_{1428}H_{2244}N_{462}S_{14}$. The heat of pasteurization damages the proteins of the egg white. A small amount of lactic acid and an aluminum salt will stabilize it and allow pasteurizing at 143°F (62°C). **Milk albumin** is made by coagulating casein. **Soybean albumin** is used to replace egg albumin in confectionery. **Synthetic egg white,** or albumin, was made in Germany from fish by extracting the soluble proteins with acetic acid, removing the fat with trichlorethylene, and hydrolyzing with sodium hydroxide. After neutralization, it is obtained as a white powder. **Fish albumin** is a good emulsifier, and can be whipped into a stiff foam for bakery products.

ALCOHOL. The common name for **ethyl alcohol,** but the term properly applies to a large group of organic compounds that have important uses in industry, especially as solvents and in the preparation of other materials. A characteristic of all alcohols is the monovalent—OH group. In the primary alcohols there is always a $\cdot CH_2OH$ group in the molecule. The secondary alcohols have a :CHOH group, and the tertiary alcohols have a distinctive :COH group. Alcohols with one OH group are called **monohydroxy alcohols;** those with more than one OH group are known as **polyhydroxy alcohols** or **polyhydric alcohols.** Another method of classification is by the terms saturated and unsaturated. The common alcohols used in industry are ethyl, methyl, amyl, butyl, isopropyl, and octyl. The alcohols vary in consistency. Methyl alcohol is like water, amyl alcohol is oily, and melissyl alcohol is a solid. Many of the alcohols are easiest made by fermentation; others are produced from natural gas or from petroleum hydrocarbons. Much of the production of ethyl alcohol is from blackstrap molasses. Alcohols, generally colorless, are similar to water in some ways and are neither alkaline nor acid in reaction.

Methyl alcohol, commonly known as **wood alcohol,** has the chemical name **methanol.** It is also referred to as **carbinol.** A colorless, poisonous liquid of the composition CH_3OH, it was originally made by the distillation of hardwoods. It is now produced chiefly by catalytic synthesis from hydrogen and carbon synthesis. It is used as a solvent in lacquers, varnishes, and shellac. On oxidation it yields formaldehyde, and is used in making the latter product for synthetic molding materials. The specific gravity of methyl alcohol is 0.795, the solidifying point is −98°C, and the boiling point is 65°C.

Solidified alcohol, marketed in tins and used as a fuel in small stoves, is a jellylike solution of nitrocellulose in methyl alcohol. It burns with a hot flame. **Sterno,** of Sterno, Inc., is this material, while **Trioxane,** employed for the same purpose, is an anhydrous formaldehyde trimer, but has the disadvantage of being water-soluble.

Butyl alcohol is a colorless liquid used as a solvent for paints and varnishes, and in the manufacture of dyes, plastics, and many chemicals. There are four forms of this alcohol, but the normal or primary butyl alcohol is the most important. **Normal butyl alcohol,** $CH_3(CH_2)_2CH_2 \cdot OH$, has a specific gravity of 0.814, and boiling point of 117°C. This form, known as **butanol,** has strong solvent power and is valued where a low evaporation rate is desired. It is also used for organic synthesis.

Fluoro alcohols are alcohols in which fluorine is substituted for hydrogen in the nonalcohol branch. They have the general composition $H(CF_2CF_2)_xCH_2OH$, with high specific gravities, 1.48 to 1.66, and high reactivity. As solvents they dissolve some synthetic resins that resist common solvents. Some of the esters are used as lubricants for temperatures to 500°F (260°C). **Acetylenic alcohols** are **methyl butynol,** $CH{:}C \cdot C(CH_3)_2OH$, with a specific gravity of 0.8672, boiling at 104°C, and used as a solvent, and **methyl pentynol,** $CH{:}C \cdot C(CH_3)_2CH_2OH$, boiling at 121°C. It is a powerful solvent. It has hypnotic qualities, and is also used for tranquilizing fish in transport.

Fatty acid alcohols, made from fatty acids or synthetically, have the general formula $CH_3(CH_2)_xOH$, ranging from the C_8 of **octyl alcohol** to the C_{18} of **stearyl alcohol.** They are easily esterified, oxidized, or ethoxilated, and are used for making cosmetics, detergents, emulsifiers, and other chemicals. **Lorol 25,** of Du Pont, is **cetyl alcohol. Elaidyl alcohol,** made from methyl oleate, is an 18-carbon alcohol. It is solid, melting at 75°F (24°C). The fatty acid alcohols vary from water-white liquids to waxy solids. The **Dytols,** of Rhom & Haas, are fatty alcohols, and the **Alfols,** of the Continental Oil Co., are straight-chain primary fatty alcohols made from ethylene and containing even numbers of carbon atoms from 6 to 18. **Polyols** are alcohols containing many hydroxyl, −OH, radicals. They react easily with isocyanates to form urethane.

ALDEHYDE. A group name for substances made by the dehydrogenation or oxidation of alcohols, such as formaldehyde from methyl alcohol. By further oxidation, the aldehydes form corresponding acids, as formic acid. The aldehydes have the radical group −CHO in the molecule, and because of their ease of oxidation are important reducing agents. They are also used in the manufacture of synthetic resins and many other chemicals. Aldehydes occur in animal tissues and in the odorous parts of plants.

Acetaldehyde is a water-white flammable liquid with an aromatic penetrating odor, used as a reducing agent, preservative, and for silvering mirrors, and in the manufacture of synthetic resins, dyestuffs, and explosives. Also called **ethanal,** it has the composition of $CH_3 \cdot CHO$, and is made by the oxidation of alcohol, by catalytic hydration of acetylene, or from ethylene. The specific gravity is 0.801 and the boiling point is 20.8°C. It is soluble in water, alcohol, and naphtha. **Paraldehyde,** $(CH_3 \cdot CHO)_3$, may be used instead of acetaldehyde in resin manufacture, has a higher boiling point, 124°C, and a higher flash point, but is not as reactive and will not reduce silver solutions to form a mirror. It is used for fulling leather. **Propion aldehyde,** or **propanal,** CH_3CH_2CHO, is made in the same way by oxidation of propyl alcohol. It has a boiling point at 48.8°C, and has reactions similar to acetaldehyde.

When acetaldehyde is condensed by reaction with a dilute alkali, it forms **acetaldol,** also called **aldol,** a viscous pale-yellow liquid of the composition $CH_3 \cdot CH(OH){:}CH_2CHO$, with a specific gravity of about 1.10, soluble in hot water and in alcohol. It is used to replace formaldehyde for synthetic resins, and for cadmium plating baths and dye baths, and for making butadiene rubber. **Paraldol,** the double molecule of aldol, is a white crystalline material melting at 82°C. When crude aldol is slightly acidified with acetic acid and heated, it yields **croton aldehyde,** also called **crotonic aldehyde** and **propylene aldehyde,** $CH_3 \cdot CH{:}CH \cdot CHO$, with a specific gravity of about 0.855 and a boiling point of 99 to 104°C. It is soluble in water, alcohol, and hydrocarbons, and is used as a solvent for resins, gums, and rubber, and in tanning leather. It has a pungent, suffocating odor, and is used in tear gases. Small quantities are sometimes used in city gas mains as a warning agent on the escape of poisonous fuel gas, as even tiny quantities will awaken a sleeping person.

Acrolein is **acrylic aldehyde,** $CH_2{:}CH \cdot CHO$, a colorless volatile liquid of specific gravity 0.8389, boiling at 52.7°C. The vapor is irritating to the eyes and nose, and the unpleasant effect of scorching fat is due to the acrolein formed. Acrolein is made by oxidation of propylene with a catalyst. It polymerizes easily, and can be copolymerized with ethylene, styrene, epoxies, and other resins to form various types of plastics. Its reactive double bond and carbonyl group make it a useful material for chemical synthesis. **Acrolein cyanohydrin,** $CH_2{:}CHCH(OH)CN$, a water-soluble liquid, boiling at 165°C, is also used to modify synthetic resins by introducing a nitrile group and a free hydroxyl into the molecular chain. It will copolymerize with ethylene and with acrylonitrile.

ALDER. The wood of several species of tree of the genus *Alnus* of the same family as the birch and beech. The **red alder** is from the *A. rubra,* or *A. oregona,* growing in the northwestern United States. The wood has a reddish-brown color and a fine even grain, is tough and resilient, can be worked easily, and takes a good polish. It has been much used for furniture, cabinetwork, and interior finish, as it rivals mahogany and walnut in appearance. **Black alder** is from the tree *A. glutinosa,* widely distributed in the Northern Hemisphere. It is of a reddish-white color, and has a smooth, fine grain, with a weight of about 35 lb/ft^3 (560 kg/m^3). It is used for plywood, cabinetwork, and toys. The wood of the alder is also used to produce smoke for curing kippered fish. The smoke is cooled to remove creosote, and is filtered. **Formosan alder** is from the *A. maritima* of Asia. The wood is light yellow streaked with reddish lines and has a fine texture.

ALKALI. A caustic hydroxide characterized by its ability to neutralize acids and form soluble soaps with fatty acids. Fundamentally, alkalies are inorganic alcohols, with the monovalent hydroxyl group —OH in the molecule, but in the alkalies this group is in combination with a metal or an ammonia group; and alkalies have none of the characteristics of alcohols. All alkalies are basic and have a pH value from 7 to 14. They neutralize acids to form a salt and water. The common alkalies are sodium hydroxide and potassium hydroxide, which are used in making soaps, soluble oils, and cutting compounds, in cleaning solutions, and for etching aluminim. All of the alkalies have a brackish taste and a soapy feel; most of them corrode animal and vegetable tissues.

ALKALI METALS. A name given to lithium, sodium, potassium, rubidium, strontium, cesium, calcium, and barium because of the basic reaction of their oxides, hydroxides, and carbonates. Carbonates of these metals are called **fixed alkalies.** The metals show a gradation in properties and increase in chemical activity with increase in atomic weights. All are silvery white and very soft. They tarnish rapidly in the air and decompose water at ordinary temperatures. In the alkali metals the electron bonding is so weak that even the impact of light rays knocks electrons free. All have remarkable affinity for oxygen. Rubidium and cesium ignite spontaneously in dry oxygen. Calcium, strontium, and barium are also called **earth metals.** Thin films of the alkali metals are transparent to ultraviolet but opaque to visible light.

ALKYD RESINS. A group of thermosetting synthetic resins known chemically as **hydroxycarboxylic resins,** of which the one produced from phthalic anhydride and glycerol is representative. They are made by the esterification of a polybasic acid with a polyhydric alcohol, and have the characteristics of homogeneity and solubility that make them especially suitable for coatings and finishes, plastic molding compounds, calking compounds, adhesives, and plasticizers for other resins. The resins have high adhesion to metals; are transparent, easily colored, tough, flexible, heat- and chemical-resistant; and have good dielectric strength. They vary greatly with the raw materials used and with varying percentage compositions, from soft rubbery gums to hard brittle solids. Phthalic anhydride imparts hardness and stability. Maleic acid makes a higher melting-point resin. Azelaic acid gives a softer and less brittle resin. The long-chain dibasic acids, such as adipic acid, give resins of great toughness and flexibility. In place of glycerol the glycols yield soft resins, and sometimes the glycerol is modified with a proportion of glycol. The resins are reacted with oils, fatty acids, or other resins such as urea or melamine, to make them compatible with drying oils and to impart special characteristics.

Since alkyd resins are basically esterification products of innumerable polybasic acids and polyhydric alcohols, and can be modified with many types of oils and resins, the actual number of different alkyd resins is unlimited, and the users' specification is normally by service requirements rather than composition. **Short-oil alkyds,** with 30 to 45% nonoxidizing oils, are used in baking enamels, while the **long-oil alkyds,** with 56 to 70% oxidizing oils, are soluble in mineral spirits and are used for brushing enamels.

Alkyd coatings are used for such diverse applications as air-drying water emulsion wall paints and baked enamels for automobiles and appliances. The properties of **oil-modified alkyd coatings** depend upon the specific oil used as well as the percentage of oil in the composition. In general, they are comparatively low in cost and have excellent color retention, durability, and flexibility, but only fair drying speed, chemical resistance, heat resistance, and salt spray resistance. The oil-modified alkyds can be further modified with other resins to produce **resin-modified alkyds.** The resin is sometimes added during manufacture of the alkyd and becomes an integral part of the alkyd, or the modifying resin is blended with the alkyd when the paint is formulated. When mixed with urea formalehyde or melamine resin, harder and more resistant baked enamels are produced. Alkyds blended with ethyl cellulose are used as tough flexible coatings for electric cable. Other resins blended with alkyds to produce special or improved properties include phenolic, rosin, vinyl, and silicone.

Alkyd plastics molding compounds are composed of a polyester resin and usually a diallyl phthalate monomer plus various inorganic fillers, depending on the desired properties. The raw material is produced in three forms—granular, putty, and glass-reinforced. As a class, the alkyds have excellent heat resistance up to about 300°F (150°C), high stiffness, and moderate tensile and impact strength. Their low moisture absorption combined with good dielectric strength makes them particularly suitable for electronic and electrical hardware such as switchgear, insulators and parts for motor controllers and automotive ignition systems. They are easily molded at low pressures and cure rapidly.

Amberlac and **Duraplex** are alkyd resins of the Rohm & Haas Co. in a wide range of formulations. The **Paraplex resins** are oil-modified soft resins used for coatings for textiles and paper, and for blending with cellulose plastics to give better adhesion on lacquers. **Paraplex, P-43HV,** however, is a **polyester-styrene** copolymer supplied as a clear liquid for laminating, molding, or casting at low pressure. **Aquaplex,** of the same company, is a group of oil-modified resins for use in water paints. **Teglac resins,** of the American Cyanamid Co., used for indoor finishes and clear-coat varnishes, are hard alkyd resins made with natural resin acids as blending agents. **Cycopol,** of the same company, is an **alkyd-styrene** copolymer resin for fast-drying enamels, giving high gloss and durability. **Beckosol 1341,** of Reichhold Chemicals, Inc., is a phenol-modified alkyd resin for fast-drying enamels. **Petrex resin,** of Hercules, Inc., is a series of alkyd resins used in lacquers, varnishes, adhesives, and inks. **Aroplaz 1248-M,** of U.S. Industrial Chemicals, Inc., is a high-gloss **phthalic alkyd resin,** soluble in mineral spirits, for industrial finishes. The **Iso Burnok 32-0,** of the T. F. Washburn Co., used as a vehicle for thixotropic paints, is a soybean oil isothalic alkyd. It has high clarity and is odorless. **Dyal,** of the Sherwin-Williams Co., is an alkyd resin for finishes, as are also **Amalite** and **Amavar,** of the American Alkyd Industries.

Plaskon Alkyd, of the Allied Chemical Corp., is a mineral-filled alkyd molding powder used for electrical parts having good arc resistance and heat resistance to 350°F (176°C). **Neolyn resins,** of Hercules, Inc., are alkyd resins produced from rosin. They are used as modifiers for nitrocellulose and for vinyl resins in lacquers and adhesives to add toughness to the film. **Cellolyn 501,** of the same company, is a lauric acid–pentaerythritol alkyd used in durable, color-stable nitrocellulose lacquers. When maleic anhydride or fumaric acid is reacted with rosin and then esterified with glycerol or other polyhydric alcohol, a series of alkyd resins is produced. **Isocyanate resins** are linear alkyds lengthened by isocyanates and then treated with a glycol or a diamine to cross-link the molecular chain. Plastics made from them are noted for good abrasion resistance.

ALLIGATOR LEATHER. A light, tough leather with platelike scales on the surface. It is made from the skins of large saurians, or lizards, of the order *Crocodilia,* abounding in muddy tropical streams. The species *Alligator mississippiensis* inhabits the swamps of the southeastern United States. Alligator leather is valued for luggage, pocketbook, and shoe leathers. It is much imitated with embossed split sheepskins, but sheepskin is soft and easily scuffed. **Lizard leather,** from the **Java ring lizard,** is another reptile leather valued for women's shoes.

ALLOY. A very old term for the admixture of a precious metal with a metal of lesser value, or with a base metal. The admixing metal, usually copper, was originally needed in gold or silver to add hardness and to prevent wear erosion of the soft metals, but it also debased the value of the coinage metal. Now the term applies more generally to any combination of metals. With the exception of the highest-purity elemental metals, most commercial metallic materials are alloys, composed of two or more different elements. Even in commercially pure grades of metals, small amounts of impurity elements are present, either from the ore or introduced during the refining process. A **metal alloy** is defined as a combination of two or more different elements (atoms), the major one of which is a metal. The metallic element present in the largest amount, by weight, is called the **base metal** or **parent metal;** other elements present are termed **alloying elements** or **alloying agents.** Most alloys are composed of two or more metallic elements. However, there are important exceptions, as in the case of steel, in which the presence of carbon atoms importantly influences steel's properties.

Structurally there are two kinds of metal alloys—single phase and multiphase. **Single-phase alloys** are composed of crystals with the same type of crystal structure. They are formed by "dissolving" together different elements to produce a **solid solution.** The crystal structure of a solid solution is normally that of the base metal. The atoms of the alloying element (solute) join the base metal (solvent) either as substitution atoms or as interstitial atoms. In the former, alloying atoms occupy some of the lattice sites normally occupied by the host atoms. In the latter, alloying atoms place themselves in between the host atoms. In substitutional solid solutions, the solute and solvent atoms are of approximately similar size; whereas in the interstitial type, the solute atoms must be small enough to fit between the atoms. The nature of solid solutions have important effect on many alloy properties. For example, strength and hardness increase with the amount of solute present, but ductility usually decreases. Electric conductivity also is generally lowered by the presence of the solute element.

In contrast to single-phase alloys, **multiphase alloys** are mixtures rather than solid solutions. They are composed of aggregates of two or more different phases. The individual phases making up the alloy are different from one another in their composition or structure. **Solder,** in which the metals lead and tin are present as a mechanical mixture of two separate phases, is an example of the simplest kind of multiphase alloy. In contrast, steel is a complex alloy composed of different phases, some of which are solid solutions. Multiphase alloys far outnumber single-phase alloys in the industrial material field, chiefly because they provide greater property flexibility. Thus, properties of multiphase alloys are dependent upon many factors, including the composition of the individual phases, the relative amounts of the different phases, and the positions of the various phases relative to one another.

When two different thermoplastic resins are blended, a **plastic alloy** is obtained. Alloying permits resin polymers to be blended that cannot be polymerized. Not all plastics are amenable to alloying. Only resins that are compatible with each other—those that have similar melt traits—can be successfully blended.

ALLOY STEEL. A general name for steels that owe their distinctive properties to elements other than carbon. Alloy steels usually take the name of the element or elements having the greatest influence on the characteristics of the alloy, regardless of the percentage of the element contained in the steel. The American Iron & Steel Institute defines alloy steel as a steel in which a minimum limit is specified or guaranteed for alloying elements. These minimum percentages are chromium, 0.25%; copper, 0.60; manganese, 1.65; nickel, 0.25; silicon, 0.60; or any amounts of titanium, tungsten, or molybdenum. Steels having casual amounts of these elements, less than those specified above, are not rated as alloy steels, although vanadium steel may be any carbon steel that has been cleansed with vanadium. Alloy steels are also marketed under a variety of trade names. Some alloy steels owe their particular properties to refinements of manufacture and selection of small amounts of alloying elements. Most tool steels and structural steels with special properties are alloy steels. High-speed steels are alloy tool steels, but are classed as a separate group. **Universal steel** is a name applied to alloy steels containing a low percentage of carbon to give them a wide range of utility. They are usually nickel-chromium steels, as these steels can be used with carbon as low as 0.15% without losing their heat-treating properties. **Low-alloy steels** are roughly defined as steels that do not have more than 5% total combined alloying elements. They are also known as **alloy constructional steels. High-alloy steels** are steels containing very large percentages of elements, usually to obtain some special property such as corrosion resistance. They are often not steels in the true sense, but are iron alloys.

ALLYL PLASTICS. A group of water-white casting plastics, related to the alkyds, produced by the polymerization of the ester of allyl alcohol or from allyl chloride, both produced from propylene. **Allyl alcohol,** $CH_2{:}CH \cdot CH_2OH$, is a colorless liquid also known as **propenol,** which can be made by heating glycerol with formic or oxalic acid. The specific gravity is 0.849, boiling point 96°C, and freezing point −129°C. **Allyl chloride,** $CH_2{:}CH \cdot CH_2Cl$, is a liquid of specific gravity 0.937, boiling point 45°C, and flash point −25°F (−30°C). **Allyl ester** is a clear syrupy liquid of specific gravity 1.26 which polymerizes with a peroxide catalyst to form allyl plastics. The liquid monomer can be poured into molds and hardened by polymerization. The polymerized castings are hard and crystal clear. Allyl plastics have a specific gravity of 1.34 to 1.40, dielectric strength of 1,275 volts per mil

(50.2 × 10^6 volts per meter), refractive index 1.57, Rockwell M hardness 116, and compressive strength 19,600 lb/in^2 (134 MPa). As the plastic has high clarity and less light dispersion than most optical glass, it is used for lenses, prisms, and reflectors. It can be colored easily with dyes, and is also used for mechanical and electrical parts. **Kriston,** of the B. F. Goodrich Chemical Co., **Allite,** of the Arol Chemical Prod. Co., and **Vibron 103,** of the Naugatuck Chemical Div., are allyl plastics. **Allymer CR-39,** of the Cast Optics Corp., is an optically clear, hard, thermosetting casting resin for clock and instrument faces, windows, and lenses. It is made from diallyl diglycol carbonate and triallyl cyanurate. The sheet material transmits 92% of ordinary light, has a high heat-distortion point, 130°C, a tensile strength of 5,500 lb/in^2 (38 MPa), a compressive strength of 22,800 lb/in^2 (156 MPa), and a Rockwell hardness of M100. It can be machined with carbide tools. It is craze-resistant, and has only half the weight of glass. **V- Lite,** of the Victor Chemical Works, is a **diallyl phenyl phosphate** monomer than polymerizes with a catalyst to form a transparent, hard, strong, flame-resistant thermosetting resin. It can also be copolymerized with the thermoplastic vinyl acetate or methyl methacrylate to produce thermosetting resins. **Selectron CR-39,** of the Pittsburgh Plate Glass Co., is an **allyl diglycol carbonate** casting resin. The cast plastic has good abrasion resistance, is resistant to the action of chemicals, and is noncrazing. It is also used to modify other plastics. **Diallyl phthalate** is a thermosetting resin cured by polymerization without water formation. The molded material, depending on the filler, has a tensile strength from 4,500 to 7,000 lb/in^2 (30 to 48 MPa), a compressive strength up to 30,000 lb/in^2 (210 MPa), a Rockwell hardness to M108, dielectric strength to 430 volts per mil (16.9 × 10^6 volts per meter), and heat resistance to 450°F (232°C). The heat-resistant asbestos-filled resin has a specific gravity of 1.70. **Dapon resin,** of the FMC Corp., is diallyl phthalate marketed as a white powder. **Diall resin,** of the Mesa Products Div., is a diallyl phthalate molding compound.

Methallyl alcohol, $CH_2{:}C \cdot CH_3 \cdot CH_2OH$, also forms esters which can be used for the production of plastics. It is a liquid boiling at 114.5°C. The allyl radical will combine with starch or sugar to form shellaclike resins. **Allyl sucrose,** made by combining sugar and allyl chloride, is a resin that produces varnishes which will withstand temperatures to 400°F (204°C), and are chemical-resistant. **Allyl starch** is a resinous material made by treating sweet-potato or grain starches with allyl chloride or allyl bromide. It is soluble in varnish solvents and is a substitute for shellac as a varnishing agent, forming an adherent, resistant film on paper, fabric, wood, or metal. it will withstand temperatures to 400°F (204°C).

ALMOND OIL. An essential oil distilled from the ground macerated kernels of the bitter almond, *Prunus amygdalus,* of the Mediterranean countries, and from the kernels of the **apricot,** *P. armeniaca.* The two oils are identical and contain the glucoside **amygdalin.** The American production is mostly from the by-product pits of the apricot-canning industry of California. The almond is a small tree closely resembling the peach. The fruit is inedible, but the seed inside the pit is marketed as roasted and salted nuts and is made into a paste for confections. Oil of bitter almonds is used in perfumery and as a flavor. For flavoring use, the poisonous hydrocyanic acid is extracted. Synthetic almond oil is **benzaldehyde,** C_6H_5CHO, a colorless volatile oil with an almond flavor, produced from benzol or from toluol, and used for producing **triphenylmethane dyes** and many chemicals.

ALPACA. A fabric made from the fine woollike hair of the alpaca, an animal of the llama family of the mountains of Bolivia, Peru, Chile, and Argentina. **Alpaca fiber** is long and fine, with a downy feel, but it does not have the strength or elasticity of fine wool, and is more closely allied to hair than to wool. There are two breeds of the alpaca animal, Huacaya and Suri, the latter having the longer and finer wool, reaching a fiber length up to 30 in (0.762 m). From 3 to 10 lb (1.4 to 4.5 kg) of fiber are obtained per animal. **Llama hair** (pronouned lyah-mah) from the llama of Bolivia is marketed as coarse alpaca. In Incan times llama wool was used by the common people for clothing, while the finer wool from the alpaca and vicuña was reserved for the upper classes. The llama is sheared every two years when the wool reaches a length of about 12 in (0.30 m). A considerable amount of stiff guard hairs occur in the fiber.

Vicuña, another animal of the llama family, is almost extinct, and the commercial vicuña cloth is made of alpaca or fine wool, or mixtures. There is a limited production of true vicuña from domestic herds in Bolivia and Peru, raised at an altitude of about 14,000 ft (4,267 m). It is the softest of all animal weaving fibers. Alpaca and vicuña cloths are used for shawls, jackets, and fine goods. Imitation **alpaca fabric** for clothing linings is a lustrous, smooth, and wiry fabric plain-woven with a cotton warp and a worsted mohair filling. When made with a rayon filling, it is called **rayon alpaca.**

ALUM. A colorless to white crystalline **potassium aluminum sulfate,** $KAl(SO_4)_2 \cdot 12H_2O$, or $[KAl(H_2O)_6]SO_4 \cdot 6H_2O$, occurring naturally as the mineral **kalunite,** and also in combination as the mineral **alunite.** It is also called **potash alum** to distinguish it from other forms. It has a sweetish taste and is very astringent. It is used as an additive in the leather and textile industries, in sizing paper, as a mordant in dyeing, in medicines as an astringent, and in baking powder. It is an important water-purifying agent. From a water solution it crystallizes out, forming positively charged particles which attract the negatively charged organic impurities, thus purifying the water as they settle out. Alum has a specific gravity of 1.757, melts in its water of crystallization at 92°C, and when heated to redness is converted to **burnt alum,** a porous, friable material which dissolves slowly in water. Alum is marketed as USP, lump, pea, nut, ground, and powdered. The **rice crystal alum** of the General Chemical Co. is from 10 to 30 mesh, and the granular is from 30 to 60 mesh.

Alumstone is a gray or pinkish massive form of alunite found in volcanic rocks. A pure variety from Italy is called **Roman alum,** or **roche alum.** The alunite of Australia is used to produce **postassium sulfate,** K_2SO_4, for fertilizer; and the residue, containing 50% alumina, is used for aluminum production. **Soda alum,** in which the potassium is replaced by sodium, occurs in the South American Andes as the mineral **mendozite.** It is more soluble than alum, but is more difficult to purify. **Filter alum,** also called **patent alum** and **aluminous cake,** used for waterworks filtration, is **aluminum sulfate,** $Al_2(SO_4)_3$, plus a varying amount of water of crystallization. The anhydrous form is used as a dehydrating agent for gases. It is a white crystalline solid readily soluble in water. When filter alum contains a slight excess of alumina, it is called basic. The **hexahydrate aluminum sulfate** of the Allied Chemical Corp. has the composition $Al_2(SO_4)_3 \cdot 6H_2O$, and is used for chemical processing.

Commerical aluminum sulfate is also called **concentrated alum,** and replaces potash alum for many uses because of its cheapness. It comes in colorless crystals having a strong astringent taste, and is used as a mordant in dyeing, in water purification, sizing papers, tanning, in printing inks, and in dry colors. For use in pickling and tanning leathers, it contains not more than 0.01% iron oxide. **Ammonia alum,** used in tanning sheepskins and fur skins, and in fireproofing and dyeing textiles, is **ammonium-aluminum sulfate,** $NH_4Al(SO_4)_2 \cdot 12H_2O$, a colorless or white crystalline powder soluble in water and, like other alums, insoluble in alcohol. It is valued for water purification because it forms chloramine, and is also used in vegetable glues and porcelain cements. For uses where an alkaline instead of an acid solution is required, **sodium phosphoaluminate** is employed. It is a white, water-soluble powder, and is a double salt containing about 70% sodium aluminate and 20 sodium orthophosphate, with the balance water. The **Alkophos** of the Monsanto Co. is **aluminum phosphate,** $AlPO_4$, a white powder used as a bonding agent for high-temperature cements, and as an alkaline flux for ceramics.

ALUMINA. The oxide of aluminum, Al_2O_3. The natural crystalline mineral is called **corundum,** but the synthetic crystals used for abrasives are designated usually as **aluminum oxide** or marketed under trade names. For other uses and as a powder it is generally called alumina. It is widely distributed in nature in combination with silica and other minerals, and is an important constituent of the clays for making porcelain, bricks, pottery, and refractories. The **alumina clay** from the large deposits in western Idaho contains an average of 28% Al_2O_3, 5.6 iron oxide, and a high percentage of titanium oxide. Such clays are used for ceramics, but the oxide, alumina, is obtained commercially chiefly by high-temperature fusing of bauxite. It is also produced from alunite with a by-product fertilizer, and obtained from the oil shales of Colorado.

The crushed and graded crystals of alumina when pure are nearly colorless, but the fine powder is white. Off colors are due to impurities. American aluminum oxide used for abrasives is at least 99.5% pure, in nearly colorless crystals melting at 2050°C. The chief uses for alumina are for the production of aluminum metal and for abrasives, but it is also used for ceramics, refractories, pigments, catalyst carriers, and in chemicals.

Aluminum oxide crystals are normally hexagonal, and are minute in size. For abrasives, the grain sizes are usually from 100 to 600 mesh. The larger grain sizes are made up of many crystals, unlike the single-crystal large grains of silicon carbide. The specific gravity is about 3.95, and the hardness is up to 2,000 Knoop. The ultrafine alumina abrasive powder of Union Carbide Corp. is of two kinds. Type A is **alpha alumina** with hexagonal crystals with particle size of 0.3 μm, density 4.0, and hardness 9 Mohs. Type B is **gamma alumina** with cubic crystals with particle size under 0.1 μm, specific gravity of 3.6, and hardness 8. Type A cuts faster, but Type B gives a finer finish. At high temperatures gamma alumina transforms to the alpha crystal. The aluminum oxide most frequently used for refractories is the **beta alumina** in hexagonal crystals heat-stabilized with sodium. It has the composition $Na_2O \cdot 11Al_2O_3$, but stablized alumina is also produced with oxides of calcium, potassium, or magnesium. The alpha and gamma alumina powders of the J. T. Baker Chemical Co. for lapping, grinding, and in dentifrices are of 99.96% purity in controlled particle sizes from 0.02 to 0.4 μm.

The aluminum oxide abrasives in all forms are sold under trade names. **Alundum,** of the Norton Co., **Aloxite,** of the Carborundum Co., and **Lionite,** of the General Abrasives Co., are aluminum oxides. **Alundum 32,** of the Norton Co., is aluminum oxide with each abrasive grain in a single crystal and, unlike the crushed and ground oxide, having the natural edges of the crystal as the cutting edges. It is about 96% Al_2O_3, and is brown. The white **Alundum 38,** of 99.5% purity, is crushed from large crystals. **Dynablast,** of the same company, has high-purity crystals agglomerated into grain sizes from 16 to 60 mesh. The 16 grit is used for metal cleaning, the 40 to 60 for granite cutting, and a finer powder, from 100 to 240 mesh, is used for glass and metal polishing.

Activated alumina is partly dehydrated alumina trihydrate, which has a strong affinity for moisture or gases and is used for dehydrating organic solvents. **Hydrated alumina** is **alumina trihydrate,** $Al_2O_3 \cdot 3H_2O$, used as a catalyst carrier. At elevated temperatures it becomes active, and the fine powder is a auxiliary catalyst. It is also used as a filler in plastics and in cosmetics, and comes in various particle sizes for use in glass and vitreous enamels to increase the strength and luster. **Hydrated alumina C-741** is this material with particles coated with stearic acid for use as a reinforcing pigment in rubber.

Activated alumina F-1, of the Aluminum Co. of America, is a porous form of alumina, Al_2O_3, used for drying gases or liquids. It will remove moisture up to 15% of the dry weight of the alumina. **Activated alumina F-6** is this material impregnated with cobaltous chloride which will change color from blue through pale pink to white with progression of the adsorption of moisture. Activated alumina is also used as a catalyst for many chemical processes.

Aluminum hydroxide is a white bulky water-insoluble fine powder of specific gravity 2.42 and refractive index 1.535, used as a base for lake pigments, for making gloss white, as a water repellent in textile and paper coatings, and as an antacid in medicine. **Boehmite,** of Du Pont, used in cosmetics, is called **colloidal alumina,** but it is an **aluminum monohydrate,** AlOOH, made by polymerizing aluminum cations in a water solution. **Baymal** of this company, used for coatings, adhesives, paints, and for making dense ceramics, is similar. The powder is in the form of tiny fibrils, 5 nm in diameter and 150 nm long. In coatings the fibrils interlock the molecules into a tight adherent film. In paints the powder acts as a **thixotropic material** to prevent sagging or running on vertical walls. The powder is both a **hydrophilic material,** that is, water-soluble, and an **organophilic material** soluble in oils and organic solvents. **Dispal,** of Continental Oil Co., is an aluminum monohydrate used as a filler in coatings and will give a hydrophilic coating on normally hydrophobic surfaces. It is in crystals 0.005 μm in size, agglomerated into particles of about 44 μm. It has very high bulking properties and is used also in cosmetics.

Alumina ceramics are the most widely used oxide-type ceramic, chiefly because alumina is plentiful, relatively low in cost, and equal to or better than most oxides in mechanical properties. Density can be varied over a wide range, as can purity—down to about 90% alumina—to meet specific application requirements. Alumina ceramics are the hardest, strongest, and stiffest of the oxides. They are also outstanding in electrical resistivity and dielectric strength, are resistant to a wide variety of chemicals, and are unaffected by air, water vapor, and sulfurous atmospheres. However, with a melting point of only 3700°F (2037°C), they are relatively low in refractoriness, and at 2500°F (1371°C), retain only about 10% of room-temperature strength. Besides wide use as electrical insulators and chemical and aerospace applications, alumina's high hardness and close dimensional tolerance capability make this ceramic suitable for such abrasion-resistant parts as textile guides, pump plungers, chute linings, discharge orifices, dies, and bearings. **Calcined alumina** is made by calcining aluminum hydroxide, and the commercial product is 99.1% Al_2O_3, with no more than 0.5% adsorbed water, and not over 0.4% silicon, iron, sodium, and titanium oxides. The powder, in particle size as small as 10 μm, is used for abrasives, glassmaking, and for refractory ceramics. The molded material has little porosity or shrinkage, and retains its strength and electrical resistance at high temperatures. **Alumina A-14,** of the Aluminum Co. of America, is a calcined alumina for electrical insulators with a particle size of 2 to 3 μm and containing not more than 0.08% soda. For electronic and nuclear ceramic parts, a 99.95% pure alumina is produced by this company. It has less than 0.01% sodium oxide. The alumina powder of Reynolds Metals Co. has an average particle size down to 1 μm and has low soda content. It is made by calcining the powder with silica sand at high temperature. Sintered parts have a service use to 3500°F (1927°C) with tensile strength to 60,000 lb/in^2 (413 MPa). **Alumina Al-200,** of the Coors Porcelain Co., for high-frequency insulators, gives a molded product with a tensile strength of 25,000 lb/in^2 (172 MPa), compressive strength of 290,000 lb/in^2 (2,000 MPa), and specific gravity of 3.36. The coefficient of thermal expansion is half that of steel, and the hardness about that of sapphire. **Alumina AD-995,** of this company, is a dense vacuum-tight ceramic for high-temperature electronic use. It is 99.5% alumina with no silica. The hardness is Rockwell N80, and dielectric constant 9.27. The maximum working temperature is 3200°F (1760°C), and at 2000°F (1093°C) it has a flexural strength of 29,000 lb/in^2 (200 MPa).

Koralox, a product of General Electric, is an alumina ceramic material formed by a selective densification process. It is used in the casting of hollow jet engine cores. These cores are then incorporated in molds into which eutectic superalloys are poured to form the turbine blades.

Tabular alumina is alumina converted to the corundum form by calcining at temperatures below the fusing point, and the tabular crystals are larger than those of calcined alumina. It comes as a fine powder or as coarse granules for making refractory ceramics and electrical insulators. **Alumina balls** are marketed in sizes from ¼ to ¾ in (0.6 to 1.9 cm) for reactor and catalytic beds. They are usually 99% alumina, having high resistance to heat and chemicals. **Alumina fibers** in the form of short linear crystals, called **sapphire whiskers,** have high strength up to 200,000 lb/in^2 (1,375 MPa) for use as a filler in plastics to increase heat resistance and dielectric properties. **Fumed alumina** powder of submicrometer size is made by flame reduction of aluminum chloride. It is used in coatings and for plastics reinforcement and in the production of ferrite ceramic magnets.

Alfrax is the trade name of the Carborundum Co. for alumina as a catalyst carrier and as a refractory. **Stupalox,** of the same company, for ring and plug gages, is alumina hot-pressed to a specific gravity of 3.95. It is tough and wear-resistant, with a transverse rupture strength of 100,000 lb/in^2 (690 MPa) and compressive strength of 450,000 lb/in^2 (3,100 MPa). **Hi-Supertite,** of Englehard Industries, Inc., is a vitreous ceramic in tube form, composed of 96% alumina, used for protecting thermocouples. It is gastight at 1450°C. **Lucalox,** of the General Electric Co., is a translucent ceramic pressed from high-purity fine alumina powder and fired at high temperature. It transmits 90% of visible light rays, and will withstand temperatures to 3600°F (1982°C). It is polycrystalline with pores removed in the firing. The transverse strength is 50,000 lb/in^2 (344 MPa), and dielectric strength 1,700 volts per mil (67 $\times$ 10^6 volts per meter). It is used for high-intensity lamps, missile nose cones, and instrument parts. **Alite,** of the U.S. Stoneware Co., used for bearings, valves, nozzles, and extrusion dies, is a white ceramic made by molding or extruding alpha alumina and sintering at 3400°F (1871°C). The bonding is by crystal growth, producing nearly a single crystal. The tensile strength is up to 32,500 lb/in^2 (224 MPa), compressive strength up to 290,000 lb/in^2 (2000 MPa), and dielectric strength 250 volts per mil (9.8 $\times$ 10^6 volts per meter). It is used for temperatures to 1600°F (871°C). **Aluminum oxide film,** or **alumina film,** used as a supporting material in ionizing tubes, is a strong, transparent sheet made by oxidizing aluminum foil, rubbing off the oxide on one side, and dissolving the foil in an acid solution to leave the oxide film from the other side. It is transparent to electrons. **Alumina bubble brick** is a lightweight refractory brick for kiln lining, made by passing molten alumina in front of an air jet, producing small hollow bubbles which are then pressed into bricks and shapes. **Alumina foam** bricks and blocks for high-temperature insulation are made by Pittsburgh Plate Glass Co. The foam has a density of 28 lb/ft^3 (448.5 kg/m^3) and porosity of 85%. The thermal conductivity at 2000°F (1093°C) is 3.7 Btu/(h)(ft^2)(°F/in) [0.002 W/(cm^2)(°C)].

ALUMINUM. Called **aluminium** in England. A white metal with a bluish tinge, symbol Al, atomic weight 26.97, obtained chiefly from bauxite. It is the most widely distributed of the elements next to oxygen and silicon, occurring in all common clays. Aluminum metal is produced by first extracting alumina (aluminum oxide) from the bauxite by a chemical process. The alumina is then dissolved in a molten electrolyte, and an electric current is passed through it, causing the metallic aluminum to be deposited on the cathode. The metal was discovered in 1727, but was obtained only in small amounts until it was reduced electrolytically in 1885. It has a specific gravity of 2.70 and a melting point of 658.7°C, is resistant to corrosion and to many chemicals, but is attacked by alkalies and hydrochloric acid. Above 400°F (204°C), however, corrosion of the metal is vigorous, and at 600°F (316°C) it converts to the oxide unless alloyed. The metal is nonmagnetic, even when highly alloyed with iron.

Pure aluminum is next to gold in order of malleability. The metal has a face-centered cubic crystal structure and is easy to cold-roll. On a volume basis aluminum has about 60% of the thermal conductivity of copper. The tensile strength of cast aluminum is 12,000 lb/in^2 (82 MPa), with elongation of 30% and Brinell hardness 30.

The chief impurities in commercial aluminum are copper, iron, and silicon, but sheet aluminum averages 99.3% pure. **Commercially pure aluminum** has a tensile strength, annealed, of 13,000 lb/in^2 (89 MPa), with elongation of 35% and Brinell hardness 23. When cold-rolled to a Brinell hardness of 44 it has a tensile strength of 24,000 lb/in^2 (165 MPa) with elongation of 5%. **High-conductivity aluminum,** called **EC aluminum,** contains 99.45% min aluminum, and has an electric conductivity 62% that of copper. The tensile strength is 12,000 lb/in^2 (82 MPa) with elongation of 23%. In the H17 hardness, with 66% cold reduction, it has a tensile strength of 20,000 lb/in^2 (137 MPa) and elongation of 2%.

Most of the aluminum used for structural products is in the form of alloys. Because of its high strength-weight factor a large part of the production of aluminum goes into transportation equipment and moving parts of machinery. It is also used for ornamental architectural work, containers, cooking utensils, chemical equipment, electric conductors, and packaging. But because the yield strength of the metal declines drastically above 375°F (190°C), it does not replace steel for large structures where there is danger of fire or locally applied heat. It extrudes easily, and is used to replace tin alloys for collapsible tubes. In powder and flake form it is used in paints and fireworks, thermit welding, and as a catalyst. **Aluminun shot,** for deoxidizing steel, comes in slightly flattened spheroids with maximum size of ½ in (1.27 cm). The reflectivity of polished aluminum is high for all wavelengths, and it is used for both heat and light reflectors.

Foamed aluminum, or **aluminum foam,** used as a core material for lightweight structures, is made by foaming the metal with zirconium hydride or other metal hydride. The released hydrogen expands the metal into a cellular structure of good strength with controlled densities from 12 to 40 lb/ft^3 (192 to 641 kg/m^3). A foam of aluminum-magnesium alloy used for insulated roofing and for building panels, with a density of 12 lb/ft^3 (192 kg/m^3), has only 7% the weight of solid aluminum. **Foamalum,** of the Foamalum Corp., is an aluminum foam. **Spangle sheet,** of the Aluminum Co. of America, is aluminum sheet with a glittering finish, used for appliances, novelties, and architecture. An aluminum alloy with a large grain is used, and the sheet is etched with an acid to cause individual grains to stand out in relief with mirrorlike facets on each of the irregularly positioned grains. The sheet reflects light in varying degrees and appears as a spangle of continuously changing patterns.

Anodized aluminum, originally called **Eloxal** in Germany, is aluminum with a hard surface of aluminum oxide produced by electrolysis with the metal as the anode. Coatings are from 0.0001 to 0.008 in (0.003 to 0.20 mm) thick, are wear-resistant, and also protect the metal from further oxidation. The coating is a nonconductor of electricity and may be used as an insulator. Oxalic acid was used in the bath for the original Eloxal process, but sulfuric and chromic acids are also used. Colors are obtained either by dyeing the coating or by adding various metals to the bath to produce the colors. The natural oxide film on aluminum is less than a millionth of an inch thick. An anodized film of 0.0004 in (0.01 mm) is necessary for corrosion and wear resistance, and for dye coloring a coating of 0.0008 in (0.02 mm) is necessary. Heavy coatings for wear and electrical resistance may be 0.006 in (0.0152 mm) thick. The wear- resistant coatings of Anodic, Inc., are 0.002 in (0.051 mm) thick, above and below the surface of the aluminum, increasing the sheet thickness 0.001 in (0.025 mm). **Alumilite** is anodized aluminum of the Aluminum Co. of America, and **Alzak** is a bright anodized sheet for reflectors with the thickness of coating less than 0.0002 in (0.005 mm), giving a light reflectance of 79%. Aluminum is difficult to electroplate by ordinary methods because it is far removed from hydrogen in the electromotive series, and water solutions always contain ionic hydrogen. Aluminum plating baths may contain lithium hydride or lithium aluminum hydride in an ethyl ether solution of anhydrous aluminum chloride, and the anodes are always of aluminum metal.

ALUMINUM ALLOY. Originally the term aluminum alloy, when not further qualified, referred to **aluminum-copper alloys** with or without small amounts of other alloying elements. The original **Duralumin,** of the Durener Metallwerke, contained 4% copper, 0.5 magnesium, and 0.5 manganese. Copper hardens and strengthens aluminum, and also gives age-hardening properties, especially when a small amount of magnesium is present. This alloy rolls well, but is subject to intercrystalline corrosion unless anodized or clad with pure aluminum. The annealed alloy has a tensile strength of 28,000 lb/in^2 (193 MPa), elongation 22%, and Brinell hardness 50.

The influence of even very small amounts of other elements on aluminum makes possible an infinite variety of alloys. The numbering system adopted by the Aluminum Assoc. for aluminum and aluminum alloys is by group classification. The 1000 series is for aluminum of over 99% purity, the last two digits indicating the minimum aluminum percentage. Thus, **aluminum 1030** is an aluminum of 99.30% min purity, and **aluminum 1085** is an aluminum of 99.85% min purity. The first-digit designations for the various aluminum-alloy groups are copper, 2; manganese, 3; silicon, 4; magnesium, 5; magnesium and silica, 6; zinc, 7; other elements, 8; an unused series, 9. The last two of the four digits identify the alloying elements in a particular alloy group. The second digit indicates modifications of the original alloy: a zero denotes the original alloy; integers 1 to 9 are assigned consecutively to alloy modifications. Experimental alloys carry the standard four-digit number preceded by X until the alloy becomes standard.

Alloys in the 1, 3, and 5 series, which owe their strength to the hardening effects of manganese and magnesium, respectively, are the non-heat-treatable classes of aluminum alloys. However, they can be strengthened by cold work. Alloys in the 2, 6, and 7 series are the heat-treatable classes. Although some of the alloys in the 4 series are heat-treatable, most of them are used only for brazing sheet and welding wire.

A standard system of letters and numbers is used to indicate the processed condition of aluminum alloys. The designation, called *temper,* follows the alloy-identification number. The word temper is used here in a broader sense than in its connection with the heat treatment of steel. In relation to aluminum alloys, temper covers the heat-treated and/or production or fabricated condition of the alloy. The letter H, for example, indicates cold work (strain hardening), and the letter T indicates the heat-treated condition. In addition, a number following the letter specifies the degree and/or combination of cold work and heat treatment. Thus the temper designation also gives a rough indication of an alloy's mechanical properties.

The finish of as-supplied wrought-aluminum materials is also designated by a standard system of letters and numerals. Finishes are classified into three major groups: mechanical finishes, chemical finishes, and coatings. Each of these groups is designated by a letter, and specific finishes in each group are identified by a two-digit number. The sequence of operations leading to the final finish can be indicated by using more than one designation.

The initial mechanical properties of non-heat-treatable aluminum alloys depend on the hardening and strengthening effect of elements such as manganese, silicon, iron, and magnesium, which are present alone or in various combinations. These alloys can be further hardened by cold work. There are three major non-heat-treatable groups—the 1000 series, the 3000 series, and the 5000 series. **Aluminum alloy 1000 series** includes **commercially pure aluminum,** which has a minimum aluminum content of 99%. The major difference between grades in the series is the level of two impurities—iron and silicon. The electric conductivity grade, designated EC, with a 99.45% aluminum content, is widely used as an electric conductor on the form of wire and busbars. The 1000-series alloys are especially noted for high electric and thermal conductivity and excellent corrosion resistance. In the annealed condition, they are relatively soft and weak. However, an annealed strength of 13,000 lb/in^2 (89 MPa) can be doubled by cold work, with a sacrifice in ductility. Typical uses of commerically pure aluminum grades are sheet metal, foil, spun ware, chemical equipment, and railroad tank cars.

Aluminum alloy 3000 series has manganese (up to about 1.2%) as the major alloying element. **Aluminum-manganese** alloys provide a moderate improvement over the 1000 series in mechanical properties without significant loss of corrosion resistance and workability. **Aluminum alloy 3004** also contains magnesium, giving additional strength improvement. Typical applications of this series are housings, cooking utensils, sheet metals, and storage tanks. Magnesium is the main alloying element in the **aluminum alloy 500** series, ranging from less than 1 up to about 5%. The **aluminum-magnesium** alloys are the strongest of the non-heat-treatable alloys, with strength increasing with magnesium content. However, high-magnesium grades are more difficult to hot-work and are susceptible to stress corrosion above 150°F (66°C). Because of the presence of lower-density magnesium, the density of these alloys is less than that of pure aluminum. They are particularly resistant to marine atmospheres and various types of alkaline solutions. Typical applications are marine hardware, building hardware, appliances, welded structures and vessels, and cryogenic equipment. Grades in the lower part of the magnesium range, known as **lurium,** take a brilliant polish and are used for reflectors and costume jewelry.

Heat-treatable aluminum alloys develop their final mechanical properties through the solid-solution hardening effects of alloying elements as well as by second-phase precipitation. Cold working is also sometimes employed to obtain optimum properties. Two treatments are generally involved in the hardening process—solution heat treatment and age or precipitation hardening. The solution-heat-treatment cycle consists of: (1) heating the alloy up to between 800 and 1000°F (427 and 540°C) and holding it at this temperature to allow the alloying elements to go into solid solution; and (2) quenching the alloy rapidly to hold the alloying element in solution. This treatment disperses the hardening element uniformly throughout the material. Age hardening, performed after solution heat treatment, is done either at room temperatures over an extended period (natural aging) or at a temperature somewhere between 240 and 450°F (115 and 230°C) over a shorter period (artificial aging). This treatment further increases strength and hardness by the precipitation of hard, second-phase particles. The second phase, an intermetallic compound, appears as a fine network on the grain boundaries. Although loss of ductility is not great, the second phase penetrates the surface aluminum-oxide layer and consequently lowers corrosion resistance.

The **aluminum alloy 2000 series** is the oldest and probably the most widely used aluminum alloy series. The principal alloying element, copper, combines with the base metal, aluminum, during heat treatment to form the hardening intermetallic compound $CuAl_2$. Copper content ranges from about 2 to 6%. Until the introduction of the 7000 series, these **aluminum-copper** alloys were the highest-strength aluminum alloys. They possess relatively good ductility but are more susceptible to corrosion than other aluminum alloys, particularly in the aged condition. Also, except for 2014 and 2219, they have limited weldability. Because of the corrosion problem, the sheet forms are often clad with commercially pure aluminum or special aluminum alloys. Because they can be heat-treated to strengths up to 75,000 lb/in^2 (517 MPa), these aluminum-copper alloys are used in structural applications. **Aluminum alloy 2014** and **aluminum alloy 2024,** the best known of the series, are widely used in the aircraft industry. **Aluminum alloy 2014** is primarily a forging grade. **Aluminum alloy 2024** was the first high-strength lightweight alloy.

Alloys in the **aluminum alloy 6000 series** contain silicon and magnesium in approximately equal amounts up to about 1.3%. Small amounts of other metals, such as copper, chromium, or lead, are also present in some of the alloys to provide improved corrosion resistance in the aged condition, or to increase strength or electric conductivity. As a group, **aluminum-silicon-magnesium** alloys in this series have the lowest strengths of the heat-treatables, but possess good resistance to industrial and marine atmospheres. Typical uses include screw machine parts, moderate-strength structural parts, furniture, bridge railings, and high-strength busbars.

The **aluminum alloy 7000 series** has the highest strength of any aluminum-alloy group. High strength is obtained by the addition of 1 to 7.5% zinc and 2.5 to 3.3% magnesium. Chromium and copper also contribute added strength, but tend to lower weldability and corrosion resistance. The wrought alloy, **aluminum alloy 7072,** with 1% zinc, is used as a coating for Alclad products rather than as a structural material. **Aluminum alloy 7001** has up to 8% zinc, about 3 copper, 2 magnesium, with small amounts of silicon, iron, manganese, chromium, and titanium. This alloy has a tensile strength of 37,000 lb/in^2 (255 MPa) with elongation of 17%, and in the T6 temper has a tensile strength of 98,000 lb/ in^2 (675 MPa) with elongation of 9%. **Aluminum alloy 7039,** for weldable tank plate, has 4% zinc and 2 to 3 magnesium. It has a tensile strength of 60,000 lb/in^2 (420 MPa) for space-systems applications where corrosion resistance is important.

Tenzaloy, of the American Smelting & Refining Co., has 8% zinc, 0.8 copper, 0.4 magnesium, and 0.1 nickel. This is similar to **ASTM alloy ZC81A.** The castings have good strength and machinability with heat treatment. The French wrought alloy known as **Zircral** contains 7 to 8.5% zinc, 1.75 to 3 magnesium, 1 to 2 copper, 0.1 to 0.4 chromium, 0.1 to 0.6 manganese, and 0.7 iron and silicon. This alloy is corrosion-resistant and has been used in aircraft construction. The tensile strength of drawn bars is about 80,000 lb/in^2 (551 MPa), with elongation 9%.

Frontier 40E alloy, of the Frontier Bronze Corp., contains 5.5% zinc, 1 iron, 0.5 magnesium, 0.5 chromium, 0.4 copper, 0.2 titanium, 0.3 manganese, and 0.3 silicon. This is essentially **ASTM alloy ZG61A.** It is corrosion-resistant, has a tensile strength of 35,000 lb/in^2 (245 MPa), elongation 4%, and Brinell hardness 70 to 80. **Aluminum alloy 7075** contains 5.1 to 6.1% zinc, 1.2 to 2 copper, 2.1 to 2.9 magnesium, 0.1 to 0.3 manganese, 0.15 to 0.4 chromium, 0.2 max titanium, 0.5 max silicon, and 0.7 max iron. In heat-treated sheet form it has a tensile strength of 77,000 lb/in^2 (530 MPa), yield strength 67,000 lb/in^2 (460 MPa), and elongation 12%.

A **superplastic aluminum alloy** has been developed than can be readily formed into complex shapes by the metal thermoforming process. It is supplied in three thicknesses: 0.60 in (1.5 cm), 0.090 in (0.229 cm), and 0.125 in (0.318 cm).

Aluminum casting alloys are available for all three major types of casting—sand, permanent-mold, and die casting. Both heat-treatable and non-heat-treatable alloys are available. Alloy compositions that do not respond to heat treatment are identified by an F following the alloy-designation number or by omission of a suffix. The heat-treatable alloys, which can be solution heat-treated and aged similar to wrought heat-treatable grades, carry the temper designations T2, T4, T5, T6, or T7. Die castings are seldom solution heat-treated because of the danger of blistering. As is true of most cast metals, the mechancial properties of aluminum castings are considerably lower than those of the wrought forms. With one or two exceptions, tensile strengths in the heat-treated condition do not exceed about 50,000 lb/in^2 (344 MPa). Ductility and hardness are also lower. **Aluminum-silicon** alloys containing 5 to 22% silicon, are the most widely used aluminum-casting alloys, primarily because of their excellent castability. They find considerable application in marine equipment and hardware because of high resistance to saltwater and saline atmospheres. They are also used for decorative parts because of their resistance to natural environments and ability to reproduce detail. They cast well even in thin sections, but the strength decreases and they become more difficult to machine as the silicon increases. Additions of copper increase the strength and improve the machinability, and also add the property of age-hardening, but decrease corrosion and wear resistance. Slight additions of magnesium also give age-hardening by the formation of Mg_2Si. The high-silicon alloys also have good thermal conductivity and a low expansion factor, and are used for engine cylinders and pistons. High-silicon casting alloys, such as **aluminum alloy 356,** with 7% silicon and 0.3 magnesium, have large needle-shaped crystals which make the alloy brittle; but adding up to 0.04% sodium refines the crystals and improves the physical properties.

Hi-Si alloy, of the Aluminum Co. of America, has 20% silicon. When heat-treated it has a hardness up to 740 Brinell. **Vanasil,** of Gillett & Eaton Plant Div., contains 22% silicon, 2.25 nickel, and 0.10 vanadium. It has a coefficient of expansion about equal to that of cast iron, and is used for pistons.

Aluminum die-casting alloys are usually high in silicon because of the casting qualities. **Alcoa 13,** with 12% silicon, has a tensile strength, die-cast, of 39,000 lb/in^2 (268 MPa) with elongation 2%. A modification of this type of high-silicon alloy is used for forgings. **Alcoa 32S,** for forgings, has 12.2% silicon, 0.9 copper, 1.1 magnesium, and 0.9 nickel. The tensile strength of the heat-treated forging is 55,000 lb/in^2 (379 MPa), elongation 9%, and Brinell hardness 120. **Alcoa A132,** for permanent-mold castings, has 12% silicon, 1.2 magnesium, 0.8 copper, and 2.5 nickel. The treated castings have a tensile strength of 47,000 lb/in^2 (324 MPa), elongation 0.5%, and hardness 125. This type of alloy retains its strength well at high temperatures, and for many uses is preferred to the sodium-modified alloys.

Tens-50, of Navan, Inc., used for permanent-mold castings to replace forgings for aircraft pylons, impeller blades, and missile fins, contains 8% silicon, 0.5 magnesium, 0.2 titanium, and 0.3 beryllium. Iron, copper, and zinc are kept at low maximums to avoid embrittlement, and the alloy is treated with 0.4% sodium. The tensile strength is 50,000 lb/in^2 (344 MPa), yield strength 44,000 lb/in^2 (303 MPa), and Rockwell hardness E88. **Reynolds alloy 357,** for high-strength castings, has 7% silicon, 0.5 magnesium, and 0.15 titanium, with the iron, copper, manganese, and zinc kept low. Castings with a T6 treatment have a tensile strength of 50,000 lb/in^2 (344 MPa) and elongation of 8%. The two most used die-casting alloys for commercial castings are **ASTM alloy S12B** and **ASTM alloy SC84B.** The first contains 12% silicon, 2 iron, 0.6 copper, 0.5 zinc, 0.35 manganese, 0.1 magnesium, 0.5 nickel, 0.15 tin, and the balance aluminum. The tensile strength is 43,000 lb/in^2 (296 MPa) and elongation 2.5%. The second has less silicon, 8.5%; more copper, about 3.5%; and more zinc, about 1%. It has a tensile strength of 46,000 lb/in^2 (317 MPa), and it machines more easily.

Aluminum-copper alloys, the earliest aluminum-casting alloys, have been largely replaced by **aluminum-copper-silicon** alloys. Copper increases strength, hardness, and machinability; and the silicon provides excellent casting properties. These alloys are especially suited to the production of castings of intricate design with large differences in section thickness or requiring pressure tightness. **Aluminum-magnesium** alloys containing up to about 10% magnesium are the most corrosion-resistant of the casting alloys. Unfortunately, they are difficult to cast, and controlled melting and pouring practices are required. Castability is improved by the addition of silicon. Heat treatment produces mechanical properties that make them attractive for such applications as automotive and aircraft parts. **Aluminum-zinc-magnesium** alloys age at room temperature to provide relatively high tensile strength. They have good machinability and corrosion resistance, but are not recommended for elevated temperatures. **Aluminum-tin** alloys, developed primarily as bearing alloys, have high load-carrying capacity and fatigue strength. Cast in sand or permanent molds, they are used for connecting rods and crankcase bearings.

ALUMINUM BRASS. A **copper-zinc** alloy containing 1 to about 2% aluminum. There are two distinct types of aluminum brass. The first is a casting brass in which a small amount of aluminum acts as a flux to eliminate impurities and give the brass greater fluidity for intricate castings. The addition of aluminum also permits the use of higher percentages of lead up to about 5%, making the castings easy to machine. The second type is wrought brass modified with aluminum, producing alloys with properties between the brasses and the aluminum bronzes. Even slight additions of aluminum, dispersed as metal in the alloy and not as oxide, improve the oxidation resistance of brasses, and brass with as little as 0.10% of aluminum will have a bright color when extruded or forged. Larger amounts increase the strength and hardness, but decrease the ductility.

A 60–40 brass containing 1% aluminum has its strength increased about 30% and its hardness about 25%. **Revalon,** of Revere Copper & Brass, Inc., contains 76% copper, 22 zinc, and 2 aluminum, and is used for condenser tubes. The soft alloy has a tensile strength of 62,000 lb/in^2 (427 MPa), elongation 50%, Rockwell B hardness 33; while the hard- drawn material has a tensile strength of 83,000 lb/in^2 (572 MPa), elongation 15%, and hardness 86. It has 22.5% of the electric conductivity of copper, and is very corrosion-resistant. **Ambraloy 927,** of the American Brass Co., has the same composition and properties. This alloy is known in England as **high-tensile brass. Ad-aluminum,** of the Chase Brass & Copper Co., has 82% copper, 15 zinc, 2 aluminum, and 1 tin. **Alcunic,** of the Scovill Mfg. Co., is a 70–30 brass modified by replacing some zinc with 2% aluminum and 1 nickel.

ALUMINUM BRONZE. A **copper-aluminum** alloy with aluminum as the chief alloying element, with or without other alloying materials. Plain additions of aluminum to copper increase the strength up to three times that of the original copper, and change the color from red to pale gold. The commercial alloys usually contain from 5 to 10% aluminum. Single-phase alpha alloys up to 7% aluminum have the structure of pure copper with the aluminum in solid solution, and the alloy is tough and ductile. At about 7.5% aluminum the structure changes from homogeneous to duplex, and the alloy becomes increasingly hard and difficult to cold- work.

All of the alloys are resistant to corrosion. They can be cast or forged, but the high-aluminum alloys are difficult to machine because of free aluminum oxide present. The duplex alloys can be hardened by quenching from a high temperature, and drawn. Aluminum bronze is used for high-strength and nonmagnetic parts, condenser tubes, and corrosion-resistant chemical equipment. The hard crystals in a soft matrix make it useful for bearings. The alloys high in aluminum are used for architectural castings to contrast in color with aluminum-silicon alloys.

Standard aluminum bronze for castings contains 8% aluminum. It has a tensile strength, annealed, of 76,000 lb/in^2 (523 MPa), elongation 55%, and Brinell hardness 125. When hardened, the tensile strength is 134,000 lb/in^2 (923 MPa), elongation 13%, and hardness 240. The density is 0.293 lb/in^3 (8110 kg/m^3) and the melting point 1940°F (1060°C). **Ambraloy 298,** of the American Brass Co., and **Revere alloy 430,** of the Revere Copper & Brass, Inc., have this composition but are wrought alloys. The hard-drawn rod has a tensile strength of 125,000 lb/in^2 (861 MPa), with elongation 5%. **Revere alloy 429** has 95% copper and 5 aluminum. In hard-drawn condenser tubes it has a tensile strength of 70,000 lb/in^2 (482 MPa), and elongation 25%. **Atlas 90,** of Ampco Metals, Inc., contains 10% aluminum, and has a tensile strength of 90,000 lb/in^2 (620 MPa) min.

Additions of iron to aluminum bronze increase the strength, refine the grain, reduce the tendency to self-anneal, and improve forging qualities. The **copper-aluminum-iron** alloys are used for cast dies, gears, and strong wear-resistant parts. They have a hardness up to 325 Brinell. **McGill metal,** of the McGill Mfg. Co., Inc., is the name of a group of these alloys. A typical analysis is 89% copper, 9 aluminum, and 2 iron. This alloy has a tensile strength up to 90,000 lb/in^2 (620 MPa), elongation 10 to 20%, and Brinell hardness 160. It is a casting metal, but can be forged, and it machines about the same as medium-carbon steel. A casting alloy used for aircraft engine parts, with 10% aluminum and 1 iron, has a tensile strength of 75,000 lb/in^2 (517 MPa) and Brinell hardness 100. This is **Lumen alloy 11-C,** of the Lumen Bearing Co., and Grade B in Federal specifications, Grade A having more iron and less aluminum. **Navy bronze 46B-186** has 7 to 9% aluminum and 2.5 to 4.5% iron. This is **SAE alloy 68,** and has a tensile strength of 80,000 lb/in^2 (551 MPa). **Daraloy 437,** of the Darling Valve & Mfg. Co., is a standard Grade A aluminum bronze used for valve stems, facings, and pressure castings. The tensile strength is 80,000 lb/in^2 (551 MPa), elongation 18%, and Brinell hardness 90. **Resistac,** of the American Manganese Bronze Co., has 9% aluminum and 1 iron. When heat-treated, the tensile strength is 90,000 lb/in^2 (620 MPa), and Brinell hardness 90. **Avialite,** a similar composition of the American Brass Co., is used for die-pressed parts and for heat-resistant parts for aircraft engines. **Ampco metal 12,** produced by Ampco Metal, has 8.5 to 9.3% aluminum and 2.5 to 3.25 iron. The annealed metal has a tensile strength of 80,000 lb/in^2 (551 MPa), elongation 40%, and Brinell hardness 131. **Ampco metal 8** is a wrought metal in sheet and rods. It contains 6.5% aluminum, 2.3 iron, 0.25 tin, 0.25 manganese, nickel, and silver, and the balance copper. The tensile strength is 82,000 lb/in^2 (565 MPa), elongation 35%, and Brinell hardness 149. A harder bronze is **Ampco metal 18-22** with 10.3 to 11% aluminum and 3 to 4.25 iron. The tensile strength is 100,000 lb/in^2 (689 MPa), elongation 6%, and Brinell hardness 223. **Ambraloy 930,** of the American Brass Co., is a wrought metal with 8% aluminum and 2.5 iron, having a tensile strength of 125,000 lb/in^2 (861 MPa). The **Tuf-Stuf alloys** of the Mueller Brass Co. have from 1 to 3% iron with 10 to 11 aluminum, but **Tuf Stuf K** contains 80% copper, 10 aluminum, 2.5 iron, 5 nickel, and 1 manganese.

Nickel is also used in aluminum bronzes, especially in those containing iron. It increases the corrosion resistance and produces dense castings suitable as hydraulic castings, but the alloys require more care in casting. **Auromet 55,** of the Aurora Metal Co., contains 76 to 80% copper, 10 to 12 aluminum, 4 to 6 iron, and 4 to 6 nickel. The wrought metal has a tensile strength of 110,000 lb/in^2 (758 MPa), elongation 2%, and Brinell hardness 250. Small additions of titanium give strength to the aluminum bronzes and act as a deoxidizer. The bronzes containing manganese are classed with the **superbronzes.** Lead is sometimes added for bearing bronzes, or for free-cutting casting alloys. It reduces the strength rapdily, and only 1.5% is needed for free cutting, though larger additions may be made to increase frictional qualities.

Calsun bronze, of the American Brass Co., has 2.5% aluminum and 2 tin. When soft, the tensile strength is 50,000 lb/in^2 (344 MPa), and elongation 30%; when hard-drawn into wire, the tensile strength is 135,000 lb/in^2 (945 MPa) and elongation 4%. **Alloy 712,** of the Bridgeport Brass Co., is a strong resilient wrought alloy used for flat springs, diaphragms, and small tubing for instruments. It is an aluminum-silicon bronze containing 3.5% aluminum, 1 silicon, and the balance copper. Rolled to spring temper it has a tensile strength of 114,000 lb/in^2 (785 MPa), elongation 3%, and Rockwell B hardness 97. The electric conductivity is 12% that of copper. When annealed, it has a tensile strength of 65,000 lb/in^2 (448 MPa) with elongation 55%.

ALUMINUM PALMITATE. One of the important **metallic soaps.** A yellow, massive salt, or a fine white powder of the composition $Al(C_{16}H_{31}O_2)_3 \cdot H_2O$, made by heating a solution of aluminum hydroxide and palmitic acid.

It is soluble in oils, alkalies, and benzol, but insoluble in water, and is used in waterproofing fabrics, paper, and leather, and in paints as a drier. In finishing leather and paper it adds to the gloss. It is also used to increase viscosity in lubricating oils. Another material of the same class is **aluminum resinate,** $Al(C_{14}H_{63}O_5)_3$, a brown mass made by heating rosin and aluminum hydroxide. **Aluminum oleate,** $Al(C_{18}H_{33}O_2)_3$, is a white salt of oleic acid used as a drier. **Aluminum stearate,** $Al(C_{18}H_{35}O_2)_3$, is a salt of stearic acid. It is repellent to water and is valued for waterproofing fabrics and as a drier, and in waterproofing concrete and stucco. It is also used to give adherence to dyes, and as a flux in soldering compounds. It is a white fluffy powder of 200 mesh, soluble in oils and in turpentine. Grades high in free fatty acid, up to 22%, do not gel readily and are useful as a flatting agent and suspending medium in paints. Grades low in free fatty acid have a thickening effect on solvents. Those with 5 or 6% are used in lubricating grease, and those with about 8% are used in paints.

ALUMINUM POWDER. Called **aluminum-bronze powder** when alloyed. The flaked powder used for paints is more properly termed **aluminum flake.** Flake powder is made by a stamping process, and used as a pigment in paints and printing inks, in silvering rubber articles, and in plastics. The powder has a high ratio of surface to volume, and the dry powder will ignite easily. All of the grades used for paint contain a major proportion of "very fine," 300 to 400 mesh, with flakes 0.000005 in (0.000127 mm) thick; and they are revolved in a drum with a lubricant to give luster and also leafing properties to form a metallic surface in the paint by capillary attraction. Aluminum powder burns with an intense heat, but the powder used in calorizing and for pyrotechnics and explosives is not flaked and polished, but consists of particles of spherical shape free from grease. This powder is called **granulated aluminum,** although this designation is also given to larger **aluminum pellets** marketed for metallurgical purposes. Aluminum powder of Union Carbide Corp., used to increase thrust in solid-fuel rockets, has a particle size from 20 to 150 μm. The ultrafine spherical powder of the National Research Corp. has particles smaller than the wavelength of visible light. The average diameter is 0.03 μm, and the maximum diameter is 0.1 μm. The high surface area gives the powder high reactivity. Aluminum powders of Aluminum Metallurgical Granules, Inc., are in the form of 99.9% pure fine granular powder to 300 mesh or fibrous "whiskers" with more than a million particles per pound. Some aluminum powders not in leafing form are also marketed for paints and enamels where a uniform dispersion rather than a concentrated surface coat of metal is desired. **Aluminum pigment 584,** of the Metals Disintegrating Co., Inc., is such a powder. **Aluminum grain,** for fast reaction in fuels and incendiaries, consists of irregularly shaped particles in sizes from 0.01 to 0.132 in (0.0254 to 0.335 cm).

Roughly a half dozen different aluminum alloy powders are being used in the production of powder metallurgy P/M parts. The major alloying elements used include copper up to 4%, magnesium up to 1%, zinc and silicon from 0.10 to 1%. One alloy contains 5.6% zinc. Aluminum P/M parts offer natural corrosion resistance, light weight, good electric and thermal conductivity. Strengths range from about 15,000 to 50,000 lb/in^2 (103 to 344 MPa), depending on composition, density, and heat treatment. In general, average fatigue limits are about half those of the wrought alloys. This is directly related to the lower density of P/M parts. Corrosion resistance, however, is not markedly affected by the porosity. Perhaps the largest use of aluminum powder is for oil-impregnated sleeve and spherical bearings.

ALUMINIZED STEEL. Also known as **alumetized steel,** or **calorized steel,** and the **Insuluminum** of the General Electric Co. The coating process consists in dip-coating and diffusing the aluminm into the steel at a temperature of about 1600°F (870°C) to form an aluminum-iron–alloy coating. It is resistant to oxidation and scaling at temperatures to about 1650°F (900°C). The process is now used for wire, sheet, and marine hardware. Variations in smoothness, brightness, depth, and color can be obtained by varying the aluminum alloy employed for the dip. Addition of silicon prevents excessive growth of the brittle FeAl alloy. **Aluminized steel** of the Sylvania Electric Products, Inc., is a cold-rolled steel which has a 0.005-in (0.127-mm) coating of an aluminum-silicon alloy rolled into the surface and then annealed in hydrogen. It has a black surface, and is used for vacuum-tube parts to give thermionic radiation and absorb stray electrons.

AMALGAM. A combination of a metal with mercury. The amalgams have the characteristic that when slightly heated they are soft and easily workable, and they become very hard when set. They are used for filling where it is not possible to employ high temperatures. A native **silver amalgam** found in South America contains 26 to 95% silver. Native **gold amalgams** are found in California and Colombia, and contain about 40% gold. Although native amalgams are chemical combinations of the metals, some of the artificial amalgams are alloys and others are compounds. **Dental amalgams** are prepared by mixing mercury with finely divided alloys composed of varying proportions of silver, tin, and copper.

Cadmium amalgam was formerly employed for filling holes in metals and was called **Evans' metallic cement.** It is a silvery-white compound of the composition Cd_5Hg_8, with about 74% mercury, the excess mercury separating out on standing. It softens at about 100°F (38°C), and can be kneaded like wax, remaining soft for a considerable time and then becoming hard and crystalline. Tin or bismuth may be added. Amalgams with an excess of cadmium are ductile and can be hammered into sheet. **Bismuth**

amalgams are lustrous, very fluid combinations of mercury and bismuth, used for silvering mirrors. They are also added to white bearing metals to make them more plastic, and to fusible alloys to lower the melting point. **Crilley metal** was a self-lubricating bearing alloy containing bismuth amalgam. The binary amalgams of mercury and bismuth are usually too fluid for ordinary use.

The usual quaternary alloy has equal parts of bismuth, mercury, tin, and lead, with the proportion of mercury increased to give greater fluidity. A bismuth amalgam containing bismuth, lead, and mercury was used for lead pencils. A **thallium amalgam,** with 8.5% thallium, which freezes at −76°F (−60°C), is used for thermometers for low readings. **Mackenzie's amalgam** is a two-part amalgam in which each part is a solid but becomes fluid when the parts are ground together in a mortar at ordinary temperatures. One part contains bismuth and mercury, and the other part contains lead and mercury. **Sodium amalgam** contains from 2 to 10% sodium. It is a silvery-white mass which decomposes water and can be used for producing hydrogen. **Potassium amalgam,** made by mixing sodium amalgam with potash, is a true chemical compound, used for amalgamating with other metals.

AMBER. A fossil resin found buried in the countries along the Baltic Sea and in Malagasy. It is employed for making varnishes and lacquers and for ornaments. The original German name for the material was **Glassa,** and in early writings it is referred to by the Greek word **elektron** and the Persian name **karaba.** It was called **vernice** by the Italian painters who used it as a varnish resin. Amber came from a coniferous tree, *Pinus succinifera,* now extinct. It is hard, brittle, and tasteless but with an aromatic odor, and dissolves in acids. It is sometimes transparent, but usually semitransparent or opaque with a glossy surface, yellow or orange in color. It takes a fine polish. When rubbed, it becomes electrically charged. Amber contains succinic acid in a complex form, and the finest specimens are known as **succinite,** although this is the name of an amber-colored garnet. The amber of Malagasy was prized for necklaces and pipe mouthpieces. It is semitransparent, wave-streaked, and honey-colored. Synthetic amber is plasticized phenol formaldehyde or other synthetic resin. **Amberoid** is reclaimed scrap amber pieces compressed into a solid, sometimes mixed with pieces of copal or other resin. It has the same uses as amber. **Amber oil,** distilled from scrap amber, is a mixture of terpenes, and is used in varnish. **Succinic acid,** $(CH_2COOH)_2$, a solid melting at 183°C, is obtained by distilling scrap amber, but is now made by fermentation of tartaric acid, or synthetically from benzene and called **butane diacid.** It is used in foods as an acidifier and taste modifier, and **succinic aldehyde** is used for making plastics. The small, highly polar molecule gives crystalline resins of high strength. The dioctyl sodium sulfo salt of succinic acid is used in the foodstuffs industry to prevent agglomeration of colloids. **Complemix,** of American Cyanamid Co., is this material. **Maleic acid,** $(HCCO_2H)_2$, formed by heating succinic acid, is made commercially from benzene and is used widely for plastics production. It can be obtained by the dehydration of **malic acid** which is a hydroxysuccinic acid. **Pomalius acid,** of Allied Chemical Corp., is a food grade of malic acid used to replace acetic acid as a more powerful acidulating agent, stabilizer, and flavor enhancer.

AMBERGRIS. A solid, fatty, inflammable substance of grayish to black color found in the intestines of some sperm whales or found floating in the ocean. **White amber,** formed by aging in the ocean, is the finest quality. When dried it pulverizes to a fine dust. Ambergris has a peculiar sweet fragrance and is highly valued as a fixative in perfumes. In Asia it is used as a spice, and in Egypt for scenting cigarettes. Lumps weigh from a few ounces up to 600 lb (272.2 kg). Synthetic ambergris, of the Swiss company Firmenick & Co., is a gamma-dihydro ionone, $C_{13}H_{22}O$, which gives the odor, with also **ambreinolide,** a gamma lactone. $C_{17}H_{28}O$, which has fixative power. **Ambropur** is a German synthetic ambergris. When the terpene alcohol **manool,** of the wood of the *Dacrydium* trees, is oxidized it yields an acetal which also has the ambergris odor.

AMETHYST. A violet or purple transparent quartz. The color is due to manganese and iron oxides, and becomes yellow on heating. It has a density of 2.65, and hardness of 7 Mohs. Amethyst is composed of alternate right- and left-hand crystals, and breaks with a rippled fracture instead of the conchoidal fracture of ordinary quartz. The crystals are doubly refractive. One of its uses is for making pivot bearings for instruments. It is the most esteemed of the quartzes for cutting into gemstones, but only deep and uniformly colored stones are used as gems and they are not common. Any large amethyst of deep and uniform color is likely to be synthetic. The chief production of natural amethyst is in Brazil and Uruguay.

AMINES. A large group of chemicals derived from ammonia, NH_3, in which one or more of the hydrogen atoms has been replaced by an organic radical. A **primary amine,** such as **methylamine,** CH_3NH_2, has one ammonia hydrogen substituted. It is used in the tanning industry for unhairing skins, and as a catalyst and solvent in the manufacture of synthetic resins. It is a gas and, like ammonia, is soluble in water and handled in water solution. It is flammable, a 40% solution having a flash point of 20°F (−7°C), and the vapors are explosive in air. A **secondary amine,** dimethylamine $(CH_3)_2NH$, is more effective for unhairing. In water solution it forms a hydrate, $(CH_3)_2NH \cdot 7H_2O$, which has a low freezing point,

$-16.8°C$. The **tertiary amine,** trimethylamine, or **secaline,** $(CH_3)_3N$, is a gas liquefying at 2.87°C. The methylamines are used widely as a source of nitrogen. **Isopropylamine,** $(CH_3)_2CNNH_2$, is used as a replacement for ammonia in many chemical processes, and as a solvent for oils, fats, and rubber. It is a clear liquid of specific gravity 0.686, boiling point 31.9°C, freezing point $-101°C$, and flash point $-15°F$ ($-25°C$). The **Sipenols,** of Alcolac Chemical Co., are a group of amines used as textile lubricants and softeners. They are clear to light yellow liquids and are dimethyl ethanol amines or dibutyl ethanol amines.

Changes in characteristics are obtainable in the amines by altering the position of the **amino group,** NH_2, in the carbon chain. **Hodag Amine 50,** of the Hodag Chemical Corp., is amino ethyl ethanolamine, $HO \cdot CH_2 \cdot CH_2 \cdot NH \cdot CH_2 \cdot CH_2 \cdot NH_2$, a clear liquid with specific gravity of 1.027. It has both primary and secondary amino groups which give versatility for production of plastics, plasticizers, agricultural chemicals, and textile wetting agents.

Fatty acid amines are used as flotation agents, oil additives to prevent sludge, and rubber-mold release agents. **Alamine,** of the General Mills Co., is **lauryl amine, palmityl amine,** or other amines of the composition RNH_2, where R is the fatty acid radical. **Delamin,** of Hercules, Inc., is a series of fatty acid amines produced from tall oil. The **Armeens,** of Armour & Co., are fatty acid amines with a wide range of uses.

Amino acids are organic compounds with both amino and carboxylic functional groups. They occur free and in combined states in nature. In the combined state, they serve as the monomers that form the carbon skeleton of protein polymers. Proteins are natural **polyamides** of various amino acids linked together by amide groups, also called **peptide linkages.** An **amide** is formed by replacing the hydroxyl group, OH, of an organic acid by an amino group to form $CONH_2$. The ease with which even complex radicals can be attached to introduce nitrogen into compounds makes the amines useful for the production of plastics, dyes, pharmaceuticals, explosives, detergents, and other chemicals.

Acetamide, also called **acetic acid amine** and **ethanamide,** is a grayish-white crystalline solid with a melting point of 77 to 81°C, specific gravity 1.139, composition $CH_3 \cdot CO \cdot NH_2$, and slight mousy odor. It is soluble in water and in alcohol. It is used as a liquid flux for soldering on painted or oily surfaces, as an antacid in lacquers and explosives, as a softening agent in glues and leather coatings, and as a nonhazing plasticizer in cellulose nitrate and acetate films. Its ability to dissolve starch and dextrine makes it useful in adhesives for waxy papers. With added corrosion inhibitors, it is used as an antifreeze, a 50% solution in water having a freezing point of $-27.5°C$.

Aramids are synthetic fibers produced from long-chain polyamides (nylons) in which 85% of the amide linkages are attached directly to two aromatic rings. The fibers are exceptionally stable and have good strength, toughness, and stiffness, which is retained well above 300°F (150°C). Two aramids are **Nomex** and **Kevlar.** They have high strength, intermediate stiffness, and are suitable for cables, ropes, webbings, and tapes. Kevlar 49, with high strength and stiffness, is used for reinforcing plastics. Nomex, best known for its excellent flame and abrasion resistance, is used for protective clothing, air filtration bags, and electrical insulation.

AMMONIA. A gas of the formula NH_3, originally called **alkaline air** and **volatile alkali** and later in water solution called **spirits of hartshorn.** It is a by-product in the distillation of coal. It is also produced from natural gas. Ammonia is readily absorbed by water, which at 60°F (16°C) takes up 683 times its own volume of the gas, forming the liquid commonly called ammonia, but which is **ammonium hydroxide,** a colorless, strongly alkaline, and pungent liquid of the composition NH_4OH with a boiling point of 38°C. At 80°F (27°C) it contains 29.4% ammonia in stable solution. It is also known as **ammonium hydrate** and **aqua ammonia,** and is used for the saponification of fats and oils, as a deodorant, for cleaning and bleaching, for etching aluminum, and in chemical processing.

Ammonia gas is used in refrigeration, in nitriding steels, and in the manufacture of chemicals. Chlorine unites with it to form **chloramines** which are used as solvents, chlorinating agents, and disinfectants. The gas does not burn in the air, but a mixture of ammonia and oxygen explodes when ignited. **Anhydrous ammonia** is the purified gas liquefied under pressure, marketed in cylinders. At 20°C the liquid has a vapor pressure of 122.1 lb/in^2 (0.04 MPa). The anhydrous ammonia used for controlled atmospheres for nitriding steel, bright annealing, and for sintering metals contains 90% NH_3, and is oxygen-free. When dissociated by heat, each pound yields 45 ft^3 (1.3 m^3) of hydrogen and 11 ft^3 (0.31 m^3) of nitrogen.

Smelling salts, sometimes referred to as ammonia, and in solution as **aromatic spirits of ammonia,** is **ammonium carbonate,** $(NH_4)_2CO_3 \cdot H_2O$, forming in colorless or white crystals. It was also called **hartshorn salts. Ammonium bicarbonate,** NH_4HCO_3, or **acid ammonium carbonate,** is a water-soluble white crystalline powder used as a source of pure ammonia and carbon dioxide and to decrease density in organic materials by creating voids such as for making foamed rubber and in the food-baking industry. It gasifies completely at 140°F (60°C). **Ammonium gluconate,** $NH_4C_6H_{11}O_7$, is a water-soluble white crystalline powder used as an emulsifier for cheese and mayonnaise and as a catalyst in textile printing.

Hydrazine, NH:NH, is a colorless liquid boiling at 113.5°C and freezing at 2°C. It is used as a propellant for rockets, yielding exhaust products of high temperature and low molecular weight. With a nickel catalyst it decomposes to nitrogen and hydrogen. It is a strong reducing agent, and is used in soldering fluxes. Reacted with citric acid, it produces the anti-tuberculosis drug **cotinazin,** which is isonicotinic acid hydrazine. It is also used as a blowing agent for foamed rubber, and for the production of plastics. For industrial applications it may be used in the form of **dihydrazine sulfate,** $(N_2H_4)_2 \cdot H_2SO_4$, a white crystalline water-soluble flake decomposing at 180°C and containing 37.5% available hydrazine. **Hydrazine hydrate,** $N_2H_4 \cdot H_2O$, is a colorless, water-miscible liquid boiling at 120.1°C, and freezing at −51.7°C. Hydrazine is made by reacting chlorine and caustic soda and treating with ammonia.

The **ammonium radical,** $-NH_4$, has the chemical reaction of an alkali metal and forms many important chemicals. **Ammonium nitrate** is made by the action of nitric acid on ammonium hydroxide. It is a colorless to white crystalline hygroscopic powder of the composition NH_4NO_3, specific gravity 1.725, melting point 170°C, decomposing at 210°C, and soluble in water, alcohol, and alkalies. It is also used in fertilizers, pyrotechnics, dental gas, insecticides, freezing mixtures, and explosives. For use as a slow-burning propellant for missiles it is mixed with a burning-rate catalyst in a synthetic-rubber binder and pressed into blocks. **Riv,** marketed by the Harrison Laboratories as a vapor-phase rust inhibitor, is ammonium nitrate. The British explosive **amatol** is a mixture of ammonium nitrate and TNT, which explodes violently on detonation. The 50–50 mixture can be melted and poured, while the 80–20 mixture is like brown sugar and was used for filling large shells. **Macite,** for treetrunk blasting, is ammonium nitrate coated with TNT, with a catalyst to make it more sensitive. **Akremite,** of the Maumee Chemical Co., is ammonium nitrate and carbon black, used as an explosive in strip mining.

Ammonium perchlorate is another explosive made by the action of perchloric acid on ammonium hydroxide. It is a white crystalline substance of the composition NH_4ClO_4, specific gravity 1.95, is soluble in water, and decomposes on heating. **Nitrogen trichloride,** NCl_3, which forms in reactions of chlorine and ammonia when there is an excess of chlorine, is a yellow oil of specific gravity 1.653, which is highly explosive. **Ammonium sulfate,** $(NH_4)_2SO_4$, is a gray crystalline water-soluble material obtained in the distillation of coal and used as a fertilizer and for fireproofing.

Ammonium chloride, or **sal ammoniac,** NH_4Cl, is a white crystalline powder of specific gravity 1.52, used in electric batteries, in textile printing, as a soldering flux, and in making other compounds. Many salts and metallic soaps are also formed in the same manner as with the alkali metals. **Ammonium vanadate,** NH_4VO_3, is a white to yellow crystalline powder used as a paint drier, in inks, as a mordant for textiles, and in pottery mixes to produce luster. The specific gravity is 2.326, and it decomposes at 210°C. **Ammonium chromate,** $(NH_4)_2CrO_4$, is a bright-yellow water-soluble granular powder used as a textile mordant, in inks, and for the insolubilization of glues. On boiling, the water solution liberates ammonia. At 180°C the powder decomposes to the dichromate. **Ammonium lactate,** $CH_3CHOHCOONH_4$, is a yellowish syrupy liquid with a slight odor of ammonia, used in leather finishing. **Ammonium stearate** is obtainable as a tan-colored waxlike solid, melting at 74°F (23°C). It can be dispersed in hot water, but above 190°F (88°C) it decomposes to ammonia and stearic acid.

AMORPHOUS METALS. Also known as **metallic glasses** and **glassy metals,** they have a noncrystalline structure that is achieved by quenching of metal-metalloid alloys. Unlike normal metals, amorphous metals have no grains or grain boundaries, thus resulting in a tight atomic network. In general they are more corrosion-resistant than crystalline metals, and they have high mechanical strength and hardness. Perhaps of greatest importance are their magnetic properties. The absence of grain boundaries allows magnetic domains to move freely as magnetic field is changed. This results in very low hysteresis and power losses, thus making these materials attractive for transformer cores. Although glasslike, amorphous metals are not brittle. However, being metastable, they cannot be hot-formed.

One class of amorphous metals, known as **Metglas** and produced by Allied Corp., contains base metals such as iron, nickel, and chromium alloyed with metalloids like carbon, phosphorous, boron, and silicon. Some of the alloys have yield strengths above 650,000 lb/in^2 (4,500 MPa) and fracture toughness better than high-strength steels. The material is produced in the form of wire, ribbon, and strip. Typical ribbon dimensions are 1½ in (4 cm) wide and between 0.001 and 0.0025 in (0.02 and 0.063 mm) thick.

AMYL ALCOHOL. A group of monohydroxy, or simple, alcohols, which are colorless liquids and have the general characteristic of five carbon atoms in the molecular chain. **Normal amyl alcohol,** $CH_3(CH_2)_4OH$, called also **fusel oil, grain oil, pentanol,** and **fermentation amyl alcohol,** has a specific gravity of 0.82 and boiling point of 137°C. It is only slightly soluble in water. It is used as a solvent for oils, resins, and varnishes; in the manufacture of amyl acetate; and in rubber vulcanization. **Secondary amyl alcohol** has a differently arranged molecule, $CH_3CHOH(C_3H_7)$. The specific gravity is 0.82 and flash point 80°F (27°C). It is used in the manufacture of secondary amyl acetate for lacquers, and in chemical manufacture. **Tertiary amyl alcohol** has the formula $(CH_3)_2C(OH)C_2H_5$, and a camphorlike odor. The specific gravity is 0.81 and boiling point 102°C. It is highly soluble in water, and soluble in alcohol and ether. It is used as a flavor, and as a plasticizer in paints, varnishes, and cellulose plastics. **Isoamyl**

alcohol, or **isobutyl carbinol,** $(CH_3)_2CHCH_2CH_2OH$, has a flash point above 80°F (27°C). It is used in pharmaceutical manufacture. **Amyl acetate,** $CH_3COOC_5H_{11}$, called **banana oil** because of its odor of bananas, is an ester made by the action of acetic acid on amyl alcohol. It is a colorless oily liquid of specific gravity 0.896 and boiling point 141°C. It is insoluble in water but soluble in alcohol. It is a good solvent and plasticizer for cellulose plastics, and is used in cellulose lacquers and adhesives. It is also used in linoleum and oilcloth, and as a banana flavor.

ANILINE. Also known as **aminobenzene, phenylamine, aminophen,** and **aniline oil** and, when first made, **krystallin** and **kyanol.** A yellowish, oily liquid of the composition $C_6H_5 \cdot NH_2$, boiling at 184.4°C, freezing at −6.2°C, and soluble in alcohol, benzene, and hydrochloric acid. The specific gravity is 1.022. It turns brown in the air, finally oxidizing into a resin. The vapor is toxic, and it is poisonous in contact with the skin, requiring protective handling. Its largest uses are in the making of dyes and rubber chemicals, but it is also used for the production of plastics, drugs, explosives, perfumes, and flavors. With nitric acid as an oxidizer it has been used as a rocket fuel. **Aniline salt** is **aniline hydrochloride,** $C_6H_5NH_2HCl$, coming in white crystalline plates of specific gravity 1.2215, melting at 198°C and soluble in alcohol.

ANNATTO. One of the chief **food colors.** It is a salmon-colored dye made from the pulp of the seeds of the tree *Bixa orellana* of the West Indies and tropical America and Africa. It contains **bixin,** $C_{25}H_{30}O_4$, a dark-red crystalline carotenoid carboxylic acid, and also **bixol,** $C_{18}H_{30}O$, a dark-green oily alcohol. It is more stable than carotene and has more coloring power. Annatto is sometimes called **bixine,** and in West Africa it is called **rocou.** It is soluble in oils and in alcohol. **Annatto paste** is used as a food color especially for butter, cheese, and margarine, but has a tendency to give a slightly mustardy flavor unless purified. It is also used as a stain for wood and silk. Water-soluble colors are made by alkaline extraction, giving orange to red shades. For coloring margarine yellow a blend of annatto and turmeric may be used. **Anattene,** of S. B. Penick & Co., is a microcrystalline powder produced from annatto, giving a range of colors from light yellow to deep orange. It comes either oil-soluble or water-soluble.

A substitute for annatto for coloring butter and margarine, having the advantage that it is rich in vitamin A, is **carrot oil** obtained from the common carrot. The concentrated oil has a golden-yellow color and is odorless and tasteless. **Carex** is a name for carrot oil in cottonseed oil solution used for coloring foods. Many of the fat-soluble coloring matters found in plant and animal products are terpenes that derive their colors from conjugated double bonds in the molecule. The yellow **carotene** of carrots and the red **lycopene** of tomatoes both have the formula $C_{40}H_{56}$, and are tetra terpenes containing 8 isoprene units but with different molecular structures. **Beta carotene,** produced synthetically from acetone by Hoffmann-La Roche, is identical with the natural food color.

A beautiful water-soluble yellow dye used to color foods and medicines is **saffron,** extracted from the dried flowers and tips of the saffron crocus, *Crocus sativas,* of Europe, India, and China. It is expensive, as about 4,000 flowers are required to supply an ounce of the dye. Saffron contains **crocin,** $C_{44}H_{70}O_{28}$, a bright-red powder soluble in alcohol. Both red and yellow colors are obtained from the orange thistlelike heads of the safflower, which are dried and pressed into cakes.

ANODE METALS. Metals used for the positive terminals in electroplating. They provide in whole or in part the source of the metal to be plated, and they are as pure as is commercially possible, uniform in texture and composition, and have the skin removed by machining. They may be either cast or rolled, with the manufacture controlled to obtain a uniform grade and to exclude impurities, so that the anode will corrode uniformly in the plating bath and will not polarize to form slimes or crusts. In some plating, as for white bronze, the anode efficiency is much higher than the cathode efficiency, and a percentage of steel anodes is inserted to obtain a solution balance. In other cases, as in chromium plating, the metal is taken entirely from the solution, and insoluble anodes are employed. Chromium-plating anodes may be lead-antimony, with 6% antimony, or tin-lead, with 7% tin. In addition to pure single metals, various alloys are marketed in anode form. The usual brass is 80% copper and 20 zinc, but other compositions are used, some containing 1 to 2% tin. **Brass anodes** are called **platers' brass. Abaloy anodes,** of the Hanson-Van Winkle-Munning Co., for silvery-white plating, are of copper, tin, and zinc. **Nickelex,** used in England as a plating undercoat for aluminum, contains 90% copper and 10 tin. **Copper anodes** for metal plating are usually hot-rolled oval-shaped bars, 99.9% pure, while those for electrotype deposits may be hot-rolled plates, electrodeposited plates, or cast plates. The **copper ball anodes** of the Udylite Corp. are forged instead of cast to give a finer and more even grain. **Zinc anodes** are 99.99% pure. **Nickel anodes** are 99+% rolled or cast in iron molds, or 97% sand-cast. **Bright nickel** anodes may have 1% or more of cobalt. **Lead anodes** have low current-carrying capacity, and may be made with a sawtooth or multiple-angled surface and ribs to provide more area and give greater throwing power. Anodes of other metals are also made with sections gear-shaped, fluted, or barrel-shaped to give greater surface area and higher efficiency. **Rhodium anodes** are made in expanded-mesh form. **Platinum anodes,** also made in mesh form, have the platinum clad on tantalum wire. Special anode metals are marketed under trade names, usually accenting the color, hardness, and corrosion resistance of the deposited plate.

ANTHRACITE. Also called **hard coal.** A variety of mineral coal found in Wales, France, and Germany, but in greatest abundance in an area of about 500 mi^2 (1,295 km^2) in northeastern Pennsylvania. It is distinguished by its semimetallic luster, high carbon content, and high specific gravity, which is about 1.70. The carbon content may be as high as 95%, but the usual fixed carbon content is from 78 to 84%. It should give 13,200 Btu/lb (72,600 kcal/kg). In theory the best grades of anthracite should have 90% carbon, 3 to 4.5 hydrogen, 2 to 5.5 oxygen and nitrogen, and only 1.7 ash. Anthracite, when pure and dry, burns without smoke or smell, and is thus preferred to bituminous coal for household furnaces. But the coal will absorb a high proportion of water, and commercial coal may be wetted down to add to the weight, thus lessening its efficiency. Hard coal is graded as anthracite and **semianthracite,** depending upon the ratio of fixed carbon to volatile matter. When the ratio is 10:1, it is anthracite.

The commercial gradings of anthracite are chiefly by size, varying from three sizes of very fine grains called silt, rice, and buckwheat, to the large size of furnace, or lump, coal. Standard ASTM sizes for anthracite are: broken, 4⅜ to 3¼ in (11.1 to 8.3 cm); egg, 3¼ to 2⁷⁄₁₆ in (8.3 to 6.2 cm); stove, 2⁷⁄₁₆ to 1⅝ in (6.2 to 4.1 cm); chestnut, 1⅝ to ¹³⁄₁₆ in (4.1 to 2.1 cm); pea, ¹³⁄₁₆ to ⁹⁄₁₆ in (2.1 to 1.4 cm); No. 1 buckwheat, ⁹⁄₁₆ to ⁵⁄₁₆ in (1.4 to 0.8 cm); No. 2 buckwheat (rice), ⁵⁄₁₆ to ³⁄₁₆ in (0.8 to 0.5 cm); No. 3 buckwheat (barley), ³⁄₁₆ to ³⁄₃₂ in (0.5 to 0.2 cm). As the coal comes from the breaker, the proportions are about 8% silt, 9 rice, 15 buckwheat, 10 pea, 24 chestnut, 23 stove, and 8 egg.

ANTIFREEZE COMPOUNDS. Materials employed in the cooling systems and radiators of internal-combustion engines to ensure a liquid circulating medium at low temperatures to prevent damage from the formation of ice. The requirements are that the compound must give a freezing point below that likely to be encountered without lowering the boiling point much below that of water, that it must not corrode the metals or deteriorate rubber connections, that it must be stable up to the boiling point, and that it must be readily obtainable commercially. Calcium chloride was early used for automobile radiators but corroded the metals. It is still used in fire tanks, sodium chromate being added to retard corrosion. Oils were also used, but the high boiling points permitted overheating of the engine, and the oils softened the rubber. Denatured **ethyl alcohol** may be used, but **methanol** is less corrosive and less expensive. A 30% solution of ethyl alcohol in water has a freezing point of about 5°F (−15°C), and a 50%

solution freezes at −24°F (−31°C). Alcohol, however, must be renewed frequently because of loss by evaporation.

Glycerol is also used as an antifreeze, a 40% solution in water lowering the freezing point to about 0°F (−18°C), and a 50% solution to −15°F (−25°C). It has the disadvantage of high viscosity, requiring forced circulation at low temperatures, but it does not evaporate easily. **Ethylene glycol** lowers the freezing point to a greater extent than alcohol and has a high boiling point so that it is not lost by evaporation, but it has a higher first cost and will soften ordinary natural rubber connections. **Acetamide** in water solution may also be used as an antifreeze. Antifreezes are sold under various trade names. **Zerone,** of Du Pont, has a methanol base, while **Zerex** has a base of ethylene glycol. **Prestone,** marketed by the Union Carbide Corp., is ethylene glycol antifreeze. **Pyro** is an antifreeze of the U. S. Industrial Chemicals, Inc. with a low freezing point. **Ramp,** of the Antara Chemical Co., is ethylene glycol with anticorrosion and antifoam agents added. **Antifreeze PFA55MB,** of the Phillips Petroleum Co., used in jet-engine fuels, is ethylene glycol monoethyl ether with 10% glycerin. **Dowtherm 209,** of Dow Chemical Co., is an antifreeze material of inhibited methoxypropanol which boils off without forming gum.

ANTIMONY. A bluish-white metal, symbol Sb, having a crystalline scale-like structure. It is brittle and easily reduced to powder. It is neither malleable nor ductile and is used only in alloys or in its chemical compounds. Like arsenic and bismuth, it is sometimes referred to as a **metalloid,** but in mineralogy it is called a **semimetal.** It does not have the free cloudlike electrons that occur in metal atoms, and thus it lacks plasticity and is a poor conductor of electricity.

The chief uses of antimony are in alloys, particularly for hardening lead-base alloys. The specific gravity of the metal is 6.62, melting point 824°F (440°C), and Brinell hardness 55. It burns with a bluish light when heated to redness in the air. Antimony imparts hardness and a smooth surface to soft-metal alloys; and alloys containing antimony expand on cooling, thus reproducing the fine details of the mold. This property makes it valuable for type metals. When alloyed with lead, tin, and copper, it forms the babbitt metals used for machinery bearings. It is also much used in white alloys for pewter utensils. Its compounds are used widely for pigments. **Antimony red** is the common name of **antimony trisulfide,** Sb_2S_3, also known as **antimony sulfide** and **antimony sulfuret,** found in the mineral stibnite, but produced by precipitation from solutions of antimony salts. It comes in orange-red crystals with a specific gravity of 4.56 and melting point 1015°F (545°C). It is used as a paint pigment, for coloring red rubber, and in safety matches. **Antimony pentasulfide,** Sb_2S_5, an orange-yellow powder, was once used for vulcanizing rubber, and it colored the rubber red.

It breaks down when heated, yielding sulfur and the red pigment antimony trisulfide.

ANTIMONY ORES. The chief ore of the metal antimony is **stibnite,** an impure form of antimony trisulfide, Sb_2S_3, containing theoretically 71.4% antimony. The usual content of the ore is 45 to 60%, which is concentrated to an average of 92% for shipment as matte. Sometimes gold or silver is contained in the ore. Stibnite occurs in slender prismatic crystals of a metallic luster and lead-gray color with a hardness of 2 Mohs. The metal is obtained by melting the stibnite with iron, forming FeS and liberating the antimony, or by roasting the ore to produce the oxide, which is then reduced with carbon. For pyrotechnic uses stibnite is liquated by melting the mineral and drawing off the metal which on cooling and solidifying is ground. Stibnite comes from China, Mexico, Japan, West Germany, Bolivia, Alaska, and the western United States.

Senarmorite, found in Mexico, Nevada, and Montana, is **antimony oxide,** Sb_2O_3, occurring in cubic crystals with a yellow color. The specific gravity is 5.2, hardness 2.5, and theoretical metal content 83.3%. **Valentinite,** also found in the same localities, has the same theoretical formula and antimony content as senarmorite, but has a rhombic crystal structure, a hardness of 3, and a specific gravity of 5.5. These oxides are used as opacifiers in ceramic enamels. **Cervantite,** found in Mexico, Nevada, and Montana, is **antimony tetraoxide,** Sb_2O_4. It has a grayish-yellow color, specific gravity of 5, and hardness of 4.5, and it contains theoretically 79.2% antimony. **Stibiconite,** from the same area, is a massive pale-yellow mineral, $Sb_2O_4 \cdot H_2O$, with a specific gravity 5.1, hardness 4.5, and antimony content 71.8%.

Kermesite, known as **red antimony,** or **antimony blend,** found in Mexico and Italy, is a mineral resulting from the partial oxidation of stibnite. The composition is Sb_2S_2O, and when pure it contains 75% antimony and 20 sulfur. It occurs in hairlike tufts, or radiating fibers of a dark-red color and metallic luster, with hardness 1.5 and specific gravity 4.5. Another sulfide ore of antimony is **jamesonite,** $Pb_2Sb_2S_5$, found in Mexico and the western United States. It has a dark-gray color, specific gravity 5.5, hardness 2.5, and contains 20% antimony. When the ore is silver-bearing, it can be worked profitably for antimony. **Stephanite** is classed as an ore of silver, but yields antimony. It is a **silver sulfantimonite,** Ag_5SbS_4, containing 68.5% silver and 15.2 antimony. It occurs massive or in grains of an iron-black color with a hardness of 2 to 2.5 and specific gravity 6.2 to 6.3. It is found in Nevada, Mexico, Peru, Chile, and central Europe. Much antimony is in lead ores and is left in the lead as hard lead. Antimony is marketed in flat cakes or in broken lumbs. The highest grade of pure refined antimony is known as **star antimony** because of the glittering spangled appearance on the surface, but starring can be done with lower grades of antimony by special cooling of the ingots. **Crude antimony** is not antimony metal, but is beneficiated ore, or ore matte, containing 90% or more of metal. High-grade antimony is +99.8% pure, Standard grade is 99 to 99.8% pure, and Chinese is 99% pure.

ANTIOXIDANT. A material used to retard oxidation and deterioration of vegetable and animal fats and oils, rubber, or other organic products. Antioxidants embrace a wide variety of materials but in general for antioxidant activity the hydroxy groups must be substituted directly in an aromatic nucleus. In the phenol group of antioxidants the hydrogen atoms must be free. In the naphthol group the alpha compound is a powerful antioxidant. Usually, only minute quantities of antioxidants are used to obtain the effect. **Ionol,** an antioxidant, or **oxidation inhibitor,** of the Shell Chemical Co., is a complex butyl methyl phenol used in gasoline, oils, soaps, rubber, and plastics. It is an odorless, tasteless, nonstaining granular powder, insoluble in water, melting at 70°C. In gasoline the purpose of an antioxidant is to stabilize the diolefins that form gums. **Norconidendrin,** an antioxidant for fats and oils, is produced from the high-phenol **confidendrin,** obtained from hemlock pulp liquor. The **Tenox antioxidants,** of Eastman Chemical Co., used for meats and poultry, are mixtures of butyl hydroxyanisole, butyl hydroxytoluene, propyl gallate, and citric acid, in solution in corn oil, glyceryl monooleate, or propylene glycol. **Tenox 2** contains 20% BHA, 6 propyl gallate, 4 citric acid, and 70 propylene glycol. **Tenox 4** contains 20% BHA, 20 BHT, and 60 corn oil. **Tenox HQ,** used to prevent rancidity in margarine, dried milk, and cooking fats, is a purified hydroquinone. **Tenamene,** of the same company, used in rubber, is a complex phenylenediamine. Most of the antioxidants for rubber and plastics are either phenols or aromatic amines. **Lead diamyldithiocarbamate (LDADC),** long used to prolong the life of hydrocarbon-based lubricants, can also be used to inhibit asphalt's cracking with age and exposure to varying climatic conditions.

A **synergist** may be used with an antioxidant for regeneration by yielding hydrogen to the antioxidant. Synergists are acids such as citric or maleic, or they may be ferrocyanides. The presence of small quantities of metallic impurities in oils and fats may deactivate the antioxidants and nullify their effect. **Phytic acid** is not only an antioxidant for oils and foodstuffs, but it also controls the metallic contaminations. It does not break down as citric acid does or impart a taste to edible oils as phosphoric acid does. It occurs in the bran of seeds as the salt **phytin,** $CaMg(C_2H_6P_2O_9)_2$, and is obtained commercially from corn steep liquor. Chemicals used to control metallic ions and stabilize the solutions are called **sequestering agents. Pasac,** of the Sanders Chemical Co., is such an agent. It is **potas-**

sium acid saccharate, $KHC_6H_8O_8$, in the form of a water-soluble white powder. **Sequelene,** of the A. E. Staley Mfg. Co., for treating hard and rusty waters, is a **sodium glucoheptonate.**

Since odor is a major component of flavor, and the development of unpleasant odors in edible fats arises from oxidation, the use of antioxidants is generally necessary, and in such use they are called **food stabilizers.** But degradation of some organic materials may not be a simple oxidation process. In polyvinyl chloride plastics the initial stage of heat degradation is a dihydrochlorination with hydrogen chloride split out of the molecular chain to give a conjugated system subject to oxidation. Materials called **stabilizers** are thus used to prevent the initial release. Traces of iron and copper in vegetable oils promote rancidity, and citric acid is used as a stabilizer in food oils to suppress this action. **Densitol,** of Abbott Laboratories, for stabilizing citrus-fruit beverage syrups, is a brominated sesame oil. It also enhances the flavor, although it has no taste.

Light stabilizers may be merely materials such as carbon black to screen out the ultraviolet rays of light. Most commercial antioxidants for foodstuffs are mixtures, and all the mixtures are synergistic with the total antioxidant effect greater than the sum of the components. **Sustane 3,** of the UOP Chemical Co., is a mixture of butylated hydroxyanisole, propyl gallate, citric acid, and propylene glycol. **Inhibitors** for controlling color in the chemical processing of fats and oils are usually organic phosphates, such as the liquids **triisooctyl phosphate** and **chloroethyl phosphate.** They are mild reducing agents and acid acceptors, and they complex with the metal salts. **Ultraviolet absorbers,** to prevent yellowing and deterioration of plastics and other organic materials, are substituted hydroxybenzophenones. The photons of the invisible ultraviolet rays of sunlight have great energy and attack organic materials photochemically. Ultraviolet absorbers are stable in this light and absorb the invisible rays. **Antirads** are antioxidants that increase the resistance of rubber or plastics to deterioration by gamma rays. Such rays may break the valence bonds and soften a rubber, or cross-link the chains and harden the rubber.

The term **corrosion inhibitors** usually refers to materials used to prevent or retard the oxidation of metals. They may be elements alloyed with the metal, such as columbium or titanium incorporated in stainless steels to stabilize the carbon and retard intergranular corrosion, or they may be materials applied to the metal to regard oxygen attack from the air or from moisture. Many paint undercoats, especially the phosphate and chromate coatings applied to steel, are corrosion inhibitors. They may contain a ferrocyanide synergist. **Propargyl alcohol,** C_2H_4CO, a liquid boiling at 115°C, is used in strong mineral acid pickling baths to prevent hydrogen embrittlement and corrosion of steel. **VPI 260,** of the Shell Chemical Co., is **dicyclohexylamine nitrite,** a white crystalline powder which sublimes to form a shield on steel or aluminum to passivate the metal and make it resistant to moisture corrosion. VPI means **vapor-phase inhibitor. VPI paper** is wrapping paper impregnated with the nitrite, used for packaging steel articles.

ANTISLIP METALS. Metals with abrasive grains cast or rolled into them, used for floor plates, stair treads, and car steps. They may be of any metal, but are usually iron, bronze, or aluminum. The abrasive may be sand, but it is more usually a hard and high-melting-point material such as aluminum oxide. In standard cast forms antislip metals are marketed under trade names. **Alumalun** is the name of an aluminum alloy cast with abrasive grains, made by the American Abrasive Metals Co. **Bronzalum** is a similar product made of bronze. The **Algrip steel** of the Alan Wood Steel Co. is steel plate ⅛ to ⅜ in (0.32 to 0.95 cm) thick, with abrasive grains rolled into one face. It is used for loading platforms and ramps.

ANTLER. The bony, deciduous horns of animals of the deer family, used for making handles for knives and other articles but now replaced commercially by plastic moldings. Antlers are true outgrowths of bone, and are not simply hardenings of tissue as are the horns of other animals. Unlike horn, antlers are solid, and have curiously marked surfaces. They are of various shapes and sizes, and are usually found on the male during the mating season, although both sexes of reindeer and American caribou possess them. They grow in from 3 to 4 months, and are shed annually.

ARGENTITE. An important ore of silver, also called **silver glance.** It has the composition Ag_2S, containing theoretically 87.1% silver. It usually occurs massive, streaked black and lead gray, with a metallic luster and a hardness of 2 to 2.5. It is found in Nevada, Arizona, Mexico, South America, and Europe. **Argyrodite** is another silver ore found in Bolivia, and is a source of the rare metal germanium. When pure, it has the composition $4Ag_2S \cdot GeS_2$, and contains 5 to 7% germanium. A similar mineral, **canfieldite,** found in Bolivia, has 1.82% germanium and some tin.

ARGOLS. Also called **wine lees.** A reddish crust or sediment deposited from wine, employed for the production of tartaric acid, cream of tartar, and rochelle salts. It is crude **potassium acid tartrate,** or **cream of tartar,** $KH(C_4H_4O_6)$. When grape fermentation is complete, the wine is drawn off and placed in storage tanks where the **lees** settle out. The amount of tartrate varies in different types of wine, from 0.1 to 1.0 pound of cream of tartar per gallon (0.01 to 0.12 g/cm^3). From wines clarified by refrigeration, as much as 1 to 3 lb/gal (0.12 to 0.36 g/cm^3) of tartrate crystallizes out. Cream of tartar is also obtained from **grape pomace,** which is the residue skins, seeds, and pulp, containing 1 to 5% tartrate. **Wine stone** is cream of tartar, 70 to 90% pure, which crystallizes on the walls of wine storage tanks. Purified cream of tartar is a colorless to white crystalline powder of specific gravity 1.956, soluble in water, and used in baking powders.

Tartaric acid is a colorless crystalline product of the composition $HOOC(CHOH)_2COOH$, which has a melting point of 170°C and is soluble in water and in alcohol. It has a wide variety of uses in pharmaceuticals, in effervescent beverages, and as a mordant in dyeing. The pods of the tamarind tree, *Tamarindus indica,* of India, contain 12% tartaric acid and 30 sugars. They are used in medicine and for bevarages under the name of **tamarind. Rochelle salts** is **potassium sodium tartrate,** $KNa(C_4H_4O_6) \cdot 4H_2O$, a colorless to bluish-white crystalline solid of specific gravity 1.79 and melting point 75°C, which is soluble in water and in alcohol. It is used in medicines and in silvering mirrors. Like quartz, it is doubly refractive, and is used in piezoelectric devices where the water solubility is not a disadvantage.

ARSENIC. A soft, brittle, poisonous element of steel-gray color and metallic luster, symbol As. The melting point is 850°C, and specific gravity 4.8. In atomic structure it is a semimetal, lacking plasticity, and is used only in alloys and in compounds. When heated in the air, it burns to **arsenious anhydride** with white odorous fumes. The bulk of the arsenic used is employed in insecticides, rat poisons, and weed killers, but it has many industrial uses, especially in pigments. It is also used in poison gases for chemical warfare. The white, poisonous powder commonly called arsenic is **arsenic trioxide,** or **arsenious oxide,** As_2O_3, also known as **white arsenic.** When marketed commercially, it is colored pink to designate it as a poison. White arsenic is marketed as Refined, + 99% pure, High-grade, 95 to 99%, and Low-grade, −95%. Arsenic is added to antimonial lead alloys and white bearing metals for hardening and to increase fluidity, and to copper to increase the annealing temperature for such uses as radiators. It is also used in lead shot to diminish cohesion, and small amounts are used as negative electron carriers in rectifier crystals.

Arsenic acid is a white crystalline solid of the composition $(H_3AsO_4)_2 \cdot H_2O$, produced by the oxidation of white arsenic with nitric and hydrochloric acids. It is soluble in water and in alcohol, has a specific gravity of 2 to 2.5, and a melting point of 35.5°C. Arsenic acid is sold in various grades, usually 75% pure, and is used in glass manufacture, in printing textiles, and in insecticides. The **arsines** comprise a large group of alkyl compounds of arsenic. They are **arsenic hydrides,** AsH_3, a colorless gas. The primary, secondary, and tertiary arsines are not basic but the hydroxides are strongly basic. The arsines are easily oxidized to **arsonic acid,**

$RAsO_3H_2$, and related acids. **Arsenic disulfide,** also known as **ruby arsenic, red arsenic glass,** and **red orpiment,** is an orange-red, poisonous powder with specific gravity 3.5, and melting point 307°C, obtained by roasting arsenopyrite and iron pyrites. The composition is As_2S_2. It is employed in fireworks, as a paint pigment, and in the leather and textile industries. Another arsenic sulfur compound used as a pigment is **orpiment,** found as a natural mineral in Utah, Peru, and central Europe. It is an **arsenic trisulfide,** As_2S_3, containing 39% sulfur and 61 arsenic. The mineral has a foliated structure, a lemon-yellow color, and a resinous luster. The specific gravity is 3.4, hardness 1.5 to 2, and melting point 300°C. Artificial arsenic sulfide is now largely substituted for orpiment, and is referred to as **king's yellow.**

ARSENIC ORES. **Arsenopyrite,** also called **mispickel,** is the most common ore of arsenic. It is used also as a source of white arsenic, and directly in pigments and as a hide preservative. The composition is FeAsS. It occurs in crystals or massive, of a silvery-white to gray-black color and a metallic luster. The specific gravity is 6.2 and hardness 5.5 to 6. Arsenic is usually not a primary product from ores, but is obtained as a by-product in the smelting of copper, lead, and gold ores. A source of white arsenic is the copper ore **enargite,** $Cu_2S \cdot 4CuS \cdot As_2S_3$, theoretically containing 48.3% copper and 19.1 arsenic. It occurs in massive form with a hardness of 3 and specific gravity of 4.45, and is gray, with a pinkish variety known as **luzonite.** The mineral is commonly intertwined with **tennantite,** $5Cu_2S \cdot 2(CuFe)S \cdot 2As_2S_3$, a mineral of a gray to greenish color. **Realgar,** known also as **ruby sulfur,** is a red or orange arsenic disulfide, As_2S_2, occurring with ores of lead and silver in monoclinic crystals. The hardness is 1.5 and specific gravity 3.55. It is used as a pigment. Another ore is **smaltite,** or **cobalt pyrites,** $CoAs_2$, occurring in gray masses of specific gravity 6.5 and hardness 5.5. It occurs with ores of nickel and copper. It may have nickel and iron replacing part of the cobalt, and is a source of cobalt, containing theoretically 28.1% cobalt.

ASBESTOS. A general name for several varieties of fibrous minerals, the fibers of which are valued for their heat-resistant and chemical-resistant properties, and are made into fabrics, paper, insulating boards, and insulating cements. The long fibers are used for weaving into fireproof garments, curtains, shields, and brake linings. The short fibers are compressed with binders into various forms of insulating boards, shingles, pipe coverings, paper, and molded products. The original source of asbestos was the mineral actinolite, but the variety of **serpentine** known as chrysotile now furnishes most of the commercial asbestos. **Actinolite** and **tremolite,** which furnish some of the asbestos, belong to a great group of widely distributed minerals known as **amphiboles** which are chiefly metasilicates of calcium and magnesium, with iron sometimes replacing part of the magnesium. They occur granular, in crystals, compact such as **nephrite,** which is the jade of the Orient, or in silky fibers as in the iron amphibole asbestos mined in the United States. This latter type is more resistant to heat than chrysotile. Its color varies from white to green and black.

Jade occurs as a solid rock and is highly valued for making ornamental objects. Jade quarries have been worked in Khotan and Upper Burma for many centuries, and large pebbles are also obtained by divers in the Khotan River. The most highly prized in China was white speckled with red and green and veined with gold. The most valued of the **Burma jade** is a grass-green variety called **Ayah kyauk.** Most jade is emerald green, but some is white and others are yellow, vermilion, and deep blue. This form of the mineral is not fibrous.

Asbestos is a hydrated metal silicate with the metal and hydroxyl groups serving as lateral connectors of the molecular chain to form long crystals which are the fibers. The formula for **chrysotile** is given as $Mg_6Si_4O_{11}(OH)_6 \cdot H_2O$. Each silicon atom in the Si_4O_{11} chain is enclosed by a tetrahedron of four oxygen atoms so that two oxygen atoms are shared by adjacent tetrahedra to form an endless chain. When the crystal orientation is perfect, the fibers are long and silky and of uniform diameter with high strength. When the orientation is imperfect, the Si_4O_{11} chain is not parallel to the fiber axis and the fibers are uneven and harsh. In chrysotile the metal connector is magnesium with or without iron, but there are at least 30 other different types of asbestos.

Chrysotile is highly fibrous, and is the type most used for textiles. The fiber is long and silky, and the tensile strength is from 80,000 to 200,000 lb/in^2 (551 to 1,378 MPa). The color is white, amber, gray, or greenish. The melting point is 2770°F (1521°C), and specific gravity is 2.4 to 2.6. Chrysotile is mined chiefly in Vermont, California, Quebec, Arizona, Turkey, and Zimbabwe. Only about 8% of the total mined is long spinning fiber, the remainder being too short for fabrics or rope. The Turkish fiber is up to ¾ in (1.9 cm) in length. Asbestos produced in Quebec is chrysotile occurring in serpentized rock in veins ¼ to ½ in (0.64 to 1.27 cm) wide, though veins as wide as 5 in (12.7 cm) occur. The fibers run crosswise of the vein, and the width of vein determines the length of fiber. **Calidria asbestos,** of Union Carbide Corp, is short-fiber chrysotile from California marketed as fibrils or pellets for use as a reinforcing agent in plastic laminates, thickening agent, and opacifier in coatings and adhesives. Chrysotile fiber has about 14% water of crystallization. At temperatures near 1800°F (980°C), it loses its water, and the dehydration has a cooling effect. Thus **chrysotile felt** is used as a heat sink in missile and space construction laminates.

Blue asbestos, from South Africa, is the mineral **crocidolite,** $NaFe(SiO_3)_2 \cdot FeSiO_2$. The fiber has high tensile strength, averaging 600,000 lb/in^2 (4,080 MPa), is heat resistant to 1200°F (650°C), and is resistant to most chemicals. The fibers are ⅛ to 3 in (0.32 to 7.6 cm) long with diameters from 0.06 to 0.1 in (0.15 to 0.25 cm). It is compatible with polyester, phenolic, and epoxy resins, and is available in all standard forms for reinforcement of plastics and insulating uses. A molding powder is made by mixing the fibers with epoxy and, after partial curing, grinding the mixture to a free-flowing powder. **Form Pack 2,** of Devcon Corp., is this fiber impregnated with Teflon for packing valves and pistons for use at temperatures to 500°F (260°C).

The classes of **cape asbestos** from South Africa are chrysotile, **amosite,** and **Transvaal blue.** Amosite has a coarse, long, resilient fiber, and is used chiefly in insulation, being difficult to spin. It comes in white and dark grades, and the fibers are graded also by length from ⅛ to 6 in (0.32 to 15.2 cm). It has a chemical resistance slightly less than crocidolite and tensile strength of 200,000 lb/in^2 (1,378 MPa). The name amosite was originally a trade name for South African asbestos, but now refers to this type of mineral. Transvaal blue is a whitish, iron-rich, **anthophyllite** $(MgFe)SiO_3$, noted for the length of its fiber. The best grades are about 1½ in (3.8 cm) long. The fibers are resistant to heat and to acids, and the stronger fibers are used for making acid filter cloth and fireproof garments. This type of asbestos is also found in the Appalachian range from Vermont to Alabama. Canadian, Vermont, and Arizona asbestos is chrysotile; that from Georgia and the Carolinas is anthophyllite.

Canadian asbestos is graded as crude, mill fibers, and shorts. Crudes are spinning fibers ⅜ in (0.95 cm) or longer, hand-cobbed. Mill fibers are obtained by crushing and screening. Shorts are the lowest grades of mill fibers. **Rhodesian asbestos** is graded in five C & GS grades separated by screen boxes. Kenya asbestos is anthophyllite, and that from Tanzania is largely amphilbole. **Nonspinning asbestos** is graded as shingle stock, ¼ to ⅜ in (0.38 to 0.95 cm); paper stock, ⅛ to ¼ in (0.32 to 0.38 cm); and shorts, 1/16 to ⅛ in (0.16 to 0.32 cm). The shorts are washed and ground for use as resistant filler in molded plastics. In England this material is known as **micro asbestos.**

Asbestos fabrics are often woven mixed with some cotton. For brake linings and clutch facings the asbestos is woven with fine metallic wire. **Asbeston,** of Uniroyal, Inc., is full asbestos fabric in various weaves and weights. The 8-oz open-weave fabric is used for cable insulation. The close-woven 16-oz cloth is used for aircraft air ducts. The 20-oz fabric is used for fireproof clothing. An **absorbent fabric** used for wiping cloths is made from 20% asbestos fiber and 80 cotton. The dyed fabric shows color only in the cotton since the asbestos does not absorb the dye. The **Colorbestos**

drapery fabric of Johns-Manville has the warp of cotton with the asbestos suronding the warp threads so that the fabric is flame- proof. It weighs 0.92 lb/yd in plain weave, and 1.1 lb in pattern. **Asbestos wick,** used for calking, is made of strands of carded long asbestos fiber twisted in the form of a soft rope, usually ¼ in (0.64 cm) in diameter. **Asbestos rope** is larger-diameter material made by twisting or braiding several strands of wick. **Caposite,** of the North American Asbestos Corp., is rope ½ to 2 in (1.3 to 5.1 cm) in diameter made of twisted rovings of long-staple asbestos covered with a braided jacket of asbestos yarn. It is used for pipe, valve, and joint insulation, and for furnace door packing. **Asbestos felt,** used for insulation, is usually made by saturating felted asbestos with asphalt, although synthetic rubber or other binder may be used. The **Pyrotex felt** of Raybestos-Manhattan, Inc., has no binder but is made from a mixture of long-fiber chrysotile and a ceramic fiber.

Asbestos shingles, for fireproof roofing and siding of houses, are normally made of asbestos fibers and portland cement formed under hydraulic pressure. They are in the natural gray color or are colored with black or green pigments. Another type of asbestos used for some insulation is **paligorskite,** known as **mountain leather,** found in Alaska. It is a complex mineral which may be an alteration product of several asbestos minerals. It absorbs moisture and is thus not suited to the ordinary uses of asbestos, but it can be reduced to a smooth pulp which is molded with a resistant plastic binder into a lightweight insulating board.

Asbestos board is a construction or insulating material in sheets made of asbestos fibers and portland cement molded under hydraulic pressure. Ordinary board for the siding and partitions of warehouses and utility buildings is in the natural mottled-gray color, but pigmented boards are also marketed in various colors. The specific gravity of asbestos board is about 2.0, and it will withstand temperatures to about 1000°F (538°C). The boards are dense and rigid, but can be worked easily with carpenters' tools. Usual thicknesses are 3/16, ¼, and ⅜ in (0.48, 0.64, and 0.95 cm). **Asbestos millboard** is also made with an organic binder with sometimes a filler of diatomaceous earth, in soft or hard grades, useful as insulation for temperatures from 800 to 1000°F (427 to 538°C). Asbestos board is marketed under trade names such as **Industal, Apac board,** and **Linabestos,** of the Keasbey & Mattison Co., and **Transite** of Johns-Manville. **Marinite,** of the same company, is a construction board useful for fire-resistant insulation, consisting of asbestos fiber, diatomaceous silica, and an inorganic binder. The density is 23 to 65 lb/ft^3 (368 to 1041 kg/ m^3). The hardest grade, Marinite 65, has a tensile strength of 750 lb/in^2 (5.3 MPa). Grade 36 has a hardness equal to that of yellow pine. **Sheetflextos,** of Keasbey & Mattison Co., is a flexible asbestos-cement wallboard in ⅛-in (0.32-cm) thickness in natural gray or waxed decorative colors. **Corrugated asbestos,**

used for roofing and siding of industrial buildings, has a thickness of ⅜ in (0.95 cm), and the usual corrugation has a pitch of 4.2 in (10.7 cm). The weight is 3.75 lb/ft^2 (18.3 kg/m^2).

Asbestos lumber is asbestos-cement board molded in the form of boards for flooring and partitions, usually with imitation wood grain molded into the surface. **Asbestos siding,** for house construction, is grained to imitate cypress or other wood and is pigmented with titanium oxide to give a clear white color. Asbestos roofing materials may also be made with asphalt or other binder instead of cement. **Fiberock,** of the Philip Carey Co., is a roofing board composed of asbestos fibers impregnated with asphalt. **Copperclad,** of Johns-Manville, is a roofing material consisting of 2-oz (57-g) Electro-sheer copper bonded to asbestos felt. **Asbestos ebony,** of this company, is an electrical panelboard made of asbestos cement bonded under high pressure with an insulating compound. It has good dielectric strength, will withstand temperatures to 300°F (149°C), and has a specific gravity of 2.05. **Ebonized asbestos,** of the Ambler Asbestos Shingle & Sheathing Co., is asbestos molded into sheets with an asphalt binder, and used for panels.

Asbestos paper is a thin asbestos sheeting made of asbestos fibers bonded usually with a solution of sodium silicate. It is strong, flexible, and white in color, and is fireproof and a good heat insulator. For covering steam pipes and for insulating walls, it is made in sheets of two or three plies. For wall insulation it is also made double with one corrugated sheet to form air pockets when in place. Thin sheets, ¹⁄₁₆ in (0.16 cm) thick, are employed for gaskets and electrical insulation. The 6-lb (2.7-kg) paper is 0.015 in (0.04 cm) thick, and the 8-lb (3.6-kg) is 0.019 in (0.05 cm) thick. Crimped asbestos paper is also used for pipe insulation to give dead air spaces between layers. **Amblerite,** of Keasbey & Mattison Co., is thin sheet made of asbestos fibers with a resilient binder, used for packing for superheated steam and chemical fittings. **Uni-Syn** sheet packing, of this company, is made by treating long asbestos fibers with synthetic rubber and felting under pressure. It is resilient and resistant to chemicals. **Prenite,** of the B. F. Goodrich Co., is neoprene-bonded asbestos sheet used for chemical packings. **Cohrlastic,** of the Connecticut Hard Rubber Co., is asbestos paper coated with white silicone rubber. The thickness is 0.024 in (0.06 cm). It has a dielectric strength of 400 volts per mil (15.8×10^6 volts per meter), withstands temperatures to 500°F (260°C), and is resistant to acids and alkalies. But the **Cohrlastic 3500,** used for high-strength flexible diaphragms for operation between −75 and 300°F (−59 and 149°C), is not asbestos, but is Orlon fabric coated with a silicone resin. **Novabestos,** of the Irvington Varnish & Insulator Co., is electrical insulating sheet, 0.003 in (0.008 cm) thick, made from asbestos fibers saturated with a silicone resin. The dielectric strength is 350 volts per mil (14.8×10^6 volts per meter). **Terratex,** of the General Electric Co., is asbestos paper with a clay binder in very thin sheets for high-voltage insulation. **Quinterra,** of Johns-Manville, is flexible asbestos paper in thickness from 0.0015 to 0.020 in (0.003 to 0.05 cm), made by a pulping process and used for insulation. Treated with silicone resin, it has high dielectric strength. **Quinterra 71** is the asbestos paper saturated with about 45% of epoxy resin in tape and sheets 0.004 in (0.01 cm) thick, used for slot lining and other insulation. The dielectric strength is 700 volts per mil (27.6×10^6 volts per meter).

ASH. The wood of a variety of species of ash trees valued for uses where strength, hardness, stiffness, and shock resistance are important. Most of the species give dense, elastic woods that polish well, but they do not withstand exposure well. The color is yellowish, which turns brown on exposure. The woods from the different species vary in their qualities, and are likely to be mixed in commercial shipments, but the general quality is high. Ash is used for quality cooperage such as tubs, flooring, veneer, vehicle parts, tool handles, bearings, and trim lumber. **American ash** and **Canadian ash,** also called **cane ash, white ash,** and **Biltmore ash,** come chiefly from the tree *Fraxinus americana* which grows over a wide area east of the Mississippi River. **Arkansas ash** is from *F. platycarpa;* **Japanese ash,** also called **tamo,** is from *F. mandschurica;* and European ash is from *F. excelsior.* **European ash** is heavier than American ash, and is tough and elastic. It is valued for hockey sticks, tennis rackets, and tool handles. Japanese ash is a close-grained wood, but browner in color.

White ash weighs 41 lb/ft^3 (657 kg/m^3) dry; **red ash,** *F. pennsylvanica,* 39 lb/ft^3 (625 kg/m^3); and **green ash,** *F. pennsylvanica lanceolata,* also called **water ash** and **swamp ash,** 44 lb/ft^3 (704 kg/m^3). This latter tree grows over the widest area throughout the states east of the Rockies, and is commercially abundant in the Southeast and Gulf states. It is a hardy tree, and has been used for farm windbreaks in the Great Plains area. All of these woods vary in tensile strength from 11,000 to 17,000 lb/in^2 (75 to 117 MPa). White ash has a compressive strength perpendicular to the grain of 2,250 lb/in^2 (16.5 MPa). **Mountain ash** and **black ash,** *F. nigra,* are also species of the American ash. The latter, also called **brown ash** and **hoop ash,** is a northern tree and was formerly used in aircraft construction. It has a specific gravity of 0.53 when oven- dried, a compressive strength perpendicular to the grain of 1,260 lb/in^2 (8 MPa), and a shearing strength parallel to the grain of 1,050 lb/in^2 (6 MPa). **Oregon ash,** *F. oregona,* is somewhat lighter and not as strong as white ash. It grows along the west coast of Canada. **Blue ash,** *F. quadrangular,* grows in the central states. **Pumpkin ash,** *F. profunda,* grows in the lower Mississippi Valley and in Florida. A wood that has similar uses to ash, for handles, levers, and machine parts, but is harder than ash, is **hornbeam.** It is from the tree

Ostrya virginiana of the eastern states of the United States. The wood is very hard, tough, and strong, but is available only in limited quantities.

ASPEN. The wood of the aspen tree, *Populus tremula,* used chiefly for match stems and for making excelsior, but also for some inside construction work. The color is yellowish, and it is tough and close-grained. The tree is native to Europe. The American aspen is from the tree *P. tremuloides,* called also **American poplar,** and from the **largetooth aspen,** *P. grandidentata.* Both species are also called poplar, and the lumber may be mixed with poplar and cottonwood. The trees grow in the lake and northeastern states and in the West. The heartwood is grayish white to light brown with a lighter-colored sapwood. It is straight-grained with a fine and uniform texture, but is soft and weak. It has a disagreeable odor when moist. The wood is used for excelsior, matches, boxes, and paper pulp. The pulp is easily bleached. **Salicin** is extracted from the bark.

ASPHALT. A bituminous, brownish to jet-black substance, solid or semisolid, found in various parts of the world. It consists of a mixture of hydrocarbons, is fusible and largely soluble in carbon disulfide. It is also soluble in petroleum solvents and in turpentine. The melting points range from 32 to 38°C. Large deposits occur in Trinidad and Venezuela. Asphalt is of animal origin, as distinct from coals of vegetable origin. Native asphalt usually contains much mineral matter; and crude **Trinidad asphalt** has a composition of about 47% bitumen, 28 clay, and 25 water. **Artificial asphalt** is a term applied to the bituminous residue from coal distillation mechanically mixed with sand or limestone. Asphalt is used for roofings, road surfacing, insulating varnishes, acid-resistant paints, and cold-molded products.

Bitumen refers to asphalt clean of earthy matter. It is obtained at Athabasca, Canada, in tar sands which are strip-mined. In general, bitumens have the characteristics that they are fusible and are totally soluble in carbon disulfide, as distinct from the **pyrobitumens,** albertite, elatarite, and coals, which are infusible and relatively insoluble in carbon disulfide. **Pyrogenous asphalts** are the residues from the distillation of petroleum or from the treatment of wurtzilite. **Asphaltite** is a general name for the bituminous asphaltic materials which are fusible with difficulty, such as gilsonite and grahamite. It is thought that **benzopyrene,** a constituent of coal-tar pitch and asphalt, will produce cancer in living tissues. This material also occurs in shale oil, soot, and tobacco smoke.

Rock asphalt, or **bituminous rock,** is a sandstone or limestone naturally impregnated with asphalt. The asphalt can be extracted from it, or it may be used directly for paving and flooring. **Kyrock** is a rock asphalt from Kentucky consisting of silica sand of sharp grains bound together with a bituminous content of about 7%. The crushed rock is used as a paving material. **Albertite** is a type of asphalt found originally in Albert County, New Brunswick, and first named **Albert coal.** It belongs to the group of asphalts only partly soluble in carbon disulfide, infusible, and designated as **carboids,** although they are true asphalts and not of vegetable origin. The commercial albertite is a type called **stellarite** from Nova Scotia. It is jet black, brittle, contains 22 to 25% fixed carbon, and yields oil and coke when distilled. It is easily lighted with a match, and burns with a bright, smoky flame, throwing off sparks. The albertite found in Utah is called **nigrite** and contains up to 40% fixed carbon. A species found in Angola is called **libollite.** These materials are weathered asphalts. **Ipsonite** is a final stage of weathered asphalt. It is black in color, infusible, only slightly soluble in carbon disulfide, contains 50 to 80% fixed carbon, and is very low in oxygen. It is found in Oklahoma, Arkansas, Nevada, and in various places in South America. The **rafaelite** found in large beds on the eastern slopes of the Andes Mountains in Argentina is a form of ipsonite.

Cutback asphalt is asphalt liquefied with petroleum distillates, used for cementing down floor coverings and for waterproofing walls. Protective coatings based on asphalt cutback form economical paints for protection against salts, alkalies, and nonoxidizing acids at temperatures up to 110°F (43°C). They are black in color, but may be pigmented with aluminum flake. They are often marketed under trade names such as **Atlastic,** of the Atlas Mineral Products Co., and **Protek-Coat,** of the Davison Chemical Co. Many corrosion-resistant coatings for chemical tanks and steel structures are asphalt solutions compounded with resins and fillers. **Perfecote,** of the Esbec Corp., for steel and concrete, contains an epoxy resin. The color is black, but it will accept a cover coat of colored plastic paint.

Modified asphalt, for laminating paper and for impregnating flooring felts, is asphalt combined with a rosin ester to increase the penetration, tack, and adhesion; but asphalt for paints and coatings may also be modified with synthetic resins. **Emulsified asphalt** is an asphalt emulsion in water solution, used for floor surfacing, painting pipes, and waterproofing concrete walls. Emulsified asphalts may be marketed under trade names such as **Elastex,** of the Truscon Laboratories, and **Ebontex,** of the Philip Carey Co. **Thermotex,** of the latter company, is an emulsified asphalt mixed with asbestos fibers, used for painting steam pipes. **Brunswick black** is a mixture of asphaltite with fatty acid pitch in a volatile solvent, used for painting roofs. **Amiesite,** of the Amiesite Asphalt Co., is asphalt mixed with rubber latex, or is a premixed asphalt with an aggregate employed for road filling. Rubbers are sometimes incorporated into paving asphalts to give resilience. The natural or synthetic rubber is mixed into the asphalt either in the form of powder or as a prepared additive. **Catalyzed asphalt**

is asphalt treated with phosphoric anhydride, P_2O_5, used for road construction to resist deterioration of the pavement from weathering. An asphalt mix developed by the Shell Chemical Co. for aircraft runways to resist the action of jet fuels is petroleum asphalt with an epoxy resin and a plasticizer. **Flooring blocks** and **asphalt tiles** are made in standard shapes and sizes from mixtures of asphalt with fillers and pigments. They are sold under many trade names, such as **Elastite** of the Philip Carey Co. and **Accotile** of the Armstrong Cork Co.

Oil asphalt, petroleum asphalt, petroleum pitch, or **asphalt oil,** is the heavy black residue left after removing the tar tailings in the distillation of petroleum. It contains 99% bitumen, is not soluble in water, and is durable. As it adheres well to metals, wood, or paper, and forms a glossy surface, it is used in roofings, or mixed with natural asphalt for paints and coatings. It is also used for roads. **Vanadiset,** of the Wilson Carbon Co., Inc., is a series of resin fractions of petroleum asphalt with small amounts of vanadium pentoxide, varying from semisolids to a brittle solid. They are used as softeners for rubber and in bitumen paints.

AVOCADO OIL. An oil obtained from the ripe, green-colored, pear-shaped fruit of the avocado, *Persea americana,* a small tree of which more than 500 varieties grow profusely in tropical America. The oil is also called **alligator pear oil.** In California, where the fruit is grown for market, it is also known as **Calavo.** The fruits weigh up to 3 lb (1.4 kg), and the seeds are 8 to 26% of the fruit. The fresh pulp contains 71% water, 20 oil, and 2.37 proteins. The seeds contain about 2% of an oil, but the avocado oil is extracted from the fruit pulp, the dehydrated pulp yielding 70% oil. In Central America the oil is extracted by pressing in bags, and the oil has been used by the Mayans since ancient times for treating burns and as a pomade. It contains 77% oleic acid, 10.8 linoleic, 6.9 palmitic, and 0.7 stearic, with a small amount of myristic and a trace of arachidic acid. It is also rich in lecithin, contains **phytostearin,** and is valued for cosmetics because it is penetrating, as lanolin is. It also contains **mannoketoheptose,** a high nonfermentable sugar. The oil has good keeping qualities and is easily emulsified. The oil-soluble vitamins are absorbed through the skin, and the oil for cosmetics is not wintered in order to retain the sterols. The specific gravity is 0.9132. Another oil used in cosmetics and for lubricating fine mechanisms is **ben oil,** a colorless to yellow oil obtained from the seeds of trees of the genus *Moringa,* notably *M. aptera, M. oleifera,* and *M. pterygosperma,* of Arabia, Egypt, India, and the Sudan. The latter species is also grown in Jamaica. The seeds contain 25 to 34% oil varying from a liquid to a solid, with specific gravity 0.898 to 0.902 and saponification value 179 to 187.

BABASSU OIL. An oil similar to coconut oil obtained from the kernels of the nut of the palm tree *Attalea orbignya* which grows in vast quantities in northeastern Brazil. There are two to five long kernels in each nut, the kernel being only 9% of the heavy-shelled nut, and these kernels contain 65% oil. A bunch of the fruits contains 200 to 600 nuts. The oil is very high in lauric acid, and is a direct substitute for coconut oil for soaps, as an edible oil, and as a source of lauric, capric, and myristic acids. The melting point of the oil is 22 to 26°C, specific gravity 0.868, iodine value 15, and saponification value 246 to 250. **Tucum oil,** usually classed with babassu but valued more in the bakery industry because of its higher melting point, is from the kernels of the nut of the palm *Astrocaryum tucuma* of northeastern Brazil. The oil is similar but heavier with melting point up to 35°C. In Colombia it is called **guere palm.**

Another similar oil is **murumuru oil,** from the kernels of the nut of the palm *A. murumuru,* of Brazil. The name is a corruption of the two Carib words marú and morú, meaning bread to eat. The oil contains as much as 40% lautic acid, with 35% myristic acid, and some palmitic, stearic, linoleic, and oleic acids. It is usually marketed as babassu oil. The **awarra palm,** *A. janari,* of the Guianas, yields nuts with a similar oil. **Cohune oil** is a white fat from the kernels of the nut of the palm **Attalea cohune** of Mexico and Central America. It is a small tree yielding as many as 2,000 nuts a year. The oil has the appearance and odor of coconut oil, and contains 46% lauric acid, 15 myristic, 10 oleic, with also stearic, capric, and linoleic acids. All of these oils yield a high proportion of glycerin. Cohune oil has a melting point of 18 to 20°C, saponification value 252 to 256, iodine value 10 to 14, and specific gravity 0.868 to 0.971. The cohune nut is much smaller than the babassu but is plentiful and easier to crack. **Curua oil** is from the nut of the palm *A. spectabilis* of Brazil. It is similar to cohune oil and is used for the same purposes in soaps and foods. **Mamarron oil** is a cream-colored fat with the odor and characteristics of coconut oil, obtained from another species of *Attalea* palm of Colombia. Another oil high in lauric acid, and similar to babassu oil, is **corozo oil,** obtained from the kernels of the nuts of the palm *Corozo oleifera* of Venezuela and Central America. **Macanilla oil** is a similar oil from the kernels of the nuts of the palm *Guilielma garipaes* of the same region. **Buri oil** is from the nuts of the palm *Diplothemium candescens* of Brazil.

BABBITT METAL. The original name for tin-antimony-copper white alloys used for machinery bearings, but the term now applies to almost any white bearing alloy with either tin or lead base. The original babbitt, named after the inventor, was made by melting together 4 parts by weight of copper, 12 tin, and 8 antimony, and then adding 12 parts of tin after fusion. It consisted, therefore, of 88.9% tin, 7.4 antimony, and 3.7 copper.

This alloy melts at 462°F (239°C). It has a Brinell hardness of 35 at 70°F (21°C), and 15 at 212°F (100°C). As a general-utility bearing metal, the original alloy has never been improved greatly, and makers frequently designate the tin-base alloys close to this composition as **genuine babbitt.**

Commercial **white bearing metals** now known as babbitt are of three general classes: tin-base, with more than 50% tin hardened with antimony and copper, and used for heavy, pounding service; intermediate, with 20 to 50% tin, having lower compressive strength and more sluggish as a bearing; lead-base, made usually with antimonial lead with smaller amounts of tin together with other elements to hold the lead in solution. These **lead-base babbitts** are lower in cost and also serve to conserve tin in times of scarcity of that metal, but they are suitable only for light service, although many ingenious combinations of supplementary alloying elements have sometimes been used to give hard, strong bearings with little tin. The high-grade babbitts, however, are usually close to the original babbitt in composition. **SAE babbitt 11,** for connecting-rod bearings, has 86% tin, 5 to 6.5 copper, 6 to 7.5 antimony, and not over 0.35 lead. A babbitt of this kind will have a compressive strength up to 20,000 lb/in^2 (137 MPa) while the high-lead alloys have only 15,000 lb/in^2 (103 MPa).

Copper hardens and toughens the alloy and raises the melting point. Lead increases the fluidity and raises the antifriction qualities, but softens the alloy and decreases its compressive strength. Antimony hardens the metal and forms hard crystals in the soft matrix which improves the alloy as a bearing metal. Only 3.5% of antimony is normally dissolved in tin. In the low-antimony alloys copper-tin crystals form the hard constituent, and in the high-antimony alloys antimony-tin cubes are also present. Alloys containing up to 1% arsenic are harder at high temperatures and are fine-grained, but arsenic is used chiefly for holding lead in suspension. Zinc increases hardness but decreases frictional qualities, and with much zinc the bearings are inclined to stick. Even minute quantities of iron harden the alloys, and iron is not used except when zinc is present. Bismuth reduces shrinkage and refines the grain, but lowers the melting point and lowers the strength at elevated temperatures. Cadmium increases the strength and fatigue resistance, but any considerable amount lowers the frictional qualities, lowers the strength at higher temperatures, and causes corrosion. Nickel is used to increase strength but raises the melting point. The normal amount of copper in babbitts is 3 or 4%, at which point the maximum fatigue-resisting properties are obtained with about 7% antimony. More than 4% copper tends to weaken the alloy, and raises the melting point. When the copper is very high, tin-copper crystals are formed and the alloy is more a bronze than a babbitt. All of the SAE babbitts contain some arsenic, ranging from 0.10% in the high-tin **SAE babbitt 10** to about 1% in the high-lead **SAE babbitt 15.** The first of these contains 90%

tin, 4.5 antimony, 4.5 copper, and 0.35 lead, while the babbitt 15 has 82% lead, 15 antimony, 1 tin, and 0.60 copper.

Because of increased speeds and pressures in bearings and the trend to lighter weights, heavy cast babbitt bearings are now little used in spite of the low cost and ease of casting. The alloys are mostly employed as **anti-friction metals** in thin facings on steel backings, the facing being usually less than 0.010 in (0.03 cm) thick, in order to increase the ability to sustain higher loads and to dissipate the heat.

Babbitts are marketed under many trade names, the compositions generally following the SAE alloy standards but varying in auxiliary constituents, the possibilities for altering the physical qualities by composition rearrangement being infinite. Some of the trade names which have been used for babbitt-type alloys marketed in ingots are: **Leantin** and **Cosmos metal** of the Lumen Bearing Co., for high-lead alloys; **stannum metal** for high-tin alloys; and **Lubeco metal** and **Lotus metal** for medium-composition alloys. **Hoo Hoo metal** and **Nickel babbitt,** of the American Brake Shoe Co., are high-tin alloys containing nickel, while **Silver babbitt** of the same company has no tin but contains a small amount of silver to aid retention of the lead and to give hardness at elevated temperatures. **Glyco** is the name of a group of lead-base alloys of Joseph T. Ryerson & Son, Inc. **Satco,** of the NL Industries, Inc., is a high-melting-point alloy for heavy service. It melts at 788°F (420°C). **Tinite,** of the Ajax Metal Co., is a tin-base metal hardened with copper. **Ajax bull,** of this company, contains 76% lead, 7 tin, and 17 antimony, modified with other elements.

BAGASSE. The residue left after grinding sugarcane and extracting the juice, employed in making paper and fiber building boards. In England it is called **megass.** The fiber contains 45% cellulose, 32 pentosan, and 18 lignin. It is marketed as dry and wet separated, and as dry fiber. The dry-separated fibers bulk 4.5 lb/ft^3 (72 kg/m^3), with 62 to 80% passing a 100-mesh screen. The dry fiber bulks 6 to 8 lb/ft^3 (96 to 128 kg/m^3), and is about 14 mesh. The fibers mat together to form a strong, tough, light, absorptive board. The finer fibers in Cuba and Jamaica are soaked in molasses and used as a cattle feed under the name of **molascuit. Celotex** is the trade name of the Celotex Corp. for wallboard, paneling, and acoustical tile made from bagasse fibers. **Ferox-Celotex** is the material treated with chemicals to make it resistant to fungi and termites. **Celo- Rock** is the trade name for Celotex-gypsum building boards. **Acousti- Celotex** is Celotex perforated to increase its sound-absorbing efficiency. In India, the Philippines, and some other countries where sugarcane is plentiful, paper is made from the bagasse. Newsprint is made with a mixture of mechanical and chemical bagasse pulp, and writing papers may be made by delignifying the bagasse and digesting with soda. **Aconitic acid,**

HOOCCH:C(COOH)CH_2COOH, occurs in bagasse and is extracted from Louisiana cane. The acid is esterified for use as a plasticizer for vinyl resins, or sulfonated for use as a wetting agent. This acid is also produced as a white powder of melting point 195°C by the dehydration of citric acid.

BALATA. A nonelastic rubber obtained chiefly from the tree *Manilkara bidentata* of Venezuela, Brazil, and the Guianas. It is similar to gutta percha and is used as a substitute. The material contains a high percentage of gums and is more tacky than rubber, but it can be vulcanized. It differs from rubber in being a transisomer of isoprene with a different polymerization. Balata has been used principally for transmission and conveyor belts and for golf ball covers. For conveyor belts heavy duck is impregnated with balata solution and vulcanized. The belts have high tensile strength, good flexibility, and wear resistance. The wood of the balata tree is used for cabinetwork and for rollers and bearings. It is called **bulletwood** in the Guianas, but this name is also applied to the wood of the gutta-percha trees of Asia. The wood is extremely hard, durable, and weighs 66 lb/ft^3 (1,057 kg/m^3). It has a deep-red color and a fine, open grain.

BALSA WOOD. The wood of large and fast-growing trees of the genus *Ochroma* growing from southern Mexico to Ecuador and northern Brazil. It is the lightest of the commercial woods and combines also the qualities of strength, stiffness, and workability. It is about one-fourth the weight of spruce, with a structural strength half that of spruce. The crushing strength is 2,150 lb/in^2 (14 MPa). The wood is white to light yellow or brownish and weighs about 8 lb/ft^3 (128 kg/m^3) from a 4-year-old tree. Wood from a 6-year-old tree weighs 10 to 12 lb/ft^3 (160 to 192 kg/m^3). Its peculiar cellular structure makes it valuable as an insulating material for refrigeration. It is also used for life preservers, buoys, floats, paneling, vibration isolators, insulating partitions, and inside trim of aircraft. The small pieces are used for model airplanes. **Balsa sawdust** may be used as a lightweight filler for plastics.

Much of the commercial wood is from the tree *O. grandiflora* of Ecuador. **Barrios balsa,** *O. concolor,* grows from southern Mexico through Guatemala and Honduras. **Limon balsa** is from the tree *O. limonensis* of Costa Rica and Panama, and **Santa Marta balsa** is the *O. obtusa* of Colombia. **Red balsa** is from the *O. velutina* of the Pacific Coast of Central America. The balsa known in Brazil as **Sumaúma** is from a kapok tree *Ceiba pentandra.* It is used for life preservers and rafts, and is quite similar to balsa. A Japanese lightweight wood used for floats, instruments, and where lightness is required is **Kiri,** from the tree *Paulownia tomentosa.* It weighs 14 to 19 lb/ft^3 (224 to 304 kg/m^3), has a coarse grain, but is strong and resists warping.

Grown as a shade tree since 1834 under the names of **paulownia** and **empress tree,** it is now common in the United States, and the wood is used as a lightweight crating lumber.

BALSAM FIR. The wood of the coniferous tree *Abies balsamea* of the northeastern states and Canada. It is brownish white in color, soft, and has a fine, even grain. It is not strong and not very durable, and is used chiefly for pulpwood and for packing boxes and light construction. The weight is 26 lb/ft^3 (417 kg/m^3). Liquid pitch comes from blisters on the outer bark. It was formerly used as a transparent adhesive. **Canada balsam,** or **Canada turpentine,** is a yellowish, viscous oleoresin liquid of pleasant odor and bitter taste, obtained from the buds of the tree. The specific gravity is 0.983 to 0.997. It is a class of turpentine, and is used as a solvent in paints and polishes, in leather dressings, adhesives, and perfumes. It is also referred to as **balm of Gilead** for medicinal and perfumery use, but the original balm of Gilead, marketed as buds, was from the small evergreen tree *Balsamodendron gileadense* of the Near East. **Southern balsam fir** is **Frazer fir,** from the tree *A. fraseri* of the Appalachian Mountains. The wood is similar to balsam fir.

BAMBOO. A genus of gigantic treelike grasses, of the order *Graminaceae,* of which the *Bambusa arundinacea* is the most common species. It grows most commonly in Indonesia, the Philippines, and southern Asia, but many species have been brought to Latin America and to the southern United States. The stems of bamboo are hollow, jointed, and have an extremely hard exterior surface. They sometimes reach more than 1 ft (0.3 m) in diameter and are often 50 ft (15 m) high, growing in dense masses. Nearly 1,000 species are known. The *B. spinosa* of the Philippines grows as much as 10 ft (3 m) in one week. Bamboo is a material which has had innumerable uses. The stalks are used for making pipes, buckets, baskets, walking sticks, fishing poles, rug-winding poles, lance shafts, window blinds, mats, arrows, and for building houses and making furniture. The weight is about 22 lb/ft^3 (352 kg/m^3). **Tonkin bamboo** is strong and flexible, and is used for making fishing poles. **Tali bamboo** of Java, *Gigantochloa apus,* is used for construction. **Betong bamboo,** *G. asper,* is one of the largest species. **Giant bamboo,** *Dendrocalamus gigantea,* of Sri Lanka, grows to a height of 100 ft (30 m). The fast-growing **eeta bamboo** is used in India as a source of cellulose for rayon manufacture.

BARITE. Sometimes spelled **baryte,** and also called **heavy spar,** and in some localities known as **tiff.** A natural **barium sulfate** mineral of the theoretical composition of $BaSO_4$, used chiefly for the production of lithopone, in chemical manufacture, and in oil-drilling muds. Mixed with synthetic rubber it is used as a seal coat for roads. For chemicals it is specified 90 to 95% pure $BaSO_4$, with not more than 1% ferric oxide. Prime white and floated grades are used for coating paper. **Baroid,** of NL Industries, Inc., used in oil wells, is barite ore crushed, dried, and finely ground. **Artificial barite, permanent white,** and **blanc fixe** are names for white, fine-grained precipitated paint grades. **Micronized barite,** for rubber filler, is a fine white powder of 400 to 1,000 mesh. Barite is widely distributed and especially associated with ores of various metals or with limestones. It occurs in crystals or massive. It may be colorless, white, or light shades of blue, red, and yellow, and transparent to opaque. Its hardness is 3 to 3.5, and its specific gravity 4.4 to 4.8. It is insoluble in water. The mineral is produced in the western states and from Virginia to Georgia. The barite of Cartersville, Ga., contains 96% $BaSO_4$, 0.6 iron, with silica, alumina, and traces of calcium, strontium, and magnesium. Large deposits of high-grade barite occur in Nova Scotia. In the West much ground crude barite is used as a drilling mud in oil wells. The white pigment marketed by the American Zinc Sales Co. under the name of **Azolite** is 71% barium sulfate and 29% zinc sulfide in 325-mesh powder. **Sunolith,** of Wishnick-Tumpeer, Inc., is a similar product. A substitute for barite for some filler uses is **witherite,** an alteration mineral of the composition $BaCO_3$, which is **barium carbonate,** found associated with barite. Precipitated barium carbonate is a white, tasteless, but poisonous powder used in rat poisons, optical glass, ceramics, and pyrotechnics, as a flatting agent in paints, and as a filler for paper. With ferric oxide it is used for making ceramic magnets. **Barium oxide,** BaO, of 99.99% purity, is made by the reduction of barite. It is used as an additive in lubricating oils.

BARIUM. A metallic element of the alkaline earth group, symbol Ba. It occurs in combination in the minerals witherite and barite, which are widely distributed. The metal is silvery white in color and can be obtained by electrolysis from the chloride, but it oxidizes so easily that it is difficult to obtain in the metallic state. Powdered or granular barium is explosive in contact with carbon tetrachloride, fluorochloromethanes, and other halogenated hydrocarbons. Its melting point is 850°C, and its specific gravity 3.78. The most extensive use of barium is in the form of its compounds. The salts which are soluble, such as sulfide and chloride, are toxic. An insoluble, nontoxic barium sulfate salt is used in radiography. Barium compounds are used as pigments, in chemical manufacturing, and in deoxidizing alloys of tin, copper, lead, and zinc. Barium is introduced into lead-bearing metals by electrolysis to harden the lead. When barium is heated to about 200°C in hydrogen gas it forms **barium hydride,** BaH_2, a gray powder which decomposes in contact with water and can be used as a source of nascent hydrogen for life rafts.

BARIUM CHLORIDE. A colorless crystalline material of the composition $BaCl_2 \cdot 2H_2O$, or in anhydrous form without the water of crystallization. The specific gravity is 3.856, and the melting point 860°C. It is soluble in water to the extent of 25% at 20°C and 37% at 100°C. In the mechanical industries it is used for heat-treating baths for steel, either alone or mixed with potassium chloride. The molten material is free from fuming and can be held at practically any temperature within the range needed for tempering steels. It is also used for making boiler compounds, for softening water, as a mordant in dyeing and printing inks, in tanning leather, in photographic chemicals, and in insecticides. **Barium chlorate,** $Ba(ClO_3)_2 \cdot H_2O$, is a colorless crystalline powder soluble in water. The melting point of the anhydrous material is 414°C. It is used in explosives as an oxygen carrier, and in pyrotechnics for green-colored light. **Barium fluoride,** BaF_2, is used in crystal form for lasers. When "doped" with uranium it has an output wavelength of 2,600 nm. Doping with other elements gives diffused wavelengths for different communication beams. **Barium cyanide,** $Ba(CN)_2$, is a poisonous, colorless, crystalline material melting at 600°C. It is marketed by the Koppers Co. as a 30% water solution for adding to cyanide plating baths, in which it removes carbonates and increases the current efficiency.

BARIUM NITRATE. Also called **nitrobarite.** A white crystalline powder of the composition $Ba(NO_3)_2$, with specific gravity of 3.24, melting at 592°C, and decomposing at higher temperatures. It is a barium salt of nitric acid obtained by roasting barite with coke, leaching out the precipitated barium sulfide, precipitating as a carbonate by the addition of soda ash, and then dissolving in dilute nitric acid. It has a bitter metallic taste and is poisonous. Barium nitrate is used in ceramic glazes, but its chief use is in **pyrotechnics.** It gives a pale-green flame in burning, and is used for green signals and flares, and for white flares in which the delicate green is blended with the light of other extremely luminous materials. It is also used as an oxygen carrier in **flare powders** and to control the time of burning of the aluminum or magnesium. **Sparklers** are composed of aluminum powder and steel filings with barium nitrate as the oxygen carrier. The steel filings produce the starlike sparks. **Barium nitrite,** $Ba(NO_2)_2$, decomposes with explosive force when heated. **Barium oxalate,** BaC_2O_4, is used in pyrotechnics as a combustion retarder.

BARLEY. The seed grains of the annual plant *Hordeum vulgare* of which there are many varieties. It is one of the most ancient of the cereal grains. The plant is hardy, with a short growing season, and can be cultivated in cold latitudes and at high altitudes, giving high yields per acre. The grains grow in a dense head with three spikelets, and the six-rowed variety has a high protein content, but has low gluten, thus making a poor breadstuff.

Pearl barley is the husked and polished grain. When used for cattle feed, barley produces lean meats. The chief industrial use is for making **malt,** for which the two-rowed varieties with low protein and thin husk are used. Malt is barley that has been germinated by moisture and then dried. Malting develops the diastase enzyme, which converts the insoluble starch into soluble starch and then into sugars. It is used for brewing beer and for malt extracts. **Caramel malt** is browned with high-temperature drying, and is used for the dark-colored **bock beer. Barley straw** is employed in Europe and Asia for making braided plaits for hats. In America it is used for packing material, especially for glassware.

BASALT. A dense, hard, dark-brown to black igneous rock, consisting of feldspar and augite and often containing crystals of green olivine. It occurs as trap or as volcanic rock. The specific gravity is 2.87 to 3, and it is extremely hard. Masses of basalt are frequently found in columns or prisms, as in the celebrated basalt cliffs of northern Ireland. It differs from granite in being a fine-grained extrusive rock, and in having a high content of iron and magnesium. Basalt is used in the form of crushed stone for paving, as a building stone, and for making rock wool. A Russian cast basalt used for electrical insulators is called **angarite.** In Germany cast basalt has been used as a building stone, for linings, and for industrial floors. It is made by melting the crushed and graded basalt, and then tempering by slow cooling. The structure of the cast material is dense with needlelike crystals, and has a hardness of 8 to 9 Mohs. **Basalt glass** is not basalt, but is pumice.

BASSWOOD. The wood of several species of lime trees, *Tilia americana, T. heterophylla, T. glabra,* and *T. pubescens,* all native to the United States and Canada. The European **limewood,** from the tree *T. cordata,* is not called basswood. The wood of the *T. glabra,* called in the eastern states the **lime tree** and the **linden,** and also the **white basswood,** *T. heterophylla,* is used for containers, furniture, and such millwood as blinds. It is soft, light in weight, and has a fine, even grain, but is not very strong or durable on exposure. The white sapwood merges gradually with the yellow-brown heartwood. The specific gravity is 0.40 when oven-dried, and the compressive strength perpendicular to the grain is 620 lb/in^2 (5.5 MPa).

BATE. Materials used in the leather industry to remove lime from skins and to make them soft and flaccid before tanning by bringing the collagen into a flaccid or unswollen condition. Since ancient times dung has been used for this purpose, and until recent years the American tanning industry imported dog dung from Asia Minor for bating leather. **Artificial bates** are now used because of their greater uniformity and cleanliness. **Boric**

acid is sometimes used for deliming, and gives a silky feel to the leather, but most bates have both a deliming and an enzyme action. **Trypsin** is a group of enzymes from the pancreatic glands of animals, and its action on skins is to dissolve the protein. They are generally used with ammonium chloride or other salt. **Oropon,** of Rohm & Haas Co., Inc., is this material carried in wood flour and mixed with a deliming salt. **Sulfamic acids** are also used as bates. The lime compounds used for dehairing are called **depilating agents.**

BAUXITE. A noncrystalline, earthy-white to reddish mineral, massive or in grains, having a composition $Al_2O_3 \cdot 2H_2O$, theoretically containing 74% alumina. It is the most important ore of aluminum, but is also used for making aluminum oxide abrasives, for refractories, white cement, and for decolorizing and filtering.

Bauxite is graded on the Al_2O_3 content. High-grade bauxite, Grade A, contains a minimum of 55% alumina and a maximum of 8 silica. Grade B contains a minimum of 50% alumina with a silica content from 8 to 16%. Chemical grades should have less than 2.5% Fe_2O_3. Grades appearing in price quotations with up to 84% alumina content are calcined and are based on the dehydrated alumina content of the ore.

Bauxite has a high melting point, 1820°C, and can be used directly as a refractory. Cement-making **white bauxite** from Greece ranks very high in alumina content. Brazilian, Arkansas, and Indian ores also contain some titanium oxide, and the Surinam ore has as high as 3% TiO_2. Two kinds of **red bauxite** are found in Italy, a dark variety containing 54 to 58% Al_2O_3, and only 2 to 4 SiO_2, but having 22 to 26% Fe_2O_3, and 2 to 3 TiO_2, and a light variety containing 60 to 66% Al_2O_3, 5 to 9 silica, 10 to 16 iron oxide, and 3 to 5 titanium oxide. The best French white bauxite contains 66 to 74% alumina, 6 to 10 silica, 2 to 4 iron oxide, and 3 to 4 titanium oxide. It is preferred for ceramic and chemical purposes, while the best grade of the red variety is used for producing aluminum, and the inferior grade for refractories and for cement manufacture. Malayan and Indonesian bauxite averages 57 to 60% Al_2O_3, 6.7 Fe_2O_3, 3 to 5 SiO_2, and 0.9 to 1 TiO_2. The large deposits on Ponape and other Pacific islands average 50 to 52% alumina, 3 to 6 silica, and 10 to 20 Fe_2O_3, but the bauxite of Hawaii contains only 35% alumina with up to 15% silica. **Phosphatic bauxite,** from the island of Trauhira off the coast of Brazil, is a cream-colored porous rock containing 31.5% alumina, 25.2 P_2O_5, 7.3 iron oxide, 6.8 silica, and 1.3 titania. **Diaspore,** $Al_2O_3 \cdot H_2O$, mined in Missouri, and **gibbsite,** $Al_2O_3 \cdot 3H_2O$, from the Guianas, are bauxites also used for refractories. Gibbsite is also called **wavellite. Filter bauxite,** or **activated bauxite,** is bauxite that has been crushed, screened, and calcined, and is usually in 20- to 60- and 30- to 60-mesh grades. It may be sold under trade names such as **Porocel** and **Floride.** It is preferred to fuller's earth for oil-refinery filtering because it can be revivified indefinitely by calcining. **Calcined bauxite** for the abrasive industry is burned bauxite and contains 78 to 84% alumina. **Laterite,** or **ferroginous bauxite,** has been used in Europe to produce alumina and iron. The laterite of Oregon contains 35% alumina, about 35 iron oxide, and about 7 silica. Low-alumina, high-silica bauxites can be lime-sintered to release the sodium aluminate which goes back into the process while the silicate goes out with the calcium, thus giving high alumina recovery with low soda loss. **Anorthosite,** an abundant **aluminum silicate** mineral containing up to 50% silica, is also used to produce aluminum. The anorthosite of Wyoming is sintered with limestone and soda ash and calcined to yield alumina and a by-product portland cement base of dicalcium silicate.

BEARING BRONZE. Any bronze used for bearing purposes, but usually referring to bronzes containing considerable lead. As an element in bearing metals lead has been called the wax of metals, forming the soft matrix or foundation for the hard crystals, but lead is not easy to keep in solution and the alloys require controlled casting. A high lead content in bronzes improves the antifriction qualities, but reduces the strength and hardness. Bronzes with about 30% lead are called **plastic bronze.** They are superior in strength to the babbitt metals, and are used for heavy mill bearings. The so-called **bearing brasses,** formerly much used for railway-car journals, are not brasses, but are lead bronzes with about 65% copper, 5 to 10 tin, and up to 30 lead. They seldom contain zinc, but the railway bearing metal called **bush metal** in England contains 72% copper, 14 tin, and 14 yellow brass ingot metal. The **journal bronze** of the U.S. Navy contains 82 to 84% copper, 12.5 to 14.5 tin, 2.5 to 4.5 zinc, and 1 lead. The **ASTM bearing bronzes** have from 70 to 85% copper, 5 to 10 tin, and 5 to 25 lead, with the zinc kept below 0.50% as an undesirable element. The 85:10:5 bronze has a tensile strength of 28,000 lb/in^2 (193 MPa), a compressive deformation limit of 18,000 lb/in^2 (124 MPa), and a Brinell hardness of 60. The 70:5:25 bronze has a tensile strength of 15,000 lb/in^2 (103 MPa), and a hardness of 40. Zinc is used, however, where higher strengths are needed, if the copper-tin is high and the zinc remains alloyed with the copper in the softer matrix. **SAE alloy 660,** and **Bunting bronze 72,** used for electric motor bearings, contain 83% copper, 7 tin, 7 lead, and 3 zinc. This alloy has a tensile strength of 34,000 lb/in^2 (234 MPa), compressive limit 22,000 lb/in^2 (151 MPa), and Brinell hardness 58. **Asarcon 773,** of Asarco, Inc., has a similar composition, but the bars and cylinders are produced by continuous casting through graphite dies from a nitrogen-atmosphere furnace to give a dense, even-grained structure. The tensile strength is 44,000 lb/in^2 (303 MPa), elongation 16%, and hardness 72.

Bearing bronzes owe much of their quality to grain structure, uniformity, and other physical properties which depend upon methods of manufacture, and they are marketed under a variety of trade names such as **Ajax metal,** of the Ajax Metal Co., and **Johnson bronze,** of the Johnson Bronze Co. **Tiger bronze,** of the National Bearings Div., American Brake Shoe Co., is a copper-tin bronze with particles of lead evenly dispersed to give low coefficient of friction. It comes in cast bars and cylinders with tensile strength 30,000 lb/in^2 (206 MPa) and Brinell hardness 53. **Magnolia isotropic bronze,** of the Magnolia Metal Co., is die-cast in bars and cylinders to produce a homogeneous structure that gives longer wear life than a sandcast bronze. Tensile strength is 30,000 lb/in^2 (206 MPa), elongation 8.5%, and Brinell hardness 70. **Promet bronze,** of the American Crucible Products Co., contains 11.5 to 15.5% lead, 4.5 to 6.5 tin, 1.25 to 2 nickel, and the balance copper. It has a fine dense structure, and needs little lubrication. The tensile strength is 33,000 lb/in^2 (227 MPa) with elongation of 8 to 13%. **Johnson bronze 40,** for sheet bearings, is **SAE bronze 795.** It contains 90% copper, 9.5 zinc, and 0.5 tin, giving a Rockwell B hardness of 65 to 71. **Johnson bronze 44** is **SAE bronze 791,** with 86 to 90% copper, 3 to 4.5 tin, 3.5 to 4.5 lead, and 3.5 to 5 zinc, having a Rockwell B hardness of 65 to 71.

An addition of small amounts of nickel to bearing bronzes helps to keep the lead in solution and improves the resistance to compression and shock. Iron up to 1% increases the resistance to pounding and hardens the bronze, but reduces the grain size and tends to segregate the lead.

A **copper-antimony alloy** used for worm gears contains 7 to 8% antimony, 1.5 to 2.5 nickel, and the balance copper. It casts well and has a triplex structure that makes a good bearing metal. The tensile strength, sand-cast, is 32,000 lb/in^2 (220 MPa), with elongation 7%.

BEARING MATERIALS. **White metals** and **bronzes** are most frequently used for machine bearings, but wood, glass, plastics, and other materials are also employed. Wood is one of the oldest of bearing materials, and is still considered an excellent material for large, low-pressure, and slow-speed bearings. The hardwoods are used, and since they absorb oil and grease little attention is required for lubrication. **Plastic bearings** are employed where electrical insulation is required. These materials have a compressive strength up to 36,000 lb/in^2 (248 MPa). Plastic bearings are not easily wetted by water, and a small amount of alkaline sodium oleate is added to the water lubricant. For food-processing equipment where lubricating oil is not desired, ball bearings are made with plastic races and steel balls. The ball bearings of the General Bearing Co. have Delrin races and Type 316 steel balls. **Nylon bearings** require no lubricant for light loads at high speed, and are suitable for textile machinery where oil might soil the fabrics. **Rubber bearings** are used where resilience is needed. Water may be used as the lubricant.

Almost any commercial metal can be used for bearings, but certain metals and alloys are particularly suited for bearings, chiefly because of the fact that a proportion of hard crystals occurs in a background, or matrix, of softer metal, thus supporting the shaft and permitting the free circulation of the lubricant. In the soft babbitt metals these crystals are formed largely by the antimony, while in the bronzes the crystals consist of a chemical compound of copper and tin. One of the metals must have wettability or affinity for lubricants because metal-to-metal contact usually results in high heat, scoring, and galling, or the tearing out of pieces of metal. To form a good bearing, one of the surfaces must slide on minute projecting irregularities. Immiscible metals that do not alloy are not as likely to gall as alloys that form welded junctions of the particles. A good bearing metal should support a load up to 1,200 lb/in^2 (10 MPa) without galling.

Cast iron is an excellent bearing metal because of the hard carbides, a soft background of iron, and considerable graphitic carbon which acts as a lubricant. Brass is not much used for bearings since zinc causes sticking, but a bearing alloy known as **Tissier's metal** was a high-copper brass with 1% arsenic to give a crystalline structure. The alloy **Hy Speed** marketed by the Buckeye Brass & Mfg. Co. contained 88% copper, 7 tin, and 5 zinc. In white metals the formation of the structure is affected by the melting times, and a well-cast alloy of inferior composition may result in a better bearing than a high-grade alloy that is poorly cast.

Aluminum bearings have high corrosion resistance against organic acids, good thermal conductivity, and higher strength than other white-metal alloys, and they are equal to babbitts in antiseizure properties. They have the disadvantage of high coefficient of expansion. Aluminum alloys for bearings must have a matrix of aluminum through which are dispersed undissolved constituents. An alloy with 4% silicon and 4 cadmium is used. Other alloys have tin, nickel, and copper or silicon. **Aluminum-tin alloys** are used for heavily loaded high-speed bearings, usually with steel backings to add strength and conserve tin. The alloys contain 20 to 30% tin. A 70–30 alloy has a Vickers hardness of 22. Adding 1% copper raises the hardness to 27 and improves the physical qualities. Usually no more than 3% copper is used. The copper is alloyed with the aluminum, but the tin remains in an almost continuous intercrystalline network in the cast metal. **Alcan Alloy AA 8280,** of Alcan Aluminum Ltd. has only 6% tin for high-duty bearings. The bearing known as **Moraine 400,** of the General Motors Corp., contains 4% silicon, 1 cadmium, and the balance aluminum. The alloy is made in strips with a steel backing, and has a thin electrodeposit bearing surface of a lead-tin-copper babbitt-type alloy. The trimetal combination offers the antifrictional advantages of a babbitt bearing, with up to 10 times the service life. **Alloy 750,** also used for automotive engine bearings, does not require a steel backing. It contains 6.5% tin, 2.5 silicon, 1 copper, 0.5 nickel, and the balance aluminum. **Magnesium bearings** are used on such products as small pumps where the die-cast metal is employed without inserts or bushings to operate with a hardened steel shaft. Alloys used are 94% magnesium and 6 tin, or 70 magnesium, 20 cadmium, and 10 lead.

Unlike metals are invariably used for shaft and bearing, and the wear is taken in the bearing. The choice of a bearing metal is usually a compromise of hardness, compressive strength, coefficient of friction, and degree of lubrication. In general, the **tin-base alloys** have low coefficients of friction and are tough and capable of withstanding shocks, but the copper bronzes are capable of withstanding heavier loads. Between the two, almost any desired combination can be obtained, depending upon the proportions of copper, tin, antimony, and lead. Load-carrying capacity is figured on the projected area surface. **Babbitts** are usually limited to 300°F (149°C) operating temperatures, and the tin and lead bronzes to 500°F (260°C). Nickel or silver in small amounts improves the hardness and strength of babbitts at higher temperatures. Some constituents of the alloys have a catalytic action on the lubricants. Tin in a bearing metal reduces the tendency of the lubricating oil to sludge, but the presence of an alkali metal may have an injurious effect on the oil. **Indium bearings,** for aircraft, are made by plating the steel backing with silver and then with lead, over which is plated the indium, and then diffused by heating. A lead-indium alloy results, which gives a strong, hard surface with good bearing properties and resistance to the corrosion of oils.

Self-lubricating bearings, or **oilless bearings,** may be sintered iron and/or bronze powders impregnated with oil or graphite, or they may be oil-impregnated wood. The iron is for low speeds at medium to heavy loads, and the bronze for high-speed bearings for appliances, motors, and power tools.

Copper sponge, used for bearings, is made by molding a mixture of copper powder and a volatile organic material, sintering to drive off the volatile matter and adhere the copper particles, and then impregnating the porous copper with lead. Molded carbon is also used for bearings where there is need for resistance to corrosive chemicals. **Purabon 5,** of the Pure Carbon Co., is a carbon-graphite mixture molded in the form of bearings and bushings for use in chemical and food equipment.

Ryertex is the name of a laminated plastic bearing material, and **Aqualite,** of the National Vulcanized Fibre Co., is the name of laminated plastic marine bearings. **Parock,** of Raybestos-Manhattan, Inc., is an oilless bearing made with graphite bonded with 20% vulcanized rubber. The coefficient of friction is 0.1 to 0.13 dry and 0.05 to 0.08 wet. **Fluoroglas,** of the

Fluorolon Laboratories, Inc., is a compound of fluorocarbon resin with glass particles and a mineral pigment, used for oilless marine bearings. The coefficient of friction is 0.04. It has high resistance to wear, and will withstand operating temperatures to 500°F (260°C). The Durometer hardness is D50 to D65. **Nolu** is an oil-impregnated **wood bearing** of the Nolu Oilless Bearing Co., and **Woodex** is a similar bearing of the Neveroil Bearing Co. These woods are normally maple, but lignum vitae may be used. **Phosphor lignum,** of the latter company, is a hardwood, oil-impregnated bearing. **Powdiron,** of the Bound Brook Oil-Less Bearing Co., is a porous iron sintered bearing that will hold up to 25% lubricant. A **microencapsulated nonsolid lubricant** is an integral part of a bearing liner material produced by Sargent Industries. The capsules of lubricant, either a liquid or paste, close to the surface break first and provide lubrication. Those further away break open progressively to give lubrication as needed.

BEECH. The wood of several species of beech trees, *Fagus atropunicea, F. ferruginea,* and *F. grandifolia,* common to the eastern parts of the United States and Canada. The wood is strong, compact, fine-grained, durable, and of a light color similar in appearance to maple. The weight is 47 lb/ft^3 (753 kg/m^3). It is employed for tool handles, shoe lasts, gunpowder charcoal, veneer, cooperage, pulpwood, and for small wooden articles such as clothespins. The beech formerly used for aircraft, *F. grandifolia,* has a specific gravity, oven-dried, of 0.66, a compressive strength perpendicular to the grain of 1,670 lb/in^2 (12 MPa), and a shearing strength parallel to the grain of 1,300 lb/in^2 (8 MPa). The wood may be obtained in large pieces, as the tree grows to a height of 100 ft (30.5 m) and a diameter of 4 ft (0.2 m). It grows from the Gulf of Mexico northward into eastern Canada. **White beech** refers to the light-colored heartwood. **Red beech** is from trees with dark-colored heartwood. The sapwood of beech is white tinged with red and is almost indistinguishable from the heartwood. The wood is noted for its uniform texture and its shock resistance.

Antarctic beech, *F. antarctica,* known locally as **rauli,** grows extensively in southern Chile. It is commonly called by the Spanish word **roble,** or oak, in South America, and is used for cooperage to replace oak. It has a coarser grain than American beech. **European beech,** *F. sylvatica,* is reddish in color, has a close, even texture, is not as heavy as American beech, but is used for tools, furniture, and small articles. **New Zealand beech,** known as red beech and **tawhai,** is from the very large tree *Nothofagus solandri.* The wood weighs 44 lb/ft^3 (705 kg/m^3), is brown in color, and has high strength and durability. **Silver beech,** of New Zealand, is *N. menziesii.* The trees grow to a height of 80 ft (24.3 m) and a diameter of 2 ft (0.61 m). The wood is light-brown, straight- grained, strong, and weighs 34 lb/ft^3 (545 kg/m^3). It is used for furniture, implements, and cooperage.

BEEF. The edible meat from full-grown beef cattle, *Bos taurus*. The meat from the younger animals that have not eaten much grass is called **veal** and is lighter in color and softer. The production of beef and beef products is one of the great industries of the world. In the industrial countries, much of the beef is prepared in organized packing plants, but also the production from city slaughterhouses is important. After slaughter and preparation of the animal, the beef is marketed in animal quarters either chilled or frozen. Fresh-killed beef from local slaughterhouses is also chilled to remove animal heat before marketing. The amount of marketable beef averages 55 to 61% of the live weight of the animal. The hide is from 5 to 7%, the edible and inedible fat and tallow are 3.5 to 7.5%, and the bones, gelatin, and glue material are 2.8 to 4.9%. From 10 to 17% of the live weight may be shrinkage and valueless materials, although the **tankage,** which includes entrails and scraps, is sold as fertilizer. **Offal** includes tongues, hearts, brains, tripe (stomach lining), livers, tails, and head, and may be from 3 to 5.5% of the live weight. The **glands** are used for the production of insulin. **Lipids** is the name for a yellow waxy solid melting at 100°C, extracted from beef spinal cord after removal of cholesterol. It contains phosphatides and complex acids, and is used in medicine as an emulsifier and anticoagulant. **Cortisone,** used in medicine, is a steroid produced from ox bile, but now made synthetically.

Canned beef, which includes **corned beef, canned hash** (beef mixed with potatoes), and various **potted meats,** is not ordinarily made from the beef of animals suitable for sale as chilled or frozen beef, but is from tough or otherwise undesirable meat animals, or from animals rejected by government inspectors as not suitable for fresh beef. In the latter case the beef canned is held at high temperature for a sufficient length of time to destroy any bacteria likely to be in the fresh meat. Federal specifications for canned corned beef require freedom from skin, tendons, and excessive fat, and a maximum content of not more than 3.25% salt and 0.2 saltpeter. Government inspection of beef for health standards is rigid, but the Federal grading of beef is little more than a rough price evaluation.

Beef extract was first made by Prof. Justus von Liebig in 1840 as a heavy concentrated paste that could be kept indefinitely. It is now made on a large scale in both paste and cube forms, and used for soups and hot beverages, but much of the extract marketed in bouillon cubes is highly diluted with vegetable protein. The so-called nonmeat beef extract is made with corn and wheat hydrolysates and yeast. Pure **nonfat beef extract** is used in the food-processing industry for soups, gravies, and prepared dishes. The extract of International Packers, Ltd., is a paste of 17% moisture content. It contains thiamine, niacin, riboflavin, pyrodoxine, pantothenic acid, vitamins B_1 and B_{12}, purine, creatine, and the nutrient proteins found only in meat. **Dehydrated beef** is lean beef dried by mechanical means into flake or powder forms. It is semicooked and, when wet with water, resumes its original consistency but has a somewhat cooked taste. Its advantage is the great saving in shipping space. Beef is also marketed in the form of **dried beef,** usually sliced and salted. **Jerked beef,** or **tasajo,** is beef that has been cut into strips and dried in the sun. It is used in some Latin American countries, but has a strong taste.

BEESWAX. The wax formed and deposited by the honey bee, *Apis mellifera.* The bees build combs for the reception of the honey, consisting of two sheets of horizontal, six-angled prismatic cells formed of wax. After the extraction of the honey, the wax is melted and molded into cakes. New wax is light yellow, but turns brown with age. It may be bleached with sunlight or with acids. It is composed largely of a complex long-chain ester, **myricil palmitate,** $C_{15}H_{31}COOC_{30}H_{61}$, and **cerotic acid,** $C_{25}H_{51}COOH$. The specific gravity is 0.965 to 0.969, and the melting point 63°C. It is easily colored with dyes, and the Germans marketed powdered beeswax in various colors for compounding purposes. Beeswax is used for polishes, candles, leather dressings, adhesives, cosmetics, molded articles, as a protective coating for etching, and as a filler in thin metal tubes for bending. It is frequently adulterated with paraffin, stearin, or vegetable waxes, and the commercial article may be below 50% pure. Standards for the Cosmetics, Toiletry, and Fragrance Assoc. require that it contain no carnauba wax, stearic acid, paraffin, or ceresin, and show no more than 0.01% ash content. Beeswax is produced in many parts of the world as a by-product of honey production from both wild and domesticated bees, the honey being used as a sweetening agent or for the making of alcoholic beverages. **Honey** varies greatly in flavor owing to the different flowers upon which the bees feed, but the chemical properties of both the honey and the wax vary little. Honey is composed largely of fructose. In the food industry small proportions are added to the sugar to enhance the flavor of cookies and bakery products. Honey, normally 82% solids, is also dehydrated to a free-flowing **honey powder** used in confectionery. Sugar may be added to raise the softening temperature and make the powder more resistant to caking. West Africa produces much wax from wild bees. Abyssinia is a large producer of beeswax, where the honey is used for making **tej,** an alcoholic drink. The ancient drink known as **mead** was a fermented honey solution. **Scale wax** is produced by removing the combs from the hives, thus forcing production of wax which is dropped in scales or particles by the bees and prevented from being picked up by a screen.

BELL METAL. A bronze used chiefly for casting large bells. The composition is varied to give varying tones, but the physical requirements are that the castings must be uniform, compact, and fine-grained. The standard is 78% copper and 22 tin, which alloy weighs 0.312 lb/in^3 (8.6 g/cm^3), is yellowish red, has a fine grain, is easily fusible, and gives a clear tone. Increasing the copper slightly increases the sonorous tone. Large bells of deeper tone are made of 75% copper and 25 tin. Big Ben, at Westminster, cast in 1856, contains 22 parts copper and 7 tin. Another bell metal, containing 77% copper, 21 tin, and 2 antimony, is harder, giving a sharper tone. An alloy for fire-engine bells contains 20% tin, 2 nickel, 0.1 silicon for deoxidation, and the balance copper. The nickel reduces the tendency to embrittlement from pounding. A bell metal marketed by the Lumen Bronze Co. contains 80% copper and 20 tin, deoxidized with phosphorus. **Silver bell metal,** for bells of silvery tone, is a white alloy containing 40% copper and 60 tin. This type of alloy, with tin contents up to as high as 60%, is also used for valves and valve seats in food machinery.

BENTONITE. A **colloidal clay** which has the property of being hydrophilic, or water-swelling, some clay absorbing as much as five times its own weight of water. It is used in emulsions, adhesives, for oil-well drilling mud, to increase plasticity of ceramic clays, and as a bonding clay in foundry molding sands. In combination with alum and lime it is used in purifying water as it captures the fine particles of silt. Because of its combined abrasive and colloidal properties it is much used in soaps and washing compounds. It is also used as an absorbent in refining oils, as a suspending agent in emulsions, and in lubricants.

Bentonite occurs in sediment deposits from a few inches to 10 ft (3 m) thick. It is stated to have been formed through the devitrification and chemical alteration of glassy igneous materials such as volcanic ash, and is a secondary mineral composed of deposits from the mineral **leverrierite,** $2Al_2O_3 \cdot 5SiO_2 \cdot 5H_2O$, crystallizing in the orthorhombic system, though some of the bentonite marketed may be montmorillonite. The finely powdered bentonite from Wyoming was originally called **wilkinite.** Wyoming bentonite is characterized by a very sticky nature and soapy feel when wet, and it is highly absorbent. Bentonites are usually light in color, from cream to olive green. Some have little swelling property, and some are gritty. The material from Otay, Calif., has been called **otaylite.** It is brownish and not as highly colloidal as **Wyoming bentonite.** Analyses of bentonites from various areas vary from 54 to 69% silica, 13 to 18 alumina, 2 to 4 ferric oxide, 0.12 to 3.5 ferrous oxide, 1 to 2.2 lime, 1.8 to 3.6 magnesia, 0.1 to 0.6 titania, 0.5 to 2 soda, and 0.14 to 0.46 potash. The material known as **hectorite** from California is lower in silica and alumina, and higher in magnesia and lime. In general, the highly colloidal bentonites contain the highest percentages of soda which have been adsorbed by the clay particles. Most crude bentonites contain impurities, but are purified by washing and treating.

Bentonites are marketed under various trade names such as **Volclay** of the American Colloid Co., **Refinite** of the Refinite Zeolite Co., and **Eyrite** of the Baroid Division, NL Industries, Inc. **Bentone,** produced in various grades by the latter is purified montmorillonite. It is a fine white powder of 200 mesh, and is used as a gelling agent for emulsion paints, adhesives, and coatings. **Bentone 18-C** is an organic compound of the material used for gelling polar organic materials such as cellulose lacquers and vinyl solutions.

BENZENE. Also called **benzol.** A colorless, highly inflammable liquid of the composition C_6H_6. It is an aromatic hydrocarbon obtained as a by-product of coke ovens or in the manufacture of gas, and also made synthetically from petroleum. Its molecular structure is the closed **benzene ring** with six CH groups in the linkage, which forms a convenient basic chemical for the manufacture of styrene and other chemicals. It is also an excellent solvent for waxes, resins, rubber, and other organic materials. It is also employed as a fuel or for blending with gasoline or other fuels. Industrially pure benzene has a distillation range from 78.1 to 82.1°C, a specific gravity of 0.875 to 0.886, and a flash point below 60°F (15.5°C). The pure nitration grade, used for nitrating and for making organic chemicals, has a 1°C boiling range starting not below 79.2°C, and a specific gravity of 0.882 to 0.886. Benzene has a characteristic order, is soluble in alcohol but insoluble in water, and all of its combinations are toxic. The terms **aromatic chemicals** and **aromatics** refer to all of the chemicals made from the benzene ring.

Nitrobenzene, $C_6H_5NO_2$, is a highly poisonous and inflammable liquid made by the action of nitric and sulfuric acids on benzene, used in soaps and cosmetics. It is called **myrbane oil** as a perfuming agent. The nitrated derivative called **benzedrine,** or **amphetamine,** originally used by wartime pilots to combat fatigue, is **phenylaminobenzine,** $C_6H_5 \cdot CH_2 \cdot CH(NH)_2 \cdot CH_3$. It is used in medicine to control obesity, but it is a stimulant to the central nervous system and is habituating. The isomer dextroamphetamine is *d*-phenylaminopropane sulfate, commonly called **Dexedrine.** It causes a rise in blood pressure and stimulates cerebral activity which lasts several hours, but it has a depressant effect on the intestinal muscles, causing loss of appetite and delayed activity of the stomach with other side effects.

Diphenyl carbonate, $(C_6H_5)_2CO_3$, is much used for the manufacture of chemicals where two benzene rings are desired. It is a white crystalline water-insoluble solid melting at 78°C. **Benzyl alcohol,** $C_6H_5CH_2OH$, is a colorless liquid soluble in water, having a boiling point of 205.2°C and a freezing point of −15.3°C. It is also called **phenylcarbinol,** and is used as a solvent for resins, lacquers, and paints. **Benzyl chloride,** $C_6H_5CH_2Cl$, is a colorless liquid of specific gravity 1.103 and boiling point 179°C, which was used as a lachrymatory gas, and is employed in the production of plastics. **Benzyl cellulose** is a thermoplastic plastic of the Imperial Chemical Industries, Ltd., produced by the action of benzyl chloride and caustic soda on cellulose. The plastic is nonflammable, resistant to acids, can be molded easily, and is produced in various grades by different degrees of benzylation. **Benzyl dichloride,** $C_6H_5CH \cdot Cl_2$, is a liquid heavier than benzyl chloride and has a higher boiling point, 212°C, but was also used as a war gas. It is also called **benzylidene chloride** and is used for producing dyestuffs.

BERYLLIUM. A steel-gray, lightweight, very hard metallic element, symbol Be, formerly known as **glucinum.** The specific gravity is 1.847, melting point 1285°C, and hardness 170 Brinell. The metal is produced in small crystalline lumps by chemical reduction, remelted, and converted to **beryllium powder** of 100 to 325 mesh, 99.6% pure. The wrought metal, obtained by powder metallurgy and hot rolling, has a tensile strength up to 90,000 lb/in^2 (620 MPa) with elongation of 15%.

Beryllium is not a tonnage metal, and the cost of finished sheet and forms ranks it almost as a precious metal. Its chief commercial use is for hardening copper and nickel for beryllium bronzes. It will also harden 18-karat gold to 300 Brinell. However, the pure metal has many uses where its high cost is permissible, chiefly in military, electronic, and atomic applications. As a structural metal, it is one-third lighter than aluminum, its stiffness-to-weight ratio is roughly six times greater than that of the ultrahigh-strength steels, and its melting point approaches that of steel. It also has excellent thermal conductivity and is nonmagnetic and a good conductor of electricity.

However, beryllium's hexagonal crystal structure combined with a high sensitivity to impurities results in an almost total lack of room-temperature ductility. This inherent brittleness seriously limits its otherwise outstanding structural service performance, as well as its fabrication. Ductility improves considerably between 390 and 750°F (200 and 400°C), but beryllium becomes brittle again above 930°F (500°C). Another limitation, its toxicity if inhaled or ingested, means that special precautionary measures are required in processing and handling, particularly when the metal is in powder or vapor form.

Most beryllium shapes and parts are produced from hot-pressed-powder block forms. Sheet, extrusion, rod, and bar are available. Beryllium foil, down to 0.0005 in (0.0013 cm) thick, is used in vacuum tubes. Wire is produced in diameters from 0.005 in (0.0127 cm) and larger down to as fine as 0.001 in (0.0025 cm). Beryllium mill products are used in specialized applications, such as nuclear systems, reentry vehicles, aircraft brakes, and satellite parts.

Because of the embrittlement problem, beryllium is most useful as an alloying element and as a composite constituent. In composites, it is used in the form of wire or particles in matrices of titanium or aluminum. For example, beryllium-wire-reinforced aluminum sheet used for pressure-bottle applications has a tensile strength of 85,000 lb/in^2 (586 MPa), a modulus of elasticity of 25 million lb/in^2 (172,250 MPa), and 5% elongation. Beryllium-reinforced titanium-alloy composites have strengths of 140,000 lb/in^2 (965 MPa) and a modulus of elasticity of 27 million lb/in^2 (186,030 MPa).

Perhaps the best-known **beryllium-aluminum composite** is **Lockalloy,** in which beryllium particles are embedded in a ductile aluminum matrix. Developed specifically for aerospace applications, one such composite with 33% aluminum has a tensile strength of 61,000 lb/in^2 (420 MPa) and a modulus of elasticity of 29 million lb/in^2 (199,810 MPa) in the extruded and annealed condition.

The low atomic number and low density of beryllium make it highly pervious even to long X-rays, the permeability to X-rays being about 17 times that of aluminum. It has also the lowest neutron cross section of any metal melting above 500°C. It is valued for X-ray windows and for atomic applications. It is a good source of neutrons when bombarded by alpha rays. It has a high capacity for conducting and absorbing heat, its specific heat factor being more than four times that of titanium, and it is thus used for heat sinks in missiles to absorb the frictional heat. The metal is nonmagnetic. It is not resistant to mineral acids, and above 900°C it is attacked by nitrogen to form **beryllium nitride,** Be_3N_2. The atom of beryllium is not spherical, and the unit molecule appears to be a flat hexagon. The structure of the metal grains is thus cryptocrystalline.

BERYLLIUM COPPER ALLOYS. **Beryllium bronze** or **beryllium copper** refers to alloys of copper and beryllium containing usually not more than 3% beryllium. The alloys are tough, and have a bronzelike crystalline structure. They are noted for their fatigue resistance, and were first employed in Germany for locomotive bearings and for springs. Beryllium bronze is now employed for springs, nonsparking tools, plastic molds, and strong mechanical parts. Its fatigue resistance is utilized in setting diamonds in drill bits. Silicon in small amounts is added to beryllium bronze to harden and strengthen the alloy further by the formation of Be_2Si. Nickel and cobalt also form chemical compounds with beryllium. A small amount of nickel refines the grain and increases the ductility. It is especially useful for this purpose in castings. A very small amount of iron, 0.25%, improves the structure and increases the hardness. A cast beryllium bronze produced by Ampco Metals, Inc., has 2.5% beryllium, and has a hardness of 325 to 375 when heat-treated. **Berylco alloy 165,** produced by the Beryllium Corp. in rods and strip, contains 1.7% beryllium. The No. 1 hard alloy has a tensile strength of 100,000 lb/in^2 (689 MPa), can be bent 180° without fracture, and has an electric conductivity 23% that of copper. **Beryldur,** of this company, is a lower-cost alloy with 1% beryllium. In thin sheet it has good formability and a tensile strength, when age-hardened, of 135,000 lb/in^2 (930 MPa). The electric conductivity is 18 to 26% that of copper. It is employed for such uses as switchgear, as it is corrosion- and erosion-resistant. Alloys for plastic molds usually have about 2% beryllium, and will harden to 365 Brinell. These alloys are also used for dies for drawing alloy steels, as they do not pick up metal on the drawing radius.

The wrought alloy called beryllium copper by the American Brass Co. has 2 to 2.5% beryllium and 0.25 to 0.50 nickel. The soft material has a tensile strength of 70,000 lb/in^2 (482 MPa), elongation 45%, and hardness 110 Brinell, while the heat-treated alloy has a tensile strength of 193,000 lb/in^2 (1,330 MPa), elongation 2%, and hardness 365 Brinell. It is valued for springs. Beryllium copper of this kind is also used for spot-welder contacts, and about 1.5% titanium may be added to give hardness stability at elevated temperatures. Wrought beryllium bronze is marketed regularly in the form of rods, wire, and strip as a product of the brass mills. A standard alloy contains 2% beryllium, 0.60 nickel plus cobalt, and the balance copper. The annealed alloy has a tensile strength of 60,000 lb/in^2 (413 MPa), and a Rockwell B hardness of 70. When precipitation-hardened, the tensile strength is 160,000 lb/in^2 (1,103 MPa), and the Rockwell C hardness is 40. Or the alloy can be cold-rolled to a hardness of Rockwell B95 with a tensile strength of 100,000 lb/in^2 (690 MPa), and then precipitation-hardened to Rockwell C45 with a tensile strength of 185,000 lb/in^2 (1,275 MPa). **Berylco 717C,** of the Beryllium Corp., has 68.8% copper, 30 nickel, and 0.5 beryllium. The heat-treated castings have a tensile strength of 118,000 lb/in^2 (813 MPa), elongation of 12%, and a hardness of Rockwell C23. **Viculoy** is the name of a group of beryllium bronzes of the Akron Bronze & Aluminum Co. for high-strength castings to replace aluminum bronze and manganese bronze for gears and marine parts.

Wrought alloys of high electric conductivity, 50% that of copper, have no more than 0.50% beryllium, with 1.5% cobalt to give a heat-treated strength up to 130,000 lb/in^2 (896 MPa). **Ampcoloy 91,** of Ampco Metals, Inc., **Tuffalloy 55,** of the Welding Sales & Engineering Co., and **Trodaloy 1,** of the General Electric Co., are **beryllium-cobalt alloys** with 2.6% cobalt, 0.40 beryllium, and the remainder copper. The cast metal has a tensile strength of 90,000 lb/in^2 (620 MPa), elongation 10%, and electric conductivity 45% that of copper. The hardness is 96 Rockwell B. The wrought metal, used for springs and tips for soldering irons, has a tensile strength of 60,000 lb/in^2 (413 MPa), and will withstand higher temperatures than regular beryllium bronze. This alloy, of the Wilber B. Driver

Co., for resistance wire, is called **Beraloy C.** The **Trodaloy 7,** of the General Electric Co., contains 0.40% chromium, 0.10 beryllium, and the balance copper. It is used for high-strength electrical parts. A high-conductivity beryllium bronze for electrical parts has 0.50% beryllium and a small amount of silver. Tensile strength is 90,000 lb/in^2 (620 MPa), and the electric conductivity is 60 to 65% that of copper.

A beryllium-copper-cobalt alloy intermediate in physical characteristics between beryllium bronze and phosphor bronze contains only 0.5% beryllium, with 2 to 3 cobalt, and the balance copper. The heat-treated alloy has a hardness of 92 to 98 Rockwell B, and an electric conductivity 50 to 60% that of copper. **Beryllium alloy 50,** of the Beryllium Corp., contains 0.35% beryllium, 1.55 cobalt, and 1 silver. The tensile strength is up to 130,000 lb/in^2 (896 MPa), and the electric conductivity is 50% that of copper. **Silvercote wire,** of the Little Falls Alloys, contains 0.5% beryllium, 2.5 cobalt, and the balance copper. The electric conductivity is 65 to 70% that of copper, and the strength and fatigue resistance are comparable with higher beryllium alloys. The wire is lightly silver-plated for easy soldering.

A **beryllium-nickel alloy** for springs has 1.9% beryllium, 0.50 manganese, and the remainder nickel. The tensile strength, cold-rolled and heat-treated, is 200,000 lb/in^2 (1,378 MPa), with elongation of 6%. It is noted for high torsional endurance. **Beryllium-nickel 260-C,** of the Brush Beryllium Co., has 2.55 to 2.8% beryllium, 0.40 carbon, and the balance nickel. The cast metal has a tensile strength up to 125,000 lb/in^2 (861 MPa) and Rockwell hardness C30. After heat treatment the strength is up to 210,000 lb/in^2 (1,447 MPa). The alloy is magnetic. Its electric conductivity is less than 3% that of copper. It is used for valves, turbine blades, and molds for glass.

BERYLLIUM ORES. Beryllium occurs widely distributed in possible recoverable quantities in more than 30 minerals, but the chief ore is **beryl,** $3BeO \cdot Al_2O_3 \cdot 6SiO_2 \cdot H_2O$. This mineral is usually in pale-yellow rhombic crystals in pegmatic dikes. The crystals are ¼ to ½ in (0.64 to 1.27 cm) in diameter, with a specific gravity of 2.63 to 2.90, and a Mohs hardness of 7.5 to 8. The ore is resistant to acid attack and requires calcining to make it reactive, though the ore of Utah, called **vitroite,** is of simpler composition and can be acid-leached. Beryl ore may contain up to 15% beryllium oxide, but most ore averages below 4%. The Indian ore contains a minimum of 12% BeO, and the beryl of Ontario has 14% BeO, or 5% metallic beryllium. The ore of Nevada contains only 1% BeO, but can be concentrated to 20%.

The secondary ores of beryllium, **bertrandite, herderite,** and **beryllonite** usually have only small quantities of BeO disseminated in the mineral, but Utah clay from Topaz Mountain, Utah, in which the bertrandite is associated with pyrolusite, fluorspar, opal, and mixed with montmorillonite and other clays, is concentrated by flotation, acid-leached, and chemically processed to 97% BeO. Other ores of beryllium are **chrysoberyl,** $BeO \cdot Al_2O_3$, and **phenacite,** which is a **beryllium silicate,** Be_2SiO_4. **Helvite,** $(MnFe)_2(Mn_2S)Be_2(SiO_2)_3$, is in cubical crystals of various colors from yellow through green to dark brown, associated with garnet and having the appearance of garnet. The specific gravity is 3.3, and the hardness 6.5.

Choice crystals of beryl, colored with metallic oxides, are cut as gemstones. **Alexandrite,** a gem variety of chrysoberyl, is emerald green in natural light but red in transmitted or artificial light. Allied Corp. produces a synthetic alexandrite for use as a tunable solid-state laser. The **emerald** is a flawless beryl-colored green with chromium. High-grade natural emeralds are found in Colombia, but occur in the United States only in North Carolina. The rose-pink, rose-red, and green beryl crystals of Malagasy, called **morganite,** are cut as gemstones, and the dark-blue stone is made by heating the green crystals. The pale blue-green crystals are **aquamarines,** and the **heliodor** is **golden beryl** from Southwest Africa. But the yellowish-green gemstone of Brazil, called **brazilianite,** is not beryl, but is a hydrous sodium-aluminum phosphate, and is softer. Crystals of chrysoberyl of lemon-yellow color found in Brazil are valued as gemstones. Synthetic emerald of the composition $3BeO \cdot Al_2O_3 \cdot 6SiO_2$ was first made in Germany by heat and pressure under the name of **Igmerald.** Synthetic emeralds are now grown from high-purity alumina, beryllia, and silica, with traces of Cr_2O_3 and Fe_2O_3 to give the green color. Synthetic beryl is used for bearings in watches and instruments.

BERYLLIUM OXIDE. A colorless to white crystalline powder of the composition BeO, also called **beryllia,** and known in mineralogy as **bromellite.** It has a specific gravity of 3.025, a high melting point, about 2585°C, and a Knoop hardness of 2,000. It is used for polishing hard metals and for making hot-pressed ceramic parts. Its high heat resistance and thermal conductivity make it useful for crucibles, and its high dielectric strength makes it suitable for high-frequency insulators. Single-crystal **beryllia fibers,** or whiskers, developed by the National Beryllia Corp., have a tensile strength above 1 million lb/in^2 (6,800 MPa).

Ceramic parts with beryllia as the major constituent are noted for their high thermal conductivity, which is about three times that of steel, and second only to that of the high-conductivity metals (silver, gold and copper). They also have high strength and good dielectric properties. Properties of typical grades of **beryllia ceramics** are: tensile strength, 14,000 lb/in^2 (96 MPa); compressive strength, 300,000 lb/in^2 (2,068 MPa); hardness (micro), 1300 Knoop; maximum service temperature, 4350°F (2400°C); dielectric strength, 5.8. Beryllia ceramics are costly and difficult to work with. Above 3000°F (1650°C) they react with water to form a volatile hydroxide. Also, because beryllia dust and particles are toxic, special handling precautions are required. Beryllia parts are used in electronic, aircraft, and missile equipment.

Beryllia is also used in ceramics to produce gastight glazes, and for this purpose was called **Degussit** in Germany. Thin films of the oxide are used on silver and other metals to protect the metal from discoloration. Very thin films are invisible, but heavier films give a faint iridescence. Two other beryllium compounds used especially in chemical manufacturing are **beryllium chloride,** $BeCl_2$, a water-soluble white powder melting at 440°C, and **beryllium fluoride,** BeF_2, melting at 800°C. Another beryllium compound, useful for high-temperature, wear-resistant ceramics, is **beryllium carbide,** Be_2C. The crystals have a hardness of 9 Mohs, and the compressed and sintered powder has a compressive strength above 100,000 lb/in^2 (690 MPa). **Berlox,** of the National Beryllia Corp., is a beryllium oxide powder in particle sizes from 80 to 325 mesh for flame-sprayed heat- and wear-resistant coatings.

BESSEMER STEEL. Steel made by blowing air through molten iron. The process, developed by Henry Bessemer in England in 1860, made possible for the first time the production of steel on a large scale. The process is now employed extensively as a preliminary step in the production of other steels, but much bar steel is made by this process. Ferromanganese and sometimes steel scrap are added to the steel when pouring into the ladle in order to regulate the content. In the blowing process the chemical action between the oxygen of the air and the molten mass increases the temperature, and the air then forms the chief fuel as the carbon is oxidized and driven off. The blowing requires only a few minutes, and the carbon is reduced to 0.04% or less. The carbon desired in the steel is then regulated by the addition of carbon to the melt. The two processes, known as acid and basic, differ in the type of refractories employed for lining the converters, and there is a difference in the resulting steel since the acid process does not remove as much sulfur and phosphorus. The acid process is used principally in the United States, and the basic process is employed in Europe where the product is called **Thomas steel.** The lining of an acid converter may be ganister or other refractory acid material. The Brassert process, invented in Austria, makes a high-quality bessemer steel by blowing oxygen through the molten steel. The oxygen must be pure, 99.5%, and nitrogen-free.

Acid **bessemer pig iron** should contain about 1% silicon, but the sulfur and phosphorus must be low. An acid bessemer steel to be free-cutting has a content of 0.09 to 0.13% phosphorus and 0.075 to 0.15 sulfur, but these amounts are too high for structural steels, so that bessemer steel is mostly employed in rod form for making screw-machine products. A good-quality acid bessemer steel contains about 0.15% carbon, 0.40 to 0.80 manganese, and up to 0.08 each of sulfur and phosphorous. For basic bessemer steel the converter has a basic lining of burned dolomite, and the basic bessemer iron has less silicon to avoid the production of much silica. The high-phosphorus pig irons of Europe are made into steel by this process, as the basic lining aids in the elimination of the phosphorus and sulfur, although lime is also added at the beginning of the blow. Low-carbon, low-manganese, basic bessemer steel, deoxidized with aluminum and called **killed steel,** is used to replace open-hearth steels for cold-forming applications. The tensile strength is 47,000 to 57,000 lb/in^2 (324 to 392 MPa), with elongations from 28 to 37%. **Synthetic steel scrap** to replace ordinary steel scrap for open-hearth steel can be made in bessemer converters by partly blowing and then casting into ingots for the furnace charge.

BIRCH. The wood of the birch trees, of which more than 15 varieties grow in the northeastern and Lake states of the United States and in Canada, and other varieties in Europe and north Asia. The birch of north Europe is called **Russian maple.** The wood of the American birches has a yellow color, is tough, strong, hard, and close-textured, and polishes well. It has a fine wavy grain sometimes beautifully figured, and can be stained to imitate cherry and mahogany. Birch is used in construction work for trim and paneling, for furniture, and for turned articles such as handles, shoe pegs, clothespins, toys, and woodenware. The lumber usually includes the wood of several species. It has a specific gravity, oven-dried, of 0.68, a compressive strength perpendicular to the grain of 1,590 lb/in^2 (10 MPa), and a shearing strength parallel to the grain of 1,300 lb/in^2 (9 MPa). **Yellow birch,** *Betula lutea,* highly prized for furniture, is now getting scarce. It is also called **silver birch** and **swamp birch.** The commercial wood includes that from the **gray birch,** *B. populifolia.* **Sweet birch,** *B. lenta,* ranks next in importance. It is called **black birch, cherry birch,** and **mahogany birch,** and may be marketed together with yellow birch. Sweet birch may also include **river birch,** *B. nigra,* but sweet birch is a heavier and stronger wood. **Paper birch,** *B. papyrifera,* is the variety known as **canoe birch** because the silvery-white flexible bark was used by the Indians in making canoes. It is also referred to as silver birch, and is much used for pulpwood. It is similar to and mixed with **white birch,** *B. alba,* the wood of which is strong, elastic, and uniform, and is much used in Vermont and New Hampshire for making spools, bobbins, handles, and toys. Yellow birch of the Canadian border reaches a height of 60 to 80 ft (18.3 to 24.4

m) and a diameter up to 2 ft (0.61 m). A 50-year-old tree has a diameter of about 15 in (0.38 m) and a height of 40 ft (12 m).

Birch oil is a viscous, yellowish, poisonous oil of specific gravity 0.956, with a characteristic birch odor, obtained by distilling **birch tar,** a product of the dry distillation of the wood of the white birch. It contains phenols, cresol, and xylenol, and is used in disinfectants and in pharmaceuticals. It is also called **birch tar oil,** and in pharmacy is known as **oil of white birch. Sweet birch oil,** also known as **betula oil,** is a lighter volatile oil distilled from the steeped bark of the *B. lenta,* or sweet birch. It contains methyl salicylate, and is used as a flavoring agent, in perfumes, in dressing fancy leathers, in cleaning solutions and soaps, and as a disinfectant to neutralize odors of organic compounds.

BISMUTH. An elementary metal, symbol Bi, sometimes occurring native in small quantities. American bismuth is obtained chiefly as a by-product in the refining of lead and copper. Foreign bismuth comes largely from the mineral bismuthinite. The metal has a grayish-white color with a reddish tinge, is very brittle, and powders easily. It is highly crystalline in rhombohedral crystals. It has few uses in its pure state. The specific gravity is 9.75, melting point 271°C, and hardness 73 Brinell. The thermal conductivity is less than that of any other metal except mercury, and it is the most diamagnetic of all the metals. It is one of the few metals that increases in volume upon solidification. It expands 3.32% when changing from the liquid to the solid state, which makes it valuable in type-metal alloys and in making small castings where sharp impressions of the mold are needed. The metal imparts to lead and tin alloys hardness, sonorousness, luster, and a lowered melting point. By regulating the amount of bismuth, it is possible to cast the alloys to fill the mold without expansion or contraction on cooling. It is used in white alloys for molds for casting plastics, and because of the property of lowering the melting point it is valued in fusible alloys and in soft solders. Very fine **bismuth wire** used for thermocouples is drawn in glass tubes, but extruded bismuth wire is marketed in diameters from 0.003 to 0.039 in (0.0062 to 0.0991 cm), in ductile enough to be wound on its own diameter without fracture.

Bismuth steel, with a very small content of bismuth, is used for transformer sheets. It has increased electrical resistance without diminished electrical permeability and lowered hysteresis. About 0.5% bismuth is used in some 18–8 stainless steels instead of selenium to add machinability without impairing the corrosion resistance. Bismuth is also used in amalgams, and is employed in the form of its salts in pigments, in pharmaceuticals as an antacid, and in many chemicals. For medicinal purposes bismuth must be completely free of traces of arsenic. The paint pigment known as **pearl**

white is **bismuth oxychloride,** BiOCl, a white crystalline powder of specific gravity 7.717, insoluble in water. Another bismuth pigment is **bismuth chromate,** $Bi_2O_3 \cdot 2CrO_3$, a water-insoluble orange-red powder. The material known in medicine as bismuth is **bismuth phosphate,** $BiPO_4$, a white powder insoluble in water.

BISMUTHINITE. An ore of the metal bismuth, found in Bolivia, Peru, central Europe, Australia, and the western United States. It is **bismuth trisulfide,** Bi_2S_3, theoretically containing 81.3% bismuth. The richest Bolivian ores contain more than 25% bismuth, and concentrates from northwest Argentina contain 40 to 48% bismuth. The mineral has a massive foliated structure with a metallic luster, a lead-gray streaked color, and a hardness of 2. The concentrated ore is roasted and smelted with carbon, and the resulting impure bismuth is refined by an oxidizing fusion. Other bismuth ores are **bismite,** or **bismuth ocher,** $Bi_2O_3 \cdot 3H_2O$, containing theoretically 80.6% bismuth, and **bismutite,** $Bi_2O_3CO_3 \cdot H_2O$, containing theoretically 78.3% bismuth, both of which are widely distributed minerals.

BITUMINOUS COAL. Also called **soft coal.** A variety of coal with a low percentage of carbon, and easily distinguished from anthracite by the property of losing moisture and breaking up into small pieces. Because of its cheapness it is the coal used most extensively for industrial fuel, but is not preferred for household use because of its smoke and odor. However, considerable of the fine or powdered coal called **slack coal** is used in making fuel briquettes. Bituminous coal is widely distributed in many countries, and is found in 28 states of the United States. Much bituminous coal is used for the production of coke, coal tar, liquid fuels, and chemicals. The bituminous coals vary in quality from near lignite to the hard grades near anthracite, called **semibituminous coal,** depending upon their geologic age. They are not true bitumens. The specific gravity of clean bituminous coal is 1.75 to 1.80. The best steam coals are the semibituminous grades from West Virginia, Virginia, Pennsylvania, and some parts of the Middle West. The latter is very compact and is extracted in large blocks, called **block coal.** Good coals for industrial use should give 13,500 to 14,500 Btu/lb (74,250 to 79,750 kcal/kg), and should have from 55 to 60% fixed carbon and 30 to 37 volatile matter. The best grades are in lumps. **Coking coals** are the higher-carbon grades low in sulfur, and with capacity to leave the residue coke in large firm lumps. **Sea coal** is a name for finely ground bituminous coal used in sand mixtures for molds for cast iron to prevent fusing of the sand to the castings.

BLACKFISH OIL. A pale-yellow waxy oil extracted from the pilot whale, porpoise, or blackfish, *Globicephala melas,* found off the North Atlantic

Coast as far south as New Jersey, and the *G. ventricosa* of other seas. The blackfish averages 15 to 18 ft (4.6 to 5.5 m) in length, with a weight of about 1,000 lb (454 kg). The oil has a saponification value of 290, iodine value 27, and specific gravity 0.929. The **dolphin oil,** of the common dolphin, *Delphinus delphis,* of all seas, is also classed as blackfish oil, as is also the oil of the killer whale, *Grampus orca,* of all seas. The oil is used as a lubricant for fine mechanisms, in cutting oils, and for treating leather. The product from the head and jaw is of the best quality, and is known as **jaw oil,** although the best grade of sperm oil is also called jaw oil. The jaw oil from blackfish does not oxidize easily, and is free-flowing at low temperatures, having a pour point of −20°F (−29°C). It consists of 71% mixed acids, of which 86% is **valeric acid,** $C_2H_5(CH_3)CHCOOH$, and 13% oleic and palmitic acids. Normal valeric acid is methyl ethyl acetic acid, and can be made by the oxidation of amyl alcohol. It is produced in the human system by the action of enzymes on amino acids. **Arginine,** of General Mills, Inc., used for treatment of shock and for reducing the toxic effects of ammonia in the blood, is a **guanidine valeric acid.** A variant of valeric acid is **levulinic acid,** a liquid boiling at 245°C, used for making synthetic resins. It is produced as a by-product in the production of furfural from corncobs, and has the composition $CH_3COCH_2CH_2COOH$ with reactive methylene groups that undergo condensation readily. It can also be produced from starches and sugars.

BOILER PLATE. Originally, a high-grade plain iron or steel plate of 9/64 in (0.36 cm) thickness or heavier, used for making steam boilers; however, the term came to mean plate of this kind for any purpose, and plates 1/4 in (0.64 cm) thick or thinner are referred to as sheet. Actually, boiler plates for boilers, tanks, and chemical equipment, and **flange plates** for dished ends, are now made in various alloy steels to incorporate high strength, corrosion resistance, and creep resistance, and also in clad steels. Ordinary boiler plate is divided into firebox, flange, and extra-soft. A plain steel **firebox plate** contains not more than 0.30% carbon, 0.30 to 0.50 manganese, and not more than 0.40 each of sulfur and phosphorus. Flange plates contain less carbon. The tensile strength is about 60,000 lb/in^2 (413 MPa). **Boiler-tube steel** may be carbon steel or alloy steels, and may be hot-rolled or cold-drawn. But the tubes are given an internal hydrostatic test and marked with the test pressure used.

Some of the **Croloy alloys** of the Babcock & Wilcox Co. are standard grades of stainless steels. **Croloy 12** is Type 410 stainless, and **Croloy 18** is Type 430 stainless. **Croloy 15–15N** will retain a tensile strength of 60,000 lb/in^2 (413 MPa) at 1200°F (649°C). It contains 15% each of chromium and nickel, 1.5 molybdenum, 1.5 tungsten, 2 manganese, 0.75 silicon, and 0.15 carbon, with about 1.2 columbium-tantalum for stabilizing.

Croloy 16–PH, for extruded shapes, has about 16% chromium, 6 nickel, up to 1 each manganese and silicon, with not over 0.12 carbon. It has a tensile strength of 220,000 lb/in^2 (1,516 MPa) and Rockwell hardness C45. **Croloy 16–1** is a low-nickel stainless steel. It has 16% chromium, 1 nickel, 1 manganese, 0.75 silicon, and a maximum of 0.03 carbon. The tensile strength is up to 95,000 lb/in^2 (654 MPa) with elongation of 28 to 38%. Stainless-steel plate is much used for chemical boilers and pressure vessels for food and chemical plants, but copper and aluminum are also employed for these purposes.

BONES. The dried bones from cattle and from Asiatic buffalo form an important item in international commerce. In general, organized packing plants do not ship much bone, but utilize it for the production of glue, gelatin, bone meal, and fertilizer. The bones shipped from packing plants are called **packer bones,** but the source of much commercial bone is from slaughterhouses, local retail meat shops, and the bones from farms and fields known as **prairie bones** and **camp bones,** the latter name being from the Argentine word campo, meaning field. American imported bone comes from Argentina, Canada, Uruguay, Brazil, South Africa, and India. **Raw bone meal** was the term for the scrap pieces and sawdust from the manufacture of these articles.

Case-hardening bone, for carbonizing steel, and **bone meal,** for animal feed, are processed to remove fats and prevent rancidity, and also steam-sterilized. **Fertilizer bone** is bone meal that is of too fine mesh for carbonizing use or for making bone black, but it is not cooked or processed, and has an analysis of about 45% ammonia and 50 phosphate of lime. Knucklebones, and also the hard shinbones, are preferred for gelatin manufacture. **Dissolved bone,** used for fertilizer, is ground bone treated with sulfuric acid, or the residue bone after dissolving out the gelatin with acid. The **steamed bone meal** produced as a residue by-product of the glue factories and soup factories is suitable only for fertilizer, but it contains only 1% ammonia. The steamed bone meal for animal feed is merely steam-sterilized raw bone.

Bone black, also called **animal charcoal** and **bone char,** is charred bone ground to a fine silky powder for use as a pigment, or as a decolorizing agent for sugar and oils. It has a deep, dense, bluish-black color valued for engraving inks of depth and tone, and to give a dull velvety-black finish to coated paper. Its covering power, however, is inferior to that of carbon black, as it has only about 10% carbon, is largely calcium phosphate, and has a very high ash content. The best blacks may be treated with acid to remove the lime salts. Federal specifications for bone black for pigment require that 97.5% shall pass through a 325-mesh screen. For filtering, the material used is from 4 to 16 mesh. The specific gravity is 2.6 to 2.8. Bone black is made by calcining ground, fat-free, dried bones in airtight retorts.

Drop black is the spent bone black from the decolorizing of sugar, which has been washed and reground for pigment use. **Ivory black,** having the same uses as bone black, is made by heating the refuse of ivory working in a closed retort and then grinding to a fine powder. **Aquablak No. 1,** of the Binney & Smith Co., is a water dispersion of bone black used to give a velvety-black color to inks, water paints, and leather finishes.

Bone oil, used in sheep dips and disinfectants, and in insecticides and fungicides, is a chemically complex oil with a pungent disagreeable odor derived as a by-product in the destructive distillation of bones. It contains nitrides, pyrroles, pyridine, and aniline.

BORAX. A white or colorless crystalline mineral used in glass and ceramic enamel mixes, as a scouring and cleansing agent, as a flux in melting metals and in soldering, as a corrosion inhibitor in antifreeze liquids, as a constituent in fertilizers, in the production of many chemicals and pharmaceuticals, and as a source of boron. Borax is a hydrous **sodium borate,** or tetraborate, first obtained from Tibet under the Persian name **borak,** meaning white. **Tincal** is the natural borax, $Na_2O \cdot 2B_2O_3 \cdot 10H_2O$, obtained originally through Iran from Tibet, and later found in quantity in the western United States. The specific gravity is 1.75, hardness 2 to 2.5, and melting point 1125°F (608°C). It contains 47.2% water, and is readily soluble in water. The borax from California and Nevada known as **colmanite** is **calcium borate,** $2CaO \cdot 3B_2O_3 \cdot 5H_2O$. That known as **kernite,** or **rasorite,** is another variety of sodium borate of the composition $Na_2O \cdot 2B_2O_3 \cdot 4H_2O$. **Ulexite,** or **boronatrocalcite,** is found in the western United States, Chile, Bolivia, and Peru, and has the composition $Na_2O \cdot 2CaO \cdot 5B_2O_3 \cdot 16H_2O$, but the ulexite of Chile and Bolivia is also mixed with sodium and calcium sulfates and sodium and magnesium chlorides. The dried mineral contains 45 to 52% boric acid. **Priceite,** or **pandermite,** chiefly from Asia Minor, is $5CaO \cdot 6B_2O_3 \cdot 9H_2O$. All of these minerals are **boron ores** suitable for producing the metal.

Borax is found in great quantities in the desert regions of the western states, and in the Andean deserts of South America, where the borate deposits are in land-locked basins at altitudes above 11,000 ft (3,353 m). The colmanite is usually associated with ulexite, shale, and clay, and is concentrated to 40% B_2O_3. Kernite ore from Kern County, California, contains about 29% B_2O_3 and is concentrated to 45%. Federal specifications for borax call for not less than 99.5% hydrous sodium borate in grades from large crystals to fine white powder. **Pyrobar** is a crystalline anhydrous borax of the American Potash & Chemical Co., produced from the brine of Searles Lake. **Puffed borax,** of this company, is an expanded borax powder with bulking density as low as 2 lb/ft^3 (32 kg/cm^3) which is very soluble and is used for detergents.

Boric acid, also called **boracic acid** and **orthoboric acid,** is a white, crystalline powder of the composition $B_2O_3 \cdot 3H_2O$, derived by adding hydrochloric or sulfuric acid to a solution of borax and crystallizing. It is also obtained from the boron ores of California, Chile, Bolivia, and Peru as one of a number of chemicals. It occurs naturally in volcanic fissures in Italy as the mineral **sassolite,** $B_2O_3 \cdot 3H_2O$. The specific gravity is 1.435 and melting point 1050°F (562°C). It is soluble in water and in alcohol. Boric acid is used as a preservative and weak antiseptic in glass and pottery mixes, in the tanning industry for deliming skins by forming calcium borates soluble in water, and as a flux in soldering and brazing. **Anhydrous boric acid** is nearly pure **boric oxide.** It is used in glass and ceramic enamels.

BORNEO TALLOW. A hard, brittle, yellowish-green solid fat obtained from the seed nuts of trees of the family *Dipterocarpaceae* of Borneo, Java, Sumatra, and Malaya. They include the *Shorea aptera, S. robusta, S. stenoptera, Hopea aspera,* and *Pentacme siamensis,* which trees also produce copals and valuable woods. The kernels of the nuts of the *S. stenoptera* contain 45 to 60% fat. It contains 39% stearic acid, 38 oleic, 21.5 palmitic, and 1.5 myristic acid. The specific gravity is 0.852 to 0.860, saponification value 185 to 200, iodine value 29 to 38, and melting point 34 to 39°C. The seeds are exported as **Sarawak illipé** nuts or **Pontianak illipé,** but they are not the same as the **illipé nuts** and **illipé tallow** from the species of *Bassia* of India. **Illipé butter,** also called **mowrah butter** and **bassia butter,** is from the seed nuts of the trees *B. longifolia* and *B. latifolia,* of India. The oil content of the seeds is 50 to 60%. The crude fat is yellow to green in color, but the refined product is colorless with a pleasant taste. It is used as a food, and in soaps and candles. **Phulwa butter,** or **Indian butter,** is from the seed nuts of the *B. butyracea* of India. It is white, stiffer than lard, and is used as a food. **Siak tallow** is from the *Palanquium oleosum* of Malaya and various species of *Dipterocarpaceae* of Indonesia. The nuts are smaller than the illipé, and are known as **calam seeds** in Sumatra. The fat is yellow to greenish, and was used as a substitute for cocoa butter. **Jaboty fat** is a similar heavy oil from the kernels of the fruit of the tree *Erisma calcaratum* of Brazil. It contains 44% palmitic acid, 28 myristic, and 4 stearic acid, and has a melting point of 40 to 46°C. The tropical forests of Brazil, Africa, and southern Asia abound in trees that produce a great variety of nuts from which useful fats and oils may be obtained.

BORON. A metallic element, symbol B, closely resembling silicon. When pure, it is in the form of red crystals or a brownish amorphous powder. Crystalline boron in lumps of 99.5% purity is extremely hard, has a metallic sheen, is not hygroscopic and resists oxidation, and is a semiconductor. It has a specific gravity of 2.31, a melting point of about 2200°C, and a

Knoop hardness of 2,700 to 3,200, equal to a Mohs hardness of about 9.3. At about 600°C boron ignites and burns with a brilliant green flame. Obtained by the electrolysis of fused boric oxide from the mineral borax, it is used mostly in its compounds, especially borax and boric acid. Minute quantities of boron are used in steels for case hardening by the nitriding process to form a boron nitride, and in other steels to increase strength and hardness penetration. In these **boron steels** as little as 0.003% is beneficial, forming an iron boride, but with larger amounts the steel becomes brittle and hot-short unless it contains titanium or some other element to stabilize the carbon. In cast steels for tools, as much as 1.5% boron may be used, but these steels cannot be forged or machined. In cast iron, boron inhibits graphitization and also serves as a deoxidizer. It is added to iron and steel in the form of **ferroboron.**

Boron 10, an isotope that occurs naturally in boron in the proportion of 10 to 12%, is 10 times more effective than lead in stopping neutrons, and is used for neutron shielding, but it does not stop gamma rays. **Boron 11,** which comprises 88 to 90% of natural boron, is transparent to neutrons. **Boron steel foil** and sheet for neutron shielding, marketed by the Chromalloy Corp. under the name of **Bo-Stan,** has the composition of a stainless steel with 2% boron evenly distributed. It is made by the sinter-wrought process from exact amounts of the metal powders, compacted under pressure, sintered at below the melting point to bond the metals, and hot-rolled. A similar foil called **Binal** is aluminum with 2% boron.

Boron fibers for reinforced structural composites are continuous fine filaments which are, themselves, composites. They are produced by vapor deposition of boron on a tungsten substrate. Their specific gravity is about 2.6, and they range from 4 to 6 mils (0.10 to 0.15 mm) in diameter. They have tensile strengths around 500,000 lb/in^2 (3,450 MPa), and a modulus of elasticity of nearly 60 million lb/in^2 (0.5 million MPa). The boron fibers are used chiefly in aluminum matrixes and in epoxy resin matrixes. Unidirectional **boron-aluminum composites** have tensile strengths ranging from 110,000 to over 200,000 lb/in^2 (758 to over 1,378 MPa). Their strength-to-weight ratio is about three times greater than that of high-strength aluminum alloys.

Boron compounds are employed for fluxes and deoxidizing agents in melting metals, and for making special glasses. Boron, like silicon and carbon, has an immense capacity for forming compounds, although it has a different valence. The boron atom appears to have a lenticular shape, and two boron atoms can make a strong electromagnetic bond, the B_2 acting like carbon but with a double ring.

Typical of a complex boron compound is the natural mineral **danburite** which occurs in orthorhombic crystals that will fuse to a colorless glass. The composition of the mineral is given as $CaB_2(SiO_4)_2$, but each silicon atom is at the center of a group of four oxygen atoms, two such groups having one oxygen atom in common. The eighth oxygen atom forms another tetrahedral group with one oxygen atom from each of the three Si_2O_7 groups, and in the center of such groups are the boron atoms. Thus, even tiny amounts of boron can affect alloys greatly. Boron is used in specialty refractories, electronics, borane fuels, and high-temperature carborane plastics.

Boron trichloride, BCl_3, is used as a catalyst. Above 12.5°C it is a gas, but it is also used as the dihydrate, a fuming liquid. **Boron trifluoride,** BF_3, is a gas used for polymerizing epoxy resins, usually in the solid form of **boron ethyl amine,** $BF_3 \cdot C_2H_5NH_2$, which releases the BF_3 at elevated temperatures. **Boron tribromide,** BBr_3, is also a highly reactive compound, used to produce boron hydrides as sources of hydrogen. **Hydrazine diboride,** $BH_3 \cdot NH_2 \cdot NH_2 \cdot BH_3$, used as a source of hydrogen in rocket fuels, is a white crystalline free-flowing powder. **Trimethyl boroxine,** $(CH_3O)_3B \cdot B_2O_3$, is a liquid used for extinguishing metal fires. Under heat it breaks down and the molten coating of boric acid smothers the fire.

Carborane plastics are produced from the boron molecule, a boron hydride of the composition $B_{10}H_{14}$, by replacing the four terminal hydrogens with two carbon atoms. The monomer, $C(HB_2H)_5C$, polymerizes at high heat and pressure to form rubbery solids which will withstand temperatures above 600°F (315°C); but the commercial plastics are usually copolymers with vinyls, silicones, or other plastics.

BORON CARBIDE. A black crystalline powder of high hardness used as an abrasive, or pressed into wear-resistant products such as drawing dies and gages, or into heat-resistant parts such as nozzles. The composition is either B_6C or B_4C, the former being the harder but usually containing an excess of graphite difficult to separate in the powder. It can be used thus as a deoxidizing agent for casting copper, and also for lapping, since the graphite acts as a lubricant. **Boroflux,** of the General Electric Co., is boron carbide with flake graphite, used as a casting flux.

The boron carbide marketed by the Norton Co. under the name of **Norbide** is B_4C, over 99% pure. It is used as a hard abrasive, and for molding. Parts molded without a binder have a compressive strength of 400,000 lb/in^2 (2,757 MPa), a tensile strength of 22,500 lb/in^2 (155 MPa), and a Knoop hardness of 2,800. The coefficient of expansion is only one-third that of steel, and the density is 2.51, or less than that of aluminum. The melting point is 4450°F (2454°C), but above 1800°F (982°C) it reacts with oxygen. The material is not resistant to fused alkalies. **Boron carbide powder** for grinding and lapping comes in standard mesh sizes to 240, and in special finer sizes to 800. The usual grinding powders are 220 and 240, and the finishing powder is 320. Boron carbide powder is also added to molten aluminum and then rolled into sheets for use in shielding against neutrons. **Boral,** of the Aluminum Co. of America, is this sheet containing 35% by weight of carbide and clad on both sides with pure aluminum. A ¼-in (0.64-cm) sheet is equal to 25 in (0.64 m) of concrete for neutron shielding.

BORON NITRIDE. A light fluffy white powder of the composition BN, used as a lubricant for high-pressure bearings, and for compacting into mechanical and electrical parts. Its X-ray pattern and platy crystal structure are almost identical with those of graphite, and it is called **white graphite.** Boron nitride has a very low coefficient of friction, but unlike carbon, it is a nonconductor of electricity, and it is attacked by nitric acid. It sublimes at 3000°C. It reacts with carbon at about 2000°C to form boron carbide. It is used for heat-resistant parts by molding and pressing the powder without a binder to a specific gravity of 2.1 to 2.25.

Sintered parts have an ivorylike appearance, a tensile strength of 3,500 lb/in^2 (25 MPa), compressive strength of 45,000 lb/in^2 (310 MPa), and dielectric strength of 1,000 volts per mil (39 $\times$ 10^6 volts per meter). They are soft, with a hardness of Mohs 2, and can be machined easily. But when compressed at very high pressure and high heat, the hexagonal crystal structure is converted into a cubic structure, and the pressed material has great hardness and strength and is stable up to about 3500°F (1925°C). This molded material is resistant to molten aluminum, is not wetted by molten silicon or glass, and is used for crucibles. Cubical crystal boron nitride equaling the diamond in hardness is produced by the General Electric Co. as an abrasive powder under the name of **Borazon.** It is produced at extremely high pressure and temperature, and the tiny crystals are reddish to black, although chemically pure crystals should be colorless to white. The cubic crystal structure has the same pattern as that of the diamond, but with greater atomic spacing. An advantage over the diamond is that it will not disintegrate at temperatures below 3500°F (1925°C). **Boron nitride fibers** are produced in diameters as small as 5 to 7 μm and in lengths to 15 in (0.38 m). The fibers have a tensile strength of 200,000 lb/in^2 (1,378 MPa). They are used for filters for hot chemicals, and as reinforcement to plastic lamination. **Boron nitride HCJ,** of Union Carbide, is a fine powder, 99% pure, used as a filler in encapsulating and potting compounds to add thermal and electric conductivity.

Another boron compound used for the production of high-temperature ceramic parts by pressing and sintering is **boron silicide,** B_4Si. It is a black, free-flowing crystalline powder. The powder offered by the Allis-Chalmers Mfg. Co. is microcrystalline, with particles about 75 μm diameter, and the free silicon is less than 0.15%. This compound normally reacts at 2190°F (1200°C) to form B_6Si and silicon, but when compacted and sintered, the ceramic forms a boron silicate oxygen protective coating, and the parts have a serviceable life in air at temperatures to 2550°F (1400°C). Molded parts have high thermal shock resistance, and can be water-quenched from 2000°F (1093°C) without shattering.

BOTANICALS. Known also as **plant extracts** and **crude drugs,** they are preparations, usually in concentrated form, obtained by chemically treating plant tissue from roots, leaves, stalks, flowers, or bark to remove odiferous, flavorful, or nutritive substances of the tissue for use in medicines, cosmetics, insecticides, and foodstuffs. Their use goes back to ancient times when they were used for herb healing and aromatic bath preparations. Over thousands of years, plant extracts have been the main ingredient in many folk remedy drugs.

Today botanicals find use in sophisticated drugs and in an immense variety of cosmetics. It is estimated that there are approximately 750,000 different species of higher plants on earth. Of these, only about 1% have been studied for possible use as botanicals.

Among the plant extracts used in cosmetics are those from cornflowers, marigolds, wild pansies, hawthorn blossoms, sage, horse chestnut, ivy, juniper, hops, mint, and chamomile. Chamomile is a popular botanical in cosmetics, such as shampoo and skin-care products.

Many of the most valuable **drug plants** contain **alkaloids** which have complex arrangements of carbon, hydrogen, oxygen, and nitrogen. Most of the alkaloids are violent poisons even in small quantities, and with few exceptions, such as quinine, are not employed as medicines except under the direction of trained physicians or pharmacists. The actual amount to produce sufficient effect without injury must be understood before use of a drug. **Pain-killing drugs,** for example, do not kill pain but diminish the conductivity of the nerve fibers, and thus prevent the brain from recording the pain. The alkaloid **hyoscine,** or **scopolamine,** a heavy liquid of the composition $C_{17}H_{21}NO_4$, causes loss of part of the normal inhibition control, and is known as **truth serum,** but what is expressed as truth by a mentally disturbed person may be only the product of imagination, and the drug should be given only by trained operators.

Drug plants are almost unlimited in number. **Belladonna,** used to relieve pain, to check perspiration, and as a dilatant, consists of the leaves, roots, flowers, and small stems of the perennial herb *Atropa belladonna* cultivated chiefly in Yugoslavia and Italy, but also grown in the United States. The plant is called **banewort** and **deadly nightshade.** The leaves are dried in the shade to retain the green color. They contain two alkaloids, **atropine** and **hyoscyamine.** Atropine is used as an antidote for military nerve gas. **Henbane** is the dried leaves and flowering tops of the ill-smelling herb *Hyoscyamus niger,* a plant of the nightshade family containing several alkaloids. It is used as a sedative. The plant is grown in southern Europe and Egypt and to some extent in the United States. It is harvested in full bloom and dried in the shade.

Stramonium is the dried leaves, flowers, small stems, and seeds of the **thorn apple** or **jimson weed,** *Datura stramonium,* an annual of the nightshade family which grows as a common weed in the United States. It is shipped from Ecuador under the name of **chamico.** It produces the alkaloids atropine and hyoscyamine, and is used as a nerve sedative, hypnotic, and antispasmodic. Under cultivation, the yield is 1,000 to 1,500 lb (453 to 680 kg) of dry leaf or 500 to 2,000 lb (227 to 906 kg) of seed per acre. In general, the yield of drug plants in cultivation is high, so that only small acreages are needed. **Aconite** is the dried root of the **monkshood,** a perennial plant, *Aconitum nepullus.* It is grown in Europe and the United States. The root contains the colorless, crystalline, extremely poisonous alkaloid **aconitine,** $C_{34}H_{47}O_{11}N$. It is used as a cardiac sedative, diaphoretic, and to relieve pain and fever. **Matrine,** obtained from the plant *Sophora angustifolia,* is a lupine alkaloid having two piperidine rings. Combined with phenol sulfonate salts under the name of **dysentol,** it is used for amoebic dysentery.

Digitalis, an important heart stimulant, consists of the leaves of the perennial **foxglove plant,** *Digitalis purpurae,* native to Europe but also grown in New England and the Pacific Northwest. The leaves contain the bitter glucoside **digitoxin,** $C_{34}H_{54}O_{11}$. **Peyote** is the root of a cactus, *Lophophora williamsii,* growing wild in the desert region of Mexico. The buttonlike tops of the root, called **mescal buttons,** contain several alkaloids, and were used by the Indians to produce a sense of well-being with visions. **Peyotina hydrochloride,** made from the buttons, is used in medicine as a heart stimulant similar to digitalis, and also as an anesthetic to the nervous system. The **glucosides,** unlike the alkaloids, contain no nitrogen. They are ethers of single sugars. The glucoside **rutin,** used for treatment of hemorrhage and for high blood pressure, is a yellow nontoxic powder extracted from flue-cured tobacco and from **Tartary buckwheat** or **rye buckwheat.** The active ingredient, **quertin,** is extracted and used to prevent hemorrhage in hypertensive persons. **Ipecac** is the rhizome of the shrub, *Cephalis ipecacuamha,* of tropical South American, chiefly Brazil. It contains **emetine,** $C_{30}H_{40}O_5N_2$, and other alkaloids, and is used as an emetic tonic, and expectorant. **Gentian,** also known as **bitter root,** is the dried root of the perennial herb *Gentiana lutea* of central Europe. It contains glucosides and is used in tonics and also in cattle feeds. **Coca** is a South American shrub from which the drug and narcotic, **cocaine,** is produced. The shrub, indigenous to Peru and Bolivia, is produced by cultivation in Java. The dried leaves when chewed have a mild narcotic effect.

Ginseng is the dried root of the plant *Panex ginseng* of China, and *P.*

quinquefolium of North America. It is used as a stimulant and stomachic. **Dog-grass root,** known in medicine as **triticum,** is the powdered dried rhizome of **Scotch grass,** *Agropyron repens,* which abounds in the meadows of the northern United States. It has a slight aromatic odor and sweet taste and is used as a diuretic. **Lobelia** is the dried leaves and tops of the small annual plant *Lobelia inflata,* known in the northeastern states as **Indian tobacco.** It contains the alkaloid **lobinine,** and is used as an emetic and antispasmodic. **Erigeron** is the herb and seeds of the **fleabane herb,** *Erigeron canadensis,* which grows wild in the north central and western states. The herb yields 0.35 to 0.65% **erigeron oil** used as an astringent and tonic. **Hamamelis** is the dried leaves of the witch hazel shrub, *Hamamelis virginiana,* of the eastern United States. It is used as a tonic and sedative, and the water extract of the leaves and twigs with 14% alcohol and 1% active ingredient is known as **witch hazel.** It is used as an external astringent. It is fragrant, and contains also tannic acid from the bark of the twigs. **Ephedrine,** or **ephedra,** is an alkaloid extracted from the dried twigs of the **mahuang,** *Ephedra sinica,* and other species of mountain shrubs of China and India. It is also grown in South Dakota. It is used as an adrenaline substitute to raise blood pressure, and also in throat medicines and nasal sprays. **Hoarhound** is an extract from the leaves and flowering tops of the small perennial herb *Marrubium vulgare* of North America, Europe, and Asia. It is used in preparations for colds, and also as a flavor in confections.

Cascara is the bark of the small **buckthorn** tree, *Rhamnus purshiana,* of the northwestern states and Canada. It was also called **chittem bark.** It was used by the Indians and given the name **cascara sagrada** by the Spaniards. The word cascara simply means bark; the word sagrada is not translated sacred but is a term applied to medicinal botanicals. Cascara is used as a laxative and tonic. The **European buckthorn,** *R. frangula,* also yields cascara. **Aloe,** used in purgative medicines, is the dried resinous juice from the leaves of the bush *Aloe vulgaris* of the West Indies and *A. perryi* of western Africa. The cut leaves are placed in troughs where the juice exudes. It is then evaporated to a viscous black mass which hardens. Both aloe and cascara contain the glucoside **emodin,** occurring also in senna and in **rhubarb.** The latter, the stalks of which are much used as a food, is known in medicines as **rheum** and is employed as a laxative and stomach tonic. **Aloin,** used in skin creams for radiation and sunburns, is an extract from the *A. vera,* a plant of the lily family growing in Florida. **Aletris root** is a botanical drug from the **stargrass,** *Aletris farinosa,* growing in the eastern United States. It is also known as **colic root,** and is used as a tonic and uterine stimulant. **Senega,** used as an emetic and stimulant, is the dried root of the **snakeroot** or **milkwort,** *Polygala senega,* a small perennial herb grown in the eastern United States. **Calumba root** is the yellow root of the woody climbing plant *Coscinium fonestratum* of Sri Lanka. It is used as a cure for tetanus. The wood, which has a bright-yellow color, is used locally as a dye. **Tanacetum** is the dried leaves and tops of the **silverweed,** *Tanacetum vulgaris* of Michigan and Indiana, used as a vermifuge. The green herb yields about 0.2% **tansy oil,** or **tanacetum oil,** which contains a terpene **tanacetene,** and borneol. It is used as an anthelmintic.

Ergot, used to stop hemorrhage, is the dried sclerotium of the fungus *Claviceps purpurea,* which develops on rye and some grasses. The purple structure that replaces the diseased rye grain contains the alkaloid **clavine,** $C_{11}H_{22}O_4N_2$. Ergot is also used to produce the alkaloids **ergotamine** and **ergonovine,** used for treating high blood pressure and migraine headache, and used for mental-disease research. It requires 1,000 lb (453 kg) of ergot to yield 1.5 oz (43 g) of ergonovine, but it is produced synthetically by the Eli Lilly & Co. from **lysergic acid,** which is synthesized from **indole propionic acid,** $C_8H_6N(CH_2)_2COOH$. **Ergot oil,** used in medicinal soaps, is obtained by extraction from the dry ergot. It was the original source of **ergosterol,** one of the most important **sterol alcohols** from plants, known as **phytosterols,** which, when irradiated, yield vitamin D. Ergot is produced chiefly in Spain, and its commercial production has not been encouraged in the United States because it is an undesirable disease on grain. Ergot also contains the amino acid **thiozine** found in blood.

Araroba, or **Goa powder,** is a brownish, bitter, water-soluble powder scraped out of the split logs of the tree *Andira araroba* of Brazil. The Carib Indians used the powder for skin diseases and it is now employed for eczema and skin infections. It contains **chrysarobin,** a complex mixture of reduction products of a complex acid contained in the wood. **Labdanum** is an oleoresin obtained in dark-brown or greenish lumps from the branches of the **rock rose,** *Cistus ladaniferus,* and other species of the Mediterranean countries. It yields a volatile oil with a powerful sweet characteristic odor which is used as a stimulant and expectorant and also as a basis for lavender and violet perfumes. It was originally obtained in Greece by combing from the fleece of sheep that browsed against the bushes. **Buchu** is the dried leaves of the South African herb *Barosma betulina* used as an anticatarrhal. **Serpasil,** of Ciba Pharmaceutical Products, Inc., is an extract from the root of the plant *Rauwolfia serpentina* of India and *R. heterophylla* of Central America. It is used as a sedative in hypertension cases, and belongs to the class of **tranquilizing agents. Rauwolfia extract** contains the two alkaloids **reserpine** and **rescinnamine,** both having hypotensive activity. They are produced by the Riker Laboratories as **Serpiloid** and **Rauwiloid,** and the product of Squibb is called **Raudixin.** A synthetic material, **meprobamate,** is a complex **propanediol dicarbamate,** and **Singoserp,** of Ciba, is a synthetic reserpine.

BOXWOOD. The wood of the **Turkish boxwood** tree *Buxus sempervirens,* native to Europe and Asia but also grown in America. It is used for rulers, instruments, engraving blocks, and inlay work. The wood is light yellow, hard, and has a fine grain and a dense structure that does not warp easily. The weight is about 65 lb/ft^3 (1,041 kg/m^3). **African boxwood,** or **cape boxwood,** comes from the tree *B. macowani,* of South Africa, and is similar to boxwood but is softer. **Kamassi wood** is a hard, fine-grained wood from the tree *Gonioma kamassi,* of South Africa, sometimes substituted for boxwood, but it does not have the straight grain of boxwood. It is valued for loom shuttles. **Coast gray boxwood,** or the **Gippsland boxwood** of New South Wales, is from the tree *Eucalyptus bosistoana.* It is a durable wood of uniform texture but it has an interlocking grain. **Maracaibo boxwood,** or **zapatero,** which comes chiefly from Venezuela, is from the tree *Casearia praecox.* It comes in straight knotless logs 8 to 10 ft (2.4 to 3 m) long and 6 to 10 in (15.2 to 25.4 cm) in diameter. The light-yellow wood has a fine uniform texture and straight grain. It replaces Turkish boxwood for all purposes except for wood engravings. The **ginkgo wood** used in China for making chessmen and chessboards is from the large tree *Ginkgo biloba* which appears to be the sole survivor of a family of trees with fernlike leaves once very abundant. The wood is white to yellowish, light in weight, fine-textured, and easy to work.

BRASS. **Copper-zinc** alloys whose zinc content ranges up to 40%. If the copper crystal structure is face-centered cubic, there will be up to 36% of zinc present. This solid solution, known as the alpha phase or **alpha brass,** has good mechanical properties, combining strength with ductility. Corrosion resistance is very good, but electric conductivity is considerably lower than in copper. When above 30 to 36% of the alloy is zinc, a body-centered-cubic crystal structure is formed, known as the beta phase, or **beta brass.** This phase is relatively brittle and high in hardness compared to the alpha phase. However, ductility increases at elevated temperatures, thus providing good hot-working properties. **Gamma brass,** with the zinc above 45%, is not easily worked, either hot or cold.

The mechanical properties of brasses vary widely. Strength and hardness depend on alloying and/or cold work. Tensile strengths of annealed grades are as low as 30,000 lb/in^2 (206 MPa), although some hard tempers approach 90,000 lb/in^2 (620 MPa). Although brasses are generally high in corrosion resistance, two special problems must be noted. With alloys containing a high percentage of zinc, dezincification can occur. The corrosion product is porous and weak. To prevent dezincification, special inhibitors—antimony, phosphorus, or arsenic—in amounts of 0.02 to 0.05% can be added to the alloy. The other problem is stress corrosion, or season cracking, which occurs when moisture condenses on the metal and accelerates corrosion.

Brass is annealed for drawing and bending by quenching in water from a temperature of about 1000°F (538°C). Simple copper-zinc brasses are made in standard degrees of temper, or hardness. This hardness is obtained by cold-rolling after the first annealing, and the degree of hardness depends upon the percentage of cold reduction. When the thickness is reduced one number of the Brown & Sharpe gage, or about 10.9% reduction, the resulting sheet is known as ¼ hard. The other grades are ½ hard, hard, extra hard, spring, and finally extra spring, which is a reduction of 10 numbers on the Brown & Sharpe gage, or about 68.7% reduction without intermediate annealing. Degrees of softness in annealed brass are measured by the grain size, and annealed brass is furnished in grain sizes from 0.010 to 0.150 mm. The ASTM standard grain sizes are: 0.015 to 0.025 mm for light anneal, 0.035 mm for drawing or rod anneal, 0.050 mm for intermediate anneal, 0.70 mm for soft anneal, and 0.120 for dead-soft anneal. Brasses with smaller grain sizes are not as ductile as with larger grain sizes, but they have smoother surfaces and require less polishing.

Even slight additions of other elements to brass alter the characteristics drastically. Slight additions of tin change the structure, increasing the hardness but reducing the ductility. Iron hardens the alloy and reduces the grain size, making it more suitable for forging, but making it difficult to machine. Manganese increases the strength, increases the solubility of iron in the alloy, and promotes the stabilization of aluminum, but makes the brass extremely hard. Slight additions of silicon increase the strength of brass, but large amounts promote brittleness, loss of strength, and danger of oxide inclusion. Nickel increases strength and toughness, but when any silicon is present, the brass becomes extremely hard and more a bronze than a brass.

There are hundreds of brasses with a bewildering array of names—some misleading such as commercial bronze, jewelry bronze and manganese bronze, which are brasses. But most brasses can be grouped into a few major classes. The **straight brasses** constitute by far the largest and most widely used group. They are binary copper-zinc alloys with zinc alloys with zinc content ranging from as low as 5% up to about 40%. Some of the common names of these alloys are: **gilding metal** (5% zinc), **commercial bronze** (10% zinc), **jewelry bronze** (12.5% zinc), **red brass** (15% zinc), **yellow brass** (35% zinc), and **muntz metal** (40% zinc). As the zinc content increases in these alloys, the melting point, density, electric and thermal conductivity, and modulus of elasticity decrease while the coefficient of expansion, strength, and hardness increase. Work hardening also increases with zinc content. These brasses have a pleasing color, ranging from the red of copper in the low-zinc alloys through bronze and gold colors to the yellow of high-zinc brasses. The color of jewelry bronze closely matches that of 14-karat gold, and this alloy and other low brasses are used in inexpensive jewelry.

The low-zinc brasses have good corrosion resistance along with moderate strength and good forming properties. Red brass, with its exceptionally high corrosion resistance, is widely used for condenser tubing. The high brasses (**cartridge brass** and **yellow brass**) have excellent ductility and high strength and are widely used for engineering and decorative parts fabricated by drawing, stamping, cold heading, spinning, and etching. Muntz metal, primarily a hot-working alloy, is used where cold working is not required.

Another group are the **leaded brasses.** These alloys have essentially the same range of zinc content as the straight brasses. Lead is present, ranging from less than 1 to 3.25%, to improve machinability and related operations. Lead also improves antifriction and bearing properties. Common leaded brasses include **leaded commercial bronze** (0.5% lead), **medium-leaded brass** (1% lead), **high-leaded brass** (2% lead), **free-cutting brass** (3.25% lead), and **hardware bronze** (1.75% lead). Free-cutting brass provides optimum machinability and is ideally suited for screw machine parts.

Another group, the **tin brasses,** are copper-zinc alloys with small amounts of tin. The tin improves corrosion resistance. Pleasing colors are also obtained when tin is added to the low brasses. Tin brasses in sheet and strip form, with 80% or more copper, are used widely as low-cost spring materials. **Admiralty brass** is a standard alloy for heat-exchanger and condenser tubing. **Naval brass** and **manganese bronze** are widely used for products requiring good corrosion resistance and high strength, particularly in marine equipment.

Casting brasses are usually made from brass ingot metal and are seldom plain copper-zinc alloys. In melting brass for casting, any over-heating causes loss of zinc by vaporization, thus lowering the zinc content. Small amounts of antimony, or some arsenic, are used to overcome this dezincification. The casting brasses are roughly divided into two classes as **red casting brass** and **yellow casting brass,** which are various compositions of copper, tin, zinc, and lead to obtain the required balance of color, ease of casting, hardness, and machining qualities.

BRASS INGOT METAL. Commercial ingots made in standard composition grades and employed for casting various articles designated as brass and bronze. They are seldom true brasses, but are composition metals intermediate between the brasses and the bronzes, and their selection for any given purpose is based on a balance of the requirements in color, strength, hardness, ease of casting, and machinability. Brass ingot metal is usually made from secondary metals, but, in general, the grading is now so good that they will produce high-grade uniform castings. In producing the ingot metal there is careful sorting of the scrap metals, and the impurities are removed by remeltings. An advantage of ingot metal over virgin metals is the ease of controlling mixtures in the foundry. The ASTM designates eight grades for brass ingot metal. No. 1 grade, the highest in copper, contains 88% copper, 6.5 zinc, 1.5 lead, and 4 tin, with only slight percentages of impurities. The No. 8 grade contains 63.5% copper, 34 zinc, 2.5 lead, and no tin. **Yellow ingot,** for plumbing fixtures, contains 65% copper, 1 tin, 2 lead, and the balance zinc. The most widely used ingot metal is the **ASTM alloy No. 2,** which is the 85:5:5:5 alloy known as **composition metal.** Yellow brass, or **yellow casting brass,** is frequently cast from **ASTM alloy No. 6,** which contains 72% copper, 22 zinc, 4 lead, and 2 tin. It has a tensile strength of 20,000 to 25,000 lb/in^2 (137 to 172 MPa), elongation 15 to 20%, and Brinell hardness 40 to 50. It is yellow in color, and makes clean, dense castings suitable for various machine parts except bearings.

BRAZIL NUTS. Also called **Pará chestnut.** The nuts of the large tree *Bertholletia excelsa* growing wild in the Amazon Valley. The trees and nuts are called **tacarí** in Brazil and **toura** in French Guiana. The tree begins to bear in 8 years and yields up to 1,000 lb (453 kg) of large round fruit pods containing 18 to 24 hard-shelled kernels which are the commercial nuts. The shelled kernels are several times the size of the peanut, and have a pleasant nutty flavor. The kernels contain 67% of a pale yellow oil of specific gravity 0.917, saponification value 192 to 200, and iodine value 98 to 106. The oil contains 51% oleic acid, 19 linoleic, 2 myristic, 3 stearic, and 12 palmitic. It is a valuable food oil, and is also a good soap oil, but is normally too high in price for this purpose as the nuts are more valued for eating.

BRAZILWOOD. The wood of the trees *Caesalpinia brasiliensis, C. crista,* and *C. echinata,* of tropical America, and *C. sappan,* of Sri Lanka, India, and Malaya. The latter species is called **sappanwood.** Brazilwood formerly constituted one of the most valuable exports from Brazil to Europe as a dyewood. It produces purple shades with a chrome mordant and crimson with alum. **Brazilwood extract** is still valued for silk dyeing, wood staining, and for inks. The wood is prized for such articles as violins and fine furniture. It has a rich bright-red color, and takes a fine, lustrous polish.

BRAZING METAL. A common name for high-copper brass used for the casting of such articles as flanges that are to be brazed on copper pipe. Federal specifications for brazing metal call for 84 to 86% copper and the balance zinc. **Brazing brass,** of the American Brass Co., has 75% copper and 25 zinc. Some alloys also contain up to 3% lead for ease of machining. It also makes the metal easier to cast.

The term brazing metal is also applied to **brazing rods** of brass or bronze used for joining metals. A common brazing rod is the 50–50 brass alloy with a melting point of 1616°F (880°C). The SAE designates this alloy as **spelter solder.** The joints made with it are inclined to be brittle. **Brazing wire,** of the Chase Brass & Copper Co., contains 59% copper and 41 zinc, while the **brazing solder,** for brazing high-zinc brasses, has 51% copper and 49 zinc. **Phos-copper,** of the Westinghouse Electric Corp., is a phosphor copper which gives joints with 98% the electric conductivity of copper. It flows at 1382°F (750°C). **Tri-Metal,** of the D. E. Makepeace Co., is a brazing metal consisting of sheet brass with a layer of silver rolled on each side. It is used for brazing carbide tips to steel tools. The silver ensures a tightly brazed joint, while the brass center acts as a shock absorbent for the cutting tool.

Brazing rods for brazing brasses and bronzes are usually of a composition similar to that of the base metal. For brazing cast iron and steel, various bronzes, naval brass, manganese bronze, or silicon bronze may be used. Brass rods may contain some silicon. Small amounts of silver added to the high-copper brazing metals give greater fluidity and better penetration into small openings.

BRICK. The most ancient of all artificial building materials, consisting of clay molded to standard shape, usually rectangular, and burned to a hard structure. In some areas bricks, known as **adobe bricks,** are still made by baking in the sun, and a modern adaptation of sun-baked brick, called **bitudobe,** has a binder of emulsified asphalt. Commercial bricks in the United States and in Europe are all of hard-burned clay. They are used for buildings, walls, and paving, and are classified apart from the bricks of fireclays used for refractories. **Brick clays** are of two general classes. The first consists of noncalcareous clays or shales composed of true clay with sand, feldspar grains, and iron compounds, which when fired become buff or salmon in color. The second class comprises calcareous clays containing up to 40% calcium carbonate, called **marls.** When fired, they are yellowish. Brick clays of the first type are widely distributed. Iron oxide in them varies from 2 to 10%, and the red color of common brick depends largely on this content. In practice, the composition of bricks varies widely, but much of the difference is also in the burning as well as in the method of pressing. **Pressed brick** is a stiff mud brick made under high pressure. It is homogeneous, and has increased density and strength. **Building brick** made by machine of ground and tempered clay has great uniformity of strength and color. Such brick is made by pressing soft, stiff, or dry. The burning is done in kilns at temperatures from 900 to 1250°C. The calcareous clays require a temperature of 1200°C to bring about chemical combination. The bricks are sorted according to hardness and color, both largely resulting from their position in the kiln. **Paving brick** is usually a hard-burned common brick. **Floor brick** is highly vitrified brick. The common hard brick for building has a crushing strength of 5,000 to 8,000 lb/in^2 (34 to 55 MPa), and weighs 125 lb/ft^3 (2,025 kg/m^3). The common standard for building-brick size is 8¼ by 4 by 2½ in (21 by 10.2 by 6.4 cm); other sizes are also used, especially 8 by 3⅞ by 2¼ in (20.3 by 9.8 by 5.7 cm). Specially sized paving bricks are 8½ by 4 by 3 or 8½ by 4 by 3½ in (21.6 by 10.2 by 7.6 or 21.6 by 10.2 by 8.9 cm). **Sand-lime bricks** for fancy walls are of sand and lime pressed in an atmosphere of steam. They are not to be confused with the sand-lime bricks used for firebrick, which are of refractory silica sand with a lime bond. Ceramic glazes and semiglazes are used on some building bricks, especially on the yellow.

BRISTLES. The stiff hairs from the back of the hog, used chiefly in making brushes. The very short bristles, rejects, and scrap pieces are used for filler in plastics. The best brush bristles do not include hair from the sides of the animals, nor the product from the fat-meat animals killed in the slaughterhouses. They come mostly from types of semiwild swine grown in cool climates, notably northern China and Russia. Bristles are in form similar to a tiny tube outwardly covered with microscopic scales and filled with a fatty substance. The so-called flag, or split end, gives the valuable paint-carrying characteristic for brushes. The taper of the bristle gives the brush stiffness at the base and resiliency toward the end. Quality varies according to the type of animal, climate, and feeding. The colors are white, yellow, gray, and black. They are graded by locality, color, and length, and in normal times the name of the place at which they are graded, such as Tsingtao, Hankow, and Chungking, is an indication of the grade. The best fibers are more than 3 in (7.6 cm) in length. The Chinese natural black bristles are sometimes sold at a premium. American bristles from Chester hogs are light in color and are of high quality. Bristles from the Duroc hog are bronze in color, are stiff, and are superior to most Chinese grades. Those from the Poland China hogs are black and stiff, but they have a crooked flag and are of poor quality. **Artificial bristles** are made from various plastics, the **nylon bristles** being of high quality and much used. **Exton,** of Du Pont, was one of the original nylon monofilament nontapered bristles. **Tynex,** of this company, now comes both tapered and level. They are more durable than natural bristles. **Casein bristles** are made by extruding an acid solution of casein, stretching the fiber, and insolubilizing with formaldehyde or other chemicals. They have good paint-carrying capacity and good wear resistance, but are dissolved by some paint solvents. **Keron bristle,** of the Rubberset Co., is produced from the protein extracted from chicken feathers. It is nearly identical in composition to natural bristle.

BROMINE. An elementary material, symbol Br. It is a reddish-brown liquid having a boiling point of 58°C. It gives off very irritating fumes, and is highly corrosive. It is one of four elements called **halogens,** a name derived from Greek words meaning salt producer. They are fluorine, chlorine, iodine, and bromine. They are all chemically active, combining with hydrogen and most metals to form **halides.** Bromine is less active than chlorine but more so than iodine. It is soluble in water. It never occurs free in nature, but is obtained by electrolysis of salt solutions. It occurs in seawater to the extent of 65 to 70 parts per million, and is extracted. It is marketed 99.7% min purity with specific gravity not less than 3.1, but dry elemental bromine, Br_2, is marketed 99.8% pure for use as a brominating and oxidizing agent. For these uses, also, bromine is available as a crystalline powder as **dibromodimethyl hydantoin,** containing 55% bromine. **Brom 55,** of McKesson & Robbins, is this material.

A pound of bromine is obtained from 2,000 gal (7,570 L) of seawater. It is also produced as a by-product from the brine wells of Michigan, and from the production of chemicals at Searles Lake, Calif., where the bromine concentration is 12 times that of seawater. It is used in the manufacture of dyes, photographic chemicals, poison gases for chemical warfare, pharmaceuticals, ethyl fluid for gasoline, disinfectants, and many chemicals. It is also employed in the extraction of gold.

BRONZE. The term bronze is generally applied to any copper alloy that has as the principal alloying element a metal other than zinc or nickel. Originally the term was used to identify copper-tin alloys that had tin as the only, or principal, alloying element. Some brasses are called bronzes because of their color, or because they contain some tin. Most commercial copper-tin bronzes are now modified with zinc, lead, or other elements.

The **copper-tin bronzes** are a rather complicated alloy system. The alloys with up to about 10% tin have a single-phase structure. Above this percentage, a second phase, which is extremely brittle, can occur, making plastic deformation impossible. Thus high-tin bronzes are used only in cast form. Tin oxide also forms in the grain boundaries, causing decreased ductility, hot workability, and castability. Additions of small amounts of phosphorus, in production of phosphor bronzes, eliminate the oxide and add strength. Because tin additions increase strength to a greater extent than zinc, the bronzes as a group have higher strength than brasses—from around 60,000 to 105,000 lb/in^2 (413 to 723 MPa) in the cold-worked high-tin alloys. In addition, fatigue strength is high.

Bronzes containing more than 90% copper are reddish; below 90% the color changes to orange-yellow which is the typical bronze color. The maximum strength is with 80% copper and 20 tin. Ductility rapidly decreases with the increase of tin up to 20%, after which it practically disappears until 80% is reached, when it again increases. Above 20% tin the alloy rapidly becomes white in color and loses the characteristics of bronze. A 90–

10 bronze weighs 0.317 lb/in^3 (8,774 kg/m^3); an 80–20 bronze weighs 0.315 lb/in^3 (8,719 kg/m^3). The 80–20 bronze melts at 1868°F (1020°C), and a 95–5 bronze melts at 2480°F (1360°C).

The family of **aluminum bronzes** is made up of alpha-aluminum bronzes (less than about 8% aluminum) and alpha-beta bronzes (8 to 12% aluminum) plus other elements such as iron, silicon, nickel, and manganese. Because of the considerably strengthening effect of aluminum, in the hard condition these bronzes are among the highest-strength copper alloys. Tensile strength approaches 100,000 lb/in^2 (689 MPa). Such strengths plus outstanding corrosion resistance make them excellent structural materials. They are also used in wear-resistance applications and for nonsparking tools. **Phosphor bronzes** have a tin content of 1.25 to 10%. They have excellent mechanical and cold-working properties and a low coefficient of friction, making them suitable for springs, diaphragms, bearing plates, and fasteners. Their corrosion resistance is also excellent. In some environments, such as salt water, they are superior to copper. **Leaded phosphor bronzes** are available with improved machinability. **Silicon bronzes** are similar to aluminum bronzes. Silicon content is usually between 1 and 4%. In some, zinc or manganese is also present. Besides raising strength, the presence of silicon sharply increases electrical resistivity. Aluminum-silicon bronze, and important modification, has exceptional strength and corrosion resistance and is particularly suited to hot working.

Gear bronze may be any bronze used for casting gears and worm wheels, but usually means a tin bronze of good strength deoxidized with phosphorus and containing some lead to make it easy to machine and lowering the coefficient of friction. A typical gear bronze contains 88.5% copper, 11 tin, 0.25 lead, and 0.25 phosphorus. It has a tensile strength up to 40,000 lb/in^2 (275 MPa), elongation 10%, and Brinell hardness from 70 to 80, or up to 90 when chill-cast. The weight is 0.306 lb/in^3 (8,470 kg/m^3). This is **SAE bronze No. 65.** A **hard gear bronze,** or **hard bearing bronze,** of the U.S. Navy, contains 84 to 86% copper, 13 to 15 tin, up to 1.5 zinc, up to 0.75 nickel, and up to 0.5 phosphorus. Hard and strong bronzes for gears are often silicon bronze or manganese bronze.

In a modified 90–10 type of bronze, the zinc is usually from 2 to 4%, and the lead up to 1%. A cast bronze of this type will have a tensile strength of about 40,000 lb/in^2 (275 MPa), an elongation 15 to 25%, and a hardness 60 to 80 Brinell, those high in zinc being the stronger and more ductile, those high in lead being weaker. Bronzes of this type are much used for general castings and are classified as **composition metal** in the United States. In England they are called **engineer's bronze.**

Architectural bronze, or **art bronze,** is formulated for color and is very high in copper. One foundry formula for art bronze of a dull-red color calls for 97% copper, 2 tin, and 1 zinc. For ease of casting, however, they are more likely to contain lead, and a **gold bronze** for architectural castings contains 89.5% copper, 2 tin, 5.5 zinc, and 3 lead. In leaded bronze the hard copper-tin crystals aid in holding the lead in solution. These bronzes are resistant to acids and are grouped as valve bronze, or as bearing bronze because of the hard crystals in a soft matrix. Federal specifications for bronze give 10 grades in a wide variation in tin, zinc, and lead. The ASTM designates five grades of **bronze casting alloys. Alloy No. 1** contains 85% copper, 10 tin, and 5 lead; **Alloy No. 5** contains 70% copper, 5 tin, and 25 lead. The British **coinage copper** is a bronze containing 95.5% copper, 3 tin, and 1.5 zinc.

BRONZE POWDER. Pulverized or powdered bronze made in flake form by stamping from sheet metal. It is used chiefly as a paint pigment and as a dusting powder for printing. In making the powder the sheets are worked into a thin foil which becomes harder under the working and breaks into small flakes. Lubricant keeps the flakes from sticking to one another. Usually stearic acid is used, but in the dusting powder hot water or nonsticky lacquers are used. The powder is graded in standard screens, and is then polished in revolving drums with a lubricant. This gives it the property of leafing, or forming a metallic film in the paint vehicle. The leaf is also called **composition leaf,** or **Dutch metal leaf,** when used as a substitute for gold leaf. **Flitters** are made by reducing thin sheets to flakes, and are not as fine as bronze powder.

The compositions of bronze powder vary, and seven alloys form the chief commercial color grades from the reddest, called pale gold, which has 95% copper and 5 zinc, to the rich gold which has 70% copper and 30 zinc. Colors are also produced by heating to give oxides of deep red, crimson, or green-blue. The powder may also be dyed in colors, using tannic acid as a mordant, or treated with acetic acid or copper acetate to produce antique finish. The color or tone of bronze powders may also be adjusted in paints by adding a proportion of mica powder. A **white bronze powder** is made from aluminum bronze, or the silvery colors are obtained with aluminum powder. The bronze powder of 400 mesh used for inks is designated as extra fine. The fine grade, for stencil work, is 325 mesh. Medium fine, for coated paper, has 85% of the particles passing through a 325-mesh screen and 15% retained on the screen. Near mesh, for paint pigment, has 30% passing through a 325-mesh screen. A 400-mesh powder has 500 million particles per gram. The old name for bronze powder is **gilding powder.** It is also called gold powder when used in cheap gold-colored paints, but bronze powders cannot replace gold for use in atmospheres containing sulfur, or for printing on leather where tannic acid would corrode the metal. **Gold pigments** used in plastics are bronze powders with oxygen stabilizers.

BROOMCORN. A plant of the sorghum family, *Holcus sorghum,* grown in the Southwest, in Illinois and Kansas, and in Argentina and Hungary. It is used for making brushes and brooms, and for the stems of artificial flowers. The jointed stems of the dwarf variety grown in the semiarid regions are 12 to 24 in (0.30 to 0.61 m) long, but the standard brush corn is up to 30 in (0.76 m) long. The fibers are yellow in color and, when dry, are coarse and hard. They are easily cleaned and readily dyed. As a brush material they have the objection that they break easily, and are therefore unsuited for mechanical brushes for hard service. **Broom root,** or **rice root,** is similar to broomcorn, and is suitable for mixing with it or for coarse brushes. It is from a type of grass, *Epicampes macroura,* of Mexico and Guatemala. The fiber is from the tough, crinkly, yellowish roots. After removal of the outer bark, the dry root is treated with the fumes of burning sulfur to improve the color. The fibers are 8 to 18 in (0.20 to 0.46 m) long. In Mexico it is called raiz de Zacaton, or **Zacaton root,** and its American name, rice, is a corruption of the Spanish word for root.

BRUSH FIBERS. Industrial brushes are made from a wide variety of fibers, varying from the fine and soft camel's hair to the hard, coarse, and brittle broomcorn. Bristles are the most commonly used, but tampico and piassava fibers are important for polishing brushes. The vegetable fibers used for brushes are tough and stiff as compared with the finer flexible and cohesive fibers used for twine and for fabrics. They may, however, come from the same plant, or even from the same leaf, as the textile fibers, but be graded out for stiffness. **Palmetto fiber** is from the **cabbage palm** tree, *Sabal palmetto,* of Florida. Whiskbrooms and brushes are made from the young leafstalks, and stiff floor sweeps from the leaves.

A fiber finer than palmetto is obtained from the twisted roots of the **scrub palmetto,** *S. megacarpa.* **Arenga fiber** is a stiff, strong fiber from the stems of the **aren palm** tree, *Arenga saccharifera,* of Indonesia. The finest grades resemble horsehair. **Kittool** is a similar strong, elastic fiber from large leaves of the palm tree, *Caryota urens,* of India and Sri Lanka. It is very resistant, and is valued for machine brushes. **Gomuti fiber** and **Chinese coir** are fibers from other species of this palm. **Bass,** or **raphia,** is a coarse fiber used for hard brushes and brooms. The heavier piassava fibers are also known as bass, but bass is from the leaves of the palm tree, *Raphia vinifera,* of West Africa. **Darwin fiber,** used for brooms and scrubbing brushes in Australia, is from the *Gahnia trifida.* **Crin** is from the leaves of a palm tree of Algeria, although the word crin originally referred to horsehair. **Crin vegetal,** or **vegetable crin,** is fiber from the leaf of the **yatay palm,** *Diplothemium littorale,* of Corrientes Province, Argentina. **Horsehair,** from the manes and tails of horses, is used for some paintbrushes.

Red sable hair is used for fine-pointed and knife-edged brushes for show-card and water-color use. It is from the tail of the **kolinsky,** *Mustela* *siberica,* of Siberia, and the pale red hair has strength and resiliency and very fine points. **Russian sable hair,** used for artists' brushes, is stronger than red sable hair, but is less pointed and is not as elastic for water painting. It is from the tail of the **fitch,** *Putorius putorius,* of Central Asia, but the so-called **fitch hair** used for ordinary flowing brushes is usually **skunk tail hair.** It is stiffer and coarser than fitch hair. **Badger hair,** also used for flowing brushes, is a resilient hair with fine points, and is from the back of the badger of Turkey and southern Russia. **Black sable hair,** used for sign-writer brushes, is not from a sable, but is the trade name for mixtures of marten hair, bear hair, and some other Siberian hairs.

Vegetable and animal fibers are not resistant to alkalies or acids and cannot be wetted with them. The artificial fibers of plastics such as nylon are resistant to many chemicals. For hard-service mechanical brushes, and for resistance to strong chemicals, brush fibers are of steel, brass, or aluminum wire. **Brush wire** for rotary-power brushes for metal brushing is soft to hard-drawn steel wire usually 0.005 in (0.013 cm) in diameter. Finer wire for soft rotary brushes is a soft steel wire 0.0025 in (0.006 cm) in diameter.

BUFFING COMPOSITIONS. Materials used for buffing or polishing metals, originally consisting of dolomitic lime with from 18 to 25% saponifiable grease as a bond. The lime acts as the abrasive, and in some compositions is partly replaced by other abrasives such as emery flour, tripoli, pumice, silica, or rouge. Harsher abrasives are used in the compositions employed for the cutting-down or buffing operations. Abrasive grains are selected for combinations of hardness, toughness, and sharpness, from the soft iron oxide to the hard and sharp aluminum oxide. Buffing compositions are usually sold under trade names for definite uses rather than by composition. **Metal polishes** for hand use are now usually liquids. The pastes, formerly known as **Putz cream** and **brash polish,** contained tripoli or pumice with oxalic acid and paraffin. The liquid polishes now generally contain finer abrasives such as pumicite or diatomite, in a detergent, together with a solvent, and sometimes pine oil or an alkali.

BUILDING SAND. Selected sand used for concrete, for mortar for laying bricks, and for plastering. Early specifications called for sand grains to be sharp, but rounded grains are now preferred because of fewer voids in the mixture. Building sand is normally taken from deposits within a reasonable haul of the site of building, and is not usually specified by analysis, but should be a hard silica sand that will not dissolve. Pure white sand for finish plaster is made by grinding limestone. Building sand is required to be clean, with not more than 3% clay, loam, or organic matter. ASTM requirements are that all grains must pass through a ⅜-in (0.95-cm) sieve, 85%

through a No. 4 sieve, and not more than 30% through a No. 50 sieve. For brick mortar all of the sand should pass through a ¼-in (0.64-cm) sieve. For plaster, not more than 6% should pass through a No. 8 sieve. **Flooring sand** for mastic flooring is a clean sand passing through a No. 3 sieve, with 7% passing through a No. 100 sieve. **Roofing sand** is a fine white silica sand. **Paving sand** is divided into three general classes: that for concrete pavements, for asphaltic pavements, and for grouting.

The United States Bureau of Public Roads requires that sand for concrete pavements should all pass through a ¼-in (0.64-cm) sieve, 5 to 25% should be retained on a No. 10 sieve, from 50 to 90% on a No. 50 sieve, and not more than 10% should pass through a No. 100 sieve. Not more than 3% of the weight should be matter removable by elutriation. For asphaltic pavements small amounts of organic matter are not objectionable in the sand. All should pass through a ¼-in (0.64-cm) sieve, 95 to 100% through a No. 10 sieve, and not more than 5% through a No. 200 sieve. **Grouting sand** should all pass through a No. 20 sieve, and not more than 5% through a No. 200 sieve. **Chat sand,** used for concrete pavements, is a by-product of zinc and lead mines. It is screened through a ⅜-in (0.95-cm) sieve.

BUILDING STONE. Any stone used for building construction may be classed as building stone. **Granite** and **limestone** are among the most ancient of building materials and are extremely durable. Two million limestone and granite blocks, totaling nearly 8 million long tons (8,128 million kg), were used in the pyramid of Giza built about 2980 B.C., the granite being used for casing. Availability, or a near supply, may determine the stone used in ordinary building, but for public buildings stone is transported long distances. Some sandstones, such as the **red sandstone** of the Connecticut Valley, weather badly, and are likely to scale off with penetration of moisture and frost. Granite will take heavy pressures and is used for foundation tiers and columns. Limestones and well-cemented sandstones are employed extensively above the foundations. Nearly half of all the limestone used in the United States in block form is **Indiana limestone.** Marble has a low crushing strength and is usually an architectural or facing stone.

Crushed stone is used for making concrete, for railway ballast, and for road making. The commercial stone is quarried, crushed, and graded. Much of the crushed stone used is granite, limestone, and **trap rock.** The latter is a term used to designate basalt, gabbro, diorite, and other dark-colored fine-grained igneous rocks. Graded crushed stone usually consists of only one kind of rock and is broken with sharp edges. The sizes are from ¼ to 2½ in (0.64 to 6.35 cm), although larger sizes may be used for massive concrete aggregate. Screenings below ¼ in (0.64 cm) are employed largely for paving. **Granite granules** for making hard terrazzo floors are marketed in several sizes, and in pink, green, and other selected colors. **Roofing granules** are graded particles of crushed rock, slate, slag, porcelain, or tile, used as surfacing on asphalt roofing and shingles. Granules have practically superseded gravel for this purpose. Black **amphibole ryolite** may be used, or gray basalt may be colored artificially for granule use. The **suzorite rock** of Quebec contains feldspar, pyroxenite, apatite, and mica, and is treated to remove the mica. **Ceramic granules** are produced from clay or shale fired and glazed with metallic salts. They are preferred because the color is uniform.

BURLAP. A coarse, heavy cloth made of plain-woven jute, or jutelike fibers, and used for wrapping and bagging bulky articles, for upholstery linings, and as a backing fabric for linoleum. Finer grades are also used for wall coverings. The standard burlap from India is largely from jute fibers, but some hibiscus fibers are used. For bags and wrappings, the weave is coarse and irregular, and the color is the natural tan. The coarse grades such as those used for wrapping cotton bales are sometimes called gunny in the United States, but **gunny** is a general name for all burlap in Great Britain. Dundee, Scotland, is the important center of burlap manufacture outside of India, but considerable quantities are made from native fibers in Brazil and other countries. Burlap is woven in widths up to 144 in (3.6 m), but 36, 40, and 50 in (0.91, 1.02, and 1.27 m) are the usual widths. **Hessian** is the name of a 9½-oz (269-g) plain-woven finer burlap made to replace an older fabric of the same name woven from coarse and heavy flax fibers. When dyed in colors, it is used for linings, wall coverings, and upholstery. **Bithess** was a name for Hessian fabric coated with bitumen used in India to spread over soft-earth areas as a seal for a top coating to form airplane runways. **Brattice cloth** is a very coarse, heavy, and tightly woven jute cloth, usually 20 oz (567 g) used for gas breaks in coal mines, but a heavy cotton duck substituted for the same purpose is called by the same name. Most burlap for commercial bags is 8, 9, 10, and 12 oz (226, 255, 283, and 340 g), feed bags being 8 oz (226 g) and grain bags 10 oz (283 g).

BUTADIENE. Also called **divinyl, vinyl ethylene, erythrene,** and **pyrrolylene.** A colorless gas of the composition $CH_2{:}CH \cdot CH{:}CH_2$ used in the production of neoprene, nylon, latex paints, and resins. The Columbian Carbon Co. is using it to produce intermediate chemicals such as **cyclododecatriene,** the prime material for nylon 6 and 12. In contact with air or oxygen it forms peroxides that are explosive, and must be safeguarded with inhibitors when shipped. Butadiene has a boiling point of −3°C, and a specific gravity of 0.6272. Commercial butadiene is at least 98% pure, with normal butane and butenes as the impurities. It contains 0.01% of phenyl beta naphthylamine or other oxidation inhibitor. It is easily produced by the dehydrogenation of butane from natural gas or petroleum, but may also be made from alcohol. **Butadiene rubber,** or **polybutadiene,** can be made with close regulation of the molecular weight to give uniform rubbers of definite characteristics. The rubbers have less resilience and a higher heat buildup than natural rubber, but they are more resistant to oxidation and abrasion, and they give much greater wear life in automotive tires. They are oil-extendable. **Plioplex 5001** is an easily dissolved emulsion of high-purity polybutadiene used for modifying polystyrene rubbers. **Amerpol CB,** or *cis*-**polybutadiene,** of Goodrich-Gulf Chemicals, is a polybutadiene rubber polymerized with a cobalt catalyst to keep the detrimental vinyl content below 1%. Carboxy-modified butadiene rubber is highly resistant to ozone, retains elasticity at subzero temperatures, and has good dielectric strength for electrical insulation. Related to butadiene is **propadiene,** a gas of the composition CH_2:C:CH_2, called also **dimethylene methane** and **alene.** It also polymerizes easily to form plastics and rubbers.

BUTTER. An edible fat made from cow's milk by curdling with bacterial cultures and churning. The production of butter is one of the large industries of the Western nations, with an annual production exceeding 10 billion pounds, 30% of which is made in the United States. Other important producers are Germany, Holland, the Scandinavian countries, Australia, New Zealand, Canada, Ireland, and Argentina. Butter is an important raw material in the bakery and confectionery industries. Federal regulations require that creamery butter shall be made exclusively from milk or cream, with or without salt and coloring matter, and shall contain not less than 80% by weight of milk fat, not over 15% moisture, and not over 2.5% salt. Butter varies greatly in color and flavor according to the feed of the animal, the processing, and the storage. The natural color is whitish in winter and yellow in summer when the animal feeds on green pasturage. Commercial butter is usually brought to a uniform yellow by coloring with annatto. Musty, garlicky, and fishy flavors may be caused by noxious weeds eaten by the animal; cheesy or yeasty flavors may be from stale cream; metallic, greasy, scorched, or alkaline flavors may be from improper processing. **Whipped butter** has 50% greater volume in the same weight and has greater plasticity for spreading.

United States grades for creamery butter range from 93 score for the best butter of fine flavor and body down to 85 score for the lowest grade having pronounced obnoxious weed flavor and defects in body, color, or salt. The grading, or scoring, of butter is done by experts. The flavor is determined by the senses of taste and smell. The flavor, body, color, and salt are rated independently, and points, or scores, are subtracted for defects. Body and texture of the butter are determined by the character of the granules and their closeness. The most common body defects are gumminess, sponginess, crumbliness, and stickiness. The most common defect in color is lack of uniformity, with waves or mottles. Defects in salting are excessive salt and undissolved salt grains. Butter held in storage at improper temperatures is likely to develop rancid or unpleasant flavors and acidity due to chemical changes, or it may absorb flavors from surrounding products. High-grade butter can be held in well-regulated cold storage for long periods without appreciable deterioration.

An important substitute for butter is **margarine. Oleomargarine** is a term still retained in old food laws, but the product is no longer manufactured. It was a compound of mutton fat with vegetable tallows and fats, invented by the French chemist Mege-Mouries. Margarine is made from a mixture of about 80% vegetable oils and 20 milk in the same manner as butter. It has a slightly lower melting point than butter, 22 to 27°C, but the melting point and a desired degree of saturation of the fatty acids can be regulated by hydrogenation of the oils. The margarine of lower melting points is used in the bakery industry, and the grades with higher melting points are for table use. From 2.5 to 4% salt is used, together with vitamins A and D, lecithin, annatto coloring, and sometimes phosphatides to prevent spattering when used for frying. **Biacetyl,** $C_4H_6O_2$, a colorless, pungent sweet liquid which gives the characteristic flavor to butter, is also added. The food value is in general higher than that of butter, but because of the competition with butter various Federal and state regulations restrict its use. **Soya butter** is made from emulsified soybean and, when fortified with butyric acid, the characteristic acid of butter, is practically indistinguishable from butter. It is, however, subject to restrictive regulations. **Butter flavors** are used in confectionery and bakery products. **Butter-Aid,** of Cino Chemical Co., is made by extracting and concentrating the esters of natural butter. It is used as a high-strength flavor in foodstuffs in the form of powder or liquid emulsion. **Butta-Van,** of the Whitehall Food Mfg. Co., is a butter flavor with vanilla. It contains butyric acid, ethyl butyrate, coumarin, vanillin, and glycerin in water solution. **Ghee butter,** used in India, is made from buffalo milk, sometimes mixed with cow's milk. It is clarified and the moisture removed by boiling and slow cooling and separating off the opaque white portion. It is light in color and granular.

Cheese is an important solid food product made from whole or skim milk. It contains all the food value of milk, including the proteins of the casein. The biotics used in the manufacture produce *n*-butyric acid, with also caproic, caprylic, and capric acids in varying amounts which produce the flavor of the various types of cheese. In the same manner **lipase enzymes** from the glands of calves and lambs are used for enhancing the flavor of food products containing milk or butterfat. The enzymes hydrolyze the butyric or other short-chained fatty acids into the glycerides. The **Lipolyzed butter** of Marschall Dairy Laboratories, Inc., is made by treating natural butterfat with the enzymes. It gives intensity and uniformity of flavor to margarine and bakery products.

CADMIUM. A silvery-white crystalline metal, symbol Cd. It has a specific gravity of 8.6, is very ductile, and can be rolled or beaten into thin sheets. It resembles tin and gives the same characteristic cry when bent, but is harder than tin. A small addition of zinc makes it very brittle. It melts at 608°F (320°C) and boils at 1409°F (765°C). Cadmium is employed as an alloying element in soft solders and in fusible alloys, for hardening copper, as a white corrosion-resistant plating metal, and in its compounds for pigments and chemicals. It is also used to shield against neutrons in atomic equipment; but gamma rays are emitted when the neutrons are absorbed, and these rays require an additional shielding of lead. The metal is marketed in small round sticks 12 in (0.31 m) long, in variously shaped anodes for electroplating, and as foil. **Cadmium foil** of the American Silver Co. is 99.95% pure cadmium and is as thin as 0.0005 in (0.013 mm). It is used for neutron shielding and for electronic applications requiring high corrosion resistance. Electrolytic cadmium is 99.95% pure. It is obtained chiefly as a by-product of the zinc industry by treating the fluc dust and fumes from the roasting of the ores. Flue dust imported from Mexico averages 0.66 ton (600 kg) of cadmium per ton (metric ton) of dust. About half the world production is in the United States. Other important producers are West Germany, Belgium, Canada, and Poland. The only commercial ore of the metal is **greenockite,** CdS, which contains theoretically 77.7% cadmium. This mineral occurs in yellow powdery form in the zinc ores of Missouri. Cadmium occurs in sphalerite to the extent of 0.1 to 1%.

Most of the consumption of cadmium is for electroplating. For a corrosion-resistant coating for iron or steel a cadmium plate of 0.0003 in (0.008 mm) is equal in effect to a zinc coat of 0.001 in (0.025 mm). The plated metal has a silvery-white color with a bluish tinge, is denser than zinc and harder than tin, but electroplated coatings are subject to hydrogen embrittlement, and aircraft parts are usually coated by the vacuum process. **Cadmium plating** is not normally used on copper or brass since copper is electronegative to it, but when these metals are employed next to cadmium-plated steel a plate of cadmium may be used on the copper to lessen deterioration. **Cadalyte,** of Du Pont, is a cadmium salt and process for cadmium plating, and **Udylite** is the name of a salt and process of the Udylite Corp.

Small amounts of cadmium added to copper give higher strength, hardness, and wear resistance, but decrease the electric conductivity. Copper containing 0.5 to 1.2% cadmium is called **cadmium copper** or **cadmium**

bronze. Hitenso is a cadmium bronze of the American Brass Co. It has 35% greater strength than hard-drawn copper, and 85% the conductivity of copper. The cadmium bronze known in England as **conductivity bronze,** used for electric wires, contains 0.8% cadmium and 0.6 tin. The tensile strength, hard-drawn, is 85,000 lb/in^2 (586 MPa), and the conductivity is 50% that of copper. **Cadmium nitrate,** $Cd(NO_3)_2$, is a white powder used for making cadmium yellow and fluorescent pigments, and as a catalyst. **Cadmium sulfide,** Cds, is used as a yellow pigment and, mixed with **cadmium selenide,** CdSe, a red powder, gives a bright-orange pigment. The sulfide is used for growing **cadmium sulfide crystals** in plates and rods for semiconductor uses. Crystals grown at 1050°C are nearly transparent, but those grown at higher temperatures are dark amber.

CAFFEINE. An alkaloid which is a white powder when it has the composition $C_8H_{10}N_4O_2$, and occurs in crystalline flakes when it has one molecule of water of crystallization. The melting point is 235°C. It is soluble in water and in alcohol. It is the most widely used of the purine compounds, which are found in plants. Caffeine stimulates physically to lessen fatigue, but in large amounts is highly toxic. Its prime use is in medicine, but most of the production is used in soft drinks. Caffeine does not normally break down in the human body, but passes off in the urine, and the effect is not cumulative, but the sarcosine, which occurs in muscles, is a decomposition product of caffeine though it normally comes from nitrogen metabolism. Caffeine is obtained from coffee, tea waste, kola nuts, or guarana by solvent extraction, or a by-product in the manufacture of non-caffeine coffees, or in the processing of coffee for the production of oil and cellulose. It is made synthetically from **dimethyl sulfate,** a volatile toxic liquid of the composition $H(CH_2)O(SO_2)O(CH_2)H$, also used for making codeine and other drugs. Synthetic caffeine is also made from urea and sodium cyanoacetate, and is equal chemically to natural caffeine.

Less than 1% caffeine is obtained from coffee, about 2 from **tea waste,** and 1.5 from kola nuts. In tea it is sometimes called **theine.** Cocoa waste contains **theobromine,** from which caffeine may be produced by adding one more methyl group to the molecular ring. The name is a deception, as there is no bromine in the molecule. Theobromine is a more powerful stimulant than caffeine. It is a bitter white crystalline powder of the composition $C_7H_8N_4O_2$, also called **dimethyl xanthine,** and used in medicine. **Guarana** contains the highest percentage of caffeine of all the beverage plants, about 3%. It comes from the seeds of the woody climbing plant *Paullinia cupana,* of the Amazon Valley. The Indians grind the seeds with water and mandioca flour and dry the molded paste with smoke. For use it is grated into hot water. **Kola nuts** are the seeds of the fruit of the large spreading tree *Kola acuminata* native to West Africa and cultivated also in tropical America, and the *K. nitida* of West Africa. The nuts of the latter tree contain the higher percentages of theobromine and caffeine. The white nuts are preferred to the pink or red varieties. **Citrated caffeine,** used in pharmaceuticals, is a white powder produced by the action of citric acid on caffeine, and contains about equal quantities by weight of anhydrous caffeine and citric acid. It is very soluble in water.

CAJEPUT OIL. A greenish essential oil distilled from the leaves of the tree *Melaleuca leucadendron* growing chiefly in Indonesia. It contains the cineole of eucalyptus oil and also the **terpinol** which is characteristic of the lilac. It has a camphorlike odor. It is used in medicine as an antiseptic and counterirritant, and in perfumes. **Naouli oil** is a similar oil from the leaves of the tree *M. viridi* of New Caledonia. **Cajeput bark,** from the same tree, is used as an insulating material in place of cork. The bark, up to 2 in (5.08 cm) thick, is soft, light, resistant, and a good insulator.

CALCITE. One of the most common and widely diffused minerals, occurring in the form of limestones, marbles, chalks, calcareous marls, and calcareous sandstones. It is a **calcium carbonate,** $CaCO_3$, and the natural color is white or colorless, but it may be tinted to almost any shade with impurities. The specific gravity is about 2.72 and Mohs hardness 3. Calcite is usually in compact masses, but **aragonite,** formed by water deposition, develops in radiating flowerlike growths often twisted erratically. **Iceland spar,** or **calc spar,** is the name for the perfectly crystallized, water-clear, flawless calcite crystals of optical grade used for the manufacture of **Nicol prisms** for polarizing microscopes, photometers, calorimeters, and polariscopes. It comes from Iceland, Spain, South Africa, and New Mexico, and some crystals have been found as large as 17 lb (7.7 kg). The common **black calcite,** containing manganese oxide, often also contains silver in proportions high enough to warrant chemical extraction of the metal.

CALCIUM. A metallic element, symbol Ca, belonging to the group of alkaline earths. It is one of the most abundant materials, occurring in combination in limestones and calcareous clays. The metal is obtained 98.6% pure by electrolysis of the fused anhydrous chloride. By further subliming it is obtained 99.5% pure. **Calcium metal** is yellowish white in color. It oxidizes easily, and when heated in the air burns with a brilliant white light. The specific gravity is 1.55, melting point 810°C, and boiling point 1440°C. Its strong affinity for oxygen and sulfur is utilized as a cleanser for nonferrous alloys. As a deoxidizer and desulfurizer it is employed in the form of lumps or sticks of calcium metal or in ferroalloys and calcium-copper. For the reduction of light-metal ores it is used in the form of the hydride. **Crystalline calcium** is also used in the form of a very reactive free-flowing powder of 94 to 97% purity, and containing 2.5% of calcium oxide with small amounts of magnesium and other impurities. The specific gravity of the powder is 1.54, and the melting point is 851°C. Natural calcium compounds, such as **dolomite,** are used directly as a flux in melting iron. Calcium is also used to harden lead, and **calcium silicide** is used in making some special steels to inhibit carbide formation.

Many compounds of calcium are employed industrially, in fertilizers, foodstuffs, and medicine. It is an essential element in the formation of bones, teeth, shells, and plants. **Oyster shells** form an important commercial source of calcium for animal feeds. They are crushed, and the fine flour is marketed for stock feeds and the coarse for poultry feeds. The shell is calcium carbonate. **Edible calcium,** for adding calcium to food products, is **calcium lactate,** a white powder of the composition $Ca(C_3H_5O_3)_2 \cdot 5H_2O$, derived from milk. **Calcium lactobionate** is a white powder that readily forms chlorides and other double salts, and is used as a suspending agent in pharmaceuticals. It contains 4.94% available calcium. **Calcium phosphate,** used in the foodstuffs industry and in medicine, is marketed in several forms. **Calcium diphosphate,** known as **phosphate of lime,** is $CaHPO_4 \cdot 2H_2O$, or in anhydrous form. It is soluble in dilute citric acid solutions, and is used to add calcium and phosphorus to foods, and as a polishing agent in toothpastes. **Calcium monophosphate** is a stable, white, water-soluble powder, $CaH_4P_2O_8 \cdot H_2O$, used in baking as a leavening agent. The anhydrous **monocalcium phosphate,** $CaH_4(PO_4)_2$, of the Victor Chemical Works, for use in prepared flour mixes, is a white powder with each particle having a coating of a phosphate that is soluble only with difficulty, to delay solution when liquids are added. **Calcium triphosphate,** $Ca_3(PO_4)_2$, is a white water- insoluble powder used to supply calcium and phosphorus to foods, as a polishing agent in dentifrices, and as an antacid. **Calcium sulfite,** $CaSO_3 \cdot 2H_2O$, is a white powder used in bleaching paper pulp and textiles, and as a disinfectant. It is only slightly soluble in water, but it loses its water of crystallization and melts at 100°C. **Calcium silicate,** $CaO \cdot SiO_2$, is a white powder used as a reinforcing agent in rubber, as an absorbent, to control the viscosity of liquids, and as a filler in paints and coatings. It reduces the sheen in coatings. **Silene EF,** of the Columbia Chemical Div., is a precipitated calcium silicate for rubber. **Micro-Cal,** of Johns-Manville, is a synthetic calcium silicate with particle size as small as 0.02 μm. It will absorb up to six times its weight of water, and 3 lb (1.36 kg) will absorb a gallon (0.0038 m^3) of liquid and remain a free-flowing powder.

Calcium metasilicate, $CaO \cdot SiO_3$, is found in great quantities as the mineral **wollastonite** near Willsboro, N.Y., mixed with about 15% of andradite. The thin, needlelike crystals are easy to crush and grind, and the impurities are separated out. The ground material is a brilliant white powder in short fibers, 99.5% passing a 325-mesh screen. It is used in flat paints, for paper coatings, as a filler in plastics, for welding-rod coatings, and for electrical insulators, tile, and other ceramics. **Calcium acetate,** $Ca(C_2H_3O_2)_2 \cdot H_2O$, is a white powder used in liming rosin and for making metallic soaps and synthetic resins. It is also called **lime acetate, acetate of lime,** and **vinegar salts.**

CALCIUM CARBIDE. A hard, crystalline substance of grayish-black color, used chiefly for the production of acetylene gas for welding and cutting torches and for lighting. It was discovered in 1892 and was widely employed for theater stage lighting and for early automobile headlights. It is made by reducing lime with coke in the electric furnace. It can also be made by heating crushed limestone to a temperature of about 1000°C, flowing a high-methane natural gas through it, and then heating to 1700°C. The composition is CaC_2, and the specific gravity is 2.26. It contains theoretically 37.5% carbon. When water is added to calcium carbide, acetylene gas is formed, leaving a residue of slaked lime. Pure carbide will yield 5.83 ft^3 (0.16 m^3) of acetylene per pound (0.5 kg) of carbide, but the commercial product is usually only 85% pure. Federal specifications require not less than 4.5 ft^3 (0.13 m^3) of gas per pound (0.5 kg).

CALCIUM CHLORIDE. A white, crystalline, lumpy or flaky material of the composition $CaCl_2$. The specific gravity is 2.512, the melting point is 772°C, and it is highly hygroscopic and deliquescent with rapid solubility in water. The commercial product contains 75 to 80% $CaCl_2$, with the balance chiefly water of crystallization. Some is marketed in anhydrous form for dehydrating gases. It is also sold in water solution containing 40% calcium chloride. Calcium chloride has been used on roads to aid in surfacing and absorb dust, and to prevent cracking from freezing. It is also used for accelerating the setting of mortars, but more than 4% in concrete decreases the strength of the concrete. It is also employed as an antifreeze in fire tanks, for brine refrigeration, as an anti-ice agent on street pavements, as a food preservative, and in textile and paper sizes as a gelling agent. Calcium chloride is obtained from natural salt brines, or as a by-product in the Solvay process.

CALCIUM-SILICON. An alloy of calcium and silicon used as a deoxidizing agent for the elimination of sulfur in the production of steels and cast irons. It is marketed as low-iron, containing 22 to 28% calcium, 65 to 70 silicon, and 5 max iron, and as high-iron, containing 18 to 22% calcium, 58 to 60 silicon, and 15 to 20 iron. It comes in crushed form, and is added to the molten steel. At the temperature of molten steel all of the calcium passes off and leaves no calcium residue in the steel. **Calcium-manganese-silicon** is another master alloy containing 17 to 19% calcium, 8 to 10 manganese, 55 to 60 silicon, and 10 iron.

CAMEL'S HAIR. The fine, tough, soft hair from the mane and back of the camel, *Camelus bactrianus,* used for artists' brushes and for industrial striping brushes. Most of the hair is produced in central Asia and Iran, and the grades preferred for brushes are from the crossbred Boghdi camel. The hair from the dromedary, also called djemel, or camel, is of poor quality. Much of the camel hair is not cut, but is molted in large patches and is picked up along the camel routes. The plucked beard hair and the coarse outerguard hair obtained in combing are the brush fibers. They are tough, silky, and resilient. The length is 5 to 8 in (12.7 to 20.3 cm). The fine body hair, or **camel wool,** which constitutes about 90% of the total fiber, is 1.5 to 2 in (3.8 to 5.1 cm) long, has a fine radiance, a pale tan color, and a downy feel. It is the textile fiber. The beard hair from the Cashmere goat is very similar to camel hair and is used for brushes. Various other hairs are used for making camel's-hair brushes, including ox-ear hair, badger hair, and sable hair.

CAMPHOR. The white resin of the *Cinnamomum camphora,* an evergreen tree with laurellike leaves, reaching a height of 100 ft (30 m). The tree occurs naturally in China and southern Japan, and is also grown in Florida. Taiwan is the center of the industry. Camphor, $C_{10}H_{16}O$, has a specific gravity of 0.986 to 0.996, and melts at 175°C. It is soluble in water and in alcohol. Camphor is used for hardening nitrocellulose plastics, but it is also used in pharmaceuticals, in disinfectants, and in explosives and chemicals. It is obtained from the trunks, roots, and large branches by steam distillation. From 20 to 40 lb (9.1 to 18.1 kg) of chips produce 1 lb (0.5 kg) of camphor. Crude camphor is pressed to obtain the **flowers of camphor** and **camphor oil.** The crude red camphor oil is fractionated into white and brown oils; the white oil is used in soaps, polishes, varnishes, cleaners, and pharmaceuticals, and the brown oil is used in perfumery. **White camphor oil** is a colorless liquid with a camphor odor and a specific gravity of 0.870 to 1.040, and is soluble in ether or chloroform. Camphor oil may also be distilled from the twigs. **Camphor sassafrassy oil** is a camphor-oil fraction having a specific gravity of 0.97. It is a sassafras tone, and is used for scenting soaps and sprays.

Borneo camphor, or **borneol,** is a white, crystalline solid obtained from the tree *Dryobalanops camphora* of Borneo and Sumatra. It is used as a substitute for camphor in cellulose plastics. It has the composition $C_{10}H_{17}OH$ and a specific gravity of 1.01, is soluble in alcohol, and sublimes at 212°C. The wood of this tree, known as **Borneo camphorwood,** or **kapur,** is used for cabinetwork. It weighs 50 lb/ft^3 (801 kg/m^3), has an interlocking grain, and a scent of camphor. It is also known as **camphorwood.**

Artificial camphor is **bornyl chloride,** $C_{10}H_{17}Cl$, a derivative of the pinene of turpentine. It has a camphor odor and the same industrial uses as camphor, but is optically inactive and is not used in pharmaceuticals. **Synthetic camphor,** made from turpentine, in refined form is equal to the natural product for medicinal use, and the technical grade is used in plastics. The camphor substitute **Lindol,** of the Celanese Corp., is **tricresyl phosphate,** or **tolyl phosphate,** $(CH_3C_6H_4)_3PO_4$, a colorless, odorless viscous liquid which solidifies at −20°C. Like camphor, it hardens cellulose nitrate and makes it nonflammable. Tricresyl phosphate is also used as an additive for gasoline to prevent buildup of carbon deposits on the spark plugs and in the engine, thus increasing power by preventing predetonation. Other uses are as a plasticizer for synthetic resins, as a hydraulic fluid, and as an additive in lubricants. It is made from petroleum and from the cresylic acid from coal. **Triphenyl phosphate,** $(C_6H_5)_3PO_4$, is also used as a substitute for camphor in cellulose nitrate, and for making coating compounds nonflammable. It is a colorless solid, melting at 49°C. **Dehydranone,** of Union Carbide Corp., Chemicals Div., is **dehydracetic acid,** $C_8H_8O_4$, a white, odorless solid with some of the properties of camphor, used in nitrocellulose and vinyl resins. **Cyclohexyl levulinate,** $CH_3CO(CH_2)_2COOC_6H_{11}$, is also used as a substitute for camphor in nitrocellulose, and also in vinyl resins and in chlorinated rubber. It is a liquid of specific gravity 1.025, boiling point 265°C, and freezing point −70°C. **Adamantine** has the odor of camphor and turpentine. It is obtained from the crude petroleum of Moravia as a stable, crystalline solid, melting at 268°C. It has the empirical formula $C_{10}H_{16}$, and the molecule has four transcyclohexane rings. **Camphorene,** $C_{20}H_{32}$, is made from turpentine by polymerizing two myrcene molecules. It is a raw material for producing geraniol and linalol.

CAMWOOD. The wood of the tree *Baphia nitida,* native to West Africa, used for tool handles and for machine bearings. It will withstand heavy bearing pressures. The wood is exceedingly hard, has a coarse, dense grain, and weighs 65 lb/ft^3 (1041 kg/m^3). It contains a red coloring matter known as **santalin,** and was once valued as a dyewood for textiles. **Barwood,** from the tree *Pterocarpus santalinus,* of West Africa, is a similar reddish hardwood containing the same dye and used for the same purposes.

CANAIGRE. A tanning material extracted from the roots of the low-growing plant *Rumex hymenosepalus* of nothern Mexico and the arid Southwest of the United States. The plant is known locally as **sour dock,** and the roots contain up to 40% tannin. The cultivated plant yields as much as 20 tons of root per acre (4.8 kg/m^2). **Canaigre extract** contains 30% tannin. It produces a firm, orange-colored leather. Canaigre was the tanning agent of the Aztec Indians, and is still extensively cultivated.

CANARY SEED. The seeds of the **canary grass,** *Phalaris canariensis,* native to the Canary Islands, but now grown on a large scale in Argentina for export, and in Turkey and Morocco for human food and for export. In international trade it is known by the Spanish name **alpiste.** It is valued as a bird food because it contains phosphates, iron, and other minerals, and is rich in carbohydrates. It is, however, low in proteins and fats, and is usually employed in mixtures. **Birdseed** is an extensive item of commerce, but the birdseed that reaches the market in the United States is usually a blend of canary seed and millet, with other seeds to give a balanced food. Canary seed is small, pale yellow in color, and convex on both sides. The term **Spanish canary seed** is applied to the choice seed regardless of origin. **Niger seed,** also valued as a birdseed, is from the plant *Guizotia abyssinica,* of the thistle, or *Compositae* family, grown in India, Africa, Argentina, and Europe. It is also known as **inga seed, rantil, kala til seed,** and **black sesame.** It is also called **gingelli** in India, although this name and **til** are more properly applied to sesame. The seed is high in proteins and fats.

CANDELILLA WAX. A yellowish amorphous wax obtained by hot water or solvent extraction from the stems of the shrubs *Pedilanthus pavonis* and *Euphorbia antisyphilitica* growing in the semiarid regions of Texas and Mexico. The plants grow to a height of 3 to 5 ft (0.9 to 1.5 m) and consist of a bundle of stalks without leaves. The stems yield 3.5 to 5% wax that consists of long-chain hydrocarbons with small amounts of esters. The wax has a specific gravity of 0.983, melting point 67 to 70°C, iodine value 37, and saponification value 45 to 65. The refined grade is purified by remelting, and contains not more than about 1% water. It is soluble in turpentine, and is used for varnishes, polishes, and leather finishes; as a substitute for carnauba wax; or to blend with carnauba or beeswax. About half the production goes into furniture and shoe polishes, but it does not have the self-polishing characteristics of carnauba wax.

CANNEL COAL. A variety of coal having some of the characteristics of petroleum, valued chiefly for its quick-firing qualities. It consists of coal-like matter intimately mixed with clay and shale, often containing fossil fishes, and probably derived from vegetable matter in lakes. It is compact in texture, dull black in color, and breaks along joints, often having an appearance similar to that of black shale. It burns with a long, luminous, smoky flame, from which it derives its old English name, meaning candle. On distillation, cannel coal yields a high proportion of illuminating gas, up to 16,000 ft^3/ton (450 m^3/ton), leaving a residue consisting mostly of ash. At low temperatures it yields a high percentage of tar oils. The proportion of volatile matter may be as high as 70%. It is found in Great Britain, and in Kentucky, Ohio, and Indiana. Cannel coal from Scotland was originally called **parrot coal,** and **boghead coal** was a streaky variety.

CARBOHYDRATES. The most abundant class of organic compounds, constituting about three-fourths of the dry weight of the plant world. They are distinguished by the fact that they contain the elements carbon, hydrogen, and oxygen, and no others. Many chemical compounds, such as alcohols and aldehydes, also have these elements only, but the term carbohydrate refers only to the starches, sugars, and cellulose, which are more properly called **saccharides.** They are best known for their use as foodstuffs, as carbohydrates compose more than 50% of all American food, but they are also used in many industrial processes. The digestible carbohydrates are the sugars and the starches. The indigestible carbohydrates are cellulose and hemicellulose, which form the chief constituents of woods, stalks, and leaves of plants, the outer covering of seeds, and the walls of plant cells enclosing the water, starches, and other substances of the plants. Much cellulose is eaten as food, especially in the leaves of vegetables and in bran, but it serves as bulk rather than as food, and is beneficial if not consumed in quantity. The digestible carbohydrates are classified as **single sugars, double sugars,** and **complex sugars,** chemically known as **monosaccharides, disaccharides,** and **polysaccharides.** The single sugars, glucose, fructose, and galactose, require no digestion and are readily absorbed into the bloodstream. The double sugars, sucrose, maltose, and lactose, require to be broken down by enzymes in the human system. **Lactose,** produced from milk solids, is a nonhygroscopic powder. It is only 16% as sweet as sugar and not as soluble, but it enhances flavor. It digests slowly. It is used in infant foods, in dairy drinks and ice cream to improve low-fat richness, in bakery products to decrease sogginess and improve browning, and as a dispersing agent for high-fat powders. **Galactose** is derived from lactose by hydrolysis. **Multisugars** are mixed sugars with the different sugars interlocked in the crystals. They dissolve rapidly to form clear solutions.

The complex sugars are the starches, dextrins, and glycogen. These require digestion to the single stage before they can be absorbed in the system. The common starches are in corn, wheat, potatoes, rice, tapioca, and sago. **Animal starch** is the reverse food of animals stored in the liver and muscles. It is **glycogen,** a sweet derivative of glycolic acid. It is not separated out commercially because it is hygroscopic and quickly hydrolyzed. **Dextran,** related to glycogen, is a **polyglucose** made up of many molecules of glucose in a long chain. It is used as an extender of blood plasma. It can be stored indefinitely and, unlike plasma, can be sterilized by heat. It is produced commercially by biotic fermentation of common sucrose sugar.

The **hemicelluloses** are agar-agar, algin, and pectin. They differ chemically from cellulose and expand greatly on absorbing water. The hemicelluloses of wood, called **hexosan,** consist of the **wood sugars,** or **hexose,** with six carbon atoms, $(C_6H_{10}O_5)_n$. They are used to make many chemicals. The water-soluble hemicellulose of the Masonite Corp., known as **Masonex** in water solution and **Masonoid** as a powder, is a by-product of the steam-exploded wood process. It is used to replace starch as a binder for foundry cores and for briquetting coal, and for emulsions. It contains 70% wood sugars, 20 resins, and 10 lignin. **Lichenin,** or **moss starch,** is a hemicellulose from moss and some seeds.

The **pentosans** are gums or resins occurring in nutshells, straw, and the cell membranes of plants. They may be classed as hemicellulose and on hydrolysis yield **pentose,** or **pentaglucose,** a sugar containing five carbon atoms. **Pectin** is a yellowish, odorless powder soluble in water and decomposed by alkalies. It is produced by acid extraction from the inner part of the rind of citrus fruits and from apple pomace. In East Africa it is obtained from sisal waste. **Flake pectin** is more soluble and has longer shelf life than the powdered form. It is produced from a solution of apple pomace containing 5% pectin by drying on steam-heated drums, and the thin film obtained is flaked to 40 mesh. Another source is **sugar-beet pulp,** which contains 20 lb (9.7 kg) of pectin per ton (907 kg).

Pectin has a complex structure, having a lacturonic acid with methanol in a glucoside chain combination. It is used for gelling fruit preserves, and the gelling strength depends on the size of the molecule, the molecular weight varying from 150,000 to 300,000. It is also used as a blood coagulant in treating hemorrhage, and for prolonging the effect of some drugs by retarding their escape through the body. **Sodium pectate** is used for creaming rubber latex, and in cosmetics and printing inks. Hemicellulose and pectin are valuable in the human system because of their ability to absorb and carry away irritants, but they are not foods in the normal sense of the term. **Oragen,** of the Consumer Drug Co., is a pectin-cellulose complex derived from orange pulp. It is used in weight-reduction diets, increasing bulk and retaining moisture, thus suppressing desire for excess food. Each of the saccharides has distinctive characteristics of value in the system, but each also in excess causes detrimental conditions.

CARBON. A nonmetallic element, symbol C, existing naturally in several allotropic forms, and in combination as one of the most widely distributed of all the elements. It is quadrivalent, and has the property of forming chain and ring compounds, and there are more varied and useful compounds of carbon than of all other elements. Carbon enters into all organic matter of vegetable and animal life, and the great branch of **organic chemistry** is the chemistry of carbon compounds. The black amorphous carbon has a specific gravity of 1.88; the black crystalline carbon known as graphite has a specific gravity of 2.25; the transparent crystalline carbon, as in the diamond, has a specific gravity of 3.51. **Amorphous carbon** is not soluble in any known solvent. It is infusible, but sublimes at 3500°C, and is stable and chemically inactive at ordinary temperatures. At high temperatures it burns and absorbs oxygen, forming the simple oxides CO and CO_2, the latter being the stable oxide present in the atmosphere and a natural plant food. Carbon dissolves easily in some molten metals, notably iron, exerting great influence on them. Steel, with small amounts of chemically combined carbon, and cast iron, with both combined carbon and graphitic carbon, are examples of this.

Carbon occurs as hydrocarbons in petroleum, and as carbohydrates in coal and plant life, and from these natural basic groupings an infinite number of carbon compounds can be made synthetically. Carbon, for chemical, metallurgical, or industrial use, is marketed in the form of compounds in a large number of different grades, sizes, and shapes; or in master alloys containing high percentages of carbon; or as activated carbons, charcoal, graphite, carbon black, coal-tar carbon, petroleum coke; or as pressed and molded bricks or formed parts with or without binders or metallic inclusions. Natural deposits of graphite, coal tar, and petroleum coke are important sources of elemental carbon. Charcoal and activated carbons are obtained by carbonizing vegetable or animal matter.

Carbon 13 is one of the isotopes of carbon, used as a tracer in biologic research where its heavy weight makes it easily distinguished from other carbon. **Carbon 14,** or **radioactive carbon,** has a longer life. It exists in the air, formed by the bombardment of nitrogen by cosmic rays at high altitudes, and enters into the growth of plants. The half-life is about 6,000 years. It is made from nitrogen in a cyclotron.

Carbon fibers are made by pyrolyzing methane at temperatures to 1500°F (843°C), and passing over a silica surface to yield masses of long and short fibers of 0.1- to 0.4-μm diameter. The carbon has a mesomorphic two-dimensional crystallite form, and the fibers are strong and flexible with a specific gravity of 1.991. Carbon fibers are produced as continuous filaments, 0.002 in (0.05 mm) in diameter, with tensile strength of 200,000 lb/in^2 (1,378 MPa). **Thornel,** of Union Carbide Corp., is a yarn made from these filaments for high-temperature fabrics. It retains its strength to temperatures above 2800°F (1538°C). **Carbon yarn,** of the Basic Carbon Corp., is 99.5% pure carbon. It comes in plies from 2 to 30, with each ply composed of 720 continuous filaments of 0.0003-in (0.0076-mm) diameter. Each ply has a breaking strength of 2 lb (0.91 kg). The fiber has the flexibility of wool and maintains dimensional stability to 5700°F (3150°C). **Ucar,** of Union Carbide Corp., is a conductive **carbon fabric** made from carbon yarns woven with insulating glass yarns with resistivities from 0.2 to 30 Ω for operating termperatures to 550°F (288°C). **Carbon wool,** of Atomic Laboratories, Inc., for filtering and insulation, is composed of pure-carbon fibers made by carbonizing rayon. The fibers, 5 to 50 μm in diameter, are hard and strong, and can be made into rope and yarn, or the mat can be activated for filter use. **Avceram RS,** of FMC Corp., is a composite rayon-silica fiber made with 40% dissolved sodium silicate. A highly heat-resistant fiber, **Avceram CS,** is woven into fabric and then pyrolyzed to give a porous interlocked mesh of **carbon silica fiber,** with a tensile strength of 165,000 lb/in^2 (1,137 MPa). **Dexsan,** of C. H. Dexter & Sons Co., for filtering hot gases and liquids, is a **carbon filter paper** made from carbon fibers pressed into a paperlike mat, 0.007 to 0.050 in (0.18 to 0.127 mm) thick and impregnated with activated carbon.

Carbon brushes for electric motors and generators, and **carbon electrodes,** are made of carbon in the form of graphite, petroleum coke, lampblack, or other nearly pure carbon, sometimes mixed with copper powder to increase the electric conductivity, and then pressed into blocks or shapes and sintered. The carbon-graphite brushes of the Pure Carbon Co. contain no metals but are made from **carbon-graphite powder** and, after pressing, are subjected to a temperature of 5000°F (2760°C), which produces a harder and denser structure, permitting current densities up to 125 A/in^2 (1,538 A/m^2). **Carbon brick,** used as a lining in the chemical-processing industries, is carbon compressed with a bituminous binder and then carbonized by sintering. If the binder is capable of being completely carbonized, the bricks are impervious and dense. **Graphite brick,** made in the same manner from graphite, is more resistant to oxidation than carbon bricks and has a higher thermal conductivity, but it is softer. The binder may also be a furfural resin polymerized in the pores. **Karbate No. 1,** of the National Carbon Co., Inc., is a carbon-base brick, and **Karbate No. 2** is a graphite brick. Karbate has a crushing strength of 10,500 lb/in^2 (72 MPa), and a weight of 110 to 120 lb/ft^3 (1,762 to 1,922 kg/ m^3). **Impervious carbon** is used for lining pumps, for valves, and for acid-resistant parts. It is carbon or graphite impregnated with a chemically resistant resin and molded to any shape. It can be machined. **Karbate 21** is a phenolic-impregnated graphite, and **Karbate 22** is a modified phenolic-impregnated graphite. Molded impervious carbon has a specific gravity of 1.77, tensile strength of 1,800 lb/in^2 (12.5 MPa), and compressive strength of 10,000 lb/in^2 (68 MPa). **Impervious graphite** has a higher tensile strength, 2,500 lb/in^2 (17.2 MPa), but a lower compressive strength, 9,000 lb/in^2 (62 MPa). The thermal conductivity is 8 to 10 times that of stainless steel. **Graphitar,** of the U.S. Graphite Co., is a strong, hard carbon molded from amorphous carbon mixed with other forms of carbon. It has high crushing strength and acid resistance, and is used for sealing rings, chemical pump blades, and piston rings. **Porous carbon** is used for the filtration of corrosive liquids and gases. It consists of uniform particles of carbon pressed into plates, tubes, or disks without a binder, leaving interconnecting pores of about 0.001 to 0.0075 in (0.025 to 0.190 mm) in diameter. The porosity of the material is 48%, tensile strength 150 lb/in^2 (1 MPa), and compressive strength about 500 lb/in^2 (3.5 MPa). **Porous graphite** has graphitic instead of carbon particles, and is more resistant to oxidation but is lower in strength.

The so-called carbons used for electric-light **arc electrodes** are pressed from coal-tar carbon, but are usually mixed with other elements to bring the balance of light rays within the visible spectrum. Solid carbons have limited current-carrying capacity, but when the carbon has a center of metal compounds such as the fluorides of the rare earths, its current capacity is greatly increased. It then forms a deep positive crater in front of which is a flame five times the brilliance of that with the low-current arc. The **sunshine carbon,** used in electric-light carbons to give approximately the same spectrum as sunlight, is molded coal-tar carbon with a core of cerium metals to introduce more blue into the light. **Arc carbons** are also made to give other types of light, and to produce special rays for medicinal and other purposes. **B carbon,** of the National Carbon Co., Inc., contains iron in the core and gives a strong emission of rays from 230 to 320 mμ units, which are the antirachitic radiations. The light seen by the eye is only one-fourth the total radiation since the strong rays are invisible. **C carbon** contains iron, nickel, and aluminum in the core, and gives powerful lower-zone ultraviolet rays. It is used in light therapy and for industrial applications. **E carbon,** to produce penetrating infrared radiation, contains strontium. **Electrode carbon,** used for arc furnaces, is molded in various shapes from carbon paste. When calcined from petroleum coke the electrodes contain only 0.2% moisture, 0.25 volatile matter, and 0.3 ash, and have a specific gravity of 2.05. The carbon is consumed both in the production of light and of furnace heat. For example, from 500 to 600 kg of carbon is consumed in producing a metric ton of aluminum.

CARBON BLACK. An amorphous powdered carbon resulting from the incomplete combustion of a gas, usually deposited by contact of the flame on a metallic surface, but also made by the incomplete combustion of the gas in a chamber. The carbon black made by the first process is called **channel black,** taking the name from the channel iron used as the depositing surface. The modern method, called the impingement process, uses many small flames with the fineness of particle size controlled by the flame size. The air-to-gas ratio is high, giving oxidized surfaces and acid properties. No water is used for cooling, keeping the ash content low. The supergrade of channel black has a particle size as low as 13 μm and a pH of 3 to 4.2. Carbon black made by other processes is called **soft black** and is weaker in color strength, no so useful as a pigment. **Furnace black** is made with a larger flame in a confined chamber with the particles settling out in cyclone chambers. The ratio of air is low, and water cooling raises the ash content. The particle surface is oily, and the pH is high.

Carbon black from clean artificial gas is a glossy product with an intense color, but all the commercial carbon black is from natural gas. To remove H_2S the sour gas is purified and water-scrubbed before burning. **Thermotomic black,** a grade made by the thermal decomposition of the gas in the absence of oxygen, is preferred in rubber when high loadings are employed because it does not retard the vulcanization, but only a small part of the carbon black is made by this process. This thermal process black has large particle size, 150 μm, and a pH of 8.5. It gives a coarse oily carbon.

The finer grades of channel black are mostly used for color pigment in paints, polishes, carbon paper, and in printing and drawing inks. The larger use of carbon black is in automotive tires to increase wear resistance of the rubber. The blacker blacks have a finer particle size than the grayer blacks, hence have more surface and absorptive capacity in compounding with rubber. Channel black is valued for rubber compounding because of low acidity and low grit content. The high pH of furnace black may cause scorching unless offsetting chemicals are used, but some furnace blacks are made especially for tire compounding. In general, the furnace black with particle sizes from 28 to 85 μm and a pH from 8 to 10, and the channel blacks with particle size of about 29 and pH of 4.8, are used for rubber. **Micronex EPC,** an impingement channel black of the Binney & Smith Co., has a particle diameter of 29 μm, and a pH of 4.8, while **Thermax MT,** a thermal process black, has a particle size of 274 μm and a pH of 7.

In rubber compounding the carbon black is evenly dispersed to become intimately attached to the rubber molecule. Fineness of the black determines the tensile strength of the rubber, the structure of the carbon particle determines the modulus, and the pH determines the cure behavior. Furnace blacks have a basic pH which activates the accelerator, and delaying-action chemicals are thus needed, but fine furnace blacks impart abrasion resistance to the rubber. Furnace black made with a confined flame with limited air has a neutral surface and a low volatility. Fineness is varied by temperature, size of flame, and time. Carbonate salts raise the pH. Most of the channel black for rubber compounding is made into dust-free pellets less than ⅛ in (0.3 cm) in diameter with a density of 20 to 25 lb/ft^3 (320 to 400 kg/m^3). **Color-grade black** for inks and paints is produced by the channel process or the impingement process. In general, carbon black for reinforcement has small particle size, and the electrically conductive grades, **CF carbon black** and **CC carbon black,** conductive furnace and conductive channel, have large particle size.

Carbon black from natural gas is produced largely in Louisiana, Texas, and Oklahoma. About 35 lb (15.9 kg) of black is avilable per 1,000 ft^3 (28 m^3) of natural gas, but only 2.2 lb (1 kg) is recovered by the channel process and 10 lb (4.5 kg) by the furnace method. By using gas from which the natural gasoline has been stripped, and by controlled preheating and combustion, as much as 27 lb (12.2 kg) can be recovered. **Acetylene black** is a carbon black made by heat decomposition of acetylene. It is more graphitic than ordinary carbon black with colloidal particles linked together in an irregular lattice structure and has high electric conductivity and high liquid-absorption capacity. Particle size is intermediate between that of channel black and furnace black, with low ash content, nonoiliness, and a pH of 6.5. It is valued for use in dry cells and lubricants. **Ucet,** of Union Carbide Corp., is in the form of agglomerates of irregular fine crystals. The greater surface area gives higher thermal and electric conductivity and high liquid absorption.

For electrically **conductive rubber,** the mixing of the black with the rubber is regulated so that carbon chain connections are not broken. Such conductive rubber is used for tabletops, for conveyor belts, and for coated filter fabrics to prevent static buildup. Carbon blacks are also made from liquid hydrocarbons, and from anthracite coal by treatment of the coal to liberate hydrogen and carbon monoxide and then high-temperature treatment with chlorine to remove impurities. The black made from anthracite has an open-pore structure useful for holding gases and liquids.

Carbon-black grades are often designated by trade names for particular uses. **Kosmovar,** of the United Carbon Co., is a black with a slight bluish top tone used as a pigment for lacquers. The specific gravity is 1.72, and mesh is 325. **Gastex** and **Pelletex** are carbon blacks of the General Atlas Carbon Co. used for rubber compounding. **Statex,** of the Columbian Carbon Co., is a colloidal furnace black for synthetic rubber compounding **Kosmos 60,** of United Carbon Co., Inc., is a furnace black of high density and structure, while **Continex FF** of this company is a finely divided furnace black. Both are used in rubber compounding, the first giving easier extrusion of the rubber, and the second giving better abrasion resistance. **Aquablak H,** of the Binney & Smith Co., is a colloidal water dispersion of channel black to give a jet-black color. **Aquablak M** is a water dispersion of furnace black to give a blue-gray tone. They are used as pigments in casein paint, inks, and leather finishes. **Liquimarl-Black,** of H. B. Taylor Co., is a stable colloidal dispersion of pure food-grade carbon black for use in coloring confectionery and for modifying colors in bakery products.

CARBON DIOXIDE. Also called **carbonic anhydride,** and in its solid state, **Dry Ice.** A colorless, odorless gas of the composition CO_2, which liquefies at −65°C and solidifies at −78.2°C. It is obtained as a by-product of distilleries, from burning lime, and from natural gas. In liquid form it is marketed in cylinders, and is used in fire extinguishers, in spray painting, in refrigeration, for inert atmospheres, for the manufacture of **carbonated beverages,** and in many industrial processes. It is also marketed as Dry Ice, a white snowlike solid used for refrigeration in transporting food products. **Cardox** is a trade name of the Cardox Corp. for liquid carbon dioxide in storage units at 30 lb/in^2 (0.21 MPa) pressure for fire-fighting equiment.

CARBON MONOXIDE. CO is a product of incomplete combustion, and is very reactive. It is one of the desirable products in synthesis gas for making chemicals, the synthesis gas made from coal containing 37% min CO. It is also recovered from top-blown oxygen furnaces in steel mills. Carbon monoxide is an intense poison when inhaled and is extremely toxic even in the small amounts from the exhausts of internal-combustion engines.

CARBON PAPER. Paper used for duplicating typewriting, pencil, or pen writing. It is made by coating the paper with a mixture of a pigment and a medium. The pigments include carbon black, Prussian blue, and organic red, or blue and green lakes. The medium is likely to be a blend of waxes and oils to give a composition of the desired consistency and melting point, but, to make a good carbon paper that will not be gummy and will not smear, a proper proportion of a high-melting, nongreasy wax, such as carnauba, must be used. Papers of special texture, preferably rag papers, are employed. Smudgeproof carbon paper has a coating of plastic lacquer.

CARBON STEEL. The **wrought carbon steels** covered here are sometimes termed **plain carbon steels.** The old shop names, **machine steel** and **machinery steel,** are still used to mean any easily worked low-carbon steel. By definition, plain carbon steels are those that contain up to about 1% carbon, not more than 1.65 manganese, 0.60 silicon, and 0.60 copper, and only residual amounts of other elements, such as sulfur (0.05% max) and phosphorus (0.04% max). They are identified by means of a four-digit numerical system established by the American Iron and Steel Institute (AISI). The first digit is the number 1 for all carbon steels. A 0 after the 1 indicates nonresulfurized grades, a 1 for the second digit indicates resulfurized grades, and the number 2 for the second digit indicates resulfurized and rephosphorized grades. The last two digits give the nominal (middle of the range) carbon content in hundredths of a percent. For example, for grade 1040, the 40 represents a carbon range of 0.37 to 0.44%. If no prefix letter is included in the designation, the steel was made by the basic open-hearth, basic oxygen, or electric furnace process. The prefix B stands for the acid Bessemer process. The letter L between the second and third digits identifies leaded steels, and the suffix H indicates that the steel was produced to hardenability limits.

For all plain carbon steels, carbon is the principal determinant of many performance properties. Carbon has a strengthening and hardening effect. At the same time, it lowers ductility, as evidenced by a decrease in elongation and reduction of area. In addition, a rise in carbon content lowers machinability and decreases weldability. The amount of carbon present also affects physical properties and corrosion resistance. With an increase in carbon content, thermal and electric conductivity decline, magnetic permeability decreases drastically, and corrosion resistance is lowered. Plain carbon steels are commonly divided into three groups, according to carbon content: Low carbon, up to 0.30%; medium carbon, 0.31 to 0.55%; and high carbon, 0.56 to 1%.

Low-carbon steels are the grades **AISI 1005** to **AISI 1030.** Sometimes referred to as **mild steels,** they are characterized by low strength and high ductility, and are nonhardenable by heat treatment except by surface-hardening processes. Because of their good ductility, low-carbon steels are readily formed into intricate shapes. Property ranges are: Tensile strength, 40,000–70,000 lb/in^2 (275–482 MPa); elongation, 25–40%; hardness, 110–150 Brinell. Cold work increases strength and decreases ductility. Where necessary, annealing is used to improve ductility after cold working. These steels are also readily welded without danger of hardening and embrittlement in the weld zone. Although low-carbon steels cannot be thoroughly hardened, they are frequently surface-hardened by various methods (carburizing, carbonitriding, and cyaniding, for example) which diffuse carbon into the surface. Upon quenching, a hard, abrasion-resistant surface is obtained.

Medium-carbon steels are the grades **AISI 1030** to **AISI 1055.** They usually are produced as killed, semikilled, or capped steels, and are hardenable by heat treatment. However, hardenability is limited to thin sections or to the thin outer layer on thick parts. Medium-carbon steels in the quenched and tempered condition provide a good balance of strength and ductility. Strength can be further increased by cold work. The highest hardness practical for medium-carbon steels is about 550 Bhn (Rockwell C55). Because of the good combination of properties, they are the most widely used steels for structural applications, where moderate mechanical properties are required. Quenched and tempered, their tensile strengths range from about 75,000 to over 150,000 lb/in^2 (517 to over 1,034 MPa).

High-carbon steels are the grades **AISI 1060** to **1095.** They are, of course, hardenable with a maximum surface hardness of about 710 Bhn (Rockwell C64) achieved in the 1095 grade. These steels are thus suitable for wear-resistant parts. So-called **spring steels** are high-carbon steels available in annealed and pretempered strips and wires. Besides their spring applications, these steels are used for such items as piano wire and saw blades. Quenched and tempered, high-carbon steels approach tensile strengths of 200,000 lb/in^2 (1,378 MPa).

Free-machining steels are low- and medium-carbon grades with additions of sulfur (0.08 to 0.13%), sulfur-phosphorus combinations, and/or lead to improve machinability. They are **AISI 1108–1151** for sulfur grades, and **AISI 1211–1215** for phosphorus-and-sulfur grades. The presence of relatively large amounts of sulfur and phosphorus causes some reduction in cold formability, weldability, and forgeability, as well as a lowering of ductility, toughness, and fatigue strength. **Calcium deoxidized steels** (carbon and alloy) have good machinability, and are used for carburized or through-hardened gears, worms, and pinions.

Low-temperature carbon steels have been developed chiefly for use in low-temperature equipment and especially for welded pressure vessels. They are low-carbon (0.20 to 0.30%), high manganese (0.70 to 1.60%), silicon (0.15 to 0.60%) steels, which have a fine-grain structure with uniform carbide dispersion. They feature moderate strength with toughness down to −50°F (−46°C).

Carbon steels having about 0.16% carbon and 0.68 manganese, but with 0.01 to 0.04% columbium to refine the grain and improve the formability and welding properties, have a tensile strength of 75,000 lb/in^2 (517 MPa), with elgonation of 27%. This type of steel is known as **coumbium steel.** They are used for shafts, forgings, gears, machine parts, and dies and gages for small runs. They forge and machine easily. Up to 0.15% sulfur, or 0.045 phosphorus, makes them free-cutting and keeps the chips from curling, but reduces the strength.

Rail steel, for railway rails, is characterized by an increase of carbon with the weight of the rail. Railway engineering standards call for 0.50 to 0.63% carbon and 0.60 manganese in a 60-lb (27-kg) rail, and 0.69 to 0.82% carbon and 0.70 to 1.0 manganese in a 140-lb (64-kg) rail. Rail steels are produced under rigid control conditions from deoxidized steels with phosphorus kept below 0.04%, and silicon 0.10 to 0.23%. Guaranteed minimum tensile strength of 80,000 lb/in^2 (551 MPa) specified, but it is usually much higher.

Sometimes a machinery steel may be required with a small amount of alloying element to give a particular characteristic and still not be marketed as an alloy steel, although trade names are usually applied to such steels. **Superplastic steels,** developed at Stanford University, with 1.3 to 1.9% carbon, fall between high-carbon steels and cast irons. They have elongations approaching 500% at warm working temperatures of 1000 to 1200°F (538 to 650°C), and 4 to 15% elongation at room temperature. Tensile strengths range from 150,000 to over 200,000 lb/in^2 (1,034 to over 1,378 MPa). The extra-high ductility is a result of a fine, equiaxed grain structure obtained by special thermal and/or mechanical deformation cycles.

CARBON TETRACHLORIDE. A heavy colorless liquid of the composition CCl_4, also known as **tetrachloromethane,** which is one of a group of chlorinated hydrocarbons. It is an important solvent for fats, asphalt, rubber, bitumens, and gums. It is more expensive than the aromatic solvents, but it is notable as a nonflammable solvent for many materials sold in solution, and is widely used as a degreasing and cleaning agent in the dry-cleaning and textile industries. Since the fumes are highly toxic it is no longer permitted in compounds for home use. It is used as a chemical in fire extinguishers such as **Pyrene,** but when it falls on hot metal it forms the poisonous gas phosgene. It is also used as a disinfectant, and because of its high dielectric strength has been employed in transformers. It was first produced in 1839 and used in Germany as a grease remover under the name of **Katharin.** Carbon tetrachloride is obtained by the chlorination of carbon bisulfide. The specific gravity is 1.595, boiling point 76°C, and freezing point 23°C. **Chlorobromomethane,** $Br \cdot CH_2 \cdot Cl$, is also used in fire extinguishers as it is less corrosive and more than twice as efficient as an extinguisher. It is a colorless, heavy liquid with a sweet odor, a specific gravity of 1.925, boiling point of 67°C, and freezing point of −65°C. It is also used as a high-gravity flotation agent.

CARNAUBA WAX. A hard, high-melting lustrous wax from the fanlike leaves of the palm tree *Copernica cerifera* of the arid region of northeastern Brazil, sometimes referred to as **Brazil wax,** or **ceara wax.** It is composed largely of **ceryl palmitate,** $C_{25}H_{51}COOC_{30}H_{61}$. The trees grow up to 60 ft (18 m) in height with leaves 3 ft (1 m) in length. The wax comes in hard, vitreous, yellowish cakes or lumps that melt at about 85°C, and have a specific gravity of 0.995. It is soluble in alcohol and in alkalies. **Olho wax** is the wax from young yellow leaves and is whitish gray. **Palha wax,** from older green leaves, is of a deeper grayish-yellow hue. In melting, water is added to the palha to make the **chalky wax.** No. 3 chalky contains up to 10% water. Olho wax without water yields the prime yellow wax. **Flora wax** is the highest quality and is clear yellow. Fully 70% of the production of carnauba goes into the manufacture of floor waxes and carbon paper. It has the property of being self-polishing in liquid floor waxes. In carbon paper it is nongreasy and nonsmearing. Other uses are in shoe polishes, in leather finishes, and for blending with other waxes in coating compounds. **Burnishing wax,** in the shoe industry, is carnauba wax blended with other waxes.

A wax quite similar to carnauba is **guaruma,** or **cauassu wax,** from the leaves of the *Calathea lutea,* a small plant with large leaves like those of the banana, growing in the lower Amazon Valley. Its melting point is 80°C. Another similar wax is from the trunk of the **wax palm,** *Ceroxylon andicola,* growing on the Andean slopes. A wax that is very similar to carnauba in properties and is more plentiful, but which contains the green leaf coloring difficult to bleach out, is **ouricury wax.** The name is also spelled **urucury** (uru, the Carib name for a shell; o means leaf). The wax is from the leaves of the palm tree *Syagrus coronata,* or *Cocos coronata,* of northeastern Brazil. Ouricury wax has a melting point of about 85°C, acid number 10.6, iodine value 16.9, and saponification value 78.8. It has the same uses as carnauba where color is not important, or it is used to blend with carnauba to increase the gloss. The nuts of the tree are called **licuri nuts,** and are used to produce **licuri oil** employed in soaps. The name **licuri wax** is sometimes erroneously given to ouricury.

Cotton wax, which occurs in cotton fiber to the extent of about 0.6%, is very similar to carnauba wax. It is a combination of C_{28} to C_{32} primary alcohols with C_{24} to C_{32} fatty acids. It has not been produced commercially. **Sugarcane wax** is a hard wax similar to carnauba occurring on the outside of the sugarcane stalk. A ton of cane contains 2 to 3 lb (1 to 1.4 kg) of wax, which concentrates in the filter press cake after clarification of the cane juice. The filter cake contains as high as 21% wax, which is solvent-extracted, demineralized with hydrochloric acid, and distilled to remove the low-molecular-weight constituents. It is used in floor and furniture polishes. The wax has a tan color, a melting point at about 176°F (80°C), and acid number 23 to 28. **Duplicane wax,** of the Warwick Wax Co., Inc., is a grade of sugarcane wax for carbon paper, and **Technicane wax** is a grade for polishes. Sugarcane wax is miscible with vegetable and petroleum waxes, and has greater dispersing action than carnauba wax. **Henequen wax,** extracted from the waste pulp of the henequen plant, has a melting point of 185°F (85°C), and is similar to carnauba. **Moss wax,** used for polishes, is extracted from Spanish moss which contains up to 4% wax. **Spanish moss** is the fiber from the plant *Tillandsia usneoides,* which grows throughout tropical and subtropical America, and along the southeastern coast of the United States, hanging from branches of trees. It is used for packing fragile articles, and for mattresses.

CARNOTITE. A mineral found in Utah and Colorado and employed as a source of uranium, radium, and vanadium. It is a vanadate of uranium and potassium, $V_2O_5 \cdot 2U_2O_3 \cdot K_2O \cdot 3H_2O$. It is found as a powder with other sands and gives them a pale-yellow color. The ore may contain 2 to 5% uranium oxide and up to 6 vanadium oxide, but it usually runs 2% V_2O_5. The vanadium is produced by roasting the ore, leaching, precipitating the oxide with acids, and sintering. The production of radium from the residue ore is a complex process, and 400 tons (362,800 kg) of ore produces only a gram of radium. **Patronite,** mined in Peru as a source of vanadium, is a greenish mineral, V_2S_9, mixed with pyrites and other materials. Carnotite ore may contain up to 2,500 parts per million of selenium and is a source of this metal.

CAROA. Pronounced car-o-áh. The fiber from the leaves of the plant *Neoglaziovia variegata* of northeastern Brazil. It is more than twice as strong as jute and is lighter in color and lighter in weight, but is too hard to be used alone for burlap. It is employed as a substitute for jute in burlap when mixed with softer fibers and also for rope, and in mixtures with cotton for heavy fabrics and suitings. Some suiting is made entirely of the finer caroa fibers. **Fibrasil** is a trade name in Brazil for fine white caroa fibers used for tropical clothing.

CARTRIDGE BRASS. One of the standard alloys of the brass mills, containing 70% copper and 30 zinc. Because of the general use of the alloy for making cartridges and for other deep drawing, the highest-purity zinc is used and all lead is excluded. It has high ductility and an attractive yellow color, and is used for deep-drawn or spun articles such as lamp bases, horns, and cornets. It brazes well and electroplates easily. Hard-rolled sheet has a tensile strength of about 80,000 lb/in^2 (551 MPa) with elongation 5%, while the annealed sheet has a strength of 49,000 lb/in^2 (337 MPa) with elongation 55%. Annealed strength is about 46,000 lb/ in^2 (317 MPa). The weight is 0.308 lb/in^3 (8.5 g/cm^3), and electric conductivity 27% that of copper. **Revere alloy No. 160,** of the Revere Copper & Brass, Inc., is this alloy. A slightly harder alloy, used for wire goods, is **eyelet brass,** containing 68% copper and 32 zinc. A grade produced by the American Brass Co. under the name of **spinning brass** contains 67% copper and 33 zinc. **Lubaloy,** of the Winchester-Western Div., contains some tin. A typical composition of the **cartridge brass strip** of the Scovill Mfg. Co. is 68.94% copper, 0.01 lead, 0.01 iron, and the balance zinc. **Primer brass** for cartridge primers may be the 70–30 alloy, or it may be cap copper.

Brass alloy 77, of the Bridgeport Brass Co., is a **mercurial brass** used for condenser tubes to resist algae growth. It is a 70–30 brass with 0.05% mercury. The mercury also inhibits dezincification. **Hi-strength brass,** of the Chase Brass & Copper Co., is a 70–30 brass processed to give a fine grain, 0.025 mm compared with 0.070 mm in regular cartridge brass. It has a tensile strength of 53,000 lb/in^2 (371 MPa), yield strength of 21,000 lb/in^2 (144 MPa), and elongation 21%. **Nebaloy,** of the New England Brass Co., is a low-cost cartridge brass for drawn and stamped products. It has 63% copper and 37 zinc, and is mill-processed to give a very fine grain.

The annealed metal has a tensile strength of 45,000 lb/in^2 (310 MPa) with elongation 40%, and the half-hard has a tensile strength of 61,000 lb/in^2 (420 MPa) with elongation 15% and Rockwell hardness of B71.

CASE-HARDENING MATERIALS. Any materials employed for adding carbon to the surface of low-carbon steels or to iron so that upon quenching a hardened case is obtained, the center of the steel remaining soft and ductile. The material may be plain charcoal, raw bone, or mixtures marketed as **carburizing compounds.** A common mixture is about 60% charcoal and 40 barium carbonate. The latter decomposes, giving carbon dioxide which is reduced to carbon monoxide in contact with the hot charcoal. If charcoal is used alone, the action is slow and spotty. Coal or coke can be used, but the action is slow, and the sulfur in these materials is detrimental. Salt is sometimes added to aid the carburizing action. By proper selection of the carburizing material the carbon content may be varied in the steel from 0.80 to 1.20%. The carburizing temperature for carbon steels is 1600°F (871°C), but for alloy steels this may vary. The articles to be carburized for case hardening are packed in metallic boxes for heating in a furnace, and the process is called **pack hardening,** as distinct from the older method of burying the red-hot metal in charcoal.

The principal liquid-carburizing material is sodium cyanide, which is melted in a pot that the articles are dipped in, or the cyanide is rubbed on the hot steel. **Cyanide hardening** gives an extremely hard but superficial case. Nitrogen as well as carbon is added to the steel by this process. Gases rich in carbon, such as methane, may also be used for carburizing, by passing the gas through the box in the furnace. When ammonia gas is used to impart nitrogen to the steel, the process is not called carburizing but is referred to as **nitriding. Tufftride,** of Kolene Corp., is a nitriding process using molten potassium cyanate with a small amount of sodium ferrocyanide in titanium-lined melting pots.

Case-hardening compounds are marketed under a wide variety of trade names. These may have a base of hardwood charcoal or of charred bone, with sodium carbonate, barium carbonate, or calcium carbonate. **Char,** of the Char Products Co., is a carburizing material in which the particles of coal-tar carbon are surrounded by an activator and covered with a carbon coating. **Accelerated Salt WS,** of Du Pont, for heat-treating baths, has a content of 66% sodium cyanide, with graphite to minimize fuming and radiation losses. For selective case hardening on steel parts a stiff paste of carburizing material may be applied to the surfaces where a carbon impregnation is desired. **Carburit,** of the Denfis Chemical Laboratories, is a carburizing paste of this kind. **Aerocarb** and **Aerocase,** of the American Cyanamid Co., are mixtures of sodium and potassium nitrates and nitrides for use in carburizing baths over a temperature range up to 1850°F (1010°C).

Chromized steel is steel surface-alloyed with chromium by diffusion from a chromium salt at high temperature. The reaction of the salt produces an alloyed surface containing about 40% chromium. **Plasmaplate** was a name given by the Linde Division to protective coatings of tungsten or molybdenum, deposited by a plasma torch which gives a concentrated heat to 30,000°F (16,650°C), but the refractory metals can now be deposited at lower temperatures by decomposition of chemical compounds. **Molybdenum pentachloride,** $MoCl_5$, is a crystalline powder which deposits an adherent coating of molybdenum metal when heated to about 900°C.

Metalliding is a **diffusion coating** process involving an electrolytic technique similar to electroplating, but done at higher temperatures [1500 to 2000°F (816 to 1093°C)]. Developed by General Electric, the process uses a molten fluoride salt bath to diffuse metals and metalloids into the surface of other metals and alloys. As many as 25 different metals have been used as diffusing metals, and more than 40 as substrates. For example, **boride coatings** are applied to steels, nickel-base alloys, and refractory metals. **Beryllide coatings** can also be applied to many different metals by this process. The coatings are pore-free and can be controlled to a tolerance of 0.001 in (0.025 mm).

CASEIN. A whitish to yellowish granular or lumpy protein precipitated from skim milk by the action of a dilute acid, or coagulated by rennet, or precipitated with whey from a previous batch. The precipitated material is then filtered and dried. Cow's milk contains about 3% casein. It is insoluble in water or in alcohol, but soluble in alkalies. Although the casein is usually removed from commercial milk, it is a valuable food accessory because it contains **methionine,** a complex mercaptobutyric acid which counteracts the tendency toward calcium hardening of the arteries. This acid is also found in the ovalbumin of egg white. Methionine, $CH_3 \cdot S \cdot CH_2CH_2CHNH \cdot COOH$, is one of the most useful of the amino acids, and is used in medicine to cure protein deficiency and in dermatology to cure acne and falling hair. It converts dietary protein to tissue, maintains nitrogen balance, and speeds wound healing. It is now made synthetically for use in poultry feeds. Some casein is produced as a by-product in the production of lactic acid from whole milk, the casein precipitating at a pH of 4.5. It is treated with sodium hydroxide to yield **sodium caseinate.**

Most of the production of casein is by acid precipitation, and this casein has a moisture content of not more than 10% with no more than 2.25% fat, and not over 4 ash. The casein made with rennet has up to 7.5% ash content, less than 1% fat, and is less soluble in alkalies. It is the type used for making plastics. **Rennet** used for curdling cheese is an extract of an enzyme derived from the stomachs of calves and lambs and is closely related to pepsin. Rennet substitutes produced from pepsin and other vegetable sources are only partial replacements and often have undesirable off-flavors. But **Sure-Curd,** of Chas. Pfizer and Co., is derived from a strain of *Endothia parasitica* and is similar to true rennet in coagulating and proteolytic properties. **Whey** is the thin sweet watery part separated out when milk is coagulated with rennet. Whey solids are used in prepared meats and other foods to enhance flavor, and in pastries to eliminate sogginess. **Tekniken,** of the Western Condensing Co., is a dry whey for use in margarine, chocolate, and cheese. **Orotic acid,** NH(CO·NH·CO·CH):C·COOH, produced synthetically, is identical with the biotic *Lactobacillus bulgaricus* of **yogurt,** the fermented milk whey used as food. It is a vitaminlike material.

Argentina and the United States are the most important producers of casein. France, Norway, and Holland are also large producers. Casein is employed for making plastics, adhesives, sizing for paper and textiles, washable interior paints, leather dressings, and as a diabetic food. **Casein glue** is a cold-work, water-resistant paste made from casein by dispersion with a mild base such as ammonia. With a lime base it is more resistant but has a tendency to stain. It is marketed wet or dry, the dry powder being simply mixed with water for application. It is used largely for low-cost plywoods and in water paints, but is not waterproof. Many gypsum wallboard cements are fortified with casein. Concentrated **milk protein,** available as **calcium caseinate** or **sodium caseinate,** is for adding proteins and for stabilizing prepared meats and bakery products. It contains eight amino acids and is high in lysine. **Sheftene,** of the Sheffield Chemical Co., is this material.

CASEIN PLASTICS. A group of thermoplastic molding materials made usually by the action of formaldehyde on rennet casein. The process was invented in 1885, and the first commercial casein plastic was called **Galalith,** meaning **milkstone.** Casein plastics are easily molded, machine easily, are nonflammable, will withstand temperatures up to 300°F (150°C), and are easily dyed to light shades. But they are soft, have high water absorption, 7 to 14%, and soften when exposed to alkalies. They are thus not suitable for many mechanical or electrical parts. They are used for ornamental parts, buttons, and for such articles as fountain-pen holders. The specific gravity of the material is 1.34, and the tensile strength is 8,000 lb/in^2 (55 MPa). They are usually marketed under trade names. Some of these are **Aladdinite,** of the Alladinite Co., Inc,; **Inda,** of the American Machine & Foundry Co.; **Erinoid,** of the Erinoid Co.; **Lactoid,** of the British Xylonite Co., Ltd.; **Lactonite,** of the British Lactonite Co.; **Ameroid,**

of the American Plastics Corp.; and **Karolith,** of the Karolith Corp. **Sicalite** is a French casein plastic. **Casein fiber** is made by treating casein with chemicals to extract the albumen and salts and forcing it through spinnerets, and again treating it to make it soft and silklike. The fiber is superior to wool in silkiness and resistance to moth attack, but is inferior in general properties. It is blended with wool in fabrics and in hat felts. **Lanitol** was an early Italian casein fiber. **Aralac fiber,** of Aralac, Inc., is a translucent, white, silky casein fiber. **Caslen,** of the Rubberset Co., is a resilient curled casein fiber used as a substitute for horsehair.

CASHEW SHELL OIL. An amber-colored, poisonous, viscous oil obtained by extraction from the by-product shells of the cashew-nut industry of India and Brazil. The cashew nut grows on the distal end of the fruit of the tree *Anacardium occidentale.* The thin-skinned, yellow, pear-shaped fruit may be eaten or used in preserves. The kernel of the seed nut, known as the **cashew nut,** is roasted and widely used as an edible nut or in confections. The kernel is crescent-shaped, and the nuts are graded by sizes from 200 per lb (0.45 kg) to 400–500 per lb (0.45 kg). On crushing, the nuts produce 45% of an edible oil, but the nuts are more valuable as a confection than for oil, and there is no commercial production of **cashew-nut oil.** One pound of shells yields 0.335 lb (0.152 kg) of cashew-nut shell oil, which contains 90% **anacardic acid,** a carboxypenta-dica-dienyl phenol, very blistering to the skin. It is used for the production of plastics, drying oils, and insulating compounds. The oil reacts with formaldehyde to give a drying oil. With furfural it produces a molding plastic. Reacted with other chemicals it forms rubberlike masses used as rubber extenders and in electrical insulating compounds. The other 10% of cashew-nut shell oil is **cardol,** a dihydroxypenta-dica-dienyl benzene. When decarboxylated the anacardic acid yields **cardanol,** a light oil liquid of the composition $C_6H_4 \cdot OH(CH_2)_6CH:CH(CH_2)_6CH_3$, with boiling point at 360°C and freezing point of about −20°C. Cardanol polymerizes with formaldehyde to form a heat-resistant, chemical-resistant, flexible resin of high dielectric strength valued for wire insulation. Small amounts of this resin also improve the chemical and electrical properties of the phenol resins. **Cardolite,** of the Irvington Varnish & Insulator Co., is a high-molecular-weight, straight-chain bisphenol derived from cashew-nut shell oil. It is used for making flexible epoxy resins, supplanting about half the normal amount of epichlorhydrin used in the resin.

CASHMERE. A fine, soft, silky fabric made from the underhair of the Cashmere goat raised on the slopes of the Himalayas in Asia. The hair is obtained by combing the animals, not by shearing, and only about 3 oz (0.09 kg) are obtained from a goat. The hair is straight and silky, but not lustrous, and is difficult to dye. The fabrics are noted for warmth, and the production now goes mostly into the making of shawls and fine ornamental garments. **Cotton cashmere** is a soft, loosely woven cotton fabric made to imitate cashmere, or it may be a cotton and wool mixture, but it lacks the fineness of true cashmere. **Cashmere hair,** used for fine paintbrushes, is from the beard of the Cashmere goat. It is similar to camel hair. **Qiviut,** the underwool of the musk ox of northern Canada, is a finer and longer fiber than cashmere, and about 6 lb (2.7 kg) may be obtained from each animal. It is shed in May or June. One pound (0.45 kg) of qiviut will make a 40-strand thread 26 miles (44 km) long. It dyes easily, and does not shrink even when boiled. It is used for fine gloves and sweaters.

CASSITERITE. Also called **tin stone.** It is the only commercial **tin ore,** and is a **tin dioxide,** SnO_2, containing theoretically 78.6% tin. It is a widely distributed mineral, but is found on a commercial scale in only a few localities, notably in Malaya, East Indies, Bolivia, Cornwall, England, Nevada, Isle of Youth, and Australia. The mineral occurs granular massive with a specific gravity of 6.8 to 7.1, a hardness of 6 to 7, and a brown to black color. It is present in the ore usually in amounts of 1 to 5%, and is found in veins, called **lode tin,** or in placer deposits. The concentrated ore averages 65 to 70% tin oxide. It is roasted to eliminate sulfur and arsenic, and then smelted in reverberatory furnaces.

CAST IRON. The generic name for a group of metals that are basically ternary alloys of carbon and silicon with iron. Included are **gray iron, ductile iron, white iron, malleable iron,** and **high-alloy iron.** The borderline between steel and cast iron is 2% carbon, which is the carbon content of saturated austenite. However, most cast irons have at least 3% total carbon, and normally the upper limit is 3.8 to 4%. Carbon is present in cast irons in two forms—as graphite, often referred to as free carbon, and as iron carbide (cementite).

The large amount of carbon and the presence of some of it as graphite are major distinguishing characteristics of cast irons' distinctive properties. Also, each of the five major cast-iron types differs from the others in the form in which carbon is present. The high carbon content makes molten iron very fluid, thus providing excellent castability. The precipitation of carbon as graphite during casting solidifcation counteracts the normal contraction of cooling metal, thus producing sound castings. The graphite also provides excellent machinability and damping qualities and adds lubrication to wearing surfaces. And, in some cast irons (white), where most of the carbon is present as iron carbide, it provides good wear resistance. Besides carbon, silicon, from 0.5 to 3.5%, is a major alloying element in cast irons. Its major function is to promote formation of graphite and to provide the desired as-cast microstructures.

The matrix structures of cast irons, where any graphite present is embedded, vary widely depending not only on casting practice and cooling rate but also on the shape and size of casting. Furthermore, it is possible to have more than one kind of matrix in the same casting. Also, the matrix structure can be controlled by heat treatment, but once graphite is formed, it is not changed by subsequent treatments. The matrix can be entirely ferritic. It differs from the ferrite found in wrought carbon steels because the relatively large amount of silicon produces a structure that makes the iron free machining. Addition of alloys can produce an acicular (needlelike) matrix. Hardening treatments yield a martensitic matrix. Other possible matrix structures are pearlite and ledeburite. Because the same composition in a cast iron can produce several different types of structure, cast irons are seldom specified by composition. Within each major type, standard grades are classified by minimum tensile strength.

Cast iron is usually made by melting pig iron and scrap in a cupola in contact with the fuel, which is normally coke. Pouring temperature, which varies with the analysis, is important, especially to prevent cold shut, which is a discontinuity in the structure caused by two streams of metal meeting, and failing to unite. With an electric furnace, scrap iron may be employed alone with carbon without pig iron, and the furnace may be operated continuously. The product is called **synthetic cast iron.**

Gun iron, formerly used for casting cannon, was a fine-grained iron of uniform texture, low in sulfur and in total carbon, made with charcoal in an air furnace.

Graphite is a weakening element in cast iron, and the high-graphite irons are desired only because of their ease of casting and machining. The lower the carbon, the stronger the cast iron. To obtain this result, steel scrap is used in the mix. Low-carbon steel of known chemical content, such as plate and rod ends and rail croppings, is used. The amount of steel varies from 15 to 60%, and the product resulting from the larger additions is called **semisteel.** Tensile strengths as high as 40,000 lb/in^2 (275 MPa) can be obtained without great reduction in the casting and machining qualities of the cast iron. Semisteel castings can be softened and made more ductile by annealing at a temperature of about 800°F (427°C), but they then lose 25 to 35% of the tensile strength. Trade names are used to designate cast iron made by special processes. **Pomoloy,** of the Pomona Pump Co., is an unalloyed cast iron with a tensile strength of 40,000 lb/in^2 (275 MPa), and hardness 215 Brinell. **DeLavaud metals,** of the U.S. Cast Iron Pipe & Foundry Co., is made by a centrifugal process in rotating steel molds. After annealing, the pipe has an outer layer of malleable iron, a center layer resembling steel, and an inner surface of gray iron. **Hi-Tem iron,** of the Bethlehem Foundry & Machine Co., is a corrosion-resistant cast iron used for processing vessels. **Hi-Tem S** is a high-manganese iron used for retorts.

High-test cast iron was originally cast iron that was superheated in the melting for pouring, poured in chilling molds, and then heat-treated, the only change in composition being to keep the silicon and manganese high. The term now means **high-strength irons** that are processed to give a careful balance of ferrite, pearlite, cementite, and carbon by the treatment, by additions of steel scrap, and by additions of nickel, chromium, and other elements that give strength to the metal by balancing the structure, but are not in sufficient quantities to classify the iron as an alloy cast iron. Tensile strengths above 50,000 lb/in^2 (344 MPa) are obtained, and all of the high-test irons are fine-grained, not spongy like gray iron. Steel scrap gives a stronger and finer structure; nickel gives ease of machining and aids in the chilling; chromium gives hardness and resistance to growth; molybdenum raises the combined carbon and adds strength and hardness. **Oxygenized iron** is high-test cast iron made by blowing air through a part of the metal and then returning the blown metal to the cupola. There is no sharp dividing line between some of these processed irons and steel, and when the combined carbon is high and the graphitic carbon is well distributed in even flakes the metal is called **graphitic steel.**

High-test cast irons are used for brake drums, cams, rolls, and high-strength parts. In many cases they are substitutes for malleable iron. They are marketed under many trade names. **Ermal** is a **pearlitic cast iron** of the Erie Malleable Iron Co. with a tensile strength up to 70,000 lb/in^2 (241 MPa). **Perlit,** of the Durson Corp., is another pearlitic cast iron. **Armite** is a synthetic cast iron of Robbins & Myers, Inc. **Jewell alloy,** of the Jewell Steel & Malleable Co., is the name of a group of high-strength and heat-resistant irons. **Ermalite,** of the Erie Malleable Iron Co., and **Wearloy,** of the Frank Foundries Corp., are high-strength, wear-resistant cast irons. **Gunite,** of the Gunite Corp., is a graphitic steel which, when quenched to a hardness of 477 Brinell, has a compressive strength of 200,000 lb/in^2 (1,378 MPa). **Arma steel,** of the Saginaw Malleable Iron Div., General Motors Corp., is a graphitic steel, or arrested malleabilized iron of high strength and shock resistance, used for connecting rods, gears, and camshafts where both high strength and bearing properties are required. **Meehanite metal,** produced under license of the Meehanite Research Institute of America, is made in a wide range of high-strength, wear-resisting, corrosion-resisting, and heat-resisting castings for dies, hydraulic cylinders, brake drums, pump parts, and gears. The normal strengths range from 35,000 to 55,000 lb/in^2 (241 to 379 MPa), compressive strengths from 135,000 to 175,000 lb/in^2 (930 to 1,206 MPa), and hardnesses from 193 to 223 Brinell. **Cylinder iron** is a general term for cast iron for engine and compressor cylinders, but also used for a variety of mechanical parts. The iron must be easily cast into a dense structure without hard spots or blowholes. Combined carbon must be sufficient to give wear resistance without brittleness, and the content of free graphite must be high enough to give a low coefficient of friction without great loss of strength.

CASTOR OIL. A light-yellow to brownish viscous oil obtained from the seed beans of the castor plant, *Ricinus communis.* In the tropics the plant grows to the proportions of a sturdy tree, but in temperate climates it is small with a poor yield. Besides its original use as a purgative in medicine, castor oil is one of the most widely used industrial vegetable oils. When pure and fresh, the oil is nearly colorless and transparent. The hot-pressed oil is brownish. It has a characteristic acrid, unpleasant taste. The specific gravity is 0.960 to 0.970, iodine value 82 to 90, saponification value 147, and solidifying point −10°C. The oil is chiefly composed of the glyceride of **ricinoleic acid,** which has a complex double-bonded molecular structure that can be polymerized easily. **Castor seeds** have the appearance of mottled colored beans and are enclosed in hard husks which are removed before crushing. The chief commercial production has been in Brazil, where two types are grown. The large Zanzibar type has seeds 16 mm long containing 30 to 35% oil, and the sanguineous type has seeds 10 mm long containing up to 60% oil. They are usually mixed in shipments, and the average yield is calculated as 0.45 lb (0.20 kg) of oil from 1 lb (0.45 kg) of beans. In the southwestern United States, dwarf disease-resistant hybrid varieties are grown that give high oil yields. Cold-pressed oil is used in medicine and lubricants, but the industrial oil is usually hot-pressed. Castor oil is used in paints, as a hydraulic oil, for treating leather and textiles, in soaps, and for making urethane resins. It increases the lathering power of soaps and their solubility in cold water. In lubricating oils and in cutting oils it has excellent keeping qualities and does not gum on exposure.

When castor oil is chemically dehydrated by removing the hydroxyl groups in the form of water by means of a catalyst, a double bond is formed giving an oil of heavy viscosity, light color, and with iodine value 116, acid value 3.5, and saponification value 191. **Dehydrated castor oil** gives a better gloss in varnishes than tung oil with a softer and less brittle film, but it has less alkali resistance than tung oil unless mixed with synthetic resin. **Sulfonated castor oil,** known as **Turkey red oil** in the textile industry, is made by treating crude hot-pressed castor oil with sulfuric acid and neutralizing with sodium sulfate. It is miscible with water and lathers like a solution of soap. It is used for the preparation of cotton fibers to be dyed, and gives clearer and brighter colors. It is also employed in soaps and in cutting compounds. Sulfonated dehydrated castor oil is used in nonalkaline water-washable skin ointments. It has a softening point of

30°C and an SO_2 content of 10%. **Synthenol,** of Spencer Kellogg & Sons, Inc., is a dehydrated castor oil for paints and varnishes. **Castung,** of the Baker Castor Oil Co., and **Isoline,** of the Woburn Degreasing Co., are dehydrated castor oils. **Copolymer 186** is a polymerized dehydrated castor oil which adds flexibility and improved general qualities to paints and outside enamels. **Mannitan drying oil,** of Atlas Chemical Industries, is an ester of dehydrated castor oil that dries faster than linseed oil and has better resistance in paints.

Hydrogenated castor oil is a hard, nongreasy, white solid melting at 82°C, used as an extender for waxes in coating compositions, and as a hard grease for making resistant lithium-type lubricating greases. Hydrogenated castor oil is odorless and tasteless, and is valued for coatings. **Castorwax,** of the Baker Castor Oil Co., **Emery S-751-R,** of the Emery Industries, Inc., and **Cenwax G,** of the W. C. Hardesty Co., Inc., are hydrogenated castor oil. In general, these materials are white, nongreasy, waxlike solids melting at about 85°C. **Primawax,** of the U.S. Cotton Oil Co., Inc., is a flaked form of hydrogenated castor oil used as a plasticizer in vinyl and cellulose plastics.

The hydrogenated ricinoleic acid, known as **hydroxystearic acid,** may also be separated and used for making waxy esters for pharmaceutical ointments, or for reacting with amines to make white, waxy solids useful as water repellents. By reacting castor oil with sodium hydroxide under heat and pressure, **sebacic acid,** $HO_2C(CH_2)_8CO_2H$, is produced. It is a powder melting at 129°C, and is a versatile raw material for alkyd resins, fibers, and heat-resistant plasticizers. It is also used for making nylon polymers and for **sebacate esters** for cold-weather lubricants, although the lower-cost azelaic and adipic acids may be substituted. Both sebacic acid and isosebacic acid are now produced synthetically from butadiene. **Isosebacic acid** is a mixture of sebacic acid with the isomers of this acid, **diethyl adipic acid** and **ethyl suberic acid.** It can replace sebacic acid for resin manufacture. Also similar in chemical properties to the ricinoleic acid of castor oil is **dimorphecolic acid,** obtained naturally from **daisy oil** from the seeds of the Cape marigold, of the genus *Dimorphotheca,* grown in California.

A substitute for castor oil in medicine is **croton oil,** a yellow-brown oil obtained from the dried ripe seeds of the small tree *Croton tiglium* of India and Sri Lanka. It has a burning taste and unpleasant odor, and is a more violent purgative than castor oil. The leaves and flowers of the tree are used like derris to kill fish. **Curcas oil** is a yellowish oil from the kernels of the seeds of the *Jatropha curcas* which grows in Central America. The kernels yield 50% oil with a specific gravity 0.920, iodine value 98 to 104, and saponification value 192. It is also a good soap oil but has an unpleasant odor. The ethyl and methyl esters of **crotonic acid** are used as monomers for flexible plastics for coatings. The acid with a composition $CH_3CH:CHCOOH$ is now made synthetically from acetylene and aldol.

CAST STEEL. Steel that has been cast into sand molds to form finished or semifinished machine parts or other articles. The general nature and characteristics of **steel castings** are, in most respects, closely comparable to wrought steels. Cast and wrought steels of equivalent composition respond similarly to heat treatment and have fairly similar properties. A major difference is that cast steel is more isotropic in structure. That is, its properties tend to be more uniform in all directions than wrought steel's properties, which generally vary, depending on the direction of hot or cold working.

Cast plain carbon steels can be divided into three groups similar to wrought steels: low-, medium-, and high-carbon steels. However, cast steel is usually specified by mechanical properties, primarily tensile strength, rather than composition. Standard classes are 60,000, 70,000, 85,000, and 100,000. Low-carbon grades, used mainly annealed or normalized, have tensile strengths ranging from 55,000 to 65,000 lb/in^2 (379 to 448 MPa). Medium-carbon grades, annealed and normalized, range from 70,000 to 100,000 lb/in^2 (482 to 689 MPa). When quenched and tempered, strength exceeds 100,000 lb/in^2 (689 MPa). Ductility and impact properties of cast steels are comparable, on average, to those of wrought carbon steel. However, the longitudinal properties of rolled and forged steels are higher than those of cast steel. Endurance-limit strength ranges between 40 and 50% of ultimate tensile strength.

Low-alloy steel castings are considered to be in the low-alloy category if their total alloy content is less than about 8%. Although many alloying elements are used, the most common are manganese, chromium, nickel, molybdenum, and vanadium. Small quantities of titanium and aluminum are also used for grain refinement. Carbon content is generally under 40%. The standard categories of low-alloy cast steels for specification purposes, in terms of tensile strength, are 65,000, 80,000, 105,000, 150,000, and 175,000 lb/in^2 (448, 551, 723, 1,034, and 1,206 MPa). For service at elevated temperatures, however, chemical compositions as well as minimum mechanical properties are often specified. **Boron cast steels** contain 0.006% max boron, which hardens and strengthens the steel without impairing the ductility and impact properties when the sulfur, phosphorus, and nitrogen contents are controlled. These steels replace some alloy steels. Alloy steels for heavy-duty castings may be chromium steel, with 0.80 to 1.10% chromium; vanadium, with 0.15 to 0.20 vanadium; chrome-vanadium; 1.50 to 3.50 nickel; nickel-chromium; or manganese, with either high or medium manganese content. One of the simplest alloy cast steels for parts subject to shock and fatigue stresses is the standard low-carbon and medium-carbon steel with 2% nickel. It is used for mining and other heavy-machinery parts, locomotive frames, and ship castings. The tensile strength is up to 85,000 lb/in^2 (586 MPa), with yield point up to 55,000 lb/in^2 (379 MPa), and elongation 25 to 32%. The nickel, with manganese up to 0.90%, gives the steel greater shock resistance at low temperatures, as ordinary steel is brittle in cold climates or when used on refrigerating equipment. A 3 to 3.50% nickel steel used for cast gears for rolling mills has a tensile strength of 110,000 lb/in^2 (758 MPa), elongation 20%, and hardness 200 Brinell as cast. A nickel-chromium-molybdenum cast steel for heavy gears contains 1.5% nickel, 0.75 chromium, 0.35 molybdenum, 0.70 manganese, and 0.35 carbon. The tensile strength is 145,000 to 160,000 lb/in^2 (999 to 1,103 MPa).

CATALYST. A material used to cause or accelerate chemical action without itself entering into the chemical combination. Catalysts are chosen for selectivity as well as activity, mechanical strength, and life. They should give a high yield of product per unit and be capable of regeneration whenever possible for economy. In the cracking of petroleum, activated carbon breaks the complex hydrocarbons into the entire range of fragments; activated alumina is more selective, producing a large yield of C_3 and C_4; and silica-alumina-zirconia is intermediate. **Contact catalysts** are the ones chiefly used in the chemical industry, and they may be in various forms. For bed reactors the materials are pelleted. **Powdered catalysts** are used for liquid reactions such as the hydrogenation of oils. **Chemical catalysts** are usually liquid compounds, especially such acids as sulfuric or hydrofluoric.

Various metals, especially platinum and nickel, are used to catalyze or promote chemical action in the manufacture of synthetics. Nitrogen in the presence of oxygen can be "fixed" or combined in chemicals at ordinary temperatures by the use of ruthenium as a catalyst. Acids may be used to aid in the polymerization of synthetic resins. Mineral soaps are used to speed up the oxidation of vegetable oils. Cobalt oxide is used for the oxidation of ammonia. Cobalt and thorium are used for synthesizing gasoline from coal. All of these are classed as **inorganic catalysts.** Sometimes more complex chemicals are employed, silicate of soda being used as a catalyst for high-octane gasoline. In the use of **potassium persulfate,** $K_2S_2O_8$, as a catalyst in the manufacture of some synthetic rubbers, the material releases 5.8% active oxygen, and it is the nascent oxygen that is the catalyst. **Sodium methylate,** also called **sodium methoxide,** $CH_3 \cdot O \cdot Na$, used as a catalyst for ester-exchange reactions in the rearrangement of edible oils, is a white powder soluble in fats but violently decomposed in water.

Aluminum chloride, $AlCl_3$, in gray granular crystals which sublime at

950°C, is used as a catalyst for high-octane gasoline and synthetic rubber, and in the synthesis of dyes and pharmaceuticals. **Antimony trichloride,** $SbCl_3$, is a yellowish solid, melting at 73.4°C, used as a catalyst in petroleum processing to convert normal butane to isobutane. This chemical is also used for antimony plating and as a cotton mordant. **Aluminosilicates** are used in fluid catalyst cracking of gasoline. They bear a negative charge even at high temperatures. **Bead catalysts** of activated alumina have the alumina contained in 3-mm beads of silica gel. **Catasil** is alumina adsorbed on silica gel, used for polymerization reactions.

Catalyst carriers are porous inert materials used to support the catalyst, usually in a bed through which the liquid or gas may flow. Materials used are generally alumina, silicon carbide, or mullite, and they are usually in the form of graded porous granules or in irregular polysurface pellets. High surface area, low bulk density, and good adherence of the catalyst are important qualities. Pellets are bonded with a ceramic that fuses around the granules with minute necks that hold the mass together as complex silicates and aluminates with no trace elements exposed to the action of the catalyst or chemicals. Catalyst carriers are usually bonded to make them about 40% porous. The pellets may be 50 mesh or finer, or they may be in sizes as large as 1 in (2.5 cm). Refractory filters known as **porous media,** used for filtering chemicals and gases at high temperatures, are essentially the same materials as catalyst carriers with ceramic bonds fired at about 1250°C, but they are usually in the form of plates or tubes, and the porosity is usually about 35%. They may be used directly as filters, or as underdrain plates for filter powders.

Sunlight or ultraviolet rays are also used as catalysts in some reactions. For example, chlorine and hydrogen combine very slowly in the dark, but combine with great violence when a ray of sunlight is turned on. **Biologic catalysts** are the **enzymes,** which are **organic catalysts** that are a form of life. They are sensitive to heat and light and are destroyed at 100°C. Enzymes are soluble in water, glycerin, or dilute saline solutions, and water must always be present for enzyme action. Their action may be stimulated or checked by other substances. When dehydrated vegetables lose their flavor by destruction of the enzymes, the flavor may be restored by adding small percentages of enzymes from the same or similar vegetables.

Enzymes have various actions. **Diastase,** found in the seeds of barley and other grains, converts starch to maltose and dextrin. **Diastase 73,** of Rohm & Haas Co., is an enzyme chemical for converting gelatinized starches to dextrose. It is **amyloglucosidase** modified to remove the bitter taste. One pound (0.45 kg) will convert 100 lb (45 kg) of starch. **Cytase,** found in seeds and fruits, decomposes cellulose to galactose and mannose. **Zymose,** found in yeast, hydrolyzes glucose to alcohol. **Thiaminase,** an enzyme which occurs in small amounts in salmon, cod, rockfish, and some other fish, destroys the vitamin thiamine, and if taken in high concentration in the human diet causes ventritional polyneuritis. **Rhozyme LA,** of the Rohm & Haas Co., is a diastatic enzyme concentrate in liquid form for desizing textiles. **Bromelin,** an enzyme used in breweries, is produced from pineapples by alcohol precipitation from the juice. **Fermcozyme** is a liquid glucose-oxidase-catalase used in carbonated beverages to remove dissolved oxygen which would combine with glucose to form gluconic acid resulting in loss of color and flavor. It is also used in egg powders to remove undesirable glucose.

Fermenting agents comprise a wide range of yeasts, bacteria, and enzymes which break down molecules to form other products. **Yeasts** are important in foodstuffs manufacture. A yeast is a fungus, and the life organisms produce carbon dioxide gas to raise doughs. These are called **leavening yeasts. Fermenting yeasts** produce alcohols by action on sugars. Many of the yeasts are high in proteins, vitamins, and minerals, and as dry, inactive powders are used to raise the nutritional values of foodstuffs. **Torula yeast,** *Torulopsis utilis,* used as an additive in processed foods, is a by-product of the sulfite paper mills, growing on the 5- and 6-carbon wood sugars. It contains more than 50% proteins, and has 10 different vitamins and 15 minerals. The dry powder is inactive, and does not cause raising in baked foods. **Prostay,** of St. Regis Paper Co., is this material.

CATECHU. An extract obtained from the heartwood and from the seed pods of the tree *Acacia catechu* of southern Asia. It is used in tanning leather, and as a dyestuff, giving brown, drab, and khaki colors. It is used in medicine as an astringent for diarrhea and hemorrhage. The name is sometimes applied to gambier, which also contains catechu tannin, $C_{15}H_9(OH)_5$. Catechu, or **cutch,** comes either as a liquid which is a water solution, or as brownish, brittle, glossy cakes. The liquid contains 25% tannin, and the solid 50%. A ton of heartwood yields, by hot-water extraction, 250 to 300 lb (113 to 136 kg) of solid cutch extract. It is a powerful astringent. When used alone as a tanning agent, the leather is not of high quality, being of a dark color, spongy, and water-absorbent. It is normally employed in mixtures. **Burma cutch** is from the *A. catechuoides.* **Indian cutch** is from the *A. sundra.* The latter is frequently adulterated with starch, sand, and other materials. **Wattle** is an extract from Australian and East African acacia, *A. dealbata,* and other species. The wattle tree is called **mimosa** in Kenya. **Wattle bark** contains 40 to 50% tannin. It gives a firm pinkish leather and is employed for sole leathers. The solid extract contains 65% tannin. **Golden wattle,** used for tanning in New Zealand, is the tree *A. pycantha.* Much wattle extract is produced in Brazil from the **black wattle. Turwar bark,** or **avarem,** used in India for tanning cattle hides, is from the tree *Cassia auricula,* and is similar to wattle.

CATGUT. String made from the intestines of sheep, used for violin strings, and for tough, durable cords for rackets and other articles. After cleansing and soaking in an alkali solution, the intestines are split, drawn through holes in a plate, cured in sulfur or other material, and graded according to size. Sheep intestines are also used for making surgical sutures, but for this purpose they are not called catgut, but simply **gut.** The sutures are encased in tubes and bombarded by electron-beam radiation for sterilization. In the meat-packing industry the intestines of sheep and goats are referred to as **casings** and are employed as the covering of sausage and other meat products. They are graded by diameter, freedom from holes, strength, color, and odor. Intestines of hogs and beef cattle are also used as casings, but they are not as edible as those from sheep.

CEDAR. A general name that includes a great variety of woods. The true cedars comprise trees of the natural order of *Coniferae,* genus *Cedrus,* of which there are three species: **Lebanon cedar,** *Cedrus libani;* **Himalayan cedar,** *C. deodora;* and **Atlas cedar,** *C. atlantica.* The differences are slight, and all of the species are sometimes classed as *C. libani.* The Himalayan cedar is also known as **deodar.** All are mountain trees, and are native to southern Europe, Asia, and northern Africa. The true cedar is yellow in color, fragrant in odor, takes a beautiful polish, and is very durable. It is used in construction work, and timbers in temples in India more than 400 years old are still in perfect preservation. The wood weighs about 36 lb/ft^3 (576 kg/m^3). Numerous species of *Cedrela* occur in tropical America, Asia, and Africa, and are also called cedar, but the wood has greater resemblance to mahogany. In the United States and Canada the name cedar is applied to woods of species of *Thuya, Juniperus,* and *Cupressus,* more properly classified as thuya, juniper, and cypress.

Spanish cedar, or **Central American cedar,** used in the United States as a substitute for mahogany in patternmaking, and for cigar boxes, furniture, carving, cabinetwork, and interior trim, is a softwood from numerous species of *Cedrela,* called in Spanish America by the name of **Cedro.** It has a light-red color sometimes beautifully figured with wavy grain, has an agreeable odor, is easily worked, seasons well, and takes a fine polish. The weight is 28 to 33 lb/ft^3 (449 to 529 kg/m^3). The trees grow to a large size, logs being available 40 in (1.02 m) square. The imports come chiefly from Central America and the West Indies, but the trees grow as far south as northern Argentina. **Paraguayan cedar** is the wood of the tree *C. braziliensis,* of Paraguay, Brazil, and northern Argentina, employed locally for cabinetwork, car building, and interior building work. It is similar in appearance to Spanish cedar but is denser, harder, and redder in color. The wood known as **southern white cedar,** and called juniper in the Carolinas, is from the tree *Chamaecyparis thyoides,* growing in the coastal belt from Maine to Florida. The heartwood is light brown tinged with pink, and the thin sapwood is lighter in color. The wood is light in weight, straight-grained, durable, and fragrant. The more plentiful **white cedar** of the West Coast, known also as **Port Orford cedar, Oregon cedar, ginger pine,** and in England as **Lawson cypress,** is from the tree *C. lawsoniana* of California and Oregon, mostly from a narrow coastal strip in Oregon to an altitude of about 5,000 ft (1524 m). Mature trees reach a height of 160 ft (49 m) and a diameter of 6 ft (1.8 m). The wood is white with a yellow tinge and a trace of red. It is rather hard and tough, with a fine straight grain, and is very durable. It has an agreeable aromatic odor, and is free from pitch. The wood is used for doors, sash, boats, matches, patterns, and where a light, strong, straight-grained wood is required. **Toon,** the wood of the tree *Cedrela toona* of India, southeast Asia, and Australia, resembles Spanish cedar but is somewhat harder, heavier, 35 lb/ft^3 (560 kg/m^3), and deeper red in color. It has a beautiful grain, and takes a high polish. It is durable, does not warp, and is used for furniture and cabinetwork.

CELLULOSE. The main constituent of the structure of plants, which, when extracted, is employed for making paper, plastics, and in many combinations. Cellulose is made up of long-chain molecules in which the complex unit $C_6H_{10}O_5$ is repeated as many as 2,000 times. It consists of glucose molecules with three **hydroxyl groups** for each glucose unit. These OH groups are very reactive, and an almost infinite variety of compounds may be made by grafting on other groups, either repetitively or intermittently, such as reaction with acetic or nitric acids to form acetates or nitrates, reaction with ethylene oxide to form hydroxyethyl cellulose, reaction with acrylonitrile to form cyanoethylated cellulose, or reaction with vinyls. Cellulose is the most abundant of the nonprotein natural organic products. It is highly resistant to attack by the common microorganisms, but the enzyme **cellulase** digests it easily, and this organism is used for making paper pulp, for clarifying beer and citrous juices, and for the production of citric acid and other chemicals from cellulose. **Takamine 4000,** of the Miles Chemical Co., is this enzyme. Cellulose is a white powder insoluble in water, sodium hydroxide, or alcohol, but it is dissolved by sulfuric acid. The highly refined insoluble cellulose with all the sugars, pectin, and other soluble matter removed is called **alpha cellulose,** or **chemical cellulose,** used for the production of chemicals. It was formerly made only from cotton linters, but is now largely made from wood pulp. **Avicel,** of the American Viscose Corp., is such a cellulose marketed as a white crystalline powder for use in foodstuffs to give body and gel stability to such products as peanut butter, cheese spreads, and prepared puddings. It forms a firm gel in water and absorbs oils easily. It is odorless and tasteless, and has no calorie content.

One of the simplest forms of cellulose used industrially is **regenerated cellulose,** in which the chemical composition of the finished product is similar to that of the original cellulose. It is made from wood or cotton pulp digested in a caustic solution. The viscous liquid is forced through a slit into an acid bath to form a thin sheet, which is then hardened and bleached. **Cellophane,** of Du Pont, is a regenerated cellulose in thin sheets for wrapping. It is transparent, dyed in colors, or embossed. It is up to 0.0016 in (0.041 mm) thick with tensile strengths from 8,000 to 19,000 lb/in^2 (55 to 130 MPa). It chars at about 375°F (190°C). The thinnest sheets, 0.0009 in (0.023 mm) in thickness, have 21,500 in^2/lb (30.8 m^2/kg). The three-digit gage system used for cellophane indicates the total film yield. Thus, 180 gage has a film yield of 18,000 in^2/lb (25.8 m^2/kg). The waterproofed material is coated with a thin film of cellulose lacquer, or the cellophane may be laminated with a film of a synthetic resin. **Cellothene,** of the Chester Packaging Products Co., is a heat-sealable film of laminated cellophane and polyethylene, the polyethylene thickness being at least 0.0005 in (0.0127 mm). Cellophane has greater transparency than polyethylene, but is not as strong nor as chemically resistant. For food packaging, the printing is done on the reverse side of the cellophane before laminating.

Purocell, of the Eastern Corp., was a highly purified and bleached cellulose produced from wood pulp and used for making high-grade writing papers. **Barcote,** of the Foote Mineral Co., is a nearly pure cellulose used in plastics or for carbonizing. It is a buff-colored, odorless powder or granular material with residual ash content of 1.6%. Some cellulose is obtained from potatoes as a by-product in the production of starch. It is pure white and is used in plastics. **Solka-Floc,** of the Brown Co., is 99.5% pure wood cellulose in the form of tough, white fibers 1 to 2 μm in diameter and 35 to 165 μm long, bulking 9 to 34 lb/ft^3 (144 to 544 kg/m^3). It is used as a filler for plastics requiring a fine surface finish and dimensional stability, such as buttons, knobs, trays, and vinyl floor tile. It is also used in welding rod coatings, in adhesives, and for cellulose chemicals. Water-soluble cellulose, or **cellulose gum,** used as a substitute for gum arabic and carob-bean flour as a stabilizer, thickener, or emulsifier, is **sodium cellulose glycollate,** or **sodium carboxymethyl cellulose,** in powder form. It is also used to increase the effectiveness of detergents. Water-soluble film is also made from this material. **Carbose,** of the Wyandotte Chemicals Corp. and **Cellocel S,** of the Dow Chemical Co., are sodium carboxymethyl cellulose. **CMC gum,** of the Dow Chemical Co., is **carboxymethyl cellulose,** used as a temporary binder for ceramic glazes. It burns out in the firing. A purified grade of this gum is used as a stabilizer in pharmaceuticals and low-acid foodstuffs. **Cellocel A** is **aluminum cellulose glycollate,** a water-soluble brownish powder used for waterproofing paper. **Natrosol,** of Hercules,

Inc., is **hydroxyethyl cellulose,** a white powder used for textile finishes and as a thickener for water-base paints. **Ethylose,** of Rayonier, Inc., is a hydroxyethyl cellulose with a low degree of substitution of ethylene oxide in the molecular chain. It is insoluble in water, but is alkali-soluble. It is used in paper coating to add gloss and water resistance. **Cellosize QP-4400,** of the Union Carbide Corporation, is a hydroxyethyl cellulose powder easily soluble in water but nongelling. It is used as a thickener in latex paints, inks, cosmetics, and pharmaceuticals. **Ceglin,** of the Sylvania Corp., is an alkali-soluble **cellulose ether** marketed as a white fibrous powder. When dissolved in a water solution of caustic soda, it forms a viscous liquid used for sizing textiles. **Sodium cellulose sulfate** is a water-soluble granular powder used as a thickener in emulsion paints, foods, and cosmetics, and for sizing paper and textiles. It produces a clear, tough, greaseproof coating. It is the sodium salt of cellulose acid sulfate produced by sulfuric acid treatment of wood pulp, with the sulfate groups in ester-type linkages on the cellulose chain. **SCS gum,** of the Tennessee Eastman Corp., used to replace gum arabic, is this material.

Ethyl cellulose is a colorless, odorless ester of cellulose resulting from the reaction of ethyl chloride and cellulose. The specific gravity is 1.07 to 1.18. It is nonflammable, very flexible, stable to light, and forms durable alkali-resistant coatings. It is used as a thin wrapping material, for protective coatings, as a hardening agent in resins and waxes, and for molding plastics. **Ethyl cellulose plastics** are thermoplastic and are noted for their ease of molding, light weight, and good dielectric strength, 400 to 520 volts per mil (15 to 20.5×10^6 volts per meter), and retention of flexibility over a wide range of temperature from −70 to 150°F (−57 to 66°C), the softening point. They are the toughest, the lightest, and have the lowest water absorption of the cellulosic plastics. But they are softer and lower in strength than cellulose-acetate plastics. **Lumarith EC,** of the Celanese Corp., is ethyl cellulose in the form of sheet, films, and molding powder. **Celcon** is a name applied by this company to acetal resin copolymers. **Hercocel E,** of Hercules, Inc., is a compounded ethyl cellulose molding powder in several formulations to give tensile strengths from 3,750 to 7,400 lb/in^2 (26 to 51 MPa), with elongation from 6 to 16%.

Ethocel is ethyl cellulose of the Dow Chemical Co., and **Stripcoat** is a solution of ethyl cellulose used for dipping automotive and aircraft replacement parts or other metal products to form a thin, waterproof protective coating to prevent corrosion. The coating strips off easily when the part is to be used. The same material is marketed by a number of other companies for the same purpose under a variety of trade names. **Methyl cellulose** is a white, granular, flaky material, which is a strong emulsifying agent, and is used in soaps, floor waxes, shoe cleaners, in emulsions of starches, glues, waxes, and fats, and as a substitute for gum arabic. It gives colorless, odorless solutions resistant to fermentation. It dissolves in cold water, but is stable to alkalies and dilute acids. In soaps it lowers the surface tension of the water and aids lathering. It is also used for tree-wound dressings, and as a moisture-conserving soil conditioner. **Colloresin,** of the General Drug Co., is methyl cellulose, and **Methocel HB,** of the Dow Chemical Co., is a hydroxybutyl methyl cellulose for use in paint removers. **Cyanoethylated cellulose** is a white fibrous solid used to produce thin transparent sheets for insulating capacitors and as carriers for luminescent phosphors. It has a high dielectric constant and low dissipation factor. A 0.002-in (0.051-mm) film has a tensile strength of 5,300 lb/in^2 (37 MPa) and is flexible. **Cyanocel,** of the American Cyanamid Co., is this material.

CELLULOSE ACETATE. An amber-colored, transparent material made by the reaction of cellulose with acetic acid or acetic anhydride in the presence of sulfuric acid. In Germany it was made by treating beechwood pulp with acetic acid in the presence of an excess of zinc chloride. It is employed for lacquers and coatings, molding plastics, rayon, and photographic film. Cellulose acetate may be the triacetate $C_6H_7O_2(OOCCH_3)_3$, but may be the tetracetate or the pentacetate, or mixtures. It is made in different degrees of acetylation with varying properties. Unlike nitrocellulose, it is not flammable, and it has better light and heat stability. It has a refractive index of 1.47 to 1.50, and a sheet ⅛ in (0.32 cm) thick will transmit 90% of the light. The specific gravity is 1.27 to 1.37, hardness 8 to 15 Brinell, tensile strength 3,500 to 8,000 lb/in^2 (24 to 55 MPa), compressive strength up to 20,000 lb/in^2 (137 MPa), elongation 15 to 80%, dielectric strength 300 to 600 volts per mil (12 to 24×10^6 volts per meter), and softening point 122 to 205°F (50 to 96°C). It is thermoplastic, and is easily molded. The molded parts or sheets are tough, easily machined, and resistant to oils and many chemicals. In coatings and lacquers the material is adhesive, tough, and resilient, and does not discolor easily. **Cellulose acetate fiber** for rayons can be made in fine filaments that are strong and flexible, nonflammable, mildewproof, and easily dyed. Standard cellulose acetate for molding is marketed in flake form. **Cellulose triacetate,** with 60 to 61.5% combined acetic acid, is more insoluble, has higher dielectric strength, and is more resistant to heat and light than other types. It is cast into sheets, and is also used for resistant coatings and for textile fibers. **Cellulose acetate film,** used for wrapping, is somewhat lighter in weight than regenerated cellulose, giving 14,500 in^2/lb (20 m^2/kg) for the 0.0015-in (0.0381-mm) film. **Kodapak,** of the Eastman Co., is cellulose acetate film for packaging.

Lumarith is a cellulose acetate of the Celanese Corp. in the form of rods, sheets, tubes, and molding powder. **Lumarith X** is a high acetyl cellulose acetate for molding. The tensile strength is up to 8,000 lb/in^2 (55 MPa), and Brinell hardness up to 12.5. **Vuelite,** of the Monsanto Chemical

Co., is transparent cellulose acetate for fluorescent light fixtures. **Cellomold** is a cellulose-acetate molding powder of F. A. Hughes & Co., Ltd., and **Celastoid** is an extrusion acetate of British Celanese, Ltd. **Tenite** is a cellulose-acetate molding material of the Tennessee Eastman Corp. **Estron** is a name adopted by this company to designate cellulose ester yarns and staple fiber. **Protectoid** is Lumarith in the form of nonflammable motion-picture film.

Cellulose acetate lacquers are sold under many trade names. They are the acetate in solvents with plasticizers and pigments. **Vimlite,** of the Celanese Corp., is a Saran screen filled with cellulose acetate. It transmits ultraviolet light, and is used for glazing. **Miramesh,** of the National Research Corp., is this material with one side coated with a film of aluminum. It is used for light diffusers and radiant-heat reflectors. **Masuron,** of John W. Masury & Son, **Nixonite,** of the Nixon Nitration Works, and **Plastacele,** of Du Pont, are cellulose-acetate materials. **Acele,** of the latter company, is a name for acetate yarns. **Celanese** is the name of cellulose acetate yarns and fabrics of the Celanese Corp. of America. **Celairese,** of the same company, is a fluffy acetate fiber used for interlinings. **Lanese** is a fine fluffy acetate fiber used to blend with wool. **Fortisan,** of the same company, is a specially processed strong acetate fiber of extreme fineness [0.0001 in (0.0025 mm) in diameter], originally developed for parachutes but now also used for fine fabrics. **Forticel,** of the same company, is a cellulose propionate plastic for injection molding. It has a flow point at 161°C, has high impact resistance, and requires less plasticizer than cellulose acetate. **Arnel,** of the Celanese Corp., is a cellulose tricetate fiber resistant to shrinkage and wrinkling in fabrics. **Arnel 60** is a cellulose acetate fiber with a circular cross section instead of the normal crenelated cross section, giving higher strength and better spinning qualities. **Hercocel A,** of Hercules, Inc., is cellulose acetate molding powder that will produce moldings with tensile strengths from 4,000 to 7,000 lb/in^2 (27 to 186 MPa), and elongations from 14 to 22%. The flow temperature is from 285 to 355°F (140 to 179°C), depending on the formulation. **Avcocel,** of the American Viscose Corp., used as a filler in plastics to increase the impact strength, is a by-product of cellulose acetate production. It contains 50% cellulose acetate and 50% white cotton.

Cellulose acetate butyrate is made by the esterification of cellulose with acetic acid and butyric acid in the presence of a catalyst. It is particularly valued for coatings, insulating types, varnishes, and lacquers. Commonly called **butyrate** or **CAB,** it is somewhat tougher and has lower moisture absorption and a higher softening point than acetate. Special formulations with good weathering characteristics plus transparency are used for outdoor applications such as signs, light globes, and lawn sprinklers. Clear sheets of butyrate are available for vacuum-forming applications. Other typical uses include transparent dial covers, television screen shields, tool handles, and typewriter keys. Extruded pipe is used for electric conduits, pneumatic tubing, and low-pressure waste lines.

Cellulose acetate propionate is similar to butyrate in both cost and properties. Some grades have slightly higher strength and modulus of elasticity. Propionate has better molding characteristics but lower weatherability than butyrate. Molded parts include steering wheels, fuel filter bowls, and appliance housings. Transparent sheeting is used for blister packaging and food containers. **Tenite III** is cellulose acetate propionate for extrusion rod and moldings of high impact strength. **Hercose C,** of Hercules, Inc., is cellulose acetate butyrate used for cable coverings and coatings. It is more soluble than cellulose acetate and more miscible with gums. It forms durable and flexible films. **Ester EAB-171,** of Eastman, is a liquid cellulose acetate butyrate for glossy lacquers, chemical-resistant fabric coatings, and wire-screen windows. It contains 17% butyl with one hydroxyl group per four anhydroglucose units. It transmits ultraviolet light without yellowing or hazing and is weather-resistant.

CELLULOSE NITRATE. Materials made by treating cellulose with a mixture of nitric and sulfuric acids, washing free of acid, bleaching, stabilizing, and dehydrating. For sheets, rods, and tubes it is mixed with plasticizers and pigments and rolled or drawn to the shape desired. The cellulose molecule will unite with from one to six molecules of nitric acid. The lower nitrates are very inflammable, but they do not explode like the high nitrates, and they are the ones used for plastics, rayons, and lacquers, although their use for clothing fabrics is restricted by law. The names **cellulose nitrate** and **pyroxylin** are used for the compounds of lower nitration, and the term nitrocellulose is used for the explosives. **Collodion** is a name given to the original solution of cellulose nitrate in a mixture of 60% ether and 40 alcohol for making fibers and film, and the name is still retained in pharmacy. The name **soluble cotton** is used to designate batches of cellulose nitrate wet with alcohol for storing for the production of lacquers, but the soluble cotton gauze, used for surgical dressings, is cotton oxidized with nitrogen dioxide.

Cellulose nitrate was first used as a plastic in England in 1855 under the name **Parkesine.** It consisted of nitrocellulose mixed with camphor and castor oil for hardening and making it nonexplosive. Later, in 1868, an improved cellulose nitrate and camphor plastic was called **Celluloid,** now the trade name of the Celanese Corp. of America for cellulose nitrate plastics. **Xylonite** was the name used in England for the nitrocellulose hardened with camphor made by Daniel Spill in 1868. The name is still used by the British Xylonite Co. for cellulose nitrate plastics. Cellulose nitrate is the toughest of the thermoplastics. It has a specific gravity of 1.35 to

1.45, tensile strength of 6,000 to 7,500 (41 to 52 MPa), elongation 30 to 50%, compressive strength 20,000 to 30,000 (137 to 206 MPa), Brinell hardness 8 to 11, and dielectric strength 250 to 550 volts per mil (9.9 to 21.7×10^6 volts per meter). The softening point is 160°F (71°C), and it is easy to mold and easy to machine. It also is readily dyed to any color. It is not light-stable, and is therefore no longer used for laminated glass. It is resistant to many chemicals, but has the disadvantage that it is inflammable. The molding is limited to pressing from flat shapes. It burns with a smoky flame, and the fumes are poisonous. Methyl or amyl alcohols are the usual solvents for the material, and various plasticizers are used, some of which aid in reducing the flammability. Camphor is the usual hardener and plasticizer, from 24 to 30% being the usual amount.

CEMENT. A material, generally in powder form, that can be made into a paste usually by the addition of water and, when molded or poured, will set into a solid mass. Numerous organic compounds used for adhering, or fastening materials, are called cements, but these are classified as adhesives, and the term cement alone means a construction material. The most widely used of the construction cements is **portland cement.** It is a bluish-gray powder obtained by finely grinding the clinker made by strongly heating an intimate mixture of calcareous and argillaceous minerals. The chief raw material is a mixture of high-calcium limestone, known as **cement rock,** and clay or shale. Blast-furnace slag may also be used in some cements. American specifications call for five types of portland cement. Type I, for general concrete construction, has a typical analysis of 63.2% CaO, 21.3 SiO_2, 6 Al_2O_3, 2.7 Fe_2O_3, 2.9 MgO, 1.8 SO_3. Type III, for use where high early strength is required, has 64.3% CaO, 20.4 SiO_2, 5.9 Al_2O_3, 3.1 Fe_2O_3, 2 MgO, 2.3 SO_3. The color of the cement is due chiefly to iron oxide. In the absence of impurities the color would be white, but neither the color nor the specific gravity is a test of quality. The specific gravity is at least 3.10. Good cement is always ground fine, with 98.5% passing a 200-mesh screen.

White cement is from pure calcite limestone, such as that found in eastern Pennsylvania. It is ground finer and used for a better class of work, but the physical properties are similar to those of ordinary cement. A typical analysis of white cement is: 65% CaO, 25.5 SiO_2, 5.9 Al_2O_3, 0.6 Fe_2O_3, 1.1 MgO, 0.1 SO_3. The white cements of France and England are made from the chalky limestones, and have superior working qualities, as they are usually ground finer. White cement is also made from inferior iron-bearing limestone by treatment with fluorspar.

Aluminous cement, or **aluminate cement,** sometimes referred to as **high-speed cement,** will set to high strength in 24 h, and is thus valued for laying roads or bank walls. It is made with bauxite, and contains a high percentage of alumina. A typical analysis is: 39.8% Al_2O_3, 33.5 CaO, 14.6 Fe_2O_3, 5.3 SiO_2, 1.3 MgO, 0.1 SO_3. **Lumnite cement,** of the Universal-Atlas Cement Co., is a cement of this type. **Accelerated cements** are intermediate cements that will set hard in about 3 days. The raw mixture for making portland cement is controlled to give exact proportions in the final product, and some quartz or iron ore may be added to balance the mix. The temperature of the rotary kiln is raised gradually to about 2650°F (1454°C). The burned clinker is then ground with a small amount of gypsum, which controls the set.

There are a number of other construction cements not classed as portland cement. **Natural cement** is made by heating to complete decarbonation, but not fusion, a highly argillaceous soft limestone. This is the most ancient of the manufactured cements, and is still called **Roman cement.** It is low in cost and will set more quickly than portland cement, but is softer and weaker. It is sometimes called **hydraulic lime.** When used for laying brick and stone it is called **masonry cement,** but ordinary **mortar** for laying brick is not this product, but is slaked lime and sand. **Cement mortar** is made with portland cement, sand, and water, with sometimes lime to aid spreading.

Oxychloride cement, or **Sorel cement,** is composed of **magnesium chloride,** $MgCl_2$, and calcined magnesia. It is strong and hard and, with various fillers, is used for floors and stucco. **Magnesia cement** is magnesium oxide, prepared by heating the chloride or carbonate to redness. When mixed with water it sets to a friable mass but of sufficient strength for covering steam pipes or furnaces. It is usually mixed with asbestos fibers to give strength and added heat resistance. The term 85% magnesia means 85% magnesia cement and 15 asbestos fibers. The cement will withstand temperatures up to 600°F (316°C).

Keene's cement, also known as **flooring cement** and **tiling plaster,** is a **gypsum cement.** It is made by burning gypsum at about 1100°F (593°C), to drive off the chemically combined water, grinding to a fine powder, and adding alum to accelerate the set. It will keep better than ordinary gypsum cement, has high strength, is white in color, and takes a good polish. **Parian cement** is similar, except that borax is used instead of alum. **Martin's cement** is made with potassium carbonate instead of alum. These cements are also called **hard-finish plaster,** and they will set very hard and white. They are used for flooring and to imitate tiling. An ancient natural cement is **pozzuolana cement.** It is a volcanic material found near Pozzuoli, Italy, and in several other places in Europe. It is a volcanic lava modified by steam or gases so that it is powdery and has acquired hydraulic properties. The chief components are silica and alumina, and the color varies greatly, being white, yellow, brown, or black. It has been employed as a construction cement since ancient times. **Trass** is a similar material found in the

Rhine district of Germany. **Santorin** is a light-gray volcanic ash with somewhat similar characteristics from the Greek island of Santorin. Artificial pozzuolana cements and trass cements are made in the United States by intergrinding pumicite, tufa, or shale with portland cement. **Slag cement** is made by grinding blast-furnace slag with portland cement. **Pozzolans** are siliceous materials which will combine with lime in the presence of water to form compounds having cementing properties. **Fly ash** is an artifical pozzolan composed principally of amorphous silica with varying amounts of the oxides of aluminum and iron and traces of other oxides. It is a fine dark powder of spheroid particles produced as the by-product of combustion of pulverized coal, and collected at the base of the stack. As an admix it improves the workability of concrete, and in large amounts its pozzolanic action adds to the compressive strength. A **fire-resistant cement,** developed by Arthur D. Little, Inc., is made of magnesium oxychlorides and magnesium oxysulfates. This inorganic resin foam cement contains 40 to 50% bond water that is released when the material is exposed to high temperatures and absorbs heat. It is said not to burn, smoke, or produce poisonous fumes when subjected to a direct flame.

CERAMICS. Ceramics, one of the three major materials families, are crystalline compounds of metallic and nonmetallic elements. The ceramic family is large and varied, including such materials as refractories, glass, brick, cement and plaster, abrasives, sanitaryware, dinnerware, artware, porcelain enamel, ferroelectrics, ferrites, and dielectric insulators. There are other materials which, strictly speaking, are not ceramics, but which nevertheless are often included in this family. These are carbon and graphite, mica, and asbestos. Also, intermetallic compounds, such as aluminides and beryllides, which are classified as metals, and cermets, which are mixtures of metals and ceramics, are usually thought of as ceramic materials because of similar physical characteristics to certain ceramics.

A broad range of metallic and nonmetallic elements are the primary ingredients in ceramic materials. Some of the common metals are aluminum, silicon, magnesium, beryllium, titanium, and boron. Nonmetallic elements with which they are commonly combined are oxygen, carbon, or nitrogen. Ceramics can be either simple, one-phase materials composed of one compound, or multiphase, consisting of a combination of two or more compounds. Two of the most common are **single oxide ceramics,** such as alumina (Al_2O_3) and magnesia (MgO), and **mixed oxide ceramics,** such as cordierite (magnesia alumina silica) and forsterite (magnesia silica). Other newer ceramic compounds include borides, nitrides, carbides, and silicides. Macrostructurally there are essentially three types of ceramics: crystalline bodies with a glassy matrix; crystalline bodies, sometimes referred to as holocrystalline; and glasses.

The specific gravities of ceramics range roughly from 2 to 3. As a class, ceramics are low tensile strength, relatively brittle materials. A few have strengths above 25,000 lb/in^2 (172 MPa), but most have less than that. Ceramics are notable for the wide difference between their tensile and compressive strengths. They are normally much stronger under compressive loading than in tension. It is not unusual for a compressive strength to be 5 to 10 times that of the tensile strength. Tensile strength varies considerably depending on composition and porosity.

One of the major distinguishing characteristics of ceramics, as compared to metals, is their almost total absence of ductility. They fail in a brittle fashion. Lack of ductility is also reflected in low impact strength, although impact strength depends to a large extent on the shape of the part. Parts with thin or sharp edges or curves and with notches have considerably lower impact resistance than those with thick edges and gentler curving contours.

Ceramics are the most rigid of all materials. A majority of them are stiffer than most metals, and the modulus of elasticity in tension of a number of types runs as high as 50 to 65 million lb/in^2 (0.3 to 0.4 million MPa) compared with 29 million lb/in^2 (0.2 million MPa) for steel. In general, they are considerably harder than most other materials, making them especially useful as wear-resistant parts and for abrasives and cutting tools.

Ceramics have the highest known melting points of materials. Hafnium and tantalum carbide, for example, have melting points slightly above 7000°F (3870°C), compared to 2600°F (1427°C) for tungsten. The more conventional ceramic types, such as alumina, melt at temperatures above 3500°F (1927°C), which is still considerably higher than the melting point of all commonly used metals. Thermal conductivities of ceramic materials fall between those of metals and polymers. However, thermal conductivity varies widely among ceramics. A two-order magnitude of variation is possible between different types, or even between different grades of the same ceramic. Compared to metals and plastics, the thermal expansion of ceramics is relatively low, although like thermal conductivity, it varies widely between different types and grades. Because the compressive strengths of ceramic materials are 5 to 10 times greater than tensile strength, and because of relatively low heat conductivity, ceramics have fairly low thermal-shock resistance. However, in a number of ceramics, the low thermal expansion coefficient succeeds in counteracting to a considerable degree the effects of thermal conductivity and tensile–compressive-strength differences.

Practically all ceramic materials have excellent chemical resistance, being relatively inert to all chemicals except hydrofluoric acid and, to some extent, hot caustic solutions. Organic solvents do not affect them. Their high surface hardness tends to prevent breakdown by abrasion, thereby retarding chemical attack. All technical ceramics will withstand prolonged heating at a minimum of 1830°F (999°C). Therefore atmospheres, gases, and chemicals cannot penetrate the material surface and produce internal reactions which normally are accelerated by heat.

Unlike metals, ceramics have relatively few free electrons and therefore are essentially nonconductive and considered to be dielectric. In general, dielectrical strengths, which range between 200 and 350 volts per mil (7.8 $\times 10^6$ and 13.8 $\times 10^6$ volts per meter), are lower than those of plastics. Electrical resistivity of many ceramics decreases rather than increases with an increase in impurities, and is markedly affected by temperature.

CERMETS. A composite material made up of ceramic particles (or grains) dispersed in a metal matrix. Particle size is greater than 1 μm, and the volume fraction is over 25% and can go as high as 90%. Bonding between the constituents results from a small amount of mutual or partial solubility. Some systems, however, such as the metal oxides, exhibit poor bonding between phases and require additions to serve as bonding agents. Cermet parts are produced by powder metallurgy (P/M) techniques. They have a wide range of properties, depending on the composition and relative volumes of the metal and ceramic constituents. Some cermets are also produced by impregnating a porous ceramic structure with a metallic matrix binder. Cermets can also be used in powder form as coatings. The powdered mixture is sprayed through an acetylene flame, and it fuses to the base material.

Although a great variety of cermets have been produced on a small scale, only a few types have significant commercial use. These fall into two main groups: oxide-base and carbide-base cermets. The most common type of **oxide-base cermets** contains aluminum-oxide ceramic particles (ranging from 30 to 70% volume fraction) and a chromium or chromium-alloy matrix. In general, oxide-base cermets have specific gravities between 4.5 and 9.0, and tensile strengths ranging from 21,000 to 39,000 lb/in^2 (144 to 268 MPa). Their modulus of elasticity runs between 37 and 50 million lb/in^2 (0.25 and 0.34 million MPa), and their hardness range is A70 to 90 on the Rockwell scale. The oxide-base cermets are used as a tool material for high-speed cutting of difficult-to-machine materials. Other uses include thermocouple-protection tubes, molten-metal-processing equipment parts, and mechanical seals.

There are three major groups of **carbide-base cermets:** tungsten, chromium, and titanium. Each of these groups is made up of a variety of compositional types or grades. **Tungsten-carbide cermets** contain up to about 30% cobalt as the matrix binder. They are the heaviest type of cermet (specific gravity is 11 to 15). Their outstanding properties include high rigidity, compressive strength, hardness, and abrasion resistance. Their modulus of elasticity ranges between 65 and 95 million lb/in^2 (0.45 and 0.65 million MPa), and they have a Rockwell hardness of about A90. They are used for gages and valve parts. Most **titanium-carbide cermets** have nickel or nickel alloys as the metallic matrix, which results in high-temperature resistance. They have relatively low density combined with high stiffness and strength at temperatures above 2200°F (1204°C). Typical properties are specific gravity, 5.5 to 7.3; tensile strength, 75,000 to 155,000 lb/in^2 (517 to 1,068 MPa); modulus of elasticity, 36 to 55 million lb/in^2 (0.25 to 0.38 million MPa); and Rockwell hardness, A70 to 90. Typical uses are gas-turbine nozzle vanes, torch tips, hot-mill-roll guides, valves, and valve seats. **Chromium-carbide cermets** contain from 80 to 90% chromium carbide, with the balance being either nickel or nickel alloys. Their tensile strength is about 35,000 lb/in^2 (241 MPa), and they have a tensile modulus of from about 50 to 56 million lb/in^2 (0.34 to 0.39 million MPa). Their Rockwell hardness is about A88. They have superior resistance to oxidation, excellent corrosion resistance, and relatively low density (specific gravity is 7.0). Their high rigidity and abrasion resistance makes them suitable for gages, valve liners, spray nozzles, bearing seal rings, bearings, and pump rotors.

Another cermet is **barium-carbonate-nickel cermet** used in higher-power pulse magnetrons. Some proprietary compositions are used as friction materials. In brake applications, they combine the thermal conductivity and toughness of metals with the hardness and refractory properties of ceramics. **Uranium-dioxide cermets** have been developed for use in nuclear reactors. Other cermets developed for use in nuclear equipment include **chromium-alumina cermets, nickel-magnesia cermets,** and **iron-zirconium-carbide cermets.**

CESIUM. Also spelled **caesium.** A rare metal, symbol Cs, obtained from the mineral **pollucite,** $2Cs_2O \cdot 2Al_2O_3 \cdot 9SiO_2 \cdot H_2O$, of Southwest Africa and Canada. The metal resembles rubidium and potassium, is silvery white and very soft. It oxidizes easily in the air, ignites at ordinary temperatures, and decomposes water with explosive violence. It can be kept only in a vacuum. The specific gravity is 1.903, melting point 28.5°C, and boiling point 670°C. It is used in low-voltage tubes to scavenge the last traces of air. It is usually marketed in the form of its compounds such as **cesium carbonate,** $CsNO_3$, **cesium fluoride,** CsF, or **cesium carbonate,** Cs_2CO_3. In the form of **cesium chloride,** CsCl, it is used on the filaments of radio tubes to increase sensitivity. It interacts with the thorium of the filament to produce positive ions. In photoelectric cells cesium chloride is used for a photosensitive deposit on the cathode, since cesium releases its outer electron under the action of ordinary light, and its color sensitivity is higher than that of other alkali metals. The high-voltage rectifying tube for changing alternating current to direct current has cesium metal coated on the nickel cathode, and has cesium vapor for current carrying. The cesium metal gives off a copious flow of electrons and is continuously renewed from the vapor. Cesium vapor is also used in the infrared signaling lamp, or **photophone,** as it gives infrared waves without visible light. **Cesium 137,** recovered from the waste of atomic plants, is a gamma-ray emitter with a half-life of 33 years. It is used in teletherapy, but the rays are not as penetrating as cobalt 60, and twice as much is required to produce equal effect.

CHALK. A fine-grained limestone, or a soft, earthy form of **calcium carbonate,** $CaCO_3$, composed of finely pulverized marine shells. The natural chalk comes largely from the southern coast of England and the north of France, but high-calcium marbles and limestones are the sources of most of the American chalk and precipitated calcium carbonate. Chalk is employed in putty, crayons, paints, rubber goods, linoleum, calcimine, and as a mild abrasive in polishes. **Whiting** and **Paris white** are names given to grades of chalk that have been ground and washed for use in paints, inks, and putty. **French chalk** is a high grade of massive talc cut to shape and used for marking. The color of chalk should be white, but it may be colored gray or yellowish by impurities. The commercial grades depend on the purity, color, and fineness of the grains. The specific gravity may be as low as 1.8.

Precipitated calcium carbonate is the whitest of the pigment extenders. **Kalite,** of the Diamond Alkali Co., is a precipitated calcium carbonate of 1-μm particle size, and **Suspenso, Surfex,** and **Nonferal** are grades with particle sizes from 5 to 10 μm. **Whitcarb RC,** of the Witco Chemical Co., for rubber compounding, is a fine-grained grade, 0.065 μm, coated to prevent dusting and for easy dispersion in the rubber. **Purecal SC,** of the Wyandotte Chemcials Corp., is a similar material. **Limeolith,** of the Kansas City Limeolith Co., **Calcene,** of PPG Industries, and **Kalvan,** of R. T. Vanderbilt Co., Inc., are precipitated calcium carbonates. A highly purified calcium carbonate for use in medicine as an antacid is **Amitone,** of Winthrop-Stearns, Inc.

CHAMOIS. A soft, pliable leather originally made from the skins of the chamois, *Antilopa rupicapra,* a small deer inhabiting the mountains of Europe but now nearly extinct. The leather was of a light-tan color, with a soft nap. All commercial chamois is now made from the skins of lamb, sheep, goat, or from the thin portion of split hides. The Federal Trade Commission limited the use of the term chamois to oil-dressed sheepskins mechanically sueded, but there are no technical precedents for such limitation. The original **artificial chamois** was made by tanning sheep-skins with formaldehyde or alum, impregnating with oils, and subjecting to mechanical sueding, but chamois is also made by various special tannages with or without sueding. Those treated with fish oils have a distinctive feel. Chamois leather will withstand soaking in hot water and will not harden on drying. It is used for polishing glass and plated metals. **Buckskin,** a similar pliable leather, but heavier and harder, was originally soft-tanned, oil-treated deerskin, but is now made from goatskins.

CHARCOAL. An amorphous form of carbon, made by enclosing billets in a retort and exposing them to a red heat for 4 or 5 h. It is also made by covering large heaps of wood with earth and permitting them to burn slowly for about a month. Much charcoal is now produced as a by-product in the distillation of wood, a retort charge of 10 cords of wood yielding an average of 2,650 gal (10,030 L) of pyroligneous liquor, 11,000 lb (4,950 kg) of gas, and 6 tons (5.4 metric tons) of charcoal. **Wood charcoal** is used as a fuel, for making black gunpowder, for carbonizing steel, and for making activated charcoal for filtering and absorbent purposes. **Gunpowder charcoal** is made from alder, willow, or hazelwood. Commercial wood charcoal is usually about 25% of the original weight of the wood, and is not pure carbon. The average composition is 95% carbon and 3 ash. It is an excellent fuel, burning with a glow at low temperatures, and with a pale-blue flame at high temperatures. Until about 1850 it was much used in blast furnaces for melting iron, and it produces a superior iron with less sulfur and phosphorus than when coke is used. **Red charcoal** is an impure charcoal made at a low temperature, and retaining much oxygen and hydrogen.

CHAULMOOGRA OIL. A brownish semisolid oil from the seeds of the fruit of the tree *Taraktogenos kurzii* and other species of Thailand, Assam, and Indonesia. It is used chiefly for skin diseases and for leprosy. A similar oil is also obtained from other genera of bushes and trees of the family *Flacourtiaceae,* and that obtained from some species of *Hydnocarpus* is superior to the true chaulmoogra oil. The tree *H. anthelminthica,* native to Thailand, is cultivated in Hawaii. This oil consists mainly of chaulmoogric and hydnocarpic acids. **Sapucainha oil,** from the seeds of the tree *Carpotroche brasiliensis,* of the Amazon Valley, contains chaulmoogric, hydnocarpic, and **gorlic acids,** and is a superior oil. **Gorliseed oil,** from the seeds of the tree *Onchoba echinata* of tropical Africa, and cultivated in Costa Rica and Puerto Rico, contains about 80% chaulmoogric acid and 10 gorlic acid. **Dilo oil** is from the kernels of the nuts of the tree *Calophyllum inophyllum* of the South Sea Islands. In Tahiti it is called **tamanu.** The **chaulmoogric acids** are cyclopentenyl compounds, $(CH)_2(CH)_2CH(CH_2)_xCOOH$, made easily from cyclopentyl alcohol.

CHEESECLOTH. A thin, coarse-woven cotton fabric of plain weave, 40 to 32 count, and of coarse yarns. It was originally used for wrapping cheese, but is now employed for wrapping, lining, interlining, filtering, as a polishing cloth, and as a backing for lining and wrapping papers. The cloth is not sized, and may be either bleached or unbleached. It comes usually 36 in (0.91 m) wide. The grade known as **beef cloth,** originally used for wrapping meats, is also the preferred grade for polishing enameled parts. It is made of No. 22 yarn or finer. For covering meats the packing plants now use a heavily napped knitted fabric known as **stockinett.** It is made either as a flat fabric or in seamless tube form, and is also used for covering inking and oiling rolls in machinery. Lighter grades of cheesecloth, with very open weave, known as **gauze,** are used for surgical dressings, and for backings for paper and maps. **Baling paper** is made by coating cheesecloth with asphalt and pasting to one side of heavy kraft or Manila paper. **Cable paper,** for wrapping cables, is sometimes made in the same way but with insulating varnish instead of asphalt. **Buckram** is a coarse, plain-woven open fabric similar to cheescloth but heavier and highly sized with water-resistant resins. It is usually of cotton, but may be of linen, and is white or in plain colors. It is used as a stiffening material, for bookbindings, inner soles, and interlinings. **Cotton bunting** is a thin, soft, flimsy fabric of finer yarn and tighter weave than cheesecloth, used for flags, industrial linings, and decorations. It is dyed in solid colors or printed. But usually the word **bunting** alone refers to a more durable, nonfading lightweight worsted fabric in plain weave.

CHEMICAL INDICATORS. Dyestuffs that have one color in acid solutions and a different color in basic or alkaline solutions. They are used to indicate the relative acidity of chemical solutions, as the different materials have different ranges of action on the acidity scale. The materials are mostly weak acids, but some are weak bases. The best known is **litmus,** which is red below a pH of 4.5 and blue above a pH of 8.3, and is used to test strong acids or alkalies. It is a natural dye prepared from several varieties of lichen, *Variolaria,* chiefly *Rocella tinctoria,* by allowing them to ferment in the presence of ammonia and potassium carbonate. When fermented, the mass has a blue color and is mixed with chalk and made into tablets of papers. It is used also as a textile dye, wood stain, and as a food colorant. **Azolitmin,** $C_7H_7O_4N$, is the coloring matter of litmus, and is a reddish-brown powder. **Orchil,** or **cudbear,** is a red dye from another species. **Alkanet,** also called **orcanette, anchusa,** or **alkanna,** is made from the root of the plant *Alkanna tinctoria* growing in the Mediterranean countries, Hungary, and western Asia. The coloring ingredient, **alkannin,** is soluble in alcohol, benzene, ether, and oils, and is produced in dry extract as a dark red, amorphous, slightly acid powder. It is also used for coloring fats and oils in pharmacy and in cosmetics, for giving an even red color to wines, and for coloring wax.

Some coal-tar indicators are: **malachite green,** which is yellow below a pH of 0.5 and green above 1.5; **phenolphthalein,** which is colorless below 8.3 and magenta above 10.0; and **methyl red,** which is red below 4.4 and yellow above 6.0. A **univeral indicator** is a mixture of a number of indicators that gives the whole range of color changes, thereby indicating the entire pH range. But such indicators must be compared with a standard to determine the pH value.

The change in color is caused by a slight rearrangement of the atoms of the molecule. Some of the indicators, such as **thymol blue,** exhibit two color changes at different acidity ranges because of the presence of more than one chromophore arrangement of atoms. These can thus be used to indicate two separate ranges on the pH scale. **Curcumin,** a crystalline powder obtained by percolating hot acetone through turmeric, changes from yellow to red over the pH range of 7.5 to 8.5, and from red to orange over the range of 10.2 to 11.8. **Test papers** are strips of absorbent paper that have been saturated with an indicator and dried. They are used for testing for acidic or basic solutions, and not for accurate determination of acidity range or hydrogen-ion concentration such as is possible with direct use of the indicators. **Alkannin paper,** also called **Boettger's paper,** is a white paper impregnated with an alcohol solution of alkanet. The paper is red, but it is turned to shades from green to blue by alkalies. **Litmus paper** is used for acidity testing. **Starch-iodide paper** is paper dipped in starch paste containing potassium iodide. It is used to test for halogens and oxidizing agents such as hydrogen peroxide.

CHERRY. The wood of several species of cherry trees native to Europe and the United States. It is brownish to light red in color, darkening on exposure, and has a close, even grain. The weight is about 40 lb/ft^3 (641 kg/m^3). It retains its shape well, and takes a fine polish. The annual cut of commerical cherry wood is small, but it is valued for instrument cases, patterns, paneling, and cabinetwork. **American cherry** is mostly from the tree *Prunus serotina,* known as the **black cherry,** although some is from the tree *P. emarginata.* The black cherry wood formerly used for airplane propellers has a specific gravity of 0.53 when oven-dried, compressive strength perpendicular to the grain 1,170 lb/in^2 (8.1 MPa), and shearing strength parallel to the grain 1,180 lb/in^2 (8.1 MPa). This tree is thinly scattered through the eastern part of the United States. The wood is light to dark reddish with a beautiful luster and silky sheen, but has less figure than mahogany. **English cherry** is from the trees *P. cerasus* and *P. avium.*

CHESTNUT. The wood of the tree *Castanea dentata,* which once grew plentifully along the Appalachian range from New Hampshire to Georgia, but is now very scarce. The trees grow to a large size, but the wood is inferior to oak in strength though similar in appearance. It is more brittle than oak, has a coarse, open grain often of spiral growth, and splits easily in nailing. The color is light brown or yellowish. It was used for posts, crossties, veneers, and some mill products. The wood contains from 6 to 20% tannin, which is obtained by soaking the chipped wood in water and evaporating. **Chestnut extract** was valued for tanning leather, giving a light-colored strong leather. The seed nuts of all varieties of chestnut are used for food and are eaten fresh, boiled, or roasted. The **European chestnut,** *C. sativa* and *C. vesca,* also called the **Spanish chestnut** and the **Italian chestnut,** has large nuts of inferior flavor. The wood is also inferior. The **horse chestnut** is a smaller tree, *Aesculus hippocastanum,* grown as a shade tree in Europe and the United States. The nut is round and larger than the chestnut. It is bitter in taste, but is rich in fats and starch and, when the saponin is removed, it produces an edible meal with an almondlike flavor used in confections in Europe. The nuts of the American horse chestnut, **buckeye,** or **Ohio buckeye,** *A. glabra,* and the **yellow buckeye,** *A. octandra,* are poisonous. The trees grow in the central states, and the dense, white wood is used for furniture and artificial limbs.

CHICLE. The coagulated latex obtained from incisions in the trunk of the evergreen tree *Achras zapota* and some other species of southern Mexico, Guatemala, and Honduras. The crude chicle is in reddish-brown pieces, and may have up to 40% impurities. The purified and neutralized gum is an amorphous white to pinkish powder insoluble in water, which forms a sticky mass when heated. The commercial purified gum is molded into blocks of 10 to 12 kg for shipment. It contains about 40% resin, 17 rubber, and about 17 sugars and starches. Under the name of **txixtle** the coagulated latex was mixed with asphalt and used as **chewing gum** by the Aztec Indians, and this custom of chewing gum has been widely adopted in the United States. Chicle is used chiefly as a base for chewing gum, sometimes diluted with gutta gums. For chewing it is compounded with polyvinyl acetate, microcrystalline wax, and flavors.

CHLORIDE OF LIME. A white powder, a **calcium chloride hypochlorite,** of the composition CaCl(OCl), having a strong chlorite odor. It decomposes easily in water and is used as a source of chlorine for cleaning and bleaching. It is produced by passing chlorine gas through slaked lime. Chloride of lime, or **chlorinated lime,** is also known as **bleaching powder,** although commercial bleaching powder may also be a mixture of calcium chloride and calcium hypochlorite, and the term **bleaches** is used for many chlorinated compounds. The dry bleaches of the Food Machinery & Chemical Corp. are chlorinated **isocyanuric acids,** the CDB-85 being a fine white powder of the composition $ClNCO_3$, containing 88.5% available chlorine. **Perchloron,** of the Pennsylvania Salt Mfg. Co., is **calcium hypochlorite,** $Ca(OCl)_2$, containing 70% available chlorine.

CHLORINATED HYDROCARBONS. A large group of materials used as solvents for oils and fats, for metal degreasing, dry cleaning of textiles, as refrigerants, in insecticides, and in fire extinguishers. They are hydrocarbons in which hydrogen atoms were replaced by chlorine atoms. They range from the gaseous methyl chloride to the solid **hexachloroethane,** CCl_3CCl_3, with most of them liquid. The increase in the number of chlorine atoms increases the specific gravity, boiling point, and some other properties. They may be divided into four groups: the methane group, including methyl chloride, chloroform, and carbon tetrachloride; the ethylene group, including dichlorethylene; the ethane group, including ethyl chloride and dichlorethane; and the propane group. All of these are toxic, and the fumes are injurious when breathed or absorbed through the skin. Some decompose in light and heat to form more toxic compounds. Some are very inflammable, while others do not support combustion. In general, they are corrosive to metals.

Chloroform, or **trichloromethane,** is a liquid of the composition $CHCl_3$, boiling point 61.2°C, and specific gravity 1.489, used industrially as a solvent for greases and resins, and in medicine as an anesthetic. It decomposes easily in the presence of light to form phosgene, and a small amount of ethyl alcohol is added to prevent decomposition. **Ethyl chloride,** also known as **monochlorethane, kelane,** and **chelene,** is a gas of the composition CH_3CH_2Cl, used in making ethyl fluid for gasoline, as a local anesthetic in dentistry, as a catalyst in rubber and plastics processing, and as a refrigerant in household refrigerators. It is marketed compressed into cylinders as a colorless liquid. The specific gravity is 0.921, freezing point −140.8°C, and boiling point 12.5°C. The condensing pressure in refrigerators is 12.4 lb (5.6 kg) at 6°F (−14°C), and the pressure of vaporization is 10.1 lb (4.6 kg) at 5°F (−15°C). Its disadvantage as a refrigerant is that it is highly inflammable, and there is no simple test for leaks. **Methyl chloride** is a gas of the composition CH_3Cl, which is compressed into cylinders as a colorless liquid of boiling point −10.65°F (−23°C) and freezing point −144°F (−98°C). Methyl chloride is one of the simplest and cheapest chemicals for methylation. In water solution it is a good solvent. It is also used as a catalyst in rubber processing, as a restraining gas in high-heat thermometers, and as a refrigerant. **Monochlorobenzene,** C_6H_5Cl, is a colorless liquid boiling at 132°C, not soluble in water. It is used as a solvent for lacquers and resins, as a heat-transfer medium, and for making other chemicals. **Trichlor cumene,** or **isopropyl trichlorobenzene,** is valued as a hydraulic fluid and dielectric fluid because of its high dielectric strength, low solubility in water, and resistance to oxidation. It is a colorless liquid,

$(CH_3)_2CHC_6H_2Cl_3$, boiling at 260°C, and freezing at −40°C. **Halane,** of the Wyandotte Chemicals Corp., used in processing textiles and paper, is dichlorodimethyl hydantoin, a white powder containing 66% available chlorine.

CHLORINATED POLYETHER. A high-priced, high-molecular-weight thermoplastic used chiefly in the manufacture of process equipment. Crystalline in structure, it is extremely resistant to thermal degradation at molding and extrusion temperatures. The plastic has resistance to more than 300 chemicals at temperatures up to 250°F (120°C) and higher, depending on environmental conditions.

Along with the mechanical capabilities and chemical resistance, chlorinated polyether has good dielectric properties. Loss factors are somewhat higher than those of polystyrenes, fluorocarbons, and polyethylenes, but are lower than many other thermoplastics. Dielectric strength is high and electrical values show a high degree of consistency over a range of frequencies and temperatures.

The material is available as a molding powder for injection-molding and extrusion applications. It can also be obtained in stock shapes such as sheet, rods, tubes, or pipe, and blocks for use in lining tanks and other equipment, and for machining gears, plugs, etc. Rods, sheet, tubes, pipe, blocks, and wire coatings can be extruded on conventional equipment and by normal production techniques. Parts can be machined from blocks, rods, and tubes on conventional metal-working equipment.

Sheet can be used to convert carbon steel tanks into vessels capable of handling highly corrosive liquids at elevated temperatures. Using a conventional adhesive system and hot gas welding, sheet can be adhered to sandblasted metal surfaces.

Coatings of chlorinated polyether powder can be applied by several coating processes. Using the fluidized bed process, pretreated, preheated metal parts are dipped in an air-suspended bed of finely divided powder to produce coatings, which after baking are tough, pinhole free, and highly resistant to abrasion and chemical attack. Parts clad by this process are protected against corrosion both internally and externally.

CHLORINATED RUBBER. An ivory-colored or white powder produced by the reaction of chlorine and rubber. It contains about 67% by weight of rubber, and is represented by the empirical formula $(C_{10}H_{13}Cl_7)_x$, although it is a mixture of two products, one having a CH_2 linkage instead of a CHCl. Chlorinated rubber is used in acid-resistant and corrosion-resistant paints, in adhesives, and in plastics,

The uncompounded film is brittle, and for paints chlorinated rubber is plasticized to produce a hard, tough, adhesive coating, resistant to oils, acids, and alkalies. The specific gravity of chlorinated rubber is 1.64, and bulking value 0.0735 gal/lb. The tensile strength of the film is 4,500 lb/in^2 (31.5 MPa). It is soluble in hydrocarbons, carbon tetrachloride, and esters, but insoluble in water. The unplasticized material has a high dielectric strength, up to 2,300 volts per mil (90.6×10^6 volts per meter). **Tornesit** and **Parlon** are chlorinated rubbers of Hercules Inc. **Paratex,** of the Truscon Laboratories, and **Roxaprene,** of the Roxalin Flexible Lacquer Co., Inc., are chlorinated-rubber coating materials. **Pliofilm,** of the Goodyear Tire & Rubber Co., is a rubber hydrochloride made by saturating the rubber molecule with hydrochloric acid. It is made into transparent sheet wrapping material which heat-seals at 105 to 130°C, or is used as a coating material for fabrics and paper. It gives a tough, flexible, water-resistant film. **Pliolite,** of this company, is a cyclized rubber made by highly chlorinating the rubber. It is used in insulating compounds, adhesives and protective paints. It is soluble in hydrocarbons, but is resistant to acids and alkalies. **Pliowax** is this material compounded with paraffin or ceresin wax. **Pliolite S-1** is this material made from synthetic rubber. **Dartex** and **Alloprene** are German chlorinated rubbers, and **Rulahyde** is Dutch. Resistant fibers have also been made from chlorinated rubbers. **Tensolite,** of the Tensolite Corp., for filter cloth, was one of these. **Betacote 95,** of Essex Chemical Corp., is a maintenance paint for chemical-processing plants which is based on chlorinated rubber. It adheres to metals, cements, and wood and is rapid-drying; the coating is resistant to acids, alkalies, and solvents.

Cyclized rubber can be made by heating rubber with sulfonyl chloride or with **chlorostannic acid,** $H_2SnCl_6 \cdot 6H_2O$. It contains about 92% of rubber hydrocarbons, and has the long straight chains of natural rubber joined together with a larger ring-shaped structure. The molecule is less saturated than ordinary natural rubber, and the material is tougher. It is thermoplastic, somewhat similar to gutta percha or balata, and makes a good adhesive. The specific gravity is 1.06, softening point 80 to 100°C, and tensile strength up to 4,500 lb/in^2 (31.5 MPa). It has been used in adhesives for bonding rubbers to metals, and for waterproofing paper.

CHLORINE. An elementary material, symbol Cl, which at ordinary temperatures is a gas. It occurs in nature in great abundance in combinations, in such compounds as common salt. It has a powerful suffocating odor, and is strongly corrosive to organic tissues and to metals. During the First World War it was used as a poison gas under the name of **Bertholite.** An important use for liquid chlorine is for bleaching textiles and paper pulp, but is also used for the manufacture of many chemicals. For bleaching, it is also widely employed in the form of compounds easily broken up. **Chlorine dioxide,** ClO_2, is a reddish-yellow gas which is more than twice as effective as a bleach as chlorine, but it is unstable and must be made just before use, from chlorine and sodium chlorite. The other two oxides of chlorine are also unstable. **Chlorine monoxide,** or **hypochlorous anhydride,** Cl_2O, is a highly explosive gas. **Chlorine heptoxide,** or **perchloric anhydride,** Cl_2O_7, is an explosive liquid. The **chlorinating agents,** therefore, are largely limited to the more stable compounds. Dry chlorines are used in cleansing powders and for detinning steel, where the by-product is tin tetrachloride.

Chlorine may be made by the electrolysis of common salt. The specific gravity of the gas is 3.214, or 2.486 times heavier than air. The boiling point is −33.6°C, and the gas becomes liquid at atmospheric pressure at a temperature of −24.48°F (−31°C). The vapor pressure ranges from 39.4 lb (17.9 kg) at 0°C to 602.4 lb (273.2 kg) at 100°C. The gas is an irritant and not a cumulative poison, but breathing large amounts destroys the tissues. Commercial chlorine is produced in making caustic soda, by treatment of salt with nitric acid, and as a by-product in the production of magnesium metal from seawater or brines. The chlorine yield is from 1.8 to 2.7 times the weight of the magnesium produced.

CHLOROPHYLL. A complex chemical which constitutes the green coloring matter of plants and the chief agent of their growth. It is obtained from the leaves and other parts of plants by solvent extraction and is used as a food color and as a purifying agent. When extracted from alfalfa by hexane and acetone, 50 tons (45.4 metric tons) of alfalfa yields 400 lb (181 kg) of chlorophyll. A higher yield is obtained in California from the cull leaves of lettuce. It is one of the most interesting of chemicals, and is a sunlight-capturing, food-making agent in plants. It has the empirical formula $C_{55}H_{72}O_5N_4Mg$, having a complex ring structure with **pyrrole,** $(CH:CH)_2NH$, as its chief building block, and a single magnesium atom in the center. It is designated as a **magnesium-porphyrin** complex. The **iron-porphyrin** complex, **hematin,** of blood, is the same structure with iron replacing magnesium. The **vanadium-porphyrin** complex of fishes and cold-blooded animals, and found also in petroleum, is the same thing with vanadium replacing the magnesium. Under the influence of sunlight and the pyrrole complex, carbon dioxide unites with water to produce formaldehyde and oxygen, and enables plant and animal bodies to produce carbohydrates and proteins. Failure of the pyrrole ring to link up with NH, connecting with sulfur instead, completely suspends the functioning of the blood.

Chlorophyll is obtained as a crystalline powder soluble in alcohol and melting at 183°C. It combines with carbon dioxide of the air to form formaldehyde which is active for either oxidation or reduction of impurities existent in the air, changing such gases to methanol, formic acid, or carbonic acid. It is thus used in household air-purifying agents. In plants, some of the formaldehyde is given off to purify the air, but most of it is condensed in the plant to form **glycolic aldehyde,** $HOCH_2CHO$, the simplest carbohydrate, and also **glyceric aldehyde,** another simple carbohydrate. Although chlorophyll is used as an odor-destroying agent in cosmetics and foods, its action when taken into the human body in quantity in its nascent state is not fully understood, and the magnesium in the complex is capable of replacing the iron in the blood complex.

The **porphyrins,** each having a nucleus of four pyrrole rings and a distinctive metal such as the magnesium of the chlorophyll of plants, are termed pigments in medicine, and the disease of unbalance of porphyrin in human blood is called porphyria. In addition to photosynthesis they have catalytic and chelating actions and may be considered as the chief growth agents in plant and animal life. For example, the **zinc porphyrin** of the eye is formed in the liver, and a lack of supply to the fluid of the eye may cause loss of vision.

Pyrrole can be obtained from coal tar and from bone oil, or can be made synthetically, and is used in the production of fine chemicals. **Pyrrolidine,** used as a stabilizer of acid materials and as a catalyst, is a water-soluble liquid, $(CH_2CH_2)_2NH$, made by the hydrogenation of pyrrole, or by treating tetrahydrofuran with ammonia. **Polyvinyl pyrrolidine,** $H_2C \cdot H_2C \cdot NH \cdot CH_2 \cdot CH_2$, is a cyclic secondary amine made from formaldehyde and acetylene. It is used as a supplementary blood plasma, and for making fine chemicals. Small amounts are added to fruit beverages such a prune juice, as a color stabilizer. It combines with the phenols which cause the oxidation, and the combination can be filtered off.

Chelating agents, used for eliminating undesirable metal ions in water solutions, for increasing color intensity in organic dyes, and for treating water and organic acids, are chemically similar to chlorophyll in having a single polyvalent metal ion surrrounded by straight-chain carboxylic or amino compounds in a ring structure, but are unlike chlorophyll in that their metal-alkyl linkage depends on electron resonance instead of electron sharing. The **Versenes,** of Versenes, Inc., are chelating agents used in agriculture to correct iron deficiency and stimulate plant growth. In medicine they are used to remove poisonous lead or mercury from the blood. The **Mullapons,** of the General Aniline & Film Corp., are chelating agents. **Potassium acid saccharate,** $HOOC(CHOH)_4COOK$, is marketed by the Sanders Chemical Co. as a chelating agent for metal cleaning and plating solutions.

CHROME-MOLYBDENUM STEEL. Any alloy steel containing chromium and molybdenum as the predominating alloying elements. Often used to refer specifically to steels in the **AISI 4100 series.** Chromium gives hardness and toughness to the steel, while molybdenum improves the forging and machining properties and increases the strength. Chrome-molybdenum, or chromium-molybdenum, steels are noted for high strength and toughness. Only small amounts of alloying elements are used in the standard steels. A chrome-molybdenum steel used for airplane tubing contains 0.80 to 1.10% chromium, 0.15 to 0.25 molybdenum, 0.40 to 0.60 manganese, and 0.25 to 0.35 carbon. It has a tensile strength of 95,000 lb/in^2 (665 MPa) min, with elongation 10% and is slightly air-hardening. It draws well, and tubes with a wall thickness of only 0.035 in (0.089 cm) are made. Molybdenum adds red hardness to steel to a greater degree than tungsten, and the amounts used in these steels are sufficient to make them slightly red-hard and air-hardening. **AISI steel 4140,** which has the same composition as the airplane tubing but with 0.40% carbon, has a tensile strength up to 260,000 lb/in^2 (1,792 MPa), elongation 8%, and Brinell hardness 490 when oil-quenched and drawn. This type of steel, with 0.30% carbon and **AISI steel 4130,** is used for structural parts where welding is to be done. The chrome-molybdenum steels marketed by the Crucible Steel Co. under the name of **Almo steel** for automotive and ordnance work have tensile strengths up to 167,000 lb/in^2 (1,151 MPa), with elongation 18%. **Croloy 2,** of the Babcock & Wilcox Co., used for boiler tubes for high-pressure superheated steam, contains 2% chromium and 0.50 max molybdenum, and is for temperatures to 1150°F (621°C). **Croloy 5** has 5% chromium and 0.50 max molybdenum, for temperatures to 1200°F (649°C) and higher pressures. **Croloy 7** has 7% chromium and 0.50 molybdenum.

Chrome-molybdenum steels, with high carbon, have great resistance to wear at high heat, and are used for die blocks for forging. **Cromo steel,** of the Michigan-Standard Alloy Casting Co., is a cast steel with a strength of 100,000 lb/in^2 (689 MPa) and elongation 15 to 25%, used for large die blocks. **Albor die steel,** of Wm. Jessop & Sons, Inc., used for dies for stamping hard metals, contains 0.90% chromium, 0.30 molybdenum, and 0.90 carbon. It is tough and deep-hardening. **Atlas No. 93,** of the Allegheny Ludlum Steel Co., is a shock-resistant steel with 0.65% chromium, 0.35 molybdenum, and 0.55 carbon.

Chrome-molybdenum steels air-harden from a relatively low temperature, about 1550°F (843°C), thus minimizing distortion, and since they are less costly than most tool steels, are often used in modified compositions for tools and strong forgings. **Lo-Air steel,** of the Cyclops Corporation, contains 1.35% molybdenum, 1 chromium, 2 manganese, 0.30 silicon, and 0.70 carbon. It hardens to about Rockwell C62, and when tempered to Rockwell C58 has a tensile strength of about 300,000 lb/in^2 (2,068 MPa) with elongation of 1%. **Hi Shock 60 steel,** of Carpenter Technology, for punches, hobs, coining dies, and shear blades, contains 1% molybdenum, 1 chromium, 2.5 copper, 0.5 nickel, 0.5 manganese, 0.3 silicon, 0.15 vanadium, and 0.68 carbon. It has high strength and toughness. **Lesco BG 41 steel,** of the Latrobe Steel Co., used to replace stainless-steel Type 440C for aircraft engine parts where high-compression stresses at temperatures above 800°F (427°C) are encountered, has 14.5% chromium, 4 molybdenum, 0.30 manganese, 0.30 silicon, and 1.10 carbon. At 800°F (427°C) the hardness is Rockwell C58, and it is above Rockwell C50 at 1000°F (538°C), while the harness of **AISI Type 440C steel** falls off rapidly at 800°F (427°C). This latter steel contains about 17% chromium, 0.75 molybdenum, 0.5 nickel, 1 manganese, 1 silicon, and 1 carbon. Chrome-molybdenum steels with carbon to 1% are used for castings for bucket lips, crusher parts, and other heavy-duty parts. They have tensile strengths up to 150,000 lb/in^2 (1,034 MPa), with elongation 12 to 14%.

CHROME-VANADIUM STEEL. Any **chromium-alloy steel** containing a small amount of vanadium which has the effect of intensifying the action of the chromium and the manganese in the steel. It also aids in the formation of carbides, hardening the alloy, and also increasing the ductility by the deoxidizing effect. The amount of vanadium is usually 0.15 to 0.25%. These steels are valued where a combination of strength and ductility is desired. They resemble those with chromium alone, with the advantage of the homogenizing influence of the vanadium. A chrome-vanadium steel having 0.92% chromium, 0.20 vanadium, and 0.25 carbon has a tensile strength of 100,000 lb/in^2 (689 MPa), and when heat-treated has a strength up to 150,000 lb/in^2 (1,034 MPa) and elongation 16%. Chrome-vanadium steels are used for such parts as crankshafts, propeller shafts, and locomotive frames. High-carbon chrome-vanadium steels are the mild-alloy tool steels of high strength, toughness, and fatigue resitance. The chromium content is usually about 0.80%, with 0.20 vanadium, and with carbon up to 1%. **Milwaloy,** of the Milwaukee Steel Foundry Co., is a casting steel. Grade 7 contains 1.5 to 1.75% chromium, 0.60 to 0.70 vanadium, and 0.30 to 0.50 carbon. It is used for high-strength parts.

Many high-alloy steels also contain some vanadium, but where the vanadium is used as a cleansing and toughening element and not to give the chief characteristics to the steel, these alloys are not classed as chrome-vanadium steel. **H.Y.C.C. steel** of the Crucible Steel Co., for example, is an oil-hardening steel of high wear resistance containing 12% chromium and 2.25 carbon with 0.25 vanadium. It is a hard, deep-hardening steel for dies. Some modified chrome-vanadium steels for high-temperature, high-strength parts are in the class of tool steels except that they may have less carbon. **Viscount 44,** of the Latrobe Steel Co., contains 5% chromium, 1 vanadium, 1.2 molybdenum, 1 silicon, 0.75 manganese, 0.40 carbon, and a small percentage of sulfides to give free machining. The tensile strength is 200,000 lb/in^2 (1,378 MPa), elongation 12%, and Rockwell hardness

C45. **Vascojet 1000,** of the Vanadium Alloy Steel Co., also for heat-resistant parts, contains 5% chromium, 1.3 molybdenum, 0.5 vanadium, and 0.4 carbon. The tensile strength is 285,000 lb/ in^2 (1,964 MPa) with elongation of 8%.

CHROMIC ACID. A name given to the red, crystalline, strongly acid material of the composition CrO_3 known also as **chromium trioxide** or as **chromic anhydride.** It is in reality not the acid until dissolved in water, forming a true chromic acid of the composition H_2CrO_4. It is marketed in the form of porous lumps. The specific gravity is 2.70, melting point 196°C. It is produced by treating sodium or potassium dichromate with sulfuric acid. The dust is irritating, and the fumes of the solutions are injurious to the nose and throat as the acid is a powerful oxidizing agent. Chromic acid is used in chromium-plating baths, for etching copper, in electric batteries, and in tanning leather. **Chromous chloride,** $CrCl_2$, is used as an oxygen absorbent and for chromizing steel. **Chromic chloride,** $CrCl_3$, is a volatile white powder used for tanning and as a mordant, for flame metallizing, and in alloying steel powders.

Chrome oxide green is a **chromic oxide** in the form of dry powder or ground in oil, used in paints and lacquers and for coloring rubber. It is a bright-green crystalline powder of the composition Cr_2O_3, with specific gravity 5.20 and melting point 1990°C, insoluble in water. The dry powder has a Cr_2O_3 content of 97% min, and is 325 mesh. The paste contains 85% pigment and 15 linseed oil. Chrome oxide green is not as bright in color as chrome green but is more permanent.

CHROMITE. An ore of the metal chromium, called **chrome ore** when used as a refractory. It is found in the United States, chiefly in California and Oregon, but most of the commercial production is in South Africa, Zimbabwe, Cuba, Turkey, the Philippines, Greece, and New Caledonia. The theoretical composition is $FeO \cdot Cr_2O_3$, with 68% chromic oxide, but pure **iron chromate** is rare. Part of the iron may be replaced by magnesium, and part of the chromium by aluminum. The silica present in the ore, however, is not a part of the molecule. Chromite is commonly massive granular, and the commercial ores contain only 35 to 60% chromic oxide. The hardness is 5.5 and the specific gravity 4.6. The color is iron black to brownish black, with a metallic luster. The melting point is about 3900°F (2149°C), but when mixed with binders as a refractory the fusion point is lowered. New Caledonia ore has 50% chromic oxide, Turkish ore averages 48 to 53%, Brazilian ore runs 46 to 48%, and Cuban ore averages only 35%. The high-grade Guleman ore of Turkey contains 52% Cr_2O_3, 14 Al_2O_3, 10.4 FeO, 4.4 Fe_2O_3, 16 magnesia, and 2.5 silica. Most of the domestic ore in the United States is low grade.

Cuban ore is rich in spinel and deficient in magnetite, and this type is adapted for refractory use even when the chromic oxide is low. Ore from Baluchistan is also valued for refractory use, as are other hard lumpy ores high in Al_2O_3 and low in iron. For chemical use the ores should have more than 45% chromic oxide and not more than 8 silica, and should be low in sulfur. Metallurgical ore should have not less than 48% chromic oxide, and the ratio of chromium to iron should not be less than 3:1. Chromite is used for the production of chromium and ferrochromium, in making **chromite bricks** and refractory linings for furnaces, and for the production of chromium salts and chemicals. For bricks the ground chromite is mixed with lime and clay and burned. Chromite refractories are neutral and are resistant to slag attack. A chrome-ore high-temperature cement marketed by the General Refractories Co. under the name of **Grefco** has a fusion point of 3400°F (1871°C).

CHROMIUM. An elementary metal, symbol Cr, used in stainless steels, heat-resistant alloys, high-strength alloy steels, for wear-resistant electroplating, and in its compounds for pigments, chemicals, and refractories. The specific gravity is 6.92, melting point 2750°F (1510°C), and boiling point 3992°F (2200°C). The color is silvery white with a bluish tinge. It is an extremely hard metal, the electrodeposited plates having a hardness of 9 Mohs. It is resistant to oxidation, is inert to nitric acid, but dissolves in hydrochloric acid and slowly in sulfuric acid. At temperatures above 1500°F (816°C) it is subject to an intergranular corrosion.

Chromium occurs in nature only in combination. Its chief ore is chromite, from which it is obtained by reduction and electrolysis. It is marketed for use principally in the form of master alloys with iron or copper. The term **chromium metal** usually indicates a pure grade of chromium containing 99% or more of chromium. A grade marketed by the Shieldalloy Corp. has 99.25% min chromium, with 0.40 max iron and 0.15 max silicon. **High-carbon chromium** has 86% min chromium and 8 to 11% carbon with no more than 0.5% each of iron and silicon. **Isochrome** is a name given by the Battelle Memorial Institute for chromium metal, 99.99% pure, made by the reduction of chromium iodide. Chromium metal is used for making alloys, particularly resistance metals, and for chemical use in which case it comes as a fine powder. Nozzles and tubing for jet engines are extruded from pure chromium powder. But chromium metal lacks ductility and is susceptible to nitrogen embrittlement, and is not used as a structural metal requiring flexural strength. **Chromium plating** is widely used where extreme hardness or resistance to corrosion is required. When plated on a highly polished metal, it gives a smooth surface that has no capillary attraction to water or oil, and chromium-plated bearing surfaces can be run without oil. For decorative purposes chromium plates as thin as 0.0002 in (0.0006 cm) may be used; for wear resistance, plates up to 0.050 in (0.127 cm) are used. Increased hardness and wear resistance in the plate are obtained by alloying 1% molybdenum with the chromium. **Alphatized steel,** of the Alloy Surfaces Co., is steel coated with chromium by a diffusion process. The deposited chromium combines with the iron of the steel and forms an adherent alloy rather than a plate. Less penetration is obtained on high-carbon steels, but the coating is harder.

CHROMIUM COPPER. A name applied to master alloys of copper with chromium used in the foundry for introducing chromium into nonferrous alloys, or to **copper-chromium alloys,** or **chrome copper,** which are one of the high-copper alloys. A chromium-copper master alloy marketed under the name of **Electromet chromium copper** by the Electro Metallurgical Sales Corp. contains 8 to 11% chromium, 88 to 90 copper, and a maximum of 1 iron and 0.50 silicon. Chrome copper in wrought form usually contains less than 1% chromium, has high impact strength, and the electric conductivity is higher than when the copper is hardened with silicon. Chrome copper with 0.50% chromium has a thermal conductivity about 85% that of high-conductivity copper, which is about twice that of aluminum alloy, or seven times that of cast iron. A wrought chrome copper of the American Brass Co. contains 0.85% chromium and 0.10 silicon, and has a tensile strength of 92,000 lb/in^2 (634 MPa) with elongation 3%. **Chromium bronze,** used for bearings, is a cast metal containing 1% chromium, 1 iron, and from 2 to 10 tin. It has high strength and good wearing qualities. **Kumium,** of the Imperial Chemical Industries, Ltd., is a copper-chromium alloy used for electrodes for spot-welding machines. It retains its hardness and conductivity at 400°C. **Copper alloy PD-135** of the Phelps Dodge Corp. is oxygen-free copper with small percentages of chromium and cadmium and is a **chromium-cadmium bronze.** The extruded metal has a tensile strength of 36,000 lb/in^2 (248 MPa) and an electric conductivity 60% that of copper. It can be cold-worked and heat-treated to strengths up to 80,000 lb/in^2 (551 MPa) and will retain its strength to above 1000°F (538°C). It is used for electric switches and contacts, high-flex wire and springs, and commutator bars.

CHROMIUM STEEL. Any steel containing chromium as the predominating alloying element may be termed chromium steel, but the name alone usually means the hard, wear-resisting steels that derive the property chiefly from the chromium content. **Straight chromium steels** refers to low-alloy steels in the **AISI 5000** series and **AISI 5100** series. Chromium combines with the carbon of steel to form a hard chromium carbide, and it restricts graphitization. When other carbide-forming elements are present, double or complex carbides are formed. Chromium refines the structure, gives deep-hardening, increases the elastic limit, and gives a slight red-hardness so that the steels retain their hardness at more elevated temperatures. Chromium steels have great resistance to wear. They also withstand quenching in oil or water without much deformation. Up to about 2% chromium may be included in tool steels to add hardness, wear resistance, and nondeforming qualities. When the chromium is high, the carbon may be much higher than in ordinary steels without making the steel brittle. Steels with 12 to 17% chromium and about 2.5 carbon have remarkable wear-resisting qualities and are used for cold-forming dies for hard metals, for broaches, and for rolls. However, chromium narrows the hardening range of steels unless balanced with nickel. Such steels also work-harden rapidly unless modified with other elements. The high-chromium steels are corrosion-resistant and heat-resistant but are not to be confused with the high-chromium stainless steels which are low in carbon, although the non-nickel Type 400 stainless steels are very definitely chromium steels. Thus, the term is indefinite but may be restricted to the high-chromium steels used for dies, and to those with lower chromium used for wear-resistant parts such as ball bearings.

Chromium steels are not corrosion-resistant unless the chromium content is at least 4%. Plain chromium steels with more than 10% chromium are corrosion-resistant even at elevated temperatures and are in the class of stainless steels, but are difficult to weld because of the formation of hard brittle martensite along the weld.

Chromium steels with about 1% chromium are used for gears and shafts. **Uma steel No. 1,** of the Republic Steel Corp., has 0.55 to 0.75% chromium, 0.35 to 0.65 manganese, and 0.15 carbon. It is a case-hardening gear steel. **Uma No. 5** has 0.85 to 1.10% chromium, 0.70 to 0.90 manganese, and 0.50 carbon. It is deep-hardening, with a tensile strength up to 135,000 lb/in^2 (930 MPa), and is used for transmission gears. This type of steel with higher carbon is deep-hardening and has high compression strength and is used for ball and roller bearings. However, **ball-bearing steel** derives many of its characteristics from the heat treatment. A good ball-bearing steel is one that has uniform undissolved carbides in a matrix of martensite and a Rockwell C hardness of 60 to 64 when quenched in oil and stress-relieved. **Teton steel,** of the Allegheny Ludlum Steel Corp., has 1.25% chromium and 1 carbon, and is used for ball bearings and hard wear-resistant parts. **Bower 315 alloy,** of Federal-Mogul Bower Bearings, Inc., for jet-engine bearings, when carbonized and heat-treated, has a Rockwell C harness of 58, and will recover its hardness after being exposed to high temperatures. It contains only 1.5% chromium, with 3% nickel, 5 molybdenum, 0.5 manganese, 0.3 silicon, and up to 0.15 carbon. **Vac-Arc Regent steel,** of the Latrobe Steel Co., for high-duty aircraft bearings, has 1.5% chromium, 1 carbon, 0.35 manganese, and 0.3 silicon. It is tempered at 400°F (204°C) to a hardness of Rockwell C60.

Chromium steels with 3 to 4.5% chromium and 0.80 to 1 carbon are used for dies for hot pressing and forging. **EB alloy,** of the Allegheny Ludlum Steel Corp., has 3.75% chromium, 0.55 vanadium, 0.70 molybdenum, and 0.65 carbon. Another hot-work steel of this company is **Potomac steel** with 5% chromium, 1.75 molybdenum, 1.25 tungsten, and 0.32 carbon. **Uniloy chrome steel** of the Cyclops Corporation, contains 4 to 6% chromium, 0.1 to 0.25 carbon, up to 0.60 manganese, with either molybdenum 0.40 to 0.60%, or tungsten 1 to 1.25%. A small amount of tungsten, or a smaller amount of molybdenum, refines the grain, increases the elastic limit, and makes the steel more resistant to corrosion at elevated temperatures. **Cyclops 62 steel,** of the same company, for shears, blanking dies, and extrusion rolls, has 3.5% chromium, 2.25 molybdenum, 1.25 vanadium, 1 silicon, 0.55 manganese, and 1 carbon. When air-cooled and tempered it has a tensile strength of 352,000 lb/in^2 (2,464 MPa) with elongation of 1%. **HWD No. 1 steel,** of Firth Sterling, Inc., for hot-forging dies, has 5% chromium, 1.55 molybdenum, 1.4 tungsten, 1 silicon, 0.3 vanadium, 0.4 manganese, and 0.35 carbon. It is air-hardening, or can be oil-quenched to Rockwell C53 and tempered to as low as Rockwell C40. A variation of this steel, **HWD No. 2 steel,** for die-casting dies and aluminum extrusion dies, has 5.25% chromium, 1.35 molybdenum, 0.5 vanadium, 1 silicon, 0.4 manganese, and 0.38 carbon. It is resistant to softening and to heat checking. Air cooling and long-time tempering to about Rockwell C48 give long wear life in die-casting dies. The **chromium-tungsten steels** are employed where resistance to hot chemicals or petroleum is required, but the steels with high tungsten are usually classed as tungsten regardless of the amount of chromium. A steel used for high-strength resistant castings is **Circle L-10** of the Lebanon Steel Foundry. It contains 5.5% chromium, 0.50 molybdenum, 0.65 manganese, 0.40 silicon, and 0.20 carbon. When heat-treated it has a tensile strength of 100,000 lb/in^2 (700 MPa) and elongation 20%. **G.S.N. steel,** of Henry Disston & Sons, Inc., is a high-chromium, high-carbon die steel. **Nilstain steel,** of the Wilber B. Driver Co., for spring wire, is an 18–8 steel with 2% manganese and 1 silicon.

A wear-resistant die steel, marketed by Carpenter Technology under the name of **Hampden steel,** contains 12.5% chromium, 0.25 nickel, 0.25 manganese, and 2.1 carbon. It is deep-hardening, has high compressive strength, and is valued for forming rolls and spinning tools. A similar-purpose tool steel with air-hardening properties is **Carpenter No. 610.** It has 12% chromium, 1.5 carbon, 0.80 molybdenum, 0.20 vanadium, and 0.30 manganese. **Huron steel,** of the Allegheny Ludlum Steel Corp., is a wear-resistant steel having 12.5% chromium, 1 vanadium, and 2 carbon. It is used for dies. **O-Hi-O** steel, of the Vanadium-Alloys Steel Co., is an air-hardening die steel having 12% chromium, 1.55 carbon, 0.85 vanadium, 0.40 cobalt, and 0.80 manganese. This company also produces **Crocar,** a wear-resistant steel with similar composition but with 2.2% carbon, and

Vasco 1741 CVM, for high-temperature bearings, pump parts and seal rings, which contain 17% chromium, 4 molybdenum, 1 vanadium, 0.30 manganese, 0.20 silicon, and 1.30 carbon. It retains a hardness of about Rockwell C60 at 1000°F (538°C). **Cromovan steel,** of Firth Sterling, Inc., contains 12% chromium, 1 vanadium, 1 molybdenum, and 1.55 carbon. It is a nondeforming, wear-resistant die steel, also used for equipment parts requiring high abrasion resistance, such as brick-mold liners. The hardness ranges from 42 to 64 Rockwell C, depending on the tempering. **Cromovan F.M.** is this steel with 0.12% sulfur to give improved machinability and better surface finish.

CINCHONA. The hard, thick, grayish bark of a number of species of evergreen trees of the genus *Cinchona,* native to the Andes from Mexico to Peru but now grown in many tropical countries chiefly as a source of **quinine.** The small tree *Remijia pendunculata* also contains 3% quinine in the bark, and quinine occurs in small quantities in other plants and fruits, notably the grapefruit. **Cinchona bark** was originally used by the Quechua Indians of Peru in powdered form and was called **loxa bark.** It derives its present name from the fact that in 1630 the Countess of Cinchon was cured of the fever by its use. In Europe it became known as **Peruvian bark** and **Jesuits' bark. Quinine** is one of the most important drugs as a specific for malaria and as a tonic. It is also used as a denaturant for alcohol as it has an extremely bitter taste. Metallic salts of quinine are used in plastics to give fluorescence and glow under ultraviolet light. Quinine is a colorless crystalline alkaloid of the composition $C_{20}H_{24}O_2N_2 \cdot 3H_20$. It is soluble with difficulty in water, and is marketed in the form of the more soluble **quinine sulfate,** a white powder of the composition $(C_{20}H_{24}O_2N_2)H_2SO_4 \cdot 2H_2O$. **Quinine bisulfate** has the same composition but with seven molecules of water. During the Second World War **quinine hydrochloride** was preferred by the Navy. It contains 81.7% quinine compared with 74% in the sulfate and is more soluble in water but has a more bitter taste. **Synthetic quinine** can be made, but is more expensive. **Atabrine,** of I. G. Farbenindustrie, is **quinacrine hydrochloride.** It is not a complete substitute, is toxic, and is a dye that colors the skin when taken internally. **Primaquine,** of Winthrop-Stearns, Inc., is an 8-aminoquinoline, and as an antimalarial is less toxic than other synthetics. In Germany, copper arsenite has been used as an effective substitute for quinine. The **maringin** of grapefruit is similar to quinine, and in tropical areas where grapefruit is consumed regularly the incidence of malaria is rare.

The bark of the tree *C. ledgeriana* yields above 7% of quinine but it is not a robust tree and in cultivation is grafted on the tree *C. succirubra* which is hardy but yields only 2 to 3% quinine. Ledgeriana trees on plantations in Mindanao and in Peru yield as high as 13% total alkaloids from the bark. Most of the world production is from the *C. officinalis* and *C. calisaya,* which are variations of *C. ledgeriana,* or **yellow bark.** The **red bark,** *C. succirubra,* is grown in India. The peak gathering of bark is 10 years after planting of the 2-year seedlings, and the trees are uprooted to obtain bark from both trunk and root. An 8-year-old tree yields 4 kg of bark, and a 25-year-old tree yields 20 kg but of inferior quality. The bark is dried and ground to powder for the solvent extraction of the alkaloids. Besides quinine the bark contains about 30 other alkaloids, chief of which are **cinchonidine, quinidine,** and **cinchonine. Totaquina** is the drug containing all the alkaloids. It is cheaper than extracted quinine, is effective against malaria, and is a better tonic. Quinidine has the same formula as quinine but is of right polarization instead of left. It is used for heart ailments. Cinchonine, $C_{19}H_{22}ON_2$, has right polarization and is 13 times more soluble in water than quinine sulfate. Cinchonidine has the same formula, but has left polarization. **Australian quinine,** or **alstonia,** is not true quinine. It is from **dita bark,** the bark of the tree *Alstonia scbolaris* of Australia, and is used as a febrifuge. It contains the water-soluble alkaloid **ditiane,** $C_{22}H_{28}O_4N_2$, and the water-insoluble alkaloid **ditamine,** $C_{16}H_{19}O_2N$. **Fagarine,** used as a substitute for quinidine for heart flutter, is extracted from the leaves of the tree *Fagara coco* of northern Argentina. **Chang shan,** used as an antimalarial in China, is the root of the plant *Dichroa Febrifuga.* It contains the alkaloid **febrifugine.**

CINNABAR. The chief ore of the metal mercury. As a pigment it was originally called **minium,** a name now applied to red lead. It is a **mercuric sulfide,** HgS, which when pure contains 86.2% mercury. The ores are usually poor, the best ones containing only about 7% mercury, and the average Italian ore having only 1.1% Hg and American ore yielding only 0.5%

The chief production is in Italy, Spain, Mexico, and the United States. Cinnabar has a massive granular structure with a hardness of 2 to 2.5, a specific gravity of about 8, and usually a dull earthy luster. It is brownish red in color, from which it derives the name **liver ore. Chinese cinnabar** is ground as a fine scarlet pigment for inks. Cinnabar is not smelted, the extraction process being one of distillation, made possible by the low boiling point of the metal. Another ore of mercury found in Mexico is **livingstonite,** $2Sb_2S_3 \cdot HgS$. It is a massive, red-streaked mineral of specific gravity 4.81 and hardness 2. **Calomel,** a minor ore in Spain, is a white crystalline mineral of the composition Hg_2Cl_2 with a specific gravity of 6.5. It is also called **horn mercury.** It is used in medicine as a purgative, but is poisonous if retained in the system. The ore found in Colorado and known as **coloradoite** is a **mercuric telluride,** HgTe. It has an iron-black color and a specific gravity of 8. **Tiemanite,** found in California and Utah, is a

mercuric selenide, HgSe, having a lead-gray color and a specific gravity of 8.2. There are more than 20 minerals classed as **mercury ores.**

CINNAMON. The thin, yellowish-brown, highly aromatic bark of the tropical evergreen laurel tree *Cinnamomum zeylanicum,* of Sri Lanka and southeast Asia. It is used as a spice, as a flavor in confectionery, perfumery, and medicine. The bark is marketed in rolls or sticks packed in bales of 112 lb (51 kg). **Cassia** is the bark from the *C. cassia* of South China. **Saigon cinnamon,** *C. loureirii,* is cinnamon, but is not as thin or as smooth a bark, and it does not have as fine an aroma and flavor. **Cassia buds** are small dried flowers of the *C. cassia,* used ground as a spice, or for the production of oil. They resemble cloves in appearance, and have an agreeable spicy odor and sweet warm taste. **Cinnamon oil, cinnamon leaf oil,** and **cassia oil** are essential oils distilled, respectively, from the bark, leaf, and bud. They are used in flavoring, medicine, and perfumery. Cinnamon oil contains about 70% **cinnamic aldehyde,** 8 to 10% eugenol, and also pinene and linalol. The specific gravity is 1.03, and the refractive index 1.565 to 1.582. The pale-yellow color darkens with age. Cinnamic aldehyde is also made synthetically. **Flasolee,** of the J. Hilary Herchelroth Co., is **amyl cinnamic aldehyde,** redistilled to remove the unpleasant odor of heptyl aldehyde, for use in perfumes. The leaf oil is used as a substitute for clove oil. About 1.9% oil is obtained from cassia buds, but it lacks the delicate fragrance of cinnamon oil. **Nikkel oil,** a bright-yellow liquid with an odor of lemon and cinnamon, is distilled from the leaves and twigs of the tree *C. laureirii* of Japan. It contains citral and cineol, and is used in perfumery. Some of the cinnamon marketed in the United States is **Padang cassia,** from the tree *C. burmannii* of Indonesia. It does not have the delicate aroma of true cinnamon.

CLAD METALS. Usually, sheet or plate metal having a face of special resistance metal welded to a base of lower-cost metal, used for making tanks, boilers, and chemical processing equipment where it is desired to have an acid-resistant or heat-resistant facing with the more easy-working and lower-cost plain steel plate. But the term also embraces other metal laminates for electrical, atomic, and other uses.

Laminated metals were used very early in the jewelry and silverware industries, and silver-clad iron was made by the Gauls by brazing together sheets of silver and iron for lower-cost products as substitutes for the Roman heavy silver tableware. An early French **duplex metal** called **doublé,** for costume jewelry, had a thin facing of a noble metal on a brass or copper base, and **Efkabimetal** was a German name for this material. **Gold shell,** used for costume jewelry, is a duplex metal with gold rolled on a rich low brass. **Abyssinian gold, talmi gold,** and other names were used for these duplex metals in traders' jewelry. **Inter-Weld metal,** of the American Silver Co., has a base metal of brass to which is silver-soldered a sheet of nickel over which is welded the gold sheet. When rolled, the gold is extremely thin, but the nickel prevents the color of the base metal from bleeding through.

Composite tool steel, used for shear blades and die parts, is not a laminated metal. The term refers to bar steel machined along the entire length and having an insert of tool steel welded to the backing of mild steel. **Clad steels** are available regularly in large sheets and plates. They are made with coatings of nickel, stainless steel, Monel metal, aluminum, or special alloys, on one or both sides of the sheet. Where heat and pressure are used in the processing, there is chemical bonding between the metals. For some uses the cladding metal on one side will be 10 to 20% of the weight of the sheet. A composite plate having an 18–8 stainless-steel cladding to a thickness of 20% on one side saves 144 lb (65 kg) of chromium and 64 lb (29 kg) of nickel per 1,000 lb (454 kg) of total plate. The coatings may also be extremely thin.

Pluramelt, of the Allegheny Ludlum Steel Corp., is composite steel with various types of stainless steels integrally bonded to a depth of 20% by a process of intermelting. **Ingaclad,** of the Ingersoll Steel & Disc Co., consists of stainless steel bonded to carbon-steel plate. **Silver-Ply** is a **stainless-clad steel** of the Jessop Steel Co., made with a stainless coating either 10 or 20% of the thickness of the plate combined with the mild steel backing by hot rolling. **Permaclad,** of the Alan Wood Steel Co., has stainless steel bonded to one side of carbon steel. **SuVeneer steel,** of the Superior Steel Corp., used for automobile bumpers, has a veneer of stainless steel bonded to spring steel. **Bronze-clad steel,** of Ampco Metal, Inc., is sheet steel with high-tensile, corrosion-resistant bronze rolled on one or both sides. The cladding is from 0.031 in (0.079 cm) up to 40% of the thickness of the sheet. It is used for tanks and chemical equipment. **Hortonclad,** of the Chicago Bridge & Iron Co., has the stainless steel or other cladding joined to the steel baseplate by a process of heating the assembly of base metal, cladding metal, and brazing material together under vacuum. Since there is no rolling, the clad thickness is uniform, and there is no migration of carbon from the steel plate to the surface of the cladding.

Titanium-clad steel, of the Lukens Steel Co., is produced by sandwichpack rolling without the use of any interlayer foil between the plate and the cladding. An atmosphere or argon gas is used during the heating, and there is no incorporation of the impurities that normally make the titanium brittle. **Nickel-clad flange steel** is also produced by this company. **Niclad,** of the Flannery Mfg. Co., has the nickel deposited on the steel by a continuous welding process. The duplex metal called **Bronze-on-steel,** of the Johnson Bronze Co., used for bearings, is made by sintering a homogeneous alloy powder of 80% copper, 10 tin, and 10 lead, to strip steel in a hydrogen atmosphere, and then rolling the strip and forming it into bearings and bushings. **Nifer,** of the Metals & Controls Corp., is **nickel-clad steel** with the nickel bonded to both sides of a carbon steel, while **Alnifer** has nickel on one side and aluminum on the other. It comes in thin gages, up to 0.010 in (0.0254 cm) for electronic uses. **Stainless-clad aluminum,** of the same company, comes in strip for automotive trim.

Stainless-clad copper is copper sheet with stainless steel on both sides, used for making cooking utensils and food-processing equipment. With stainless steel alone, heat remains localized and causes sticking and burning of foodstuffs. Copper has high heat conductivity, is corroded by some foods, and has an injurious catalytic action on milk products. Thus, the stainless-clad copper gives the conductivity of copper with the protection of stainless steel. The internal layer of copper also makes the metal easier to draw and form. **Rosslyn metal,** of the American Cladmetals Co., is this material. **Ferrolum,** of Knapp Mills, Inc., is sheet steel clad with lead to give protection against sulfuric acid in tanks and chemical equipment. **Copper-clad steel** usually has a cladding of copper equal to 10% of the total thickness of the sheet on each side of a soft steel. But **Conflex,** of the Metals & Controls Corp., has the copper laminated to a hardenable carbon steel so that spring characteristics can be given by heat treatment of the finished parts. The electric conductivity is 30% that of solid copper.

Brass-clad steel, used for making bullet jackets and shell cases, consists of 90–10 brass on one side of a low-carbon-steel sheet, with the brass equal to 20% of the weight of the sheet. **Bronco metal,** of the Metals & Controls Corp., is copper-strip coated on both sides with 25% by weight of phosphor bronze. The bronze gives good resiliency for springs, and the material has an electric conductivity 55% that of solid copper.

Coppered steel wire is produced by wet-drawing steel wire which has been immersed in a copper sulfate or copper-tin sulfate solution. The tin gives a brass finish or a white finish, depending on the proportion of tin. **Fernicklon,** of the Kenmore Metals Corp., is nickel-coated wire for instrument use, made by nickel-plating steel or copper rod and then drawing into wire. **Copper-clad steel wire,** marketed by the Copperweld Steel Co., for line wires, screens, and staples, has an electric conductivity 40% that of an equal section of pure copper, and a tensile strength 250% higher than that of copper. **Copperply wire,** of the National-Standard Co., has either 5 or 10% by weight of copper electroplated on hard-drawn or annealed steel wire in 5 to 36 B&S gage. The conductivity of the 10% coated wire is 20% that of copper wire, or 23% when low-carbon soft wire is used. It is employed for electric installations where high strength is needed. **Nickel-clad copper wire** is used where an electric conductor is required to resist oxidation at high temperatures. It is made by inserting a copper rod into a nickel tube and drawing. **Kulgrid,** of Sylvania Electric Products, Inc., is a nickel-clad copper wire for lead-in wires. The cladding is 28% of the total weight, and the electric conductivity is 70% that of solid copper. The tensile strength of the hard-drawn wire is 85,000 lb/in^2 (586 MPa) and it resists oxidation at high tempertures.

Feran, a German duplex metal, was made by passing strips of aluminum and iron through rolls at a temperature of 350°C and then cold-rolling to sheet. **Alclad,** of the Aluminum Co. of America, is an aluminum-clad aluminum alloy, with the exposed pure aluminum giving added corrosion resistance, and the aluminum-copper base metal giving strength. The German **Lautal** with pure aluminum rolled on is called **Allautal. Zinnal** is a German aluminum sheet with tin cladding on both sides, while **Cupal** is a copper-clad aluminum sheet. **Copper-clad aluminum** is regularly available in sheet, strip and tubing. The copper is rolled on as a coating equal to 5% of the total thickness on each side, or 10% on one side, with a minimum thickness of copper of 0.001 in (0.003 cm). It gives a metal with good working characteristics, and high electric and heat conductivity. **Alcuplate,** of the Metals & Controls Corp., is aluminum with copper bonded to both sides, used for stamped and formed parts where good electric conductivity and easy soldering in combination with light weight are desired. **Alsiplate,** of this company, has silver bonded to both sides of aluminum sheet. **Alfer,** of the Metals & Control Corp., is **aluminum-clad steel.** The aluminum cladding is 10% of the total thickness on each side. It comes in strips of thin gages. **Aliron,** of this company, is a five-ply metal in very thin gages for radio-tube anode plates. It has a core of copper amounting to 40% of the thickness, with a layer of iron and a layer of aluminum on both sides. The copper gives good heat dissipation, and the iron-aluminum compound formed when the metal is heated makes it highly emissive. **Aluminum-clad wire** for electric coils is copper wire coated with aluminum to prevent deterioration of the enamel insulation caused by copper oxide. The wire of the Westinghouse Electric Corp. is made by silver-plating the wire bar to improve adhesion, inserting the bar in a thin-walled aluminum tube, and drawing to size. The aluminum skin is about 0.0025 in (0.006 cm) thick.

CLAY. Naturally occurring sediments that are produced by chemical actions resulting during the weathering of rocks. Often clay is the general term used to identify all earths that form a paste with water and harden when heated. The **primary clays** are those located in their place of formation. **Secondary clays** are those that have been moved after formation to other locations by natural forces, such as water, wind, and ice. The U.S. Department of Agriculture distinguishes clay as having small grains, less than 0.002 mm in diameter, as distinct from **silt** with grains from 0.002 to

0.05 mm. Most clays are composed chiefly of silica and alumina. Clays are used for making pottery, tiles, brick, and pipes, but more particularly the better grade of clays are used for pottery and molded articles not including the fireclays and fine porcelain clays. **Kaolins** are the purest forms of clay. The clayey mineral in all clays in kaolinite, or minerals closely allied, as **anauxite,** $Al_2O_3 \cdot 3SiO_2 \cdot 2H_2O$, and **montmorillonite,** $Al_2O_3 \cdot 4SiO_2 \cdot 2H_2O$, the latter having an expanding lattice molecular structure which increases the bond strength of ceramic clays. When the aluminum silicates are in colloidal form, the material is theoretically true clay, or **clayite.** Some clays, however, derive much of their plasticity from colloids of organic material, and since all clays are of secondary origin from the weathering or decomposition of rocks, they may vary greatly in composition. Hardness of the clay depends on the texture as well as on the cohesion of the particles. Plasticity involves the ability of the clay to be molded when wet, to retain its shape when dry, and to have the strength to withstand handling in the green or unfired condition. The degrees of plasticity are called fat, rich, rubbery, and waxy; or the clays may be termed very plastic, which is waxy; sticky plastic, medium plastic, and lean, which is nonplastic. Clays that require a large amount of water for plasticity tend to warp when dried. Those that are not easily worked may be made plastic by **ceramic binders** such as alkaline starch solutions, ammonium alginate, or lignin. For making pressed or cast whiteware, methyl cellulose is used as binder for the clay. It gives good binding strength, and it fires out of the ceramic with an ash residue of only 0.5%.

Clays with as much as 1% iron burn red, and titanium increases this color. Yellow ochers contain iron as a free hydrate. Most clays contain quartz sand and sometimes powdered mica. **Calcareous clays** are known as **marls.** Pyrites burn to holes in the brick bordered by a ring of magnetic iron oxide, and a clay should be free of this mineral. Limestone grains in the clay burn to free lime which later slakes and splits the ceramic. Most of the common brick clays are complex mixed earths likely to have much undesirable matter that makes them unsuitable for good tile, pipe, or pottery. **Kingsley clay** of Georgia, used for artware, wall tile, dishes, and refractories, has only 0.4% iron oxide, 0.15 Na_2O, 0.1 K_2O, and 0.05 CaO. It contains about 45% silica, 40 alumina, and 1.15 titanium oxide. The **seito ware** of Japan is made with the **Gaerome clay** found near Nagoya. It is a granite with quartz particles, and when used with a high percentage of zirconium oxide produces ceramics of close density and brilliant whiteness. Alumina clay of western Idaho contains on a dry basis 28.7% alumina, 5.6 iron oxide, and a high percentage of titania.

CLOVES. The dried flower buds of the evergreen tree *Caryophillus aromaticus,* grown chiefly in Zanzibar, but also in Malagasy, East Africa, and

Indonesia. The buds yield 15 to 19% of a pungent yellowish essential oil, **clove oil,** also called **caryophil oil** and **amboyna.** It contain 85% **eugenol,** and also the terpene **clovene,** $C_{15}H_{24}$. Clove oil is used in medicine as an antiseptic, in toothpastes, in flavoring, and for the production of artificial vanilla. Eugenol is a viscous phenol-type liquid. It is also the basis for carnation-type perfumes. **Clove buds** are chiefly valued as a highly aromatic spice. Lower-grade **Zanzibar cloves** containing only about 5% oil are used in the strootjes cigarettes of Indonesia, in a mixture of 75% tobacco and 25 cloves. Ground clove is also an efficient antitoxidant, and is sometimes used in lard and pork products. The clove tree attains a height up to 40 ft (12.2 m), bearing in 7 or 8 years, and continuing to bear for a century, yielding 8 to 10 lb (3.6 to 4.5 kg) of dried cloves annually. Clove stems are also aromatic, but contain only 5 to 6% oil of interior value. Clove was one of the most valued spices of medieval times. It grew originally only on five small islands, the Moluccas, in a volcanic-ash soil, and was carried by Chinese junks and Malayan outriggers to India from whence the Arabs controlled the trade, bringing the tree also to Zanzibar. The *Victoria* of Magellan's fleet returned to San Lucar with 26 tons (24 metric tons) of cloves, enough to pay for the loss of the other four ships and the expenses of the voyage around the world.

COAL. A general name for a black mineral formed of ancient vegetable matter, and employed as a fuel and for destructive distillation to obtain gas, coke, oils, and coal-tar chemicals. Coal is composed largely of carbon with smaller amounts of hydrogen, nitrogen, oxygen, and sulfur. It was formed in various geological ages and under varying conditions, and occurs in several distinct forms. **Peat** is the first stage, followed by lignite, bituminous coal, and anthracite, with various intermediate grades. The mineral is widely distributed in many parts of the world. The value of coal for combustion purposes is judged by its fixed carbon content, volatile matter, and lack of ash. It is also graded by the size and percentage of lumps. The percentage of volatile matter declines from peat to anthracite, and the fixed carbon increases. A good grade of coal for industrial power-plant use should contain 55 to 60% fixed carbon and not exceed 8% ash. The Btu value should be 13,500 to 14,000/lb (74,250 to 77,000 kcal/kg). Finely ground coal, or **powdered coal,** is used for burning in an air blast like oil, or may be mixed with oil. Coal in its natural state absorbs large amounts of water and also, because of impurities and irregular sizes, is not so efficient as a fuel as the **reconstructed coal** made by crushing and briquetting lignite or coal and waterproofing with a coating of pitch. **Anthracite powder** is used as a filler in plastics. **Carb-O-Fil,** of the Shamokin Filler Co., is powdered anthracite in a range of particle sizes used as a carbonaceous filler. It has a plasticizing effect. It can also be used to replace carbon black in phenolic resins.

Increasing amounts of coal are being used for the production of gas and chemicals. By the hydrogenation of coal much greater quantities of phenols, cresols, aniline, and nitrogen-bearing amines can be obtained than by means of by-product coking, and low grades of coal can be used. The finely crushed coal is slurred to a paste with oil, mixed with a catalyst, and reacted at high temperature and pressure. **Synthesis gas,** used for producing gasoline and chemicals, is essentially a mixture of carbon monoxide and hydrogen. It is made from low-grade coals. The pulverized coal is fed into a high-temperature reactor with steam and a deficiency of oxygen, and the gas produced contains 40% hydrogen, 40 carbon monoxide, 15 carbon dioxide, 1 methane, and 4 inert materials. It is also made by passing steam through a bed of incandescent coke to form a **water gas** of about equal proportions of carbon monoxide and hydrogen. It is also made from natural gas.

COATED FABRICS. The first coated fabric was a rubberized fabric produced in Scotland by Charles Macintosh in 1823 and known as **Mackintosh cloth** for rainwear use. The cloth was made by coating two layers of fabric with rubber dissolved in naphtha and pressing them together, making a double fabric impervious to water. **Rubberized fabrics** are made by coating fabrics, usually cotton, with compounded rubber and passing between rollers under pressure. The vulcanized coating may be no more than 0.003 in (0.008 cm) thick, and the resultant fabric is flexible and waterproof. But most coated fabrics are now made with synthetic rubbers or plastics, and the base fabric may be of synthetic fibers, or a thin plastic film may be laminated to the fabric.

Coated fabrics now have many uses in industrial applications, and the number of variations with different resins and backing materials is infinite. They are usually sold under trade names, and are used for upholstery, linings, rainwear, bag covers, book covers, tarpaulins, outerwear, wall coverings, window shades, gaskets, and diaphragms. Vinyl-type resins are most commonly used, but for special purposes other resins are selected to give resitance to wear, oils, or chemicals. The coated fabric of Reeves Bros., Inc., called **Reevecote,** for gaskets and diaphragms, is a Dacron fabric coated with Kel-F fluorocarbon resin. An industrial sheeting of the Auburn Mfg. Co. is a cotton fabric coated with urethane rubber. It is tough, flexible, and fatigue-resistant, and gives 10 times better wear resistance than natural rubber.

Vinyl-coated fabrics are usually tough and elastic, and are low in cost, but unless specially compounded are not durable. Many plastics in the form of latex or emulsion are marketed especially for coating textiles. **Rhoplex WN-75** and **WN-80** are water dispersions of acrylic resins for this purpose. **X-Link 2833,** of the National Starch & Chemical Corp., is a vinyl-acrylic copolymer. Coatings cure at room temperature, have high heat and light stability, give softness and flexibiilty to the fabric, and withstand repeated dry cleaning. **Polectron,** of the General Aniline & Film Corp., is a water emulsion of a copolymer of vinyl pyrrolidone with ethyl acrylate. It forms an adherent, tough, and chemical-resistant coating. **Geon latex,** of the B. F. Goodrich Co., is a water dispersion of polyvinyl chloride reson. Polyvinyl chloride of high molecular weight is resistant to staining, abrasion, and tearing, and is used for upholstery fabrics. The base cloth may be of various weights from light sheetings to heavy ducks. They may be embossed with designs to imitate leather. The **Boltaflex cape vinyl,** of the Bolta Co., is a rayon fabric coated with a vinyl resin embossed with a leatherlike grain. It has the appearance, feel, and thickness of a split leather and, when desired, is impregnated with a leather odor.

One of the first of the upholstery fabrics to replace leather was **Fabrikoid,** of Du Pont. It was coated with a cellulose plastic, and was in various weights, colors, and designs, especially for automobile seating and book covers. **Armalon** is twill or sateen fabric coated with ethylene plastic for upholstery, and **Pontan** is a rubberized fabric made to imitate colored leathers. For some uses, as for draperies or industrial fabrics, the fabric is not actually coated, but is impregnated, either in the fiber or in the finished cloth, to make the fabric water-repellent, immune to insect attack, and easily cleaned. **Tontine** is a plastic-impregnated fabric for window shades. The **Fairprene fabrics,** also of Du Pont, are cotton fabrics coated with chloroprene rubber or other plastics. **Corfam,** of the same company, used as a leather substitute, is a nonwoven sheet of urethane fibers reinforced with polyester fibers, with a porous texture. The fabric can be impregnated or coated.

Pliosheen, of the Goodyear Tire & Rubber Co., is rayon fabric coated with a Chemigum type of synthetic rubber. **Terson voile,** of the Athol Mfg. Co., for umbrellas, rainwear, and industrial linings, is a sheer-weight rayon coated with a vinyl resin. It weighs 2 oz/yd^2 (0.07 kg/m). **Vynside,** of the Columbia Mills, Inc., for book covers, is a heavy buckram cotton fabric coated with a vinyl resin in colors. **Lantuck fabric,** of the Wellington Sears Co., is a nonwoven fabric with a vinyl coating which permits deeply embossed patterns for upholstery. Coated fabrics may also be napped on the back, or coated on the back with a flock, to give a more resilient backing for upholstery. **Kalistron,** of the U.S. Plywood Corp., has such a flock backing.

Impregnated fabrics may have only a thin, almost undetectable surface coating on the fibers to make them water-repellent and immune to bacterial attack, or they may be treated with fungicides or with flame-resistant chemicals or waterproofing resins. **Stabilized fabrics,** however, are not waterproofed or coated, but are fabrics of cotton, linen, or wool that have been treated with a water solution of a urea-formaldehyde or other thermosetting resin to give them greater resiliency with resistance to creasing and resistance to shrinking in washing. **Shrinkproof fabrics** are likewise not coated fabrics, but have a light impregnation of resin that usually remains only in the core of the fibers. The fabric retains its softness, texture, and appearance, but the fibers have increased stability. Various resin materials are marketed under trade names for creasproofing and shrinkproofing fabrics, such as **Lanaset,** a methylomelamine resin of the American Cyanamid Co., and **Synthrez,** a methylourea resin of Synthron, Inc.

Under the general name of **protective fabrics,** coated fabrics are now marketed by use characteristics rather than by coating designation since resin formulations vary greatly in quality. For example, the low-cost grades of vinyl resins may be hard and brittle at low temperatures and soft and rubbery in hot weather, and thus unsuitable for all-weather tarpaulins. Special weaves of fabric are used to give high tear strength with light weight, and the plastic may be impregnated, coated on one side or both, bonded with an adhesive or electronically bonded, or some combination of all of these. Flame resistance and static-free qualities may also be needed. Many companies have complete lines to meet definite needs. The **Coverlight** fabrics of Reeves Bros., Inc., which come in many thicknesses and colors, are made with coatings of neoprene, Hypalon, or vinyl chloride resin, with weights from 6 to 22 oz/yd^2 (0.18 to 0.67 kg/m^2) and widths up to 72 in (1.8 m). The **H.T.V. Coverlight** is a high-tear nylon fabric with specially formulated vinyl coating. The 22-oz (0.62-kg) grade for such heavy-duty, all-weather uses as truck-trailer covers and concrete-curing covers remains flexible at temperatures down to −50°F (−46°C).

COBALT. A white metal, Co, resembling nickel but with a bluish tinge instead of the yellow of nickel. Although allied to nickel, it has distinctive differences. It is more active chemically than nickel. It is dissolved by dilute sulfuric, nitric, or hydrochloric acids, and is attacked slowly by alkalies. The oxidation rate of pure cobalt is 25 times that of nickel. Its power of whitening copper alloys is inferior to that of nickel, but small amounts in nickel-copper alloys will neutralize the yellowish tinge of the nickel and make them whiter. The metal is diamagnetic like nickel, but has nearly three times the maximum permeability. Like tungsten, it has the property of adding red-hardness to cutting alloys. It also hardens alloys to a greater extent than nickel, especically in the presence of carbon, and can form more chemical compounds in alloys than nickel.

Cobalt has a specific gravity of 8.756, melting point at 1493°C, hardness 85 Brinell, and electric conductivity about 16% that of copper. The tensile strength of pure cast cobalt is 34,500 lb/in^2 (238 MPa), but with 0.25% carbon it is increased to 62,000 lb/in^2 (427 MPa). The metal is employed in cutting alloys and tool steels, in magnet alloys, in high-permeability alloys, and as a catalyst; and its compounds are used as pigments and for producing many chemicals. The metal has two forms: a close-packed hexagonal crystal form, which is stable below 417°C, and a cubic form stable at higher temperatures to the melting point. Cobalt has valences of 2 and 3, while nickel has only a valence of 2.

The natural cobalt is **cobalt 59,** which is stable and nonradioactive, but the other isotopes from 54 to 64 are all radioactive, emitting beta and gamma rays. Most have very short life except **cobalt 57** which has a half-life of 270 days, **cobalt 56** with a half-life of 80 days, and **cobalt 58** with a half-life of 72 days. **Cobalt 60,** whith a half-life of 5.3 years, is used for radiographic inspection. It is also used for irradiating plastics, and as a catalyst for the sulfonation of paraffin oils since the gamma rays cause the reaction of sulfur dioxide and liquid paraffin. Cobalt 60 emits gamma rays of 1.1- to 1.3-MeV energy which gives high penetration for irradiation. The decay loss in a year is about 12%, the cobalt changing to nickel.

Cobalt metal is marketed in rondels, or small cast slugs, in shot and anodes, and as a powder. Powders with low nickel content for making cobalt salts and catalysts are produced by Sherritt-Gordon Mines, Ltd., in particle size down to 1 μm. About one-quarter of the supply of cobalt is used in the form of oxides and salts for driers, ceramic frits, and pigments. **Cobalt carbonyls** are used for producing **cobalt powder** for use in powder metallurgy, as catalysts, and for producing cobalt chemicals. **Dicoblat octacarbonyl,** $Co_2(CO)_8$, or **cobalt tetracarbonyl,** is a brownish powder melting at 51°C, and decomposing at 60°C to **tetracobalt dodecacarbonyl,** $(CoCO_3)_4$, a black powder which oxidizes in the air.

The best-known **cobalt alloys** are the **cobalt-base superalloys.** The desirable high-temperature properties of low creep, high stress rupture, and high thermal-shock resistance are attributed to cobalt's allotropic change to a face-centered cubic structure at high temperatures. Most of these superalloys contain around 20% chromium for good oxidation resistance. An important group of these alloys is known as **Stellites.** They contain chromium and various other elements such as tungsten, molybdenum, and silicon. These extremely hard alloy carbides in a fairly hard matrix give excellent abrasion and wear resistance and are used as hard-facing alloys and for jet engine parts. **Eatonite,** of the Eaton Manufacturing Co., used for valve facings to withstand corrosive action of hot gases and antiknock fuels is a nickel-cobalt-tungsten alloy.

The interesting properties of cobalt-containing permanent, soft, and constant-permeability magnets are a result of the electronic configuration of cobalt and its high curie temperature. In addition, cobalt in well-known Alnico-magnet alloys decreases grain size and increases coercive force and residual magnetism.

Cobalt is a significant element in many **glass-metal seal alloys** and **low-**

expansion alloys. One alloy with 54% cobalt and 36 nickel, known as **Invar,** has a thermal coefficient of expansion of nearly zero over a small temperature range. Another alloy with 57 to 63% cobalt, named **Co-Elinvar,** is characterized by an invariable modulus of elasticity over a wide temperature range and low expansion. **Cobalt-chromium alloys** are used in dental and surgical applications because they are not attacked by body fluids. Alloys named **Vitallium** are used as bone replacements and are ductile enough to permit anchoring of dentures on neighboring teeth. They contain about 65% cobalt.

Cobalt is a necessary material in human and animal metabolism, and is used in fertilizers in the form of **cobaltous carbonate,** $CoCO_3$, in which form it is easily assimilated. This form occurs in nature in the mineral **cobalt spar** and mixed with magnesium and iron carbonates. **Cobaltous citrate,** $Co(C_6H_5O_7) \cdot 2H_2O$, is a rose-red powder soluble in water, used in making pharmaceuticals. **Cobaltous fluorosilicate,** $CoSiF_6 \cdot H_2O$, is an orange-red water-soluble powder used in toothpastes. It furnishes fluorine and silica as well as cobalt. **Cobaltous hydroxide,** $Co(OH)_2$, has a high cobalt content, 61.25%, is stable in storage, and is used for paint and ink driers, and for making many other compounds. **Cobaltous chloride,** $CoCl_2$, a black powder, is an important cobalt chemical. It is also used as a **humidity indicator** for silica gel and other desiccants. As the **desiccant** becomes spent, the blue of the cobaltous chloride changes to the pink color of the hexahydrate, but when the material is regenerated by heating to drive off the moisture, the blue reappears.

Cobalt metal may be obtained from the sulfur and arsenic ores by melting and then precipitating the cobaltous hydroxide powder which is high in cobalt, has high stability in storage, and is readily converted to the metal or the oxide or used directly for driers and other applications. The chief **cobalt ores** are cobalite and smaltite. **Cobalite,** or **cobalt glance,** from Ontario and Idaho, is a sulfarsenide, CoAsS, and occurs with **gersdorffite,** NiAsS. Another sulfide is **linnaeite,** Co_3S_4, containing theoretically 58% cobalt, but usually containing also nickel and iron. Cobalt is also found with pyrites as the mineral **bieberite,** which is **cobaltous sulfate,** $CoSO_4 \cdot 7H_2O$, but combined with iron sulfate. Some cobalt is extracted from the iron pyrites of Pennsylvania, the concentrated pyrite containing 1.41% cobalt, 42 iron, and 0.28 copper. **Erythrite** is a hydrous cobalt arsenate occurring in the smaltite deposits of Morocco. **Skutterudite** also occurs in Morocco. It is a silvery-gray brittle mineral of the composition $(CoNiFe)As_3$, with a specific gravity of 6.5 and hardness of 6.

Asbolite, an important ore in Shaba and in New Caledonia, is a soft mineral, hardness 2 Mohs, consisting of varying mixtures of cobaltiferous manganese and iron oxides. A number of minerals classed as **heterogenite,** black in color and containing only cobalt and copper, occur in copper deposits, especially in Shaba. Among these are **mindigite,** $2Co_2O_3 \cdot CuO \cdot 3H_2O$, and **trieuite,** $2Co_2O \cdot CuO \cdot 6H_2O$. **Carrollite,** $CuS \cdot Co_2S_3$, a steel-gray mineral with a specific gravity of 4.85 and hardness of 5.5, is an important ore in Zimbabwe. The copper ores of the Zaire and Zimbabwe form one of the chief sources of commerical cobalt. Some of the metal is exported as **white alloy,** containing 40% cobalt, 9 copper, and the balance iron. Cobalt occurs naturally in many minerals, and the metal may be considered as a by-product of other mining. Small quantities are produced regularly as a by-product of zinc production in Australia, although the cobalt content of the concentrate is only 0.015%. Some cobalt is obtained from the lead and zinc ores of Missouri. Its relative scarcity is a matter of cost of extraction.

COBALT OXIDE. A steel-gray to blue-black powder employed as a base pigment for ceramic glazes on metal, as a colorant for glass, and as a chemical catalyst. It gives excellent adhesion to metals and is valued as an undercoat for vitreous enamels. It is the most stable blue, as it is not changed by ordinary oxidizing or reducing conditions. It is also one of the most powerful colorants for glass, 1 part in 20,000 parts of a batch giving a distinct blue color. Cobalt oxide is produced from the cobalt-nickel and pyrite ores, and the commercial oxide may be a mixture of the three oxides. **Cobaltous oxide,** CoO, is called **gray cobalt oxide** but varies in color from greenish to reddish. It is the easiest to reduce to the metal, and reacts easily with silica and alumina in ceramics. **Cobaltic oxide,** Co_2O_3, occurs in the mixture only as the unstable hydrate, and changes to the stable **black cobalt oxide,** or **cobalto-cobaltic oxide,** Co_3O_4, on heating. Above about 900°C this oxide loses oxygen to form cobaltous oxide.

Cobalt dioxide, CoO_2, does not occur alone, but the dioxide is stable in combination with other metals. The blue-black powder called **lithium cobaltite,** $LiCoO_2$, is used in ceramic frits to conserve cobalt, since the lithium adds fluxing and adherent properties. The pigment known as **smalt,** and as **royal blue** and **Saxon blue,** is a deep-blue powder made by fusing cobalt oxide with silica and potassium carbonate. It contains 65 to 71% silica, 16 to 21 potash, 6 to 7 cobalt oxide, and a little alumina. It is used for coloring glass and for vitreous enameled signs, but does not give good covering power as a paint pigment. **Thenaud's blue** is made by heating together cobalt oxide and aluminum oxide. **Rinmann's green** is made by heating together cobalt oxide and zinc oxide.

COBALT STEELS. Cobalt is a much rarer and more expensive metal than nickel, but it has powerful influencing properties in steel and iron alloys, and is used in tool steels, cutting alloys, and magnet steels. Small amounts decrease the impact resistance. The cobalt steels retain hard carbides at high temperatures, and the high-speed steels containing cobalt are harder than the regular tungsten steels and can be operated at higher temperatures, but since they are not plastic at ordinary forging temperatures they require higher heats for forging and more care in heat treatment. A typical super-high-speed steel is **Braecut steel,** of the Braeburn Alloy Steel Corp. It contains 12% cobalt, 6.25 molybdenum, 5.25 tungsten, 4.25 chromium, 2.25 vanadium, and 1.15 carbon. It develops a hardness of Rockwell C70. Another alloy called a "balanced" high-speed steel, for drills, cutters, and lathe tools, is **Rex 49 steel,** of the Crucible Steel Co. of America. It has 5% cobalt, 6.75 tungsten, 4.25 chromium, 3.75 molybdenum, 2 vanadium, 0.45 manganese, 0.3 silicon, and 1.1 carbon. It hardens by oil quenching to Rockwell C68. The qualities of hardness and heat resistance are developed in the cobalt steels best when considerable amounts of chromium are present.

Cobalt increases residual magnetism in steel and increases the coercive magnetic force, and more than 30% of the cobalt supply is used in magnetic alloys. But the magnetic and permeability alloys may have little or no iron and are high alloys rather than steels, as is also the case with the cobalt alloys used for heat-resistant spring wire, although many such high alloys used for high-temperature applications are called steels. **Ferrovac WD65,** of the Crucible Steel Co. of America, for such uses as aircraft bearings to operate at temperatures to 900°F (482°C), contains 5% cobalt, 15 chromium, 4 molybdenum, 2.5 vanadium, 2.25 tungsten, and 1.12 carbon. It is air-hardening, and will harden to Rockwell C65. **Alloy MA-18NiCoMo,** of the International Nickel Co., Inc., contains 7% cobalt, 18 nickel, 5 molybdenum, 0.5 titanium, and not over 0.05 carbon. When heat-treated it has a yield strength of 300,000 lb/in^2 (2,068 MPa) and retains high strength and resistance to stress corrosion at elevated temperatures. Many tool steels and some low-alloy steels also contain cobalt. **Unimach UCX2,** of the Cyclops Corporation, is a low-alloy steel which can be heat-treated to a tensile strength of 270,000 lb/in^2 (1,861 MPa), with elongation of 5% and Rockwell hardness of C50, and which is not notch-sensitive. It contains 1% cobalt, 1.1 chromium, 1 silicon, 0.7 manganese, 0.25 molybdenum, 0.15 vanadium, and 0.39 carbon. When annealed it is easily deep-drawn.

COCAINE. An alkaloid derived from the leaves of the **coca** shrub. It is used as a local anesthetic and as a narcotic. It is habit forming. In small and moderate doses it is stimulating and increases physical energy. Depression usually follows. Continued heavy use of cocaine has debilitating effects on the nervous system and can lead to insanity. Cocaine crystallizes from alcohol and is readily soluble in ordinary solvents except water. In the manufacture of cocaine, the alkaloids of coca leaves are hydrolyzed to **ecgonine.**

COCHINEAL. A dyestuff of animal origin, which before the advent of coal-tar dyes was one of the most important coloring materials. Cochineal is the female of the *Coccus cacti,* an insect that feeds on various species of cactus of Mexico. The insects have no wings, and at the egg-laying season are brushed off the plants, killed by boiling, and dried. They are dark reddish brown in color. Cochineal contains 10 to 20% pure coloring matter, carminic acid, mostly in the eggs, from which the **carmine red,** $C_{11}H_{12}O_7$, is obtained by boiling with mineral acid. Carmine red produces brilliant lake colors of various hues with different metals. Commercial cochineal may be adulterated with starch, kaolin, red lead, or chrome lead. The brilliant red pigment known as **carmine lake** is made by precipitating a mixture of cochineal and alum. Salmonella-free cochineal in water solution is now used in foods to give a reddish-purple color. A species of cochineal insect that feeds on the leaves of the **tamarisk** tree, *Tamarix manifera,* produces **manna,** a viscous, white, sweet substance composed mostly of sugars. It forms in small balls and falls usually in May to July. When dry, it is hard and stable, and is a good food. It is native to the Near East.

COCOA BEANS. The seed beans from the large fruit pods of the **cacao tree,** *Theobroma cacao,* native to Mexico, and *T. leiocarpum,* native to Brazil. The tree was cultivated in Mexico from ancient times, and the beans were used by the Aztecs to produce a beverage called **choclatl** which contained the whole substance of the fermented and roasted bean flavored with vanilla. Cocoa beans are now produced in many countries, and the United States imports them from about 40 countries. Ghana, Nigeria, and Brazil are noted producers. The flavor and aroma vary with soil and climate, and differences in curing methods also produce differences in the beans, so that types and grades are best known by the shipping ports and districts in which they grow. **Mico coca** is wild cocoa of Central America. The beans are smaller and are noted for fine flavor. Cocoa beans are shipped dried but not roasted. They are roasted just before use to develop the flavor, to increase the fat content, and to decrease the tannin content. The hard shells are removed, and the roasted seeds are ground and pressed to produce **bitter chocolate,** generally known as **chocolate liquor. Sweet chocolate** is made by adding sugar and flavoring, usually vanilla. **Cocoa,** for beverage purposes, is made by removing about 60% of the fatty oil from chocolate by hydraulic pressing and powdering the residue, to which is usually added ground cocoa shells. The removed fatty oil is **cocoa butter,** used for bakery products, cosmetics, and pharmaceuticals. A hundred pounds of cocoa beans yields 48 lb (21.8 kg) of **chocolate powder,** 32 lb (14.5 kg) of cocoa butter, and 20 lb (9.1 kg) of waste. Also an artificial cocoa butter is made by fractionating palm kernel oil. **Pakena,** a substitute cocoa butter of the Dura Commodities Corp., contains 53% lauric acid, 21.5 myristic, 12 palmitic, 8 oleic, 3.5 stearic, and 2 capric acids. Besides fat, chocolate contains much starch and protein and has high food value, but is not as stimulating as the cocoa since the alkaloid is largely contained in the waste and shells. These contain 1 to 1.5% theobromine and are used for the synthetic production of caffeine. The chocolate is used in the manufacture of confectionery, chocolate bars, bakery products, and flavoring syrups. **Microfine cocoa,** of the Cook Chocolate Co., used for bakery products, is ground to 325 mesh, and contains from 9 to 16% cocoa butter. **Postonal** is a German substitute for cocoa butter for pharmaceuticals. It is a polymerized ethylene oxide containing chemically combined castor oil.

Cocoa powder, used in the United States for beverages and for adding chocolate flavor to foodstuffs, as distinct from the sweet chocolate used in Latin countries for beverages, was originally made from the shells, but is now made from the residue cake after extraction of the chocolate liquor and the pressing out of the cocoa butter. It is widely used as a flavor for cakes and confectioneries. Sugar makes the powder easily soluble in water; **instant cocoa** is cocoa powder processed with about 70% sugar and sometimes also with nonfat milk powder. The fat content of commercial cocoa powders ranges from 6 to 22% with a color range from light brown to reddish black. **Breakfast cocoa** is the high-fat grade. Cocoa powder is usually acidic with the pH as low as 3.3, but **Dutch cocoa,** for nonacid foods, is **stabilized cocoa** with the pH raised to as high as 9.0 by treatment with solutions of sodium or potassium carbonate.

COCOBOLA. The wood of the hardwood tree *Dalbergia retusa,* of Central America, also known as **Honduras rosewood.** It is a beautiful wood, extremely hard, and very heavy with a weight of 75 to 85 lb/ft^3(1,202 to 1,362 kg/m^3). It has orange and red bands with dark streaks and takes a fine polish. The thick sapwood is hewn off before shipment, and the heartwood logs are usually not more than 18 in (45.7 cm) in diameter. The wood is used for canes, turnery, inlaying, scientific-instrument cases, and knife handles. **Cocos wood,** also called **cocoawood** and **West Indian ebony,** used chiefly for inlaying, is from the tree *Brya ebenus* of tropical America. The sapwood is light yellow, and the heartwood is brown streaked with yellow. The grain is dense and even, and the wood is hard and tough.

COCONUT OIL. The oil obtained from the thick kernel or meat adhering to the inside of the shell of the large nuts of the palm tree *Cocos nucifera* growing along the coasts of tropical countries. The tree requires salt air, and inland trees do not bear fruit unless supplied with salt. The name coco is the Carib word for palm. **Copra** is the dried meat of the coconut from which the oil is pressed. The dried copra contains 60 to 65% oil. It is an excellent food oil, and is valued as a shortening for crackers, but its use for margarine has declined. It is also valued for soaps because of its high lathering qualities due to the large percentage of lauric and myristic acids, though these acids are irritating to some skins. It is also employed as a source of lauric acid, but lauryl alcohol is now made synthetically. Coconut oil was once the chief illuminating oil in India, and the oil for burning was exported under the name of **Cochin oil.** This oil was cold-pressed and filtered and was water-clear. Coconut oil has a melting point of 27 to 32°C, specific gravity 0.926, saponification value 251 to 263, and iodine value 8 to 9.6. It contains 45 to 48% lauric acid, 17 to 20 myristic, 10 capric, 5 to 7 palmitic, up to 5 stearic, and some oleic, caprylic, and caproic acids.

In sun-drying coconut meat to make copra there is a loss of some of the sugars and other carbohydrates, and some proteins. The oil from copra contains more free fatty acid than that from fresh dried coconut and is rancid, requiring neutralization, decolorization, and deodorization. The meal and cake are also dirty and rancid but are useful for animal feed or fertilizer. **Dehydrated coconut** meat gives a better yield of oil and is not rancid. The **copra cake** of India is called **poonac.** The chief production of copra and coconut oil is in southern Asia, Indonesia, the Philippines, and in the South Sea Islands. About 5,000 coconuts are required to produce a metric ton of copra, and the average yield of crude oil is 63%. The stearine separated from crude coconut oil by the process of wintering is known as **coconut butter** and is used in confectionery. It has a melting point of 27 to 32°C. Hydrogenated coconut oil is a soft solid with a melting point of 45°C. **Desiccated coconut,** produced by oven drying or dehydration of the fresh coconut meat, is used shredded as a food and also powdered in many bakery products as a food and stabilizer. It has high food value, containing not less than 60% oil, 15 carbohydrates, 14 cellulose, 6 to 7 protein, and various mineral salts and considerable vitamin B. It is easily digested and has antitubercular value, but its characteristic coconut flavor is not universally liked and its use is largely confined to confections.

COFFEE. The seed berries, or beans, of the **Arabian coffee** tree, *Coffea arabica,* the **Liberian coffee,** *C. liberica,* and the **Congo coffee,** *C. robusta,* of which the first species furnishes most of the commercial product. The coffee bean contains the alkaloid caffeine used in medicine as a stimulant and in soft drinks, but most of the commercial coffee beans are used for the preparation of the beverage coffee, with small quantities for flavoring.

The alkaloid is stimulating and is harmless in small amounts as it does not break down in the system and is easily soluble in water and thus carried off rapidly, but in large quantities at one time it is highly toxic. Coffee contains niacin, and also rubidium and other metallic salts useful in small quantities in the human system.

The Arabian coffee plant is a small evergreen tree first introduced to Europe through Arabia. The first plants were brought to America in 1723, and the trees are now grown in most tropical countries. It requires a hot, moist climate, but develops best at higher altitudes. There are numerous varieties, and the coffee beans also vary in aroma and taste with differences in climate and cultivation. The Liberian and Congo species, grown on the west coast of Africa, are more hardy plants, but the coffee is different in aroma and is used only for blending. **Mocha coffee** and **Java coffee** are fragrant varieties of Arabian coffee. The fruits are small fleshy berries containing two greenish seeds. They are dried in the sun, or are pulped by machine and cleaned in fermenting baths and dried in ovens or in the sun. After removal of the skin from the dried beans they are graded and shipped as green beans. The general grades are by shipping ports or regions with numbered grades or qualities. Coffee is always roasted for use. This consists in a dry distillation with the formation of new compounds which produce the flavor and aroma. The **caffeic acid** in coffee is a complex form of cinnamic acid which changes readily to a complex coumarin. **Coffee-Captan,** of Cargille Scientific, Inc., is alpha furfuryl mercaptan, one of the essential constituents in the aroma of freshly roasted coffee. It is a water-white liquid used in masking agents, and is also a vulcanizer for rubber. **Coffee flavor,** made synthetically for adding to coffee blends, is furfural mercaptan. The **mercaptans** are thioalcohols, or sulfur alcohols, which have compositions resembling those of the alcohols but react differently to give **mercaptals** with aldehydes and **mercaptols** with ketones and produce various flavors from offensive to pleasant.

Brazilian coffee is the base for many blends, though the average quality is not high. In blends, **Medellin coffee** from Colombia is used for rich flavor, Mexican **Coatepec** for winey flavor, El Salvador for full body, Costa Rican for fragrance, and Arabian mocha for distinctive flavor. Some coffees, such as Guatemalan, which have a full body and rich flavor are used without blending, though trade-named coffees are usually blends because of the lack of quantity of superior types. **Powdered coffees,** commonly known as **instant coffee,** are produced by evaporating coffee brew. To drink, it is only necessary to add hot water. **Chicory,** which is used extensively in Europe for blending with coffee, is the dried, roasted, and ground root of the perennial plant *Cichorium intybus,* native to Europe. From 5 to 40% chicory may be used in some blends of coffee. It gives a taste preferred by some. **Caffeine-free coffee** brands have the alkaloid removed by solvent extraction and the tannic acid neutralized to improve digestibility.

COIR. A fiber by-product of the coconut industry. The fiber is retted from the outer husks, hammered with wooden mallets, and then combed and bleached. The coarse and long fibers are used for brushmaking; the finer and curly fibers are spun into coir yarn used for mats, cordage, and coarse cloths. In the West Indies it is mixed with sisal and jute to make coffee-bag cloth. In the Philippines it has been used with cement to make a hard-setting, lightweight board for siding. In India **coir fiberboard** is made by bonding with shellac, pressing, and baking. The boards are hard and have a good finish. Coir is easily dyed. The Sri Lankan **coir yarn** is sold in two quality grades, **Kogalla** and **Colombo**, with subdivisions according to the thickness and texture. The yarn is properly called coir, and the harsh brush fiber is best known as **coconut fiber.** Coir yarn averages 330 m/kg. The Indian yarn is in 450-yd (411-m) lengths tied into bundles. A hundred nuts yield 17 or 18 lb (7.7 or 8.2 kg) of fiber. **Coconut shell,** a by-product of the copra industry, is used for making activated charcoal and for **coconut shell flour** used as a filler in molded plastics. It has a composition similar to walnut shell, being chiefly cellulose with about 30% lignin, 17 pentosan, and 5 methoxyl.

COKE. The porous, gray, infusible residue left after the volatile matter is driven out of bituminous coal. The coal is heated to a temperature of 1200 to 1400°C, without allowing air to burn it, and the volatile matter expelled. The residue, which is mainly fixed carbon and ash, is a cellular mass of greater strength than the original coal. Its nature and structure make it a valuable fuel for blast furnaces, burning rapidly and supporting a heavy charge of metal without packing. Soft, or bituminous, coals are designated as coking or noncoking, according to their capacity for being converted into coke. Coal low in carbon and high in ash will produce a coke that is friable and not strong enough for furnace use, or the ash may have low-melting constituents that leave glassy slag in the coke. Coke is produced in the beehive and by-product ovens, or is a by-product of gas plants. One ton (907 kg) of coal will yield an average of 0.7 ton (635 kg) of coke, 11,500 ft^3 (325 m^3) gas, 12 gal (45 L) tar, 27 lb (12 kg) ammonium sulfate, 50 gal (189 L) benzol, 0.9 gal (3.4 L) toluol and naphtha, and 0.5 lb (0.2 kg) naphthalene, but the product yield varies with the temperature. When steel production is low and coking ovens are run at lower temperature with a longer cycle, the yield of naphthalene is low.

The fixed carbon of good coke should be at least 86%, and sulfur not more than 1%. The porosity may vary from 40 to 60%, and the apparent specific gravity should not be less than 0.8. **Foundry coke** should have an ignition point about 1000°F (538°C), with sulfur below 0.7%, and the pieces should be strong enough to carry the burden of ore and limestone. Coke suitable for foundry use is also made from low-grade coals by reducing them to a semicoke, or char, and briquetting, but **semicoke** and **smokeless fuel** are generally coals carbonized at low temperatures and briquetted for household use. These fuels are sold under trade names such as **coalite** and **Carbolux,** and they are really by-products of the chemical industry since much greater quantities of liquids and more lighter fractions in the tar are obtained in the process.

Pitch coke, made by distilling coal tar, has a high carbon content, above 99%, with low sulfur and ash, and is used for making carbon electrodes. **Petroleum coke** is the final residue in the distillation of petroleum, and forms about 5% of the weight of the crude oil. With the sand and impurities removed it is about 99% pure carbon, and is used for molded carbon products. **Calcined coke** is petroleum coke that has been calcined at 2400°F (1316°C) to remove volatile matter. It is used for electrodes. **Carbonite** is a natural coke found in England and in Virginia. It is a cokelike mineral formed by the baking action of igneous rocks on seams of bituminous coal.

COLD-MOLDED PLASTICS. This is the oldest group of plastics materials and they were introduced into the United States in 1908. The materials fall into two general categories: inorganic or refractory materials, and organic or nonrefractory materials.

Inorganic cold-molded plastics consist of asbestos fiber filler and either a silica-lime cement or portland cement binder. Clay is sometimes added to improve plasticity. The silica-lime materials are easier to mold although they are lower in strength than the portland cement types.

In general, advantages of these materials include high arc resistance, heat resistance, good dielectric properties, comparatively low cost, rapid molding cycles, high production with single cavity molds (thus low tool cost), and no need for heating of mold. On the other hand, they are relatively heavy, cannot be produced to highly accurate dimensions, are limited in color, and can be produced only with a relatively dull finish. They have been used generally for arc chutes, arc barriers, supports for heating coils, underground fuse shells, and similar applications.

Organic cold-molded plastics consist of asbestos fiber filler materials bound with bituminous (asphalt, pitches, and oils), phenolic, or melamine binders. The binder materials are mixed with solvents to obtain proper viscosities, then thoroughly mixed with the asbestos, ground and screened to form molding compounds. The bituminous-bound compounds are lowest in cost and can be molded more rapidly than the inorganic compounds;

the phenolic and melamine-bound compounds have better mechanical and electrical properties than the bituminous compounds and have better surfaces as well as being lighter in color. Like the inorganic compounds, organic compounds are cold-molded, followed by oven curing.

Compounds with melamine binders are similar to the phenolics, except that melamines have greater arc resistance, lower water absorption, are nontracking, and have higher dielectric strength.

Major disadvantages of these materials, again, are relatively high specific gravity, limited colors, and inability to be molded to accurate dimensions. Also, they can be produced only with a relatively dull finish.

Compounds with bituminous binders are used for switch bases, wiring devices, connector plugs, handles, knobs, and fuse cores. Phenolic and melamine compounds are used for similar applications where better strength and electrical properties are required.

An important benefit of cold-molded plastics is the relatively low tooling cost usually involved for short-run production. Most molding is done in single-cavity molds, in conventional compression-molding presses equipped for manual, semiautomatic, or fully automatic operation.

The **water-fillable plastics** used to replace wood or plaster of paris for ornamental articles such as plaques, statuary, and lamp stands, and for model making, are thermosplastic resins that cure to closed-cell lattices that entrap water. The resin powders are mixed with water and a catalyst and poured into a mold without pressure. They give finer detail than plasters, do not crack or chip, and are light in weight, and the cured material can be nailed and finished like wood. Water content can be varied from 50 to 80%.

COLD-ROLLED STEEL. Almost any steel can be cold-rolled, and some high-alloy and stainless steels are specially rolled to obtain mirrorlike finishes for special purposes. But the old term cold-rolled steel still applies generally to low-carbon, open-hearth steel that has been worked into strips, sheet, or bars by cold rolling, giving a good finish and a grain oriented in one direction. Bar stock is usually drawn to final dimensions through dies, and may be called **cold-finished steel.**

The carbon content of cold-rolled steel is usually from 0.08 to 0.12% with managanese from 0.30 to 0.80%. After the regular hot rolling has been completed, the steel is annealed and pickled, and then passed cold through finishing rolls which smooth and polish the surfaces and increase the hardness and tensile strength. Only a slight reduction is made by cold-rolling sheet steel. For the making of cold-rolled strip steel the slabs after hot rolling are sheared to length, then hot-rolled into strip which is wound on a coil. The coils are recoiled to loosen the scale, pickled, rolled in breakdown mills, and annealed. The cold rolling is then accomplished until the desired hardness is obtained. It is usually in four tempers, but sometimes in six, from No. 1, hard, to No. 6, dead soft. Dead-soft steel is for severe drawing and cupping work. It has a minimum tensile strength of 37,500 lb/in^2 (258 MPa) and elongation of 40%. Medium-soft, or quarter-soft, is for forming or light drawing. The minimum tensile strength is 42,500 lb/in^2 (293 MPa) with elongation 20%. Medium-hard, or half-hard, is for bending at sharp right angles across the grain. The tensile strength is 50,000 lb/in^2 (344 MPa). Hard-rolled steel is for flat work and easy punching. The tensile strength is 55,000 lb/in^2 (379 MPa) and elongation 5%. The average tensile strength of the normalized sheet is 50,000 lb/in^2 (344 MPa). **Cold-drawn steel** is bar or rod steel that has been finished by cold drawing through dies. Cold drawing doubles the yield point of hot-rolled bars, and imparts a high finish. The tensile strength of commercial, low-carbon, cold-drawn steel is 70,000 lb/in^2 (482 MPa) and elongation 15%.

COLUMBITE. An ore of the metal columbium. Its composition varies and may be $FeO \cdot Cb_2O_5$ or $(FeMn)Cb_2O_6$, or it may also contain tungsten and other metals. It is produced chiefly in Nigeria and marketed on the basis of the Cb_2O_5 content. But columbium occurs more usually in combination with tantalum. Concentrates generally average 44 to 70% Cb_2O_5 and 0.4 to 7% Ta_2O_5. The combined mineral known as **columbotantalite,** mined in South Dakota, Idaho, and the Congo, is marketed on the basis of the total $Ta_2O_5 \cdot Cb_2O_5$ content, and as the tantalum increases and the specific gravity increases, the mineral is called tantalite. The black mineral is associated with pegmatite, and some crystals are up to a ton in weight. Columbite concentrates contain about 60% **columbium pentoxide,** Cb_2O_5.

COLUMBIUM. An elementary metal, symbol Cb. It is also called **niobium,** with the symbol Nb. It occurs in the minerals columbite and tantalite and, as it closely resembles tantalum, is difficult to separate from it. Columbium has a fine yellowish-white color, a specific gravity of 8.57, a melting point of 2415°C, a tensile strength of 48,000 to 59,000 lb/in^2 (330 to 406 MPa), annealed, and up to 130,000 lb/in^2 (896 MPa) in drawn wire. The electric conductivity is one-eighth that of copper. It is ductile and malleable when pure, but slight amounts of impurities harden the metal. When free of nitrogen, oxygen, and hydrogen it is extremely ductile, and a 3-in (7.6-cm) bar can be rolled into foil 0.0005 in (0.0127 mm) thick without annealing. The metal can be machined readily with a cutting lubricant. It is insoluble in most acids and not easily attacked by alkalies. The metal oxidizes in the air, but the film of oxide retards further oxidation, and the normal rate is only about 5% that of molybdenum. However, about 750°F (399°C) there is severe oxidation caused by nucleation and growth of a porous columbium pentoxide which keeps a continuously refreshed surface of metal exposed to corrosion. At high temperatures, also, the metal absorbs gases. Its gas-absorbent properties are greater than those of tantalum, and it is thus used in high-vacuum tubes. The color of the metal is more attractive than that of tantalum, and it is used for jewelry. Columbium is the only metal, other than gold and silver, that can reasonably meet the characteristics needed for a **coinage metal:** intrinsic value, aesthetic value, durability, and widespread availability to prevent a national monopoly. Specimen coins have been made, but usage requires an act of Congress and international acceptance.

Although costly, columbium is not a rare metal, and it is estimated to be 50% more abundant in nature than lead. It is used in stainless steels in small amounts to inhibit intergranular corrosion. In chromium steels it reduces air hardening, increases impact strength, reduces creep, shortens the annealing time, and improves oxidation resistance. Small amounts are added to brass and copper alloys to aid retention of temper hardness at elevated temperatures. **Ferrocolumbium,** for adding columbium to steel, is marketed as an alloy containing 50 to 60% columbium, 7 silicon, and the balance iron.

Columbium alloys have some use for high-temperature parts in turbines and missiles. An alloy of 80% columbium, 10 titanium, and 10 molybdenum retains a tensile strength of 11,000 lb/in^2 (75 MPa) at 2500°F (1371°C). A **columbium-tungsten alloy,** with 11% tungsten, 3 molybdenum, and 2 hafnium, for instrument use, retains a tensile strength of 28,000 lb/in^2 (193 MPa) at 2000°F (1093°C). **Fansteel 80,** of the Fansteel Metallurgical Corp., is a **columbium-zirconium alloy** for missile and aircraft parts. It has a density of 8.6 and a melting point at 4350°F (2399°C), and retains a tensile strength of 18,000 lb/in^2 (124 MPa) at 2000°F (1093°C). **Fansteel 82** is an alloy of columbium, zirconium, and tantalum. It has a density of 10.26 and a melting point of 4550°F (2510°C). The tensile strength is 55,000 lb/in^2 (379 MPa), and it retains a strength of about 30,000 lb/in^2 (206 MPa) at 2000°F (1093°C) with high oxidation resistance. **Alloy FS-85,** of the same company, for aerospace parts, contains 28% tantalum, 10.5 tungsten, 0.9 zirconium, and the balance columbium. The melting point is 4695°F (2590°C). It retains a tensile strength of 15,000 lb/in^2 (103 MPa) above 2500°F (1371°C) and retains ductility at −320°F (−196°C). **Columbium alloy Cb-65,** of the Union Carbide Corp., contains 7.5% titanium and 0.75 zirconium It has a tensile strength of 37,000 lb/in^2 (255 MPa) at 1800°F (982°C). It can be hot-extruded, forged, or cold-worked. It has high oxidation resistance and a low neutron cross section. **Haynes alloy Cb-752** contains 10% tungsten and 5 zirconium. At ordinary temperatures it has a tensile strength of 125,000 lb/in^2 (861 MPa) with good ductility, and it retains a tensile strength of 50,000 lb/in^2 (344 MPa) at 2000°F (1093°C). This alloy comes as **columbium foil**

in a thickness of 0.002 in (0.051 mm) for high-temperature honeycomb structures. **Haynes alloy Cb-753** contains 5% vanadium; 1.25 zirconium; small amounts of carbon, nitrogen, and oxygen; and the balance columbium. Thin sheets retain a tensile strength of 27,400 lb/in^2 (189 MPa) up to 2200°F (1204°C).

Columbium selenide, $CbSe_2$, is more electrically conductive than graphite and forms an adhesive lubricating film. It is used in powder form with silver, copper, or other metal powders for self-lubricating bearings and gears. Columbium also comes in the form of **columbium oxide,** Cb_2O_5, a white powder melting at 1520°C, and as **potassium columbate,** $4K_2O \cdot 3Cb_2O_5 \cdot 16H_2O$. **Columbium ethylate,** $Cb(OC_2H_5)_5$, has a melting point of 6°C. It is used for producing thin dielectric films and for impregnating paper for dielectric use. Other such **metal alcoholates** are **columbium methylate,** $Cb(OCH_3)_5$, with a melting point of 53°C, and the **tantalum alcoholates** of the same formulas. **Columbium carbide,** CbC, is an extremely hard crystalline powder, which can be molded with a metal binder and sintered for use in cutting tools. The melting point is about 3800°C. It is made by sintering columbium powder and carbon in a hydrogen furnace.

COMPOSITION METAL. Also called **composition brass,** although it does not have the characteristics of a true brass. A general name for casting alloys that are in a mid-position between the brasses and the bronzes. The most widely used standard composition metal is **ounce metal,** containing 85% copper, 5 zinc, 5 tin, and 5 lead, which derived its name from the fact that originally 1 oz (0.03 kg) each of the white metals was added to 1 lb (0.45 kg) of copper. It makes a good average bearing metal and, as it gives a dense casting that will withstand liquid pressures, is also used for valves, pumps, and carburetor parts. It casts well, machines easily, and takes a good polish, so that it is widely employed for mechanical castings. It has about the same coefficient of expansion as copper, and can thus be used for pipe fitting. **ASTM alloy No. 2** is this metal. It contains 84 to 86% copper, 4 to 6 zinc, 4 to 6 tin, and 4 to 6 lead, and may also contain up to 0.75% nickel and small amounts of iron, either as intentional additions to increase strength or as impurities. The minimum tensile strength is 26,000 lb/in^2 (179 MPa), yield point 12,000 lb/in^2 (82 MPa), and elongation 15%. Well-cast alloys may have strengths as high as 32,000 lb/in^2 (220 MPa), with elongation 20% or higher, and Brinell hardness 50 to 59. The weight is 0.31 lb/in^3 (5 kg/m^3). This alloy is also called **red casting brass, hydraulic bronze,** and **steam brass,** and has also been used for forgings, producing parts with a tensile strength of 33,000 lb/in^2 (227 MPa) and elongation 25%.

In the high-copper red casting-brass series, for any given content of copper and zinc, the higher the ratio of tin to lead, the stronger but less ductile the alloy. The higher the content of zinc, the more ductile the alloy. For cast pipe fittings, the alloy may have 80 to 86% copper, 4 to 15 zinc, 2 to 6 lead, and 3 to 6 tin. This type of alloy is called **valve bronze,** and when the copper content is higher it is called **valve copper.** It should cast readily without cracks, checks, or porous spots. The **M bronze** of the U.S. Navy, for valves, contains 86 to 91% copper, 6.25 to 7.25 tin, 1.5 to 5 zinc, 1 to 2 lead, and not over 0.25 iron. It has a tensile strength of 34,000 lb/in^2 (234 MPa) and elongation 17%. It will withstand continuous temperatures up to 500°F (260°C), while the 85:5:5:5 bronze can be used for temperatures only to 400°F (204°C). **ASTM alloy No. 1,** designated as high-grade red casting brass for general castings, contains 85% copper, 1.5 lead, 6.5 tin, and 4 zinc. It has a tensile strength of 36,000 lb/in^2 (248 MPa), elongation 25%, and Brinell hardness 50 to 60.

Nickel is added to composition metals for hydraulic and steam castings to densify the alloy and make the lead more soluble in the copper. One company uses an alloy containing 84.5% copper, 5 lead, 7 zinc, 2.5 tin, and 1 nickel for casting injectors and lubricator parts. The nickel is added to the melt in the form of nickel shot which contains 5 to 7% silicon to deoxidize the metal and increase the hardness. For heavy high-pressure hydraulic castings as much as 5% silicon may be added to alloys containing nickel, giving strengths above 40,000 lb/in^2 (275 MPa). The alloys for machinery bearings usually contain higher proportions of tin or lead, or both, and are classified as high-lead bronze, but **Johnson bronze No. 44,** for bearings, contains 88% copper, 4 tin, 4 lead, and 4 zinc. The **hardware bronze** used for casting hardware and automobile fittings to be highly polished and plated is likely to be a true copper-zinc brass or a leaded brass with only a small amount of lead. **Oreide bronze,** a term still used in the hardware industry, was the metal employed for carriage and harness hardware. It contains 87% copper and 13 zinc, and polishes to a golden color. The hardware bronze of the Chase Brass & Copper Co. contains 86% copper, 12.25 zinc, and 1.75 lead. Aluminum, even in small amounts, is not considered a desirable element in the red casting brasses as it decreases the ductility and requires more care in casting.

CONCRETE. A construction material composed of portland cement and water combined with sand, gravel, crushed stone, or other inert material such as expanded slag or vermiculite. The cement and water form a paste which hardens by chemical reaction into a strong stonelike mass. The inert materials are called **aggregates,** and for economy no more cement paste is used than is necessary to coat all the aggregate surfaces and fill all the voids. The concrete paste is plastic and easily molded into any form or troweled to produce a smooth surface. Hardening begins immediately, but precautions are taken, usually by covering, to avoid rapid loss of moisture since the presence of water is necessary to continue the chemical reaction and increase the strength. Too much water, however, produces a concrete that is more porous and weaker. The quality of the paste formed by the cement and water largely determines the character of the concrete.

Proportioning of the ingredients of concrete is referred to as designing the mixture, and for most structural work the concrete is designed to give compressive strengths of 2,500 to 5,000 lb/in^2 (16 to 34 MPa). A rich mixture for columns may be in the proportion of 1 volume of cement to 1 of sand and 3 of stone, while a lean mixture for foundations may be in the proportion of 1:3:6. Concrete may be produced as a dense mass which is practically artificial rock, and chemicals may be added to make it waterproof, or it can be made porous and highly permeable for such use as filter beds. An air-entraining chemical may be added to produce minute bubbles for porosity or light weight. Normally, the full hardening period of concrete is at least seven days. The gradual increase in strength is due to the hydration of the tricalcium aluminates and silicates. Sand used in concrete was originally specified as roughly angular, but rounded grains are now preferred. The stone is usually sharply broken. The weight of concrete varies with the type and amount of rock and sand. A concrete with traprock may weigh 155 lb/ft^3 (2,483 kg/m^3). Concrete is stronger in compression than in tension, and steel bars or mesh are embedded in structural members to increase the tensile and flexural strengths. In addition to the structural uses, concrete is widely used in precast units such as block, tile, sewer and water pipe, and ornamental products.

Concrete blocks may be made from cement, sand, and gravel, or from cement and sand alone. For insulating purposes they may be made with cement and asbestos fibers. **Careystone,** of the Philip Carey Co., is cement with asbestos pressed into blocks, into corrugated slabs for roofing and siding, or into sheathing and wallboard. **Reinforced concrete** is a combination of concrete with a steel internal structure generally composed of rods or metal mesh. The strength of the concrete is thus greatly increased, and it is used for buildings, bridges, telegraph poles, roads, and fences. **Nonslip concrete,** for steps, is made by applying aluminum oxide grains, sizes 3 to 60 mesh, to the concrete before it hardens.

Insulating concrete and lightweight concretes are made by special methods, or by the addition of spongy aggregates. Slag may be used for this purpose. **Aerocrete,** of the Aerocrete Corp., is a porous lightweight concrete produced by adding aluminum powder to the cement. The reaction between the aluminum flakes and the lime in the cement forms hydrogen bubbles. **Durox,** of the U.S. Durox Co., produced as lightweight blocks, panels, and wall units, is a **foamed concrete** made from a mixture of sand, lime, cement, and gypsum, with aluminum powder which reacts to produce $3CaO \cdot Al_2O_3$ and free hydrogen which generates tiny bubbles. The set material contains about 80% cells and has only about one-third the weight of ordinary concrete with a compressive strength of 1,000 lb/in^2 (6 MPa). **Acid-resistant concrete,** developed by the Dutch firm of Ocrietfabrick, and called **Ocrete,** is made by passing the well-dried concrete products through a treatment tunnel containing **silicon tetrafluoride** gas, SiF_4, which converts the free lime to calcium fluoride. In the center of the concrete parts where moisture still remains, silicic acid is formed and fills the pores. The parts have increased density and are more wear-resistant than the original concrete.

Many prepared aggregates are used for special-purpose concretes. **Haydite** is a lightweight aggregate made by kiln-burning shale to produce a material of expanded cellular structure. Haydite concrete weighs below 100 lb/ft^3 (1,602 kg/m^3), but is not as strong as gravel concrete. **Superock** and **Waylite** are trade names for expanded aggregate made by treating molten slag with water or steam. **Microporite** is a German aggregate made by steam-treating ground silica and lime. **Calicel,** of the Keasbey & Mattison Co., is a lightweight spongy aggregate made by fusing silicates of lime and alumina and cooling to produce a stone of cellular structure. **Fluftrok,** of the Fluftrok Corp., is a lightweight aggregate made by heating obsidian in a kiln. The rock expands to 16 times its original volume, forming a porous material. Mixed with about 10% portland cement it is made into building blocks that are light and strong. A **conductive concrete,** known as **Marconite,** produced by Marconi Communication Systems, England, can be used for radio frequency grounding of TV, radio, and computer systems. The special aggregate can be added to the concrete mix to provide predetermined resistivity values.

CONDUCTIVE (ELECTRICAL) PLASTICS AND ELASTOMERS. Polymers made electrically conductive by the addition of carbon black nickel, silver, or other metals. Volume resistivities of plastics and rubbers, which normally are in excess of 10^8 Ω/cm can be lowered to between 10^{-1} and 10^6 Ω/cm by addition of conductive materials. **Carbon black** is the most widely used filler. The relationship of carbon black loading and volume resistivity is not proportioned. Up to a 25% loading, conductivity significantly increases, but it falls off sharply thereafter. Generally, the addition of carbon black lowers the polymer's mechanical properties. However, the use of carbon fibers to enhance conductivity improves mechanical properties.

Polyethylene and **polyvinyl chloride** resins loaded with carbon black are perhaps the most widely used conductive plastics. Plastics often made conductive by adding up to 30% carbon fiber are **polysulfone, polyester, polyphenylene sulfide, nylon 6/6, ethylene tetrafluoroethylene,** and **vinylidene fluoride-polytetrafluoroethylene.**

While **silicone** is the most widely used base polymer for **conductive rubber,** other rubbers frequently used in compounding conductive elastomers include SBR, EPDM, TPR, and neoprene.

Another type of **electrically conductive polymers** are materials that are doped with either electron donors, such as alkali metal ions and iodine, or electron acceptors, such as arsenic pentafluoride. Also referred to as **organic conductors,** their conductivity is about one-hundredth that of copper. The most widely used are **polyacetylene, polyparaphenylene,** and **polyparaphenylene sulfide.**

Polyacetylene, used in the form of foil for battery electrodes, has an energy storage density comparable to that of a lead-acid automobile battery, but can deliver 20 to 25 times the current. By stretching the foil, the fibers of which the foil is composed conduct electricity preferentially in one direction. Environmental stability, especially water sensitivity, is a problem with these materials. It can be improved by encasing them in other plastics. Another problem is that these polymers are difficult to form. Polyacetylene is insoluble and infusible, polyparaphenylene can be formed only by sintering, while polyparaphenylene can be melt-processed. **Phthalocyanines** can also be made electrically conductive by doping them with an electron acceptor, such as iodine, bromine, and charge-transfer salts.

CONDUCTORS. A term usually applying to materials, generally metals, employed for conducting electric current, though heat conductors and sound conductors have important uses. Silver is the best conductor of electricity, but copper is the most commonly used. The conductivity of pure copper is 97.6% that of silver. The electric conductivity of metals is often expressed as a percentage of the electric conductivity of copper, which is arbitrarily set at 100%. Tough-pitch copper is the standard conductivity metal, and it is designated as the International Annealed Copper Standard (IACS).

Because of the low conductivity of zinc, the brasses have low current-carrying capacity, but are widely used for electric connections and parts because of their workability and strength. The electric conductivity of aluminum is only 63% that of copper, but it is higher than that of most brasses. **Copper wire** for electric conductors in high-temperature environments has a plating of heat-resistant metal. **Aluminum wire,** usually with a steel core, is used for power transmission because of the long spans possible. Steel has a conductivity only about 12% that of copper, but the current in a wire tends to travel near the surface, and the small steel core does not reduce greatly the current-carrying capacity. Aluminum is now much used to replace brass in switches and other parts. Aluminum wire for electric equipment is usually commercially pure aluminum with small amounts of alloying elements such as magnesium which give strength without reducing the conductivity. Plastics, glass, and other nonconductors are given conductive capacity with coatings of transparent lacquer containing metal powder, but **conductive glass** usually is made by spraying on at high temeprature an extremely thin invisible coating of tin oxide. Coated glass panels are available with various degrees of resistivity.

CONTACT METALS. Metals used for contact points or surfaces for electric apparatus. The qualities required are high electric conductivity, corrosion resistance, and wear resistance. Because of its superior electric conductivity and corrosion resistance, silver is preferred where great wear resistance is not needed. Fine silver, of 99.9% purity, is used widely as a contact facing, but where greater hardness is required coin silver of 10% copper content is used. Nickel brass is harder, is lower in cost, and has good resistance to corrosion and wear, and the conductivity is sufficiently high for pushbuttons. Nickel silvers are now used for low-cost electric contacts and conductors although the conductivity is only about 8% that of copper. They have high tensile strengths, are corrosion resistance without plating, and are easily worked. **Silver-zinc alloy,** with 75% silver, is used for telephone jacks. **Spring contacts** have 92.5% silver, 7 copper, and 0.5 tin. **Radio contacts** contain 72 to 92.5% silver and 7.5 to 28% copper. **Internally oxidized alloys,** such as **silver-cadmium,** give good resistance to arc corrosion. The alloy is heated in an oxidizing atmosphere to disperse cadmium oxide in the silver matrix. Higher hardness is obtained in the alloy with small amounts of copper, nickel, tin, or manganese. For current breakers, a high percentage of tungsten carbide may be added.

Where high resistance to arcing is required tungsten is employed, but the conductivity is low. However, for high-speed equipment it is more important to have a heat-resistant metal than to have high electric conductivity. An ordinary low-melting contact metal will last only about 25,000 contacts, while a telegraph relay may operate 6 million times a day. For relays and signal instruments, platinum hardened with iridium, palladium, osmium, or rhenium may be used. Telephone contacts may be pure palladium, although this metal has only 20% of the conductivity of platinum. Sensitive relays are of **platinum-palladium alloy** with 25% palladium. **Ignition contacts** are of **platinum-osmium alloy** with 35% osmium. For low contact pressure and high corrosion resistance an **osmium-rhodium alloy** with 35% rhodium may be used. It has high hardness.

Silver-tungsten alloys, made by powder metallurgy, combine the conductivity of silver with the arc resistance of tungsten, and are used for contacts. **Silver-tungsten alloys** of the Gibson Electric Co. contain 20 to 90% silver with hardnesses 80 to 100 Rockwell B and good conductivity. For use in electric-discharge drilling tools the alloys contain only small amounts of silver. **Tungsten Alloy EDM,** of Firth Sterling, Inc., is such an alloy. **Gibsiloy M-12** is a **silver-molybdenum alloy** containing a high percentage of molybdenum, used for circuit breakers subject to high arcing. **Gibsiloy KA alloys** are alloys of silver with cadmium oxide to prevent welding and sticking of switches and relays. The conductivity is from 75 to 80% that of copper. **Silver-graphite alloys,** with low percentages of graphite, are used for contacts that require rubbing or sliding lubrication and for brushes and collector rings. **Aeralloy,** of the H. A. Wilson Co., is a **platinum-ruthenium alloy** with high ruthenium content, used for aircraft magneto contacts. It is hard and gives long wear life. **Wilco No. 6 alloy,** for automotive magneto contacts, is platinum hardened with ruthenium. **Gibsiloy UW-8,** used for contacts of oil-immersed motor starters, is a copper-tungsten alloy.

COOLANTS. Liquids employed for quenching steels in heat treating, although this term is also used to designate the cutting oils used on machines to cool the work and improve the cutting. When water is used for the normal water-hardening steels, it may be modified with soda or other material to give a less drastic and more uniform cooling. A water bath containing 5% sodium hydroxide gives uniform rapid cooling. Oils are used in cooling or quenching baths for many alloy steels, as they remove the heat from the steel more uniformly and not as suddenly as water. **Quenching oils** are usually compounded, although fish oils alone are sometimes employed. Fish oils, however, have offensive odors when heated. Vegetable oils alone are likely to oxidize and become gummy. Animal oils become rancid. Lard and palm oils give low cooling rates, while cottonseed, neatsfoot, and fish oils give more rapid cooling. Mineral oils compounded with fish, vegetable, or animal oils are sold under trade names and vary considerably in their content. **Oil-quenching baths** are usually kept at a temperature of not over 150°F (66°C) by providing cooling pipes. **Tempering oils** differ from quenching oils only in that they are compounded to withstand temperatures up to about 525°F (274°C).

COPAL. A general name for fossil and other hard resins found in nearly all tropical countries and used in making varnishes and lacquers, adhesives, and coatings, though now largely replaced by synthetic resins. Copals are distinguished by their solubility in chloral hydrate. All of the copals are also soluble in alcohol, linseed oil, and turpentine. The hardest varieties come from Africa. **Zanzibar copal,** from the tree *Trachylobium verrucosum,* or from species no longer existent, is one of the hardest of the varnish resins, with a melting point of 240 to 360°C, compared with 180 to 200°C for **Congo copal** from Guinea. **Madagascar copal** is from the tree *Hymenaca verrucosa,* and is darker than Zanzibar. **Gum benguela** is a semifossil resin from the tree *Gulbourtia copaifera* of West Africa. The melting point is 170°C. Many species of trees of the genus *Hymenaca* of tropical America furnish copals. **Animi gum,** or **gum Zanzibar,** is from the stem of the plant, *H. coubarii* of Zanzibar and East Africa. It belongs to the group called **East African copals,** but is distinguished from other copals by its solubility in alcohol. The specific gravity is about 1.065, and melting point 245°C. The **Brazilizan copal** known as **jutahycica resin** is from the jatahy tree which is plentiful in the Amazon Valley. **Jatabó** and **trapucá resins** are fossil copals from species of *Hymenaca* of the state of Bahia, Brazil. **Congo gum,** chiefly from the tree *Copaifera demensi,* is the most insoluble of the natural resins, but after thermal processing, it is soluble in a wide range of solvents. The specific gravity of copals is from 1.04 to 1.13. The colors vary from white through yellow, red, brown, to brownish black.

The commercial copals are classed in five groups: East African, West African, Manila, East Indian, and South American. The name copal is applied in Indonesia to the resin of the tree *Agathis alba,* closely related to the kauri pine. The types include **Manila copal, Loba,** and **Boea.** In Malaya the tree has been classed as *Dammara orientalis* and the copal is known as **white dammar.** In the Philippines the tree is called **almacido,** and the gum, Manila copal. There are seven grades of Manila copal, from No. 1 pale, scraped chunks, to the No. 7 dust. Hard copal is harder than dammar, and has a higher melting point, but the hardness of the resin depends greatly upon the seasoning time in the ground. The semihard and soft copals are produced directly from the trees by tapping. The melting point of copal from *A. alba,* collected 1 day after tapping, averages 85°C, compared with 105°C when collected 3 months after tapping. **Fossil copal,** or **copalite,** of high quality, is obtained by separation from the low-grade coals of Utah, which contain about 5%. The copal has an amberlike appearance of light yellow to red color, with a specific gravity of 1.02 to 1.06, melting point 165°C, and hardness about the same as that of Congo gum.

COPPER. One of the most useful of the metals, and probably the one first used by man. It is found native and in a large number of ores. Its apparent plentifulness is only because it is easy to separate from its ores and is often a by-product from silver and other mining. Copper has a face-centered cubic crystal structure. It is yellowish red in color, tough, ductile, and malleable, gives a brilliant luster when polished, has a disagreeable taste and a peculiar odor. It melts at 1083°C and boils at 2310°C. The specific gravity is 8.91, and weight 0.321 lb/in^3 (5 kg/m^3). It is the best conductor of electricity next to silver, with a conductivity 97% that of silver. The coefficient of expansion is 0.000017 per degree Celsius. The tensile strength of cast copper is from 17,000 to 20,000 lb/in^2 (117 to 137 MPa) with elongation 49 to 50%. Annealed wrought copper has a strength of 32,000 lb/

in^2 (220 MPa) with elongation 56%, while cold-drawn copper has a tensile strength of 56,000 lb/in^2 (386 MPa) with elongation 6%. The **busbar copper** used in electrical devices has a tensile strength up to 40,000 lb/in^2 (275 MPa). Copper does not have the ductility of brass for metalworking, but does not work-harden as rapidly as brass. Pure copper is difficult to cast, as the molten metal absorbs oxygen, forming oxides. Copper is used for electric conductors; for making brasses and bronzes; for sheathing, fittings, and pipe; and for cast articles. Small amounts of copper are added to some steels to give corrosion resistance.

Copper is marketed in three general grades: electrolytic, lake, and casting. **Secondary copper** is a term used to designate copper recovered from smelting scrap and old copper alloys. Commercial wrought copper in bars, wire, sheets, and rods is marketed as **electrolytic tough pitch, oxygen-free copper, phosphorized copper,** and **arsenical copper.**

Electrolytic copper has a purity not less than 99.9%. Electrolytic tough pitch, or **high-conductivity copper,** deoxidized with phosphorus without residue and annealed, has an electrical resistance of 0.67879 $\mu\Omega/\text{in}^3$ (16 $\mu\Omega/\text{cm}^3$) at 20°C, which is taken as 100% conductivity. This copper has the disadvantage of becoming brittle when heated and of not giving a high finish.

Much of the copper marketed for commercial use as copper contains slight amounts of silicon or other hardener, but even as little as 0.40% arsenic or other impurity will reduce the electric conductivity drastically. Oxygen-free copper is 99.9% pure, has high conductivity, is not subject to brittleness, and will withstand much cold working. Phosphorized copper contains residual phosphorus. It has high strength, higher hardness and resistance to corrosion than other copper, but has lower conductivity. **Amphos,** of American Metal Climax, Inc., is phosphorized copper which has enough phosphorus to raise the softening point, giving better workability. It comes in billets for producing high-strength wrought metals. Anaconda condenser tubes of arsenical copper contain 0.25% arsenic.

Silver-bearing copper is 99.9% pure, carrying 8 to 30 oz (0.2 to 0.9 kg) of silver per ton (907 kg). The silver raises the annealing temperature, and the metal is used for high-speed motors, commutators, and semiconductor cases. **Amsil,** of American Metal Climax, Inc., with 15 oz (0.4 kg) of silver per ton (907 kg), has the electric conductivity at room temperature of pure copper. **Lake copper,** from the Lake Superior region, is a silver-bearing copper having varying amounts of silver up to about 30 oz (0.9 kg) per ton (907 kg).

Hard-drawn wire or sheet arsenical copper has a tensile strength of 60,000 lb/in^2 (413 MPa) while the annealed material has a strength of 32,000 lb/in^2 (220 MPa) and elongation 45%. **Cast copper** has only 80 to 90% the conductivity of wrought copper. A special grade of copper having high ductility, high conductivity, and fatigue resistance is made without melting by converting electrolytic cathode copper directly into rods and strips by rolling at elevated temperature in a reducing atmosphere. The Phelps Dodge Corp. produces this copper under the name of **PDCP Copper. Electro-sheet copper** is thin sheet copper produced by electrodeposition. It is marketed by the American Brass Co. in roll sheets of 0.0013 to 0.0094 in (0.3302 to 0.2388 mm) thick, and is used for roofing and dampproofing. **Rocan copper** is a sheet copper of Revere Copper & Brass, Inc., having high strength and resistance to corrosion fatigue. It contains 0.50% arsenic. It is used for roofing and leaders. **Roofing copper** is hot-rolled soft copper sheet in 14- to 32-oz (0.4- to 0.9-kg) weights, but **cornice copper** is cold-rolled to a hard temper. The 16-oz (0.5-kg) sheet is used for gutters and leaders. **Braziers' copper** is a term used to designate heavy sheets of copper weighing from 1.5 to 6 lb/ft^2 (7.3 to 29 kg/m^2) used for coppersmiths' work. **Coppersmiths' copper** is hot-rolled, soft-temper, heavy sheets up to ½ in (1.27 cm). **Copper foil** is sheet copper less than 0.005 in (0.012 cm) in thickness. **Free-cutting copper** is deoxidized copper containing up to 0.70% tellurium, marketed in rods for making screw-machine products.

Cupaloy, of the Westinghouse Electric Corp., is a nearly pure copper containing small amounts of silver and chromium, the chromium forming a hard crystalline structure and the silver acting as a stabilizer. When temper-hardened the wrought alloy has a tensile strength of 70,000 lb/ in^2 (482 MPa), elongation 15%, and Rockwell B hardness 80 to 85. It has 85% the electric conductivity of copper. It is used for welding electrodes, commutator bars, and strong electrical parts. The **copper-silver alloy** developed by the Army Signal Corps for high-strength electric conductors contains 6.5% silver. Wire drawn to a tensile strength of 160,000 lb/in^2 (1,103 MPa) has a conductivity 70% that of copper. When drawn to a tensile strength of 116,000 lb/in^2 (799 MPa), it has a conductivity 85% that of copper. Fine-gage **copper-silver wire** contains 94% copper and 6 silver, and has a tensile strength of 150,000 lb/in^2 (1,034 MPa) and a conductivity 70% that of copper. **Copper-iron alloy,** with 12,5% iron, has a tensile strength of 150,000 lb/in^2 (1,034 MPa) and conductivity 50% that of copper. For this alloy magnesium is used as a deoxidizer in pouring the ingots at high temperature. **Switch copper,** of the Revere Copper & Brass, Inc., is electrolytic copper in shaped bars of close tolerance and burnished finish. It is used for switchblades and electrical parts. The minimum tensile strength is 36,000 lb/in^2 (248 MPa), elongation 15% min, and Rockwell B hardness 35 to 65.

Copper powder is usually chemically reduced copper in noncrystalline form. It is used in a liquid vehicle for copper coating, or for sintering. The powder produced electrolytically is flaked, suitable for pigment but not for sintering. Copper powder is usually 98.3% pure, and all particles pass through a 350-mesh screen. Copper powder in flake form used for paint and ink pigment is produced by electrolysis. The fine grains of large surface area give high green strength and a uniformly dense structure in the sintered molding. **Copper shot** is copper in the form of round globules, used chiefly in alloying gold and silver. **Leaded copper** is copper in commercial rods and shapes containing a small amount of lead to make it free-machining. **Sulfur copper** is copper containing about 0.3% sulfur, marketed in rods for the production of screw-machine products. It is free-cutting, but does not machine as easily as tellurium copper, and has higher electric conductivity, about 96% that of standard copper. Copper is also marketed in the form of master alloys such as **copper aluminum,** which is an alloy of 50% copper and 50 aluminum melting at 1070°F (577°C), used for making aluminum alloys.

COPPER ACETATE. Also known as **crystals of Venus.** A dark-green crystalline poisonous powder of the composition $Cu(CH_3COO)_2 \cdot H_2O$, of specific gravity 1.882 and melting point 115°C. It is soluble in water and in alcohol. It is used as a pigment in paints, lacquers, linoleum, inks, and for making artificial verdigris or patina on copper articles. It is also used as a catalyst in making phthalic anhydride plastics. When used for mildew-proofing cotton cloth, the copper precipitates out to form the **waxate,** or copper soap coating. **Verdigris** is an old name for basic copper acetate as a blue-green pigment, but the name is now usually applied to the bluish-green corrosion crust on copper. The greenish-brown crust known as **patina,** formed on bronze, is esteemed as a characteristic of antiquity. It is a basic sulfate of copper, usually with oxides of tin, copper, and lead. Another green copper paint pigment is **copper carbonate,** also called **artificial malachite.** It is a poisonous powder of the composition $CuCO_3 \cdot Cu(OH)_2$, made by adding sodium carbonate to a solution of copper sulfate. The specific gravity is 3.7. It is insoluble in water. As a pigment it is also named **mineral green, Bremen Green,** and **mountain green.**

COPPER-ALLOY POWDER. There is a rather large range of compositions of copper-alloy powders available, including brass, bronze, and copper-nickel powders. **Brass powders** are the most widely used for powder metallurgy (P/M) structural parts. Conventional grades are available with zinc content from around 10 to 30%. Sintered brass parts have tensile strengths up to about 35,000 and 40,000 lb/in^2 (241 to 275 MPa), and elongations of from 15 to around 40%, depending on composition, design, and processing. In machinability, they are comparable to cast- and wrought-brass stock of the same composition. Brass P/M parts are well suited for applications requiring good corrosion resistance, and where free-machining properties are desirable. **Copper-nickel powders,** or **silver-nickel powders,** contain 10 to 18% nickel. Their mechanical properties are rather similar to the brasses, with slightly higher hardness and corrosion resistance. Because they are easily polished, they are often used in decorative applications. **Copper powders** and **bronze powders** are used for filters, bearings, and electrical and friction products. However, bronze powders are relatively hard to press to densities that give satisfactory strength for structural parts. The most commonly used bronze powder contains 10% tin. The strength properties are considerably lower than iron-base and brass powders, being usually below 20,000 lb/in^2 (137 MPa).

COPPER ORES. There are about 15 copper ores of commercial importance, and these are widely distributed in almost all parts of the world. More than 40 countries produce copper on a commercial scale. The average copper content of ores, however, is usually low, and copper would be an expensive metal if it were not for the valuable by-products: silver, gold, nickel, and other metals. About 80% of the ores in the United States contain only 1.17 to 1.57% copper and are concentrated before smelting. The direct smelting ores average from 4.3 to 6.2% copper. The most important ore of copper is **chalcopyrite,** also known as **copper pyrites** and **yellow copper ore.** It occurs widely distributed, associated with other minerals, and may carry gold and silver. It is the chief copper ore in many parts of the United States, Canada, Chile, Africa, England, and Spain. Chalcopyrite is a sulfide of copper and iron, $CuFeS_2$, containing theoretically 34.5% copper. It usually occurs massive, with a hardness of 3.5 and a specific gravity of 4.2. The color is brass yellow, with greenish-black streaks. To obtain the copper, the ore is first smelted with enough sulfur to combine with all of the copper, producing a matte which is a mixture of CuS_2 and FeS together with impurities. Air is then blown through the molten matte in a reverberatory furnace, converting the iron sulfide to oxide and the sulfur to sulfur dioxide. The remaining copper is cast into pigs which are called **blister copper,** owing to its blistered appearance. Blister copper contains 96 to 99% copper, with various metals and arsenic and sulfur. It is not used commercially, but is refined in furnaces or electrolytically. The **cement copper** shipped from Cyprus contains about 51% copper.

Chalcocite is another important ore found in Montana, Arizona, Alaska, Peru, Mexico, and Bolivia. It is a **cuprous sulfide,** Cu_2S, containing theoretically 79.8% copper. It usually occurs massive, but crystals are also found. The hardness is 2.5 to 3, and the specific gravity 5.5. It has a shining lead-gray color. But the emerald-green platy mineral **chalcolite** is a **copper-uranium mica,** $CuO \cdot 2UO_3 \cdot P_2O_5 \cdot 8H_2O$, with a high percentage of uranium oxide, U_3O_8. **Tennantite,** or **gray copper ore,** found in Colorado, Wyoming, and Montana, has the composition $3Cu_2S \cdot As_2S_3$, with iron and antimony. When much of the arsenic is replaced by antimony it is called **tetrahedrite. Azurite,** also called **blue copper carbonate** and **chessylite,** is found with other copper ores. It is a basic carbonate of copper, $Cu(OH)_2 \cdot 2CuCO_3$, occurring in azure-blue crystals. **Malachite,** or **green copper ore,** is an important carbonate ore, $Cu(OH)_2 \cdot CuCO_3$, containing theoretically 57.4% copper. It has a bright-green color, specific gravity 3 to 4, and hardness 3.5 to 4. **Cuprite,** or **red copper ore,** is a **cuprous oxide,** Cu_2O, containing theoretically 88.8% copper. It occurs usually massive, but sometimes in crystals. The specific gravity is 6, and the hardness 3.5 to 4. The color may be various shades of red, with an adamantine luster in the clear crystalline form, or a dull earthy luster in the massive varieties. Cuprite is found in the copper deposits in Arizona, and is one of the ores in Chile, Peru, and Bolivia.

Bornite, also known as **horseflesh ore, peacock ore,** and **variegated ore,** is an important ore of copper widely distributed and mined in Chile, Peru, Canada, and the United States. It occurs in massive form, having a bronze color that turns purple on exposure. The compositon is Cu_5FeS_4, having theoretically 63.3% copper. It has a metallic luster and a hardness of 3. **Chrysocolla** is a highly refractory ore of copper occurring in the oxidized parts of copper veins of Arizona and New Mexico. It is a hydrous **copper silicate** of the composition $CuSiO_3 \cdot 2H_2O$. It occurs in compact masses with a specific gravity of 2 to 2.4 and a hardness of 2 to 4. The color is green to bluish. It was used as a green pigment by the ancient Greeks. Large reserves of this ore occur in The Gambia and other copper regions of Africa, and it it treated by high-temperature methods to obtain the copper. **Atacamite** is an ore found in Bolivia, Arizona, and Australia. It is a copper chloride with copper hydroxide, $CuCl_2 \cdot 3Cu(OH)_2$, generally found in confused crystalline aggregates, fibrous or granular. The hardness is 3 to 3.5, specific gravity 3.75, and the color may be various shades of green. The unique copper ores of Japan, called **kuromono,** are complex sulfide-sulfate replacement minerals.

Much **native copper** metal occurs in the Lake Superior region, particularly in Michigan, but it occurs irregularly and not in continuous veins. The Ontonagon boulder of native copper in the National Museum, weighing 3 tons (2.7 metric tons), came from Michigan. A mass of native copper found in 1847 was 10 ft (3 m) long and weighed 6 tons (5.4 metric tons). The largest ever found weighed 18 tons (16.3 metric tons).

COPPER OXIDE. There are several oxides of copper, but usually the term refers to **red copper oxide,** or **cuprous oxide,** Cu_2O, a reddish crystalline powder formed by the oxidation of copper at high temperatures. It also occurs naturally in cuprite ore. The specific gravity is 6.0 and the melting point 1235°C. It is insoluble in water but soluble in acids and alkalies. It is used in coloring glass and ceramics red, in electroplating, and in alternating-current rectifiers. **Rextox,** of the Westinghouse Electric Corp., is copper upon which a layer of copper oxide has been formed. Electric current will flow easily from the oxide to the copper, but only with difficulty from the copper to the oxide. It may be used for transforming alternating current into pulsating direct current. **Black copper oxide,** or **cupric oxide,** CuO, is a brownish-black amorphous powder of specific gravity 6.4 and melting point 1065°C. It is used for coloring ceramics green or blue. In its natural ore form it is called **tenorite. Copper hydroxide,** formed by the action of an alkali on the oxides, is a poisonous blue powder of the composition $Cu(OH)_2$ and specific gravity 3.37. It is used as a pigment.

COPPER STEEL. Steel containing up to 0.25% copper and very low in carbon, employed for construction work where mild resistance to corrosion is needed and where the cost of the higher-resistant chromium steels is not warranted. It is employed in sheet form for culverts, ducts, pipes, and for such manufacturing purposes as washing-machine boilers. The alloy steels containing considerable copper for special purposes are not classed as copper steels. The copper neutralizes the corroding influence of the sulfur in the steel, and also aids in the formation of a fine-grained oxide that retards further corrosion. Copper is not added to unalloyed high-carbon steels because it causes brittleness and hot-shortness. Since the carbon content of copper steel is usually very low, the material is in reality a **copper iron.** Unless balancing elements, especially nickel, are present, more than 0.2% copper in steel may cause rolling defects. Molybdenum in small quantities may also be added to give additional corrosion resistance, and the percentage of carbon may be raised to 0.40% when about 0.05% molybdenum is added. **Toncan iron,** of the Republic Steel Corp., has this composition, and has a tensile strength of 40,000 to 48,000 lb/in^2 (275 to 330 MPa), elongation 32 to 40%, and weight 0.283 lb/in^3 (7.8 g/cm^3).

The copper-bearing iron specified for culverts by the ASTM contains not less than 0.20% copper and not more than 0.10% carbon, manganese, phosphorus, sulfur, and silicon as impurities.

COPPER SULFATE. Also called **bluestone** and **blue vitriol.** An azure-blue crystalline lumpy material of the composition $CuSo_4 \cdot 5H_2O$ and specific gravity 2.286. It is soluble in water and insoluble in alcohol. When heated, it loses its water of crystallization and melts at 150°C. It is used for wet electric-battery solutions, for copperplating, in dyestuffs, in germicides, in coppering steel, and in various chemical processes. In its natural form, called **chalcanthite,** it is a rare mineral found in arid regions and deposited from the water in copper mines. It is produced as a by-product in copper refineries, or by the action of sulfuric acid on copper or copper oxide.

CORAL. A shiny, hard, calcareous material valued for jewelry, buckles, beads, and novelties. It is a growth composed of the skeletons of *Corallium nobile* and other species of aquatic protozoa. The structures are built up by these creatures into forms like leafless trees or shrubs, fans, mushrooms, or cups. **White coral** is common and not used commercially. The most valuable is the **red coral,** a twiglike species that grows about 12 in (30 cm) high with thin stems. **Pink coral** and **black coral** are also valued. Red and pink corals come from the Indian Ocean and off the coast of northeastern Africa. Black coral is from southeastern Asia. The red and black varieties are very hard and take a beautiful polish. The pink is softer, with a more delicate appearance, and is used for beads. The rate of growth of coral is very slow. The gleaming white sand of tropical beaches called **coral sand** is usually not coral, but consists of the disintegrated limy skeletons of the seaweed *Halimeda opuntia.*

CORDAGE. A general term for the flexible string or line of twisted fibers used for wrapping, baling, power transmission, and hauling. **Cordage fibers** are any materials used for making ropes, cables, twine, and cord. In general, the cordage fibers are hard compared with those used for weaving into fabrics, but cotton and some other soft fibers are used for cord. **Twine** is cordage less than 3/16 in (0.48 cm) in diameter, and composed of two or more rovings twisted together. **Rope** is cordage made by twisting several yarns into strands, and then twisting the strands into a line. A **cable** is a strong rope, usually referring to the large sizes of special construction. **Cord** is an indefinite term for twine, but is, more specifically, the soft cotton twines used for wrapping. The term **string** is used for the weak cotton cords used for wrapping light packages. **Seaming twines** are made of flax fibers. **Seine twine** is a three-strand cotton twine with 2 to 56 plies per strand. Most of the **binder twine** is made from sisal, but **Indian twine** is made from jute. Ramie fiber is used for marine twines. **Binder twine** has 15 turns per foot (49 turns per meter) and 500 ft/lb (336 m/kg). **Baler twine,** for heavier work, has 12 turns per foot (39 turns per meter) and 125 ft/lb (84 m/kg). Before the advent of synthetics, about half of American strong cordage was from Manila hemp, and about 30% from sisal. Manila help is very resistant to seawater. Sisal is used for the cheaper grades of rope, but it absorbs water easily. True hemp is considered a superior fiber for strong ropes. Untarred **hemp rope** is used for elevator cables, and tarred hemp is employed for ship cables. **Marine rope,** used by the Navy, was formerly true hemp, then Manila hemp, and is now often synthetic fiber. Most industrial rope has at least three strands, each strand having at least two yarns, and may be hard lay, medium lay, or soft lay. Twisting may be S twist or Z twist, conforming approximately to the shape of these letters. Cable twist has the twists alternating in each successive operation. Hawser twist, to give greater strength and resilience, has the plies twisted SSZ.

Cordage fibers are also obtained from a wide variety of plants. Generally, after the fibers are retted, the softer and finer fibers are separated out for use in weaving into fabrics and the harder and coarser fibers are marketed as cordage fibers. **New Zealand hemp,** or **New Zealand flax,** is a strong cordage fiber obtained from the leaves of the swamp lily, *Phormium tenax,* grown in New Zealand and Argentina. The fibers are white, soft, and lustrous. One variety of the plant reaches a height of 16 ft (4.9 m) and the other variety 6 ft (1.8 m). **Olona fiber**, grown in Hawaii and used locally for fishnets, is from the nettle plant, *Touchardia latifolia.* The bast fibers of the bark of the slender branches are soft and flexible, are very water-resistant, and have a tensile strength three times that of Manila hemp. **Gravatá** is a Brazilian name for the very long and resistant fibers from the leaves of the pineapple plant, *Ananas sagenaria.* The leaves of this species are up to 7 ft (2.1 m) in length. The fiber known as **widuri** of Indonesia is bast fiber from the tree *Calotropis gigantea* which yields the madar kapok. It has great strength and is resistant to seawater. It is used for ropes and fishnets. **Agel fiber** is from the stems and leaves of the **gebang palm** of the Celebes where the various grades are used for sailcloth, rope, and fishnets; the coarser fibers are woven into Bangkok hats. The fibers from the leafstalks are fine and white. **Caraguatá** is a strong, highly resistant fiber from the plant *Bromelia balansea* of Paraguay. It is employed by the Indians for making hammocks, and is now used for cordage and burlap fabrics.

Synthetic fibers are also used for cordage. **Nylon rope** is about twice as strong as **Manila rope,** is lighter, and because of its property of stretching rapidly but recovering slowly it makes a desirable rope for lifting and towing, giving a smooth, shock-absorbing pull. Nylon ropes are used for pulling airplane gliders and for tugboat lines. **Mylar rope,** of U.S. Plastics Rope, Inc., is made by slitting Mylar film, and stretching and spinning the strands. A three-strand rope of 1-in (2.5-cm) diameter has a breaking strength of 18,000 lb/in^2 (124 MPa), compared to 9,000 lb/in^2 (62 MPa) for Manila rope of the same size. Moisture absorption is less than 0.3%. Elongation at 50% of breaking strength is about 4.75%. **Saran rope,** made by the Plymouth Cordage Co. for chemical-plant use, is formed of three strands of vinylidene chloride monofilament. The breaking strength is 70% that of Manila rope, and it is flexible and chemical-resistant, but it is not recommended for temperatures above 170°F (77°C). The **M-cord** of the same company is a strong wrapping twine made with a core of Manila fiber wrapped with a tough, smooth paper. Nylon and some other plastics have a tendency to fray in cordage, and may be coated with polyvinyl butyral to give abrasion resistance. **Chemclad** is **rayon cordage** coated with polyvinyl chloride. The nylon rope of Rochester Ropes, Inc., is steel-wire rope with an extruded coating of nylon in various colors, used for automotive brake cables, aircraft control cables, and luggage handles. **Glass rope**, woven from continuous filaments of glass fiber, is used for chemical and electrical applications where resistance to chemicals or electrical insulation is needed. It is strong, but is expensive and has low flexing strength. It comes in diameters from ¼ to ¾ in (0.64 to 1.90 cm). **Fiberglas cordage,** of the Owens-Corning Fiberglas Co., is marketed in diameters from 1/64 to ⅛ in (0.04 to 0.32 cm) and made of continuous filament or staple glass fibers. The ⅛-in (0.32-cm) untreated continuous-filament cord has a breaking strength of 258 lb (116.5 kg). **Newbroc,** of Hitemp Wires, Inc., is chemical-resistant and heat-resistant thread and cord made with continuous-filament glass fiber impregnated with Teflon plastic, in diameters from 0.0046 to 0.076 in (0.12 to 0.19 cm). It remains flexible at subzero temperatures and is used for lacings and for sewing canvas. The 0.020-in (0.05-cm) fiber has a tensile strength of 70 lb (31.6 kg). Cordage made with high-modulus polyethylene fiber has high tensile strength and elasticity, and is used for tugboat hawsers.

CORE OILS. Liquid binders used for sand cores in foundry work. The binder should add strength to the core, should bake to a dry bond, should not produce much gas, and should burn out after the metal is poured, so that the sand core will collapse. Linseed oil is considered one of the best binders, but it is usually expensive and may be mixed with cheaper vegetable oils or with mineral oil. In some cases fish oil or rosin is also used. Molasses, dextrin, or sulfite liquor may be included in prepared core oils. The specifications of the American Foundrymen's Association call for 50% raw linseed oil, 25 H grade rosin, and 25 water-white kerosene, with no fish oil. A good core oil should have a specific gravity of 0.9368 max, flash point 165 to 200°F (73 to 93°C), Saybolt viscosity 155 min, and iodine number 154, and should be of light color. However, any drying oil or semidrying oil can be used to replace all or part of the linseed oil. Perilla and corn oils are used, and core oils of linseed- and soybean-oil mixtures have good strength. The liquor from sulfite pulp mills contains lignin and is used as a core binder. **Glutrin,** of the Robeson Process Co., is a core oil with sulfite liquor. **Truline** is a resinous binder in a powder form marketed by Hercules Inc. **Uformite 580,** of the Rohm & Haas Co., is a **core binder** especially for aluminum sand cores. It is a modified urea-formaldehyde resin which bakes in the core at 325 to 375°F (162 to 190°C), and will break down in the core at temperatures above 450° F (232°C). **Cycor 191,** of the American Cyanamid Co., is a urea-formaldehyde resin in water solution for sand cores for short-cycle baking in an electronic oven. **Dexocor** and **Kordex,** of the Corn Products Sales Co., are dextrin binders.

CORK. The thick, spongy bark of a species of oak tree, *Quercus suber,* grown in Spain, Portugal, Italy, Algeria, Morocco, Tunisia, and to a limited extent the United States. It is used for bottle stoppers, insulation, vibration pads, and floats for rafts and nets. The scrap cuttings are used for refrigerator insulation, packing for the transportation of fruits, and the manufacture of linoleum and pressed products. When marketed as **granulated cork,** this material usually comes in sizes of ½ in (1.27 cm) and No. 8 mesh. Cork is also used natural or in the form of pressed composition for gaskets, oil retainers, roll coverings, polishing wheels, and many other articles. The material has a cellular structure with more than 50% of the volume in air cells. The cell structure is peculiar, and each cell is in contact with 14 neighboring cells, and because of lack of capillarity it does not absorb moisture. When dried, cork is light, porous, easily compressed, and very elastic. It is one of the lightest of solid substances, the specific gravity being 0.15 to 0.20. It also has low thermal conductivity. Charring begins at 250°F (121°C), but it ignites only with difficulty in contact with flame. The cork tree grows to a height of about 30 ft (9 m). After it has attained the age of about 25 years it can be barked in the summer, and this barking is repeated every 8 or 10 years. The quality of the bark improves with the age of the tree, and with proper barking, a tree will live for 150 years or more. The thickness of the bark varies from ½ to 2 in (1.27 to 5.08 cm). **Cork bark** is shipped in bales of 170 lb (77.1 kg), and **cork wastes** in bales of 148 lb (67.1 kg).

Brazilian cork is the bark of the tree *Angico rayado,* called **pao santo bark,** and also from the trees *Piptadenia incuriale* and *P. colomurina.* The bark has a cellular structure and, when ground, has the appearance of a low grade of true cork, but is softer. It is suitable for insulation. A substitute for cork for insulation packings and acoustical panels is marketed by the Palmetex Corp. under the name of **Palmetex.** It is the compressed pith from the internal fibers of the sawtooth palm, *Cerano repens,* of the eastern Gulf states. It has lower conductivity than cork, but without a binder it is more friable. **Corkboard** is construction board made by compressing granulated cork and subjecting it to heat so that the particles cement themselves together. It is employed for insulating walls and ceilings against heat and cold, and also as a sound insulator. **Cork tile** is corkboard in smaller, regularly shaped blocks for the same purposes. The natural gum in the cork is sufficient to bind the particles, but other binders may be used. Corkboard produced by the Armstrong Cork Co. is marketed in sheets in thicknesses from 1 to 6 in (2.5 to 15.2 cm). The weight is from 6 to 10 lb/ft^3 (96 to 160 kg/m^3), depending upon the amount of compression and the binder. The heat conductivity is about one-third that of wood. Corkboard retains the properties of cork, being cellular, and without capillarity or tendency to absorb moisture. **Novoid corkboard,** of the National Cork

Co., is made with both large and small granules tightly packed to leave only small air spaces. **Joinite,** of L. Mundet & Son, Inc., is a corkboard for use under machinery to deaden vibration and noise. **Corkoustic** is a sound-absorbent corkboard of the Armstrong Cork Co. for walls and ceilings. It has a sound-absorbing coefficient of 0.30 compared with 0.032 for brick walls. **Linotile,** of the same company, is a resilient tile made of powdered cork, oxidized oils, and color pigments.

CORN. One of the most important food grains of the world for both human and animal consumption, but also used industrially for the production of starch, glucose, alcohols, alcoholic beverages, and corn oil. Corn was unknown to Europe before the discovery of America, where it was one of the chief foods of the Indians from Canada to Patagonia. In Europe and in foreign trade it is known by its original name of **maize,** and the Incan name of **choclo** still persists in South America for the grains on the cob. In Great Britain corn means all hard grains including wheat, and the American term corn is an abbreviation of the name **Indian corn.** In South Africa it is called **mealies.** Corn is the seed grain of the tall leafy plant *Zea mays,* of which there are innumerable varieties of subspecies. It grows in temperate climates and in the high elevations of the tropics where there is a warm growing season without cold nights, but high commercial yields are limited to areas where there is a combination of well-drained friable soil, plenty of moisture, few cloudy days, and no night temperatures below 66°F (19°C) during the growing season of 4 months. Corn is an unnatural plant, with seeds not adapted for natural dispersal; it does not revert to a wild species. It is a product of long selection. No wild plants have ever been found, but it is believed to have been a cultivated selection from the grass **teosinte** of Mexico. About half the world production of corn is in the United States and Argentina, but large amounts are also grown in southern Europe and northern India.

Confectionery flakes, used as an additive and conditioner in candy, cookies, and pastries, is a bland, yellowish, flaky powder made from degerminated yellow corn. It contains 8% protein, and is pregelatinized to require no cooking. The **pregelatinized corn flour** of the General Foods Corp., used to improve texture, binding qualities, and flavor of bakery products, is a cream-colored powder which hydrates in cold water and needs no cooking. It contains 82% starch, 9 protein, 1 corn oil, and 8 moisture, and is a food ingredient rather than an additive, although it may replace 10% of the wheat flour. In the corn belt of the United States, 40% of the corn grown is used for hog feed, while in the dairy belt the hogs are fed on skim milk, buttermilk, and whey, and most of the corn is fed to poultry or shipped commercially.

Corn grains grow in rows on a cob enclosed by leafy bracts. They are high in starch and other food elements, and form a valuable stock feed especially for hogs and poultry. Nearly 90% of the commercial corn in the United States is for animal feed. But corn is one of the cheapest and easiest sources of starch, and much of the Argentine corn is used for starch and glucose.

Sweet corn is a type of **soft corn,** *Z. saccharata,* cultivated for direct eating and for canning. There are about 70 varieties grown widely on farms, but not cultivated for industrial applications. **Popcorn,** *Z. everta,* has very hard, small, elongated oval grains which, when heated, explode into a white, fluffy, edible mass without further cooking. It was used by the Indians as a food for journeys, and is now grown for food and confections. The corns cultivated for stock feeding and for starch and glucose are varieties of **flint corn,** *Z. indurata,* and **dent corn,** *Z. indentata.* Flint corn has long cylindrical ears with hard smooth grains of various colors. Dent corn has larger and longer ears which are tapering, with white or yellow grains. About 300 varieties of dent corn are grown in the corn belt of the United States, while the Argentine corn is largely flint varieties which yield high starch. Much of the corn grown in the United States is **hybrid corn.** This is not a species, but consists of special seed stocks produced by crossing inbred strains. It is resistant to disease and gives high yields. The **waxy corn** grown in Iowa produces a starch comparable with the root starches. In the wet milling of corn for the production of cornstarch, the germ portion of the grain is separated as a by-product and used for the extraction of **corn oil,** or **maize oil.** The germ contains 50% oil which is a bright-yellow liquid of specific gravity 0.920 to 0.925, iodine value 123. It contains 56% linoleic, 7 palmitic, 3 stearic, and the balance mainly oleic acid. About 1.75 lb (0.80 kg) of oil per bushel of corn is obtained by crushing the germ, and another 1.4% is obtained by solvent extraction. About 1% of oil remains in the **corn oil meal** marketed as feed. Corn oil is used as an edible oil as a substitute for olive oil and in margarine, and also in soaps, belt dressings, core oils, and for vulcanizing into factice. Corn syrups and glucose are produced directly from the starchy corns. **Zein,** of the Corn Products Refining Co., is a protein extracted from corn. It is dissolved in alcohol to form a lacquerlike solution which will dry to a hard, tough film. It is used as a substitute for shellac and is more water-resistant than shellac. **Zein G210** is a water solution of prolamine protein extracted from corn gluten, used to produce hard, tough, grease-resistant coatings, and for formulating polishes and inks. **Corn tassels** are used for livestock and poultry feed. They are a rich source of vitamins. About 270 lb (122 kg) of dry tassels are produced per acre. **Cornstalks** contain up to 11% sugars, usually about 8% sucrose, and 2 other sugars, but little sugar is produced commercially from this source, the stalks being used as cattle feed. **Corncobs** are used to produce **cob meal** for feeds, and also processed to produce lignin, xylose, furfural, and dextrose. **Korn-Kob,** of Kube-Kut, Inc., is granular corn cob used as an abrasive material for finishing metal parts in tumbling barrels. It is tougher than maple and will not absorb water as wood granules do.

Kafir corn is a variety of sorghum grass not related to true corn. The plant is a tall annual with a stalk similar to corn but with smaller leaves and long cylindrical beardless heads containing small round seed grains. It is widely grown in tropical Africa, and a number of subvarieties are grown on a limited scale in Kansas, Texas, and Oklahoma. The grain is similar in composition to corn, but has a peculiar characteristic flavor. It is used as flour in bread mixtures, and in biscuit and waffle flour.

CORROSION-RESISTANT CAST ALLOYS. The name usually refers to the chemical-resistant chromium or chromium-nickel cast ferrous alloys. They are also referred to as **cast stainless steels.** Many of these alloys are also heat-resistant, or resistant to scaling at high temperatures, and there is no real dividing line between the corrosion-resistant and the heat-resistant alloys.

These cast ferrous high alloys are specified by a designation code established by the Alloy Casting Division of the Steel Founders' Society of America. They are used for continuous or intermittent service in corrosive environments at temperatures less than 1200°F (649°C). They have a minimum of 8% alloy content. In general, corrosion resistance of corresponding cast and wrought alloys is comparable. At room temperature, all corrosion resistant grades tolerate food products, oxidizing salts and acids, and ordinary water. As temperatures and concentrations of corrodants increase, the choice of grade narrows. The **iron-chromium cast steels** are comparable to the 400 series of wrought stainless steels. Some of the grades are martensitic and hardenable by heat treatment, and others are ferritic and virtually nonhardenable. The martensitic grades have their best corrosion resistance when fully hardened. The ferritic grades are normally supplied in the annealed condition. The iron-chromium grades are generally highly resistant to oxidizing solutions and are used for parts and equipment in chemical plants processing nitric acid and nitrates. De-aerated or reducing conditions are unfavorable to them. In general, the ferritic grades have greater resistance to most corrosive environments than the martensitic grades.

Iron-chromium-nickel cast steels are austenitic grades, generally comparable to the 300 series of wrought stainless steels. Because austenitic steels undergo no change in phase, heat treatment has only a minor effect on mechanical properties. However, these alloys must be properly heat-treated to ensure complete solution of carbides for maximum corrosion resistance. Within this group there are grades suitable for parts exposed to strong, hot, weakly oxidizing solutions such as sulfurous, sulfuric, acetic, and phosphoric acids. Three grades are suitable for handling hot chlorides and hydrochloric and hydrofluoric acids.

CORUNDUM. A very hard crystalline mineral used chiefly as an abrasive, especially for grinding and polishing optical glass. It is **aluminum oxide,** Al_2O_3, in the alpha, or hexagonal, crystal form, usually containing some lime and other impurities. It is found in India, Burma, Brazil, and in the states of Georgia and the Carolinas, but most of the commercial production is in South Africa. The physical properties are theoretically the same as for synthetic alpha alumina, but they are not uniform. The melting point and the hardness are generally lower because of impurities, and the crystal structure also varies. The hexagonal crystals are usually tapered or barrel-shaped, but may be flat with rhombohedral faces.

The Hindu word corundum was originally applied to gemstones. The ruby and the sapphire are corundum crystals colored with oxides. **Oriental topaz** is yellow corundum containing ferric oxide. **Oriental emerald** is a rare green corundum, but it does not have the composition of the emerald, and the use of the name is discouraged in the jewelry industry. The clear-colored crystals are sorted out as gemstones, and the premium ore is the large-crystal material left after sorting. Some material is shipped in grain. The crude ore is washed, crushed, and graded. There are four grades of abrasive corundum shipped from South Africa: Grade A is over 92% Al_2O_3, Grade B is 90 to 92%, Grade C is 85 to 90%, and Grade D is under 82%. In the United States most of the natural corundum used for optical-glass grinding is in sizes from 60 to 275 mesh, while the grain sizes for coarse grinding and snagging wheels are 8 to 36 mesh. Corundum is now largely replaced by the more uniform manufactured aluminum oxide, and even the name **synthetic corundum,** or the German name **Sintercorund,** is no longer used.

COSMETICS. Substances applied to the outer surface of the body for enhancing appearance and/or for improving the condition of the skin. Most cosmetics also contain odorants and perfume oil. **Face powders** are composed of white pigments having high covering power, such as titanium oxide and zinc oxide; pigments, such as iron oxide and talc (hydrated magnesium silicate), to import slip; and adhesion-promoting ingredients, such as zinc or magnesium stearate. **Rouges** for the face, which contain many of the ingredients present in face powders, are produced in pressed powder or paste form. The coloring agents are usually water-insoluble bright red lakes and the binder is an oil, lanolin, or gum tragocanth. The ingredients of lipstick are principally a vehicle of castor oil and a mixture of waxes, such as beeswax, carnauba wax, candililla wax, lanolin, butyl stearate, and spermaceti. A great variety of other substances are used for special effects. The color ingredients are usually lakes.

Mascaras, used on eyelashes, are made of an oil-soluble soap base, such as triethanolamine stearate; waxes; and color pigments, such as carbon blacks, iron oxide, and ultramarine blue.

Nail polishes, or **nail lacquers,** are made of a nitrocellulose, gum resins, and plasticizers dissolved in a mixture of solvents. For color and opacity, lakes and a substance like titanium oxide are also present.

Although produced in great variety, most **skin creams,** or **cold creams,** are emulsions composed of oils, water, beeswax, and borax. A typical cold cream contains spermaceti, beeswax, oil of lemon or mineral oil, borax, and rose water. **Handcreams** and **hand lotions** for protection against chapping are emulsions formed from a soap, an oil, and glycerine. Other ingredients that can be present include water.

The active ingredients in **astringents,** sold by the name of **skin bracers** or **aftershave lotions,** are witch hazel or alcohol. Often they contain 50% water by volume. **Refiners** are astringents containing aluminum salts that when applied to the skin cause slight swelling, which in turn causes the pores to look smaller for a brief period of time. **Clarifiers** are liquids containing such chemicals as bromelin, resorcinol, or a salicylate, which remove the skin's top layer of dead cells and give the skin a fresher appearance. **Facial masks,** consisting of various "clay" minerals, such as bentonite and kaolin, produce a tight film over the skin upon drying, causing the skin pores to become smaller. Paint-on–peel-off masks use polyvinyl alcohol or vinyl pyrrolidone to form the dry film.

Suntan lotions are formulated to protect the skin against damage from excessive exposure to sunlight. They generally are composed of ingredients similar to those in other skin creams. In addition, however, substances that screen out ultraviolet radiation are present.

Deodorants are of two different types. **Antiperspirants** use zinc and/or aluminum salts that have an astringent action to block the pores through which perspiration is secreted. Other deodorants prevent the bacterial decomposition of the perspiration that produces unwanted odors. These **antibacterial deodorants** contain germicides, such as hexachlorophene.

Bath salts are generally composed of sodium sesquicarbonate or sodium phosphates dissolved in alcohol along with some color and perfume oil. **Bubble bath preparations** contain foaming agents such as sulfated alcohols or sulfated glyceryl monolaurate. In one type of **bath oil,** perfume oils are mixed with an agent such as polyoxyethylene sorbitan monolaurate, which disperses the oil in the water. In another type of bath oil, the perfume is dissolved in a low-viscosity oil.

Shampoos for washing hair are composed of one or more detergent materials. **Soaps** derived from coconut oil are the most widely used because they are high in detergency, are excellent foaming agents, and are resistant to precipitation by hard water. In recent years increasing use has been made of **synthetic detergents,** such as sulphated castor oil, sulphated lauryl alcohol, and sulphated glyceryl monolaurate.

Hair rinses and **hair conditioners** are intended to restore the hair to its natural condition after shampooing or the use of various treatments. The **acid rinses** remove scum left by the shampoo and restore the hair's acid pH to its previous level. The **conditioning rinses,** which restore the hair's natural oily coating, contain stearalkonium chloride. Also included may be such ingredients as an alkali, an emollient of oil or fatty substance, thickeners, humectants, and fragrances.

Hair sprays coat the hair with a film that makes the hair strands stick together. Available as lotions, gels, and sprays, they contain a synthetic resin such as vinyl pyrrolidone dissolved in alcohol and water.

COTTON. The white to yellowish fiber of the calyx, or blossom, of several species of plants of the genus *Gossypium* of the mallow family. It is a tropical plant, and the finest and longest fibers are produced in hot climates, but the plant grows well in a belt across southeastern United States and as far north as Virginia. It requires a growing season of about 200 days with an average summer temperature about 75°F (24°C) and a dry season during the time of ripening and picking. Cotton was used in India and China in most ancient times, was described in Greece as a vegetable wool of India, but was not used in Europe until the early Middle Ages. All of the Asiatic species are short-staple, and the long-staple cottons are from species cultivated by the American Indians. Cotton has a wide variety of uses for making fabrics, cordage, and padding, and for producing cellulose for plastics, rayon, and explosives.

There are many species and varieties of the plant, yielding fibers of varying lengths, coarseness, whiteness, and silkiness. Cotton fiber contains 88 to 96% cellulose (dry weight), together with protein, pectin, sugars, and 0.4 to 0.8% wax. Ordinary treatment does not remove the wax. When the wax is removed by ether extraction, the fiber is stronger but is harsh and difficult to spin. The most noted classes are Sea Island, Egyptian, American upland, Brazilian, Arabian, and Nanking. **Sea Island cotton,** *G. barbadense,* was native to the West Indies, and named when brought to the islands off the American coast. It is grown best in hot moist climates, and is the longest, finest, and silkiest of the fibers. Its length varies from 1¼ to 2½ in (3.18 to 6.35 cm), but it is cream-colored. **Egyptian cotton,** grown in Egypt and the Sudan, came originally from Peruvian seed. **Peruvian cotton,** *G. acuminatum,* is long-staple, silky, has strength and firmness, but is brownish in color. The **tanguis cotton** from Peru is valued for fine English fabrics.

Egyptian cotton, or **maco cotton,** is next in quality to Sea Island. The long staple is from 1⅛ to 1⅜ in (2.86 to 3.49 cm), and the extra-long staple is over 1⅜ in (3.49 cm). It has a fine luster and great strength. It also has a remarkable twist, which makes a strong, fine yarn. It is used chiefly in yarns for the production of fine fabrics, thread, and automobile-tire fabrics. **American-Egyptian cotton** is grown in Arizona. The fiber has an average length of 1⅝ in (4.13 cm), and has the same uses as the Egyptian. **Upland cotton,** *G. hirsutum,* is the species originally grown by the Aztecs of Mexico. It is whiter than Egyptian or Sea Island cotton, and is the easiest and cheapest to grow. There are 1,200 named varieties of this plant. The short-staple upland has a fiber under 1⅛ in (2.86 cm) in length, and can be spun only into coarse and medium yarns, but it is the most widely grown of cottons in the United States. Long-staple upland is from 1⅛ to 1⅜ in (2.86 to 3.49 cm) in length. The common grades of cotton fiber in the United States vary in diameter from 0.0006 to 0.0009 in (0.0152 to 0.0229 mm). Sea Island cotton fiber is as fine as 0.0002 in (0.005 mm), compared with 0.001 in (0.025 mm), for the coarse Indian cotton. The cotton of India, China, and the Near East is from *G. herbaceum,* and the fiber is short, ⅜ to ¾ in (0.95 to 1.91 cm), but strong.

Cotton linters removed from the cottonseed after ginning are from 0.04 to 0.6 in (0.10 to 1.5 cm) long. The first cuts, or longer fibers, are used for upholstery and for mattresses, and amount to 20 to 75 lb (9 to 34 kg) per ton (907 kg) of seed. The second-cut short fibers vary from 125 to 180 lb (57 to 82 kg) per ton (907 kg) of seed, and are called **hull fiber.** The No. 1 grade of long linters is spinnable, and can be used for mixing with cotton for yarns. This grade is also used for making absorbent cotton. The short hull fiber is cleaned and processed to produce **chemical cotton,** which is a pure grade of alpha cellulose used for making rayon, nitrocellulose, and plastics. Chemical cotton is marketed as loose pulp in bales, and as sheet pulp with the sheet stacked in bales of 200 or 400 lb (91 or 181 kg), or with the continuous sheet in rolls. Formerly, cotton linters were considered as the only source of pure cellulose for making nitrocellulose explosives, but pure alpha cellulose from wood is now used for this purpose.

Chaco cotton, grown in Argentina, is from Louisiana seed, and probably 70% of total world cotton is now grown from American upland seed although it varies in characteristics because of differences in climate and soil. Cotton is shipped in bales of 478 lb (216 kg) each. **Cotton yarn** is put up in 840-yd (768-m) hanks, and the number, or count, of cotton yarn indicates the number of hanks to the pound. No. 10 cotton yarn, for example, has 10 hanks, or 8,400 yd/lb (16,933 kg/m).

Mercerized cotton, developed in 1851 by John Mercer, is prepared by immersing the yarn in a stretched condition in a solution of sodium hydroxide, washing, and neutralizing with dilute sulfuric acid. Mercerized yarns have a silky luster resembling silk, are stronger, have less shrinkage, and have greater affinity for dyes. The fabrics are used as a lower-cost substitute for silk, or the yarns are mixed with silk.

Absorbent cotton is cotton fiber that has been thoroughly cleaned and has had its natural wax removed with a solvent such as ether. It is very absorbent and will hold water. It is marketed in sterilized packages for medical use. **Cotton batting** is raw cotton carded into matted sheets, and usually put up in rolls to be used for padding purposes. **Cotton waste,** used in machine shops for wiping under the general name of waste, is usually in mixed colors, but the best grades are generally all white, of clean soft yarns and threads without sizing. It is very oil-absorbent. **Comber waste** consists of the lengths of fiber up to 1 in (2.5 cm), and is not sold with the waste from yarns, but is sent to mills that produce cheap fabrics. **Cotton fillers,** used as reinforcing materials in molding plastics to replace wood flour or other fibers, are made by cutting cotton waste or fabric pieces into short lengths. **Filfloc,** of the Rayon Processing Co., is cotton flock for this purpose; **Fabrifil** is cotton fabric cut into small pieces; **Cordfil** is cotton cord cut into very short pieces. These fillers give greater strength to the molded product than wood flour. **Acetylated cotton** is a mildew-proof cotton made by converting part of the fiber to cellulose acetate by chemical treatment of the raw fiber. **Aminized cotton** is produced by reacting the raw cotton with aminoethyl sulfuric acid in an alkaline solution. Amino groups are chemically combined with the cellulose of the fiber, which gives ion-exchange properties and good affinity for acid wool dyes, and also absorption of metallic waterproofing agents. **Cyanoethylated cotton** is produced by treating the fibers with acrylonitrile, and caustic and acetic acid. The acrylonitrile reacts with the hydrogen of the hydroxyl groups, forming cyanoethyl ether groups in the fiber. The fibers retain the original feel and appearance, but have increased heat strength, better receptiveness to dyes, and strong resistance to mildew and bacteria attack. Another method of adding strength, chemical resistance, and dyeing capacity to cotton fibers is by treating them with anhydrous monoethylamine. It forms an amine-cellulose complex instead of the hydrogen bond. Since cotton is nearly pure cellulose, many chemical variations can be made, and even some dyes may alter the fiber.

COTTON FABRICS. Cotton cloth is made in many types of weave and many weights, from the light, semitransparent **voile,** made of two-ply, hard-twisted yarn, and **batiste,** a fine, plain-woven fabric, to the coarse and heavy canvas and duck. They may have printed designs, as in **calico,** which is highly sized; or yarn-dyed plain stripes, plaids, or checks, as in **gingham;** or woven figures, as in **madras. Muslin,** a plain white fabric widely used for garments, filtering, linings, and polishing cloths, has a downy nap on the surface. The full-bleached cloth is usually of finer yarns than the unbleached. Cheaper grades are usually heavily sized, and the sizing is removed in washing. **Crinoline** is an open-weave fabric of coarse cotton yarn, and is heavily sized to give stiffness. It was originally made as a dress fabric of horsehair and linen. It is now used for interlinings, and as a supporting medium where a stiff, coarse fabric is needed. **Wigan** is similar to crinoline, but is more closely woven. **Percale** is a softer fabric similar to calico but with a higher yarn count. **Swiss** is a plain-woven, fine, thin muslin, stiff and crisp. **Dotted Swiss** is a very thin, transparent, plain-woven cotton with colored swivel or lappet woven dots. It is sized stiff and crisp. **Dimity** is a plain-woven, sheer fabric with ribs in the form of corded checks or stripes. It comes in white or colors. **Organdy** is a plain-woven, thin, transparent, crisp fabric stiffened with shellac or gum, usually in delicate color shades. All of these are plain-woven. **Poplin** is a lateral-ribbed fabric, often mercerized. It is heavier than broadcloth. **Rep** has a rib produced by heavy warp yarns. **Crash** is a rough-texture fabric with effects produced by novelty yarns. **Charmeuse** in the cotton industry designates a soft, fine, satin-weave fabric of Egyptian cotton used industrially as a lining material. **Chambray** is a plain-woven, lightweight cotton similar to gingham but with no pattern and a dyed warp and white filling. It is used for linings, shirtings, and dresses. **Cotton damask** is a type of jacquard-figured fabric having warp sateen figures in a filling sateen ground, or vice versa. The surface threads of the figures lie at right angles to those in the ground so that the light is diffusely reflected, causing them to stand out in bold relief. The fabric is usually of coarse or medium yarns, 15's to 30's, bleached and finished to imitate linen. **Cotton crepe** is a cotton fabric having a pebbled surface. The pebble is produced with sulfonated oil, lauric acid ester oil, or other soluble oil which is washed off after the treatment. When the word crepe is used alone, it usually signifies silk crepe. **Domet** is a warp-stripe cotton fabric similar to flannel, used for apparel linings. **Venetian** is a highly mercerized, stout, closely woven fabric with the yarn in reverse twist. It is used as a lining for hats, pocketbooks, and luggage. **Cottonade** is a coarse heavy cotton fabric made to look like woolens and worsteds in weave and finish, and is used for men's suit linings. **Eiderdown** is a cotton fabric of knitted soft-spun yarns, heavily napped on one or both sides. It is used for shoe and glove linings. **Tarlatan** is a thin cotton fabric with a net weave, heavily sized, used for linings. **Cambric** was originally a fine, thin, hard-woven linen but is now a strong cotton fabric of fine weave and hard-twist yarn. It was used as **varnished cambric** and **varnished cloth** with a coating of insulating varnish or synthetic resin. The strength exceeded that of the older **varnished silk** but was less than **varnished rayon.** A 0.003- to 0.008-in (0.076- to 0.203-mm) thick fabric of Irvington

Varnish & Insulator Co. made from high-tenacity rayon has a dielectric strength of 1,000 volts per mil (39.4 $\times$ 10^6 volts per meter).

Strex, developed by Uniroyal, Inc., is an elastic full-cotton fabric that has 100% elongation without the use of rubber. It is made from yarn that has a twisting like a coiled spring. The fabric is used for surgical bandages, gloves, and wearing apparel. **Glass cloth** is a name given to cotton fabric made of smooth, hard-twisted yarns which do not lint. It is used for wiping glass, but is now largely replaced by silicone-treated soft papers. It may be of the type known as **sponge cloth,** which is a twill fabric of **nub yarn** or honeycomb effect, or it may be of **terry cloth,** which has a heavy loop pile on one or both sides. Another **wiping cloth** for glass and instruments where a lint-free characteristic is important is made with a cotton warp and a high-tenacity rayon filling. It is strong, soft, and absorbent. The **Nun-Lint cloth** of Rittenbaum Bros., Inc., for polishing glass and fine instruments, is a nonwoven fabric made by binding the cotton fibers with a plastic.

Twill is a fabric in which the threads form diagonal lines. **Tackle twill,** used for football uniforms, is also used in olive-drab color for Army parachute troop uniforms. It is a strong, snag-resistant fabric having a right-hand twill with a rayon warp and combed cotton filling. It is 8.5 oz/yd^2 (0.29 kg/m^2), 180-lb (82-kg) warp, and 80-lb (36-kg) filling. **Cavalry twill** is not a cotton cloth, but is of worsted or rayon twill woven with a diagonal raised cord. It is similar to **gabardine** except that gabardine has a single cord and cavalry twill has a double cord. **Bedford cord** has the cord running lengthwise, and the cord is more pronounced than in cavalry twill. These three are usually woolen fabrics, but **parade twill** is a mercerized cotton fabric of combed two-ply yarns, with the fabric vat-dyed in tan. It is employed for work clothing. **Byrd cloth** is a wind-resistant fabric made originally for Antarctic use. It has a close-twill weave with about 300 threads per in. It is soft and strong, and comes in light and medium weights. **Sateen** is fabric made with a close-twill weave of mercerized cotton in imitation of satin. The wind-resistant sateen used for military garments is a 9 oz/yd^2 (0.30 kg/m^2) cotton fabric in satin weave with two-ply yarn in warp and filling. The thread count is 112 ends per in, 68 picks per in. The fabric is singed, mercerized, and given a water-repellent finish. **Foulard** is a highly mercerized twill-woven cotton with a silky feel. It is plain or printed, and is used for dresses or sportswear. **Cotton duvetyn** is a twill-woven, mercerized cotton fabric with a fine nap that gives it a soft velvety feel. It is much used for apparel linings and pocket linings. **Brilliantine** is a lightweight fabric with a cotton warp and a twilled worsted filling, yarn-dyed. It is used for apparel linings.

Balloon cloth is a plain-woven cotton fabric used originally as a base material in making coated fabrics for the construction of balloons, but now used in many industries under the same name. The various grades differ in weight, thread count, and strength. Grade HH, having 120 threads per inch in each direction, is most widely used. A Navy fabric has a weight of 2.05 oz/yd^2 (0.07 kg/m^2) and a tensile strength of 38 lb/in^2 (0.26 MPa) in each direction. When several layers are built up and rubberized or plastic-coated, they may be on the bias, and the outside layer coated with aluminum paint to reduce the heat absorption. **Gas cell fabric** is a single-ply, coated balloon cloth. **Airplane cloth,** formerly used for fabric-covered training planes, is a plain-woven cotton fabric of two-ply combed yarns mercerized in the yarn. It is usually 4 oz/yd^2 (0.14 kg/m^2), but wide fabrics may be 4.5 oz/yd^2 (0.15 kg/m^2). The cotton is 1⅛ in min staple, and the threads per inch are 80 to 84.

COTTONSEED OIL. One of the most common of the vegetable oils, used primarily as a food oil in salad oils, margarine, cooking fats, and for sardine packing. It also has a wide industrial use in lubricants, cutting oils, soaps, quenching oils, and in paint oils, although soybean oil is used as a more abundant substitute. The hydrogenated oil is widely used as a cooking grease. Its food value is lower than that of lard, but it is often preferred because it is odorless and does not scorch. Cottonseed oil is expressed from the seed of the cotton plant, *Gossypium,* and is entirely a by-product of the cotton industry, its production depending upon the cotton crops. The yield of seed is 890 lb (403 kg) per 478-lb (217-kg) bale of cotton, and 100 lb (45 kg) of seed yields 15.5 lb (7 kg) of oil. When the seeds are crushed whole, the oil is dark in color and requires careful refining. The American practice is to hull the seeds before crushing. The oil is colorless, nearly odorless, and has a specific gravity of 0.915 to 0.921. Upland cottonseed contains about 25% oil, which has 40% linoleic, 30 oleic, and 20 palmitic acids. The residue is caked and sold as **cottonseed meal** for cattle feed and fertilizer. About 900 lb (408 kg) of meal and from 450 to 620 lb (204 to 281 kg) of hulls are obtained per short ton (0.9 metric ton) of seed, the yield of hulls varying inversely with the yield of linters. The American oil has an iodine value up to 110, and a saponification value of 192 to 200. Egyptian and Indian oils are inferior in color, and the Indian oil has a fishy odor and a fluorescence. **Cottonseed stearin** is the solid product obtained by chilling the oil and filtering out the solid portion. It has an iodine value between 85 and 100, and consists largely of palmitin. It is used for margarine, soap, and as a textile size. Winter-yellow cottonseed oil is the expressed oil after the stearin has been removed.

COTTONWOOD. The wood of the large trees *Populus monilifera, P. deltoides,* and other species of the United States and Canada. It is a soft wood of a yellowish-white color and a fine, open grain. It is sometimes called

poplar, or **Carolina poplar,** and **whitewood.** The weight is about 30 lb/ ft^3 (480 kg/m^3). The wood is easy to work, but is not strong and warps easily. It is used for packing boxes, paneling, and general carpentry. The *P. deltoides,* or **eastern cottonwood,** used in paneling, has a specific gravity when kiln-dried of 0.43, a compressive strength perpendicular to the grain of 650 lb/in^2 (4.5 MPa), and a shearing strength parallel to the grain of 660 lb/in^2 (4.6 MPa). This wood comes from the lower Mississippi Valley. **Black cottonwood** is from the large tree *P. trichocarpa,* of the Pacific Coast. The wood is used for boxes, excelsior, and pulpwood. It has a light color, uniform texture, and fairly straight grain. **Swamp cottonwood,** *P. heterophylla,* also called **river cottonwood,** grows in the Mississippi and Ohio River valleys. **Balsam poplar** is from the tree *P. balsamifera,* of the northeastern states. It is a soft weak wood used chiefly for containers and for making excelsior. The tree also goes under the Algonquin name of **tacamahac.** The wood may be marketed as cottonwood even when mixed with aspen. It is an excellent paper-pulp material. The name cottonwood is also applied to the wood of the tree *Bombax malabaricum,* native to India, which produces kapok. The wood is white in color, soft, and weighs about 28 lb/ ft^3 (448 kg/ m^3). It is much softer than cottonwood.

CREOSOTE. Also called **dead oil** and **pitch oil.** A yellowish poisonous oily liquid obtained from the distillation of coal tar. It has the odor of carbolic acid, a specific gravity of 1.03 to 1.08, and a boiling point of 200 to 300°C. The crude **creosote oil** is used as a wood preservative and as a harsh disinfectant. Creosote is also obtained in the distillation of pinewood tar, and is then a yellowish liquid with a smoky odor, a mixture of phenols and derivatives. Creosote oil contains **acridine,** a dibasic pyridine, used as an insecticide, and is also the source of other complex heterocyclic ring compounds. The distillation of wood also produces charcoal, gas, and **methyl acetate,** a sweet-smelling liquid of the composition $CH_3COO \cdot CH_3$, and boiling point 54°C, used as a solvent.

Cresol, also known as **cresylic acid** and as **methyl phenol,** obtained in the distillation of coal tar, is a mixture of three isomers of cresol, $CH_3 \cdot C_6H_4 \cdot OH$, and **xylenol,** $(CH_3)_2 \cdot C_6H_3 \cdot OH$. The crude material is a brownish-yellow liquid solidifying at 11°C. It is used for making plastics, in ore flotation, in refining petroleum, in soap-emulsion cutting oils as a disinfectant, and in medicine as a strong antiseptic such as **Lysol,** which is a 50% solution of cresols in liquid soap. It is also used in the production of other chemicals. **Orthocresol** is a colorless solid with a melting point of 30°C and a boiling point of 191.5°C. It is soluble in alcohol, but only slightly soluble in water. It is used in the manufacture of cumerones, disinfectants, and fumigants, and as a plasticizer. **Metacresol** is a yellow liquid freezing at 12°C and boiling at 202.8°C. It is used in the manufacture of photographic developers, nitrocresols, disinfectant soaps, printing inks, paint and varnish removers; as a preservative in leathers, glues, and pastes; in the reclaiming of rubber; and in making synthetic resins, perfumes, and pharmaceuticals. **Paracresol** is a colorless solid melting at 36°C and boiling at 202.5°C. It is the least soluble of the cresols. It is used in the manufacture of cresotinic acid dyes, disinfectants, and pharmaceuticals.

CRYOLITE. A mineral of the composition Na_3AlF_6, found in commercial quantities in Greenland, and used as a flux in the electric production of aluminum, in the making of special glasses and porcelain, as a binder for abrasive wheels, and in insecticides. One ton (907 kg) of cryolite is used for flux for 40 tons (36,280 kg) of aluminum. For glass batches 30 lb (14 kg) of cryolite is equivalent to 22.7 lb (10 kg) soda ash, 16.3 lb (7 kg) fluorine, and 11 lb (5 kg) aluminum hydrate. It acts as a powerful flux because of its solvent power on silicon, aluminum, and calcium oxides. In opal and **milky glasses** it forms a complex AlF_6 anion, retaining the alumina and preventing loss of the fluorine. Cryolite occurs in masses of a vitreous luster, colorless to white, with a hardness of 2.5. It fuses easily. **Kryolith,** of the Pennsylvania Salt Co., is cryolite of 98 to 99% purity, and **Kyrocide** is a grade of 90% purity. The latter is the dust from the natural ore, and is used as an insecticide. Synthetic cryolite is made by reacting fluorspar with boric acid to form fluoroboric acid, and then reacting with hydrated alumina and sodium carbonate to form cryolite and regenerate boric acid.

CRYPTOSTEGIA RUBBER. Rubber obtained from the leaves of two species of perennial vines native to Malagasy, *Cryptostegia grandiflora* and *C. madagascariensis.* The former was grown in India, and the rubber was known as **palay rubber.** It was brought to Mexico and Florida as an ornamental plant and now grows extensively in Mexico and the West Indies. The maximum rubber content is found in the leaves 3½ months old, at which time it is 2 to 3% of the dry weight of the leaf. There is also about 8% resin in the leaf, which must be separated from the rubber as it makes it soft and tacky. The *C. madagascariensis* contains less rubber, but the leaves of hybrid plants grown from both species give increased yields of rubber. The hybrid does not come true to type from seed, and is propagated from cuttings. When extracted and separated from the resin, cryptostegia has the same uses as ordinary hevea rubber.

Another plant that yields rubber from the leaves is the desert **milkweed,** *Asclepias erosa, A. subulata,* and other species growing in the dry regions of southwest United States. The short and slender leaves are produced only on the young stems and the gathering season is short. The dry leaves are ground, and the rubber is obtained by solvent extraction. The average rubber content is about 2%, but as much as 12% has been obtained from some species of wild plants. As with guayule and cryptostegia, a considerable amount of resin is extracted with the rubber. **Goldenrod rubber** is extracted similarly from the leaves of the goldenrod, the dry leaves containing as much as 7% rubber mixed with resin. The species which contains the most rubber is *Solidago leavenworthii.* It does not occur in the plant as a latex, but is in isolated globules in the cells, mostly in the leaf. The **milk bush,** *Euphorbia tirucalli,* of Cuba and Jamaica, also produces rubber of good elasticity, but the crude latex from the bush causes skin blisters, and the extraction requires special treatment.

Dandelion rubber is the gum latex extracted from the roots of the Russian dandelion, which, when separated from the contained resin, has practically the same characteristics as the rubber from the hevea tree. Dandelion rubber, from various species of the genus *Taraxacum,* chiefly the plants known as **kok sagyz, tau sagyz,** and **crim sagyz,** native to Turkmen, is produced in Russia. The plant is grown only on a small scale in the United States and Canada. The roots, which extend 15 to 20 in (38 to 51 cm) into the ground, contain up to 10% rubber after the plant has passed the first-year flowering period. The normal yield is about 6% rubber with also considerable resin. The dry roots also contain a high percentage of inulin.

CUMERONE. A colorless oily liquid of the composition C_8H_6O, used chiefly in making synthetic resins. It occurs in the fractions of naphtha between 165 and 175°C. It has a specific gravity of 1.096, is insoluble in water, and is easily oxidized. Another similar product is **indene,** C_9H_{10}, a colorless liquid of specific gravity 0.993, boiling at about 182°C, obtained from coal tar. When oxidized it forms phthalic acid, and with sulfuric acid it polymerizes readily. It is a bicyclic ring compound with an active double bond and methylene group in the five-membered ring fused to the benzene nucleus. It can be reacted with butadiene to form an **indene-butadiene rubber** of superior properties. All of the **cumenes** are variants of benzene.

The **indene resins** are classed with the cumerone resins, but they are lighter in color and are used in varnishes. The simple polymer, or diindene resin, is a crystalline solid melting at about 58°C. The polyindene resins are made by polymerizing indene with ultraviolet light and oxygen. The **cumerone resins,** which are polymers of $C_6H_4 \cdot O \cdot CH{:}CH$, made by the action of sulfuric or phosphoric acid on cumerone, are very soluble in organic solvents, and are used in lacquers, waterproofing compounds, molding, and adhesives. The specific gravity of the molded resins is 1.05 to 1.15. They have high dielectric strength. **Paracumerone,** also called **paraindene** and **cumar gum,** is a synthetic resin which is a copolymer of cumerone and indene. The grades vary from a soft gum to a hard brown solid, with melting points from 5 to 140°C. Varnishes made with it are resistant to alkalies. **Nevindene,** of the Neville Co., is a cumerone-indene resin of specific gravity 1.08 and melting point 10 to 160°C, used for compounding with rubber and synthetics. **Nevilloid C-55** is a **cumerone-indene resin** in water emulsion for coatings. It forms cohesive translucent films of slightly tacky nature. Blended with melamine resin it forms a clear and hard film. **Cumar** is the name of a cumerone-indene resin of the Barrett Co., but the name cumar has been applied to a range of pale-yellow to reddish-brown coal-tar resins which are polymers of indene, cumerone, and other compounds, with melting points from 45 to 160°C. They are used in rubber compounding to increase tensile strength and tear resistance. **Piccoumaron resins** of the Pennsylvania Industrial Chemical Corp. are paracumerone-indene thermoplastic resins produced by the polymerization of unsaturates in coal-tar oils. They vary from light liquids to tacky solids with melting points from 10 to 120°C. The colors vary from pale yellow to reddish brown. They are resistant to alkalies and are used in paints and waterproofings for concrete, and in adhesives for floor tile.

CUPRONICKEL. Any alloy of copper and nickel, as nickel and copper form solid solutions in all proportions. Nickel and copper are found naturally alloyed in many copper mines, and as the two metals are difficult to separate, the cupronickel alloys were among the first alloys used. They are ductile and malleable, are very corrosion-resistant, and have a maximum hardness at about 50% nickel. A 15% nickel alloy can be rolled from a thickness of 1⅝ in (4 cm) to about 0.040 in (0.10 cm) without intermediate annealing. Nickel whitens copper and gives the alloys a characteristic pinkish-white color sometimes called yellow, which can be whitened by the addition of small amounts of cobalt. At least 10% nickel is needed to obtain a nickel-white color, and this amount gives high corrosion resistance. Cupronickel with 2.5% nickel is used for the driving bands of shells, 15% nickel for bullet jackets and condenser tubes, and 25% for coinage. The higher-nickel alloys are used for resistance wire, for corrosion-resistant equipment, and for parts where strength, toughness, and a white color without plating are required. For the latter use, however, the alloys contain other elements. When color is important, the alloys should not contain zinc because of the change in color by dezincification. Aluminum is also used as a whitener and hardener. An old formula for a silvery-white alloy for silverware, called **minargent,** was 100 parts copper, 70 nickel, 6 antimony, and 2 aluminum. The standard cupronickels are 5, 15, 20, and 30% nickel.

Supernickel is a name given by the American Brass Co. to the 70–30 alloy for condenser tubes. This alloy and the 80–20 alloy are also marketed by the Revere Copper & Brass, Inc., for condenser tubes. The 80–20 alloy is used to replace nickel-silver for electrical springs. It has strength and flexural endurance comparable to Monel metal. The 70–30 alloy has a tensile strength, annealed, of 49,000 lb/in^2 (337 MPa) with elongation 50%, and when hard-drawn the strength is 75,000 lb/in^2 (517 MPa) with elongation 5%. The weight is 0.323 lb/in^3 (5.174 kg/m^3). The white alloy of the Scovill Mfg. Co. known as **Adnic,** used for valve diaphragms and parts for chemical equipment, has 70% copper, 29 nickel, and 1 tin. When hard-drawn it has a tensile strength of 113,000 lb/in^2 (779 MPa) and elongation 10%. The hot-rolled rod has a strength of 65,000 lb/in^2 (448 MPa) and elongation 45%. This is essentially the same alloy marketed by the Revere Copper & Brass, Inc., as **admiralty metal. Revere alloy 508,** originally developed by the British Non-Ferrous Metals Research Association for condenser tubes, contains 88.5% copper, 10 nickel, and 1.5 iron. **Cufenloy 30,** of the Phelps Dodge Copper Products Corp., for heat-exchanger tubes to operate at temperatures to 800°F (427°C) and pressures to 8,000 lb/in^2 (55 MPa), contains 29.1% nickel, 0.5 iron, 0.35 manganese, and the balance copper. The tensile strength is 77,000 lb/in^2 (530 MPa), and at 1050°F (566°C) the strength is 43,000 lb/in^2 (296 MPa). This type of alloy is also used for castings, giving a cast strength of 65,000 lb/in^2 (448 MPa) with elongation of 20%.

Small amounts of silicon added to cupronickel form a nickel silicide that gives a **hardenable copper,** which is the ancient tool material of the Incas. But, since the nickel is usually low in these alloys, they are not classed with the cupronickels. An alloy of this type marketed by the American Brass Co. under the name of **Tempaloy** contains 95% copper, 4 nickel, and 1 silicon. It is hard and strong, and can be forged or cold-rolled. Annealing by heating throws the nickel silicide into solid solution and produces a soft and ductile metal, which can be hardened again by holding at a temperature of 450°C for several hours. The tensile strength is from 50,000 to 150,000 lb/in^2 (344 to 1,034 MPa) with hardness above 200 Brinell. Small amounts of manganese and iron may be added to cupronickel to harden and strengthen the alloy. **Davis metal,** of the Chapman Valve Mfg. Co., for valves and fittings, contains 67% copper, 29 nickel, 2 iron, 1.5 manganese, and 0.5 carbon and silicon. The castings are hard and corrosion-resistant. **Cufenium,** a white alloy used as a base metal in tableware, contains 72% copper, 22 nickel, and 6 iron. Some chromium may be added for further corrosion resistance. **Everbrite,** of the American Manganese Bronze Co., has 60% copper, 30 nickel, 3 iron, 3 silicon, and 3 chromium. It is white in color, has a tensile strength, cast, of 75,000 lb/in^2 (517 MPa), elongation 14%, and Brinell hardness 170. **Cataract metal,** of the Niagara Falls Smelting & Refining Corp., is the name of a series of cupronickel alloys containing small amounts of other elements. **Copper-nickel alloy** is the name used for the 75–25 and 50–50 cupronickel slabs, ingots, and shot used for adding nickel to brasses. Federal specifications for copper-nickel alloy call for a minimum of 65% copper and 25 nickel with other elements allowable.

The high nickel-cupronickel alloys are very corrosion-resistant at high temperatures, and are used for electrical-resistance wire and strip where the temperature does not exceed 1100°F (593°C). **Lucero,** of the Driver-Harris Co., is a 70–30 alloy, and **Copel** is a 55–45 alloy for resistance wire. The 55–45 alloy has an electrical resistivity of 294 $\Omega/(\text{mil}\cdot\text{ft})^3$, and tensile strength, cold-drawn, of 140,000 lb/in^2 (965 MPa). Because of its constant-temperature coefficient of resistance, it was given the name of **Constantan** when used in low-temperature pyrometers. It has a high thermoelectric effect with either copper or iron, but cannot be used for high-temperature pyrometers as the melting point is 1290°C. **Advance metal,** of the Driver-Harris Co., and **Cupron,** of the Wilber B. Driver Co., are names for this alloy for pyrometer use. A **copper-nickel-tin alloy,** developed by Bell Labs for springs, relays, and connectors, has the tin uniformly dispersed in a copper-nickel matrix. The alloy is about 15% stronger than copper-beryllium and about 50% stronger than phosphor bronze.

CURUPAY. The wood of the tree *Piptadenia cebil,* native to Argentina, Paraguay, and Brazil. In northern Argentina and Paraguay it is also known under the Guarani name **cevil.** The wood is very hard and heavy, weighing 74 lb/ft^3 (1,185 kg/m^3), and it has a reddish color and a handsome, wavy grain. It is used as an ornamental hardwood and is much employed locally for construction. Another wood of the same order is **angico,** from the *Angico rigida* of Brazil, also known as **queenwood;** the lighter-colored wood is called **angico vermelho,** or **yellow angico.** It is very hard, with a dense close grain, a reddish-brown color, and weight of 70 lb/ft^3 (1,121 kg/m^3). It is employed where a heavy hardwood is required, and in cabinetmaking.

CUTTING ALLOYS. Usually of complex Co-Cr-W-Fe-Si-C composition, used for lathe and planer tools for cutting hard metals. They form a class distinct from the cemented carbides which are not true alloys, from the refractory hard metals which are chemical compounds, and from the cobalt high-speed steels which are high in iron and usually have less carbon. The hardness is inherent in the alloy, and is not obtained by heat treatment as with the cobalt steels. Cutting alloys are cast to shape and are usually marketed in the form of tool bits and shear blades. Complex alloys, however, may have heat-transition points at which the metal complexes change structure, limiting the range of use.

Since the development of balanced super-high-speed steels and cermet-type cutting tools, these alloys with a high proportion of the scarcer cobalt have lost their importance as cutting alloys, and, because of their high corrosion, heat, and wear resistance, are used chiefly for weld-facing rods and heat-corrosion applications. One of the earliest of the alloys, called **Cooperite,** was based on nickel. The first of the commercial cobalt cutting alloys was **Stellite,** of the Haynes Stellite Co., in various composition grades and with trade names such as **J-metal** and **Star J-metal.** The hardest alloy, with a Rockwell hardness to C68, contained about 45% cobalt, 32 chromium, 17 tungsten, 1.5 iron, 1.5 silicon, and up to 2.7 carbon. The tensile strength is above 100,000 lb/in^2 (689 MPa), and compressive strength about 325,000 lb/in^2 (2,240 MPa). It is silvery white in color. The **Delloy** of the Penn Rivet Corp. contained a somewhat similar composition. Other similar alloys were **Speedaloy, Rexalloy, Crobalt,** and **Borcoloy,** the latter two containing also boron for added wear resistance. This type of alloy is now also used in **surgical alloys** for surgical tools and dental plates since they are not attacked by body acids and set up no electromotive currents. To make them more workable for this purpose, they usually contain a higher content of cobalt, 60% or more, with a smaller amount of molybdenum instead of tungsten, and with less carbon and silicon.

CUTTING OILS. Oils used on cutting tools and on work being machined to aid in the cutting action. The chief object of the oil is to act as a lubricant between the tool and the work, decreasing the friction. These oils, therefore, are usually heavy oils or compounds distinguished from the thin solutions of soluble oils employed for flooding the work with the object of keeping it cool. However, the cutting oil also carries off much of the heat and enables the tool to stand up longer under the cutting action. There is no sharp dividing line between the cutting oils and the coolants which are designated in the mechanical industries as **soluble oils** because they are generally used in the water solution. These soluble oils also have some lubricating action, but the old coolant known as **soda water** in machine shops is a solution of soap and soda ash in water.

Lard oil is a good cutting oil, but is seldom used alone because of its cost. It is mixed with mineral and vegetable oils, or with hydrogenated oils. The **mineral-lard oils** are used for machining copper alloys or for turning steel where a good surface finish is required. Lard oil mixed with kerosene is used for cutting aluminum and for Monel metal. Ordinary **mineral oils** are used for light cutting operations and combine lubrication and cooling. For cutting brass, an emulsion of oil in soapy solution is often employed. Carbon tetrachloride mixed with turpentine has been used for cutting hard steels. For screw cutting, a paraffin oil mixed with a vegetable oil may be used, but for cutting fine threads a heavy oil mixed with white lead is preferred.

In general, the soluble oils are made by treatment of an oil with sodium hydroxide or other alkali; they emulsify easily because of the formation of sodium oleate and sodium palmitate; and they can be mixed with water in all proportions. Federal specifications for compounded soluble oil permit a maximum of only 10% water, with no separation of the water in 24 h at

75°F (23.5°C). Cutting oils, or **cutting compounds,** are marketed under many trade names. They may vary widely in composition, and contain rosin or rosin oil to improve cutting action, cresol or other disinfectants, corrosion inhibitors, or antioxidants to prevent rancidity, but they are generally based on sulfonated oils, or **sulfurized oils.** The sulfur adds film strength to the oil and gives better penetration of the cutting tool. **Cleartex, Sultex,** and **Transultex** are grades of sulfurized oils of Texaco., Inc. **Sulchlor,** of the Carlisle Chemical Works, is a modified sperm oil with sulfur and chlorine in the molecule. It is used for high-pressure cutting. **Aquadag,** of the Acheson Colloids Co., is a solution of about 22% of colloidal graphite in water. **Antisep,** of E. H. Houghton & Co., is a high-sulfur oil concentrate with an antiseptic and a rust inhibitor.

CYPRESS. A number of different woods are called cypress, but when the name is used alone it is likely to refer to the wood of the **Italian cypress,** *Cupressus sempervirens,* native to the Mediterranean countries but now grown in the Gulf states and in California. This wood is light in weight, soft, light brown in color, and has a pleasant aromatic odor. It is very durable, and is used for furniture, chests, doors, and general construction. The **citrus wood,** or **citron board,** from which the massive dining tables of ancient Rome were made, were heavy plates of the wood of this tree cut across the trunk near the roots to show a variegated grain. The wood was cut in Mauritania. **Arizona cypress,** *C. arizonica,* is a smaller tree, and the wood is used chiefly for fence posts. The wood usually referred to in the eastern United States as cypress, and also as **marsh cypress, red cypress, bald cypress, yellow cypress, gulf cypress,** and **southern cypress,** is from the coniferous tree *Taxodium distichum;* the **pond cypress** is from *T. ascendens,* of the southeastern states. Southern cypress grows along the coast from Delaware to Mexico, especially in Florida and the lower Mississippi Valley. The red cypress is along the coast and the yellow is inland, the coastal types being darker in color. The trees are sometimes very old, reaching a height of 120 ft (37 m) in 800 years. The wood is yellowish red or pink in color, and is moderately hard with an open grain. The weight is about 32 lb/ft^3 (513 kg/m^3). It is very durable, and is valued for shingles, tanks, boatbuilding, or construction where resistance to weather exposure is needed. The wood called yellow cypress on the West Coast, also known as **Sitka cypress, Alaska cedar,** and **yellow cedar,** is from the tree *Chamaecyparis nootkaensis,* or *Cupressus sitkaënsis,* growing on the Pacific Coast from Alaska to Oregon. The trees reach 6 ft (2 m) in diameter and 120 ft (37 m) in height in 500 years. The heartwood is bright yellow, and the sapwood slightly lighter. The wood has a fine, uniform, straight grain, is light in weight, moderately hard, easily worked and polished, shock-resistant, and durable. It is used for furniture, boatbuilding, and interior finish.

Monterey cypress, *C. macrocarpa* of California, is one of the chief trees planted on reforestation projects in New Zealand.

DAMMAR. Also written **damar.** The resin from various species of trees of the genus *Shorea, Balanocarpus,* and *Hopea,* but the name is also applied to the resins of other trees, especially from the *Agathis alba,* the source of Manila copal. There is no dividing line between the dammars and the copals, and dammar may be considered as a recent or nonfossil **copal,** the Malay word damar meaning simply a gum. The best and hardest dammars are from deposits at the bases of the trees, which are then the seasoned or fossil resins like the copals. Dammar is obtained by tapping the trees and collecting the solidified gum after several months. It is used in varnishes, lacquers, adhesives, and coatings. The usual specific gravity is 1.04 to 1.12, and the melting point up to 120°C. The average grade of dammar does not have a melting point much higher than 100°C. Dammar is a spirit varnish resin, gives a flexible film, but is softer and less durable than the copals. It is noted for its complete solubility in turpentine. It is also soluble in alcohol, and the Batavia and Singapore dammars are soluble in chlorinated compounds and in hydrocarbons. Dammar is classified according to color and size, the best grades being colorless and in large lumps. The high-grade pale-colored dammars from Batavia and Sumatra, including the so-called **cat's-eye dammar,** are from species of *Hopea.* Most of the **white dammar** equivalent to Manila copal comes from Malagasy. It is semihard to hard, and is used in paints where resistance to wear is required, as in road-marking paints, but is not as hard as Congo copal. In general, the true dammars are from the *Shorea* and *Balanocarpus,* and they are inferior in hardness to the fossilized resins approaching the copals. The *Shorea* resins are usually dark in color. The **Malayan black dammar, dammar hitam,** is from a species of *Balanocarpus.* The plentiful **dammar penak** is from the Malayan tree *B. heimii,* which also yields the important wood known as **chengal** used for furniture and boatbuilding. **Black dammar** is from the tree *Canarium strictum,* of India, and comes in black, brittle lumps, easily ground to powder. The reddish **dammar sengai** is also from a species of *Canarium.* These are types of elemi. **Dewaxed damar,** for making colorless, glossy lacquers, is highly purified dammar in xylol solution.

DEGRADABLE PLASTICS Plastics that are decomposed by any of three mechanisms—biodegradation, solubility, and photodegradation. **Biodegradable plastics** are those that are susceptible to being assimilated by microorganisms, such as fungi and bacteria, through enzyme action. The assimilating action requires heat, oxygen, and moisture. For all practical purposes almost all synthetic polymers are immune to enzyme attack. Only **aliphatic polyesters** and **urethanes** derived from **aliphatic ester diols** and low-molecular-weight (under 500) unbranched **polyethylene** derivatives can be assimilated. Certain mutant soil microorganisms, when inoculated into resistant types of polymers in waste disposal areas, have increased the degradability of the polymers. Union Carbide Corp. has formulated **polycaprolactone resins** which are biodegradable in contact with a nutrient soil environment. They are not attacked by airborne spores.

The solubility of **water-soluble plastics** varies with formulations, molecular weight, and temperature. **Hydroxypropyl cellulose** is insoluble in water above 115°F (46°C). Below this temperature, when immersed in water, it quickly forms a slippery gel on the outer surface. The gel layer must dissolve and wash away before further dissolving takes place. **Polyethylene oxides** are soluble in water above 150°F (66°C). They are nontoxic, eatable but nonnutritive, nonchloric, and wash through plumbing without damage or clogging. They are resistant to grease, oil, and petroleum hydrocarbons. The solubility of **polyvinyl chloride** depends on degree of alchoholization. Thus, completely alcoholized grades are hot-water-soluble and cold-water-soluble. Partially alcoholized types (about 87%) are soluble in both hot and cold water.

Photodegradable plastics are sensitive to ultraviolet light. Energy in the form of photons breaks down the bonds between the carbon and hydrogen atoms and oxygen-reactive free radicals are formed. The free radicals react with oxygen in the environment to produce peroxide and hydroperoxides that decompose further to produce carbonyl groups, hydroxyl groups, water, and carbon dioxide. The best photodegradable materials are the linear, nonaromatic, molecular structured plastics. Unvulcanized **syndiotactic polybutadiene** is typical. It is degradable under direct sunlight in periods ranging from one week to more than one year. Additives such as pigments, ultraviolet accelerators and promoters, and ultraviolet absorbers and antioxidants promote ultraviolet degradation in **polyethylenes, polystyrenes, polypropylenes, polybutadienes, polybutylenes, ABS,** and **polyvinyl chloride.**

DENATURANTS. Materials used chiefly for mixing with ethyl alcohol to be employed for industrial purposes to prevent the use of the alcohol as a beverage and to make it tax-free under the Tax Free Industrial Alcohol Act. The qualities desired in a denaturant are that its boiling point should be so close to that of the alcohol that it is difficult to remove by ordinary distillation, and that it should be ill-tasting. Some of the denaturants are poisonous and cause death if the alcohol is taken internally. The usual denaturants are **methyl alcohol, pyridine, benzene, kerosene,** and **pine oil.** One or several of these may be employed, but denaturants must be approved by the Bureau of Internal Revenue. Completely **denatured alcohol** is a term used to designate alcohol containing poisonous denaturants, and these are employed only for antifreeze, fuels, and lacquers, but not in contact with the human body. Special denatured alcohol is alcohol containing denaturants authorized for special uses, such as pine oil for hair tonics. Many approved denaturants are marketed under trade names. **Denol** is the name of a mixture of primary and secondary aliphatic higher alcohols. **Agadite** is a compounded petroleum product. **Hydronol** is a hydrogenated organic product. Denaturants are also used in imported oils that are permitted entry at lowered tax rates for industrial use so that they cannot be diverted for edible use. Rapeseed oil, for example, is denatured with brucine.

DERRIS. The root of various species of vines of the bean family, *Derris uliginosa, D. elliptica,* and *D. trifoliata,* growing in Indonesia. It is imported as crude root, and marketed as a fine powder of 200 mesh for use as an insecticide diluted with dusting clay to a rotenone content of 1%, or as a spray in kerosene or other liquid. The root contains **rotenone,** a colorless, odorless, crystalline solid poison of complex composition, $C_{22}H_{22}O_6$, and melting point 163°C. The value of rotenone as an insecticide is that it is highly toxic to cold-blooded animals, including insects and worms, and nonpoisonous to warm-blooded animals. It is widely used as an agricultural insecticide as it is harmless to birds. It is about 30 times more toxic to cutting worms than lead arsenate.

Rotenone is also found in many other plants, and when separated has the same toxic power. **Cubé** is the root of the vine *Lonchocarpus utilis,* of Peru, containing rotenone and used for the same purposes as derris. **Timbó,** also known as **urucú** and as **tingi** and **conambi,** is the root of the vine *L. urucu,* of Brazil, also containing rotenone and used in the same manner. **Barbasco** is a name applied to timbó and all other fish-killing plants of the Orinoco Valley. The Caribs used the root either in shredded or in extract form for catching and killing fish. A cubic foot of root will poison an acre of water without harming the fish as food. The tubers of the wild yam called barbasco yield **diosgenin,** a steroidal used in the synthesis of steroids, which are oxidized to produce cortisone. Other plants of the same family are **nicou, nekoe,** and **haiari** of the Guianas, and rotenone sometimes goes under the name of **nicouline.** The high yield of rotenone from Indonesian derris, up to 12%, is due to careful selection and propagation in cultivation, the semiwild roots of South America sometimes containing only about 2%. The Brazilian government standard for timbó is 4% rotenone content. From 1 to 4% rotenone is also obtained from the long leathery shoots of the perennial weed *Tephrosia virginiana,* known as **devil's shoestring,** growing in Texas. Piperonyl butoxide is sometimes mixed with rotenone to give greater insect-killing power.

DETERGENTS. Materials which have a cleansing action like soap. Although soap itself is a detergent, as are also the sodium silicates and the phosphates, the term usually applies to the synthetic chemicals, often referred to as **detergent soaps** or **soapless soaps,** which give this action. The detergents may be the simple sulfonated fatty acids such as turkey-red oil; the **monopole soaps,** or highly sulfonated fatty acids of the general formula $(SO_2OH)_xR \cdot COONa$; or the **gardinols,** which are sulfonated fatty alcohols.

All of the synthetic detergents are **surface-active agents,** or **surfactants,** with unsymmetrical molecules which concentrate and orient at the interface of the solution to lower interfacial tension. They may be **anion-active agents,** with a positive-active ion; **cation-active agents,** with a negative-active ion; or **nonionic agents.** Most of the household detergents are anion-active and are powders. Most of the nonionics are liquids, and are useful in textile processing since they minimize the difference in dye affinity of various fibers. The cationics have lower detergency power and are usually skin irritants, but they have disinfectant properties and are used in washing machines and dairy cleansers. They are called **invert soaps** by the Germans. The synthetic detergents do not break down in the presence of acids or alkalies, and they do not form sludge and scum, or precipitate salts in hard waters like soap. They do not form quantities of suds like some soaps, but suds contribute little to cleansing, and are not desirable in automatic washing equipment. **Textile softeners** are different from surface-active agents. They are chemicals that attach themselves molecularly to the fibers, the polar, or charged end, of the cation orienting toward the fiber, with the fatty tails exposed to give the softness to the fabric. **Arqued 2HT,** of Armour & Co., is a distearyldimethyl ammonium chloride for this purpose. A special-purpose surfactant used for dispersing oil slicks on the sea is **Dispersol,** of Imperial Chemical Industries. It is a polyethanoxy dissolved in isopropyl alcohol. It is soluble in oil but not in water. It agglomerates the oil into small blobs that are scattered by the winds and eventually destroyed by marine organisms.

Synthetic detergents have now largely replaced soaps for industrial uses. They are employed in textile washing, metal degreasing, paper-pulp processing, and industrial cleansing. They are also used in household cleansers, soapless shampoos, and toothpastes. **Biodegradable detergents** are those which can be chemically disintegrated by bacteria so that the discharged wastes do not contaminate the ground waters. **Millox,** of Millmester Onyx Corp., is a group of biodegradable detergents made by the reaction of sucrose and fatty acids with a linking of ethylene oxide. This type of detergent is more powerful than petroleum-based detergents. **Millox 120** is made from the fatty acids of coconut oil, and **Millox 180** is from tallow. The detergents produced from straight-chain paraffinic hydrocarbons derived from petroleum cracking are **alkyl aryl sulfonates,** R·Ar·SO_3Na, or **alkylbenzene sulfonates.** These detergents do not break down in the wastes and therefore do tend to contaminate the ground waters. The detergent characteristics vary with the number of carbon atoms in the alkyl chain and the arrangement of atoms in the chain. Detergency increases to a maximum at 12 to 15 atoms and then decreases. These detergents are 10 times as bulky as soda ash, but can be mixed with alkaline or phosphate cleaners.

The detergents are more efficient than toilet soaps, but tend to leave the skin with an alkaline hardness. Lecithin may be used in detergent bars to reduce tackiness, and starch may be used for hardening. **Nytron,** of the Solvay Process Co., is a **sodium sulfonate** derived from petroleum hydrocarbons. It is a buff-colored powder. **Surfax 1288,** of E. F. Houghton & Co., is an **aryl sulfopropionate** with only slight detergent power, used in textile processing for rewetting and as a leveling agent for dye baths. **Clavenol,** of the Dexter Chemical Corp., is a **polyethylene glycol** condensate of the nonionic class.

The **Ultrawets** of the Atlantic Refining Co., **Kamenol D** of the Kamen Soap Products Co., **Oronite detergent** of the Standard Oil Co., **Kreelon** of the Wyandotte Chemicals Corp., **Parnol** of Jacques Wold & Co., **Wicamet** of Wica Chemicals, Inc., and **Santomerse** of the Monsanto Chemical Co. are alkyl aryl sulfonates. This type of chemical is available in powder, bead, or paste forms, and one molecule in 40,000 molecules of water gives good detergency. It is effective in hard water or in acid and alkaline solutions. **Sulframin E,** of the Ultra Chemical Works, Inc., is this material in liquid form.

Superonyx, of the Onyx Oil & Chemical Co., is a modified **sodium alkyd sulfate** and is a neutral detergent and dye assistant for processing textiles. **Maprosyl 30,** of the same company, is called a **modified soap.** It has the detergent and emollient properties of soap but does not form scum as soap does, and does not cause skin irritation as many detergents do. Unlike soap, it is soluble in highly alkaline solutions, and unlike most detergents, it has high foaming qualities. It is a **sodium lauroyl sarcosinate** produced from fatty acids, and may also be in the form of stearoyl, linoleyl, or derivatives of other fatty acids. The **sarcosine** is **methyl glycine,** $CH_3NHCH_2CO_2H$, an amino acid occurring in small amounts in animal muscle, but now made synthetically. It is a decomposition product of caffeine. **Lauryl pyridium chloride** is also a soaplike detergent. It is a tan-colored semisolid with a soapy feel and with germicidal properties. It is used for textile washing.

The **Pluronics** of the Wyandotte Chemicals Corp. are nonionic detergents produced from polyoxypropylene glycol, ethylene oxide, and ethylene glycol. When the ethylene oxide content is 70% the detergent is a solid which can be flaked. It is formulated with alkyl sulfonate and sodium carboxymethyl cellulose for laundry work. Somewhat similar chemicals to the detergents are used as **dispersing agents** for latex, paper coatings, dyestuffs, and agricultural sprays. **Daxad 11,** of the Dewey & Almy Chemical Co., is a polymerized salt of alkyl naphthalene sulfonic acid. Its action is to impart an electric charge to each particle, giving a repelling action to space the particles and prevent agglomeration or settling. It increases fluidity and permits a higher solids content in dispersions without increasing the viscosity.

DEXTRIN. Also called **amylin.** A group of compounds with the same empirical formula as starch $(C_6H_{10}O_5)_x$, but with a smaller value of x. The compounds have strong adhesive properties and are used as pastes, particularly for envelopes, gummed paper, and postage stamps, for blending with gum arabic, in pyrotechnic compositions, and in textile finishing. Dextrin is a white, amorphous, odorless powder with a sweetish taste. It dissolves in water to form a syrupy liquid, and is also distinguished from starch by giving violet and red colors with iodine. Dextrin is made by moistening starch with a mixture of dilute nitric and hydrochloric acids, and then exposing to a temperature of 100 to 125°C. Dextrin varies in grade chiefly owing to differences in the type of starch from which it is made. **British gum** is a name given to dextrins that give high tack for paste use. **Cartonite,** of Paisley Products, Inc., is a liquid solution of a converted dextrin used as an adhesive in box-sealing machines. It is also marketed as a brown water-soluble powder. **Koldrex,** of the A. E. Staley Mfg. Co., is a formulated dextrin which dissolves easily in cold water to produce stable liquid adhesives of uniform viscosity. It is produced by combining dextrin with borax, preservatives, and defoamers, and then spray-drying the mixture into powder. The **borated dextrin** of the National Starch and Chemical Corp., for automatic packaging machines, has high initial tack and good adhesion. It gives 400 sealings per minute.

DIAMOND. A highly transparent and exceedingly hard crystalline stone of almost pure carbon. When pure, it is colorless, but it often shows tints of white, gray, blue, yellow, or green. It is the hardest known substance, and is placed as 10 on the Mohs hardness scale. But the **Mohs scale** is only an approximation, and the hardness of the diamond ranges from 5,500 to 7,000 on the **Knoop scale,** compared with 2,670 to 2,940 for boron carbide, which is designated as 9 on the Mohs scale. The diamond always occurs in crystals in the cubical system, and has a specific gravity of 3.521 and a refractive index of 2.417. Carbon is normally quadrivalent in flat planes, but in the diamond the carbon atoms are arranged in face-centered lattices forming interlocking tetrahedrons and also hexagonal rings in each cleavage plane.

The diamond has been valued since ancient times as a gemstone, but it is used extensively as an abrasive, for cutting tools, and for dies for drawing wire. These **industrial diamonds** are diamonds that are too hard or too radial-grained for good jewel cutting. **Jewel diamonds** have the formation in regular layers, while industrial diamonds are grown in all directions. Technically these are called **feinig** and **naetig. Ballas diamonds,** valued for industrial drilling, are formed with the crystallization starting from one central point. The stones thus formed do not crack in the tool as easily as those with layer formation. Stones for diamond dies are examined in polarized light to determine the presence of internal stresses. They are then drilled normal to the rhombic dodecahedron plane with cleavage planes parallel to the die-hole axis to obtain the greatest die-service life. The stones for industrial purposes are also the fragments and the so-called **bort** which consists of the cull stones from the gem industry including stones of radiating crystallization that will not polish well. Bort also includes a cryptocrystalline variety of diamond in brown, gray, or black colors, known as **black diamonds, carbonados,** or **carbons,** found in Brazil in association with gem diamonds. The carbons have no cleavage planes, are compact, and thus offer greater resistance to breaking forces. The carbons vary greatly in quality and hardness. Some rare natural diamonds of South America contain small amounts of aluminum and other elements which give stability to the crystal above the normal disintegrating temperature. These diamonds are not suitable as gemstones but are efficient semiconductors.

The value of diamonds is based on the gem value and is determined by color, purity, size, and freedom from flaws. The weight is measured in carats. Diamond splinters as small as $\frac{1}{500}$ carat may be cut and faceted. Small diamonds are sieved into straight sizes, and the tinted stones are separated. Then each stone is examined for cut, brilliance, and degree of perfection, and diamond merchants who sell by grade are meticulously careful of their reputation for uniform judgment. The most valued gems are blue-white. A faint straw color detracts from the value, but deep shades of yellow, red, green, or blue are prized. The largest diamond found in Brazil, the **Vargas diamond,** was a flawless stone weighing 726.6 carats. It was cut into 23 stones. The famous **Kohinoor diamond** weighed originally 793 carats, and the **Jonkers diamond** from South Africa was a blue-white stone weighing 726 carats. The **Cullinan diamond,** or Star of Africa, measured 4 by 2.5 by 2 in (10 by 6 by 5 cm), and weighed 3,106 carats. The annual world production of natural diamonds reaches as high as 28 million carats, or about 6 tons (5.4 metric tons), of which 5 tons (4.5 metric tons) are industrial diamonds. An average of 250 tons (22.7 metric tons) of ore is processed to obtain 1 carat. In Angola the average find is 0.14 carat per cubic meter of ore.

Most of the diamonds come from South Africa, Brazil, India, and Zaire. The average diamond content of the Bushimaie deposits of Zaire is 5 or 6 carats per meter. The diamonds are associated with pebbles of flint, jasper, agate, and chalcedony, but diamonds usually occur in **kimberlite,** an intrusive rock with the appearance of granite but with a composition similar to basalt plus much olivine. It occurs in South Africa, North Carolina, and Arkansas. Diamonds are formed at very high pressures and heat, and since at ordinary pressure the diamond disintegrates into graphite at 1600°F (871°C), the natural diamonds could not have been released until the temperature of the rock was below that point. The stones found in the beach sands of Southwest Africa and in sandstone in Brazil are not native to the sand, but were washed into it after scattering from the exploded rock. Diamonds have been found irregularly in Arkansas since their discovery in 1906. The average weight of the Arkansas diamonds is less than 1 carat, with the largest 40.22 carats. Some diamonds are found in the Appalachian region, the largest, from West Virginia, weighing 34.46 carats. Few of the American diamonds are of gem quality, but they are of full hardness.

Synthetic diamonds are produced from graphite at pressures from 800,000 to 1.8 million lb/in^2 (5,512 to 12,402 MPa) and temperatures from 2200 to 4400°F (1204 to 2427°C). A molten metal catalyst of chromium, cobalt, nickel, or other metal is used, which forms a thin film between the graphite and the growing diamond crystal. Without the catalyst much higher pressures and temperatures are needed. The shape of the crystal is controllable by the temperature. At the lower temperatures cubes predominate, and at the upper limits octahedra predominate; at the lower temperatures the diamonds tend to be black, while at higher temperatures they are yellow to white. The synthetic diamonds produced by the General Electric Co. are up to 0.01 carat in size, and are of industrial quality comparable with natural diamond powders. The synthetic blocky powder, Grade MBG-11, is harder and tougher than natural diamond powder and is used for cutting wheels. Synthetic diamonds have been produced in a flame at about 7000°F (3871°C), but the time element in a torch flame is short, and the crystal particle size is extremely small. With high pressure, larger crystals can be grown at lower temperatures. **Diamond crystals** are used as temperature-sensitive resistant elements in thermistors for ovens and furnaces and for temperature control in chemical processing. They are used at temperatures from −200°C up to 650°C.

Diamond dust is a powder obtained by crushing the fragments of bort, or from refuse from the cutting of gem diamonds. It is used as an abrasive for hard steels, for cutting other stones, and for making **diamond wheels** for grinding. Grit sizes for grinding wheels are 80 to 400. The coarsest,

No. 80, for fast cutting, has about 1,400 particles per square inch in the face of the bonded wheel. The National Bureau of Standards designates nine grades of **diamond powder** from No. ¼ to No. 60. The No. ¼ has a particle size of ½ μm, equivalent to a mesh of 60,000. It is used for metallographic polishing. The No. 60 has a particle size of 35 to 85 μm, equivalent to a mesh of 230, and is for rough polishing. The Elgin National Watch Co. grades diamond powder in 10 graduated sieves with from 1,600 to 105,625 openings per inch (630 to 41,585 openings per centimeter). The finest sieve grade is then placed in pure water, and as the heavier particles sink first, at intervals the water suspensions are removed to segregate the particles into absolute uniformity. These **Dymo diamond powders,** with particle size to ½ μm for fine polishing, are marketed in various colors for easy selection of grade. **Hyprez,** of the Engis Equipment Co., is also accurately graded diamond powder for fine polishing. **Swarf** is diamond powder obtained by collecting the dust from diamond-wheel grinding operations and separating the diamond dust chemically. The **blocky diamond powder** of Englehard Industries, Inc., used for metal-bonded grinding wheels and saws for stone and ceramics, contains no perfect cubes, but has about 75% of blocks having equal measurement of all three dimensions, 20% rectangles with one side shortened or lengthened, and 5% needles, flats, and slivers. The powder comes in grade sizes from 16–20 mesh to 270–325 mesh, and gives high cutting efficiency. **Abrasive RVG-D,** of General Electric, for use in resin-bonded wheels for high-speed grinding and polishing of tungsten carbide, is a diamond powder of irregularly shaped friable grains coated with a copper-base alloy to give lubricity and heat dissipation. Grain sizes are from 40–60 to 325–400 mesh.

DIATOMACEOUS EARTH. A class of compact, granular, or amorphous minerals composed of hydrated or **opaline silica,** used as an abrasive, for filtering, in metal polishes and soaps, as a filler in paints and molding plastics, for compacting into insulating blocks and boards, and in portland cement for fine detail work and for waterproofing. It is formed of fossil diatoms in great beds and is not earthy. In mineralogy it is called **diatomite,** and an old name for the ground powder is **fossil flour.** Tripoli and kieselguhr are varieties of crystalline diatomite.

The American production of the mineral is mainly in Oregon, California, Washington, Idaho, and Nevada. After mining, the material is crushed and calcined. When pure it is white; with impurities it may be gray, brown, or greenish. The powder is marketed by fineness and chemical purity. The density is usually 12 to 17 lb/ft^3 (192 to 272 kg/m^3). Its high resistance to heat, chemical inertness, and dielectric strength, and the good surface finish it imparts make it a desirable filler for plastics. For insulating purposes, bricks or blocks may be sawed from the solid or molded from the crushed material, or it may be used in powdered form. **Diatomite block** has a porosity of 90% of its volume and makes an excellent filter. **Celite,** of Johns-Manville, is a 325-mesh uncalcined, amorphous diatomaceous earth for portland-cement mixtures, paper finishes, and for use as a flatting agent in paints. **Sil-O-Cel,** of the same company, is diatomaceous earth in powder or in insulating block to withstand temperatures to 1600°F (371°C). **Superex,** of the same company, is calcined diatomite powder bonded with asbestos fibers to resist temperatures up to 1900°F (1038°C). **Dicalite,** of the Philip Carey Co., is a fine diatomite powder weighing 8 to 8.5 lb/ft^3 (128 to 136 kg/m^3) loose, and 15 to 17 lb (7 to 8 kg) tamped, used for heat-insulating cement or as insulation for walls. **Compressible insulation,** to absorb the expansion stresses known as drag stress, between the firebrick and the steel shell of metallurgical furnaces, may be of diatomaceous silica. **Superex SG,** for blast furnaces, is a composite block with Superex on the hot side and a blanket of fine spun-glass fibers on the cold side. It recovers to 97% of its original thickness in cooling from a temperature of 1900°F (1038°C) and a compression of 10%.

DIE-CASTING METAL. Any alloy employed for making parts by casting in metal molds, or dies, in pressure-casting machines as distinct from other permanent-mold castings where no pressure is used. The pressures may be as high as 25,000 lb/in^2 (172 MPa) to give a uniform dense structure and smooth finish in castings of intricate design and varying section thicknesses. The cost of equipment, including heat-resistant dies, thus limits the process economically to high-production quantities. Temperature limits on the tools restrict the process to nonferrous metals. A characteristic of the castings, also, is that they must have a draft of at least 2° on all sides to give rapid ejection from the die.

Zinc-base alloys, with melting points below 800°F (427°C), constitute the largest single tonnage of die castings, but the high-melting alloys, based on aluminum, magnesium, or copper, with melting points to about 1600°F (871°C), are used for a very wide range of mechanical parts. **Lead-base alloys** and low-melting alloys of lead, zinc, tin, and bismuth are also cast in steel or bronze dies, including **slush castings** in which excess metal is poured out after a skin of metal on contact with the die has set, leaving a hollow casting. But these are cast without pressure, and are classed as permanent-mold castings rather than die castings, although the composition of the alloys may be essentially the same.

Die-casting alloys are standardized into relatively small groups with symbol numbers under the zinc, aluminum, magnesium, and copper alloys; but the possibilities of variation, particularly with the minor ingredients, are infinite, and trade names and company numbers are often used to designate uniform quality standards for particular uses.

DIE STEELS. Any tool steels used for dies. Originally, the term meant high-grade carbon steel with about 0.90 to 1.1% carbon that could be hardened to retain a keen cutting edge and not shrink or warp greatly in the hardening. Unalloyed tool steels are still widely used for many types of dies. They have the advantage of low cost and ease of machining. Also, they are easy to harden and temper, have high impact strength, and have good fatigue endurance. They are used for cold-header dies, stamping and forming dies and punches, shear blades, small tools, sledges, and for dowels, bushings, shafts, and other machine parts. Often, the die steels now have small amounts of vanadium or other elements, not enough to rate them as alloy steels but sufficient to give them added physical properties. The modern die steels are made to a high degree of uniformity, with the manganese and silicon adjusted to give depth of hardening and uniformity of grain structure. Typical of this class of steel is **H-9 Extra Hard,** or **cold-heading die steel,** of Carpenter Technology. It contains 0.90% carbon, 0.40 manganese, and 0.40 silicon, and is used for coining dies, cold-heading dies, and knurls. A water-hardening die steel is **Special A.S.V. steel** of Firth Sterling, Inc., for punching, forming, swaging, and threading dies. It has 1 to 1.10% carbon, 0.30 manganese, 0.30 silicon, with sulfur and phosphorus below 0.015%. This is a shallow-hardening steel, with a very hard surface ranging from Rockwell C66 to C56 according to temper, and a tough, shock-resistant interior.

The **low-alloy die steels** are distinguished from the low-alloy construction steels in that the small percentages of alloying elements are carefully balanced to give certain physical characteristics. They may be classed with the carbon tool steels, but are given trade names. Very small amounts of vanadium refine the grain and give finer cutting edges. As little as 0.30% chromium in the steel increases the pearlite and also refines the graphite, while free, hard carbides are formed with about 0.60% chromium. A typical steel of this kind is **Artdie steel** of the Columbia Tool Steel Co., used for high-production jewelry dies. It contains 0.95% carbon, 0.30 chromium, 0.20 vanadium, 0.25 manganese, and 0.25 silicon. **Silvan Star steel,** of Firth Sterling, Inc., is called a carbon steel, but it contains 0.20% vanadium, with 1.03 carbon, 0.25 manganese, and 0.25 silicon. It is suitable for a wide range of dies and tools. It is easily hardened by water quenching to Rockwell C67, and tempered in hot oil or fused salts to Rockwell C62 for keen hardness or to C58 for tough swaging dies.

Small amounts of tungsten or molybdenum add toughness and wear resistance. **SAE steel 6150** and **SAE steel 6195,** with about 1% chromium, 0.15 vanadium, 0.45 tungsten, and up to 1% carbon, are used for dies that require a high degree of hardness, toughness, and fatigue resistance. **Albany steel,** of the Allegheny Ludlum Steel Corp., is a steel of this type, and **Alhead steel,** of the same company, for cold-heading dies, has 1.5%

tungsten, 1.5 cobalt, and 1 carbon. **Break-die steel,** of the Crucible Steel Co., has 0.20% molybdenum, 1 chromium, 0.70 manganese, 0.30 silicon, with only 0.35 carbon. **Hedervan steel,** of the Latrobe Steel Co., is a 0.90-carbon cold-heading steel with up to 3.5% vanadium carbide particles to increase wear resistance. **BR-3 die steel,** of the same company, for long-run blanking and forming dies, and to withstand the high abrasion in such uses as brick molds, contains 4.5% vanadium in the form of hard carbides, with 5.25 chromium, 1.10 molybdenum, 0.70 manganese, 0.30 silicon, and sufficient carbon, 2.8, to form the carbides. Tools can be operated without cracking at a hardness of Rockwell C65.

The possibilities of variation of physical characteristics of die tool steels by even slight changes in the alloying elements and in the treatment are unlimited, and usually there are many hundreds of these steels offered commercially at any one time, the methods of marketing being on performance in particular uses rather than on composition. Die steels also include alloy steels classified by characteristic types, as nondeforming steel or hot-work steel. They also include the prehardened steel blocks used for making plastic molds, die-casting dies, forging dies, and dies for pressing ceramics. These **die-block steels** usually contain a high percentage of chromium or vanadium with high carbon to form hard carbides, but they may be low-alloy, low-carbon steels for easy die sinking or hobbing, to be face-hardened by carburizing or nitriding. **Ottawa 60 steel,** of the Allegheny Ludlum Steel Corp., contains 12% vanadium, 1 chromium, 1 molybdenum, and 3.25 carbon. **Airdi 225 steel,** of the Crucible Steel Co., is an air-hardening die-block steel with 12% chromium, 1 molybdenum, 0.25 vanadium, and 2.25 carbon. **Prestem 5M21 steel,** of the Heppenstall Co., for die-casting dies and high-pressure press dies, is prehardened to Rockwell C36 to C45. It contains 3.38% molybdenum, 3 nickel, 0.65 manganese, 0.30 silicon, 0.15 chromium, and 0.20 carbon. **Cromoco die steel,** of Firth Sterling, Inc., for forming dies, swages, and metal shears, has 12% chromium, 1 cobalt, 1 molybdenum, and 1.6 carbon. **Select B-FM steel,** of the Latrobe Steel Co., is an air-hardening 15% chromium steel with sulfur added for free machining. **Lo-Air steel,** of the Cyclops Corporation, for dies, punches, and shear blades, is an air-hardening steel containing 2% manganese, 1.35 molybdenum, 1 chromium, 0.30 silicon, and 0.70 carbon. **Multimold steel,** of the Bethlehem Steel Co., for die-casting dies, which has some red hardness, is a low-alloy steel with 0.80 chromium, 0.30 molybdenum, and 0.35 carbon. But **Thermold AV steel,** of the Cyclops Corporation, is an **ASTM H13 steel,** with 5% chromium, 1.40 vanadium, 1 silicon, 0.40 manganese, and 0.35 carbon. **Thermold 75,** for aluminum casting dies, has less chromium and vanadium with an addition of molybdenum and cobalt. **Pressurmold steel,** of the Braeburn Alloy Steel Corp., is also a modified chromium precipitation-hardening steel. These steels are tempered to less than Rockwell C55, but can be nitrided when high hardness is wanted. **Badger steel,** of the Latrobe Steel Co., is a nonshrinking die steel with 1.20% manganese, 0.50 tungsten, 0.50 chromium, 0.30 silicon, and 0.94 carbon. It is tempered to Rockwell C63 for dies. **MGR punch steel** of the same company is tempered to Rockwell C63 for punches. It contains 5% chromium, 1.20 tungsten, 1.20 molybdenum, 0.95 silicon, 0.30 manganese, and 0.55 carbon, and is shock-resistant.

DISINFECTANTS. Materials used for killing germs, bacteria, or spore, and thus eliminating causes of disease or bad odors in factories, warehouses, or in oils and compounds. The term **antiseptic** is employed in a similar sense in medicine, and the term **germicide** is often used for industrial disinfectants. Some disinfectants are also used as preservatives for leather and other materials, especially chlorine and chlorine compounds. **Phenol** is one of the best-known disinfectants, and the germ-killing power of other chemicals is usually based on a comparison with it. Practically all bacteria are killed in a few minutes by a 3% solution of phenol in water, but phenol has the disadvantage of being irritating to the skin. Industrial disinfectants are usually sold as concentrates to be diluted to the equivalent of a 3 to 5% solution of phenol.

Too large a proportion of disinfectants in oils, solutions, or the air may be injurious to workers, so the advice of health officials is ordinarily obtained before general use. Creosote oil and cresylic acid are employed in emulsions in disinfecting sprays and dips, but continuous contact with creosote may be injurious. **Formaldehyde** has high germicidal power, and is used for hides and leather, and some air sprays may contain chemicals such as chlorophyll which unite with moisture in the air to produce formaldehyde. But formaldehyde is not generally recommended for odor control as it is an **anesthetizer.** It desensitizes the olfactory receptors so that the individual is no longer able to detect the odor. **Masking agents,** which introduce a stronger, more pleasant odor, are likewise not a recommended method of disinfecting. They do not destroy the undesirable odor, and may permit raising the total odor level to unhealthy proportions. Elimination of odors requires chemicals that neutralize or destroy the cause of the odors without causing undesirable effects.

The **silver** ion is an effective cleanser of water that contains bacteria which produce sulfur-bearing enzymes, and silver sterilization is done with silver oxides on activated carbon, or with organic silver compounds. The safe limit of silver in water for human consumption is specified by the U.S. Health Service as 10 parts per billion, and as with many other disinfectants the use requires competent supervision. Antiseptic atmospheres may be produced by spraying chloramine T, iodine, or argyrol. **Chloramine T** is a white crystalline powder of the composition $CH_3C_6H_4SO_2NClNa \cdot 3H_2O$, soluble in water and in organic solvents. Besides its use as an antiseptic and germicide it is also employed as an oxidizing and chlorinating agent. **Thymol** is used as a disinfectant in ointments, mouthwashes, soaps, and solutions. Condensation products of thymol with other materials are also used. **Thymoform,** $C_{21}H_{28}O_2$, made by condensing thymol with formaldehyde, is a yellow powder used as an antiseptic dusting powder. **Thymidol** is an antiseptic made by condensing thymol with menthol. **Dihydroxyacetic acid,** $CH(OH)_2COOH$, and its sodium salt, both white powders, are used in cosmetics, pharmaceuticals, and in coatings for food-wrapping papers, as they are nontoxic and do not irritate the skin. **Hexylresorcinol,** $CH_3(CH_2)_5C_6H_3(OH)_2$, is a more powerful antiseptic than phenol and is not injurious to the skin or tissues. **Caprokol,** of Sharpe & Dohme, is hexylresorcinol. The antiseptic throat lozenge of this company, known as **Sucret,** has a base of sugar and glucose, with hexylresorcinol and a flavor. **Pinosylvine** is a natural antiseptic extracted from the heartwood of the pine tree, where it protects the tree against decay and insects. It is related chemically to resorcinol, and its germ-killing power is 30 times that of phenol. **Ceresan M,** of Du Pont, is a powder designated as **ethyl mercury toluene sulfonanilide,** used for disinfecting seeds to protect against soil-borne plant diseases. **Pittside,** of the Columbia Chemical Div., used as an industrial germicide, is a stabilized **calcium hypochlorite,** in water-soluble granules. Disinfectants sold under trade names are usually complex chemicals, and may be chlorinated or fluorinated phenyl compounds not harmful to the skin.

DISPERSION-STRENGTHENED METALS. Particulate composites in which a stable material, usually an oxide, is dispersed throughout a metal matrix. The particles are less than 1 μm in size, and the particle volume fraction ranges from only 2 to 15%. The matrix is the primary load bearer while the particles serve to block dislocation movement and cracking in the matrix. Therefore, for a given matrix material, the principal factors that affect mechanical properties are the particle size, the interparticle spacing, and the volume fraction of the particle phase. In general, strength, especially at high temperatures, improves as interparticle spacing decreases. Depending on the materials involved, dispersion-hardened alloys are produced by either powder metallurgy, liquid metal, or colloidal techniques. They differ from **precipitation-hardened alloys** in that the particle is usually added to the matrix by nonchemical means. Precipitation-hardened alloys derive their properties from compounds that are precipitated from the matrix through heat treatment.

There are a rather wide range of dispersion-hardened-alloy systems. Those of aluminum, nickel, and tungsten, in particular, are commercially significant. **Tungsten thoria,** a lamp-filament material, has been in use for more than 30 years. And **dispersion-hardened aluminum** alloys are in wide commercial use today. Known as **SAP alloys,** they are composed of aluminum and aluminum oxide and have a unique combination of good oxidation and corrosion resistance plus high-temperature stability and strength considerably greater than that of conventional high-strength aluminum alloys. Another dispersion-hardened alloy is **TD nickel,** a dispersion of thoria in a nickel matrix. The alloys are three to four times stronger than pure nickel at 1600 to 2400°F (871 to 1316°C). Other metals that have been dispersion-strengthened include copper, lead, zinc, titanium, iron, and tungsten alloys.

DIVI-DIVI. The dried seed pods of the tree *Caesalpinia coriaria,* native to tropical America, employed in tanning leather. Most of the divi-divi is produced in Colombia, the Dominican Republic, and Venezuela. It is used chiefly in blends with other tannins to increase acidity, to give a light color to the leather, and to plump and soften the leather. The pods are about 3 in (7.6 cm) long, and contain up to 45% pyrogallol tannin. They require to be kept from fermentation, which develops a red coloring matter. The best pods are the thickest and lightest in color, and they are used to replace gambier, valonia, and myrobalans. The commercial extract contains 25% tannin. **Algarobilla,** from the pods of the *C. brevifolia,* of Chile, is a similar tanning agent. **Cascalote** is from the pods of the tree *C. cacolaco* of Mexico, and is the standard tanning material of Mexico. It is also used to replace quebracho for oil-well-drilling mud. **White tan,** or **tari,** is from the pods of the *C. digyna* of the Far East. **Tara,** or **Bogotá divi-divi,** also called **cevalina,** is from the pods of the tree *C. tinctoria* of Colombia and Peru. The pods contain 32% tannin, and 1,000 lb (454 kg) of tara pods produce 500 lb (227 kg) of **tara powder.** The material makes a soft leather, and is used to replace sumac.

DOGWOOD. A heavy hardwood noted for its ability to stay smooth under long-continued rubbing. Its outstanding use is for shuttles for weaving. The texture is fine and uniform. Other uses of the wood are for small pulleys, golf-club heads, mallet heads, jewelers' blocks, skate rollers, and bobbins. There are 17 known varieties of the plant in the United States, only four of which grow to tree size. The **white dogwood** is *Cornus florida;* the **Pacific dogwood** is *C. nuttalli; rough-leaf dogwood* is *C. asperfolia;* and **blue dogwood** is *C. alternifolia.* Dogwood grows widely throughout the eastern states. **Turkish dogwood** was formerly imported for shuttles, as was also the **Chinese dogwood,** or **kousa,** *C. kousa.*

DOLOMITE. A type of limestone employed in making cement and lime, as a flux in melting iron, as a lining for basic steel furnaces, for the production of magnesium metal, for filtering, and as a construction stone. It is a carbonate of calcium and magnesium of the composition $CaCO_3 \cdot MgCO_3$, differentiated from limestone by having a minimum of 45% $MgCO_3$. It occurs widely distributed in coarse, granular masses or in fine-grained compact form known as **pearl spar.** The specific gravity is 2.8 to 2.9 and hardness 3.5 to 4. It is naturally white, but may be colored by impurities to cream, gray, pink, green, or black. For furnace linings it is calcined, but for fluxing it is simply crushed. The raw dolomite, marketed by Basic Refractories, Inc., for open-hearth steelmaking, is washed crushed stone in ⅝-in (1.6-cm) size. When calcinated at a temperature of about 3100°F (1704°C), dolomite breaks down to MgO and CaO, and it is limited to about 3000°F (1649°C) as a refractory. **Calcined dolomite** used in Germany as a water-filter material under the name of **magno masse** is in grain sizes 0.5 to 5.0 mm. Dolomite for the production of magnesia, some of which is cut as building marble, contains 10 to 20% magnesia, 27 to 33 lime, 1 to 12 alumina, 40 to 46 carbonic acid, 1 to 5 silica, and 0 to 3 iron oxide. The dolomite found in huge deposits in Oklahoma contains 30.7% CaO, 21.3 MgO, and only very small amounts of silica, alumina, and iron oxide. For the production of magnesium metal, calcined dolomite and ferrosilicon are brought to a high temperature in a vacuum and the magnesium is driven off as a vapor. In the ceramic industry dolomite is sometimes called **bitter spar** and **rhombic spar.**

DOUGLAS FIR. The wood of the tree *Pseudotsuga taxifolia,* of the northwestern United States and British Columbia. It is sometimes called **Oregon pine, Douglas pine, Douglas spruce, red fir, fir, yellow fir,** and **Puget Sound pine.** The wood of young trees with wide growth rings is reddish brown and is the type called red fir, though the true red fir is from the large tree *Abies magnifica* of California and Oregon, the lumber of which is called **golden fir,** and the wood of which is used also for paper pulp. The wood of older trees of slower growth with narrow rings is usually yellowish brown and is called yellow fir. Both woods may come from the same tree. The narrow-ringed wood is stronger and heavier. Douglas fir averages below longleaf pine in weight, strength, and toughness, but above loblolly pine in strength and toughness, though below it in weight. The grain is even and close, with resinous pores less pronounced than in pitch pine. It is a softwood, and is fairly durable. The weight is 34 lb/ft^3 (545 kg/m^3). The compressive strength perpendicular to the grain is 1,300 lb/in^2 (9 MPa); the shearing strength parallel to the grain is 810 lb/in^2 (5.5 MPa).

Douglas fir is used for general construction and millwork, plywood, boxes, flooring, and where large timbers are required. It is also used for pulping and yields kraft paper of high folding endurance but low bursting strength. The fibers are large. The trees grow to great heights, the average being 80 to 100 ft (24 to 30 m). The stand is estimated at more than 450 billion bd ft (1 billion m^3) or about one-fourth of all timber in the United States. **Douglas fir bark** contains from 7.6 to 18.3% of a catechol tannin, the bark of young trees yielding the higher percentages. It is suitable for tanning heavy leathers, and yields a pliable, light-colored leather. **Silvacon 383,** of the Weyerhaeuser Timber Co., is Douglas fir bark in flaky corklike granules used in flooring and acoustical tile. **Silvacon 490** is the bark as a reddish powder used in dusting powders and paints. **Silvacon 508** is hard, spindle-shaped small fibers from the tissue of the bark, used as a filler for plastics and in asphalt and fibrous paints. **Douglas-fir bark wax** is a hard glossy wax extracted from the bark of the Douglas fir, and is a partial replacement for carnauba wax. A ton of bark yields 150 lb (68 kg) of wax by solvent extraction with 150 lb (68 kg) of tannin and 10 lb (4.5 kg) of quercetin as by-products.

DRIERS. Materials used for increasing the rapidity of the drying of paints and varnishes. The chief function of driers is to absorb oxygen from the air and transfer it to the oil, thus accelerating its drying to a flexible film. They are in reality catalyzers. Excessive use of driers will destroy the toughness of the film and cause the paint to crack. Solutions of driers are called liquid driers; it is in this form that **paint driers** are most used. Certain oils, such as tung oil, have inherent drying properties and are classed as drying oils but not as driers. Driers may be oxides of metals, but the most common driers are metallic salts of organic acids. **Manganese acetate,** $(CH_3COO)_2Mn \cdot 4H_2O$, is a common paint drier. It is a pinkish crystalline powder soluble in water and in alcohol. **Sugar of lead,** used as a drier, is **lead acetate,** $Pb(CH_3COO)_2 \cdot 3H_2O$, a white crystalline powder with a faint acetic acid odor, also used as a mordant in textile printing. It is also known as **plumbous acetate** and **Goulard's powder. Lead oleate,** $Pb(C_{18}H_{33}O_2)_2$, is a drier made by the action of a lead salt on oleic acid. It is also used for thickening lubricants. **Lead linoleate,** $Pb(C_{18}H_{31}O_2)_2$, is a drier made by adding litharge to linseed oil, and heating. Lead and manganese compounds together act more effectively as driers than either alone. **Lead resinate** adds toughness of film as well as drying power. **Zinar,** of Newport Industries, Inc., is a **zinc resinate** with 5.6% zinc content. **Cobalt octoate,** which has about 12% cobalt in combination with hexoic acid, is used as a drier. **Cobalt driers** are twice as rapid in drying power as manganese driers, but too rapid drying often makes a wrinkled film which is desirable for some finishes but not for others.

Naphthenate driers are metallic salts made with naphthenic acids instead of fatty-oil acids. They are usually more soluble in paint solvents, and since the naphthenic acids can be separated into a wide range of molecular weights by distillation, a wider variety of characteristics can be obtained. **Sodium naphthenate,** with 8.6% metal content, and **potassium naphthenate,** with 13.1%, are powders that are good bodying agents and emulsifiers as well as driers. **Tin naphthenate,** with 20% tin, may be added to lubricating oils as an antioxidant. **Mercuric naphthenate,** with 29% mercury, retards the growth of bacteria and mold when added to finishes. **Barium naphthenate,** with 22.6% barium, has binding and hardening properties, and is used in adhesives and in linoleum. **Uversols,** of the Harshaw Chemical Co., are naphthenic acid salts of aluminum, calcium, cobalt, lead, manganese, or zinc, in liquid form for use as paint driers, wetting agents, and catalysts. **Octoic driers,** of the Witco Chemical Co., are metallic salts made with ethylhexoic acid, and the metal content is lower than that of driers made with naphthenic acids. They are light in color, have no odor, and have high solubility. The **Octasols,** of the Harshaw Chemical Co., are ethylhexoic acid metal salts. **Drying agents** for resin coatings and inks may act by oxidation or other chemical reaction. **Sulfur dichloride,** S_2Cl_2, speeds the drying action of coatings and inks formulated with alkyd, urea, or melamine resins, and such inks dry almost instantly.

DRILL ROD. Tool-steel round rod made to a close degree of accuracy, generally not over or under 0.0005-in (0.0127-mm) the diameter size, and usually polished. It is employed for making drills, taps, reamers, punches, or for dowel pins, shafts, and rollers. Some mills also furnish square rods to the same accuracy under the name of drill rod. Common drill rod is of open-hearth high-carbon steel hardened by quenching in water or in oil. The usual commercial sizes are from 1½ in (3.8 cm) in diameter down to No. 80, which is 0.0135 in (0.343 cm) in diameter. The usual lengths are 1 to 3 ft (0.3 to 0.9 m). The sizes are by the standard of drill gages, with about 200 different diameters. The carbon content is usually from 0.90 to 1.05%, with 0.25 to 0.50 manganese, 0.10 to 0.50 silicon, and a maximum of 0.04 phosphorus or sulfur. It also comes in high carbon with from 1.50 to 1.65% carbon and 0.15 to 0.35 manganese. Drill rod can also be obtained regularly in high-speed steels, and in special alloy steels for dowel pins. **Needle wire** is round tool-steel wire used for making needles, awls, and latch pins. It comes in coils, in diameters varying by gage sizes from 0.010 to 0.105 in (0.025 to 0.267 cm). **Needle tubing** for surgical instruments and radon implanters is stainless-steel tubing 0.014 to 0.203 in (0.036 to 0.516 cm) in diameter in 6-ft (1.8-m) lengths. The **hypodermic tubing** of J. Bishop & Co. is hard-drawn stainless-steel tubing 0.008 to 0.120 in (0.020 to 0.304 cm) outside diameter, with wall thicknesses from 0.004 to 0.012 in (0.010 to 0.304 cm), in 2-ft (0.6-m) lengths, with a fine finish. **Capillary tubing** is also stainless steel, but comes in lengths to 200 ft (61 m), with outside diameters from 0.060 to 0.125 in (0.152 to 0.318 cm). The inside bore can be had in various diameters from 0.006 to 0.025

in (0.015 to 0.064 cm) for the 0.060-in (0.152-cm) tubing, and from 0.010 to 0.024 in (0.025 to 0.061 cm) for the 0.125-in (0.318-cm) tubing. **Stud steel** is an English name for round bar steel made to close limits and hardened and descaled, used for heavy pins and studs. **Pin bar** is small-diameter rod of case-hardening steel used for dowel pins. **Drill steel,** for mine and quarry drills, comes in standard rounds, octagons, squares, and cruciform bars, solid or hollow, usually in carbon steel.

DRYING OILS. Vegetable oils which are easily oxidized by exposure to the air and thus suitable for producing a film in paints and varnishes, and known as **paint oils.** The best drying oils are those which contain the higher proportions of unsaturated acids, in which oxidation causes polymerization of the molecules. The drying of an oleoresinous varnish takes place in two stages. First, the reducer or solvent evaporates, leaving a continuous film composed of gums and drying oil. The drying oil is then oxidized by exposure, leaving a tough, hard skin. This oxidation is hastened by driers, but the drying oil itself is responsible for the film. The drying power of oils is measured by their **iodine value,** as their power of absorbing oxygen from the air is directly proportional to their power of absorbing iodine. Linseed oil is the most common of the drying oils, though tung oil and oiticica oil are faster in drying action. Linseed oil alone will take about 7 days to dry, but can be quickened to a few hours by the addition of driers. Linseed oil and other oils may be altered chemically to increase the drying power.

Conjugated oils are oils that have been altered to give conjugated double bonds in place of isolated double bonds in the molecules of the fatty acids. **Conjulinol,** of the Woburn Degreasing Co., is a drying oil of this class made from linseed oil. The iodine value is 154, and the drying time is greatly reduced. Normally, soybean oil is not classed as a drying oil although it may be blended with drying oils for paint use. But by chemical alteration it can be given good drying power. **Conjusoy** is a drying oil made by conjugation of soybean oil. The iodine value is 128, and the drying time is about half that of boiled linseed oil.

Castor oil, which has poor drying properties, is dehydrated to form a good drying oil. Other methods are used to alter oils to increase the drying power, notably polymerization of the linoleic and some other acids in the oils; or oils may be fractionated and reblended to increase the percentage of acids that produce drying qualities. The **Admerols,** of the Archer-Daniels-Midland Co., comprise a series of drying oils made by treating linseed or soybean oil with butadiene, styrene, or pentaerythritol. **Kel-X-L oil,** of Spencer-Kellogg & Sons, is a modified linseed oil with an iodine value up to 170, used as a substitute for tung oil in quick-drying varnishes. **Kellin,** of the same company, is a quick-drying blended oil with a linseed-oil base, while **Kellsoy** is a similar oil with a soybean-oil base. **Cykelin,** of the same company, is a quick-drying oil made by treating linseed oil with **cyclopentadiene,** $(CH:CH)_2 \cdot CH_2$, a low-boiling liquid obtained from coal tar or from cracking petroleum. **Cykelsoy** is another drying oil made by treating soybean oil with cyclopentadiene. **Dorscolene,** of Dorward & Sons Co., is a drying oil made from fractionated and blended fish oils. The German substitute drying oil known as **Resinol** was a liquid obtained by the distillation of the heavy fractions of the benzolated oils derived from scrubbing coke-oven gas. **Resigum** is the final residue in the distillation of tar-oil benzol which has been washed with sulfuric acid, caustic soda, and water. It contains a maximum naphthalene content of 5%. It is miscible with resins or copals, and with vegetable oils, and makes a good paint without other drying oils. Synthetic drying oil, of Shell Development Co., is **glycerol allyl ether** derived from propylene gas obtained in cracking petroleum. **C oil,** of the Standard Oil Development Co., is a heavy, sticky liquid with a butadiene base. In paints it gives high adhesion to metals and masonry, and produces a smooth, hard, glossy coating with good chemical resistance.

Although the great volume of drying oils is produced from linseed, soybean, tung, oiticica, castor, and fish oils, many other oils have drying properties and are used in varying quantities. **N'gart oil** is from the seed nuts of a climbing plant of Africa, and is equal in drying power to linseed oil. **Lallemantia oil,** obtained from the seeds of the *Lallemantia iberica*, of southeastern Europe and Asia, resembles linseed oil in physical properties. **Isano oil,** obtained from the kernel of the nut of the *Ongokea klaineana* of tropical Africa, is a pale-yellow viscous oil that has little drying power, but when heat-treated sets up an exothermic action to produce a varnish oil. **Anda-assu oil,** also used in Brazil for paints, is from the seeds of the plant *Joannesia princeps.* The seeds yield 22% of a clear yellow oil with an iodine value of 142 which is bodied by heating. **Manketti oil** is a varnish oil with about two-thirds the drying power of linseed oil. It is a light-yellow viscous oil from the seed nuts of the tree *Ricinodendron rautanenii,* of southwest Africa.

Chia-seed oil is a clear amber-colored oil extracted from the seeds of the plant *Salvia hispanica* of Mexico. It has a higher drying value than linseed oil. The seeds yield about 30% oil, which contains 39% linolenic acid, 45 linoleic, 5 palmitic, 2.7 stearic, with some arachidic, oleic, and myristic acids. The specific gravity is 0.936, iodine value 192, and acid value 1.4. The seeds scatter easily from the pods and are difficult to collect.

DUCK. A strong, heavy cotton fabric employed for sails, awnings, tents, heavy bags, shoe uppers, machine coverings, and where a heavy and durable fabric is needed. It is woven plain, but with two threads together in the warp. It is made in various weights, and is designated by the weight in ounces per running yard 22 in (0.6 m) wide. It is marketed unbleached, bleached, or dyed in colors, and there are about 30 specific types with name designations usually for particular uses such as **sailcloth.** When woven with a colored stripe, it is called **awning duck. Russian duck** is a fine variety of **linen duck.** Large quantities of cotton duck are used for making laminated plastics and for plastic-coated fabrics, and it is then simply designated by the weight. **Belt duck,** for impregnated conveyor and transmission belts, is made in loosely woven soft ducks and in hard-woven, fine-yarn hard fabric. The weights run from 28 to 36 oz (0.80 to 1.02 kg). **Conveyor belting** for foodstuff plants is usually of plastic fabric for cleanliness. The **Transilon** of Extremultus, Inc., is a belting of good strength and flexibility to operate over small-diameter rollers. It is made of nylon fabric faced on both sides with polyvinyl chloride sheet. It may have a variety of surface finishes such as tetrafluoroethylene.

Hose duck, for rubber hose, is a soft-woven fabric of plied yarns not finer than No. 8, made in weights from 10 to 24 oz (0.28 to 0.68 kg). The grade of duck known as **elevator duck** for conveyor belts is a hard-woven 36-oz (1.02-kg) fabric. **Plied-yarn duck** is used for Army tents instead of flat duck as it does not tear easily and does not require sizing before weaving. **Canvas** is duck of more open weave. The term is used loosely in the United States to designate heavy duck used for tarpaulins, bags, sails, and tents. But more properly it is a heavy duck of square mesh weave more permeable than ordinary duck, such as the canvas used for paintings and for embroidery work. The word duck is from the Flemish doeck meaning cloth, originally a heavy linen fabric. The word canvas is from the Latin cannabis, originally a coarse, heavy hempen cloth for tents. **Osnaburg cloth** is a heavy, coarse, plain-woven fabric used for wrapping and bailing, and for inside sacks for burlap flour bags. It is made from lower grades of short-staple cotton and from waste. In colored checks and stripes it is used for awnings.

Drill is a stout, twilled cotton fabric used for linings and where a strong fabric lighter than duck is required. It differs from duck also in that it has a warp-flush weave that brings more warp than filling to the face of the cloth. It comes unbleached, bleached, or piece-dyed, or may be yarn-dyed. It is made in various weights, and is designated in ounces per yard, the same as duck. Tan-colored drill is called **khaki. Denim** is a heavy, twill-woven, warp-flush fabric usually lighter in weight than drill. The warp is yarn-dyed. The filling is made with one black and one white yarn. It is much used for worker's clothing, and the light weights for sportswear are called **jean.** Denim is also used industrially where a tough fabric is needed. **Art denim,** in plain colors or woven with small figures, is used for upholstery.

DUCTILE IRON. Ductile irons, which have been in commercial use less than 50 years, have the basic compositions of gray irons, but differ in the shape of the graphite particles. In contrast to the flat flakes present in gray irons, the graphite in ductile irons is nodular or spheroidal. Because of this distinctive feature, ductile irons have also been called **nodular irons** and **SG (spheroidal) irons.** In both structure and some properties, ductile irons resemble steel. They are sometimes considered a high-silicon steel with the addition of graphite. The spherical graphite nodules have a minimum effect on the matrix, which can be varied considerably by foundry practice from completely pearlitic to totally ferritic. Also, by hardening heat treatments, martensitic or banitic structures can be produced to give high strength and hardness. Alloying elements also influence structure as they do in steels. For example, nickel strengthens the matrix without forming carbides, and molybdenum increases strength, hardness, and hardenability.

The ASTM classifies and codes ductile irons according to mechanical properties. A three-number designation is used to specify minimum tensile strength, yield strength, and elongation. For example, the designation 60:40:18 indicates the following minimums: 60,000 lb/in^2 (413 MPa) tensile strength, 40,000 lb/in^2 (275 MPa) yield strength, and 18% elongation. There are five such ASTM standard grades: Types 60:40:18 and 65:45:12 have maximum toughness and machinability and are used for valve and pump bodies, gear housings, and farm-machine parts. Type 80:55:06 has maximum strength in the as-cast condition, and, in general, good creep strength. It can be surface-hardened for use in applications requiring resistance to wear and good strength, as in mining machinery and automotive and diesel engine parts. Also, its heat resistance is an advantage in moderate high-temperature service. Type 100:70:03 is usually normalized and tempered to develop a combination of strength, ductility, and wear resistance. It also responds well to surface hardening and is used for high-strength gears and automotive and machine components. Type 120:90:02 provides maximum strength and wear resistance and is used for pinions, gears, rollers, and slides. There is also a **heat-resistant nodular iron** that contains more silicon than the other grades. It has the highest resistance to scaling, but has the lowest room-temperature ductility. It is used chiefly for parts in metal and glass-producing machinery and furnaces.

High-alloy ductile iron grades, also referred to as **austenitic ductile iron,** are highly alloyed, with nickel ranging from 18 to as high as 36%, and chromium running between 1.5 and 5%. Their relationship to the standard ductile irons is similar to that of wrought stainless to low-alloy steels. They were developed primarily for corrosion- and heat-resistant applications. Some of the grades have elevated-temperature strengths comparable with that of 18–8 stainless steel. These austenitic grades are nonmagnetic and readily welded. Typically, they are used in the chemical industry for impeller, pumps, and valves; in paper mills; in saltwater service; and for parts requiring dimensional stability. A **nickel-molybdenum ductile iron** is another alloy grade with a tough bainite and martensitic structure. It has a tensile strength of over 200,000 lb/in^2 (1,378 MPa).

DYESTUFFS. Materials employed for giving color to textiles, paper, leather, wood, or other products. They may be either natural or artificial. Many chemicals will stain and color other materials, but a product is not considered a dye unless it will impart a distinct color of some permanence to textiles. The natural dyestuffs may be mineral, animal, or vegetable, but the artificial dyes are derived mainly from coal-tar bases. **Tyrian purple,** from various Mediterranean snails, was in ancient times the most noted of the animal dyestuffs. **Cochineal** and **kermes** are other animal dyes. One of the earliest metallic or mineral dyestuffs was called **iron buff.** It was made by allowing pieces of iron to stand in a solution of vinegar to corrode. Fabrics that had been dipped in this solution were rinsed in a solution of wood ashes. **Mineral dyes** now include ocher, chrome yellow, and Prussian blue. **Vegetable dyes** may be water solutions of woods, barks, leaves, fruits, or flowers. The buff and brown textile colors of early New England were made by boiling fresh green butternuts in water, while a dark-red dye was made by boiling the common red beet in water. The yellow to red colors known by the Algonquin name of **puccoon** were from the orange-red juice of the root of the **bloodroot,** a perennial of the poppy family. Vegetable dyes now include brazilwood, barwood, sappanwood, fustic, logwood, madder, henna, saffron, annatto, indigo, and alkanet. The **camphire** of the ancients mentioned in the Bible and Koran was a reddish-orange dyestuff made by grinding to a paste the red, sweet-scented spikes of the small cypress tree, *Lawsonia inermis,* of Egypt and the Near East. It was used by Eastern and Roman women to stain the fingernails, and is now used under the name of **henna** for dyeing leather and as a hair dye. It gives various shades from yellowish to red or brown. **Argol,** a brilliant red used extensively until replaced by synthetic dyes, is from the **orchilla,** a lichen found in the Canaries and Near East. It was used to produce the brilliant colors of the medieval **Florentine cloth. Chinese green, buckthorn bark,** or **lokao,** is the powdered bark of the buckthorns, *Rhamorus globosa* and *R. utilis,* of China and Russia. It is used in dyeing silk and cotton. **Weld,** from the plant *Reseda luteola* of Europe, produces a very bright-yellow color with an alum mordant. With indigo it produces green. **Woad** is the dried fermented leaves of the plant *Isatis tinctoria* of Europe. It gives a blue color, but is now little cultivated.

Synthetic dyes are mostly coal-tar or aniline colors. They are more intense, brighter, faster, and generally cheaper than natural dyestuffs. There are thousands of these dyes, classified by a color index rather than by their chemical composition. The dyes are complex chemicals, but they usually contain characteristic groups of atoms so that the color or change in color can be predicted. They are generally marketed under trade names or company numbers. The chemical names refer to the forms of atom groupings. The **azo dyes,** with an $\cdot N:N\cdot$ linkage, constitute about half of the production. **Azobenzene,** $C_6H_5\cdot N:N\cdot H_5C_6$, is made from nitrobenzene, and is in crystalline red plates melting at 68°C. This may be converted to **hydrazobenzene,** $C_6H_5NHNHC_6H_5$, a solid of camphorlike odor melting at 131°C. **Substituted azo dyes** constitute a class containing OH and NH groups, made by coupling amines or phenols with the salts. The azo dyes are in general poisonous, but are sometimes used in restricted quantities to color foodstuffs. Some are poisonous in contact with the skin, such as **xylyazonaphthol** and the sodium salt of **sulfophenylazo,** designated by the Food and Drug Administration as **Red No. 32** and **Orange No. 1,** and proscribed for use in coloring lipstick and oranges.

The other three important classes are: **anthraquinone dyes, indigoid dyes,** and **thioindigoid dyes,** the latter being **sulfur dyes.** The sulfur dyes may be made by treating the organic compounds with sodium sulfide. They are fast to washing and to light, but the range of color is limited, and their use is generally limited to fibers where a strongly alkaline bath is tolerable.

Some of the synthetic dyes will color animal fibers well and not vegetable fibers, or vice versa, while some will color all fibers. Some, called **direct dyes,** can be dyed direct, while others require a mordant. Some are permanent, or fast, while others are water-soluble and will fade when the fabric is washed, or some may not be light-fast and will fade when exposed to light. Direct dyes usually have a weak OH bond between the nitrogen in the dye and the fiber. In **reactive dyes** the dye reacts with the fiber to produce both an OH and an oxygen linkage, the chlorine combining with the hydroxyl to form a strong ether linkage. Such dyes are fast and very brilliant. **Acid dyes** contain a carboxylic or sulfonic acid group and operate best in an acid bath. **Basic dyes** have an amino group, and are marketed as salts. **Vat dyes** are insoluble and are applied in the soluble colorless form and then reduced or oxidized to color. They usually have an anthraquinone structure, and are solubilized by the reducing agent, a hydroxyl group, OH, diffusing into the fiber where it is fixed.

Color carriers, used to aid adherence of dyes to synthetic fibers, are usually chemicals that act as swelling agents to open the fiber structure, such as phenylphenol, benzoic acid, or dichlorobenzoic acid. The **Ketosol 75,** of Union Carbide, is 75% methylphenyl carbinol and 25% acetophenone. Monochlorobenzene, used as a color carrier for Dacron fiber, acts to promote a concentrated layer of dye solution around the fiber. **Ring-dyed fiber** is a synthetic fiber not receptive to dyes that has been passed through a bath of a receptive plastic before dyeing. The dye then adheres to the coated surface and encases the fiber.

EBONY. A hard, black wood valued for parts subject to great wear, and for ornamental inlaying. It is the wood of various species of trees of the ebony family, *Ebenaceae,* although the name is also applied to some woods of the genus *Dalbergia,* family *Leguminosae.* **Black ebony,** from the tree *Diospyros dendo* of West Africa, and ebony, from the tree *D. melanoxylos,* of India, are the true ebonies. Black ebony has a black heartwood with brownish-white sapwood. It is next to lignum vitae in hardness, has a fine, open grain, and weighs 78 lb/ft^3 (1,250 kg/m^3). It is used for inlaying, piano keys, and turnery. The ebony of India is also extremely hard, with a fine and even grain. The heartwood is black with brownish streaks. **Marblewood,** or **Andaman marblewood,** is an ebony from the tree *D. kurzii* of India and the Andaman Islands. The wood is black with yellowish stripes. It has a close, firm texture, is hard, takes a fine polish, and weighs 65 lb/ft^3 (1,041 kg/m^3). **Marble ebony** is another species from Malagasy. The ebony from Japan, called **kaki,** is from the tree *D. kaki.* It has a black color streaked with gray, yellow, and brown. The grain is close and even, and the wood is very hard, but the weight is less than that of African ebony. Ebony wood is shipped in short billets, and is graded according to the color and the source, as Niger, Macassar, or Cameroon. **Green ebony** is a name sometimes given to the cocoswood of the West Indies. **Artificial ebony,** formerly composed of asphaltic compounds, is now usually molded plastics. **Partridgewood,** a heavy blackish wood used for fine inlay work, is **acapau,** from the large tree *Voucapapoua americana* of the Amazon Valley. It is valued in Brazil for furniture because of its resistance to insect attack.

ELASTOMERS. **Synthetic rubbers,** often referred to as **rubbers,** are hydrocarbon, polymeric materials similar in structure to plastic resins. The difference between plastics and elastomers is largely one of definition based on the property of extensibility, or stretching. The American Society for Testing and Materials defines an elastomer as "a polymeric material which at room temperature can be stretched to at least twice its original length and upon immediate release of the stress will return quickly to approximately its original length." Some grades of plastics approach this rubberlike state, for example, certain of the polyethylenes. Also, a number of plastics have elastomer grades, such as the olefins, styrenes, fluoroplastics, and silicones. As indicated above, the major distinguishing characteristic of elastomers is their great extensibility and high-energy storing capacity. Unlike many metals, for example, which cannot be strained more than a fraction of 1% without exceeding their elastic limit, elastomers have usable elongations up to several hundred percent. Also, because of their capacity for storing energy, even after they are strained several hundred percent, virtually complete recovery is achieved once the stress is removed.

Up until World War II, almost all rubber was natural. During the war, synthetic rubbers began to replace the scarce **natural rubber,** and, since that time, production of synthetics has increased until now their use far surpasses that of natural rubber. There are thousands of different elastomer compounds. Not only are there many different classes of elastomers, but individual types can be modified with a variety of additives, fillers, and reinforcements. In addition, curing temperatures, pressures, and processing methods can be varied to produce elastomers tailored to the needs of specific applications.

In the raw-material or crude stage, elastomers are thermoplastic. Thus **crude rubber** has little resiliency and practically no strength. By a vulcanization process in which sulfur and/or other additives are added to the heated crude rubber, the polymers are cross-linked by means of covalent bonds to one another, producing a thermoset-like material. The amount of cross-linking which occurs between the sulfur (or other additive) and the carbon atoms determines many of the elastomer's properties. As cross-linking increases, resistance to slippage of the polymers over one another increases, resilience and extensibility decreases, and the elastomer approaches the nature of a thermosetting plastic. For example, hard rubbers, which have the highest cross-linking of the elastomers, in many respects are similar to phenolics. In the unstretched state, elastomers are essentially amorphous because the polymers are randomly entangled and there is no special preferred geometrical pattern present. However, when stretched, the polymer chains tend to straighten and become aligned, thus increasing in crystallinity. This tendency to crystallize when stretched is related to an elastomer's strength. Thus, as crystallinity increases, strength also tends to increase.

There are roughly 20 major classes of elastomers; we cannot do much more here than identify them and highlight the major characteristics of each group. Two basic specifications provide a standard nomenclature and classification system for these classes. The ASTM standard D1418 categorizes elastomers into compositional classes. A joint ASTM-SAE specification (ASTM D2000/SAE J200) provides a classification system based on material properties. The first letter indicates specific resistance to heat aging, and the second letter denotes resistance to swelling in oil.

Styrene-butadiene elastomers, sometimes also called **Buna S, SBR,** and **GR-S,** are copolymers of butadiene and styrene. They are similar in many ways to the natural rubbers, and were the first widely used synthetics. They top all elastomers in volume of use, chiefly because of their low cost and use in auto tires. A wide range of property grades are produced by varying the relative amounts of styrene and butadiene. For example, styrene content varies from as low as 9% in low-temperature resistant rubbers to 44% in rubbers with excellent flow characteristics. Those grades with styrene content above 50% are by definition considered plastics. Carbon black is sometimes added also as it substantially improves processing and abrasion resistance. SBR elastomers are similar in properties to natural rubber. They are non-oil-resistant and are generally poor in chemical resistance. Although they have excellent impact and abrasion resistance, they are somewhat below natural rubber in tensile strength, resilience, hysteresis, and some other mechanical properties. The largest single use is in tires. Other applications are similar to those of natural rubber.

Neoprene, also known as **chloroprene,** was developed in the 1930s, and has the distinction of being the first commercial synthetic rubber. It is chemically and structurally similar to natural rubber, and its mechanical properties are also similar. Its resistance to oils, chemicals, sunlight, weathering, aging, and ozone is outstanding. Also, it retains its properties at temperatures up to 250°F (121°C), and is one of the few elastomers that does not support combustion, although it is consumed by fire. In addition, it has excellent resistance to permeability by gases, having about one-fourth to one-tenth the permeability of natural rubber, depending on the gas. Although it is slightly inferior to natural rubber in most mechanical properties, neoprene has superior resistance to compression set, particularly at elevated temperatures. It can be used for low-voltage insulation, but is relatively low in dielectric strength. Typical products made of chloroprene elastomers are heavy-duty conveyor belts, V belts, hose covers, footwear, brake diaphragms, motor mounts, rolls, and gaskets. **Butyl rubbers,** also referred to as **isobutylene-isoprene elastomers,** are copolymers of isobutylene and about 1 to 3% isoprene. They are similar in many ways to natural rubber, and are one of the lowest-priced synthetics. They have excellent resistance to abrasion, tearing, and flexing. They are noted for low gas and air permeability (about 10 times better than natural rubber), and for this reason make a good material for tire inner tubes, hose, tubing, and diaphragms. Although butyls are non-oil-resistant, they have excellent resistance to sunlight and weathering, and generally have good chemical resistance. They also have good low-temperature flexibility and heat resistance up to around 300°F (149°C); however, they are not flame-resistant. They generally have lower mechanical properties such as tensile strength, resilience, abrasion resistance, and compression set, than the other elastomers. Because of their excellent dielectric strength, they are widely used for cable insulation, encapsulating compounds, and a variety of electrical applications. Other typical uses include weather stripping, coated fabrics, curtain wall gaskets, high-pressure steam hoses, machinery mounts, and seals for food jars and medicine bottles.

Isoprene is **synthetic natural rubber.** It is processed like natural rubber, and its properties are quite similar although isoprene has somewhat higher extensibility. Like natural rubber, its notable characteristics are very low hysteresis, low heat buildup, and high tear resistance. It also has excellent flow characteristics, and is easily injection-molded. Its uses complement those of natural rubber. And its good electrical properties plus low moisture absorption make it suitable for electrical insulation. **Polyacrylate elastomers** are based on polymers of butyl or ethyl acrylate. They are low-volume-use, specialty elastomers, chiefly used in parts involving oils (especially sulfur-bearing) at elevated temperatures up to 300°F (149°C) and even as high as 400°F (204°C). A major use is for automobile transmission seals. Other oil-resistant uses are gaskets and O rings. Mechanical properties such as tensile strength and resilience are low. And, except for recent new formulations, they lose much of their flexibility below −10°F (−23°C). The new grades extend low-temperature service to −40°F (−40°C). Polyacrylates have only fair dielectric strength, which improves, however, at elevated temperatures.

Nitrile elastomers, or **NBR rubbers,** known originally as **Buna N,** are copolymers of acrylonitrile and butadiene. They are principally known for their outstanding resistance to oil and fuels at both normal and elevated temperatures. Their properties can be altered by varying the ratio of the two monomers. In general, as the acrylonitrile content increases, oil resistance, tensile strength, and processability improve while resilience, compression set, low-temperature flexibility, and hysteresis characteristics deteriorate. Most commercial grades range from 20 to 50% acrylonitrile. Those at the high end of the range are used where maximum resistance to fuels and oils is required, such as in oil-well parts and fuel hose. Low-acrylonitrile grades are used where good flexibility at low temperatures is of primary importance. Medium-range types, which are the most widely used, find applications between these extremes. Typical products are flexible couplings, printing blankets, rubber rollers, and washing machine parts. Nitriles as a group are low in most mechanical properties. Because they do not crystallize appreciably when stretched, their tensile strength is low, and resilience is roughly one-third to one-half that of natural rubber. Depending on acrylonitrile content, low-temperature brittleness occurs at from −15 to −75°F (−26 to −60°C). Their electrical insulation quality varies from fair to poor. **Polybutadiene elastomers** are notable for their low-temperature performance. With the exception of silicone, they have the lowest brittle or glass transition temperature, −100°F (−73°C), of all the elastomers. They are also one of the most resilient, and have excellent abrasion resistance. However, resistance to chemicals, sunlight, weathering, and permeability by gases is poor. Some uses are shoe heels, soles, gaskets, and belting. They are also often used in blends with other rubbers to provide improvements in resilience, abrasion resistance, and low-temperature flexibility.

Polysulfide elastomer, also known as **Thiokol,** is rated highest in resistance to oil and gasoline. It also has excellent solvent resistance, extremely low gas permeability, and good aging characteristics. Thus it is used for such products as oil and gasoline hoses, gaskets, washers, and diaphragms. Its major use is for equipment and parts in the coating production and application field. It is also widely applied in liquid form in sealants for the aircraft and marine industries. Thiokol's mechanical properties, including strength, compression set, and resilience, are poor. Although Thiokol is poor in flame resistance, it can be used in temperatures up to 250°F (121°C). **Ethylene-propylene elastomers,** or **EPR rubber,** are available as copolymers and terpolymers. They offer good resilience, flexing characteristics, compression-set resistance, and hysteresis resistance, along with excellent resistance to weathering, oxidation, and sunlight. Although fair to poor in oil resistance, their resistance to chemicals is good. Their maximum continuous service temperature is around 350°F (177°C). Typical applications are electrical insulation, footwear, auto hose, and belts. **Urethane elastomers** are copolymers of diisocyanate with their polyester or polyether. Both are produced in solid gum form and viscous liquid. With tensile strengths above 5,000 lb/in^2 (34 MPa) and some grades approaching 7,000 lb/in^2 (49 MPa), urethanes are the strongest available elastomers. They are also the hardest, and have extremely good abrasion resistance. Other notable properties are low compression set, and good aging characteristics and oil and fuel resistance. The maximum temperature for continuous use is under 200°F (93°C), and their brittle point ranges from −60 to −90°F (−51 to − 68°C). Their largest field of application is for parts requiring high wear resistance and/or strength. Typical products are forklift truck wheels, airplane tail wheels, shoe heels, bumpers on earth-moving machinery, and typewriter damping pads. **Chlorosulfonated polyethylene elastomer,** commonly known as **Hypalon,** contains about one-third chlorine and 1 to 2% sulfur. It can be used by itself or blended with other elastomers. Hypalon is noted for its excellent resistance to oxidation, sunlight, weathering, ozone, and many chemicals. Some grades are satisfactory for continuous service at temperatures up to 350°F (177°C). It has moderate oil resistance. It also has unlimited colorability. Its mechanical properties are good but not outstanding, although abrasion resistance is excellent. Hypalon is frequently used in blends to improve oxidation and ozone resistance. Typical uses are tank linings, high-temperature conveyor belts, shoe soles and heels, seals, gaskets, and spark plug boots. **Epichlorohydrin elastomers** are noted for their good resistance to oils, and excellent resistance to ozone, weathering and intermediate heat. The homopolymer has extremely low permeability to gases. The copolymer has excellent resilience at low temperatures. Both have low heat buildup, making them attractive for parts subjected to repeated shocks and vibrations. **Fluorocarbon elastomers,** fluorine-containing elastomers, like their plastic counterparts, are highest of all the elastomers in resistance to oxidation, chemicals, oils, solvents, and heat—and they are also the highest in price. They can be used continuously at temperatures over 500°F (127°C) and do not support combustion. Their brittle temperature, however, is only −10°F (−23°C). Their mechanical and electrical properties are only moderate. Unreinforced types have tensile strengths of less than 2,000 lb/in^2 (13 MPa) and only fair resilience. Typical applications are brake seals, O rings, diaphragms, and hose. The **phosphonitrile plastics** and elastomers have high elasticity and high temperature resistance. They are derived from **chlorophosphonitrile,** or **phosphonitrilic chloride,** $P_3N_3Cl_3$, which has a hexagonal ring of alternating atoms of phosphorus and nitrogen with the chlorine atoms attached. In the plastic monomers the chlorine atoms are replaced by other groups. The **PN Polymers,** of Horizons, Inc., contain OCH and C_3F_7 groups and are synthetic rubbers of high oxidation and chemical resistance. The tensile strength is up to 700 lb/in^2 (4.9 MPa) with elongation up to 600%. The material is used for such applications as gaskets and cratings for severe service. A **phosphonitrilic fluorocarbon elastomer,** of the Firestone Tire and Rubber Co., is a semiorganic **phosphazene polymer.** It remains flexible and serviceable at temperatures from −70 to 350°F (−57 to 177°C), and is highly resistant to oils and solvents over that temperature range. **Viton** is a **vinylidene fluoride hexafluoropropylene tetrafluorethylene** copolymer of Du Pont with extra high resistance to solvents, hydrocarbons, steam, and water. Elastomers of the same type produced by 3M have the trade name **Fluorel. Perfluoroelastomer** is a specialty, costly, fully chlorinated elastomer of Du Pont that resists commercial solvents, bases, and jet fuels. **Kalrez** can be used at continuous service temperatures of up to 550°F (288°C) and at temperatures up to 650°F (343°C) for short time periods. Firestone's **phosphonitrilic fluoroelastomer (PNF200)** has improved low-temperature flexibility and resistance to heat aging and fuels. Its maximum continuous service temperature is 350°F (176°C).

Chlorinated polyethylene elastomers are produced by substitution of chlorine for hydrogen on a high-density polyethylene chain resulting in a fully saturated structure with no double or triple bonds. The elastomer requires the catalytic reaction of a peroxide for curing. Thus, most molded parts are black. Five grades of **CPE polymers** are produced, differing principally in chlorine content. The higher chlorine content grades have best oil and fuel resistance, tear resistance, gas impermeability, and hardness.

Those with lower chlorine content have lower viscosities, better low-temperature properties, and improved resistance to heat and compression set. **Silicone elastomers** are polymers composed basically of silicon and oxygen atoms. There are four major elastomer composition groups. In terms of application, silicone elastomers can be divided roughly into the following types: general-purpose, low-temperature, high-temperature, low-compression-set, high-tensile–high-tear, fluid-resistant, and room-temperature vulcanizing. All silicone elastomers are high-performance, high-price materials. The general-purpose grades, however, are competitive with some of the other specialty rubbers, and are less costly than the fluorocarbon elastomers. Silicone elastomers are the most stable group of all the elastomers. They are outstanding in resistance to high and low temperatures, oils, and chemicals. High-temperature grades have maximum continuous service temperatures up to 600°F (316°C); low-temperature grades have glass transition temperatures of −180°F (−118°C). Electrical properties, which are comparable to the best of the other elastomers, are maintained over a temperature range from −100°F (−73°C) to over 500°F (260°C). However, most grades have relatively poor mechanical properties. Tensile strength runs only around 1,200 lb/in^2 (8 MPa). However, grades have been developed with much improved strength, tear resistance, and compression set. **Fluorosilicone elastomers** have been developed which combine the outstanding characteristics of the fluorocarbons and silicones. However, they are expensive and require special precautions during processing. A unique characteristic of one of these elastomers is its relatively uniform modulus of elasticity over a wide temperature range and under a variety of conditions. Silicone elastomers are used extensively in products and components where high performance is required. Typical uses are seals, gaskets, O rings, insulation for wire and cable, and encapsulation of electronic components.

ELEMI. A soft, sticky, opaque resin with a pleasant odor, obtained from the **Pili tree,** *Canarium luzonicum,* of the Philippines, and employed for giving body and elasticity to lacquers and in lithographic inks. In medicine it is used in ointments. It contains **dipentene,** $C_{10}H_{18}$, which is called **limonene** from its lemonlike odor, and is known as **cajeputene** when obtained from cajeput. **Limonene dioxide,** or **dipentene dioxide,** a colorless liquid of the composition $C_{10}H_{12}O_2$, is a valuable synthetic chemical for making epoxy resins and for cross-linking acrylic and other resins. Elemi also contains a related terpinene oil, **phellandrene.** Substitute elemi resins are obtained from various trees of the family *Burseraceae* of tropical Africa and America. The pili trees are hacked or stripped, and the resin collects on the bark, a tree yielding about 5 lb (2.3 kg) per year. **West Indian elemi** is from the tree *Dacryodes hexandra* of the West Indies. **Nauli gum** is elemi from the tree *C. commune* of the Solomons. **Elemi oil,** obtained by distilling elemi, is a colorless liquid of specific gravity 0.87 to 0.91, used in perfumes and in medicines. It has an aniselike odor.

ELKSKIN. The commercial name for soft, pliable, and durable leather made from the bundled rawhides known as kips, or from overgrown calf by a special tanning process and impregnation with oils. It is used chiefly for children's shoe uppers and for pocketbooks. A heavier elkskin, or **elk leather,** for sport shoes and boots, is made from cowhides by the same treatment. Elkskin, like chamois, dries out to its original softness after wetting. **Smoked elk** is elk leather dyed cream-colored to imitate the original leather of elks, which was smoked over a wood fire.

ELM. The wood of several species of the elm tree, of the eastern United States and Canada, and northern Europe. The wood of the **American elm,** or **white elm,** *Ulmus americana,* has a fine grain, has a weight of about 40 lb/ft^3 (641 kg/m^3), is hard and tough, and is whitish brown in color. It is the best known commercially of the six species grown in the United States. The American elm is not a forest tree, but is grown as a shade and ornamental tree. It does not grow in the mountains. The trees sometimes reach a diameter of 6 ft (1.8 m) and a height of 100 ft (30 m). The tough, durable wood is valued for ax handles, and for parts requiring a combination of strength, bending qualities, and ability to withstand rough usage. The wood of this tree, and also of the rock elm, was formerly used for superstructures of naval ships because it did not sharp-splinter like oak. It was also the favorite wood for hubs and spokes of heavy wagon wheels. **Rock elm,** or **hickory elm,** *U. thomasii,* is also native to the United States and Canada. It has a very fine, close grain and is slightly heavier. It is sometimes called **cork elm,** although this name applies to the **wahoo,** or **winged elm,** *U. alata,* of the southeastern states, because of the corky appearance of the twigs. The winged elm is grown as a shade tree, but the wood was valued for vehicle parts. **English elm,** *U. procera,* has a straight trunk and rounded crown more like the oaks. **Chinese elm,** *U. parvifolia,* has small leaves and is very resistant to disease. **Slippery elm** is a smaller forest tree, *U. fulva,* of the northeastern United States. Considerable lumber came from this tree under the name of **red elm.** The inner bark of the tree is mucilaginous with a sweet taste and characteristic odor. It was used by the Indians as a chewing gum and as a poultice for skin infections. The dried and powdered bark is now used in medicine for skin infections and for the throat. It contains **ulmic acid,** or **geic acid,** $C_{20}H_{14}O_6$.

EMERY. A fine-grained, impure variety of the mineral corundum, with the fine crystals of aluminum oxide embedded in a matrix of iron oxide.

It usually contains only 55 to 75% Al_2O_3. The specific gravity is 3.7 to 4.3, and the hardness about 8. It occurs as a dark-brown granular massive mineral. It is used as an abrasive either ground into powder or in blocks and wheels. In the natural block material the grains are irregular, giving a varying grinding performance. The grains are graded in sizes from 220 mesh, the finest, to 20 mesh, the coarsest. **Emery paper** and cloth are usually graded from 24 to 120 mesh, and the grains are glued to one side of 9- by 11-in (23- by 28-cm) sheets. **Flour of emery** is the finest powder, usually dust from the crushing. **Emery cake** as made for buffing and polishing is now not likely to be made of emery, but a graded combination of aluminum oxide and iron oxide, with a higher percentage of the hard aluminum oxide for buffing, and higher iron oxide for polishing. It is furnished in various grades of fineness, with grains of 120 to 200 mesh, or flour sizes, F, FF, and FFF. Emery takes its name from Cape Emery, on the Island of Naxos.

EMULSIFYING AGENTS. Materials used to aid in the mixing of liquids that are not soluble in one another, or to stabilize the suspension of nonliquid materials in a liquid in which the nonliquid is not soluble. The suspension of droplets of one liquid in another liquid in which the first liquid is not soluble is called an **emulsion.** The emulsion of oil and water, used in machine shops as a cutting lubricant and work coolant, may be made with soap as the emulsifying agent. The emulsifying agent protects droplets of the dispersed medium from uniting and thus separating out. The oil itself may be treated so that it is self-emulsifying. Sulfonated oils contain strong negatively charged ester sulfate groups in the molecule, and do not react and conglomerate with the molecules of a weakly charged liquid. They will thus form emulsions with water without any other agent.

In emulsions of a powder in a liquid an emulsifying agent called a **protective colloid** may be used. This is usually a material of high molecular weight such as gelatin, and such materials form a protective film around each particle of the contained powder. **Saponin** and **starches** are commonly used thus as **suspending agents.** For the suspension of drug materials in pharmaceutical mixtures, **gum arabic** or **tragacanth** may be used. Starches, egg albumin, and proteins are common emulsifying agents for food preparation. **Alginates** are among the best suspending agents for a wide range of emulsions because of the numerous repelling charges in the high molecular weight and the irregular configuration of the chain, but when added to protein-containing liquids such as many foodstuffs, the similar conditions of the algin and the protein molecules cause a neutralization reaction and a precipitation of the agglomerated particles. Suspending agents generally increase the viscosity of the liquid, and with high concentrations of some gums or resins the water molecules may be completely encased in the resin lattice as a semisolid or water-filled plastic.

Sucrose esters, used as emulsifiers for foods, cosmetics, and drugs, are made from sugar and palmitic, lauric, or other fatty acids. The monoesters are soluble in water and in alcohol, and the diesters are oil-soluble. **Sorbester,** of Howards of Ilford, Ltd., for emulsifying fatty foodstuffs, is a diester of sucrose. **Sucrodet D-600,** of the Millmaster Chemical Corp., is a white, tasteless, and odorless powder made from sugar and palmitic acid. **Myrj 45,** of the Atlas Chemical Industry, is polyoxyethylene stearate. **Propylene laurate** is a light, high-boiling liquid that is self-emulsifying in water, and is employed in foodstuffs and pharmaceuticals to stabilize the mixtures. The sodium salt of **ursolic acid** is a strong emulsifying agent for oil-in-water mixtures. The acid is a complex triterpene obtained from the skins of the cranberry.

Some solid materials may be suspended indefinitely in liquids if ground to such a fineness that the electronegative mutually repelling force, or **zeta potential,** of the particles is greater than the force of gravity. Silica, for example, has only a feeble electronegativity, and if the particles are below about 1 μm in size they will give a permanent suspension in water. These finely ground solids are used as **thickening agents** for paints and coatings. Bentonite is thus used in adhesives and paints. Thickening agents may add other properties such as better adhesion or strengthening the film. Some long-chain chemicals used as emulsifying agents in cutting oils also give antirust properties. **Thickening agent ASE-95,** of the Rohm & Haas Co., has a powerful thickening action on water-base emulsions which can be halted at any desired viscosity by neutralizing the acidity. It is an acrylic copolymer of 20% solids containing an organic acid with a pH of 3. When added to the emulsion to be thickened, the solids dissolve in minute particles, and the process is stopped at the desired viscosity by adding an alkali.

ENAMEL. A coating which upon hardening has an enameled or glossy face. **Pottery enamels, ceramic enamels,** or **ceramic coatings,** and **vitreous enamels** are composed chiefly of quartz, feldspar, clay, soda, and borax, with saltpeter or borax as fluxes. The quartz supplies the silica, and such enamels are fusible glasses. In acid-resisting enamels alkali earths may be used instead of borates. To make enamels opaque, opacifiers are used. They may be tin oxide for white enamel, cobalt oxide for blue, or platinum oxide for gray. Enamel-making materials are prepared in the form of a powder which is called **frit.** The frit-making temperature is about 2400°F (1316°C), but the enamel application temperatures are from 1400 to 1600°F (760 to 871°C). Each succeeding coat has a lower melting point than the one before it so as not to destroy the preceding coat. It must also have about the same coefficient of expansion as the metal to prevent cracking. Enamels for aluminum usually have a high proportion of lead oxide to lower the melting point, and enamels for magnesium may be based on lithium oxide. Some enamels for low-melting metals have the ceramic frit bonded to the metal with monoaluminum phosphate at temperatures as low as 400°F (204°C).

The mineral oxide coatings fused to metals are often called **porcelain enamels,** but they are not porcelain, and the term vitreous enamel is preferred in the industry, although ceramic-lined tanks and pipe are very often referred to as **glass-lined steel.** The composition varies greatly, one company having more than 3,000 formulas. Vitreous enameled metals are used for cooking utensils, signs, chemical tanks and piping, clock and instrument dials, and siding and roofing. Ground coats are usually no more than 0.004 in (0.010 cm) thick, and cover coats may be 0.003 to 0.008 in (0.006 to 0.020 cm) thick. The hardness ranges from 150 to 500 Knoop. Thick coatings on thin metals are fragile, but thin coatings on heavy metals are flexible enough to be bent. Standard porcelain-type enamel has a smooth glossy surface with a light reflectance of at least 65% in the white color, but pebbly surfaces that break up the reflected image may be used for architectural applications.

High-temperature coatings may contain a very high percentage of zirconium and will withstand temperatures to 1650°F (899°C). **Refractory enamels,** for coating superalloys to protect against the corrosion of hot gases to 2500°F (1371°C), may be made with standard ceramic frits to which is added boron nitride with a lithium chromate or fluoride flux. Blue undercoats containing cobalt are generally used to obtain high adhesion on iron and steel but some of the **enameling steels** do not require an undercoat, especially when a specially compounded frit or special flux is used. When **sodium aluminum silicate,** $Na_2O \cdot Al_2O_3 \cdot 6SiO \cdot xH_2O$, is used instead of borax, a white finish is produced without a ground coat. **Mirac,** of the Pemco Co., is a white enamel which gives good adhesion direct to steel. Enamels containing titanium oxide will adhere well to steels alloyed with a small amount of titanium. **Ti-Namel,** of the Inland Steel Co., is an enameling steel containing titanium.

Many trade names are applied to vitreous enamels and to enameled metals. **Vitric steel,** of the Republic Stamping & Enameling Co., is an enameled corrugated sheet steel for construction. **Majolica** is an old name for marblelike enamels made by mixing enamels of different colors, but **mottled graywear** is made with cobalt oxide on steel that has a controlled misting on the surface. **Cloisonné enamel** is an ancient decorative enamel produced by soldering thin strips of gold on the base metal to form cells into which the colored enamel is pressed and fused into place. It requires costly hand methods, and is now imitated in synthetic plastics under names such as **Enameloid.**

The word enamel in the paint industry refers to glossy **varnishes** with pigments or to paints of oxide or sulfate pigments mixed with varnish to give a glossy face. They vary widely in composition, in color and appearance, and in properties. As a class, enamels are hard and tough, and offer good mar and abrasion resistance. They can be formulated to resist attack by the most commonly encountered chemical agents and corrosive atmospheres. Because of their wide range of useful properties, enamels are one of the most widely used organic finishes in industry, and are especially used as household appliance finishes. **Japan** is a name applied to black baking enamels. The same finish in other colors would not be called japan, as the original oriental lacquer was always black. Japan consists of a pigment, a gum, a drying oil, and a reducer, the same as any oil enamel. It is always baked, which drives off the solvent and fuses the gum into a uniform vitreous layer. Japans give a tough durable finish to small machine parts, but require expensive between-coat rubbing, and are now replaced by synthetic baking finishes. The modified phenolmelamine and alkyd-melamine synthetic resins produce tough and resistant enamel coatings. Quick-drying enamels are the cellulose lacquers with pigments. **Fibrous enamel,** used for painting roofs, is an asphalt solution in which asbestos fibers have been incorporated. When of heavy consistency and used for calking metal roofs, it is called **roof putty.**

EPOXY RESINS. A class of synthetic resins characterized by having in the molecule a highly reactive **oxirane** ring of triangular configuration consisting of an oxygen atom bonded to two adjoining and bonded carbon atoms. They are usually made by the reaction of epichlorohydrin with phenol compounds, but epoxidation is also done by the oxidation of a carbon-to-carbon double bond with an organic peracid such as peracetic acid. **Epichlorohydrin** is produced from allyl chloride, and is a colorless liquid with a chlorine atom and an epoxide ring. The **dipoxy resins** made by the oxidation of olefins with peracetic acid have higher heat resistance than those made with bisphenol. Epoxidation is not limited to the making of plastic resins, and **epoxidized oils,** usually epoxidized with peracetic acid, are used as paint oils and as plasticizers for vinyl resins.

Epoxy resins are generally more costly than many other resins, but, because of their unusual combinations of high mechanical and electrical properties, they are important, especially for such uses as adhesives, resistant coatings, and for encapsulation of electronic units. The resins are thermosetting and inert. For encapsulation, they cast easily with little shrinkage. They have very high adhesion to metals and nonmetals, heat resistance from 350 to 500°F (177 to 260°C), dielectric strength to 550 volts per mil (22 volts per meter), and hardness to Rockwell M110. The tensile strength may be up to 12,000 lb/in^2 (82 MPa), with elongation to 2 to 5%, but some resilient encapsulating resins are made with elongation to 150% with lower tensile strengths. The resins have high resistance to common solvents, oils, and chemicals.

An unlimited variety of epoxy resins is possible by varying the basic reactions with different chemicals or different catalysts, or both, by combination with other resins, or by cross-linking with organic acids, amines, and other agents. To reduce cost when used as laminating adhesives they may be blended with furfural resins, giving adhesives of high strength and high chemical resistance. Blends with polyamides have high dielectric strength, mold well, and are used for encapsulating electrical components. The **Epon resins** of the Shell Chemical Corp. are epoxy resins made with epichlorohydrin and bisphenols. **Kem-Krete,** of the Sherwin-Williams Co., a coating for ceramic blocks, and **Erkopon,** a coating for tanks, of the Earl Paint Corp., are based on **ethoxyline resins,** or polyaryl ethylene oxide condensates, made from epichlorohydrin and bisphenol with an ethylenediamine catalyst. By using a polyamide curing agent an epoxy can be made water-emulsifiable for use in water-based paints. The **water-soluble epoxy** of General Mills, Inc., is a reaction product of **Genepoxy M195,** a bisphenol epichlorohydrin epoxy resin, and **Versamid 265-WR70,** an ethylene glycol ether acetate. **Resin X-3441,** of the Dow Chemical Co., is an epoxy resin with 19% bromine in the molecule. It is flame-resistant. Another grade, with 49% bromine, is a semisolid, used for heat-resistant adhesives and coatings. **Oxiron resins,** of the FMC Corp., are **epoxidized polyolefins.** They have five or more reactive epoxy groups along each molecule of the chain instead of the usual two terminal epoxy groups on each molecule. With dibasic acids or anhydrides they form strong, hard resins of high heat resistance; or resins of lower viscosity are made for laminating and casting. **Novolac DEN 438,** of the Dow Chemical Co., is an epoxy resin with a high distortion point, 570°F (299°C), made by the reaction of epichlorohydrin with a phenol-formaldehyde resin with an anhydride catalyst. As an adhesive for laminates it gives very high strength at elevated temperatures. **Narmco adhesive,** of Whittaker Corp., is an **epoxy adhesive** with a sheer strength of 5,500 lb/in^2 (38 MPa). Epoxies can be copolymerized with other resins. **Epoxy-acrylate resin,** used for glass-fiber laminates, combines the resistance and adhesiveness of the epoxy with the fast cure and strength of the acrylate. **Epoxy resin ERRA-0300,** of Union Carbide, has cyclopentyl oxide terminal groups instead of diglycidyl ether. The yield strength at 392°F (200°C) is 18,200 lb/in^2 (123 MPa), and it has a heat deflection temperature of 434°F (223°C). **Aracast epoxy resins,** of Ciba-Geigy Corp., are produced by a reaction of **hydantoin** with epichlorohydrin. Hydantoin is a nitrogen-containing heterocyclic compound. They have high mechanical properties, good dielectrical characteristics, and ultraviolet light resistance. They retain light transmission properties after thermal aging of several thousand hours at 302°F (150°C).

Novoloids are fibers containing at least 85%, by weight, cross-linked **novalac epoxies. Kynol** is a novoloid noted for its exceptionally high temperature resistance. At 1920°F (1049°C) the fiber is virtually unaffected. The fiber also has high dielectric strength and excellent resistance to all organic solvents and nonoxidizing acids.

A family of one-component epoxy resins, named **Arnox,** are produced by General Electric Co. Suitable for compression, transfer, injection molding, filament winding, and pultrusion, they cure rapidly at temperatures of 250 to 350°F (121 to 177°C). The compression and transfer molding grade is a black, mineral-filled compound. The injection molding grade is a pelletized glass-fiber-reinforced compound with a shelf life of 9 to 12 months below 80°F (27°C).

ESSENTIAL OILS. **Aromatic oils** found in uncombined form in various parts of plants and employed for flavors, perfumes, disinfectants, medicines, stabilizers; for masking undesirable odors; and as raw materials for making other products. They are usually the esters upon which the odiferous properties of the plants depend, and are called essential oils because of their ease of solubility in alcohol to form essences. They are also called **volatile oils,** although this term is sometimes also applied to the light and volatile distillates from petroleum. The essential oils are of four general classes: the **pinenes** or **terpenes** of coniferous plants, containing carbon and hydrogen of the empirical formula $C_{10}H_{16}$, such as **oil of turpentine; oxygenated oils** containing carbon, hydrogen, and oxygen, such as **oil of cassia; nitrogenated oils** containing carbon, hydrogen, oxygen, and nitrogen, such as **oil of bitter almonds; sulfurated oils** containing carbon, hydrogen, and sulfur, such as **oil of mustard.**

Although fixed vegetable oils are obtained by expression, the essential oils are obtained by distilling the buds, flowers, leaves, twigs, or other parts of the plant. **Rose oil** is found only in the flowers. **Orange oil** and **lemon oil** are from the flowers and the fruits, but are of different compositions. **Sweet birch oil** and **cinnamon oil** are from the bark. **Valerian** and **calamus** are only in the roots, while **sandalwood oil** and **cedar oil** are only in the wood. Sometimes the essential oil is not in the plant, but is developed when the plant is macerated with water. The **alpha pinene** extracted from turpentine is used for paints and varnishes as it has a high evaporation rate. It is a water-white liquid of pleasant odor boiling at 163°C. It is also used in the synthesis of camphor. **Pinic acid** is a complex carboxycyclobutane acetic acid produced from alpha pinene. Its esters are used for synthetic lubricants. **Balsams** are solid or semisolid resinous oils, and are mixtures of resins with cinnamic or benzoic acid, or both, with sometimes another volatile oil. They are obtained from a variety of trees, and are used in antiseptics, perfumes, flavors, and in medicine.

Some of the essential oils contain alkaloids which have a physiological effect. **Wormwood oil,** distilled from the dried leaf tops of the perennial herb *Artemisia absinthium,* native to southern Europe but also grown in the United States, is used in medicine for fevers, and also for flavoring the liqueur absinthe. The drug **santonin,** used for worm treatment for animals, is an alkaloid extracted from the unopened flower heads of the **Levant wormseed,** *A. cina,* of the Near East, but **wormseed oil,** or **Baltimore oil,** used for the same purpose, is an essential oil containing the alkaloid **ascoridole.** It is distilled from the seeds and leaf stems of the annual plant *Chenopodium anthelminticum,* grown in Maryland.

ESTERS. Combinations of alcohols with organic acids, which form several important groups of commercial materials. The esters occur naturally in vegetable and animal oils and fats as combinations of acids with the alcohol glycerin. The natural fats are usually mixtures of esters of many acids, coconut oil having no less than 14 acids. Stearic, oleic, palmitic, and linoleic acid esters are the common bases for most vegetable and animal fats, and the esters of the other acids such as linolenic, capric, and arachidic give the peculiar characteristics of the particular fat, although the physical characteristics and melting points may be governed by the basic esters. Esters occur also in waxes, the vegetable waxes being usually found on the outside of leaves and fruits to protect them from loss of water. The waxes differ from the fats in that they are combinations of monacids with monohydric, or simple, alcohols, rather than with glycerin. They are harder than fats and have higher melting points. Esters of still lower molecular weights are also widely distributed in the essential oils of plants where they give the characteristic odors and tastes. All of the esters have the characteristic formula ArCOOR or RCOOR, where R represents an **alkyl group,** and Ar an **aryl group,** that is, where R is a univalent straight-chain hydrocarbon having the formula C_nH_{2n+1}, and Ar is a univalent benzene ring C_6H_5. In the esters of low molecular weight which make the odors and flavors, the combination of different alcohols with the same acid yields oils of different flavor. Thus the ester methyl acetate, CH_3COOCH_3, is **peppermint oil;** amyl acetate, $CH_3COOC_5H_{11}$, is banana oil; and isoamyl acetate, $CH_3COO(CH_2)_3(CH_3)_2$, is **pear oil.** Esters are used as solvents, flavors, perfumes, waxes, oils, fats, fatty acids, pharmaceuticals, and in the manufacture of soaps and many chemicals. Ester liquid lubricants have good heat and oxidation resistance at high temperatures and good fluidity at low temperatures. They are widely used in jet aircraft.

Ester alcohols are intermediates that require less acid for esterification. **Texanol,** of Eastman Chemical Co., has both a hydroxy group and an ester linkage with the empirical formula $C_{12}H_{24}O_3$. It produces a wide range of chemicals and compounds with low, −57°C, pour point.

ETCHING MATERIALS. Chemicals, usually acids, employed for cutting into, or etching, the surface of metals, glass, or other material. In the metal industries they are called **etchants.** The usual method of etching is to coat the surface with a wax, asphalt, or other substance not acted upon by the acid; cut the design through with a sharp instrument; and then allow the acid to corrode or dissolve the exposed parts. For etching steel, a 25% solution of sulfuric acid in water or a ferric chloride solution may be used. For etching stainless steels a solution of ferric chloride and hydrochloric acid in water is used. For high-speed steels, brass, or nickel, a mixture of nitric and hydrochloric acids in water solution is used, or nickel may be etched with a 45% solution of sulfuric acid. Copper may be etched with a solution of chromic acid. Brass and nickel may be etched with an acid solution of ferric chloride and potassium chlorate. For red brasses, deep etching is done with concentrated nitric acid mixed with 10% hydrochloric acid, the latter being added to keep the tin oxide in solution and thus retain a surface exposed to the action of the acid. For etching aluminum a 9% solution of copper chloride in 1% acetic acid, or a 20% solution of ferric chloride may be used, followed by a wash with strong nitric acid. Sodium hydroxide, ammonium hydroxide, or any alkaline solutions are also used for etching aluminum. Zinc is preferably etched with weak nitric acid, but requires a frequent renewal of the acid. Strong acid is not used because of the heat generated, which destroys the wax coating. A 5% solution of nitric acid will remove 0.002 in (0.005 cm) of zinc per minute, compared with the removal of over 0.005 in (0.013 cm) per minute in most metal-etching processes. Glass is etched with hydrofluoric acid or with **white acid.** White acid is a mixture of hydrofluoric acid and ammonium bifluoride, a white crystalline material of the composition $(NH_4)FHF$. **Sodium chlorate** may be used as the electrolyte in producing **chemical finishes.** The process in which the metal is removed chemically to give the desired finish as a substitute for mechanical machining is called **chemical machining.**

ETHER. The common name for **ethyl ether,** or **diethyl ether,** a highly volatile, colorless liquid of the composition $(C_2H_5)_2O$ made from ethyl alcohol. It is used as a solvent for fats, greases, resins, and nitrocellulose, and in medicine as an anesthetic. The specific gravity is 0.720, boiling point 34.2°C, and freezing point −116°C. Its vapor is heavier than air and is explosive. Actually, ether is a more general term, and an ether is an **alkyl oxide** with two alkyl groups joined to an oxygen atom. The ethyl ether would thus be expressed as $C_2H_5 \cdot O \cdot C_2H_5$, and there are many ethers. **Butyl ether,** $(C_4H_9)_2O$, has a much higher boiling point, 140°C; is more stable; and is used as a solvent for gums and resins. **Isopropyl ether,** $(CH_3)_2CHOCH(CH_3)_2$, is a by-product in the manufacture of isopropyl alcohol from propylene. It has a higher boiling point than ethyl ether, 69°C; lower solubility in water; and is often preferred as an extractive solvent. **Methyl ether,** or **dimethyl ether,** also known as **wood ether,** is a colorless gas of the composition $(CH_3)_2O$, with a pleasant aromatic odor. The boiling point is −23.5°C. The specific gravity is 1.562 or, as a liquid compressed in cylinders, 0.724. It is used for fuel, as a welding gas, as a refrigerant, and for vapor-pressure thermometers. **Hexyl ether,** $C_6H_{13}OC_6H_{13}$, has a high boiling point, 226.2°C; very low water solubility; and a specific gravity of 0.7942. It is stable and not volatile, with a flash point of 170°F (77°C). It is used in foam breakers, and in chemical manufacture where anhydrous properties are desired. A low-boiling chemical used as an extractive solvent and for plastics because of its stability in alkalies and its high water solubility is **methylal,** $CH_3OCH_2OCH_3$. It is a water-white liquid boiling at 42.3°C. Ether reacts slowly with the oxygen of the air to form highly explosive and poisonous compounds, so that long-stored ether is dangerous for use as an anesthetic.

ETHYL ALCOHOL. Also called **methyl carbinol,** and **ethanol** when made synthetically. It is the common beverage alcohol, which when denatured for nonbeverage purposes is called **industrial alcohol.** About 90% of the ethyl alcohol used in the United States is denatured. Ethyl alcohol is a colorless liquid with a pleasant odor but burning taste. The composition is CH_3CH_2OH, specific gravity 0.79, boiling point 78.5°C, and freezing point −117.3°C. It mixes with water in all proportions and takes up moisture from the air. It burns with a bluish flame and high temperature, yielding carbonic acid and water. The ignition temperature is 965°F (518°C). It is one of the best solvents, and dissolves many organic materials such as gums, resins, and essential oils, making solutions called **essences.**

Alcohol is sold by the proof gallon, a 100 proof containing 50% alcohol by volume and having a specific gravity of 0.7939. The term **alcohol,** alone, refers to 188 to 192 proof. High-purity alcohol, **grain alcohol,** and pure ethyl alcohol are terms for 190 proof. **Absolute alcohol,** or **anhydrous alcohol,** is 200 proof, free of water. **Methylated spirits** is a term first used in England to designate the excise-free mixture of 90% ethyl alcohol and 10 wood alcohol for industrial use. Denatured ethyl alcohol, made unsuitable for beverage purposes, may be marketed under trade names such as **Synasol** of Union Carbide. **Solox,** of U.S. Industrial Chemicals, Inc., consists of 100 parts 190-proof alcohol, 5 ethyl acetate, and 1 gasoline, used for lacquers, fuel, and as a solvent. **Neosol,** of the Shell Chemical Corp., is 190-proof ethyl alcohol denatured with four parts of a mixture of tertiary butyl alcohol, methyl isobutyl ketone, and gasoline.

Ethyl alcohol is used as a solvent in varnishes, explosives, extracts, perfumes, pharmaceuticals, as a fuel, as a preserving agent, as an antifreeze, and for making other chemicals. Up to 15% of alcohol can be used in gasoline motor fuels without change in the carburation. The German motor fuel **Monopolin** was a mixture of absolute alcohol and benzene. Ethyl alcohol is classed as a poison when pure, but is employed as a beverage in many forms. In small quantities it is an exhilarant and narcotic. In all countries large amounts of **beverage alcohol** are made from starches, grains, and fruits, retaining the original flavor of the raw material and marketed directly as wines, whiskies, and brandies. But **synthetic wines** are made by fermenting sugar and adding vegetable extracts to supply flavor and bouquet. No methyl alcohol or fusel oil is produced in the process. Alcohol is produced easily by the fermentation of sugars, molasses, grains, and starch. It is also made cheaply by hydrating ethylene produced by the cracking of petroleum hydrocarbons. In Europe it is also made from the waste liquor of pulp mills by fermentation of the wood sugar. **Sulfite pulp liquor** contains 1.8% fermentable hexose sugar. It is also made directly from wood waste by fermenting the wood sugar molasses.

A substitute for ethyl alcohol for solvent purposes and as a rubbing alcohol is **isopropyl alcohol,** or **isopropanol,** a colorless liquid of the composition $(CH_3)_2CHOH$, boiling point 82°C, and produced by the hydration of propylene from cracked gases. It is also used as a stabilizer in soluble oils, and in drying baths for electroplating. **Petrohol,** of the Enjay Chemical Co., is isopropyl alcohol. **Trichloroethanol,** $CCl_3 \cdot CH_2OH$, is a viscous liquid with an ether odor, boiling at 150°C and freezing at 13°C, slightly soluble in water, used for making plasticizers and other chemicals. The spent grain from alcohol distilleries, called **stillage,** is dried and marketed as livestock feed, and is a better feed than the original grain because of the high concentration of proteins and vitamins, with the starch removed. The **leaf alcohol** which occurs in fruits and many plants is a hexene alcohol. It is made synthetically for blending in synthetic flavors and for restoring full flavor and fragrance to fruit extracts.

ETHYL SILICATE. A colorless liquid of the composition $(CH_2H_5)_4SiO_4$, used as a source of colloidal silica in heat-resistant and acid-resistant coatings and for moldings. The specific gravity is 0.920 to 0.950. It is a **silicic acid ester,** with a normal content of 25% available silica, though the **tetraethyl orthosilicate** has 27.9% available silica, and the **Ethyl silicate 40** of Union Carbide has 40% silica. The latter is a brown liquid. Water hydrolyzes ethyl silicate to alcohol and **silicic acid,** H_4SiO_2, which dehydrates to an adhesive amorphous silica. For molding, the ester is mixed with silica powder, and for such products as bearings, wood flour may be incorporated to absorb and retain the lubricating oil. Ethyl silicate solutions are employed for the surface hardening of sand molds and graphite molds for special casting. Silicic acid ester paints are used to harden and preserve stone, cement, or plaster, and for coating insulating brick. They are resistant to heat and to chemical fumes. **Kieselsol,** a German material for clarifying wine and fruit juices by precipitation of the albumin, is a 15% water solution of silicic acid.

ETHYLENE. Also called **ethene.** A colorless, inflammable gas, $CH_2:CH_2$, produced in the cracking of petroleum. Ethylene liquefies at −154.8°F (−68.2°C). It was first produced in Holland by dehydrating ethyl alcohol with sulfuric acid, and is now made from cracking petroleum or by breaking down alcohol by catalytic action. It was originally employed for enriching illuminating gas to give it a more luminous flame, and was called **olefiant gas** because it formed an oil, ethylene dichloride, called **Dutch liquid,** when treated with chlorine. Ethylene is now used to produce ethyl alcohol, acrylic acid, and styrene, and it is the basis for many types of reactive chemicals. **Ketene,** for example, used as a reactant in connecting polymers to improve physical properties of the plastics, has the basic formula $H_2C:C:O$, which is ethylene-modified by substituting oxygen for two of the hydrogens. **Butyl ethyl ketene** of Eastman Chemical Products, Inc., for modifying compounds with active double bonds or active hydrogens, is a yellow liquid of specific gravity 0.826, having the composition $(C_4H_9)(C_2H_5):C:C:O$. **Calorene** is ethylene in pressure cylinders for flame cutting. When burned with oxygen, it gives a flame lower in temperature than acetylene, and it is more stable in storage. For making resins and waxes, and for solvent use, it may be employed in the form of **ethylene diamine,** $NH_2CH_2CH_2NH_2$, a colorless liquid of specific gravity 0.968, boiling at about 120°C. **Ethylene imene,** C_3H_7N, is a very reactive chemical useful for making a wide range of products. It is a water-white liquid of specific gravity 0.79, boiling at 66°C, soluble in water and in common solvents. The **imene ring** in the molecule has two carbon atoms and a nitrogen atom forming a triangle. The ring is stable with basic chemicals, but is strongly reactive to acid compounds, opening at the carbon-nitrogen bond to receive hydrogen. By acid catalyzation and control with alkaline solutions to avoid violent simultaneous opening of the two carbon bonds, the material can be polymerized or made to receive other chemical groups.

Trichlorethylene is a heavy colorless liquid of pleasant odor of the composition $CHCl:CCl_2$, also known as **westrosol.** Its boiling point is 87°C and its specific gravity 1.471. It is insoluble in water and is unattacked by dilute acids and alkalies. It is not inflammable and is less toxic than tetrachlorethane. Trichlorethylene is a powerful solvent for fats, waxes, resins, rubber, and other organic substances, and is employed for the extraction of oils and fats, for cleaning fabrics, and for degreasing metals preparatory to plating. The freezing point is −88°C, and it is also used as a refrigerant. It is also used in soaps employed in the textile industry for degreasing. **Tri-Clene** is a trade name of Du Pont for trichlorethylene, marketed for dry cleaning. **Triad** and **Perm-A-Clor** are trade names of the Detroit Rex Products Co. for trichlorethylene stabilized with a basic organic stabilizer that prevents breakdown of the solvent in degreasing metals.

Ethylene resins are a class of synthetic resins which range from grease-like liquids in the low molecular weights to waxlike materials at molecular weights from about 4,000 to 10,000 to tough white solids at molecular weights above about 12,000, which are thermoplastic resins melting at 210 to 235°F (99 to 112°C). In the ethylene molecule the two carbon atoms, each of which has two attached hydrogen atoms, are linked together with a straddle bond of the Nos. 1 and 2 electrons of the carbons, which normally form the hexagonal carbon ring. This type of **double bond** is not double in a mechanical sense and is termed a **reactive bond,** that is, a bond that can be broken readily to receive other attachments.

Polyox resins, of Union Carbide, are white granular powders of water-soluble **ethylene oxide plastics** with a wide range of molecular weights for films, fibers, and molded articles. **Polyox film** has a tensile strength of 1,800 to 2,400 lb/in^2 (12.4 to 16.5 MPa), with elongation from 100 to 2,000%, and heat seals at temperatures from 170 to 265°F (77 to 129°C). It is used for packaging soaps, detergents, and chemicals to be added in measured amounts without removing the package. The plastic has high adhesive strength, and is also used for adhesives where water solubility is wanted. **Radel,** of this company, is an oriented polyethylene oxide for film and molded parts. The film heat-seals at 135 to 160°F (57 to 71°C) and comes in thicknesses from 0.0015 to 0.010 in (0.004 to 0.025 cm). The opaque film has biaxial orientation which allows heat printing and increases its water solubility.

ETHYLENE GLYCOL. Also known as **glycol** and **ethylene alcohol.** A colorless syrupy liquid, CH_2OHCH_2OH, with a sweetish taste, very soluble in water. It has a low freezing point, −25°C, and is much used as an antifreeze in automobiles. A 25% solution has a freezing point of − 5°F (−20.5°C), without appreciably lowering the boiling point of the water. It has the advantage over alcohol that it does not boil away easily, and permits the operation of engines at much higher temperatures than with water, giving greater fuel efficiency. It is also used for the manufacture of acrylonitrile fibers, and as a solvent for nitrocellulose. It is highly toxic in contact with the skin.

Diethylene glycol, $C_4H_{10}O_3$, is a water-white liquid boiling at 244°C, used as an antifreeze, as a solvent, and for softening cotton and wool fibers in the textile industry. A 50% solution of diethylene glycol freezes at −28°C. **Cellosolve,** $C_2H_5OCH_2CH_2OH$, of Union Carbide, is the monoethyl ether of ethylene glycol. It is a colorless liquid boiling at 135.1°C, and is a powerful solvent used in varnish removers, cleaning solutions, and as a solvent for paints, varnishes, plastics, and dyes. **Carbitol,** of the same company, is an ether of diethylene glycol of the composition $CH_3CH_2OCH_2CH_2OCH_2CH_2OH$, used as a solvent for oils, dyes, resins, and gums. The boiling point is 201.9°C and freezing point −75°C. **Propylene glycol,** or **propanediol,** $CH_3 \cdot CHOH \cdot CH_2OH$, is a colorless and odorless liquid boiling at 188°C, used in cosmetics and perfumes; in flavoring extracts as a humectant, wetting agent, and color solvent; and in baked foods to maintain freshness. **Methyl carbitol,** with one less CH_2 group, is also a high boiling solvent for gums and resins, and **carbitol acetate** is used as a high-boiling solvent for cellulose acetate. **Glycol diformate,** $HCOOCH_2CH_2OOCH$, used as a solvent for cellulose acetate and nitrocellulose, is a colorless liquid soluble in water, alcohol, and ether. It hydrolyzes slowly, liberating formic acid.

EUCALYPTUS. A tree genus of several hundred species native to Australia, but now grown in many parts of the world. It is known as **gumwood** in southern United States. The **blue gum,** which attains a height of 300 ft (91 m), is grown on the West Coast of the United States. The wood has a pale straw color and is hard and tough. It has a twisted grain and shrinks and warps easily, but is very durable. The weight is about 50 lb/ ft^3 (801 kg/m^3), being heavier than southern gum. **Salmon gum,** from *E. salmonophloria,* has a salmon-red color, is dense and hard, and has a fine, open grain. It is superior and has a great variety of uses. The weight is about 60 lb/ft^3 (961 kg/m^3). **Red gum,** from *E. calophylla,* has a yellowish-red color, is strong, tough, and weighs about 45 lb/ft^3 (721 kg/ m^3). The grain is fine, but has gum veins intersecting. Other species of gumwood are marketed under the names of **York gum, blackbutt, tuart,** and **Australian red mahogany.**

Three Australian timbers—**jarrah, karri,** and **ironbark**—are members of the *Eucalyptus* genus. Jarrah resembles karri so closely that it is difficult to distinguish one from the other. Both are dark red in color and similar in weight and appearance. Ironbark is heavier than either of these and is more gray in color. Also, it is nearly always severely surface checked, a characteristic which does not detract significantly from its strength. It is very strong, having a modulus of rupture in bending of 27,100 lb/in^2 (187 MPa), whereas jarrah and karri run about 16,000 and 19,000 lb/in^2 (110 and 130 MPa) respectively. Ironbark and jarrah are rated "very durable" by the British Forest Products Research Laboratory. Karri is rated only "moderately durable."

The **wandoo tree,** *E. redunca,* of Western Australia, the wood of which is known as **redunca wood,** has a high percentage of pyrogallol tannin. The solid extract is called **myrtan,** and it produces a solid, firm sole leather lighter in color than that from chestnut. The wood of the *E. saligna,* of South Africa, is hard, and has a fine, even, interlocking grain which makes it strong in all directions. It is used in the United States for small turned articles, saw handles, and paintbrush handles. It has a reddish tinge. **Blackbutt** is from the trees *E. pilularis, E. patens,* and some other species native to Australia, but now grown in other countries. It is used as a substitute for oak, but tends to warp and crack.

Eucalyptus oil, obtained from the dried leaves of the *E. globulus,* is used in pharmaceuticals for nose and throat treatment. It is the source of **cineole,** also called **eucalyptole.** From 3 to 4% oil is obtained from the leaves. It is a pungent yellowish oil. This type of eucalyptus oil contains **phellandrine,** used in Australia as an antiknock agent in gasoline. **Eucalyptus dives oil,** from the leaves of the Australian tree *E. dives,* contains 92 to 94% **piperitone,** and is used in the manufacture of menthol. The yield is about 50% **levomenthol** with a melting point of 33 to 35°C. It lacks the odor of USP menthol of which only 15% can be produced from this oil. Much eucalyptus oil is produced in Chile. More than 300 species of eucalyptus trees are known, and each produces a different type of oil.

EXPANDED METAL. Sheet metal that has been slit and expanded to form a mesh, which is used for reinforced-concrete work or plaster wall construction, and also for making grills, vents, and such articles as trays, where stiffness is needed with light weight. The expanded metal has greater rigidity than the original metal sheet, and also permits a welding of the concrete or plaster through the holes. It is made either with a plain diamond-shaped mesh, or with rectangular meshes. One type is made by slitting the sheet and stretching the slits into the diamond shape. The other variety is made by pushing out and expanding the metal in the meshes so that the flat surface of the cut strand is nearly at right angles to the surface of the sheet. Expanded metal is made from low-carbon steel, iron, or special metals, in sheets from 8 to 12 ft (2.4 to 3.7 m) in length and 3 to 6 ft (0.9 to 1.8 m) in width, in several thicknesses. It is also marketed as **metal lath,** usually 96 in (2.4 m) long, and 14 to 18 in (0.4 to 0.5 m) wide. Expanded metal of the U.S. Gypsum Co. is made of stainless steel and aluminum alloys in various thicknesses with openings from ½ to 1½ in (1.27 to 3.81 cm). **Rig-**

idized metal, or **textured metal,** is thin sheet that is not perforated, but has the designs rolled into the sheet so that the rigidity of the sheet is increased 2 to 4 times. Thus, extremely thin sheets of stainless steel can be used for novelties, small mechanical products, and paneling. Rigidized steel, of the Rigidized Metals Corp. and previously known as **Rigid-Tex steel,** is made in many ornamental designs and also comes in vitreous enameled sheets for paneling. **Crimp metal,** of the American Nickeloid Co., has various embossed designs in either raised or depressed ridges rolled into the polished side of the metal. **Perforated metals** are sheet metals with the perforations actually blanked out of the metal. They are marketed in sheets in carbon steel, stainless steel, or Monel metal, with a great variety of standard designs. Those with round, square, diamond, and rectangular designs are used for screens and for construction. **Agaloy,** of the Agaloy Tubing Co., is perforated metal made into tube form.

EXPANSIVE METAL. An alloy which expands on cooling from the liquid state. The expansive property of certain metals is an important characteristic in the production of accurate castings having full details of the mold such as type castings. The alloys are also used for proof-casting of forging dies, for sealing joints, for making duplicates of master patterns, for holding die parts and punches in place, and for filling defects in metal parts or castings. Antimony and bismuth are the metals most used to give expansion to the alloys. **Lewis metal,** one of the original expansive alloys, had one part of tin and one of bismuth, and melted at 138°C. **Matrix alloy** and **Cerromatrix,** of the Cerro Copper and Brass Co., contains 48% bismuth, 28.5 lead, 14.5 tin, and 9 antimony. The melting point is 248°F (120°C), tensile strength 13,000 lb/in^2 (91 MPa), and Brinell hardness 19. **Cerrobase,** of this company, is another alloy balanced to give the exact impression of the mold without shrinkage or expansion in cooling. It is harder than lead, and melts at 255°F (123°C).

EXPLOSIVES. A material which, upon application of a blow, or by rise in temperature, is converted in a small space of time into other compounds more stable and occupying much more space. Commercial explosives are solids or liquids that can be instantaneously converted by friction, heat, shock, or spark into a large volume of gas, thereby developing a sudden rise in pressure which is utilized for blasting or propelling purposes. **Gunpowder** is the oldest form of commercial or military explosive, but this has been replaced for military purposes by more powerfully acting chemicals. **Smokeless powder** was a term used to designate nitrocellulose powders as distinguished from the smoky black gunpowder. **Blasting powders** are required to be relatively slow acting to have a heaving or rending effect. **Military explosives** used as propellants must not give instantaneous detonation, which would burst the gun, but are arranged to burn slowly at first and not reach a maximum explosion until the projectile reaches the muzzle. This characteristic is also required in explosives used for the explosive forming of hard metals. The more rapid-acting **high explosives** are generally used for bombs, torpedoes, boosters, and detonators. The **detonators** are extremely sensitive explosives, such as the fulminates, set off by a slight blow but too sensitive to be used in quantity as a charge. The **booster explosives** are extremely rapid but not as sensitive as the detonators. They are exploded by the detonators and in turn set off the main charge of explosive. Some explosives such as **nitroglycerin** can be exploded by themselves, while others require oxygen carriers or carbon carriers mixed with them. Other requirements of explosives are that they should not react with the metal container, should be stable at ordinary temperatures, and should not decompose easily in storage or on exposure to air.

Shaped charges of high explosive give a penetrating effect, known as the Monroe effect, used in armor-piercing charges. A solid mass of explosive spends itself as a flat blast, but with a conical hole in the charge, and having the open end facing the target, a terrific piercing effect is generated by the converging detonation waves coming from the sides of the cone. This effect drives a jet of hot gases through the steel armor. **Permissible explosives** are explosives that have been passed by the Bureau of Mines as safe for blasting in gaseous or dusty mines. Most of the permissibles are of ammonium nitrate or gelatin base. **Wet-hole explosives,** for oil-well and mining operations, may be ammonium nitrate in plastic containers, or various combinations in containers. **Nitramex 2H,** of Du Pont, is **TNT-ferrosilicon-ammonium nitrate** in a metal can. **Lox,** used in mines and quarries, is an explosive consisting of a paper cartridge filled with carbon black or wood pulp soaked in liquid air. It cannot be tamped as it is very sensitive. It is fired by electric detonators. **Cardox,** an explosive used in coal mining, consists of liquid carbon dioxide in a steel cylinder with aluminum powder. The powder is fired by an electric spark, heating and gasifying the carbon dioxide. **Picric acid,** or **trinitrophenol,** $C_6H_2(OH)(NO_2)_3$, a lemon-yellow crystalline solid melting at 248°F (120°C), is a powerful explosive used in shells, and because of its persistant color also used as a dyestuff. It is called **melanite** by the French, **lyddite** by the English, and **schimose** by the Japanese. It is made by treating phenol with sulfuric and nitric acids, or can be produced by treating acaroid resin with nitric acid. It reacts with metals to form dangerous explosive salts, so that the shells must be lacquered. **Cressylite,** used for shells, is a mixture of picric acid and trinitrocresol. It has a lower melting point.

Explosive D, or **dunnite,** made by the neutralization of picric acid with ammonium carbonate, is **ammonium picrate,** $C_6H_2(NO_2)_3ONH_4$. It forms orange-red needles that explode when heated to 300°C, but is not highly sensitive to friction. It is used as a bursting charge in armor-piercing shells. **Trinitrotoluene,** or **trinitrotoluol,** $C_6H_2(CH_3)(NO_2)_3$, also commonly known as **TNT** and also called **trotyl** and **tolite,** is the principal constituent of many explosives. It resembles brown sugar in appearance, it melts at 80°C, and the fumes are poisonous even when absorbed through the skin. Its detonation velocity is 23,000 ft/s (7,010 m/s). It is thus not as powerful as picric acid, but it is stable, not hygroscopic, and does not form unstable compounds with metals. It is safe in handling because it does not detonate easily, but is exploded readily with mercury fulminate, and is used for shrapnel, hand grenades, mines, and depth bombs. TNT is made by the nitration of toluol with nitric and sulfuric acids. The intermediate product, **dinitrotoluol,** is employed with hexanitrodiphenylamine for torpedoes. **Hexanitrodiphenylamine,** $(NO_2)_3C_6H_2 \cdot NH \cdot C_6H_2(NO_2)_3$, is a powder that explodes with great violence. It is highly poisonous, and causes painful blisters and inflammation. The commercial explosive **sodatol** is made by mixing TNT with nitrate of soda.

Trinitroaniline, $(NO_2)_3C_6H_2NH_2$, commonly known as **TNA,** is derived from aniline by nitration, and is one of the strongest of the high explosives. It is a yellowish-green crystalline powder melting at 215°C. It stains the skin yellow but is not poisonous. It is more sensitive to shock than TNT and is more costly. **Trinitroanisol,** used in Japanese Baka planes, has the composition $C_6H_2OCH_3(NO_2)_3$. It is about equal to TNT in power, and has the advantage that it does not attack metals.

Tetryl, or **pyronite,** $(NO_2)_3C_6H_2N(NO_2)CH_3$, is a nitro derivative of benzene. It is a yellow crystalline powder that melts at 130°C and explodes when heated to 186°C. It is more sensitive to shock than TNA, and has a higher rate of detonation than TNT. It is too sensitive to be used as a shell filler, and is employed as a booster and in commercial explosives to replace mercury fulminate for detonators. The high explosive **RDX** is **cyclotrimethylene trinitroamine,** and has a detonation velocity of 27,500 ft/s (8,382 m/s). It is used in bombs, torpedoes, mines, and rockets, but is very sensitive to shock and is mixed with waxes or plasticizers to reduce sensitivity. **PETN** is **pentaerythritol tetranitrate,** with a detonation velocity of 26,500 ft/s (8,077 m/s). **Pentolite** is a 50–50 mixture of TNT and PETN with less sensitivity and a detonation velocity of 25,000 ft/s (7,620 m/s). It is used as a booster. When aluminum powder is added to high explosives, the brisance, or blast effect, is increased. A powerful explosive used during the Second World War contained 40% RDX, 40 TNT, and 20 aluminum powder. Various combinations of high explosives are now used in thin sheet form for **explosive welding** of laminated metals.

FABRICS. **Woven fabrics** and **knit fabrics** are composed of webs of fiber yarns. The yarns may be of either filament (continuous) or staple (short) fibers. In knit fabrics, the yarns are fastened to each other by interlocking loops to form the web. In woven fabrics, the yarns are interlaced at right angles to each other to produce the web. The lengthwise yarns are called the warp, and the crosswise ones are the filling (or woof) yarns.

The many variations of woven fabrics can be grouped into four basic weaves. In the **plain weave fabric,** each filling yarn alternates up and under successive warp yarns. With a plain weave, the most yarn interlacings per square inch can be obtained for maximum density, "cover," and impermeability. The tightness or openness of the weave, of course, can be varied to any desired degree. In **twill weave fabrics,** a sharp diagonal line is produced by the warp yarn crossing over two or more filling yarns. **Satin weave fabrics** are characterized by regularly spaced interlacings at wide intervals. This weave produces a porous fabric with a smooth surface. Satins woven of cotton are called **sateen.** In the **leno weave fabrics,** the warp yarns are twisted and the filling yarns are threaded through the twist openings. This weave is used for meshed fabrics and nets.

Because the variety of woven fabrics is endless, we can only briefly outline here the way woven textiles are characterized or specified. Generally, specifications include the type of weave; the thread count, both in warp and fillings; whether the yarn is filament or staple; the crimp, in percent; the twist per inch; and the yarn numbers for warp and fill. Over the years a rather unsystematic fabric-designation system has evolved. For example, some fabrics, such as twills and sateens, are designated by width in inches, number of linear yards per pound, and number of warp and filling threads per inch. Other fabrics are identified by width, ounces per linear yard, and warp and filling count.

While the largest single use of woven fabrics is, of course, wearing apparel, they are used in many other areas: in mechanical applications such as machine and conveyor belting, for filtration, for packaging, and as reinforcement for plastics and rubber.

FAT LIQUORS. Oil emulsions used in tanneries for treating tanned leather to lubricate the fibers, increase the flexibility, and improve the finish. There are two general types of fat-liquor emulsions: acid and alkaline. The acid group includes sulfonated oils and some soluble-oil combinations. Alkaline types are emulsions of oils with soaps or alkalies. Leather may be treated first with an alkaline liquor and then with an acid, or borax or soda ash may be added to sulfonated oils to produce alkaline liquors. For suede and white leathers, egg-yolk emulsions may be used. The oils employed in emulsions may be sperm, cod, or castor oil, and those that are neutral have a neatsfoot-oil base. The soaps are usually special for the tannery trade. Prepared fat liquors are marketed under trade names. **Tanners' greases,** used for sponging or milling onto the leather, are also trade-named mixtures of waxes, sulfonated oils, and soaps.

FATS. Natural combinations of glycerin with fatty acids, some fats having as many as 10 or more different fatty acids in the combination. They are derived from animal or vegetable sources, the latter source being chiefly the seeds or nuts of plants. Fats in a pure state would be odorless, tasteless, and colorless, but the natural fats always contain other substances that give characteristic odors and tastes. Fats are used directly in foods and also in the making of various foodstuffs. They are also used in making soaps, candles, and lubricants, and in the compounding of resins and coatings. They are also distilled or chemically split to obtain the fatty acids.

Fats are most important for food, containing more than twice the fuel value of other foods. They are also important carriers of glycerin necessary to the human system. Metabolism, or absorption of fats into the system, is not a simple process, and is varied with the presence of other food materials. The fats with melting points above 45°C are not readily absorbed into the system. The heavy fats are called tallow. Lack of certain fats, or fatty acids, cause skin diseases, scaly skin, and other conditions. Some fatty acids are poisonous alone, but in the glyceride form in the fats they may not be poisonous but beneficial. Fats can be made synthetically from petroleum or coal. **Edible fats** were first made synthetically by the Germans in wartime by the hydrogenation of brown coal and lignite and then esterifying the C_9 to C_{16} fractions of the acids. But the world resources of natural fats are potentially unlimited, especially from tropical nuts, forming a cheap source of fatty acids in readily available form.

FATTY ACIDS. A series of **organic acids** deriving the name from the fact that the higher members of the series, the most common ones, occur naturally in animal fats, but fatty acids are readily synthesized, and the possible variety is almost infinite. All of these acids contain the **carboxyl group** ·COOH. The acids are used for making soaps, candles, and coating compounds; as plasticizers; and for the production of plastics and many chemicals. The hydrogen atom of the group can be replaced by metals or alkyl radicals with the formation of salts or esters, and other derivatives such as the halides, anhydrides, peroxides, and amides can also be made. The **neoacids,** in general, have the formula R(COOH) in which the R is substituted methyl or other groups. Some of the fatty acids can be polymerized to form plastics. Various derivatives of the acids are used as flavors, perfumes, driers, pharmaceuticals, and antiseptics. Certain fatty acids, such as oleic and stearic, are common to most fats and oils regardless of their source, while others, such as arachidic and erucic, are characteristic only of specific fats and oils.

Saturated acids are acids that contain all the hydrogen with which they can combine, and they have the type formula $C_nH_{2n+1}COOH$. They have high melting points. **Unsaturated acids,** such as oleic, linoleic, and linolenic, are liquid at room temperature, and are less stable than saturated acids. **Fatty acid glycerides** in the form of animal and vegetable fats form an essential group of human foods. Fats of the highly unsaturated acids are necessary in the metabolism of the human body, the glycerides of the saturated acids such as palmitic being insufficient alone for food.

Polyunsaturated acids of the linoleic type with more than one double bond lower blood cholesterol, but saturated acids with no double bonds do not. **Arichidonic acid** with four double bonds lowers blood cholesterol greatly. It is manufactured in the body from linoleic acid if vitamin B_6 is present. **Linoleic acid,** $C_{18}H_{32}O_2$, the characteristic unsaturated food acid, has two double bonds. **Linolenic acid,** $C_{18}H_{30}O_2$, found in linseed oil, has three double bonds.

The names of the fatty acids often suggest their natural sources, though commercially they may be derived from other sources or made synthetically. **Butyric acid,** $CH_3CH_2CH_2 \cdot COOH$, is the characteristic acid of butter. Also called **butanoic acid** and **ethylacetic acid,** it is made synthetically as a colorless liquid with a strong odor and completely soluble in water. With alcohols it forms butyrates of pleasant fruity odors used as flavors. The cellulose esters of butyric acid are used in lacquers, and have good water resistance and easy solubility in hydrocarbons. The acid is also used as a starting point for fluoro rubbers.

Some acids, such as linoleic, are found in greater amount in cold-climate products, while some other acids are found in most abundance in hot-climate products. **Lauric acid,** or **dodecanoic acid,** $CH_3(CH_2)_{10}COOH$, occurs in high percentage in the oil of the coconut and other kernels of tropical palm nuts. It is a saturated acid much lower in carbon and hydrogen than linoleic acid, and is a semisolid melting at 44°C. It is one of the chief constituents of coconut oil that gives sudsing properties to soaps. It is also used for making detergents and plasticizers, and as a modifier for waxes in coatings and polishes. **Neo-Fat 12,** of Armour & Co., is 95% pure lauric acid. The ester of lauric acid is used for treating cotton fabrics to give a pebbly surface. **Lauralene** is a lauric acid of the Beacon Co. with an acid value of 324 and saponification number of 366. **Methyl laurate** is often preferred to lauric acid for all the uses. It is a stable, noncorrosive, water-white liquid. **Methyl esters** of other acids are similarly used. **Methyl stearate** is an economical compounding agent for rubbers, waxes, and textile coatings. **Myristic acid,** $CH_3(CH_2)_{12}COOH$, is a hard crystalline solid melting at 58°C, obtained from coconut oil. It is soluble in alcohol and is compatible with waxes and oils. It is used in cosmetics, and will produce high-lathering soaps that are not irritating to the skin as are the coconut-oil soaps. **Neo-Fat 14,** of Armour Industrial Chemical Co., is myristic acid 94% pure.

Caprylic acid, $CH_3(CH_2)_6COOH$, obtained from coconut oil, has a melting point of 11°C, acid number of 382, and iodine value of 1. It is used in cosmetics, as a fungicide, and in the manufacture of pharmaceuticals. **Capric acid,** or **decanoic acid,** $CH_3(CH_2)_8COOH$, obtained from coconut oil, is a bad-smelling white crystalline solid melting at 31.5°C with an acid value of 321. It is used for making esters for perfumes and flavors. **Neo-Fat 10** is capric acid 92% pure, containing 5% lauric acid and 3 caprylic acid. **Aliphat 2** and **Aliphat 3,** of the General Mills Co., are caprylic acid and capric acid, respectively. **Caproic acid,** or **hexanoic acid,** $CH_3(CH_2)_4COOH$, occurs in coconut and palm kernel oils, but is produced synthetically on a large scale for the manufacture of hexylresorcinol, hexylphenols, flavors, and high-boiling plasticizers. It is a liquid boiling at 203°C and has a goatlike odor from which it derives its name. **Oenanthic acid,** or **heptoic acid,** is a homolog of caproic acid with one more carbon atom. When polymerized with lactam it gives a nylon stronger and more flexible than ordinary nylon 6. **AB fatty acid,** of E. F. Drew & Co., Inc., used for soaps, is composed of the acids from coconut oil distilled to remove most of the low fractions to improve color and odor. It contains 60% lauric acid, 18 myristic, 7 palmitic, 7 oleic, 3 linoleic, 3 capric, and 1 each of stearic and caprylic. Some fatty acids that occur only occasionally in small amounts in vegetable oils are made synthetically. **Undecylenic acid,** $CH_2{:}CH(CH_2)_8COOH$, is a highly reactive acid of this kind used for making synthetic resins, fungicides, and perfumes. The **Duomeens,** of the Armour Industrial Chemical Co., are alkyl trimethylenediamines derived from fatty acids, and are used as pigment dispersants, metalworking lubricants, and flotation agents. They have the general formula $RNHCH_2CH_2NHH$, where R is the alkyl group from the fatty acid. **Duomeen C** is from coconut oil, **Duomeen S** is from soybean oil, and **Duomeen O** is from oleic acid. The **lactams** and **lactones,** used in making plastics, form a wide range of amino-fatty acid ring compounds. They are produced from fatty acids.

FEATHERS. The light fluffy outgrowth or plumage of birds. The industrially important feathers are those from the duck, goose, chicken, and ostrich. Radiantly colored feathers from many other types of birds are used for ornamental and art purposes. An important featherwork art exists in Mexico as a development of the Aztec featherwork. **Down** is the soft feathers of young birds or the soft undergrowth of adult birds, used as a stuffing material. **Eiderdown,** from the eider duck, is highly valued as an insulation in sleeping bags. In Iceland the female duck plucks the down from her breast to line the nest, and this down is gathered commercially after the birds are hatched.

The midrib and quill of **chicken feathers** are made into protein plastic, and the fluffy barbs are used as stuffing, but many of the feathers are processed directly into protein. The inedible protein is used for making brush bristles and insulating fiber, or is split into edible proteins for poultry feed.

Ostrich feathers, from the domesticated ostriches of Argentina, South Africa, and Australia, are used for ornamental purposes, hats, and dusting brushes. The ostrich has 24 feathers on each wing, some as long as 25 in (0.6 m), and the grade depends upon the color and the length. Male ostrich feathers are black. The female feathers are a soft gray, with white feathers in the wings and tail. The life of the ostrich is 50 to 75 years, and the feathers begin to be clipped at the age of 10 months. **Ostrich eggs,** which weigh 4 lb (1.8 kg) and are laid every other day, are a valuable food by-product.

FELDSPAR. A general name for a group of abundant minerals used for vitreous enamels, pottery, tile, and glass in fertilizers; in fluxes; for roofing granules; and as an abrasive in soaps and cleaning compounds. Ground feldspar is also used for extinguishing magnesium fires, as it melts and gives a smothering action. There are many varieties of feldspar, but those of greatest commercial importance are the **potash feldspars, orthoclase** or **microline,** $K_2O{:}Al_2O_3 \cdot 6SiO_2$, the **soda feldspar, albite,** $Na_2O{:}Al_2O_3 \cdot 6SiO_2$, and the **calcium feldspar, anorthite,** $CaO{:}Al_2O_3 \cdot 2SiO_2$. Orthoclase and microline have the same composition but different crystal structures. Anorthite crystals occur in many igneous rocks, and are white, gray, or reddish in color. **Aplite,** used as a flux for ceramics, has more silica and less alumina. **Japanese aplite** has 77.6% silica, 12.8 alumina, 3.7 K_2O, and 3.9 Na_2O, with small amounts of calcia, magnesia, and iron oxide. Orthoclase is called **suntone. Adularia** is a pure form of orthoclase with only a little sodium. Pieces with an opalescent sheen are called **moonstone,** and are used as gemstones. This stone is white with a bluish adularescence caused by the action of light on the laminations. The hardness is 6 to 6.5 Mohs, but the cleavage in two directions makes it fragile. The blue opalescent moonstone of New Mexico is **sanidine,** a quartz mineral. **Amazon stone,** or **amazonite,** is a beautiful green microline found in Italy, Malagasy, and Colorado, and used as a gemstone. The Amazone stone of Virginia has bluish-green and white streaks, and was formerly shipped to Germany for cutting into ornamental objects. The colors of feldspar are from mineral oxides and impurities and are white, gray, yellow, pink, brown, and green. Albite is generally white, while microline is more often green.

All of the chemical components of feldspar are glassmaking materials. In making glass about 150 lb (68 kg) are used to each 1,000 lb (454 kg) of sand. But the mineral in its natural occurrence varies widely in composition even in the same mine, and thus must be controlled chemically to obtain uniform results in glass and ceramic enamels. It occurs in pegmatite dikes associated with quartz, mica, tourmaline, garnet, and spodumene. The mineral is ground to a uniform size, from 80 to 140 mesh, and shipped in bags. Crude unground feldspar is also marketed in bulk. The melting point varies from 1185 to 1490°C, but the preferred range is 1250 to 1350°C. The hardness is 6 to 6.5, and the index of refraction is 1.518 to 1.588, the lowest being orthoclase and the highest anorthite. The specific gravity is 2.44 to 2.62 for orthoclase and microline, and 2.6 to 2.8 for anorthite. Tennessee and North Carolina feldspar has about 70% SiO_2 and 17 Al_2O_3, with 9 to 11 K_2O, and 2 to 3 Na_2O. New England feldspar is lower in silica and higher in potash. **Potash spar** from New York and New Jersey has about 12% K_2O, and is suited for glass and pottery. **Soda spar,** with about 7% Na_2O, is preferred for ceramic enamels. **Cornwall stone,** from England, is a kaolinized feldspar with about 2% CaO. A similar stone from North Carolina is called **Carolina stone.** Aplite is a ceramic fluxing stone found in Virginia and used chiefly to supplement feldspar to provide more alkalies. It is a white massive material of feldspars and other minerals, containing 60% silica, 24 alumina, 6 calcia, 6 sodium oxide, and 3 potassium oxide. Another feldspar material is **alaskite,** a feldspar and quartz mixture from North Carolina. It is classed as a pegmatitic granite. Ground feldspar for enamels is sometimes called **glass spar. Dental spar** is specially selected potash feldspar used in making artificial teeth.

FELT. A fabric of wool, fur, hair, or synthetic fibers made by matting the fibers together under pressure when thoroughly soaked or steam-heated. The matting may also be accomplished by blowing the wet fibers under a powerful air blast and then pressing. The animal fibers mat together, owing to minute scales on their surface. Cotton and other vegetable fibers do not have the property of felting, but a percentage of vegetable or synthetic fibers may be incorporated to vary the characteristics of the felt. So great is the felting property of wool that only 20% is needed in mixtures.

Wool felt can be composed of 100% virgin wool, or a combination of synthetic fibers and reused wool. The top grade, which has a density of 0.26 g/cm^3, is used where high strength, purity, and fineness are needed. It is made of the best grades of wool, which are usually white. Wool felts are produced in sheet and roll form. Standard widths are the minimum widths of trimmed felt or the width between pinholes of untrimmed felt. All widths of 60 and 72 in (152 and 183 cm) refer to felt made in roll form. All sizes of 36 by 36 in (91 x 91 cm) refer to felt made in sheet form and have a tolerance of 1½ in (3.8 cm). Special sizes are available.

Synthetic fiber felts are composed of such fibers as polyester, nylon, Teflon, polypropylene, and acrylic. They are made by a needle loom process that simulates the natural entanglement of wool fiber.

New and reworked wool and noils are mixed with cotton, rayon waste, ramie, jute, casein fiber, and other fibers. Cotton decreases the density and prevents voids in the felt. **Kapok** gives lower thermal and sound conductance, and **insulating felt** may contain a high percentage of kapok. Felt is made of staple fibers of about 1½ in (3.8 cm) in length, and noils of ¾ to 1 in (1.9 to 2.5 cm). Longer fibers tend to mat. Shorter fibers lack depth of penetration to give necessary strength. Since most of the wool used is secondary or waste, all grades are employed, from the fines to the coarse carpet wools. Grading is by characteristic symbols. Thus, felt No. 26R1 is a felt with a specific gravity of 0.26 in roll form of first quality. But, although true felt is based on wool, most of the roll and sheet felt is now produced from synthetic fibers mechanically or chemically bonded, and they have the chemical resistance and physical properties of the particular synthetic fiber. **Needled felt** is a fabric made of natural or synthetic fibers physically interlocked by the action of a needle loom. It may also be treated chemically, or by heat or moisture, for special effects.

Felt is one of the earliest manufactured materials. It is now used for insulation, sound and vibration absorption, padding and lining in instrument cases, hats, in roofing, and where a soft resilient fabric is needed. Although the best hat felts are made with nutria or beaver fur, vast quantities of rabbit furs or mixed furs and wool are used. **Hair felt** is made of cattle hair, and is used for insulating cold-water pipes and refrigerating equipment, and for cushioning and padding. The **Ozite felt** of the American Felt Co. is an all-hair felt. Felt comes in thicknesses from ¼ to 2 in (0.6 to 5.1 cm), the ¼ in (0.6 cm) weighing 4 oz/ft^2 (18g/cm^2). The **K felt** of this company is made to Army-Navy specifications and weighs 3.24 lb/(yd^2)(in) [0.69 kg/(m^2)(cm)] of thickness. It has a tensile strength of 12 lb/in^2 (0.08 MPa) and compressive strength of 3 lb/in^2 (0.02 MPa) at 50% deflection. It is for sound and thermal insulation. **Filtering felts,** for filtering gases and liquids, are usually made from various synthetic fibers to meet specific chemical-resistance requirements. **Teflon felt,** of the American Felt Co., for filtering hot strong acids and alkalies, is made from fluorocarbon fibers. Because of the high chemical and physical properties of this fiber it is called **dragon fur** in the felting industry. It also has a low friction coefficient, and is repellent to sticky materials, giving high filtering efficiency and easy cleaning. The **Scottfelt,** of Scott Paper Co., is a **foam filter** made from urethane foam compressed under heat and pressure to 1/15 of its original thickness. The tensile strength is increased from 35 to 270 lb/ in^2 (0.2 to 1.9 MPa), and the porosity may be graduated up to 1,340 per linear in.

Cattle- and goat-hair felts are also used for glass polishing. **Baize** is an old name for a thin woolen felt used for desk and table tops, box linings, and for bases of instruments. Its name is derived from the fact that it was originally bay, or brown, in color, but the industrial baize is now usually green. The name is now used to designate a plain-woven, loose, cotton or woolen fabric with a short, close nap, in plain colors for the same purposes. **Feltex,** of the Philip Carey Corp., is an asphalt-saturated felt for roofing, and **Mica-kote** is a heavy felt coated with asphalt and finished with mica flakes, used for roofing. **Unisorb,** of the Felters Co., is a heavy felt in blocks and sheets for isolation pads under machinery to absorb vibration. **Slaters' felt** is a tarred sheathing felt used in building construction, usually in 25- and 30-lb (11.3- and 13.6-kg) rolls. **Fire felt,** of Johns-Manville, is made of asbestos fibers felted into sheets, blocks, or shapes, for boiler and furnace insulation. **Slatekote,** of the same company, is a heavy felt saturated with asphalt and coated with colored crushed slate, used for roofing. The term **roofing felt** is also applied to the thick asphalt-impregnated papers used for that purpose, and **papermakers' felt** is the woven wool or part-wool belting used in papermaking.

FERRIC OXIDE. The red **iron oxide,** Fe_2O_3, found in abundance as the ore **hematite,** or made by calcining the sulfate. It has a dark-red color and comes in powder or lumps. The specific gravity is 5.20, and melting point about 1550°C. It is used as a paint pigment under such names as **Indian red, Persian red,** and **Persian Gulf oxide.** In cosmetics and in polishing compounds it is called rouge. The Persian red oxide from the Island of Hormuz contains from 60 to 90% Fe_2O_3, and is marketed on a 75% basis. **Brown iron oxide** is made from ferrous sulfate and sodium carbonate and is not a pure oxide, though its chemical formula is given as Fe_2O_3. It is also called **iron subcarbonate,** and is used in making green glass, in paints, and in rubber.

The names **metallic red** and **metallic brown** are applied to pigments from Pennsylvania ores containing a high percentage of **red iron oxide. Venetian red** is a name for red iron oxide pigments mixed with various fillers. The **Tuscan red** pigments are red iron oxide blended with up to 75% of lakes, but may also be barium sulfate with lakes. **Ferric oxide pigments** make low-priced paints, and are much used as base coats for structural steel work. The natural oxides come chiefly from Alabama, Tennessee, Pennsylvania, Iran, and Spain.

The **Mapico colors** of the Binney & Smith Co. are **iron oxide pigments** refined under controlled conditions to give uniformity free of other mineral impurities. **Mapico red** and **Mapico crimson** contain 98% Fe_2O_3, the balance being almost entirely material lost on ignition or water-soluble impurity. The red oxide has a spheroidal particle shape, while the crimson has an acicular, or needle-shaped, particle. **Mapico lemon yellow** contains 87% Fe_2O_3, with 11.85% ignition loss. The particles are acicular and are only half the size of the crimson particles, being only 0.1 to 0.8 μm. **Mapico brown** contains 93.1% Fe_2O_3 and 5% FeO. Its particles are cubical and of sizes from 0.2 and 0.4 μm. **Mapico black** contains 76.3% Fe_2O_3 and 22.5% FeO, with a cubical particle shape. The **Auric brown** of Du Pont, used for giving light-fast shades to paper, is a hydrated ferric oxide ground to an extremely fine particle size.

Yellow iron oxide, known also as **ferrite yellow** and **Mars yellow,** used as a paint pigment, is $Fe_2O_3 \cdot 3H_2O$ plus from 2 to 12% calcium sulfate. It is made by precipitating ferrous hydroxide from iron sulfate and lime and then oxidizing to the yellow oxide. **Black ferric oxide, ferroferric oxide,** or **magnetic iron oxide,** is a reddish-black amorphous powder, $FeO \cdot Fe_2O_3 \cdot H_2O$. It is used as a paint pigment, for polishing compounds, and for decarbonizing steel. The finely ground material used as a pigment is called **magnetic black,** and when used for polishing it is called black rouge. **Hammer scale** is the iron oxide Fe_3O_4, formed in the hot rolling or forging of steel, and is used for decarbonizing steel by packing the steel articles in the scale and raising to a high temperature. It is very hard, 5.5 to 6.5 Mohs, and is also used as an abrasive.

FERROCHROMIUM. A high-chromium iron master alloy used for adding chromium to irons and steel. It is also called **ferrochrome.** It is made from chromite ore by smelting with lime, silica, or fluorspar in an electric furnace. **High-carbon ferrochrome,** of Union Carbide, contains 66 to 70% chromium in grades of 4.5, 5, 6, and 7% carbon. It is used for making tool steels, ball-bearing steels, and other alloy steels. It melts at about 1250°C. It is marketed as crushed alloy in sizes up to 2 in (5 cm), and as lump alloy in lumps up to about 75 lb (34 kg). **Low-carbon ferrochrome** of this company and the Vanadium Corp. of America contains 67 to 72% chromium, in grades of 0.00, 0.10, 0.15, 0.20, 0.50, 1, and 2% carbon. It is used for making stainless steels and acid-resistant steels. **Simplex ferrochrome,** of Union Carbide, contains as little as 0.01% carbon, It comes in pellet form to dissolve easily in the steel, and is used for making low-carbon stainless steels. Low-carbon ferrochrome is also preferred for alloy steel mixtures where much scrap is used because it keeps down the carbon and inhibits the formation of hard chromium carbides. The various grades of ferrochromium are also marketed as **high-nitrogen ferrochrome,** with about 0.75% nitrogen for use in making high-chromium cast steels which would normally have a coarse crystalline structure. The nitrogen refines the grain and increases the strength. **Foundry-grade ferrochrome,** for making cast irons, contains 62 to 66% chromium and 5 carbon. The **V-5 Foundry alloy** of the Vanadium Corp. has about 40% chromium, 18 silicon, and 9 manganese. It is used for ladle additions to cast iron to give uniform structure and increase the strength and hardness. Addition of 1% of the alloy to a cast iron of 3.40% total carbon, with resultant balance of 1.30% silicon, 0.60 manganese, and 0.35 chromium, gives a dense iron of good hardness.

FERROMANGANESE. A master alloy of manganese and iron used for deoxidizing steels, and for adding manganese to iron and steel alloys and bronzes. Manganese is the common deoxidizer and cleanser of steel, forming oxides and sulfides that are carried off in the slag. Ferromanganese is made from the ores in either the blast furnace or the electric furnace. Standard ferromanganese has 78 to 80% manganese. British ferromanganese contains about 7% carbon, but the content in the American alloy is usually 5 to 6.5%. **Low-carbon ferromanganese** is also marketed containing 0.10 to 1% carbon. Low-phosphorus ferromanganese contains less than 0.10% phosphorus. The alloys are marketed in lumps to be added to the furnace. **Spiegeleisen** is a form of low-manganese ferromanganese with from 15 to 30% manganese and from 4.5 to 5.5 carbon. The German name, meaning mirror iron, is derived from the fact that the crystals of the fractured face shine like mirrors. Spiegeleisen has the advantage that it can be made from low-grade manganese ores, but the quantity needed to obtain the required proportion of manganese in the steel is so great that it must be premelted before adding to the steel. It is used for making irons and steels by the bessemer process. Grade A spiegeleisen, of Union Carbide, has 19 to 21% manganese, and 1 silicon; Grade B has 26 to 28% manganese and 1 silicon. The melting point is from 1950 to 2265°F (1066 to 1240°C).

FERROPHOSPHORUS. An iron containing a high percentage of phosphorus, used for adding phosphorus to steels. Small amounts of phosphorus are used in open-hearth screw steels to make them free-cutting, and phosphorus is also employed in tinplate steels to prevent sticking together of the plates in annealing. Ferrophosphorus is made by melting phosphate rock together with the ore in making the pig iron. The phosphorus content is about 18% and is chemically combined with the iron. Another grade, made in the electric furnace and containing 23 to 25% phosphorus, is used for adding phosphorus to bronzes. A master alloy for adding selenium to steels to give free-machining qualities to steel, particularly the stainless steels, is **ferroselenium.** A typical ferroselenium, of the American Smelting & Refining Co., contains about 52% selenim and 0.90 carbon.

FERROSILICON. A high-silicon master alloy used for making silicon steels, and for adding silicon to transformer irons and steels. It is made in the electric furnace by fusing quartz or silica with iron turnings and carbon. It is marketed in various grades with from 15 to 90% silicon. The silicon forms a chemical combination with the iron, but the alloys having more than about 30% silicon are fragile and unstable. The silicon also causes the carbon to be excluded in graphite flakes. The alloys of high silicon content are called **silicon metal.** One producer markets two grades, 15 and 45% silicon, while another has 15, 50, 75, 85 and 90% grades. Grades with silicon from 80 to 95% are marketed for use where small ladle additions are made for producing high-silicon steels, and also for producing hydrogen by reaction with caustic soda. The alloy is marketed in lumps or crushed form. Silicon is often added to steels in combination alloys with deoxidizers or other alloying elements. **Ferrosilicon aluminum,** containing about 45% silicon and 12 to 15 aluminum, is a more effective deoxidizer for steel than aluminum alone. It is also used for adding silicon to aluminum casting alloys. **Silvaz,** of Union Carbide, is a ferrosilicon aluminum containing also vandium and zirconium. The alloy serves as a deoxidizer, fluxes the slag inclusions, and also controls the grain size of the steel. **Simanal,** of the Ohio Ferro Alloy Corp., is a **deoxidizing alloy** containing 20% each of silicon, aluminum, and manganese. **Alsifier,** of the Vanadium Corp., contains 40% silicon, 20 aluminum, and 40 iron. The aluminum and silicon are in the form of an aluminum silicate which forms a slag that is eliminated during the teeming of the steel. **Alsimin** is a Swiss ferrosilicon aluminum with 50% aluminum. **Silicon aluminum** is a master alloy for adding silicon to aluminum alloys, and it does not contain iron. A 50–50 silicon aluminum, of Alloys & Products, Inc., has a melting point of 1920°F (1049°C), but is soluble in aluminum at 1275°F (690°C). It comes in pyramid waffle form for breaking into small lumps.

FERROTITANIUM. A master alloy of titanium with iron used as a purifying agent for irons and steel owing to the great affinity of titanium for oxygen and nitrogen at temperatures above 800°C. The value of the alloy is as a cleanser, and little or no titanium remains in the steel unless the percentage is gaged to leave a residue. The **ferrocarbon titanium** is made from ilmenite in the electric furnace, and the carbon-free alloy is made by reduction of the ore with aluminum. Ferrotitanium comes in lumps, crushed, or screened. **High-carbon ferrotitanium,** of the Vanadium Corp., has 17% titanium and 7 carbon. It is used for ladle additions for cleansing steel. **Low-carbon ferrotitanium** has 20 to 25% titanium, 0.10 carbon, 4 silicon, and 3.5 aluminum. It is used as a deoxidizer and as a carbide stabilizer in high-chromium steels. **Graphidox,** of the same company, has 10% titanium, 50 silicon, and 6 calcium. It improves the fluidity of steel, increases machinability, and adds a small amount of titanium to increase the yield strength. The **Grainal alloys** of this company, for controlling alloy steels, have various compositions. Grade No. 6 has 20% titanium, 13 vanadium, 12 aluminum, and 0.20 boron. **Tam alloy No. 78,** of the Titanium Alloy Mfg. Div., contains 15 to 18% titanium, 7 to 8 carbon, with low silicon and aluminum. It is used in cast iron and steels. **Tam alloy No. 35** has 18 to 21% titanium, and only 3.5 to 4.5 carbon. Its melting point is 2750°F (1510°C). Ferrotitaniums with 18 to 22% titanium are used for making fine-grained forging steels. **Carbotam,** of the same company, contains 16 to 17% titanium, 2.5 to 3 silicon, 6.5 to 7.5 carbon, 1.5 to 2 boron, and less than 1 calcium. It is used for cast steels to contain boron for high hardness. **Manganese titanium** is used as a deoxidizer for high-grade steels and for nonferrous alloys. A common grade contains 38% manganese, 29 titanium, 8 aluminum, 3 silicon, 22 iron, and no carbon. **Nickel titanium** is used for hard nonferrous alloys. The low-iron grade contains 15% titanium, 5 aluminum, 4 silicon, 1 iron, and 75 nickel. **Thermocol,** of the Vanadium Corp., is a **ferrocolumbium** for adding columbium to steel. It contains 53% columbium and 0.15 max carbon. It has an exothermic reaction which prevents chilling of the molten metal.

FERROUS SULFATE. Also called **iron sulfate** and **green vitriol.** It is a green crystalline material of the composition $FeSO_4 \cdot 7H_2O$. It occurs naturally as the mineral **melanterite** and is a by-product of the galvanizing and tinning industries. The specific gravity is 1.898, the melting point is 64°C, and it is soluble in water. On exposure to the air it becomes yellowish because of the formation of basic iron sulfate, and on heating to 140°C it becomes a white powder, $FeSO_4 \cdot H_2O$, which also occurs as the mineral **szomolnokite.** Ferrous sulfate, under the name of **copperas,** is an important salt in the ink industry to give color permanence to the inks. It is also employed in water purification, as a disinfectant, in polishing rouge, as a mordant in dyeing wool, and in the production of pigments. **Ferric sulfate** is a grayish amorphous powder of the composition $Fe_2(SO_4)_3 \cdot 9H_2O$, or $Fe_2(SO_4)_3$. The specific gravity of the hydrous is 2.1, and of the anhydrous 3.097. It is very soluble in water, and is used as a pigment, as a mordant in dyeing, for etching aluminum and steel, and as a disinfectant. **Ferrisul,** of the Monsanto Co., is anhydrous ferric sulfate used for speeding the action of metal pickling baths and for descaling boilers. In etching steel, the action of anhydrous ferric sulfate is 30 times more rapid than sulfuric acid.

FERTILIZERS. Materials added to the soil to supply plant food either directly or by chemical reaction with the soil. The preparation of fertilizers is now one of the major industries, and the commercial fertilizers include nitrates, phosphates, potash salts, calcium salts, and mixtures. They may also include the materials which regulate the acidity of the soil for better plant production, such as lime, and the materials which act as **soil conditioners,** i.e., **synthetic mulches,** such as methyl cellulose or polymeric plasticlike organic chemicals. **Plant regulators** are fertilizers containing selected metals or minerals for specific plant foods, and are applied either in the soil or to the plant.

Chemicals used as fertilizers must not be of such a nature as to kill earthworms. It is stated that a minimum of at least 50,000 earthworms per acre are needed for invigorating and loosening the soil. Also, millions of bacteria are in every pound of good soil, and millions of ants, bugs, and invertebrates in every acre perform a tremendous pattern of interdependent chemical conversion. Thus, fertilizers should not contain drastic chemicals that make the soil sterile.

Much barnyard manure is employed as fertilizer, but does not enter the commercial mixed fertilizers except the dried and ground sheep and cow manures. Much local fertilization is also done by the planting and plowing under of legumes that bring nitrogen from the air and also serve as soil conditioners. Conditioning of the soil, for the retention of moisture and to prevent hard-caking so that plants may take deep root and have the needed elements readily available, is a necessary part of fertilization. Decayed vegetable matter, or peat moss, may thus be added to the soil as **humus.** These materials also often add plant foods to the soil. **Fersolin,** of the Timber Engineering Co., is such a material produced by heating sawdust with a catalyst below the charring temperature to convert the cellulose to lignin and humus. It is usually mixed with fertilizers to give greater plant yields. **Merloam,** a soil conditioner of the Monsanto Co., is a vinyl acetate–maleic acid compound.

Chilean nitrate, phosphate rock, and **potash** are the chief natural minerals used as fertilizers. Nitrogen is needed in most soils, and phosphorus is a necessary ingredient in soils. Large quantities of **muriate of potash** are used in fertilizers to supply K_2O, while vast quantities of hydrated lime are employed to supply MgO and to reduce the acidity of some soils. Potassium, calcium, and sodium are also supplied in combination forms especially with ammonia to yield nitrogen. Ammonium sulfate yields both nitrogen and sulfur. **Ground gypsum** is a source of sulfur trioxide for cotton, tobacco, grapes, and some other crops. It also helps to liberate soluble potash and stimulates growth of nitrogen-fixing bacteria in the soil. **Calcium cyanamid** is employed as a fertilizer to yield nitrogen and calcium. **Crude urea** is now also used as a fertilizer, and has the nitrogen in the same form as in the natural guanos and manures. **Uramon,** of Du Pont, is a urea fertilizer in the form of a dark-brown powder easily soluble in water and yielding 42% nitrogen, equivalent to 51% ammonia. It also contains calcium and phosphorus. **Ureaform,** developed by the U.S. Department of Agriculture, is a hygroscopic powder made by reacting urea with a small amount of formaldehyde and crushing. It may also be mixed with ammonium nitrate. **Fertilizer pellets** that resist the bleaching action of rains and release nitrogen slowly are made in granules from ammonium sulfate with a binder of asphalt and wax.

Superphosphate, or **phosphate fertilizer,** is made by treating the phosphate rock with sulfuric or nitric acid, reacting with ammonia to neutralize the acid and add nitrogen, and then adding potash salts. The final ground product contains 12% each of nitrogen, phosphoric acid, and potash. Or it may be produced by digesting the rock with ammonium sulfate yielding ammonium phosphate and gypsum. The ammonium sulfate is then treated with sulfuric acid to yield 70% phosphoric acid and ammonium sulfate. The German fertilizer **Nitrophoska** is a nitrate-phosphate-potash made by treating phosphate rock with nitric acid, neutralizing with ammonia, and then granulating with potassium salts. The **calcium nitrate tetrahydrate** which is precipitated off is also used as fertilizer. The calcium nitrate used in Europe is produced by treating phosphate rock with nitric acid. It is highly alkaline and efficient in release of nitrogen, but is very hygroscopic and sets up in lumps. Fish meal, castor pomace, cottonseed meal, soybean meal, copra cake, and other residues from oil pressing are used as commercial fertilizers. Tankage from the meat-packing plants is also an important fertilizer material. **Whale guano,** from South Georgia and Newfoundland, was a mixture of whale-meat meal and bone meal. Ground bone meal is used in fertilizers to give phosphorus, calcium, and other mineral salts to the soil. Some plants require boron, and borax is applied as a fertilizer to some soils. Many vegetable products obtain their coloring and some characteristic properties from small quantities of copper, manganese, rubidium, iodine, and other elements that do not occur in all soils. Boron is necessary for sunflower growth, iron is necessary for pineapples, molybdenum is needed for cauliflower, and cobalt oxide is necessary in the soil to prevent salt sickness in cattle. Lack of manganese in the soil also causes yellow spot on leaves of tomatoes and citrus fruits. For use as a fertilizer, **manganese sulfate,** $MnSO_4$, comes as a water-soluble powder of porous spherical particles. Most plants require minute quantites of zinc to promote formation of **auxin,** a complex butyl-cyclopentene ring compound needed for growth of plants. However, plants require balanced feeding, and indiscriminate use of fertilizers is often injurious. Too much manganese in the soil, for example, may cause necrosis, or inner bark rot, on apple trees, or excess of some common fertilizers may cause abnormal growth of stalk and leaves in plants.

FIBERBOARD. Heavy sheet material of fibers matted and pressed or rolled to form a strong board, used for making containers and partitions, and for construction purposes. Almost any organic fiber may be used, with or without a binder. The softboards are made by felting wood pulp, wood chips, or bagasse, usually without a binder. **Masonite,** of the Masonite Corp., is produced from by-product wood chips reduced to the cellulose fibers by high steam pressure. The long fibers and the lignin adhesive of the wood are retained, and no chemicals are used in pressing the pulp into boards. **Masonite quarter board,** for paneling, is made in boards ¼ in (0.64 cm) thick. **Presdwood** is a grainless grade made by compressing under hydraulic pressure, and is dense and strong.

These types belong to the class known as **hardboard,** in the processing of which the carbohydrates and soluble constituents of the original wood are dissolved out and the relative proportion of lignin is increased, resulting in a grainless, hard, stiff, and water-resistant board free from shrinkage. The density of most hardboards is greater than 1.0, and the modulus of rupture is from 5,000 to 15,000 lb/in^2 (34 to 103 MPa). The lignin acts as a binder for the fibers, but some hardboards are made harder and more resistant by adding a percentage of an insoluble resin. The usual weights range from 50 to 65 lb/ft^3 (801 to 1,041 kg/m^3), but with added resin binder the weight may be as much as 70 lb (31.8 kg), and **densified hardboard,** made with high pressure, is 85 lb (38.6 kg) or above. Hardboards have uniform strength in all directions and have smooth surfaces. Tensile strengths are up to 7,700 lb/in^2 (53 MPa) and compressive strength of 26,000 lb/in^2 (179 MPa).

Irradiated wood is natural wood impregnated with resins of low molecular weight and irradiated with gamma rays from cobalt 60 which cross-links the resin molecules and binds them to the fibers of the wood. The resins add strength and hardness without changing the grain structure and color of the wood. Maple, impregnated with 0.5% by weight of methyl methacrylate and irradiated, is three times as hard as the natural wood but can be worked with ordinary tools. **Particleboards,** made with wood particles, have lower density, about 40 lb/ft^3 (641 kg/m^3), and have greater flexibility but lower strength than hardboard. The process is not limited to the making of boards. **Wood particles** are also used for low-cost molded parts, with up to 90% wood particles and the balance urea, phenolic, or melamine resin. Birch or maple particles are preferred. These **Granuplast moldings** are made with low heat and pressure to densities of 45 to 85 lb/ft^3 (721 to 1,362 kg/m^3).

Wood molding powder is made by the same method of treating wood fibers with steam pressure and hydrolyzing the hemicellulose, leaving the lignin free as a binder. Hardboard is used for counter tops, flooring, furniture, and jigs and templates. **Forall,** of the Forest Fiber Products Co., is a light-colored hardboard in thicknesses from ⅜ to ¾ in (1 to 1.9 cm), made by compressing Douglas fir free of bark. It is grain-free, and will not split or splinter. **Presdply,** of the Masonite Corp., has surfaces of grainless hard Presdwood and a core of soft plywood that will hold screws.

Hardwood, of the Elmendorf Corp., is a hardboard made from hardwood waste compressed into sheets under heat and hydraulic pressure. The surface is hard with a high polish. A hardboard, developed by the Scottish Cooperative Wholesale Society and called **heatherwood,** is made by pulping heather and pressing into boards with a synthetic resin binder. **Heather,** or **heath,** is a small flowering shrub, *Ericaceae tetralix* and *E. cinerea,* which grows profusely in Great Britain. **Wonderwood,** of the Wonderwood Corp., is a development of the **Novopan** made in Switzerland. It is made by pulping chipped waste wood and compressing with a resin binder. **Tensilite 300,** of the J. P. Lewis Co., is made of pulp combined with nitrile rubber and a phenolic resin and pressed into sheets. The specific gravity is 1.35, dielectric strength 600 volts per mil (23.6×10^6 volts per meter), and compressive strength 32,000 lb/in^2 (220 MPa). It is suitable for mechanical and electrical applications as well as for paneling. The **Forest hardboard** of the Forest Fiber Products Co. is made of chipped wood that is pulped and mixed with synthetic resin and wax and then hydraulically pressed. The board has a smooth glossy face, and is suitable for making furniture and toys and for paneling. **Prespine,** of the Curtis Co., Inc., is a paneling board of lower hardness and density made by mixing 5 to 15% phenolic resin with sawdust and wood chips and pressing at only 200 lb/in^2 (1.4 MPa). These hardboards are used for many construction parts, but the lighter and less dense fiberboards are preferred for insulation and some construction uses. **Granite board,** of National Starch and Chemical Corp. is a strong, nonsplintering **building board** made from fine particles of eastern white pine molded under pressure with a resin binder to a density about equal to natural wood. It has an acoustical value higher than that of natural wood. **Kimflex board,** of the Kimberly-Clark Corp., is a lightweight, pliable fiberboard used for shoe counters. It is made from balsawood pulp, using rubber latex as a binder. **Electrite,** of the West Virginia Pulp & Paper Co., is a wood fiberboard of high strength and high dielectric strength for electrical panels. **Temlok,** of the Armstrong Cork Co., is a fiberboard made from pinewood fibers impregnated with resin and compressed into building boards and tiles. **Temwood,** of the same company, is a lightweight board of wood fibers hydraulically pressed into grainless boards in hard and semihard grades. **Temboard** is a decorative wood fiberboard used for interior paneling. **Veneer fiberboard,** of the Elmendorf Co., is made by cutting veneer waste into fibers of 0.010- to 0.015-in (0.025- to 0.038-cm) thickness with strand lengths from 1 to 8 in (2.54 to 20.32 cm). The flat side of the strand is edge-grained, and when felted the broad surface lies parallel to the faces of the board. From 10 to 20% phenolic resin is used as a fiber binder. **Disfico board,** of the Diamond State Fibre Co., for making trunks and boxes, is made of pressed jute and hemp fibers. It comes in sheets in plain colors. **Thermax,** of the Northwest Magnesite Co., is an insulating board made of shredded wood fibers with a fire-resistant cement.

FIBERS. By definition (ASTM) a **fiber** has a length at least 100 times its diameter or width, and its length must be at least 0.2 in (0.5 cm). Length also determines whether a fiber is classified as staple or filament. **Filaments** are long and/or continuous fibers. **Staple fibers** are relatively short, and, in practical applications, range from under 1 to 6 in (2.5 to 15.2 cm) long (except for rope, where the fibers can run to several feet). Of the natural fibers, only silk exists in filament form, while synthetics are produced as both staple and filaments.

The internal, microscopic structure of fibers is basically no different from that of other polymeric materials. Each fiber is composed of an aggregate of thousands of polymer molecules. However, in contrast to bulk plastic forms, the polymers in fibers are generally longer and aligned linearly, more or less parallel to the fiber axis. Thus fibers are generally more crystalline than are bulk forms.

Also in contrast to bulk forms, fibers are not used alone, but either in assemblies or aggregates such as yarn or textiles or as a constituent with other materials, such as in composites. Also, compared with other materials, the properties and behavior of both fibers and textile forms are more critically dependent on their geometry. Hence fibers are sometimes characterized as tiny microscopic beams, and, as such, their structural properties are dependent on such factors as cross-sectional area and shape, and length. The cross-sectional shape and diameter of fibers vary widely. Glass, nylon, Dynel, and Dacron, for example, are essentially circular. Some other synthetics are oval, while others are irregular and serrated round. Cotton fibers are round tubes, and silk is triangular.

Fiber diameters range from about 10 to 40 μm (0.01 to 0.04 mm) in diameter. Because of the irregular cross section of many fibers, it is common practice to specify diameter or cross-sectional area in terms of fineness, which is defined as a weight-to-length or linear density relationship. One exception is wool, which is graded in micrometers. The common measure of linear density is the denier, which is the weight in grams of a 9,000-m length of fiber. Another measure is the tex, which is defined as grams per 1 km. A millitex is the number of grams per 1,000 km.

Of course, the linear density, or denier, is also directly related to fiber density. This is expressed as the denier/density value, commonly referred to as denier per unit density, which represents the equivalent denier for a fiber with the same cross-sectional area and a density of 1.

The cross-sectional diameter or area generally has a major influence on fiber and textile properties. It affects, for example, yarn packing, weave tightness, fabric stiffness, fabric thickness and weight, and cost relationships. Similarly, the cross-sectional shape affects yarn packing, stiffness, and twisting characteristics. It also affects the surface area, which in turn determines the fiber contact area, air permeability, and other properties.

FIBER-REINFORCED PLASTICS. A broad group of composite materials composed of fibers embedded in a plastic resin matrix. The short designation for these materials is **FRP.** In general, they have relatively high strength-to-weight ratios and excellent corrosion resistance compared to metals. They can be formed economically into virtually any shape and size.

In size, FRP products range from tiny electronic components to large boat hulls. In between these extremes, there are a wide variety of FRP gears, bearings, bushings, housings, and parts used in all product industries.

Glass is by far the most-used fiber in FRPs. **Glass-fiber-reinforced plastics** are often referred to as **GFRP** or **GRP. Asbestos fiber** has some use, but is largely limited in applications where maximum thermal insulation or fire resistance is required. Other fibrous materials used as reinforcements are paper, sisal, cotton, nylon, and Kevlar. For high-performance parts and components, more costly fibers, such as boron and graphite, can be specified.

Although a number of different plastic resins are used as the matrix for reinforced plastics, thermosetting **polyester resins** are the most common. The combination of polyester and glass provides a good balance of mechanical properties as well as corrosion resistance, low cost, and good dimensional stability. In addition, curing can be done at room temperature without pressure, thus making for low processing-equipment costs. For high-volume production, special **sheet-molding compounds** are available in continuous sheet form. Resin mixtures of thermoplastics with polyesters have been developed to produce high-quality surfaces in the finished molding. The common thermoplastics used are acrylics, polyethylenes, and styrenes.

Other glass-reinforced thermosets include phenolics and epoxies. **GRP phenolics** are noted for their low cost and good overall performance in low-strength applications. Because of their good electrical resistivity and low water absorption, they are widely used for electrical housings, circuit boards, and gears. Since epoxies are more expensive than polyesters and phenolics, **GRP epoxies** are limited to high-performance parts where their excellent strength, thermal stability, chemical resistance, and dielectric strength are required.

Initially, GRP materials were largely limited to thermosetting plastics. Today, however, more than 1,000 different types and grades of **reinforced thermoplastics** are commercially available. Leaders in volume use are nylon and the styrenes. Others include sulfones and ABS. Unlike thermosetting resins, GRP-thermoplastic parts can be made in standard injection-molding machines. The resin can be supplied as pellets containing chopped glass fibers. As a general rule, a GRP thermoplastic with chopped fibers at least doubles the plastic's tensile strength and stiffness. Glass-reinforced thermoplastics are also produced as sheet materials for forming on metal stamping equipment.

FILTER FABRICS. Any fabric used for filtering liquids, gases, or vapors, but, because of the heat and chemical resistance usually required, are generally of synthetic or metal fibers. Weave is an important consideration.

Plain weave permits maximum interlacings, and in a tight weave gives high impermeability to particles. Twill weave has lower interlacings in sharp diagonal lines, and gives a more selective porosity for some materials. Satin weave has fewer interlacings, is spaced widely but regularly, and is used for dust collection and gaseous filtration.

Fibers are chosen for their particular chemical resistance, heat resistance, and strength. Dacron has good acid resistance except for concentrated sulfuric or nitric acid. It can be used to 325°F (162°C). High-density polyethylene has good strength and abrasion resistance, and its smooth surface minimizes clogging of the filter, but it has an operating temperature only to 230°F (110°C). Polypropylene can be used to 275°F (134°C). Nylon gives high strength and abrasion resistance. It has high solvent resistance, but low acid resistance. Its operating limit is about 250°F (121°C). Teflon is exceptionally resistant to a wide variety of chemicals. It can be operated above 400°F (204°C), and its waxy, nonsticking surface prevents clogging and makes it easy to clean, but the fiber is available only in single-filament form.

FILTER SAND. A natural sand employed for filtration, especially of water. Much of the specially prepared filter sand comes from New Jersey, Illinois, and Minnesota, and is from ocean beaches, lake deposits, and sandbanks. The specifications for filter sand require that it be of fairly uniform size, free from clay and organic matter, and chemically pure, containing not more than 2% combined carbonates. The grain sizes are specified in millimeters, the most common being from 0.35 to 0.65 mm. Very fine sand clogs the filter. **Greensand,** produced from extensive beds in New Jersey, is used as a water softener. It is a type of marl classed as **zeolite,** and consists largely of **glauconite,** which is a greenish granular mineral containing up to 25% iron, with a large percentage of silica and some potash and alumina. Synthetic zeolite is a **sodium alumina silicate,** $Na_2O \cdot Al_2O_3 \cdot 6SiO_2 \cdot xH_2O$, made by reacting caustic soda with bauxite to form sodium aluminate and then reacting with sodium silicate. In addition to filtering, the greensand softener extracts the calcium and magnesium from the water. It is regenerated for further use by passing common salt brine through it. The **Zeolex** of J. M. Huber Corp. is **sodium silicoaluminate** in extremely fine powder form. Up to 2% is added to such products as dried egg-yolk powder to prevent caking and to keep dry ingredients in automatic food-processing equipment free-flowing.

Molecular sieves are synthetic crystalline zeolites whose molecules are arranged in a crystal lattice so that there are a large number of small cavities interconnected by smaller pores of uniform size, the network of cavities and pores being up to 50% of the volume of the crystal. The sieves consist of three-dimensional frameworks of SiO_4 and AlO_4 tetrahedra.

Electrovalence of each tetrahedron is balanced by the inclusion in the crystal of a metal cation of Na, Ca, or Mg. Firing in a kiln drives out the water, and by exchanging the sodium ion for a smaller or larger ion the pore openings can be varied from 0.2 to 1.2 nm. For gasoline upgrading, 0.4-nm openings are used, while 1.0-nm openings serve for removing oil vapor or hydrogen sulfide from gas. The mean path required for oxygen and nitrogen molecules is about 0.1 μm. **Filter plates,** for filtering acids and oils, are porous fused alumina with pores from 0.09-to 0.30-mm diameter. **Zeolon,** of the Norton Co., is a zeolite with a crystal structure known as **nordenite.** The pore diameter is about 1.0 nm. **Zeolite 4A,** of Linde, for chemical separations, will pass molecules no larger than 0.4 nm.

FIRECLAY. Clays that will withstand high temperatures without melting or cracking, used for lining furnaces, flues, and for making firebricks and lining tiles. Common fireclays are usually **silicate of alumina.** Theoretically these clays contain 45.87% alumina and 54.13 silica, but in general they contain considerable iron oxide, lime, and other impurities. Most of the American clays are from New Jersey, Kentucky, Pennsylvania, Ohio, and Missouri. They are largely $Al_2O_3 \cdot SiO_2$, with CaO, Fe_2O_3 and TiO_2. Those low in iron oxide, lime, magnesia, and alkalies are chosen. The clays are grouped as low-duty, intermediate, high-duty, and super-duty. The low-duty has low alumina and silica with high impurities, and is limited to a temperature of 1600°F (871°C). Standard types are good for temperatures of 2400 to 2700°F (1316 to 1482°C), and the super-duty to temperatures of 2700 to 3000°F (1482 to 1649°C). Kiln-burned clay should have a balanced proportion of coarse, intermediate, and fine grain sizes. Clays with an excess of silica are also used. The German **Klingenberg clay** used for crucibles has about 60% silica. The term **refractory clay** embraces nearly all clays having a melting point above 160°C. But the clays alone are likely to shrink and crack, and they may be mixed with other clays, sand, or graphite. **Firebrick** is made in various shapes and sizes and is usually white or buff in color. Common firebrick from natural clays will melt at from 2800 to 3100°F (1538 to 1704°C).

Insulating firebrick is made with fireclay and a combustible material, such as sawdust, which burns out to leave a porous structure. The weight is 1.25 to 4 lb (0.57 to 1.81 kg) per brick compared with 8 lb (3.6 kg) for regular firebrick. Firebrick containing more than 47.5% alumina is not classed as fireclay brick but as **alumina brick.** The **SL firebrick** of Johns-Manville is a **kaolin firebrick** with 62% silica, and can be used up to 2300°F (1260°C). Spalling is a common failure of fireclay brick, but high-duty should show little spalling under long soaking at 1650°F (899°C) or alternating periods of heating and cooling at higher temperatures. **Alamo brick**

and **Varnon brick,** of the Harbison-Walker Refractories Co., are high-duty firebrick. **Kaosil firebrick,** of the same company, is designated as a **semisilica firebrick.** It is made from low-alkali siliceous kaolin of Pennsylvania, rotary-fired at high temperature. The nominal composition is 75.6% silica, 21.8 alumina, 1.7 titania, 0.5 iron oxide, 0.27 magnesia, 0.15 lime, and 0.10 alkalies. The brick can be used in a soaking heat of 2700°F (1482°C), is resistant to spalling and to fluxing by alkalie slags, and has high load-carrying ability. **Korundal,** of the same company, is a corundum-mullite brick for temperatures to 3425°F (1880°C). It contains 91% corundum alumina, 8 silica, and less than 1% iron oxide, lime, magnesia, and alkalies. It melts at 2020°C, converting all the mullite to corundum, but slow cooling returns the brick to the original mixture.

Some other materials used in making firebrick are chromite, bauxite, diatomaceous earth, and magnesite, or the artificial materials silicon carbide and aluminum oxide, but brick made of these are designated by the name of the material or by trade names. **Chromite brick** will withstand temperatures up to 3700°F (2038°C), and **magnesia brick** up to 3900°F (2149°C), while silicon-carbide brick without a clay binder will withstand heats to 4000°F (2204°C). **Firecrete,** of Johns-Manville, is a lightweight refractory consisting of calcined high-alumina clay used for furnace doors and floors. It will withstand temperatures of 2400°F (1316°C) in continuous operation. **Insuline,** of the Quigley Co., is a calcined fireclay in small cellular particles. In insulating brick it is called **Insulbrix,** and as a lightweight concrete it is known as **Insulcrete. Allmul firebrick,** of the Babcock & Wilcox Co., for glass furnaces, is mullite with no free silica.

FIRE EXTINGUISHERS. Materials used for extinguishing fires, usually referring to chemicals in special containers rather than the materials, like water, used in quantity for cooling and soaking the fuel with a noncombustible liquid. There are three general types of fire extinguishers: those for smothering, such as carbon dioxide; those for insulating the fuel from the oxygen supply, such as licorice and protein foams, which also includes mineral powders that melt and insulate metallic fires; and chemicals which react with the combustion products to terminate the chain reaction of combustion, such as **bromotrifluoromethane,** $CBrF_3$, a nontoxic colorless gas liquefied in cylinders. **Freon FE 1301,** of Du Pont, is this chemical, while **Freon 13B1** of the same company is **monobromotrifluoromethane** gas pressurized with nitrogen. The relative effectiveness of extinguishers varies with the type of fuel in the fire, but on an average, with bromotrifluoromethane taken as 100%, dibromodifluoromethane would be about 67%, the dry chemical sodium hydrogen carbonate 66, carbon tetrachloride 34, and carbon dioxide about 33. **Fire-Trol,** of the Arizona Agrochemical Co., is ammonium sulfate and **attapulgic clay** for fighting forest fires. It is sprayed in a water solution, and the slurry mixture coats the trees to stop fire spread.

FISHERY PRODUCTS. Fisheries constitute one of the largest industries of the world. In addition to its use as food, fish is important as the source of fatty oils, animal feeds, fertilizer, vitamin products, fish flour, protein powders, pearl essence, and skins. More than half of all species of vertebrates (animals with backbones) are fish, and more than 40,000 kinds of fish have been classified, varying from the small goby, weighing less than 0.01 oz (0.28 g) to the whale shark, sometimes weighing more than 20 tons (18,140 kg). Extreme shapes vary from the snakelike eel to the sea horse, but the **marine herring,** *Clupea herengus,* is designated by the U.S. Fish and Wildlife Service as the typical **fish** because of its abundance; its lack of extremes in form, size, and structure; and its water efficiency due to its streamlined shape and fin arrangement.

Ocean perch, *Sebastes marinas,* is the chief commercial fish caught in the New England area, and **menhaden,** *Brevoortia tyrannus,* constitutes the chief catch of the Atlantic coasts. The red **snapper,** *Lutianus aya,* and the red **grouper,** *Epinephelus morio,* constitute the major part of the catch in the Gulf of Mexico. **Smelt,** *Oamerus mordax,* planted in the Great Lakes in 1912, constituted 97% of the commercial catch of fish in Lake Erie before the recent pollution of the lake.

One third of all fish and frozen packaged fish in the United States is ocean perch, or **rosefish,** compared with 24% for haddock. **Rockfish,** very low in oil and sodium and high in protein, is important in the frozen-fish industry of the West Coast, reaching a tonnage about half that of the total salmon. With these processed fish, only about 25% of the total weight of the fish is packaged, the remainder being used for oils, feeds, and fertilizer. The variety of **tuna fish** known as **skipjack,** *Katsuwonus pelamis,* is the most important commercial fish of the Pacific. It is migratory, ranging from California to Japan and the Philippines. Known as **aku** in Hawaii, it abounds in that area, but is also caught in warm areas of the Atlantic. The **albacore tuna** likewise ranges over the Pacific and Indian Oceans and into the Atlantic.

The usual types of fish processed for marketing as **smoked fish** are mackerel, mullet, sturgeon, catfish, and flounder. The **sturgeon** of the Caspian Sea attains a length of 30 ft (9.1 m) and a weight of 4,000 lb (1,814 kg). It is valued for meat, liver vitamins, isinglass, oil, skins, and **caviar,** the latter being the **roe,** or eggs, also obtained from shad and some other large fish. In the processing of frozen fish, an average of only 33% of the whole fish is shipped as edible packaged fish, but the amount of fillets taken from cleaned fish may be as high as 55% for pollock and 70% for large, 8-lb

(3.61-kg) steelhead trout. **Industrial fish,** for the production of oil, animal feeds, and fertilizer, consists usually of mixed small fish, but in some areas there is no sorting, and fish inedible to humans may be included.

The **fish scales** which are removed from processed edible fish are high in edible proteins, and are also used in animal feeds. Scales from the pollock contain 70% protein. **Fish solubles** usually consist of a 50% solids concentration of the residue liquor, known as **stickwater,** from processing plants and canneries. In addition to protein, it is rich in vitamins B, G, and B_{12}. It is mixed with alfalfa leaf meal for animal feed. **Fishskins** have a close texture and are impervious, but sharkskin is the only fishskin of commercial importance.

Fish meal is produced from whole fish or from the residue of processed food fish. Whole fish is ground and cooked below 212°F (100°C) to avoid loss of protein, and the oil solvent is extracted. About 2% oil is retained in the meal, but for the manufacture of fish flour this residue oil is removed by alcohol extraction. Prior to the extraction of the oil, which may be up to 16%, fish meal contains up to about 23% protein, and up to about 30% minerals, including calcium, phosphorus, iron, and copper. The proteins have all the essential amino acids to supplement cereal foods for poultry raised by commercial methods where the birds lack access to normal feeding.

Fish flour is a low-cost source of protein, and is used for enriching flours, baby foods, sauces, and prepared soups, and for adding proteins to breads, cakes, pastries, and other bakery products. It is prepared from fish meal by refining and deodorizing. It is an additive rather than a flour; it does not thicken soups, and in bakery products it does not have the elastic and extensible properties inherent in cereal flours. In the food industries it is called **animal protein concentrate,** and the high-protein grade contains 95% animal protein. It is also high in calcium and phosphorus, and contains thiamin, niacin, and riboflavin. An edible **fish-protein concentrate** may now be made directly from finely ground industrial fish. Oils, fats, fatty acids, and lipid-containing materials are extracted by a solvent such as isopropyl alcohol. About 20% by weight of raw fish is recovered as a dry, odorless, tasteless powder which can be mixed with a variety of foods to upgrade the human diet. Fish flour has negligible contents of carbohydrates, and no more than 0.4% fat, but **Viking egg white,** an odorless gray powder made in Germany from whitefish, is a soluble albumin used as a substitute for egg white in bakery goods.

Fish oils are obtained by boiling the fish and skimming off the oil, or by solvent extraction from the fish meal. The crude oil has a brownish color and an offensive odor, but it is usually decolorized and deodorized. Oil content of fish varies from 0.5 to about 16%, depending on the type of fish, the season, and the area. Fish in cold waters tend to have more oil than those in warm waters. There is only a small difference in the composition of oils from different species. They usually contain 20 to 30% of saturated acids and 70 to 80% unsaturated acids. The average specific gravity is about 0.930. Much of the commercial oil is from the cod, herring, menhaden, sardine, and salmon. **Japan fish oil** consists of a mixture of sardine and herring oils. Fish oil is of the nondrying class, and is used for lubricants, leather dressings, soaps, and heat-treating oils, but is also used for blown oils or for fractionating for use in paints and in plastics.

FLAX. A fiber obtained from the flax, or linseed, plant, *Linum usitatissimum,* used for making the fabrics known as linens, and for thread, twine, and cordage. It is valued because of its strength and durability. It is finer than cotton, and very soft, and the fibers are usually about 20 in (50.8 cm) long. Flax consists of the **bast fibers,** or those in the layer underneath the outer bark, which are of fine texture. The plants are pulled up by the roots, retted, or partly decayed, scraped, and the fibers combed out and bleached in the sun. For the best European flax the preparation is entirely by hand. The important centers of flax preparation are in Russia, central Europe, Italy, Ireland, France, and Egypt. Some flax is also grown in the United States. The plants that are grown for the oil seed yield a poor fiber and are not employed to produce flax.

FLINT. An opaque variety of **chalcedony** or nearly pure amorphous **quartz** which shows no visible structure. It is deposited from colloidal solution and is an intimate mixture of quartz and opal. It contains 96 to 99% silica, and may be colored to dull colors by impurities. Thin plates are translucent. When heated, it becomes white. Flint is finely crystalline. It breaks or chips with a convex, undulating surface. The hardness is 7, and the specific gravity is 2.6. It was the prehistoric utility material for tools, and was later used with steel to give sparks on percussion. **Gun flints** are still made from a type of flint mined at Brandon, England, for special uses. **Lydian stone,** or **touchstone,** was a cherty flint used for testing gold. Flint is now chiefly used as an abrasive, and in pottery and glass manufacture. **Flint paper** for abrasive use contains crushed flint in grades from 20 to 240 mesh, usually coated on one side of 70- or 80-lb paper. Flint is also used in the form of grinding pebbles. **Potters' flint,** used for mixing in ceramics to reduce the firing and drying shrinkage and to prevent deformation, is ground flint of about 140 mesh made from white **French pebbles. Bitstone** is a name used in the ceramic industry for calcined flint chips ground to the size of wheat, employed for sprinkling on the bottom of the saggers so that the ware will not stick in firing. **Hornstone** is a flint with chalcedony inclusions. It splinters rather than chips, and is not used for abrasives.

FLUORINE. An elementary material, symbol F, which at ordinary temperatures is an irritating pale-yellow gas, F_2. Fluorine gas is obtained by the reduction and electrolysis of fluorspar and cryolite. It has a density of 1.69 and a boiling point of −187°C, and it solidifies at −223°C. It is used in the manufacture of fluorine compounds. It combines violently with water to form hydrofluoric acid, and it also reacts strongly with silicon and most metals. **Liquid fluorine,** at temperatures below −367°F (−221°C), is used as an oxidizer for liquid rocket fuels. In combustion, a pound of fluorine produces a pound of hydrogen fluoride which is highly corrosive. Fluorine is one of the most useful of the halogens.

The gas **sulfur hexafluoride,** SF_6, resembles nitrogen in its inactivity. It is odorless, colorless, nonflammable, nontoxic, and five times as heavy as air. It is used as a refrigerant, as a dielectric medium in high-voltage equipment, as an insecticide propellant, and as a gaseous diluent. It remains stable to 800°C. **Aluminum fluoride,** AlF_3, is a white crystalline solid used in ceramic glazes and for fluxing nonferrous metals. Other metallic fluorides are marketed for special purposes. **Silver difluoride,** AgF_2, is a blackish powder used as a fluorinating agent; it contains about 26% fluorine. **Chlorofluorine gas,** Cl_3, is a violent fluorinating agent, and is used for the fluorination of some metals otherwise difficult to separate, such as uranium.

FLUOROCARBONS. Compounds of carbon in which fluorine instead of hydrogen is attached to the carbon atoms. They range from gases to solids. When not less than two fluorine atoms are attached to a carbon atom they are very firmly held, and the resulting compounds are stable and resistant to heat and chemicals. Fluorocarbons may be made part hydrocarbon and part fluorocarbon, or may contain chlorine. The fluorocarbons used as plastic resins may contain as much as 65% fluorine and also chlorine, but are very stable. **Liquid fluorocarbons** are used as heat-transfer agents, hydraulic fluids, and fire extinguishers. Benzene-base fluorocarbons are used for solvents, dielectric fluids, lubricants, and for making dyes, germicides, and drugs. Synthetic lubricants of the fluorine type consist of solid particles of a fluorine polymer in a high-molecular-weight fluorocarbon liquid.

FLUOROPLASTICS. Also termed **fluorocarbon resins** and **fluorine plastics.** A group of high-performance, high-price engineering plastics. They are composed basically of linear polymers in which some or all of the hydrogen atoms are replaced with fluorine, and are characterized by relatively high crystallinity and molecular weight. All fluoroplastics are natural white and have a waxy feel. They range from semirigid to flexible. As a class, they rank among the best of the plastics in chemical resistance and elevated-temperature performance. Their maximum service temperature ranges up to about 500°F (260°C). They also have excellent frictional properties and cannot be wet by many liquids. Their dielectric strength is high and is relatively insensitive to temperature and power frequency. Mechanical properties, including tensile creep and fatigue strength, are only fair, although impact strength is relatively high.

There are three major classes of fluoroplastics. In order of decreasing fluorine replacement of hydrogen, they are **fluorocarbons, chlorotrifluoroethylene,** and **fluorohydrocarbons.** There are two fluorocarbon types: **tetrafluoroethylene (PTFE** or **TFE)** and **fluorinated ethylene propylene (FEP).** PTFE is the most widely used fluoroplastic. It has the highest useful service temperature, 500°F (260°C), and chemical resistance. FEP's chief advantage is its low-melt viscosity, which permits it to be conventionally molded.

Teflon, of Du Pont, is a tetrafluoroethylene of specific gravity up to 2.3. The tensile strength is up to 3,500 lb/in^2 (23.5 MPa), elongation 250 to 350%, dielectric strength 1,000 volts per mil (39.4 $\times$ 10^6 volts per meter), and melting point 594°F (312°C). It is water-resistant and highly chemical-resistant. **Teflon S** is a liquid resin of 22% solids, sprayed by conventional methods and curable at low temperatures. It gives a hard, abrasion-resistant coating for such uses as conveyors and chutes. Its temperature service range is up to 400°F (204°C). **T-film,** of Eco Engineering Co., is thin Teflon film for sealing pipe threads. **Teflon fiber** is the plastic in extruded monofilament, down to 0.01 in (0.03 cm) in diameter, oriented to give high strength. It is used for heat- and chemical-resistant filters. **Teflon tubing** is also made in fine sizes down to 0.10 in (0.25 cm) in diameter with wall thickness of 0.01 in (0.03 cm). **Teflon 41-X** is a collodial water dispersion of negatively charged particles of Teflon, used for coating metal parts by electrodeposition. **Teflon FEP** is **fluorinated ethylenepropylene** in thin film, down to 0.0005 in (0.001 cm) thick, for capacitors and coil insulation. The 0.001-in (0.003-cm) film has a dielectric strength of 3,200 volts per mil (126 $\times$ 10^6 volts per meter), tensile strength of 3,000 lb/in^2 (20 MPa), and elongation of 250%.

Chlorotrifluoroethylene (**CTFE** or **CFE**) is stronger and stiffer than the fluorocarbons and has better creep resistance. Like FEP and unlike PTFE, it can be molded by conventional methods.

The fluorohydrocarbons are of two kinds: **polyvinylidene fluoride** (**PVF$_2$**) and **polyvinyl fluoride** (**PVF**). While similar to the other fluoroplastics, they have somewhat lower heat resistance, and considerably higher tensile and compressive strength.

Except for PTFE, the fluoroplastics can be formed by molding, extruding, and other conventional methods. However, processing must be carefully controlled. Because PTFE cannot exist in a true molten state, it cannot be conventionally molded. The common method of fabrication is by compacting the resin in powder form and then sintering.

Whitcon, of the Whitford Chemical Corp., is **fluorocarbon powder** of 1-μm particle size for use as a dry lubricant or for incorporation into rubbers, plastics, and lubricating greases. **Plaskon Halon,** grade G80, of Allied Chemical Co., is a tetrafluoroethylene powder of fine particle size, 25 to 30 μm, for molded parts that have tensile strengths to 6,500 lb/in^2 (44.5 MPa).

Fluorothene plastic, of Union Carbide, has the formula $(CF_2 \cdot CFCl)_n$, differing from Teflon in having one chlorine atom on every unit of the polymer chain, replacing the fourth fluorine atom. It is transparent, and molded parts have a specific gravity of 2.1, a tensile strength of 9,400 lb/in^2 (65 MPa), and high dielectric strength, and it will withstand temperatures to 300°F (149°C). **KEL-F,** of Minnesota Mining & Mfg. Co., is **chlorotrifluoroethylene** used for moldings, gaskets, seals, liners, diaphragms, and coatings. The molded parts have high chemical resistance. The compressive strength is 30,000 lb/in^2 (210 MPa), but it can be heat-treated to increase the compressive strength to 80,000 lb/in^2 (560 MPa). The tensile strength of the molded material is 5,000 lb/in^2 (34 MPa), but oriented fibers have tensile strength to 50,000 lb/in^2 (344 MPa). **Fluorocarbon rubber** produced by this company for tubing, gaskets, tank linings, paints, and protective clothing has a tensile strength of 3,000 lb/in^2 (20 MPa), elongation of 600%, heat resistance to 400°F (204°C), and high resistance to oils and chemicals. It is a saturated fluorocarbon polymer containing 50% fluorine. **Aclar,** of the Allied Chemical Corp., is chlorotrifluoroethylene transparent packaging film which is exceptionally resistant to oils and chemicals, has a moisture-barrier efficiency 400 times that of polyethylene film, has good strength to 390°F (199°C), and retains its flexibility to −300°F (−184°C). It is also used for wire covering.

FLUORSPAR. Also called **fluorite.** A crystalline or massive granular mineral of the composition CaF_2, used as a flux in the making of steel, for making hydrofluoric acid, in opalescent glass, in ceramic enamels, for making artificial cryolite, as a binder for vitreous abrasive wheels, and in the production of white cement. It is a better flux for steel than limestone, making a fluid slag, and freeing the iron of sulfur and phosphorus. About 5.54 lb (2.5 kg) of fluorspar is used per ton of basic open-hearth steel.

Fluorspar is mined in Illinois, Kentucky, Nevada, and New Mexico. American ore usually runs 35 to 75% CaF_2, but high-grade ore from Spain and Italy contains up to 98%. The specific gravity is 3.18, hardness 4, and the colors light green, yellow, rose, or brown. When ground, the color is white. The melting point is 1650°F (899°C). The usual grades for fluxing are smaller than ½ in (1.27 cm) and contain 85% min CaF_2, with 5 max SiO_2.

High-grade fluorspar for ceramic frit has 95 to 98% CaF_2, 3 max SiO_2, and 0.12 max Fe_2O_3, and is known as No. 1 ground. **Acid spar** is a grade used in making hydrofluoric acid. It contains over 98% CaF_2 and 1 max SiO_2, and is produced by flotation. It is also used for making refrigerants, plastics, and chemicals, and for aluminum reduction. **Optical fluorspar** is the highest grade but is not common. **Fluoride crystals** for optical lenses are grown artificially from acid-grade fluorspar. Pure **calcium fluoride,** Ca_2F_6, is a colorless crystalline powder used for etching glass, in enamels, and for reducing friction in machine bearings. It is also used for ceramic parts resistant to hydrofluoric acid and most other acids. **Calcium fluorite** has silicon in the molecule, $CaSiF_6 \cdot 2H_2O$, and is a crystalline powder used for enamels. The clear rhombic fluoride crystals used for transforming electric energy into light are **lead fluoride,** PbF_2.

FLUX. A substance added to a refractory material to aid in its fusion. A secondary action of a flux, which may also be a primary reason for its use, is as a reducing agent to deoxidize or decompose impurities and remove them as slags or gases. In soldering, a flux may serve to remove oxides from the surface to be soldered. Materials such as charcoal or impure boron carbide used to cover baths of molten metals may also be considered as fluxes. **Fluxing stone** is a common term for the limestone or dolomite used in the melting of iron. About 900 lb (408 kg) of limestone are employed for every long ton of pig iron produced in the blast furnace. If iron ore were reduced without a basic flux, the silica and alumina would unite with the iron oxides to form double silicates of iron and alumina, and there would be a heavy loss of iron. With the addition of limestone, the silica and alumina, having strong affinity for the lime and magnesia, form compounds that contain very little iron. These compounds form a liquid slag which floats on the surface of the molten iron and can be removed readily. The flux also removes sulfur and phosphorus from the iron. Some iron ores contain sufficient lime carbonate to be almost self-fluxing. Lime is more effective as a flux than limestone, but is more expensive. The action of the blast furnace is first to convert the limestone to lime. Upon being heated to 1525°F (829°C), limestone breaks down to lime, which then begins fusion with silica to form the slag at about 2600°F (1427°C). Limestone for use as flux must be fairly pure, or additional undesirable compounds will be formed. For brass, bronze, or soft white metals, resins may be used, and the covering flux may be charcoal, salt, or borax. Cryolite is a flux for aluminum and for glass.

Fluxing alloys for brasses and bronzes are phosphor tin, phosphor copper, or silicon copper. They deoxidize the metals at the same time that alloying elements are added. For tinning steel, palm oil is used as a flux. For ordinary soldering, zinc chloride is a common flux. Tallow, rosin, or olive oil may also be used for soldering. Acetamide is used for soldering painted metals. For silver solders, borax is a common flux. For soldering stainless steel, the borax is mixed with boric acid, or pastes are made with zinc chloride and borax. Borax may also be used as a welding flux. **White flux** is a mixture of sodium nitrate and nitrite, and is a strong oxidizer used for welding.

Welding fluxes for high-temperature welding are usually coated on the rod and contain a deoxidizer and a slag former. **Lithium fluoride,** LiF, is a powerful flux with the fluxing action of both lithium and fluorine, and it gives a low-melting liquid slag. Deoxidizers may be ferromanganese or silicomanganese. **Slag formers** are titanium dioxide, magnesium carbonate, feldspar, asbestos, or silica. Soluble silicate is a binder, while cellulose may be used for shielding the arc.

FOAM MATERIALS. Materials with a spongelike, cellular structure. They include the well-known sponge rubber, plastic foams, glass foams, refractory foams, and a few metal foams. Ordinary chemically blown **sponge rubber** is made up of interconnecting cells in a labyrinthlike formation. When made by beating latex, it may show spherical cells with the porous walls perforated by the evaporation of moisture. It is also called **foam rubber.** Special processes are used to produce celltight and gastight **cellular rubber** which is nonabsorbent. **Unicel ND,** of Du Pont, used as a blowing agent for sponge rubber, is dinitropentamethyl tetramine. It is mixed into the rubber, and in the presence of the rubber acids it is decomposed, liberating gas during the vulcanization to form small cells. The cellular rubber of Uniroyal, Inc., produced in sheets of a density of 3.5 to 12 lb/ft^3 (56 to 192 kg/m^3) for refrigeration insulation, is made with a chemical that releases nitrogen gas to produce innumerable microscopic cells during the molding. The **Rubatex** of the Rubatex Div., Great American Industries, Inc., is this type of cellular rubber. It comes in sheets of any thickness for gaskets, seals, weather stripping, vibration insulation, and refrigerator insulation.

In the form of insulation board for heat and cold, nitrogen-filled rubber has a thermal conductivity of 0.237 Btu/(h)(°F) [1.343 $W/(cm^2)$(°C)] temperature difference per inch of thickness. The 5.5-lb/ft^3 (88-kg/m^3) board has a crushing strength of 33 lb/in^2 (0.23 MPa). Under the name of **Royal insulation board** it is marketed in thicknesses from ¾ to 1½ in (1.9 to 3.8 cm), but most of the so-called sponge rubbers are not made of natural rubber but are produced from synthetic rubbers or plastics, and may be called by a type classification, such as **urethane foam,** or marketed under trade names. Some of these foams may be made by special processes, and some of the materials are marketed in liquid form for use as foamed-in-place insulation. The urethane foams expanded with fluorocarbon gas are

95% gas in closed microscopic pores. They are rigid, have low weight, and are used in sheets with metal, paper, felt, or other facings for wallboard and roofing, and in thermal insulation. The contained gas has very low thermal conductivity. **Phenolic foam** is made by incorporating sodium bicarbonate and an acid catalyst into liquid phenol resin. The reaction liberates carbon dioxide gas, expanding the plastic.

Bubblfil, of Du Pont, is **cellular cellulose acetate** expanded with air-filled cells to densities from 4 to 9 lb/ft^3 (64 to 144 kg/m^3), for use as insulation and as a buoyancy material for floats. It is tough and resilient. **Strux,** of the Strux Corp., is **cellulose acetate foam,** made by extruding the plastic mixed with barium sulfate in an alcohol-acetone solvent. When the pressure is removed, it expands into a light, cellular structure. **Foamex,** of Firestone, is a foam rubber made from synthetic latex in several density grades. It is stronger than that made from natural rubber and is flexible at very low temperatures. **Polystyrene foam** is widely used for packaging and for building insulation. It is available as prefoamed board or sheet, or as beads that expand when heated. Densities range from 1 to almost 5 lb/ft^3 (16 to 80 kg/m^3).

Styrofoam, of Dow Chemical Co., is polystyrene expanded into a multicellular mass 42 times the original size. It has only one-sixth the weight of cork, but will withstand hot water or temperatures above 170°F (77°C), as it is thermoplastic. It is used for cold-storage insulation, and is resistant to mold. **Ensolite,** of Uniroyal, Inc., and **Vinylaire,** of the Dura Flex Co., are foamed vinyl plastisols. **Plastic-Cell** of the Sponge Rubber Products Co., is an expanded polyvinyl chloride molded to densities from 2 to 12 lb/ft^3 (32 to 192 kg/m^3). It is used for floats, buoys, and insulation. **Vynafoam,** of the Interchemical Corp., used for gaskets, seals, and refrigerator doors, is a vinyl plastisol which is sprayed on and cured by heat into a foamed texture with increase in volume of 400%. It is white in color, and has an insulating K value of 0.36, or about the same as felt. **Permafoam,** of the Hudson Foam Plastics Corp., has only half the weight of foamed rubber with greater strength and high resistance to oxidation. It is a foamed polyester resin, is odorless, flame-resistant, and resistant to oils and solvents. It is used for upholstery and insulation.

Silicone foam, of Dow Corning Corp., used for insulation, is silicone rubber foamed into a uniform unicellular structure of 8 to 24 lb/ft^3 (128 to 384 kg/m^3) density. It will withstand temperatures above 600°F (316°C). For structural sheets the rubber is foamed between two sheets of silicone glass laminate. **Spongex,** of the Sponge Rubber Products Co., is a foamed silicone rubber for cushioning and damping.

Vinyl foams are widely used. They are made from various types of vinyl resins with the general physical properties of the resin used. Open-cell vinyl foam contains interconnecting voids and is very flexible. It is made by mechanical foaming by absorption under pressure of an inert gas in a vinyl plastisol. It is used for furniture and transport seating. Closed-cell vinyl foam contains separate discrete voids. It is made by chemical foaming, using chemical blowing agents. It is used for impermeable insulation and marine floats.

An **epoxy foam** of the Shell Chemical Co. comes as a powder consisting of an Epon epoxy resin mixed with diaminodiphenyl sulfone. When the powder is placed in a mold or in a cavity, and heat is applied, it foams to fill the space and cures to a rigid foam. The foam has a density of 16 lb/ft^3 (256 kg/m^3), tensile strength of 360 lb/in^2 (2.5 MPa), compressive strength of 710 lb/in^2 (4.9 MPa), and it will withstand temperatures to 500°F (260°C). **Lockfoam H-602,** of the Nopco Chemical Co., is a pour-in-place urethane foam that expands with a fluorocarbon to a density of 2 lb/ft^3 (32 kg/m^3). The insulation K factor is 0.13 Btu (137 J), and it retains its properties at subzero temperatures when used for freezer insulation. **Expandoform,** of the Armstrong Cork Co., and **Polyfoam,** of the General Tire & Rubber Co., are polyether-based urethanes that expand to a density of 2 lb/ft^3 (32 kg/m^3) into stable, rigid foam of good strength for refrigerator insulation. **Carthane foam,** of the Carwin Co., used in aircraft and missile sandwich structures, is a polymethylene polyphenyl isocyanate that is infusible and withstands temperatures to 900°F (482°C) before beginning to carbonize. **Hetrofoam,** of the Hooker Chemical Corp., is a urethane-type foam with a high chlorine content that makes it fire-resistant. The thermal conductivity K factor is 0.10, and it is used for refrigerator and building insulation. **Thurane foam board,** of the Dow Chemical Co., is a foamed urethane in the form of strong, rigid boards ¾ to 16 in (1.9 to 41 cm) thick, with densities from 1.80 to 2.30 lb/ft^3 (29 to 39 kg/m^3).

Glass foam is used as thermal insulation for buildings, industrial equipment, and piping. **Ceramic foams** of alumina, silica, and mullite are used principally for high temperature insulation. **Aluminum foam** is a **metal foam** that has found appreciable industrial use as a core material in sandwich composites. **Foamed zinc,** of Foamalum Corp., is a lightweight structural metal with equal strength in all directions, made by foaming with an inert gas into a closed cell structure. It is used particularly for shock and vibration insulation. **Foaming agents** for metals are essentially the same as those used for plastics. They are chemical additives that release a gas to expand the material by forming closed bubbles. Or they may be used to cause froth as in detergents or fire-fighting foams.

FOIL. Very thin sheet metal used chiefly for wrapping, laminating, packaging, insulation, and electrical applications. **Tinfoil** is higher in cost than some other foils, but is valued for wrapping food products because it is not poisonous; it has now been replaced by other foils such as aluminum.

Ordinary tinfoil is made in thicknesses from 0.00024 in (0.006 mm) to 0.00787 in (0.200 mm), the former having 16,037 in^2/lb (22.7 m^2/kg) and the latter 432 in^2/lb (0.612 m^2/kg). Tinfoil for radio condensers has 14,500 in^2/lb (20.6 m^2/kg). An English modified tinfoil, which has greater strength and is nontoxic in contact with foods, contains 8.5% zinc, 0.15 nickel, and the balance tin. It can be rolled to thinner sheets.

Lead foil, used for wrapping tobacco and other nonedible products, is rolled to the same thickness as tinfoil, but because of its higher specific gravity has less coverage. The thinnest has 10,358 in^2/lb (14.7m^2/kg), and the thickest 279 in^2/lb (0.4m^2/kg). Lead foil has a dull luster, but it may be modified with some tin and other elements to give it a brighter color. **Stainless steel foil** is produced in thicknesses from 0.002 to 0.015 in (0.005 to 0.038 cm) for laminating and for pressure-sensing bellows and diaphragms. Type 302 steel foil, for facing, comes in a thickness of 0.003 in (0.008 cm) highly polished, in rolls 24 in (61 cm) wide. **Steel foil** is carbon steel coated with tin and produced by the U.S. Steel Corp. in sheets as thin as 0.001 in (0.003 cm) with widths to 40 in (102 cm). It is easily formed and soldered and is used for strong packaging and laminating.

Aluminum foil has high luster, but is not as silvery as tinfoil. The thin foil usually has a bright side and a matte side because two sheets are rolled at one time. Aluminum foil also comes with a satin finish, or in colors or embossed designs. It is made regularly in 34 thicknesses, from 0.006 mm, having 43,300 in^2/lb (61m^2/kg), to 0.200 mm having 1,169 in^2/lb (1.7m^2/kg), but can be made as thin as 60,000 in^2/lb (85m^2/kg). The most-used thickness is 0.00035 in (0.0089 mm), with 29,300 in^2/lb (42m^2/kg). For electrical use the foil is 99.999% pure aluminum, but foil for rigid containers is usually **aluminum alloy 3003,** with 1 to 1.5% manganese, and most other foil is of **aluminum alloy 1145,** with 99.45% aluminum. The tear resistance of thin aluminum foil is low, and it is often laminated with paper for food packaging. **Trifoil,** of Tri-Point Industries, Inc., is aluminum foil coated on one side with Teflon and on the other with an adhesive. It is used as coatings for tables and conveyors or liners in chemical and food plants. Since polished aluminum reflects 96% of radiant heat waves, this foil is applied to building boards or used in crumpled form in walls for insulation. **Alfol** is a name applied to crumpled aluminum foil for this purpose by the British National Physical Laboratory. Aluminum foil cut in tiny strips has been used for scattering from aircraft to confuse radar detection. A bundle of 6,000 such strips weighs only 6 oz (0.2 kg) and scatters widely.

Aluminum yarn, for weaving ribbons, draperies, and dress goods, is made from aluminum foil, 0.001 to 0.003 in (0.003 to 0.008 cm) thick, by gang-slitting to widths from 0.0125 to 0.125 in (0.0317 to 0.3175 cm) and winding the thread on spools. **Gold foil** is called **gold leaf** and is not normally classed as foil. It is used for architectural coverings and for hot-embossed printing on leather. It is made by hammering in books, and can be made as thin as 0.0000033 in (0.0000083 cm), a gram of gold covering 5.184 in^2 (3.4m^2). Usually, gold leaf contains 2% silver, and copper for hardening.

Metal film, or **metal foil,** for overlays for plastics and for special surfacing on metals or composites, comes in thicknesses from flexible foils as thin as 0.002 in (0.005 cm) to more rigid sheets for blanking and forming casings for intricately shaped parts. **Hot stamping foils** are decorative foils on a disposable carrier film applied to parts by means of a heated die. The carrier is usually polyester or cellophane. There are many types of these foils, including **metallic pigment foils, printed foils,** and **vacuum metallized foils. Composite metal films,** marketed by Composite Sciences, Inc., come in almost any metal or alloy, such as a film of tungsten carbide in a matrix of nickel alloy for wear-resistant overlays. They are made by rolling and sintering 325-mesh powder. The foils may be cemented to the substrate with an epoxy resin, but for operating temperatures above 400°F (204°C), brazing or welding is needed. **Metal overlays** give a smoother and more uniform surface than is usually obtained by flame-spray or chemical deposition.

FOODSTUFFS. A great group of materials employed for human consumption, while those employed for feeding animals are called **feeds.** Foodstuffs are derived mainly from vegetable and animal life, but some, like common salt, are produced from mineral sources, and some may be entirely synthesized. Foodstuffs are intended primarily for the maintenance and growth of the body, and technically could embrace drugs which are taken primarily for their physiological effects, and cosmetics that feed the skin and hair. Tobacco, used also for its physiological effect, is classed with foodstuffs in government statistics.

Proteins, carbohydrates, and fats are called the essential foodstuffs, and about 1.5 lb (0.7 kg) total of these are considered as needed daily for the human system. But these alone are inadequate for metabolism. Almost every element is required directly, or indirectly as catalysts, in the building up and maintenance of the innumerable highly complex chemical compounds in the human body, and the form in which they are taken into the body is of great importance. Iodine, for example, is an essential element in the thyroid gland, with minute amounts also required in every cell of the body, but if taken directly into the system in even small amounts is an intense poison.

Some complex chemical compounds, required for proper health, cannot be synthesized in the human body and must be taken in through the eating of foodstuffs containing them. First-class proteins are essential for full health, but they are not synthesized in the human body and are not available in vegetable products, so that they must be obtained from the eating of meats from animals that synthesize them. About one-third of the protein required daily should come from animal sources. Fish and shellfish can supply the essential animo acids, but abnormally large quantities would be necessary.

Concentrated foods are now used in vast quantities for military supplies and for prepared food mixes. They may be spray-dried powders, comminuted dehydrated vegetables or fruits, or freeze-dried cooked foods. For example, dehydrated onions can be handled in automatic metering equipment for adding to soup mixes or meat dishes, and 1 lb (0.5 kg) of the dried onions equals 8 lb (3.6 kg) of fresh onions. For mixes, cooking times are balanced by partial pressure cooking or puffing. Celery is cross-cut in fine flakes and dehydrated. Rehydration gives 25 to 1 volume. For use uncooked it requires only immersement in cold water. **Concentrated wine,** used in flavoring foods, is made by evaporating sherry or other wines. One ounce (0.03 kg) is equal to 8 oz (0.23 kg) of wine, but it contains no alcohol. **Banana crystals,** for flavoring bakery products and milk-based beverages, is a light-tan crystalline powder made by vacuum dehydration of bananas, the fruit of the tall treelike herbaceous plant, *Musa paradisiaca,* native to southern Asia but now grown in all tropical countries. There are more than 300 varieties of the plant. The powder contains 50% banana solids and 50% corn-syrup solids, while the natural banana contains 25% solids and 75% water. The powder crystals are easily soluble in water, reconstituting to 80% banana solids and 20% corn-syrup solids, giving a true banana flavor and the natural food value. For food manufacturers bananas are also marketed in canned form, either sliced or mashed.

For efficient high-production processing of bakery, confectionery, and other food products, most types of fruits and flavors are now available in the forms of dry powders, nuggets, pastes, solutions, or cooked or semicooked sections. Advantages are economy in shopping and storing, ease of handling in automatic equipment, and uniformity of quality. For example, **fruit nuggets** consisting of foamed particles are easily metered in automatic machines. Nuggets of berries contain no seeds. For dry-mix packaging the flavors may be in capsule.

Calorific value is only an imperfect measurement of food value. The trace percentages of mineral compounds and such chemical compounds as vitamins are of vital importance to proper metabolism and health. **Therapeutic foods** are mixtures, usually in powder form to be mixed in water, to give a balance of proteins, fats, and carbohydrates for food satisfaction without overeating. They may be designed for reducing excessive weight, for adding weight, or as a soft diet for invalids or babies, and are usually intended to contain sufficient calorific value for a rounded diet. But it is extremely difficult to obtain all the essential ingredients in a synthetic food. Iron is a common essential, readily absorbed from many natural foods, but as an additive for deficient systems it must be in chemical compounds that yield ferrous iron to the system. Trace quantities of zinc in enzymatic proteins are necessary to catalyze the reaction of carbonic anhydrase to sustain life, and are also needed in the liver and in the eyes. Other metals, such as cobalt and rubidium, are also necessary, and the highly complex processes are as yet only imperfectly understood.

FORMALDEHYDE. Also called **methylene oxide.** A colorless, poisonous gas of the composition HCHO, boiling at −21°C. It is very soluble in water, and is marketed as a 40% solution by volume, 37% by weight, under the name of **formalin.** The commercial formalin is a clear colorless liquid with a specific gravity of 1.075 to 1.081. When shipped by tank cars it contains 11 to 12% methanol, or 6 to 7% when shipped in drums, as a stabilizer to prevent precipitation of polymerized formaldehyde. The material is obtained by oxidation from methyl alcohol. It is used in making plastics, as a reducing agent, as a disinfectant, and in the production of other chemicals. **Trioxane** is polymerized formaldehyde, or a ring compound of anhydrous formaldehyde, $(HCHO)_3$. It is marketed as colorless crystals of a pleasant ether-alcohol odor, with a specific gravity of 1.17, melting point of 62°C, and boiling point of 115°C. It is used as a source of dry formaldehyde gas, as a tanning agent, and as a solvent. It ignites at 113°C and burns with a hot, odorless flame, and is used in tablet form to replace solidified alcohol for heating.

Glyoxal, of Union Carbide, is **dialdehyde,** or **ethanedial,** CHO · CHO, marketed as a water solution as a substitute for formaldehyde for resin manufacture and as a hardening and preserving agent, and for treating rayon fabrics. It is faster in action and has less volatility and odor than formaldehyde. It is a light-yellow liquid boiling at 50°C. **Paraformaldehyde,** $(CH_2O)_3$, also called **paraform,** is a white amorphous powder used instead of formaldehyde where a water solution is not desirable. It is used as a catalyst and hardener for the resorcinol and some other synthetic resins, and as an antiseptic. **Formamide,** $H(CO)NH_2$, is a clear, viscous water-soluble liquid with a faint odor of ammonia, boiling at 210°C. It is used as a solvent for metal chlorides and inorganic salts, and for lignin, glucose, or cellulose, and as a softener for glues. **Formol** is a trade name for a solution of formaldehyde in methanol and water used as an antiseptic. **Formcel,** of the Celanese Corp., is a solution of formaldehyde in either methanol or butanol, to replace formalin where a water-free solution is desired. **Glutar aldehyde,** $O{:}HC(CH_2)_3 \cdot CH{:}O$, made from acrolein, has a reaction similar to formaldehyde, and is used as a cross-linking agent in plastics, and for insolubilizing starches, casein, and gelatin.

Hexamine is a white, crystalline powder used chiefly for the manufacture of synthetic resins in place of formalin and its sodium hydroxide catalyst. It is formed by the action of formaldehyde and ammonia. It is hexamethylene tetramine, $(CH_2)_6N_4$, melting at 280°C, and is very stable when dry. It is readily soluble in water and in alcohol. It is also known as **formin, ammonio formaldehyde, urotropin, crystogen, aminoform,** and **cystamine.** In pharmacy it is called **methenamine.** In the presence of an acid it yields formaldehyde and is used in medicine as an internal antiseptic. It is also used as an accelerator for rubber.

FORMIC ACID. Also called **methanoic acid** and **hydrogen carboxylic acid.** Formic acid is the simplest of the organic acids, with the composition HCOOH, and was originally distilled from red ants, receiving its name from the Latin name for ants. It is made synthetically, or is obtained from the black liquor of sulfite paper mills where it occurs as a sodium salt. It is a pungent colorless liquid of specific gravity 1.22, boiling point 101°C, and freezing point 8.4°C, soluble in water and in alcohols. It is an easily oxidized reducing agent, and small amounts will blister the skin and give the stinging sensation of ant or bee bites or nettle stings, all of which are caused by formic acid. The acid has greater reducing action than acetic acid, and is thus used in textile furnishing, especially in the chrome dyeing of wool. It is also used in leather processing as a dye-bath exhausting agent. Other uses are as a food preservative, in electroplating, as a germicide, as a fermentation assistant in brewing, and as a coagulant for rubber. **Methyl formate,** $HCOOCH_3$, is a white volatile liquid boiling at 31.75°C, with a pleasant ester odor, soluble in water. It hydrolyzes to form methanol and formic acid, and is used as an intermediate, but it is also a good solvent for cellulose esters and for acrylic resins. **Protan,** of Hercules Inc., is a **sodium formate.**

FUEL BRIQUETTES. Also termed **coal briquettes.** Various-shaped briquettes made by compressing powdered coal, usually with an asphalt or starch binder, but sometimes as **smokeless fuel** without a binder. They are sometimes also made waterproof by coating with pitch or coal tar. They have the great advantage over raw coal that they do not take up large amounts of water as coal does and thus have uniformity of firing. Fuel briquettes are made from anthracite screenings usually mixed with bituminous screenings as the bituminous coals require no binders. The usual forms of the briquettes are pillow-shaped, cubic, cylindrical, ovoid, and rectangular, and the usual size is not over 5 oz (0.14 kg). The term **packaged fuel** is used for cube-shaped briquettes wrapped in paper packages, used for hand firing in domestic furnaces. The **coal briquettes** of the Blaw-Knox Co. are made with anthracite dust and a small amount of Pocahontas-type bituminous coal with an asphalt binder. The cakes are 3 by 3 by 3 in (7.6 by 7.6 by 7.6 cm) wrapped in kraft paper. **Coal logs,** of the Coal Logs Co., Inc., are briquetted smokeless fuel which consists of high-temperature coke made by carbonizing Utah low-grade coal, yielding 20 to 40 gal (76 to 151 L) of tar as a by-product per ton of coal. **Charcoal briquettes** for home fuels are charcoal powders pressed with a starch binder.

FUEL OIL. Distillates of petroleum or shale oil used in diesel engines and in oil-burning furnaces. True fuel oils are the heavier hydrocarbons in kerosene, but the light or distillate oils are used largely for heating, and the heavy or residual oils for industrial fuels. In some cases only the light oils, naphtha and gasoline, are distilled from petroleum, and the residue is used for fuel oil, but this is wasteful of the lighter oils. Commercial grades of **furnace oil** for household use and **diesel oil** for trucks may be low grades of kerosene. Fuel oil of low specific gravity requires preheating to obtain complete automization. At 10°Bé the minimum temperature to atomize fuel oil is 300°F (149°C), but at 40°Bé a temperature of only 40°F (4°C) is required. Fuel oils used in oil burners are 28 to 32°Bé, and have a Btu content of 142,000/gal (537,470/L), completely atomizing at 90°F (32°C). **Gas oil,** which receives its name from its use to enrich fuel gas and increase the luminance of the flame, is also used as a fuel in oil engines. It is the fraction of petroleum distilling off after kerosene, or above about 300°C. It is brownish in color and has a specific gravity of about 0.850. Ignition temperatures of crude oils vary from 715 to 800°F (379 to 427°C) in air at atmospheric pressure. **Bunker C oil,** for diesel engines, is a viscous black oil of specific gravity 1.052 to 0.9659 with a flash point above 15°F (−9°C). The National Bureau of Standards lists six grades of fuel oils with flash points from 100 to 200°F (38 to 93°C). For general comparison of oil and coal, with comparable fire and boiler equipment, 4.5 bbl of Bunker C oil equal 1 net ton of low-ash, highly volatile West Virginia coal.

FUELS. The term normally covers a wide range, since innumerable organic materials can be used as fuel. Coal, oil, or natural gas, or products derived from them, are the basic industrial fuels, but other materials are basic in special situations, such as sawdust in lumbering areas and bagasse in sugarcane areas. **Fuel conversion factors** for industrial fuels are based on the relation to the fuel value of a metric ton of coal having a colorific value of 12,000 Btu/lb (12.6×10^6 J). Thus, a ton of lignite is equal to 0.3 ton (0.3 metric tons) of coal; a ton of crude oil is equal to 1.4 tons (1.3 metric tons) of coal; 1,100 ft^3 (31 m^3) of natural gas is equal to 1.33 tons of coal; and 1,000 kWh of hydroelectricity is equal to 0.4 ton (0.4 metric ton) of coal.

But modern technical reference to fuels generally applies to **high-energy fuels** for jet engines, rockets, and special-use propellants, and the comparisons of these fuels are in terms of **specific impulse,** which is the thrust in pounds per pound of propellant per second. The molecular weight of the products produced by the reaction of a fuel must be extremely low to give high specific impulse, that is, above 400. Hydrogen gives a high specific-impulse rating, but it has very low density in the liquid state and other unfavorable properties, so that it is usually employed in compounds. The initial impulse of a rocket is proportional to the square root of the combustion temperature of the fuel. **Hydrogen fuels** reacted with pure oxygen produce temperatures above 5000°F (2760°C), and some fuels may react as high as 9000°F (4982°C). Aluminum powder or lithium added to hydrogen increases the efficiency. **Boron fuels** in general release 50% more thermal energy than petroleum hydrocarbons. The first Saturn space rocket had kerosene-liquid oxygen in the first stage and liquid hydrogen-oxygen in the following three stages. **Solid rocket fuels,** designed for easier handling, have a binder of polyurethane or other plastic. **Fuel oxidizers,** for supplying oxygen for combustion, may be ammonium, lithium, or potassium perchlorates. In solid fuels, oxidizers make up as much as 80% of the total.

A **monopropellant** high-energy fuel is a chemical compound which, when ignited under pressure, undergoes an exothermic reaction to yield high-temperature gases. Examples are nitromethane, methyl acetylene, ethylene oxide, and hydrogen peroxide. Gasoline oxidized by hydrogen peroxide gives a specific impulse of 248, while pentaboranes under pressure and oxidized with hydrogen peroxide give a specific impulse of 363. **ASTM fuel A,** for jet engines, is isooctane, and **ASTM fuel B** is isooctane and toluene. **Turbine jet fuel,** or **JP fuel,** for commercial aircraft is a special grade of kerosene. American specifications are for a flash point of 125°F (51°C), but for military planes the flash point may be 110°F (43°C) using additions of methane or naphthene. The specific heat of standard commercial turbine jet fuel is 0.47 Btu/(lb)(°F) (2.5 kcal/kg) compared with 3.39 Btu (3,578 J) for hydrogen gas fuel. The **naphthalenes,** such as **decahydronaphthalene,** have high thermal stability, and they have a high density which gives high thermal energy per unit volume. **Naphthene,** C_nH_{2n}, is a general term for nonaromatic cyclic hydrocarbons in petroleum.

Hydrazine with liquid oxygen has a specific impulse of 282, and with liquid fluorine as an additive the specific impulse is 316. The rocket fuel **hydyne** is a 60–40 mixture of unsymmetrical dimethylhydrazine and diethylene triamine. Other liquid fuels may be this dimethylhydrazine with nitric acid, or with nitrogen tetroxide, or with liquid oxygen. **Alkyl boranes** are used in rockets, but they leave a deposit and they exhaust boric acid in a dense cloud. **Diborane,** B_2H_6, is the simplest compound of boron and hydrogen. It is a gas which burns with high flame speed and high heat. It is used to produce the **boron hydrides** such as **decarborane,**

$B_{10}H_{14}$, or $HH(H \cdot BB \cdot H)_5HH$, employed for boron fuels and for making boron plastics.

Sodium boronhydride, a white crystalline solid of composition $NaBH_4$, made by reacting sodium hydride with methyl borate, is also used to produce the boranes for fuels. **Triethyl borane,** $(C_2H_5)B$, used for jet fuels and as a flame-speed accelerator, is a colorless liquid. It is spontaneously flammable, its vapors igniting with oxygen. Any element or chemical which causes spontaneous ignition of a rocket fuel is called a **hypergolic material.**

Chemical radicals are potential high-energy fuels, as the recombining of them produces high specific impulses. But chemical radicals normally exist only momentarily, and are thus not stable materials and, in general, are not commercial fuels. **Ion propellants** operate on the principle that like charges repel each other, and the fuel is an ion-plasma jet actually formed outside of the engine. The original fuel is a metal such as cesium from which electrons can be stripped by passing the vapor through a hot screen, leaving positive cesium ions, which are formed into a beam and exhausted from the thrust jet to be electronically neutralized in the ionized plasma.

FULLER'S EARTH. A soft, opaque clay with a greasy feel used as a filtering medium in clarifying and bleaching fats, greases, and mineral and vegetable oils. It absorbs the basic colors in the organic compounds. It is also used as a pigment extender and a substitute for talcum powder. It was formerly much used in the textile industry as a fuller for woolen fabrics, cleansing them by absorbing oil and grease. It is a hydrated compound of silica and alumina. It may contain 75% silica, 10 to 19 alumina, 1 to 4 lime, 2 to 4 magnesia, and sometimes ferric oxide. The usual color is greenish white to greenish brown. The rose-colored fuller's earth from Florida is a variation of **montmorillonite,** $(MgCa)O \cdot Al_2O_3 \cdot 4SiO_2$. The Florida fuller's earth marketed by the Floridian Co. under the name of **Floridin** is a grayish-white material graded by sizes from B, which is 16 to 30 mesh, to XXX, which will pass 90% through a 200-mesh screen. **Florex** is this material processed by extrusion to increase the absorption capacity. A typical analysis gives 58.1% silica, 15.43 alumina, 4.95 iron oxide, 2.44 magnesia, with small amounts of CaO, Na_3O, and K_2O. The specific gravity is 2.3. **Florigel** is the hydrated material which forms viscous suspensions in water, and is used to replace bentonite as a filler for soaps, and for clarifying liquids. **Diluex** is finely powdered Florida fuller's earth used as a diluent or carrier for insecticides. **Activated clay,** or **bleaching clay,** for bleaching oils, may be acid-treated fuller's earth, or it may be bauxite or kaolin. **Florite,** of the Floridian Co., is activated bauxite. It is a red, granular, porous material of 20 to 60 mesh.

FULMINATES. Explosives used in percussion caps and detonators because of their sensitivity. They may be called **cap powder** in cartridge caps and detonators when used for detonating or exploding artillery shells. **Mercury fulminate,** $Hg(CNO)_2$, a gray or brown sandy powder, is the basis for many detonating compositions. It is made by the action of nitric acid on mercury and alcohol, and is 10 times more sensitive than picric acid. It may be mixed with potassium chlorite and antimony sulfide for percussion caps. The fulminates may be neutralized with a sodium thiosulfate. The **azides** are a group of explosives containing no oxygen. They are compounds of hydrogen or a metal and a monovalent N_3 radical. **Hydrogen azide,** HN_3, or **axoic acid,** and its sodium salt are soluble in water. **Lead azide,** $Pb(N_3)_2$, is used as a substitute for fulminate detonators. It is much more sensitive than mercury fulminate and in large crystals is subject to spontaneous explosion, but it is precipitated to suppress crystal formation and to form a free-flowing powder less sensitive to handling. **Lead bromate,** $Pb(BrO_3)_2 \cdot H_2O$, is in colorless crystals which will detonate if mixed with lead acetate. For primer caps, a substitute for mercury fulminate is a mixture of lead styphnate, lead triagoacetate, lead nitrate, and lead sulfocyanate. **Lead styphnate,** used as a detonator, is made by the sulfonation and nitration of resorcinol to form **styphnic acid,** or **trinitroresorcinol,** $(NO_2)_3C_6H(OH)_2$. This powder is treated with magnesia and with a lead nitrate solution to form the lead styphnate powder. **Azoimide,** or **iminazoic acid,** $HN \cdot N_2$, is an extremely explosive colorless gas liquefying at 37°C which can be used in the form of its salts. **Silver azoimide,** $AgN \cdot N_2$, is highly explosive, and **barium azoimide,** BaN_6, explodes with a green flash. **Cyanuric triazide** is a powerful explosive made by reacting cyanuric chloride with sodium azide. **Lead nitrate,** a white crystalline water-soluble powder of the composition $Pb(NO_3)_2$, is used in match heads and explosive compositions. It is also employed as a mordant in dyeing and printing, and in paints.

FURFURAL. Also known as **furfuraldehyde, furol,** and **pyromucic aldehyde.** A yellowish liquid with an aromatic odor, having a composition $C_4H_3O \cdot CHO$, specific gravity 1.161, boiling point 161.7°C, and flash point 132°F (56°C). It is soluble in water and in alcohol but not in petroleum hydrocarbons. On exposure it darkens and gradually decomposes. Furfural occurs in different forms in various plant life and is obtained from cornstalks, corncobs, straw, oat husks, peanut shells, bagasse, and rice. Furfural is used for making synthetic plastics, as a plasticizer in other synthetic resins, as a preservative, in weed killers, and as a selective solvent especially for removing aromatic and sulfur compounds from lubricating oils. It is also used for the making of butadiene, adiponitrile, and other chemicals.

Various derivatives of furfural are also used, and these, known collectively as **furans,** are now made synthetically from formaldehyde and acetylene which react to form butyl nedole. This is hydrogenated to butanediol, then dehydrated to tetrahydrofuran. Furan, or **tetrol,** C_4H_4O, used for plastics manufacture, is a colorless liquid boiling at 32°C. **Methyl furan,** or **sylvan,** $C_4H_3O \cdot CH_3$, is a colorless liquid boiling at 62°C. **Tetrahydrofuran,** $(CH_2)_4O$, is a water-white liquid boiling at 66°C, having strong solvent powers on resins. It reacts with carbon monoxide to form adipic acid, and is also an intermediate for the production of other chemicals. **Furfuryl alcohol,** $C_4H_3O \cdot CH_2OH$, a yellow liquid with a brinelike odor, boiling at 176°C, and flash point at 167°F (75°C), is used as a solvent for nitrocellulose and for dyes, and for producing synthetic resins. It is made by hydrogenation of furfural. **Furfuryl alcohol resins,** made by reacting with an acid catalyst, are liquid materials that are low in cost and highly chemical-resistant. They are much used for protective coatings, tank linings, and chemical-resistant cements. They are dark in color. The **Alkor cement** of the Atlas Mineral Products Co. is a furfuryl alcohol solution which produces coatings resistant to chemicals and to temperatures to 380°F (193°C). **Furacin,** of the Norwich Pharmaceuticals, is a nitro-furfural semicarbazone, a yellow crystalline powder made by nitrating furfural and reacting it with hydrazine hydrate. It is used as a bacterial treatment for wounds and burns. **Furoic acid,** or **pyromucic acid,** $C_4H_3O \cdot COOH$, is a colorless crystalline powder soluble in water. It is **furan carboxylic acid** used for making pharmaceuticals, flavors, and resins.

Tetrahydrofurfuryl alcohol, $(CH_2)_3OCH \cdot CH_2OH$, is the usual starting point for making furfuryl esters, ethers, and straight-chain compounds, and it is also a high-boiling solvent for gums, resins, and dyes. It is a liquid of specific gravity 1.064, boiling at 177.5°C, and is soluble in water. **Furfural acetone,** $C_4H_3O \cdot CH{:}CHCOCH_3$, is a reddish-brown liquid boiling at 229°C. Furfural, when treated with aniline at 150°C, forms an insoluble black **furfural-aniline resin** used in resistant protective coatings and enamels. **Furfural-acetone resin,** or **furfuracetone,** is a transparent elastic resin made by the reaction of furfural and acetone in the presence of an alkali. Furfural also polymerizes with phenol to form **furfuralphenol resins** that are self-curing. They have high heat, chemical and electrical resistance, and have excellent adhesion to metals and other materials, making them adaptable for chemical and electrical coatings. The resins have high gloss, but a very dark color. The **Tygon resins** of the U.S. Stoneware Co. are furfural resins used for brush application as protective coatings for such purposes as chemical tank linings. They cure by self-polymerization, will withstand temperatures to 350°F (177°C), and are resistant to acids, alkalies, alcohols, and hydrocarbons. **Furafil,** of the Quaker Oats Co., is a by-product material containing modified cellulose, lignin, and resins, used as an extender for phenolic plywood glues, as an additive for phenolic molding resins, and as a binder for foundry sand molds. Under the name of **Fur-Ag,** it is used as a conditioner and anticaking agent in fertilizer mixtures. The material is a dark-brown, absorbent powder.

Furane plastics have high adhesion and chemical resistance, but they do not have high dielectric strength, and are black or dark in color. They are used for pipe, fittings, and chemical-equipment parts, and for adhesives and coatings. **Eonite** is produced from **Durez 16470,** a furfural alcohol resin of the Hooker Chemical Co. The pipe will resist hot acids and alkalies to 300°F (149°C), is strong, and does not sag in long lengths. **Furfural-ketone resin** is used to blend with epoxy laminating resins to reduce cost and improve the properties.

FUSIBLE ALLOYS. Alloys having melting points below the boiling point of water, 100°C. They are used as binding plugs in automatic sprinkler systems, for low-temperature boiler plugs, for soldering pewter and other soft metals, for tube bending, and for casting patterns and many ornamental articles and toys. They are also used for holding optical lenses and other parts for grinding and polishing. They consist generally of mixtures of lead, tin, cadmium, and bismuth. The general rule is that an alloy of two metals has a melting point lower than that of either metal alone. By adding still other low-fusing metals to the alloy a metal can be obtained with almost any desired low melting point. The original **Newton's alloy** contained 50% bismuth, 31.25 lead, and 18.75 tin. **Newton's metal,** used as a solder for pewter, contained 50% bismuth, 25 cadmium, and 25 tin. It melts at 203°F (95°C), and will dissolve in boiling water. **Lipowitz alloy,** another early metal, contained 3 parts cadmium, 4 tin, 15 bismuth, and 8 lead. It melts at 158°F (70°C), is very ductile, and takes a fine polish. It was employed for casting fine ornaments, but now has many industrial uses. A small amount of indium increases the brilliance and lowers the melting point 1.45°C for each 1% of indium up to a maximum of 18%. **Wood's alloy,** or **Wood's fusible metal,** was patented in 1860 and was the first metal used for automatic sprinkler plugs. It contained 7 to 8 parts bismuth, 4 lead, 2 tin, and 1 to 2 cadmium. It melts at 160°F (71°C), and this point was adopted as the operating temperature of sprinkler plugs in the United States; in England it is 155°F (68°C). The alloy designated as **Wood's metal** by the Cerro Corp. contains 50% bismuth, 25 lead, 12.5 tin, and 12.5 cadmium. It melts at 158°F (70°C). An early alloy for tube bending contained 50% bismuth, 16.7 lead, 13.3 tin, and 10 cadmium. It melts at 158°F (70°C), and can be easiy removed from the tube after bending by dipping in boiling water or by applying steam.

Cerrobend, or **Bendalloy,** of the Cerro Corp., is a fusible alloy for tube bending which melts at 160°F (71°C). **Cerrocast** is a **bismuth-tin alloy** with pouring range of 280 to 338°F (138 to 170°C), and shrinkage of only 0.0001 in/in (0.0025 cm/cm), used for making pattern molds. **Cerro-safe,** or **Safalloy,** is a fusible metal used for toy-casting sets as the molten metal will not burn wood or cause fires. Alloys with very low melting points are sometimes used for this reason for pattern and toy casting. A fusible alloy with a melting point at 60°C contains 26.5% lead, 13.5 tin, 50 bismuth, and 10 cadmium. These alloys expand on cooling and make accurate impressions of the molds. **Boiler-plug alloys** have been made under a wide variety of trade names with melting points usually ranging from 212 to 342°F (100 to 172°C). **D'Arcet's alloy,** melting at 200°F (93°C), contained 50% bismuth, 25 tin, and 25 lead. **Lichtenberg's alloy,** melting at 198°F (92°C), contained 50% bismuth, 30 lead, and 20 tin. **Guthrie's alloy** has 47.4% bismuth, 19.4 lead, 20 tin, and 13.2 cadmium. **Rose's alloy** contained 35% lead, 35 bismuth, and 30 tin. **Homberg's alloy,** melting at 251°F (121°C), contained 3 parts lead, 3 tin, and 3 bismuth. **Malotte's metal,** melting at 203°F (95°C), had 46% bismuth, 20 lead, and 34 tin. The variation of these different alloys was largely because of the relative cost of the different alloying metals at various times. Fusible metals have also been used in strip forms to test the temperature of steels for heat-treating. The **Temperite alloys** of the Cornish Wire Co. were for this purpose with melting points between 300 and 625°F (149 and 329°C) in steps of 25°F (−4°C). The **Tempil pellets** of the Tempil Corp. are alloy pellets made with melting points in steps of 12.5, 50, and 100°F (−10.8, 10, and 37.7°C) for measuring temperatures from 113 to 2500°F (45 to 1371°C). The **Semalloy metals** of Semi-Alloys cover a wide range of fusible alloys with various melting points. **Semalloy 1010** with a melting point at 117°F (47°C) can be used where the melting point must be below that of thermoplastics. It contains 45% bismuth, 23 lead, 19 indium, 8 tin, and 5 cadmium. **Semalloy 1280,** for uses where the desired melting is near the boiling point of water, melts at 204°F (96°C). It contains 52% bismuth, 32 lead, and 16 tin.

FUSTIC. Known also as **Cuba wood.** The wood of the tree *Chlorophora tinctoria* of tropical America, used for cabinetmaking and as a dyewood. It has a yellow color, is very hard, and has a fine, open grain. The weight is about 41 lb/ft^3 (657 kg/m^3). The liquid extract of the wood produces the yellow dyestuff **morin,** $C_{15}H_1i_1O_7$, and the red dye **morindone,** $C_{15}H_{11}O_5$. **Fustic extracts** are mordant dyes and give colors from yellow to olive with various mordants. Morin is used also as an indicator to detect aluminum, with which it develops a green fluorescence. **Young fustic,** or **Hungarian yellow wood,** is a yellow dyewood from the *Rhus cotinus.* **Osage orange,** called **bois d'arc,** is the bright orange wood of the bush *Maclura pomifera* growing in the swamplands of Texas and Oklahoma. It has a high tannin content and is used in the textile and leather industries for orange-yellow and gold colors and to blend with greens. As a tanning agent it may be blended with quebracho and chestnut extracts.

GALLIUM. An elementary metal, symbol Ga, originally called **austrium.** It is silvery white, resembling mercury in appearance but having chemical properties more nearly like aluminum. It melts at 85.6°F (30°C) and boils at 2912°F (1600°C), and this wide liquid range makes it useful for high-temperature thermometers. Like bismuth, the metal expands on freezing, the expansion amounting to about 3.8%. Pure gallium is resistant to mineral acids, and dissolves with difficulty in caustic alkali. It forms many salts at different valences. The weight is only about half that of mercury, having a specific gravity of 5.9. Commercial gallium has a purity of 99.9%. In the molten state it attacks other metals, and small amounts have been used in tin-lead solders to aid wetting and decrease oxidation, but it is expensive for this purpose. **Gallium-tin alloy** has been used when a low-melting metal was needed. It is also used as an electron carrier in silicon semiconductors, and crystals of **gallium arsenide,** GaAs, are used as semiconductors. Gallium arsenide can be used in rectifiers to operate to 600°F (316°C). The material has high electron mobility. This material in single-crystal bars is produced by Monsanto for lasers and modulators. **Gallium selenide,** GaSe, **gallium triiodide,** GaI_3, and other compounds are also used in electronic applications.

Gallium exists in nature in about the same amount as lead, but it is widely dissipated and not found concentrated in any ore. It is found in small amounts associated with zinc ores and is recovered from smelter flue dust. In Germany it is produced as a by-product of copper smelting. It is also a minor constituent in the mineral sphalerite to the extent of 0.01 to 0.1%, and it also occurs in almost all aluminum ores in the ratio of 50 to 100 grams of gallium per ton of aluminum. In the United States it is a by-product of aluminum production. About 1 oz (0.03 kg) of gallium is obtained commercially per ton (0.9 metric ton) of bauxite.

GALLS. Tanning materials obtained from the **nutgalls,** or **gall nuts,** from the oaks of Europe and the Near East and from the sumac of China and Japan. Nutgalls are plant excrescences caused by the punctures of insects. They contain 50 to 70% tannins, and are the richest in tannin of all the leather-tanning materials. The tannin is also valued for ink-making and in medicine for treating burns. **Green galls,** or **Aleppo galls,** are obtained from the twigs of the **Aleppo oak,** *Quercus infectoria,* a shrub of the Near East. Those of blue color are the best quality, with green second, and the white of inferior grade. **Chinese galls,** from species of *Rhus,* are in the form of irregular roundish nuts which enclose the insect. They show no vegetable structure but have a dense resinous fracture and are very high in tannin.

The product known as **gall** in the pharmaceutical industry is an entirely unrelated material. It is **beef gall,** or **ox bile,** a bitter fluid from the livers of cattle. It is used for steroid production, and also in the textile industry for fixing dyes and in soaps for washing dyed fabrics. **Steroids,** or **hormones,** made from ox bile, have a great number of possible combinations that have an influence on the behavior of the human cellular system. They are based on a 4-ringed, 17-carbon cyclopentamorphen anthrene nucleus, and arranging the side group in different ways gives compounds with distinct physiological properties. **Cortisone,** made by moving the oxygen atom of the steroid nucleus from the 12th to the 11th position, is one of the many steroids. Steroids are not synthesized from the more plentiful cholesterol, from the **stigmastrol** of vegetable oils, or from the **sapogenins** of plants.

GALVANIZED STEEL AND IRON. Iron or steel sheets coated with zinc and used for roofing, sheathing, culverts, and for making articles that require rust-resistance without painting. The most common sheets are of low-carbon steel, but the copper-bearing iron and steels are also used to give added corrosion resistance. The common dipped sheets are annealed, pickled, cold-rolled to increase the polish, and then dipped in molten zinc. They have a characteristic spangle which in thick coats is likely to peel and flake off under distortion. Sheets are graded by perfection of coating, gage, and size, as primes, seconds, and gray-coated. Primes are the perfect sheets. **Galvanized sheets** are either plain or corrugated. They come usually in lengths of 6, 7, and 8 ft (1.8, 2.1, and 2.4 m) and widths of 24, 26, 28, 32, 34, and 36 in (61, 66, 71, 81, 86, and 91 cm). The thickness is by gage sizes, usually from No. 14 to No. 30. The No. 14 sheet carries a 2-oz (0.06-kg) zinc coating.

In hot-dip galvanizing the molten zinc reacts with the steel to form a zinc-iron alloy which is brittle. Addition of a small amount of aluminum to the zinc retards formation of the brittle alloy, and control of temperature also reduces the formation of the alloy and produces a more ductile coating. Small amounts of magnesium added to the zinc greatly improve the corrosion resistance in marine atmospheres. With an alloy 97% zinc and 3 magnesium, about 0.05% by weight is added to the coating. **Zinc alloy steel,** of the Inland Steel Co., is **zinc-coated steel** sheet with the coating fused in and alloyed with the steel so that there is no spangle. The sheet can be spun or drawn without cracking the coating. **Gal-Van-Alloy** is the trade name of this company for the galvanized sheet. **Bethcon Jetcoat,** of the Bethlehem Steel Co., is high-temperature hot-dip galvanized steel sheet in which the zinc is chemically combined to leave no free zinc. It has a uniform gray color without spangles. It is easily welded, and is more corrosion-resistant and takes paint better than ordinary galvanized steel.

Electrogalvanized steel, a zinc-coated steel made by continuous electroplating, has a uniform homogenous deposit of pure zinc tightly bonded to the steel, and has a smooth, shiny surface that does not peel or flake. **Weirzin,** of the Weirton Steel Co., is zinc-electroplated sheet steel. **Zincgrip,** of the Armco Steel Corp., is zinc-coated sheet steel for deep drawing. The coating does not flake or peel in the drawing dies. **Galv-Weld,** of Morton L. Clark Enterprises, is a low-melting zinc alloy in stick form for regalvanizing welded joints. **Coronized steel** refers to steel parts with a zinc-nickel coating. The steel parts are first nickel-plated, and then coated with zinc and heat-treated at about 700°F (371°C) to give a smooth, semilustrous, nickel-zinc surface.

Galvanized wire is steel wire coated with zinc for rust resistance either by dipping in molten zinc or by continuous electroplating, and is marketed plain, twisted, or barbed. **Barbed wire** has double-pointed wire barbs held in the twisted wire. **Bethanized steel,** of the Bethlehem Steel Co., is wire electrogalvanized and then passed through a die which polishes and hardens the coating.

GAMBIER. Also spelled **gambir.** A tanning and dyeing material extracted from the leaves and twigs of the shrubs, *Uncaria gambier, U. dacyoneuro,* and other species of India, Malaysia, and the East Indies. It is similar to catechu, and is also called **white cutch, yellow cutch, cube cutch,** and **tara japonica.** The cubes of extract are brittle, have a dull-gray color, and a bitter astringent taste. The material contains **catechin,** $C_{15}H_{14}O_6$, a yellow astringent dye also found in catechu and kino resin. Catechin is soluble in hot water and in alkaline solutions and is neutral with no acid properties. Gambier also contains **catechutannic acid,** a reddish tannin. The liquid extract contains 25% tannin, and the cube gambier has 30 to 40%. Gambier is used in tanning leather and in dyeing to give yellow to brown colors. It produces the **cutch brown** on cotton fabrics. It has excellent fastness as a dye and is used in shading logwood and fustic, or as a mordant for fixing basic dyes. In tanning it gives a soft porous leather, has less astringency than other tannins, and is employed for tanning coat leathers and for blending with other tannins. Gambier is also used in boiler compounds and in pharmaceuticals. **Plantation gambier,** or **Singapore cube,** is refined clear gambier in small square cubes, while ordinary gambier may have dark patches or a black color with a fetid odor. **Gambier bulat** is round gambier, and **gambier papu** is in long black sticks, both used locally for chewing.

GARNET. A general name for a group of minerals varying in color, hardness, toughness, and method of fracture, used for coating abrasive paper and cloth, for bearing pivots in watches, for electronics, and the finer specimens for gemstones. Garnets are trisilicates of alumina, magnesia, calcia, ferrous oxide, manganese oxide, or chromic oxide. The general formula is $3R''O \cdot R'''_2O_3 \cdot 3SiO_2$, in which R'' is Ca, Mg, Fe, or Mn, and R''' is Al, Cr, or Fe. There are thus 12 basic types of garnets, but sodium and titanium may also occur in the crystals, replacing part of the silicon, and there are also color varieties. Asterism in the garnet, because of the isometric crystal, appears as 4-, 8-, or 12-ray stars instead of the 6 rays of the ruby.

Hardness of garnet varies from the Mohs 6 of grossularite to 7.5 of almandite and rhodolite, a hard garnet having a Knoop hardness of 1,350. The specific gravity is 3.4 to 4.3, with a melting point about 1300°C. The color is most often red, but it may be brown, yellow, green, or black. **Spanish garnet** is pale pink. High-iron garnets have the lowest melting points and fuse to a dark glass, while the high-chromium garnets have high melting points. Garnets occur in a wide variety of rocks in many parts of the world. The immense alluvial schist deposits of the Emerald Creek area of Idaho contain about 10% garnet. **Almandite,** $3FeO \cdot Al_2O_3 \cdot 3SiO_2$, forms crystals of a fine deep-red color with a hardness of about 7.5, which fuses to a glassy mass at 1315°C. It is the most common garnet and is produced chiefly in New York state. It is crushed and graded for coating abrasive paper and cloth. The choice crystals are used as gemstones. **Cape ruby,** from South Africa, is a red almandite. **Pyrope** has the composition $3MgO \cdot Al_2O_3 \cdot SiO_2$. It has a deep-red color to nearly black, and hardness of 6.5 to 7.5. The nearly transparent or translucent crystals are selected for gemstones. **Rhodolite,** a pale-rose to purple variety of garnet, is a mixture of pyrope and almandite. These two garnets are found in the eastern states. The garnet of North Carolina is a by-product of kyanite mining, the ore containing 10% garnet, 15 kyanite, and 30 mica.

Andradite has the composition $3CaO \cdot Fe_2O_3 \cdot 3SiO_2$; and the colors are yellow, green, or brown to black, with a hardness of 6.5. The orange-yellow variety is called **hessenite;** the yellow-green is **topazolite;** the green is **demantoid;** and the black variety is called **melanite.** The **Uralian emeralds** are choice green crystals of demantoid from the Ural Mountains. **Uvarovite** has the composition $3CaO \cdot Cr_2O_3 \cdot 3SiO_2$, and is emerald green. **Grossularite,** $3CaO \cdot Al_2O_3 \cdot 3SiO_2$, or $Ca_3Al_2(SiO_4)_3$, may be white, yellow, or pale green. **Succinite** is an amber variety; **romanzovite** is brown; and **wiluite** is pale green. **Spessartite,** $3MnO \cdot Al_2O_3 \cdot 3SiO_2$, is brown to red, the yellowish-brown variety being **rothoffite.** The sodium garnet is **lagoriolite,** and a variety of calcium spessartite is called **essonite.**

The best **abrasive garnets** come from almandite, although andradite and rhodolite are also used. Garnet is crushed, ground, and separated and graded in settling tanks and sieves. Hornblende is a common impurity and is difficult to separate, but good-quality abrasive garnet should be free of this softer mineral. **Garnet-coated paper** and cloth are preferred to quartz for the woodworking industries, because garnet is harder and gives sharper cutting edges, but aluminum oxide is often substituted for garnet. The less expensive quartz is sometimes colored to imitate garnet. The grades of garnet grains used on **garnet paper** and cloth range from No. 5, the coarsest, which is about 15 mesh, to 7/0, the finest, which is about 220 mesh. The paper used as a backing is a kraft of 50 to 70 lb (23 to 32 kg) weight, or a manila stock. The usual size is 9 by 11 in (23 by 28 cm). The cloth is usually in two weights, the lightweight being used as a flexible rubbing-down material. Some garnet is made into wheels by the silicate process, but vitrified wheels are not made because of the low melting point of garnet.

Synthetic garnets for electronic applications are usually **rare-earth garnets** with the general structure of grossularite, but with a rare-earth metal substituted for the calcium, and iron substituted for the aluminum and the silicon. **Yttrium garnet** is thus $Y_3Fe_2(FeO_4)_3$. **Yttrium-aluminum garnets** of 3-mm diameter are used for lasers. **Gadolinium garnet** is marketed by Semi-Elements, Inc., for microwave use. **Gadolinium gallium garnet** made from gadolinium oxide and gallium oxide is used for computer bubble memories.

GASKET MATERIALS. Any sheet material used for sealing joints between metal parts to prevent leakage, but gaskets may also be in the form of cordage or molded shapes. The simplest gaskets are waxed paper or thin copper. A usual requirement is that the material will not deteriorate by the action of water, oils, or chemicals. Gasket materials are usually marketed under trade names. **Felseal,** of the Felt Products Mfg. Co., consists of sheets of paper or fiber, 0.010 to 0.125 in (0.025 to 0.318 cm) thick, coated with Thiokol to withstand oils and gasoline. **Corbestos,** of the Dana Corp., to resist high heat and pressure, consists of sheet metal coated with graphited asbestos, the sheet metal being punched with small tongues to hold the asbestos. **Chrome lock,** of the Products Research Co., is felt impregnated with zinc chromate to prevent corrosion and electrolysis between dissimilar metal surfaces.

Foamed synthetic rubbers in sheet form, and also plastic impregnants, are widely used for gaskets. Some of the specialty plastics, selected for heat resistance or chemical resistance, are used alone or with fillers, or as binders for fibrous materials. **Haveg 16075,** of the Haveg Industries, Inc., a gasketing sheet to withstand hot oils and superoctane gasolines, is based on **Viton,** of Du Pont, a copolymer of vinylidene fluoride and hexafluoropropylene. It contains about 65% of fluorine, has a tensile strength of 2,000 lb/in^2 (13 MPa) with elongation of 400%, and will withstand operating temperatures to 400°F (204°C) with intermittent temperatures to 600°F (316°C). Another synthetic of this company, used for gaskets of high chemical resistance at temperatures to 350°F (177°C), is **Hypalon CSM-60,** a chlorosulfonyl polyethylene, made by reacting polyethylene with sulfur and chlorine. It has a tensile strength of 2,250 lb/in^2 (15.5 MPa), elongation of 30%, and Durometer hardness of 60. The square rope-type packing of the Marlo Co., called **Graphlon C,** has a core of braided asbestos fibers impregnated with Teflon, over which is a braided jacket of Teflon fiber, with an outer jacket of graphite fabric. It has a low coefficient of friction, is self-lubricating, has high chemical resistance, and withstands service temperatures to 675°F (357°C).

GASOLINE. Known in England as **petrol.** A colorless liquid hydrocarbon obtained in the fractional distillation of petroleum. It is used chiefly as a motor fuel, but also as a solvent. Ordinary gasoline consists of the hydrocarbons between C_6H_{14} and $C_{10}H_{22}$, which distill off between the temperatures 69 and 174°C, usually having the light limit at **heptane,** C_7H_{16}, or **octane,** C_8H_{18}. The octane number is the standard of measure of detonation in the engine. **Motor fuel,** or the general name gasoline, before the wide use of high-octane gasolines obtained by catalytic cracking, meant any hydrocarbon mixture that could be used as a fuel in an internal-combustion engine by spark ignition without being sucked in as a liquid and without being so volatile as to cause imperfect combustion and carbon deposition. These included also mixtures of gasoline with alcohol or benzol.

Gasanol, used in the Philippines, contained only 20% gasoline, with 5 kerosene and 75 ethyl alcohol, while the German **Dynakol** contained 70% gasoline with alcohol and benzol. The **Dynax motor fuel** contained a methanol-benzol blend. In Brazil, alcohol produced from excess sugar is mixed with gasolines, and at times all commercial automotive fuels there contain alcohol. But gasolines containing as much as 15% alcohol require special carburation. In the United States alcohol was not normally used in gasoline fuels until the oil shortages of the 1970s, after which its use increased. **Gasohol** is the trade name of the University of Nebraska Agriculture Products Industrial Utilization Committee for a blend of 90% unleaded gasoline and 10% (by volume) anhydrous ethanol. The **ethanol** is derived mostly from corn.

The common commercial gasolines in the United States had an upper limit at 225°C, and an average specific gravity of 0.75, with the **aromatic-free gasolines** having a specific gravity of 0.718. **Aviation gasoline** formerly had a boiling range below 150°C, but aviation gasolines are now the high-octane cracked and treated gasolines. Straight-run distillation yields gasoline octane numbers from about 40 to 60, and the yield is 20 to 30% of the petroleum. In the heat-cracking process the heavy hydrocarbons of the petroleum are fragmentated and converted into lighter hydrocarbons of gasoline range. The octane number of the gasoline then ranges from 55 to 70, and the yield is higher. With catalytic cracking the octane number can be brought up to 100 or higher, and the yield proportionately increased. **Cracked gasolines** may be rich in the olefins, C_nH_{2n}, or ethylene series, which have antiknock properties but which polymerize and form resins with high heat. They are stabilized with antioxidants. Filtering straight-run gasoline with bauxite removes sulfur impurities that cause knocking and raises the octane number. The octane number and the antiknock qualities are improved by slight additions of tetraethyl lead, but high-quality gasolines that do not need lead additions are produced by catalytic cracking. Small amounts of tricresyl phosphate may also be added to motor gasoline to give more uniform combustion and eliminate knock. All lead compounds are toxic, and the exhaust fumes from **leaded gasoline** may remain in the atmosphere for long periods with injury to animals and vegetation. The federal government's Clean Air Act, passed in 1970, requires the use of **unleaded gasoline** in all motor vehicles built in 1975 or later. To attain the minimum of 91 octane in unleaded gas, as required by the federal government and automobile manufacturers, a manganese compound, called **MMT,** and the high-octane boosters, reformates and alkylates, replace the formerly used **tetraethyl lead.**

Alkylated gasoline is made by adding **neohexane,** which is produced by combining isobutane and ethylene, or by adding other alkylates. **Alkylates** are produced by reacting butylenes plus propylenes and amylenes with isobutane and an acid catalyst. They are clean-burning fuels with high octane rating. As much as 35% may be added to a premium grade of gasoline to improve the sensitivity and raise the aviation performance number to above 115.

Synthetic gasoline was first produced in Germany by the Bergius process of hydrogenation of powdered lignite at high temperatures and pressures to produce gasoline, an intermediate oil, and a heavy oil. But gasolines are also produced from the high-volatile bituminous coal. The low-grade fuel bituminous coals contain a high proportion of **anthraxylon,** or **vitrain,** which is that part of bituminous coal consisting of undisintegrated parts of trees and plants. This structure of coal has less carbon and can be liquefied more easily to produce gasoline and oils. A low-grade bituminous, such as Vigo No. 4 vein, with 65% anthraxylon, gives a liquefaction to gasoline and oils of more than 96%, while a Pennsylvania medium-volatile fuel coal gives only 79% liquefaction and requires a greater consumption of hydrogen. **Natural gasoline** consists of the liquid hydrocarbons from C_5 upward, extracted from natural gas and separated from propane and other higher fractions. It has high vapor pressure, and is used for blending with motor fuels. However, under controlled production, gasolines from the various sources are the same.

Gasoline gel, used in incendiary bombs, is gasoline made into a thick gel with a chemical thickener. It adheres to the surface where it strikes and will produce a temperature of 3000°F (1649°C) for 10 min. **Kerosene gel,** similarly made, is used to dissolve tight formations and increase the flow in oil wells. The gel known as **napalm** is sodium palmitate, but may also be an aluminum soap made with an oleic, naphthenic, and coconut fatty acid mixture. The same principle is used for making quick-firing gels for commercial power boilers, but diesel oil is used instead of gasoline. The motor fuel known as **triptane** is **trimethyl butane.** In automotive engines it is free from knock so that higher compressions may be employed. In blends with 100-octane gasoline it increases the power output about 25% in aviation engines.

GELATIN. A colorless to yellowish water-soluble tasteless colloidal hemicellulose obtained from bones or from skins and used as a dispersing agent, sizing medium, coating for photographic films, and stabilizer for foodstuffs and pharmaceutical preparations. It is also flavored for use as a food jelly, and is a high-protein, low-calorie foodstuff. While albumin has a weak, continuous molecular structure that is cross-linked and rigidized by heating, gelatin has an ionic or hydrogen bonding in which the molecules are brought together into large aggregates, and it sets to a firmer solid. Gelatin differs from glue only in the purity. **Photographic gelatin** is made from skins. **Vegetable gelatin** is not true gelatin, but is algin from seaweed.

Collagen is the gelatin-bearing protein in bone and skins. The bone is dissolved in hydrochloric acid to separate out the calcium phosphate and washed to remove the acid. The organic residue is called **osseine** and is the product used to produce gelatin and glue. About 25% of the weight of the bone is osseine, and the gelatin yield is about 65% of the osseine. One short ton (0.9 metric ton) of green bones, after being degreased and dried, yields about 300 lb (136 kg) of gelatin. When skins are used, they are steeped in a weak acid solution to swell the tissues so that the collagen may be washed out. The gelatin is extracted with hot water, filtered, evaporated, dried, and ground or flaked.

GERMANIUM. A rare elementary metal, symbol Ge. It is a grayish white crystalline metal of great hardness, being 6.25 Mohs. The specific gravity is 5.35, and melting point 958°C. It is resistant to acids and alkalies. The metal is trivalent and will form chain compounds like carbon and silicon. It gives greater hardness and strength to aluminum and magnesium alloys, and as little as 0.35% in tin will double the hardness. It is not used commonly in alloys, however, because of its rarity and great cost. It is used chiefly in the form of its salts to increase the refraction of glass, and as metal in rectifiers and transistors. A **germanium-gold alloy,** with about 12% germanium, having a melting point at 359°C, has been used for soldering jewelry.

Germanium is obtained as a by-product from flue dust of the zinc industry, or it can be obtained by reduction of its oxide from the ores, and is marketed in small irregular lumps. Metal of 99.9+ purity for electronic use is made by passing an ingot slowly through an induction heater so that the more soluble impurities pass along through the molten zones and are cut off at the end of the ingot. **Germanium crystals** are grown in rods up to 1⅜ in (3.49 cm) in diameter for use in making transistor wafers. High-purity crystals are used for both P and N semiconductors. They are easier to purify and have a lower melting point than other semiconductors, specifically silicon.

The chief ore of the metal is **germanite,** which is a copper ore found in southwest Africa. Germanite contains no less than 20 elements. Together with about 45% copper and 30 sulfur, it contains 6 to 9% germanium, and 1 gallium, with various amounts of iron, zinc, lead, arsenic, silica, titanium, tungsten, molybdenum, nickel, cobalt, manganese, cadmium, and carbon. **Renierite,** found in the Congo, is a **germanium sulfide** containing up to 7.8% germanium in the ore. The rare lead-silver ore, **ultrabasite,** found in central Europe, also contains germanium. Many other metal ores, such as lepidolite, sphalerite, and spodumene, contain small amounts of germanium, so that it has an indirect use, especially in ceramics. As much as 1.6% **germanium oxide,** GeO, occurs in some English coals.

GILSONITE. A natural asphalt used for roofing, paving, floor tiles, storage-battery cases; in coatings; and for adding to heavy fuel oils. It is also referred to as **Utah coal resin.** It is a lustrous, black, almost odorless, brittle solid, having a specific gravity of 1.10. Gilsonite is one of the purest asphalts and has high molecular weight. It is soluble in alcohol, turpentine, and mineral spirits. The mineral was named for Samuel Gilson, who discovered it about 1875. The very pure gilsonite called **uintahite** takes its name from the Indian word "uintah" meaning mountain. The mineral in Utah occurs in vertical veins up to 20 ft (6.1 m) wide and 1,400 ft (427 m) deep, sometimes 8 miles (13.5 km) long. The selects come from the center of the veins, are very pure, have high solubility in naphtha, and have fusing points from 270 to 310°F (132 to 154°C). But most commercial gilsonite is now melted and regraded. **Elaterite** and **wurtzilite** are similar asphalts found in Utah, used chiefly for acid-resisting paints. **Grahamite,** or **glance pitch,** is another pure asphalt found in large deposits in Oklahoma, and in Mexico, Trinidad, and Argentina. It is used in insulation and molding materials and in paints. **Gilsonite dust,** used for foundry cores, is a by-product of Utah mining. **Manjak** is a variety of grahamite from Cuba, Barbados, and Trinidad, used for insulation and varnishes. It is the blackest of the asphalts, has a higher melting point than gilsonite, but is usually not as pure. **Millimar,** of R. T. Vanderbilt Co., Inc., is processed gilsonite in 20-mesh powder used for rubber compounding. **Gilsulate,** of the American Gilsonite Co., is an insulation grade of gilsonite. **Millex,** of Cary Chemicals, Inc., is a dark-brown gilsonite melting at 250°F (121°C), used for blending in synthetic rubbers and vinyl resins to improve processing. **Gilsonite coke** is a high-grade, low-sulfur coke produced by the American Gilsonite Co. from Utah gilsonite, 630 tons (571 metric tons) of gilsonite yielding 250 tons (227 metric tons) of coke, 1,300 bbl of gasoline, and 330 bbl of fuel oil. **Insulmastic,** of the Pittsburgh Coke & Chemical Co., is a solution of gilsonite with mica flakes and asbestos fibers, used for protective coatings.

GINGER. The most important spice obtained from roots, and one of the first Oriental spices known in Europe. It is the prepared root of the perennial herb *Zingiber officinale,* grown in India, China, Indonesia, and Jamaica. The roots are pale yellow and contain starch, an essential oil, and a pungent oleoresin, **gingerin.** The first crop is the best; the product from the regrowth is called **rhatoon ginger** and is inferior. Ginger is employed as a spice and condiment, in flavoring beverages and confections, and in medicine as a digestive stimulant and carminative. It has a cooling effect on the body. **Preserved ginger** is made by peeling off the thick scaly skin of the boiled roots and boiling in a sugar solution. **Dried ginger** is prepared by drying the peeled roots in the sun. When the roots have been boiled in lime water before peeling, **black ginger** is produced. **White ginger** is made by bleaching the roots.

Other plants of the ginger family are also used as spices and flavors. **Angelica** is from the perennial herb *Angelica archangelica,* of Syria and Europe. All parts of the plant are aromatic. **Angelica oil,** distilled from the fruit, is used in perfumes, flavors for vermouth and bitters, and in medicine. **Candied angelica** consists of the stems steeped in syrup. It is bright green in color, has an aromatic taste, and is used for decorating confections. **Galangal** is from the roots of the perennial herb *Languas officinarum,* of China. It has an aromatic odor and pungent taste similar to a mixture of ginger and pepper. It is used chiefly as a flavor, but also in medicine. **Turmeric,** highly popular in India and Malaya as an ingredient of curries, is from the rhizomes of the perennial plant *Curcuma longa,* of southeast Asia and Indonesia. When used as a dye it is called **India saffron.** The roots are cleaned and dried in the sun. It is very aromatic, with a pungent bitter taste. **Curry powders** are not turmeric alone, but mixtures of turmeric with pepper, cumin, fenugreek, and other spices. Turmeric is used to flavor and color foodstuffs, and as a dyeing agent for textiles and leather and in wood stains. The natural dye is reddish brown, and gives a yellowish color to textiles and foods. It is also used as a chemical indicator, changing color with acidity or alkalinity. **Zedoary** consists of the dried roots of the perennial plant *C. zedoaria*, grown in India. **Zedoary oil** is a viscid liquid of reddish color with an odor resembling ginger and camphor. It is used in flavoring, in medicine, and in perfumery.

GLASS. An amorphous solid made by fusing silica (silicon dioxide) with a basic oxide. Its characteristic properties are its transparency, its hardness and rigidity at ordinary temperatures, its capacity for plastic working at elevated temperatures, and its resistance to weathering and to most chemicals except hydrofluoric acid. The major steps in producing glass products are: (1) melting and refining, (2) forming and shaping, (3) heat-treating, and (4) finishing. The mixed batch of raw materials, along with broken or reclaimed glass, called **cullet,** is fed into one end of a continuous-type furnace where it melts and remains molten at around 2730°F (1499°C). Molten glass is drawn continuously from the furnace and runs in troughs to the working area, where it is drawn off for fabrication at a temperature of about 1830°F (999°C). When small amounts are involved, glass is melted in pots.

There are a number of general families of glasses, some of which have many hundreds of variations in composition. It is estimated that there are over 50,000 glass formulas. The **soda-lime glasses** are the oldest, lowest in cost, easiest to work, and the most widely used. They account for 90% of the glass used in the United States. They are composed of silica, sodium oxide (soda), and calcium oxide (lime). These glasses have only fair to moderate corrosion resistance and are useful at temperatures up to about 860°F (460°C) annealed and 480°F (249°C) in the tempered condition. Thermal expansion is high and thermal shock resistance is low compared with other glasses. These are the glass of ordinary windows, bottles, and tumblers.

Bottle glass is a simple soda-lime glass, the greenish color being due to iron impurities. The lime acts as a flux, and the calcium silicate produced gives the glass a chemical stability which quartz and fused silica do not have. The brilliance and sparkle in glass for bottles for food and drug containers are obtained by adding a small amount of barium. Small amounts of cerium will absorb ultraviolet light without distorting the passage of visible light. Cerium also acts as a **glass decolorizer** by changing the crystal structure of the iron impurities, energizing it from the normal divalent which is bluish to the trivalent which is colorless. **Crown glass** for windows is a hard, white soda-lime glass high in silica, a typical composition being 72% SiO_2, 13 CaO, and 15 Na_2O. It derives its name from the circular crowning method of making the sheets, but it is highly transparent and will take a brilliant polish. **Hard glass,** or **Bohemian glass,** for brilliant glassware, is a potash-lime glass with a high silica content, the potash glasses in general not being as hard as the soda glasses, and having lower melting points. The artistry is largely responsible for the quality of Bohemian glass.

Lead glasses or **lead-alkali glasses** are produced with lead contents ranging from low to high. They are relatively inexpensive, and are noted for high electrical resistivity and a high refractory index. Corrosion resistance varies with lead content, but they are all poor in acid resistance compared with other glass. Thermal properties also vary with lead content. The coefficient of expansion, for example, increases with lead content. High-lead grades are the heaviest of the commercial glasses. As a group, lead glasses are the lowest in rigidity. They are used in many optical components, for neon-sign tubing, and for electric light-bulb stems.

Borosilicate glasses, which contain boron oxide, are the most versatile of the glasses. They are noted for their excellent chemical durability, for resistance to heat and thermal shock, and for low coefficients of thermal expansion. There are six basic kinds. The low-expansion type is best known as **Pyrex** or **Kimax.** The low-electrical-loss types have a dielectric-loss factor second only to fused silica and some grades of 96% silica glass. Sealing types, including the well-known **Kovar,** are used in glass-to-metal sealing applications. Optical grades, which are referred to as crowns, are characterized by high light transmission and good corrosion resistance. Ultraviolet-transmitting and laboratory-apparatus grades are the two other borosilicate glasses.

Because of this wide range of types and compositions, borosilicate glasses are used in such products as sights and gages, piping, seals to low-expansion metals, telescope mirrors, electronic tubes, laboratory glassware, ovenware, and pump impellers.

Aluminosilicate glasses are roughly three times more costly than the borosilicate types, but are useful at higher temperatures and have greater thermal-shock resistance. Maximum service temperature in the annealed condition is about 1200°F (649°C). Corrosion resistance to weathering, water, and chemicals is excellent, although acid resistance is only fair compared with other glasses. Compared with 96% silica glasses, which they resemble in some respects, they are more easily worked and are lower in cost. They are used for high-performance power tubes, traveling-weave tubes, high-temperature thermometers, combustion tubes, and stove-top cookware.

The **Vicor glass** is a silica glass made from a soft alkaline glass by leaching in hot acid to remove the alkalies, and then heating to 2000°F (1093°C) to close the pores and shrink the glass. The glass will withstand continuous temperatures to 1600°F (871°C) without losing its strength or clarity. It will withstand temperatures to 1800°F (982°C) but becomes cloudy and opaque. The glass has high thermal shock resistance. **Filter glass** is flat, porous glass sheets or disks to replace high-alloy metals for filtering chemicals. The filter disks of Ace Glass, Inc., come in five porosities, from A, with pore diameters from 145 to 175 μm, to E, with pore diameters of 4 to 8 μm.

Phosphate glass, developed by the American Optical Co., will resist the action of hydrofluoric acid and fluorine chemicals. It contains no silica, but is composed of P_2O_5 with some alumina and magnesia. It is transparent and can be worked like ordinary glass, but it is not resistant to water. **Fluorex glass,** marketed in rods and tubes by the Haverford Glass Co., is a phosphate glass containing 75% P_2O_5 and less than 0.5 silica. It is decomposed by alkalies.

Sodium-aluminosilicate glasses that are chemically strenghened are used in premium applications, such as aircraft windshields. Molten salt baths are used in the strengthening process. **Corning code 0315 glass,** a chemically strengthened glass, has the highest modulus of rupture of any commercially available glass.

Industrial glass is a general name usually meaning any glass molded into shapes for product parts. The lime glasses are the most generally used because of low cost, ease of molding, and adaptability to fired colors. For such uses as light lenses and condenser cases, the borosilicate heat-resistant glasses may be used. **Glass flake** is produced by cooling rapidly very thin sheets of glass which shatter into fine flakes 0.0003 in (0.0008 cm) thick. It is used for reinforcement in plastics to permit higher filler loadings than with glass fiber. It also gives higher strength and rigidity, and dielectric strengths as high as 3,000 volts per mil (118×10^6 volts per meter) in plastics with 65% glass flake filler. The flake used as a substitute for mica flake in paints and plastics comes in particles 1 to 4 μm thick and $\frac{1}{32}$ in (0.079 cm) in diameter. In paints it produces a tilelike appearance. **Carboglas 1600** is a coating material of a chemically resistant polyester solution containing glass flake. When sprayed, a 0.030-in (0.076-cm) coating will contain 120 layers of the thin glass flake.

Filmglas, of the Owens-Corning Corp., is glass flake in tiny platelets. The 10- to 18-mesh flake is 2 μm thick. **Powdered glass,** used as a filler in plastics and coatings, is made by grinding broken scrap glass.

Cer-Vit, of Owens-Illinois Glass Co., consists of tiny ceramic crystals in a vitreous matrix and is used for molded electronic parts. The thermal expansion can be varied from zero to small positive or negative values. **Glass spheres** and **ceramic spheres** are thin-walled hollow balls of glass or oxide ceramics used as fillers and strengthening agents in plastics and lightweight composites.

Fused silica glass is 100% silicon dioxide. If it occurs naturally, the glass is known as **fused quartz** or **quartz glass.** There are many types and grades of both glasses, depending on the impurities present and the manufacturing method. Because of its high purity level, fused silica is one of the most transparent glasses. It is also the most heat-resistant of all glasses; it can be used at temperatures up to 1650°F (899°C) in continuous service, and to 2300°F (1260°C) for short-term exposure. In addition, it has outstanding resistance to thermal shock, maximum transmittance to ultraviolet light, and excellent resistance to chemicals. Unlike most glasses, its modulus of elasticity increases with temperature. However, because fused silica is high in cost and difficult to shape, its use is restricted to such specialty applications as laboratory optical systems and instruments and crucibles for crystal growing. Because of a unique ability to transmit ultrasonic elastic waves with little distortion or absorption, fused silica is also used in delay lines in radar installations.

Flint glass for windows is a highly transparent soda-lime quartz glass. **Lustraglas,** of the American Saint Gobain Corp., is a highly transparent flat-drawn quartz glass.

Thermopane, of the Libbey-Owens-Ford Glass Co., is a heat- and sound-insulating window glass made with two panes of glass separated by a metal bonded to the edges of the glass, leaving an insulating layer of dehydrated air between the two glass sheets. The highly refractive flint glass used in the **rhinestones** for cheap jewelry and for the **paste diamonds,** which derive their name from their softness as compared with the diamond, contains lead. It has an index of refraction of 1.67, but lacks the double refraction and regular molecular arrangement of true gem crystals. Lead is also used in the **crystal glass** for cut glassware, and the brilliance from the lead is higher in the potash glasses than the soda. **English crystal,** which is a potash glass, contains as high as 33% lead oxide. It has a high clarity and brilliancy, but is soft. Leaded glasses are heavy, a crystal glass having a specific gravity of 6.33 compared with 2.25 for a borate glass.

Ninety-six percent silica glasses are similar to fused silica. Although less expensive than fused silica, they are still more costly than other glasses. Compared to fused silica, they are easier to fabricate, have a slightly higher coefficient of expansion, about 30% lower thermal stress resistance, and a lower softening point. They can be used continuously up to 1470°F (799°C). Uses include chemical glassware and windows and heat shields for space vehicles.

The boric oxide glasses, or **borax glasses,** are transparent to ultraviolet rays. The so-called **invisible glass** is a borax glass surface treated with a thin film of sodium fluoride. It transmits 99.6% of all visible light rays, thus casting back only slight reflection and giving the impression of invisibility.

Ordinary soda and potash glasses will not transmit ultraviolet light. Glass containing 2 to 4% ceric oxide absorbs ultraviolet rays, and is also used for X-ray shields. Glass capable of absorbing high-energy X-rays or gamma rays may contain tungsten phosphate, while the glass used to absorb slow neutrons in atomic-energy work contains cadmium borosilicate with fluorides. The shields for rocket capsule radio antennas are made of 96% silica glass. It is transparent to radio signals, and will withstand temperatures above 900°F (482°C). **Fluorescent glass** for mercury-vapor discharge tubes contains ceric oxide. **Kromex glass** of the MacBeth-Evans Glass Co., used for gasoline-dispensing pumps, is a glass made to stop the passage of ultraviolet rays.

Optical glass is a highly refined glass; it is usually a flint glass of special composition, or made from rock crystal, used for lenses and prisms. It is cast, rolled, or pressed. In addition to the regular glassmaking elements, silica and soda, optical glass contains barium, boron, and lead. The high-refracting glasses contain abundant silica or boron oxide. A requirement of optical glass is transparency and freedom from color. Traces of iron make the glass greenish, while manganese causes a purple tinge. First-quality optical glass should contain a minimum of 99.8% SiO_2. Borax is used in purifying and in increasing the strength and brilliance of the glass. Besides the control of chemical composition, careful melting and cooling are necessary to obtain fine transparency, followed by intense polishing. The pouring temperature is about 1200°C. The best optical glass has a transparency of 99%, compared with 85 to 90% for ordinary window glass. A **borate glass** with lanthanum and tantalum oxides but no silica is used for airplane-camera lenses and for eyepieces for wide-angle field glasses. It has a high refractive index and a low dispersion. The **telescope mirror glass** of Owens-Illinois, called **Cer-Vit optical glass,** has near-zero expansion and no deflection of light rays. **Beryllium fluoride glass** of the American Optical Co. is made by substituting beryllium fluoride for silicon dioxide. It has a low refractive index with low color dispersion, and light travels faster through it than through ordinary optical glass. It has the disadvantage that it is hygroscopic.

Plate glass is any glass that has been cast or rolled into a sheet, and then ground and polished. But the good grades of plate glass are, next to optical glass, the most carefully prepared and the most perfect of all the commercial glasses. It generally contains slightly less calcium oxide and slightly more sodium oxide than window glass, and small additions of agents to give special properties may be added, such as agents to absorb ultraviolet or infrared rays, but inclusions that are considered as impurities are kept to a minimum. The largest use of plate glass is for storefronts and office partitions. Plate glass is now made on a large scale on continuous machines by pouring on a casting table at a temperature of about 1000°C, smoothing with a roller, annealing, and then setting rigidly on a grinding table and grinding to a polished surface. Normally, the breaking stress of a glass with a ground surface is much less than that of blown or pressed glass, but highly polished plate glass with the surface flaws removed may have double the breaking strength of average pressed glass. The chief advantage of plate glass, however, is that the true parallelism of the ground surfaces eliminates distortion of objects seen through the glass. **Herculite** is a glass of the PPG Industries that withstands temperatures up to 650°F (343°C) without cracking. **Carrara structural glass** of this company is made like plate glass and ground to true plane surfaces. It is made in many colors in thicknesses from 11/32 to 1½ in (0.86 to 3.81 cm) or laminated to give various color effects. It is used for storefronts, countertops, tiling, and paneling. The weight of the 11/32 in (0.86 cm) is 4.5 lb/ft^2 (72 kg/m^3). It does not craze, check, or stain like tile. **Spandrelite,** of this company, is an ornamental structural plate glass made by fusing ceramic color to a plate glass. It is used for cladding buildings.

Conductive glass, employed for windshields to prevent icing and for uses where the conductive coating dissipates static charges, is plate glass with a thin coating of stannic oxide produced by spraying glass, at 900 to 1300°F (482 to 704°C), with a solution of stannic chloride. Coating thicknesses are 50 to 550 nm, and will carry current densities of 6 W/in^2 (9,300 W/m^2) indefinitely. The coatings are hard and resistant to solvents. The light transmission is 70 to 88% that of the original glass, and the index of refraction is 2.0, compared with 1.53 for glass. **Electrapane,** of the Libbey-Owens-Ford Co., and **Nesa,** of PPG Industries, are conductive glasses.

Transparent mirrors are made by coating plate glass on one side with a thin film of chromium. The glass is a reflecting mirror when the light behind the glass is less than in front, and is transparent when the light intensity is higher behind the glass. **Photosensitive glass** is made by mixing submicroscopic metallic particles in the glass. When ultraviolet light is passed through the negative on the glass, it precipitates these particles out of solution, and since the shadowed area of the negative permit deeper penetration into the glass than the highlight areas, the picture is in three dimensions and in color. The photograph is developed by heating the glass to 1000°F (538°C). **Photochromic glass** becomes dark under ultraviolet light and regains full transparency when the rays are removed. In sunlight, rearrangement of the oxides forms dark spots that impede the light rays.

Metal salts are used in glass for coloring as well as controlling the characteristics. Manganese oxide is added to most glass to neutralize iron oxide, but an excess colors the glass violet to black. **Jena blue glass** gets its color and fluorescence from a mixture of cobalt oxide and ceric oxide. **Ruby glass,** of a rich red color, is produced with selenium and cadmium sulfide, or with copper oxide. It is also produced with **purple of cassius,** or **gold-tin purple,** a brown powder which is a mixture of the yellow **gold chloride,** $AuCl_3$, and the dark-brown tin oxide. **Copper-ruby glass** has a greenish tinge and is suitable for automobile taillights, but signal lens glass is made with selenium, cadmium sulfide, and arsenious oxide, which give a distinct ruby color with heat treatment. **Amber glass** is made with controlled mixtures of sulfur and iron oxide that give tints varying from pale yellow to ruby amber. Amber glass is much used for medicine bottles to prevent entry of harmful light rays. The **actinic glass** used for skylights and factory windows has a yellow tint that softens and diffuses the light and impedes passage of heat rays. **Neophane glass** is glass containing neodymium oxide which reduces glare and is used in yellow sunglasses, or small amounts may be used in windshield glass. **Opticolor glass,** of the Chicago Dial Co., containing neodymium oxide for television tube filters, transmits 90% blue, green, and red light rays, and only 10% of the less desirable yellow rays. The pictures are bright with improved color contrasts. **Opalescent glass,** or **opal glass,** used for light shades, tabletops, and cosmetic jars, has structures that cause light falling on them to be scattered, and they thus are white or translucent. They owe their properties to tiny inclusions with different indexes of refraction, such as fluorides, sulfides, or oxides of metals. **Alabaster glass** has inclusions of larger dimensions in lower numbers per unit than opal glass, and it shows no colors, whereas opalescent glass appears white by reflected light but shows color images through thin sections. Opal glass may contain lipidolite, a mineral which contains various metals.

Monax glass, of the MacBeth-Evans Glass Co., is a white diffusing glass for lamp shades and architectural glass. **Glass blocks** for translucent units in factory walls are made from types of opalescent glass. **Insulex,** of the Owens-Illinois Glass Co., is a translucent glass brick of this kind. **Cellulated glass,** or **foamed glass,** is **expanded glass** in the form of blocks and sheets used for thermal insulation of walls and roofs. It is made by heating pulverized glass with a gas-forming chemical at the flow temperature of the glass. The glass expands and forms hollow cells which may comprise up to 90% of the total volume. The density is usually about 10 lb/ft^3 (160 kg/m^3). It has a compressive strength of 100 lb/in^2 (0.7 MPa), a flexural strength of 75 lb/in^2 (0.5 MPa), and will retain rigidity to 800°F (427°C). **Low-melting glass,** for encapsulating electronic components, is made by adding selenium, thallium, arsenic, or sulfur to give various melting points from 260 to 660°F (127 to 349°C). These glasses can be vaporized and condensed as thin films. Some are insulators, some are semiconductors, and all are chemical-resistant.

Polarized glass, for polarizing lenses, is made by the American Optical Co. by adding minute crystals of tourmaline or peridote to the molten glass, and stretching the glass while still plastic to bring the axes of the crystals into parallel alignment. **Florentine glass** does not refer to a mixture but to an ornamental glass made by casting on an embossed bed, or by rolling with a roll upon which the designs are cut.

GLASS CERAMICS. A family of fine-grained crystalline materials made by a process of controlled crystallization from special glass compositions containing nucleating agents. They are sometimes referred to as **nucleated glass, devitrified ceramic,** or **vitro ceramics.** Since they are mixed oxides, different degrees of crystallinity can be produced by varying composition and heat treatment. Some of the types produced are cellular foams, coatings, adhesives, and photosensitive compositions.

Glass ceramics are nonporous and generally either opaque white or transparent. Although not ductile, they have much greater impact strength than commercial glasses and ceramics. However, softening temperatures are lower than those for ceramics, and they are generally not useful above 2000°F (1093°C). Thermal expansion varies from negative to positive values depending on composition. Excellent thermal-shock resistance and good dimensional stability can be obtained if desired. These characteristics are used to advantage in "heat-proof" skillets and range tops. Like chemical glasses, these materials have excellent corrosion and oxidation resistance. They are electrical insulators and are suitable for high-temperature, high-frequency applications in the electronics field.

Pyroceram, of the Corning Glass Works, is a hard, strong, opaque-white nucleated glass with a flexural strength to above 30,000 lb/in^2 (206 MPa), a density of 2.4 to 2.62, a softening point at 2460°F (1349°C), and high thermal shock resistance. It is used for molded mechanical and electrical parts, heat-exchanger tubes, and coatings. **Macor,** by the same company, is a machinable glass ceramic. **Pyroceram balls,** made by the Hartford-Universal Co., are used to replace steel balls in bearings and valves. They have the hardness of hardened tool steel with only one-third the weight, are corrosion-resistant, have a low coefficient of expansion, and will withstand temperatures to 2200°F (1204°C). **Nucerite,** used by the Pfaudler Co. for lining tanks, pipes, and valves, is nucleated glass. It has about four times the abrasion resistance of a hard glass, will withstand sudden temperature differences of 1200°F (649°C), and has high impact resistance. It also has high heat-transfer efficiency.

Polychromic glass is a glass ceramic that reproduces colors much like photographic film in a thin layer or in thick sections of glass up to 1 in (2.5 cm) thick. Before processing, the glass is colorless and transparent. The colored images are produced when activated by ultraviolet light and heat.

GLASS FIBER. Fine flexible fibers made from glass are used for heat and sound insulation, fireproof textiles, acid-resistant fabrics, retainer mats for storage batteries, panelboard, filters, and electrical insulating tape, cloth, and rope. Molten glass strings out easily into threadlike strands, and this **spun glass** was early used for ornamental purposes, but the first long fibers of fairly uniform diameter were made in England by spinning ordinary molten glass on revolving drums. The original fiber, about 0.001 in (0.003 cm) in diameter, was called **glass silk** and **glass wool,** and the loose blankets for insulating purposes were called **navy wool.** The term navy wool is still used for the insulating blankets faced on both sides with flameproofed fabric, employed for duct and pipe insulation and for soundproofing.

Glass fibers are now made by letting the molten glass drop through tiny orifices and blowing with air or steam to attenuate the fibers. The usual composition is that of soda-lime glass, but it may be varied for different purposes. The glasses low in alkali have high electrical resistance, while those of higher alkali are more acid-resistant. They have very high tensile strengths, up to about 400,000 lb/in^2 (2,757 MPa).

The standard glass fiber used in glass-reinforced plastics is a borosilicate type known as **E-glass.** The fibers spun as single glass filaments, with diameters ranging from 0.0002 to 0.001 in (0.0005 to 0.003 cm), are collected into strands that are manufactured into many forms of reinforcement. E-glass fibers have a tensile strength of 500,000 lb/in^2 (3,445 MPa). Another type, **S-glass,** is higher in strength, but because of higher cost, its use is limited to advanced, higher-performance products.

Staple glass fiber is usually from 0.00028 to 0.00038 in (0.0007 to 0.0009 cm), is very flexible and silky, and can be spun and woven on regular textile machines. **Glass fiber yarns** are marketed in various sizes and twists, in continuous or staple fiber, and with glass compositions varied to suit chemical or electrical requirements. **Vitron yarn** of Glass Fibers, Inc., is a plied low-twist yarn for braided insulation for wire. **Glass sewing thread** has a high twist. The minimum breaking strength of the 0.014-in (0.036-cm) thread is 12 lb (5 kg).

Halide glass fibers are composed of compounds containing fluorine and various metals such as barium, zirconium, thorium, and lanthanum. They appear to be promising for fiber-optic communication systems. Their light-transmitting capability is many times better than that of the best silica glasses now being used.

Glass fiber cloth and **glass fiber tape** are made in satin, broken twill, and plain weaves, the satin-wear cloth being 0.007 in (0.02 cm) thick weighing 7 oz/yd^2 (0.24 kg/m^2). **Lagging cloth** for high-temperature pipe insulation is 20 oz/yd^2 (0.68 kg/m^2). **Glass insulating sheet,** for electrical insulation, is a tightly woven fabric impregnated with insulating varnish, usually in thicknesses from 0.005 to 0.012 in (0.013 to 0.030 cm). **Glass fabric** of the Soule Mill for varnishing is 0.001 in (0.003 in) thick and weighs 0.81 oz/yd^2 (0.03 kg/m^2).

Glass cloth of plain weave of either continuous fiber or staple fiber is much used for laminated plastics. The usual thicknesses are from 0.002 to 0.023 in (0.005 to 0.058 cm) in weights from 1.43 to 14.7 oz/yd^2 (0.05 to 0.50 kg/m^2). Cloth woven of monofilament fiber in loose rovings to give better penetration of the impregnating resin is also used. **Glass mat,** composed of fine fibers felted or intertwined in random orientation, is used to make sheets and boards by impregnation and pressure. **Fluffed glass fibers** are tough twisted glass fibers. For filters and insulation the felt withstands temperatures to 1000°F (538°C). **Chopped glass,** consisting of glass fiber cut to very short lengths, is used as a filler for molded plastics. Translucent corrugated building sheet is usually made of glass fiber mat with a resin binder. All of these products, including chopped fiber, mat, and fabric preimpregnated with resin, and the finished sheet and board, are sold under a wide variety of trade names. Glass fiber bonded with a thermosetting resin can be performed for pipe and other insulation coverings. **Glass fiber block** is also available to withstand temperatures to 600°F (316°C). **Glass filter cloth** is made in twill and satin weaves in various thicknesses and porosities for chemical filtering. **Glass beltings,** for conveyor belts that handle hot and corrosive materials, is made with various resin impregnations. Many synthetic resins do not adhere well to glass, and the fiber is sized with vinyl chlorosilane or other chemical.

GLASS SAND. Sands employed in glassmaking. They are all screened, and usually washed, to remove fine grains and organic matter. The grain standards of the American Ceramic Society specify that all should pass through a No. 20 screen, between 40 and 60% should remain on a No. 40 screen, between 30 and 40% should remain on a No. 60 screen, between 10 and 20% on a No. 80 screen, and not more than 5% should pass through a No. 100 screen. Sand for first-quality optical glass should contain 99.8% SiO_2, and maximums of 0.1 Al_2O_3 and 0.02 Fe_2O_3. Third-quality flint glass may contain only 95% SiO_2 and as high as 4 Al_2O_3. Only in the eighth- and ninth-quality amber glasses is the content of Fe_2O_3 permitted to reach 1%. **Potters' sand** is usually a good grade of glass sand of uniform grain employed for packing to keep the ware apart.

GLUCOSE. A syrupy liquid of the composition $CH_2OH(CHOH)_4CHO$, which is a monosaccharide, or simple sugar, occurring naturally in fruits and in animal blood, or made by the hydrolysis of starch. It is also produced as a dry white solid by evaporation of the syrup. Glucose is made readily from cornstarch by heating the starch with dilute hydrochloric acid, which is essentially the same process as occurs in the human body. Commercial glucose is made from cornstarch, potato starch, and other starches, but in Japan it is also produced from wood. Glucose is only 70% as sweet as cane sugar and has a slightly different flavor. It is used in confectionery and other foodstuffs for blending with cane sugar and syrups to prevent crystallization on cooling and because it is usually cheaper than sugar. It is used in tobacco and inks to prevent drying, and in tanning as a reducing agent. The name glucose is usually avoided by the manufacturers of edible products, because of prejudices against its substitution for sugar, but in reality it is a simple form of sugar easily digested. It is used in medicine as a blood nutrient and to strengthen heart action, and may be harmful only in great excess. When free from starch, it is called **dextrose.** It is also marketed as corn syrup, but corn syrup is not usually pure glucose, as it contains some dextrine and maltose. The **maltose,** or **malt sugar,** in the combination has the empirical formula $C_{12}H_{22}O_{11}$. When hydrolyzed in digestion, it breaks down easily to glucose. It is produced from starch by enzyme action. When purified, it is transparent and free of malt flavor. It is not as sweet as the sucrose of cane sugar, but is used in confectionery and as an extender of cane sugar. **Dry corn syrup** is in colorless glasslike flakes. It is made by instantaneous drying and quick cooling of the syrup. **Sweetose,** of the A. E. Staley Mfg. Co., is a crystal-clear enzyme-converted corn syrup used in confectionery to enhance flavor and increase brightness.

Glucose derived from grapes is called **grape sugar.** The glucose in fruits is called **fruit sugar, levulose,** or **fructose.** This is dextroglucose, and when separated out is in colorless needles which melt at 104°C. It is used for intravenous feeding and is absorbed faster than glucose. It is also used in low-calorie foods, and in honey to prevent crystallization. It is normally expensive, but is made synthetically. It can be made from corn and is superior to corn syrup as a sweetening agent. **Isomerose,** of Clinton Corn Processing Co., is a corn syrup that is treated with an enzyme to convert part of the dextrose into fructose. It contains 14% frutose and has double the sweetness of sucrose. **Corn sugar** is also a solid white powder, being glucose with one molecule of water of crystallization. When the refined liquor is cooled, the corn sugar crystallizes in a mother liquor known as **hydrol,** or **corn-sugar molasses.** This molasses is screened and washed off, and marketed for livestock feed. The crystalline monohydrate sugar is known as **cerelose. Methyl glucoside,** made from corn glucose, is a white crystalline powder melting at 164°C. It has the composition $C_7H_{13}O_6$, with four esterifiable hydroxyls, and is used in making tall oil esters and alkyd resins. **Ethyl glucoside,** $C_8H_{16}O_6$, is marketed as a colorless syrup in water solution with 80% solids and a specific gravity of 1.272. It is noncrystallizing, and is used as a humectant and plasticizer in adhesives and sizes.

GLUE. A cementing material usually made from impure gelatin from the clippings of animal hides, sinews, horn and hoof pith, from the skins and heads of fish, or from bones. The term **animal glue** is limited to **hide glue,**

extracted bone glue, and **green bone glue.** Fish glue is not usually classed with animal glue, nor is casein glue. The vegetable glues are also misnamed, being classed with the mucilages. Synthetic resin glues are more properly classed with adhesive cements. Animal glues are **hot-work glues** which are applied hot and which bind on cooling. Good grades of glue are semitransparent, free from spots and cloudiness, and are not brittle at ordinary temperatures. **Bone glue** is usually light amber in color; the strong hide and sinew glues are light brown. The stiffening quality of glue depends upon the evaporation of water, and it will not bind in cold weather. Glues made from blood, known as **albumin glues,** and from casein are used for some plywood. However, they do not have the strength of the best grades of animal glue, and are not resistant to mold or fungi. **Marine glue** is a glue insoluble in water, made from solutions of rubber or resins, or both. The strong and water-resistant plywoods are now made with synthetic resin adhesives.

Animal glue has been made since ancient times, and is now employed for cementing wood, paper, and paperboard. It will not withstand dampness, but white lead or other material may sometimes be added to make it partly waterproof. Casein glues and other protein glues are more water-resistant. **Soybean glue** is made from soybean cake and is used for plywood. It is marketed dry. It has greater adhesive power than other vegetable glues, or pastes, and is more water-resistant than other vegetable pastes. Hide glue is used in the manufacture of furniture, abrasive papers and cloth, gummed paper and tape, matches, and print rollers. The bone glues are used either alone or blended in the manufacture of cartons and paper boxes. Green bond glue is used chiefly for gummed paper and tape for cartons. In making bone glue the bones are crushed, the grease extracted by solvents, and the mineral salts removed by dilute hydrochloric acid. The bones are then cooked to extract the glue. Glues are graded according to the quality of the raw material, the method of extraction, and the blend.

There are 16 grades of hide glue and 15 grades of bone glue. Those with high viscosity are usually the best. Most glue is sold in ground form, but also as flake or pearl. Glues for such uses as holding abrasive grains to paper must have flexibility as well as strength, obtained by adding glycerin. The **animal protein colloid** of Swift & Co. is a highly purified bone glue especially adapted for use as an emulsifier, and for sizing, water paints, stiffening, and adhesives. Hoof and horn-pith glue is the same as bone glue, and is inferior to hide glue. **Fish glue** is made from the jelly separated from fish oil, or from solutions of the skins. The best fish glue is made from Russian isinglass. Fish glues do not form gelatin well and are usually made into **liquid glues** for photo mounting, gummed paper, household use, and for use in paints and sizes. The liquid glues are also made by treating other glues with a weak acid. Pungent odors indicate defective glue. The glues made from decomposed materials are weak. Preservatives such as sulfur dioxide or chlorinated phenol may be used. The melting point is usually about 140°F (60°C).

GLYCERIN. A colorless, syrupy liquid with a sweet, burning taste, soluble in water and in ethyl alcohol. It is the simplest **trihydroxy alcohol,** with the composition $C_3H_5(OH)_3$. It has a specific gravity of 1.26, a boiling point at 220°C, and a freezing point at 17°C. It is also called **glycerol,** and was used as a lotion under the name of **sweet oil** for more than a century after its discovery in 1783.

Glycerin occurs as **glycerides,** or combinations of glycerin with fatty acids, in vegetable and animal oils and fats, and is a by-product in the manufacture of soaps and in the fractionation of fats, and is also made synthetically from propylene. Coconut oil yields about 14% glycerin, palm oil 11%, tallow 10%, soybean oil 10%, and fish oils 9%. It does not evaporate easily and has a strong affinity for water, and is used as a moistening agent in products that require to be kept from drying, such as tobacco, cosmetics, foodstuffs, and inks. As it is nontoxic, it is used as a solvent in pharmaceuticals, as an antiseptic in surgical dressings, as an emollient in throat medicines, and in cosmetics. Since a different type of group can replace any one or all of the three **hydroxyl groups,** (OH), a large number of derivatives can be formed, and it is thus a valuable intermediate chemical, especially in the making of plastics. It is also used as a plasticizer in resins, and to control flexibility in adhesives and coatings. An important use is in nitroglycerin and dynamite. In water solutions the freezing point is lowered, reaching −60°F (−51°C), the lowest point, at 37% of water, and it is thus valuable as an antifreeze. **Glycerogen,** a German substitute made by the hydrogenation of wood hexose, is not pure glycerin, but also contains glycols and other hexyl alcohols.

GOLD. A soft, yellow metal, known since ancient times as a precious metal on which all material trade values are based. It is so chemically inactive that it is found mostly in the native state. It is found widely distributed in all parts of the world. It is employed chiefly for coinage, ornaments, jewelry, and for gilding. Gold is extracted by crushing the ores and catching the metal with quicksilver. About 25% of the gold produced in the United States is placer gold and about 5% is a by-product of the copper industry. The average gold recovered from ore in western United States is 0.2 oz/ton (0.006 kg/metric ton) of ore; that in Alaska is 0.044 oz/ton (0.001 kg/metric ton); and the low-grade carbonaceous ore of Nevada contains about 0.3 oz/ton (0.009 kg/metric ton). **Native gold** is usually alloyed with silver, placer gold being the purest. The natural alloy of gold and silver was known as **electrum,** and under the Egyptian name of **Asem** was thought to be an elementary metal until produced as an alloy by the Romans.

Among most civilized nations gold has always been the standard upon which trade values were set, even when gold itself was not used. For more than 25 centuries, until the extensive use of the precious metals for industrial purposes, gold retained a 15½- or 16-to-1 value with silver, the only other metal meeting the tests for a **coinage metal.** These tests are: that it must have an intrinsic value to the people as a whole, as for ornamentation; that it must be readily workable but highly corrosion-resistant and permanent; that while reasonably scarce, it must be available in all parts of the worldmit cannot be a monopoly of any nation or group of nations. Thus, the value of gold (and silver) is regulated by law in all countries, and only gold and silver pieces are true coinage, those of other metals being merely **tokens,** and paper money being merely promissory notes.

Gold is the most malleable of the metals, and can be beaten into extremely thin sheets. A gram of gold can be worked into leaf covering 6 ft^2 (0.6 m^2), and only 0.0000033 in (0.0000084 cm) thick, or into a wire 1.5 miles (2.5 km) in length. **Cast gold** has a tensile strength of 20,000 lb/in^2 (137 MPa). The specific gravity is 19.32, and the melting point 1943°F (1062°C). It is not attacked by nitric, hydrochloric, or sulfuric acid, but is dissolved by aqua regia, or by a solution of azoimide, and is attacked by sodium and potassium cyanide plus oxygen. The metal does not corrode in the air, only a transparent oxide film forming on the surface. Its reflectivity of ultraviolet and visual light rays is low, but it has high reflectivity of infrared and red rays, and is thus valued for plating some types of reflectors. For radiation-control coatings for spacecraft, **gold flake** is used. Gold of 14-karat purity is used in the form of tiny laminar platelets with overlap to form a film. **Gold flake No. 14,** of the Metz Refining Corp., has particles 10 μm in diameter and 0.5 μm thick. The particles of No. 16 are 30 μm in diameter and 1 μm thick. Gold-plated grid wires in electronic tubes give high conductivity and suppress secondary emission. **Gold-gallium** and **gold-antimony** alloys of the J. M. Ney Co., for electronic uses, come in wire as fine as 0.005 in (0.013 cm) in diameter, and in sheet as thin as 0.001 in (0.003 cm). The maximum content of antimony in workable gold alloys is 0.7%. A gold-gallium alloy with 2.5% gallium has a resistivity of 90 Ω per cir mil ft ($15 \times 10^{-8}\ \Omega \cdot m$), and has a tensile strength of 55,000 lb/in^2 (379 MPa) with elongation of 22%. **Gold powder** and **gold sheet,** for soldering semiconductors, are 99.999% pure. The gold wets silicon easily at a temperature of 700°F (371°C). Chemically reduced powder is amorphous and comes in particle sizes from 1 to 8 μm. **Atomized gold powder** has spherical free-flowing particles from 30 to 400 mesh. The "gold" powder used in some paints and plastics is bronze powder. The gold coating on spacecraft for radiation control must be hard for resistance to meteorite particles. The metallizing powder contains up to 20% platinum or palladium with a glass frit.

Because of its softness, gold is almost always alloyed with other metals, usually copper, silver, or nickel, and graded on a basis of degrees of fineness in 1,000 parts, or on the basis of karat gold value, pure gold being 24 karats. **Green gold,** used in making jewelry, is an alloy of gold, silver, and copper, graded from 14 to 18 karats. The 18-karat green gold contains 18 parts gold and 6 silver, with no copper. The 15-karat grade contains 15 parts gold, 8 silver, and 1 copper. The 14-karat grade contains 14 parts gold, 8 ¼ silver, and 1 ¾ copper. **Coinage gold** in the United States is 90% gold and 10 copper. In England it is 91.66% gold and 8.33 copper, and this alloy is called **standard gold.** In Australia 8.33% silver is used instead of the copper, and the **gold-silver alloy** is called **Australian gold. Dental gold** is a term for a wide range of wrought and cast alloys with usually from 65 to 90% gold, 5 to 12 silver, and frequently platinum and sometimes palladium. A very small amount of iridium may also be used for hardening. Colors vary from white to yellow, and such alloys are also used for jewelry and acid-resistant plates. Gold is easily electroplated on other metals from cyanide solutions in controlled thicknesses from 0.000005 to 0.005 in (0.000013 to 0.013 cm). **Sodium gold cyanide,** used for gold-plating solutions, is a water-soluble yellow powder of the composition $NaAu(CN)_2$, containing 46% gold. **Gold plate** is thus much used for jewelry, ornaments, and for chemical resistance, but the gold is so soft that thin plates wear off easily. But small amounts of other metals harden the gold and also give a color range from red and pink gold to lemon yellow. For plating electrical contacts, alloys with 1 to 6% nickel are used. The hardness of gold doubles with each 2% increase in nickel content, but the electrical resistance increases. Hard gold plate is a **gold-indium plate.** The process of the Indium Corp. of America is to plate the gold and indium successively and then alloy by diffusion at 330°F (166°C). Precipitation hardening takes place, giving a hard, wear-resistant coating.

South Africa is the most important producer of gold. Other important producers are Canada, the United States, and Australia, but gold is produced in 90 other countries. The ore of gold known as **calaverite,** found in Colorado, California, and west Australia, is a **gold telluride,** $AuTe_2$, occurring in monoclinic crystals, while the variety known as **krennerite** is Au_8Te_{16}, in orthorhombic crystals. They are found with the **gold-silver telluride** called **sylvanite,** $(AuAg)Te_2$, a silvery-white granular mineral of specific gravity 9.35 and hardness 2.5, which is easily fusible.

GRANITE. A coarse-grained, igneous rock having an even texture, and consisting largely of quartz and feldspar with often small amounts of mica and other materials. There are many varieties. Granite is very hard, compact, and takes a fine polish, showing the beauty of the crystals. It is the most important building stone, and is also used as an ornamental stone. An important use is also for large rolls in pulp and paper mills. **Granite surface plates,** for machine-shop layout work, are made in sizes up to 30 by 72 in (76 by 183 cm) and 10 in (25 cm) thick, ground and highly polished to close accuracy. It is extremely durable, and since it does not absorb moisture as limestone and sandstone do, it does not weather or crack as these stones do. The colors are usually reddish, greenish, or gray. **Rainbow granite** may have a black or dark-green background with pink, yellowish, and reddish mottling, or it may have a pink or lavender background with dark mottling. The weight is 170 lb/ft^3 (2,723 kg/m^3), the specific gravity 2.72, and the crushing strength is from 23,000 to 32,000 lb/in^2 (158 to 220 MPa). The most notable granite quarries are in northern New England. **Mount Airy granite** from North Carolina is light gray in color, and is a **biotite** containing feldspar, quartz, and mica. It is somewhat lighter in weight and of lower crushing strength than **Maine granite.** The hard composite igneous rock **diabase,** called **black granite,** is used by the Bahn Granite Surface Plate Co. for making precision parallels for machine-shop work. **Unakite** is a granite of Virginia and North Carolina with a mosaic of red, green, and other colors. It is used as an ornamental building stone. **Balfour pink granite** of North Carolina contains 72% silica, 14.1 alumina, 2 soda, 6 potash, 1.2 iron oxide, 0.12 titanium oxide, 0.20 manganese oxide, and 0.36 lime. The granite known as **pegmatite,** of which there are vast quantities, contains beryllium in the form of beryl as a minor constituent.

GRAPHITE. A form of carbon. Also called **plumbago.** It was formerly known as black lead, and when first used for pencils was called **Flanders' stone.** A natural variety of elemental carbon having a grayish-black color and a metallic tinge.

Natural graphite comes chiefly from Mexico, India, Sri Lanka, and Malagasy. In the United States it is found in Alabama. It occurs in veins in rocks, and always contains some impurities. The high-grade lump graphite of Sri Lanka, often preferred for crucibles, is found in closely foliated masses in underground veins in gneiss interbedded with limestone. It occurs in two forms: foliated and amorphous. **Foliated graphite** is used principally for crucibles and lubricants, and amorphous for lead pencils, foundry facings, electric brush carbons, molded parts, and paint pigments. It is infusible, subliming at 6700°F (3704°C), but oxidizing above 600°C. It is a good conductor of heat and electricity, is resistant to acids and alkalies, and is readily molded. **Molded graphite** is usually made by mixing calcined petroleum coke with a binder of coal-tar pitch, pressing, baking in an inert atmosphere, and then heating to above 3500°F (1927°C) to promote crystal growth. Molded parts increase in strength with increasing temperature up to about 4500°F (2482°C). The specific gravity of graphite is 2 to 2.5. The hardness is 1 to 2, sometimes less than 1, and it has a decidedly greasy feel. It is a good lubricant, especially when mixed with grease, but the natural graphite contains silica and other abrasive materials so that the artificial graphite is preferred for lubricants.

Graphite is marketed in grades by purity and fineness. No. 1 flake should contain at least 90% graphitic carbon. Mexican **amorphous graphite** contains 80%. The best Malagasy graphite contains 95% carbon, while the powders may be as low as 75%. **Crystalline graphite** and **flake graphite** are synonymous terms for material of high graphite content as distinguished from amorphous. Some natural graphite, useful for paints, contains as little as 35% graphite carbon, but high-grade graphite, suitable for crucibles and nuclear reactors, is made from low-grade ores containing 20% carbon by flotation, purification at high heat, and pressing into blocks. Ultrapure graphite, for nuclear reactors, is graphitized at temperatures to 5400°F (2982°C) to free it of silicon, calcium, aluminum, and manganese, and treated with a Freon gas to eliminate boron and vanadium. In general, **artificial graphite** made at high temperatures in the electric furnace is now preferred for most uses because of its purity.

Recrystallized graphite is produced by a proprietary hot-working process which yields recrystallized or "densified" graphite with specific gravities in the 1.85 to 2.15 range, as compared with 1.4 to 1.7 for conventional graphites. The material's major attributes are a high degree of quality reproducibility, improved resistance to creep, a grain orientation that can be controlled from highly anisotropic to relatively isotropic, lower permeability than usual, absence of structural macroflaw, and ability to take a fine surface finish.

Graphite fibers are produced from organic fibers such as rayon. The fiber or textile form (for example, fabric, yarn, or felt) is graphitized at temperatures up to 5400°F (2982°C). The resulting fibers are high-purity (99.9%) graphite, with extremely high individual fiber strengths. They run about one-third of a mil in diameter, and have a specific gravity of from 1.7 to 2.0. When used in composites, they are generally made into yarn containing some 10,000 fibers. Depending on the precursor fiber, their tensile strength ranges from 200,000 to nearly 500,000 lb/in^2 (1,378 to 3,445 MPa); and their modulus of elasticity is from 28 million to 75 million lb/in^2 (0.2 million to 0.5 million MPa). **Graphite fiber-epoxy composites** provide exceptionally high strength and stiffness, and because of their light weight are finding increasing use for golf club shafts and tennis racquet frames. **PT graphites** are graphite fibers impregnated or bonded with an organic resin (such as furfural) and then carbonized. The result is a graphite-reinforced carbonaceous material with a high degree of thermal stability. The composite has a low density (0.93 to 1.2 specific gravity), and what is reported to be the highest strength-to-weight ratio of any known material at temperatures in the 4000 to 5000°F (2204 to 2760°C) range.

Extremely fine particles of pure artificial graphite, or **colloidal graphite,** will remain in suspension indefinitely, and are marketed in distilled water, oils, or solvents, under trade names. **Mexacote,** of the U.S. Graphite Co., is colloidal graphite powder to be mixed with water for spraying on sand molds. When a solution of colloidal graphite in alcohol is sprayed on machine bearings, the alcohol evaporates to leave a thin coating of graphite as a lubricant. **Prodag,** of Acheson Industries, Inc., is a solution in water for foundry facings. **Dag dispersion 154,** of the same company, is colloidal graphite in ethyl silicate used to produce black coatings on glass. **Dag 440** is **graphite powder** in silicone resin for use as a resistance coating for continuous temperatures up to 500°F (260°C). A 0.002-in (0.005-cm) coating has a volume resistivity of 100 Ω/in^2 (0.065 Ω/m^2). **Grafita** and **Grafene,** of the U.S. Graphite Co., are grease and oil solutions of colloidal graphite for producing lubricating films. **Acheson 1127,** of Acheson Colloids, is a **release agent** for spraying on molds in aluminum die-casting machines. It is a water solution of graphite powder. It prevents adherence of the metal to the die and also gives better flow of metal because of reduced friction.

For making lead pencils, amorphous graphite is mixed with clay and fired, the amount of clay determining the hardness of the pencil. **Flexicolor pencils,** of Koh-I-Noor, Inc., have a plastic binder to give pliable strength to the pencil. **Polo Graphite,** of the Pure Oil Co., is a graphite powder with maximum particle size of 0.001 in (0.003 cm) used for molded parts. The parts have a smooth surface and a tensile strength to 6,000 lb/in^2 (41 MPa). **Graphite carbon raiser** is a term given to graphite powder used for adding to molten steel to raise the carbon content. Molded **graphite brushes** for motors and generators may have metal powders mixed with the graphite to regulate the conductivity or may be molded from carbon-graphite powder especially heat-treated. **Graphited metals,** used for bearings and bushings, are made by molding metal powders with graphite and sintering, and may contain up to about 45% graphite evenly dispersed in the metal matrix to act as a lubricant. Or powdered oxides of the metals may be used, and these oxides are reduced in the sintering to give greater porosity for oil retention. **Genelite,** of the General Electric Co., is a porous **graphited bronze** made with oxides of copper, tin, and lead, with graphite powder. It will absorb about 20% by volume of oil. It has a compressive strength of 50,000 lb/in^2 (344 MPa), and a tensile strength of 8,000 lb/in^2 (55 MPa). **Durex bronze,** of the General Motors Corp., made by reducing the oxides with graphite under pressure, will take up 29% by volume of oil. Graphited metals, with the matrix of bronze or of babbitt, are marketed in the forms of rods and bushings under a variety of trade names, such as **Gramix,** of the U.S. Graphite Co.;

Graphex, of the Wakefield Bearing Corp.; and **Ledaloyl,** of the Johnson Bronze Co. **Iron-bonded graphite,** developed by the Ford Motor Co. for oilless bearings, is made by powder metallurgy with a content of 40 to 90% graphite and the balance iron powder and a calcium-silicon powder. The calcium-silicon overcomes the normal low compatibility of the iron and graphite, resulting in a strong, nonbrittle material.

The **supergraphite,** of the National Carbon Co., used for rocket casings and other heat-resistant parts, is recrystallized molded graphite. It will withstand temperatures to 5500°F (3038°C). **Pyrolytic graphite,** developed by the General Electric Co., is an oriented graphite. It has high density, with a specific gravity of 2.22, has exceptionally high heat conductivity along the surface, making it very flame-resistant, is impermeable to gases, and will withstand temperatures to 6700°F (3704°C). It is made by deposition of carbon from a stream of methane on heated graphite, and the growing crystals form with thin planes parallel to the existing surface. The structure consists of close-packed columns of graphite crystals joined to each other by strong bonds along the flat planes, but with weak bonds between layers. This weak and strong electron bonding is characteristic of a semimetal, giving the laminal structure. The material conducts heat and electricity many times faster along the surface than through the material. The flexural strength is 25,000 lb/in^2 (172 MPa) compared with less than 8,000 lb/in^2 (55 MPa) for the best conventional graphite. At 5000°F (2760°C) the tensile strength is 40,000 lb/in^2 (275 MPa). Sheets as thin as 0.001 in (0.003 cm) are impervious to liquids or gases. It is used for nozzle inserts and reentry parts for spacecraft. **Boron Pyralloy**, of High Temperature Materials, Inc., for atomic shielding, is this material with an addition of boron.

GRAVEL. A natural material composed of small, usually smooth, rounded stones or **pebbles.** It is distinguished from sand by the size of the grain, which is usually above ¼ in (0.64 cm), but gravel may contain large stones up to 3 in (7.62 cm) in diameter, and some sand. It will also contain pieces of shale, sandstone, and other rock materials. Gravel is used in making concrete for construction, and as a loose paving material. Commercial gravel is washed to remove the clay and organic matter, and screened. **Pea gravel** is screened gravel between ¼ and ½ in (0.64 to 1.27 cm) in diameter. It is used for surfacing with asphalt, or for roofing. Gravel is sold by the cubic yard or by the ton, and is shipped by weight. **Bank-run gravel,** with both large and small material, weighs about 3,000 lb/yd^3 (1,779 kg/m^3).

GRAY CAST IRON. The most common, least costly, and most widely used of the cast irons. It also holds the honor of being the cheapest of all engineering metals. The distinguishing microstructural characteristic is the presence of graphite flakes in a matrix of, usually, ferrite, pearlite, and austenite. The graphite flakes occupy about 10% of the total volume of the metal so that gray irons have a lower density than steels. Alloying produces a broader range of properties than is possible in unalloyed types. Common elements added are chromium, copper, nickel, molybdenum, and vanadium. Standard gray irons are classified according to tensile strengths into nine main classes from 20,000 to 65,000 lb/in^2 (137 to 448 MPA), in increments of 5,000 lb/in^2 (34 MPa). A complicating factor is that the tensile strength varies with casting section thickness. Thus strengths of class 20 [20,000 lb/in^2 (137 MPa)], for example, can range from 13,000 to 40,000 lb/in^2 (89 to 275 MPa), depending on the thickness of the test bar. This strength increase occurs as section thickness decreases because of the faster cooling rates in thinner sections.

Gray cast irons possess the highest fluidity of the ferrous casting metals and are thus well suited to the production of relatively intricate and thin-walled castings. In addition, solidification shrinkage is low, ranging from 0 to about 2%, compared to about 5% for steel and malleable iron. Also, the machinability is superior to virtually all grades of steel. Gray irons are notably low in tensile strength but high in compressive strength—about three to four times the equivalent values of tensile strength. Their compressive strengths are higher than those of most nonferrous materials and are about equal with those of non-heat-treated low-alloy steels. They retain their strength properties down to below −300°F (−184°C) and up to about 800°F (427°C). Unlike that of most ferrous metals, the modulus of elasticity of gray irons is not constant. It ranges from 12 million to 20 million lb/in^2 (82,700 to 138,000 MPa). It decreases with increasing strain and varies with the specific grade of iron. Because their impact resistance is lower than that of most other cast irons, gray irons are not suitable for applications where extreme overloading might be encountered. However, they have good damping capacity, which increases as the amount of flake graphite increases.

Gray irons have excellent wear resistance because of the presence of graphite. Hardened gray iron is used to obtain maximum wear properties. Although not generally considered a corrosion-resistant material, gray irons offer better corrosion resistance than most carbon steels. The rusting action forms a relatively adherent protective coating that offers fairly good resistance to the atmosphere, soil, acids, and alkalies. For example, some cast-iron water and gas pipes have been in service for more than 100 years. Because of gray irons' low cost and excellent castability, they are produced in a wide size range for a great variety of parts and components. The largest single use may be for automotive cylinder blocks. Huge machine and equipment bases, large gears, heavy compressors, diesel engine castings, heavy rolls, high-pressure cylinders, press and crusher frames, flywheels, and brake drums are other common uses.

Gray irons containing over about 3% in alloying elements are classified as **alloy cast irons.** The alloying elements are silicon, nickel, chromium, copper, and aluminum, either singly or in combination. One type of **high silicon iron** contains 14 to 17% silicon. Other grades contain 4 to 6%. The 14 to 17% grades are used for handling corrosive liquid chemicals. Their shock resistance is poor and they are virtually unmachinable. The 4 to 6% silicon irons have excellent resistance to scaling and growth in use in temperatures up to 1650°F (899°C), but their thermal shock resistance is low.

There are a number of **austenitic gray-iron alloy** grades that contain 18% or more of nickel, up to 4% chromium, and up to 7% copper. Because they have good scaling and growth resistance up to about 1500°F (816°C), and are much tougher and more resistant to thermal shock than silicon- and chromium-alloy white irons, they are used where heat, corrosion, and wear resistance are important. For example, traditionally these alloy irons have served as stove tops as well as for heavy-duty industrial products such as flood gates, seawater valves, turbocharger housings, and caustic pumps and valves.

High-aluminum irons, which contain 18 to 25% aluminum, have excellent scaling resistance along with good corrosion resistance in the presence of hot acids. However, brittleness and the difficulty in producing sound castings limits their use. Small amounts of tungsten increase the tensile and transverse strength of cast iron, but reduce its hardness. **Tungsten cast iron,** with 1.2% tungsten, has a tensile strength of about 40,000 lb/in^2 (275 MPa). Case hardening of cast iron is done by using powdered ferromanganese on the mold surface. **Nitrided cast iron,** hardened by nitrogen treatment to give great wear resistance, usually has a special composition. A typical composition for the **Nitricastiron,** as given by the Electro Metallurgical Co., is 2.9% total carbon, 0.6 combined carbon, 2.7 silicon, 0.75 manganese, 1.3 chromium, 1 aluminum, 0.16 vanadium, 0.25 molybdenum, with sulfur and phosphorus below 0.07. It can be hardened to 800 to 1,000 Brinell to a depth of 0.004 to 0.006 in (0.010 to 0.015 cm).

GREENHEART. The wood of the tree *Octotea rodioei* of Guyana, especially valued for shipbuilding, dock timbers, planking, and lock gates because of its resistance to fungi and termites. It also goes under the name of **Demerara greenheart,** and under its native name of **bibiru.** The tree grows to a height of 120 ft (37 m), with diameter of 2 to 3 ft (0.6 to 0.9 m), clear of branches for 50 to 70 ft (15 to 21 m). The wood is very strong and hard with good wearing qualities. The average weight is 62 lb/ft^3 (993 kg/m^3) with 12% moisture. The specific gravity, oven-dried, is 0.80. The heartwood is light olive to nearly black, and the sapwood pale yellow to greenish.

Surinam greenheart is the wood **bethabara,** from the tree *Tecoma leucoxylon* of Surinam and French Guiana. It is distinguished from greenheart by the yellow deposits in the pores, and is not as resistant as greenheart.

Manbarklak, the wood of the tree *Eschweilera corrugata,* of Surinam, is reddish brown in color, weighs 76 lb/ft^3 (1,218 kg/m^3) is equal to greenheart for marine construction, but is scarcer. **Angelique,** the wood of the tree *Dicorynia paraensis* of the Guianas and lower Amazon, is very resistant to fungi and insect attack, and is used as a substitute for greenheart in marine construction. It is hard but not as heavy as greenheart, weighing 50 lb/ft^3 (801 kg/m^3). It has an olive-brown color with reddish patches. **African greenheart** is **dahoma,** the wood of the large tree *Piptadena africana* of the west coast of Africa. It is yellowish brown in color, weighs 50 lb/ft^3 (801 kg/m^3), but is not as resistant to marine borers as greenheart. The wood of a species of tonka bean tree of Panama, known as **almendro,** *Coumarouna panamensis,* is very resistant to marine borers and is used as a substitute for greenheart. It is yellowish brown in color and is very heavy. The almondlike seeds of the tree are used as a food.

GRINDING PEBBLES. Hard and tough rounded small stones, usually of flint, employed in cylindrical mills for grinding ores, minerals, and cement. Pebbles from Greenland, marketed usually through Denmark and known as **Danish pebbles,** are of great hardness and toughness. Newfoundland also supplies these pebbles. Quantities of **flint pebbles** also come from Denmark for use in tube mills. They are smooth, round pebbles formed by the washing of the sea on the chalk cliffs, and come from the islands off the Danish coast. Danish pebbles are graded in seven sizes, according to French standards, from No. 0 which is 1⅛ to 1¾ in (3 to 4 cm) to size E which is 4 to 5 in (10 to 13 cm). Small pebbles, ½ in (1.27 cm) in diameter, are used for polishing iron castings by tumbling. American grinding pebbles are chiefly from Minnesota, Ohio, Nevada, and from the beaches of California. **Quartzite pebbles** are produced in Nova Scotia and Saskatchewan. **Quartz pebbles** from Alabama are 99% silica and low in iron, but they do not wear as well as true flint pebbles. Granite, rhyolite, and andesite pebbles are also used for grinding. The **granite pebbles** produced in North Carolina from feldspar or other mineral processing are broken into cubes and wet-milled to remove the edges and corners. They are graded in sizes from 1½ to 5 in (3.8 to 12.7 cm). Because of greater uniformity of size and hardness, manufactured abrasive materials are now generally preferred to natural pebbles.

The **tumbling abrasives,** for use in tumbling barrels, come in aluminum oxide or silicon carbide preformed balls, cubes, triangles, or cylinders of various sizes to conform to the parts being tumble-polished. Typical are the **aluminum oxide balls,** ¾ to 2 in (1.9 to 5.1 cm) diameter, pressed to a uniform density and marketed under the name of **Starrlum** by the American Refractories & Crucible Corp. **Porcelain grinding balls** are made of high-grade resistant porcelain, and are marketed in stock sizes for grinding and polishing. They have the advantage over flint pebbles of greater uniformity. The material is dense, and is 80% heavier and 50% harder than flint.

GRINDSTONES. **Sandstones** employed for grinding purposes. Grindstones are generally used for the sharpening of edged tools, and do not compete with the hard emery, aluminum oxide, and silicon carbide abrasive wheels which are run at high speeds for rapid cutting. Grindstones are quaried from the sandstone deposits and made into wheels usually ranging from 1 to about 6 ft (0.3 to about 1.8 m) in diameter, and up to 16 in (41 cm) in thickness. They are always operated at low speeds because of their inability to withstand high centrifugal stresses. The grades vary from coarse to fine. Good grindstones have sharp grains, without an excess of cementing material that will cause the stone to glaze in grinding. The texture must also be uniform so that the wheel will wear evenly. The hard silica grains are naturally cemented together by limonite, clay, calcite, quartz, or mixtures. Too much clay causes crumbling, while too much calcite results in disintegration in the atmosphere. An excess of silica results in a stone that is too hard.

GUAIACUM OIL. An essential oil distilled from the wood of the guayacan tree of Paraguay, used in medicine, soaps, and perfumes. It is light gray in color, and is solid at temperatures below 45°C. The odor is that of a combination of tea leaves and roses. It is also called **guaiacwood oil.** The wood yields 5 to 6% of the oil. **Guaiac gum,** also called **guaiacum,** is gum resin of the true lignum vitae trees for use in varnishes, as a chemical indicator, and to prevent rancidity and loss of flavor in preserved and dehydrated foods. It is an effective antioxidant, although its chief use is in medicine as a stimulant and laxative. The resin comes in greenish-brown tears. **Azulene** is a blue dye extracted from guaiacum oil, from eucalyptus oil, and from some balsams. The azulenes derive names from the source, as **guaiazulene,** but they all have the empirical formula $C_{15}H_{18}$, the molecule having two rings and five double bonds. The synthetic material is known as **vetivazulene.** It comes in cobalt-blue crystals melting at 90°C.

GUANIDINE NITRATE. A white granular powder of the composition $(H_2N \cdot CNH \cdot NH_2) \cdot HNO_3$, produced from cyanamide and ammonium nitrate, and employed in making flashless propellants and pharmaceuticals, for coating blueprint paper to speed up development time, and in photographic fixing baths. It melts at 206 to 212°C and is strongly acid. **Guanidine carbonate** is a white nonhygroscopic granular powder of the composition $(H_2N \cdot CNH \cdot NH_2)_2 \cdot H_2CO_3$. It decomposes at 198°C. It is soluble in water, and has an alkaline reaction. Guanidine carbonate is used in the manufacture of pharmaceuticals, as an emulsifier in soaps, and in photographic developing solutions. Other derivatives of guanidine, such as **diphenyl guanidine,** are used as plasticizers and accelerators for rubber.

Guanidine, also called **carbamidene,**, is a white crystalline hygroscopic powder of the composition $HN{=}C(NH_2)_2$, produced from ammonium thiocyanate, chloropicrin, or cyanogen chloride for use in plastics, pharmaceuticals, and other chemicals. It is a strong base and forms salts with acids even as weak as carbonic. It is classified generally with the single-carbon group of chemicals based on **cyanogen,** N :CC:N, which itself is a poisonous colorless gas. The chemistry of this vast interrelated group is highly complex, including urea and the amino acids, and the **purines**, which are the basic cyclic organic compounds of which **uric acid** is the best known. The purines also include caffeine and **xanthine,** found in plants, and **guanine,** found in guano and in the fish scales used for imitation pearl. **Kelzan,** of the Kelco Co., used as a thickening and stabilizing agent, is synthetic **Xantham gum** made by fermentation of glucose. It is nontoxic and soluble in water. Xantham gum is a polysaccharide of **xanthic acid** occurring in many plants.

GUAVA. The fruit of the tree *Psidium guajava* of the myrtle family to which the clove, allspice, and eucalyptus belong. Because of its content of acids, sugars, pectin, and vitamins, it is valued in the food industries for blending in jellies, preserves, and beverages. The pulp is also used in ice-cream manufacture. Vast quantities of guava are used for hard jellies in Latin America. The fruit has more than 10 times the vitamin C content of the orange and retains it better. It also contains vitamins A and B_1 and 11.6% carbohydrates. The tree is native to America from Mexico to Peru and is also grown extensively in Brazil and the West Indies. It was one of the original Aztecan and Incan fruits, and is still known under the Carib name of **guayaba** in Brazil and Argentina. There are about 150 varieties, and the large seedy fruit has a fragrant aroma and distinctive sweet flavor.

GUAYULE A perennial plant, *Parthenium argentatum,* of the *Compositae* family, grown in the semiarid regions of northern Mexico and southern California as a source of rubber. The plants are hardy woody shrubs that mature into the highest rubber content in 7 years. They contain in the dry state up to 22% guayule rubber, in all parts except the leaves. The plant is uprooted in 3 to 5 years and is crushed and pulverized in mills, and the rubber extracted by flotation. The guayule rubber contains from 20 to 25% resin so that it is suitable only for blending or for cements unless deresinated. Natural crude guayule is softer than hevea rubber, owing to the content of natural resins which act as plasticizers. In the low-sulfur compounds it remains permanently tacky, and is thus valued for use as a coating adhesive for the permanently tacky binding tape known as Scotch tape. When deresinated by extraction of the resin with acetone or other solvent, the rubber is suitable for all the uses of hevea rubber. The by-product **guayule resin** is used in plastics. From 400 to 1,000 lb (181 to 454 kg) of rubber are produced per acre under cultivation.

GUM. The wood of the tree *Liquidambar styraciflua,* of the United States and Mexico. It is called **red gum** and **sweet gum.** In England it is known as **California red gum** although the gumwood of California is from a eucalyptus tree. In Europe, also, the term **satin walnut** is used for the heartwood and **hazel pine** for the sapwood. Gum has a reddish-brown color, is soft with a fine, close grain, and weighs about 40 lb/ft^3 (641 kg/m^3). It is used for furniture, veneer, inside trim, cooperage, and the making of pulp for book paper. The timber is cut mostly in the southern states, especially in Louisiana, Mississippi, and Arkansas. The trees have a height 80 to 100 ft (24 to 30 m) and average 1½ to 3 ft (0.5 to 0.9 m) in diameter with a straight clear trunk. Red gum is from the heartwood of mature trees and is reddish brown. **Sap gum** comes from the outer portion of logs or from young trees and is white tinged with pink. Nearly 25% of all the hardwood used in the United States is red gum. It has an interlocking grain which gives a fine appearance in veneers, but has a tendency to warp.

Gum is graded according to standards of the National Hardwood Lumber Association from firsts through selects to No. 3B common. Local names for red gum are **southern gum, sycamore gum, bilsted,** and **star-leafed gum. Cotton gum,** or **tupelo,** of Louisiana, is from the tree *Nyssa aquatica.* It is also known as **water tupelo, tupelo gum, swamp gum, sour gum. Black gum,** or **black tupelo,** also of the southern states is *N. sylvatica.* **Swamp tupelo,** also called **water gum** and **swamp black gum,** is from the tree *N. biflora.* **Ogeche tupelo** is from the tree *N. ogecha,* and is not common. It is called **gopher plum, wild limetree,** and **sour tupelo.** Black tupelo grows from New Hampshire to central Texas, but water tupelo is found chiefly along the coasts and river valleys of the South. The shipments of woods are usually mixed. Tupelo woods from the various species of *Nyssa* are fine-textured but with large pores. The heartwood is brownish gray, and the sapwood is grayish white. They are tough and difficult to split, having an interlocking grain, and find wide use for such articles as mallets, toilet seats, furniture, and bottle cases.

GUM ARABIC Also called **acacia gum.** The gum exudation of the small tree *Acacia arabica,* and various other species of acacia trees of Africa. **Kordofan gum,** or **hashab gum,** is a variety from the Red Sea area and forms the chief export of the Sudan. It is obtained by tapping the wild tree *A. verek,* and is of high quality. **Sennaar gum** is gum arabic exported from

Arabian ports on the Red Sea. **Gum senegal,** from the *A. senegal,* comes from the dry regions of northwest Africa. **Gum talha, talco gum,** or **talh gum,** is a brittle and low grade of gum arabic from the North African acacia, *A. stenocarpa.* Gum arabic is used for adhesives, for thickening inks, in textile coatings, and in drug and cosmetic emulsions. As a binder in pharmaceutical tablets the powder acts as a **disintegrating agent** to make the tablets easily soluble in water. In confectionery glazes it prevents crystallization of the sugar.

To obtain the gum the trees are wounded and the sap is allowed to run out, forming in yellowish, transparent lumps. It is also marketed as a white powder of 120 mesh, soluble in water, but insoluble in alcohol. Gum arabic is a mixture of calcium, magnesium, and potassium salts of **arabic acid,** in a complex of the saccharides **arabinose, galactose, rhamnose** or **mannomethylose,** and the open-chain **glucuronic acid.** It has a molecular weight of 240,000, and an acid reaction. For drug uses gums are selected, blended, and ground to a powder of uniform characteristics. **Arabasan,** of the Tetroid Co., Inc., is a spray-dried blended gum of this kind. It is a white powder, colorless in solution. **Larch gum,** or **galactan gum,** leached from chips of the western larch, is a copolymer of arabinose and galactose and is very similar to gum arabic. **Stractan,** of Stein Hall and Co., is this material in white powder form. It is used as a stabilizer and binder in coatings, adhesives, pharmaceuticals, and in low-calorie foodstuffs. **Tamarind seed gum** is from the pod beans of the tamarind tree, *Tamarindus indica,* of India. It is a white to tan, water-soluble powder used as a low-cost alternate for gum arabic. It is a polysaccharide but differs from arabic chemically by containing 12% tartaric acid and 30% sugars. It is used also in medicine and for beverages under the name of tamarind.

Various synthetic water-soluble gums and emulsifiers are now used as replacements for gum arabic in drugs, cosmetics, adhesives, and foodstuffs. The **Bemul** of the Beacon Co. is a glyceryl monostearate, soluble also in alcohol. The water-soluble **Kelzan gum** of the Kelco Co. is made from glucose. The wood of the gum arabic tree is the **satinwood** of the Near East, valued since ancient times for its great durability. It is light in weight, hard, close-grained, and has an orange-brown color that darkens with age. But the satinwood of Brazil, known also as **setim,** is a yellowwood from the large tree *Aspidosperma eburneum,* used for inlays.

GUNMETAL The name for a casting bronze containing 88% copper, 10 tin, and 2 zinc. It was originally used for small cannon but is now used where the golden color and strong noncrystalline structure are desired. It casts and machines well, and is suitable for making steam and hydraulic castings, valves, and gears. It has a tensile strength of 32,000 to 45,000 lb/in^2 (226 to 310 MPa), with elongation 15 to 30%. The specific gravity is

8.7, weight 0.315 lb/in^3 (8,719 kg/m^3), and Brinell hardness 65 to 74. This alloy is the same as the **G bronze** of the U.S. Navy. In England it is called **admiralty gunmetal,** and is specified as **BES No. 383** for sand castings. **Gunmetal ingot,** of H. Kramer & Co., may have the zinc replaced by 2% lead. Such an alloy is easier to machine but has less strength. **Modified gunmetal** contains lead in addition to the zinc. It is used for gears and for bearings. A typical modified gunmetal by William H. Barr, Inc., contains 86% copper, 9.5 tin, 2.5 lead, and 2 zinc. It has a tensile strength up to 40,000 lb/in^2 (275 MPa), elongation 15 to 25%, Brinell hardness 63 to 72, and weight 0.31 lb/in^3 (8,580 kg/m^3).

GUNPOWDER. Also known as **black powder.** An explosive extensively used for blasting purposes and for fireworks. It was introduced into Europe prior to 1250, and was the only propellant used in guns until 1870. It is now superseded for military uses by high explosives. Black powder deteriorates easily in the air from the absorption of moisture. It is a mechanical mixture of potassium nitrate, charcoal, and sulfur, in the usual proportions of 75, 15, and 10. More saltpeter increases the rate of burning; additional charcoal decreases the rate. A slow-burning powder for fireworks rockets may have only 54% saltpeter and 32 charcoal. Commercial black powder comes in grains of graded sizes and is glazed with graphite. The grain sizes are known as **pebble powder,** large-grain, fine-grain, **sporting powder, mining powder, Spanish spherical powder,** and **cocoa powder.** The potential energy of gunpowder is estimated at 500 ft·tons/lb (305,000 kg·m/kg), but the actual gun efficiency is less than 10% of this. A temperature of about 2100°C is produced by the explosive. Gunpowder is the slowest-acting of all the explosives, and has a heaving, not a shattering, effect. Hence, it is effective for blasting and breaking up stone. **Blasting powder** is divided by Du Pont into two grades: A and B. The **A powder** contains saltpeter; the **B powder** contains nitrate of soda. The other ingredients are the usual sulfur and charcoal. B powder is not so strong or water-resistant as A powder, but is cheaper and is extensively used. **Pellet powder** is blasting powder made up in cylindrical cartridges for easier use in mining. **White gunpowder** is a powder in which the saltpeter is replaced by potassium chlorate. It is very sensitive and explodes with violence. It is used only for percussion caps and for fireworks, **Lesmok powder,** used in 22-caliber cartridges, is composed of 15% black powder and 85 nitrocellulose.

GURJUN BALSAM. Also known as **wood oil,** and sometimes called **East Indian copaiba.** An oleoresin obtained from various species of the *Dipterocarpus* tree, about 50 varieties of which grow in India, Burma, Sri Lanka, and the Malay Peninsula. It is a clear liquid with a greenish fluorescence.

The specific gravity is 0.955 to 0.965. It is soluble in benzene. Gurjun balsam is used in lacquers and varnishes that are capable of resisting elevated temperatures. The burmese trees form two groups yielding products known as kanyin and in. **Kanyin oils** are brown in color, while the **In oils** are whitish and heavier. Gurjun balsam may consist of either or both of these products. Commercial **gurjun oil** is obtained by steam distillation of the balsam, and has a specific gravity of 0.900 to 0.930. It is soluble in alcohol. **Copaiba balsam** is a resin obtained from the **copaifera tree,** a species of *Dipterocarpus,* of South America. **Maracaibo copaiba** and **Para copaiba** are the principal varieties. They are dark yellow or brown in color, and are soluble in alcohol. The resin is used as a plasticizer, in varnishes, tracing paper, and in pharmacy. The specific gravity is 0.940 to 0.990.

GUTTA PERCHA. A gum obtained by boiling the sap of species of trees of the order *Sapotaceae,* chiefly *Palaquium gutta* and *P. oblongifolia,* native to Borneo, New Guinea, and Malaya. It is grayish white, very pliable, but not elastic like rubber. It is harder and a better insulator than rubber. Gutta percha, like rubber, will vulcanize with sulfur and form a hard material. It is used for mixing with rubber, but its chief use was in the covering of electric cables. It has a greater pliability than rubber for the given hardness required in cable insulation. This property with its greater resistance to water makes it valuable for golf balls and dental fillings. It molds easily at 180°F (82°C). It is also employed like balata for impregnating driving belts, and for washers and valve seats, and in adhesives. Gutta percha is often imported as mixtures with inferior guttas from other trees. **Gutta soh,** for example, is a mixture from Singapore often colored red artificially. **Gutta siak** is a low-grade gutta from Sumatra. It has a reddish color, and is lightly elastic. **Gutta sundik** is from the tree *Payena leerii* of Malaya and Indonesia. It is white in color, and is mixed with gutta percha. Another gutta used to adulterate gutta pecha is **gutta hangkang,** from the *Palaquium leiocarpum* of Borneo. It is slightly reddish. **Gutta jangkar** is a low-grade red gutta from Sarawak. **Gutta susu** of Indonesia is a gray-colored material, slightly elastic. With increasing use of synthetic resin insulating materials, gutta percha has become of only secondary importance.

GYPSUM A widely distributed mineral which is a hydrated **calcium sulfate,** $CaSO_4 \cdot 2H_2O$, used for making building plaster, wallboard, tiles, as an absorbent for chemicals, as a paint pigment and extender, and for coating papers. Natural gypsum of California, containing 15 to 20% sulfur, is used for producing ammonium sulfate for fertilizer. Gypsum is also used to make sulfuric acid by heating to 2000°F (1093°C) in an air-limited furnace. The resultant calcium sulfide is reacted to yield lime and sulfuric acid. Raw gypsum is also used to mix with portland cement to retard the set. Compact massive types of the mineral are used as building stones. The color is naturally white, but it may be colored by impurities to gray, brown, or red. The specific gravity is 2.28 to 2.33, and the hardness 1.5 to 2. It dehydrates when heated to about 190°C, forming the hemihydrate $2CaSO_4 \cdot H_2O$, which is the basis of most gypsum plasters. It is called **calcined gypsum,** or when used for making ornaments or casts is called **plaster of paris.** When mixed with water, it again forms the hydrated sulfate that will solidify and set firmly owing to interlocking crystallization. Theoretically, 18% of water is needed for mixing, but actually more is necessary. Insufficient water causes cracking. Water solutions of synthetic resins are mixed with gypsum for casting strong waterproof articles. **Palestic,** of the Palestic Corp., is gypsum mixed with a urea-formaldehyde resin and a catalyst. It expands slightly on hardening, and thus gives a good impression of the mold. The tensile strength of the molded material is 1,100 lb/in^2 (7 MPa), and the compressive strength is 12,000 lb/in^2 (82 MPa).

Calcium sulfate without any water of crystallization is used for paper filler under the name of **pearl filler,** but is not as white as the hydrated calcium sulfate called **crown filler.** The paint pigment known as **satin spar** is a fibrous, silky variety of gypsum, and is distinct from the pigment called **satin white,** made by precipitation of aluminum sulfate with lime. **Terra alba** is an old name for ground gypsum as a paint filler. The anhydrous calcium sulfate in powder and granular forms will absorb 12 to 14% of its weight of water, and is used as a drying agent for gases and chemicals. It can then be regenerated for reuse by heating. **Drierite,** of the W. A. Hammond Drierite Co., is this material. The mineral mined at Billingham, England, as **anhydrite** is **anhydrous calcium sulfate.** It is used for producing sulfur, sulfur dioxide, and ammonium sulfate. The **Hydrocal** of the U.S. Gypsum Co., used as a filler for paints and plastics, is $CaSO_4 \cdot \frac{1}{2}H_2O$. Much calcined gypsum, or plaster of paris, is used as **gypsum plaster** for wall finish. For such use it may be mixed in lime water or glue water, and with sand. Because of its solubility in water it cannot be used for outside work. **Neat plaster,** for walls, is the plaster without sand. When the term **plaster** alone is used, it generally refers to gypsum plasters, but the **calcium plaster** made from spent fuller's earth has greater workability and better water resistance than gypsum plaster.

Plasterboard, or **gypsum wallboard,** consists of sheets or slabs of gypsum mixed with up to 15% of fibers, employed as a fire-resistant material for walls, ceilings, or partitions, but most of the **wallboard** used in dwelling houses is gypsum board faced on both sides and edged with paper. It usually comes ½ in (1.27 cm) thick, 4 by 8 ft (1.2 by 2.4 m), weighing 2 lb/ft^2 (10 kg/m^2). **Macoustic,** of the National Gypsum Co., is a lightweight gypsum acoustical plaster. **Grain board** is a fireproof gypsum board with an imitation wood-grain surface used for walls. The hard-finish plasters for **flooring plaster** may contain alum or other materials. **Scott's cement** is a plaster made by grinding lime with calcined gypsum. It sets rapidly. **Mack's cement** is a hard cement made of dehydrated gypsum to which is added a small percentage of calcined sodium sulfate and potassium sulfate. It sets quickly and has good adhesion. It is used for walls, and for floors when mixed with sand.

Patterns, models, and molds of plaster of paris have their strength raised and are made water-resistant by impregnating the dried and hardened plaster with a synthetic resin, particularly the furane resins which cure at low temperatures without pressure. The ordinary casting plaster of 100 parts solids and 60 parts water, called **hydrocal,** is slightly acid, and must be treated with an acid-catalyzing resin, while the low-expansion plaster of 100 parts solids and 45 water, called **hydrostone,** is alkaline and treated with alkaline-catalyzed resin. Impregnation of plaster-of-paris castings with resin raises the compressive strength from 2,000 up to 9,000 lb/in^2 (13 to 62 MPa), and the flexural strength from 700 to 3,500 lb/in^2 (5 to 24 MPa). Special liquid resins are marketed for this purpose. The **Plaspreg** of Furane Plastics, Inc., is a furane solution to which a catalyst is added before use.

A crystalline variety of gypsum, known as **selenite,** occurs in transparent crystals and usually splits in thin laminations. A fine-grained, marblelike variety, called **alabaster,** is employed in ornamental building work, and for lamps, vases, and novelties. Much alabaster is produced in Colorado. **Travertine,** which resembles alabaster but is grained like wood, is a water-deposited calcium carbonate. The **Italian travertine** is notable as a decorative building stone, but is also found in Georgia, Montana, and California. It is an ancient building material, and the great Colosseum at Rome was built of this stone.

HAFNIUM. An elementary metal, symbol Hf. It occurs in nature in about the same amount as copper, but is sparsely disseminated and is costly to extract. All zirconium minerals contain some hafnium, but the two metals are so similar chemically that separation is difficult. All zirconium chemicals and alloys may contain some hafnium, and hafnium metal usually contains about 2% zirconium. The melting point is higher than that of zirconium, about 4000°F (2204°C), and heat-resistant parts for special purposes have been made by compacting hafnium powder to a density of 98%. The metal has a close-packed hexagonal structure. The electric conductivity is about 6% that of copper. It has excellent resistance to a wide range of corrosive environments. Because it has a high thermal-neutron-capture cross section and excellent strength up to 1000°F (538°C), hafnium is useful in unalloyed form in nuclear reactors.

Hafnium oxide, or **hafnia,** HfO_2, is a better refractory ceramic than zirconia, but is costly. The inversion of the crystal from the monoclinic to the tetragonal occurs at 3100°F (1704°C) with an expansion of only 3.4%, compared with 2000°F (1093°C) and an expansion of 7.5% in zirconia. **Hafnium carbide,** HfC, produced by reacting hafnium oxide and carbon at high temperature, is obtained as a loosely coherent mass of blue-black crystals. The crystals have a hardness of 2910 Vickers, and a melting point of 4160°C. It is thus one of the most refractory materials known. Heat-resistant ceramics are made from **hafnium titanate** by pressing and sintering the powder. The material has the general composition $_x(TiO_2)\cdot n(HfO_2)$, with varying values of x and n. Parts made with 18% titania and 82% hafnia have a density of 0.26 lb/in^3 (7,197 kg/ m^3), a melting point at about 4000°F (2204°C), a low coefficient of thermal expansion, good shock resistance, and a rupture strength above 10,000 lb/in^2 (68 MPa) at 2000°F (1093°C).

HAIR The fibrous covering of the skins of various animals, used for making coarse fabrics and for stuffing purposes. It is distinguished from wool in having no epidermal scales. It cannot be spun readily, although certain hairs, such as camel hair, are noted for great softness and can be made into fine fabrics. **Horsehair** is from the manes and tails, and is used as a brush fiber and for making haircloth. It is largely imported from China and Argentina, cleaned and sorted. The imported hair from live animals is more resilient than domestic hair from dead animals. **Brushhair** is usually cut to 3 to 5½ in (7.6 to 14 cm) long, but tail hair for making **curled hair** for weaving comes in lengths up to 30 in (76 cm). **Cattle hair** is taken from slaughtered animals in packing plants. The body hair is used as a binder in plaster and cements, for hair felt, and for stuffing. The tail hair is used for upholstery, filter cloth, and stuffing. The **ear hair** is used for brushes. It has a strong body and a fine tapered point suitable for poster brushes. In the brush industry it is known as **ox hair. Camox** is a mixture of squirrel hair and ox hair to combine the fineness of squirrel hair with the springiness of ox hair for one-stroke brushes and flowing brushes.

Artificial horsehair, or **monofil,** was a single-filament cellulose acetate fiber, used for braids, laces, hairnets, rugs, and pile fabrics. **Crinex** is a German artificial horsehair made of cuproammonium filament and used as a brush fiber. **Haircloth** is a stiff, wiry fabric with a cotton or linen warp and a filling of horsehair. It is elastic and firm, and is used as a stiffening and interlining material. The colors are black, gray, and white. The fabric is difficult to weave and disintegrates easily, as the hairs cannot be made into a single strand and must be woven separately. **Press cloth,** used for filtering oils, was made from human hair, which has high tensile strength, resiliency, and resistance to heat. The hair came from China, but filter fabrics are now made from synthetic fibers. **Rabbit hair** from Europe and Australia is used for making felt hats and is referred to as **rabbit fur,** although it does not felt like wool. The white rabbit hair known as **Angora wool** is from the Angora rabbit of France and Belgium, called **Belgian**

hare. The hairs are clipped or plucked four times a year when they are up to 3 in (8 cm) long. They are soft and lustrous, dye easily to delicate shades, and are used for soft wearing apparel. Because of its fluffiness and hairy characteristics, the wool is difficult to spin, and is usually employed in mixtures.

HEAT INSULATORS. Materials having high resistance to heat rays, or low heat conductivity, used as protective insulation against either hot or cold influences. The materials are also called **thermal insulators.** Insulators for extremely high external temperatures, as on aerospace vehicles, are of ablative materials. Efficiency of heat insulators is measured relatively by the Btu/(hr) (ft^2)(°F/ft), known as the K factor. The thermal conductivity of air and gases is low, and the efficiency of some insulators, especially fibrous ones, is partly due to the air spaces. On the other hand, the thermal conductivity of a porous insulator may be increased if water is absorbed into the spaces.

A wide variety of materials are used as thermal insulators in the forms of powder or granules for loose fill, blanket batts of fibrous materials for wall insulation, and in sheets or blocks. Although metals are generally high heat conductors, the polished white metals may reflect as much as 95% of the heat waves, and make good **reflective insulators.** But for this purpose the bright surface must be exposed to air space. Aluminum has a high K factor, up to 130, but crumpled aluminum foil is an efficient thermal insulator as a fill in walls. Wool and hair, either loose or as felt, with a K factor of 0.021, are among the best of the insulators, but organic materials are usable only for low temperatures, and they are now largely replaced by mineral wool or ceramic fibers. Mineral wool has a low K factor, 0.0225. The **Tipersul,** of Du Pont, is a **potassium titanate fiber** used loose or in batts, blocks, or sheets. Its melting point is 2500°F (1371°C), and it will withstand continuous temperatures to 2200°F (1204°C). Another ceramic fiber, called **Fibrox,** for the same purpose, is a **silicon oxycarbide,** SiCO, in light fluffy fibers.

Magnesia or asbestos, or combinations of the two, is much used for insulation of hot pipelines, while organic fibrous materials are used for cold lines. **High-heat insulators,** for furnaces and boilers, are usually made of refractory ceramics such as chromite. For intermediate temperatures, expanded glass, such as the **Foamglas** of the Pittsburgh-Corning Corp., may be used. Foamed glass blocks will withstand heats to 1000°F (538°C), and the blocks have a crushing strength of 150 lb/in^2 (0.7 MPa). Some rigid materials of good structural strength serve as structural parts as well as insulators. **Roofinsul,** of Johns-Manville, used for roof decks, is a lightweight board compressed from wood fibers. **Ludlite board,** of the Allegheny Ludlum Steel Co., for paneling, is thin stainless steel backed with a magnesia-asbestos composition. Insulators in sheets, shapes, and other forms are sold under a great variety of trade names. **Dry-Zero,** of the Dry-Zero Co., for refrigerator insulation, consists of kapok batts encased in fiberboard. The French cold-storage insulation known as **Isotela** consists of pads of matted coir. The **balsam wool,** of the Wood Conversion Co., is wood fibers chemically treated to prevent moisture absorption.

HEAT-RESISTANT CAST ALLOYS. Cast alloys intended for service in which the metal temperature will exceed 1200°F (649°C). Although some cast heat-resistant alloys are available in compositions similar to wrought heat-resistant or stainless steels or other alloys, it is necessary to differentiate between them. Cast alloys are made to somewhat different chemical specifications than wrought alloys. Therefore the heat-resistant cast alloys are identified by a designation code set up by the Alloys Casting Division of the Steel Founders' Society.

Heat-resistant cast alloys have both high strength and good chemical stability at temperatures between 1200 and 2200°F (649 and 1204°C). Although not immune to corrosive media, their rate of corrosion is low compared to low-alloy steels and cast irons. There are three groups of heat-resistant grades, according to composition and microstructure.

Iron-chromium cast steels contain 8 to 30% chromium and under 7% nickel. They are predominantly ferritic and therefore have relatively low hot strength. They are seldom used in critical load-bearing parts at temperatures above 1200 or 1400°F (649 or 760°C). They have excellent resistance to oxidation and to sulfur-containing atmospheres. **Iron-chromium-nickel cast steels** contain 18 to 32% chromium and 8 to 22% nickel, with the chromium content always higher than the nickel. Being partially or completely austenitic, they have greater high-temperature strength and ductility than the ferritic grades. They are characterized by good high-temperature strength, hot and cold ductility, and resistance to oxidizing and reducing conditions. They are particularly suited for parts that operate in atmospheres high in sulfur. They also have good weldability and generally good machinability. **Nickel-chromium-iron cast steels** contain 10 to 20% chromium and 30 to 70% nickel. They are austenitic, and the nickel predominates. If it were not for their relatively high cost and the problem of corrosion in high-sulfur atmospheres, these alloys could be used for practically all applications up to 2100°F (1149°C). They have good hot strength, carburization resistance, and thermal-fatigue resistance. Therefore they are widely used for load-bearing parts subject to large temperature differentials and cyclic heating. Although their resistance to high-sulfur atmospheres is low, they will withstand reducing and oxidizing atmospheres.

The alloys are generally marketed on the basis of the maximum service temperature rather than on the content. An alloy of 67% nickel, 16 chromium, 12 iron, and 1 manganese has a tensile strength of 64,000 lb/in^2

(441 MPa), which drops to 30,000 lb/in^2 (206 MPa) at about 1500°F (816°C), but will resist oxidation for long periods at 1800°F (982°C). The alloys are sold under trade names such as **Hoskins alloys** of the Hoskins Mfg. Co., **Fahrite** of the Ohio Steel Foundry Co., **Cromax** and **Veriloy** of the Driver-Harris Co., **Accoloy** of the Alloy Engineering & Casting Co., the **Q-Alloys** of the General Alloys Co., **Amsco alloy** of the Amsco Div., Abex Corp. Grades of these alloys may be modified with manganese, aluminum, silicon, titanium, and other elements to increase heat resistance, lower the coefficient of expansion, add creep resistance, and increase strength, or to increase resistance to hot corrosive chemicals.

HEAT-TRANSFER AGENTS. Liquids or gases used as intermediate agents for the transport of heat or cold between the heat source and the process, or for dissipating heat by radiation. Water, steam, and air are the most common heat-transfer agents, but the term is usually applied only to special materials. Air can be used over the entire range of industrially important temperatures, but it is a poor heat-transfer medium. Water can be used only between its freezing and boiling points, unless high pressures are employed to keep the water liquid. A liquid agent should have a wide liquid range, be noncorrosive and nontoxic, and have low vapor pressure to minimize operational loss.

Gallium, with a freezing point at 85.6°F (29°C) and boiling point at 3600°F (1982°C), offers an exceptionally wide liquid range, but it is too costly for ordinary use, and the liquid metal also attacks other metals. Mercury is used for heat transfer, but is costly and toxic, and at 1200°F (649°C) it exerts a vapor pressure of 500 lb/in^2 (3.4 MPa). **Anisol** is a methyl phenyl ether of the composition $C_6H_5OCH_3$. It freezes at −37°C and boils at 154°C, has low vapor pressure, and is used for heat transfer although its chief use is as a solvent for plastics and for recrystallization processes. **Aroclor 1248,** of Monsanto, can be used for temperatures up to 550°F (288°C) and, like anisol, is easily pumped at room temperature. It is a chlorinated biphenyl and is noncombustible.

Brine solutions of sodium or calcium chlorides are used for heat transfer for temperatures down to −6°F (−21°C), but are corrosive to metals. Molten sodium and potassium salts are used for temperatures from 600 to 1400°C, but are corrosive to metals. The **sodium-potassium salt,** NaK, called **Nack,** is also highly corrossive. The salt known as **Hitec,** which is a 50–50 mixture of sodium nitrite and potassium nitrate, melts at 282°F (139°C), remains liquid at high temperatures, and has no appreciable vapor pressure at 1200°F (649°C). **Tetraryl silicate** remains liquid between −40 and 700°F (−40 and 371°C), but is costly for most uses.

HEAVY ALLOY. A name applied to **tungsten-nickel alloy** produced by pressing and sintering the metallic powders. It is used for screens for X-ray tubes and radioactivity units, for contact surfaces for circuit breakers, and for balances for high-speed machinery. The original composition was 90% tungsten and 10 nickel, but a proportion of copper is used to give a lower sintering temperature and to give better binding as the copper wets the tungsten. Too large a proportion of copper makes the product porous. In general, the alloys weigh nearly 50% more than lead, permitting space saving in counterweights and balances, and they are more efficient as gamma-ray absorbers than lead. They are highly heat-resistant, retain a tensile strength of about 20,000 lb/in^2 (137 MPa) at 2,000°F (1093°C), have an electric conductivity about 15% that of copper, and can be machined and brazed with silver solder.

An alloy of 90% tungsten, 7.5 nickel, and 2.5 copper has a tensile strength of 135,000 lb/in^2 (930 MPa), compressive strength 400,000 lb/in^2 (2,757 MPa), elongation 15%, Rockwell hardness C30, and weight 0.61 lb/in^3 (16,885 kg/m^3). **Kenertium,** of Kennemetal, Inc., has this composition. **Hevimet,** of the General Electric Co., and **Mallory 1000,** of P. R. Mallory, Inc., are similar metals. **Mallory 3000** is in the form of rolled sheet for radiation shielding. The tensile strength of the sheet is 195,000 lb/in^2 (1,344 MPa), with a Rockwell hardness of A63. **Fansteel 77 metal,** of Fansteel, Inc., contains 89% tungsten, 7 nickel, and 4 copper. The density is 16.7, tensile strength 85,000 lb/in^2 (586 MPa), elongation 17%, and Brinell hardness to 280. The coefficient of expansion is low, 0.0000065 (cm)(cm)/°C. **Heavy metal powder,** of the Astro Alloys Corp., for making parts by powder metallurgy, is prealloyed with the tungsten in a matrix of copper-nickel to prevent settling out of the heavy tungsten.

HELIUM. A colorless, odorless, elementary gas, He, with a specific gravity of 0.1368, liquefying at −268.9°C, freezing at −272.2°C. It has a valency of zero and forms no electron-bonded compounds. It has the highest ionization potential of any element. The lifting power of helium is only 92% that of hydrogen, but it is preferred for balloons because it is inert and nonflammable, and is used in weather balloons. It is also used instead of air to inflate large tires for aircraft to save weight. Because of its low density, also, it is used for diluting oxygen in the treatment of respiratory diseases. Its heat conductivity is about six times that of air, and it is used as a shielding gas in welding, and in vacuum tubes and electric lamps. Because of its inertness helium can also be used to hold free chemical radicals which, when released, give high energy and thrust for missile propulsion. When an electric current is passed through helium it gives a pinkish-violet light, and is thus used in advertising signs. Helium can be obtained from atmospheric nitrogen, but comes chiefly from natural gas, the gas of Texas yielding 0.94%, with some gases yielding as much as 2%. It also occurs in the mineral **cleveite.** Helium is transported as a liquid in trucks and tube trailers. Helium marketed by Matheson Gas Products Co. for use in semiconductor production and where noncontaminating atmospheres are needed is 99.9999% pure.

HEMLOCK The wood of the coniferous tree *Tsuga mertensiana,* of northeastern United States. This species is also called **mountain hemlock,** and is now scarce. **Eastern hemlock,** *T. canadensis,* was formerly a tree common from eastern Canada to northern Alabama. In the southern area it is called **spruce pine,** and in the northern area **hemlock spruce.** The wood is coarse with an uneven texture, splintering easily. The trees are up to 80 ft (24 m) in height and up to 3 ft (0.9 m) in diameter. It is used for paper pulp, boxes and crates, and inferior lumber. **Western hemlock,** *T. heterophylla,* is a wood produced in abundance from Alaska to northern California. It is known also as **West Coast hemlock, hemlock spruce, Prince Albert fir, gray fir, Alaskan pine,** and **western hemlock fir.** Trees 100 years old are about 20 in. (0.5 m) in diameter and 140 ft (43 m) high. The wood is light in color, with a pinkish tinge, light in weight, moderately soft, and straight grained. It is nonresinous and free from resin ducts, but black knots are frequent. The select grades of the lumber are free from knots and suitable for natural and paint finishes. The wood is used for general construction, boxes, woodenware, and pulpwood. The lumber often comes mixed with Douglas fir. It is easy to work but does not plane smooth like pine. It has frequent dark streaks from heart rot, common in old trees. **Hemlock-bark extract** is obtained from the bark of the eastern hemlock, and is an important tanning material. **Western hemlock bark** is not in general use for tanning, but the bark contains 22% tannin. The extract is used with resorcinol-formaldehyde or other resins as **cold-setting adhesives** for plywood. They are strong and water-resistant. **Adhesive HT-120,** of Raynier, Inc., is hemlock-bark extract modified with a phenol resin.

HEMP. A fiber from the stalk of the plant *Cannabis sativa,* valued chiefly for cordage, sacking, packings, and as a fiber for plastic filler. In normal times it is grown principally in southern Russia, central Europe, the Mediterranean countries, and Asia, but during the Second World War was extensively cultivated in the United States. The fiber, which is obtained by retting, is longer than that of the flax plant, up to 75 in (2 m), but is coarser and not suitable for fine fabrics, although the finest and whitest fibers are sorted out in Europe and used in linen fabrics. It is also more difficult to separate the fiber and to bleach. It is stronger, more glossy, and more durable than cotton, and has been used for toweling and coarse fabrics to replace the heavy linen fabrics. It is high in alpha cellulose, containing about 78%. **Hemp rope** was once the chief marine cordage, but it has been replaced largely by rope of abaca which is lighter and more water-resistant.

Hemp contains a toxic alkaloid, and in India the stalks are chewed for the narcotic effect. The drug, known in medicine as **cannabis,** is called **marijuana** when smoked in cigarettes. Cannabis is an exhilarator and painkiller, and is used in medicine as a depressive antidote, but in excess the drug causes hallucinations. The plant's resin, which in a fully ripe cultivated plant covers the flowers and top leaves, contains the active ingredients. The least potent grade is **bhang,** derived from the tops of uncultivated plants. **Ganja** is the product of select, cultivated plants. The most potent and highest-grade version of the drug, called **charas,** is made from the resin alone and is the only grade which may properly be referred to as **hashish. Synthetic cannabis,** or **synhexyl,** is a pyrahexyl more powerful in action than the natural material. **Hempseed oil,** used in paints and varnishes, is made by pressing the seed of the hemp plant. It has a specific gravity of 0.926 and an iodine value of 148. **Oakum,** used for seam calking, is made from old hemp ropes pulled into loose fiber and treated with tar, usually blended with some new tow. Some grades may have sunn or jute fibers. It comes in balls or in rope form. **Marine oakum** is made entirely from new tow fibers.

HERRING OIL. A fish oil obtained by extraction from several species of fish of the herring family, *Clupeidae.* The **sardine** is the smaller fish of this family. The **Norwegian herring,** *Clupea harengus,* or **sea herring,** is the sardine of Maine, eastern Canada, and the North Sea. The herring is an abundant fish, but it is objectionable as a food because of the quantity of sharp bones. In the very small sardines the bones are soft and edible when cooked. In Norway the oil is produced by boiling the whole fish, pressing, and separating the oil from the water centrifugally. A process used in the United States is to grind the whole fish into liquid form, remove the oil, and condense the remaining solution until it is 50% solids, which is marketed as homogenized condensed fish for use as poultry feed.

In California and western Canada the sardine is a much larger fish, the **pilchard,** *Sardinia coerulea,* usually about 8 in (20 cm) long. The pilchard, or **California sardine,** once constituted about 25% of the entire fish catch of the United States by weight, but since 1948 the number of sardines in California waters has decreased greatly. The oil yield is about 30 gal/ton (125 L/metric ton) of fish, but much of the sardine oil is a by-product of the canning industry. The oil content of herring is 10 to 15% of the total weight of the fish, being low in the 1-year-old fish and reaching a peak in the third year. The fish builds up its oil in the summer. In winter the herring tends to stay close to the bottom, or at great depths, and uses up much of its oil. Commercially, the yield of herring oil is from 3 to 5 gal (1.4 to 19 L) per 250-lb (113-kg) bbl of raw fish. Much of the fish oil of South Africa is from the pilchard, *C. sagax.* In France, Spain, and Portugal the

European pilchard, *C. pilchardus,* and the *C. sardinus* are used. The oil from the latter has a high iodine value. In Norway, the **sprat,** *C. sprattus,* is also used. The **Japanese herring** is the *C. pallasi.* **Herring oil,** or **sardine oil,** is employed as a quenching oil in heat-treating, either alone or mixed with other oils; in soaps; printing inks; lubricants; and for finishing leather. It is also fractionated to use in blends for paint oils. Herring oil contains 25% of **clupanodonic acid,** $C_{21}H_{35}COOH$, 20% **arachidonic acid,** $C_{19}H_{31}COOH$, 18% **palmitoleic acid,** $C_{15}H_{29}COOH$, 13 linoleic, 9 oleic, 8 palmitic, and 7 myristic. The specific gravity is 0.920 to 0.933, iodine value 123 to 142, and saponification value 179 to 194. It can be made clear and odorless by hydrogenation. Sardine oil contains some stearic acid, higher percentages of palmitic and linoleic acids, and less of the other acids. **Pilchard oil** is quite similar, but has less oleic acid. Both of these oils contain about 15% of **tetracosapolyenoic acid,** a 24-carbon acid, also occurring in herring oil.

HICKORY. The wood of the **shagbark hickory** tree, *Carya ovata,* and other species of the walnut order. It is prized as a wood for ax, pick, and other tool handles, and for other items where resiliency and shock absorption are necessary. The color of the thick sapwood is white, and the heartwood is reddish brown. It has a fine, even, straight grain, and is tough and elastic, having 30% greater strength than white oak and twice the shock resistance, although not as durable. The weight is 45 to 52 lb/ ft^3 (721 to 833 kg/m^3). The chief producing states are Arkansas, Louisiana, Mississippi, Tennessee, and Kentucky, but the trees grow from New Hampshire to Texas. A mature shagbark tree 200 to 300 years old averages 100 ft (30 m) high and over 2 ft (0.6 m) in diameter. For handle manufacture the white wood and the red wood are considered equal in physical properties, and both possess the smooth feel required for handles. The average specific gravity when kiln-dried is 0.79, compressive strength perpendicular to the grain 3,100 lb/in^2 (21 MPa), and shearing strength parallel to the grain 1,440 lb/in^2 (10 MPa). There are more than 30 species of hickory, including the pecan trees. Besides the *C. ovata* three other species are important for the commercial wood: the **shellbark hickory,** *C. laciniosa,* also called **kingnut;** the **pignut hickory,** *C. glabra,* also called **black hickory** and **bitternut;** and the **mockernut hickory,** *C. alba,* also called **ballnut, hognut,** and **white hickory.** The kernels of the nuts of all species are edible, although some are bitter and astringent. The pecan hickories include the **pecan,** *C. illinoensis,* the **water hickory,** *C. aquatica,* and **nutmeg hickory,** *C. myristicaeformis.* The pecan trees are cultivated widely in the southern states for the nuts. **Pecan nuts** are widely used in confectionery and bakery products, but they become off-taste rapidly unless sprayed with an antioxidant.

HIGH BRASS. Sometimes called **common brass,** and formerly known as **market brass.** The most common of all the commercial wrought brasses. The usual mill standard is 65% copper and 35 zinc, and grades containing from 66 to 70% copper are referred to as **deep-drawing brass.** High brass is marketed in sheets, rolls, and strips, and is used largely for drawing, forming, and spinning. In the hard tempers it is used for parts made by blanking, forming, and bending. It is a cold-working material and is not suitable for hot working. The 65–35 brass marketed by the American Brass Co. under the name of **yellow brass** has a tensile strength of 45,000 lb/in^2 (310 MPa) and elongation 60% when soft, and a strength of 76,000 lb/in^2 (523 MPa) and elongation 5% when hard-rolled. The weight is 0.306 lb/in^3 (8,470 kg/m^3), melting point 930°C, and coefficient of expansion 0.0000106. Bar stock of high brass, for turned parts, called **high-brass bar,** invariably contains some lead, up to about 3%, as the unleaded brass is tough and the turned chips do not break easily. The most commonly used **die-casting brass** is a modified 60–40 yellow brass. The nominal composition is 57% copper, 1.5 lead, 0.25 tin, with sometimes small amounts of iron, silicon, or manganese, and the balance zinc. This is **ASTM alloy Z30A.** The tensile strength is 45,000 lb/in^2 (310 MPa), elongation 10%, and melting point about 1575°F (857°C). It is more difficult to cast than aluminum alloys.

The alloy listed in Federal specifications as **commercial brass** for wrought shapes actually covers the brasses from muntz metal to high brass, and is leaded. It contains 60 to 65% copper, with lead permissible up to 3.75%. The government specifications for commercial brass for castings are equally broad. **Butt brass,** for hinges, has 64% copper, 35 zinc, and 1 lead. The term **etching brass** refers to the temper rather than to the composition. It is a high-brass sheet in quarter-hard or half-hard temper used for nameplates and dials. **Bobierre's metal** is an old name for 63–37 high brass. This alloy is called in England **basis brass,** and **BES No. 265. Bristol brass** and **Prince's metal** are old names for high brass. **Spring brass** is usually a high brass with 66 to 72% copper rolled 8 numbers hard, but may be modified with tin or other elements and called **spring bronze.** The tensile strength, hard-rolled, is about 68,000 lb/in^2 (468 MPa), and elongation 11%; spring wire may be 100,000 lb/in^2 (689 MPa). SAE specifications require that a spring wire be capable of bending 180° around a wire of its own diameter without cracking. **Pin metal,** for common pins, has 62% copper and 38 zinc, and is not annealed from the working. **Shim stock** is usually a soft high brass in thicknesses of 0.001 and 0.002 in (0.003 and 0.005 cm) for shims and gaskets.

HIGH-LEAD BRONZE. Bronze alloys containing high percentages of lead to give a soft matrix metal for bearing use, as distinct from bronzes containing small amounts of lead to make them free-machining. The first high-lead bronze was invented in England in 1870 under the name of **Dick's bronze,** and was used on British railways. In 1892 C. B. Dudley in the United States produced **ExB metal** containing 77% copper, 15 lead, and 8 tin. It is still used as a car bearing metal and called **car brass.** A common type of **leaded bronze** used for bearings is the 80:10:10 mixture; this alloy is also known as **ordnance bronze.** It has a Brinell hardness of 58, a tensile strength of 30,000 lb/in^2 (206 MPa), and when deoxidized with phosphorus has a dense structure. Lead does not alloy well with copper unless a catalyzer is present, the copper absorbing only about 3.5% lead. It also tends to sweat out at a temperature of 327°C. High-lead bronze is now deoxidized with phosphorus, or contains small amounts of nickel, arsenic, or some other element to aid in holding up the lead.

The alloys containing tin are true bronzes, and are not as difficult to cast as the copper-lead alloys. Some lead bronzes also contain antimony, which gives them a crystalline structure useful for bearings. They are easy to cast. **Retz alloy** and **Reith alloy** contained about 75% copper, 10 each of tin and lead, and 5 antimony. **Cyprus bronze** contained 65% copper, 30 lead, and 5 tin. **Allen red metal** is a copper-lead alloy with 50% lead and a small amount of sulfur to hold the lead in solution. **Allen's metal** was an early alloy containing 40% lead, 55 copper, and 5 tin. Part of the lead was put in the form of galena ore or lead sulfide. **Johnson bronze No. 25,** used for high-speed bearings, contains 75% copper, 19 lead, 5 tin, and 1 nickel. High-lead bronzes are resistant to acids and, when used for casting chemical machine parts, are called **antiacid bronze.** The name **lead bronze** is used in England for an alloy of 75% copper, 1 to 2 tin, and the balance lead.

A bronze for motor bearings, **Johnson alloy 29,** of the Johnson Bronze Co., is **SAE alloy 67,** containing 78% copper, 15 lead, and 7 tin. **Lubrico,** of the Buckeye Brass Mfg. Co., contains 75% copper, 20 lead, and 5 tin. **Sabeco metal,** of the Fredericksen Co., has 21% lead and 9 tin. **Sumet bronze,** of the Sumet Corp., is the trade name of a group of bearing bronzes in grades from the softest, with 28% lead and Brinell hardness 30 to 33, to the hardest, with 17.5% lead and hardness of 58 to 62 Brinell. **Arctic bronze,** of the National Bearing Metals Corp., is the name of leaded bearing bronzes chill-cast in metal molds to give fine grain structure. **Bearium,** of the Bearium Metals Corp., refers to a group of high-lead bronzes containing 17.5 to 28% lead and about 10 tin. The softest grade, with a hardness of 35 Brinell, has a compressive limit of 7,800 lb/in^2 (54 MPa). **Durbar bronze,** of the Buffalo Die Casting Corp., has 24% lead and 4 tin. **Durbar hard bronze** has 10% tin and 20 lead. **Monarch metal,** of the Monarch Alloy Co., is a high-lead bearing bronze in various compositions. It is melted in sealed crucibles and water-cooled in pouring, resulting in a fine dispersion of the lead. **Tri-Alloy,** of the Ford Motor Co., is a high-lead alloy for crankshaft bearings for heavy loads and high speeds. It contains 35 to 40% lead, 4.5 to 5 silver, and the balance copper. The alloy is cast on strip steel and formed into bearings. **Copper-lead alloys** are now made also by sintering together copper powder and copperplated lead powder. They can thus be made in any proportion of lead.

HIGH-SPEED STEEL. A general name for high-alloy steels which retain their hardness at very high temperatures and are used for metal-cutting tools. All high-speed steels are based on either tungsten or molybdenum (or both) as the primary heat-resisting alloying element. The property of red-hardness is the ability of the steel to retain the hard carbides at the high cutting temperatures up to about 1750°F (954°C). Tungsten adds red-hardness to steel; chromium gives deep-hardening and strength; vanadium adds hardness, refines the grain, and improves the cutting edge. Molybdenum has a much more pronounced effect than tungsten on red-hardness, but alone it tends to make the steel more brittle and also subject to decarbonization. Cobalt gives red-hardness, but the steels are more brittle and difficult to forge.

The tungsten steels form the oldest class and are an outgrowth of the older mushet steels. **Mushet steel,** or **air-hardening steel,** was invented in England in 1868. It contained 5 to 8% tungsten, 2.5 manganese, 1.5 silicon, and very high carbon. **Taylor-White steel,** marketed in 1900, contained 8% tungsten, 1.8 chromium, 1.15 carbon, 0.18 manganese, and 0.25 silicon. In 1900 the Bethlehem Steel Co. brought out the tungsten-chromium-vanadium steel now known as 18:4:1. Because of its balance of red-hardness, toughness, and cutting edge, this formula, and the English preference for a harder 14:4:2 steel, still remains the basic standard for tungsten high-speed steel, although most modern steels are now modified. **Star-Zenith,** of Carpenter Technology, is an 18:4:1 steel with the alloying elements in slight excess of standards. **ML steel,** of the Allegheny Ludlum Steel Co., has the vanadium increased to 1.85% and an addition of 0.50 molybdenum. It gives a keener edge.

The first **molybdenum high-speed steel** was **Mo SH steel** of the Sanderson Steel Co., marketed in 1898 and called **self-hardening steel.** Later, research by the Army Ordnance Department to substitute molybdenum for the imported metal tungsten resulted in the **Watertown Arsenal steel** containing 9.5% molybdenum, 4 chromium, 1.25 to 2 tungsten, 0.90 to 1.5 vanadium, up to 0.40 manganese and 0.50 silicon, and 0.80 carbon. **Motung steel** contained about 8% molybdenum, 1.5 tungsten, 4 chromium, 1.25 vanadium, up to 0.50 each of manganese and silicon, and up to 0.85 carbon. **Van Lom steel,** of Vasco, had up to 10% molybdenum, about 4 chromium and 4 vanadium, and up to 1.2 carbon. The early molybdenum steels required careful heat treatment to prevent decarburization, and later steels contained less molybdenum and more tungsten. **Bethlehem 66 steel,** of the Bethlehem Steel Co., had about 5.5% tungsten, 5 molybdenum, 4 chromium, 1.5 vanadium, and 0.80 carbon. **Unicut steel,** of the Cyclops Corp., had 6.25 tungsten, 6.25 molybdenum, 4 chromium, 2.4 vanadium, and 1 carbon. **Twin Mo steel,** of H. Boker & Co., Inc., is a 6:6:2 tungsten-molybdenum steel. **Speed Star steel,** of the Carpenter Steel Co., contains 5.5% tungsten, 4.25 molybdenum, 4 chromium, 15 vanadium, and 0.80 carbon. It is a general-purpose tool steel with a fine grain and surface hardness of Rockwell C65. **Star-Mo M-2 steel,** of Firth Sterling, is typical of the steels designed to combine the desirable tool qualities of both the tungsten and the molybdenum steels. It contains 6.40% tungsten, 5 molybdenum, 4 chromium, 2 vanadium, and 0.83 carbon. It weighs 6% less than the 18:4:1 types of steel, is easily machined at the annealed hardness of less than 240 Brinell, and is hardened to about Rockwell C65. The high-speed steels are sometimes modified for use as casting alloys for such uses as jet engine parts. **Haynes alloy 589,** of Union Carbide, for high-heat abrasion resistance in moving parts, contains 17% chromium, 16 molybdenum, 2 vanadium, with up to 4% carbon, 1.5 silicon, and maximum 0.5 manganese.

The **cobalt high-speed steels** are the "super" steels for high production, but they are balanced in composition for particular service and usually require a specified heat treatment. Those with high tungsten and cobalt have high red-hardness and can be run at higher speeds and feeds. Those with high molybdenum have a lower red-hardness and a narrower heat-treating range. **Circle C steel,** of Firth Sterling, has 18.5% tungsten, 4.5 chromium, 2 vanadium, 1 molybdenum, 9 cobalt, and 0.77 carbon, while **Super Hi-Mo steel** of this company has 1.8 tungsten, 8.5 molybdenum, 4 chromium, 5 cobalt, 1.2 vanadium, and 0.80 carbon. **Congo steel,** of the Braeburn Steel Corp., for cutting hard materials, has 12% cobalt, 4 tungsten, 5 molybdenum, 4 chromium, 1.4 vanadium, and 0.78 carbon.

Rex AAA steel, of the Crucible Steel Co., is an 18:4:1 steel modified with 5% cobalt and 0.50 molybdenum. **Red Cut Cobalt steel** of Vasco, **Co-Co steel** of the Colonial Steel Co., and **Maxite steel** of the Columbia Tool Steel Co. were also 18:4:1 steels modified with cobalt and molybdenum. **Vasco Supreme steel** of Vasco has 5% vanadium instead of the usual 1%, with 5% cobalt, 4.75 chromium, 12.5 tungsten, 0.25 manganese, and 1.5 carbon, and has good wear life at cutting speeds double those for standard 18:4:1 steel. **XDH steel,** of Firth Sterling, is an 18:4:1 steel with low carbon, 0.55%, to give greater shock resistance for dies and punches.

General-purpose **cutting steels** are usually tungsten-molybdenum steels. **Crusader XL steel,** of the Latrobe Steel Co., which hardens to Rockwell C66 and gives good red-hardness and wear resistance, contains

6% tungsten, 6 molybdenum, 4.1 chromium, 3.2 vanadium, and 1.20 carbon, while **Electrite Corsair XL** is a variation of this steel with added sulfur to make it easier machining for form tools. **Electric Star steel,** of this company, for machining abrasive castings, contains 5.5% tungsten, 4.5 molybdenum, 4.5 chromium, 4 vanadium, 0.25 silicon, 0.25 manganese, and 1.28 carbon. **STM steel,** of the Simonds Saw & Steel Div., for saws, knives, and chisels, has up to 9.5% molybdenum, 1.5 tungsten, 3.75 chromium, 2 vanadium, and from 3 to 8 cobalt, with 0.80 to 1 carbon. **Amotun steel,** of the Atlantic Steel Co., for dies and taps, is quite similar with 6% cobalt, **Rex M2S,** of Crucible Steel, is a grade of M2 high-speed cutting steel with 1% carbon to give a Rockwell hardness of C65 with only a slight decrease in impact strength.

For high wear resistance when cutting at red heats, some of the steels have higher percentages of tungsten and cobalt. **Imperial Major steel,** of A. Milne & Co., has 13% cobalt and 22 tungsten. **Gray Cut Cobalt steel,** of Vasco, has 20.5% tungsten, 12.25 cobalt, 4.25 chromium, 1.3 vanadium, 0.60 molybdenum, and 0.80 carbon. The high-speed steels are sold under a great variety of trade names in rods, bars, flats, and tool shapes. Among these are **Supremus steel** and **Jessco steel** of the Jessop Steel Co., **Milvan steel** of A. Milne & Co., **Rex, Champion,** and **Peerless** of the Crucible Steel Co., **Blue Chip, Van Chip,** and **Circle M** of Firth Sterling, **Clarite steel** of the Columbia Tool Steel Co., and **Panther steel** and **DBL steel** of the Allegheny Ludlum Steel Co.

HOBBING STEELS. Used for making plastic molds by pressing the hard model into the steel, they may be plain carbon steel for case hardening, or they may be low-alloy steels. **Press E-Z steel,** of the Jessop Steel Co., is a hobbing steel containing 0.40 to 0.50% carbon and 0.15 to 0.20 manganese with no alloying elements. It is marketed annealed to 100 Brinell, and when carburized and hardened has a Rockwell C hardness of 60. **AISI steel 3110** is a hobbing steel containing 0.50% manganese, 0.60 chromium, and 1.25 nickel. It takes deep impressions, and when hardened has high core strength. **AISI steel 3312** is a hobbing steel with 3.5% nickel and 0.5 chromium. After air hardening, the core strength is 160,000 lb/in^2 (1,103 MPa) and core hardness 360 Brinell. **Speed Alloy,** of W. J. Holliday & Co., is a mold steel that machines easily and gives deep hardening and high core strength when hardened. It is a 0.30% carbon steel with some chromium and molybdenum. Tensile strength, annealed, is 107,000 lb/in^2 (737 MPa), elongation 26%, and Brinell hardness 207. These materials, however, are low-alloy steels rather than plain carbon steels.

HOLLY. The wood of the tree *Ilex aquifolium,* and several other species of *Ilex,* or holly tree, native to Europe, and the tree *I. opaca,* of southeastern United States. It is valued as a wood for inlaying because of its white color, its fine, close grain, and its ease of staining to imitate ebony. It is hard, and the weight is 47 lb/ft^3 (753 kg/m^3). It is also used for scientific and musical instruments, model boats, and sporting goods.

HORN. The excrescent growth, or horns, from the heads of certain animals, notably beef cattle. Horn is used for making handles and various articles. The quality depends largely upon the size and age of the animal from which it comes, the No. 1 grade being the large steer horns, and the No. 2 those below 40 lb (18 kg) per hundred. Horns occur on the head in pairs and are hollow, growing on a core of pithy bone. The horns are split by saws, soaked to make them flexible, and then flattened under pressure. **Horn meal,** made from the bone refuse, is sold largely as fertilizer. **Horn pith,** extracted by boiling the horns, is used for glue and for gelatin.

HOT-DIE STEELS. A general name for alloy steels that resist shock and retain their hardness when used in operations involving the forging, shearing, or punching of metals at relatively high temperatures. They are also called **hot-work steels.** These are two general types, one being chromium steel with around 5% chromium, and the other tungsten steel with 8 to 10% tungsten, but they are usually modified with other elements and often do not differ greatly from the general purpose high speed steels.

The plain chromium steels are oil-hardening. They develop a high hardness and are deep-hardening with high impact value, and are used for dies for compressive action, as for header machines. The tungsten steels are air-hardening, have high impact value and superior service life at high heats, but, unless modified, are not hard enough for cutting dies. Both types of steel now usually have additions of vanadium, molybdenum, nickel, and other elements to give added physical qualities. The commonly used 3 to 5% chromium complex steels are air-hardening and have the alloy content adjusted to give a balance of hardness, toughness, and resistance to heat checking. **LPD die steel,** of the Latrobe Steel Co., for forging and extrusion dies, has 5% chromium, 1.6 molybdenum, 1.3 tungsten, 1 silicon, 0.30 vanadium, 0.30 manganese, and 0.35 carbon. **VDC die steel,** for die-casting dies, has 5.25% chromium, 1.2 molybdenum, 1 vanadium, 1 silicon, 0.40 manganese, and 0.40 carbon. Both of these steels harden to Rockwell C55.

A tough and wear-resistant air-hardening steel for swaging and forming dies and for knives is **Airque V,** of the Braeburn Alloy Steel Div. It has 5.25% chromium, 1.15 molybdenum, 1 vanadium, 0.50 manganese, 0.30 silicon, and 1.25 carbon. It hardens to Rockwell C65. **HWD 3 steel,** of Firth Sterling, for forging dies and aluminum die-casting dies, has a quite similar composition but with only 0.40 carbon. It hardens to Rockwell

C52, but can be nitrided for greater surface hardness. **Potomac M steel,** of the Allegheny Ludlum Steel Corp., is also similar and may be either air- or oil-hardened. **Hot-work steel B-47** of the same company, for extrusion and forging dies and hot-punch tools, is more complex. It has 4.25% each of chromium, tungsten, and cobalt, 2.25 vanadium, 0.40 molybdenum, 0.35 manganese, 0.25 silicon, and 0.40 carbon. It retains a tensile strength of 178,000 lb/in^2 (1,246 MPa) at 800°F (427°C). **Crescent steel** and **La Belle steel,** of the Crucible Steel Co., and **C.Y.W. steel,** of Firth Sterling, are 3.5 to 3.9% chromium steels modified to require only easy heat treatment to give a good balance of physical properties.

The tungsten steels are generally suited for more severe service, but are more costly. **Mohawk steel,** of the Allegheny Ludlum Steel Co., has 14% tungsten, 3.5 chromium, 0.70 vanadium, and 0.50 carbon, while **Atlas steel** has 10% tungsten, 3.5 chromium, 0.45 vanadium, and 0.35 carbon. **Peerless A steel,** of the Crucible Steel Co., has 9% tungsten, 3.25 chromium, and 0.25 vanadium. **Vasco Marvel steel,** of Vasco, contains 9.25% tungsten, 3.5 chromium, 0.40 vanadium, and 0.35 carbon, while **Vasco Extrude Die steel,** for hot-extrusion dies, has 15.5% tungsten, 3 molybdenum, 4 chromium, and 2 nickel. **Excelo steel,** of the Carpenter Steel Co., for hot shears, has 2.5% tungsten, 1.5 chromium, 0.35 vanadium, and 0.55 carbon, but **T-K steel,** for swaging dies, has 9% tungsten, 3.5 chromium, 0.40 vanadium, and 0.35 carbon, and for higher heat resistance **D.Y.O. steel** has 14.5% tungsten, 4 chromium, and 0.50 vanadium. **Hot-work 8 steel,** of the Bethlehem Steel Co., is molybdenum steel with 8.5% molybdenum and 1.5 vanadium. It is tough, shock-resistant, abrasion-resistant, and resists heat checking. General-purpose, easily heat-treated steels with low tungsten and chromium and some vanadium are not highly heat-resistant, but are suitable for die-casting dies and, because of their resistance to shock, are also used for upsetting dies. **Tungo steel,** of the Colonial Steel Co., and **Par-Exc steel,** of Vasco, are examples of these.

Die block steel refers to hot-work steels furnished in finished blocks for forging and die-casting dies, heat-treated to a temper permitting machining but not requiring further heat treatment. **Hardtem steel** and **C55 steel,** of the Heppenstall Co., are such steels. They eliminate the possibility of warpage in the heat treating of the finished die. For general-purpose use in plants not equipped with extensive heat-treating facilities, low-alloyed tungsten-chromium steels are produced that have a good balance of hardness, toughness, and heat resistance for both hot forging and heading and for cold work. **JS steel,** of Firth Sterling, has 2.5% tungsten, 1.4 chromium, 0.25 vanadium, 0.30 manganese, and 0.50 carbon. It can be oil-hardened over a wide temperature range. When tempered at 800°F (427°C), it has a hardness of Rockwell C49, but can be readily pack-hardened to give a surface hardness of Rockwell C65. Where high hardness and compressive strength are needed, high-chromium, air-hardening steels modified with small percentages of other elements are also used as general-purpose hot-work steels. **Olympic FM die steel,** of the Latrobe Steel Co., is such a steel. It contains 12% chromium, 0.90 vanadium, 0.75 molybdenum, 0.50 manganese, 0.30 silicon, 1.5 carbon, plus sulfides to make it readily machinable.

HYDROCHLORIC ACID. Also called **muriatic acid,** and originally called **spirits of salt.** An inorganic acid used for pickling and cleaning metal parts; producing of glues and gelatin from bones; manufacturing chlorine, pharmaceuticals, dyes, and pyrotechnics; for tanning, etching, and reclaiming rubber; and treating oils and fats. It is a water solution of **hydrogen chloride,** HCl, and is a colorless or yellowish fuming liquid, with pungent, poisonous fumes. The specific gravity of the gas is 1.269, the solidifying point −112°C, and boiling point − 83°C. It is made by the action of sulfuric acid on sodium chloride, or common salt. The commercial acid is usually 20°Bé, equaling 31.45% HCl gas, and has a specific gravity of 1.16. Other grades are 18 and 22°Bé. **Fuming hydrochloric acid** has a specific gravity of 1.194, and contains about 37% hydrogen chloride gas. Hydrochloric acid is shipped in glass carboys. Anhydrous hydrogen chloride gas is also marketed in steel cylinders under a pressure of 1,000 lb/in^2 (6 MPa) for use as a catalyst. The boiling point is 85.03°C. The acid known as **aqua regia,** used for dissolving or testing gold and platinum, is a mixture of 3 parts hydrochloric acid and 1 nitric acid. It is a yellow liquid with suffocating fumes.

HYDROCYANIC ACID. Also called **prussic acid, formonitrile,** and **hydrogen cyanide.** A colorless, highly poisonous gas of the composition HCN. The specific gravity is 0.697, the liquefying point is 26°C, and it is soluble in water and in alcohol. It is usually marketed in water solutions of 2 to 10%. It is used for the production of acrylonitrile and adiponitrile and for making sodium cyanide. It is also employed as a disinfectant and fumigant, as a military poison gas, and in mining and metallurgy in the cyanide process. It is so poisonous that death may result within a few seconds after it is taken into the body. It was used as a poison by the Egyptians and Romans, who obtained it by crushing and moistening peach kernels. It is produced synthetically from natural gas. The French war gas known as **vincennite** was hydrocyanic acid mixed with stannic chloride. **Manganite** was a mixture with arsenic trichloride. **HCN discoids,** of the American Cyanamid Co., are cellulose disks impregnated with 98% hydrocyanic acid, used for fumigating closed warehouses.

HYDROFLUORIC ACID. A water solution of **hydrogen fluoride,** HF. It is a colorless, fuming liquid, highly corrosive and caustic. It dissolves most metals except gold and platinum, and also glass, stoneware, and organic material. The choking fumes are highly injurious. It is widely used in the chemical industry, for etching glass, and for cleaning metals. In cleaning iron castings it dissolves the sand from the castings. The specific gravity of the gas is 0.713, and liquefying point −19°C. Hydrofluoric acid is made by treating calcium fluoride or fluorspar with sulfuric acid. It is marketed in solution strengths of 30, 52, 60, and 80%. The anhydrous material, HF, is used as an alkylation catalyst. **Hydrobromic acid,** HBr, is a strong acid which reacts with organic bases to form bromides that are generally more reactive than chlorides. The technical 48% grade has a specific gravity of 1.488.

HYDROGEN. A colorless, odorless, tasteless elementary gas. With an atomic weight of 1.008, it is the lightest known substance. The specific gravity is 0.0695, and its density ratio in relation to air is 1:14.38. It is liquefied by cooling under pressure, and its boiling point at atmospheric pressure is −252.7°C. Its light weight makes it useful for filling balloons, but, because of its flammable nature, it is normally used only for signal balloons, for which use the hydrogen is produced easily and quickly from hydrides. Hydrogen produces high heat, and is used for welding and cutting torches. For this purpose it is used in atomic form rather than the usual H_2 molecular form. Its high thrust value makes it an important rocket fuel. It is also used for the hydrogenation of oil and coal, for the production of ammonia and many other chemicals, and for **water gas,** a fuel mixture of hydrogen and carbon monoxide made by passing steam through hot coke.

Hydrogen is so easily obtained in quantity by the dissociation of water and as a by-product in the production of alkalies by the electrolysis of brine solutions that it appears as a superabundant material, but its occurrence in nature is much less than many of the other elements. It occurs in the atmosphere to the extent of only about 0.01%, and in the earth's crust to the extent of about 0.2%, or about half that of the metal titanium. However, it constitutes about one-ninth of all water, from which it is easily obtained by high heat or by electrolysis.

Hydrogen has three isotopes. **Hydrogen 2,** called **deuterium,** occurs naturally in ordinary **hydrogen 1** to the extent of one part in about 5,000. Deuterium has one proton and one neutron in the nucleus, with one orbital electron revolving around. A gamma ray will split off the neutron, leaving the single electron revolving around a single proton. The physicist's name for hydrogen 1 is **protium.** Deuterium is also called **double-weight hydrogen. Deuterium oxide** is known as **heavy water.** The formula is H_2O, but with the double-weight hydrogen the molecular weight is 20 instead of the 18 for ordinary water.

Heavy water is used for shielding in atomic reactors as it is more effective than graphite in slowing down fast neutrons. It is also made with oxygen 17 and oxygen 18. Chemicals for special purposes are also made with hydrogen 2. The **deuterated benzene** of Ciba, Ltd., is made with deuterium, and the formula is expressed as C_6D_6. **Hydrogen 3** is **triple-weight hydrogen,** and is called **tritium.** It has two neutrons and one proton in the nucleus, and is radioactive. It is a beta emitter with little harmful secondary ray emission, which makes it useful in self-luminous phosphors. It is a solid at very low temperatures. **Liquid hydrogen** for rocket fuel use is made from ordinary hydrogen. It is required to be within 0.00001% of absolute purity. This material has a boiling point at −423°F (−253°C), and the weight is 0.6 lb/gal (0.07 kg/L). With a chamber pressure of 300 lb/in^2 (2 MPa), the specific impulse is 375. The **Hydripills** of Metal Hydrides, Inc., used for producing small quantities of hydrogen, are tablets of a mixture of sodium borohydride and **cobalt chloride,** $CoCl_2$. When water is added, the chloride reacts to produce hydrogen from both the borohydride and the water. **Gelled hydrogen** for rocket fuel is liquid hydrogen thickened with silica powder.

HYDROGENATED OILS. Vegetable or fish oils that have been hardened or solidified by the action of hydrogen in the presence of a catalyst. Partial hydrogenation also clarifies and makes odorless some oils. The solidifying process is carried on to any desired extent, and these oils have a variety of uses. For mechanical uses they are employed in cutting oils, and in place of palm oil in tinplate manufacture. By hydrogenation the fatty acids, such as oleic acid, are converted into stearic acid. Peanut oil, coconut oil, and cottonseed oil can thus be made to have the appearance, taste, and odor of lard, or they can be made like tallow. **Lard compound,** previous to the passage of the Food and Drug Act of 1906, was cottonseed oil mixed with oleostearin from beef tallow. It was later sold under trade names, but has now been replaced by hydrogenated oils under the general name of **shortenings,** and under trade names such as **Crisco.** Hydrogenated oils have lower iodine values and higher melting points than the original oils.

IMPREGNATED WOOD. Also called **compressed wood** or **densified wood.** Many types are forms of laminated wood. **Compreg,** developed by the U.S. Forest Products Laboratory, consists of many layers of 1/16-in (0.16-cm), rotary cut, yellow-birch plies bonded with about 30% resin under a pressure of 600 to 1,500 lb/in^2 (4 to 9 MPa). The specific gravity is 1.22 to 1.37, and tensile strength 43,000 to 54,000 lb/in^2 (296 to 372 MPa), depending upon the resin and the molding pressure. **Impreg,** developed by this laboratory for use in making patterns and models, is produced from 1/16-in (0.16-cm) laminations of mahogany impregnated with a low-molecular-weight phenolic resin and bonded under pressure into a uniform solid of good dimensional stability. **Flaypreg,** another member of this group of wood products, is hard, strong, dense, and low in cost. It is made from wood flakes, usually fir or spruce, impregnated with resin, machine-felted, and pressed. The specific gravity is 1.39, and the water absorption is only 0.44% compared with 1.46 for Compreg. It is used for making gears, cams, patterns, and tabletops. **Delwood,** of the Gisholt Machine Co., is molded of wood chips, chopped glass fiber, and a binder of polyester resin. It has the strength of hard maple, and takes nails and screws better than wood. It is used for shoe lasts, picture frames, and furniture. **Pregwood,** of the Formica Co., is a wood laminate impregnated with a phenolic resin and cured into a hard sheet. But **Impreg weldwood,** of the United States Plywood Corp., has the wood plies impregnated only to a short depth before compressing so that it remains a true plywood. It has higher strength and is more resistant than an ordinary resin-bonded plywood. The **Sprucolite,** of the Sprucolite Corp., for bearings, rolls, gears, and pulleys, is cross-laminated like plywood with thin sheets of western spruce, but is **plywood block** impregnated with resin and subjected to high pressure to make it dense and hard. Its weight is about 35% that of cast iron. A similar English laminated wood, called **Hydulignum,** consists of thin birch veneers impregnated with vinyl formal resin and compressed into a dense board with a specific gravity of 1.31 and a tensile strength of 45,000 lb/in^2 (310 MPa). Still another type of building board, **Dylite,** of the Koppers Co., has a core of polystyrene plastic jacketed on both sides with plywood or gypsum.

The material developed by the U.S. Forest Products Laboratory under the name **Staypak** is made by compressing veneered softwood containing no resin except that used to bond the veneers. It has a smooth satiny finish, a specific gravity of 1.3 to 1.4, and about double the tensile and flexural strengths of birch. The color is darker than the original wood because of the flow of lignin. **Hiden,** of the Parkwood Corp., is a synthetic hardwood of about the same density and hardness as lignum vitae. It is made of birch veneers impregnated with phenolic resin and compressed to 30% of the original thickness and cured. The board are used for tabletops, cutlery handles, sheet-metal forming dies, and textile picker sticks.

Wood impregnated with **polyethylene glycol** is known as **Peg.** This treatment is used for walnut gunstocks for high-quality rifles and for tabletops. This impregnant can be used to reduce checking of green wood during drying. Wood can also be vacuum-impregnated with certain liquid vinyl monomers and then treated by radiation or catalyst heat systems, which transform the vinyl into a plastic. **Methyl methacrylate,** or **acrylic,** is a common resin used to produce this type of product, known as **wood-plastic combinations** or **WPC.** A principal commercial use of this modified wood is as parquet flooring and for sporting goods such as archery bows. It is produced in squares about 5½ in (14 cm) on a side from strips about ⅞ in (2.2 cm) wide and 5⁄16 in (0.8 cm) thick. It has a specific gravity of 1.0.

WPC material resists indentation from rolling, concentrated, and impact loads better than white oak. This is largely due to improved hardness, which is increased 40%. Abrasion resistance is no better than white oak. Arco Chemical Co. produces **acrylic-impregnated wood,** which if cut, sanded, and polished, exhibits the original surface finish.

INDIGO. Once the most important of all vegetable dyestuffs and valued for the beauty and permanence of its color. Commercial blue indigo is obtained from the plants *Indigofera tinctoria* and several other species, of India and Java, and the plant *Isatis tinctoria,* of Europe, by steeping the freshly cut plants in water, and after decomposition of the glucoside **indican,** $C_{14}H_{17}O_6N$, the liquid is run into beating vats where the indigo separates out in flakes which are pressed into cakes. About 4 oz (0.1 kg) of indigo are produced from 100 lb (45 kg) of plants. **Indigo red,** or **indirubin,** $C_{16}H_{10}N_2O_2$, is a crimson dyestuff obtained in the proportion of 1 to 5% in the manufacture of indigo. **Indigo white** is obtained by reducing agents and an alkali. Another product obtained in the manufacture of synthetic indigo is **indole,** a white crystalline solid with a melting point of 52°C. In concentrations it has a powerful, disagreeable odor, but in extreme dilution has a pleasant floral odor and is used in many perfumes. It occurs naturally in oils of jasmine, neroli, orange blossom, and others, and is made synthetically as **benzopyrrole. Skatole** is made by adding a methyl group to the No. 3 position of the indole ring. It is a solid melting at 95°C, and is found as a decay product of albumin in animal excrement. It has an overpowering fecal odor, and the synthetic material is used as a fixative in fine perfumery. **Oxindole,** or **hydroxyindole,** is a lactam of aminophenyl acetic acid, easily made synthetically, and is the basis for the production of a wide variety of chemicals.

INDIUM. A silvery-white metal with a bluish cast, whiter than tin. It has a specific gravity of 7.31, tensile strength of 15,000 lb/in^2 (103 MPa), and elongation 22%. It is very ductile and does not work-harden, as its recrystallization point is below normal room temperature, and it softens during rolling. The metal is not easily oxidized, but above its melting point, 155°C, it oxidizes and burns with a violet flame.

Indium was first found in zinc blende, but is now obtained as a by-product from a variety of ores. Because of its bright color, light reflectance, and corrosion resistance, it is valued as a plating metal, especially for reflectors. It is softer than lead, but a hard surface is obtained by heating the plated part to diffuse the indium into the base metal. It has high adhesion to other metals. When added to chromium plating baths it reduces brittleness of the chromium.

In spite of its softness small amounts of indium will harden copper, tin, or lead alloys and increase the strength. About 1% in lead will double the hardness of the lead. In solders and fusible alloys it improves wetting and lowers the melting point. In lead-base alloys a small amount of indium helps to retain oil film and increases the resistance to corrosion from the oil acids. Small amounts may be used in gold and silver dental alloys to increase the hardness, strength, and smoothness. Small amounts are also used in silver-lead and silver-copper aircraft-engine bearing alloys. **Lead-indium alloys** are highly resistant to corrosion, and are used for chemical-processing equipment parts. **Gold-indium alloys** have high fluidity, a smooth lustrous color, and good bonding strength. An alloy of 77.5% gold and 22.5 indium, with a working temperature at about 500°C, is used for brazing metal objects with glass inserts. **Silver-indium alloys** have high hardness and a fine silvery color. A silver-indium alloy of the Westinghouse Electric Corp., used for nuclear control rods, contains 80% silver, 15 indium, and 5 cadmium. The melting point is 1375°F, (746°C), tensile strength 42,000 lb/in^2 (289 MPa), and elongation 67%, and it retains a strength of 17,600 lb/in^2 (120 MPa) at 600°F (316°C). It is stable to irradiation, and is corrosion-resistant in high-pressure water up to 680°F (360°C). The thermal expansion is about six times that of steel.

Indium sulfate, used for plating, has three forms. The normal sulfate is $In_2(SO_4)_3 \cdot 9H_2O$; the acid salt is $In_2(SO_4)_3 \cdot H_2SO_4 \cdot 7H_2O$; and the basic salt is $In_2O(SO_4)_2 \cdot 6H_2O$. **Indium oxide** is an amorphous yellow powder of the composition In_2O_3, and specific gravity 7.179, used to give a beautiful yellow color to glass. The color may be varied from light canary to dark tangerine-orange. **Indium monoxide,** InO, is black, and is not stable.

INDUSTRIAL JEWELS. Hard stones, usually **ruby** and **sapphire,** used for bearings and impulse pins in instruments and for recording needles. **Ring jewels** are divided into large and small. The large rings are about 0.050 in (0.127 cm) in diameter and 0.012 in (0.030 cm) in thickness with holes above 0.006 in (0.015 cm) in diameter. Ring jewels are used as pivot bearings in instruments, timepieces, and dial indicators. From 2 to 14 are used in a watch. **Vee jewels** are used in compasses and electrical instruments. **Cup jewels** are used for electric meters and compasses. **End stones** are flat undrilled stones with polished faces to serve as end bearings. **Pallet stones** are rectangular impulse stones for watch escapements. **Jewel pins** are cylindrical impulse stones for timepiece escapement. In making **bearing jewels** the synthetic sapphire boules are split in half, secured to wooden blocks, and then sawed to square blanks. These are then rounded on centerless grinding machines, and flat-ground to thickness by means of copper wheels with diamond powder. **Quartz bearings** are made from fused quartz rods. A notch is ground in the end of the rod, then polished and cut off, repeating the process for each bearing. Quartz has a hardness of only 7 Mohs, while the synthetic ruby and sapphire have a hardness of 8.8, but quartz has the advantage of low thermal expansion.

The making of industrial jewels was formerly a relatively small, specialized industry, and a national stockpile of cut jewels was maintained for wartime emergencies. But the process of slicing and shaping hard crystals for semiconductors and other electronic uses is essentially the same, and stones of any required composition and cut to any desired shape are now regularly manufactured.

INGOT IRON. Nearly chemically pure **iron** made by the basic open-hearth process and highly refined, remaining in the furnace 1 to 4 h longer than the ordinary time, and maintained at a temperature of 2900 to 3100°F (1593 to 1704°C). In England, it is referred to as **mild steel,** but in the United States the line between iron and steel is placed arbitrarily at about 0.15% carbon content. Ingot iron has as low as 0.02% carbon. It is obtainable regularly in grades 99.8 to 99.9% pure iron. Ingot iron is cast into ingots and then rolled into plates or shapes and bars. It is used for construction work where a ductile, rust-resistant metal is required, especially for tanks, boilers, enameled ware, and for galvanized culvert sheets. The tensile strength, hot-rolled, is 48,000 lb/in^2 (330 MPa), elongation 30%, and Brinell hardness 82 to 100. Dead soft, the tensile strength is 38,500 lb/in^2 (270 MPa), elongation 45%, and Brinell hardness 67. **Armco ingot iron,** of Armco Steel Corp., is 99.94% pure, with the carbon 0.013 and the manganese 0.017%. It is used as a rust-resistant construction material, for electromagnetic cores, and as a raw material in making special steels. The specific gravity is 7.858, and melting point 1530°C. **Enamelite** is a sheet iron especially suited for vitreous enameling, produced by the Sharon Steel Co. Ingot iron may also be obtained in grades containing 0.25 to 0.30% copper, which increases the corrosion resistance. Iron of very low carbon content may also be used for molds and dies which are to be hobbed. The iron is quite plastic under the hob and is then hardened by carburizing. **Plastiron,** of Henry Disston & Sons, Inc., is such an iron.

INK. Colored liquids or pastes used for writing, drawing, marking, and printing. Black writing inks usually contain gallotannate of iron which is obtained by adding an infusion of nutgalls to a solution of ferrous sulfate. Good **writing ink** is a clear, filterable solution, not a suspension. It must flow easily from the pen without clogging, give a smooth, varnishlike coating, and adhere to inner fibers of the paper without penetrating through the paper. It must have an intense color that does not bleach out. Ink is essentially a pigment in a liquefying and adhesive medium, but the iron-gallotannate writing inks develop their full color by chemical action and become insoluble in the outer fibers of the paper. For the proper development of the black color in gallotannate inks a high percentage of iron is needed, and this requires a liberal use of acid, which will tend to injure the paper. It is thus usually the practice to reduce the amount of iron and bring up the color with dyes or pigments. Gums or adhesive materials may also be added.

Carbon inks are composed of lampblack or carbon black in solutions of gums or glutenous materials. **India ink** is a heavy-bodied drawing ink. The original India ink, or **Chinese ink,** was made with a jet-black carbon pigment produced by burning tung oil with insufficient air. The pigment was imported into Europe in compressed sticks known as **indicum.** India ink was originally only black ink, but the name is now used also for colored heavy drawing inks made with various mineral pigments. **Marking inks** are usually solutions of dyes that are fast to laundering, but they may also be made with silver salts which develop full color and stability by application of heat. **Fountain-pen ink** is not a special-composition ink, but is a writing ink free of sediment and tendency to gum. It usually contains tannic, gallic, and hydrochloric acids with a pH above 2 to avoid corrosion. **Permanent inks** contain dissolved iron, not over 1%, to avoid sludge. **Ball-point ink** is usually a paste and is a true solution with 40 to 50% dye concentration. It must be stable to air, noncorrosive, and a good lubricant. An **encaustic ink** is a special writing ink that will penetrate the fibers of the paper and set chemically to make erasure difficult, but an **indelible ink** for textiles is a marking ink. **Invisible-writing inks,** or **sympathetic inks,** are inks that remain invisible until the writing is brought out by the application of heat or with another chemical which develops the color. They are made with sal ammoniac or salts of metals. **Magnetic ink** for use on bank checks to permit mechanical processing contains 50 to 70% of a magnetic powder smaller than 5-μm particle size. The powder may be hydrogen-reduced iron, carbonyl iron, or electrolytic iron.

Printing inks are in general made with carbon black, lampblack, or other pigment suspended in an oil vehicle, with a resin, solvent, adhesive, and drier. But there are innumerable modifications of printing inks to meet different conditions of printing and varieties of surfaces. The oil or chemical vehicles are innumerable, and the pigments, resin strengtheners and gloss formers, adhesives, tackifiers, and driers vary greatly to suit the nature and surface coating of the base material. Oils may dry by oxidation, polymerization, absorption, or solvent evaporation, and resins may be used to add gloss, strength, hardness, and color fastness, or to increase the speed of drying. It is estimated that there are about 8,000 variables in an ink, and thus printing ink is a prescription product for any given job. They are not normally purchased on composition specification, but on ability to meet the requirements of the printing. **Aremco-Mark 530,** of Aremco Products, Inc., is a black **ceramic ink** that withstands temperatures to

2000°F (1093°C). These inks are ceramic solutions that must be heat-cured. **Flexographic ink** and **rotogravure ink** may be made with cellulose-acetate-propionate ester resin which is soluble in alcohol and in other resins. When used with urea it cross-links to form a permanent thermoset film.

INSECTICIDES. Chemicals, either natural or synthetic, used to kill or control insects, particularly agricultural pests. They are also referred to as **pesticides.** Of about 800,000 known species of insects, half feed directly on plants, retarding growth of the plant and causing low yields and inferior crops. The production of insecticides is now one of the important branches of the chemical industry, and increasing quantities are used, but the specification and use of insecticides require much skill because of the cumulative toxic effect in the earth and in animal or plant life. Indiscriminate use may destroy honey bees and other useful insects, produce sterility of soils by killing worms and anaerobic life, and poison the waters of lakes and streams. DDT, for example, is highly valuable for the control of malaria and other insect-borne diseases, but its uncontrolled use as an insecticide has been disastrous to wildlife.

Insecticides are generally classed as **stomach poisons** and **contact poisons.** Stomach poisons include **calcium arsenate,** a white powder of the composition $Ca_3(AsO_4)_2$, which constitutes about half of all insecticides used, and also paris green, lead arsenate, and white arsenic. An **antimetabolite** is not a direct poison, but acts on the insect to stop the desire for food so that the insect dies from starvation. Dimethyltriazinoacetamilide, used against corn-ear worms, is such a chemical. Contact poisons include rotenone dust, sulfur dust, and nicotine sulfate solution.

A **larvicide** is a chemical, such as chloropicrin, used to destroy fungi and nematodes in soils, and insect eggs and organisms in warehouses. Chemicals used against fungi and bacteria are called **fungicides** and **bactericides,** and those used to control plant diseases caused by viruses are called **viricides.** None of these are properly classed as insecticides, but are often used with them. **Herbicides** are used to kill weeds, usually by overstimulating cell growth. **Aminotriazole,** used as a weed killer, may cause cancer in animal life. Most of the pest-control chemicals are cumulative toxic poisons. **Benzene hexachloride** destroys bone marrow, and all of the chlorinated hydrocarbons affect the liver. **Deodorants** may have an insect-kill action, but are usually chemicals such as chlorophyll which combine with impurities in the air to eliminate unpleasant odors.

Insecticides may be solids or liquids, and the solids may be applied as a fine powder, usually in dilution in a dusting powder, or the powder may be suspended in a liquid carrier. Usually, the proportion of poison mixed with a mineral powder is no more than 5%. The mineral carrier, or **dusting**

powder, for this purpose should be gritless and inert to the insecticide. Ordinary **dusting clay** is a light, fluffy, air-floated kaolin, or it may be finely ground soft limestone. **Sodium fluorosilicate,** Na_2SiF_6, comes as a fine white powder for this purpose. Calcium sulfate is also used as a carrier but has itself a poisoning effect. One of the oldest of solid insecticides still used, either dry or in liquid sprays, is **Bordeaux mixture,** made by reacting copper sulfate with lime, giving a product with an excess of hydrated lime. Liquid carriers for insecticides may be kerosene or other petroleum hydrocarbons, or they may be liquid chemicals that have toxic properties, but they must be chosen to avoid deleterious effects, such as the yellowing of papers and organic materials in warehouses or archives, or the injuring of plants from active chlorine in some chemicals.

Some materials, such as citronella oil, used as mosquito repellents in households, have little or no value as insecticides for the eradication of mosquitoes in important applications such as at military sites or mining and lumbering camps. The **aerosol bomb** employed during the Second World War contained 3% DDT, 2 to 20 pyrethrum concentrate, 5 cyclohexanone, 5 mineral oil, and the balance a carrier gas. **Dimethyl phthalate,** a liquid of the composition $C_6H_4(CO_2CH_3)_2$, is a mosquito repellent having an effect lasting 1½ h in the open air. **Thiourea** is used to kill mosquito larvae in water and is harmless to fish.

The insecticide called **DDT** is dichlorodiphenyltrichloroethane, $C_6H_3Cl_2(C_6H_4 \cdot CH_2CCl_3)$, used effectively during the Second World War against flies, mosquitoes, body lice, and agricultural pests. It has no noxious odor, but it is cumulative and in concentration is toxic to man and to other warm-blooded animals. Oil paint containing 0.5 to 5% DDT kills flies on walls painted with it. **Chlordane** is a liquid of the composition $C_{10}H_6Cl_8$. It is a powerful insecticide.

Sabadilla is used in cotton sprays. It is also known as **cevadilla,** or **Indian barley,** and consists of the dried ripe poisonous seeds of the plant *Veratrum sabadilla* of the lily family growing in Central America, of which there are about 20 known species in Central and South America. The seeds contain **veratric acid,** from which is derived **veratraldehyde,** or **vetraldehyde,** a crystalline solid of the composition $(CH_3O)_2C_6H_3 \cdot CHO$, which gives the heliotrope flavor to the vanilla of Samoa and to some synthetic vanilla from coniferin. The alkaloid poison is extracted with a hydrocarbon solvent, and when the extract is used as an insecticide in combination with a synthetic, it gives greatly increased toxicity. The powdered seeds are also used as an agricultural insecticide dust which has greater staying power than pyrethrum. The cresols in various forms are also used as insecticides. **Dinitroorthocresol,** a yellow crystalline material melting at 83.5°C, is used in fruit-tree sprays. Sodium antimony lactophenate, known as **salp,** is an effective insecticide against chewing insects. Some insecticides are sprayed on the ground or on the foliage to be absorbed into the plant, poisoning the insect that feeds on the plant.

A **microencapsulated insecticide,** consisting of **pyrethroids** inside a plyurea shell, is produced by 3M Co. The pyrethroids become active when a biological synergist permeates the shell wall. Unprotected, pyrethroids normally decompose in several minutes to several hours.

Sodium fluoride, or **fluorol,** NaF, is a water-soluble white powder used as a wood preservative as well as an insecticide and vermin poison, although this material is better known as an industrial chemical. Single crystals of it are used for windows for ultraviolet and infrared equipment as it transmits these rays. When wood is treated with an alkaline water solution of acrylonitrile ethylates the cellulose fibers are cyanoethylated and the wood becomes resistant to the attack of enzymes and fungi. Wood treated with pyradine and acetic anhydride is given dimensional stability as well as resistance to insect and fungi attack.

A **fumigant** is a liquid, powder, or gas which kills insects, worms, or burrowing animals by toxic fumes. For general use a fumigant should not be injurious to grains or stored foodstuffs. **Repellents** are fumigants used for driving out, rather than killing, insects. However, some repellents contain naphthalene, rotenone, or other materials having toxic properties, and these are insecticides rather than fumigants. **Methyl bromide,** or **bromomethane,** CH_3Br, a gas with a liquefying point at 4.6°C, is an effective fumigant not injurious to grains. Methyl bromide is also used for fumigating clothing warehouses and does not shrink or wrinkle woolen fabrics. **Dihydroacetic acid** is used on dried fruits in storage to prevent decomposition. It acts in the nature of a fungicide.

INSULATORS. Any materials that retard the flow of electricity, and are used to prevent the passage or escape of electric current from conductors. No materials are absolute nonconductors; those rating lowest on the scale of conductivity are therefore the best insulators. An important requirement of a good insulator is that it not absorb moisture which would lower its resistivity. Glass and porcelain are the most common line insulators because of low cost. Pure silica glass has an average dielectric strength of 500 volts per mil (20×10^6 volts per meter), and glass-bonded mica about 450 volts per mil (17.7×10^6 volts per meter), while ordinary porcelain may be as low as 200 volts per mil (8×10^6 volts per meter), and steatite about 240 volts per mil (9.4×10^6 volts per meter). Slate, steatite, and stone slabs are still used for panelboards, but there is now a great variety of insulating boards made by compressing glass fibers, quartz, or minerals with binders, or standard laminated plastics of good dielectric strength may be used. **Vulcoid,** of the Budd Co., and **Vulcabeston,** of Johns-Manville, are typical. For slots and separators, natural mica is still valued because of its heat resistance, but, because of the irregular quality of natural mica and the difficulty of handling the small pieces, it has been largely replaced by synthetic mica paper, polyester sheet, or impregnated papers or fabrics. The impregnated fish paper of the Spaulding Fibre Co., called **Armite,** comes in thicknesses down to 0.004 in (0.010 cm), and has a dielectric strength of 500 volts per mil (20×10^6 volts per meter).

Synthetic rubbers and plastics have now replaced natural rubber for wire insulation, but some aluminum conductors are insulated only with an anodized coating of aluminum oxide. Wires to be coated with an organic insulator may first be treated with hydrogen fluoride, giving a coating of copper fluoride on copper wire and aluminum fluoride on aluminum wire. The thin film of fluoride has high dielectric strength and heat resistance. The AIEE classification for wire insulation is by heat resistance, from **Class O insulation,** for temperatures to 195°F (90°C), to **Class C insulation,** for temperatures above 355°F (179°C).

Insulating oils are mineral oils of high dielectric strength and high flash point employed in circuit breakers, switches, transformers, and other electric apparatus. An oil with a flash point of 285°F (140°C) and fire point of 310°F (154°C) is considered safe. A clean, well-refined oil will have high dielectric strength, but the presence of as low as 0.01% water will reduce the dielectric strength drastically. The insulating oils, therefore, cannot be stored for long periods because of the danger of absorbing moisture. Impurities such as acids or alkalies also detract from the strength of the oil. Since insulating oils are used for cooling as well as for insulating, the viscosity should be low enough for free circulation, and they should not gum. **Askarel** is an ASTM designation for **insulating fluids** which give out only nonflammable gases if decomposed by an electric arc. They are usually chlorinated aromatic hydrocarbons such as trichlorobenzene, but fluorinated hydrocarbons are also used. They have high dielectric strength, and a dielectric constant below 2. **Insulating gases** are used to replace air in closed areas to insulate high-voltage equipment. Sulfur hexafluoride for this purpose, with a density of 0.755, has a dielectric strength 2.35 times that of air. The insulating oil, fluids, and gases are generally referred to as **dielectrics,** although this term embraces any insulator.

Insulation porcelain, or **electrical porcelain,** is not usually an ordinary porcelain except for common line insulators. For such uses as spark plugs they may be molded silica, and for electronic insulation they may be molded steatite or specially compounded ceramics, more properly called **ceramic insulators.** Insulation porcelains compounded with varying percentages of zirconia and beryllia have a crystalline structure, and have good dielectric and mechanical strengths at temperatures as high as 2000°F (1093°C). These porcelains may have some magnesia, but are free of silica. However, **zircon porcelain** is made from zirconium silicate, and the molded and fired ceramic is equal to high-grade steatite for high-frequency insulation. **Vitrolain,** of the Star Porcelain Co., is an electrical porcelain of high strength and density with porosity of only 0.25%. **Thyrite,** of the General Electric Co., is a porcelain that possesses the property of being an insulator at low potentials and a conductor at high potentials. It is used for lightning arresters. The German **Hartporzellan,** or **hard porcelains,** are specially compounded ceramics having good resistance to thermal shock. The material called **Nolex** by the Naval Ordnance Laboratory is made by hot-molding finely powdered synthetic fluorine mica. The molded parts are practically pure mica. They can be machined, have high dimensional precision because they need no further heat treatment, and have high dielectric strength. Beryllia is a valued insulator for encapsulation coatings on heat-generating electronic devices as it has both high electrical resistivity and high heat conductivity.

IODINE. A purplish-black, crystalline, poisonous elementary solid, chemical symbol I, best known for its use as a strong antiseptic in medicine, but also used in many chemical compounds and war gases. In tablet form it is used for sterilizing drinking water, and has less odor and taste than chlorine for this purpose. It is also used in cattle feeds. Although poisonous in quantity, iodine is essential to proper cell growth in the human body, and is found in every cell in a normal body, with larger concentration in the thyroid gland.

The Chilean production of iodine is a by-product of the nitrate industry. In Scotland, Norway, and Japan it is produced by burning seaweed and treating the ashes. A ton of seaweed produces about a pound of iodine. It is also produced from salt brines and from seawater, and in California from the wastewaters of oil wells, the brine containing 65 parts of iodine per million. The lump iodine from this source is 99.9% pure. As much as 1,000 tons (907 metric tons) of iodine is present in a cubic mile of seawater. The specific gravity of iodine is 4.98, melting point 114.2°C, and boiling point 184°C. It is insoluble in water, but is soluble in alcohol, ether, and alkaline solutions. **Tincture of iodine,** a 7% alcohol solution of iodine in a 5% solution of potassium iodide, is used in medicine as a caustic antiseptic. As an antiseptic, iodine has the disadvantage that it burns and stains the skin. **Vodine,** of the Lanteen Laboratories, Inc., is a 2% oil-and-water emulsion of iodine containing also lecithin. It does not burn, and the faint stain washes off easily. An **iodophor** is a chemical containing iodine which is released on contact with organic material. **I-O-Dynamic,** of Field Bros., Inc., is a detergent containing iodine. **PVP iodine** of General Aniline & Film (GAF Corp.) is iodine combined with polyvinyl pyrrolidine to give a product that retains the germ-killing properties of iodine without the toxic and burning effects. **Wescodyne,** of the West Disinfecting Co., is another nonburning and nonstaining iodine. **Clearodine,** of the U.S. Peroxide Corp., is a water-soluble iodine in powder form for disinfectant purposes. In water solution it releases a colorless **hypoiodous acid,** IHO. It has a higher bacteriological effect than ordinary iodine and does not stain or irritate. **Iodine cyanide,** ICN, an extremely poisonous colorless crystalline material soluble in water, is used as a preservative for furs and museum specimens. Alfa Inorganics produces the acid as **iodic acid** HIO_3, and a stable **iodine pentoxide,** I_2O_5, is also marketed. The iodine atom is very regular with a valence of 7, but having three spheron pairs in opposite polarity which can be broken to give valences of 1, 3, 5, and 7. A wide range of compounds are made for electronic and chemical uses.

IRIDIUM. A grayish-white metal of extreme hardness, symbol Ir. It is insoluble in all acids and in aqua regia. The melting point is 2443°C, and the specific gravity is 22.50. The annealed metal has a hardness of 172 Brinell. Iridium is found in the nickel-copper ores of Canada, pyroxinite deposits of South Africa, and platinum ores of Russia and Alaska. It occurs naturally with the metal osmium as an alloy, known as **osmiridium,** 30 to 60% osmium, used chiefly for making fountain-pen points and instrument pivots. Iridium is employed as a hardener for platinum, the jewelry alloys usually containing 10%. With 35% iridium the tensile strength of platinum is increased to 140,000 lb/in^2 (965 MPa). **Iridium-rhodium alloys** are used for high-temperature thermocouples. Electrodeposition is difficult, but coatings have been obtained with a fused sodium cyanide electrolyte or with an iridium salt and organic compound in solution and, after application, volatilizing the vehicle and reducing the compound to the metal. Iridium-clad molybdenum or tungsten is produced by swaging an iridium tube onto the base metal and drawing at 600°C. Iridium plating is used on molybdenum to protect against oxidation at very high temperatures. Above 600°C iridium tarnishes, and above 1000°C it forms a volatile oxide. It gives a bright ductile plate with a Vickers hardness of 170. It is also used as a catalyst. It is resistant to most molten metals except copper, aluminum, zinc, and magnesium. **Iridium wire** is used in spark plugs as it resists attack of leaded aviation fuels. **Iridium-tungsten alloys** are used for springs operating at temperatures to 800°C. Iridium intermetallic compounds such as Cb_3Ir_2, Ti_3Ir, and $ZrIr_2$ are superconductors. Iridium is multivalent with most of its compounds formed in its trivalent state, as $IrCl_3$. The metal is sold by the troy ounce, a cubic inch of the metal weighing 11.82 troy oz (0.4 kg).

IRON. One of the most common of the commercial metals. It has been in use since the most remote times, but it does not occur native except in the form of meteorites, and early tools of Egypt were apparently made from nickel irons from this source. The common iron ores are magnetic pyrites, magnetite, hematite, and carbonates of iron. To obtain the iron the ores are fused to drive off the oxygen, sulfur, and impurities. The melting is done in a blast furnace directly in contact with the fuel and with limestone as a flux. The latter combines with the quartz and clay, forming a slag which is readily removed. The resulting product is crude pig iron which requires subsequent remelting and refining to obtain commercially pure iron. A short ton (0.9 metric ton) of ore, with about 1,000 lb (454 kg) of coke and 600 lb (272 kg) of limestone, produces an average of 1,120 lb (508 kg) of pig iron. Sintered iron and steel are also produced without blast-furnace reduction by compressing purified iron oxide in rollers, heating to 2200°F (1204°C), and hot-strip rolling. The final cold-rolled product is similar to conventional iron and steel.

Originally, all iron was made with charcoal, but because of the relative scarcity of wood and the greater expense, charcoal is now seldom used in the blast furnace. **Charcoal iron** has less sulfur and phosphorus than iron made with coke, and cast iron made from it has a dense structure and a tendency to chill. **Elverite,** of the Fuller Lehigh Co., is a charcoal type of cast iron which gives a hard chill with a soft, gray-iron core. **Charcoal pig iron** was formerly imported from Sweden and Norway, and was used for such purposes as car wheels, magnet cores, and for making high-grade steels for boiler tubes. **Stora** was a name for Swedish charcoal iron used for making malleable iron.

Iron is a grayish metal, which until recently was never used pure. It melts at 1525°C and boils at 2450°C. Even very small additions of carbon reduce the melting point. It has a specific gravity of 7.85. All commercial irons except ingot iron and electrolytic iron contain perceptible quantities of carbon, which affect its properties. Iron containing more than 0.15% chemically combined carbon is termed steel. When the carbon is increased to above about 0.40%, the metal will harden when cooled suddenly from a red heat. Iron, when pure, is very ductile, but a small amount of sulfur, as little as 0.03%, will make it **hot-short,** or brittle at red heat. As little as 0.25% of phosphorus will make iron **cold-short,** or brittle when cold. Iron forms carbonates, chlorides, oxides, sulfides, and other compounds. It oxidizes easily and is also attacked by many acids.

Because pure iron is allotropic, it can exist as a solid in two different crystal forms. From subzero temperatures up to 1670°F (910°C), it has a body-centered cubic structure and is identified as alpha (α) iron. Between 1670 and 2552°F (910 and 1400°C), the crystal structure is face-centered cubic. This form is known as gamma (γ) iron. At 2552°F (1400°C) and up to its melting point of 2802°F (1540°C), the structure again becomes body-centered cubic. This last form, called delta (δ) iron, has no practical use. The transformation from one allotropic form to another is reversible.

Thus, when iron is heated to above 1670°F (910°C), the alpha body-centered cubic crystal changes into face-centered cubic crystals of gamma iron. When cooled below this temperature, the metal again reverts back to a body-centered cubic structure. These allotropic phase changes inherent in iron make possible the wide variety of properties obtainable in ferrous alloys by various heat-treating processes.

Electrolytic iron is a chemically pure iron produced by the deposition of iron in a manner similar to electroplating. Bars of cast iron are used as anodes and dissolved in an electrolyte of ferrous chloride. The current precipitates almost pure iron on the cathodes which are hollow steel cylinders. The deposited iron tube is removed by hydraulic pressure or by splitting, and then annealed and rolled into plates. The iron is 99.9% pure, and is used for magnetic cores and where ductility and purity are needed. Highly refined nearly pure iron is marketed by the Westinghouse Electric Corp. under the name of **Puron** in rod form for use for spectroscopic and magnetic standards. It contains 99.95% iron, with only 0.005 carbon, 0.003 sulfur, and less than 0.001 phosphorus. By high-temperature hydrogen annealing, the carbon can be reduced to 0.001%, bringing the purity to 99.99%. **Iron whiskers** are single-crystal pure iron fibers, 0.00004 in (0.0001 cm) in diameter, for electronic uses. The tensile strength is as high as 500,000 lb/in^2 (2,825 MPa).

Iron powder, as originally produced in Sweden, is made by reducing iron ore by the action of carbon monoxide at a temperature below the melting point of the iron and below the reduction point of the other metallic oxides in the ore. In the United States it is made by the reduction of iron oxide mill scale, or by electrolysis of steel borings and turnings in an electrolyte of ferric chloride. Iron powders are widely used for pressed and sintered structural parts, commonly referred to as **P/M parts.** Pure iron powders are seldom used alone for such parts. Small additions of carbon in the form of graphite and/or copper are used to improve performance properties.

Iron-copper powders contain from 2 to 11% copper. Small amounts of graphite are sometimes added. Copper increases strength properties, improves corrosion resistance, and tends to increase hardness, but it lowers ductility somewhat. Densities of around 6 g/cm^3 are most common for iron-copper parts, although densities approaching 7.0 can be achieved. Strengths range between 30,000 and 100,000 lb/in^2 (206 and 680 MPa), depending on density and heat treatment.

Iron-carbon powders contain up to 1% graphite. When pressed and sintered, internal carburization results and produces a carbon-steel structure, although some free carbon remains. In general, **iron-carbon steel P/M parts** have densities of around 6.5 g/cm^3. However, densities of over 7.0 are used to produce higher mechanical properties. These carbon-steel

P/M parts have higher strength and hardness than those of iron, but they are usually more brittle. As-sintered strengths range from about 35,000 to 70,000 lb/in^2 (241 to 482 MPa) depending on density. By heat treatment, strengths up to 125,000 lb/in^2 (861 MPa) are achieved. The mechanical properties of ferrous powder parts can be considerably improved by impregnating or infiltrating them with any one of a number of metallic and nonmetallic materials, such as oil, wax, resins, copper, lead, and babbitt.

P/M forgings are produced by hot- or cold-forming specially shaped metal preforms in mechanical or hydraulic presses. Most P/M forgings are made with iron or steel powders. **Aluminum powders** and **titanium powders** can also be used.

Soft **electrolytic iron powder** for making sintered molded parts has 99% min iron, 0.04 max carbon, a hydrogen loss of 0.9 max, and 0.1 max of other impurities. Up to 40% passes through a 325-mesh screen. The apparent density is 2.5, and the final density of the molded part is 7.2 to 7.7. **Magnetic iron powder,** used for electrical cores of high permeability, is made by reducing iron oxide with hydrogen. For this purpose it must be free of carbon and sulfur. A 10- to 40-μm powder with an apparent density of 2.75 will compress to a density of 6.85 at 150,000 lb/ in^2 (1,034 MPa). **Carbonyl iron powder,** used for magnetic cores for high-frequency equipment and for medical application of iron, is metallic iron of extreme purity and in microscopic spherical particles. It is made by the reaction of carbon monoxide on iron ore to give liquid **iron carbonyl,** $Fe(CO)_5$, which is vaporized and deposited as a powder.

HVA iron powder, produced electrolytically in Germany for making strong, high-density parts by powder metallurgy, contains a maximum of 0.02% silicon, 0.05 carbon, 0.06 manganese, 0.01 phosphorus, and 0.01 sulfur. The standard powder has up to 35% of the grains under 0.06 mm, up to 30% from 0.10 to 0.06 mm, and a maximum of 5% over 0.20 mm. The sponge grade, for making porous parts, has a greater percentage of smaller particles, with up to 50% under 0.06 mm. Parts made with electrolytic iron powder to a density of 7.5 have a tensile strength of about 40,000 lb/in^2 (275 MPa) with elongation of 30%.

Reduced iron, used for special chemical purposes, is a fine gray amorphous powder made by reducing crushed iron ore by heating with hydrogen. **Nu-Iron,** of the U.S. Steel Corp., is iron powder made by reducing Fe_2O_3 with hydrogen at 1300°F (704°C) to FeO, and then at 1100°F (593°C) to iron to prevent sintering of the particles. **H-iron,** of the Bethlehem Steel Co., is a **hydrogen-reduced iron.** Iron powder made by this process is usually pyrophoric, and is made nonpyrophoric by heating in a nonoxidizing atmosphere. **Ferrocene,** of the Ethyl Corp., used as a combustion-control additive in fuels, and in lubricants and plastics for heat stabilization and radiation resistance, is a **dicyclopentadienyl iron,**

$(C_5H_5)_2Fe$. It is a double-ring compound with the iron atom between parallel planes and symmetrically bound to the five carbon atoms of each ring. It is an orange-yellow powder melting at 173°C, but resisting pyrolysis to 470°C. **Iron 55** is a very pure radioactive iron produced in minute quantities by cyclotronic bombardment of manganese and used in medicine for increasing the red blood pigment, **hemoglobin,** in the human system. **Busheled iron** is an inferior grade of iron or steel made by heating bundles of scrap iron or steel in a furnace and then forging and rolling into bars. It is not uniform in composition or in the welding, and is used only in isolated places or in wartime.

IRON ORES. Iron-bearing minerals from which iron can be extracted on a commercial scale. The chief iron ores in order of importance are hematite, magnetite, limonite, and siderite. More than 90% of the iron ores mined in the United States are **red hematite,** Fe_2O_3, containing theoretically 70% iron, but usually not over 60%. The districts include the Lake Superior region and northern Alabama. It is also the ore from the Furness district in England and parts of Spain and Germany. The color is various shades of reddish brown, and the structure is usually earthy. The variety known as **kidney ore** is columnar with a fibrous appearance; **specular hematite** has a brilliant luster and foliated structure. The specific gravity is 4.8 to 5.3. **Brown hematites** contain from 35 to 55% iron. Ores containing more than 50% iron are considered high grade. Lake Superior ores are now averaging only 52% Fe, with 8 silica. This region also has large reserves of **taconite,** a ferroginous chert which is an alteration product from **greenalite,** or ferrous silicate. Taconite eventually leaches and becomes a hematite by the loss of silica, but the Mesaba taconite is a very hard gray-green sedimentary chert in the form of a compact silica rock with 20 to 35% iron. It cannot be used directly in a blast furnace but is crushed to powder, concentrated to 65% iron, and pelletized for use. **Rubio iron ore** of Spain is classed both as limonite and brown hematite. It is a **hydrated ferric oxide,** $Fe_2O_3 \cdot H_2O$, brown-streaked with a silica gangue. It contains 77.4% ferric oxide, 9.2 silica, 1.76 alumina, and 1.1 manganese oxide.

The **hematite ores** are preferred for the bessemer process because of their freedom from phosphorus and sulfur. **Natural iron** is the percentage of iron in the ore before drying; **dry iron** is the percentage of iron in the ore after drying at 212°F (100°C). Low-grade hematite and limonite ores can be concentrated by passing the ground ore through a reducing gas at high temperature, which causes a part of the iron in each particle to become magnetic and thus capable of being separated. Low-grade ores of 30% iron are also ground to a fine powder and separated from the gangue by flotation, concentrated to 60% iron, and pelleted with a bentonite binder. **Self-fluxing iron ore** is concentrated iron ore combined with limestone and pelleted.

Magnetite, or **magnetic iron ore,** is found in northern New York, in New Jersey, and in Pennsylvania. It has the composition $FeO \cdot Fe_2O_3$, containing theoretically 72.4% iron but usually only about 62%. Magnetite may also contain some nickel or titanium. The specific gravity is 5.18, the melting point is 1540°C, the color is iron black with a metallic luster, and the material is strongly magnetic. The natural magnet known as **lodestone** is magnetite.

Siderite and carbonate ores are used in Great Britain, Germany, and Russia. Much of these ores is not considered commercial in the United States, but the **dogger iron ore** of Germany contains as high as 35% iron. **Siderite** is the chief iron ore in Great Britain. It is found in Staffordshire, Yorkshire, and Wales, and in the United States in Pennsylvania and Ohio. It is an **iron carbonate,** $FeCO_3$, containing theoretically 48.2% iron. It usually occurs granular or compact and earthy. Its specific gravity is 4.5 to 5, and the hardness is 3.5 to 4. The color is light to dark brown, with a vitreous luster. It often is impure, with a mixture of clay materials or forming stratified bodies with coal formations. It is also known as **carbonate ore, ironstone,** and as **spathic iron ore.** Impure forms mixed with clay and sands are called **clay ironstones,** or **black band.** The ironstone and blackband ores are the important ores of England and Scotland, but there is only a slight usage of carbonate ore in the United States. The **minette ore** of Luxembourg and Germany averages about 26% iron, 22 CaO, 11 silica, and 2 $CaOSiO_2$, with 0.54 phosphorus and 0.06 vanadium. The ore mined in Norway and Sweden is very pure, and it is the ore used for the **dannemora iron** made with charcoal as a fuel.

Limonite, also called **brown hematite, brown ore, bog iron ore,** and **shot ore** when in the form of loose rounded particles, was the common iron ore of early New England. It is a mineral of secondary origin formed by the water solution of other iron minerals. Its composition is $2Fe_2O_3$ $3H_2O$, containing theoretically 59.8% iron, but usually 30 to 55%. It occurs earthy or in stalactitic forms of a dark-brown color. The specific gravity is 3.6 to 4. Limonite is found in Alabama, Tennessee, and Virginia, and is also an English and German ore. **Goethite** is a minor ore of iron of the composition $Fe_2O_3 \cdot H_2O$, found in the Lake Superior hematite deposits and in England. It is yellowish brown in color with a specific gravity of 4.3. It is also called **turgite.** The largest known deposits of high-grade iron ores are in Brazil and extend into Venezuela. The ore in southwestern Brazil is with deposits of manganese. The hematite of Minas Gerais contains 57 to 71% iron. Large deposits of high-grade iron ores are also found in the Labrador-Quebec regions of Canada, in India, and in the South Africa-Zimbabwe area. The nearest approach to a native iron is the iron-nickel mineral **awaruite,** $FeNi_2$, found in gravel in New Zealand and Alaska, and **josephinite,** $FeNi_3$, found in serpentine in Oregon.

IRON PYRITE. A mineral sometimes mined for the zinc, gold, or copper associated with it, but chiefly used for producing sulfur, sulfuric acid, and copperas. It is an **iron disulfide,** FeS_2, containing 53.4% sulfur. It often occurs in crystals, also massive or granular. It is brittle, with a hardness of 6 to 6.5 and specific gravity 4.95 to 5.1. The color is brass yellow, and it is called **fool's gold** because of the common error made in detection. In the Italian method of roasting pyrites, the oxygen combines with the sulfur to form sulfur dioxide, and also with the iron. In a yield of 100,000 tons (90,700 metric tons) of concentrated sulfuric acid there is obtained a by-product of 50,000 tons (45,350 metric tons) of 65% iron ore. **Pyrite** is found in rocks of all ages associated with different minerals. The pyrites mined in Missouri, known as **marcasite,** also used for gemstones, have the formula FeS, and the gem specimens have a yellow color with a greenish tinge. **Sulfur pyrites,** found in great quantity in Shasta County, Calif., are roasted to produce sulfuric acid, and the residue is used for making cement and ferric sulfate. The distillation of 370 tons (336 metric tons) of pyrite of Ontario yields 62 tons (56 metric tons) of sulfur and 252 tons (229 metric tons) of iron-oxide sinter, the latter being used in steel mills. In Japan pyrite is chlorinated, producing sulfur, copper, gold, silver, and zinc, and the residue of sintered pellets of Fe_2O_3 is used in blast furnaces.

IRON SHOT. An abrasive material made by running molten iron into water. It is employed in tumbling barrels and also in the cutting and grinding of stones. The round sizes between No. 30 and No. 20 are used for the shot peening of mechanical parts to increase fatigue resistance. **Peening shot** is marketed by SAE numbers from 6 to 157, the size numbers being the diameter in thousandths of an inch. Grit numbers in shot for grinding are from No. 10 to No. 200; the No. 10 is 0.0787 in (0.200 cm); and the No. 200 is 0.0029 in (0.007 cm) in diameter. **Steel grit** is made by forcing molten iron through a steam jet so that the metal forms into small and irregular pieces and large globules are crushed into steel grit in sizes from No. 8 to No. 80. It is preferred to sand for sandblasting some materials. **Steelblast,** of the Steelblast Abrasives Co., is this material for tumbling and sandblasting.

Tru-Steel shot, of the American Wheelabrator & Equipment Corp., is steel shot which has been heat-treated to give toughness to prevent breakdown into fragments. **Cutwire shot,** of the Precision Shot Co., is made by cutting SAE 1065 steel wire into short pieces. **Pellets shot,** of Pellets, Inc., is steel wire of a hardness of Rockwell 50 cut to lengths equal to the diameter of the wire, ranging from 0.028 to 0.41 in (0.071 to 1.041 cm). **Kut-**

Steel is a similar abrasive of the Steelblast Abrasives Co. **Permabrasive,** of the National Metal Abrasive Co., is a treated **malleable iron shot** to give a soft resilient body with a hard exterior. All of these materials for metal cleaning are termed **blasting shot.**

IRONWOOD. The name for several varieties of wood, which may refer to any exceedingly hard wood that is used for making bearings, gears, tool handles, or parts of machinery. In the United States ironwood is most likely to refer to **hackia,** the wood of the hackia tree, *Ixora ferrea,* of the West Indies and of tropical South America, or it may refer to the wood of the quebracho tree. Hackia is brown in color, has a coarse, open grain, and is very hard and tough. The weight is about 55 lb/ft^3 (881 kg/m^3). It is also used for furniture. The Burmese tree, *Mesua ferrea,* furnishes the wood **gangaw,** which is also known as ironwood. It is a tough, extremely hard wood of a rose-red color weighing 70 lb/ft^3 (1,121 kg/ m^3). Innumerable varieties of hardwoods occur in the tropical jungles, many unnamed or known only by native names. Many have rich colors, shading from yellow, orange, and red to purplish-black, and in beautiful grains which can be varied by the angle of cutting. They are generally commercialized only on a small scale for use in furniture, cabinetwork, inlays, and panel facings.

Ekki, *Lophira alata procera,* sometimes called **African ironwood** or **azobe,** comes from West Africa. It is an excellent timber for piling, wharf and dock construction, bridge planking, ties, and all heavy timber structures. The wood is a chocolate-brown color, sometimes verging on dark red, and has a speckled surface caused by yellowish deposits in the pores. It is so heavy that the timbers will not float in water.

ISINGLASS. A gelatin made from the dried swimming bladders, or **fish sounds,** of sturgeon and other fishes. **Russian isinglass** from the sturgeon is the most valued grade, and is one of the best of the water-soluble adhesives. It is used in glues and cements, and in printing inks. It is also used for clarifying wines and other liquids. It is prepared by softening the bladder in water and cutting it into long strips. These dry to a dull-gray, horny, or stringy material. Isinglass is also known as **ichthyocolla. Ichthyol** is an entirely different material. It is a reddish-black syrupy liquid with a peculiar odor and burning taste, distilled from an asphaltic shale found in the Austrian Tyrol. It is used as an antiseptic astringent. The name isinglass was also used for mica when employed as a transparent material for stove doors.

IVORY. The material that composes the tusks and teeth of the elephant. It is employed mostly for ornamental parts, art objects, and piano keys, although the latter are now usually a plastic which does not yellow. The color is the characteristic ivory white, which yellows with age. The specific gravity is 1.87. The best grades are from the heavy tusks weighing more than 55 lb (25 kg), sometimes 6 ft (1.8 m) long. The ivory from animals long dead is a gray color and inferior. The softer ivory from elephants living in the highlands is more valuable than the hard and more brittle of the low marshes. The west coasts of Africa, India, and southern Asia are the chief sources of ivory. The tusks of the hippopotamus, walrus, and other animals, as well as the fossil mammoth of Siberia, also furnish ivory, although of inferior grades. **Odontolite** is ivory from fossil mammoths of Russia. Ivory can be sawed readily, and is made into thin veneers for ornamental uses. It takes a fine polish. Artificial ivory is usually celluloid or synthetic resins.

Vegetable ivory, used for buttons and small articles, is from the **ivory nut,** called **tagua nut** in Colombia and Ecuador, and **jarina** in Brazil. The ivory nut is the seed of the low-spreading palm tree, *Phytelephas macrocarpa,* which grows in tropical America. The nuts are about 2 in (5 cm) in diameter, growing in clusters and encased in shells. They have a fine white color and an even texture. They can be worked easily, and they harden on exposure to the air. They take dyes readily and show fine polished colors. A similar nut called the **dom nut** is used in Eritrea for making buttons and novelties.

JAPAN WAX. Sometimes called **sumac wax** and **Japan tallow.** A vegetable fat used for extending beeswax, and in candles and polishes. It occurs between the kernel and the outer skin of the berries of plants of the genus *Rhus,* which grow in Japan and in California. The fat, which is misnamed wax, is extracted by steaming and pressing the berries, and then refining. The crude wax is a coarse greenish solid, but the bleached wax is in cream-colored cakes which darken to yellow. Japan wax melts at 51°C. The specific gravity is about 0.975. It contains chiefly palmitic acid in the form of glyceryl palmitate with also stearic and oleic acids, and a characteristic acid, **japonic acid,** which is also found in catechu. The saponification value is 220. It is sometimes adulterated with common tallow. **Lac** is the gum exudation of the wood of the same sumac plants, notably *R. vernicifera,* of Japan and Korea. It is used as a drying oil in clear lacquers and baking enamels. Lac is distinct from shellac of the lac insect.

JEWELRY ALLOYS. An indefinite term which refers to the casting alloys used for novelties to be plated and to the copper and other nonferrous sheet and strip alloys for stamped and turned articles. The base-metal jewelry industry, as distinct from the precious-metal industry, produces costume jewelry, trinkets, souvenirs, premium and trade goods, clothing accessories known as notions, and low-cost religious goods such as medals and statuettes. The soft white alloys of this type are not now as important as they were before the advent of plastics, and articles now produced in metals are likely to be made from standard brasses, nickel brasses, and cupronickels, but laminated and composite sheet metals are much used to eliminate plating. When the base metals are clad or plated with gold, they are called **gold-filled metal** if the gold alloy is 10-karat or above and the amount used is at least 5% of the total weight. When the coating is less than 5% it is called **rolled-gold plate.** Formerly, a great variety of trade names were used for jewelry alloys. **Argentine metal** was a silvery alloy for casting statuettes and small ornaments. It contained 85.5% tin and 14.5 antimony, and produced silvery-white, hard, and clean-cut castings. An alloy known as **Alger metal** contained 90% tin and 10 antimony. A harder but more expensive white alloy known as **Warnes metal** contained 10 parts tin, 7 nickel, 7 bismuth, and 3 cobalt. **Fahlum metal,** used for stage jewelry, contained 40% tin and 60 lead. When faceted, it makes highly reflective brilliants. **Rosein** was a white metal for jewelry and ornamental articles containing 40% nickel, 30 aluminum, 20 tin, and 10 silver. When polished it has a high white luster without plating. **Mock silver** was an aluminum alloy containing about 5% silver and 5 copper. **Argental** was a rich low brass whitened and strengthened with about 5% cobalt. **Kuromi,** a Japanese white jewelry alloy, is copper whitened with tin and cobalt. **Argent français,** or **French silver,** is cupronickel containing considerable silver. **Platinoid** was a nickel-silver-tungsten alloy with the small amount of tungsten added to the melt in the form of phosphor tungsten. **Proplatinum,** another substitute for platinum, was a nickel-silver-bismuth alloy. **Nuremberg gold,** with a nontarnishing golden color, contained 90% copper, 7.5 aluminum, and 2.5 gold.

JOJOBA WAX. Also called **jojoba oil,** pronounced ho-hó-bah. A colorless, odorless liquid wax obtained by pressing or solvent extraction from the seed beans of the evergreen shrub *Simmondsia californica* growing in the semiarid region of southwest United States and northern Mexico. The beans contain 50% oil which has high-molecular-weight acids and alcohols and no glycerin and is a true wax. It is used as a substitute for sperm oil in lubricants, in leather dressing, and also in cosmetics. **Jojoba alcohol,** extracted from the bean, is a complex alcohol used as an antioxidant to prevent deterioration of fats.

JUJUBE. The oval fruit of the small spiny tree *Ziziphus jujuba* of the buckthorn family growing in dry alkaline soils. It is native to northern China, but is also grown in the Mediterranean countries and in the southwestern United States. It is also known as the **Chinese date,** or **tsao.** The fruit has high food value, being higher in sugars than the fig, 65 to 75%, and higher in protein than most fruits, 2.7 to 6%, but is very acid, containing up to 2% acid. Some varieties develop butyric acid in ripening. Jujube is used for flavoring, and in confections, preserves, and sweet pickles.

JUNIPER. The wood of the juniper tree *Juniperus virginiana* growing in the eastern United States from Maine to Florida. It is also called **red cedar, red juniper,** and **savin.** The heartwood has a bright to dull-red color, and the thin sapwood is nearly white. The wood is light in weight, soft, weak, and brittle, but is durable. It is used for chests, cabinets, and closet lining because of its reputed value for repelling moths. It was formerly employed on a large scale for lead-pencil wood, and was known as **pencil cedar,** but it is now scarce, and other woods are used for this purpose, notably **incense cedar,** *Libocedrus decurrens,* of California and Oregon, a close-grained brown wood with a spicy resinous odor. **Rocky Mountain juniper** is from a medium-sized tree of the Mountain states, *J. scopulorum,* valued for fence posts and lumber. **African pencil cedar** is from the tree *J. procera,* of eastern Africa. It is harder and heavier and less fragrant than incense cedar. There are more than 40 species of juniper. The fruit or **juniper berries** of the common varieties are used in flavoring gin. **Cade oil** is a thick brownish essential oil, of specific gravity 0.950 to 1.055, distilled from the wood of the **European juniper** and used in antiseptic soaps. It is also called **juniper tar oil.** The juniper oil known as **savin oil** is distilled from the leaves and tops of the **juniper bush,** *J. sabina,* of North America and Europe. It is used in medicine as a diuretic and vermifuge, and also in perfumes. **Cedarwood oil,** used in perfumes and soaps, is distilled from the sawdust and waste of the eastern red cedar. The red heartwood contains 1 to 3% oil.

JUTE. A fiber employed for making burlap, sacks, cordage, ropes, and upholstery fabrics. It is obtained from several plants, of which **Corchorus capsularis** of India is the most widely cultivated, growing in a hot, steaming climate. This fiber in Brazil is called **juta indiana,** or **Indian jute.** Most of the commercial jute comes from Bengal. The plant grows in tall slender stalks like hemp, and the fiber is obtained by retting and cleaning. The fiber is long, soft, and lustrous, but is not as strong as hemp. It also loses its strength when damp, but is widely used because of its low cost and because of the ease with which it can be spun. Its chemical composition is intermediate between hemp and kapok. It contains 60% alpha cellulose, 15 pentosan, 13 lignin, and 4.5% of **uronic anhydride** which is also a constituent of kapok. The crude fiber may be as long as 14 ft (3.6 m), but the commercial fibers are from 4 to 8 ft (1.2 to 2.4 m). The butts, or short ends of the stalks, and the rough fibers are used for paper stock. **Jute**

paper, used for cement bags, is a strong paper made of these fibers usually mixed with old rope and old burlap in the pulping. It is tan in color. **Jute fiber** is also used for machine packings. **Rel-Kol,** of the Reliance Packing Co., is jute fiber treated with synthetic rubber and braided into square sections.

Brazilian jute is fiber from plants of the mallow family, *Hibiscus kitaibelifolius,* or *H. ferox,* cultivated in the state of Saõ Paulo, Brazil. It is also called **juta paulista.** The fiber is long, strong, resilient, and durable. It is used for making burlap for coffee bags. The fiber known as **pacopaco** is from the *H. cannabinus,* of Bahia and Minas Gerais. It is used for cordage, burlap, and calking. In Indonesia it is called **Java jute.** In India it is called **Bimlipatam jute** and **Deccan hemp,** and is mixed with *Corchorus* fibers as jute and in burlap. As a substitute for hemp it is known as **sunee,** or **brown Indian hemp.** The species known as **amaniurana** is from the *H. furcellatus* of the Paraguay River Valley. The fibers are very soft and silky and are made into sacks and bags. The species *H. sabdariffa,* known as **vinagreira,** is used as a substitute for jute. In the East Indies it is called **roselle fiber** and has characteristics similar to India jute, but is lighter in color. In El Salvador it is called **kenaf fiber** and is used for coffee bags. **Brazilian hemp** is from the *H. radiatus,* and is stronger than true hemp. It is cultivated in Bahia. **Aramina fiber** is from the very long stalks of a plant *Urena lobata,* of the mallow family, of Brazil. In the north of Brazil it is known as **carrapicho.** The fiber is used for cordage, twine, and burlap fabrics. In Cuba this fiber is known as **white malva** and **bowstring hemp. Cuba jute** is from the *Sida rhombifolia.* Both Cuba jute and aramina fiber belong to the mallow family, *Malvaceae,* and in Cuba and Venezuela are also known as **malva fiber.** The *U. lobata* is also grown in Zaire where it is known as **Congo jute.**

Pitafiber, used in Colombia and Central America for coffee bags, is from a plant of the pineapple genus, *Ananas magdalenae.* The fiber is a light cream color, lustrous, very long, and finer and more flexible than most hard fibers, so that it is useful for ropes, twines, and fabrics, although most of the fibers of this botanical class are brush fibers. It has excellent resistance to salt water. The word pitameans yellow or reddish in the Indian language, and this name is also used for grades of yucca and other fibers. **Tucum fiber** is from the leaves of the oil palm *Astrocaryum tucuma* of Brazil. The fibers are long, flexible, water-resistant, and durable. They are used for ropes, hammocks, and marine cordage, but are not classed as burlap fiber. Another fiber of great length and noted for resistance to insect attack is **curana,** from the stalks of the plant *Bromelia sagenaria* of northeast Brazil. There are two chief grades: the white and the roxo or purple.

KAOLIN. Also called **China clay.** A pure form of hydrated **aluminum-silicate clay.** There are three distinct minerals, **kaolinite, nicrite,** and **dick-**

ite, all having similar composition. The formula for kaolin is usually given as $Al_2O_3 \cdot 2SiO_2 \cdot 2H_2O$, but is also expressed as $Al_2Si_2O_5(OH)_4$. It occurs in claylike masses of specific gravity 2.6, and of a dull luster. Kaolin is used for making **porcelain** for **chinaware,** and **chemical porcelain** for valves, tubes, and fittings; as a refractory for bricks and furnace linings; for electrical insulators; as a pigment and filler in paints; as a filler in plastics; and as an abrasive powder.

In firebricks kaolin resists spalling. Its melting point is 3200°F (1760°C), but this lowers with impurities. The color of all varieties is white, but inferior grades burn to a yellow or brownish color, and it should be free of iron. Porcelain made from kaolin is fired at about 2300°F (1260°C), but the upper service limit of the products is only about 500°F (260°C) since it has a low heat-transfer rate and low thermal shock resistance. Porcelain parts have a specific gravity of 2.4 to 2.9, a hardness of 7.5 Mohs, and a compressive strength from 60,000 to 90,000 lb/in^2 (413 to 620 MPa).

When kaolin is employed as an inert colloidal pigment in paints, it is called **Chinese white.** The powder is hydrophobic and cannot be wet by water, but it has good compatibility in oils and many organic solvents. As an extender in plastics and rubbers it reduces absorption of moisture and increases dielectric strength. Kaolin is a decomposition product of granite and feldspar, and its usual impurities are quartz, feldspar, and mica, which can be washed out. The **aluminum silicate RER-45,** of the Georgia Kaolin Co., is purified kaolin ground to a fineness of 0.2- to 4.5-μm particle size. It is used in paints, coatings, and plastics. The **modified kaolin** of the Minerals & Chemical Corp., for rubber and plastics, has the composition $(Al_2O_3)_3(SiO_2)_2$, and the particles are in thin flat plates averaging 0.55 μm.

The **Cornwall kaolin** of England and the **Limoges kaolin** of France are the best known. English China clays contain little or no iron oxide, and the yellow clays contain only organic materials that can be bleached out. The best grade of English clay is used for coating and filling paper. **Cornish clay,** known as **China stone,** is used for the best grades of porcelain glazes. Cheaper grades of kaolin, called **mica clay,** are used for earthenware glazes, and as an absorbent in oil purifying. The clay of Kentucky and Tennessee, known as **ball clay,** occurs in massive beds of great purity. It has high plasticity and good bonding strength, and is light in color when fired. It is used for high-grade porcelain and for wall tiles. Impure varieties of kaolin, called **kaolonic earth,** are used for refractories. **Kaolin fiber** of extreme fineness, with average diameter of 3 μm, is made from kaolin containing about 46% alumina, 51 silica, and 3 iron and titanium oxides. It will withstand continuous temperatures to 2000°F (1093°C). **Kaowool,** of the Babcock & Wilcox Co., is kaolin fiber in the form of insulating blankets of ¼ to 3 in (0.64 to 7.62 cm) thickness. **Kaowool paper** is made from the fibers compressed to thicknesses up to 0.08 in (0.20 cm) with or without a binder. It will withstand temperatures to 2000°F (1093°C) and is used for filters, separators, and gaskets. The fibers may also be compressed into **kaowool blocks** of 4 in (10 cm) thickness.

Halloysite has about the same composition as kaolinite but contains more alumina and water. It occurs with kaolinite. In association with alunite in Arkansas it is called **newtonite.** Some varieties, such as **glossecollite,** are waxlike, and with an electron microscope the grains appear as tubular structures. A variety of halloysite is marketed in France as a fine gray powder for use as a filler in rubber latex compounds. **Indianaite,** or **allophane,** is an impure halloysite. It is a white, waxy clay found in Indiana, and is used for pottery. The Indiana halloysite used for refractories is called **malinite. Bone clay** is a pure kaolin from feldspar and granite. It makes a strong porcelain. But **bone china** is a name given to high-grade English pottery made with China clay and 25% calcined bone. **Fresh clay** is formed from feldspar and quartz. It gives resilience to the porcelain. Clays can be mixed to give desired characteristics.

Micronized clay is pure kaolin ground to a fineness of 400 to 800 mesh, used as a filler in rubber. **Dixie clay,** of the R. T. Vanderbilt Co., is ground kaolin of 300 mesh, used as a stiffening or reinforcing agent in rubber and adhesives. **Osmose kaolin** is kaolin deposited by electroosmosis from an alkaline solution to eliminate iron and other impurities and raise the alumina ratio. As a fine powder it is used for making electrical insulators and synthetic mica, and for cosmetics. **Osmo,** of E. Fougera & Co., is a finely ground kaolin for cosmetics. **Alumina flake,** of the Aluminum Flake Co., is a white, flaky kaolinic clay from Missouri, ground for use in paints, adhesives, and rubber. The **tile kaolin** of the Georgia Kaolin Co. is selectively mined and water-washed to obtain low iron and titania content. It has high plasticity and fires white with good translucency. The mean particle size is about 1.2 μm with 65% below 2 μm. The pH is 4 to 6.5. **Quickcast kaolin** of the same company is a coarse powder of iron-free Georgia kaolin with 200-mesh particle size. It is used for mixing with other clays to increase the casting and drying times and to improve whiteness in tile and ceramic products.

KAPOK. A silky fiber obtained from the seed pods of the silk-cotton trees of the genera *Ceiba, Bombax, Chorisia,* and *Ochroma,* now grown in most tropical countries. It is employed for insulation and fine padding work. It is extremely light and resilient. The chemical constituents are the same as those of jute, but the proportions are different. Kapok is low in alpha cellulose, 43%, and high in pentosans, 24%, lignin, 15%, and uronic anhydride, 6.6%. Most of the commercial kapok normally comes from Java and is from the *Ceiba pentandra.* The tree was brought originally from Brazil where it was known as **samaúma.** It grows to a height up to 100 ft (30 m), with diameters to 10 ft (3 m), making the picking of the kapok pods difficult. The fibers are long, white, and silky, similar in appearance to cotton, but are too brittle for spinning. It is also known as **Java cotton** and **Illiani silk.** A silky cotton somewhat similar but inferior to kapok is **madar,** from the shrub *Calotropis gigantea* of India and the East Indies. Another species, **akund,** is from the *C. procera.* Both fibers are sometimes mixed with kapok, but are less resilient. They are known in the East Indies as **vegetable silk. Red silk cotton,** known as **simal,** is from the large tree *Bombax ceiba* or *B. malabaricum,* of India. The fiber is reddish. **Shilo fiber** is from the tree *B. ellipticum.* **White silk cotton** is from the tree *Cochlospermum gossypium* of India. It is quite similar to kapok. The kapok of Ecuador is from various species of the tree *Chorisia.* The fiber from the balsa trees, of the genus *Ochroma* of Central America, Colombia, and Ecuador, is dark in color. **Mexican kapok** is from the *Bombax palmeri* and from the *Ceiba schottii* and *C. acuminata,* the fibers from the two latter being more buoyant than the Java fiber. The kapok from the Lower Amazon region is from the *C. samaum̃a,* and the **pochote fiber** of El Salvador is from the *C. aesculifolia.* The **balsa fiber** of Central America is from the tree *O. velutina.* **Paina kapok** is from a ceiba tree of Brazil. **Kapok oil** is a semidrying oil obtained from the seeds of the kapok tree. It is used in margarine and for soaps.

A substitute for kapok for sound and heat insulation, and as stuffing for life jackets, cushions, furniture, and toys, is *typha,* or **cattail fiber.** It is the fluffy fiber from the cylindrical flowers of the *Typha latifolia* which grows in swamps throughout the temperate climates of North America. As an insulating material the fiber has about 90% the efficiency of wool. **Milkweed floss,** used also as a substitute for kapok, is the bluish-white silky fiber from the seed pods of the common milkweed, *Asclepias syrica,* of the eastern United States. It is not grown commercially.

KARAKUL. The curly lustrous fur pelts of newly born lambs of the karakul sheep originally of Afghanistan and Siberia, but now grown extensively in southwest Africa for the fur. The wool of the older sheep is clipped for carpet wool, but the fur pelts of the young animals are highly valued for coats and garment trim. The best qualities are those with small curl and medium curl to the fur. The fur skins are also graded by color, the gray and flora being most valued and the brown the least valued. **Persian lambskin** is a name for pelts of small tight curl.

KAURI GUM. A fossil gum dug from the ground in New Zealand and New Caledonia, used in varnishes and enamels to increase the body and increase the elasticity and hardness. It is also used in adhesives, and the lower grades of chips are used in linoleum. It was first known as **New Zealand gum,** and first spelled cowrie, although **cowrie** is the name of a genus of mollusk shells found in the Indian Ocean and formerly used as money in China. Kauri is a product of kauri tree exudations buried for long periods, but it also comes from the conifer tree *Agathis australis.* There is little extraction of the gum from the present kauri forests, whose wood is employed for lumber, but some **bush gum** is obtained by collecting the deposits in the forks of branches. **Range gum** is found in clay deposits, and some is transparent. **Swamp gum** is brown in color and varies from hard to friable. The fossil gum has a specific gravity of 1.05, a melting point of 182 to 232°C, and is soluble in turpentine, benzol, and alcohol. The **kauri tree** grows to a height of 100 ft (30 m) and great diameters, and yields a yellowish-brown, straight-grained wood free from knots and much prized as a useful softwood. It weighs 36 lb/ft^3 (577 kg/m^3). Mottled and figured **kauri pine** is used as a cabinet wood.

KERMES. A brilliant red natural dyestuff similar in color to cochineal, having a beautiful tone and being very fast. It is one of the most ancient dyes, but is now largely replaced by synthetic dyestuffs. Kermes is an insect found on the **kermes oak** tree, *Quercus coccifera,* of southern Europe and North Africa. The body of the animal is full of a red juice, and the coloring matter, **kermesic acid,** $C_{18}H_{12}O_9$, is separated out in brick-red crystals. It has only about one-tenth the coloring power of cochineal. **Garouille** is the root bark of the kermes oak. It contains 18 to 32% tannin, and is used for making sole leather. The color is darker than oak tannin.

KEROSENE. Originally the name of illuminating oil distilled from coal, and also called **coal oil.** It is now a light, oily liquid obtained in the fractional distillation of petroleum. It distills off after the gasoline, and between the limits of 174 and 288°C. It is a hydrocarbon of the composition $C_{10}H_{22}$ to $C_{16}H_{34}$, with a specific gravity between 0.747 and 0.775. Commercial kerosene may be as high a distillate as 325°C, with a corresponding higher specific gravity up to 0.850, but in states where it is distinguished from gasoline in the tax laws it is more sharply defined. In Pennsylvania, kerosene is defined as having a flash point above 46°C, with not over 10% distillable at 175°C and not over 45% at 200°C. Kerosene is employed for illuminating and heating purposes, and as a fuel in internal-combustion engines, and for turbine jet fuels. The heaviest distillate, known as **range oil,** is sufficiently volatile to burn freely in the wick of a heating range, but not so volatile as to be explosive. It is nearly free from odor and smoke. **Deodorized kerosene,** used in insect sprays, is a kerosene highly refined by treatment with activated earth or activated carbon. **Fialasol,** of Fred Fiala, Inc., is a **nitrated kerosene** used as a solvent for the scouring of wool. It is less flammable and has a slower rate of evaporation than kerosene, and it is odorless.

KHAYA. A class of woods from trees of the genus *Khaya,* growing chiefly in tropical West Africa and known commercially as **African mahogany.** The woods closely resemble mahogany, but they are more strongly figured than mahogany, are slightly lighter in weight, softer, and have greater shrinkage. The pores are larger, and the wood coarser. The wood is used for furniture and store fixtures, musical instruments, and paneling. It is not as suitable for patterns as mahogany. African mahoganies are marketed under the names of the shipping ports, as the shipments from the various ports usually differ in proportion to the different species cut in the region. The chief wood of the genus, from which the native name Khaya derives, is known commercially as **dry zone mahogany,** *K. senegalensis.* It is also known as **kail** and **oganwo.** It grows from Gambia to Angola on the west coast and eastward to Uganda. The heartwood is dark reddish brown, and the thick sapwood is grayish to pinkish red. The most favored commercial wood is that of the **red khaya,** or **red mahogany,** *K. ivorensis,* known locally as **dukuma** and **dubini.** This wood is highly figured, with interlocking grain, and when quartered shows a ribbon figure with alternate light and dark stripes. It comes chiefly from the Ivory Coast, Ghana, and Nigeria. **Sassandra mahogany** is chiefly this species. **Duala mahogany** is chiefly **white mahogany,** *K. anthoteca,* known also as **diala, krala,** and **mangona.** The wood is lighter in color but tinged with red. **Big-leaf mahogany,** *K. grandifolia,* has a reddish-brown color. Much African mahogany is cut into veneer, and the standard thickness for the face veneer is $^{1}/_{28}$ in (0.09 cm).

Gaboon mahogany and **Port Lopez mahogany** are chiefly **okume wood,** from the tree *Aucoumea klaineana* of the Guinea coast. The tree belongs to the family *Meliaceae* to which khaya belongs, and the wood resembles African mahogany but is lighter in weight and softer. It is light pinkish brown in color. It is shipped chiefly to Europe where it is used for furniture, chests, boxes, and boats. **Cola mahogany,** from the Ivory Coast and Ghana, is **niangon,** *Tarrietia utilis.* The heartwood is light reddish brown, and the wood shows a herringbone figure on the quartered surface. It is heavier than khaya, and the pores are larger and more numerous. **Cherry mahogany,** or **makore,** is a plentiful wood on the Ivory Coast, Ghana, and in Nigeria. It is from the tree *Minusops heckelii.* The wood is dark reddish brown without figure. It is heavier than khaya and is finer in texture.

KIESELGUHR. A variety of tripoli, or **infusorial earth,** obtained in Germany, and employed chiefly as an absorbing material. It is also used as an abrasive, as a heat insulator, for making imitation meerschaum, and as an absorbing material for nitroglycerin in making dynamite. Kieselguhr is very absorbent and will hold 75% of its own weight of sulfuric acid. It is insoluble in water. Its desirable characteristics as an insulator are closed cells and very high porosity, giving low density and low thermal conductivity. It is also used as a catalyst carrier in chemical processing. Kieselguhr from Oberhole, Germany, has 88% silica, 0.1 alumina, 8.4 water, and the remainder organic matter. **Randanite,** found at Clermont-Ferrand, France, is similar to kieselguhr but has a gray color. **Moler earth,** from the Jutland Peninsula of Denmark, is similar to kieselguhr and is made into insulation bricks. **Nonpareil insulating brick,** of the Armstrong Cork & Insulating Co., is made of pulverized kieselguhr mixed with ground cork, molded into brick form, and dried. The cork is burned out, leaving small air pockets to increase the insulating effect. The bricks will withstand temperatures up to 1000°C; the heat transmission is lower than for natural kieselguhr.

KINO RESIN. Also known as **gum kino.** The red exudation of the tree *Pterocarpus marsupium,* of India and Sri Lanka, and of *P. erinaceus,* of West Africa, formerly much used for colored varnishes and lacquers and used in throat medicines. **Bengal kino,** or **butea gum,** from the tree *Butea frondosa,* is now limited to medicinal use, as is also the **Australian red kino** from species of eucalyptus. Kino belongs to the group of red resins known as **dragon's blood** when used in spirit varnishes for musical instruments and furniture, but it is now replaced by synthetic colors. The dragon's blood resins from the East Indies are from the ripe fruit of various species of the tree *Daemonorops.* The resin is separated out by boiling and is shipped in small oval drops or long cylindrical sticks. It is used in fine lacquers and varnishes. The dark-red dye known as **red sandalwood** is the boiled-down juice of another kino tree, *P. santelinus,* of India. Still another *Pterocarpus* tree of southern India produces the wood **padouk,** valued for furniture, cabinetwood, and veneer. The heartwood is red with black stripes. It is hard and takes a high polish.

KYANITE. A natural **aluminum silicate,** $Al_2O_3 \cdot SiO_2$, used as a refractory especially for linings of glass furnaces and furnaces for nonferrous metals. Molded or cast ceramic parts have a nearly zero thermal expansion up to 2300°F (1260°C). It is a common mineral but occurs disseminated with other minerals and is found in commercial quantities and grades in only a few places. Most of the world production has been in eastern India, but high-grade kyanite is now obtained from Kenya. It is also mined in North Carolina, Georgia, Virginia, and California. The related minerals sillimanite and dumortierite are mined in western United States. The specific gravity is 3.56 to 3.67, and the hardness is 6 to 7 Mohs. Kyanite of 97 to 98% purity is obtained by flotation, but gravity concentrates rarely exceed 90%. Low-grade kyanite ore from California, containing 35% kyanite and much quartz, is used for ceramics. Kyanite is usually marketed ground to finenesses from 35 to 325 mesh. Low-grade kyanite is used in glassmaking as a source of alumina to increase strength and chemical and heat resistance. Aluminum silicate minerals are widely distributed in nature combined in complex forms, and the aluminum silicate extracted from them is called synthetic. **Kyanite powder,** produced from Florida beach sands, has round single-crystal grains of kyanite and sillimanite in nearly equal proportions. The beach sands are screened and graded and marketed as granules of various sizes for compacting and sintering into refractory ceramic parts. The smooth rounded grains give bonding properties superior to those of crushed kyanite. **Cerox ceramic,** of the Babcock & Wilcox Co., is a synthetic aluminum silicate in molded shapes for electronic and furnace parts. It is marketed in grades containing 50 to 90% alumina. **Cerox 200,** with a melting point at 3290°F (1800°C), contains 64.3% alumina, 32.4 silica, 2.1 titania, and 1 iron oxide. It is a hard, white, nonporous ceramic which is spall- and wear-resistant.

The mineral known as **staurolite** is a vitreous, translucent, red-brown complex aluminum silicate, $Fe_2Al_9O_7(SiO_4)_4OH$. The orthorhombic crystals occur quite commonly as twins crossing at nearly 90° in cruciform shape and are cut as gemstones. **Zoisite** is a complex calcium aluminum silicate occurring worldwide, usually in the mineral **saussurite,** which is altered feldspar. A variety of zoisite named **tanzanite** is a dark-blue gemstone found in Tanzania. **Tourmaline** is a boron aluminum silicate of wide occurrence and usually black or green in color. But pink crystals of gem quality are found in Malagasy, yellow in Sri Lanka, and dark green in Brazil. A colorless variety found in Malagasy is called **achroite.** The fine red tourmaline, **rubellite,** is scarce. Tourmalines from the pegmatite veins of New England, Pennsylvania, and California are found in all colors. Tourmaline, from Brazil, has good piezoelectric properties, and can be used for frequency control and pressure transducers, but is more expensive than quartz crystal. It is used in submarines to measure depth. Tourmaline is also a **pyroelectric.** When heated, an electric charge is produced, with the crystal becoming positive and negative at opposite ends, the poles reversing themselves on cooling.

LACQUER. Originally the name of an **Oriental finish** made with lac. The original **Chinese lacquer** was made with the juice of the **lac plant,** *Rhus vernicifera,* mixed with oils. The juice is milky but it darkens with age, and the lacquer is a glossy black. Later, the name referred to transparent coatings made with shellac, and to glossy pigmented spirit varnishes. Still later, it referred to quick-drying finishes made with nitrocellulose or cellulose acetate. The word lacquer has now come to mean glossy, quick-drying finishes that are dried by evaporation of the solvents or thinners in which a resin (plastic) vehicle is dissolved.

The true lacquers were made with copals and other natural resins, a pigment, a softener or plasticizer, and one or more volatile solvents, with the time of drying controlled by the evaporation of the solvents. Too rapid drying may cause the cloudiness called blushing, because of the absorption of moisture from the cooling caused by the rapid evaporation. Various resins impart different characteristics. Dammar gives high gloss and hardness. Kauri gives hardness and wear resistance. The softeners are usually the amyl, ethyl, and butyl phthalates. The usual solvents are anhydrous alcohol, ethyl, acetate, benzol, and toluo. But modern lacquers may contain nitrocellulose or cellulose esters, or they may be made with acrylic, melamine, or other synthetic resins. **Shellac** is often referred to as a **spirit lacquer.**

For industrial work, lacquers are usually sprayed, and the feature is quick-drying. **Brushing lacquers** are also quick-drying, but they have a longer drying period to prevent streaks and lumps in the application. Lacquers are harder and tougher than enamels, but not as elastic, and they are more expensive. They are usually not as solvent-resistant or as weather-resistant, and are generally not suited for exterior work. A good lacquer requires no buffing, and retains the original gloss well. The word lacquer is also used to describe a highly transparent varnish used to produce a thin protective film on polished or plated metals to preserve their luster. Lacquers are sold under a variety of trade names. Some early cellulose lacquers were marketed as **Duco, Agateen, Zapon, Brevolite,** and **Zeloctite,** but solutions of methyl acrylate or other crystal-clear resins may now be used. **Bronzing liquids** are the clear lacquer-base mediums marketed ready for incorporation of bronze or aluminum powders. **Cable lacquers** are clear, black, or colored lacquers prepared from synthetic resins, and are characterized by high dielectric strength, resistance to oils and heat, and by their ability to give tough, flexible coatings. **Chromate-protein films** may be used instead of clear lacquers for the protection of metal parts. The parts are dipped in a solution of casein, albumin, or gelatin, and, after drying, dipped in a weak chromic acid solution. The thin, yellowish film is hard and adherent and will withstand temperatures to 300°F (149°C).

LAMPBLACK. A soot formed by the smudge process of burning oil, coal tar, resin, or other carbonaceous substances in an insufficient supply of air, the soot being allowed to settle on the walls or floors of the collecting chambers. Lampblack is practically pure carbon, but inferior grades may contain unburnt oil. It is chemically the same as carbon black made from gas but, since it may contain as high as 2.5% oil, it is not generally used in rubber. The particle size is large, 65 to 100 μm, and the pH is low, 3 to 3.5. However, some **amorphous carbon,** made from crude oil by spraying the oil and air into a closed retort at 3000°F (1649°C) to obtain partial combustion, is equal to carbon black for many uses. Lampblack is used in making paints, lead pencils, metal polishes, electric brush carbons, crayons, and carbon papers. It is grayish black in color, and is flaky and granular. The color is not as intensely black as carbon black. One pound (0.5 kg) occupies from 200 to 230 in^3 (3,278 to 3,770 cm^3). For use as a pigment for japan the powder should pass through a 325-mesh screen. **Lampblack oil** is a coal-tar product marketed for making lampblack.

LANCEWOOD. The wood of the tree *Guatteria virgata,* of tropical America. It is used as a substitute for boxwood, and for fine work where a uniform, tough, durable wood is needed. The wood is yellowish in color and has a fine, close, smooth grain. The weight is 52 to 63 lb/ft^3 (833 to 1,009 kg/m^3). It is very hard and elastic. **Yaya** is a name given in the Honduras trade to lancewood. **Degami lancewood,** or **degami wood,** is a yellowish wood with a fine, dense grain, from the *Calycophyllum candidissimum* of the West Indies. **Burma lancewood,** used in India for implements, is a strong, straight-grained, heavy wood from the large tree *Homalium tomentosum.* It has a light-brown color and weighs 60 lb/ft^3 (961 kg/m^3).

LARCH. The wood of the coniferous tree *Larix occidentalis,* of the northwestern United States and southwestern Canada. It is also called **western larch, western tamarack, mountain larch, Montana larch,** and **hackmatack.** The wood is heavier than most softwoods, having a specific gravity of 0.48 and weight of 36 lb/ft^3 (577 kg/m^3). It is fine textured and straight grained. In strength and hardness it ranks high among the softwoods, but shrinks and swells more than most softwoods. The heartwood is reddish brown and the narrow sapwood is yellowish. It finishes well, but splits easily. Butt logs contain galactan gum, which darkens the color. Larch is used for bridge timbers, flooring, paneling, and general construction. The trees reach a diameter of 5 ft (1.5 m) and a height up to 200 ft (61 m). Shipments of western larch and Douglas fir mixed are known commercially as **larch fir.** Larch is also the name given to the **tamarack** tree of New England, *L. larincina.* **European larch,** *L. europea,* is an important wood in Russia and some other countries.

LARD. The soft white fat from hogs. It is used chiefly as a shortening in bakery and food products, and as a cooking grease. The inedible grades are used for the production of lard oil and soaps, and for splitting into fatty acids and glycerin. The types of edible lard for use in the United

States are defined in regulations of the Department of Agriculture. **Steam-rendered lard** is made by applying steam directly to the fats in a closed container. Open-kettle rendered lard is made by applying steam to the outside of the kettle. The **neutral lard** used for making margarine is produced by applying hot water in place of steam. **Rendered pork fat** is an edible material that does not meet the specifications for lard. Average production of edible lard is about 24 lb (11 kg) per animal.

Leaf lard is from around the kidneys and intestines and is the best edible grade. **White grease** is an inedible lard from the kidneys and the back. **Yellow grease** is the residue from the parts of the hog remaining after the parts yielding white grease are separated. USP grade of lard is called **adeps** in pharmacy. It is purified internal fat of the abdomen, and is a soft unctuous solid of bland taste. It is used in **benzoated lard** for ointments. **Modified lard,** of Armour & Co., is a **plastic lard** which is plastic at lower temperatures and also retains its body at higher temperatures than ordinary hydrogenized shortenings. It is made by treating with sodium methylate which re-forms the esters in random orientation to give the soft plastic texture.

Lard oil is an oil expressed from lard by subjecting it to hydraulic pressure. Prime or first-grade lard oils are nearly colorless, or greenish, and have little odor. The commercial oils vary from the clear sweet oil to the acid and offensive-smelling brown oils. The oils contain oleic, stearic, and palmitic acids. They are used in cutting and in lubricating oils, sometimes in illuminating oils. They may be adulterated with cottonseed oil or blown oils. The flash point of pure lard oil is 480°F (249°C), saponification value 192, and specific gravity 0.915. **Mineral lard oil** is a mixture of refined mineral oil with lard oil, the fatty content being 25 to 30%. The flash point is about 300°F (149°C). **Petrofac,** of the Sun Oil Co., is a lard-oil substitute made entirely from petroleum. **Lardine** is an old name for blown cottonseed oil used in lubricants.

LATEX. The milklike juice of the rubber tree, now much used instead of the cured crude rubber for many rubber applications such as adhesives, rubber compounds, and rubber powder. The properties of latex vary with the type of tree, age of tree, method of tapping, and climate. Latex from young trees is less stable than from older trees. Intensive tapping of the trees results in less rubber content, which may vary from 20 to 50%. For shipping, a preservative and anticoagulant is added to the latex, usually ammonia or sodium sulfate. Concentrated 60% latex is a stable liquid of creamlike consistency. **Heveatex** is the trade name of latex of various grades of the Heveatex Corp. **Latex foam** is a cellular sponge rubber made by whipping air into latex, pouring into molds, and vulcanizing. **Aircell,** of the B. F. Goodrich Co., is a latex foam made from natural rubber latex.

Artificial latex is a water dispersion of reclaimed rubber. **Dispersite** was a trade name of Dispersions Process, Inc., for water dispersions of crude or reclaimed rubbers, produced by swelling and dissolving the rubber in an organic solvent, treating with an organic acid or with ammonia, and emulsifying. It resembles latex, but is softer and more tacky, and is used for adhesives. **Lotol** and **Revertex** are brands of latex. The Revertex process consists in the concentration of latex by evaporation in the presence of protective colloids. Latex foams are now usually made by incorporating a chemical which releases a gas to form the cells.

The term latex now also refers to water dispersions of synthetic rubbers and of rubberlike plastics. **Neoprene latex** is a water dispersion of neoprene rubber, and has the dispersed particles smaller than those of natural latex, giving better penetration in coating paper and textiles. **Naugatex J-8174,** made by Uniroyal, Inc., is a butadiene-styrene latex with 68% solids for producing foamed rubber, and **Nitrex 2614** is a nitrile rubber latex. **Nubun,** of the Firestone Tire & Rubber Co., is **Buna S latex. Adiprene L,** of Du Pont, is a urethane rubber suspension, and **Nelco 70-A,** of the Allied Chemical Corp., for use in paints and spar varnishes, is a 70% solids suspension of urethane resin in mineral spirits. **Pliolite 5352,** of the Goodyear Tire & Rubber Co., for producing foams, is a synthetic-natural isoprene rubber latex. **Latex water paints** are now usually made with synthetic rubber or plastic dispersions.

LAUAN. The wood of trees of several genera of the Philippines, Malaya, and Sarawak, known in the American market as **Philippine mahogany.** The woods resemble mahogany in general appearance, weight, and strength, but the shrinkage and swelling with changes in moisture are greater than in the true mahoganies. The lauan woods are used for furniture and cabinet woods, paneling, and boatbuilding. The lauans belong chiefly to the genus *Shorea,* and the various species have local or common names. The so-called dark-red Philippine mahogany is **tangile,** *S. polysperma,* and **red lauan,** *S. negrosensis.* Tangile is also called **Bataan mahogany,** and has the closest resemblance to true mahogany of all the species. The thick sapwood is light red, and the heartwood dark brownish red. It has greater tendency to warp than mahogany. Red lauan has larger pores, but is favored for boat construction because of the large sizes available. **Tiaong,** from the *S. teysmanniana,* resembles tangile but is lighter and softer. **Almon,** from the tree *S. eximia,* is harder and stronger than red lauan or tangile, but is coarser in texture and less lustrous.

White lauan is from a different genus of the same family and is *Pentacme contorta.* It has about the same mechanical properties as tangile, but is gray with a pinkish tint. **Mindanao lauan,** *P. mindanensis,* is quite similar but is lighter and softer. **Mayapis,** from the tree *Shorea palosapis,* is coarser in texture than tangile and is subject to warping and checking like red lauan. It is intermediate in color, and the light-colored wood is marketed as white lauan, while the dark wood is sold as red lauan. **Yellow lauan** is from the trees **kalunti,** *S. kalunti,* **manggasinoro,** *S. philippensis,* and **malaanonang,** *S. polita.* Yellow lauan is yellowish in color and has lower strength and greater warpage than other lauans. **Bagtikan,** from the tree *Parashorea malaanonan,* is reddish gray in color, not lustrous, but is heavier and stronger than the other lauans. Sometimes mixed with Philippine mahogany is the wood known as **lumbayan,** from an entirely different family of trees. It is from the tree *Tarrietia javanica.* The thick sapwood is light gray in color and the heartwood reddish. The weight and strength are about equal to tangile but the pores are larger. When marketed separately, it is a more valuable wood than the lauans for furniture manufacture. The reddish woods from Sarawak, Sumatra, and Malaya known as **meranti,** or **morenti,** are *Shorea* species of the lauan types. In the East Indies morenti is used for barrels, casks, and tanks for palm oil. Similar woods from North Borneo are called **seraya,** or known as **Borneo cedar** or **Borneo mahogany.** The *Shorea* trees yield Borneo tallow and dammar. **Merawan** is a wood from various species of trees of the genus *Hopea* of Malaya. It is valued for furniture and interior work. Much of the so-called mahogany normally shipped from the Philippines is **Apitong,** the wood of the tree *Dipterocarpus grandiflorus,* also grown in Borneo and Malaya. The wood weighs 44 lb/ft^3 (705 kg/m^3).

LAVA. A name given to ceramic material used for molding gas-burner tips, electrical insulating parts, nozzles, and handles. It may be calcined talc, steatite, or other material. As produced by the American Lava Corp., it is molded from **magnesium oxide,** and it is hardened by heat treatment after shaping and cutting. It is baked at 2000°F (1093°C). The compressive strength is from 20,000 to 30,000 lb/in^2 (138 to 207 MPa). It will resist moisture and has high dielectric strength. Rods as small as 0.020 in (0.05 cm) in diameter can be made. **Alsimag** is the trade name of lava produced by this company from ground talc and sodium silicate, but **Alsimag 602** is phosphate-bonded steatite talc, used for thyratron tubes and as a substitute for mica in receiving tube spacers. **Porcelava** is a similar ceramic produced by Burgess & Co., for electric heating appliances. **Isolantite,** of the Isolantite Co. of America, Inc., is a **steatite** ceramic molded with a binder and then vitrified. It can be machined or threaded before firing and, by allowing for the contraction, parts can be made with great accuracy. The specific gravity of the vitrified material is 2.5, hardness 8 to 9 Mohs, and crushing strength 80,000 to 120,000 lb/ in^2 (552 to 828 MPa), and the dielectric strength of ⅛-in (0.32-cm) thickness is 40,500 V. **Lavalloy,** of the Lava Crucible Co., is a high-strength ceramic made from a mixture of mullite and alumina. **Lavolain,** of the Star Porcelain Co., is a strong, dense, heat-resistant ceramic for electrical insulation.

LEAD. A soft, heavy, bluish-gray metal, symbol Pb, obtained chiefly from the mineral galena. It surface-oxidizes easily, but is then very resistant to corrosion. It is soluble in nitric acid but not in sulfuric or hydrochloric, and is one of the most stable of the metals. Its crystal structure is cubic face-centered. It is very malleable, but it becomes hard and brittle on repeated melting because of the formation of oxides. The specific gravity of the cast metal is 11.34, and that of the rolled is 11.37. The melting point is 621°F (327°C), and boiling point 2777°F (1523°C). The tensile strength is low, that of the rolled metal being about 3,600 lb/in^2 (25 MPa) with elongation of 52% at normal temperatures, but at low temperatures the strength is greatly increased. At −40°F (−40°C) it is about 13,000 lb/in^2 (89 MPa), with elongation of 30%. The coefficient of expansion is 0.0000183, and the thermal conductivity is 8.2% that of silver. The electric conductivity is only 7.8% that of copper. When used in storage batteries, the metal is largely returned as scrap after a period and is remelted and marketed as **secondary lead,** as is also that from pipes and cable coverings.

Lead alloys easily with tin and zinc, and forms many commercial alloys, including solders and bearing metals. Although the amount of lead going into chemicals other than white lead, red lead, and tetraethyllead is relatively small, there are about 200 lead chemicals that have a wide variety of industrial uses. Lead and its compounds are poisonous, and are not used in contact with the skin or near foodstuffs. **Tetraethyllead,** also called **ethyl lead,** is a liquid of the composition $Pb(C_2H_5)_4$, used in a blend with halogens as an additive in gasoline to increase octane number and resistance to knock. The compound is so volatile that it distributes widely in the atmosphere from exhaust fumes, and the lead is not recovered. It is highly toxic, and gasolines containing it are dyed for identification. Various other compounds, such as cyclopentadiene manganese carbonyl, can be used as antiknock agents. **Methyl lead,** $(CH_3)_4Pb$, a liquid boiling at 110°C, is more effective then tetraethyllead in some respects, and is used in mixtures.

Commercial lead is sold in pigs weighing 100 lb (45 kg). Seven grades of pig lead are marketed. **Corroding lead,** 99.94% pure, is for making white lead. **Chemical lead,** 99.90% pure, with some silver and copper, is the **undesilverized lead** from Missouri ores. Soft undesilverized lead is 99.93% pure and is used for storage batteries, sheet, and for pipes and cable coverings. Chemical lead has a hardness of 80 Rockwell B compared with 60 for the purer corroding lead. **Desilverized lead** is produced from the silver-lead ores of the Rocky Mountain states and is at least 99.85% pure. **Common lead A** is desilverized and is 99.85% pure, while **common**

lead B is 99.73% pure. **Acid lead** is made by adding copper to fully refined lead, and **copper lead** is produced in the same way. These two grades are 99.90 and 99.85% pure. The percentages given for the seven grades of lead are ASTM minimums. **Work lead** is the pig lead from the blast furnaces before the silver is extracted. About 60% of the world production of lead is in the United States, Australia, Mexico, Germany, and Canada, though the ores are widely distributed and many other countries produce lead.

Lead wool is lead in a shredded form used for calking. The strands are 0.005 to 0.015 in (0.013 to 0.038 cm) in diameter, and come in ropes ⅝ to ¾ in (1.59 to 1.90 cm) in diameter. **Blue lead** is a term meaning all lead products such as pipe and shot that have not been changed chemically in manufacture. Blue lead is also a name for a basic sulfate, a by-product of the smelting of lead ores obtained by collecting and filtering the smoke from the furnace. It is a mixture of the products of the partial combustion of the ore and coal. It consists of lead sulfate, lead sulfide, lead sulfite, lead oxide, zinc oxide, and carbon. It is used in base-coat paints for steel.

Sheet lead is produced by cold-rolling, and is designated by weight. A sheet of lead 1⁄64 in (0.38 cm) thick weighs approximately 1 lb/ft^2 (4.9 kg/m^2), and is called a 1-lb (0.5-kg) sheet. Thus 4-lb (1.8-kg) sheet lead is 1⁄16 in (0.16 cm) thick, and 8-lb (3.6-kg) sheet is ⅛ in (0.32 cm) thick. As a sound barrier in building construction an 0.008-in (0.02 cm) thickness of lead is equal to ¾ in (1.9 cm) of fir plywood for sound absorption.

Lead has a high capacity for the capture of neutrons and gamma rays and is used for **radiation shielding** in the form of sheet lead or as metal powder in ceramic mortars and blocks, paints, and in plastic composite structures. **DS Lead,** of Sherrit Gordon Mines, Ltd., is a **dispersion-strengthened lead** containing up to 1.5% lead monoxide evenly distributed through the structure. The oxide combines chemically with the lead, doubling the strength and stiffness of the metal, but increasing its brittleness. It is used for chemical piping and fittings. A **neoprene-lead fabric,** of Raybestos-Manhattan, Inc., is a neoprene fabric impregnated with lead powder. It has a radiation shielding capacity one-third that of solid lead sheet. It comes in thicknesses of 1⁄32 to ¼ in (0.08 to 0.64 cm), and its flexibility makes it suitable for protective clothing and curtains. **Shielding cements** for X-ray and nuclear installation shielding are **metallic mortars** containing a high percentage of lead powder with ceramic oxides as binders and other elements for selective shielding. They are mixed with water to form plasters or for casting into sections and blocks. The formulation varies with the intended use for capture, attenuation, or dissipation of neutrons, gamma rays, and other radiation. **Shielding paints** are blended in the same manner. **Chemtree 82,** of the Chemtree Corp., is a lead powder with an organic binder soluble in water for application as a paint or molded into brick. The cured material is 95% lead with linear attenuation about 60% that of lead. **Chemtree 1-6-18,** designed for neturon beam attenuation, has 78% lead and a high hydrogen content together with the neutron-capturing isotopes lithium-6 and boron-10.

Battery-plate lead for the grid plates of storage batteries contains 7 to 12% antimony, 0.25 tin, and small amounts of arsenic and copper. **Silvium alloy,** for positive-plate grids, contains silver. **Lead-coated copper,** used for roofing and for acid-resistant tanks, is 16-oz (0.45-kg) copper sheet, coated on both sides with lead, made with either a rough or smooth finish. **Leadtex,** of Revere Copper & Brass, Inc., is a lead-coated sheet copper. **Roofloy,** of this company, is a strong, stiff, creep-resistant sheet lead used for flashings and gutters. It contains 99.7% lead, 0.20 tin, 0.015 calcium, and 0.012 magnesium. **Tea lead,** used for wrapping tea, is lead with 2% tin. **Tinsel lead** contains 1.5% antimony and up to 4% tin. **Shot lead** is lead hardened with 2 to 6% antimony and a small amount of arsenic. **Frangible bullets,** which shatter on striking a target surface and are used for aerial gunner practice, are made of lead powder and synthetic resin powder pressed into shape.

Antimonial lead is an alloy containing up to 25% antimony with the balance lead, used for storage-battery plates, type metal, bullets, tank linings, pipes, cable coverings, bearing metals, roofing, collapsible tubes, toys, and small cast articles. The alloy is produced directly in the refining of some lead ores, but much is also made by adding antimony to soft lead. The alloy is also known as **hard lead,** and in England is called **regulus metal.** Much of the lead used in the United States is in the form of antimonial lead. The antimony hardens the lead and increases the tensile strength. The usual alloy contains 4 to 8% antimony and has about twice the tensile strength of pure lead. Up to about 0.10% arsenic stabilizes and hardens the alloy, and from 0.25 to 1% tin may be added to improve the casting properties. Antimonial lead, of NL Industries, Inc., for chemical linings, contains 6 to 8% antimony, weighs 0.398 lb/in^3 (11,016 kg/m^3), and melts at 475 to 555°F (245 to 290°C). An antimonial lead with 6% antimony has a tensile strength of 4,100 lb/in^2 (27 MPa) with elongation of 47%, as rolled. After heat treatment at 455°F (235°C) and aging one day, the tensile strength is 12,600 lb/in^2 (85 MPa) with elongation of 3%.

A hard lead with 10% antimony and 90% lead has a tensile strength of 8,800 lb/in^2 (61 MPa), elongation 17%, Brinell hardness 17, and melting point 486°F (252°C). **Cable lead,** or **sheathing lead,** used to cover telephone and power cables to protect against moisture and mechanical injury, contains about 1% antimony. The alloy for collapsible tubes usually contains 2% antimony. Antimonial lead may be used for machine bearings, but for this purpose it usually contains considerable tin and is classed with the babbitt metals. **Hoyt metal,** used for bearings, contains 6 to 10% antimony. Alloys containing from 70 to 90% lead, 5 to 20 antimony, and 2 to 20 tin have been used for railway-car bearings under the name of **lining metal. Dandelion metal,** used for locomotive crosshead linings, had 72% lead, 18 antimony, and 10 tin.

LEAD ALKALI METALS. Lead hardened with small amounts of alkali metals used chiefly as bearing metals. Alloys of this type are also called **tempered lead** and **alkali lead.** The original alkali lead, known as **Bahnmetall,** was made under a German patent. It contains 0.73% calcium, 0.04 lithium, 0.55 sodium, and the balance lead, and is made by electrolysis of the fused alkali salts, using a molten lead cathode. **Calcium-lead alloy** is made in the same manner, and contains up to 1% calcium. The calcium forms a chemical compound, Pb_3Ca, with part of the lead, and the crystals of this compound are throughout the lead matrix. The Brinell hardness is 35, and the compressive strength 25,000 lb/in^2 (172 MPa), being thus superior in strength to the high-tin-bearing metals. But the metal is difficult to melt without oxidation, and is easily corroded. The melting point is 370°C, and it retains its bearing strength at more elevated temperatures than babbitts. **Calcium lead,** with about 0.04% calcium, is used as cable sheathing to replace antimonial lead, giving greater fatigue resistance. It is also used for grids in storage batteries, and has a lower rate of self-discharge than antimonial lead. The small amount of lithium in Bahnmetall is intended to prevent the corrosion set up by the calcium. It also increases the compressive strength to about 30,000 lb/in^2 (206 MPa). **Mathesius metal,** another German alloy, contains calcium and strontium to form Pb_3Sr crystals. **Frary metal,** of NL Industries, Inc., is a lead-calcium-barium alloy containing about 1.5% of the alkali metals. These alloys give low friction loss at low temperatures, but they have higher coefficients of friction than the lead-tin alloys at temperatures above 65°C. A European lead-calcium-barium alloy known as **Ferry metal** contained 0.25% mercury to decrease the coefficient of friction. Another bearing alloy, **Noheet metal,** contained some antimony and tin in addition to the alkali metals. **B metal,** a German alloy, contains 0.73% calcium, 0.66 sodium, 0.05 lithium, and 0.03 potassium.

LEAD ORES. The chief ore of the metal lead is **galena,** a lead sulfide, PbS, containing theoretically 86.6% lead. The ore, however, contains many other minerals and usually carries only 4 to 11% lead. It is concentrated by gravity methods to contain 40 to 80% galena. Galena has a bright metallic luster, streaked gray color, a specific gravity of about 7.5, and hardness of 2.75. It frequently contains silver and sometimes cadmium, bismuth, and copper. The lead is obtained from the concentrated ore by roasting to remove the sulfur, and smelting. The **ingot lead** from the blast furnace contains silver, copper, zinc, and other impurities. It is refined and desilverized. Southern Missouri is the chief source of galena in the United States.

The abundant lead ores **cerussite** and **anglesite** are secondary minerals formed by the oxidation of galena. Cerussite is a lead carbonate, $PbCO_3$, found in crystals or in granular crystalline aggregate or massive. Its color is white to gray, transparent to opaque. The hardness is 3 to 3.5, and the specific gravity 6.55. Anglesite is usually found in the oxidized portions of lead veins associated with galena and other minerals. It is a common mineral occurring in many localities. Anglesite is a **lead sulfate** of the composition $PbSO_4$, containing 68% lead. It occurs in crystals, massive or granular. Its hardness is 2.75 and specific gravity 6.12 to 6.39. The color may be white or pale shades of yellow or blue, or it may be colorless.

Basic lead sulfate, $PbSO_4 \cdot PbO$, called **lanarkite,** is in white crystals with a specific gravity of 6.92. It is formed by the action of heat and air on galena. It is used as a paint pigment combined with zinc oxide. **Alamosite** is **lead metasilicate,** $PbSiO_3$, and is in white crystals. It can be produced by fusing litharge and silica, and is used in glazing ceramics and in fireproofing textiles. **Barysilite** is **lead orthodisilicate,** $Pb_2Si_2O_7$, a white solid of specific gravity 6.6. Other rarer minerals are **clausthalite,** which is **lead selenide,** PbSe, occurring in lead-gray cubical crystals of specific gravity 8.1, **attoite,** which is **lead telluride,** PbTe, occurring in tin-white cubical crystals of specific gravity 8.16, and **lead tungstate,** $PbWO_4$, called **lead wolframite,** and occurring as **raspite** in monoclinic crystals, or as **stolzite** in colorless tetragonal crystals. Lead tungstate is manufactured as a yellow powder with a specific gravity of 8.235, and it is used as a paint pigment.

Vanadinite is a minor ore of lead and a source of vanadium. It is a mineral of secondary origin found in the upper oxidized parts of lead veins. It is found in Arizona, New Mexico, Mexico, and Spain. It has the composition $9PbO \cdot 3V_2O_5 \cdot PbCl_2$, with phosphorus and arsenic sometimes replacing part of the vanadium. It occurs in reddish crystals and globular forms and as incrustations. The specific gravity is 7, and hardness 3.

LEAD PIGMENTS. Chemical compounds of lead used in paints to give color. They are to be distinguished from the lead compounds such as lead oleate, used as driers for paints. **White lead** is the common name for **basic lead carbonate,** the oldest and most important lead paint pigment, and also used in putty and in ceramics. It is a white, amorphous, poisonous powder of the composition $2PbCO_3 \cdot Pb(OH)_2$. It is insoluble in water, and decomposes on heating. The specific gravity is 6.7. It is made from metallic lead, and is marketed dry, or mixed with linseed oil and turpentine in paste form. **Lead carbonate,** $PbCO_3$, is used as a pigment in the same way as the basic compound, but it discolors more easily. Basic lead sulfate, called **sub-**

limed white lead, makes a fine white pigment. Commercial sublimed white lead contains 75% lead sulfate, 20 lead oxide, and 5 zinc oxide. Commercial white lead may be mixed with lithopone, magnesium oxide, antimony oxide, witherite, or other materials.

The basic **silicate white lead,** $3PbO \cdot 2SiO_2 \cdot H_2O$, is made by fusing silica sand with litharge and hydrating by ball-milling with water. It has corrosion-inhibiting properties, and is used in metal-protective paints for underwater service. It is also used in ceramics and as a stabilizer in vinyl plastics. **Basic lead silicate** is made from silica and litharge with sulfuric acid as a catalyst. The material has a core of silica with a surface coating of basic lead silicate, giving a pigment of lower weight per unit of volume but retaining the activity of the silicate white lead. **Chrome yellow,** or **Leipzig yellow,** is **lead chromate** of the composition $PbCrO_4$, and comes in yellow, poisonous crystals. The specific gravity is 6.123. It is insoluble in water and decomposes at 600°C. **Basic lead chromate,** $2PbO \cdot CrO_3$, is red in color and is used for anticorrosive base coats for steel. **American vermilion,** also called **Chinese scarlet** and **chrome red,** is basic lead chromate made from white lead. **Orange mineral** is the red oxide, Pb_3O_4. **Cassel yellow,** or **Montpelier yellow,** is oxychloride of lead. Mixtures of white lead and heavy spar are known as **Venetian white. Dutch white** is composed of three parts of sulfate and one part of carbonate. **Lead thiosulfate,** PbS_2O_3, is a white insoluble powder used chiefly in matches. All of the lead compounds are poisonous when absorbed through the skin or taken internally. **Lead sulfide,** PbS, is used as a feeler in missiles as it is very sensitive to heat rays.

Litharge is the yellow **lead monoxide,** PbO, also called **massicot.** It is a yellow powder used as a pigment and also in the manufacture of glass, and for the fluxing of earthenware. An important use is as a filler in rubber. With glycerin, litharge is used as a plumbers' cement. The specific gravity is 9.375. Litharge is produced by heating lead in a reverberatory furnace and then grinding the lumps. For storage-battery use the black oxide, or suboxide of lead, is now largely substituted. **Lead dioxide,** or **lead peroxide,** PbO_2, comes as a brown crystalline powder, and is used in making matches, dyes, and pyrotechnics, as a mordant, and as an oxidizing agent. The natural mineral is called **plattnerite.** Litharge, or **lead oxide,** converts to Pb_3O_4 at 800°F (427°C).

LEADED HIGH BRASS. An alloy containing approximately 65% copper, from 0.5 to 3.5% lead, and the remainder zinc. It is one of the standard grades of brass, and is also called **free-cutting brass,** or **high-speed brass.** It is easier to machine than high brass but is less ductile. It is used especially for cupped, drawn, or formed parts on which a clean thread must be cut. The lead is present in finely dispersed particles, and the property of free cutting is gained at the expense of its drawing capacity, but the material is mostly employed in the form of rods for screw-machine work or in sheets for blanking. The free-cutting brass rod of the Bridgeport Brass Co. contains 61.5% copper, 35.5 zinc, and 3 lead. All impurities, including iron, are below 0.25%, since as little as 0.50% iron will harden brass 20 points Rockwell. The free-cutting brass of the Chase Brass & Copper Co. has the same amount of copper, but has 3.5% lead. A brass of this class will have a tensile strength of 57,000 lb/in^2 (392 MPa), elongation 25%, and Brinell hardness up to 110. **Ledrite brass** is the name of the Bridgeport Brass Co. for leaded brass containing 60 to 63% copper and 2.5 to 3.75 lead. The alloy known as **architectural bronze,** used for extruded moldings and for forgings, contains 57% copper, 40 zinc, 0.25 tin, and 2.75 lead. The free-machining alloy known as **arsenical bronze** is a leaded high brass modified with other elements. A typical analysis is copper, 56.5%; zinc, 39; lead, 0.70; nickel, 2; iron, 1.20; and arsenic, 0.60. The tensile strength is 65,000 to 87,000 lb/in^2 (448 to 599 MPa), elongation 11 to 40%. It is both wear- and corrosion-resistant.

LEADED STEEL. A free-machining steel containing about 0.25% lead. Lead does not alloy with iron, but when a stream of finely divided lead is shot at the stream of molten steel to the ingot mold the lead is distributed in the steel in tiny particles and strings. The lead gives free machining without imparting to the steel the unfavorable characteristics given by sulfur or phosphorus. There is little weakening of the physical properties of the steel. In cutting, the lead acts as a lubricant, reducing friction of the cutting tool, and aiding in producing a smooth finish. **Ledloy,** of the Inland Steel Co., contains 0.15 to 0.30% lead in the regular SAE grades of steel. **La-Led steel** of the LaSalle Steel Co. is a low-carbon leaded free-machining bessemer screw steel marketed in cold-finished rounds. **Super La-Led steel** contains 0.25% lead with up to 0.50 sulfur. **Rycut 40 steel,** of Joseph T. Ryerson & Son, Inc., is a chrome-molybdenum AISI steel 4140 containing 0.15 to 0.35% lead. It can be machined 50% faster than unleaded steel of the same composition. The tensile strength is 129,000 to 200,000 lb/in^2 (889 to 1,378 MPa) with elongation of 12 to 21%, depending on the tempering. **Ledloy 170,** of the same company, is leaded cold-drawn seamless tubing. It is a 1015 analysis steel with 0.15 to 0.35% of lead added.

LEATHER. The skin or hides of animals, cured by the chemical action of tannins. Leather is used for belting, gaskets, shoes, jackets, handbags, linings, coverings, and a variety of other products. The action of tannins precipitates the protein of the hide, changes its colloidal structure, and makes it more pliable and capable of resisting decay. The process of tanning hides consists essentially in soaking them in solutions of the tanning material after they have been unhaired in caustic lime. This soaking may be for weeks or months, after which the hides are washed, oiled, and rolled. Rapid tanning in strong solutions gives inferior leathers.

The quality of leather depends upon the type of animal, its physical condition, the care used in taking off the hide, the method of preserving the hide before tanning, and the care used in tanning. Leather is made from many kinds of skins, including sheep, goat, deer, alligator, seal, and shark, although the bulk of commercial leather is made from **cattle hides,** often referred to as **cowhide** or **steerhide.** Commercially, the term **hides** refers to the skins of full-grown beef cattle. Leather from larger animals such as cowhide and steerhide are split into layers and divided into three grades. **Top-grain leather** is the top or outside layer of the hide and is the best quality. It has close fibers, and is more flexible, durable, and attractive than the other top grades. Top grain is most often used natural and is seldom embossed. **First-split leather** comes from the layer next to the top layer and is second to top grain in quality. It is not as flexible or durable as top grain and is sometimes coated and embossed to resemble the better-grade leather. **Second-split leather,** the inner layer of the hide, is lowest in quality, and generally considered waste.

Kip skins and **calfskins** are the names given to the hides of the younger animals. Calfskins, from animals that have not eaten grass, produce a softer leather than cattle hides. Hides of other animals besides the beef cattle, *Bos taurus,* are usually designated with the name of the animal, as in **horsehide.** The hides of smaller animals are designated as **skins,** either tanned or untanned, as in **pigskin.** Quantities of wild pigskins, known as **capivara skins,** are imported from Colombia.

Kid is leather made from the skins of young goats, but commercial **kidskin** leather is now from both young and old animals, and the term **goatskin** is not liked in the American trade. It is thin and has a fine, close-grained texture, with tiny groups of pores, and the leather is soft and pliable but strong and nonscuffing. Kid is usually chrome-tanned and dyed to many colors. It has a natural lustrous surface which is heightened by glazing or polishing. The leather is sorted by grain or size of pores, weights, and sizes into 10 grades. India, Argentina, Brazil, South Africa, and northwest Africa are large producers of goatskins. About 70% of the goat and kid leather of the world is tanned in the United States although this country produces less than 1% of the skins. Kidskin is used for shoes, gloves, pocketbooks, jerkins, and pads and linings. The term **Vici kid,** used in the shoe industry, was the name originally given to chrome-tanned leather. The best kidskins come from arid regions, and these are used for the fine French kid leather. In the glove trade the term **chevreau** is used to designate young goats that have never browsed, while **chevrette** refers to the small skins of the older kids that have eaten grass. **Capeskin** is goatskin from South Africa. **Glazed kid** is made by pressing a seasoning agent into the leather and then ironing the dry leather with a glass cylinder.

Buffalo hides are from the domesticated **water buffalo** of southern Asia and the Philippines. They are heavy, with a course grain, and are used to make heavy, rough leather for buffing wheels and heavy boots, and also for **rawhide** mallets and gears. Rawhide is a treated but untanned form of leather.

Belting leather is usually made from salted hides free from cuts and scratches, and is either oak- or chrome-tanned. It is then stuffed with oils or tallows. Belting leather weighs about 0.035 lb/in^3 (969 kg/m^3) and has an ultimate strength of about 3,800 lb/in^2 (26 MPa). Belting is now sold by the thickness instead of the weight because the leather is easily weighted by the addition of heavy impregnations. But because of the use of direct drives and fabric belts impregnated with resins or rubbers, leather has lost its former importance for power transmission. **Lace leather,** for splicing machinery belting, was originally **porpoise leather** tanned with alum and impregnated with cod oil, making a tough, pliable, and strong leather, but it is now a semitanned, oil-treated cowhide or calfskin. Lacing for fastening sporting goods is now usually made of plastics, such as the **Cotalace** of Freydberg Bros.-Strauss, Inc. **Helvetia leather,** for light belts and lacings, was weak-tanned with gambier and oil-impregnated. **Belt dressings** are compounds of waxes and oils to maintain flexibility, and rosin or chemicals to improve grip.

Leather was once used widely for packings and gaskets but, because it is dissolved by mineral oils and becomes brittle at high temperatures, it has now been largely replaced for such uses by compounded materials. **Fibrated leather,** of the Armstrong Cork Co., for gaskets and packing, is shredded leather impregnated with a protein binder. It is resilient, resistant to oils, and will withstand temperatures to 300°F (149°C). It comes in sheets.

The **wax calf leather,** once popular for men's shoe uppers, was a leather heavily stuffed with grease and wax. It sheds water and is wear-resistant, but oily leathers are no longer desired in shoes. **Patent leather** is an old name for glossy burnished black leather for shoes and handbags. It cracks easily, and plastic finishes are now usually employed. **Helios leather,** of the Wilmington Enameling Co., has a base coat of linseed-oil gel on a chrome-tanned leather, and a finish coat of urethane plastic, giving 20 times more flex life than patent leather. **Ostrich leather,** once valued for handbags, is marked with tiny rosettes where the quills are extracted. It is now imitated in embossed calfskin. **Kangaroo leather** is a strong, supple, and durable leather made from the skins of the Australian kangaroo, used chiefly for shoe uppers and gloves. The skins measure from 2 to 12 ft^2 (0.19 to 1.11

m^2) in area, and the small ones are known as **wallaby.** The fibers have an interwoven structure, and the leather does not scuff easily. It takes a brilliant polish. It is rated as the strongest shoe leather per unit of weight. **Cordovan** is a tough, smooth, close-grained leather made from the hind quarters of horsehides. It takes a beautiful polish and is used for boots and fancy articles.

Upholstery leather is very thin, finely finished leather once valued for upholstery, seats, and coverings for various articles, but now largely replaced by plastic-coated fabrics. It consists of split hides, tanned to a soft, even texture, and usually dyed in colors. **Chrome tanned leather** is softer and stronger than ordinary leather tanned with barks or quebracho. In splitting, the full hide thickness of about ¼ in (0.64 cm) can be split into three or four thicknesses. After splitting, the leather is retanned and "nourished" with cod oil. Hand buffs are top grains with the top of the grain snuffed off. The second split of 3/64 in (0.11 cm) is called deep buff, and has an artificial grain put on. The third split is called No. 2 split. What remains is called a slab and is unsuited for upholstery leather. Splitting with four cuts gives buffing, machine buff, No. 1 split, No. 2 split, and slab. Upholstery leather is finished by japanning, by coating with lacquer, by dyeing with aniline, or by combinations of the last with either of the first two methods. **Spanish leather,** used for upholstery, is made by tanning in strong quebracho liquor which draws the grain and gives a slightly wrinkled appearance.

Leather dust consists of the light, fluffy fibers blown from the buffing and sueding wheels in tanneries. It is used as a covering material for **artificial suede** and **sueded fabrics,** and as a filler in adhesives and calking compounds. In Europe it is extensively collected and separated by grade and color, but the American material generally lacks uniformity as it is a mixture of many varieties of leather with different tannages and colors. **Leather flour** is ground and graded leather, uniform in color and in mesh, used as a filler and for sueded fabrics.

Pulped scrap leather is made into stiff board or flexible sheets, and may be called **leather board, fiber leather, synthetic leather,** or **artificial leather.** Leather board and sheet is used for luggage, shoe heels, shoe linings, gaskets, and clutch linings. **Cotton-leather,** of the Southern Friction Materials Co., is made from two to six plies of cotton fabric impregnated with a thermosetting resin, calendered and surface-ground to resemble leather. It is flexible and resistant to oil and water, and is used for shoe soles, buffing wheels, and machinery mountings. But for many of the former large uses of leather, including luggage and some types of shoes, leather has been replaced by impregnated laminated fabrics or fibers sheeted with a binder.

LECITHIN. A light-brown, soft, salvelike material of bland taste and odor obtained from soybeans by solvent and steam extraction. It is also obtained from cottonseed, and occurs in small amounts in animal and plant cells. Lecithin is a **glyceride** of complicated chemical structure containing an amine, phosphoric acid, and choline. It bleaches to a faint gold color, and is insoluble in water but soluble in oils and in alcohol. At about 150°F (66°C) it melts to an oil that will disperse readily in chocolate and other foods as an antioxidant to prevent spoilage, as a stabilizer and emulsifier, and as a dietetic. It is also employed in casein paints to regulate viscosity, in printing inks to impart flow, as a softener and wetting in some other products, and as a curing accelerator in synthetic rubbers. The **choline,** $CH_2OHCH_2N(CH_3)_3OH$, is used in medicine as a heart stimulant. **Emultex R,** of the A. E. Staley Mfg. Co., is a lecithin concentrate produced from soybeans. It is used as a stabilizer in water-based paints to prevent pigment migration and settling in storage. **Sta-Sol Plus,** of the same company, is a concentrated grade of soybean lecithin containing 68% active ingredients. It is a low-moisture emulsifier for coatings and paints.

LEMONGRASS OIL. A pale-reddish essential oil distilled from various species of *Cymbopogon* grass of the East Indies, Malaya, and tropical America. It contains about 75% **citral,** $C_9H_{15}CHO$, which is an aldehyde of several essential oils. The chief constituents are **citronellal,** $C_{10}H_{18}O$, and geraniol. It also contains some **ionone,** $C_{13}H_{20}O$, a terpene used as a base for violet perfumes. The oil has a powerful lemonlike odor. The plant is a tall, rapid-growing grass. Three tons of the East Indian grass yield 24 bottles of oil, each bottle being 25 oz (0.74 L) by volume. **Florida lemongrass** yields 6.3 lb (2.9 kg) of oil per ton of grass. The residue **lemongrass pulp** in Florida is dehydrated and mixed with cane molasses for cattle feed. The oil is soluble in 70% alcohol. It is used as a stabilizer and perfume in soaps and in cosmetics. The two chief grades are **East Indian lemongrass,** from *C. flexuosus,* and **West Indian lemongrass,** from *C. citratus.* The latter contains less citral and is less soluble, but partly owing to a difference in the method of distillation. **Citronella oil** is a yellow oil with a pleasant odor, distilled from the grass *C. nardus* of Sri Lanka, Java, and India. **Java citronella** is called **serah grass.** It contains high percentages of geraniol and citronellal. It is used as a stabilizer and perfume in soaps, as an insect repellent, and for the production of **synthetic menthol** and of geraniol.

Java citronella oil is highest in citronellal, which yields about 20% levo menthol or about 12% USP menthol. Sri Lanka oil is highest in geraniol, being sold on the basis of 85%, and is used in soaps and insecticides. The rose perfume **rhodinal** is obtained from **Réunion geranium oil,** or **geranium Bourbon.** It is an isomer of **citronellol,** the alcohol of citronellal, and can be made synthetically by isomerizing citronellol with ultraviolet light. **Palmarosa oil,** called **Turkish geranium oil** because the original shipments were through Turkey, is from the **motia grass,** *C. martini,* of India and Java. It contains 75 to 90% geraniol. Another variety of *C. martini,* known as **sofia grass,** sofia meaning mediocre, yields the product called **ginger grass oil,** inferior to palmarosa, and containing less geraniol.

LEUCAENA. A leafy tropical evergreen found principally in Mexico. Its deep taproot enables it to live on moisture far below the surface. The plants also fix their own nitrogen from air in the soil. Because its foliage is rich in nitrogen, it can be used as fertilizer and also as cattle feed. The trees yield high amounts of wood and the stumps regrow rapidly.

LICORICE. The sweet roots of a group of plants of the order *Fabraeceae.* The common licorice of the Near East and Spain is *Glycyrrhiza glabra,* and of Italy, *G. echinata.* Certain other species,,all native to the Mediterranean countries, are also employed. The roots are crushed and boiled in water, and the juice strained and evaporated. Finely ground root is added to aid in hardening, but starch or other adulterant may be added. In medicine it is known as **glycyrrhiza,** and it has an estrogenic, or female-hormone, action. The dark-colored extracted juice of the roots contains a glucoside, which will not ferment. The **glycyrrhizin** of MacAndrews & Forbes Co. is an ammoniated form of the tri-terpenoid glycoside. It is a sweetening agent used in foodstuffs to reduce the amount of sugar and also to intensify flavor. Licorice also contains a saponin, or froth-producing substance. In fire extinguishers it is sold under trade names. An insulating building board, made with a proportion of the residue fibers after extraction of the juice, is produced by the National Gypsum Co. It is resistant to insect attack. Licorice is employed in confectionery, medicines, and tobacco, as a flavoring, and in beverages as a frothing agent, but the saponins, in excess, are toxic.

LIGNIN. A colorless to brown crystalline product recovered from paper-pulp sulfite liquor, and used in furfural plastics, as an extender of phenol in phenolic plastics, as a corrosion inhibitor, in adhesives and coatings, as a natural binder for compressed-wood products, and for the production of synthetic vanilla when it contains coniferin. It is also used as a fertilizer, providing humus and organic material and some sulfur to the soil. For this use it may be mixed with phosphates. Lignin is coprecipitated with natural and synthetic rubbers to produce stronger and lighter-weight products. When lignin is incorporated into nitrile rubber by coprecipitation the rubber has higher tensile and tear strengths and greater elongation than with an equal loading of carbon black. **Indulin,** of the West Virginia Pulp & Paper Co., is a pine wood lignin used as an extender for rubber latex to improve coating strength, and in phenolic resins to decrease cost. It is also used as a sequestering agent. About 100,000 tons (90,700 metric tons) of lignin occurs in the liquor from 1 million tons (907,000 metric tons) of sulfate-process paper pulp.

The melting point of lignin is 250 to 275°C. It is a complex material, $C_{41}H_{32}O_6$, with four hydroxyl and four methoxyl groups, pine lignin containing 15% **methoxyl.** The contained **lignoceric acid** is a C_{24} fatty acid melting at 84.7°C. Lignin occurs in the waste liquor as a sodium compound, and is precipitated by acidifying. For making laminates the lignin from soda-process aspen is preferred to spruce kraft lignin because of lower melting point and better flow. The lignin known as **tomlinite** is obtained from kraft pulp liquor by treating with CO_2, separating the black liquor, and acidifying with sulfuric acid to yield the water-insoluble brown powder. **Lignin binder,** for surfacing roads, is waste liquor fromsulfite mills containing about 5% lignin. **Lignosol,** of Lignosol Chemicals, Ltd., is a concentrated lignin liquor or powder. **Spruce extract** is the liquor from sulfite spruce pulping used for mixing with quebracho for tanning leather, or in combination with syntans. The tanning agent in the liquor is **lignosulfonic acid. Alkali lignin** is produced by acidulation of the black liquor from the soda process. It is used as the organic expander in the negative plates of storage batteries, as an emulsifier for asphalt to reduce cold flow, and as a binder in paperboard products. It may also be hydrogenated to produce methane, methanol, and tar acids.

Lignosulfonates produced from the waste liquor of sulfate pulping of soft wood are used to reduce the viscosity of slurries and oil-well muds, as foam stabilizers, as extenders in glues, cements, and synthetic resins, and as a partial replacement for quebracho in tanning leather. **Orzan A,** of the Crown Zellerbach Corp., is **ammonium lignosulfonate,** while **Orzan S** is **sodium lignosulfonate. Toranil,** of the Lake States Yeast Corp., is **calcium lignosulfonate.** It comes as a tan powder or as a brown viscous solution. These lignosulfonates have the wood sugars and the sulfur dioxide removed, and will polymerize with heat.

LIGNITE. Also called **brown coal.** A variety of organic mineral of more recent age than coal, occurring in rocks of tertiary age, and intermediate in composition between wood and coal. It is widely distributed over Europe, and found in many parts of the world. Freshly cut lignite often contains a large quantity of water, up to 40%, and is sometimes also high in ash. When dried it breaks up into fine lumps and powder. Dry lignite contains 75% carbon, 10 to 30 oxygen, and 5 to 7 hydrogen. It kindles easily but burns with a low calorific power and a smoky flame. In retort gas production, lignite loses its gas in half the time required for gas removal from bituminous coal, with temperature of 1270°F (688°C), compared with 1655°F (901°C) for bituminous. California and Arkansas lignites are processed to extract montan wax, **sap brown dye,** and **Van Dyke brown** pigment. The color of lignite varies from brown to black, and the lower grades of brown lignite show the woody structure. The **pitch coal** is brownish black, breaks with a pitchlike fracture, and shows no woody structure. Lignite is briquetted by crushing and pressing with a binder under heat. Belgian **lignite briquettes** have a binder of 8% asphalt, with 2 flour to assist binding. The reserves of lignite in the United States are estimated at 1,000 billion tons (907 billion metric tons), and form a future fuel for use in powdered form, or for distillation of oils and coal-tar products. Texas lignite is carbonized at 950°F (510°C) to produce a solid fuel known as **char.** From 7,000 tons (635 metric tons) of lignite the yeild is 3,200 tons (2,902 metric tons) of char and 2,300 lb (2,086 metric tons) of by-product tar. **Lignite tar** is a mixture of hydrocarbons with nitrogen, oxygen, and sulfur complexes. The tar distillate contains 74% neutral oil, 23 tar acids, 3 tar bases, and 0.9 sulfur. The neutral oils are cracked for motor fuels. Other fractions are used for producing phenol, pyridine, and epoxy resins. **Jet** is a hard, compact, black lignite found in Colorado, Utah, and in Yorkshire, England. The Utah jet occurs in the form of flattened-out trees, with the entire tree metamorphosed into jet. It was formerly much used for making buttons and ornamental articles, but is now largely replaced by plastics, and the less brittle black onyx is substituted for jewelry.

LIGNUM VITAE. The wood of the **guayacum** trees, *Guaiacum offinale* and *G. sanctum,* of tropical America, but the commercial shipments of lignum vitae are also likely to contain other species. The wood of the **guayacan** tree of Brazil and Paraguay is also called by this name. The best quality of the true lignum vitae comes in logs up to 18 ft (5.5 m) long and up to 3 ft (0.9 m) in diameter, and originates mostly from Cuba, Haiti, Yucatan, Dominican Republic, and the west coast of Mexico and Central America, but species of the trees grow as far south as Paraguay. It is very hard, heavy, and tough. The color is brown to greenish black. The grain is very finely twisted, and the wood has a greasy feel, containing 3% of natural resin. The specific gravity is 1.14, with a weight of 80 lb/ft^3 (1,282 kg/m^3) and a crushing strength of 10,500 lb/in^2 (71 MPa). The wood is used in places where extreme hardness is needed, such as for pulley blocks. It is also used for rollers, mallet heads, handles, novelties, bearings, and furniture. In machine bearings it withstands pressures up to 4,000 lb/in^2 (27 MPa), and is used for propeller-shaft bearings in steamships. **Palo santo,** *Bulnesia sarmenti,* known as guayacan, of Argentina and Paraguay, and also **vera,** *B. arborea,* of Colombia and Venezuela, are closely related to lignum vitae and were formerly classified as *Guaiacum.* The latter is called **Maracaibo lignum vitae,** and the former is known as **Paraguayan lignum vitae.**

The trees are larger than true lignum vitae, but the wood is not as hard. **Philippine lignum vitae** is a hard, heavy wood with interwoven grain, from the tree *Xanthestemon verdugonianus* of the Philippines. **Yellow guayacan,** deriving its name from the yellow flowers, is from the *Tecoma guayacan,* an important timber tree of Panama and Central America. The wood is very hard and strong, but neither this wood nor the Philippine wood has the self-lubricating bearing properties of the oily true lignum vitae. **Ibera vera,** of Argentina and Paraguay, is also called guayacan locally. It is from the tree *Caesalpinia melanocarpa,* and resembles lignum vitae in hardness, strength, and durability, but is lacking in the oily gum resin. It is a valuable ornamental construction wood.

LIME. A **calcium oxide,** CaO, chemically known as **calcia,** occurring abundantly in nature, chiefly in combination with carbon dioxide as calcium carbonate, in limestone, marble, chalk, coral, and shells. Lime is employed in mortars and cements, as a flux in iron melting, in many chemical processes, as an absorbent, and for liming acid soils. It is obtained by heating limestone in a furnace or kiln to about 1000°F (538°C) to burn out the carbonic acid gas. The residue is called **quicklime** or **caustic lime.** Pure quicklime is white and amorphous or crystalline. The specific gravity is 3.2 and melting point 4660°F (2578°C). **Chemical lime,** used in the chemical industries and for water softening, is a high-calcium lime with minimum impurities. A typical analysis is 97.9% CaO, 0.43 silica, 0.45 iron and aluminum oxides, and 0.52 magnesium oxide. The **Kemidol quicklime** marketed by the U.S. Gypsum Co. for glass manufacture is a free-flowing fine powdered lime made from Ohio dolomite. When dolomite quicklime is hydrated, no more than 2 or 3% of the magnesium oxide is converted to the hydroxide, and the lime has a high neutralizing value, useful for neutralization of chemical solutions. **Oyster shells** have about the same calcium carbonate content as the best grade of high-calcium limestone, with low impurities, and they are used for making quicklime.

Commercial limes for building purposes contain about 94% calcium oxide, some calcium carbonate, and less than 0.50% magnesia. Water causes the lime to slake with much heat, leaving a white powder, $CaO \cdot H_2O$. High-calcium limes slake rapidly, expand greatly on slaking, and are the strongest. Limes with much magnesia slake slowly, but magnesia produces the slip that makes easier working. The so-called **lean limes** contain considerable silica, alumina, and iron oxide, and are slow-slaking and difficult to work. Lime is marketed in lumps or ground to 20 mesh, and as mill run. **Hydrated lime** is made by grinding quicklime, slaking the powder with water, and sifting to a fine powder. It is easier to handle and is a more reliable product than ordinary lime. High-grade hydrated lime will have a fineness so that 98% will pass through a 100-mesh sieve and will contain not over 2% magnesia. Some grades contain less than 0.50% magnesia, and 98% pass through a 200-mesh sieve. The pure hydrated lime for the chemical industry has 98.2% $Ca(OH)_2$, or 74.4% CaO, and is a fine air-float powder. This is **calcium hydroxide,** or **lime hydrate,** and is also used in paints, for dehairing hides, and in medicine. **Hydraulic lime** that will set under water is a hydrated lime with more than 10% silica. The **Grappier cement** of France and Belgium is made from limestones that contain 20 to 22% silica and 2 alumina. **Lime mortar,** made of a mixture of hydrated lime, sand, and water, will have a compressive strength up to 400 lb/in^2 (3 MPa). **Soda lime,** used for freeing air of carbon monoxide, as a chemical purifying agent, and in gas masks, is a mixture of hydrated lime with sodium hydroxide.

LIME OIL. An essential oil expressed from the rind of the **lime,** the fruit of the small tree *Citrus aurantifolia,* native to India but now grown in many tropical and subtropical countries. The fruit is smaller than the lemon, and is greenish yellow in color. Montserrat in the West Indies, Italy, and Florida are the chief commercial centers. **Lime juice,** from the pulp of the fruit, is used in flavoring and to give freshness to perfumes, in beverages, and was early used to prevent scurvy, but contains only about 25% as much vitamin C as orange juice.

Of more commercial importance is **lemon oil,** a bright-yellow oil of pleasant odor expressed from the rind of the lemon, the fruit of the tree *C. limonia,* grown commercially in the Mediterranean countries, California, Florida, and the West Indies. About 1,000 lemons are required to produce 1 lb (0.5 kg) of oil. The oil is high in citral with also limonene, terpineol, and citronellal. It is used in flavors, soap, and perfumery. Synthetic lemon oil, used in confectionery and beverages, contains no terpenes. It has a much stronger flavor than the natural oil. **Lemon juice** is from the pulp. It contains 5% citric acid, and is used in beverages and as a bleaching agent. **Lemon juice powder,** prepared by spray-drying from a solution that encases each particle, is a free-flowing, nonhygroscopic powder having long storage life. It is used for foodstuff flavoring.

Citric acid produced from lemons, limes, and pineapples is a colorless, odorless, crystalline powder. It has the chemical composition $(CO_2HCH_2)_2C(OH)COOH$. Its specific gravity is 1.542, and its melting point is 153°C. It is used in medicine, flavoring extracts, and in soft drinks, but is toxic in excess. It is also used in inks, etching, and as a resist in textile dyeing and printing. It is a good antioxidant and stabilizer for tallow and other fats and greases, but is poorly soluble in fats. **Tenox R,** of Eastman Chemical Products, Inc., a soluble antioxidant, consists of 20% citric acid with 60% propylene glycol and 20% butylated hydroxyanisol. Citric acid is also used as a preservative in frozen fruits to prevent discoloration in storage. Its salt, **sodium citrate,** is a water-soluble crystalline powder used in soft drinks to give a nippy saline taste, and is also used in plating baths. Citric acid is also produced by the fermentation of blackstrap molasses.

LIMESTONE. A general name for a great variety of **calcite rocks.** Immense quantities of limestone are used as flux in the melting of iron, for the manufacture of lime, and as a building stone. In a broad sense limestone includes dolomite, marble, chalk, or any mineral consisting largely of $CaCO_3$. When the proportion reaches 45% and the limestone is in the double carbonate, $CaCO_3 \cdot MgCO_3$, it is called **dolomite.** On calcination, limestone yields lime, CaO. **Portland stone,** of England, consists of fossils cemented together with lime, and the **coquina rock** of Florida is made up entirely of shells, mostly the tiny **coquina shell,** *Donax variabilis.*

The **Indiana limestone,** valued for building, is a noncrystalline massive rock with aggregate filler, and matrix of 98% calcium carbonate in gray and buff colors. It is an **oolite** with small round grains resembling fish roe. The weight is 135 to 150 lb/ft^3 (2,163 to 2,403 kg/m^3), with compressive strength from 4,000 to 7,000 lb/in^2 (27 to 48 MPa). The mill blocks are from 8 to 12 ft (2.4 to 3.7 m) in length, and are 3 ft 6 in (1 m), or 4 ft 4 in (1.3 m), square. **Bangor limestone,** from Alabama, is also an important building stone. It is an oolite with 97% $CaCO_3$. **Dolomitic limestone** from Minnesota is also used for building. **Kasota stone** from Minnesota is a recrystallized variety resembling marble and coming in yellow and pink. It contains 49% $CaCO_3$ and 39% $MgCO_3$ with silica. Its crushing strength is more than double that of Indiana limestone. It is used largely for interior work. **Mankato stone** contains slightly higher $CaCO_3$ with about 4.5% alumina in addition to silica. It weighs 155 lb/ft^3 (2,483 kg/m^3), and has a compressive strength of 13,500 lb/in^2 (92 MPa). Desirable buff and yellow coloring in limestones comes from small amounts of iron oxides, but iron sulfides produce weather staining, and are not desired in building stone. **Sohnhofen stone,** or **lithographic stone,** produced at Sohnhofen, Germany, is a limestone with an exceptionally homogeneous fine grain, very low in impurities, and with a Mohs hardness of 3. It occurs in layers 2 to 6 in (5 to 15 cm) thick, with partings of gray clay. Limestone is widely distributed, and a very large proportion of all crushed and broken stone used for construction purposes in the United States is limestone. **Agricultural limestone,** known as **agstone,** used for liming soils, is stone of high calcium carbonate content ground to a fineness of 200 to 325 mesh. A grade for chemical use is specified with a minimum of 97.5% $CaCO_3$, and maximums of 0.30 silica, 0.20 alumina, and 0.07 iron.

LINEN. A general name for the yarns spun from the fiber of the variety of flax plant cultivated for its fiber, or for the cloth woven from the yarn.

Linen yarns and fabrics have been made from the earliest times, and the ancient Egyptian linen fabrics were of exceeding fineness containing 540 warp threads per inch, not equaled in Europe until the twentieth century. Ireland, Belgium, and France are the principal producers of linen. Linen yarns are used for the best grades of cordage, and linen fabrics are employed industrially wherever a fine, even, and strong cloth is required. **Linen fabrics** are sold under a wide variety of trade names. They are graded chiefly according to the fineness of the yarns and the class of weave and may contain some fine hemp fibers. **Lisle** was formerly a fine, hard linen thread, made at Lille, France, but is now a fine, smooth yarn made of long-staple cotton spun tightly in a moist condition. **Tow yarns** are the coarsest linen yarns, used for making **crash,** a coarse, plain-woven fabric used for towels or covers. **Chintz** is a plain-woven linen fabric in brightly colored designs. It is also made in cotton as **cotton chintz.** The name is from the Hindu word meaning color. **Damask** is a jacquard reversible woven linen fabric for table linen. It is now also made in cotton, silk, or rayon.

LINOLEUM. A general name for floor-covering material consisting of a mixture of ground cork or wood flour, rosin or other gum, blown linseed oil, pigments, with sometimes a filler such as lithopone, on a fabric backing of burlap or canvas, rolled under pressure to give a hard, glossy surface. It is pigmented or dyed in plain colors, or printed with designs. The cork is usually in about 50-mesh particles. Linoleum was patented in 1863 and displaced a more expensive floor covering known as **Kamptulicon,** which was made of rubber, cork, rosin, and oil. Because of its low cost, linoleum still has a wide usage, but it is not durable and has a tendency to crack, and it has been largely replaced by sheeting compounded with synthetic resins marketed under trade names and by tile. **Battleship linoleum,** a very heavy grade in plain colors or mosaics, is described in Federal specifications as made with oxidized linseed oil, fossil or other resins or oxidized rosin, and an oleoresinous binder, mixed with ground cork, wood flour, and pigments, processed on a burlap back. Because of its resiliency, battleship linoleum is still valued for high-grade flooring, and is coated with an adherent, wear-resistant lacquer based on cellulose ester and alkyd resins.

LINSEED OIL. This oil is the most common of the drying oils, and is widely used for paints, varnishes, linoleum, printing inks, and soaps. It is obtained by pressure from the seeds of the flax plant, *Linum usitatissimum,* which is cultivated for oil purposes. It is sometimes referred to as **flaxseed,** though this name properly belongs only to the seed of the flax plant for producing flax fiber. The varieties producing linen fiber do not yield much seed. The seed contains up to 40% oil, and Argentine seed averages 38.5%, but the usual conversion factor is 34%. The linseed plant can stand high temperatures and drought, but is sensitive to cold, while the flax fiber plant needs a cool and humid climate. In normal times 80% of the world's linseed is grown in the Mesopotamia region of Argentina and Uruguay, and much of the remainder in northern India and southern Russia, while the flax plant is grown in northern Europe and the Baltic region. When prices are high, much linseed is grown in the United States and Canada. The commercial oil is hot-pressed and has a bitter taste, but a cold-pressed oil is used in Russia for food purposes. The residue, **linseed cake,** is used for cattle feed and fertilizer. The oil imported from Holland under the name of **Haarlem oil** is linseed oil mixed with sulfur and turpentine. It is used in the southern states as a diuretic medicine. Linseed oil contains about 48% linoleic and 34 linolenic acids. The specific gravity is 0.925 to 0.935. It is a yellowish oily liquid with a peculiar odor and a bland taste. It is soluble in turpentine, ether, and benzene. It dries with a distinct gloss and makes a hard film.

The best **Baltic oil** is used as a standard in measuring the drying power of other oils. Genuine linseed oil has an iodine value of at least 170, and the best approaches 190. The linseed grown in cooler climates from the same type of plant generally yields oil of higher iodine value than that grown in warm climates. Baltic oil has an iodine value of 190 to 200, though this high value is from the type of plant as well as from the climate, the European plant being the flax plant yielding less seed. Oils from seeds grown in different areas vary widely in the acid content. North Dakota oil contains 26 to 33% linolenic acid, while the **Punjab oil** of India has as high as 54% of this acid. The **puntis oil** used in Pakistan to extend linseed oil is a fish oil from the puntis, *Barbus stigma,* caught off the Indian coast.

For varnish use, linseed oil may be used as **boiled linseed oil,** prepared by heating to not over 600°F (316°C) in a closed container, or by heating with oxidizing driers such as the salts of lead or manganese. When prepared with driers, it is called **bung oil. Stand oil,** also known as **lithographic oil,** is linseed oil heated for several hours without blowing, at a temperature of 550 to 650°F (288 to 343°C). It has the consistency of honey and is used in oil enamel paints. Blown oils and boiled oils are not greasy like the original oil. **Linoxyn** is a trade name for blown linseed oil. The purity and adulteration of linseed oil for paint and varnish use are controlled by state laws. The law of the state of Ohio, which is typical, defines boiled linseed oil as prepared from pure, raw linseed oil heated to a temperature of 225°F (107°C), and incorporating no more than 4% by weight of drier, and with specific gravity at 60°F (16°C) of not less than 0.935 and not greater than 0.945. **Esskol** and **Solinox,** of Textron, Inc., are hydrogen-treated linseed oils used as substitutes for tung oil and castor oil. **Keltrol L,** of this company, is **styrenated linseed oil** made by reacting linseed oil with styrene. It is used for paints, in which it dries rapidly to a hard, tough, and alkali-resistant film. The heavy-bodied linseed oils of ADM Chemicals are oxidized oils with specific gravities of 0.980 to 0.990, and iodine numbers from 210 to 230. **Linopol,** of the Sherwin-Williams Co., is a polymerized linseed oil in water emulsion for use in latex paints.

LITHIUM. The lightest of all metals, with a specific gravity of 0.534. It is abundant, and is found in more than 40 minerals, but is obtained chiefly from lepidolite, spodumene, and salt brines. The dried crude concentrate from the flotation cells contains about 20% Li_2O. It may be extracted by solar evaporation from the brines of underground lakes in Nevada and other locations. It occurs also in seawater in lesser concentrations. North Carolina has immense reserves of lithium ores.

Lithium melts at 186°C and boils at 1370°C. It is unstable chemically and burns in the air with a dazzling white flame when heated to just above its melting point. The metal is silvery white but tarnishes quickly in the air, and a **lithium nitride,** Li_3N, is formed. The metal is kept submerged in kerosene. Lithium resembles sodium, barium, and potassium, but has a wider reactive power than the other alkali metals. It combines easily with oxygen, nitrogen, and sulfur to form low-melting-point compounds which pass off as gases, and is thus useful as a deoxidizer and degasifier of metals. In glass the small ionic radius of lithium permits a lithium ion coupled with an aluminum ion to displace two magnesium ions in the spinel structure. **Lithium cobaltite,** $LiCoO_2$, and **lithium zirconate,** Li_2ZrO_3, are also used in ceramics. **Lithium carbonate,** Li_2CO_3, is a powerful fluxing agent for ceramics, and is used in low-melting ceramic enamels for coating aluminum. It is used in medicine to treat mental depression.

Lithium metal, 99.4% pure, is produced by the reduction of **lithium chloride,** LiCl. The salts of lithium burn with a crimson flame, and lithium chloride is used in pyrotechnics. It is also used for dehumidifying air for industrial drying and for air conditioning, as it absorbs water rapidly. It is also employed in welding fluxes for aluminum and in storage batteries. The anode is lithium, the cathode is a **lithium-tellurium alloy,** and the electrolyte is a molten bath of lithium salts at 800°F (427°C). **Lithium ribbon,** of Foote Mineral Co., for high-energy battery use, is 99.96% pure metal in continuous strip form, 0.02 in (0.05 cm) thick. It comes on spools packed dry under argon. An anyhdrous form of **lithium hexafluoroarsenate** powder is used as the anode in dry batteries.

Because of the low weight, lithium compounds give the highest content of hydrogen, oxygen, or chlorine. **Lithium perchloride,** $LiClO_4$, is a stable solid used as a source of oxygen in rockets and flares, with lithium chloride as a residue. One cubic foot yields 91 lb (41 kg) of oxygen. On a volume basis it has 29% more oxygen at normal temperature than liquid oxygen at its boiling point. **Lithium hydride,** LiH, a white or gray powder of specific gravity 0.82 and melting point 680°C, is used for the production of hydrogen for signal balloons and floats. A l-lb (0.5-kg) can of hydride when immersed in water will liberate 45 ft^3 (1.3 m^3) of hydrogen gas. It is more stable to heat than sodium hydride, and it provides molecular hydrogen, not atomic hydrogen. **Lithium aluminum hydride,** or **lithium alanate,** $LiAlH_4$, is used in the chemical industry for one-step reduction of esters without heat. Lithium metal is very sensitive to light, and is also used in light-sensitive cells. **Lithium 7,** which comprises 94% of natural lithium, has low neutron absorption. The other isomer, **lithium 6,** has high neutron absorption, and is used in nuclear reactors.

LITHIUM ALLOYS. Lithium is soluble in most commercial metals only to a slight extent. It is a powerful deoxidizer and desulfurizer of steel, but no lithium is left in the **lithium-treated steel.** In stainless steels it increases fluidity to produce dense castings. Cast iron treated with lithium has a fine grain structure and increased density with high impact value. Not more than 0.01% remains in the casting when treated with lithium-copper. In magnesium alloys the tensile strength is increased greatly by the addition of 0.05% lithium. The solid solubility of lithium in lead is not over 0.09%, but lithium refines the grain structure of the lead, increasing the strength, and it hardens the lead by the formation of a compound, Pb_3Li_2. **Lithium-treated lead** is called **alkali lead,** and is used for machine bearings. Lithium up to 15% is added to magnesium to make alloys.

Lithium-copper master alloys consist of a group of foundry alloys containing usually 90, 95, or 98% copper with the balance lithium, used for deoxidizing and degasifying nonferrous alloys. Lithium combines easily in the molten bath with oxygen, hydrogen, nitrogen, sulfur, and the halides. The compounds formed are stable, of a nonmetallic nature, of low melting points, and volatilize easily so as to pass off as vapors at the pouring temperature of the metals. **Lithium copper** is a high-conductivity, high-density copper containing a minute quantity of residual lithium, 0.005 to 0.008%, made by treating copper with a 50–50 lithium-calcium master alloy. The conductivity of lithium copper is 101.5% IACS. The tensile strength is 31,500 to 36,500 lb/in^2 (216 to 251 MPa), with elongation 60 to 72%. The wrought metal is tougher than phosphorized copper, and has exceptional deep-drawing properties. The case metal has a density of 8.92. Lithium is an excellent desulfurizer for nickel alloys. From 1 to 7% calcium may also be included in lithium-copper master alloys. **Lithium-calcium alloys** usually contain 30 to 50% lithium, with the balance calcium. They are silvery white in color, with a metallic luster, and are hard and brittle. The melting range is 230 to 260°C. They must be kept in tight containers under kerosene. The alloys are used for treating steel, cast iron, or nickel where no residual copper is to be left. **Copper-manganese-lithium** contains 60 to 70% copper, 27 to 30 manganese, 0.5 to 5 lithium and sometimes 5 to 7 calcium. **Copper-silicon-lithium** contains 80 to 84% copper, 10 to 11 silicon, 2.5 to 10 lithium, and sometimes 2.5 calcium.

LITHIUM ORES. One of the chief ores is **lepidolite** which also carries more rubidium than any other known mineral, containing from a trace to 3% **rubidium oxide,** Rb_2O. It may also have as much as 0.77% **cesium oxide.** It is a **lithia mica,** $LiF \cdot KF \cdot Al_2O_3 \cdot 3SiO_2$, occurring in small plates together with muscovite. It is the most widespread of the lithium minerals, being found in various parts of the Unites States, Canada, northern Zimbabwe, South Africa, India, China, Japan, Russia, and Germany. The hardness is 2.5 to 4, and specific gravity 2.8. It has a pearly luster, and color pink and lilac to grayish white. It is insoluble in acids. Lepidolite is employed as a source of lithium compounds, and of the metals rubidium and cesium. It is also used in making opal and white glasses. **Glassmaker's lepidolite** contains 4% Li_2O; West African lepidolite usually contains 3.75%. **Amblygonite,** plentiful in Sweden, has the formula $Al_2O_3 \cdot 2LiF \cdot P_2O_5 \cdot Li_2O$, and contains 4.24 to 5.26% lithium. The amblygonite from South Dakota is sold on the basis of 8 to 9% Li_2O; and the ore from southern Zimbabwe contains 9% Li_2O, 48% P_2O_5, and 34% Al_2O_3. Other lithium ores are **lithiophilite,** $Li_2O \cdot 2MnO \cdot P_2O_5$, containing 4.6 to 5% lithium; **chryolithionite,** $3LiF \cdot 3NaF \cdot 2AlF_3$, with 5.35% lithium; **petalite,** $Li_2O \cdot Al_2O_3 \cdot 8SiO_2$, with 1.4 to 2.26% lithium; and **manandonite,** $7Al_2O_3 \cdot 2LiO{:}2B_2O_3 \cdot 6SiO_2$, with 2.13% lithium.

LITHOPONE. Also known under various trade names: **Ponolith, Sunolith, Beckton white, Zincolith, Sterling white,** and others. A white pigment consisting of barium sulfate and zinc sulfide. A standard lithopone is 66% barium sulfate and 34 zinc sulfide. High-strength lithopones contain about 50% zinc sulfide, which is one of the whitest pigments. **Titanated lithopone** contains a percentage of titanium dioxide. **Tidolith** is a titanated lithopone of the United Color & Pigment Co. having 85% lithopone and 15 titanium oxide. **Cadmolith** is the trade name of the Glidden Co. for cadmium red and cadmium yellow lithopones used as pigments for plastics, as they are chemical-resistant and nonbleeding. Commercial lithopone is a fine white powder used in the manufacture of paints, inks, oilcloth, linoleum, and rubber goods. For paints the powder should pass through a 325-mesh screen. The ground paste should contain 76 to 80% pigment and 20 to 24 linseed oil. As a paint pigment, lithopone has good hiding power and is lower in cost than other whites, but is not as durable for outside use as white lead or zinc white. It is one of the most-used white pigments for interior work. **Albalith** is a 70–30 lithopone of the New Jersey Zinc Co., also used in rubber goods, inks, and paper.

LOCUST. The wood of the locust tree, *Robinia pseudoacacia,* also known as **acacia, false acacia, black locust,** and **red locust.** The tree is native to North America, but is also grown in Europe. The wood is strong and durable, with a weight of 43 to 52 lb/ft^3 (689 to 833 kg/m^3). Its hardness is about the same as ash, and the strength, flexibility, and shock resistance are greater than oak. The grain is coarse, but the surface is lustrous and satiny. Locust is used for furniture, wheel spokes, posts, crossties, and in construction. **Honey locust** is a lighter and weaker wood from the tree *Gleditsia triacanothos.* The name locust is also applied to the wood of the tree *Hymensea courbaril* of tropical America. This wood has a brownish color, with an open grain, and takes a beautiful polish. The wood of the **Australian locust,** *Acacia melanoxylon,* known as **Australian blackwood** and **Tasmanian blackwood,** and employed for cabinetwork, is reddish brown to black in color and beautifully grained. It is similar in durability and appearance to rosewood, but lighter in weight. Various species of true acacia trees furnish the tannins catechu and wattle. The **silver wattle** of New Zealand, *A prominens,* used for ax handles and fruit cases, is hard and tough. The black locust has clusters of very fragrant white flowers, and it is now widely grown as a shade tree and for shelterbelts in the eastern area of the United States.

LOCUST BEAN GUM. Also called **locust bean flour,** and **carob flour.** A tasteless and odorless white powder obtained by milling the bean kernels of the locust trees of tropical America, Africa, and the Mediterranean countries, notably *Cerafonia siliqua* of Cyprus, Syria, and Spain. The **carob bean,** or pod, contains 6 to 10 hard seeds, which are the **locust beans,** the bean in the pod averaging 9% of the weight. In Gambia the locust bean is called **netto.** When dissolved in water and boiled, it produced an adhesive, transparent jelly, which dries into a colorless, strong, elastic film. It contains galactose and mannose in a complex polymer, and is a polysaccharide or complex sugar. It is used for coating textiles and also as a thickener and binder in glues, pastes, and latex, in leather finishes, and in sizings for yarn and paper. The flour dissolves in cold water and swells in warm water. It is edible and is also used in jellies and bakery products. The dried pods are used also in flavoring dog biscuit and tobacco. In the Near East they are eaten like candy, and are also used as cattle and horse feed. The pods also yield **tragasol** gum, which is used as a textile size and in leather tanning. The production of carob beans in Spain is large, and fully 90% of the crop is used locally for livestock feed.

The **algaroba tree,** *Prosopis chilensis,* growing in semiarid regions of Mexico and Central and South America, called also **mesquite,** and in Hawaii called **keawe,** furnished an important stock feed from the pods and beans, which are similar to the locust. An acre of mesquite produces four times as much food for beef cattle as an acre of corn. The wood of the mesquite also contains up to 1% pyrogallol tannin, valuable for tanning leather. **Guar gum,** used as a replacement for carob bean gum and gum arabic in foods and pharmaceuticals, is a water-soluble, odorless, and tasteless white powder obtained from the endosperm of the seed of the guar plant, *Cyamopsis tetragonoloba,* grown in Pakistan as cattle feed and cultivated in Texas and Arizona. The gum is a polysaccharide with a straight chain of mannose united having one galactose group on every other mannose unit. There is a 3:1 ratio of mannose to galactose compared with a 4:1 ratio in locust bean gum. It also contains about 6% protein, about the same as in locust bean gum. **Guar flour** is a nearly white guar gum. Its high swelling properties in cold water make it suitable as a disintegrating agent in medical tablets. It has more than six times the thickening power of starch, and is used for upgrading starches. **Jaguar gum,** of Stein, Hall & Co., is guar gum. **Jaguar 315,** of this company, and **Guartec,** of General Mills, Inc., are derivatives of guar gum that gel in either acid or alkaline solutions, and will form stiff gels with as much as 99% water content. They are polymers of mannose and galactose.

LOGWOOD. An extract obtained from the wood of the tree *Haematoxylon campechianum,* of tropical America, used as a black dye or as a darkening agent in browns and grays. The wood yields 15% extract. The coloring matter, **hematine,** $C_{16}H_{12}O_6$, forms brownish-red crystals and is produced only in the aged wood or by oxidation of the white extract of fresh wood. **Logwood extract,** or hematine, is marketed in crystals, solid extract, or water extract.

LOOFA SPONGE. A yellowish, porous, skeletal, fibrous body obtained by retting the fruit or seed pods of the tropical edible cucumber, *Luffa cylindrica* and *L. acutangula,* obtained in India, Japan, and the Caribbean countries. It is also called **vegetable sponge,** although it is harder and coarser than sponge. The paper-thin skin is easily removed when the pod dries. The skinned sponge is washed to remove the slimy interior, and when redried the seeds are shaken out. The seeds are used to produce a food oil similar to olive oil but colorless and tasteless. The sponges vary in size from 8 to 24 in (20 to 61 cm) long. The plants grow rapidly, giving four annual crops in Brazil, and a hill of 3 seeds can produce 30 sponges. Loofa sponges are used chiefly for filters in feed tanks of ships. A stiff curly fiber used for making hats is obtained from the product by further retting.

LOW-ALLOY STEELS. Also known as **alloy constructional steels.** They are roughly defined as steels that do not have more than 5% total combined alloying elements. One or more of the following elements are present: manganese, nickel, chromium, molybdenum, vanadium, and silicon. Of these, nickel, chromium, and molybdenum are the most frequent. They are designated by a numerical code prefixed by AISI (American Iron and Steel Institute) or SAE (Society of Automotive Engineers). The last two digits show the nominal carbon content. The first two digits identify the major alloy element(s) or group. For example, 2317 is a nickel-alloy steel with a nominal carbon content of 0.17%

Whereas surface hardness attainable by quenching is largely a function of carbon content, the depth of hardness depends in addition on alloy content. Therefore a principal feature of low-alloy steels is their enhanced hardenability compared to plain carbon steels. Like plain carbon steels, however, low-alloy steels' mechanical properties are closely related to carbon content. In heat-treated, low-alloy steels, the alloying elements contribute to the mechanical properties through a secondary hardening process that involves the formation of finely divided alloy carbides. Therefore, for a given carbon content, tensile strengths of low-alloy steels can often be double those of comparable plain carbon steels.

A majority of low-alloy steels are produced in surface hardening (carburizing) and through hardening grades. The former are comparable in carbon content to low-carbon steels. Grades such as 4023, 4118, and 5015 are used for parts requiring better core properties than obtainable with the surface-hardening grades of plain carbon steel. The higher-alloy grades, such as 3120, 4320, 4620, 5120, and 8620, are used for still better strength and toughness in the core.

Most through, or direct, hardening grades are of medium-carbon content and are quenched and tempered to specific strength and hardness levels. Alloy steels also can be produced to meet specific hardenability limits as determined by end quench tests. Identified by the suffix H, they afford steel producers more latitude in chemical composition limits. The **boron steels,** which contain very small amounts of boron, are also **H steels.** They are identified by B's after the first two digits of their designation.

A few low-alloy steels are available with high carbon content. These are mainly **spring-steel** grades 9260, 6150, 5160, 4160, and 8655, and **bearing steels** 52100 and 51100. The principal advantages of low-alloy spring steels are their high degree of hardenability and toughness. The bearing steels, because of their combination of high hardness, wear resistance, and strength, are used for a number of other parts, in addition to bearings.

A specialty group of **high-strength low-alloy (HSLA) steels** contain 0.05 to 0.33% carbon, 0.2 to 1.65 manganese, and small additions of other elements (such as chromium, columbium, copper, molybdenum, and nickel), which dissolve in a ferritic-matrix structure to provide high strength and corrosion resistance. Their cost and strength falls between that of structural carbon steels and quenched-and-tempered steels. They are available in most commercial forms and are usually supplied and used in the as-rolled condition, although they are also available in other conditions. Most of these **HSLA steels** after a short period of exposure to air develop a pebbly, rusty-looking surface film which markedly reduces the corrosion rate. Because this protective film turns to a deep purple-brown after several years, HSLA steels are being used as exposed structural members in large buildings. Other advantages are their high ratios of yield to tensile strength, and good toughness, abrasion resistance, and weldability.

HSLA steels are covered by several different specifications. The American Society for Testing and Materials (ASTM), for example, classifies them into six groups based on chemical composition and mechanical properties. The SAE covers 12 grades with emphasis on toughness and fabricability. HSLA steels have minimum yield points between 40,000 and 70,000 lb/in^2 (275 and 482 MPa), and ultimate tensile strengths from 60,000 to 85,000 lb/in^2 (413 to 586 MPa). In general, the fatigue endurance limit for 10^6 cycles is about 50 to 60% of their ultimate tensile strength. While room-temperature toughness is roughly comparable to other carbon steels, their lower transition temperature gives them better notch toughness at low temperatures. HSLA steels are more resistant to abrasion than structural carbon steels. Of all HSLA steels, the intermediate manganese grades are the best in this respect. Most grades of HSLA are two to eight times more resistant to atmospheric corrosion than plain carbon steels, due to the presence of such elements as nickel, copper, phosphorus, and chromium. In fresh-and saltwater environments, however, there is little improvement in corrosion resistance over plain carbon steels. In general, the cold formability of HSLA steels is not as good as that of plain carbon steels, but conventional techniques can be used. Most HSLA steels are readily welded by conventional arc, gas, and resistance welding processes. In a group of specialty HSLA steels, small additions of rare-earth metals and controlled cooling change the shape and distribution of sulfide inclusions from stringerlike type to small dispersed globules. This improves cold formability markedly.

As contrasted to the HSLA steels, **quenched-and-tempered steels** are usually treated at the steel mill to develop optimum properties. Generally low in carbon, with an upper limit of 0.2%, they have minimum yield strengths from 80,000 to 125,000 lb/in^2 (551 to 861 MPa). Some two dozen types of proprietary steels of this type are produced. Many are available in three or four different strength or hardness levels. In addition, there are several special abrasion-resistant grades. Mechanical properties are significantly influenced by section size. Hardenability is chiefly controlled by the alloying elements. Roughly, an increase in alloy content counteracts the decline of strength and toughness as section size increases. Thus specifications for these steels take section size into account. In general, the higher-strength grades have endurance limits of about 60% of their tensile strength. Although their toughness is acceptable, they do not have the ductility of HSLA steels. Their atmospheric-corrosion resistance in general is comparable, and in some grades, it is better. Most quenched-and-tempered steels are readily welded by conventional methods.

Dual-phase steels are HSLA steels that combine the strength of conventional HSLA steels with the formability of low-carbon steels. To attain this combination of properties, they are specially heat-treated, and sometimes temper-rolled to produce a two-phase microstructure—ferrite and martensite. A typical dual-phase steel before forming has 21 to 27% total elongation and around 65,000 lb/in^2 (448 MPa) yield strength. Fully strengthened, yield strengths range up to 110,000 lb/in^2 (758 MPa) depending on the grade.

Low-carbon ferritic steels, which were developed by Inco, Ltd., are low-alloy steels containing nickel, copper, and columbium. They are precipitation-hardened and have yield strengths from 70,000 to 100,000 lb/in^2 (482 to 689 MPa) in sections up to ¾ in (1.9 cm). They possess excellent welding and cold-forming characteristics. A major use of these steels has been for vehicle frame members. Atmospheric corrosion resistance is roughly three or four times that of carbon steels.

LOW-EXPANSION ALLOYS. Alloys, usually of nickel and iron, which have a very low coefficient of expansion at ordinary temperatures, used chiefly for instrument parts and for lead wires to be sealed in glass or other material. The **coefficient of expansion** of a metal is not a quality of the metal but merely a comparative measure of volume change with heat-energy application. This increase in volume is caused by change in shape of the molecule and distance apart of the atoms. Chemically uncombined iron, tungsten, and nickel normally have no molecular shape changes in the solid stage. For low-expansion alloys, elements that cause molecular changes in the solid are avoided. The alloy first developed in France under the name of **Invar** has an expansion so low for measuring purposes, under ordinary conditions, that it is generally taken as zero. This is about 0.0000004 per degree Celsius, but above 120°C the coefficient rises. Invar is used for the measuring guides of accurate instruments and for parts for watches. A typical Invar as produced by Inco, Ltd., contains 54% cobalt. This product may be used for parts and piping whenever strength is required at low temperatures. Such an alloy is also very resistant to corrosion. **Nivar** is a similar alloy. **Nivarox,** a Swiss alloy for watch and clock springs of exceptional accuracy, has 37% nickel, 8 chromium, 54 iron, 0.85 manganese, 0.90 beryllium, 1 titanium, 0.20 silicon, and 0.10 carbon. Another Swiss alloy noted for corrosion resistance as well as high fatigue resistance is **Contracid.** It has 60% nickel, 15 chromium, 7 molybdenum, 15 iron, 2 manganese, 0.5 silicon, and 0.5 to 0.75 beryllium. **Nicol,** of the National-Standard Co., for instrument springs, contains 40% cobalt, 20 chromium, 15 nickel, 7 molybdenum, 2 manganese, 0.04 beryllium, 0.15 carbon, and the balance iron. When heat-treated it has a tensile strength of 368,000 lb/in^2 (2,536 MPa), and a Vickers hardness of 700. It has an electrical resistance of 600 Ω/cir mil ft (100×10^{-8} Ω·m), retains its elasticity to 600°F (316°C), and is highly corrosion-resistant.

Elinvar, used for chronometer balances and springs for gages and instruments, has low thermal expansion and almost invariable modulus of elasticity. Elinvar for hair springs for watches contains 33 to 35% nickel, 53 to 61 iron, 4 to 5 chromium, 1 to 3 tungsten, 0.5 to 2 manganese, 0.5 to 2 silicon, and 0.5 to 2 carbon. **Elinvar Extra,** of the Hamilton Watch Co., contains 42% nickel, 5.5 chromium, 2.5 titanium, about 0.5 each of manganese, aluminum, and silicon, 0.06 max carbon, and the balance iron. It is nonmagnetic, corrosion-resistant, and has low thermal expansion. **Elgiloy,** of Elgiloy Co., originally a **watch spring alloy,** is superior to carbon and stainless steel for coil and flat springs, and is marketed in wire, rod, and strip for many applications requiring a combination of high fatigue strength, heat-corrosion resistance, dimensional stability, and nonmagnetic characteristics. The alloy contains 40% cobalt, 20 chromium, 15.5 nickel, 7 molybdenum, 2 manganese, 0.15 carbon, 0.04 beryllium, and the balance iron. The tensile strength is 368,000 lb/in^2 (2,536 MPa), modulus of elasticity 29.5 million, and Rockwell C hardness 58. The physical properties depend upon cold work and heat treatment at 900°F (482°C), and service temperatures are thus below this point. **Platinite,** a French alloy with a coefficient of expansion about the same as that of glass, contains 46% nickel and 54 iron. It is used in light bulbs. **Dilver** and **Adr** are also French low-expansion, nickel-iron alloys. **Super-Invar** is a Japanese product containing 5% cobalt to replace an equal amount of nickel. It has a nearly zero coefficient of expansion at ordinary temperatures. **Nilvar,** of the Driver-Harris Co., used for instrument parts, measuring tapes, and connections through glass, is in various grades with about the same coefficient of expansion as glass or with the zero expansion of Invar. **Fernico,** of the General Electric Co., is an alloy used for sealing vacuum tubes.

Low-expansion alloy 42, of Carpenter Technology, is an Invar with 42% nickel and 0.10 carbon, used for temperatures above 400°F (204°C). It has an electrical resistance of 430 Ω/cir mil ft (71×10^{-8} Ω·m). The hot-rolled material has a tensile strength of 82,000 lb/in^2 (574 MPa), and elongation 30%. It can be hardened by cold work. **Sylvania 4 alloy,** of

Sylvania Electric Products, Inc., is in the form of wire with the same expansion as glass. It has 42% nickel, 5.65 chromium, with small amounts of manganese, silicon, carbon, and aluminum, and the balance iron. The wire is fired at 2300°F (1260°C) in a hydrogen atmosphere to give a coating of Cr_2O_3 which seals it to the glass. **Sealmet HC-4,** of the Allegheny Ludlum Steel Co., is similar.

The **sealing alloy** of the Wilbur B. Driver Co., called **Niron 52,** contains 52% nickel and the balance iron. **Rodar,** of this company, for sealing hard, heat-resistant glass, has 29% nickel, 17 cobalt, 0.3 manganese, and the balance iron. **Niromet 46,** used with vitreous enamels, has 46% nickel and 54 iron. The electrical resistivity is 275 Ω. **Dumet,** of the Westinghouse Electric Corp., is a composite metal for lead-in wires for electric lamps and electric tubes. It is made of a nickel-steel core with a thin coating of brass and an outer shell of copper tubing. This composite rod is then drawn into wire of diameters 0.010 to 0.040 in (0.025 to 0.101 cm). The radial expansion is 0.000009 per degree Celsius, which is close to that of soft glass. The tensile strength is up to 79,000 lb/in^2 (544 MPa), with elongation 18 to 26%. **Kovar,** of the Stupakoff Ceramic & Mfg. Co., is an alloy of 29% nickel, 17 cobalt, 0.3 manganese, and the balance iron, having an expansion the same as that of heat-resistant glass for all temperatures up to the softening point of the glass. It is used for making gastight window assemblies, instrument and radio-tube sealing, and for metal attachments to glass. When the metal is heated to a red heat and applied to the glass, the metal oxide formed dissolves in the glass and gives a gastight seal.

LUBRICATING GREASE. Usually a compound of a mineral oil with a soap, employed for lubricating machinery where the speed is slow or where it would be difficult to retain a free-flowing oil. The soap is one that is made from animal or vegetable oils high in stearic, oleic, and palmitic acids. The lime soaps give water resistance, or a mineral soap may be added for this purpose. Aluminum stearate gives high film strength to the grease. All of these greases are more properly designated as **mineral lubricating grease.** Originally, **grease** for lubricating purposes was hog fat or the inedible grades of lard, varying in color from white to brown. Some of these greases were stiffened with fillers of rosin, wax, or talc, which were not good lubricants. The stiffness of such a grease should be obtained with a mineral soap. ASTM specifications for heavy journal bearing grease require 45% soap content. About 2% calcium benzoate increases the melting point. Mineral lubricating grease may contain from 80 to 90% mineral oil and the remainder a lime soap. Federal specifications prescribe 85% mineral oil. Chemicals may be added to improve the physical properties of grease. **Oronite GA-10,** of the Oronite Chemical Co., for example, is a sodium salt of terephthalic acid used as a gelling agent in high-temperature greases. It adds water resistance and stabilizes against emulsion. **Ortholeum 300,** marketed as a brown flaky powder by DuPont, is a mixture of complex amines, and small amounts added to a grease will give high heat stability. **Braycote 617,** of the Bray Oil Co., is a synthetic grease for rockets subject to both heat and cold. It is a mixture of perfluorotrialkylamines gelled with tetrafluoroethylene. It comes as a translucent lardlike semisolid with a boiling point at 230°C. The lubricating grease known as **trough grease,** used in food plants for greasing trays, tables, and conveyors, contains no mineral oil and is edible.

Lime greases do not emulsify as readily as those made with a soda base, and are thus more suitable for use where water may be present. **Hard grease** flows at a temperature of about 90°C; medium grease flows at 75 to 80°C. Paraffin wax, sometimes added, is an adulterant and not a lubricant. **Graphite grease** contains 2 to 10% amorphous graphite, and is used for bearings, especially in damp places. Federal specifications call for 2 to 3% graphite. For large ball and roller bearings a low-lime grease is used, sometimes mixed with a small percentage of graphite. **Cylinder grease** is made of about 85% mineral oil or mineral grease and 15% tallow. Compounded greases are also marketed containing animal and vegetable oils, or are made with blown oils and compounded with mineral oils. The fatty acids in vegetable and animal oils, however, are likely to corrode metals. Tannin holds graphite in solution; in the gear grease sold under the name **Gredag** by the Acheson Colloids Co. a graphite-tannin mixture is used. **Metaline,** of the R. W. Rhoades Metaline Co., Inc., is a compound of powdered antifriction metal, oxide, and gums, which is packed in holes in the bearings to form self-lubricating bearings. **Lead-Lube grease,** of Knapp Mills, Inc., has finely powdered lead metal suspended in the grease for heavy-duty lubrication.

Sett greases are mixtures of the calcium soaps of rosin acids with various grades of mineral oils. They are low-cost semisolid greases used for lubricating heavy gears or for greasing skidways. Clay fillers may be added to improve the film strength, or copper or lead powders may be incorporated for heavy load conditions. **Solidified oil** is also a name given to grease made from lubricating oil with a soda soap and tallow, used for heavy bearings. **Cup grease** is made with soda soaps and light lubricating oils. Greases made with potash and soda soaps tend to form soap fibers when water is present. A metallic soap that contains no fibers is called a **neat soap,** and gives a smooth grease. Greases made with lithium stearate have good water and high-temperature resistance, and have a buttery texture. **Alrania grease,** of the Shell Oil Co., is a grease of this type. Fatty acids used for grease making may be hardened by hydrogenation to remove polyunsaturated acids. The greases have greater resistance to heat discoloration, and do not gum or become rancid. **Lubrex 45,** of W. C.

Hardesty, Inc., is a hardened fatty acid of this type. **Slushing oil,** for use in protecting machine parts from corrosion in shipping or storage, is usually a low-melting grease preferably compounded with a waxy fat such as lanolin. **Paralan,** of the American Lanolin Corp., is such a slushing oil having a lanolin base.

LUBRICATING OILS. Oils used for lubricating the bearing parts of machinery. They are usually the heavy distillates following kerosene in the fractional distillation of petroleum, between the temperatures 253 and 317°F (123 and 158°C). They are separated into grades, light, medium, and heavy, depending upon the molecular weight. They are also classed as pale, when yellow to reddish in color, and dark, when brownish black. The flash points range from 300 to 600°F (149 to 316°C), and the specific gravities usually from 0.860 to 0.940. **Neutral oils** are light oils obtained by distillation without cracking, and they will not emulsify in contact with water as do the paraffin oils. They are thus desirable for crankcase lubrication and in circulating systems. Lubricating oils may be bleached with acid, and they may be mixed with vegetable or animal oils. The ideal of lubrication is to obtain a full fluid film with little clearance between the moving surfaces so that the shaft rotates on a film of oil. Hydrodynamic lubrication with pressure gives this condition. Only a boundary lubrication with contact merely on the bottom is obtained when the clearance is too great, the oil viscosity too low, the load too heavy, or the speed is too slow, so that the film does not support the shaft.

Animal oils are more greasy than mineral oils, but they are acid. Vegetable oils are greasy and have more oiliness, but they oxidize easily and are also acid. They are likely to gum in use unless an antioxidant is employed. Vegetable and animal oils add the property of adhesion to the lubricating oil, but in no case should any element be added to an oil that will cause emulsification. Federal specifications for marine engine oil call for 15 to 20% blown refined rapeseed or peanut oil. This lubricating oil has a flash point of 350°F (177°C). **Steam cylinder oil** has 5 to 10% fatty acid vegetable oils, and the flash point is 450°F (232°C). **Absorbed oil** is a name of a combination oil of E. F. Houghton & Co. which acts as both film and lubricant. **Amlo** is a trade name of a mineral oil refined wax-free, used for low-temperature lubrication. The silicones are now often used to replace lubricating oils for very high and very low temperature conditions, but in general the lubricating value is not high.

Antioxidants used in oils to reduce oxidation and minimize sludging and acid formation are usually tin compounds such as tin dioxide, tin tetraphenyl, and tin ricinoleate. Tin dust alone also has an inhibitory action. Detergents are compounded in lubricating oils for internal-combustion engines in order to prevent and break down carbon and sludge deposits.

High percentages of animal or vegetable oils may be added to lubricating oils for use on textile machinery. They are called **stainless oils** for this purpose, since such oils wash out of the textile more easily than mineral oils. They also give lower coefficients of friction. The high lubricating qualities of the vegetable oils without the disadvantage of gumming can be obtained with mineral oils by the addition of an **oiliness agent** such as the **cetyl piperidine ricinoleate.** The **EP lubricants** (extreme pressure) for heavy-duty gear lubrication are made with a high-quality oil compounded with a lead-sulfonated soap.

For extreme high pressure and high temperatures where oils and greases oxidize, **molybdenum disulfide,** MoS_2, is used alone or mixed with oils or silicones. It is a fine black powder, available in particle sizes as small as 0.75 μm, which adheres strongly to metal surfaces, gives a low coefficient of friction, and will permit operation up to 750°F (399°C), but it has an acid reaction and is corrosive to metals. Molybdenum disulfide resembles graphite but is twice as dense. The sulfur attaches itself with a weaker electron bond on one side than the other, forming laminal plates or scales in the molecular structure which tend to split off and give the sliding or lubrication action. Molybdenum disulfide may be used as a filler in nylon gears and bearings to reduce friction. It also increases the flexural strength of the plastic. **Molysulfide,** of the Climax Molybdenum Co., is this material. **Tungsten disulfide** is also used as a lubricant in the same way as molybdenum disulfide. The electron bond of sulfur to tungsten is stronger than that to molybdenum, and it is thus more stable at high temperatures. The tungsten disulfide of Sylvania Electric Products, Inc., is a crystalline, gray-black powder with particle size from 1 to 2 μm. **Molykote,** of the Alpha Corp., and **Liqui-Moly,** of the Lockrey Co., are molybdenum-disulfide lubricants. **Dry-film lubricants** are usually graphite or molybdenum sulfide in a resin or volatile solution. They are sprayed on the bearing surface, and the evaporation of the solvent leaves an adherent thin film on the bearing. **Selenium disulfide,** SeS_2, will retain its lubricating qualities at temperatures to 2000°F (1093°C) and is useful for lubrication under vacuum because of its low emission of gas. Other materials used as **dry lubricants** are **tantalum disulfide,** TaS_2, **tantalum diselenide,** $TaSe_2$, **titanium ditelluride,** $TiTe_2$, and **zirconium diselenide,** $ZrSe_2$.

Hydraulic fluids for the operation of presses must lubricate as well as carry the pressure. They are mostly mineral oils, but chemicals are used where high temperatures are encountered, such as in die-casting machines. **Lindol HF-X,** of the Celanese Corp. of America, is a flame-resistant hydraulic fluid with a tricresyl phosphate base. **Skydrol,** of the Monsanto Chemical Co., for aircraft hydraulic systems, is an oily ester produced from petroleum gas. The ignition point is 1050°F (566°C), and it will operate at temperatures as low as −40°F (−40°C). The **Fluorolube oils** of the

Hooker Chemical Corp. are polymers of trifluorovinyl chloride fractionated to provide grades from a colorless low-viscosity oil to an opaque heavy grease. They have high lubricating values, are resistant to acids and alkalies, and have an operating range from 300°C down to very low subzero temperatures. **Hydraulic fluid QF-6-7009,** of the Dow Chemical Co., for closed systems operating from −25 to 550°F (−32 to 288°C), is a diphenyl didodecyl silane. **Refrigeration oils,** for lubricating refrigerating machinery, are mineral oils refined to remove all moisture and wax. **Ansul oil,** of the Ansul Chemical Co., is an oil of this class which will remain stable at temperatures as low as −70°F (−57°C).

The nature of the bearing metals often has an effect upon the action of the lubricating oil. In highly alloyed metals some elements act as catalyzers to oxidize the oil, or the acids or moisture in the oils may act to break down the metal. In lead-bearing metals free magnesium causes disintegration of the lead in contact with moisture. The alkali-lead metals also tend to dissolve in contact with animal or fish oils. Normally, however, none of the white bearing metals are attacked by the animal and vegetable oils used for lubrication unless there are perceptible amounts of a freely oxidizing element present. Graphite adds to the effectiveness of a lubricating oil and can be held in suspension with a tannin. **Graphite lubricants** are used where continuous lubrication is difficult, for running in, for springs, or for bearings where heavy films are desired. The **Dag Lubricants** and**Dag Dispersants** of Acheson Colloids Co. comprise a large group of lubricants, lubricant coatings, and mold partings consisting of graphite or molybdenum sulfide in oils, resins, or solvents, usually applied by spray application. **Glydag** is a solution of 10% graphite in glycerin, **Castordag** is graphite in castor oil, **Glydag B** is graphite in butylene glycol, and **Dag Dispersion 223** is molybdenum disulfide in an epoxy resin. **Neolube,** of Huron Industries, is graphite in alcohol. With these, the carrier liquid evaporates, leaving a film of graphite on the bearing. **Polyphenyl ether** lubricants are highly radiation-resistant. They lubricate after absorbing gamma-ray doses that solidify mineral oils. They are used as specialty lubricants under extreme high-temperature conditions.

LUMINOUS PIGMENTS. Pigments used in paints to make surfaces visible in the dark and in coatings for electronic purposes. They are used for signs, watch and instrument hands, airfield markings, and signals. They are of two general classes. The so-called permanent ones are the **radioactive paints,** which give off light without activation, and the **phosphorescent paints,** or **fluorescent paints,** which require activation from an outside source of light. The radioactive paints contain a radioactive element that emits alpha and beta rays which strike the phosphors and produce visible light. Radium, sometimes used for paints for watch hands, gives a greenish-blue light, but it emits dangerous gamma rays. Also, the intense alpha rays of radium destroy the phosphors quickly, reducing the light. Strontium 90 gives a yellow-green light and has a long half-life of 25 years, but it emits both beta and gamma rays and is dangerous. **Tritium paints,** with a tritium isotope and a phosphor in the resin-solvent paint base, have a half-life of 12.5 years and require no shielding. The self-luminous phospors for clock and instrument dials contain tritium, which gives off beta rays with only low secondary emission so that the glass or plastic covering is sufficient shielding. Other materials used are **krypton 85,** with a half-life of 10.27 years, **promethium 147,** with a half-life of 2.36 years, and **thallium 204,** with a half-life of 2.7 years.

Fluorescent paints depend upon the ability of the chemical to absorb energy from light and to emit it again in the form of photons of light. This variety usually has a base of calcium, strontium, or barium sulfide with traces of other metal salts to improve luminosity, and the vehicle contains a moistureproof gum or oil. Temporary luminous paints may be visible for long periods after the activating light is withdrawn. A paint activated by 5 min of exposure to sunlight may absorb sufficient energy for 24 hr of luminosity. **Luminous wall paints** used for operating rooms to eliminate shadows are made by mixing small amounts of zinc or cadmium sulfide into ordinary paints. After being activated with ultraviolet rays, they will give off light for an hour and a half.

Phosphorescent paints are lower in cost than radioactive paints, and may be obtained in various colors. In general, the yellow and orange phosphorescent pigments are combinations of zinc and cadmium sulfides, the green is zinc sulfide, and the violet and blue pigments are combinations of calcium and strontium sulfides. They are marketed in powder form to be stirred into the paint or ink vehicle since mixing by grinding lowers the phosphorescence. The natural minerals are not used, as the pigments must be of a high degree of purity, as little as a millionth part of iron, cobalt, or nickel killing the luminosity of zinc sulfide. These phosphorescent pigments are called **phosphors,** but technically they are incomplete phosphors; and copper, silver, or manganese is coprecipitated with the sulfide as activators or to change the color of the emitted light. The metals that are used as activators are called **phosphorogens,** and their atoms diffuse into the lattice of the sulfide. For fluorescent screens the phosphors must have a rapid rate of extinguishment so that there will be no time lag in the appearance of the events. For television, electron microscope, and radar screens the phosphors must cease to flow $\frac{1}{50}$ sec after withdrawal of excitation. They must also be of very minute particle size so as not to give a blurred image. For a white television screen, mixtures of blue zinc sulfide with silver and yellow zinc-beryllium silicates are used. For color television the screen is completely covered with a mixture of various colored phosphors, especially rare-earth metal combinations. For scintillation counters for gamma-ray-detection phosphors the pulses should be of longer duration, and for this purpose crystals of cadmium or cadmium tungstate are used.

Fluorescent fabrics for signal flags and luminescent clothing are impregnated with fluorescent chemicals which can be activated by an ultraviolet light that is not seen with the eye. Some fluorescent paints contain a small amount of luminous pigment to increase the vividness of the color by absorbing the ultraviolet light and emitting it as visible color. **Fluorescein,** made from phthalic anhydride and resorcinol, has the property of fluorescence in a solvent. Since cellulose acetate will keep it in a permanently solvent state, acetate rayon is used as the carrier fabric. Signal panels are distinguishable from a plane at great heights even through a haze, and at night they give a brilliant glow when activated with ultraviolet rays. Fluorescent paints for signs may have a white undercoat to reflect back the light passing through the semitransparent pigment. In passing through the color pigments the shorter violet and blue wavelengths are changed to orange, red, and yellow hues, and the reflected visible light is greater than the original light. **Uranine,** the sodium salt of fluorescein, is used by fliers to mark spots in the ocean. One pound (0.5 kg) of uranine will cover an acre (4,047 m^2) of water to a brilliant yellowish green easily seen from the air. One part of uranine is detectable in 16 million parts of water. The **luminous plastic** marketed by the U.S. Radium Corp. for aircraft markings is coated on the inside with radioactive material to give visibility in the dark. The **fluorescent plastic** of the Rohm & Haas Co. is acrylic sheet containing a fluorescent dye. Lettering or designs cut from the sheet will glow brightly in the dark after exposure to light. It is used for direction signs and decorative panels. **Spot-Lite Glo,** of Conrad Precision Industries, is a **phosphorescent frit** for incorporating into ceramics for luminous signs. It contains a zinc sulfide that is stable at high heats and has a long afterglow.

Whitening agents, optical whiteners, or **brightening agents,** used to increase the whiteness of paper and textiles, are fluorescent materials that convert some of the ultraviolet of sunlight into visible light. The materials are colorless, but the additional light supplied is blue, and it neutralizes yellow discolorations and enhances the whiteness. They were first developed in Germany and called **blankophors. Ultrasan,** the first of the German blankophors, was a 1,3,5-triazine derivative. The **M.D.A.C.** of the Carlisle Chemical Works, Inc., is a methyl diethyl aminocoumarin of the empirical formula $C_{14}H_{17}O_2N$. It comes in tan-colored granules melting at 70°C, soluble in water and in acid solutions. It gives a bright-blue fluorescence in daylight and adds whiteness to fabrics and makes colors more vivid. As little as 0.001% added to soaps, detergents, or starches is effective for wool and synthetic fibers, but it is not suitable for cotton. It is also used to overcome yellow casts in varnishes and plastics, and in oils and waxes. **Solium,** of Lever Bros., used in detergents, is a whitener of this type. The **DAS triazines,** triazinyl diaminostilbene disulfonic acid, used with naphthyl triazole, is effective for cotton and rayon. Luminous materials also occur in nature as organic compounds. The luminous material of the firefly, **luciferin,** is a water-soluble protein.

LUTES. Adhesive substances, usually of earthy composition, deriving the name from the Latin, *lutum,* meaning mud. A clay cement was used by the Romans for cementing iron posts into stone. Although they often contain a high percentage of silica sand or clays, the active ingredient is usually sulfur. They may also contain other reactive ingredients such as lead monoxide or magnesium compounds. **Plumber's lutes** are used for pipe joints and seams and for coating pipes to withstand high temperatures. Plaster of paris mixed with a weak glue will withstand a dull-red heat. **Fat lute** is pipe clay mixed with linseed oil. **Spence's metal** is the name of an old lute for pipe jointing. It was made by introducing iron disulfide, zinc blende, and galena into melted sulfur. It melts at 320°F (160°C) and expands on cooling. It makes a good cast joint which is resistant to water, acids, and alkalies. It is not a metal, but is a mixture of sulfur with metallic oxides. **Sulfur cements,** or lutes, usually have fillers of silica or carbon to improve the strength. They are poured at about 235°F (113°C). They form a class of acid-proof cements used for ceramic pipe connections. Modern lutes for very high heat resistance do not contain elemental sulfur. Industrial lutes are used for sealing in wires and connections in electrical apparatus, and are compounded to give good bonding to ceramics and metals. A lute cement for adhering knife blades to handles is composed of magnesium acid sulfates, calcined magnesia, with fine silica sand or powder. The term **sealant** generally refers to a wide range of mineral-filled plastics formulated with a high proportion of filler for application by troweling or air gun.

MADDER. Formerly the most important dyestuff with the exception of indigo. It is now largely replaced by the synthetic mauve dye alizarin. It was grown on a large scale in France and the Near East and was known by its Arabic name **alizari** and by the name **Turkey red.** Madder is the ground root of the plant *Rubia tinctorum,* which has been stored for a time to develop the coloring matter, the orange-red **alizarin,** $C_{14}H_{18}O_4$, which is a dihydroxyanthraquinone, a powder melting at 289°C. It occurs in a madder root as the glucoside, **ruberythric acid,** $C_{26}H_{28}O_{14}$, but is now made synthetically from anthracene. Its alkaline solution is used with mordants to give **madder lakes.** With aluminum and tin it gives **madder red,** with calcium it gives blue, and with iron it gives violet-black. **Purpurin,** $C_{14}H_{18}O_5$, is also obtained from madder, but is now made synthetically. Madder gives fast colors.

MAGNESIA. A fine white powder of **magnesium oxide,** MgO, obtained by calcining magnesite or dolomite and refining chemically. It is used in pharmaceuticals, in cosmetics, in rubbers as a scorch-resistant filler, in soaps, and in ceramics. It requires 6.5 tons (5.9 metric tons) of dolomite to yield 1 ton (0.9 metric ton) of pure magnesia powder. Particle size of the powder is 0.5 μm. For chemical uses it is 99.7% min purity with no more than 0.06% iron oxide and 0.08 calcium oxide, and the magnesia for electronic parts has a maximum of 0.03% iron oxide and 0.0025% boron. This powder is converted from magnesium hydroxide. **Maglite,** of Whittaker, Clark & Daniels, Inc., used for rubbers, is produced from seawater. **Magox magnesia,** of Basic Chemicals, is 98% pure MgO extracted from seawater. It comes in particle sizes to 325 mesh in high- and low-activity grades for rubber, textile, and chemical uses. A very pure magnesia is also produced by reducing magnesium nitrate.

Magnesia ceramic parts, such as crucibles and refractory parts, are generally made from magnesia that is usually electrically fused and crushed from the large cubic crystals. The crystals have ductility and can be bent. The particle size and shape are easily controlled in the crushing to fit the needs of the molded article. Pressed and sintered parts have a melting point at about 5070°F (2765°C), and can be employed to 2300°C in oxidizing atmospheres or to 1700°C in reducing atmospheres. The material is inert to molten steels and to basic slags. **Magnafrax 0340,** of the Carborundum Co., is magnesia in the form of plates, tubes, bars, and disks. The material has a density of 3.3, and a thermal conductivity twice that of alumina. Its vitreous structure gives it about the same characteristics as a single crystal for electronic purposes. **Magnorite,** of the Norton Co., is fused magnesia in granular crystals with a melting point of 2800°C, used for making ceramic parts and for sheathing electric heating elements. **K-Grain magnesia,** of Kaiser Aluminum and Chemical Corp., is 98% magnesia, containing no more than 0.4% silica. The magnesia ceramic, of the Corning Glass Co., is 99.8% purity. The cast, pressed, or extruded parts when high-fired have a fine-grained dense structure with practically no shrinkage and a flexural strength of 15,500 lb/in^2 (106 MPa).

MAGNESITE. A white to bluish-gray mineral used in the manufacture of bricks for basic refractory furnace linings and as an ore of magnesium. The ground, burned magnesite is a light powder, shaped into bricks at high pressure, and baked in kilns. Magnesite is a magnesium carbonate, $MgCo_3$, with some iron carbonate and ferric oxide. Magnesite releases carbon dioxide on heating, and forms magnesia, MgO. When heated further it forms a crystalline structure known as **periclase,** which has a melting point of 5070°F (3076°C) and specific gravity of 3.58. The mineral periclase occurs in nature but is rare. A crystalline form is called **breunnerite.** The fused magnesia made in the arc furnace is actually synthetic periclase. The synthetic material is in transparent crystals up to 2 in (5 cm), which are crushed to powder for thermal insulation and for making refractory parts. Magnesite in compact earthy form or granular masses has a vitreous luster, and the color may be white, gray, yellow, or brown. The hardness is 3.5 to 4.5, and the specific gravity about 3.1. The American production of crude magnesite is in Nevada, Washington, and California.

The product known as **dead-burned magnesite** is in the form of dense particles used for refractories. It is produced by calcining magnesite at a temperature of 1450 to 1500°C. **Caustic magnesite** is a product resulting from calcination at 700 to 1200°C, which leaves from 2 to 7% carbon dioxide in the material and gives sufficient cementing properties for use as a refractory cement. Beluchistan magnesite has 95 to 98% $MgCO_3$, with 0.5 to 1% iron oxide. Manchurian dead-burned magnesite has 90.9% magnesia with 4 silica, and some iron oxide and alumina.

Magnesite for use in producing magnesium metal should have at least 40% MgO, with not over 4.5 CaO and 2 FeO. **Brucite,** a natural hydrated magnesium oxide found in Ontario, contains a higher percentage of magnesia than ordinary magnesite and is used for furnace linings. Austrian magnesite has from 4 to 9% iron oxide, which gives it the property of fritting together more readily. Magnesite is a valued refractory material for crucibles, furnace brick and linings, and high-temperature electrical insulation because of its basic character, chemical resistance, high softening point, and high electrical resistance. Its chief disadvantage is its low resistance to heat shock. Magnesite brick and refractory products are marketed under a variety of trade names, such as **Ritex,** of the General Refractories Co., and **Ramix,** of Basic Dolomite, Inc. It is also used as a covering for hot piping. The German artificial stone called **Kunststein** is magnesite.

MAGNESIUM. A silvery-white metal, symbol Mg, which is the lightest metal that is stable under ordinary conditions and produced in quantity. It is the sixth most abundant element, and is only 64% the weight of aluminum and 23% the weight of iron. By volume, 1 lb (0.5 kg) of magnesium produces five times as much rolled metal as 1 lb (0.5 kg) of copper. It was originally called **magnium** by Sir Humphry Davy. The specific gravity is 1.74, melting point 651°C, and boiling point 1120°C. It has a tensile strength when cast of 14,000 lb/in^2 (96 MPa). The rolled metal has a tensile strength of 25,000 lb/in^2 (172 MPa), and elongation 4%. The strength is somewhat higher in the forged metal. Magnesium has a close-packed hexagonal structure that makes it difficult to roll cold, and its narrow plastic range requires close control in forging. Repeated reheating causes grain growth. It is thus usually hot-worked, and at 600°F (316°C) it can be deep-drawn better than other common metals. It is the easiest to machine of the metals. Its heat conductivity is half that of aluminum, and it has high damping capacity. Electrolytic magnesium is usually 99.8% pure, and the metal made by the ferrosilicon-hydrogen reduction process may be 99.95% pure.

Magnesium is immune to alkalies, it is resistant to seawater, and its sound-damping characteristics make it valuable for diaphragms, resonators, and shielding. Because of its flammability at high temperatures, it cannot be used safely for parts subject to temperatures above 450°F (232°C). The metal is produced regularly in sheet. **Thinstrip magnesium,** of the Burgess Battery Co., is commercially pure magnesium sheet in thicknesses from 0.005 to 0.010 in (0.013 to 0.025 cm) for die forming and for paneling. **Magnesium sheet** can be drawn in dies without folding or wrinkling. For better corrosion resistance the metal may be given a thin coating of zinc by dipping in a pyrophosphate-zinc solution. Electroplating the metal is done by first depositing a thin zinc plate followed by a copper flash coating as a base for the nickel or other plate. **Anodized magnesium** is produced by immersing in a solution of ammonium fluoride and applying a current of 120 V. The fluoride film has a thickness of only 0.0001 in (0.0003 cm), but it removes cathodic impurities from the surface of the magnesium, giving greater corrosion resistance and better adhesion of paints.

Magnesium is valued chiefly for parts where light weight is needed, but it is not ordinarily used alone. It is a major constituent in many light aluminum alloys, and superlight alloys can be made by alloying magnesium with lithium. **Photoengraving plates** made of commercially pure magnesium, or of slightly alloyed metal, are easier to etch than zinc, lighter in weight, and resistant to wear. As a facing and shielding material in building construction, the light weight of magnesium gives high coverage, 1 lb (0.5 kg) of 0.005-in (0.013-cm) sheet covering 22.2 ft^2 (2 m^2).

The pure metal ignites easily, and even when alloyed with other metals the fine chips must be guarded against fire. In alloying, it cannot be mixed directly into molten metals because of flashing, but is used in the form of master alloys. The metal is not very fluid just above its melting point, and casting is done at temperatures considerably above the melting point so that there is danger of burning and formation of oxides. A small amount of beryllium added to magnesium alloys reduces the tendency of the molten metal to oxidize and burn. The solubility of beryllium in magnesium is only about 0.05%. As little as 0.001% lithium also reduces fire risk in melting and working the metal. Molten magnesium decomposes water so that greensand molds cannot be used, as explosive hydrogen gas is liberated.

For the same reason water sprays cannot be used to extinguish magnesium fires. The affinity of magnesium for oxygen, however, makes the metal a good deoxidizer in the casting of other metals.

Magnesium is produced commercially by the electrolysis of a fused chloride, or fluoride obtained either from brine or from a mineral ore, or it can be vaporized from some ores. Much of the magnesium produced in the United States is from brine wells of Michigan, whose brine contains 3% $MgCl_2$, and from seawater. From seawater the magnesium hydroxide is precipitated, filtered, and treated with hydrochloric acid to obtain a solution of magnesium chloride from which the metal is obtained by electrolysis. The magnesium ion content of seawater is 1,270 parts per million. One cubic mile (4.8 km^3) of seawater contains up to 12 million lb (5.4 million kg) of magnesium. In production, 1 lb (0.5 kg) of magnesium is obtained from 100 gallons (379 L) of seawater. Magnesium is also obtained from dolomite by extracting the oxide by reacting the burned dolomite with crushed ferrosilicon in a sealed retort and filtering the vapor in a hydrogen atmosphere. The dolomite of Ohio averages 20% magnesium oxide, and the metal is obtained 99.98% pure in a solid, dense, crystalline mass which is then melted with a flux and poured into ingots. Magnesium metal and ferrosilicon are produced from the olivine of Washington state. In Russia, magnesium is produced from the mineral carnallite. In Germany it is produced from dolomite, carnallite, magnesite, and the end lyes of the potash industry. The metal can also be produced from serpentine, olivine, and other siliceous ores by heating the powdered ore in a vacuum retort and driving off the metal as a vapor which is condensed.

Magnesium powder, for pyrotechnic and chemical uses, is made by reducing metallic magnesium into particles in the shape of curly shavings to give maximum surface per unit of weight. It is produced in four grades: Cutting powder, Standard powder, Special specification, and Fireworks powder. **Cutting powder** is finely cut shavings in a matted condition, made from magnesium of 99.8% purity. **Standard powder** is loose powder in fineness from 10 to 200 mesh. **Fireworks powder** is 100 mesh. The speed of ignition increases rapidly with decreasing particle size. A 200-mesh powder is used for **flashlight powder,** and a 30- to 80-mesh for more slowly burning flares. For flares, magnesium gives a brilliant light of high actinic value. **Incendiary powder,** for small-arms incendiary ammunition, is magnesium powder mixed with barium peroxide. **Ophorite** is an English name for magnesium powder and potassium perchlorate used as an igniter for incendiaries. The material known as **goop,** used in fire bombs, is a rubbery mixture containing magnesium powder coated with asphalt, gasoline, and chemicals.

A wide variety of **magnesium chemicals** is used for applications where the magnesium may be the desirable element or as the chemical carrier for another element. **Magnesium nitrate,** a colorless crystalline powder of the composition $Mg(NO_3)_2 \cdot 6H_2O$, is made from magnesite and used in dry colors and pyrotechnics, and to produce magnesia. **Magnesium methoxide,** $Mg(OCH_3)_2$, is a white powder used for drying alcohol to produce absolute alcohol, and also for producing stable alcohol gels for use as solid alcohol fuels. The gels are made by adding water to an alcohol solution of the magnesium methoxide. **Magnesium fluoride,** MgF_2, or **sellaite,** is a pale-violet, crystalline powder that melts at 1396°C. It has a very low refractive index and is used on lenses and instrument windows to eliminate reflection. To apply the coating, the fluoride is dissolved in **dimethyl formamide,** $(CH_3)_2NCHO$, and mixed with an essential oil to wet the glass surface. The applied coating is fired at 500°C to leave a coating of pure magnesium fluroide about one-quarter light wavelength thick.

MAGNESIUM ALLOYS. Although magnesium is plentiful in nature, and is the lightest in weight of the structural metals, the use of magnesium alloys has been limited to applications where only moderate temperatures and low stresses are encountered. Alloying has been based largely on the conventional methods for iron and aluminum, but the molecular structure of the metal is very different, and requires selective alloying based on chemical combinations.

Present magnesium alloys, employing additions of aluminum, manganese, silicon, and silver, have a wide range of use in applications where temperatures do not exceed 300°F (149°C) and dynamic stresses are light. The alloys have good sound-damping qualities, have a high strength-weight factor, have high stiffness, and are resistant to alkalies. They are used for housings for machinery, hand tools, and office equipment, and for such items as ladders, loading ramps, marine hardware, and nonstressed lightweight structural parts.

Standard ASTM magnesium alloys are designated by letter symbols indicating the chief alloying elements with following numbers showing approximate percentage of the contained elements. Aluminum is represented as A, manganese as M, silicon as S, zirconium as K, thorium as H, silver as Q, and rare earths as E. Thus, **magnesium alloy AZ92A** contains about 9% aluminum and 2 zinc. The terminating letters signify physical treatments such as annealed or aged.

In general, the magnesium alloys have a strength-weight factor more than 10 times that of steel, and are easily machined. The AZ92A casting alloy has a tensile strength of 14,000 lb/in^2 (96 MPa) with elongation of 2%. In wrought form this alloy has a tensile strength of 50,000 lb/in^2 (344 MPa) with elongation of 8%.

Wrought alloys usually contain aluminum and manganese; forgings contain aluminum, manganese, and zinc; and die-casting alloys contain aluminum, manganese, and silicon. Aluminum up to 8% refines the grain, increasing hardness, strength, and rigidity; but with higher percentages the alloys become brittle unless modified with other elements. Small amounts of manganese give added corrosion resistance and higher strength. Silicon combines with the magnesium to give a dispersion of the compound Mg:Si:Mg, increasing the strength and heat resistance. The rare-earth metals are usually added as oxides, giving increased stability by a dispersion of high-energy particles.

Zinc, zirconium, lithium, and beryllium form their unit molecules in platelets similarly to magnesium, and the alloys form distinctive classes. **Magnesium alloy ZK60A** has a specific gravity of 1.83, melting point of 1175°F (634°C), tensile strength of 50,000 lb/in^2 (344 MPa), and elongation of about 8%. **Magnesium alloy HK31A** has a tensile strength of 31,000 lb/in^2 (213 MPa), but retains a tensile strength of 23,000 lb/ in^2 (158 MPa) at 500°F (260°C).

The **magnesium-lithium alloys** may contain up to 15% lithium and are called lithium alloys. The attractive feature is the light weight; they were originally developed for aerospace and military uses such as housings, vehicle protective armor, and ammunition containers. **Lithium alloy L14,** produced at Frankford Arsenal, containing 14% lithium and the balance magnesium, has a tensile strength of 14,500 lb/in^2 (102 MPa) with 30% elongation. When modified with 1% aluminum, 1 zinc, and 0.5 silicon, the die-cast metal has a tensile strength of 36,000 lb/in^2 (248 MPa) with elongation of 1%.

The **lithium alloy LA141A** marketed by Brooks & Perkins, Inc., has 14% lithium and 1 aluminum. It has a density of 0.045 lb/in^3 (1,245 kg/ m^3), a melting point of 1075°F (579°C), a tensile strength of 19,000 lb/ in^2 (130 MPa), and elongation of 10%. In general, the use of lithium alloys is limited to service temperatures below 300°F (149°C), and they are very plastic at elevated temperatures unless stabilized with high-energy molecular combinations.

Magnesium alloys may be marketed under company designations or trade names, but they generally follow the standard compositions with slight variations. **Magnesium alloy AM240,** of the Aluminum Co. of America, for permanent-mold castings, has up to 12% aluminum and 0.10 manganese. It has a tensile strength of 20,000 lb/in^2 (140 MPa) with elongation of 2%. **Revere alloy M,** for extruded bars, has up to 2% manganese, while **Revere alloy JS-1,** for sheet and plate, contains 5% aluminum, 1 zinc, and 0.25 manganese. This alloy has a tensile strength of 40,000 lb/in^2 (275 MPa) with elongation of 15%. **Eclipsalloy 56,** of Bendix, has about 9% aluminum, 2 zinc, 0.30 silicon, 0.10 manganese, 0.05 copper, and 0.01 nickel. The tensile strength is about 34,000 lb/in^2 (238 MPa), elongation 6%, and Brinell hardness 60. **Dowmetal AZ91C** is a casting metal of the Dow Chemical Co. with a similar composition but with less zinc. **Dowmetal**

F, for extruded rod, has 4% aluminum and 0.30 manganese. It has a tensile strength of 42,000 lb/in^2 (289 MPa). One of the earliest aircraft structural magnesium alloys was called **Electron.** The British Air Ministry specification of this alloy was for 1.3 to 1.7% manganese and the balance magnesium, with a tensile strength of 45,000 lb/in^2 (310 MPa). **Dowmetal M** is this alloy. **Dowmetal H** of Dow Chemical and **Mazlo alloy AM26S** of Aluminum Co. of America are of this type. **ASTM alloy AZ80X,** for forgings, is a modification of the **C metal** used in England for die castings. **Magnesium alloy ZA124** is a die-casting alloy of NL Industries containing 12% zinc, 4 aluminum, and 0.3 manganese. It has high creep and corrosion resistance. **Magnesium alloy AZ88,** of the same company, with 8.25% aluminum, 8.25 zinc, and 0.2 manganese, has excellent castability and can be used for complex die castings. C metal contains up to 9% aluminum, up to 1.5 manganese, with slight additions of silicon, copper, and nickel. The tensile strength is about 40,000 lb/in^2 (275 MPa) with elongation of 12%.

Magnesium-nickel is a master alloy of magnesium and nickel used for adding nickel to magnesium alloys and for deoxidizing nickel and nickel alloys. A magnesium-nickel marketed by Alloys & Products, Inc., contains about 50% of each metal, is silvery white in color, and is furnished in round bar form. **Magnesium-Monel** contains 50% magnesium and 50 Monel metal. Alloys of magnesium with nickel, Monel, zinc, copper, or aluminum, used for deoxidizing nonferrous metals, are called **stabilizer alloys.**

MAGNESIUM CARBONATE. A white, insoluble powder of the composition $MgCO_3$, containing also water of crystallization. The specific gravity is 3.10. It is made by calcining dolomite with coke, slaking with water, saturating with carbonic acid gas, and crystallizing out the magnesium carbonate. It is employed as an insulating covering for steam pipes and furnaces, for making oxychloride cement, in boiler compounds, and as a filler for rubber and paper. **Montax,** of the R. T. Vanderbilt Co., used as a filler, is a mixture of hydrated magnesium carbonate and silica powder. Magnesium carbonate is a good heat insulator because of the great number of microscopic dead-air cells in the material. The insulating material known as 85% magnesia has a density of 12 lb/ft^3 (192 kg/m^3), a thermal conductivity of [2.5 W/(m·°C)] at 100°F (38°C), and a conductivity of 0.46 Btu (485 J) at 400°F (204°C). As an insulating pipe covering it is usually mixed with abestos fibers to give structural strength. Federal specifications for magnesia call for the hydrated magnesium carbonate, $4MgCO_3 \cdot Mg(OH)_2 \cdot 5H_2O$, combined with 10% asbestos fibers, for use in heat-insulating blocks. The hydrated carbonate is a fine white powder called **magnesia alba levis,** slightly soluble in water, and used in medicine.

MAGNESIUM SULFATE. A colorless to white, bitter-tasting material occurring in sparkling needle-shaped crystals of the composition $MgSO_4 \cdot$

$7H_2O$. The natural mineral is called **epsomite,** from Epsom Spa, Surrey, England. In medicine it is called **epsom salt.** It is used in leather tanning, as a mordant in dyeing and printing textiles, as a filler for cotton cloth, for sizing paper, in water-resistant and fireproof magnesia cements, and as a laxative. It can also be obtained in the anhydrous form, $MgSO_4$, as a white powder. The specific gravity of the hydrous material is 1.678, and of the anhydrous 2.65. It occurs naturally as deposits from spring waters, and is also made by treating magnesite with sulfuric acid. In Germany it is produced from the mineral **kieserite,** $MgSO_4 \cdot H_2O$, which is abundant in the Strassfurt district, and is used as a source of sulfuric acid and magnesium. The magnesium sulfate from the waters at Seidlitz, Bohemia, was called **Seidlitz salt,** but **Seidlitz powder** is now a combination of Rochelle salt and sodium bicarbonate. **Synthetic kieserite** is made from the olivine of North Carolina. The mineral **langbeinite,** found in the potash deposits at Carlsbad, N.M., is a potassium-magnesium-sulfate containing 22% potassium oxide and 18 magnesia and is used in the production of potassium sulfate and magnesium metal. **Sulpomag** is langbeinite with the halite and clay washed out. When the magnesium of epsomite is replaced by zinc, the mineral is called **goslarite,** and when replaced by nickel, it is called **morenosite,** a green mineral occurring in nickel mines.

MAGNET WIRE. Insulated wire for the winding of electromagnets and for coils for transformers and other electrical applications. Since compactness is usually a prime consideration, **high-conductivity copper** is used in the wire, but where weight saving is important, aluminum wire may be used and has the advantage that an extremely thin anodized coating of oxide serves as the insulation either alone or with a thin varnish coating. Square or rectangular wires may be used, but ordinary magnet wire is round copper wire covered with cotton and an enamel, in OD sizes from No. 40, AWG [0.0071 in (0.018 cm)], to No. 8 [0.1380 in (0.351 cm)]. **Vitrotex,** of the Anaconda Wire & Cable Co., is a magnet wire coated with a resin enamel and covered with alkali-free glass fiber. It will withstand temperatures to 130°C, and the glass fibers dissipate heat rapidly. **Silotex,** of this company, has a silicon resin and glass-fiber insulation and will withstand temperatures to 300°C. Various types of synthetic resins are used as insulation to give high dielectric strength, heat and abrasion resistance, and flexibility to withstand bending of the wire without cracking. Heat resistance is usually designated by the AIEE class standards. **Formvar magnet wire** has a coating of vinyl acetal resin with dielectric strength of 1,000 volts per mil (39.4×10^6 volts per meter). The General Electric wire coating called **Alkanex** is a modified glycerol terephthalate polyester resin for operating tempertures to 155°C. **Bondar coating,** of Westinghouse, is an epoxy-modified polyester amide for temperatures to 155°C. **Carthane**

8063, of the Carwin Co., is a liquid urethane resin for flexible and abrasion-resistant coating on magnet wire.

MAGNETIC MATERIALS. Metallic and ceramic materials that become magnetized when placed in a magnetic field. All magnetic materials can be classified into two broad groups—soft magnetic materials and hard magnetic materials. **Soft magnetic materials,** sometimes called **electromagnets,** do not retain their magnetism when removed from a magnetic field. **Hard magnetic materials,** sometimes referred to as **permanent magnets,** retain their magnetism when removed from a magnetic field. Cobalt is the major element used for obtaining magnetic properties in hard magnetic alloys.

The common soft magnetic alloys are iron, iron-silicon alloys, and nickel-iron. **Irons** are widely used for their magnetic properties because of their relatively low cost. Typical **iron-silicon magnetic alloys** contain 1, 2, 5, and 4% silicon. There are about six types of nickel-irons, sometimes called **permeability alloys,** used in magnetic applications. For maximum magnetostriction the two preferred nickel contents are 42 and 79%. Additions of molybdenum give higher resistivities, and additions of copper result in higher initial permeability and resistivity.

Magnet steels are a major class of hard magnetic materials. **Tungsten magnet steels** contain 5 to 6% tungsten and about 0.65 carbon. They can be hot-forged, and machined when annealed. They are hardened in water and then tempered in hot oil. With increased carbon the coercive force increases, but the maximum induction and the residual magnetism decrease. The coercive force of a 6% tungsten steel is about 75 Oe (5,963 A/m). Molybdenum may be used instead of tungsten, but is employed usually only with other elements. The cobalt-chromium steels with some molybdenum have very high tensile strengths. **Comol** has 17% molybdenum, 12 cobalt, and 71 iron. **Indalloy** and **Remalloy** have similar compositions: about 20% molybdenum, 12 cobalt, and 68 iron. **Chromindur** has 28% chromium, 15 cobalt, and the remainder iron, with small amounts of other elements that give it improved strength and magnetic properties. In contrast to Indalloy and Remalloy, which must be processed at tempertures as high as 2280°F (1250°C), Chromindur can be cold-formed.

Chromium magnet steels are less expensive, and contain up to 5% chromium (usually 3), and about 1 carbon. A standard grade contains 2.25 to 4% chromium, 0.45 manganese, and 0.95 carbon. The magnetic properties are similar to those of the tungsten steels, and they have the same difficulty in hardening because of breakage when the carbon is high. **Tuncro,** of the Allegheny Ludlum Steel Co., is a tungsten-chromium magnet steel, and **Armat** is a magnet steel of the Jessop Steel Co. **Cobalt magnet steels** contain from 18 to 60% cobalt, part of which may be replaced by the less expensive chromium, or some tungsten may be used. Some cobalt magnet steels contain 1.5 to 3% chromium, 3 to 5 tungsten, 0.50 to 0.80 carbon, with high cobalt. **Alfer magnet alloys,** first developed in Japan to save cobalt, were **iron-aluminum alloys. MK alloy** had 25% nickel, 12 aluminum, and the balance iron, close to the formula Fe_2NiAl. It is age-hardening, and has a coercive force of 520 Oe (41,340 A/m) and maximum energy product of 1.35 million G·Oe (10,746 T·A/m). **Oerstit 400,** used by the Germans during the Second World War because it gave high coercive force in proportion to weight, contained 22% cobalt, 16 nickel, 8 aluminum, 4 copper, and the balance iron. **Cunife,** of Indiana General Corp., is a nickel-cobalt-copper alloy that can be cast, rolled, and machined. It is not magnetically directional like the tungsten magnets, and thus gives flexibility in design. The weight is 0.300 lb/in^3 (8,304 kg/m^3), the electric conductivity is 7.1% that of copper, and it has good coercive force. **Cunife 1** contains 50% copper, 21 nickel, and 29 cobalt. **Cunife 2,** with 60% copper, 20 nickel, and 20 iron, is more malleable. This alloy, heat-treated at 1100°F (593°C), is used in wire form for permanent magnets for miniature apparatus. It has a coercive force of 500 Oe (39,750 A/m). **Hipernom,** of Westinghouse Electric Corp., is a high-permeability nickel-molybdenum magnet alloy containing 79% nickel, 4 molybdenum, and the balance iron. It has a curie temperature of 860°F (460°C) and is used for relays, amplifers, and transformers.

In the **Alnico alloys,** hard magnetic alloys made by General Electric Co., a precipitation hardening occurs with AlNi crystals dissolved in the metal and aligned in the direction of magnetization to give greater coercive force. This type of magnet is usually magnetized after setting in place. **Alnico 1** contains 63% iron, 20 nickel, 12 aluminum, and 5 cobalt. The alloy is cast to shape, is hard and brittle, and cannot be machined. The coercive force is 400 Oe (31,800 A/m). **Alnico 2,** a cast alloy with 17% nickel, 12.5 cobalt, 10 aluminum, 6 copper, and the balance iron, has a coercive force of 560 Oe (44,520 A/m). The cast alloys have higher magnetic properties, but the sintered alloys are fine-grained and stronger. **Alnico 4** contains 12% aluminum, 28 nickel, 5 cobalt, and the balance iron. It has a coercive force of 700 Oe (55,650 A/m), or 10 times that of a plain tungsten magnet steel. **Alnico 8,** of the Crucible Steel Co., has 35% cobalt, 34 iron, 15 nickel, 7 aluminum, 5 titanium, and 4 copper. The coercive force is 1,450 Oe (115,275 A/m). It has a hardness of Rockwell C59. The magnets are cast to shape and finished by grinding. **Hyflux Alnico 9,** of the same coercive force, has an energy product of 9.5 million G·Oe (75,620 T·A/m). The magnets of this material, made by Indiana General Corp., are cylinders, rectangles, and prisms, usually magnetized and oriented in place. The **Alnicus magnets,** of the U.S. Magnet & Alloy Corp., are alnico-type alloys with the grain structure oriented by directional solidification in the casting which increases the maximum energy output. **Ticonal, Alcomax,** and **Hycomax** are alnico-type magnet alloys produced in Europe. Various other alloys of high coercive force have been developed for special purposes. **Silmanal,** with 86.75% silver, 8.8 manganese, and 4.45 aluminum, has a coercive force of 6,300 Oe (500,850 A/m), but a low flux density. **Platinax,** with 76.7% platinum and 23.3 cobalt, has a coercive force of 2,700. **Bismanol,** developed by the Naval Ordnance Laboratory, is a **bismuth-manganese alloy** with 20.8% manganese. It has a coercive force of 3,600 Oe (286,560 A/m), but oxidizes easily. **Cobalt-platinum,** as an intermetal rather than an alloy, has a coercive force above 4,300 Oe (341,850 A/m) and a residual induction of 6,450 G (0.645 T). It contains 76.8% by weight of platinum and is expensive, but is used for tiny magnets for electric wristwatches and instruments. **Placovar,** of the Hamilton Watch Co., is a similar alloy that retains 90% of its magnetization flux up to 650°F (343°C). It is used for miniature relays and focusing magnets. **Ultramag,** of the Mallory Metallurgical Co., is a **platinum-cobalt** magnet material with a coercive force of 4,800 Oe (381,600 A/m). The curie temperature is about 500°C, and it has only slight loss of magnetism at 350°C, whereas cobalt-chromium magnets lose their magnetism above 150°C. The material is easily machined. **Alloy 1751,** of Engelhard Industries, is a cobalt-platinum intermetal with a coercive force of 4,300 Oe (341,850 A/m), or of 6,800 Oe (540,600 A/m) in single-crystal form. The metal is not brittle and can be worked easily. It is used for the motor and index magnets of electric watches.

Soft magnetic ceramics, also referred to as **ceramic magnets, ferromagnetic ceramics,** and **ferrites (soft)**, were originally made of an iron oxide, Fe_2O^3, with one or more divalent oxides such as NiO, MgO, or ZnO. The mixture is calcined, ground to a fine powder, pressed to shape, and sintered. Ceramic and intermetal types of magnets have a square hysteresis loop and high resistance to demagnetization, and are valued for magnets for computing machines where a high remanence is desired. A ferrite with a square loop for switching in high-speed computers contains 40% Fe_2O_3, 40% MnO, and 20% CdO. Some intermetallic compounds, such as **zirconium-zinc,** $ZrZn_2$, which are not magnetic at ordinary temperatures, become ferromagnetic with properties similar to ferrites at very low temperatures, and are useful in computers in connection with subzero superconductors. Some compounds, however, are the reverse of this, being magnetic at ordinary temperatures and nonmagnetic below their transition temperature point. This transition temperature, or **Curie point,** can be arranged by the compounding to vary from subzero temperatures to above 212°F (100°C). **Chromium-manganese-antimonide,** $Cr_xMn_{2x}Sb$, is such a material. Chromium-manganese alone is ferromagnetic, but the antimonide has a transition point varying with the value of $_x$.

Vectolite is a lightweight magnet of Indiana General made by molding and sintering ferric and ferrous oxides and cobalt oxide. The weight is 0.114 lb/in_3 (3.2 g/cm^3). It has high coercive force, and has such high electrical resistance that it may be considered as a nonconductor. It is very brittle, and is finished by grinding. **Magnadur,** of the Ferrox Corp., was made from barium carbonate and ferric oxide, and has the formula $BaO(Fe_2O_3)_6$. **Indox,** of Indiana General, and **Ferroxdure,** of the Phillips Research Laboratories, are similar. This type of magnet has a coercive force to 1,600 Oe (127,200 A/m), with initial force to 2,600 Oe (206,700 A/m), high electrical resistivity, high resistance to demagnetization, and light weight, with specific gravity from 4.5 to 4.9. **Ferrimag,** of the Crucible Steel Co., and **Cromag,** of Henry L. Crowley & Co., Inc., are ceramic magnets. Strontium carbonate is superior to barium carbonate for magnets but is more costly. **Lodex magnets,** of General Electric Co., are extremely fine particles of iron-cobalt in lead powder made into any desired shape by powder metallurgy.

Ceramic permanent magnets are compounds of iron oxide with oxides of other elements. The most used are **barium ferrite, oriented barium ferrite,** and **strontium ferrite. Yttrium-iron garnets (YIG)** and **yttrium-aluminum garnets (YAG)** are used for microwave applications.

Flexible magnets are made with magnetic powder bonded to tape or impregnated in plastic or rubber in sheets, strip, or forms. **Magnetic tape** for recorders may be made by coating a strong, durable plastic tape, such as a polyester, with a **magnetic ferrite powder.** For high-duty service, such as for spacecraft, the tape may be of stainless steel. For recording heads the ferrite crystals must be hard and wear-resistant. **Ferrocube,** of Ferrox Corp., is manganese zinc. The tiny crystals are compacted with a ceramic bond for pole pieces for recorders. **Plastiform,** of the Leyman Corp., is a barium ferrite bonded with rubber in sheets and strips. **Magnyl,** of the Applied Magnetic Corp., is vinyl resin tape with the fine magnetic powder only on one side. It is used for door seals and display devices.

Rare-earth magnetic materials, used for permanent magnets in computers and signaling devices, have coercive forces up to 10 times that of ordinary magnets. They are of several types. **Rare-earth-cobalt magnets** are made by compacting and extruding the powders with a binder of plastic or soft metal into small precision shapes. They have high permanency. **Samarium-cobalt** and **cesium-cobalt magnets** are cast from vacuum melts and, as made by Bell Laboratories, are chemical compounds, $SmCo_5$ and $CeCo_5$. These magnets have intrinsic coercive forces up to 28,000 Oe (2.2 million A/m). The **magnetooptic magnets** produced by IBM for memory systems in computers are made in thin wafers, often no more than a spot in size. These are ferromagnetic ceramics of **europium-chalcogenides.** Spot-size magnets of europium oxide only 4 μm in diameter perform reading and writing operations efficiently. Films of this ceramic less than a wavelength in thickness are used as memory storage mediums.

Magnetic fluids consist of solid magnetic particles in a carrier fluid. When a magnetic field is applied, the ultramicroscopic iron oxide particles become instantly oriented. When the field is removed, the particles demagnetize within microseconds. Typical carrier fluids are water, hydrocarbons, fluorocarbons, diesters, organometallics, and polyphenylene ethers. Magnetic fluids can be specially formulated for specific applications such as damping, sealing, and lubrication.

MAHOGANY. A name applied to a variety of woods. All of the true mahogany, however, comes from trees of the family *Meliaceae,* but of various genera and species. The tropical cedars, Spanish cedar and Paraguayan cedar, belong to this family. Mahogany, of the tree *Swietenia mahogani* and other species of *Swietenia,* is obtained from Mexico to as far south as northern Argentina. The Central American has the best reputation and is frequently referred to under the Spanish name **caoba.** The mahogany from Cuba and Santo Domingo has a close grain and beautiful color and is valued for furniture. The so-called **Horseflesh mahogany** from Cuba is from the **sabicu tree. Baywood** is an English name originally applied to a superior, straight-grained mahogany from the shores of the Bay of Honduras. **Colombian mahogany** is the wood of the tree *Cariniana pyriformis* of northern South America. It resembles mahogany but is heavier and harder.

The wood of the mahogany tree is obtainable in large logs. It has a reddish color of various shades. The grain is often figured, and it has a high luster when polished. It seasons well, does not warp easily, and is prized for furniture and cabinetwork and for small patterns in foundry work. The weight varies from 32 to 42 lb/ft^3 (513 to 673 kg/m^3), and the hardness and closeness also vary. The beautiful curled-grain woods are from selected forks of the trees. The mahogany formerly used for airplane propellers, and used also for small boats and boat trim, is either African *Khaya* or American *Swietenia,* with average specific gravity kiln-dried of 0.50. The compressive strength is up to 1,760 lb/in^2 (12 MPa) perpendicular to the grain, and the shearing strength 860 lb/in^2 (6 MPa) parallel to the grain.

Australian red mahogany is from the tree *Eucalyptus resinifera* of Australia. It is hard, durable, of a dark-red color, with a coarse, open grain. **Crabwood,** used as a substitute for mahogany, is the wood of the **carapa tree,** *Carapa guianensis,* of Brazil and the Guianas. It has a deep-reddish-brown color with a coarse grain and weighs 40 lb/ft^3 (641 kg/m^3). This tree produces the seed nuts from which **carapa fat** is pressed and used for soap, candles, and as an edible fat. **Oleo vermelho,** from the tree *Myrospermum erythoxylum* of Brazil, is a fine-grained reddish cabinet wood similar to mahogany. The specific gravity is 0.954. It has an agreeable odor. **Cameroon mahogany** is from the tree *Bassia toxisperma* of West Africa. The kernels of the nuts of the tree yield about 60% of a yellowish-white semisolid oil known as **djave butter,** or **adjab butter,** used in Europe for soapmaking. It is used locally for food by steaming off the traces of hydrocyanic acid.

MALLEABLE IRON. A high-tensile-strength white cast iron produced by a long heat treatment of white chilled castings. Iron for malleable iron is usually melted in the reverberatory furnace, which gives it greater strength and ductility than iron melted in the cupola in contact with the fuel. The iron contains 1 to 1.5% silicon and is cooled rapidly to produce white iron. The castings seldom exceed 10 lb (4.5 kg) in weight. After casting, the parts are packed in annealing pots and subjected to an increasing temperature to about 1650°F (899°C), for a period of 48 to 60 h. They are then cooled slowly with temperature decreasing 8 to 10° F (4 to 5°C) per hour to below 1275°F (690°C). The resulting iron has the carbon in regular tiny particles instead of flakes as in gray cast iron. The ordinary American malleable iron called **blackheart iron** has a white shell and a dark core, the outside being completely decarbonized. It is tough and more ductile than the coarse-grained **whiteheart iron** made at higher annealing temperatures.

The iron for malleable iron must have enough silicon to promote graphitization of the iron carbide at sustained high temperature, and sufficient manganese to offset the stabilizing effect of sulfur.

Ferritic malleable irons are composed of a ferrite matrix interspersed with nodules of carbon. As a class, they have high toughness and ductility and excellent machinability. There are two standard grades: 32510 and 35018. Of the two, **malleable iron grade 32510** has the higher carbon content, ranging from 2.30 to 2.65%. The high carbon makes this grade extremely fluid and therefore suitable for thin and/or intricate castings. Sections as thin as ⅓₂ in (0.08 cm), in small areas, can be successfully cast. This grade is mainly used for automotive parts, agricultural implements, and electrical products. **Malleable iron 35018** has lower carbon content (2 to 2.45%), slightly higher strength, and somewhat higher ductility than grade 32510. However, in casting properties and annealability, the two grades are about equal. Typical applications of grade 35018 are railroad, high-pressure, and oil-field castings. An iron known as **quick malleable iron,** used for automotive engine castings, has 2.2% carbon, 1.5 silicon, 0.30 to 0.60 manganese, and 0.75 to 1 copper. The copper also increases the tensile strength. For ease of machining, malleable iron normally should have 2.60% or more of carbon, although less carbon is desirable for strength and ductility. An average composition is carbon, 2%; silicon, 0.60 to 1.10; manganese, under 0.30; phosphorus, under 0.20; and sulfur, 0.06 to 0.15, with 1% copper for corrosion resistance. Malleable iron is used for castings for implements, pipe fittings, building hardware, and small machine parts requiring strength. The tensile strength is a minimum of 50,000 lb/in^2 (344 MPa), and the Brinell hardness 115, with elongation 18%, but higher strengths may be obtained with small controlled additions of alloying elements. Chromium gives a stabilizing effect, and increased amounts of manganese may be used, with also a small amount of molybdenum. But the white irons made with considerable contents of silicon, manganese, nickel, plus heat treatment, are classed as **austenitic cast iron** and not as malleable iron.

The material known as **pearlitic malleable iron** is malleable iron made by controlled heat treatment which develops a matrix of pearlite with temper carbon nodules and from 0.3 to 0.9% combined carbon distributed in the matrix. It has tensile strengths from 65,000 lb/in^2 (448 MPa) with 10% elongation up to 100,000 lb/in^2 (689 MPa) with 2% elongation. The weight averages about 8% less than for forged steel. It machines more easily than steel, gives a fine finish, has good wear resistance and fatigue resistance, and has better damping than steel. It is used to replace steel forgings for such parts as universal-joint yokes, gears, and small crankshafts. The hardness is Brinell 200 to 260.

The so-called **certified malleable irons** are made to ASTM specifications. Other general names, such as **shockproof iron,** are used to designate irons with controlled upper levels of silicon and manganese, with elongation to 5%, increased strength, and hardness to 200 Brinell. Malleable irons are also sold under many trade names, such as **Belmalloy** and **Flecto metal.**

The **Promal,** of the Link-Belt Co., and **Supermal,** of the Jeffrey Mgf. Co., used for such parts as chain links, are specially processed to give high fatigue resistance with hardness to 170 Brinell. The **Z metal,** of the Ferrous Metals Corp., is a pearlitic malleable iron with tensile strength of 70,000 lb/in^2 (482 MPa), hardness 180 Brinell, and elongation to 18%. The **Cu-Z metal** contains about 1% copper, and other grades contain copper and molybdenum to add corrosion resistance, refine the grain, and increase the density, giving high strength and ductility.

MANGANESE. A metallic element, symbol Mn, found in the minerals manganite and pyrolusite, and with most iron ores and traces in most rocks. Manganese has a silvery-white color with purplish shades. **Distilled manganese,** with no iron, and with carbon and silicon not over 0.006% total, has a fine silvery-gray luster very resistant to corrosion. It is brittle but hard enough to scratch glass. The specific gravity is 7.42, melting point 1245°C, and weight 0.268 lb/in^3 (7,418 kg/m^3). It decomposes water slowly. It is not used alone as a construction metal. The electrical resistivity is 100 times that of copper or three times that of 18--8 stainless steel. It also has a damping capacity 25 times that of steel, and can be used to reduce the resonance of other metals.

Manganese is used in the steel industry as a deoxidizer and as a hardener, and nearly all steel now contains some manganese. For this purpose it is used largely in the form of ferromanganese. Manganese is also added to steel in considerable amounts for the production of wear-resistant alloy steels. **Manganese metal,** for adding manganese to nonferrous alloys, is marketed in crushed form containing 95 to 98% manganese, 2 to 3 max iron, 1 max silicon, and 0.25 max carbon, but for the controlled addition of manganese to nonferrous metals and to high-grade steels, high purity, 99.9% plus, electrolytic manganese metal is now used. Manganese metal has very high sound-absorbing properties, and **copper-manganese alloys** with high percentages of manganese are used as **sound-damping alloys** for thrust collars for jackhammers and other power tools.

Electrolytic manganese can be produced from low-grade ores by electrochemical methods, 99.9% pure metal. The material produced by the Electro Manganese Corp. from high-grade ores is designated as **electromanganese.** It comes in chips about 1/16 in (0.16 cm) thick in sizes larger than 1 in (2.54 cm) square. It is 99.97% min pure metal of high purity, of 150 to 325 mesh, employed for pyrotechnic and metallurgical uses. **Manganese tablets,** for use in steelmaking, are made by pressing electrolytically reduced powder in an inert atmosphere and then coating the tablets with ammonium chloride to prevent oxidation. **Manganese carbonyl,** used for vapor deposition of manganese coatings, is a yellow crystalline solid of the composition $Mn_2(CO)_{10}$, melting at 154°F (68°C) and soluble in common organic solvents.

MANGANESE ALLOY. A series of nonferrous alloys containing manganese, copper, and nickel. They can be forged readily and machined, and then age-hardened at 500 to 1000°F (260 to 538°C) to a hardness of 400 to 500 Brinell with tensile strengths of 200,000 lb/in^2 (1,378 MPa) and high wear resistance. **Chace alloy 720**, of the W. M. Chace Co., contains 60% copper, 20 manganese, and 20 nickel. It is soft and ductile, but can be hardened to 400 Vickers. The tensile strength of the forged metal varies from 98,000 to 220,000 lb/in^2 (675 to 1,516 MPa) depending upon the hardness. The specific gravity is 8.25. It is used for high-strength, corrosion-resistant parts and for springs. **Wyndaloy 720,** of the Wyndale Mfg. Co., is this metal in forgings. These alloys are also noted for their high electrical resistivity, the 20% manganese alloy having a resistivity of 300 to 500 Ω/cir mil ft, and a conductivity of only 2 to 3.5% IACS. It is thus used for resistor wire and strip.

The alloys have a high coefficient of thermal expansion, and the **high expansion alloy** with 72% manganese, 18 copper, and 10 nickel has the highest expansion of any of the common strong alloys. A high-manganese alloy, used for rheostat resistors and electrically heated expansion elements, is **Chace alloy 772.** It contains 72% manganese, 18 copper, and 10 nickel. It can be machined, drawn, and stamped. The cold-rolled metal has a hardness of 220 Vickers and a tensile strength of 115,000 lb/in^2 (792 MPa), with elongation 6.5%. It is nonmagnetic, has a heat conductivity of only 12% that of steel, and a vibration damping constant 25 times that of steel. The electrical resistivity is 1,050 Ω/cir mil ft. Manganese alloy **resistor wire** is produced by the Driver-Harris Co.

MANGANESE-ALUMINUM. A **hardener alloy** employed for making additions of manganese to aluminum alloys. Manganese lowers the thermal conductivity of aluminum but increases the strength and increases the contraction. Manganese up to 1.5% is used in aluminum alloys when strength and stiffness are required. A manganese-aluminum marketed by the Niagara Falls Smelting & Refining Corp. contains 25% manganese and 75 aluminum. **Manganese-boron** is another alloy used for deoxidizing and hardening bronzes. It contains 20 to 25% boron, with small amounts of iron, silicon, and aluminum. For deoxidizing and hardening brasses, nickel bronze, and cupronickel, **manganese copper,** or **copper manganese,** may be used. The alloys used contain 25 to 30% manganese and the balance copper. The best grades of manganese copper are made from metallic manganese and are free from iron. For nickel bronzes and nickel alloys the manganese copper must be free of both iron and carbon, but grades containing up to 5% iron can be used for manganese bronze. Grades made from ferromanganese contain iron. Manganese copper is usually marketed in slabs with notched sections, or as shot. It has a lower melting point than metallic manganese and is thus more easily dissolved in the brass or bronze. The 30% alloy melts at about 1600°F (871°C).

MANGANESE BRONZE. A brass containing iron and manganese which, because of its hardness and crystalline structure, is called a bronze. It casts more easily than aluminum bronze and is used for propeller blades, valve stems, engine frames, and machinery parts requiring high strength and resistance to seawater. Manganese is a deoxidizer in the alloys, but in excess, usually up to 3.5%, it hardens and strengthens the alloy, increases the solubility of the iron in the brass, and acts to stabilize the aluminum when this metal is used. Manganese has nearly the whitening power of nickel in copper alloys. **Turbadium bronze** was an old name used by the British Admiralty for manganese bronze containing 50% copper, 44 zinc, 1 iron, 1.75 manganese, 2 nickel, and 0.5 tin, used for casting propellers and marine parts. The original **Turbiston's bronze** contained 55% copper, 41 zinc, 1 aluminum, 2 nickel, and 1 iron. ASTM and Federal specifications for manganese bronze call for 55 to 60% copper, 38 to 42 zinc, up to 3.5 manganese, up to 1.5 tin, 1.5 aluminum, and up to 2 iron. This alloy has a minimum tensile strength of 65,000 lb/in^2 (448 MPa) and elongation 30% as cast; the wrought metal has a strength of 72,000 lb/in^2 (496 MPa).

The manganese bronze marketed as **Amcoloy 666** by Ampco Metals, Inc., in extruded round rods for gears, cams, shafts, and bushings, contains 2.5% manganese, 1 aluminum, 0.7 silicon, 0.40 max lead, 57 to 60 copper, and the balance zinc. The tensile strength is 80,000 to 88,000 lb/in^2 (551 to 606 MPa), with elongation 12 to 18%, and hardness Rockwell B87 to B91. Even very small amounts of lead decrease the strength of manganese bronze and lower the ductility. Phosphor copper is sometimes added to make the metal easier to pour. The most popular manganese bronze in the automotive and aviation industries for castings contains little manganese. It has 57% copper, 40.5 zinc, 1 aluminum, 1 iron, and 0.5 manganese. It is tough and wear-resistant, and has a tensile strength of 65,000 lb/in^2 (448 MPa) min. Manganese bronze has high shrinkage, 3/16 in/ft (1.5 cm/m), and large fillets are necessary between changes in section thickness. **Manganese-tin alloy,** used in England as a substitute for nickel silver, is a white alloy containing 16% manganese, 8 tin, and the balance copper. The tensile strength is 57,000 lb/in^2 (392 MPa), with elongation of 48%, but when cold-worked the strength is increased to 103,000 lb/in^2 (710 MPa) with elongation of 2%.

A **super manganese bronze,** used for aircraft engine parts, contains 69% copper, 20 zinc, 2 manganese, 2.5 iron, and 6.5 aluminum. It has a tensile strength of 110,000 lb/in^2 (758 MPa) and Brinell hardness 225 when heat-treated. A **manganese-aluminum brass,** under the name of **Hy-Ten-Sl,** marketed by the American Manganese Bronze Co., contains 66% copper, 19 zinc, 10 aluminum, and 5 manganese. The castings have a tensile strength up to 105,000 lb/in^2 (723 MPa), elongation 15%, and Brinell hardness 175. It is also made in wrought forms. The alloys known as **manganese casting brass** are usually Muntz metal containing a small amount of manganese. The original **manganese brass,** patented in 1876 under the name of **Parsons' alloy,** contained 56% copper, 41.5 zinc, 1.2 iron, 0.7 tin, 0.1 manganese, and 0.46 aluminum. An alloy used by one automotive company has 58% copper, 40 zinc, and 2 of a master alloy consisting of tin, iron, manganese, and aluminum. The lead content is not permitted to exceed 0.15%. The tensile strength is 70,000 lb/in^2 (490 MPa) and elongation 20%. It is used as a substitute for malleable iron or drop-forged steel. **Lumen manganese brass,** of the Lumen Bearing Co., is a 60–40 brass with 3% of the copper replaced with 1% iron and some manganese, tin, and aluminum.

MANGANESE ORES. Manganese is a widely dispersed metal, occurring in many ores and in many parts of the world. The ores are used largely for producing ferromanganese, but some low-grade ores are reduced electrolytically to the metal, and the oxide ores are used directly in dry batteries, glassmaking, and in the chemical industry. **Pyrolusite** is the most important manganese ore. It is a **manganese dioxide,** MnO_2, with a black color and a metallic luster. The specific gravity is 4.75 and hardness 2 to 2.5. It is mined in various parts of Europe, Australia, Brazil, Argentina, Ghana, Cuba, India, Canada, and the United States. It is valued for glass manufacture, and when used as a decolorizer for glass, pyrolusite has been called **glassmakers' soap.**

Some of the high-grade Montana and Ghana ores are used for batteries. **Battery-grade manganese** must be free of lead, copper, iron, and other impurities which are electronegative to zinc, and which would decrease the potential and the life of the dry cell. Battery manganese must also have the oxygen readily available, and should be poorly crystallized and consist of the gamma oxide known as **cryptomelane,** or a black pseudoamorphous powder. Pyrolusite normally has an orthorhombic crystal structure, but also occurs pseudoamorphic, or as **psilomelane,** a colloidal form of the oxide, and it is this material which is separated as the battery grade. But the natural material is an alteration product in the ore veins, and is irregular in quality. High-grade battery manganese of uniform purity is manufactured from low-grade ores by leaching the crushed ore with sulfuric acid and precipitating the heavy metal sulfides with barium sulfide, aerating to oxidize the iron and sulfur, and electrolyzing the solution to obtain MnO_2 on the graphite anodes. For use as a dry-cell depolarizer, it is ground so that 65% passes a 100-mesh screen. Synthetic manganese dioxide made by electrolysis of the sulfate or by chemical reduction of the carbonate shows an irregularly shaped amorphous structure under a microscope. It is more reactive and more uniform than the natural material, and gives a longer battery life with a smaller quantity. High-grade battery manganese is also made by reacting manganese sulfate with sodium chlorate at 200°F (93°C) in the presence of sulfuric acid, and the synthetic manganese oxides are now preferred for battery use.

The ore known as **bog manganese,** also called **wad,** is an impure mixture of MnO_2 and MnO, together with other oxides. It is a soft, friable mineral of black or brown color and is an impure psilomelane. The wad ore of Arkansas, used for making manganese sulfate, contains 15 to 50% manganese. **Manganblend,** or **alabandite,** is a natural **manganese sulfide,** MnS, and is an iron-black mineral with a specific gravity of 4 and a Mohs hardness of 3.5 to 4. This material in ground form is marketed by the Foote Mineral Co. for making amber glass. It is stable and produces a clear amber color without muddiness.

Rhodochrosite, found in several parts of the United States and in central Europe, is a **manganese carbonate,** $MnCO_3$, with usually some iron replacing part of the manganese. It has a rose-red to dark-brown color, with a vitreous luster, specific gravity 3.5, and hardness 3.5 to 4.5. It usually has a massive cleavable structure. **Manganite,** found with other manganese minerals and with iron, is an iron-black mineral of the composition $Mn_2O_3 \cdot H_2O$, containing theoretically 62.4% manganese. It is found in Germany, England, and in the Lake Superior region of the United States. **Hausmannite,** found in Washington state, is another hydrated oxide, $Mn_3O_4 \cdot 2H_2O$, containing theoretically 62.26% manganese. It is used for coating welding rods. **Bementite,** from the same area, is a hydrated silicate, $8MnO \cdot 7SiO_2 \cdot 5H_2O$, containing 40.8% manganese. The high silica makes it difficult to use. **Rhodonite,** $MnO \cdot SiO_2$, found in Colorado, contains 35% manganese so tightly bound chemically that it is difficult to separate by ordinary methods. It is vaporized with a high-intensity arc, and the simple oxides, MnO_2 and SiO_2, then condense.

In general, an ore for ferromanganese should contain at least 35% manganese. Much of the American ore contains only 5 to 10% manganese. Arkansas ores are low-grade, with as little as 18% manganese, and high-grade, with more than 70%. The low grades of Montana ore are concentrated by a nodulizing process up to 58% manganese. Manganese is also extracted from low-grade ores by a chemical process of leaching the pulverized ore with acid, treating with calcium chloride to remove calcium sulfate, and then with limestone, and filtering off the iron oxide. In Germany, low-grade ore is made into ferromanganese by first smelting to spiegeleisen and then treating part in an acid and part in a basic converter before mixing. The three grades of ore designated by the Metals Reserve Co. are: High grade, with 48% min manganese and 7 max iron; Low grade A, with 44% min manganese and 10 max iron; and Low grade B, with 40% min manganese and no maximum on iron. Chemical-grade manganese ore should have 80 to 90% MnO_2, equivalent to 51% min manganese, and not more than 2% iron. Indian ore is classed as First grade when it has 50% min manganese, Second grade with 48 to 50%, and Third grade with 45 to 48%. The battery-grade ore from Papua averages 86% manganese dioxide. The ore of Zaire averages 50% manganese and 4.5 min iron. By the Nossen process ores with as low as 11% Mn are converted to either metallurgical or battery-grade manganese dioxide. The process consists in leaching with nitric acid, evaporating the filtered manganese nitrate, and then decomposing in heated drums to form MnO_2, HNO, and some NO_2.

MANGANESE STEEL. All commercial steels contain some manganese which has been introduced in the process of deoxidizing and desulfurizing with ferromanganese, but the name was originally applied only to steels containing from 10 to 15% manganese. Steels with from 1.0 to 1.5% manganese are known as **carbon-manganese steel, pearlitic manganese steel,** or **intermediate manganese steel. Medium manganese steels,** with manganese from 2 to 9%, are brittle and are not ordinarily used, but steels with 2% or more of manganese modified with small amounts of molybdenum and chromium form an important class of high-strength, air-hardening, nondeforming tool steels. Typical of this class is **Lo-Air die steel,** which contains 2% manganese, 1.35 molybdenum, 1 chromium, 0.30 silicon, and 0.70 carbon. It hardens to Rockwell C58, with a tensile strength of 300,000 lb/in^2 (2,068 MPa) and elongation of 1%. The original **Hadfield manganese steel** made in 1883 contained 10 to 14.5 manganese and 1 carbon.

Manganese increases the hardness and tensile strength of steel. In the absence of carbon, manganese up to 1.5% has only slight influence on iron; as the carbon content increases, the effect intensifies. Air hardening becomes apparent in a 0.20 carbon steel with 1.5% manganese, and in a 0.35 carbon steel with 1.4% manganese. The manganese steels used for dipper teeth, tractor shoes, and wear-resistant castings contain 10 to 14% manganese, 1 to 1.4 carbon, and 0.30 to 1 silicon. The tensile strength is up to 125,000 lb/in^2 (861 MPa), elongation 45 to 55%, weight 0.286 lb/in^3 (7,916 kg/m^3), and Brinell hardness, when heat-treated, of 185 to 200. Cold working hardens this steel, and dipper teeth in service will work-harden to a hardness up to 550 Brinell.

High-manganese steels are not commercially machinable with ordinary tools, but can be cut and drilled with tungsten carbide and super-high-speed tools. The austenitic steels, with about 12% manganese, are exceedingly abrasion-resistant and harden under the action of tools. They are nonmagnetic. The coefficient of expansion is about twice that of ordinary steel. Various trade names are used to designate the high-manganese steels. **Rol-man steel,** marketed by the Manganese Steel Forge Co., contains 11 to 14% manganese and 1 to 1.4 carbon, and has a tensile strength of 160,000 lb/in^2 (1,120 MPa) and elongation up to 50%. **Amsco steel,** of the American Manganese Steel Co., contains 12 to 13% manganese and 1.2 carbon. The tensile strength is 125,000 lb/in^2 (861 MPa), and it will work-harden to 500 Brinell. **Tisco steel,** of the Taylor-Wharton Iron & Steel Co., has up to 15% manganese, and is used for rails and crossovers where high resistance to abrasion is needed. **Timang,** of this company, is a high-manganese steel made in the form of wire for rock screens. A German stainless type of steel, made without nickel, has 125% manganese. It is called **Roneusil steel. High-manganese steels** are brittle when cast and must be heat-treated. For castings of thin sections or irregular shapes where the drastic water quenching might cause distortion, nickel up to 5% may be added. The **manganese-nickel steels** have approximately the same characteristics as the straight manganese steels. Nickel is also used in high-manganese steel wire, and the hard-drawn wire has strengths up to 300,000 lb/in^2 (2,068 MPa). **Manganal** is a hot-rolled plate steel of high strength and wear resistance marketed by Joseph T. Ryerson & Son, Inc. It contains 11 to 13% manganese, 2.5 to 3.5 nickel, and 0.60 to 0.90 carbon. The tensile strength is 150,000 lb/in^2 (1,034 MPa). Pearlitic **nickel-manganese steel** contains only 1.25% manganese with 1.25 nickel. It has high yield point and ductility. A **manganese-aluminum steel,** developed by the Ford Motor Company, has 30% manganese, 9 aluminum, 1 silicon, and 1 carbon. Its tensile strength is 120,000 lb/in^2 (840 Mpa) with elongation of 18%, but it work-hardens rapidly, and when cold-rolled and heat-aged the tensile strength is 300,000 lb/in^2 (2,068 MPa) with a yield strength of 290,000 lb/in^2 (1,999 MPa). This alloy forms a special type of stainless steel, with high resistance to oxidation and sulfur gases to 1400°F (760°C).

Structural steels with 0.50% carbon and from 1 to 2 manganese have tensile strength above 90,000 lb/in^2 (620 MPa). **Martinel steel,** or **Martin elastic limit steel,** of Alfred Holt & Co., was an early English steel of this type. **D-steel,** developed by the British Admiralty for warship construction, contains 1.1 to 1.4% manganese, 0.33 carbon, and 0.12 silicon. The tensile strength is 96,000 lb/in^2 (661 MPa) and elongation 17%. Penn Central rails have 1.30 to 1.60% manganese and 0.65 carbon. **Man-Ten steel,** of the U.S. Steel Corp., is a medium-carbon, medium-manganese structural steel, with corrosion resistance twice that of carbon steel. The tensile strength of the steel is 75,000 lb/in^2 (517 MPa), with elongation of 20%. Steels containing 1.30 to 1.90% manganese replace more expensive alloy steels for automotive parts. Most mills now list these steels as special alloy machinery steels; those containing about 0.10% sulfur are designated as **manganese screw stock.**

The **SAE steel X1330** and **X1340** are of this type. **E.Z. Cut steel** plate, of Joseph T. Ryerson & Son, Inc., is a free-machining steel for molds, gears, and machine parts. It has 0.14 to 0.21% carbon, 1.15 to 1.4 manganese, and 0.17 to 0.23 sulfur. The tensile strength is 65,000 lb/in^2 (455 MPa) and elongation 30%, but when carbonized and water-quenched the tensile strength is 100,000 lb/in^2 (689 MPa). **Max-El No. 4 steel,** of the Crucible Steel Co., is a pearlitic manganese steel with a small amount of chromium and 0.75% carbon, used for spring collets and called **collet steel.** Slight amounts of chromium will increase the strength and hardness of the intermediate manganese steels. The **tank car steel** M-128, of Lukens Steel Co., is a **manganese-vanadium steel** with 0.25% carbon, up to 1.5 manganese, and 0.02 or more vanadium. It has a minimum tensile strength of 81,000 lb/in^2 (558 MPa) with elongation of 18%. This type of steel with up to 1.75 manganese is used for forgings.

MANGROVE. An extract from the bark of the mangrove tree, *Rhizophora mangle,* of Venezuela and Colombia, the **red mangrove,** *R. racemosa,* of Nigeria, the **East African mangrove,** *R. mucronata,* and other species of

Africa, the East Indies, southern Asia, and tropical America, used for tanning leather. In Java, it is called **baku bark.** The East African bark contains 22 to 38% tannin, and the Nigerian bark contains 15 to 29%, usually at the low level. The South American barks range in tanning content from 5 to 45%. In the **Brazilian mangrove,** the leaves contain 24%. The solid extract marketed in blocks contains 62 to 63% solids and 53 to 54 tannin. The liquid extract contains 25 to 35% tannin. Red mangrove contains a red coloring matter which is objectionable in tanning, but can be decolorized with albumin. **White mangrove** produces a pale pinkish-brown leather, fairly soft and of firm texture. Mangrove from East Africa is called **mangrove cutch,** and is sometimes erroneously referred to as wattle. The bark is sold in fibrous form and in pieces.

MANILA HEMP. A fiber obtained from the leaf stalks of the **abaca** plant, *Musa textilis,* a tree of the banana family growing in the Philippines, the East Indies, and Central America. It is employed for rope and cordage and is the strongest of the vegetable fibers. The fibers are also very long, from 4 to 8 ft (1.2 to 2.4 m), and do not stiffen when wet. It is thus valued for marine cordage. The best grades are light in weight, soft and lustrous, and white in color. The finest fibers, called **lupis fibers,** are used in the East for weaving into cloth. The plant grows to a height of 20 to 30 ft (6.1 to 9.1 m), with huge leaves characteristic of the banana. Each successive layer of leaves toward the stalk yields fibers that are lighter in color, higher in strength, and of finer texture than those outside. There are 15 grades.

MAPLE. The wood of maple trees native to the United States and Canada which include 13 species in the United States. Of these the **sugar maple,** *Acer saccharum,* is the most plentiful and the most important. Other names for this tree and wood are **hard maple** and **rock maple.** The wood is tough and hard, close-grained, and does not splinter easily. The heartwood is light reddish brown, and the wide sapwood is white. The wood has an average specific gravity when kiln-dried of 0.67, compressive strength perpendicular to the grain of 2,170 lb/in^2 (15 MPa) and shearing strength parallel to the grain of 1,520 lb/in^2 (11 MPa). **Black maple,** *A. nigrum,* is similar to sugar maple and is marketed with it. The **broadleaved maple,** also known as **bigleaf maple** and **Oregon maple,** is *A. macrophyllum,* and is the only species native to the western states. **Silver maple,** *A. saccharinum,* grows most extensively in the middle states. It is also called **soft maple, white maple, river maple,** and **swamp maple. Box elder,** *A. negundo,* grows over the northern states east of the Rocky Mountains. The **red maple** is *A. rubrum,* and the **vine maple** is *A. circinnatum.* The wood of the maples may be white or yellowish to brownish, and is close-grained and hard. It often has a curly, twisted grain. The weight is about 40 lb/ft^3 (641 kg/m^3). The wood of the soft maples is not as heavy or as strong as that of the sugar maple. Maple is used for furniture, cabinetwork, flooring, rollers, measuring rules, forming dies, shoe heels and lasts, and where a hard, fine-grained wood is needed. **Rose maple,** used in Australia for cabinetwork and paneling, is not a maple but is from the tree *Cryptocarya ethyroxylon.* The pinkish-brown wood has a wavy grain, weighs 45 lb/ft^3 (721 kg/m^3), is hard, and has a fragrant odor.

Maple sugar, used in confectionery, and in sweetening agents as **maple syrup,** is the boiled-down sap of the sugar maple tree, harvested by tapping the tree in the early spring. The sugar contains the calcium salt of succinic acid. The ratio of sap to sugar is 40:1, and an average production is 2 lb (1 kg) per tree, or 20 qt (19 L) of sap from a 15-in (0.4-m) tree. Maple sugar is produced chiefly in Vermont, New Hampshire, New York, and Canada. **Maple flavor** is made artificially by the reaction of an alpha amino acid with a reducing saccharide.

MARBLE. A compact crystalline limestone used for ornamental building, for large slabs for electric-power panels, and for ornaments and statuary. In the broad sense, marble includes any limestone that can be polished, including **breccia, onyx,** and others. Pure limestone would naturally be white, but marble is usually streaked and variegated in many colors. **Carrara marble,** from Italy, is a famous white marble, being of delicate texture, very white and hard. In the United States the marbles of Vermont are noted and occur in white, gray, light green, dark green, red, black, and mottled. A typical white **Vermont marble** slightly mottled with gray is 99% pure carbonates with only slight amounts of manganese and aluminum oxides and organic matter. But about 60% of American marble is quarried in Tennessee and Georgia. It is highly crystalline, and is colored white, gray, bluish, or pink. The 56-ton (51-metric ton) block in the tomb of the Unknown Soldier at Arlington, Va., is **yule marble** from Colorado. **Alabama marble** is a pure, low-porosity material of good statuary grade, mostly white. Much of the **Tennessee marble** is marked with stylolites of zigzag colored bands, and is used for floor tile. The Victoria pink and Cumberland pink marbles of Tennessee have low porosity and high compressive strength, about 17,000 lb/in^2 (117 MPa). The marbles of southern Uruguay are famous for great variety of beautiful colors, and they occur in immense blocks.

Marble has a specific gravity of about 2.72 and weighs about 170 lb/ ft^3 (2,723 kg/m^3) with compressive strengths from 8,000 to above 15,000 lb/ in^2 (55 to 103 MPa). It will ordinarily withstand heat up to 1200°F (649°C) without injury. **Translucent marble** is selected and processed marble, semitransparent to light. **Statuary marble** is always selected by experts who have had long experience in cutting. It must be of a single shade, and be free of hard or soft spots, iron inclusions, and other defects. **Marble chips** are irregular small graded pieces of marble marketed for making artificial stone. It is a by-product of marble quarrying. **Marble flour,** or **marble dust,** is finely ground chips used as a filler or abrasive in hand soaps and for casting. **Marbelite** is an **artificial marble** used for casting statues and small ornamental articles. It is made by heating potassium alum in water and adding about 10% heavy spar, and then casting in rubber molds. Marble dust may be added. **Exsilite,** of the Thermo American Fused Quartz Co., is a synthetic onyx marble in slabs as large as 2 by 3 ft (0.6 by 0.9 m), and up to 3 in (7.6 cm) thick. It is made by fusion of pure silica with mineral colors incorporated at high temperatures.

MASTIC. The gum exudation of the tree *Pistacia lentiscus,* called **Chios mastic,** and from the *P. cabulica,* called **Bombay mastic,** both small evergreens native to the Mediterranean countries. In ancient times the resin was highly valued for artists' paints and coating lacquers, adhesives, and for incense, dental cements, and as a chewing gum from which use it derives its name. Because of high cost, its use is now largely limited to fine art paints and lacquers and as an astringent in medicine.

Mastic is obtained by making an incision in the tree, a tree yielding 6 to 11 lb (2.7 to 5.0 kg) annually. There are two general grades, the purer resin adhering to the tree, and the resin collecting on the ground. It is easily soluble in turpentine but is more expensive than many other natural resins, and is used for high-grade pale varnishes for artwork. The name mastic is also erroneously applied to asphalt when used in calking or adhesive compounds.

MAURITIUS HEMP. The fiber obtained from the fleshy leaves of the plant *Furcraea gigantea,* of Mauritius, Nigeria, and Ghana, used for rope and cordage. The product from West Africa is often erroneously termed sisal. The plant belongs to the lily family. Similar fibers are obtained from other species, notably *F. foetida,* of Brazil. The *F. gigantea* is also grown in Brazil under the name of **piteira.** Each plant yields about 40 leaves annually, from 10 to 12 ft (3.0 to 3.7 m) in length, each leaf giving 35 g of fiber. The plant *F. cabuya* produced the ancient cordage fiber of the Mayas. The term **cabuya,** which means cordage, is applied to the fibers of the several species growing through Central America, the West Indies, and northern South America to Ecuador. The fibers of the *F. cabuya* of Costa Rica are up to 100 in (2.5 m) long. The leaves yield up to 3.5% of their green weight in dry fiber, which is coarser than henequen but is used for coffee-bag fabric. The fibers of the cabuya of Ecuador, *F. andina,* are not as long. They are used extensively for burlap for bagging. **Fique fiber,** of Colombia, used for rope and for coffee bags, is from the leaves of the *F. macro-*

phylla. The leaves are longer than those of henequen, and the fiber is finer and more lustrous.

MEERSCHAUM. A soft, white or gray, claylike mineral of the composition $3SiO_2 \cdot 2MgO \cdot 2H_2O$, used for making pipes and cigar holders, but also employed for making various other articles, as it can be cut easily when wet and will withstand heat. When fresh, the mineral absorbs grease and makes a lather; its German name means seafoam. It is used as a filler in soaps in Germany. The hardness is about 2 and the specific gravity 1.28. Most of the commercial meerschaum comes from Asia Minor; the mines at Eskisehir have been worked for 20 centuries. A little is produced in New Mexico and some in Spain. **Artificial meerschaum** is made from meerschaum shavings, kieselguhr, and from silicates of aluminum, calcium, and magnesium.

MELAMINE RESIN. A synthetic resin of the alkyd type made by reacting melamine with formaldehyde. The resin is thermosetting, colorless, odorless, and resistant to organic solvents. It is more resistant to alkalies and acids than urea resins, has better heat and color stability, and is harder. The melamine resins have the general uses of molding plastics, and also are valued for dishes for hot foods or acid juices and they will not soften or warp when washed in hot water. **Melamine,** a trimer of cyanamide, has the composition $(N{:}C \cdot NH_2)_3$.

It may be made by reacting urea with ammonia at elevated temperatures and pressure. It has a specific gravity of 1.56 and melting point 354°C. Melamine alone imparts to other resins a high gloss and color retention. The melamine resins have good adhesiveness but are too hard for use alone in coatings and varnishes. They are combined with alcohol-modified urea-formaldehyde resins to give coating materials of good color, gloss, flexibility, and chemical resistance. **Uformite MU-56,** of the Resinous Products & Chemical Co., is a urea-modified melamine-formaldehyde resin used for coatings and varnishes. **Melmac 1077,** of the American Cyanamid Co., is a melamine-formaldehyde molding resin with cellulose filler. It has a tensile strength of 7,500 lb/in^2 (51 MPa) and dielectric strength 325 volts per mil (12.8×10^6 volts per meter). **Melmac 592** has a mineral filler. It has a dielectric strength of 400 volts per mil (16×10^6 volts per meter) and will withstand temperatures to 300°F (149°C). **Melurac 300** is a melamine-urea-formaldehyde resin with a lignin extender used as an adhesive for water-resistant plywood. **Melmac 483** is a phenol-modified melamine-formaldehyde resin solution used for laminating fibrous materials. **Melmac 404** is a highly translucent melamine-formaldehyde resin for molding high-gloss buttons. **Lanoset** is a methylol-melamine made by alkylating a melamine-formaldehyde resin with methyl alcohol. It is used for shrinkproofing woolen fabrics. **Resimene 812,** of Monsanto, is a colorless melamine-formaldehyde resin powder that can be dissolved in water or ethyl alcohol, for impregnating paper or fabrics, or for laminating.

MENHADEN OIL. An oil obtained by steaming or boiling the fish *Brevoortia tyrannus,* caught along the Atlantic Coast of the United States. It was first called **porgy oil,** the Maine name for the fish. Other names for fish are **whitefish, fatback,** and **mossbunker.** The fish, when fully grown, are 12 to 15 in (30 to 38 cm) long, weighing about 1 lb (0.5 kg). They yield up to 15% oil, although fish from warm southern waters yield less oil. In May the fish migrate north to the New England coast, and they return south to below the Carolinas in November. An annual catch of 1.5 billion fish yields 10.2 million gal (38.6 million L) of oil and 103,000 tons (93,421 metric tons) of meal. Menhaden is not a desirable food fish because of its oily nature. The oil contains 27% oleic acid, 20 arachidonic, 16 clupanodonic, 17 palmitoleic, 7 myristic, and 1 stearic acid. It has an iodine number of 140 to 180, and a specific gravity of 0.927 to 0.933. It is used for dressing leather, mixing in cutting oils, and for making paint oils. It is also hydroxylated with acetic acid and used for making polyisocyanate and alkyd resins. Menhaden oil polymerizes easily, and the drying power is good, but it does not give an elastic film as do the vegetable oils. Its strong odor is due to the clupanodonic acid ester. The residue **fish meal** is sold for poultry feed and fertilizer. The meal is not as rich in vitamins A and D as that from some other fish, but as much as 15% can be used in poultry feed without producing a fishy taste in the eggs.

MERCURY. Also called **quicksilver.** A metallic element, symbol Hg. It is the only metal that is a liquid at ordinary temperatures. Mercury has a silvery-white color and a high luster. Its specific gravity is 13.596. The solidifying point is −40°F (−40°C), and its boiling point is 622°F (328°C). It does not oxidize at ordinary temperatures, but when heated to near its boiling point it absorbs oxygen and is converted into a red crystalline powder, **mercuric oxide,** HgO, used as a pigment in marine paints. Mercury is derived chiefly from the mineral cinnabar. Spain, Italy, Russia, Mexico, and the western United States are the chief producers. The metal is marketed in steel flasks holding 75 lb (34 kg). European flasks hold 76 lb (34.4 kg). It is used for separating gold and silver from their ores, for coating mirrors, as an expansive metal in thermometers, in mercury-vapor lamps, in tanning, in batteries, for the frozen-mercury molding process, mercury-vapor motors, as a circulating medium in atomic reactors, in amalgams, and in its compounds for fungicides, pharmaceuticals, paint pigments, and explosives. The black **mercurous oxide,** Hg_2O, is used in skin ointments. **Mercuric chloride,** or **corrosive sublimate,** $HgCl_2$, is an extremely poisonous, white crystalline powder soluble in water and in alcohol, used as a wood preservative, as an insecticide and rat poison, in tanning, as a mordant, and as a caustic antiseptic in medicine. **Vermilion red,** one of the oldest paint pigments, is red **mercury sulfide,** HgS, made directly by heating mercury and sulfur. It is a brilliant water-insoluble red powder of specific gravity 8.1. Because of its expense, it is often mixed with other red pigments. **Mercurochrome,** $C_{20}H_8O_6Na_2Br_2Hg$, is a green crystalline powder which gives a deep-red solution in water, and is used as an antiseptic. Mercury forms a vast number of compounds, all of which are poisonous and some of which are explosive. **Mercury 203** is radioactive.

METALLIC MATERIALS. About three-quarters of the elements available can be classified as **metals,** and about half of these are of at least some industrial or commercial importance. Although the word *metal,* by strict definition, is limited to the pure metal elements, common usage gives it wider scope to include **metal alloys.** While pure metallic elements have a broad range of properties, they are quite limited in commercial use. Metal alloys, which are combinations of two or more elements, are far more versatile and for this reason are the form in which most metals are used by industry.

Metallic materials are crystalline solids. Individual crystals are composed of unit cells repeated in a regular pattern to form a three-dimensional crystal-lattice structure. A piece of metal is an aggregate of many thousands of interlocking crystals (grains) immersed in a cloud of negative valence electrons detached from the crystals' atoms. These loose electrons serve to hold the crystal structures together because of their electrostatic attraction to the positively charged metal atoms (ions). The bonding forces, being large because of the close-packed nature of metallic crystal structures, account for the generally good mechanical properties of metals. Also, the electron cloud makes most metals good conductors of heat and electricity.

Metals are often identified as to the method used to produce the forms in which they are used. When a metal has been formed or shaped in the solid, plastic state, it is referred to as a **wrought metal.** Metal shapes that have been produced by pouring liquid metal into a mold are referred to as **cast metals.**

There are two families of metallic materials—ferrous and nonferrous. The basic ingredient of all **ferrous metals** is the element iron. These metals range from cast irons and carbon steels, with over 90% iron, to specialty iron alloys, containing a variety of other elements that add up to nearly half the total composition.

Except for commercially pure iron, all ferrous materials, both irons and steels, are considered to be primarily **iron-carbon alloy** systems. Although the carbon content is small (less than 1% in steel and not more than 4% in cast irons) and often less than other alloying elements, it nevertheless is the predominant factor in the development and control of most mechanical properties.

By definition, metallic materials that do not have iron as their major ingredient are considered to be **nonferrous metals.** There are roughly a dozen nonferrous metals in relatively wide industrial use. At the top of the list is aluminum, which next to steel is the most widely used structural metal today. It and magnesium, titanium, and beryllium are often characterized as **light metals** because their density is considerably below that of steel.

Copper alloys are the second nonferrous material in terms of consumption. There are two major groups of copper alloys: **brass,** which is basically a binary-alloy system of copper and zinc, and **bronze,** which was originally a copper-tin-alloy system. Today, the bronzes include other copper-alloy systems.

Zinc, tin, and lead, with melting points below 800°F (427°C), are often classified as **low-melting alloys.** Zinc, whose major structural use is in die castings, ranks third to aluminum and copper in total consumption. Lead and tin are rather limited to applications where their low melting points and other special properties are required. Other low-melting alloys are bismuth, antimony, cadmium, and indium.

Another broad group of nonferrous alloys is referred to as **refractory metals.** Such metals as tungsten, molybdenum, and chromium, with melting points above 3000°F (1649°C), are used in products that must resist unusually high temperatures. Although nickel and cobalt have melting points below 3000°F (1649°C), they are often considered as refractory metals also, because of their excellent heat resistance and their use in heat-resistant alloys.

Finally, the **precious metals** and **noble metals** families have the common characteristic of high cost. In addition, they have in common unusually high corrosion resistance and generally high density.

METALLIC SOAP. A term used to designate compounds of the fatty acids of vegetable and animal oils with metals other than sodium or potassium. They are not definite chemical compounds like the alkali soaps, but may contain complex mixtures of free fatty acid, combined fatty acid, and free metallic oxides or hydroxides. The name distinguishes the water-insoluble soaps from the soluble soaps made with potash or soda. Metallic soaps are made by heating a fatty acid in the presence of a metallic oxide or carbonate, and are used in lacquers, leather and textiles, paints, inks, ceramics, and grease. They have the properties of being driers, thickening agents, and flattening agents. They are characterized by ability to gel in solvents and oils, and by their catalytic action in speeding the oxidation of vegetable oils.

When made with fatty acids having high iodine values, the metallic soaps are liquid, such as the oleates and linoleates, but the resinates and tungstates are unstable powders. The stearates are fine, very stable powders. The fatty acid determines the physical properties, but the metal determines the chemical properties. Aluminum stearate is the most widely used metallic soap for colloid products. **Aluminum soaps** are used in polishing compounds, in printing inks and paints, for waterproofing textiles, and for thickening lubricating oils. The resinates, linoleates, and naphthanates are used as driers, the lead, cobalt, and manganese being the most common. **Calcium rosinate** is an insoluble soap of the composition $Ca(C_{44}H_{62}O_4)_2$ produced by boiling rosin with lime water, and filtering. It is also called **calcium abiotate,** and is a yellow powder of high molecular weight, 1349.06, used as a paint drier and for waterproofing. **Calcium linoleate** is a white amorphous powder of the composition $Ca(C_{18}H_{31}O_2)_2$ used in paints and in waterproofing. It is insoluble in water but soluble in alcohol. **Calcium stearate,** $Ca(C_{18}H_{35}O_2)_2$, is a white fluffy powder of 250 mesh. It is used as a flatting agent in paints, for waterproofing cements and stucco, as a lubricant for rubber and plastic molds, as a softening agent in lead pencils, and in drawing compounds for steelwire drawing. The calcium stearate of the Mallinckrodt Chemical Works, used in food and pharmaceutical emulsions, is an air-float powder of 325 mesh.

Barium stearate, $Ba(C_{18}H_{35}O_2)_2$, is a waxy nontacky white powder with molecular weight 703 and melting point 140 to 150°C. It is used as a dry lubricant for molding plastics, greaseless bearings, wax compounding, and wire drawing. **Chemactant PFC-5,** of Chemactant, Inc., is **barium lanolate,** a soft waxy soap, 25% barium and 75 lanolin acids, used as an additive for paints and coatings to improve adhesion and pigment dispersion. **Strontium stearate,** $Sr(C_{18}H_{35}O_2)_3$, is a white powder with molecular weight 654 and melting point 130 to 140°C. It is partly soluble and gels in benzol, mineral spirits, and hydrocarbons. It is used in grease and wax compounding and in crimson flares and signals. **Chromium stearate** is a dark-green powder of the composition $Cr(C_{18}H_{35}O_2)_3$ which melts at 95 to 100°C. It is used in ceramics and plastics, in plastic waxes and greases, and as a catalyst. **Manganese stearate** is a pink powder of the composition $Mn(C_{18}H_{35}O_2)_2$, with a melting point 100 to 110°C. It is used in wax and grease compounding. **Cerium stearate,** $Ce(C_{18}H_{35}O_2)_2$, is a very inert, waxy white powder melting at 100 to 110°C, used as a catalyst and in waterproofing compounds. **Nickel stearate** is a green powder, $Ni(C_{18}H_{35}O_2)_2$, with a melting point 150 to 160°C, soluble in aromatic hydrocarbons, and forming gels with petroleum oils. It is used in lubricants, waterproofing compounds, leather finishes, and as a flux in nickel welding.

Dibasic **lead stearate,** $2PbOPb(C_{18}H_{35}O_2)_2$, is a soft white powder of specific gravity 2.02, insoluble in water or mineral spirits. It is used in greases, cutting oils, and as a heat and light stabilizer in vinyl plastics. **Lithium stearate** is a white, odorless powder melting at 217°C. It is used in machinery greases and as an oil-soluble emulsifying and dispersing agent in cosmetics. The **lithium soap greases** are very adhesive to the bearings and are heat-resistant. The **Stanolith greases** of the Standard Oil Co. of Indiana are lithium soap greases containing an oxidation inhibitor. **Uni-Temp grease,** of Texaco, is a lithium soap grease made with a synthetic hydrocarbon instead of an oil. It has uniform lubricating qualities between −100 and +300°C.

METHANE. Also known as **marsh gas,** in coal mines as **firedamp,** and chemically as **methyl hydride.** A colorless, odorless gas, CH_4, employed for carbonizing steel, in the manufacture of formaldehyde, and as a starting point for many chemical compounds. The molecule has no free electrons and is the only stable **carbon hydride,** though it reacts easily on the No. 1 and No. 2 electrons of the carbon to form the hexagonal molecule called the **benzene ring.** It may thus be considered as the simplest of the vast group of **hydrocarbons** derived from petroleum, coal, and natural gas. Methane occurs naturally from the decomposition of plant and animal life, and is also one of the chief constituents of illuminating gas. It is made synthetically by the direct union of carbon or carbon monoxide with hydrogen. It is also produced by the action of water on **aluminum carbide,** a gray, massive substance of the composition Al_4C_3. Methane has a specific gravity of 0.560 and, since it is much lighter than air, it is easily diffused in it. In air the gas is highly explosive, although the gas alone is not explosive. It liquifies at −258.6°F (−161°C). **Pintsch gas,** once used widely for car lighting, contained up to 60% methane and was made by spraying petroleum oil into a hot retort. This type of gas, under the name of **oil gas,** or **carbohydrogen,** and containing as high as 85% hydrogen, gives a low-temperature flame used for flame-cutting torches. **Nitromethane** is nitrated methane. It is a liquid explosive more powerful than TNT. Methane is also the chief constituent of **natural gas** from oil fields. Natural gas contains usually at least 75% methane, although some Pennsylvania gas contains 98.8%, and some gas from Kentucky as low as 23%. A typical pipeline gas containing gas from several fields and freed of carbon dioxide, hydrogen sulfide, and water vapor has 78% methane, 13 ethane, 6 propane, 1.7 butane, 0.6 pentane, and some higher gases. **Re-formed gas,** used for copper refining, contains 86% methane, and is free of H_2S and from the higher homologs of CH_4. Natural gas has an energy value of 1,035 Btu/ft^3 (26,023 J/m^3), almost double that of manufactured gas from coal, but **synthetic gas,** or oil gas, made from crude oil, can be had with an energy value equivalent to that of natural gas. **Sour gas** is natural gas with more than one grain of H_2S per 100 ft^3 (2.8 m^3). This hydrogen sulfide is removed to eliminate the odor before being piped. The propane, butane, and heavier hydrocarbons may also be removed for the production of chemicals. Much of the American production of natural gas is in Texas and Louisiana, but there are large reserves in California and other areas and throughout the Canadian plains area.

MICA. Known originally as **Muscovy glass.** A group of **silicate** minerals with monoclinic crystals which break off easily into thin, tough scales, varying from colorless to black. **Muscovite** is the common variety of mica, and is called **potash mica,** or **potash silicate,** $H_2KAl_3(SiO_4)_3$. It has superior dielectric properties, and is valued for radio capacitors. **Phlogopite** is **magnesium mica,** $H_2KMgSi(SiO_4)_3$, and is distinguished from muscovite by its decomposition in sulfuric acid. It is also called **amber mica.** It is superior to muscovite in heat resistance, but is softer and has a brownish-yellow color. It comes from India, Canada, Malagasy, and Tanzania, while the chief producers of muscovite are India, Brazil, and Argentina. The peculiar crystal structure of phlogopite, and the almost infinite number of chemical combinations in which it can be produced, has made it an attractive mica for synthetic production which is generally replacing the natural product. The crystal structure is a repetition of four-layer units, and the layers are in ionic thicknesses of indefinite extent. It consists of extremely thin sheets of strongly bonded silica tetrahedra with a weak ionic bond joining the sheets.

The **chromium mica** known as **fuchsite** comes only in small emerald-green flakes. The rare greenish **vanadium mica** known as **roscoelite** is usually in fine scales, and where there are considerable amounts, as in the sandstones of Colorado, Utah, and Arizona, it is most valuable as a source of vanadium, as it contains 1.5 to 3.5% vanadium. **Margarite** is a yellow or purple **calcium mica,** $CaO(SiO_2)_2 \cdot (Al_2O_3) \cdot H_2O$, and is a transition product. The chief uses of mica are as an electrical insulator, a heat insulator, and a filler in plastics and insulating materials.

The value of sheet mica depends greatly upon the absence of staining, especially from iron inclusions which decrease the electrical insulating value. Most stains are black from the magnetite or other iron oxide. Reddish stains are usually red iron oxide. The brown-colored micas containing much iron are valueless as electrical insulators. While selected high-grade mica is an excellent insulating material, natural mica is, in general, unsuited for economic high-production use because of the difficulty of handling the small irregular pieces and the great wastage of time and material in the selection of uncontaminated pieces. India has been the largest producer of the highest grades of phlogopite, and about 80% of this production comes from Bihar, but less than 1% of Bihar trimmed block mica is suitable for such uses as condenser film. Argentine mica runs 80% stained, 15 semiclear, and only 5 clear.

The specific gravity of mica is 2.7 to 3.1, and the hardness from 2 to 3. Dielectric strength is not an accurate measurement of the electrical quality of mica, and it is measured by the **Q value,** or reciprocal of the **power factor,** the power factor being the measure of the loss of electric energy in a capacitor in which mica is the dielectric. High-quality capacitor mica should have a minimum Q value of 2,500, or a power factor of 0.04% at a frequency of 1 mHz. **Ruby mica** is the finest grade of **Indian mica** for electric capacitor use. **Madras mica** is greenish, and is not high-quality electric mica. Indian mica is graded as No. 5, first-quality and second-quality bookform; No. 6, loose with powder; extra loose, first-quality and second-quality; special loose, second-quality; and No. 6A loose,third-quality. **Argentine mica** has eight size grades from No. 6, in sheets of 1 to 3 in^2 (6.5 to 19.3 cm^2) to AA, or Special, in sheets of 48 in^2 (310 cm^2) or more.

Mica is marketed as cut or uncut block, sheet, splittings, and ground. **Block mica** is a deceptive term, since the pieces are not blocks. While Indian mica has been mined sometimes in blocks as large as 15 ft (4.6 m), and in sheets as large as 24 by 30 in (61 by 76 cm), the commercial block mica is usually no more than 0.030 in (0.076 cm) thick, and American importers designate block mica as pieces not less than 0.007 in (0.018 cm) thick with a minimum usable area of 1 in^2 (6.5 cm^2). **Mica splittings** have a maximum thickness of 0.0012 in (0.003 cm) and a minimum usable area of ¾ in (1.9 cm). About 85% of the mica imported into the United States is in thin splittings. **Film mica** is split from the best qualities of block mica to thicknesses from 0.0012 to 0.004 in (0.003 to 0.010 cm). Importers recognize 11 quality grades of block mica from the densely stained to the clear, all by visual inspection. Splittings, which are usually only 1 or 2 in (2.5 or 5.1 cm) in diameter, can be cemented together for use, but it is a costly operation. The original **micanite** made in India consisted of small pieces cemented together with shellac. **Built-up mica** is made by bonding the pieces with a synthetic resin and compressing at high temperature to give uniform thickness. Small pieces and the scrap from manufacture are made into mica powder for use as a filler in plastics and paints, or the powder is chemically cleaned and magnetically separated for making reconstituted sheet.

Because of the difficulty and cost of handling, natural mica sheet has now become almost obsolete for most uses. Sheets and strips made from mica flake can be handled in automatic machines, and are of a uniform quality not found in natural mica. **Reconstituted mica** without a binder is made by sheeting mica flakes by papermaking methods and submitting the sheet to high pressure and temperature to unite the flakes by recrystallization. Natural mica powder processed to give high purity may be used, but synthetic mica flake is preferred because of the ability to select a composition to suit the requirements for dielectric strength and heat resistance. In general, also, the absence of hydroxyl ions in synthetic mica gives higher heat resistance.

Mica sheet with a ceramic binder has been called **ceramoplastic insulation.** The **Synthamica 202,** of the Mycalex Corp., is made with a heat-resistant grade of synthetic mica flake with a glass bond. It comes in continuous strips 3 in (7.6 cm) wide in thicknesses from 0.002 to 0.007 in (0.005 to 0.018 cm). The dielectric strength is 1,000 volts per mil (39×10^6 volts per meter) operating temperature to 1800°F (982°C), and tensile strength 10,000 lb/in^2 (689 MPa). **Mykroy sheet,** of Electronic Mechanics, Inc., is glass-bonded mica in heavier sheets for panels and structural parts of electronic equipment. **Mica paper** and **mica mat** are usually made with an organic binder to give flexibility. One of the earliest mica papers, called **Watsonite,** was ground natural mica, dehydrated by heating, and sheeted with a resin binder. Many of the mica papers have superior dielectric strength, but the heat resistance is limited to that of the binder. **Crystal M,** of the Minnesota Mining & Mfg. Co., is a thin, flexible mica paper with a melting point at about 1900 °F (1038°C). The termal conductivity is from 0.30 to 1.5 Btu (317 to 1583 J), and it is used for fire-resistant thermal insulation. Mica paper made with about 90% mica flake and 10% epoxy resin has a dielectric strength of 1,300 volts per mil (51×10^6 volts per meter), power factor of 0.012, and flexural strength of 40,000 lb/in^2 (275 MPa). **Isomica,** of the Mica Insulator Co., is mica paper with an epoxy binder. **Transformer-grade mica,** for Class H insulation, may be made with a silicone binder. **Flexi-Mica,** of the Spruce Pine Mica Co., is mica bonded with a silicone resin. The 0.002-in (0.006-cm) sheet has a dielectric strength of 800 volts per mil (31×10^6 volts per meter) and a tensile strength of 10,000 lb/in^2 (68 MPa).

The first **synthetic mica** as made by Siemens & Halske was produced by melting a mixture of 11.6% aluminum oxide, 32.6 magnesium oxide, 30.7 kieselguhr, and 25.1 K_2SiF_6. The synthetic mica developed by the Bureau of Mines is a **fluorine-phlogopite mica** produced by calcinating a mixture of quartz, bauxite, and magnesite to drive off the carbon dioxide, adding potassium fluorosilicate, and melting at 1400°C. As the furnace cools, the mica crystals grow from a seed at the bottom of the crucible. The mica has the composition $K_4Mg_{12}Al_3O_{40}F_8$, and is similar to a natural mica in which the hydroxyl radical has been replaced by fluorine. **Fluorine mica** has superior heat resistance and dielectric strength, but it is harder and not as flexible as natural phlogopite mica. The number of combinations that can be produced in the fluorine type of phlogopite mica alone is very large. Potassium can be replaced by sodium, rubidium, calcium, cesium, strontium, barium, thallium, or lead. Magnesium can be replaced by iron, cobalt, nickel, manganese, titanium, copper, or zinc. Aluminum can be replaced by iron, manganese, vanadium, boron, or beryllium. **Boron phlogopite** as made synthetically is soft and flexible, with a melting point at 1150°C. Almost any desired combination of characteristics of heat resistance, electrical resistivity, and chemical resistance can be obtained within wide ranges in synthetic micas by varying the composition. However, sheet mica made synthetically embraces the same costly procedures of splitting and handworking of the relatively small sheets as for natural mica, so that synthetic mica is normally made as flake for reconstituting into sheet and strip of uniform purity and thickness.

Plastics are often heavily filled with mica powder and marketed under trade names such as **Micabond** and **Lamicoid** in the form of sheets, tubes, and molded parts, but they are distinct from the **mica ceramics** molded with an inorganic binder and having generally higher physical properties. One of the first of these, called **Mycalex,** produced by the Mycalex Corp. and the General Electric Co., was composed of the ground mica and lead borate heated together to the softening point of the borate, 675°C, and compressed while plastic. Part of the mica combines to form a lead borosilicate. Such a molding has good strength, water resistance, resistance to arcing, and a low coefficient of expansion. **Supramica 620BB,** of the Mycalex Corp., has a lead borate binder. The molded parts have a specific gravity of 3.8, a flexural strength of 12,000 lb/in^2 (82 MPa), hardness of Rockwell M110, and an operating temperature to 1200°F (649°C). Mica ceramics are made in various grades. Molded parts made with **boron mica** have a power factor below 0.07, and those of **barium mica** have power factors as low as 0.03. The traverse strengths of the mica ceramics are as high as 10,000 lb/in^2 (68 MPa) at 400°C.

The first wartime German **mica substitute,** called **Glushartgewebe,** was made by impregnating a very fine high-alkali glass-fiber fabric with an alcohol solution of osmose kaolin and a synthetic resin and compressing at high temperature. The General Electric **Terratex** was made from bentonite and asbestos fiber impregnated with ethyl silicate. Other substitute micas made from bentonite were **Alsifilm** and **Diaplex,** in thin, transparent hard sheets. **Tissuglas,** of the Amflex Products Dept., American Machine & Foundry Co., is made of extremely fine glass fibers matted on a papermaking machine into sheets 0.0006 to 0.012 in (0.0015 to 0.0304 cm) thick, in continuous rolls up to 38 in (97 m) wide. It withstands temperatures to 1200°F (649°C) and has higher dielectric strength than natural mica. **Fiberfilm,** of the same company, for such uses as capacitors and transformers, comes in thicknesses from 0.0008 to 0.0017 in (0.0020 to 0.0043 cm), and has a dielectric strength of 4,000 volts per mil (158×10^6 volts per meter) and an operating temperature to 250°C. It is made with fine glass fiber bonded with tetrafluoroethylene resin. It is stiffer and stronger than ordinary plastic film. **Glass paper** developed by the Naval Research Laboratory is made from borosilicate glass flake bonded with alkyd, phenolic, or silicone resin. Where temperatures are not high, various plastic films are used for slot insulation. **Anilite,** of the National Vulcanized Fibre Co., is a phenolic impregnate in sheets as thin as 0.004 in (0.010 cm). **Kynor,** of Pennwalt Corp., is vinylidene fluoride in thin sheet. The dielectric strength is 1,280 volts per mil (50.4×10^6 volts per meter), and heat distortion temperature 340°F (171°C).

MICA POWDER. Mica in very small flakes used as a filler in plastics, in paints, in roofing, and asphalt shingles, and for making glass-bonded mica. When produced by grinding the small scrap pieces from mica workings, it is known as **ground mica.** The ground mica from Canada is **phlogopite.** It comes in 20, 60, 120, and 150 mesh. Mica in paints helps to bond the film and prevent cracking, acting similarly to aluminum leaf. Mica powder for plastic filler and paints is produced directly from mica schists by froth flotation. The powder recovered averages more than 50% of sizes smaller than 200 mesh and 20% between 150 and 200 mesh. The recovered mica equals in every way the powder made by grinding mica scrap. **Sericite mica,** or **damourite,** from North Carolina kaolin workings, and in pockets in the fireclay deposits of Adelaide Co., Australia, is a type of potash mica related to muscovite but softer. It occurs as a finely divided powder with a talclike feel. Mixed with aluminum powder it produces a finish paint superior to aluminum alone. It has some activating properties, and is a substitute for zinc oxide in rubber. It is also used as a filler in plastics, and replaces graphite as a foundry core and mold wash. The recovery from North Carolina clay is 10% of the gross weight of the clay. **Water-ground mica** is ground to a fineness of 90% through a 325-mesh screen. It is for paint and rubber use. The mica powder **Micalith G,** of the General Mining Assoc., is sericite mica from Pennsylvania washings with 0.5% graphite embedded in the mica crystals. The graphite improves the wetting and dispersion in paints. **Micronized mica** is a powder of a fineness of 400 to 1,000 mesh, used as a filler. **Mica flake,** used in the manufacture of shingles and roofing, is washed from pegmatite deposits, but the mica flake used for molding into mica ceramic electric insulators is ground phlogopite scrap, or from various compositions of synthetic mica.

MICROSPHERES. Spherical particles used in plastics and other materials as fillers and reinforcing agents. They are made of glass or ceramics or resins. There are two different kinds of **glass microspheres**—solid and hollow. Solid spheres, made of soda-lime glass, range in size from 4 to 5,000 μm in diameter and have a specific gravity of about 2.5. Hollow glass microspheres have densities ranging from 5 to 50 lb/ft^3 (80 to 801 kg/m^3) and diameters from 20 to 200 μm. The strength of standard hollow glass spheres in terms of hydrostatic pressure required to reduce volume of the spheres by 10% is about 220 lb/in^2 (1.5 MPa). A stronger microsphere of 3M Co., known as **B40BX spheres,** withstands a hydrostatic pressure of 2,200 lb/in^2 (15 MPa) for a 10% volume loss. The spheres in plastics improve tensile, flexural, and compressive strength and lower elongation and water absorption. They also serve as thermal and sound insulators. **Plastic microspheres** are used mostly in the production of syntactic foams. **Polyvinylidene chloride microspheres** are excellent resin extenders and are used in sandwich construction of boat hulls. **Epoxy microspheres** are used as low-density bulk fillers for plastics and ceramics, and were developed for use in submerged deep-water floats. They can withstand hydrostatic pressures of 10,000 lb/in^2 (68 MPa). **Phenolic microspheres** filled with nitrogen are used for production of polyester foams and syntactic epoxy foams. **Polystyrene microspheres** are also used to produce syntactic foams.

MILK. The secretion of the mammary glands of mammals. Commercial milk in the United States is almost entirely from the cow. Besides its use as a food for direct consumption and in bakery products, it is employed for making cheese, butter, casein, ice cream, lactic acid, and lactose chemicals. The production of cow's milk in the United States from registered dairy animals averages about 6,100 lb (2,767 kg) per cow, but the output from cows under the Dairy Herd Improvement Assoc. is 8,000 lb (3,629 kg) per year per animal. The composition of milk is 87.34% water, 3.75 butterfat, 3 caseinogen, 0.4 lactalbumin, and 0.75 mineral salts. It also contains vitamins A, B, and C. The vitamin A content is low during the winter months when the animals do not feed on green grasses. The ultraviolet rays of sunlight may destroy the fat-digesting enzyme and the vitamins B and C in milk unless milk is protected in opaque bottles or in cartons.

Much of the American production of milk is used in foodstuffs manufactured in the form of **dried milk,** or **milk powder,** which is made largely from **skim milk,** that is, milk from which the fatty **cream** has been separated. One gallon (3.8 L) of milk yields 1.2 lb (0.54 kg) of dried milk. **Powdered milk** produced by spray drying in an air stream has 90% of its particles larger than 75 μm and will disperse more easily than a powder of fine particles. Milk powders processed above 180°F (82°C) may have a cooked taste, but powders with no off-taste are produced by evaporation at 95°F (35°C) and a quick dry at 160°F (71°C). **Milk proteins** may be separated from milk and used for enriching foodstuffs. **Crest 6S,** of the Crest Foods Co., is a milk protein powder containing 60% soluble protein and 26% lactose, with calcium and phosphorus. It also improves the whiteness and waterholding ability of milk powders. **Condensed milk,** formerly used widely where refrigeration was not available, is now used in confectionery to replace cream. It is very sweet, and contains 8.5% milk fats, 19.5 milk protein solids, 42 sugar, and 30 water.

While milk has many useful applications in cookery and commercial baking, it is essentially a natural baby food to build up the original low-calcium soft bones and to provide calcium for bones and teeth. Too much milk in the diet of adults may give an excess of calcium in the arteries and also cause cirrhosis or fatty growth of the liver. It may also cause an excess of lactic acid. The **Savortex** milk powder of the Western Dairy Products, Inc., for use in sausage and other comminuted meat products, has sodium substituted for a part of the calcium by ion-exchange methods. It is less greasy, and gives a smoother texture to the meats.

Pasteurized milk is milk that has been heated to kill disease organisms. Sterilization at 285°F (140°C) for a few seconds leaves better flavor and color than heating at low temperature for long periods. **Homogenized milk** is milk that has been treated by sonic vibration to break up the fat globules and distribute them evenly in the liquid. In natural milk the fat and proteins are in colloidal solution. Under the U.S. Food and Drug Act milk must contain not less than 3% milk fat and 8.5 solids not fat. In most countries the handling of milk is regulated by laws, as it is easily contaminated.

Cow's milk for direct consumption and for industrial use is produced on a large scale in the United States, Canada, Europe, Argentina, New Zealand, and Australia. **Kumiss,** sweetened cow's milk fermented with yeast, is used as a nutrient. **Soy milk** is a water solution of soybean solubles combined with sugar, calcium phosphate, irradiated sesame oil, and vitamins. It has the approximate composition of cow's milk and is chemically almost indistinguishable from it. It is used in foodstuffs, but is not permitted to be marketed under the name of milk. **Sesa-Lac,** of John Kraft Sesame Corp., is a **milk substitute** used by food processors in bakery products, creamed desserts, and beverages. It is made by crushing hulled sesame seed, homogenizing, and spray drying into a cream-colored powder. It has good nutritional value and has a pleasant sweet flavor. Another **imitation milk** is made from sodium caseinate, coconut oil, or other vegetable oils.

Lactic acid derives its name from the fact that it was originally obtained from milk. It has a wide industrial use and is now produced synthetically from sulfite pulp liquor, or by fermentation of carbohydrates. **Lactic acid** is a liquid of the composition $CH_3CHOHCOOH$, but a carbon may attach to four different groups, and the acid can thus exist in four forms. The dextrorotary isomer which occurs in milk is optically active. This form, called **sarcolactic acid,** is also produced naturally in the human muscles and joints by exercise to give the normal feeling of tiredness and induce sleep, but large amounts of lactic acid taken into the body are injurious. The dextrorotary acid made by the fermentation of sucrose is used in beverages and foodstuffs to minimize undesirable odors, and in animal feeds to improve the efficiency of protein utilization. **Polylactic resins,** made by the heat reaction of lactic acid with castor oil or other fatty oil, are soft, elastic resins used to produce tough, water-resistant coatings. Many useful chemicals are made from lactic acid. **Lactonitrile,** $HO(CH_3)CHCN$, is a colorless liquid of specific gravity 0.9834, which has the reactions of an alcohol and those of a nitrile.

MILLET. The very small seeds of a number of grasses. It is one of the most ancient of food grains, and is an important grain in Asia, being used as a food by a third of the population of the world. Nearly 40 million acres (161,876 million m^2) are cultivated to this grain in India alone. The plant is drought-resistant, but will not withstand frost. The seeds are high in phosphate, protein, minerals, and oil. In the United States millet seed is used as a birdseed, and the plant is grown for pasturage and silage. **German millet,** also known as **foxtail millet** and **Hungarian grass,** is from the grass *Setaria italica.* The seed contains phosphates and many minerals and vitamins A and E. It is an important food in Europe and Asia, but in North America it is a forage crop. **White millet,** or **proso millet,** from the grass *Pancium miliaceum,* is one of the richest grains in food value, but is employed in the United States only as a birdseed. It is much used in Russia as a food. The millet from the plant *Sorghum vulgare* is a staple food in India. **Sanwa millet,** from the *Echinochloa frumentacea,* is an important food in Japan. The plant will produce as many as eight forage crops per year.

MILLSTONE. Any stone employed for grinding paint, cement, grain, or minerals. Millstones are made from sandstone, basalt, granite, or quartz conglomerate. **Burrstone** is a millstone made from chalcedony silica of cellular texture, usually yellowish in color. The stone is used as a building stone. **Esopus stone** is a conglomerate of this type from Ulster County, New York. The most noted burrstones for grinding grains are the European stones cut in wedge-shaped blocks 6 in (15.2 cm) long, 3½ to 4 in (8.9 to 10.2 cm) wide, and 2 to 2½ in (5.1 to 6.4 cm) thick bound together with iron hoops. The French stones are chalcedony quartz, creamy white, with a cellular structure, the cavities formerly occupied by fossil shells. German burrstones are basaltic lava. Millstones vary greatly in sharpness of grain and size of grain, and thus synthetic stones of even texture are often preferred. **Chaser stones** are very large stones run on edge in mills for grinding minerals.

Pulpstones are blocks of sandstone cut into wheels and used for grinding, chiefly for the grinding of wood pulp in paper manufacture. The American pulpstones are produced in Ohio and West Virginia. The sandstones for pulpstones must be uniform in texture, have sharp grains, have medium hardness, and be composed of even quartz grains of which 85% will be retained on a 150-mesh screen, and 90% on a 200-mesh screen. The cementing material may be siliceous, calcareous, or argillaceous, but must be firm enough to hold the stone together when working under pressure, and soft enough to wear faster than the quartz grains and prevent glazing. The standard diameter of pulpstones is 54 in (137 cm) and width of face 27 in (69 cm). The stones are aged or seasoned from 1 to 2 years before use. Aging is quickened by heating the stones to about 180°F (82°C) in a closed room and cooling slowly. Large pulpstones for paper mills are now made of silicon carbide or aluminum oxide in fitted sections.

MINERAL WOOL. A fibrous material employed as a heat insulator in walls or as a sound insulator. It was first obtained as a natural product from volcanic craters in Hawaii and was known as **Pele's hair.** It is made by mixing stone with molten slag from blast furnaces and blowing steam through it. Slags from copper and lead furnaces are also used. **Slag wool** is made from slag without the rock. A lead slag containing 30 to 50% calcium and magnesium oxides makes a mineral wool that will withstand temperatures to 1500°F (816°C) when made into blocks or boards. Mineral wool usually consists of a mass of fine, pliant, vitreous fibers, which are incombustible and a nonconductor of heat. **Rock wool** is made by blowing molten rock in the same manner, and is more uniform than common mineral wool, with physical qualities depending, however, on the class of rock used. The rock wool marketed by Johns-Manville under the name of **Banrock** is made from high-silica limestone and is used for insulating oven walls for temperatures up to 1000°F (538°C). **Zerofil** is a nonsweating, low-temperature insulating material of the same company consisting of rock wool coated with asphalt. The rock wool of the Philip Carey Co., known as **Rocktex,** has a heat conductivity of 0.22 Btu/ (h)(ft^2)(°F) [0.69 W/(m^2)(°C)] difference in temperature. **Rock cork** is a name for a low-temperature insulating material made of rock wool molded in sheet form with a waterproof binder, used for walls in cold-storage rooms. **Mineral-wool board,** of the Armstrong Cork Co., is a moisture-resistant board in thicknesses of 1, 1½, 2, 3, and 4 in (2.5, 3.8, 5.1, 7.6, and 10.2 cm) for cold-storage insulation. The thermal conductivity is 0.31 to 0.33 Btu [0.98 to 1.04 W/ (m^2)(°C)] at 90°F (32.2°C). **Mono-Block,** of the Baldwin-Hill Co., is rock wool made into standard blocks and slabs by a felting process. It is used for both cold and heat insulation to temperatures up to 1600°F (871°C).

The weight is 18 lb/ ft^3 (288 kg/m^3). The heat conductivity at 200°F (93°C) is 0.325 Btu [1.0W/(m^2)(°C)].

Granulated mineral wool is the fiber milled into pellets of about ½ in (1.25 cm) in diameter. The pellets can be poured into a space for insulation, giving a density of 4 to 6 lb/ft^3 (64 to 96 kg/m^3), or they may be pressed into insulating blocks. **Supertemp block,** of the Eagle Picher Lead Co., is such a block compressed with a mineral binder and having tiny dead-air pockets. The thermal conductivity is 0.50 Btu [1.5 W/ (m^2)(°C)] and the service temperature is 1700°F (927°C). The block insulation of the Harbison-Walker Refractories Co., for backing up firebrick, is mineral wool compressed with a mineral binder to a density of 21 lb/ft^3 (336 kg/ m^3). The service temperature is 1900°F (1038°C). **Rockwool quilt** consists of felted fibers stitched between layers of kraft paper for wall insulation.

Selected minerals, to give various characteristics, may be used to make rock wools. **Wollastonite,** found in California and New York, is melted to produce a rock wool. It is also ground to a white acicular powder of 30 to 350 mesh for use as a filler in molded plastics. The grains are minute fibers. **Ceramic fiber** differs from mineral wool in that it is made of special composition. **Aluminum-silicate fiber** is made by melting alumina and silica and passing the molten material through a stream of high-pressure air which produces a fluffy mass of extremely fine fibers up to 3 in (7.62 cm) long. The melting point of the fiber is about 3000°F (1649°C), but the maximum usable temperature is about 2300°F (1260°C) as the fibers devitrify at higher temperatures. The fibers are used for thin insulating paper, panelboard, filler for plastics, chemical-resistant rope, or woven into fabrics. Aluminum silicate fabrics of 15 to 75 oz/yd (0.5 to 2.3 kg/m) are used for filters, gaskets, belting, insulation, and protective clothing. **Fiberfrax,** of the Carborundum Co., is an aluminum silicate fiber. **Dyna-Flex,** of Johns-Manville, is a **ceramic felt** made from aluminum silicate chromia fibers. It will withstand continuous temperatures to 2700°F (1482°C).

MOHAIR. The long, lustrous fleece of the **Angora goat,** *Capra angorensis,* important as a textile fiber because of its luster, length, strength, and spinning qualities. **Mohair fabrics** are used for upholstery material for hard service, and valued for summer wearing apparel, linings, and plushes. Army **necktie cloth,** notable for durability and noncreasing, has a cotton warp and a mohair filling. Mohair commonly contains shorter fibers that are coarse and difficult to dye. These are known as **kemp** and sometimes comprise 18% of the fiber. They are removed by combing. Mohair has a natural curl but no crimp, and does not felt as wool does. The **mohair fiber** used for fine paint brushes is usually from the kid and is of white color with a silklike luster and a slippery feel, and the scales are not sharply defined as in wool. The staple length is 5 to 8 in (12.7 to 20.3 cm), but the Turkish fiber is up to 10 in (25.4 cm) long. American production is mostly in Texas. The average weight of the clip is 5.3 lb (2.4 kg) per animal. South Africa produces much mohair.

MOLD STEEL. A term that generally refers to the steels used for making molds for plastics. Mold steel should have a uniform texture that will machine readily with die-sinking tools. It must have no microscopic porosity, and must be capable of polishing to a mirrorlike surface. When annealed, it should be soft enough to take the deep imprint of a hob, and when hardened it must be able to withstand high pressure without sinking and have sufficient tensile strength to prevent breakage of the small mold sections. It should not warp under ordinary heat-treating processes. For most plastic molding a case-hardening, low-carbon, low-alloy steel is used, but it must be of tool-steel quality and not of structural-steel grade. For hot sinking a typical analysis is 0.10% carbon, 0.50 manganese, 0.60 chromium, and 1.25 nickel. Such a steel has an annealed hardness of 120 Brinell, but will give a hardened case of Rockwell C62 and core of C21, with a tensile strength up to 120,000 lb/in^2 (827 MPa). For molds that are machined and not hobbed, a higher carbon content is used which increases the strength of the steel.

For severe pressures or resistance to abrasion, high-carbon manganese steels with 1.60% manganese may be used, and for high hydraulic pressures, alloy steels may be employed. When electrolytic methods of die sinking are used, the machining ability of the steel is not important, and it can be selected for its use characteristics. The **Cascade die steel,** of the Latrobe Steel Co., has 4.1% nickel, 1.2 aluminum, 0.20 vanadium, 0.25 chromium, 0.30 manganese, 0.30 silicon, and 0.20 carbon. Precipitation hardening gives this steel uniform hardness, and it polishes to a fine finish.

A carburizing steel for plastic molds is **SAE steel 3312,** containing 3.5% nickel and 1.2 chromium. **Carpenter 158 steel** has 3.5% nickel, 1.5 chromium, 0.40 manganese, and 0.10 carbon. It gives a surface hardness of Rockwell C62 with a core strength of 165,000 lb/in^2 (1,137 MPa). **Samson Extra,** of this company, is a deep-hobbing steel of good machinability and high core strength. It contains 0.10% carbon, 0.30 manganese, 0.20 silicon, and 2.30 chromium. When water-quenched to a hardness of Rockwell C65, it has a tensile strength of 120,000 lb/in^2 (827 MPa). **Hoballoy,** of the Crucible Steel Co., is a steel of this type with 0.10% carbon. **Plastalloy,** of Henry Disston & Sons, Inc., is a low-carbon, low-alloy, fine-grained steel for molds to be hobbed. **Speed-Cut steel,** of the Vanadium Alloys Steel Co., used for die-casting molds, will air-harden slightly. It contains 1.12% chromium, 0.85 manganese, 0.50 molybdenum, 0.30 silicon, 0.40 carbon, and is free of sulfur and phosphorus. When machined at the heat-treated condition up to 325 Brinell, it requires no further heat treatment, but can be annealed to 170 Brinell for hobbing.

MOLYBDENUM. A silver-white metal, symbol Mo, occurring chiefly in the mineral molybdenite but also obtained as a by-product from copper ores. About 90% of all molybdenum is produced in the United States, the remainder coming from Chile, Mexico, Peru, and Norway. The metal has a specific gravity of 10.2 and a melting point of 4750°F (2621°C). It is ductile, softer than tungsten, and is readily worked or drawn into very fine wire. It cannot be hardened by heat treatment, but only by working. The rolled metal has a tensile strength up to 260,000 lb/in^2 (1,792 MPa) with Brinell hardness 160 to 185. Above a temperature of 1400°F (760°C) the metal forms an oxide that evaporates as it is formed, but the resistance to corrosion is high, and molybdenum heating elements can be used to 3000°F (1649°C). The electric conductivity is 34% that of copper. The thermal expansion is low, and the heat conductivity is twice that of iron. It is one of the few metals that has some resistance to hydrofluoric acid. While unalloyed molybdenum in the soft state has a tensile strength of 97,000 lb/in^2 (668 MPa) with elongation of 42%, small amounts of other elements will harden and strengthen it greatly. **Molybdenum-titanium alloy,** with only 0.5% titanium, has a tensile strength of 132,000 lb/in^2 (910 MPa), and at 1600°F (871°C) retains a strength of 88,000 lb/in^2 (606 MPa). This alloy and one modified with an addition of 0.8 zirconium are used for tubing which will retain its shape to near its melting point.

Molybdenum, 99.95% pure, is used for support members in radio and light bulbs, heating elements for electric furnaces, arc-resistant electric contacts, and high-temperature structural parts for jet engines and missiles. Sheet is available as thin as 0.001 in (0.003 cm), and wire as fine as 0.004 in (0.010 cm). **Molybdenum ribbon** which retains its flexibility after high temperatures is produced by National Research Corp. As thin as 0.0016 in (0.0041 cm), it can be bent or twisted into small knots. **Molybdenum sheet,** of the General Electric Co., in thicknesses to 0.375 in (0.953 cm) can be deep-drawn and can be bent double without cracking. Molybdenum is used as a flame-resistant and corrosion-resistant coating for other metals, and may be arc-deposited. **Spraybond,** of the Metallizing Engineering Co., is molybdenum wire for this purpose. But protective coatings of molybdenum can be deposited by vapor deposition of **molybdenum pentachloride** reduced with hydrogen chloride. The pentachloride melts at 194°C and boils at 268°C, but a temperature of 850°C is used for deposition.

Molybdenum carbonyl, $MoCO_6$, may also be used for this purpose. Various molybdenum compounds are used for pigments, catalysts, and in chemical manufacture. **Sodium molybdate** in the anhydrous form, Na_2MoO_4, is used as a dry powder in fertilizers, and the sulfides and selenides of molybdenum are used in lubricants.

Most of the molybdenum used is employed in alloys or in alloy steels. For the latter purpose it is used in the form of **ferromolybdenum,** which is made directly from the ore by reduction with carbon, lime, and silicon, and adding iron. A standard grade of ferromolybdenum contains 50 to 60% molybdenum with up to 2.5% carbon, and is marketed on the basis of the contained molybdenum. **Calcium molybdate,** $CaMoO_4$, is also used for adding molybdenum to steels. It contains about 60.7% MoO_3, with the balance lime. The specific gravity is 4.35. **Molyte,** of the Molybdenum Corp., is calcium molybdate with a flux. It is heavier and will sink in the molten steel. Also, briquettes of **molybdic oxide,** containing 70 to 75% molybdenum trioxide and 12 carbon, are used for alloying steel. **Molybdenum-chromium,** produced by the Shieldalloy Corp., for making nonferrous alloys, contains 68 to 72% molybdenum, 28 to 32 chromium, with 0.50 max iron. **Climelt,** of the Climax Molybdenum Co., is a **molybdenum-tungsten alloy** for making high-temperature structural parts. It has a melting point at 5100°F (2816°C). It contains 70% molybdenum and 30 tungsten. The tensile strength is 121,000 lb/ in^2 (834 MPa) with elongation of 26%, and it retains a strength of 65,000 lb/in^2 (448 MPa) at 1800°F (982°C).

MOLYBDENUM CAST IRON. Cast iron containing small amounts of molybdenum added to the iron as ferromolybdenum or as calcium molybdate. Molybdenum in iron is not a carbide former or a graphitizer. It goes into direct solid solution and refines the matrix, increasing the strength, toughness, and wear resistance. Usually the manganese is increased when molybdenum is added, and the irons are more uniform in structure than plain cast iron. A plain molybdenum cast iron with 0.65% molybdenum has a tensile strength of 44,000 lb/in^2 (303 MPa) and Brinell hardness 223, which can be raised by heat treatment to above 60,000 lb/in^2 (413 MPa) with a hardness above 300 Brinell. Greater hardness can be obtained in the iron with the addition of small amounts of chromium. A chrome-molybdenum cast iron with 0.50% chromium, 0.50 molybdenum, and 1.25 manganese added to the cupola, resulting in an iron of 3.1% total carbon, 0.30 molybdenum, 0.33 chromium, and 0.75 manganese, has a tensile strength of 48,000 lb/in^2 (330 MPa), Brinell hardness 240, and an increase of about 40% in transverse strength over the plain iron. Raising the amount of molybdenum increases the hardness, but the machinability is kept by the addition of nickel or copper. **Nickel-molybdenum cast irons** have hardnesses as high as 300 Brinell without massive carbides that interfere with machining. An iron with 0.80% each of molybdenum and nickel, without chromium, has a tensile strength of 50,000 lb/in^2 (344 MPa) and a hardness of 270 Brinell. It has great uniformity, and can be cast in combined thin and heavy sections.

MOLYBDENUM ORES. Molybdenite is the chief ore of the metal molybdenum. It is a **molybdenum disulfide,** MoS_2, containing 60% molybdenum, occurring in granite, gneiss, and granular limestone. **Molybdenite** resembles graphite in appearance, with a lead-gray color, metallic luster, and greasy feel. The hardness is 1, and specific gravity 4.75. It is infusible. The American production of molybdenite is from Colorado, New Mexico, and Nevada, but about half of the molybdenum is a by-product of copper mining. **Wulfenite,** another important ore, is a **lead molybdate,** $PbMoO_4$, and occurs in lead veins with other ores of lead. It is found in Utah, Nevada, Arizona, and New Mexico. Wulfenite occurs in crystals and also massive granular. The specific gravity is 6.7 and hardness 3. Its color is yellow, orange, gray, red, or white. **Molybdite,** another ore, is a hydrous ferric molybdate of the composition $Fe_2O_3 \cdot 3MoO_3 \cdot 7H_2O$. It occurs either crystalline massive or as an earthy powder. It is yellowish, with a specific gravity of 4.5 and a hardness of about 1.5. Molybdenum ores are converted into ferromolybdenum or into calcium molybdate for use in adding molybdenum to steel. **Molybdenum trioxide,** MoO_3, the most important molybdenum compound for chemical manufacture, is made by heating molybdenite in air. It is a white crystalline powder which melts at 795°C. It is an electrical insulator, but becomes a conductor when molten.

MOLYBDENUM STEEL. Next to carbon, molybdenum is the most effective hardening element for steel. It also has the property, like tungsten, of giving steel the quality of red-hardness, requiring a smaller amount for the same effect. It is also used in hot-work steels, and to replace part of the tungsten in high-speed steels. It is added to heat-resistant irons and steels to make them resistant to deformation at high temperatures and to creep at moderate temperatures. It increases the corrosion resistance of stainless steels at high temperatures. Molybdenum in small amounts also increases the elastic limit of steel, reduces the grain size, strengthens the crystalline structure, and gives deep-hardening. It goes into solid solution, but when other elements are present it may form carbides and harden the steel, giving greater wear resistance. It also widens the heat-treating range in tool steels. As it decreases the temper brittleness of aluminum steels, small amounts are added to nitriding steels.

Plain **carbon-molybdenum steels** are easier to machine than other steels of equal hardness. **Molybdenum structural steels** usually have from 0.20 to 0.75% molybdenum. **SAE steels 4130** and **4140** contain about 1% chromium and 0.20 molybdenum, and are high-strength forging steels for such uses as connecting rods. **Allenoy,** used by the Allen Mfg. Co. for hollow-head screws, is **SAE 4150 steel.** SAE steels 4615 to 4650 have no chromium but contain about 1.75% nickel and 0.25 molybdenum. **SAE 4615** is used for molds and dies to be hobbed. It is easily worked. The alloying elements increase the rate of carbon pickup in carburizing, and the steel has a core hardness of 280 Brinell after hardening. **SAE 4650 nickel-molybdenum steel** is used for forming dies and, when hardened and drawn to a hardness of 435 Brinell, has a tensile strength of 215,000 lb/in^2 (1,482 MPa). **Hyten M steel,** of Wheelock, Lovejoy & Co., and **Durodi steel,** of A. Finkl & Sons, are high-strength nickel-chromium-molybdenum steels. Up to 3% molybdenum is used in stainless steels for cast parts for hot-oil and chemical equipment. **Lebanon 22-XM steel** has 19.5% chromium, 9 nickel, and 3 molybdenum. **Welmet,** of the Welland Electric Steel Foundry, is a similar steel.

The old **Damascus steel** and **Toledo steel** were molybdenum steels, the molybdenum being in the original ore. **Damascene steel** is a name referring to the wavy marks made on blades by forging and was not necessarily a molybdenum steel. But the original **Wootz steel,** or **Indian steel,** of this type, contained small percentages of aluminum incorporated in some unknown manner. Wootz steel was made in the crucible, although the crucible method was not used in Europe until 1740.

MONEL METAL. A natural alloy produced directly from Canadian bessemer matte obtained by reduction of the nickel ore, but now usually made by alloying. The average composition is nickel, 67%; copper, 28; iron, manganese, silicon, and other elements, 5. The alloy may be cast, rolled, or forged, and can be annealed after cold working. It is resistant to corrosion and to the action of many acids, and will retain its bright nickel-white surface under ordinary conditions. The melting point is about 2460°F (1349°C) and weight 0.318 lb/in^3 (8,802 kg/m^3). The tensile strength is 65,000 lb/in^2 (448 MPa) with elongation up to 50%, depending on the treatment. Cast metal may have tensile strengths to 100,000 lb/in^2 (689 MPa). The alloys are used for chemical and marine parts where corrosion resistance and white color are important. The synthetic alloys are widely known by letter symbols, but International Nickel Co. now uses numbers for regular grades. **Monel alloy 400,** the cold-roll-hardening alloy, is similar to K Monel. **Mond metal** was an early Monel with 4% manganese.

S Monel is an alloy of Monel metal with 3.75% silicon. It is used for valves and for castings subject to wear and corrosion. It has a hardness of 275 to 390 Brinell when heat-treated. **K Monel** is a modification of Monel metal that will age-harden. The nominal composition is 66% nickel, 29 copper, 2.75 aluminum, 0.9 iron, 0.50 silicon, 0.75 manganese, 0.50 titanium, and 0.15 carbon. The soft alloy, with a hardness of 145 Brinell, will have a hardness above 300 Brinell when heat-treated, and a tensile strength up to 160,000 lb/in^2 (1,103 MPa). It will retain its hardness up to 700°F (371°C) and is suitable for high-pressure steam valves. **R Monel** contains 67% nickel, 30 copper, and 0.35 sulfur. The cold-drawn rods have a tensile strength of 90,000 lb/in^2 (620 MPa), elongation 25%, and Brinell hardness 180. **H Monel** is a sandcasting alloy containing 63% nickel, 31 copper, 3 silicon, 2 iron, and 0.75 manganese. The tensile strength is 100,000 lb/in^2 (689 MPa), elongation 15%, and Brinell hardness 210. **Z Monel** contains 98% nickel, and is really a high-strength nickel. **Ebonized nickel** is Monel metal in commercial forms with a lustrous ebony finish obtained by an oxidizing process.

MORDANT. A substance used in dyeing for fixing the color. A mordant must have an affinity for the material being dyed and be chemically reactive; that is, the molecule must have free electrons that combine with the dyestuff. The vegetable fibers, such as cotton and linen, frequently require mordants. Basic aniline dyes require a mordant for cotton or rayon, and the water-soluble acid dyes need a mordant for vegetable fibers. The mordant may be applied first, usually in a hot solution, or simultaneously with the dye. **Mordant dyes** are dyes chemically able to accept and hold a metal atom, the metal being added in the mordant. The metal atom is held to the oxygen atoms and to one of the nitrogen atoms in each chromophoric group of the dye.

Besides fixing the color, mordants sometimes also increase the brilliancy of the dye, particularly when chelation occurs. Common mordants are **alum** and **sodium bichromate,** but salts of aluminum and other metals are used. **Sodium stannate,** used both as a mordant and for fireproofing textiles, is a gray-white water-soluble powder of the composition $Na_2SnO_3 \cdot 3H_2O$ made by treating tin oxide with caustic soda. **Chromium acetate,** used as a mordant for chrome colors, and to produce khaki shades with iron solutions, is a grayish-green powder of the composition $Cr(C_2H_3O_2)_2 \cdot H_2O$. It is soluble in water. In gilding, the term mordant is used to mean a viscous or sticky substance employed to make the gold leaf adhere, but such a material does not have the chemical action of a mordant.

MOTHER OF PEARL. The hard, brilliant-colored internal layer of the **pearl oystershell** and of certain other marine shellfish. It is employed for knife handles, buttons, inlay work, and other articles. Large oysters of the Indian Ocean, especially off Sri Lanka and in the Persian Gulf, furnish much of the mother of pearl. Other producers are Australia and the lands bordering the Coral Sea. The large Hawaiian pearl oyster is *Pinctada galtsoff.* The iridescent appearance is due to the structure of the nacre coating, and the shells are also called **nacre.** Mother of pearl is brittle but can be worked with steel saws and drills using a weak acid lubricant. The shells are thick and heavy, and disks cut from them are split for buttons. A beautiful pink nacre occurs on the inner surface of the **conch shell,** *Strombus gigas,* a sea snail which grows in great abundance off the Caicos Islands near Haiti. There are about 56 species, and they grow up to 12 in (30 cm) long and 8 in (20 cm) wide. The shells were formerly much used for making cameos. **Mussel shell,** from fresh-water mussels of the Mississippi River, are also called **pearl shell,** but they do not have the iridescence of mother of pearl. There is a large production in Iowa, used for buttons. The waste shell from the manufacture is crushed and marketed for poultry feed. **Pearl essence,** used for making **imitation pearls** and in plastics and lacquers, is a motley-silver compound extracted from the scales of fish, notably the sardine, herring, and alewife. Only a few types of fish have iridescent scales. The chief constituent is guanine, a chemical related to caffeine. One ton (0.907 metric ton) of scale is produced from 100 tons (90.7 metric tons) of Pacific herring, and this yields 1 lb (0.45 kg) of essence. **Nacromer,** of the Mearl Corp., is pearl essence. It gives high luminosity and iridescence to lacquers. **Synthetic pearl essence** may be ground nacreous shells in a liquid vehicle, or it may be produced chemically. Compounds of basic lead carbonate and lead monohydrogen phosphate have multiple reflectivity and give an optical effect resembling that of mother of pearl. **H-scale,** of the Celanese Corp., and **Ko-Pearl,** of the Ultra Ray Pearl Essence Corp., are synthetic pearl essences.

MUCILAGE. A sticky paste obtained from linseed and other seeds by precipitation from a hot infusion, and used as a light adhesive for paper and as a thickening agent. It contains arabinose, glucose, and galactose, and is easily soluble in water. Mucilage as a general name also includes **water-soluble gums** from various parts of many plants and has the same uses. It is the stored reserve food of plants. There are two types: the cell-content mucilage, which acts as a disorganization product of some of the carbohydrates, and membrane mucilage, which acts as a thickening agent to the cell wall. Membrane mucilage occurs in the acacias, algae (seaweeds), linun (flax plants), ulmus (elms), and astragalus. When it is collected in the form of exudations from the trees, it is called a gum. **Cherry gum,** from the *Prunus cerasus,* is this type of water-soluble gum, as is also **medlar gum,** from a small tree *Mespilus germanica,* of the same family as the cherry, grown in Europe and the Near East.

MULLITE. A mineral found originally in the Isle of Mull and employed as a refractory material for firebrick and furnace linings. The natural material occurs as fused argillaceous sediment inclusions in the mineral **buchite,** but it is rare and is produced artificially. It can be made by decomposing sillimanite or kyanite. **Artificial mullite,** or **synthetic mullite,** a ceramic material made by a prolonged fusing in the electric furnace of a mixture of silica sand or diasphoric clay and bauxite, has the composition $3Al_2O_3 \cdot 2SiO_2$. It is valued as a refractory because it does not soften below its high melting point, 3290°F (1810°C), and its interlocked grain will withstand continuous temperature changes without spalling. The bricks are resistant to flame and to molten ash, and have a low, uniform coefficient of thermal expansion and a heat conductivity only slightly above that of fireclays. Normally, mullite has very fine crystals which change form and become enlarged after prolonged heating, making the product porous and permeable. For stable high-temperature refractories the mullite is prefused to produce larger crystals. At very high temperatures mullite tends to decompose to form corundum and alkali-silicate minerals of lower heat resistance. Mullite is also made by burning silica clay and alumina at a very high temperature, and is used for making spark plugs, chemical crucibles, and extruding dies. For spark-plug cores it is fired at a temperature of 1450°C and aged before use. The tensile strength is above 9,000 lb/in^2 (62 MPa), or double that of porcelain, and it has high dielectric strength. The hardness is 6 to 7 Mohs.

Sillimanite, as well as **andalusite** found in California, has the composition $Al_2O_3 \cdot SiO_2$, with the same melting point and specific gravity of 3.20. **Diaspore andalusite** from Mono County, Calif., is used for spark plugs. **Andalusite crystals** of fine brilliance and of brick-red to yellow-green colors, found in Brazil, are used as gemstones. Sillimanite is decomposed to mullite and silica when heated above 1550°C. **Dumortierite,** produced in Nevada, has the composition $8Al_2O_3 \cdot B_2O_3 \cdot 6SiO_2 \cdot H_2O$. Deep-blue crystals of this mineral from the Colorado River resemble lapis lazuli and are cut as gemstones. A high-grade electric porcelain material marketed by Champion Sillimanite, Inc., under the name of **Champion sillimanite,** is a mixture of andalusite and dumortierite. **Shamra** is the trade name of a mullite refractory produced by the Mullite Refractories Co. **Mullfrax** is a mullite refractory of the Carborundum Co. **Tervex,** of Du Pont, is **foamed mullite** with a uniformly latticed honeycomb structure for light-weight, heat-resistant structural parts.

MUNTZ METAL. A yellow brass containing 60% copper and 40 zinc, invented in 1832 by George F. Muntz. In England it is called **yellow metal.** It is also known as **malleable brass,** and is now a standard product of the brass mills. Muntz metal has a tensile strength, when annealed, of 52,000 lb/in^2 (358 MPa), elongation 50%, and Rockwell B hardness 30. When hard-rolled, the strength is 85,000 lb/in^2 (586 MPa). The weight is 0.3015 lb/in^3 (8,304 kg/m^3). It is used for sheathing, marine fittings, bolts, and parts exposed to corrosion. Wrought Muntz metal is frequently modified with small amounts of tin to give greater hardness and corrosion resistance, and is then called **naval brass** or **naval bronze.** The usual composition is 60% copper, 39.25 zinc, and 0.75 tin. The **Roman bronze** of Revere Copper & Brass, Inc., has this composition, and the naval brass of the same company has 60% copper, 39 zinc, 0.75 tin, with 0.2 lead to make it easy to machine. Roman bronze has a tensile strength, annealed, of 60,000 lb/in^2 (413 MPa), and elongation 45%. When hard-drawn, the strength is 82,000 lb/in^2 (565 MPa), and elongation 20%. The leaded naval brass is slightly lower in strength.

Federal specifications for **naval brass rod** call for 0.5 to 1.5% tin and 0.20 lead. **Chamet bronze,** of the Chase Brass & Copper Co., has this composition. The naval brass marketed by the American Brass Co. under the name of **Tobin bronze** contains 59 to 60% copper, 0.5 to 1 tin, and the balance zinc. In rod form it has a fine-grained structure. The tensile strength of the soft rod is 52,000 lb/in^2 (358 MPa), and of the hard-drawn bar 67,000 lb/in^2 (461 MPa). The electric conductivity is 25% that of copper. A **forging bronze** used for corrosion-resistant marine parts, fittings, and hardware has 58.5% copper, 39.2 zinc, 1 tin, and 1 iron. It forges easily, and produces parts with a tensile strength from 60,000 to 75,000 lb/in^2 (413 to 517 Mpa). Bronzes with considerable amounts of tin are difficult to hot-work because of tin sweat, and because low-melting impurities such as lead and bismuth segregate in the grain boundaries.

Forging brass is usually Muntz metal containing up to 1.75% lead, with frequently some iron. With the copper content below 56% the brass is brittle; with copper above 63% the wear on dies is high. **Brass forgings** made in smooth dies are tough and compact and need no polishing, being simply pickled in a nitric acid bath to bring out the color. Forging brass in England is called **hot-stamping brass.** The British standard, **BES 218,** calls for 58% copper, 40.5 zinc, and 1.5 lead. The 60–40 brass modified with lead is called **extruding brass. Relleum brass,** of the Mueller Brass Co., for forgings, has 59% copper, 34 zinc, and 2 lead.

Muntz metal is also modified with small amounts of iron and manganese. **Delta metal,** first marketed by Alexander Dick in 1883, was a 60–40 brass modified with iron, manganese, and a small amount of lead. An earlier brass known as **Aich's metal** had 60% copper, 38.2 zinc, and 1.8 iron. Somewhat similar casting brasses were called **Gedge's metal** and **Sterro metal,** and a forging alloy was called **Mach's metal.** Iron above 0.35% forms a separate iron-rich constituent which is stable and gives hardness and high strength to the alloys. The addition of manganese helps to absorb the iron and also hardens the alloy. These alloys were also called **high-strength brass,** and were employed for such uses as hydraulic cylinders and marine forgings, but are now largely replaced by manganese bronze or aluminum bronze. **Durana metal** was a forging brass with 1.5% iron and 1.5 aluminum.

MUSIC WIRE. A high-grade, uniform steel originally intended for strings for musical instruments, but now employed for the manufacture of spiral springs. It is the highest grade of **spring wire** and is made of acid open-hearth steel or electric steel, free from slag or dirt, and low in sulfur and phosphorus. For springs it is usually **SAE steel 1085** reduced about 80% in 8 or 10 drawing passes, but piano wire may have as many as 40 draws from a No. 7 rod. The tensile strength of spring wire, when hard-drawn, is from 225,000 to 400,000 lb/in^2 (1,551 to 2,757 MPa), but it should be tough enough to bend 180° flat upon itself without cracking, or wind into a close helix with inside diameter 1 to 1½ times the diameter of the wire. The wire 0.187 in (0.475 cm) in diameter has an ultimate strength of 230,000 lb/in^2 (1,585 MPa); 0.015 wire has a strength of 400,000 lb/in^2 (2,757 MPa). The wire is usually marketed in gage sizes according to Washburn & Moen and the music-wire gages. Wire below 0.034 in (0.086 cm) in diameter (No. 15 gage) is furnished on reels. Larger sizes are in coils. Wire for springs is from No. 00, which is 0.0085 in (0.0216 cm) in diameter, to No. 36, which is 0.102 in (0.259 cm). The smallest size of music wire is 0.003 in (0.008 cm) in diameter. **Piano wire,** intended for piano strings but much used for springs, is a cold-drawn, high-quality wire formerly only drawn from Swedish billets or rods but now made from American steels. Piano wire ranges in diameter from 0.03 to 0.065 in (0.076 to 0.165 cm). The tensile strength is from 350,000 to 400,000 lb/in^2 (2,413 to 2,757 MPa). **Supertensile steel wire,** of the American Steel & Wire Co., for A strings of guitars, is a high-carbon steel made by long cold working and heat treatment. It has a tensile strength of 460,000 lb/in^2 (3,170 MPa), permitting a vibrational load of 37.7 lb (17 kg) on a 0.010-in (0.025-cm) wire. **Tire cord wire,** of this company, used for automobile tire plies and extra-thin conveyor belts, has a tensile strength of 350,000 lb/in^2 (2,413 MPa). The 0.0059-in (0.0149-cm) wire is stranded into a ⅓₂-in (0.079-cm) yarn and coated with rubber to make the ply. **High tensile wire,** of the National-Standard Co., in diameters of 0.005 in (0.0125 cm) and finer, used as a reinforcement in high-strength plastics, contains 0.90% carbon and has a tensile strength of 575,000 lb/in^2 (3,963 MPa). **Stitching wire,** for wire-stitching machines, is not as hard-drawn, and has a tensile strength of about 290,000 lb/in^2 (1,999 MPa).

While the modulus of elasticity of music wire is 30 million lb/in^2 (0.2 million MPa) compared with no more than 28 million lb/in^2 (0.2 million Mpa) for highly alloyed steels, the plain carbon steels cannot be used for springs operating at elevated temperatures, and the general-purpose stainless steels, such as Type 302, are limited to operation at about 500°F (260°C). **Alloy NS-A286,** of the National-Standard Co., a precipitation-hardening austenitic alloy used for springs for jet engines and gas turbines, contains about 25% nickel, 15 chromium, 2 titanium, 1.5 manganese, 1.25 molybdenum, 0.75 silicon, 0.3 vanadium, 0.08 max carbon, and not over 0.01 boron. The tensile strength of the wire with 30% reduction is 198,000 lb/in^2 (1,365 MPa) at 1350°F (732°C). **Alloy NS-355,** for highly stressed springs in jet engines, has a nominal composition of 15.65% chromium, 4.38 nickel, 2.68 molybdenum, 1 manganese, 0.32 silicon, 0.12 copper, 0.124 nitrogen, and 0.142 carbon. The 0.004-in (0.010-cm) wire has a tensile strength of 500,000 lb/in^2 (3,446 MPa). The modulus of elasticity is 29.3 million lb/in^2 (0.2 million MPa) at normal temperatures and 24.6 million lb/in^2 (0.17 million MPa) at 800°F (427°C).

MUSTARD. A pungent yellowish powder produced from the seeds of the **black mustard,** *Brassica nigra,* and the **white mustard,** *B. alba,* an annual plant of the turnip family grown in most temperate-climate countries. The ground seeds are treated with water, and the enzyme action yields a complex sulfur compound with a sharp taste and pungent action. It is used as a condiment, and in medicine as a counterirritant and emetic. It is one of the most ancient of the condiments and has a stimulating effect on the salivary glands. The product from the black mustard is more pungent than that from the white, and ground mustard may be a mixture of the two to give the desired pungency. **Mustard oil** is a volatile oil obtained from the seeds. It is a powerful vesicant and has a pungent aromatic odor. Also called **horseradish essence,** it is used in medicine and in flavors. Mustard seed which yields on expression the fixed **mustard-seed oil** used extensively in India as a food oil and in lubricants is *B. juncea.* It is closely related to rape oil, containing about 50% erucic acid. Another variety of *Brassica* of India yields **jamba oil,** which contains 46% erucic acid and is a substitute for rape oil for lubricants. Synthetic mustard oil is **allyl isothiocyanate.** It is a nasal irritant and lacrimator and small quantities are added to solvent glues as a sniffing deterrent.

MYROBALAN EXTRACT. A liquid or solid extract made from the dried, unripe fruit of several species of *Terminalia,* especially *T. chebala* of India and China. The *Phyllanthus emblica* of India also yields the fruit. The dried fruit, which resembles a plum, contains 30 to 40% tannin in the pulp. The fruit is graded and marketed as **myrobalans** chiefly on the basis of color, the lighter the color the higher the grade. The best grades of fruit are oval, pointed, and solid in structure. Inferior grades are round and spongy. The **bimlies** from Madras are rated best. Liquid myrobalan extract contains 25 to 30% tannin, and solid extract contains about 53%. It is used in tanning light leathers, and gives a quick tan. The natural acidity of myrobalan plumps the leather, but when used alone on heavy hides it makes a porous leather. It is used with other tannins.

NAPHTHA. A light, colorless to straw-colored liquid which distills off from petroleum between 70 to 90°C. The specific gravity is from 0.631 to 0.660, or slightly higher, ranging from C_6H_{14} to C_7H_{18}. The lightest of the distillates used as solvents for fats, rubber, and resins approaches petroleum ether; the heaviest distillates approach benzin and gasoline, and are used for fuel. **Benzin** is a light distillate ranging from C_8H_{18} to C_9H_{20}, with specific gravity from 0.635 to 0.660, and boiling-point limits from 120 to 150°C. **Petroleum ether,** or **petroleum spirits,** ranges from C_5H_{12} to C_6H_{14} and is distilled off between 40 and 60°C. Benzine is used as a solvent, as a cleaner, and in lighters. **Varnoline** and **white spirit** are old names for petroleum ether used as a general-purpose solvent, especially for varnishes.

The name naphtha is also applied to heavier distillates from petroleum and to various grades of light oils obtained in the distillation of coal tar. They are widely used because of their low volatility and safety in handling. These include **solvent naphtha,** having a specific gravity of 0.862 to 0.872, with a boiling point below 160°C, and **heavy naphtha,** a dark liquid of specific gravity between 0.925 and 0.950, and boiling point between 160 to 220°C. **High-flash naphtha** for solvent purposes is a petroleum fraction with a specific gravity within the range of the gasolines and a boiling point from 150 to 200°C. Various trade names are given to the light petroleum distillates used as solvents and in paints and varnishes, such as **Naphtholite. Solvesso,** of Humble Corporation, is a hydrogenated distillate in various grades with specific gravities from 0.797 to 0.937. **Stoddard solvent** is a standardized fraction of petroleum, or naphtha, used in dry cleaning or as a solvent. It is water-white in color, has a flash point above 100°F (38°C), and consists of the distillation fraction not over 410°F (210°C) with 50% below 350°F (177°C). **HiSolve VM,** of the Pennsylvania Industrial Chemical Corp., is a petroleum naphtha with an initial distillation point at 100°C, and an end distillation at 135°C. It is used as a thinner in paints, as an ink solvent, and in paint removers. **HiFlash naphtha,** of the same company, is a coal-tar solvent for rubber, for dry cleaning, and for varnish thinning. The distillation range is 145 to 185°C, and the specific gravity is 0.860.

NAPHTHALENE. Also called **tar camphor.** A white solid of the composition $C_{10}H_8$, which is one of the heavy distillates from coal tar, and may also be obtained from natural gas. Crude naphthalene has a melting point of 70 to 78.5°C. The pure crystalline flakes melt at 79.4°C and boil at

218°C, but vaporize slowly at room temperature. Naphthalene burns with a smoky flame, is soluble in benzene and in hot alcohol, but not in water. Refined naphthalene comes in balls, flakes, and pellets, largely for use as an insect repellent, but most of the material is sold in technical grades for use in making dyestuffs, synthetic resins, coatings, tanning agents, and Celluloid. **Naphthalene crystals** are very transparent to fluorescent radiation, and light produced in thick layers can escape and reach a photo surface. Low pulses can be obtained from the absorption of beta and gamma rays in naphthalene. It is thus used in photomultiplier tubes as a gamma-ray detector.

The **Halowax** and **Halowax oil** of the Koppers Co. consist of refined fractionated chloronated naphthalene ranging from low-viscosity oils through waxlike solids to hard resinlike solids. The melting range is 40 to 180°C. They are chemically stable, resistant to acids and alkalies, have high dielectric strength, are nonflammable, and are used as solvents, for waterproofing and fireproofing fabrics, and for electric-cable coatings. **Monoamyl naphthalene,** used to give plasticity to synthetic resins at low temperatures is an amber liquid, $C_{10}H_7C_5H_{11}$, specific gravity 0.96, and boiling point 279°C. On hydrogenation, naphthalene yields the solvents Tetralin and Decalin. **Tetralin,** $C_{10}H_{12}$, of Du Pont, has a specific gravity of 0.975 and boiling point 206°C. It is a clear liquid that is a good solvent for fats, oils, and resins. It will burn with a bright flame and can also be used as a fuel. The flash point is 78°C. **Decalin,** $C_{10}H_{18}$, is a liquid with specific gravity of 0.884 and boiling point 190°C.

NEATSFOOT OIL. A pale-yellow, inedible oil obtained by boiling the feet and shin bones of cattle in water, skimming the oil from the surface, and filtering. It was formerly highly valued for leather dressing and as a lubricating oil. For high-grade, cold-test lubricating oil for fine instruments the stearin is pressed out. The oil contains 67% oleic acid, 17 palmitic, 9 palmitoleic, 3 stearic, 1 myristic, 1 myristoleic, and 2 arachidonic and clupanodonic. The specific gravity is 0.916, iodine value 70, and saponification value 197.

NICKEL. A silvery-white metal with a yellowish cast first isolated in 1751, but used in alloy with copper since ancient times. Nickel has a specific gravity of 8.84, melts at 2646°F (1450°C), and is magnetic up to 680°F (360°C). It is marketed in grains or powder, in electrolytic sheets, blocks, shot, and in malleable forms. The metal is resistant to corrosion and to most acids except nitric. The electric conductivity is 16% that of copper. The tensile strength of hard-rolled sheet is 115,000 lb/in^2 (792 MPa), elongation 5%, and Rockwell B hardness 100. The tensile strength when annealed is 70,000 lb/in^2 (482 MPa), elongation 45%, and hardness 60. Commercially pure wrought nickel contains some cobalt which alters its physical properties. The nominal composition of Canadian nickel is 99.4% nickel including cobalt, 0.2 manganese, 0.1 copper, 0.1 carbon, 0.15 iron, and 0.05 silicon. The most commonly used commercial nickel is 99% pure, and is called **A nickel.** The very pure nickel used for sintered plates in nickel-cadmium batteries is produced by vaporizing nickel carbonyl and depositing the nickel as a powder. High-purity thin nickel strip, 99.9% pure, produced by Metals for Electronics, Inc., is rolled by a continuous process from metal powder. Commercially pure nickel resists the corrosive action of caustic soda and other alkalies even at temperatures of 600°F (316°C). It forms a thin coating of black oxide, which halts further corrosion. **Nickel 200,** of International Nickel, will corrode no more than 0.0001 in (0.0003 cm) per year from a 50% concentration of caustic soda.

Nickel is difficult to cast when pure as it absorbs oxygen and also dissolves carbon and sulfur. **Z nickel,** of the International Nickel Co., is a 98% nickel with alloying elements. The tensile strength, hot-rolled, is 105,000 lb/in^2 (723 MPa), with 15 to 35% elongation. When cold-drawn, the tensile strength is 120,000 to 175,000 lb/in^2 (827 to 1,206 MPa), elongation 15 to 25%, and Brinell hardness 220 to 340. The electric conductivity is 12% that of copper. **D nickel** contains 95.2% nickel and 4.5 manganese. In the annealed state it has a tensile strength of 75,000 lb/in^2 (517 MPa), elongation 40%, and Brinell hardness 140. **Duranickel,** of this company, used for springs, bellows, valve disks, and instrument parts, is an alloy containing 93.7% nickel, 4.4 aluminum, 0.5 silicon, 0.35 iron, 0.30 manganese, 0.17 carbon, and 0.05 copper. The hot-rolled metal has a tensile strength of 105,000 lb/in^2 (723 MPa) with elongation of 40%. When age-hardened, the tensile strength is 177,000 lb/in^2 (1,220 MPa) with elongation of 16%, Brinell hardness of 375, and electric resistivity of about 270. For springs it has high fatigue resistance and is not affected by heat to 550°F (288°C). A modification of the alloy having higher electric conductivity is called **Permanickel.** Nickel has its greatest use in alloys, particularly in alloy steels, cupronickel, nickel brasses, and German silver. It is also used to make white gold, 15% nickel changing the color of gold to white. Nickel steels have high strength and resistance. Nickel is also used in coinage alloys and in commercial heat-resistant and corrosion-resistant alloys.

Nearly 80% of the world's nickel production comes from Ontario, Canada, and much of the remainder from the garnierite ores of New Caledonia. The standard ASTM grades of virgin nickel are **Electrolytic,** containing 99.5% nickel; **X shot,** containing 98.9; **A shot,** with 97.75; and **Ingot,** with 98.5. But electrolytic nickel is available 99.95% pure, including not over 0.04 cobalt. **Nickel shot,** for adding nickel to iron in the ladle, is a master alloy of nickel and iron containing 50% nickel. **Sponge nickel** is a porous form of nickel made by leaching with caustic soda an alloy of 50%

nickel and 50 aluminum. The friable alloy is ground to 140 mesh before dissolving out the aluminum. **Raney nickel,** of the Raney Catalyst Co., is sponge nickel used as a hydrogenation catalyst. Dry **reduced nickel,** used as a catalyst for the hydrogenation of vegetable oils, is made by precipitating nickel hydroxide or nickel carbonate onto kieselguhr and reducing with hydrogen at high temperature. One lb (0.5 kg) of nickel is required to hydrogenate about 3,500 lb (1,588 kg) of oils, and the contaminated spent nickel is not recovered.

The **nickel powders** of Sherritt-Gordon Mines, Ltd., are 99.9% pure, graded in particle sizes from 1 μm to 50 mesh. E grade is used for vacuum tubes and semiconductor devices. **Nickel briquettes** of this company are pillow-shaped, compacted S-grade powder for alloying. The nickel used for the **nickel-cadmium battery** is **nickel hydroxide** NiO(OH). With cadmium as the negative plate and potassium hydroxide as the electrolyte it gives 1.35 volts per cell.

NICKEL ALLOYS. Any alloy containing nickel as the base metal, or as the chief alloying element. Nickel goes into solution in copper in all proportions and continually raises the melting point of copper alloys. In brasses and bronzes, nickel is used for the color effect and for toughening and strengthening the alloys. Nickel is employed in both ferrous and nonferrous alloys to produce heat-resistant and acid-resistant metals. Nickel-manganese alloys are used for electric-resistance wire. **Cold-resistant alloys** for use at subzero temperatures where ordinary steel would be brittle may be steels with a high proportion of nickel. **AMF alloy,** for liquid-air valves, was an early French alloy containing about 55% nickel, 3 manganese, 0.4 carbon, and the balance iron.

A corrosion-resistant nickel-base alloy of the Waukesha Foundry Co., for use as a bearing metal against stainless-steel shafts, and called **Waukesha metal,** contains about 80% nickel, 8 tin, 6 zinc, 4 lead, and 2 manganese. The tensile strength is 92,000 lb/in^2 (634 MPa), with elongation of 9% and Brinell hardness to 200. A group of **acid-resistant alloys** produced by the Burgess-Parr Co. under the names of **Illium** and **Parr metal** are complex alloys. **Illium G** contains 56% nickel, 22.5 chromium, 6.4 molybdenum, 6.5 iron, 6.5 copper, 1.25 manganese, 0.65 silicon, and 0.20 carbon. The density is 0.31 lb (0.14 kg), the melting point is about 2400°F (1316°C), and tensile strength of the cast metal is 68,000 lb/in^2 (468 MPa) with elongation of 7.5% and Brinell hardness 170. **Illium R** contains 64% nickel, 22 chromium, 5 molybdenum, 6 iron, 2.5 copper, 0.3 manganese, 0.15 silicon, and 0.05 carbon. This is a wrought alloy, and the tensile strength, annealed, is 113,000 lb/in^2 (779 MPa) with elongation of 54%. The cold-worked metal with 20% reduction and a hardness of Brinell 280 has a tensile strength of 142,000 lb/in^2 (979 MPa) with elongation of 23%.

A group of **nickel-chromium-molybdenum alloys** marketed by the Union Carbide Corp. under the name of **Hastelloy** in cast and wrought forms has high strength and resistance to chemicals at elevated temperatures. **Hastelloy A** contains 60% nickel, 20 molybdenum, and 20 iron. It has a specific gravity of 8.8, tensile strength, annealed, of about 115,000 lb/in^2 (792 MPa), and Brinell hardness 97. It casts easily and resists most acids except nitric. **Hastelloy B** has 28% molybdenum, 1 max chromium, 5.5 iron, 1 max manganese, 1 silicon, 0.12 carbon, and the balance nickel. The tensile strength, cast, is 80,000 lb/in^2 (551 MPa) with elongation of 8%. This alloy is resistant to molten chloride salts. **Hastelloy C** is resistant to nitric acid and to free chlorine. The nominal composition is 17% molybdenum, 16.5 chromium, 6 iron, 4 tungsten, 1 each of manganese and silicon, 0.15 max carbon, and the balance nickel. The tensile strength of the wrought metal is 130,000 lb/in^2 (896 MPa), with elongation of 45%, and that of the castings is 80,000 lb/in^2 (551 MPa) with elongation of 8%. **Hastelloy D** contains about 4% copper, 9 silicon, 1 manganese, 1 max each of chromium and iron, and 0.12 max carbon, with the balance nickel. It is resistant to hot sulfuric acid as it becomes coated with a resistant sulfate film. The tensile strength is 38,000 lb/in^2 (262 MPa). **Hastelloy X** is a high-temperature alloy to replace cobalt alloys. The nominal composition is 45% nickel, 22 chromium, 9 molybdenum, 0.15 carbon, and the balance iron. It comes in sheets, rods, and wire, and the annealed metal can be formed. At room temperature the tensile strength is 71,000 lb/in^2 (489 MPa), and at 1200°F (649°C) it is 52,000 lb/in^2 (358 MPa) with elongation of 16%. It has high oxidation resistance, and has been used for ceramic pressure molds at 2200°F (1204°C). **Hastelloy R-235** is a wrought alloy with a tensile strength of 167,000 lb/in^2 (1,151 MPa) and elongation of 43% that retains a strength of 90,000 lb/in^2 (620 MPa) at 1500°F (816°C). Its strength is derived from the formation of Ni_3Al and $Ni_3(AlTi)$ compounds. It contains 15.5% chromium, 5.5 molybdenum, 10 iron, 2.5 cobalt, 1 silicon, 1 manganese, 2 aluminum, 2.5 titanium, and 0.16 carbon, with the balance nickel.

An age-hardening strip metal, **Ni-Span C alloy,** developed by the International Nickel Co., contains 5.5% chromium, 2.5 titanium, 0.03 carbon, and the balance nickel. The cold-worked metal has a tensile strength of 200,000 lb/in^2 (1,378 MPa), elongation 7%, and Brinell hardness 395. **Inconel W,** of this company, used for springs at temperatures to 1000°F (538°C), has 70% nickel, 5 to 9 iron, 0.4 to 1 aluminum, 2 to 2.75 titanium, 17 chromium, 1 manganese, 0.70 silicon, 0.50 copper, 0.08 carbon, and 0.01 sulfur. The annealed metal has a tensile strength up to 160,000 lb/in^2 (1,103 MPa) with elongation 15 to 35%, while the spring-hard metal has a tensile strength up to 275,000 lb/in^2 (1,896 MPa) with elongation 2 to 5%.

A series of **paramagnetic alloys,** called **Nitinol,** developed by the Naval Ordnance Laboratory, are intermetallic compounds of nickel and titanium rather than **nickel-titanium alloys.** The compound TiNi contains theoretically 54.5% nickel, but the alloys may contain Ti_2Ni and $TiNi_3$ with about 50 to 60% nickel. The TiNi and nickel-rich alloys are paramagnetic, with a permeability value of 1.002, compared with the unity value of a vacuum. A 54.5% nickel alloy has a tensile strength of 110,000 lb/in^2 (758 MPa) with elongation of about 15%, and hardness of Rockwell C35. The alloys close to the TiNi composition are ductile and can be cold-rolled. The nickel-rich alloys are hot-rolled. They can be hardened by heat treatment to give hardnesses to Rockwell C68 and tensile strengths to 140,000 lb/in^2 (965 MPa). This class of alloy can also be modified with small amounts of silicon or aluminum, forming complex intermetallic compounds that can be solution-treated.

The Nitinols, with nickel content ranging from 53 to 57%, are known as **memory alloys** because of their ability to "remember," or return to, a previous shape upon being heated. This unusual behavior stems from a diffusionless transformation of the alloy. These shape-memory alloys have excellent fatigue strength, and damping around room temperature is reported to be one of the highest ever measured in a metal.

The alloy known as **nickel-aluminum,** containing about 20% nickel, is a master alloy for adding nickel to aluminum alloys or for making aluminum bronzes. **Nickel-aluminum bronze,** used for dies, molds, cast propellers, and valve seats, is usually an 8 to 10% aluminum bronze with an addition of nickel to give increased strength, corrosion resistance, and heat resistance to about 750°F (399°C). But **nickel-aluminum alloys** are used for coil springs. **Duranickel,** with 93.7% nickel and 4.4 aluminum, is age-hardening, and the cold-drawn wire has a tensile strength of 225,000 lb/in^2 (1,551 MPa) and is corrosion-resistant to 650°F (343°C). **Tin-nickel alloy,** with 65% tin and 35 nickel, is used only for electroplating. It deposits the chemical compound SnNi, giving a bright-silvery-white, corrosion-resistant plate.

Nickel alloy powders are used for flame-sprayed coatings for hard surfacing and corrosion resistance. **Metco 14E,** of Metco, Inc., is an alloy powder containing 14% chromium, 3.5 silicon, 2.75 boron, 4 iron, 0.60 carbon, with the balance nickel. The alloy is self-fluxing and gives an extremely hard coating. **Colmonoy 72,** of Wall Colmonoy Corp., is a similar alloy powder but with the addition of a high percentage of tungsten. Coatings have a melting point of 1950°F (1066°C) and retain hardness and wear resistance at high temperatures.

NICKEL BRONZE. A name given to bronzes containing nickel, which usually replaces part of the tin, producing a tough, fine-grained, and corrosion-resistant metal. A common nickel bronze containing 88% copper, 5 tin, 5 nickel, and 2 zinc has a tensile strength of 48,000 lb/in^2 (330 MPa), elongation 42%, and Brinell hardness 86 as cast. When heat-treated or age-hardened, the tensile strength is 87,000 lb/in^2 (599 MPa), elongation 10%, and Brinell hardness 196. Small amounts of lead take away the age-hardening quality of the alloy, and also lower the ductility. But small amounts of nickel added to bearing bronzes increase the resistance to compression and shock without impairing the plasticity. A bearing bronze of this nature, U.S. Navy 46B22, for heavy loads at high speeds, contains 15 to 20% lead, 73 to 80 copper, 5 to 10 tin, and 1 nickel. In the **leaded nickel-copper** of American Brass, which contains 1% nickel, 1 lead, 0.2 phosphorus, with the balance copper, a nickel phosphide is dispersed in the alloy by heat treatment, giving a machinability of 80% that of a free-cutting brass. The tensile strength is 85,000 lb/in^2 (586 MPa), elongation 5%, and electric conductivity 55% that of copper.

For decorative bronze parts, nickel is used to give a white color. In the hardware industry the old name **Chinese bronze** was used for these white alloys. At least 10% nickel is needed to give a white color. This amount also gives corrosion resistance to the alloy. When more than 15% nickel is used, the bronzes are difficult to machine unless some lead is added. Hardware and plumbing fixtures of these alloys do not require plating. **Eclipse bronze** is a **white bronze** of Sargent & Co. **M-M-M alloy,** of Manning, Maxwell & Moore, Inc., used for pressure valves for super-heated steam, contains 60 to 65% nickel, 24 to 27 copper, 9 to 11 tin, and small amounts of iron, silicon, and manganese. **Mercoloy,** of the Merco Nordstrom Valve Co., is a white valve bronze containing 60% copper, 25 nickel, 10 zinc, 1 tin, 2 lead, and 2 iron. It has a tensile strength of 44,000 lb/in^2 (303 MPa), with elongation 20%.

Ni-Vee bronzes of International Nickel are copper-base alloys containing 5% nickel and 5 tin. Lead and zinc may be added for economy, machinability, and to give bearing qualities. The amount of lead or zinc is shown by the number designation. **Ni-Vee Z2** contains 88% copper, 5 nickel, 5 tin, and 2 zinc. It has a tensile strength of 50,000 lb/in^2 (344 MPa), and Brinell hardness of 85. It is used for gears, cams, and construction-castings. **Ni-Vee L10** contains 80% copper, 5 nickel, 5 tin, and 10 lead. The tensile strength is 35,000 lb/in^2 (241 MPa), elongation 10%, and Brinell hardness 80. It is used for bearings and for acid-resistant castings. **Ni-Vee L5Z5** has a tensile strength of 40,000 lb/in^2 (275 MPa), elongation 20%, and Brinell hardness 80. It is used for valves and pump castings. A wide variety of copper-base alloys containing 10 to 40% nickel, and generally classed as nickel bronzes, are marketed under trade names such as **Aterite** and **Alcumet,** and used for hardware, plumbing fixtures, and where a white color and good corrosion resistance are required. Some of these alloys may also contain iron, aluminum, or manganese for added strength and hardness. **Nickel-silicon bronze** of American Brass, called **Cunisil 837,** contains 1.9% nickel, 0.60 silicon, and the balance copper. It is corrosion-resistant and has a tensile strength of 90,000 lb/in^2 (620 MPa), and the rod has a machinability rating 40% that of free-cutting brass. **Nironze 635,** of Bridgeport Brass, has a similar composition. These alloys, used for bolts, screws, and marine hardware, can be precipitation-hardened to Rockwell B95, with a yield strength of 90,000 lb/in^2 (620 MPa) and elongation of 12%. The electric conductivity is 30% that of copper.

NICKEL CAST IRON. A high-strength alloy cast iron containing nickel. Nickel, like silicon, assists the graphite formation and the carbide decomposition, and therefore reduces chill and acts to eliminate hard carbide spots, chilled edges, and mottled areas. About 1% nickel is equivalent to 0.5% silicon, but the effect of nickel is progressive; unlike silicon, it does not make the iron brittle. Nickel in amounts from 0.5 to 10% will also progressively harden cast iron. A gray cast iron which will have a Brinell hardness of 174 is raised to 350 by the addition of 9% nickel. Since the nickel promotes the formation of the graphite in fine crystals, the iron has a high resistance to wear. Tensile strengths up to 65,000 lb/in^2 (448 MPa) are obtained in these irons. Nickel obstructs the passage of electric currents in iron, and iron with 5% nickel is used for resistance grids. High-nickel iron is also nonmagnetic.

Most nickel irons contain small amounts of chromium to increase the chilling power and refine the grain. It also increases the strength and hardness. **Ni-Tensyl iron** is the trade name of International Nickel for a nickel cast iron made by adding to the melt a graphitizer of nickel-silicon to cause partial decomposition of the combined carbon. With 1.75% nickel this iron has high strength and a hardness of 320 Brinell. The silicon content is about 1.40% and the chromium about 0.35%. An iron of this type used for heavy-duty drawing dies is **Paralloy** of the Youngstown Foundry & Machine Co. As cast, it has a hardness up to 300 Brinnell, and a tensile strength to 50,000 lb/in^2 (344 MPa). When heat-treated, the hardness is up to 550 Brinell with tensile strength to 40,000 lb/in^2 (275 MPa). **Ryanite,** of the Allyne-Ryan Foundry Co., for dies, is a similar iron. **Lectrocast,** of the Detroit Gray Iron Foundry, used for automobile body dies, has 2.75% nickel and 0.70 chromium. **Tensloy,** of the Ensign Foundry Co., and **Novite,** of the Novo Engine Co., have about 1.5% nickel and 0.50 chromium. **Mitchalloy A,** of the Robert Mitchell Co., Ltd., has 2.5% nickel and 0.90 chromium. **Cariron, Bryiron,** and **Diamite** are **nickel-chromium irons,** but alloy irons now usually conform to ACI specifications.

High-nickel-chromium irons are used for pump and compressor parts handling hot liquids and may contain up to 30% nickel and 5 chromium.

Pyrocast is a high-test nickel-chromium cast iron of the Pacific Foundry Co. **Niresist,** developed by International Nickel, is an alloy cast iron containing Monel metal with chromium and manganese. A typical analysis range is nickel, 12 to 15%; copper, 5 to 7; chromium, 1.25 to 4; manganese, 1 to 1.5; silicon, 1 to 2; and total carbon, 2.75 to 3.1. The tensile strength is 20,000 to 35,000 lb/in^2 (137 to 241 MPa), with a Brinell hardness 130 to 170. It can be chilled to a hardness of 350 to 400. It has a low coefficient of expansion, 0.00001 per degree Fahrenheit (0.00002 per degree Celsius), or about the same as aluminum-silicon alloys used for pistons. **Nogroth metal** is such an iron. **Nicrosilal,** developed by the British Cast Iron Research Assoc., contains 5% silicon, 17 nickel, and 3 chromium. It has high-heat distortion resistance, and is used for heat-resistant castings.

Vanadium in nickel irons adds strength and wear resistance. **Vanick** is the name of a **nickel-vanadium cast iron** of the Malleable Iron Fittings Co. Nickel cast irons and nickel-chromium cast irons are marketed under many trade names, such as **Frankite,** of the Frank Foundries Corp.; **Alco Ni-Iron,** of the American Locomotive Co.; **Elfur iron,** of the Cramp Brass & Iron Foundry; **Tylerite,** of the W. S. Tyler Co.; **Domite,** of the Dominion Wheel & Foundry Co.; and **Maxtensile,** of the Farrel-Birmingham Co. The nickel-chromium cast irons may also contain molybdenum for greater hardness and wear resistance. **Ni-Chillite,** of the Mackintosh-Hemphill Co., is a nickel-chromium-molybdenum chilled cast iron for heavy rolls. **Mocasco iron,** of the Motor Casting Co., is a wear-resistant iron capable of being cast into thin sections without chill. **Mocasco 30,** for cylinders, has 1 to 1.35% nickel, 0.25 to 0.30 chromium, and 0.75 molybdenum. **Strenes metal** is the name of the Advance Foundry Co. for a nickel-chromium-molybdenum cast iron for heavy dies. **Durite** is a wear-resistant iron of the Birdsboro Steel Foundry & Machine Co. **Ironite,** of the Kinite Corp., is a chromium-nickel-vanadium cast iron used for cams, gears, and wear-resistant parts. **Low-expansion iron,** for dies, gages, and machine parts held to close tolerances, is nickel cast iron. **Minovar,** of the International Nickel Co., is such an iron with only one-third the expansion of ordinary cast iron.

NICKEL-CHROMIUM ALLOYS. Alloys of nickel and chromium employed as heat-resistant metals, for resistance wires, and as corrosion-resistant metals for chemical machinery. An alloy of 80% nickel and 20 chromium will withstand temperatures up to 2100°F. **Chromel A,** of the Hoskins Mfg. Co., has this composition. The resistivity is 650 Ω/cir mil ft, tensile strength 120,000 lb/in^2 (827 MPa), and melting point 1420°C. **Chromel C,** with 60% nickel, 16 chromium, and 24 iron, has an electrical resistivity 10% higher than Chromel A at high temperatures, but its oxidation resistance is limited to 1700°F (927°C). It is used for electric-appliance heating wire. **Chromel D** contains 35% nickel, 18.5 chromium, and the balance iron. Its resistivity at room temperature is 600 Ω/cir mil ft, but its hot resistivity is greater than Chromel A. It is for temperatures not over 1200°F (649°C).

For general-purpose castings for use at temperatures to 1800°F (982°C) the **HT alloy** of the Alloy Casting Institute has 35% nickel, 15 chromium, and 1.5 columbium. For lower temperatures, from 1600 to 1800°F (871 to 982°C), the **HK alloy** has 20% nickel, 25 chromium, and 0.40 max carbon. **Misco metal,** of the Michigan-Standard Alloy Casting Co., contains 35% nickel and 10 chromium, with the balance iron. **Misco C** contains 29% chromium, 9 nickel, 0.55 manganese, 0.60 silicon, and 0.25 carbon. **Miscrome** is resistant to nitric acid. It contains 28% chromium with no nickel. **Centricast alloys,** of the same company, are these corrosion-resistant alloys centrifugally cast for pipes. A group of alloys for heat-resistant parts is marketed by the Electro-Alloys Div., American Brake Shoe Co., under the name **Thermalloy. Thermalloy 72,** for temperatures to 2000°F (1093°C), has 12 to 14% chromium and 58 to 63 nickel. **Thermalloy 40** is a general-use machinable alloy to resist temperatures from 1200 to 1500°F (649 to 816°C). It contains about 26% chromium and 12 nickel. **Chemalloy F32M,** of the same company, is primarily for parts to resist chemical action. It has 26 to 29% chromium, 3 to 6 nickel, and 1 to 1.5 molybdenum. It has high strength at elevated temperatures, is resistant to sulfur atmospheres, and is used for parts for refineries and pulp mills.

A group of heat-resistant and acid-resistant nickel-chromium alloys is sold by the Michigan-Standard Alloy Casting Co. as Standard alloy. **Standard alloy HR-3,** a cast alloy for heat-treating equipment, contains 37% nickel, 17 chromium, 1.5 silicon, 1.2 manganese, and 0.50 carbon. At room temperature it has a tensile strength of 68,000 lb/in^2 (468 MPa) with elongation of 10%, and at 1800°F (982°C) the tensile strength is 12,000 lb/in^2 (82 MPa) with elongation of 26%. It has a Brinell hardness of 187. It is austenitic, and is not affected by heat treatment. It resists air oxidation to 2000°F (1093°C) and is resistant to tempering salts and carburizers. A lower-cost metal for general heat resistance to 2100°F (1149°C) is **Standard alloy HR-6,** which is **AISI alloy Type 309** containing 25% chromium and 12 nickel with 1 manganese, 1.25 silicon, and 0.40 carbon. Another group is marketed by the Cooper Alloy Foundry Co. under the names of **Sweetaloy** and **Cooper alloys. Cooper alloy S-16** has 14 to 20% chromium and 65 nickel; **Cooper alloy S-21A** is this combination with 3% molybdenum and 1.5 silicon. **Cooper alloy V2B** has high resistance to acids, high strength and wear resistance, and welds easily. It contains about 19.25% chromium, 10 nickel, 3 silicon, 3.25 molybdenum, 2 copper, 0.60 manganese, 0.15 beryllium, and 0.07 max carbon. The hardness as cast is 302 Brinell, but it can be quench-annealed for machining. The tensile strength is 150,000 lb/in^2 (1,034 MPa), with elongation of 3%. **Cimet,** of the Driver-Harris Co., used for mine pump parts, has 26% chromium, 10 nickel, and the balance iron. **Gridnic alloys,** of this company, are nickel-chromium alloys used for radio grids. **Nirex,** of the same company, has 80% nickel, 14 chromium, and 6 iron. In annealed sheet form it has a tensile strength of 90,000 lb/ in^2 (620 MPa) with elongation 50%. **Chromel No. 502** has 36% nickel, 20 chromium, and the balance iron. It is used for furnace fixtures, either rolled or cast.

Inconel, developed by the International Nickel Co., used for dairy and food equipment, contains 79.5% nickel, 13 chromium, 6.5 iron, 0.08 carbon, and slight amounts of copper, manganese, and silicon. The tensile strength of the hot-rolled plate is 100,000 to 140,000 lb/in^2 (689 to 965 MPa) with elongation 20 to 40%, and Brinell hardness 160 to 240. Heat-treating boxes of this alloy give furnace service life of 12,000 h at 2000°F (1093°C). **Inconel X,** a wrought alloy for gas turbine blades, contains 70 to 73% nickel, 15 chromium, 2.5 titanium, 0.70 to 1.20 columbium, 0.40 to 1 aluminum, 5 to 9 iron, 0.30 to 1 manganese, 0.5 max silicon, 0.20 max copper, and 0.08 max carbon. It has a tensile strength of 160,000 lb/in^2 (1,103 MPa), retains 90% of its strength at 900°F (482°C), and 80% up to 1100°F (593°C). The specific gravity is 8.3, and the melting range is 2540 to 2600°F (1393 to 1449°C). It can be age-hardened to 400 Brinell, but loses its hardness at 1500°F (816°C). **Incoloy T,** of this company, has high corrosion resistance to above 1600°F (871°C), but loses three-quarters of its strength at that temperature. It contains about 32% nickel with cobalt, 20 chromium, 1.5 manganese, 1 silicon, 1.25 titanium, with maximums of 0.50 copper and 0.10 carbon. The tensile strength is above 80,000 lb/in^2 (551 MPa). **Inconel 718,** for use at subzero temperatures, has up to 55% nickel with contained cobalt, up to 21% chromium, 5 columbium plus tantalum, 3 molybdenum, and 1 each of cobalt and titanium.

Nickel-chromium-iron alloys, with silicon and other elements forming complex alloys, are employed for acid-resistant and corrosion-resistant castings and wrought metals for high temperatures. **Durimet,** of the Duriron Co., Inc., is marketed in various grades normally containing 19.5 to 23% nickel, 18 to 22 chromium, 2.75 to 3.75 silicon, 0.50 to 0.75 manganese, 1 to 1.5 molybdenum, and 0.25 to 0.45 copper. **Durimet 20** contains 29% nickel, 20 chromium, 1.75 molybdenum, 3.5 copper, 1 silicon, and 0.07 max carbon. The heat-treated castings have a tensile strength of 65,000 to 75,000 lb/in^2 (448 to 517 MPa), elongation 35 to 50%, and Brinell hardness 120 to 150. It is especially resistant to sulfuric and mixed acids. This alloy is produced as **Esco 20** by the Electric Steel Foundry Co., **Isocast 20** by Empire Steel Castings, Inc., **Aloyco 20** by the Alloy Steel Products Co., **Fahrite C-20** by the Ohio Steel Foundry Co., **Misco 20** by the Michigan Steel Castings Co., and **Utilloy 20** by the Utility Electric Steel

Foundry. **Carpenter stainless 20** is this alloy in wrought forms with slightly less copper and more molybdenum. The tensile strength of the wrought bar is 85,000 lb/in^2 (586 MPa), elongation 35 to 50%, and hardness 150 to 180 Brinell. **Carpenter stainless 20Cb-3** contains 34% nickel, up to 1% columbium plus tantalum, and is for high resistance to hot acids. **Durimet FA-20,** of the Cooper Alloy Foundry Co., has 20% chromium, 29 nickel, 4 copper, 3.5 molybdenum, 1 silicon, 1 manganese, with the carbon kept below 0.07 to prevent intergranular corrosion. The tensile strength is 75,000 lb/in^2 (517 MPa), elongation 40%, and Brinell hardness 150. It is resistant to sulfuric acid and to strong, hot alkalies and chlorates. **Donegal alloy DC-50,** of the Donegal Mfg. Corp., used for pump parts, has corrosion resistance about equal to an 18–8 steel. It contains 16.5% chromium, 4 nickel, 4 copper, and 0.05 carbon. When hardened, it has a tensile strength of 135,000 to 210,000 lb/in^2 (930 to 1,447 MPa), with elongation of 6 to 15%, and hardness of 375 to 440 Brinell.

Chemalloy N4, of the Electro-Alloys Div., used as a stain-resistant metal for food equipment, contains 30% nickel, 5 chromium, 5 silicon, and the balance iron. It will resist oxidation to 1400°F (760°C). **Pyromet 860,** of Carpenter Technology, used for turbine blades and aircraft engine parts, contains 43% nickel, 14 chromium, 6 molybdenum, 4 cobalt, 3 titanium, 1 aluminum, 0.1 carbon, and the balance iron. The tensile strength is up to 180,000 lb/in^2 (1,241 MPa), and it retains good strength up to about 1400°F (760°C). **Rezistal,** of the Crucible Steel Co., is marketed in many grades resistant to heat, acids, and corrosion. They include stainless steels and special-composition nickel-chromium alloys. They are also produced in supersoft sheet form, 0.012 to 0.018 in (0.030 to 0.046 cm) thick, for flashings, gutters, and expansion joints in building construction. They provide high strength and durability but may be cut with ordinary tools. **Rezistal 2600,** formerly known as **Atha's 2600 alloy,** contains 22% nickel, 8 chromium, 1.75 silicon, 1 copper, 0.70 manganese, and 0.25 carbon. It is nonmagnetic, resistant to acids, and is easily machinable. **RA 330 alloy,** of Rolled Alloys, Inc., contains 35% nickel, 19 chromium, 1.25 silicon, 2 manganese, 0.50 copper, and 0.06 carbon. It is austenitic, and has a tensile strength of 85,000 lb/in^2 (586 MPa) with elongation of 40%. At 1200°F (649°C) the tensile strength is 59,000 lb/in^2 (406 MPa), and at 1800°F (982°C) it is 10,500 lb/in^2 (71 MPa).

NICKEL-CHROMIUM STEEL. Steel containing both nickel and chromium, usually in a ratio of 2 to 3 parts nickel to 1 chromium. The 2:1 ratio gives great toughness, and the nickel and chromium are intended to balance each other in physical effects. The steels are especially suited for large sections which require heat treatment because of the deep and uniform hardening power. Hardness and toughness are the characteristic properties of these steels. Nickel-chromium steel containing 1 to 1.5% nickel, 0.45 to 0.75 chromium, and 0.38 to 0.80 manganese is used throughout the carbon ranges for case-hardened parts and for forgings where high tensile strength and great hardness are required.

Low nickel-chromium steels, but with more carbon, from 0.60 to 0.80%, are used for drop-forging dies and other tools. **R.D.S. steel,** of Carpenter Technology, is an oil-hardening, tough tool steel containing 1.75% nickel, 1 chromium, 0.50 manganese, and 0.75 carbon. An industrial steel of this company, **Carpenter steel 5-317,** contains the same amounts of nickel and chromium, but 0.50% carbon and less manganese. It is oil-hardening, and is used for gears, shafts, and shock-resisting parts. The tensile strength is up to 295,000 lb/in^2 (2,033 MPa). **Samson steel,** of this company, used for machinery parts for severe service, contains 1.25% nickel, 0.60 chromium, 0.50 manganese, and various amounts of carbon. The 0.40 carbon steel, when heat-treated, has a tensile strength of 240,000 lb/in^2 (1,654 MPa) and Brinell hardness 440. **Beaver steel,** of the Colonial Steel Co., used for water-quenched forging-die blocks, contains 1.5% nickel, 0.75 chromium, 0.60 manganese, and 0.55 carbon. **Colona steel,** of the same company, used for oil-quenched forging dies, has the same nickel and chromium content, but somewhat more manganese and carbon. **Nikro-Trimmer steel,** of the same company, used for hot trimming dies, has higher carbon, 0.85%, but only small amounts of nickel and chromium, 0.30 to 0.55% of each. **Simplex steel,** of the Crucible Steel Co., used for forgings, has 1.25% nickel and 0.60 chromium. In the low-carbon, case-hardening grades, for gears, it has a tensile strength of 90,000 lb/in^2 (620 MPa). **SAE steel 3330,** containing 3.5% nickel and 1.5 chromium, when oil-tempered has a tensile strength of 205,000 lb/in^2 (1,413 MPa) and elongation 13%. **Quality steel** is the trade name of Quality Steels, Ltd., for nickel-chromium steels in various grades.

Nickel-chromium steels may have temper brittleness, or low impact resistance, when improperly cooled after heat treatment. A small amount of molybdenum is sometimes added to prevent this brittleness. **Encem steel,** of W. T. Flather, Ltd., is a molybdenum steel of this type. **Miraculoy,** of the Sivyer Steel Castings Co., contains 1.25% nickel, 0.65 chromium, 0.40 molybdenum, and 1.55 manganese. The tensile strength is 115,000 lb/in^2 (792 Mpa) with elongation 18% and Brinell hardness 275. **Miscoloy No. 60,** of the Michigan Steel Casting Co., is another steel of this type. A nickel-chromium **coin steel,** used by the Italian government for coins, was a stainless type containing 22% chromium, 12 nickel, and some molybdenum.

Low-carbon nickel-chromium steels are water-hardening, but those with appreciable amounts of alloying elements require oil quenching. Air-hardening steels contain up to 4.5% nickel and 1.6 chromium, but are brittle unless tempered in oil to strengths below 200,000 lb/in^2 (1,378 MPa). The alloy known as **Krupp analysis steel** contains 4% nickel and 1.5 chromium. The steel under the name of **Millaloy,** used by Doelger & Kirsten, Inc., for heavy shear blades, is of this analysis with 0.40% carbon. Blades hardened to 530 Brinell have an ultimate strength of 312,000 lb/in^2 (2,151 MPa) and elongation 11%. **Nikrome** is a nickel-chromium steel of Joseph T. Ryerson & Son, Inc. **Nikrome M** contains 2.25% nickel, 2 chromium, and 0.45 molybdenum, with 0.40 carbon. It is characterized by high strength and uniform hardening, and can be machined up to 450 Brinell. **H.T.M. steel,** of this company, is a 2% nickel steel with chromium and molybdenum. **Nykrom** is a steel of W. T. Flather, Ltd. **Ohioloy** is a nickel-chromium steel of the Ohio Steel Foundry Co. All of the nickel-chromium steels require special heat treatment to bring out their best qualities. The nickel-chromium steels used for high strength at elevated temperatures in rockets and aircraft are variations of the stainless-type alloys. **Unitemp 212,** of the Universal-Cyclops Steel Co., which has a tensile strength of 187,000 lb/in^2 (1,289 MPa), and will retain a strength of 108,000 lb/in^2 (744 MPa) at 1400°F (760°C) with high corrosion resistance, contains 25% nickel, 16 chromium, 4 titanium, 0.08 boron, 0.50 columbium, 0.50 aluminum, 0.05 zirconium, 0.50 manganese, 0.50 silicon, and 0.15 carbon.

NICKEL-COBALT ALLOYS. A group of alloys containing 20 to 30% cobalt and 70 to 80 nickel. Nickel has a yellowish cast and cobalt a blue cast; alloys of the two metals are almost pure silver in luster and resemble silver. They are expensive, because of the high cost of cobalt, but the two metals are codeposited in an electroplating bath to form an alloy deposit on iron, steel, or nonferrous metals. The alloy is harder than either nickel or cobalt alone, and is also more corrosion-resistant. Another type of nickel-cobalt complex alloy is **Konal,** developed by the Westinghouse Electric Corp., as a heat-resistant and acid-resistant alloy. It contains 73% nickel, 17.5 cobalt, 6.5 iron, 2.5 titanium, and 0.2 manganese. At a temperature of 600°C this alloy has a tensile strength of 66,000 lb/in^2 (455 MPa). At ordinary temperatures the strength is 100,000 lb/in^2 (689 MPa). It was developed for radio-tube filaments. **Alloy C** of M & T Chemicals is a 50–50 alloy of nickel and cobalt used for plating.

NICKEL COPPER. An alloy of nickel and copper employed for adding nickel to nonferrous alloys. A 50–50 nickel copper has a melting point of 2330°F (1277°C) and dissolves readily. **Nickel-copper shot** of this composition is marketed for ladle additions to iron and steel. The copper-nickel master alloy designated in Federal specifications contains 60% nickel, 33 copper, 3.5 manganese, and may contain up to 3.5 iron. It is used for alloying high-strength bronzes. **Copper-nickel alloy** is also used for special purposes. A grade of **Thermalloy,** of the Electro Alloys Co., used as a temperature-sensitive magnetic metal for magnetic shunts in watt-hour meters, has 66.5% nickel, 30 copper, and 2 iron. The permeability falls off with increase in temperature and compensates for errors due to temperature changes.

NICKEL-MOLYBDENUM IRON. A group of alloys used for high acid resistance. They may contain up to 40% molybdenum, which takes the place of the chromium used in the more common corrosion-resistant alloys. The most usual alloy in this class contains about 20% iron, 20 molybdenum, 60 nickel, and small amounts of carbon. This alloy is very resistant to hydrochloric and sulfuric acids, but for high general acid resistance the iron content should be below 10%. Iron adds hardness and stiffness to the alloys, but decreases the acid resistance. Manganese improves the workability, but more than 3% decreases the acid resistance. The alloys cast easily, and the 20:20:60 alloy is readily machinable. It can be hot-rolled into sheet, or cold-rolled. The melting point is 1300°C, and the weight is 0.315 lb/in^3 (8,719 kg/m^3). The tensile strength, forged, is 118,000 lb/in^2 (813 MPa) and Brinell hardness 207. This alloy is resistant to all acids except nitric. **Chemalloy H1,** of the electro-Alloys Div., for severe acid resistance such as for valve and pump parts, contains 21% molybdenum, 21 iron, and the balance nickel.

NICKEL-MOLYBDENUM STEEL. An alloy steel which is most used in compositions of 1.5% nickel and 0.15 to 0.25 molybdenum, with varying percentages of carbon up to 0.50%. These steels are characterized by remarkably uniform properties, are readily forged and heat-treated. Molybdenum produces toughness in the steels, and in the case-hardened steel gives a tough core. Roller bearings are made of this class of steel. A steel used by the Ingersoll Steel & Disc Co. for hand-shovel blades contains an average of 0.45% carbon, 0.50 manganese, 1.35 nickel, and 0.40 molybdenum. When hot-rolled and heat-treated, it has a tensile strength of 240,000 lb/in^2 (1,654 MPa) and elongation 6%. **Superalloy steel,** of the same company, is **SAE steel 3160.** A 5% nickel steel with 0.30% carbon and 0.60 molybdenum has a tensile strength of 175,000 to 230,000 lb/in^2 (1,206 to 1,585 MPa) with elongation 12 to 22%, depending upon the heat treatment. Molybdenum is more frequently added to the steels containing also chromium, the molybdenum giving air-hardening properties, reducing distortion, and making the steels more resistant to oxidation.

NICKEL ORES. Nickel occurs in minerals as sulfides, silicates, and arsenides, the most common being **pyrrhotite,** or **magnetic pyrites,** a sulfide of iron of the formula $Fe_{1-x}S$, where $_x$ is between 0 and 0.2. When $_x$ is zero the mineral is called **troilite.** Pyrrhotite has nickel associated with the iron sulfide. The ore of Copper Cliff, Ontario, is calcined to remove the sulfur, and the nickel is removed, leaving a fine magnetite which is pelletized and fired to give an iron concentrate of 68% iron. The chief sulfide ore deposit at Sudbury, Ontario, contains sulfides of iron, nickel, and copper, and small amounts of other elements, and some of the matte after removal of the iron and sulfur is used as Monel metal without separating the natural alloy. The extensive ore deposits at Lynn Lake, Manitoba, yield an ore averaging 1.74% nickel and 0.75 copper. The **garnierite,** or **noumeite,** of New Caledonia is a nickel silicate containing also iron and magnesium. It is amorphous and earthy, of an apple-green color, with a specific gravity of 2.2 to 2.8, and hardness of 3 to 4. The ore contains about 5% nickel, and is smelted with gypsum to a matte of sulfides of nickel and iron, the sulfur coming from the gypsum. This is then bessemerized, and the matte crushed, roasted to oxide, and reduced to nickel. The material exported from New Caledonia under the name of **fonte** is a directly smelted cast iron containing about 30% nickel.

A minor ore of nickel called **millerite,** occuring in Europe and in Wisconsin, is a **nickel sulfide,** NiS, containing theoretically 64.7% nickel. It is found usually in radiating groups of slender crystals with a specific gravity of 5.6, hardness 3.5, and of a pale-yellow color and metallic luster. **Niccolite,** NiAs, is a minor ore containing theoretically 43.9% nickel, usually with iron, cobalt, and sulfur. It is found in Canada, Germany, and Sweden. The mineral occurs massive, with a specific gravity of 7.5, hardness 5 to 5.5, and a pale-copper color. Nickel is also produced as a by-product from copper ores.

NICKEL SILVER. A name applied to an alloy of copper, nickel, and zinc, which is practically identical with alloys known in the silverware trade as **German silver. Packfong,** meaning **white copper,** is an old name for these alloys. The very early nickel silvers contained some silver and were used for silverware. **Wessell's silver** contained about 2%, and **Ruolz silver** about 20%. **Baudoin alloy,** a French metal, contained 72% copper, 16 nickel, 1.8 cobalt, 2.5 silver, and the balance zinc, but the white jewelry alloys called **Paris metal** and **Lutecine** contained about 2% tin instead of the silver. The English nickel silver known as **Alpaca,** used as a base metal for silver-plated tableware, had about 65% copper, 20 zinc, 13 nickel, and 2 silver. Such an alloy takes a fine polish, has a silvery-white color, and is resistant to corrosion.

Nickel silver is made in regular grades of 5 to 30% nickel, with up to 70% copper, and the balance zinc. Nickel whitens brass and makes it harder and more resistant to corrosion, but the alloys are more difficult to cast because of shrinkage and absorption of gases. They are also subject to fire cracking and are more difficult to roll and draw than brass. The most common nickel silver is the 18% nickel alloy with 55 to 65% copper, and the balance zinc. The high-copper grades are used for parts where there is much fabricating. As a spring material, with 55% copper, this alloy has a tensile strength of 110,000 lb/in^2 (758 MPa), and Brinell hardness 160, when cold-rolled. Telephone-equipment springs normally contain 55% copper, 18 nickel, and 27 zinc. **Copper alloy 766** is a nickel silver widely used for electric contacts, connectors, and current-carrying springs. It contains 56.5% copper, 12 nickel, and 31.5 zinc. It has a tensile strength up to 120,000 lb/in^2 (827 MPa) and does not need plating to resist corrosion even in marine atmospheres. Contacts made with this material by Elco Corp. have a resistance of 0.007 Ω and electric conductivity of 7.6% IACS. **Scovill nickel silver** for soft-temper extrusions has 13% nickel, 43 copper, and 44 zinc. It contains alpha grains in an amorphous beta matrix. **Benedict metal,** of the American Brass Co., originally had 12.5% nickel, with 2 parts copper to 1 zinc, but the alloy used for hardware and plumbing fixtures contains about 57% copper, 2 tin, 9 lead, 20 zinc, and 12 nickel. The cast metal has a strength of 35,000 lb/in^2 (241 MPa) with elongation of 15%. The white alloy known as **dairy bronze,** used for casting dairy equipment and soda-fountain parts, has 63% copper, 4 tin, 5 lead, 8 zinc, and 20 nickel. The higher-nickel alloys have a more permanent white finish for parts subject to corrosion. **Ambrac 854,** of the American Brass Co., is a wrought metal with 65% copper, 30 nickel, and 5 zinc. **Pope's Island white metal,** of this company, used for jewelry, had 67% copper, 19.75 nickel, and 13.25 zinc. **Victor metal** is an alloy of 50% copper, 35 zinc, and 15 nickel, used for cast fittings. It is a white metal with a yellow shade. It casts easily and machines well.

For threaded parts and for casting metals, the nickel silvers usually contain some lead for easier machining. Federal specifications for casting metals call for 65% copper, 20 nickel, 4 tin, 5 lead, and 6 zinc. **Leaded nickel silver,** for making keys, has 46.5% copper, 10 nickel, 40.75 zinc, and 2.75 lead. The extruded metal has a tensile strength of 80,000 lb/in^2 (551 MPa), elongation of 33%, and Rockwell hardness B82. **White nickel brass,** for cast parts for trim, is a standard 18% nickel alloy with or without lead. **Silveroid,** an English alloy for this use, is a cupronickel without zinc. An English alloy for tableware, under the name of **Newloy,** contains 35% nickel, 64 copper, and 1 tin. The **stainless nickel** used for silverware by Viners, Ltd., has 30% nickel, 60 copper, and 10 zinc, and is deoxidized with manganese copper, using borax as a top flux.

A number of other alloys of copper, nickel, and zinc are termed **nickel brass. Nickel-silicon brass** contains very small percentages of silicon, usually about 0.60%, which forms a **nickel silicide,** Ni_2Si, increasing the strength and giving heat-treating properties. Rolled nickel-silicon brass, containing 30% zinc, 2.5 nickel, and 0.65 silicon, has a tensile strength of 114,000 lb/in^2 (785 MPa). **Imitation silver,** for hardware and fittings, was a nickel brass containing 57% copper, 25 zinc, 15 nickel, and 3 cobalt. The bluish color of the cobalt neutralizes the yellow cast of the nickel and produces a silver-white alloy. **Silvel** was a nickel brass containing 67.5% copper, 26 zinc, and 6.5 nickel, with sometimes a little cobalt. Nickel brass is an alloy used where white color and corrosion resistance are desired.

Seymourite was an alloy of 64% copper, 18 nickel, and 18 zinc, produced by the Seymour Mfg. Co. It had a white color and was corrosion-resistant. **Nickeline,** used by the Yale & Towne Mfg. Co. for hardware, had 58 to 60% copper, 16.5 nickel, 2 tin, and the remainder zinc. It had high strength, a white color, and casts well. **Nickelene** is an old name applied to nickel brass of various compositions, but an alloy patented in 1912 under this name had 55% copper, 12.5 nickel, 20.5 zinc, 10 lead, and 2 tin. Most of these alloys have good casting qualities, but they do not machine easily unless they contain some lead. Up to 2% lead does not affect the color and does not decrease the strength greatly.

NICKEL STEEL. Steel containing nickel as the predominating alloying element. The first nickel steel produced in the United States was made in 1890 by adding 3% nickel in a bessemer converter. The first nickel-steel armor plate, with 3.5% nickel, was known as **Harveyized steel.** Small amounts of nickel steel, however, had been used since ancient times, coming from **meteoric iron.** The nickel iron of meteorites, known in mineralogy as **tacnite,** contains about 26% nickel.

Nickel added to carbon steel increases the ultimate strength, elastic limit, hardness, and toughness. It narrows the hardening range but lowers the critical range of steel, reducing danger of warpage and cracking, and balances the intensive deep-hardening effect of chromium. The nickel steels are also of finer structure than ordinary steels, and the nickel retards grain growth. When the percentage of nickel is high, the steel is very resistant to corrosion. Above 20% nickel the steel becomes a single-phase austenitic structure. The steel is nonmagnetic above 29% nickel, and the maximum value of permeability is at about 78% nickel. The lowest expansion value of the steel is at 36% nickel. The percentage of nickel employed usually varies from 1.5 to 5%, with up to 0.80% manganese. The bulk of nickel steels contain 2 and 3.5% nickel. They are used for armor plate, structural shapes, rails, heavy-duty machine parts, gears, automobile parts, and ordnance.

The standard ASTM **structural nickel steel** used for building construction is an open-hearth steel containing 3.25% nickel, 0.45 carbon, and 0.70 manganese. This steel has tensile strength from 85,000 to 100,000 lb/in^2 (586 to 689 MPa) and a minimum elongation 18%. An automobile steel used by one of the larger companies contains 0.10 to 0.20% carbon, 3.25 to 3.75 nickel, 0.30 to 0.60 manganese, and 0.15 to 0.30 silicon. When heat-treated, it has a tensile strength up to 80,000 lb/in^2 (551 MPa) and an elongation 25 to 35%. Forgings for locomotive crankpins containing 2.5% nickel, 0.27 carbon, and 0.88 manganese have a tensile strength of 83,000 lb/in^2 (572 MPa), elongation 30%, and reduction of area 62%. A **nickel-vanadium steel,** used for high-strength cast parts, contains 1.5% nickel, 1 manganese, 0.28 carbon, and 0.10 vanadium. The tensile strength is 90,000 lb/in^2 (620 MPa) and elongation 25%. **Univan steel,** of the Union Steel Casting Co., for high-strength locomotive castings, is a nickel-vanadium steel of this type. **Unionaloy steel,** of this company, is an abrasion-resistant steel.

The Federal specifications for 3.5% nickel-carbon steel call for 3.25 to 3.75% nickel, and 0.25 to 0.30 carbon. This steel has a tensile strength of 85,000 lb/in^2 (586 MPa) and elongation 18%. A hot-rolled 3.5% nickel medium-carbon steel, **SAE steel 2330,** when oil-quenched develops a strength up to 220,000 lb/in^2 (1,516 MPa), and Brinell hardness from 223 to 424, depending upon the drawing temperature. Standard 3.5 and 5% nickel steels are regular products of the steel mills, though they are often sold under trade names. Steels with more than 3.5% nickel are too expensive for ordinary structural use. Steels with more than 5% nickel are difficult to forge, but the very high content nickel steels are used when corrosion-resistant properties are required. **Nicloy,** of the Babcock & Wilcox Tube Co., used for tubing to resist the corrosive action of paper-mill liquors and oil-well brines, contains 9% nickel, 0.10 chromium, 0.05 molybdenum, 0.35 copper, 0.45 manganese, and 0.20 silicon, and 0.09 max carbon. The heat-treated steel has a tensile strength of 110,000 lb/in^2 (758 MPa), with elongation 35%. The **cryogenic steels,** or **low temperature steels,** for such uses as liquid-oxygen vessels, are usually high-nickel steels. **ASTM steel A-353,** for liquid-oxygen tanks at temperatures to −320°F (−196°C), contains 9% nickel, 0.85 manganese, 0.25 silicon, and 0.13 carbon. It has a tensile strength of 95,000 lb/in^2 (654 MPa) with elongation of 20%. **Lukens Nine Nickel,** for temperatures down to −320°F, contains 9% nickel, 0.80 manganese, 0.30 silicon, and not over 0.13 carbon. It has a minimum tensile strength of 90,000 lb/in^2 (620 MPa) with elongation of 22%.

NICKEL SULFATE. The most widely used salt for nickel-plating baths, and known in the plating industry as **single nickel salt.** It is easily produced by the reaction of sulfuric acid on nickel, and comes in pea-green water-soluble crystalline pellets of the composition $NiSO_4 \cdot 7H_2O$, of a specific gravity of 1.98, melting at about 100°C. **Double nickel salt** is **nickel ammonium sulfate,** $NiSO_4 \cdot (NH_4)_2 \cdot SO_4 \cdot 6H_2O$, used especially for plating on zinc. To produce a harder and whiter finish in nickel plating, **cobaltous sulfamate,** a water-soluble powder of the composition $Co(NH_2SO_3)_2 \cdot 3H_2O$, is used with the nickel sulfate. **Nickel plate** has a normal hardness of Brinell 90 to 140, but by controlled processes file-hard plates can be obtained from sulfate baths. **Micrograin nickel,** of Metachemical Processes Ltd., with a grain diameter of 0.00002 in (0.00005 cm), is such a hard plate.

Other nickel salts are also used for nickel plating. **Nickel chloride,** $NiCl_2 \cdot 6H_2O$, is a green crystalline salt which, when used with boric acid, gives a fine grained, smooth, hard, strong plate. It requires less power, and the bath is easy to control. **Nickel carbonate,** $2NiCO_3 \cdot 3Ni(OH)_2 \cdot 4H_2O$, comes in green crystals not soluble in water, but soluble in acids and in solutions of ammonium salts. **Nickel carbonyl,** $Ni(CO)_4$, used for nickel plating by gas decomposition, is a yellow volatile liquid. It is volatilized in a closed vessel with hydrogen as the carrier, and the nickel is deposited at about 350°F (177°C). It will adhere to glass and wood as well as to metals. The material is a strong reducing agent, and is explosive when mixed with oxygen. **Nickel nitrate,** $(NiNO_3)_2{:}6H_2O$, used in electric batteries, comes in thin flat flakes.

NITRIC ACID. Also called **aqua fortis** and **azotic acid.** A colorless to reddish fuming liquid of the composition HNO_3, having a wide variety of uses for pickling metals, etching, and in the manufacture of nitrocellulose, plastics, dyestuffs, and explosives. It has a specific gravity of 1.502 (95% acid) and a boiling point of 86°C, and is soluble in water. Its fumes have a suffocating action, and it is highly corrosive and caustic. **Fuming nitric acid** is any water solution containing more than 86% acid and having a specific gravity above 1.480. Nitric acid is made by the action of sulfuric acid on sodium nitrate, or purified Chilean saltpeter, and condensation of the fumes. It is also made from ammonia by catalytic oxidation, or from the nitric oxide produced from air. The acid is sold in various grades, depending on the amount of water. The strengths of the commercial grades are 38, 40, and 42°Bé, containing 67.2% acid. C.P., or reagent grade, is 43°Bé, with 70.3% acid, very low in iron, arsenic, or other impurities. It is usually shipped in glass carboys. **Anhydrous nitric acid** is a yellow fuming liquid containing the unstable anhydride, **nitrogen pentoxide,** N_2O_5. It is violently reactive and is a powerful nitriding agent. The dark-red fuming liquid known as **nitrogen tetroxide,** N_2O_4, is really a concentrated water solution of nitric acid, as this oxide is an unstable polymer of NO_2. It is used as an oxidizer for rocket fuels, as it contains 70% oxygen. **Mixed acid,** or **nitrating acid,** is a mixture of nitric and sulfuric acids used chiefly in making nitrocellulose and nitrostarch. Standard mixed acid contains 36% nitric and 61 sulfuric, but other grades are also used.

NITRIDING STEELS. Low- and medium-carbon steels with combinations of chromium and aluminum or nickel, chromium, and aluminum.

Nitriding consists of exposing steel parts to gaseous ammonia at about 1000°F (538°C) to form metallic nitrides at the surface. The hardest coatings are obtained with aluminum-bearing steels. Nitriding of stainless steel is known as **Malcomizing.** After nitriding, these steels have extremely high surface hardnesses of about 92 to 95 Rockwell N. The nitride layer also has considerable resistance to corrosion from alkali, the atmosphere, crude oil, natural gas, combustion products, tap water, and still salt water. Nitrided parts usually grow about 0.001 to 0.002 in (0.003 to 0.005 cm) during nitriding. The growth can be removed by grinding or lapping, which also removes the brittle surface layer. Most uses of nitrided steels are based on resistance to wear. The steels can also be used in temperatures as high as 1000°F (538°C) for long periods without softening. The slick, hard, and tough nitrided surface also resists seizing, galling, and spalling. Typical applications are cylinder liners for aircraft engines, bushings, shafts, spindles and thread guides, cams, and rolls.

A composition range of **Nitralloy** steel is 0.20 to 0.45% carbon, 0.75 to 1.5 aluminum, 0.9 to 1.8 chromium, 0.4 to 0.70 manganese, 0.15 to 0.60 molybdenum, 0.3 max silicon. Nitralloy is marketed by various steel companies. **Nitrard,** of the Firth-Sterling Steel Co., is also the name of a nitriding steel. Nitralloy steel is used for tools, gages, gears, and shafts. Unlike the soft core of ordinary case-hardened steels, it will have a tough core with high hardness. **Nitralloy 135** contains 0.35% carbon, 0.55 manganese, 0.30 silicon, 1.20 copper, 1 aluminum, 0.20 molybdenum, and has a tensile strength, hardened, of 138,000 lb/in^2 (951 MPa) with elongation of 20% and Brinell hardness of 280. **Nitralloy N** is similar but with about 3.5% nickel, higher chromium, and less carbon. It gives a Brinell hardness of 415.

Carbonitrided steel is produced by exposing the steel at about 1500°F (816°C) in a carbon-nitrogen atmosphere and then quenching in oil. The depth of the case depends on the length of time of treatment. The surface is harder and more wear-resistant than carbon case-hardened steel.

NITROCELLULOSE. A compound made by treating cellulose with nitric acid, using sulfuric acid as a catalyst. Since cotton is almost pure cellulose, it was originally the raw material used, but alpha cellulose made from wood is now employed. The cellulose molecule will unite with from 1 to 6 molecules of nitric acid. **Trinitrocellulose,** $C_{12}H_{17}O_7(NO_3)_3$, contains 9.13% nitrogen, and is the product used for plastics, lacquers, adhesives, and Celluloid. It is classified as cellulose nitrate. The higher nitrates, or **pyrocellulose,** are employed for making explosives. It was originally called **guncotton,** and the original United States government name for the explosive was **Indurite,** from the Indian Head Naval Powder Factory. It was called **cordite** in England. The nitrated cellulose is mixed with alcohol and ether, kneaded into a dough, and squeezed through orifices into long multitubular strings which are cut into short cylindrical grains. Solid grains become smaller as they burn, so that there would be high initial pressure and then a decreasing pressure of gases. When the multitubular grains burn, the surface becomes greater and thus there is increasing pressure. **FNH powder,** or **flashless powder,** is nitrocellulose which is nonhygroscopic and which contains a partially inert coolant, such as potassium sulfate, to reduce the muzzle flash of the gun. **Ballistite** is a rapid-burning, double-base powder used in shotgun shells and as a propellant in rockets. It is composed of 60% nitrocellulose and 40 nitroglycerin, made into square flakes 0.005 in (0.013 cm) thick or extruded in cruciform blocks.

NITROGEN. An element, symbol N, which at ordinary temperatures is an odorless and colorless gas. The atmosphere contains 78% nitrogen in the free state. It is nonpoisonous and does not support combustion. Nitrogen is often called an inert gas, and is used for some inert atmospheres for metal treating and in light bulbs to prevent arcing, but it is not chemically inert. It is a necessary element in animal and plant life, and is a constituent of many useful compounds. Lightning forms small amounts of nitric oxide from the air which is converted into nitric acid and nitrates, and bacteria continuously convert atmospheric nitrogen into nitrates. Nitrogen combines with many metals to form hard nitrides useful as wear-resistant metals. Small amounts of nitrogen in steels inhibit grain growth at high temperatures, and also increase the strength of some steels. It is also used to produce a hard surface on steels. Nitrogen has five isotopes, and **nitrogen 15** is produced in enrichments to 95% for use as a tracer.

Most of the industrial use of nitrogen is through the medium of nitric acid, obtained from natural nitrates or from the atmosphere. Fixation of nitrogen is a term applied to any process whereby nitrogen from the air is transferred into nitrogen compounds, or **fixed nitrogen,** such as nitric acid or ammonia. The first step is by passing air through an electric arc to produce **nitric oxide,** NO, a heavy, colorless gas, which oxidizes easily to form **nitrogen dioxide,** NO_2, a brown-colored gas of a disagreeable odor. This oxide reacts with water to form nitric acid. Or, atmospheric nitrogen can be converted to the oxide by irradiation of the compressed heated air with uranium oxide. **Calcium cyanamide,** $CaCN_2$, made by reacting atmospheric nitrogen with calcium carbide, is used as a fertilizer and as a chemical raw material. The chemical radical **cyanamid,** or **hydrogen cyanamide,** $H_2N \cdot C \cdot N$, is marketed as a stable, colorless 50% aqueous concentrate. The nitrogen-containing gas **Drycolene,** of the General Electric Co., used for furnace atmospheres for sintering metals, contains 78%

N_2, 20% CO, and 2% H_2. It is produced by burning hydrocarbon gases and air, removing the moisture, and passing through incandescent charcoal to convert the CO_2 and residual moisture to CO and H_2. Nitrogen liquefies at about −195°C, and solidifies at about −210°C. Nitrogen gas occupies 696 times as much space as the **liquid nitrogen** used in surgery.

NITROGLYCERIN. A heavy, oily liquid known chemically as **glyceryl trinitrate** and having the empirical formula $C_3H_5(NO_3)_3$. It is made by the action of nitric acid on glycerin in the presence of sulfuric acid. It is highly explosive, detonating upon concussion. Liquid nitroglycerin when exploded forms carbonic acid, CO_2, water vapor, nitrogen, and oxygen; 1 lb (0.5 kg) is converted into 156.7 ft^3 (4.4 m^3) of gas. The temperature of explosion is about 628°F (330°C). For use as a commercial explosive it is mixed with absorbents, usually kieselguhr or wood flour, under the name of **dynamite.** Cartridges of high density explode with greater shattering effect than those of low density. By varying the density and also the mixture of the nitroglycerin with ammonium nitrate, which gives a heaving action, a great diversity in properties can be obtained.

Dynamites are rated on the percentage, by weight, of the nitroglycerin that they contain. A 25% dynamite has 25% by weight of nitroglycerin, and has a rate of detonation of 11,800 ft/s (3,597 m/s). The regular grades contain from 25 to 60%. **Ditching dynamite** is the 50% grade. It has a rate of detonation of 17,400 ft/s (5,304 m/s), and will detonate sympathetically from charge to charge along a ditch line. **Extra dynamite** has half of the nitroglycerin replaced by ammonium nitrate. It is not so quick and shattering, and not as water-resistant, but is lower in cost. It is used for quarrying, for stump and boulder blasting, and for highway work. A 50% extra dynamite has a detonation rate of 10,800 ft/ s (3,292 m/s). **Hercomite** and **Hercotol** are extra dynamites of Hercules, Inc., while **Durox** is an ammonium dynamite of Du Pont, and **Agritol,** a low-velocity dynamite also of Du Pont, is a low-density ammonium dynamite for stump blasting.

Gelatin dynamite is made by dissolving a special grade of nitrocotton in nitroglycerin. It has less fumes, is more water-resistant, and its plasticity makes it more adaptable for loading solidly in holes for underground work. It is marketed as straight gelatin or as **ammonium gelatin,** called **gelatin extra.** The gelatin dynamites come in grades from 20 to 90%. All have a detonation rate of 8,500 ft/s (2,591 m/s), but modified high-pressure gelatin has rates to 19,700 ft/s (6,005 m/s). These, however, produce large amounts of fumes and are not for use in mines or confined spaces. **Blasting gelatin,** called **oil-well explosive,** is a 100% dense and waterproof gelatin with the appearance of crude rubber, and having a detonation rate of 8,500 ft/s (2,591 m/s). **Gelamite** and **Hercogel** are gelatin **blasting dynamites** of Hercules, Inc., although the **Bituminite,** of this company, is a slow permissible ammonium nitrate dynamite for coal mines. **Gelobel** is a gelatin dynamite, and **Monobel** is an ammonium dynamite marketed by Du Pont for mine blasting. The **Gelodyn explosive** of the Atlas Powder Co. is a combination of ammonium gelatin dynamite that is plastic, gives a shattering effect, and does not produce excessive fumes. It is used for construction blasting. **Amocol,** of this company, is a blasting explosive composed of grained ammonium nitrate mixed with ground coal. The double-base solid propellant for rockets, known as **ballistite,** is nitroglycerin-nitrocellulose. With potassium perchlorate as an oxidizer it gives a specific impulse of 180 to 195. It leaves plumes of white smoke. Dynamite is also sometimes used for explosive metal forming as it releases energy at a constant rate regardless of confinement, and produces pressures to 2 million lb/in^2 (1,378 MPa). For bonding **metal laminates** a thin sheet, or film, of the explosive is placed on top of the composite, and the progressive burning of the explosive across the film produces an explosive force downward and in vectors that produces a microscopic wave, or ripple, in the alloyed bond that strengthens the bond but is not visible on the laminated sheet.

NONDEFORMING STEEL. Also called **nonshrinking steel.** A group of alloy steels which have the characteristic that they do not easily deform, or go out of shape, when heat-treated. This property makes them suitable for making dies, gages, or tools that must be accurate. The usual nondeforming steel contains from 1 to 1.75% manganese, with or without chromium or other alloying elements. The carbon content is the same as tool steels of the same grade. The phosphorous, sulfur, and silicon impurities are kept as low as possible. The steels are oil-hardening, and do not have the tough core of ordinary tool steels. They have low resistance to shock and are thus not suited for bending or forming dies, except when they have additional alloying elements. But, except for such uses as blanking dies and reamers, the characteristic of nondeformation is not a leading specification in the procurement of a tool steel, and many alloyed tool steels are hardenable at relatively low temperatures and are nondeforming. The principle involved in the nondeforming quality is to obtain a steel so balanced in ingredients that there will be no serious phase changes, i.e., change in shape of the molecules, during the heating and cooling cycle. Structural steels for such uses as casement windows and doors must be nondeforming to assure close fit, and for this purpose many low-alloy steels are considered nondeforming.

Hi Wear 64 steel, of Carpenter Technology, a wear-resistant steel for feed rolls, molds for compacting powdered metals, and for blanking and forming dies for abrasive materials, contains 4% tungsten, 2 manganese, 1 molybdenum, 0.90 chromium, 0.25 silicon, and 1.50 carbon. When air-hardened from 1550°F (843°C) and tempered at 350°F (177°C), a 1-in (2.54-cm) piece will return to within 0.0005 in (0.0013 cm) of its original length, and it does not deform. The nondeforming steel marketed by the Cyclops Steel Co., under the name of **Wando,** contains 1.05% manganese, 0.50 chromium, 0.50 tungsten, and 0.95 carbon. It is employed for making intricate dies, taps, and other tools. **Mansil die steel,** of Henry Disston & Sons, Inc., is an oil-hardening steel containing 1.15% manganese, 0.90 carbon, 0.50 tungsten, and 0.50 chromium. It is deep-hardening, and is used for blanking dies, broaches, reamers, and gages. **Mangano,** of the Latrobe Electric and Steel Co., contains about 1.60% manganese, 0.20 chromium, and 0.95 carbon. **Stentor,** of Carpenter Steel, is a nondeforming tool steel containing 1.6% manganese, 0.25 silicon, and 0.90 carbon. It hardens at a low temperature range, 1400 to 1440°F (760 to 782°C), which aids in avoiding warpage. **Vega steel,** of the same company, is an air-hardening nondeforming steel that hardens uniformly at low temperatures even in large sections. It contains 2% manganese, 1.35 molybdenum, 1 chromium, 0.30 silicon, and 0.70 carbon. It hardens to Rockwell C61 from a temperature of 1550°F (843°C). **A-H steel,** of the Bethlehem Steel Co., also has high manganese, 2%, with 1.5 chromium, 1 molybdenum, and 1 carbon. It machines easily when annealed, and air-hardens to Rockwell C62. **Exl-Die steel,** of the Columbia Tool Steel Co., and **Saratoga steel,** of the Allegheny Ludlum Steel Co., have about 1.15% manganese, 0.50 chromium, 0.50 tungsten, and 0.90 carbon. **Amcoh steel,** of A. Milne & Co., called **hollow-die steel,** for ring dies and drawing dies, also has tungsten. It contains 1.25% manganese, 0.50 chromium, 0.50 tungsten, 0.95 carbon, and is oil-hardening. **Hargus steel,** of the Ziv Steel & Wire Co., for blanking dies, reamers, and gages, has 1% manganese, 0.35 nickel, and 1 carbon. A hollow-die steel of the Timkin Roller Bearing Co., called **Graph-Mo steel,** has 1% manganese, 1.29 silicon, 0.25 molybdenum, and 1.45 carbon. The high carbon with silicon gives free graphite to aid machining when annealed, and provides hard carbides for wear resistance when hardened. **Truform steel,** of the Jessop Steel Co., is a high-manganese, oil-hardening steel for cutters and dies. **Deward steel,** of the Allegheny Ludlum Steel Co., contains 1.55% manganese, 0.90 carbon, and 0.30 molybdenum. **Paragon steel,** of the Crucible Steel Co., has about this amount of manganese with 0.50 to 0.75% chromium and 0.25 vanadium. **Kiski steel,** of Braeburn, is similar.

Many of the high-chromium, wear-resistant steels are designated as nondeforming. **Airvan steel,** of Firth Sterling, Inc., used for heavy-duty blanking and forming dies, contains 5.25% chromium, 1.15 molybdenum, 0.25 vanadium, and 1 carbon. When hardened by air cooling, it has a hardness of Rockwell C64 with a compressive strength of 570,000 lb/in^2 (3,927 MPa), and when tempered at about 950°F (510°C) to remove strains, the hardness is Rockwell C59 with added toughness. **Chromovan steel,** of this company, is a nondeforming, wear-resistant steel for coining dies and thread-rolling dies. It has 12.5% chromium, 1.6 carbon, 0.80 molybdenum, and 1 vanadium. **A-H5 steel,** of Bethlehem Steel, is a die steel that has high resistance to distortion and is also wear-resistant and shock-resistant. It contains 1% carbon, 5.25 chromium, 1.1 molybdenum, 0.25 vanadium, and 0.60 manganese. It anneals to 212 Brinell and air-hardens to Rockwell C62. **Carpenter No. 484 steel** is an air-hardening steel for dies and tools. It has 5% chromium, 1 molybdenum, 0.20 vanadium, 0.70 manganese, 0.20 silicon, and 0.95 carbon. **Carpenter No. 883** is a tougher tool steel for forging dies and bulldozer dies. It has 5% chromium, 1.35 molybdenum, 0.45 vanadium, 1.10 silicon, 0.35 manganese, and 0.40 carbon. **Airtem steel** of the Lehigh Steel Co., used for blanking and forming dies, has 5.25% chromium, 1.25 molybdenum, 1 silicon, 0.50 manganese, 0.30 vanadium, and 1.30 carbon. It is also shock-resistant. **Air-kool steel,** of the Crucible Steel Co. of America, used for blanking and trimming dies and gages, has 5.25% chromium, 1.15 molybdenum, 0.50 vanadium, and 0.95 carbon. **Sagamore steel,** of Allegheny Ludlum, has 5% chromium, 1 molybdenum, 0.25 vanadium, and 1 carbon. **Ontario steel,** of the same company, is a steel for greater resistance to abrasion. It contains 12% chromium, 0.80 molybdenum, 0.25 vanadium, and 1.5 carbon.

NONMAGNETIC STEEL. Steel and iron alloys used where it is important that no magnetic circuits be set up or magnetic effects be induced. **Manganese steel** containing 14% manganese is nonmagnetic and casts readily but is not machinable. **Nickel steel** containing high nickel is also nonmagnetic. Many mills regularly produce nonmagnetic steels containing from 20 to 30% nickel. **Manganese-nickel steels** and **manganese-nickel-chromium steels** are nonmagnetic and may be arranged to combine desirable features of the nickel and manganese steels. A nonmagnetic steel of the Jessop Steel Co. has 10.5 to 12.5% manganese, 7 to 8 nickel, and 0.25 to 0.40 carbon. The electrical resistance is 70 $\mu\Omega/cm^3$ compared with 5 to 7 for brass. It has low magnetic permeability and low eddy-current loss, can be machined readily, and work-hardens only slightly. The tensile strength is 80,000 to 110,000 lb/in^2 (551 to 758 MPa), elongation 25 to 50%, and specific gravity 8.02. It is austenitic and cannot be hardened. The 18–8 austenitic chromium-nickel steels are also nonmagnetic. The nonmagnetic alloy used for watch gears and escapement wheels is not a steel but is a **cupronickel-manganese** containing 60% copper, 20 nickel, and 20 manganese. It is very hard, but can be machined with diamond tools.

NONSHATTERING GLASS. Also referred to as **shatterproof glass, laminated glass,** or **safety glass,** and when used in armored cars it is known as

bulletproof glass. A material composed of two sheets of plate glass with a sheet of transparent resinoid between, the whole molded together under heat and pressure. When subjected to a severe blow, it will crack without shattering. The first of these was a German product marketed under the name of **Kinonglas,** which consisted of two clear glass plates with a cellulose nitrate sheet between, and was first used for protective shields against chips from machines. Nonshattering glass is now largely used for automobile and car windows. The original cellulose nitrate interlining sheets had the disadvantage that they were not stable to light and became cloudy. Cellulose acetate was later substituted. It is opaque to actinic rays and prevents sunstroke but has the disadvantage of opening in cold weather, permitting moisture to enter between the layers. The acrylic resins are notable for their stability in this use; in some cases they are used alone without the plate glass, especially for aircraft windows. Polyvinyl acetal resins, as interlinings for safety glass, are weather-resistant and will not discolor. Polyvinyl butyral is much used as an interlayer, but in airplane glass at temperatures about 150°F (66°C) it tends to bubble and ripple. Silicone resins used for this purpose will withstand heats to 350°F (177°C), and they are not brittle at subzero temperatures. **Silastic Type K,** of the Dow Chemical Co., is such a silicone resin used as an interlayer. **Flexseal,** of the Pittsburgh Plate Glass Co., is a laminated plate glass with a vinyl resin interplate with an extension for sealing into the window frame. It will withstand pressure of 20 lb/in^2 (0.14 MPa), with a ⅛-in (0.32-cm) plastic interplate, and is used for aircraft windows. **Duplate** is the trade name of the Duplate Corp. for a nonshattering glass. Standard bulletproof glass is from 1½ in (3.81 cm), 3 ply, to 6 in, 5 or more ply.

NONWOVEN FABRIC. In the most general sense, fibrous-sheet materials consisting of fibers mechanically bonded together either by interlocking or entanglement, by fusion, or by an adhesive. They are characterized by the absence of any patterned interlooping or interlacing of the yarns. In the textile trade, the terms nonwovens or **bonded fabrics** are applied to fabrics composed of a fibrous web held together by a bonding agent, as distinguished from felts, in which the fibers are interlocked mechanically without the use of a bonding agent. There are three major kinds of nonwovens based on the method of manufacture. **Dry-laid nonwovens** are produced by textile machines. The web of fibers is formed by mechanical or air-laying techniques, and bonding is accomplished by fusion-bonding the fibers or by the use of adhesives or needle punching. Either natural or synthetic fibers, usually 1 to 3 in (2.5 to 7.6 cm) in length, are used. **Wet-laid nonwovens** are made on modified papermaking equipment. Either synthetic fibers or combinations of synthetic fibers and wood pulp can be used. The fibers are often much shorter than those used in dry-laid fabrics, ranging from ¼ to ½ in (0.64 to 1.27 cm) in length. Bonding is usually accomplished by a fibrous binder or an adhesive. Wet-laid nonwovens can also be produced as composites, for example, tissue-paper laminates bonded to a reinforcing substrate of scrim. **Spin-bonded nonwovens** are produced by allowing the filaments emerging from the fiber-producing extruder to form into a random web, which is then usually thermally bonded. These nonwovens are limited commercially to thermoplastic synthetics such as nylons, polyesters, and polyolefins. They have exceptional strength because the filaments are continuous and bonded to each other without an auxiliary bonding agent. Fibers in nonwovens can be arranged in a great variety of configurations that are basically variations of three patterns: parallel or unidirectional, crossed, and random. The parallel pattern provides maximum strength in the direction of fiber alignment, but relatively low strength in other directions. Cross-laid patterns (like wovens) have maximum strength in the directions of the fiber alignments and less strength in other directions. Random nonwovens have relatively uniform strength in all directions.

NUTMEG. The brown, round, wrinkled seed of the plumlike fruit of the evergreen tree *Myristica fragrans,* native to the Moluccas but now grown extensively also in Grenada. The bright-red aril covering of the seed is called **mace.** The trees average about 20 lb (9 kg) of kernels per year, but a large tree may bear as many as 10,000 nutmegs annually. The average yield in Grenada is taken as 1,500 lb (680 kg) of green nutmegs per acre per year, giving 720 lb (327 kg) of dry sound nutmegs and 150 lb (68 kg) of mace per acre. The nutmeg tree grows best on tropical islands at a height of 500 to 1,500 ft (152 to 457 m) above sea level. It begins to bear at 6 years, and will bear for a century. The ripe fruit splits, and the seeds fall to the ground. Nutmeg is a delicately flavored spice for foodstuffs, but in large amounts is highly toxic. Mace has a finer but weaker flavor and is used as a savory, but the **oleoresin mace** of Fritzche Bros., Inc., a dark-brown liquid produced from mace, gives a lasting spicy nutmeg flavor, and is used as a substitute for nutmeg oil. **Nutmeg butter** is a solid yellow fat obtained from the rejected nutmegs of the spice trade. To obtain the fat the kernels are roasted and ground before extraction. The nutmeg contains about 40% of the fat. It is used chiefly in ointments. **Nutmeg oil** is an essential oil extracted from the nutmeg and used in medicine, in flavoring tobacco, and in dentifrices. It is also called **myristica oil** and is high in **myristicin,** a yellow poisonous oil of the composition $C_3H_5 \cdot C_6H_2(O_2CH_2)OCH_3$. It is now synthesized from pine oil.

NUX VOMICA. The seeds of the ripe fruit of the deciduous tree *Strychnos nux vomica* of India, Ceylon, and Australia, used as the source of the alkaloids strychnine and brucine. The powdered seed may also be used. The fruits contain three to five hard grayish seeds which yield 1 to 1.25% strychnine alkaloid and about the same amount of brucine. **Strychnine** is an odorless, crystalline, intensely bitter powder of the composition $C_{21}H_{22}N_2O_2$ with a very complex multiring molecular structure. It is a spinal stimulant and in quantity is a violent convulsive poison. It is used in proprietary and prescription medicines of the tonic class, and also in rat poisons. For medicinal use it is employed mostly in the form of *strychnine sulfate* which is easily soluble in water. **Brucine** is a bitter crystalline alkaloid of the composition $C_{23}H_{26}N_2O_4$ with similar characteristics but much less active. It is **dimethoxystrychine.** It is also used as a denaturant for rapeseed oil and other industrial oils. The woody vine **woorali,** *S. toxifera,* of the Amazon and Orinoco valleys, from which the arrow poison **curare** was obtained, contains strychnine and **curine,** a benzyl isoquinoline alkaloid. Curare inactivates the motor nerves without affecting the sensory and central nervous system, and is used in medicine as a local anesthetic. The synthetic **Mytolon,** of Winthrop-Stearns, is used as a more potent and safer substitute. It is a complex diethylaminopropylaminobenzoquinone benzyl chloride in the form of red crystals.

NYLON. A group of **polyamide resins** which are long-chain polymeric **amides** in which the amide groups form an integral part of the main polymer chain, and which have the characteristic that when formed into a filament the structural elements are oriented in the direction of the axis. Nylon was originally developed as a textile fiber, and high tensile strengths, above 50,000 lb/in^2 (344 MPa), are obtainable in the fibers and films. But this high strength is not obtained in the molded or extruded resins because of the lack of oriented stretching. When **nylon powder** that has been precipitated from solution is pressed and sintered, the parts have high crystallinity and very high compressive strength, but they are not as tough as molded nylon.

Nylons are produced from the polymerization of a dibasic acid and a diamine. The most common one of the group is that obtained by the reaction of adipic acid with hexamethylenediamine. The nylon molding and extruding resin of Du Pont, **Elvamide 8042,** formerly **Zytel 42,** has a tensile strength of 12,500 lb/in^2 (106 MPa) with elongation above 100%, a flexural strength of 13,800 lb/^{2} (95 MPa), hardness of Rockwell R118, a flow temperature of 480°F (249°C) and a dielectric strength of 350 volts per mil (14×10^6 volts per meter).

All of the nylons are highly resistant to common solvents and to alkalies, but are attacked by strong mineral acids. Molded parts have light weight, with a specific gravity about 1.14, good shock-absorbing ability, good abrasion resistance, very low coefficient of friction, and high melting point, up to about 482°F (250°C). A disadvantage is their high water absorption and the resulting dimensional changes in moldings in service. They are much used for such parts as gears, bearings, cams, and linkages. The electrical characteristics are about the same as those of the cellulosic plastics. As a wire insulation, nylon is valued for its toughness and solvent resistance. **Nylon fibers** are strong, tough, and elastic, and have high gloss. The finer fibers are easily spun into yarns for weaving or knitting either alone or in blends with other fibers, and they can be crimped and heat-set. The **Nyloft fiber** of the Firestone Tire & Rubber Co., used for making carpets, is nylon staple fiber, lofted, or wrinkled, to give the carpet a bulky texture resembling wool. **Caprolan tire cord,** made from nylon 6 of high molecular weight, has the yarn drawn to four or five times its original length to orient the polymer and give one-half twist per inch. **Nylon film** is made in thicknesses down to 0.002 in (0.005 cm) for heat-sealed wrapping, especially for food products where tight impermeable enclosures are needed. **Nylon sheet,** for gaskets and laminated facings, comes transparent or in colors in thicknesses from 0.005 to 0.060 in (0.013 to 0.152 cm). **Nylon monofilament** is used for brushes, surgical sutures, tennis strings, and fishing lines. Filament and fiber, when stretched, have a low specific gravity down to 1.068, and the tensile strength may be well above 50,000 lb/in^2 (344 MPa). Nylon fibers made by condensation with oxalic acid esters have high resistance to fatigue when wet.

Nylon 6 is made from **caprolactam,** which has the empirical formula $(CH_2)_5NH \cdot C{:}O$, with a single 6-carbon ring. Molded parts have a tensile strength of 11,700 lb/in^2 (79 MPa), elongation 70%, dielectric strength 440 volts per mil (17.3×10^6 volts per meter) and melting point 420°F (216°C). **Nylon foam,** or **cellular nylon,** for lightweight buoys and flotation products, is made from nylon 6. The foam is produced by Du Pont in slabs, rods, and sheets. Densities range from 1 to 8 lb/ft^3 (16 to 128 kg/m^3). The low densities are flexible, but the high-density material is rigid with a load-carrying capacity about the same as that of balsa wood. **Nylon 6/10** is tough, relatively heat-resistant, and has a very low brittleness temperature. It absorbs about one-third as much moisture as type 6 and half as much as type 6/6. **Nylon 9** is made from soybean oil by reacting with ozone. It has better water resistance than other nylons and is used for coatings. **Nylon 11,** originally marketed in France as **Rilsan,** is a polycondensation product of **aminoundeconoic acid** which is made by a complex process from the recinoleic acid of castor oil. This type of nylon has superior dimensional stability and is valued for injection moldings. **Nylon 12** is a similar plastic. It is a lauro lactam synthesized from butadiene. It has a low water absorption and good strength and stability, and it is used for packaging film, coatings for metals, and moldings. **Nylon 4** is a polypyrrolidine used for textile fibers. The molecular chain has more amide groups than do the chains of other nylons, and its ability to absorb moisture is about the same as that of cotton. Fabrics made from it do not have the hot feel usual with other synthetic fibers and they have better pressability and are free of static. **Alrac Nylon,** of Radiation Research Corp., is nylon 4. Moldings of **nylon 6/6** have a specific gravity of 1.14 and have a tensile strength of 11,500 lb/in^2 (78 MPa). It is used for gears and mechanical parts, and its physical properties may be further increased by adding glass fibers or spheres as fillers. **Nylon copolymers** of types 6 and 6/6 are flexible materials with extra-high impact resistance, even at −40°F (−40°C), and good heat resistance.

Nylon products are marketed under many trade names. **Nylatron G,** of the Polymer Corp., is graphite-impregnated nylon in rods and strip for making gears, bearings, and packings. **Nylasint,** of this company, for bearings, is sintered nylon impregnated with oil. **Flalon,** of Burgess-Berliner Assoc., is a **nylon flannel** of 15-denier crimped fibers carded on both faces. It has the appearance of cotton flannel, but is superior in heat resistance and wear resistance. **Raynile,** of Hewitt-Robbins, Inc., is a rayon-nylon fabric with nylon traverse threads. It is flexible and has about twice the strength of cotton fabric. It is used for conveyor belts. **Fiberthin,** of the U.S. Rubber Co., is a thin waterproof fabric used to replace heavier tarpaulins for protective coverings. It is woven of nylon and coated with plastic. It weighs 5 oz/yd^2 (0.17 kg/m^2) and has a tensile strength of 175 lb/in (31 kg/cm) of width. **Facilon,** of Sun Chemical Corp., a caprolan nylon fabric impregnated with vinyl resin, is used for facing wall panels and flexible floor coverings. It comes in colors and embossings.

OAK. The wood of a large variety of oak trees, all of the natural order *Cupuliferae,* genus *Quercus.* European oak, under various names, such as **Austrian oak** and **British oak,** are from two varieties of the tree *Q. robur.* The wood is light brown in color, with a coarse, open grain, firm texture, and weight of about 45 lb/ft^3 (720 kg/m^3). **American red oak** is from the tree *Q. rubra* or *Q. falcata.* It is also called **black oak,** although black oak is from the *Q. velutina,* and the **red oak** of the Lake states is *Q. borealis.* The heartwood is reddish brown, and the sapwood whitish. **Southern red oak** of the Gulf Coast, a valued wood for furniture and cabinetwork, is the **shumard oak,** *Q. shumardii,* also known as **Schneck oak** and **Texas oak. Nuttall oak,** *Q. nuttallii,* of the lower Mississippi Valley, is also called red oak. American **white oak** is from the tree *Q. alba* of the eastern states. The heartwood is brown, and the sapwood white. The grain of these species is coarse, but the texture is firm. **Post oak,** of the southern states, is *Q. stellata.* **Chestnut oak,** of the Appalachian range, is *Q. montana,* but this name is also applied to the **chinquapin oak,** *Q. muehlenbergii,* a large tree which grows profusely over a wide area of the eastern half of the United States, and was early valued for railroad ties and heavy construction timbers. **Overcup oak,** *Q. lyrata,* is an important tree from New Jersey to Texas. **Scarlet oak,** of Pennsylvania, is *Q. coccinea.* **Western white oak,** *Q. garryana,* has a more compact texture and straighter grain. **Spanish oak,** *Q. oblongifolia,* is native to California and New Mexico. The grain is finer and denser. American oaks are widely distributed in the United States and Canada. There are more than 400 varieties of oak on the North American continent. An enormous stand of oak in Costa Rica is made up of immense trees of **copey oak,** *Q. copeyensis,* the trees being up to 8 ft (2.4 m), in diameter with clean boles to 80 ft (24.4 m) to the first limb. The wood has a hardness between that of white and live oaks, and the bark has a high content of tannin.

Oak is used for flooring, furniture, cask staves, and where a hard, tough wood is needed. For cabinetwork the boards are variously sawed at angles and quarters to obtain grain effects known as **quartered oak. Fumed oak** is not a kind of oak, but a finish produced by the action of ammonia vapor. **Butt oak,** or **pollard oak,** also known as **burwood,** is the wood of the decapitated **European oak** trees, *Q. pedunculata* and *Q. sessiliflora,* of Great Britain. A pollard tree is one whose head has been cut for ornamental purposes. The growth in height is permanently arrested and innumerable branches shoot out from the trunk, which produce humps, or burrs, with the grain of the wood running in all directions. **Burr oak** is valued for ornamental work. Burr oak of the northern and central United States is not a pollard oak but is a name for the tree *Q. macrocarpa.* The commercial red and white oaks have an average specific gravity when kiln-dried of 0.69. The compressive strength perpendicular to the grain is 1,870 lb/in^2 (13 MPa) with shearing strength parallel to the grain of 1,300 lb/in^2 (9 MPA).

The woods often called oaks in the Southern Hemisphere are not true oaks. **Australian oaks** are from a variety of trees, and **Chilean oak** is from a species of beech. **Beef oak,** of Australia, is a hard, heavy, brownish wood from the tree *Grevillea striata.* It has an irregular grain. **She oak** is from the Australian tree *Casuarina stricta,* and **swamp oak** is from *C. suberosa.* These woods are lighter in weight than oak. **Silky oak,** used for cabinetwork, is a brownish wood that has a uniform texture and can be quartersawn to show attractive figuring. It is from the tree *Cardwellia sublimis* of Australia.

Oak extract, which is an important tanning material for the best grades of heavy leather, is chiefly from the bark of the **swamp chestnut oak,** *Q. prinus,* but also from the white oak and red oak. The **tanbark oak** of California is the tree *Lithocarpus densiflora.* The extract of the **scarlet oak,** *Q. coccinea,* is dark in color and is known as **quercitron extract.** The bark of the tanbark oak yields 10 to 14% tannin, but the extract contains 25 to 27% tannin. **Quercetin** is a complex phenyl benzyl pyrone derived from oak bark and from Douglas fir bark. It is an antioxidant and absorber of ultraviolet rays, and is used in rubber, plastics, and in vegetable oils. **Valonia** consists of the acorn cups of the oak *Q. aegilops* of Asia Minor and the Balkans. **Smyrna valonia** contains 32 to 36% tannin which produces a light-colored, lightweight leather with a firm texture and bloom. When used alone, however, valonia makes a brittle leather, and is thus always used in blends. Valonia is marketed as cups or as extract, the latter containing about 60% tannin.

OATS. An important grain which is the seed of the tall plant *Avena sativa.* The grain is surrounded by a hull, and grows in many spikelets as a spreading or one-sided panicle inflorescence. It can be grown farther north than any other grain except rye, and on poor soils. Although it is one of the most nutritious of grains, most of the oats grown in the United States are used for animal feed. **Rolled oats** and **oatmeal** are used as cereal foods and for some bakery products, but the grain is not suitable for breadmaking. **Oat hulls** are used for the production of furfural and other chemicals. The largest production of oats is in the United States and Russia, but large quantities are produced in Canada, western Europe, and Argentina. It is the chief grain crop of Scotland. The yield per acre (4,047 m^2) in the United States is about 30 bu (1 m^3) but it is twice that figure in Great Britain. Oats are often called by the Spanish name **avena** in international trade. **Turkish oats,** cultivated in central Europe, are from the species *A. orientalis.* **Horse gram,** used as a substitute for oats in India, is from the plant *Dolichus bifloris.* The **gram,** from the *Cicer arientinum,* is an important food grain in India.

OCHRE. A compact form of earth used for paint pigments and as a filler for linoleum. It is an argillaceous and siliceous material, often containing compounds of barium or calcium, and owing the yellow, brown, or red colors to hydrated iron oxide. The tints depend chiefly upon the proportions of silica, white clay, and iron oxide. Ochres are very stable as pigments. They are prepared by careful selection, washing, and grinding in oil. They are inert, and are not affected by light, air, or ordinary gases. They are rarely adulterated, because of their cheapness, but are sometimes mixed with other minerals to alter the colors. **Chinese yellow** and many other names are applied to the ochres. **Golden ochre** is ochre mixed with chrome yellow. **White ochre** is ordinary clay. A large part of the American ochre is produced in Georgia. **Sienna** is a brownish-yellow ochre found in Italy and Cyprus. The material in its natural state is called **raw sienna. Burnt sienna** is the material calcined to a chestnut color. Indian red and Venetian red are hematite ochres.

Vandyke brown is a deep-brown pigment made originally from lignitic ochre from Cassel, Germany. It was named after the Dutch painter Van Dyck, and is also called **Cassel brown, Cassel earth,** and **Rubens brown.** It is also obtained from low-grade coals of Oklahoma and California. Imitation Vandyke brown is made from a mixture of lampblack, yellow ochre, and iron oxide. **Cologne earth** is a Vandyke brown made from American clays which are mixtures of ochre, clay, and bituminous matter, roasted to make the color dark. **Yellow ochre** and **brown ochre** are limonite, but yellow iron oxide is made in Germany by the aeration of scrap iron in the presence of copperas. **Umber** is a brown siliceous earth colored naturally with iron oxides and manganese oxide. It comes chiefly from Italy and Cyprus. For use as a pigment it is washed with water and finely ground. It is inert and very stable. **Cyprus umber** is a rich coffee-brown color and as a pigment has good covering qualities. It is a modified marl with impregnations of iron and manganese. **Burnt umber** is redder in color than umber, and is made by calcining the raw umber. **Caledonian brown** and **Cappagh brown** are varieties of umber found in Great Britain.

OILCLOTH. A fabric of woven cotton, jute, or hemp, heavily coated with turpentine and resin compositions, usually ornamented with printed patterns, and varnished. It was employed chiefly as a floor covering, but a light, flexible variety having a foundation of muslin is used as a covering material. This class comes in plain colors or in printed designs. It was formerly the standard military material for coverings and ground protection, but has been replaced by synthetic fabrics. **Oilskin** is a cotton or linen fabric impregnated with linseed oil to make it waterproof. It was used for coverings for cargo and for waterproof coats, but has now been replaced by coated fabrics. **Oiled silk** is a thin silk fabric impregnated with blown linseed oil which is oxidized and polymerized by heat. It is waterproof, very pliable, and semitransparent. It was much used for linings, but has now been replaced by fabrics coated with synthetics.

OILS. A large group of fatty substances which are divided into three general classes: vegetable oils, animal oils, and mineral oils. The **vegetable oils** are either fixed or volatile oils. The fixed oils are present in the plant in combined form, and are largely glycerides of stearic, oleic, palmitic, and other acids, and they vary in consistency from light fluidity to solid fats. They nearly all boil at 500 to 600°F (260 to 316°C), decomposing into other compounds. The volatile, or essential, oils are present in uncombined form and bear distillation without chemical change.

Seed oils, or **oilseeds,** obtained from various plant seeds, are fatty acids of varying chain lengths containing hydroxy, keto, epoxy, and other functional groups. The oils are chemically very pure. Among important uses of these oils are for polymers, surface coatings, plasticizers, surfactants, and lubricants. The seeds of the **Chinese tallow tree** are coated with a semisolid fat. An oil similar to linseed oil is inside the kernel. The oil can be used as a substitute for cocoa butter and for fatty acids in cosmetics.

Fish oils are thick, with a strong odor. Vegetable and animal oils are obtained by pressing, extraction, or distillation. Oils that absorb oxygen easily and become thick are known as drying oils and are valued for varnishes, because on drying they form a hard, elastic, waterproof film. Unsaturation is proportional to the number of double bonds, and in food oils these govern the cholesterol depressant effect of the oil. Oils and fats are distinguished by consistency only, but waxes are not oils. **Mineral oils** are derived from petroleum or shale and are classified separately. The most prolific sources of vegetable oils are palm kernels and copra. About 2,500 lb (1,134 kg) of palm oil is produced per acre (4,047 m^2) annually, and the yield of coconut oil per acre (4,047 m^2) from plantation plantings is 1,200 lb (544 kg). This compares with 350 lb (159 kg) of oil per acre (4,047 m^2) from peanuts and 200 lb (91 kg) per acre (4,047 m^2) from soybeans. Under comparable aggressive plantation work, from 10 to 20 times more palm and coconut oil can be produced per acre than peanut or soybean oil. Babassu oil is almost chemically identical with coconut oil, and vast quantities of babassu nuts grow wild in northeast Brazil.

Blown oils are fatty oils that have been oxidized by blowing air through them while hot, thereby thickening the oil. They are mixed with mineral oils to form special heavy lubricating oils, such as marine engine oil, or are employed in cutting oils. They are also used in paints and varnishes, as the drying power is increased by the oxidation. The flash point and the iodine value are both lowered by the blowing. The oils usually blown are rapeseed, cottonseed, linseed, fish, and whale oils. The **blown fish oils** of the Archer-Daniels-Midland Co., used for paints, enamels, and printing inks, are preoxidized and destearinized, and have specific gravities from 0.980 to 1.025. **Crystol oils,** of this company, are kettle-boiled fish oils for paints.

OILSTONE. A fine-grained, slaty silica rock used for sharpening edged tools. The bluish-white and opaque white oilstones of fine grain from Arkansas are called **novaculite,** and received their name because they were originally used for razor sharpening. They are composed of 99.5% chalcedony silica and are very hard with a fine grain. Novaculite is a deposit from hot springs. It is fine-grained, and the ordinary grades are employed for the production of silica refractories. Arkansas oilstones are either hard or soft and have a waxy luster. They are shipped in large slabs or blocks, or in chips for tumbling barrel finishing. **Washita oilstone,** from Hot

Springs, Ark., is a hard compact white stone of uniform texture. **Ouachita stones** come in larger and sounder pieces but are coarser than the Arkansas. **Water-of-Ayr stone,** also known as **Scotch hone,** is a fine sandstone used with water instead of with oil. **Artificial oilstones** are also produced of aluminum oxide. **India oilstone** was originally blocks of emery, but the name now may refer to aluminum oxide stones.

OITICICA OIL. A drying oil obtained from the kernels of the nuts of the tree *Licania rigida* of northeastern Brazil. The oil contains about 80% **licanic acid,** which, like the **eleostearic acid** of tung oil, gives a greater drying power than is apparent from the iodine value. The specific gravity is 0.944 to 0.971, saponification value 187 to 193, and iodine number 142 to 155. The properties as a varnish oil are much like those of tung oil, both producing wrinkled films when applied pure, and both lacking high gloss. **Cicoil** is a name for a treated oiticica oil with improved qualities. The oiticica nuts are 1 to 2 in (2.5 to 5.1 cm) long with the kernel about 60% of the nut, yielding about 60% oil. The average yield per tree is 350 lb (159 kg) of nuts, but a full-grown tree may yield 10 times that amount. Another species of the tree, *L. crassifolia,* of Surinam, yields a similar oil. Mexican oiticica is from the nuts of another species and is called **cacahuanache oil.** The kernels yield 69% of light-colored heavy oil.

OLEFIN COPOLYMERS. A group of polyolefin plastics. They are thermoplastics, and are also referred to as **polyolefin copolymers.** The principal olefin copolymers are the polyallomers, ionomers, and ethylene copolymers. The **polyallomers,** which are highly crystalline, can be formulated to provide high stiffness and medium impact strength, moderately high stiffness and high impact strength, or extra-high impact strength. Polyallomers, with their unusually high resistance to flexing fatigue, have "hinge" properties better than those of polypropylenes. They have the characteristic milky color of polyolefins; they are softer than polypropylene but have greater abrasion resistance. Polyallomers are commonly injection-molded, extruded, and thermoformed, and they are used for such items as typewriter cases, snap clasps, threaded container closures, embossed luggage shells, and food containers.

Ionomers are nonrigid plastics characterized by low density, transparency, and toughness. Unlike polyethylenes, density and other properties are not crystalline-dependent. Their flexibility, resilience, and high molecular weight combine to provide high abrasion resistance. They have outstanding low-temperature flexural properties, but should not be used at temperatures above 160°F (71°C). Resistance to attack from organic solvents and stress-cracking chemicals is high. Ionomers have high melt strength for thermoforming and extrusion coating, and a broad temperature range for blow molding and interjection molding. Representative ionomer parts include injection-molded containers, housewares, tool handles, and closures; extruded film, sheet, electrical insulation, and tubing; and blow-molded containers and packaging.

There are four commercial **ethylene copolymers,** of which ethylene vinyl acetate (EVA) and ethylene ethyl acrylate (EEA) are the most common.

Ethylene vinyl acetate, or **EVA copolymers,** approach elastomers in flexibility and softness, although they are processed like other thermoplastics. Many of their properties are density-dependent, but in a different way from polyethylenes. Softening temperature and modulus of elasticity decrease as density increases, which is contrary to the behavior of polyethylene. Likewise, the transparency of EVA increases with density to a maximum that is higher than that of polyethylenes, which become opaque when density increases above around 0.935 g/cm^3. Although EVA's electrical properties are not as good as those of low-density polyethylene, they are competitive with vinyl and elastomers normally used for electrical products. The major limitation of EVA plastics is their relatively low resistance to heat and solvents, the Vicat softening point being 147°F (64°C). EVA copolymers can be injection-, below-, compression-, transfer-, and rotationally molded; they can also be extruded. Molded parts include appliance bumpers and a variety of seals, gaskets, and bushings. Extruded tubing is used in beverage vending machines and for hoses for air-operated tools and paint spray equipment.

Ethylene ethyl acrylate, or **EEA copolymer,** is similar to EVA in its density-property relationships. It is also generally similar to EVA in high-temperature resistance, and like EVA it is not resistant to aliphatic and aromatic hydrocarbons as well as chlorinated versions thereof. However, EEA is superior to EVA in environmental stress cracking and resistance to ultraviolet radiation. As with EVA, most of EEA's applications are related to its outstanding flexibility and toughness. Typical uses are household products such as trash cans, dishwasher trays, flexible hose and water pipe, and film packaging.

Two other ethylene copolymers are **ethylene hexene,** or **EH copolymer,** and **ethylene butene,** or **EB copolymer.** Compared with the other two, these copolymers have greater high-temperature resistance, their useful service range being between 150 and 190°F (66 and 88°C). They are also stronger and stiffer, and therefore less flexible, than EVA and EEA. In general, EH and EB are more resistant to chemicals and solvents than the other two, but their resistance to environmental stress cracking is not as good.

OLEIC ACID. Also called **red oil, elaine oil, octadecenoic acid,** and **rapic acid,** although the latter is a misnomer based on a former belief that it was the same as the erucic acid of rapeseed. It occurs in most natural fats and oils in the form of the glyceride, and is obtained in the process of saponification or by distillation. Much of this acid is obtained from lard and other animal fats, but **Emery 3758-R,** of Emery Industries, is produced from soybean or other vegetable sources. It is an oily liquid with a specific gravity of 0.890, boiling at 286°C. Below about 14°C it forms colorless needles. It is a complex acid of the composition $CH_3 (CH_2)_7CH{:}CH(CH_2)_7COOH$, and if heated to the boiling point of water it reacts with oxygen to form a complex mixture of acids, including a small percentage of acetic and formic acids. When it is hydrogenated in food fats it converts to stearic acid. When reacted with potassium hydroxide it is converted into an acetate and a palmitate. It is also readily converted to pelargonic and other acids for making plastics. Oleic acid is a basic foodstuff in the form of the glyceride, and the acid has a wide use for making soaps, as a chemical raw material, and for finishing textiles. In soluble oils and cutting compounds it forms **sodium oleate,** $C_{17}H_{33}COONa$. The two commercial grades of oleic acid, yellow and red, are known as distilled red oil and saponified. They may be sold under trade names. **Alcholein 810,** of Arnold, Hoffman & Co., Inc., is a clear, distilled red oil used for textile treating. **Monoenoic acid** is a modified isomer of oleic acid which produces soaps that are nonirritating to the skin. It is used in cosmetics. **Hydrofol C-18,** of the Archer-Daniels-Midland Co., is this acid.

OLIVE OIL. A pale greenish oily liquid extracted from the ripe fruit of the olive tree, *Olea europaea,* a small evergreen grown largely in the Mediterranean countries but also in California and Argentina. The fruits are eaten ripe (purple) and green. They are rich in oil, and vast quantities are crushed for oil. The oil contains 69 to 85% oleic acid, 7 to 14 palmitic acid, 4 to 12 linoleic acid, with some stearic, arachidic, and yristic acids. The specific gravity is 1.912, iodine value 85, and saponification value 190. The best grades of the oil are used for food chiefly as a salad and cooking oil, and in canning sardines, but some is used in the manufacture of castile soaps. The industrial oil consists of the **olive oil foots** obtained in the third pressing or in the last extraction with carbon bisulfide, and is used for finishing textiles, degumming silk, and in soaps. **Florence oil** is a grade of Italian olive oil. In Italy olive oil is also known as **Lucca oil. Synthetic olive oil,** or **olive-infused oil,** is used as a foodstuff. It is made from highly refined corn oil by infusing the corn oil with about 20% of a paste made of finely ground, partly dehydrated ripe olives ground with a small amount of corn oil. The olive-infused oil has the flavor of olive oil, and also contains carotene, or vitamin A, contained in the olive pulp. Other fractionated oils reblended to give high oleic acid content are also used as substitutes for olive oil. **Olevene,** of Jacques Wolf & Co., is a sulfonated synthetic oil used instead of olive oil for treating textiles.

OLIVINE. A translucent mineral, usually occurring in granular form, employed as a refractory. The formula is usually given as $(Mg \cdot Fe)_2 \cdot SiO_4$, but it is a solid solution of **forsterite,** $2MgO \cdot SiO_2$, and **fayalite,** $2FeO \cdot SiO_2$. The fayalite lowers the refractory quality, but forsterite is not found alone. The mineral is also called **chrysolite,** and the choice green stones used as gems are called **peridot. Dunite** deposits in Washington and North Carolina carry up to 90% olivine which has only 5 to 15% fayalite. It is olive green in color, vitreous, with a hardness of 6.5 to 7 and a specific gravity of 3.3 to 3.5. As a refractory it is neutral up to about 1600°C but may then react with silica. The fayalite fuses out at 2700°F (1482°C), making the material porous and subject to attack by iron oxide. Although the name olivine indicates a green color, not all is green. Dunite takes its name from Dun Mountain of New Zealand, dun being the Irish and Scottish word for reddish brown. The melting point of forsterite is 3470°F (1912°C). When used mixed with chrome ore, the low-fusing elements form a black glass which presents a nonporous face. Some refractory material marketed as forsterite may be olivine blended with magnesite, or may be serpentine treated with magnesite. **Forsterite firebrick** in the back walls of basic open-hearth steel furnaces gives longer life than silica brick but only two-thirds that of chrome-magnesite brick. Forsterite refractories are usually made from olivine rock to which MgO is added to adjust the composition to $2MgO \cdot SiO_2$. **Monticellite,** $CaMgSiO_4$, may also occur with forsterite. They are also made by synthetic mixtures of MgO and silica. The thermal expansion of olivine is lower than that of magnesite. **Olivine sand** is substituted for silica sand as a foundry sand where silica is expensive. There are large deposits of olivine in the Pacific Northwest. When olivine is used as a foundry sand, it is noted that the heat-resisting qualities decrease with particle size. Olivine contains from 27 to 30% magnesium metal, and is also used to produce magnesium by the electrolysis of the chloride. **Magnesium-phosphate** fertilizer is made by fusing olivine with phosphate rock at 1600°C, tapping off the iron, and spray-cooling and crushing the residue. It contains 20% citric acid-soluble phosphate, 14% MgO, 29% CaO, and 23% SiO_2, and is useful for acid soils.

ONYX. A variety of chalcedony silica mineral differing from agate only in the straightness of the layers. The alternate bands of color are usually white and black, or white and red. Onyx is artificially colored in the same way as agate. It is used as an ornamental building stone, usually cut into slabs, and for decorative articles. **Onyx marble** is limestone with impurities arranged in banded layers. American onyx comes largely from Arizona, California, and Montana. **Mexican onyx** is banded limestone obtained from stalactites in caves. These materials are cut into such articles as lamp stands. **Argentine onyx** is a dark-green or a green-yellow translucent stone of great decorative beauty. In the United States it is called **Brazilian onyx** and is used for bookends, lamp bases, inkstands, and ornaments. **Opalized wood** is an onyxlike petrified wood from Idaho. It is cut into ornaments.

OPACIFIERS. Materials used in ceramic glazes and vitreous enamels primarily to make them nontransparent, but opacifiers may also enhance the luster, control the texture, promote craze resistance, or stabilize the color of the glaze. An opacifier must have fire resistance so as not to vitrify or decrease the luster. Tin oxide is a widely used white opacifier, and up to 3% also increases the fusibility of the glaze or enamel. Titanium oxide adds scratch hardness and high acid resistance to the enamel. It also increases the flow, making possible thinner coats which minimize chipping. Opacifiers may also serve as the pigment colors. Thus, cobalt oxide gives a blue color, and platinum oxide gives a gray. Lead chromate gives an attractive red color on glazes fired at 900°C, but when fired at 1000°C the lead chromate decomposes and a green chromium oxide is formed. If the glaze is acid, the basic lead chromate is altered and the color tends toward green. **Lufax 77A,** of the Rohm & Haas Co., is a crystalline zirconia which provides nuclei for the formation of zirconia crystals from the molten enamel, adding gloss and opacity and stabilizing the color on the blue side. Antimony oxide as an opacifier gives opaque white enamels of great brilliance but is expensive and poisonous. The **zirconium opacifiers** have a wide range of use from ordinary dishes to high-heat electrical porcelain and sanitary-ware enamels. The amount of zirconium oxide used is a minimum of 3%. The opacifiers may be in prepared form with lead oxide or other materials to give particular characteristics. **Opax,** of the Titanium Alloy Mfg. Co., is a zirconium oxide with small percentages of silica, sodium oxide, and alumina. It is used for hard-glaze dinnerware and wall-tile glaze. **Zircopax** is zirconium silicate, $ZrSiO_4$, with 33.5% silica in the molecule. It gives color stability and craze resistance. **Superpax,** of the same company, is a finely milled zirconium silicate powder with an average particle size less than 5 μm. In white ceramic glazes very small amounts will give opacity. The **Ultrox opacifiers** of M & T Chemicals, Inc., are refined zirconium silicates. **Ultrox 1000W,** for maximum opacity and whiteness, has 65% ZrO_2 and 35% SiO_2 with particle size of 0.5 μm. Lead oxide is used to lower the melting point of a glaze. Matte effects are obtained by adding barium oxide, magnesia, or other materials to the opacifier.

OPEN-HEARTH STEEL. Steel made by the process of melting pig iron and steel or iron scrap in a lined regenerative furnace, and boiling the mixture with the addition of pure lump iron ore, until the carbon is reduced. The boiling is continued for a period of 3 to 4 hr. The process was developed in 1861 by Siemens in England. The furnaces contain regenerative chambers for the circulation and reversal of the gas and air. The fuels used are natural gas, fuel oil, coke-oven gas, or powdered coal. Both the acid- and the basic-lined open-hearth furnaces are used, but most steel made in the United States is basic open hearth. Ganister is used as a lining in the acid furnaces, and magnesite in the basic.

An advantage of the open-hearth furnace is the ability to handle raw materials that vary greatly and also to employ scrap. Iron low in silicon requires less heating time. The duplex process consists in melting the steel in an acid bessemer furnace until the silicon, the manganese, and part of the carbon have been oxidized, and then transferring to a basic open-hearth furnace where the phosphorus and the remainder of the carbon are removed. Open-hearth steel is of uniform quality, and is produced in practically all types.

OPIUM. The dried juice from the unripe capsules of the **poppy plant,** *Papaver somniferum,* cultivated extensively in China, India, and the Near East, but also growing wild in many countries. The opium poppy is an annual with white flowers. After the petals drop off, the capsules are cut and the juice exudes and hardens. The crude opium is a brownish mass. It contains about 20 alkaloids which are useful in medicine. Opium alone is a powerful narcotic, but the material is usually processed and the alkaloids are employed separately or in combinations for their particular effects.

Morphine, $C_{17}H_{19}NO_3 \cdot H_2O$, a white powder melting at 253°C, is the most important of the opium alkaloids. It is a powerful narcotic and pain-killer. It has a complex five-ring molecular structure which can be synthesized from the three-ring **phenanthrene,** $C_{14}H_{10}$, an isomer of anthracene occurring in coal tar. **Codeine,** a white powder melting at 247°C, is a methyl ether of morphine, and is a pain-killer less drastic than morphine. It is much used in cough medicines. **Dionine** is **ethyl morphine,** and is also an important drug. **Heroin** is **diacetyl morphine.** It is a powerful narcotic, but its use is prohibited in the United States. **Colchicine,** $C_{22}H_{25}NO_6$, is a complex three-ring alkaloid used as a gout remedy. Its action is to quicken the release of **heparin** from intestinal cells, which decomposes fat in the blood and prevents blood clotting. It is chemically similar to morphine, but has the acetyl amino group in a different position.

Laudanum is an alcohol solution of opium. **Amidone** is a German synthetic morphine. It is a diphenyl dimethylaminoheptanone, is stronger than morphine as a pain-killer, and, like morphine, is an exhilarant and is habit-forming. The English drug **Heptalgin** is a similar morphine substitute. **Poppy-seed oil** is a colorless to reddish-yellow liquid of specific gravity of about 0.925 and iodine number 157 used as a drying oil in artists' varnishes. The cold-pressed white oil is used locally as an edible oil. The very dark grades are used in soaps and in paints. The oil from the seed does not contain opium.

ORE. A metal-bearing mineral from which a metal or metallic compound can be extracted commercially. Earths and rocks containing metals that cannot be extracted at a profit are not rated as ores. Ores are named according to their leading useful metals. The ores may be oxides, sulfides, halides, or oxygen salts. A few metals also occur native in veins in the minerals. Ores are usually crushed and separated and concentrated from the **gangue** with which they are associated, and then shipped as **concentrates** based on a definite metal or metal oxide content. The metal content to make an ore commercial varies widely with the current price of the metal, and also with the content of other metals present in the ore. Normally, a sulfide copper ore should have 1.5% copper in the unconcentrated ore but, if gold or silver is present, an ore with much less copper is workable, or, if the deposit can be handled by high-production methods, a mineral of very low metal content can be utilized as ore. Low-grade lead minerals can be worked if silver is recoverable, and low grade manganese minerals become commercial when prices are high. Thus, the term ore is only relative, and under different economic conditions, minerals that are not considered ores in one country may be much used as ores in another.

OSMIUM. A platinum-group metal noted for its high hardness, about 400 Brinell. The heaviest known metal, it has a high specific gravity, 22.65, and a high melting point, 4890°F (2698°C). The boiling point is about 9900°F (5468°C). Osmium has a close-packed hexagonal crystal structure, and forms solid-solution alloys with platinum, having more than double the hardening power of iridium in platinum. However, it is seldom used to replace iridium as a hardener except for fountain-pen tips where the alloy is called **osmiridium.** The name osmium comes from the Greek word meaning odor, and the tetroxide formed is highly poisonous. Osmium is not affected by the common acids, and is not dissolved by aqua regia. It is practically unworkable, and its chief use is as a catalyst.

OXALIC ACID. Also known as **ethane diacid.** A strong organic acid of the composition HO_2CCOOH, which crystallizes as the ortho acid $(Ho)_3CC(OH)_3$. It reduces iron compounds, and is thus used in writing inks, in stain removers, and in metal polishes. When it absorbs oxygen it is converted to the volatile carbon dioxide and to water, and it is used as a bleaching agent, as a mordant in dyeing, and in detergents. Oxalic acid occurs naturally in some vegetables, notably **Swiss chard,** and is useful in carrying off excess calcium in the blood. The acid is produced by heating sodium formate and treating the resulting oxides with sulfuric acid, or it can be obtained by the action of nitric acid on sugar, or strong alkalies on sawdust. It comes in colorless crystals with a specific gravity of 1.653, melting at 101.5°C, and soluble in water and in alcohol. **Oxamide,** $(CONH_2)_2$, is a stable anhydrous derivative with a high melting point, 419°C. It is a white crystalline powder used in flameproofing and in wood treatment. **Potassium ferric oxalate,** $K_3Fe(C_2O_4)_3$, is stable in the dark, but is reduced by the action of light, and is used in photography.

OXYGEN. An abundant element, constituting about 89% of all water, 33% of the earth's crust, and 21% of the atmosphere. It combines readily with most of the other elements, forming their oxides. It is a colorless and odorless gas and can be produced easily by the electrolysis of water, which produces both oxygen and hydrogen, or by chilling air below −300°F (−184°C), which produces both oxygen and nitrogen. The specific gravity of oxygen is 1.1056. It liquefies at −113°C at 59 atm. **Liquid oxygen** is a pale-blue, transparent, mobile liquid. As a gas, oxygen occupies 862 times as much space as the liquid. Oxygen is one of the most useful of the elements, and is marketed in steel cylinders under pressure, although most of the industrial uses are in the form of its compounds. An important direct use is in welding and metal cutting, for which it should be at least 99.5% pure.

Oxygen is the least refractive of all gases. It is the only gas capable of supporting respiration, but is harmful if inhaled pure for a long time. **Ozone** is an allotropic form of oxygen with three atoms of oxygen, O_3. It is formed in the air by lightning, or during the evaporation of water, particularly of spray in the sea. In minute quantities in the air it is an exhilarant, but pure ozone is an intense poison. It has a peculiar odor, which can be detected with 1 part in 20 million parts of air. Ozone is a powerful oxidizer, capable of breaking down most organic compounds, and bleaching vegetable colors. **Liquid ozone** explodes violently in contact with almost any organic substance. It is bright blue in color, and is not attracted by a magnet, although liquid oxygen is attracted. Ozone absorbs ultraviolet rays, and a normal blanket in the upper ozonosphere at heights of 60,000 to 140,000 ft (18,288 to 42,672 m), with 1 part per 100,000 of air, shields the earth from excess short-wave radiations from the sun. As an oxidizer in the rubber industry, ozone is known as **activated oxygen.** It is used widely as a catalyst in chemical reactions. It is made commercially by bombardment of oxygen with high-speed electrons.

Oxygen for bleaching and oxidizing purposes may be obtained from compounds that readily yield the gas, such as the liquid **hydrogen peroxide,** H_2O_2, or the granular solid **sodium peroxide,** Na_2O_2. The C.P. grade of hydrogen peroxide is a colorless liquid with 90% H_2O_2 and 10 water. The specific gravity is 1.39. It contains 42% active oxygen by weight, and one volume yields 410 volumes of oxygen gas. Grades for oxidation and bleaching contain 27.5 and 35% H_2O_2. It is also used as an oxidizer for liquid fuels. A variety of chemicals is used for providing oxygen for chemical reactions. These are known as **oxidizers** or **oxidants,** and they may be peroxides or superoxides which are compounds with the oxygen atoms singly linked. They break down into pure oxygen and a more stable reduced oxide. Sodium peroxide is used in submarines to absorb carbon dioxide and water vapor and to give off oxygen to restore the air. To provide oxygen in rockets and missiles, **lithium nitrate,** $LiNO_3$, with 70% available oxygen, and **lithium perchlorate,** $LiClO_4 \cdot 3H_2O$, with 60% available oxygen, are used. Another rocket fuel oxidizer which is liquid under moderate pressure and is easily stored is **perchloryl fluoride,** ClO_3F, normally boiling at −52°F (−47°C).

Albone, of Du Pont, is hydrogen peroxide, and **Solozone** is sodium peroxide with 20% available oxygen. **Ingolin** was a German name for hydrogen peroxide used in rockets. **Liquid air** was used in the first V-2 rockets, with alcohol, potassium permanganate, and hydrogen peroxide. Liquid air is used in the chemical industry, and for cold-treating. It is atmospheric air liquefied under pressure, and contains more than 20% free oxygen. The boiling point is −310°F (−190°C), and 1 ft_3 (0.028 m^3) makes 792 ft^3 (22 m^3) of free air.

Tetrabutyl hydroperoxide, an organic peroxide, is a powerful oxidizing agent used as an accelerator in curing rubbers, as a drying agent in oils, paints, and varnishes, and as a combustion aid for diesel fuel oils. The commercial 60% solution in water has a boiling point of 82°C and specific gravity of 0.859. **Urea peroxide,** $(Co \cdot NH_2)_2O_2$, is a white crystalline material with 16% by weight of active oxygen, used in bleaching, polymerization, and in oxidation processes. **Magnesium peroxide,** MgO_2, **calcium peroxide,** CaO_2, and **zinc peroxide,** ZnO_2, are stable white powders insoluble in water, containing, respectively, 14.2, 13.6, and 7.4% active oxygen. They are used where the oxidation is required to be at high temperatures. **Uniperox,** of the Union Oil Co. of California, is a peroxide of the composition $C_7H_{13}OOH$, made from petroleum fractions. At low temperatures it is stable, but at 110°C the decomposition is exothermic and rapid. It is used as a diesel fuel additive to raise the cetane number, and also as a polymerization catalyst for synthetic resins.

OZOKERITE. Also known as **mineral wax,** and as **earth wax.** A natural paraffin found in Utah and in central Europe, and used as a substitute or extender of beeswax, and in polishes, candles, printing inks, crayons, sealing waxes, phonograph records, and insulation. Ozokerite is a yellowish to black greasy solid, melting at 55 to 110°C and having a specific gravity of

0.85 to 0.95. It is soluble in alcohol, benzol, and naphtha, but not in water. The wax occurs in rocks, which are crushed, and the wax is melted out. The wax is then refined by boiling, treating with an alkali, and filtering. The refined and treated ozokerite is called **ceresin** and is white to yellow in color and odorless. The melting point is up to 142°F (61°C). It is used for waxed paper, polishes, candles, and compounding.

A similar wax, called **montan wax,** or **lignite wax,** is produced in Germany from lignite. Montan wax is white to dark brown in color, and has a melting point of 80 to 90°C, usually 83 to 85°C. The wax is obtained from the powdered lignite by solvent extraction with a mixture of benzene and ethyl alcohol and subsequent removing of the bitumen by oxidation with chromic acid. The brown coals of Oklahoma and Texas also contain as much as 13% montan wax. **IG wax S** is extracted and purified montan wax, but **IG wax V** is a synthetic substitute consisting of the octadecyl ether of vinyl alcohol, $C_{17}H_{34}CH_2 \cdot O \cdot CH \cdot CHOH$. Montan is valued for leather finishes, polishes, phonograph records, insulation compounds, and as a hard wax in candles. The mineral waxes are sold in white, waxy cakes, or in flakes.

PAINT. A general name sometimes used broadly to refer to all types of **organic coatings.** However, by definition, paint refers to a solution of a pigment in water, oil, or organic solvent, used to cover wood or metal articles either for protection or for appearance. Solutions of gums or resins, known as varnishes, are not paints, although their application is usually termed painting. Enamels and lacquers, in the general sense, are under the classification of paints, but specifically the true paints do not contain gums or resins. **Stain** is a varnish containing enough pigment or dye to alter the appearance or tone of wood in imitation of another wood, or to equalize the color in wood. It is usually a dye rather than a paint.

Enamel paint is an intimate dispersion of pigments in either a varnish or a resin vehicle, or in a combination of both. Enamels may dry by oxidation at room temperature and/or by polymerization at room or elevated temperatures. They vary widely in composition, in color and appearance, and in properties. Although they generally give a high-gloss finish, some give a semigloss or eggshell finish and still others give a flat finish. Enamels as a class are hard and tough and offer good mar- and abrasion-resistance. They can be formulated to resist attack by the most commonly encountered chemical agents and corrosive atmospheres, and have good weathering characteristics.

Because of their wide range of useful properties, enamels are probably the most widely used organic coating in industry. One of their largest areas of use is as coatings for household appliances—washing machines, stoves, kitchen cabinets, and the like. A large proportion of refrigerators, for example, are finished with synthetic baked enamels. These appliance enamels are usually white, and therefore must have a high degree of color and gloss retention when subjected to light and heat. Other products finished with enamels include automotive products; railway, office, sports, and industrial equipment; toys; and novelties.

House paint for outside work consists of high-grade pigment and linseed oil, with a small percentage of a thinner and drier. The volatile thinner in paints is for ease of application, the drying oil determines the character of the film, the drier is to speed the drying rate, and the pigment gives color and hiding power. Part or all of the oil may be replaced by a synthetic resin. Many of the newer house paints are water-base paints.

Paints are marketed in many grades, some containing pigments extended with silica, talc, barytes, gypsum, or other material; fish oils, or inferior semidrying oils in place of linseed oil; and mineral oils in place of turpentine. Metal paints contain basic pigments such as red lead, ground in linseed oil, and should not contain sulfur compounds. Red lead is a rust inhibitor, and is a good primer paint for iron and steel, though it is now largely replaced by chromate primers. White lead has a plasticizing effect which increases adhesion. It is stable and not subject to flaking. Between some pigments and the vehicle there is a reaction which results in progressive hardening of the film with consequent flaking or chalking, or there may be a development of water-soluble compounds. Linseed oil reacts with some basic pigments, giving chalking and flaking. Fading of a paint is usually from chalking. The composition of paints is based on relative volumes since the weights of pigments vary greatly, although the custom is to specify pounds of dry pigment per gallon of oil.

Bituminous paints are usually coal tar or asphalt in mineral spirits, used for the protection of piping and tanks, and for waterproofing concrete. For line pipe heavy pitch coatings are applied hot, but a bitumen primer is first applied cold. The **Bitumastic primer** of the Koppers Co., Inc., for such purpose, is refined coal-tar pitch in a quick-drying solvent. The bituminous paints have poor solvent resistance, but have high outdoor weathering resistance. **Battery paint** is usually asphalt or gilsonite in a petroleum solvent. It forms a heavy, acid-resistant, and water-resistant coating. Ordinary **aluminum paint** is made with aluminum flake in an oil varnish or in a synthetic lacquer. In lacquers the powder does not leaf, and the paint dries to a hard, metallic surface with a frosted effect. Aluminum paints will reflect 70% of the light rays, and they are used for painting tanks, but where high resistance is needed, especially for industrial atmospheres, the paints have a synthetic resin base. For painting chimneys and ovens, aluminum paints consist of aluminum flake in a silicone resin, and they resist heats to 1000°F (538°C). When aluminum is used in asphalt paints for tanks and roofing, the aluminum pigment leaf comes to the surface to form a reflective shield. The so-called **heat-resistant paints** are usually aluminum pigment in a silicone resin. The heat resistance comes more from the reflective power of the aluminum than from the actual melting point of the resin. Sericite mica flake is sometimes mixed with aluminum flake to give a different color tone. The **Opal-Glo paint** of the Sherwin-Williams Co. contains a small amount of opaque aluminum particles to give a three-dimensional opalescent glow without the metallic sheen of a flaked powder. The **Lumiclad paint** of the Asbestos Mfg. Corp., for roofing, is aluminum flake and asbestos powder in an oil-resin vehicle. **Calibrite,** of Clairmont Polychemical Corp., is a borited aluminum powder which retains the silvery color of aluminum in the paint.

Lead powder may be incorporated in paints as a protection against gamma rays. **Leadoid paint** is an English paint of this kind. **Ceramic paints** are refractory oxides or carbides in a soluble silicate vehicle, but they are generally only temporary repair coatings. But the **Pyromark paint** of the Tempil Corp. has the color pigment in a silicone vehicle which is converted by applied heat into an inorganic silica film which will withstand temperatures to 2500°F (1371°C). **Intumescent paints,** which bubble and swell to form an insulating barrier to protect the base material from fire damage, may contain borax or a percentage of an intumescent resin. **Resyn 1066,** of the National Starch & Chemical Corp., is such an additive. It is a high-solids emulsion of a vinyl resin. **Masonry paints** may have a silicone resin base for water resistance, but they may also be made with synthetic rubbers and be designated for special purposes such as **traffic paints, road-marking paints,** and **pool paints.** Road-marking paints were formerly made with Manila copal, but they are now made with synthetics. **Imron,** of Du Pont, to withstand heavy traffic on industrial building floors, is based on a urethane resin. It dries quickly without a catalyst, and is resistant to greases and cutting oils. **Pliolite AC,** of the Goodyear Tire & Rubber Co., used in road-marking and pool paints, is a styrene-acrylate copolymer resin. It needs no catalyst for curing, and has high adhesion to concrete. The **Mobilzinc 7 paint** of the Socony Paint Products Co., for use on iron and steel, withstands temperatures to 750°F (399°C) and severe weathering conditions. It contains zinc silicate up to 82% by weight and bonds chemically with the metal. **Foliage paints** are made with a base of vinyl acrylic resin. They are used for coloring Zoyzia grass lawns in winter and for other turfs. **Vitalon,** of Mallinckrodt Chemical Works, is such a colorant.

Antifouling marine paints contain soaps of copper, arsenic, and mercury to inhibit action of marine organisms. A paint produced by International Paint Co. consists of a controlled-release biocide of organotin and copper in a copolymer base of methylmethacrylate and tributyltin, with oxylene as the solvent. Known as **SPC,** the paint film is made smoother by water friction, resulting in lower frictional resistance of a ship's hull as it passes through water.

Water-base paints consist essentially of finely divided ingredients, including plastic resins, fillers, and pigments, suspended in water. An organic medium may also be involved. There are three types of water-base coatings: **emulsion coatings** or **latexes, dispersion coatings,** and **water-soluble coatings.** Emulsions, or latexes, are aqueous dispersions of high-molecular-weight resins. Strictly speaking, latex coatings are dispersions of resins in water, whereas emulsion coatings are suspensions of an oil phase in water.

Emulsion and latex coatings are clear to milky in appearance, have low gloss, excellent resistance to weathering, and good impact resistance. Their chemical and stain resistance varies with composition. Dispersion coatings consist of ultrafine, insoluble resin particles present as a colloidal dispersion in an aqueous medium. They are clear or nearly clear. Their weathering properties, toughness, and gloss are roughly equal to those of conventional solvent paints.

Water-soluble types, which contain low-molecular-weight resins, are clear finishes, and they can be formulated to have high gloss, fair to good chemical and weathering resistance, and high toughness. Of the three types, they handle and flow most like conventional solvent coatings.

The simplest **water paints** consist of gypsum or whiting with some zinc oxide, with water as the vehicle and glue for adhesion. **Calcimine** is an old name for wall paint made with whiting and glue and some linseed oil and water colors. **Whitewash** may be merely quicklime and water, or may be slaked lime, salt, whiting, and glue. These materials are still used for interior painting of farm buildings where low cost is the prime factor. **Casein paints** consist of pigments and extenders in a casein solution. Interior paints and enamels are now mostly water paints with a vehicle of a latex water solution of a synthetic resin. The resin may be an acrylic emulsion, a styrene-acrylic, or water-dispersible polyesters and alkyds, or water-soluble epoxies. They can be applied to wet surfaces, and they cure rapidly to water and chemical-resistant films. The acrylic resin emulsions are valued because of their ability to produce pastel shades and their good flow and leveling properties for one-coat application.

Paint removers, for removing old paint from surfaces before refinishing, are either strong chemical solvents or strong caustic solutions. In general, the more effective they are in removing the paint quickly, the more damaging they are likely to be to the wood or other organic material base. The hiding power of a paint is measured by the quantity which must be applied to a given area of a black and white background to obtain nearly uniform complete hiding. The hiding power is largely in the pigment, but when some fillers of practically no hiding power alone, such as silica, are ground to microfine particle size, they may increase the hiding power greatly. Paint making is a highly developed art, and the variables are so many and the possibilities for altering the characteristics by slight changes in the combinations are so great that the procurement specifications for paints are usually by usage requirements rather than by composition.

PALLADIUM. A rare metal found in the ores of platinum, symbol Pd. It resembles platinum, but is slightly harder and lighter in weight and has a more beautiful silvery luster. It is only half as plentiful but is less costly. The specific gravity is 12.10 and the melting point is 1554°C. The Brinell hardness of the annealed metal is 40, with a tensile strength of 27,000 lb/in^2 (186 MPa), and that of the hard-drawn metal with 60% reduction is 100 with a tensile strength of 50,000 lb/in^2 (344 MPa). It is highly resistant to corrosion and to attack by acids, but, like gold, it is dissolved in aqua regia. It alloys readily with gold, and is employed in some white golds. It alloys in all proportions with platinum, and the alloys are harder than either of the constituents.

Although palladium has low electric conductivity, 16% that of copper, it is valued for its resistance to oxidation and corrosion. Palladium-rich alloys are widely used for low-voltage electric contacts. **Palladium-silver alloys,** with 30 to 50% silver, for relay contacts, have 3 to 5% the conductivity of copper. A palladium-silver alloy with 25% silver is used as a catalyst in powder or wire mesh form. A **palladium-copper alloy** for sliding contacts has 40% copper with a conductivity 5% that of copper. Many of the palladium salts, such as **sodium palladium chloride,** Na_2PdCl_4, are easily reduced to the metal by hydrogen or carbon monoxide, and are used in coatings and electroplating. A **palladium-iridium alloy** with 20% iridium has a Brinell hardness of 140, and can be work-hardened to Brinell 260 with a tensile strength of 190,000 lb/in^2 (1,309 MPa). A **palladium-nickel alloy** with 20% nickel has a hardness of Brinell 200, and can be rolled to a hardness of Brinell 360 with a tensile strength of 170,000 lb/in^2 (1,172 MPa). **Palladium alloys** are also used for instrument parts and wires, dental plates, and fountain-pen nibs. Palladium is valued for electroplating as it has a fine white color which is resistant to tarnishing even in sulfur atmospheres. **Palladium leaf** is palladium beaten into extremely thin foil and used for ornamental work like gold leaf. Hydrogen forms solid solutions with palladium, forming **palladium sponge** which has been used for gas lighters. **Palladium powder** is made by chemical reduction and has a purity of 99.9% with amorphous particles 0.3 to 3.5 μm in diameter. Atomized powder has spherical particles of 50 to 200 mesh and is free-flowing. The powders are used for coatings and parts for a service temperature to 2300°F (1260°C). **Palladium flake** has tiny laminar platelets of average diameter of 3 μm and thickness of 0.1 μm. The particles form an overlapping film in coatings.

PALM OIL. An oil obtained from the fleshy covering of the seed nuts of several species of palm trees, chiefly *Elaeis guineensis,* native to tropical Africa, but also grown in Central America. The tree attains a height of about 60 feet (18 m) and the nuts occur in large bunches similar to dates. The fruit is of an elongated ellipse shape, about 1½ in (3.8 cm) long, enclosing a single kernel. The fleshy part carries about 65% oil, which is a semisolid fat. The iodine value is about 55, and the saponification value 205. West African palm oil has four grades: edible, with 11% max free fatty acid; soft, with 18% max; semihard, with 35% max; and hard, with more than 35%. The high-grade edible oil is from unfermented fruits. Fresh palm oil has an agreeable odor and a bright-orange color, but the oil often has a rancid stench and is of varying colors. The oil is used as a fluxing dip in the manufacture of tinplate, for soaps, candles, margarine, and for the production of palmitic acid. About 10% by weight of the palm oil is recovered as by-product glycerin in making soaps or in producing the acid.

Palm oil contains 50 to 70% **palmitic acid,** $C_{15}H_{31}COOH$, which in the form of glyceride is an ingredient of many fats. When isolated, it is a white crystalline powder of specific gravity 0.866, and melting point 65°C, soluble in hot water. It is used in soaps, cosmetics, pharmaceuticals, food emulsifiers, and in making plastics. The **Neo Fat 16,** of the Armour Chemical Co., is 95% pure palmitic acid, with 4% stearic and 1 myristic acids. This is a powder with an acid value of 220 and a saponification value of 221. But the **Greco 55L,** of A. Gross, Inc., which is a white crystalline solid, for cosmetics and soaps, is 50% palmitic acid with the balance stearic acid.

The oil from the kernel of the palm nut, known as **palm kernel oil,** is different in characteristics from palm oil. It contains about 50% lauric acid, 15 myristic acid, 16 oleic acid, and 7 palmitic acid, together with capric and caprylic acids found in coconut oil, while palm oil is very high in palmitic and oleic acids. The specific gravity is 0.873, iodine number 16 to 23, saponification value 244 to 255, and melting point 24 to 30°C. The American species of palm oil is from the dwarf tree *E. melanococca* growing from Mexico to Paraguay and called **noli palm** in Colombia. The pulp of the nuts yields 30% of an oil similar to African palm oil. The tall **Paraguayan palm** *Acrocomia sclerocarpa* has the fruit also in bunches, and the pulp yields 60% of oil similar to palm oil.

PAPAIN. The dried extract, or enzyme, obtained from the fruit and sap of the papaya tree, *Carica papaya,* of tropical America, East Africa, and Asia. It is marketed as a dry, friable powder, and has a complex structure. It is a **proteolytic agent,** which splits proteins, and it also contains a lipase which accelerates the hydrolysis of fatty acid glycerides, and it contains an antibacterial. The latex from the fruit is dried by low heat, since temperatures above 70°C destroy the enzymes. Papain is used in beer and other beverages to remove protein haze, in medicine as a digestive aid and in combination with urea and chlorophyll to promote the healing of wounds, as a meat tenderizer, in degumming silk, and treating textiles.

The **papaya** tree grows to a height of 25 ft (7.6 m) without branches, and is crowned with large leaves. The melonlike fruit grows out from the trunk, and has orange-colored flesh. It is eaten raw like a melon, but as it spoils rapidly it is not easily shipped. The papaya is called **pawpaw** in Florida. **Meat tenderizers** marketed by Papaya Industries, Inc., in powder and liquid forms, are papain with or without seasoning spices. They are applied to the meats before or during cooking. Papain is also injected into the beef animals 10 min before slaughtering. The enzyme spreads throughout the circulatory system of the animal, remaining in the meat and tenderizing it during the cooking cycle. It also inhibits discoloration of the meat in aging. **Pro-Ten,** of Swift & Co., is a solution of papain for this purpose. **Augment,** of the Calgon Corp., is a mixture of papain powder and sodium chloride for use in a dip solution for tenderizing meats.

PAPER. The name given to cellulose made into paste form from plant sources and rolled into thin sheets, used as a material for writing, printing, and wrapping. It may be considered as a thin felting of fibers bonded by a water-soluble cellulose formed on the fiber surfaces, to which a coating material such as clay may be added with starch or other sizing material. Most papers are less than 0.006 in (0.152 cm) thick, but the dividing line between paper and **paperboard** is taken as 0.012 in (0.030 cm). Properties of papers are controlled by the following variables: (1) type and size of fiber; (2) pulp processing method; (3) web-forming operation; and (4) treatments applied after the paper has been produced.

The original **Egyptian paper,** known as **papyrus,** was made from the stems of the rush *Cyperus papyrus* growing along the Nile. It was made in sheets, sometimes as long as 130 ft (33 m). The Chinese process of papermaking from hemp and linen rags was brought to the Near East when the Arabs took Samarkand in A. D. 704. The papers used in Medieval Europe were **charta damascena,** from the Arab factory at Damascus, and **charta bombycina,** from the factory at Bombyce near Antioch, both sold in reams (from the Arabic word razmah). **Greek parchment,** used in later medieval times, was made from cotton. **Aztec paper,** called **amatl,** was made from the inner bast fibers of species of wild fig trees, *Ficus.* The fibers were felted into sheets and beaten with a ribbed mallet. The thin white sheets used for writing were then polished with a curved stone celt which closed the pores and smoothed the sheets. Some very large sheets were made for folding into books. The fibers of the yellow fig tree, *F. petolaris,* were made into a yellow paper used for coloring for decorations.

There are many varieties and grades of paper, depending upon the source of the cellulose and the method of manufacture. Wood is a lignified form of cellulose, and the wood is chipped and cooked with chemicals to dissolve out the lignin. The material so treated is known as **chemical wood pulp** to distinguish it from **mechanical wood pulp** used for making wallboard and **newsprint paper,** the latter requiring some chemical pulp to give fiber and strength. There are four processes for producing chemical pulp: the sulfite, with calcium sulfite; the soda, with sodium carbonate; the sulfate, with sodium sulfate; and the magnesium bisulfite. Hardwoods are cooked in a soda-ash solution and sulfited. The bleaching of pulp is done with chlorine dioxide which oxidizes lignin to water-soluble colorless compounds without reducing the strength of the cellulose.

Cellulose fiber papers, made from wood pulp, constitute by far the largest number of papers produced. A great many of the engineering papers are produced from draft or sulfate pulps. The term kraft paper is used broadly today for all types of sulfate papers, although it is primarily descriptive of the basic grades of unbleached sulfate papers, where strength is the chief factor and cleanliness and color are secondary. Kraft can be altered by treatments to produce various grades of condenser, insulating, and sheathing papers.

Book paper is usually a mixture of sulfate and soda pulp, tha latter process producing a bulky pulp. **Wrapping paper** is a strong, coarse paper made usually from mixed pulps. **Manila paper** is a strong wrapping paper originally made from Manila hemp, but the name is now applied to any strong chemical wood pulp or mixed paper of a slightly buff, or Manila, color. **Clupak paper,** of the West Virginia Pulp & Paper Co., is a tough extensible kraft paper for bags and wrappings, made by compressing the plastic web of paper on a rubber blanket in the papermaking machine. The paper is soft but strong, and stretches 10% in any direction. **Absorbent paper,** such as for **blotting paper** and **filter paper,** is made from spongy bulky fibers, such as poplar, or is loosely felted fiber. The **Kimtowels,** of the Kimberly-Clark Corp., to replace cotton waste for machine cleaning, are made from bulky, specially treated pulp. The paper can be saturated with oil or solvent for cleaning purposes. The **Netone filter paper** of the National Filter Media Co. is a 60-lb kraft paper impregnated with neoprene to give chemical resistance. This type of paper has 3 to 10 times the strength of cellulose papers, and it is also used for electrical insulation. But the **Permalex paper** of the Rogers Corp., for electrical insulation, is a kraft paper in which the cellulose fibers have been treated to replace hydroxyl groups with cyanoethyl groups. The paper has high tensile strength, is more heat-resistant, and has higher dielectric strength than ordinary paper. The **X-Crepe paper,** of the Cincinnati Industries, Inc., used as a substitute for burlap for bags, as a barrier paper, and as a reinforcement for laminated plastics, is a heavy, soft-texture kraft paper that is creped and cross-creped to produce a material that is stretchable in varying degrees from 15 to 60%. It has a bursting strength to 260 lb/in^2 (1.8 MPa) **Balancing paper,** used with the core material in structural plastic laminates to prevent warping, is heavy kraft paper impregnated with a phenol resin.

Cotton is nearly pure cellulose and makes an excellent paper material. Old cotton rags are thus scoured and used for papermaking. Linen rags are also used and produce a fine grade of **writing paper.** The best quality writing and **printing papers** are 50, 75 or 100% rag papers. **Bond paper** is a hard-finished writing paper made from spruce, which has a long fiber. Highly rolled and coated printing papers are called **supercalendered papers.** They are used for printing fine-screen halftones. In England this paper is called **art paper. Tuf-Flex T,** of the Martin Cantine Co., is a tough flexible printing paper impregnated with an acrylic resin and clay coated which takes color printing readily. **Offset paper** for offset printing is given the required porosity without affecting physical properties by coating with an alkali-swellable resin. Fine linen **ledger paper** is made with 100% white rags. Good-quality bond **typewriter paper** may have 80% white rags. These papers are sold by weight per ream, a ream usually consisting of 500 sheets of a specified size in inches. **Watermarked paper** can be made in various ways, but the simplest method is by printing the mark with a solution of castor oil in methyl alcohol. Papers are generally described in terms of basic weight, which is the weight in pounds per 3,000 ft^2. Standard-weight papers are 90 to 105 lb; lightweight papers are 60 to 65 lb.

Drawing paper is a heavy paper, usually white or buff color, employed for making drawings. For mechanical drawings the buff color is preferred as it is easier on the eyes and not so readily soiled. Drawing papers are smooth or rough, the smooth being hot-pressed. Good grades of drawing paper should permit considerable erasure without destroying the appearance. Buff detail paper for pencil use is made slightly rough or grained. High-grade paper for ink work is extra-hard-sized and coated. Drawing paper is marketed in rolls of widths from 30 to 72 in (76 to 183 cm) and in standard sheets varying from cap, 17 by 13 in (43 by 33 cm), to antiquarian, 52 by 31 in (132 by 79 cm). **Tracing cloth** is made from thin, fine cotton or linen fabric, of plain weave, heavily sized and glazed on one side. It is used for making tracings in ink and is quite transparent. It can also be obtained with the glaze on both sides. Tracing cloth is usually marketed in rolls of 24 yd (22 m). The sizing of ordinary tracing cloth is easily soluble in water, and will therefore not withstand wetting, but special grades are made with impervious resin coatings. Plastic-treated papers are now made that have high strength and better transparency than tracing cloth while retaining the drafting qualities of a fine paper. **Tracing paper** is usually a good grade of hard transparent tissue paper in sheets and rolls, in white or buff colors. The **Ozatrace paper** of the General Aniline & Film Corp., for tracings and for maps, is a 16-lb, 100% rag-textured vellum paper with a transparent resin added in the pulp. It takes ink without feathering, and pencil lines may be erased as many as 10 times on the same spot. **Tracing paper PTM-173,** of the Frederick Post Co., is made of rag paper stock with microscopic pores filled with a synthetic resin, roll-pressed to give an evenly textured surface. **Vindure paper**, of George Vincent, Inc., is 100% rag paper processed to give transparency, dimensional stability, and water resistance.

Granite paper is made by the addition of colored fibers to the pulp or by adding several shades of dyed pulp to the regular stock. **Oatmeal paper,** used chiefly for wallpaper, has a flaky finish produced by washing a solution of wood flour over the sheet on the forming wire in the paper machine. The wood flour may be natural or dyed in colors. **Cartridge paper** is 50- to 80-lb Manila paper, waxed on one side, originally used for muzzle-loading cartridges, but now employed where a stiff, waterproof material is needed. **Glassine** is a transparent thin paper used for envelope windows and for sanitary wrapping. It is made of sulfite pulp subjected to long-continued beating and supercalendered. **Glassoid** is a more highly finished transparent paper. **Onionskin paper** is a lightweight highly finished transparent writing paper made transparent by hydration of the pulp in the beaters. Transparent papers are now often made water-resistant and stronger by adding a synthetic resin to the pulp. **Albanene tracing paper,** of Keuffel & Esser Co., is a thin rag paper treated with a transparent synthetic resin. It takes ink well, and erases easily. **Silicone tissue,** for wiping glass, is soft tissue paper treated with silicone resin. **Tissue paper** is a very thin, almost transparent paper. It may be loosely felted to give absorbent qualities, or it may have a hard, smooth surface for wrapping paper. **Detergent paper,** for washing windows, is a soft paper impregenated with a detergent. **Keel,** of the Kee-Lox Mfg. Co., is a paper of this type.

Crepe paper has many consumer and industrial uses. Creping imparts stretch, strength, bulk, conformability, and texture similar to that found in fabrics. Creping consists of forming small pleats, or folds, with a blade. **Cross-creped paper** is made by creping in two opposite directions.

To make paper smooth-surfaced and resistant to the spreading of inks, adhesive sizing materials are used together with inert fillers such as China clay which give body, weight, opacity, and added strength to the paper. The usual coating adhesives are starches and proteins. The proteins, such as casein, are more uniform than starch, but give a more brittle film. Starch films are not water-resistant unless the treated starches are used.

Waterproof paper was formerly paper treated with a copper-ammonium solution and hot-rolled, or was paper coated with rubber latex to which had been added a creaming agent such as a metallic soap, but various synthetic resins are now incorporated in the sizing or mixed in the pulp. High-styrene butadiene latex gives a flexible and glossy film for printing papers. Acrylic latex also gives a strong, glossy, and flexible coating. Polyvinyl acetate is also used for printing papers. **Scriptite 31,** of the Monsanto Chemical Co., used to give a tough, water-impervious surface to offset papers, is a methylated methylol melamine resin which forms a molecular link with the protein of the coating. **Wet-strong paper** is usually specially processed paper in which the water resistance is due to the processing and interlocking of the fibers as well as to impregnation with a small amount of melamine, urea-formaldehyde, or other resin. It is used for maps, documents, and wrapping. **Resistall,** of the L. L. Brown Paper Co., is a paper of this type. The **Anti-adhesive paper** of the Central Paper Co., for interleafing sticky materials and for box linings, is a kraft paper treated with a silicone resin. The **Kastek paper** of the Plastic Film Corp., for waterproof wrapping, is 30- to 100-lb paper with a very thin film of polyethylene or vinyl resin bonded to the surface. The **washable wallpaper** of Richard E. Thibaut, Inc., has a 0.00088-in (0.0022-cm) Lamarith cast film laminated to the paper.

Capacitor paper, used as a dielectric in capacitors, is made from Swedish spruce sulfate pulp, is highly purified, and is nearly transparent. It is extremely thin, 0.00015 to 0.0004 in (0.00038 to 0.00101 cm), but is strong and tough. **Insulating paper,** commonly called **varnished paper,** is a standard material for insulation of electric equipment. It is usually bond or kraft paper coated on both sides with black or yellow insulating varnish. The thicknesses are 0.002 to 0.020 in (0.005 to 0.050 cm) with dielectric strengths of 500 to 2,000 volts per mil (20 to 80 volts per meter). Special insulating varnishes of high dielectric strength are now marketed for this purpose. **Cyanoethylated paper,** used in condensers, is a thin paper treated with acrylonitrile which improves the electrical insulating properties. **Quinorgo,** of Johns-Manville, is a group of asbestos papers containing 80% chrysotile fibers bonded with organic resins. The resins are varied to give differing physical properties. The papers come in thicknesses from 0.003 to 0.015 in (0.007 to 0.038 cm) and have high dielectric strength. **Laminating paper,** for making laminated plastics, is a white or brown paper of uniform basis weight and uniform internal structure capable of having a controlled resin pickup. It usually comes in thicknesses from 0.004 to 0.020 in (0.010 to 0.050 cm). **Nibro-Cell,** of the Brown Co., is such a paper. **Flame-proofed paper** is paper treated with ammonium sulfate and ammonium and sodium phosphates. Paper and nonwoven textile fabrics may be treated with a **fire-retardant agent** such as **FireTard,** of

National Starch and Chemical Corp., an emulsion of vinylidene chloride copolymer with an equal amount of antimony trioxide. **Metallized paper,** of Smith Paper, Inc., used for capacitors, is a lacquer-coated kraft paper with a thin layer of zinc deposited on one side. **Vaculite,** of the Vaculite Corp., used for packaging and as a barrier paper, is **aluminum-coated paper** produced by vacuum metallizing. It has the appearance of bright aluminum, but has the flexibility and physical properties of paper.

Building paper, used for sheathing houses, is a heavy kraft paper, plain or rosin-sized. Specially treated building papers are also marketed under trade names. The **barrier paper** of the Presstite Engineering Co., for lining storage rooms, is kraft paper saturated with gilsonite, asphalt, and wax. It is odorless and black in color. **Weatherite,** of Johns-Manville, is a kraft building paper treated with a black waterproofing. **Copperskin** is an insulating construction material made by facing 1-oz (0.03-kg) electro-sheet copper on one or two plies of heavy building paper impregnated with bitumen. **Cop-O-Top,** of the Chase Brass & Copper Co., and **Copper-kote,** of the Cheney Co., are similar materials. **Sisalkraft,** of the Sisalkraft Co., is a waterproof building paper made with sisal fibers. **Fibreen,** of this company, is a tough, strong, flexible, waterproof paper used for wrapping bundles of steel and other heavy products. It is made of two layers of kraft paper reinforced with two crossed layers of sisal fibers embedded in asphalt, and the whole combined under heat and pressure. **Brownskin,** a waterproof sheathing paper of the Angier Corp., is high-strength building paper impregnated with a bituminous compound and crimped to give it stretch and resiliency. **Burlap-lined paper,** for heavy wrapping, has 4- to 10-oz (0.ll- to 0.28-kg) burlap laminated to heavy kraft paper with asphalt as the binder. It is waterproof. **Papier mâché** is comminuted paper made into a water paste with an adhesive binder and molded. It was formerly widely used for toys, dishes, and novelties, but dishes and novelties now made of paper stock are produced directly from the wood pulp and are more uniform and stronger. Kraft paper impregnated with phenolic resin laminated in blocks and heavy sheet is used for short-run tooling. **Panelyte 750,** of Thiokol Chemical Corp., is the material with a compressive strength of 28,000 lb/in² (193 MPa). **Laminated paperboard** is made by laminating together plies of paper about ¹⁄₁₆ in (0.16 cm) thick. Density runs between 30 and 37 lb/in² (0.21 and 0.25 MPa). It is made in two general qualities, an interior and a weather-resistant quality. The main differences between these types are in the kind of bond used to laminate the layers together and in the amount of sizing used in the pulp stock from which the individual layers are made. Laminated paperboard is regularly manufactured in thicknesses of ³⁄₁₆, ¼, and ⅜ in (0.48,, 0.64, and 0.95 cm) for construction uses. For industrial uses, such as furniture and automotive liners, ⅛ in (0.32 cm) thickness is common.

There are three major types of **inorganic fiber papers: Asbestos papers,** the most widely used, are nonflammable, resistant to elevated temperatures, and have good thermal insulating characteristics. They are available with or without binders and can be used for electrical insulation or for high-temperature reinforced plastics. **Fibrous glass papers** can be used to produce porous and nonhydrating papers. Such papers are used for filtration and thermal and electrical insulation, and are available with or without binders. High-purity **silica glass papers** are also available for high-temperature applications. **Ceramic fiber (aluminum silicate) papers** provide good resistance to high temperatures, low thermal conductivity, and good dielectric properties, and can be produced with good filtering characteristics.

Synthetic organic fiber papers consist of synthetic fibers, synthetic pulp, or plastic film. Plastics used include polyethylene, nylon, acrylic, and polypropylene. In general, synthetic papers have greater dimensional stability and tear resistance than conventional natural fiber papers.

PAPER PLANTS. Cellulose for papermaking is obtained from a wide variety of plant life, made directly into **paper pulp,** or obtained from old rags which were originally made from vegetable fibers. Animal fibers incorporated into some papers are fillers for special purposes and not papermaking materials. The papyrus of Egypt was made from a reed, but the **baobab** of India was from the bark of the tree *Adamsonia digitata.* **Rice paper** of China came from the *tetrapanax papyriferum,* but the so-called rice paper used for cigarettes in the United States is made from flax fiber. **Cigarette paper** is also made from ramie and sunn hemp. The distinction between cigarette paper and the tissue paper used for wrapping is that it must be free of any substance that would impart a disagreeable flavor to the smoke and it must be opaque and pure white, must burn at the same rate as tobacco, and must be tasteless.

Wood pulp is now the most important papermaking material. Spruce is the chief wood used for the sulfite process, but hemlock and balsam fir are also used. Aspen and other hardwoods are used in the soda process, and also southern pine. White fir is readily pulped by any process, but western red cedar is high in lignin content, about 30%, and reduced with difficulty by the sulfate process to a dark-colored pulp. It is pulped by the kraft process. Its fibers are fine and short, yielding a paper of high bursting strength. Normally, the **pulpwoods** of the West Coast are western hemlock, white fir, and Sitka spruce, leaving the Douglas fir to the lumber mills. The same species of trees grow in Alaska and British Columbia as in Oregon and Washington, but Douglas fir decreases to the north and hemlock and spruce become more abundant but with smaller trees. A stand of spruce in Canada at the age of 80 years yields about 18 tons (16 metric tons) of pulp per acre, while a stand of pine in the southern states at 24 years of age yields about the same amount. Western hemlock, balsam, and spruce are the chief pulpwoods of Canada. Pines are used extensively in the United States, especially for kraft paper, paperboard, and book paper. More then 50% of all pulpwood used in the United States is now from the southern states, and about 10% of this is salvage from lumber mills. But, in general, special methods are used for pulping pine since conventional sulfite liquor does not free the fibers as the phenolic compounds in the heartwood condense to form insoluble compounds.

The kraft paper made by Rayonier, Inc., and called **Fibrenier,** is sulfate-pulped from a mixture of 50% western hemlock, 25 western red cedar, and 25 Douglas fir. The fir has a coarse fiber which gives high tear strength; cedar has a long thin fiber which gives a smooth surface; hemlock is abundant and used as a filler. Poplars are also used for pulpwood, and the Scott Paper Co. uses fast-growing scrub alder. Newsprint made from hardwoods has a bursting strength 20% higher than that made from softwoods, and the brightness value is higher, but the pulping of hardwoods is usually a more involved chemical process.

In England fine printing papers are made by the soda process from **esparto** grass. It gives a soft, opaque, light paper, although the cellulose content is less than 50%. Esparto is the plant *Stipa tenacissima* of the dry regions of North Africa. In Tunisia it is called **alfa.** It grows to a height of about 3 ft (1 m), with cylindrical stem. The fine, light fibers, about ½ in (1.27 cm) long, are from the leaves. Some grades of cardboard and some newsprint are made from straw. **Deluwang paper** of the East Indies is made from the scraped and beaten bark of the **paper mulberry** tree, *Broussonetia papyrifera.* It is an ancient industry in Java, and the paper is used for lamp shades and fancy articles. Under the name of **tapa cloth** the sheets were dyed and used as a muslinlike fabric by the Polynesians. The strips are welded together by overlapping and beating together the wet material. Bagasse is of increasing importance as a papermaking material in the sugar-growing areas of the world.

PARAFFIN. A general name often applied to paraffin wax, but more correctly referring to a great group of hydrocarbons obtained from petroleum. Paraffin compounds begin with methane, CH_4, and are sometimes called the methane group. The compounds in the series have the general formula C_nH_{2n+2}, and include the gases methane and ethane, and the products naphtha, benzine, gasoline, lubricating oils, jellies, and the common paraffin. The name paraffin indicates little affinity for reaction with other substances. In common practice the name is limited to the waxes that follow petroleum jelly in the distillation of petroleum. These waxes melt at from 40 to 60°C, and consist of the hydrocarbons between C_{22} and C_{27}; the refined waxes may range up to 90°C. They burn readily in the air. Paraffin occurs to some extent in some plant products, but its only commercial source is from natural petroleum. **Chlorinated paraffin** is a pale to amber-colored, odorless, soft wax or viscous oil of specific gravity 0.900 to 1.50. It is flame-resistant, and is used in treating paper and textiles. **Chlorowax,** of the Diamond Alkali Co., is a chlorinated paraffin for adhesive, fire-resistant, and water-resistant compounds. It is a cream-colored powder containing 69 to 73% chlorine, and is insoluble in water but soluble in organic solvents. A water-dispersible form of this wax is **Delvet 65,** a white, viscous liquid containing 65% solids. **Clorafin 42,** of Hercules, Inc., is an amber-colored viscous liquid chlorinated paraffin containing 42% chlorine. **Cereclor,** of the Chemical Mfg. Co., Inc., is a similar product. **Clorafin 70** is a yellow solid containing 70% chlorine and softening at 90 to 100°C. The former is used as a plasticizer in resins and for coatings; the latter is for flameproofing and waterproofing textiles.

Paraffin oil is drip oil from the wax presses in the process of extracting paraffin wax from the wax-bearing distillate in the refining of petroleum. The oil is treated, redistilled, and separated into various grades of lubricating oils from light to heavy. They may be treated and bleached with sulfuric acid, and neutralized with alkali. When decolorized with acid and sold as filtered, they are brilliant liquids, but are not suitable in places where they may be in contact with water, since the sulfo compounds present cause emulsification. The specific gravities of paraffin oils are between 21 and 26°Bé. **Triton oil** is a 100% pure paraffin oil produced by the Union Oil Co. of California.

PARAFFIN WAX. The first distillate taken from petroleum after the cracking process is known as wax-bearing, and is put through a filter press and separated from the oils. The wax collected on the plates is called **slack wax,** and contains 50% wax and 50 oil. This is chilled to free it from oil. The yellow wax is filtered to make a white semitranslucent refined wax, which is odorless and tasteless. For large-scale operations, solvent methods of wax extraction are used. Paraffin wax is soluble in ether, benzine, and essential oils. **Match wax** has a melting point of 105 to 112°F (40 to 44°C); white crude wax, 111 to 113°F (43 to 44°C); yellow crude, 117 to 119°F (46 to 48°C); and special white, 124 to 126°F (51 to 52°C). The refined waxes are in various melting-point ranges from 115 to 136°F (45 to 58°C) and are used for coating paper and for blending in coating and impregnating compounds. They are also used in candles and other products. The refined paraffin wax used for molded goods and for rubber compounding is a white solid having a melting point of 122°F (50°C) and a specific gravity 0.903.

Borneo wax has a very high melting point and a hard crystalline structure which makes it valuable for coatings and for high-quality candles. By treatment of the waxes from American petroleums to remove the low-melting constituents, a similar wax is obtained having branched-chain molecules and a fine crystalline structure. This is known as **microcrystalline wax. Aristowax,** of the Union Oil Co. of California, is a treated wax of this kind with melting points from 145 to 165°F (63 to 74°C). **Petrolite wax,** of the Petrolite Corp., is a microcrystalline wax with melting point at 195°F (90°C). **Sunwax,** of the Sun Oil Co., is a microcrystalline wax in two grades, a brown with melting point of 175°F (79°C) and a yellow with melting point of 185°F (85°C). **Warcosine wax,** of the Warwick Wax Co., Inc., is a white microcrystalline wax melting at about 153°F (67°C), while **Fortex wax** has a melting point at about 195°F (90°C). Microcrystalline wax does not emulsify easily like carnauba wax, but when oxidized with a catalyst it is emulsifable and the melting point is raised so that it can be used in hard, self-polishing floor waxes. **Cardis wax** and **Polymekon wax,** of this company, have melting points at 198 and 250°F (92 and 121°C). **Petronauba D,** of the Bareco Oil Co., is an oxidized wax with a melting point of 192.2°F (89°C) used as a partial replacement for carnauba. Microcrystalline waxes are also compounded with polyethylene and other materials to increase strength, flexibility, and other properties. The paraffin waxes are sold under many trade names. **Arwax,** of the American Resinous Chemicals Corp., may contain butyl rubber or polyethylene. **Advawax 2575,** of the Advance Solvents & Chemical Corp., for paper coatings, contains polyisobutylene. **Santowax,** of the Monsanto Chemical Co., is a high-melting microcrystalline wax. **Wax tailings** is a name for the distillate that comes from petroleum after the wax-bearing distillate is removed. It contains no wax, but at ordinary temperatures looks like beeswax. It is very adhesive and is employed in roofings and for waterproof coatings.

Synthetic paraffin wax, called **Ruhr wax,** is made in Germany from lowgrade coals and other hydrocarbon sources. The waxes, with molecular weights of 900 to 1600, are white flakes, water white when melted. They are odorless, free of sulfur and aromatics, with ash content below 50 ppm. The melting points are between 105 and 126°C. They are used as additives in paper coatings and printing inks and for mixing with refined paraffin.

PARCHMENT. Originally, goatskin or sheepskin specially tanned and prepared with a smooth hard finish for writing purposes. It was used for legal documents, maps, and fancy books, being more durable than the old papers. The extremely thin high-quality parchment that was used for documents and handmade books was rubbed with pumice and flattened with lead. Parchment now is usually **vegetable parchment.** It is made from a base paper of cotton rags or alpha cellulose called **waterleaf** which contains no sizing or filling materials. The waterleaf is treated with sulfuric acid which converts a part of the cellulose into a gelatinlike amyloid. When the acid is washed off, the amyloid film hardens on the fibers and in the interstices of the paper. The strength of the paper is increased, and it will not disintegrate even when fully wet. The paper now has a wide usage in food packaging as well as for documents as a competitor of the resin-treated wet-strong papers. The wet strength and grease resistance are varied by differences in acid treatment and subsequent sizing. **Patapar,** of the Paterson Parchment Paper Co., is a vegetable parchment marketed in many grades. Parchment papers are also waterproofed by dipping in the solution of copper hydroxide and ammonium hydroxide known as **Schweitzer's reagent,** and then hot-rolling. **Vellum** is a thick grade of writing paper made from high-grade rag pulp pebbled to imitate the original calfskin parchment called vellum.

PASTEBOARD. A class of thick paper used chiefly for making boxes and cartons, and for spacing and lining. It may be made by pasting together several single sheets, but more usually by macerating old paper and rolling into heavy sheets. It may also be made of straw, certain grasses, and other low-cellulose paper materials, and is then known as **strawboard**. Colloquially, the term pasteboard applies to any paper-stock board used for making boxes, including the hard and stiff boards made entirely from pulp, and the term pasteboard is not liked in the paper industry. The bulk of packaging boards are now pulp boards treated with resin and are called **carton boards. Cardboard** is usually a good quality of chemical pulp or rag pasteboard used for cards, signs, or printed material, or for the best-quality boxes. Ivory board, for art printing and menu cards, is a highly finished cardboard clay-coated on both sides. **Bristol board** is a high-class white cardboard, supercalendered with China clay, or it may be made by pasting together sheets of heavy ledger paper, but the name is also applied to any high-grade printing or drawing board over 0.06 in (0.15 cm) thick. **Index bristol** is always made solid on a Fourdrinier machine to prevent splitting in use or warping. The original board made in Bristol, England, was made in this way. **Jute board,** used for folding boxes, is a regular product of the paper mills, and is a strong solid board made of kraft pulp. **Chipboard** is a cheap board made from mixed scrap paper, used for boxes and book covers. When made with a percentage of mechanical wood pulp, it is called **pulpboard.** A heavy rope-pulp paper or board, usually reddish in color and used for large expansion filing envelopes, is called **paperoid.**

PEANUT OIL. Also known as **groundnut oil.** A pale-yellow oil with a distinctive nutty taste and odor, obtained from the pressing of the seed kernels of the peanut, a legume of the genus *Arachis,* of which there are many species. It was native to Brazil, brought to Africa in slave ships and thence to the United States. It is now grown in many countries. The Spanish peanut, cultivated in temperate climates, has small seeds, while the common variety, *A. hypogaea,* known in the United States as the **Virginia peanut,** has pods up to 1½ in (3.8 cm) long with seeds twice the length of the Spanish varieties. The **Spanish peanut,** however, is easily grown and gives a high yield per acre. The **Brazilian peanut,** *A. nambyquarae,* has pods up to 3 in (7.5 cm) in length. Vast quantities of peanuts are roasted and marketed as food nuts or for confections or ground to make edible **peanut butter.** The best grades of cold-pressed oils are marketed as edible oils, but the oil is also used industrially for soaps, in diesel-engine fuels, and for blending in lubricating and varnish oils. The **arachidic acid,** $CH_3(CH_2)_{18}COOH$, contained in the oil to the extent of 4%, however, makes a hard soap. The oil also contains 52 to 65% oleic, 21 to 25 linoleic, besides palmitic, stearic, and lignoceric acids. The specific gravity is from 0.916 to 0.922, saponification value 189 to 196, and iodine value 83 to 101. The oil known as **arachis oil,** or as **Katchung oil** when imported from the Orient, is from the peanut *A. hypogaea.* It is used in lubricating, for varnishes, and for softening leather. **Peanut meal,** left after extraction of the oil, is sold as stock feed, or that from the final extraction of the inedible oil is used for fertilizer. **De-oiled peanuts** are marketed as low-calorie, nonfat food. Most of the oil is removed by solvent extraction, but the nuts retain the high-protein value, color, and flavor. The calorific value is reduced about 80%.

PEAT. An earthy mass formed by the rapid accumulation of quick-growing mosses and plants, and valued as a fuel in countries where fuels are expensive. Large quantities are used for fuel in Finland, Switzerland, Ireland, and some other countries. In Russia large amounts are used to produce fuel gas and for processing the tar into chemicals. In the United States it is used for fertilizer, insulation, and packing. The dried **peat moss** is used for making insulating board. Peat bogs, or beds, are found mainly in moist districts in temperate climates. The top layers are only slightly decayed, are brown in color, and are of low specific gravity. But at greater depths peat is nearly black and is very compact. In the peat of southern New Jersey there are layers of trees as large as 5 ft (1.5 m) in diameter, buried for centuries.

The reserves of peat are very large in the states bordering on the Great Lakes, those in the state of Minnesota alone being estimated at 7 billion tons (6.35 billion metric tons). Fresh peat often contains as high as 80% moisture and must be dried before use. Wicklow dried peat contains about 71% volatile matter, 27% being fixed carbon, and 28% coke. The calorific value of peat is about 5,000 Btu (5.3 million J). It is sometimes semicarbonized and made into fuel briquettes. **Charred peat** is peat that has been subjected to a temperature to cause partial decomposition. It is marketed as fertilizer. Peat is also distilled, yielding mainly gas and a high percentage of tar. **Peat wax,** extracted from peat in England, is a hard wax with characteristics similar to those of montan wax, and is a substitute for it.

PENTAERYTHRITOL. A tetrahydric alcohol of the composition $HOH_2C \cdot C(CH_2OH)_3$ produced by the condensation of formaldehyde and acetaldehyde. It is a white, crystalline solid melting at about 262°C. The commercial grade is 85 to 90% pure. It is employed for the production of explosives, plastics, drying oils, and chemicals. Pentaerythritol will combine with the fatty acids of vegetable oils to form esters that are superior to linseed oil as drying oils. Combined with the fatty acids of linseed oil, it will give a drying oil that will dry completely in 6 h compared with 16 h for a bodied linseed oil, and as a varnish oil it gives higher gloss and greater water resistance. **Synthetic waxes** are also made by combining pentaerythritol with long-chain, saturated, fatty acids, and these waxes have higher melting points than beeswax or carnauba wax, but do not have the natural gloss of carnauba. **Pentawax 177,** of the Heyden Chemical Co., is a **pentaerythritol stearate.** It is a light-brown wax melting at 53°C, used for coating paper, in printing inks, and in cosmetics. **Pentamull 126** of this company is an ester of pentaerythritol and oleic acid. It is an amber-colored oil used as an emulsifying agent. **Pentex** is a technical grade of pentaerythritol containing 85% monopentaerythritol and the balance higher polymers. It produces fast-drying and glossy varnishes. The **Pentalyn resins** of Hercules, Inc., which are employed to replace copals in varnishes, are pentaerythritol esters of rosin. **Pentalyn 802A** is a phenol-modified pentaerythritol rosin ester for gloss printing inks and traffic paints. It is pale in color, and has high resistance to chemicals and wear.

PEPPER. One of the oldest and most important of the spices. **Black pepper,** the common household spice, is the ground, dried, unripe fruit of the evergreen shrub or vine, *Piper nigrum,* of India and Southeast Asia. There are two grades of Indian pepper, **Alleppey** and **Tellicherry;** the latter is bolder and heavier and the more expensive grade. The fruits are small, berrylike drupes. They change in ripening from green to bright red to yellow. When dry the unripe berries are reddish brown or black. The vine comes into full production in 3 years and lives for 20 to 30 years. A vine yields 5 to 10 lb (2.3 to 4.5 kg) of pepper. **White pepper,** preferred for the preparation of commercial foods, is from the nearly ripe berries. It has a yellow to gray color. Pepper is used as a condiment and stimulates the flow of gastric juices. White pepper is not as pungent as black pepper. Commercial pepper is often a blend of two kinds. **Pepper oil,** used for flavoring, is a yellowish essential oil of specific gravity 0.873 to 0.916 with a pepperlike odor and flavor but not pungent like pepper. It is extracted from the common pepper berries. **Pepperoyal,** of the Griffith Laboratories, is pepper flavor extracted from black pepper and converted to minute soluble globules that disperse easily in foodstuffs. Soluble pepper, of Fritzsche Bros, used for food processing, is a liquid solution of black-pepper oleoresin from which there is no precipitation of piperine crystals during processing or cold storage. The synthetic **piperidine,** $C_5H_{11}N$, is a colorless liquid with an odor resembling pepper. It yields crystalline salts, and it occurs in natural pepper in combination with **piperic acid** in the form of the alkaloid **piperine,** $C_{17}H_{19}NO_3$, which is the chief active constituent of pepper. **Piperazine,** made synthetically, is used in medicine as an anthelmintic and as an intermediate for pharmaceuticals. It is a six-membered heterocyclic ring compound with two nitrogen atoms in the para position. **B-Cap,** of the Evans Chemetic Co., is a cinnaylidene acetoyl piperide, a synthetic with a pepperlike flavor used in prepared foods.

Long pepper, esteemed in some countries for preserves and curries, is more aromatic and is sweeter than common pepper. It is from the tiny fruits of the climbing plant *P. retrofractum* of Malaya and the *P. longum* of India and Indonesia. **Ashanti pepper,** of western Africa, is from the vine *P. guineense,* also known as **Guinea pepper,** although this name is applied to **grains of paradise,** the pungent peppery seeds of the perennial herb *Aframomum melegueta* of West Africa, which are used for flavoring and in medicine. The seeds are also called **alligator pepper** and **melegueta pepper,** and are used as pepper in Europe, but the plant is of the ginger family. **Cubeb** is the dried unripe fruit of the climbing vine *P. cubeba* of India, Indonesia, and the West Indies. The berries resemble those of black pepper, but have a strong, peculiar odor and a bitter aromatic taste. Cubeb is used in medicine and cigarettes. **Cubeb oil** is from the berries, which yield 10 to 16% of the pale-green oil with a pepperlike odor. It is used in perfumery and in soaps. **Paprika** is the ground dried fruit of the *Capsicum annuum* of Europe and America. When full, red, ripe pods are used, and the seeds, cores, and stems are removed, a uniform maximum red color is produced. Yellow pods give low red value. Paprika is used as a condiment. **Chili pepper** is from the smaller podlike berries of species of *Capsicum* which grow as small trees or shrubs. It is a tropical plant. The ground fruits as a condiment are known as **red pepper,** or **cayenne pepper.** In medicine it is known as **capsicum,** and is used as a carminative and as a source of vitamin P. It is also used in soft drinks in place of ginger. The Samoan beverage known as **kava** is made by steeping in water the ground root of a species of pepper plant, *Piper methysticum.* It has a peppery flavor. But the Kava of Borden, Inc., is not a pepper but an instant coffee powder processed to neutralize the coffee acids.

PEPPERMINT. An oil distilled from the perennial herb *Mentha piperita,* which grows in the temperate climates of America, Europe, and Asia. The oil has a pleasant odor and a persistent cooling taste, and is valued as a flavor and in soaps, toothpastes, perfumes, and pharmaceuticals. The oil contains **menthol,** $C_{10}H_{20}O$, which is extracted for use as an antiseptic, in perfumery, and in medicine for colds and as an antispasmodic and anodyne. **Japanese peppermint** is from the *M. arvensis* grown extensively in Japan, Brazil, and the United States. The oil is less fragrant, and is used for the production of menthol as it has a higher menthol content. The plant is propagated from roots and grows to a height of 2 to 3 ft (0.6 to 1 m). It is cut when it blooms and partly cured like hay. The crude oil is obtained by steam distillation; the menthol is obtained by freezing and recrystallization, with a yield of 50% menthol crystals to total crude oil. The residue oil is called **cornmint oil.** It retains the peppermint flavor and is used in perfumery and flavoring. **Spearmint** is from the *M. viridis,* grown largely in Michigan. The oil is sharper in odor and taste and used chiefly in chewing gums. **Pennyroyal oil** is distilled from the dried leaves and tops of the small annual plant *Hedeoma pulegiodes* or *M. pulegium,* which grows in the eastern United States. The oil is a counterirritant and is used in liniments. It is also used in insect repellents and for the production of menthol. The plant yields 0.7% oil, and the oil will yield 65% menthol with a melting point of 33 to 35°C, or 40% of 42°C menthol. **Horsemint oil** is from the plant *M. canadensis,* used for the production of thymol. Menthol and menthol substitutes are also synthesized from coal tar. **Cyclonol,** of W. J. Bush & Co., Inc., is a derivative of cyclohexanol. It lacks one H and one CH of the structural formula of menthol, but has the characteristic odor and cooling effect. **Levomenthol,** produced synthetically by the Glidden Co., is used as a replacement for natural menthol.

PERFUME OILS. Volatile oils obtained by distillation or by solvent extraction from the leaves, flowers, gums, or woods of plant life, although a few are of animal origin. Perfumes have been used since earliest times, not only for aesthetic value, but also for antiseptic value and for religious purposes. Simple perfumes usually take their name from the name of the plant, but the most esteemed perfumes are blends, and the blending is considered a high art. It is done by tones imparted by many ingredients. Some oils with repugnant odors have an attractive fragrance in extreme dilution and a persistence which is valued in blends and for stabilization. Some oils with heavy odors, such as **coumarin,** are used in dilution to give body. Since many of the odors come directly from esters, aldehydes, or ketones, they can be made synthetically from coal-tar hydrocarbons and alcohols. Synthetics are now most used in perfumes, although some natural odors have not yet been duplicated synthetically, and about 30,000 aromatics have been developed. A perfume may contain 50 components, sometimes as high as 300, and the average perfume manufacturer employs about 3,000 components. Some of the chemicals are not odors, but give lasting qualities or enhance odor. Some are used as fixatives or blending agents. **Hydroquinone dimethyl ether,** $C_8H_{10}O_2$, has an odor of sweet clover but is used as a fixative in other perfumes.

In general, the aldehyde odors are fugitive, and some become acid in the presence of light or oxygen. Ketones are more stable. Esters are usually stable, but some are saponified in hot solution and cannot be used for soaps. Some esters, made from complex high alcohols, are used to give a fresh top note to floral perfumes. **Linalyl acetate,** produced from citral, is an example. Acid perfumes neutralize free alkali and cannot be used in soaps. Phenol odors alter the color of soaps, and the odor may also become disagreeable.

Some odors are never extracted from the flowers, but are compounded. **Crab apple,** for example, which is a peculiarly sweet odor, is compounded of 16 oils, including bois de rose, ylang-ylang, nutmeg oil, jasmine, musk, heliotropin, coumarin, and others. **Wisteria** is the honeylike odor of the mauve and white flowers of the climbing plant *Wisteria sinensis.* The oil is never extracted but is compounded from geranium, Peru balsam, benzoin, bois de rose, and synthetics. Some oils such as **lavender,** from the flowers of the *Lavandula vera,* have no value when used alone but require skillful blending to develop the pleasant odor. Apple and peach odors are **allyl cinnamate.** Synthetic **rose** is the ester of phenyl ethyl alcohol made from benzene and ethylene oxide. Although the natural rose odor is readily extracted, it is more expensive.

Fixatives are used for the finer perfumes. They are essential oils that are less volatile and thus delay evaporation. The animal oils, such as musk and civet, are of this class, and also the balsam oils. Some evil-smelling distillates from chemical manufacture may also be used as fixatives. Musk is from the male musk deer of Tibet. It is one of the most expensive materials. Synthetic musk is as powerful as the natural. The synthetic musk of Du Pont is called **Astrotone. Musk ambrette** is made from metacresol. **Ambrette oil** has a strong musklike odor distilled from the **musk seed,** or **amber seed,** of the plant *Hibiscus abelmoschus* of Ecuador, India, and Egypt. **Civet** is an odorous yellow fluid from the civet cat of tropical Asia and Ethiopia. **Civettone** is a liquid with a clean odor and easily soluble in alcohol, distilled from civet. **Patchouli oil** is one of the best fixatives for heavy perfumes. It is a powerfully odorous viscous liquid obtained by distilling the fermented leaves of the shrub *pogostemon patchouli* of India, China, and the Philippines. The odor resembles sandalwood. **Cassie** is a valuable oil with an odor similar to violet obtained by maceration in oil of the flowers of the shrub *Acacia farnesiana* of the Mediterranean countries and the West Indies. It is used to scent pomades and powders. **Versilide,** of Givaudan-Delawana, Inc., is a cyclic ketone synthetic musk that is very stable in soaps and cosmetics and does not discolor.

Attar of rose is one of the most ancient and popular of perfume oils. The name is derived from the Persian attar, and is sometimes incorrectly given as **ottar** but with the same French pronunciation. The finest attar of rose is from Bulgaria, where it is distilled from the flowers of the **damask rose,** *Rosa damascena.* The fresh oil is colorless, but turns yellowish green. About 20,000 lb (9,072 kg) of flowers are needed to make 1 lb (0.5 kg) of essence, and it is so valuable that it is usually adulterated with geraniol or synthetic rose. In France the oil is obtained from the *R. centifolia.* **Rose water** is the scented water left after distillation, or is made by dissolving attar in water. The **Otto of baronia** of Australia is a high-grade **rose oil.**

Geranium oil is obtained from the leaves or flowers of the *Pelargonium graveolens* of the Mediterranean countries and other species of geranium. It is used as an adulterant or substitute for rose oils in perfumes and soaps. **Zdravetz oil** is a geranium oil from the *P. macnorhijum* of Cyprus, used in rose bouquet and lavender perfumes. Many geranium and rose oils are derived from geraniol obtained from citronella and other oils. A synthetic rose-geranium is **diphenyl methane,** $(C_6H_5)_2CH_2$, a colorless solid melting at about 25°C. **Benzophenone** is also used for rose-geranium perfumes. It is a **diphenyl ketone,** $C_6H_5(CO)C_6H_5$, melting at 47°C. This material is also used for making fine chemicals. **Geraniol,** from citral, is a colorless liquid with a sweet, delicate rose odor. **Vetiver,** a very sweet-scented oil used in high-grade perfumes and in medicine, is distilled from the roots of the **khuskhus** plant, *Vetiveria zizamoides,* native to India but produced chiefly in Java, Réunion, and Haiti. **Opopanox,** used in incense and in medicine, is an oleoresin from the roots of the *pastinaca opopanox* of the Orient and British Somaliland. **Frankincense,** used in incense and perfumes, and in medicine under the name of **olibanum,** is a gum resin from the tree *Boswellia carterii* of the Sudan and Somaliland. It comes in hard yellow grains. **Kiounouk,** used as a fixative, is a clear yellowish semiliquid obtained from olibanum. **Mecca balsam,** used in oriental types of perfume, is a greenish oleoresin from the plant *Commiphora opobalsamum* of Arabia. It has the odor of rosemary. **Rosemary** is an oil distilled from the fresh flowering tops of the sweet-smelling evergreen shrub *Rosemarinus officinalis* of the Mediterranean countries. It is used in eau de cologne, soaps, and medicine. **Jasmine oil,** a highly valued perfume material, is from the fragrant flowers of the shrub *Jasminum grandiflorum,* a species of jasmine grown in southern France especially for perfume. The oil is extracted from the fresh flowers by enfleurage. A synthetic **jasmine-rose oil,** which also has a **peach-apricot flavor** and a sweet taste, is **benzyl propionate,** $CH_3CH_2COOCH_2C_6H_5$. It is a liquid boiling at 220°C, and is used in perfumes and as a flavor. **Ylang-ylang,** or *cananga oil,* is a valuable essential oil from the flowers of the tree Canangium odorata, cultivated in Indonesia, Malagasy, and the Philippines. No more than 150 lb (68 kg) of flowers are obtained from a tree, but about 400 lb (181 kg) are required to produce 1 lb (0.5 kg) of oil. It contains linalol and geraniol. Another oil that rivals ylang-ylang in fragrance is **champaca oil,** from the flowers of the large tree *Michelia champaca,* of southern Asia.

Lavender oil, used with rosemary in eau de cologne, and also as **lavender water** in a mixture of the oil in water and alcohol, is obtained from the flowers of the shrub *Lavandula officinalis* of southern Europe. The dried flowers are fragrant and are used in sachets. **Spike lavender** is an inferior oil from the plant *L. latifolia* of France and Spain. It is used in perfumes and sometimes as a food flavor. **Herbandin,** of the UOP Chemical Co., is a synthetic ester made from petroleum and used as a replacement or extender for natural lavender oil. It has a pronounced lavender odor. Espantone is a synthetic ketone with a spike lavender odor. **Bay oil,** or **myrcia oil,** used in the toilet alcohol known as **bay rum,** and also in perfumes, is distilled from the leaves of the bay tree, *Pimenta acris,* of the West Indies, 60 lb (27 kg) of leaves yielding 1 lb (0.5 kg) of oil, and 1 gal (0.004 m^3) of bay oil being used to 100 gal (0.4 m^3) of rum to make bay rum. It contains eugenol, and has a spicy odor.

Carnation oil is obtained by solvent extraction or by enfleurage from the flowers of the *Dianthus caryophyllus,* of which there are more than 2,000 varieties grown in the Mediterranean countries. The less highly cultivated plants give the richest perfumes. **Violet oil** is derived by solvents or maceration in hot oils from the flowers of the blue and purple varieties of *Viola odorata.* **Synthetic violet** is made from **ionone,** $C_{13}H_{20}O$, derived from lemongrass oil. The ionones are made synthetically by condensation of citral with acetone. They are monoenol and dienol butones. **Velvione,** of Rhodia, Inc., is such an ionone with a powerful violet odor. The true violet odor is **irone**, a complex seven-ring compound. It can be obtained from the iris root and is one of the most odoriferous materials obtained from plants. **Orris** is the dry root of the *Iris florentian,* and the powdered root is used in violet powders and as a flavor. **Oak moss** was one of the perfumes of ancient Egypt. It is obtained from the lichen *Evernia prunastri* and *E. furfuracea* growing on oak and spruce trees of southern Europe. The resinous extract has the odor of musk and lavender. It is used as a fixative in perfumes of the poppy type. **Rue oil,** used for sweet pea perfume, is distilled from the plants *Ruta graveolans* of France and *R. montana* of Algeria.

As with jewelry, the manufacture of perfumes is normally classed as a luxury industry. But the vital test of essentiality comes in wartime, and these aesthetic materials are always considered as essential to the public morale, and hence basic. Even under desperate wartime conditions France never stopped the manufacture of perfumes, and during the life struggle of England in the Second World War the restrictive regulations placed upon perfume manufacture had to be abandoned quickly because of public pressure. Wartime restrictions placed on the imports of perfume oils into the United States during the Second World War were immediately abandoned.

PERILLA OIL. A light yellow oil obtained from the seeds of the plant *Perilla ocimoides* of China and Japan, and employed in varnishes, core oils, printing ink, and linoleum. It dries to a harder, tougher, glossier film than linseed oil. The specific gravity is about 0.935, iodine value 200, and saponification value 191. It contains 41 to 46% linolenic acid, 31 to 42 linoleic, and 3 to 10 oleic acid. In Japan it is called **egoma oil.** The raw oil tends to form globules, but this is overcome by boiling or blowing. The blown oil is rapid-drying and is more weather-resistant than linseed oil.

PERMEABILITY ALLOYS. A general name for a group of nickel-iron alloys with special magnetic properties. These soft magnetic materials possess a magnetic susceptibility much greater than iron. An early alloy determined by experiment was made up theoretically of 78.5% nickel, 21.5 iron, but with other elements approximately as follows: carbon 0.04%; silicon, 0.03; cobalt, 0.37; copper, 0.10; and manganese, 0.022. It is produced by the Western Electric Co., sometimes with chromium or molybdenum, under the name of **Permalloy,** and is used in magnetic cores for apparatus that operates on feeble electric currents, and in the loading of submarine cables. It has very little magnetic hysteresis.

Supermalloy, developed by the Bell Laboratories for transformers, contains 79% nickel, 15 iron, 5 molybdenum, and 0.5 manganese, with total carbon, silicon, and sulfur kept below 0.5%. It is melted in vacuum, and poured in an inert atmosphere. It can be rolled to a thinness of 0.00025 in (0.00064 cm). The alloy has an initial permeability 500 times that of iron. **Supermendur,** of the Westinghouse Electric Corp., contains 49% iron, 49 cobalt, and 2 vanadium. It is highly malleable, and has very high permeability with low hysteresis loss at high flux density. **Duraperm,** of the Hamilton Watch Co., is a high-flux magnetic alloy containing 84.5% iron, 9.5% silicon, and 6 aluminum. **Perminvar,** of the Western Electric Co., is an alloy containing 45% nickel and 25 cobalt, intended to give a constant magnetic permeability for variable magnetic fields. **A-metal** is a nickel-iron alloy containing 44% nickel and a small amount of copper. It is used in transformers and loudspeakers to give nondistortion characteristics when magnetized.

Permenorm, first produced in Germany as **Orthonel** and developed at the U.S. Naval Ordnance Laboratory, contains 50% nickel and 50 iron. It is subjected to drastic rolling in one direction followed by a heat treatment to obtain chemical combination of the nickel and iron. It has a grain orientation which, when subjected to a magnetic field, produces a square hysteresis loop indicating instantaneous magnetization. It is used as a core material and in magnetic amplifiers. **Deltamax,** of the Arnold Engineering Co., is this material. **Alfenol,** developed by the Naval Ordnance Laboratory, contains no nickel, but has 16% aluminum and 84 iron. It is brittle and cannot be rolled cold, but can be rolled into thin sheets at a temperature of 575°C. It is lighter than other permeability alloys, and has superior characteristics for transformer cores and tape-recorder heads. A modification of this alloy, called **Thermenol,** contains 3.3% molybdenum without change in the single-phase solid solution of the binary alloy. The permeability and coercive force are varied by heat treatment. At 18% aluminum the alloy is practically paramagnetic. The annealed alloy with 17.2% aluminum has constant permeability.

Aluminum-iron alloys with 13 to 17% aluminum are produced in sheet form by the Metals & Controls Corp. for transformers and relays. They have magnetic properties equal to the 50–50 nickel-iron alloys and to the silicon-iron alloys, and they maintain their magnetic characteristics under changes in ambient temperature.

Conpernik, of the Westinghouse Electric Corp., contains equal amounts of nickel and iron with no copper. It is called constant-permeability nickel, and has little permeability variation. It differs from **Hipernik** of the same company only in heat treatment. When heat-treated, Hipernik has higher permeability than silicon transformer steel up to flux density of about 10 kilogausses. Both alloys are used in transformer cores. **Hipernik V** is grain-oriented, giving a square hysteresis loop, with a low degree of coercive force, 0.15 Oe (12 A/m), and a high ratio of residual magnetism to saturation. It comes in thin strip for use in instruments and computer elements. **Hiperco,** of this company, contains 35% cobalt, 64 iron, and 1 chromium. It is ductile and easy to roll to extremely thin strip for instrument parts. The heat-treated metal has high permeability and low hysteresis loss, but in this condition it has low tensile strength and is brittle. **Vicalloy,** developed by the Bell Laboratories, is a high-permeability alloy containing 36 to 62% cobalt, 6 to 16 vanadium, and the balance iron. It is cast and hot-swaged, then drawn into wire or tape as fine as 0.002 in (0.005

cm). It retains high permanent magnetism, with coercive forces to 250 Oe (19,895 A/m) and residual flux of 8,000 G, and it retains its magnetism to 600°F (316°C).

Iron-nickel permeability alloys are used as loading by wrapping a continuous layer around the full length of the cable. When nickel-copper alloys are used, they are employed as a core for the cable. **Magnetostrictive alloys** are iron-nickel alloys which will resonate when the frequency of the applied current corresponds to the natural frequency of the alloy. They are used in radios to control the frequency of the oscillating circuit. **Magnetostriction** is the stress that occurs in a magnetic material when the induction changes. In transducers it transforms electromagnetic energy into mechanical energy. **Temperature-compensator alloys** are iron-nickel alloys with about 30% nickel. They have the characteristics that they are fully magnetic at −20°F (−29°C) but lose the magnetic permeability in proportion to rise in temperature until at about 130°F (54°C) they are nonmagnetic. Upon cooling they regain permeability at the same rate. They are used in shunts in electrical instruments to compensate for errors due to temperature changes in the magnets.

PERSIMMON WOOD. The wood of the common persimmon trees, *Diospyros virginiana,* of the southeastern United States, and the **black persimmon,** *D. texana,* of western Texas. It is used for shuttles, golf-stick heads, and tools, and takes a fine polish. The tree belongs to the ebony family, of which there are listed about 160 species. The persimmon is a small tree not over 50 ft (15 m) high and 12 in (30 cm) in diameter. The wood is very hard, strong, and compact, and retains its smoothness under long rubbing. The sapwood is light brown in color, 2 to 5 in (5 to 13 cm) wide, and the heartwood is black but very small, the thickness of a pencil, for example, in a 6-in (15 cm) tree. The weight is 49 lb/ft^3 (785 kg/m^3). The fruit is cultivated for food, and there are only limited quantities of the wood. A fine wood used for fancy articles is **olive wood,** from the olive fruit tree of California. It is yellow with beautifully streaked dark lines.

PETITGRAIN. An essential oil obtained by distillation of the leaves and small twigs of the **bitter orange** tree, *Citrus aurantium,* native to tropical Asia, but now grown in other countries. In Spain, it is known as the **Seville orange.** Paraguay is the chief producer of high-grade petitgrain, which is one of the best fixatives for fine perfumes and is also used in flavoring extracts. One pound (0.5 kg) of petitgrain is obtained from 100 to 150 lb (45 to 68 kg) of leaves. The fruits of the tropical bitter orange are large and of the finest golden appearance, but the pulp is very acid. The juice is used only for blending in orange drinks. It contains a dilactone, **limonin,** which gives it a bitter taste. The rind is used in **marmalade** and **candied orange peel.** An essential oil distilled from the rind is known as **curaçao,**

and is used in perfumery and in **curaçao liqueur.** The flowers are very fragrant, and from them **neroli oil** is distilled. Neroli is used in perfumery blends and for mixing with synthetic perfumes. **Neroli Portugal** is inferior, and comes from the sweet orange *C. sinensis* by extraction. **Orange oil,** obtained by expression from the ripe rind of the sweet orange, is a less valuable oil. **Bergamot oil** is from the rind of the fruit of the small spiny tree *C. bergamia* of Italy. It has a soft sweet odor, and is used in perfume blends and in soaps. The golden-yellow pear-shaped fruit has an acid inedible pulp.

The **bioflavonoids,** used in cold remedies, are obtained from the white pulp, or **albedo,** of citrous fruits. They are alkaline-soluble crystalline compounds, variations of **chromone,** a **benzpyrone,** and **flavone,** the **phenyl benzpyrone.** The pressed product from the pulp is acidified, crystallized, and dried. **Citroid,** of the Grove Laboratories, is a product of this kind. About 150 distinct chemicals have been produced from citrous fruits. The bioflavonoids are six-membered, double-ring compounds. Some are isolated and used directly, as the **naringin** from grapefruit peel, which is an effective substitute for quinine. Others are synthesized easily. The chromone and flavone are really the parent substances of many natural vegetable dyes, drugs, and tannins, and are readily convertible to these materials.

PETROLATUM. A jellylike substance obtained in the fractional distillation of petroleum. Its composition is between $C_{17}H_{36}$ and $C_{21}H_{44}$, and it distills off above 303°C. It is also called **petroleum jelly.** It is used for lubricating purposes and for compounding with rubber and resins. When highly refined for the pharmacy trade, it is used as an ointment. The specific gravity is from 0.820 to 0.865. It is insoluble in water but readily soluble in benzine and in turpentine. For lubricating purposes it should be refined by filtration only and not with acids, and should not be adulterated with paraffin. The melting point should be between 115 and 130°F (46 and 54°C). Petrolatum of Grade O, used as a softener in rubber, is a pale-yellow, odorless semisolid of specific gravity 0.84 and melting point 115 to 118°F (46 to 48°C). **Sherolatum** is petrolatum jelly of the Sherwood Petroleum Co., Inc., and **Vaseline** is petroleum jelly of the Chesebrough Mfg. Co. But the trade-named petroleum jellies for pharmaceutical uses may be compounded with other materials.

PETROLEUM. A heavy, liquid, flammable oil stored under the surface of the earth, and originally formed as the by-product of the action of bacteria on marine plants and animals. It consists chiefly of carbon and hydrogen in the form of hydrocarbons, including most of the liquids of the **paraffin series,** C_5H_{12} to $C_{16}H_{34}$, together with some of the gases, CH_4 to C_4H_{10}, and most of the solids of the series from $C_{17}H_{36}$ to $C_{27}H_{56}$. It also contains hydrocarbons of other series. While petroleum is used primarily for the production of fuels and lubricating oils, it is one of the most valuable raw materials for a very wide range of chemicals. The name **petrochemicals** is used in a general sense to mean chemicals derived from petroleum, but it does not mean any particular class of chemicals. Sulfur and helium are by-products from petroleum working.

Petroleums from different localities differ in composition, but tests of oils from all parts of the world give the limits as 83 to 87% carbon, 11 to 14 hydrogen, with sulfur, nitrogen, and oxygen in amounts from traces to 3%. Mexican and Texan oils are high in sulfur. The crude oil is split by distillation into naphtha, gasoline, kerosene, lubricating oils, paraffin, and asphalt. It may also be split by cracking, that is, by subjecting it to violent heating in the absence of air. This process yields a higher proportion of volatile products because of the breaking down of the more complex molecules by the high heat. Liquefied petroleum gases, including **propane, butane, pentane,** or mixtures, are marketed under pressure in steel cylinders as **bottled gas.** Propane, $CH_3CH_2CH_3$, is used in cook stoves. Butane, which has an additional CH_2 group, is used to enrich illuminating and heating gas. Propane and butane gases have heating values from 2,800 to 3,000 Btu/ft^3 (1 billion to 1.1 billion J/m^3). **Liquid gas** is also used for internal-combustion engines, as a solvent, and for making many chemicals. **Pyrogen** is the trade name of gas obtained during the process of recovering gasoline from natural gas. It is marketed in cylinders for use in flame cutting torches. **Road oil,** used on dirt roads, is a heavy-residue oil from the refineries.

Certain highly refined oils used in medicine as laxatives and used in other specialized applications are referred to as **mineral oils. White mineral oil** is petroleum highly refined to color. Russian and Rumanian oils with high naphthene content were particularly suitable for this refining, and the oil was called **Russian oil.** Pennsylvania paraffin-base oils and Texas asphaltic-base oils are difficult to refine to color. The American white Russian oil is refined from mid-Continent and Gulf Coast oils which contain high naphthenes. It is used as a laxative, as a carrier of many drugs, in textile spinning, as a plasticizer for synthetic resins, and for sodium dispersions where the alkali metal would normally react with any impurities. **Klearol** and **Blandol,** of L. Sonneborn Sons, Inc., are viscosity grades of white mineral oil.

Petroleum from Baku was used from ancient times for lighting purposes, and the Bolivian oil was used in the sixteenth century for burning. The first commercial wells in the United States were opened in 1859 at Titusville, Pa. The chief production of petroleum is in Mexico, the United States, Russia, Rumania, Asia Minor, Peru, and northern South America, but large reserves exist in many other places. About half the world production of crude petroleum is in the Western Hemisphere, but the largest single field is in the Persian Gulf area. Only about 1% of the production of crude oil is from the Pennsylvania field, but much of the motor lubricating oil is from this oil. Many undersea deposits of petroleum are known, but some are at great depth; for example, the water at the Rockall Plateau, northeast of Ireland, is 5,000 ft (1,524 m) deep with 10,000 ft (3,048 m) of sediment above the oil deposit.

PEWTER. A very old name for tin-lead alloys used for dishes and ornamental articles, but now referring to the use rather than to the composition of the alloy. Tin was the original base metal of the alloy, the ancient **Roman pewter** having about 70% tin and 30 lead, although iron and other elements were present as impurities. Pewter, or **latten ware,** of the sixteenth century contained as high as 90% tin, and a strong and hard **English pewter** contained 91% tin and 9 antimony. This alloy is easily cold-rolled and spun, and can be hardened by long annealing at 225°C and quenching in cold water and tempering at 110°C. Pewter is now likely to contain lead and antimony, and very much less tin; when the proportion of tin is below about 65%, the alloys are unsuited for vessels to contain food products, because of the separation of the poisonous lead. Antimony is also undesirable in food containers because of its poisonous nature, but when the tin content is low, antimony is needed to make the alloy susceptible to polishing. Ordinary pewter, with 6% antimony, 1.5 copper, and the balance tin, has a Vickers hardness of 23. With the addition of 1.5% bismuth, the hardness is 29. Best pewter, used for high-class articles, contains 100 parts tin, 8 antimony, 2 bismuth, and 2 copper. Triple has 83 parts tin, 17 antimony, or some lead to replace part of the antimony. Pewter should have a peculiar bluish-white luster when polished. It can be spun easily. Pewters containing much lead are dark in color and must be plated.

Britannia metal is a type of latten ware which also usually contains copper. The color is silvery white with a bluish tinge, or with a yellowish tinge if the copper is high. The ordinary composition is quite similar to some babbitts, 89% tin, 7.5 antimony, and 3.5 copper. It takes a fine polish and does not tarnish easily. It is easily worked by stamping, rolling, or spinning, or may be cast. Some zinc may be used in the casting alloys. English Britannia metal has 94% tin, 5 antimony, and 1 copper. **Hanover white metal** contains 87% tin, 7.5 antimony, and 5.5 copper. **Dutch white metal** has 81.5% tin, 8.5 antimony, and 9.5 copper. **Queen's metal** is a Britannia metal with a small amount of zinc. A typical composition is 88.5% tin, 7 antimony, 3.5 copper, and 1 zinc. The zinc helps to strenghten the alloy. **Minofor** is another grade containing up to 9% zinc and up to 1 iron. When zinc is used in these metals, the antimony is lowered because both metals tend to make the alloy brittle. **Ashberry metal,** for tableware, contains some nickel and aluminum. **Tutanic metal,** a German utensil alloy, had 88 to 92% tin, up to 3 copper, and 6 to 7.5 lead. **Ludensheid plate** had 72% tin, 24 antimony, and 4 copper, with a trace of lead to increase fluidity. **Wagner's alloy** and **Koeller's alloy** were names for utensil alloys, the latter containing some bismuth to improve the casting.

PHENOL. Also known as **carbolic acid.** A colorless to white crystalline material of sweet odor, having the composition C_6H_5OH, obtained from the distillation of coal tar and as a by-product of coke ovens. It is also made by alkylating benzene with propylene to form **cumene,** which is **isopropylbenzene,** $C_6H_5CH(CH_3)_2$, and then oxidizing to cymene hydroperoxide and finally splitting to form phenol and acetone. Or, it can be made by oxidation of toluene and then a catalytic conversion to phenol. It is also produced by hydrogenation of the lignin from sulfite waste liquor. It is a valuable chemical raw material for the production of plastics, dyes, pharmaceuticals, syntans, and other products.

Phenol melts at about 43°C and boils at 183°C. The pure grades have melting points of 39, 39.5 and 40°C. The technical grades contain 82 to 84% and 90 to 92% phenol. The crystallization point is given as 40.41°C. The specific gravity is 1.066. It dissolves in most organic solvents. By melting the crystals and adding water, liquid phenol is produced, which remains liquid at ordinary temperatures. Phenol has the unusual property of penetrating living tissues and forming a valuable antiseptic. It is also used industrially in cutting oils and compounds and in tanneries. The value of other disinfectants and antiseptics is usually measured by comparison with phenol. The **phenol coefficient** of a disinfectant is the ratio of the dilution required to kill the Hopkins strain of typhoid bacillus in a specified time compared with the dilution of phenol required for the same organism under the same conditions and time. However, phenol is poisonous and gives dangerous burns on the skin, so that as a disinfectant it must be used with caution.

Phenol is a very reactive and versatile compound. In coal tar it occurs with many homologs, and many of the complex chemicals occurring in vegetable life are homologs or complexes of phenol. It can be easily nitrated, sulfonated, or halogenated. The hydroxyl hydrogen is readily replaceable by strong bases to produce **phenolates,** and these are readily convertible to ethers, such as **diphenyl ether,** or **diphenyl oxide,** $(C_6H_5)_2O$, which has a boiling point at 259°C and is used as a heat-transfer medium. Ortho and para compounds are obtained by direct substitution, and these are called **substituted phenols. Thiophenol,** C_6H_5SH, also called **mercaptobenzene** and **phenyl mercaptan,** is an evil-smelling liquid boiling at 172°C, used for making dyes and pharmaceuticals. One of the most important applications of phenol is its condensation with formaldehyde to produce synthetic resins, and these resins can be varied greatly by alterations in the phenol. The term **phenols** is a class name for a wide variety of materials, as distinct from the normal phenol.

The Koppers Co. markets **alkylated phenols** for the production of tough and oil-soluble phenol-formaldehyde resins. Tertiary **butyl phenol,** $C_{10}H_{14}O$, comes in flakes melting at 98°C. Tertiary **amyl phenol,** $C_{11}H_{10}O$, is also in flakes melting at 90°C. **Nonyl phenol** is a mixture of monoalkyl phenols with side chains of random-bracketed alkyl radicals on the molecule. It is a liquid of specific gravity 0.94, boiling at 300°C, used for making oil-soluble phenolic resins and lubricating oil additives. **Phenyl phenol** is used in cosmetics to control odors and prevent bacterial deterioration, and also as a preservative in paper, paints, and leather. **Bisphenol** is a hydroxy phenol, $C_{15}H_{16}O$, with high reactivity, used for producing modified phenolic resins and epoxy resins, as an antioxidant for oils, and as a stabilizer for resins. **Santobrite,** of the Monsanto Chemical Co., is chlorinated phenol, or **sodium pentachlorophenol,** used as a preservative. **Biolite** is a formulation of Santobrite used for slime and mildew control in laundries and in paper-pulp plants. Para **nitrophenol,** made by treating phenol with cold dilute nitric acid, is used as a preservative for leather goods to prevent mold growth. **Lorothiodol,** of the Hilton-Davis Chemical Co., is a sulfurized chlorinated phenol. It is a more powerful antiseptic than phenol, is odorless and nontoxic, and is used in soaps. **Pentaphen,** of Sharples Chemicals, Inc., is a para tertiary amyl phenol, $C_5H_{11}C_6H_4OH$, used for making pale-colored, oil-soluble resins by condensation with aldehydes.

A wide range of related materials is produced from **diphenyl,** also called **biphenyl** and **phenylbenzene.** It has the composition $(C_6H_5)_2$, is a liquid below 71°C, and boils at 255°C. It is a stable compound and is used as a heat-transfer medium. But it is made easily by heating benzene to eliminate H_2 and, because of its low cost, is a valuable intermediate chemical. Like phenol, it can be modified easily to produce innumerable compounds. **Tetramethylbenzene,** or **durene,** for example, is a 10-carbon aromatic liquid used for making **pyromellitic acid** and its anhydride for the production of polyester resins. Many other useful acids are also produced.

PHENOL-FORMALDEHYDE RESIN. A synthetic resin, commonly known as **phenolic,** made by the reaction of phenol and formaldehyde, and employed as a molding material for the making of mechanical and electrical parts. It was the earliest type of hard, thermoset synthetic resins, and its favorable combination of strength, chemical resistance, electrical properties, glossy finish, and nonstrategic abundance of low-cost raw materials has continued the resin, with its many modifications and variations, as one of the most widely employed groups of plastics for a variety of products. The resins are also used for laminating, coatings, and casting resins.

The reaction was known as early as 1872 but was not utilized commercially until much later. A condensation product of 50 parts phenol and 30 parts 40% formaldehyde made under an English patent of 1905 was called **Resinite,** and was originally offered as a substitute for Celluloid. Various modifications were made by other inventors. **Redmanol** was one of the first of the American products by the Bakelite Corp. **Juvelite** was made in Germany by condensing the phenol and formaldehyde with the aid of mineral acids, and **Laccain** was made under an English patent by using organic acids as catalysts. A Russian phenol resin, under the name of **Karbolite,** employed an equal amount of naphthalenesulfonic acid, $C_{10}H_7SO_3H$, with the formaldehyde.

The hundreds of different phenolic molding compounds can be divided into six groups on the basis of major performance characteristics. General-purpose phenolics are low-cost compounds with fillers such as wood flour and flock, and are formulated for noncritical functional requirements. They provide a balance of moderately good mechanical and electrical properties, and are generally suitable in temperatures up to 300°F (149°C). Impact-resistant grades are higher in cost. They are designed for use in electrical and structural components subject to impact loads. The fillers are usually either paper, chopped fabric, or glass fibers. Electrical grades, with mineral fillers, have high electrical resistivity plus good arc resistance, and they retain their resistivity under high-temperature and high-humidity conditions. Heat-resistant grades are usually mineral-or glass-filled compounds that retain their mechanical properties in the 375 to 500°F (190 to 260°C) temperature range. Some of these, such as **phenylsilanes,** provide long-term stability at temperatures up to 550°F (288°C). Special-purpose grades are formulated for service applications requiring exceptional resistance to chemicals or water, or combinations of conditions such as impact loading and a chemical environment. The chemical-resistant grades, for example, are inert to most common solvents and weak acids, and their alkali resistance is good. Nonbleeding grades are compounded specially for use in container closures and for cosmetic cases.

The resins are marketed usually in granular form partly polymerized for molding under heat and pressure, which complete the polymerization, making the product infusible and relatively insoluble. They may also come as solutions, or compounded with reinforcing fillers and pigments. The tensile strength of a molded part made from a simple phenol-formaldehyde resin may be only about 6,000 lb/in^2 (41 MPa), with a specific gravity of 1.27 and dielectric strength of about 450 volts per mil (17.7×10^6 volts per meter). Reinforcement is needed for higher strength, and with a wood-flour filler the tensile strength may be as high as 10,000 lb/ in^2 (68 MPa). With a fabric filler the flexural strength may be 15,000 lb/in^2 (103 MPa), or 18,000 lb/in^2 (124 MPa) with a mineral filler. The specific gravity is also raised, and the mineral fillers usually increase the dielectric strength.

Proper balance of fillers is important, since too large a quantity may produce brittleness. Organic fillers absorb the resin and tend to brittleness and reduced flexural strength, although organic fibers and fabrics generally give high impact strength. Wood flour is the most usual filler for general-service products, but prepared compounds may have mineral powders, mica, asbestos, organic fibers of macerated fabrics, or mixtures of organic and mineral materials. The **Resinox** resins, of Monsanto, are in grades with fibrous and mineral fillers, and **Moldarta,** of the Westinghouse Electric Corp., is in various grades. **Bakelite** was the original name for phenol plastics, but trade names now usually cover a range of different plastics, and the types and grades are designated by numbers.

The specific gravity of filled phenol plastics may be as high as 1.70. The natural color is amber, and, as the resin tends to discolor, it is usually pigmented with dark colors. Normal **phenol resin** cures to single-carbon methylene groups between the phenolic groups, and the molded part tends to be brittle. Thus, many of the innumerable variations of phenol are now used to produce the resins, and modern phenol resins may also be blended or cross-linked with other resins to give higher mechanical and electrical characteristics. Furfural is frequently blended with the formaldehyde to give better flow, lower specific gravity, and reduced cost. The alkylated phenols give higher physical properties. **Phenol-phosphor resin** is a phenol resin modified with phosphonitrilic chloride. When cured the resin contains 15% phosphorous, 6 nitrogen, and less than 1 chlorine. The tensile strength is 7,000 lb/in^2 (48 MPa) and it will withstand continuous temperatures to 500°F (260°C). The **Flexiphen 160** resin, of the Koppers Co., has some of the single-carbon methylene linkages replaced by hydrocarbon chains, giving 30% higher flexural strength with 5% lower specific gravity. **Resinox 3700,** of the Monsanto Chemical Co., is a mineral-filled phenolic resin of high arc resistance and high dimensional stability for electrical parts. **Synvar,** of the Synvar Corp., and **Durez,** of the Hooker Chemical Corp., come in a number of grades.

Phenol resins may also be cast and then hardened by heating. The cast resins usually have a higher percentage of formaldehyde and do not have fillers. They are poured in syrupy state in lead molds and hardened in a slow oven. **Crystallin,** of the Crystallin Products Corp., **Phenalin,** of Du Pont, and **Catalin,** of the American Catalin Corp., are cast phenol plastics. **Ivoricast,** of West Coast Enterprises, is a shock-resistant cast phenolic plastic with wood-flour filler which cures at low heat. **Prystal** is the name of water-clear Catalin, and **Bois glacé** is **Catalin-coated wood** for desk tops. **Fiberlon** is a cast phenol plastic of the Fiberloid Corp. **Marblette** is a cast plastic of the Marblette Corp.

PHONOLITE. Also known as **clinkstone.** An aluminum-potassium-silicate mineral used in the production of glass, and in Germany for the production of aluminum. Phonolite is a variety of feldspar. It varies greatly in composition, the best of the Eifel Mountains mineral containing 20 to 23¼% alumina, 7 to 9 K_2O, 6 to 8 Na_2O, and 50 to 52 silica. A variety of this mineral, **nepheline,** from the Kola Peninsula, is used in Russia to produce aluminum, with soda and potash as by-products. **Nepheline syanite** from Peterborough Co., Ontario, Canada, is used in the ceramic industry in pottery, porcelain, and tile to increase translucency and reduce warpage and crazing. From 3 to 5% added to structural clay increases the mechanical strength. As a substitute for potash feldspar in wall tiles, it increases the fluxing action and lowers the fusing point. **Agalmatolite,** a name derived from the Greek words meaning **image stone,** is the massive form of phonolite from which the Chinese carve figures and bas-reliefs. It has a soft greasy feel, and varies in color.

PHOSGENE. The common name for **carbonyl chloride,** $COCl_2$, a colorless, poisonous gas made by the action of chlorine on carbon monoxide. It was used as a poison war gas, called **D-stoff** by the Germans and **collongite** by the French. But it is now used in the manufacture of metal chlorides and anhydrides, pharmaceuticals, perfumes, isocyanate resins, and for blending in synthetic rubbers. It liquefies at 8.2°C, and solidifies at −118°C. It is decomposed by water. When chloroform is exposed to light and air, it decomposes into phosgene. One part in 10,000 parts of air is a toxic poison. For chemical warfare it is compressed into a liquid in shells. **Lacrimite,** also a poison war gas, is thiophosgene mixed with stannic chloride. **Diphosgene,** $ClCOOC \cdot Cl_3$, called **green cross, superpalite,** and **perstoff,** is an oily liquid boiling at 128°C. It is an intense lachrymator, has an asphyxiating odor, and is a lung irritant.

PHOSPHOR BRONZE. Originally called **steel bronze** when first produced at the Royal Arsenal in Vienna. It was 92–8 bronze deoxidized with phosphorus and cast in an iron mold. It is now any bronze deoxidized by the addition of phosphorus to the molten metal. It may or may not contain residual phosphorus in the final state. Ordinary bronze frequently contains cuprous oxide formed by the oxidation of the copper during fusion. By the addition of phosphorus, which is a powerful reducing agent, a complete reduction of the oxide takes place. Phosphor bronzes have excellent mechanical and cold-working properties and a low coefficient of friction, making them suitable for springs, diaphragms, bearing plates, and fasteners. In some environments, such as salt water, they are superior to copper. Phosphor-bronze casting metals for bearings usually contain lead. ASTM Grade B phosphor bronze has 4.0 to 5.5% tin, 2.5 to 4 lead, and 0.03 to 0.25 phosphorus. A foundry alloy known as **standard phosphor bronze** contains about 80% copper, 10 tin, and 10 lead, with about 0.25

phosphorus. When chill-cast, it is fine-grained and has a tensile strength up to 33,000 lb/in^2 (227 MPa), Brinell hardness 65, and weight 0.325 lb/in^3 (8,996 kg/m^3). One large automotive company lists this metal as **phosphor casting bronze.** The phosphor-bronze ingot metal of this composition marketed by H. Kramer & Co. contains 0.50 to 1% phosphorus, part of which is absorbed in the melting and casting. The residual phosphorus is 0.25%.

Anaconda phosphor bronze, a wrought metal marketed by the American Brass Co.,is in grades containing from 5 to 10% tin and 90 to 95 copper. The 90–10 grade for springs has high resistance to fatigue. This is ASTM Grade D wrought phosphor bronze. When soft it has a tensile strength of 65,000 lb/in^2 (448 MPa) with elongation of 65%, and in spring hardness the tensile strength is 125,000 lb/in^2 (861 MPa) with elongation of 4%. **Duraflex** is the name of this company for hard-rolled strip and wire phosphor bronze for springs.

Seymour phosphor bronze, of the Seymour Mfg. Co., used for springs, has 95% copper, 4.75 tin, and 0.25 phosphorus. This is **ASTM spring bronze,** Grade A. The Grade B has 7.75% tin with higher physical properties. For coil and flat springs, these alloys have service temperatures to 225°F (107°C). **Carobronze,** of Wrought Bearing Metals, Inc., has 91.2% copper, 8.5 tin, and 0.3 phosphorus. It is produced in hard-drawn tubes and rods to give greater wear resistance and higher impact value than cast bronze for bearings. **Corvic bronze,** of the Chase Brass & Copper Co., has 98.2% copper, 1.5 tin, and 0.30 phosphorus. The tensile strength of the spring material is 95,000 lb/in^2 (654 MPa), and the electric conductivity is 42% that of copper. **Telnic bronze** of this company contains 1% nickel, 0.20 phosphorus, 0.5 tellurium, and the balance copper. The phosphor bronze marketed by Revere Copper & Brass, Inc., in sheet, plate, and strip contains 1.25 to 10% tin with not over 0.05% residual phosphorus. The soft material with hardness to Rockwell B75 has tensile strengths from 40,000 to 73,000 lb/in^2 (275 to 503 MPa), while the hard metal, with Rockwell B hardness from 75 to 104, has tensile strengths from 90,000 to 129,000 lb/in^2 (620 to 889 MPa). **Phosphor bronze rod,** for producing screw-machine parts, contains 88% copper, 4 tin, 4 zinc, and 4 lead. It is free-cutting and has high strength. Hard-drawn phosphor-bronze wire may have a tensile strength exceeding 120,000 lb/in^2 (827 MPa). **White phosphor bronze,** for bearings, contains 72% lead, 12 antimony, 15 phosphor tin, and 1 copper, and is not a bronze.

PHOSPHOR COPPER. An alloy of phosphorus and copper, used instead of pure phosphorus for deoxidizing brass and bronze alloys, and for adding phosphorus in making phosphor bronze. It comes in 5, 10, and 15% grades and is added directly to the molten metal. It serves as a powerful deoxidizer, and the phosphorus also hardens the bronze. Even slight additions of phosphorus to copper or bronze increases the fatigue strength. Phosphor copper is made by forcing cakes of phosphorus into molten copper and holding under until the reaction ceases. Phosphorus is soluble in copper up to 8.27%, forming Cu_3P, which has a melting point of 707°C. A 10% phosphor copper melts at 850°C, and a 15% at 1022°C. Alloys richer than 15% are unstable. Phosphor copper is marketed in notched slabs or in shot. In Germany **phosphor zinc** was used as a substitute to conserve copper. **Metallophos** is a name for German phosphor zinc containing 20 to 30% phosphorus. The name phosphor copper is also applied to commercial copper deoxidized with phosphorus and retaining up to 0.50% phosphorus. The electric conductivity is reduced about 30%, but the copper is hardened and strengthened. **Phosphor tin** is a master alloy of tin and phosphorus used for adding to molten bronze in the making of phosphor bronze. It usually contains up to 5% phosphorus and should not contain lead. It has an appearance like antimony, with large glittering crystals, and is marketed in slabs. Federal specifications call for 3.5% phosphorus, and with not over 0.50% impurities.

PHOSPHORIC ACID. Also known as **orthophosphoric acid.** A colorless, syrupy liquid of the composition H_3PO_4 used for pickling and rustproofing metals, for the manufacture of phosphates, pyrotechnics, and fertilizers, as a latex coagulant, as a textile mordant, as an acidulating agent in jellies and beverages, and as a clarifying agent in sugar syrup. The specific gravity is 1.834, melting point 42.35°C, and it is soluble in water. The usual grades are 90, 85, 75%, technical 50%, and dilute 10%. As a cleanser for metals, phosphoric acid produces a light etch on steel, aluminum, or zinc, which aids paint adhesion. **Deoxidine,** of the American Chemical Paint Co., is a phosphoric acid cleanser for metals. **Nielite D,** of Nielco Laboratories, is phosphoric acid with a rust inhibitor, used as a nonfuming pickling acid for steel. **Phosphoric anhydride,** or **phosphorus pentoxide,** P_2O_5, is a white, water-soluble powder used as a dehydrating agent and also as an opalizer for glass. It is also used as a catalyst in asphalt coatings to prevent softening at elevated temperatures and brittleness at low temperatures. **Granusic,** of the J. T. Baker Chemical Co., is this material in granular form for removing water from gas streams.

PHOSPHORUS. A nonmetallic element, symbol P, widely diffused in nature, and found in many rock materials, in ores, in the soil, and in parts of animal organisms. Commercial phosphorus is obtained from phosphate rock by reduction in the electric furnace with carbon, or from bones by burning and treating with sulfuric acid. **Phosphate rock** occurs in the form of land pebbles and as hard as rock. It is plentiful in the Bone Valley area of Florida, and also comes from Tennessee, Idaho, and South Carolina.

Vast quantities are mined in Morocco and Tunisia. Large deposits are found on many of the Pacific Islands, the Christmas Island resources being estimated at 30 million tons (27 million metric tons) and those on Nauru at 100 million tons (91 million metric tons). It is a calcium phosphate high in P_2O_5. The mineral **apatite,** widely distributed in the Appalachian range, in Idaho, Brazil, and French Oceania, is also a source of phosphorus, containing up to 20% P_2O_5, with iron oxide and lime. The Egyptian rock contains 62 to 70% tricalcium phosphate. The aluminocalceous phosphate rock of Senegal is treated to obtain a very soluble fertilizer known as **phosphal.** Florida hard phosphate rock contains 80% phosphate of lime. A ton of phosphorus is obtained from 7.25 tons (6.58 metric tons) of rock, requiring 30 lb (14 kg) of electrodes and 11,850 kW of electricity. The Tennessee Valley Authority produces about 8 tons (7 metric tons) of **expanded slag** for each ton (0.9 metric tons) of phosphorus produced from the phosphate rock. The slag from the smelter is run onto a forehearth at about 2000°F (1093°C) and treated with water, high-pressure steam, and air. The expanded slag formed is crushed to ⅜-in (0.95-cm) size, bulking 50 lb/ft^3 (801 kg/m^3). It is called **TVA slag** and is used for making lightweight concrete blocks. The **superphosphate** used for fertilizers is made by treating phosphate minerals with concentrated sulfuric acid. It is not a simple compound, but may be a mixture of **calcium acid phosphate,** $CaHPO_4$, and calcium sulfate. **Nitrophosphate** for fertilizer is made by acidulating phosphate rock with a mixture of nitric and phosphoric acids, or with nitric acid and then ammoniation and addition of potassium or ammonium sulfate. Natural rock phosphate in finely ground form is also used as a fertilizer for legume crops, but the untreated natural rock is not readily soluble, and is thus not as quick-acting as a fertilizer.

There are two common forms of phosphorus, yellow and red. The former, also called **white phosphorus,** P_4, is a light-yellow waxlike solid, phosphorescent in the dark and exceedingly poisonous. Its specific gravity is 1.83 and it melts at 44°C. It is used for smoke screens in warfare and for rat poisons and matches. **Yellow phosphorus** is produced directly from phosphate rock in the electric furnace. It is cast in cakes of 1 to 3 lb (0.45 to 1.36 kg) each. **Red phosphorus** is a reddish-brown amorphous powder, having a specific gravity of 2.20 and a melting point of 725°C. Red phosphorus is made by holding white phosphorus at its boiling point for several hours in a reaction vessel. Both forms ignite easily. **Phosphorus sulfide,** P_4S_3, may be used instead of white phosphorus in making matches. **Phosphorus pentasulfide,** P_2S_5, is a canary-yellow powder of specific gravity 1.30, or solid of specific gravity 2.0, containing 27.8% phosphorus, used in making oil additives and insecticides. It is decomposed by water.

Phosphorus is an essential element in the human body, a normal person having more than a pound of it in the system, but it can be taken into the system only in certain compounds. **Nerve gases** used in chemical warfare contain phosphorus which combines with and inactivates the choline sterase enzyme of the brain. This enzyme controls the supply of the hormone which transmits nerve impulses, and when it is inactivated the excess hormone causes paralysis of the nerves and cuts off breathing. **Organic phosphates** are widely used in the food, textile, and chemical industries. **Tributyl phosphate,** for example, is a colorless liquid, used as a plasticizer in plastics and as an antifoaming agent in paper coatings and textile sizings. Flour and other foodstuffs are fortified with **ferric phosphate,** $FePO_4 \cdot 2H_2O$. **Iron phosphate** is used as an extender in paints. **Tricalcium phosphate,** $Ca_3(PO_4)_2$, is used as an anticaking agent in salt, sugar, and other food products and to provide a source of phosphorus. The tricalcium phosphate used in toothpastes as a polishing agent and to reduce the staining of chlorophyll has the formula $(10CaO \cdot H_2O \cdot 3P_2O_5)3H_2O$, and is a fine white powder. **Dicalcium phosphate,** used in animal feeds, is precipitated from the bones used for making gelatin, but is also made by treating lime with phosphoric acid made from phosphate rock. **Diammonium phosphate,** $(NH_4)_2HPO_4$, is a mildly alkaline, white crystalline powder used in ammoniated dentifrices, for pH control in bakery products, in making phosphors and to prevent afterglow in matches, and for flameproofing paper.

For manufacturing operations, phosphorus is generally utilized in the form of intermediate chemicals, but the phosphorus marketed by the American Agricultural Chemical Co., for use in doping semiconductors and in electroluminescent coatings, is 99.9999% pure. **Phosphorus tricloride,** PCl_3, is an important chemical for making phosphites. It is a colorless liquid boiling at 76°C. It decomposes in water to form phosphorus and hydrochloric acid. **Phosphorus oxychloride,** $POCl_3$, is a very reactive liquid used as a chlorinating agent and for making organic chemicals. In water it decomposes to form phosphoric and hydrochloric acids. **Phosphorus thiochloride,** $PSCl_3$, is a yellowish liquid containing 18.5% phosphorus and 18.6% sulfur. It is used for making insecticides and oil additives.

PHTHALIC ANHYDRIDE. A white cystalline material of the composition $C_6H_4(CO)_2O$, with a melting point of 130.8°C, soluble in water and in alcohol. It is made by oxidizing naphthalene, or it is produced from orthoxylene derived from petroleum. It is used in the manufacture of alkyd resins, for the production of dibutyl phthalate and other plasticizers, dyes, and many chemicals. **Chlorinated phthalic anhydride** is also used as a compounding medium in plastics. It is a white, odorless, nonhygroscopic stable powder containing 50% chlorine. It gives higher temperature resistance and increased stability to plastics. **Niagathal** is a chlorinated phthalic anhydride of the Niagara Alkali Co. **Tetrahydrophthalic anhydride** is a white crystalline powder with a molecular weight of 152.1, melting at 100°C, used to replace phthalic anhydride where a lighter color is desired. It is produced by condensing butadiene with maleic anhydride. In synthetic resin coatings it gives higher adhesion. **Terephthalic acid** may be obtained as a by-product in the production of phthalic anhydride from petroleum. It has a long-chain alkyl group having an amide linkage on one end and a methyl ester on the other. It is used for producing textile fibers. Its sodium salt is used as a gelling agent for high-temperature lubricating greases for uses to 600°F (316°C). It forms fine **crystallites** of soft flexible fibers in the grease. **Oronite GA10** is this material. **Isophthalic acid,** made by oxidation of ethyl benzene and orthoxylene, produces alkyd paint resins of greater heat stability than phthalic anhydride. **Maleic anhydride,** $(CHCO)_2O$, is a white crystalline solid used to replace phthalic anhydride in alkyd resins to increase the hardness for baking enamels and to resist yellowing.

PIASSAVA. Also called **Pará grass** and **monkey grass.** A coarse, stiff, and elastic fiber obtained from a species of palm tree, *Leopoldinia piassaba,* of Brazil, used for making brushes and brooms. The plant has long beards of bristlelike fibers, which are combed out and cut off the young plants. These fibers sometimes reach a length of 4 ft (1.2 m). The soft, finer fibers are made into cordage, and the coarser ones are used for brushes. Piassava is very resistant to water. The fiber for brush manufacture is separated into three classes, the heavy fibers being known as **bass,** the medium as **bassine,** and the fine as **palmyra.** The bass is used for heavy floor sweeps. The fiber of the palm *Attalea funifera,* which grows in the state of Bahia, Brazil, and is also called piassava, is a harder and stronger material than the piassava of Amazonas. It is used for marine cordage, and is resistant to salt water. A substitute for piassava is **acury,** from the leaves of the palm *A. phalerata* of Matto Grosso. It is used for cordage and brushes, and the coarser fibers are used for brooms.

PICKLING ACIDS. Acids used for pickling, or cleaning castings or metal articles. The common pickling bath for iron and steel is composed of a solution of sulfuric acid and water, 1 part acid to 5 to 10 parts water being used. This acid attacks the metal and cleans it of the oxides and sand by loosening them. For pickling scale from stainless steels a 25% cold solution of hydrochloric or sulfuric acid is used, or hydrofluoric acid with the addition of anhydrous ferric sulfate is used. Hydrofluoric acid solutions are sometimes used for pickling iron castings. This acid attacks and dissolves away the sand itself. For bright-cleaning brass a mixture of sulfuric acid and nitric acid is used. For a matte finish the mixed acid is used with a small amount of zinc sulfate. Copper and copper alloys can be pickled with sulfuric acid to which anhydrous ferric sulfate is added to speed the action, or sodium bichromate is added to the sulfuric acid to remove red cuprous oxide stains. Brass forgings are pickled in nitric acid to bring out the color. Since all of these acids form salts rapidly by the chemical action with the metal, they must be renewed with frequent additions of fresh acid. The French pickling solution known as **framanol,** used for aluminum, is a mixture of chromium phosphate and triethanolamine. The latter emulsifies the grease and oil, and the aluminum oxide film is dissolved by the phosphoric acid, leaving the metal with a thin film of chromic oxide.

The temperature of most pickling is from 140 to 180°F (60 to 82°C). An increase of 20°F (−7°C) will double the rate of pickling. **Acid brittleness** after pickling is due to the absorption of hydrogen when the acid acts on iron, and is reduced by shortening the pickling time. **Inhibitors** are chemicals added to reduce the time of pickling by permitting higher temperatures and stronger solutions without hydrogen absorption. **Hibitite,** of Monsanto, is a brown liquid of the composition $C_{27}H_{45}NO_{10}S_2$, used as a metal pickling inhibitor. Addition of a small amount of 2% tincture of iodine to a 5% sulfuric acid solution gives a 95% retardation of acid attack on steel without decreasing the rate of dissolution of the rust. In plating baths, **fluoboric acid,** HBF_4, has high throwing power and has a cleansing effect by dissolving sand and silicides from iron castings and steel surfaces. It is a colorless liquid with specific gravity of 1.33 and decomposes at 130°C. **Pennsalt FA-42,** of Pennwalt, is this material. It is a 42% solution of fluoboric acid for pickling and for control of acidity of plating baths.

Phosphoric acid is employed in hot solution as a dip bath for steel parts to be finished to a rough or etched surface. It leaves a basic iron phosphate coating on the steel which is resistant to corrosion and gives a rough base for the finish. **Coslettized steel** is steel rustproofed by dipping in a hot solution of iron phosphate and phosphoric acid. **Parkerized steel** is rustproofed steel treated in a bath of iron and manganese phosphates. **Bonderized steel** is steel treated with phosphoric acid and a catalyst to give a rough, tough, rust-resistant base for paints. **Granodized steel** is produced with zinc phosphate. In general, the coatings left on steel by phosphate treatments are extremely thin, not over 0.0002 in (0.0005 cm). The iron-manganese coatings are black, and the iron-zinc-phosphate coatings are gray.

PIG IRON. The iron produced from the first smelting of the ore. The melt of the blast furnace is run off into rectangular molds, forming, when cold, ingots called pigs. Pig iron contains small percentages of silicon, sulfur, manganese, and phosphorus, besides carbon. It is useful only for resmelting to make cast iron or wrought iron. Pig iron is either sand-cast or machine-cast. When it is sand-cast, it has sand adhering and fused into the surface, giving more slag in the melting. Machine-cast pig iron is cast in steel forms and has a fine-grained chilled structure, with lower melting point. Pig irons are classified as bessemer or nonbessemer, according to whether the phosphorus content is below or above 0.10%. There are six general grades of pig iron: **low-phosphorus pig iron,** with less than 0.03% , used for making steel for steel castings and for crucible steelmaking; **bessemer pig iron,** with less than 0.10% phosphorus, used for bessemer steel and for acid open-hearth steel; **malleable pig iron,** with less than 0.20%, used for making malleable iron; **foundry pig iron,** with from 0.5 to 1%, for cast iron; **basic pig iron,** with less than 1%, and low-silicon, less than 1%, for basic open-hearth steel; and **basic bessemer,** with from 2 to 3%, used for making steel by the basic bessemer process employed in England.

Since silicon is likely to dissolve the basic furnace lining, it is kept as low as possible, 0.70 to 0.90%, with sulfur not usually over 0.095%. Pig irons are also specified on the content of other elements, especially sulfur. The sulfur may be from 0.04 to 0.10%, but high-sulfur pig iron cannot be used for the best castings. The manganese content is usually from 0.60 to 1%. Most of the iron for steelmaking is now not cast but is carried directly to the steel mill in car ladles. It is called **direct metal.** Foundry pig iron is graded by the silicon content, No. 1 having from 2.5 to 3%, and No. 3 from 1.5 to 2% silicon. **Silvery iron** is a name for pig iron of high silicon content because of its silvery fracture. **Puddling iron** is a grade of pig iron used for making wrought or puddled iron in a puddling furnace. A requirement is that the silicon be low, with manganese 0.5 to 1%.

Chateaugay iron is a low-phosphorus pig iron produced from New York state **magnetite ore.** The original ore as mined contains about 28% iron. The standard analysis of the pig iron is total carbon 4%, silicon 0.75 to 4.0, sulfur 0.03 max, phosphorus 0.03 max, manganese 0.10 to 0.15. Chateaugay iron is used for casting rolls, gears, and machine parts. **Norskalloy** is a name for pig iron produced from Norwegian ores containing anadium and titanium. The standard grade contains from 4 to 4.5% total carbon, 0.5 to 1.5 silicon, 0.20 manganese, 0.20 to 0.25 phosphorus, 0.30 to 0.40 vanadium, and 0.40 to 0.80 titanium. From 15 to 20% Norskalloy pig is added to mixtures where vanadium is required. **Mayari iron** is pig iron made from Cuban ores which contain vanadium and titanium, or is pig iron made to duplicate the Cuban iron. These irons are considered especially suitable for heavy rolls or high-grade castings. Mayari pig, as marketed by the Bethlehem Steel Co., contains 1.60 to 2.50% chromium, 0.80 to 1.25 nickel, 0.25 to 2.25 silicon, 0.10 to 0.20 titanium, 0.05 to 0.08 vanadium, 3.8 to 4.5 total carbon, 0.60 to 2 manganese, under 0.05 sulfur, and under 0.10 phosphorus. **Nikrofer** is a German pig iron from Greek ore that is similar in composition.

PIGMENT. A substance, usually earthy or clayey, which when mixed with oil or other adhesive carrier and a solvent forms a paint. Pigments usually give body as well as color to the paint, and the paint **hiding power** is measured by comparison tests when in the form of a mixed paint. If the hiding power of lithopone is taken as 100, the hiding power of zinc sulfide is 240, and that of titanium dioxide is 400. **Color standards** are distinct from hiding power. Pure mangnesium oxide is used as the standard for the measurement of whiteness. The chemical interaction must also be considered in pigments. For example, zinc oxide increases wear life and mildew resistance in paints, but may tend to react and cause blisters. The use of pigments is not confined to paints. In ceramics, their primary choice is by color, but they usually also add other physical properties to the ceramic. In plastics they add body and strength, as distinct from dyes which do not add body.

A pigment is distinct from a **filler** in that it must retain its opacity when wet. White powdered quartz, used sometimes as a filler, is not a good pigment as it becomes glassy when wet. Fillers that retain their opacity are called **extenders,** or **auxiliary pigments,** and the final mixed pigment is called a **reduced color.** But an extender pigment, such as silica, that does not have good hiding power in itself will increase the hiding power and depth of color of a pigment if the extender is of such fine particle size that it will be dispersed in the voids between the pigment particles. Extremely fine silica will also cement itself chemically to lead pigments and add wearing qualities. The **Hi-Sil** of the Pittsburgh Plate Glass Co. is a silica with a particle size of 0.025 μm. As a pigment for rubber, it adds strength and wear resistance to the rubber. **Extender RX-2022,** of the Chas. Pfizer & Co., is iron phosphate in the form of white powder for use in paints. It has low hiding power, with refractive index of 1.7, but is a corrosion inhibitor and is used in undercoats. **Antimony trioxide** is an opaque yellow powder used in plastics and coatings as a flame-retardant pigment and opacifier. **Firemaster HHP,** of Michigan Chemical Corp., is this material. **Thermoguard S,** of M & T Chemicals, is antimony trioxide as a 99.8% pure white powder with average particle size less than 1 μm.

Pigments are mostly of mineral origin, the vegetable pigments such as logwood and the animal pigments such as cochineal being ordinarily classified as dyestuffs. Bone black, however, is an example of an animal pigment, and vine black is a typical vegetable pigment. **Vine black,** a fine pigment for inks, was originally made by charring grapevine stems, and was known as **Frankfurt black,** but similar pigments are now made from fruit pits, nut shells, or wood pulp, and are called **vegetable black.** Pigments are also produced by dyeing clays with aniline dyestuffs. These are called **lakes. Dutch yellow,** for example, is a **yellow lake** made by adsorption of a yellow dye such as quercitron by an inert material such as calcium carbonate. Various chemicals such as copper acetate and potassium acetate are used as pigments. **Potassium acetate,** CH_3COOK, is a white powder also used in making crystal glass. Pigments should be ground fine enough so that all of the powder will pass through a 325-mesh screen.

Natural pigments include ochre, umber, ground shale, hematite, and sienna. **Sepia** is a dark-brown pigment originally made from the black, inky secretion found in an internal ink sac of the *Sepia* mollusk. It is used in India inks of sepia or dark-brown color and was formerly employed directly as a writing ink. The **red-ochre** pigments are the natural red iron oxides of high oxide content. The yellows, or siennas, are the oxides mixed with considerable clay. The browns, or umbers, have manganese present in the clays. **Terre verte,** or **Verona green,** of Cyprus, is a fine blue-green earth valued highly in the Middle Ages as a pigment. It contains 53% silica, 26 ferric oxide, 16 potash, and some magnesia, manganese, and other impurities. **Manganese green,** or **Cassel green,** is **barium manganate,** $BaMNO_4$, a green poisonous powder of specific gravity 4.85, insoluble in water. **Mineral green,** or **Scheele's green,** is **copper arsenite,** $CuHAsO_3$. It is a light green amorphous powder used in paints and in textile printing. It is also used in medicine. **Orange pigments,** from yellow to brilliant red, with high tinting strength and great fastness, are made with mixtures of **lead chromate,** $PbCrO_4$, **lead molybdate,** $PbMoO_4$, and **lead sulfate,** $PbSO_4$. Pink to maroon is obtained in ceramic enamels with **calcium stannate,** $CaSnO_3 \cdot 3H_2O$, a white crystalline powder that loses its water at 350°C. **Gloss white,** used as a reduced-color white pigment, for surface-coating pulp papers, and in printing inks, is a coprecipitation product consisting of 75% blanc fixe and 25 aluminum hydrate.

The most important yellow is **chrome yellow,** but it fades easily. However, a pigment's light-fastness and tinting effect depend on the crystal structure as well as on the chemical composition. Normal **lead chromate** has a monoclinic crystal form, and it gives the strongest and most light-fast of the chrome yellows. Coprecipitated lead chromate and lead sulfate are orthorhombic and greenish in hue, giving primrose and lemon yellows poorer in light-fastness and rust-inhibiting action. **Strontium chromate,** $SrCrO_4$, gives a lemon yellow of good light-fastness. **Yellow ochre** is inferior as a color but durable. **Cadmium yellow** is cadmium sulfide and is a brilliant, permanent pigment but expensive. **Cadmium selenide** produces a bright **cadmium red** which when mixed with the sulfide produces **cadmium orange.** The yellow pigment called **mosaic gold,** or **artificial gold,** is **stannic sulfide,** SnS_2, used in gilding and bronzing paints. **Stannous sulfide,** SnS, used for incorporation in bearings, is a black crystalline material melting at 880°C. Cadmium red and cadmium orange are produced by calcining selenium with cadmium sulfide. These **cadmium sulfoselenide** pigments give brilliant colors. Cadmium pigments are used in **camouflage paints** to give greater reflection of infrared rays. A building painted with a green containing cadmium has the same reflection as grass or leaves, and is indistinguishable in aerial photographs. An ancient lemon-brown pigment is **bistre.** It was obtained from the collected chimney soot of wood fires and much used by the old masters. It is very durable in watercolors. The most important green is **chrome green,** which is chrome yellow mixed with Prussian blue; it gives a good color but no permanence.

Ultramarine is the most important blue. It is used in paints and inks and also as **bluing** for whitening paper, textiles, and organic materials by neutralizing the yellow cast. It is an ancient pigment, formerly made by grinding **lapis lazuli,** an azure-blue gemstone which is a sodium silicate sulfide. Ultramarine is now made by calcining a mixture of aluminum silicate and sodium sulfide, and has the empirical formula $Na_7Al_6Si_6O_{24}S_2$. It is a deep-blue, water-soluble powder of 325 mesh, often marketed as a linseed-oil paste. **Egyptian blue,** a chemical-resistant pigment, is a double silicate of calcium and copper, $CaO \cdot CuO \cdot 4SiO_2$. It was used by the ancients, and paintings 1,900 years old still retain the color. It is now made by fusing powdered quartz, chalk, copper oxide, and sodium carbonate. **Cobalt blue** is a good color but is expensive. **Prussian blue,** or **Chinese blue,** is **ferric ferrocyanide,** $Fe_4[Fe(CN)_6]_3$, a blue amorphous powder. It is made by combining iron chloride and potassium ferrocyanide. **Celestial blue** is the light-blue pigment made by extending Prussian blue with barytes. **Milori blue,** used for coloring matches, inks, lacquers, and soaps, is ferric ferrocyanide with gypsum or barium sulfate. **Vermilion pigment** is mercury sulfide, which gives a fine color and is permanent, but it is expensive. High-grade blacks are usually **lampblack, bone black,** and **ivory black,** but may be extended with graphite. **Spanish black** is a name used in old texts for the black pigment made by burning cork. It is light and of soft texture. **Mineral black,** or **slate black,** is made by grinding black slate. **Metronite** is a white mineral composed of magnesium and calcium carbonates and magnesium silicate, used as a paint filler and extender. The pigment known as **Plessy's green** is **chromium phosphate,** $CrPO_4$, a bluish-green powder insoluble in water.

The **chemical colors** known as **phthalocyanines** give high tinting strength and resistance to deterioration by high-temperature baking. They are used for paints, inks, and plastics, and are available as dry colors, in oil- or water-dispersible pastes, and in resin chips for plastics. They are chelated metallic salts of **tetrabenzoporphyrazine,** which is made from phthalamide or ammonium phthalic anhydride in the presence of iron, nickel, or copper salts. **Monastral blue,** of Du Pont, is **copper phthalocyanine,** a salt in which the copper is held in a chelate ring complex with four nitrogen atoms. The green is made by chlorination. **Fastolux pigments,** of the Sun Chemical Corp., are phthalocyanines in fine particle size with a complete range of blue and green colors. The red and bluish-red colors of Du Pont are linear **quinacridones** made from terephthalic acid.

The alpha crystal has a bluish-red color, the beta crystal has an intense violet color, and the gamma material has a true red color. The crystal structure can be controlled, and combinations give a range of brilliant, nonbleeding red and violet shades. The **Mercadium colors,** made by Hercules Inc., are compounds of the sulfides of mercury and cadmium to give colors from light orange to dark maroon. All of these pigments give permanence, light and chemical resistance, and a very high tinting strength. The tinting strength of the blue is 15 times that of ultramarine blue, and the green is superior to conventional pigments in brightness and permanence. **Heliogen blue** and **Heliogen green,** of the General Dyestuff Corp., are phthalocyanine pigments, and **Ramapo blue** and **Ramapo green** are the pigments with a barium resinate extender.

A reactive **fungicidal pigment** used in ship paints and antifouling paints is **cupric oxide hydrated,** a fine dark-brown powder of the composition $4CuO \cdot H_2O$. **Copper quinolinolate** is also used in fungicidal paints. **Cunilate,** of the Scientific Oil Co., is this material. A yellow pigment used in anticorrosive and blister-resistant metal primers is **zinc tetroxychromate,** $4Zn(OH)_2 \cdot ZnCrO_4$. **Metallic pigments** are most frequently bronze powder or aluminum powder. They are used to increase light reflectivity as well as for appearance. **Stainless-steel flake** for pigment is marketed by Charles Hardy, Inc., as a fine powder in a paste with stearic acid and a solvent. Added to a clear lacquer or varnish it gives a hard silvery coating resistant to corrosive fumes. **Aluminum flake** gives high heat reflectivity as well as light reflectivity, and is used in silicone-based paints for high heats. Aluminum powder gives iridescent effects when dispersed in vinyl compounds. **Vinylum,** of the Argus Chemical Laboratory, is such a powder in a vinyl copolymer.

PINE. The wood of coniferous trees of the genus **Pinus,** of which there are 37 species in the United States. The **white pine,** or **northern white pine,** *P. strobus,* grows widely in Canada and in the northeastern United States. The trees are 80 to 100 ft (24 to 30 m) high, with trunks 3 to 9 ft (1 to 3 m) in diameter, reaching full size in 80 years. The wood is soft, straight-grained, and free from rosin. The heartwood is light brown, and the sapwood white. It is the chief wood for pattern making and is also extensively used for cabinetwork and general carpentry. **Cork pine** is a name for the clear, soft, white pieces used for patterns. The white pine is now scarce in New England, and red spruce is used in its place. **Yellow pine** is a name for the wood of the **longleaf** or **longstraw pine** tree, *P. palustris,* of the southeastern states, and **shortleaf pine,** *P. echinata,* of the southeast and middle western states, also called **North Carolina pine** and **rosemary pine.** The leaves, called needles or straws, of the longleaf pine are up to 18 in (46 cm) in length. The longleaf pine tree furnishes the best grades of yellow pine and also is the chief source of turpentine. It is also called **Georgia pine, southern pine, hard pine,** and **hill pine.**

Slash pine, also known as **Cuban pine** and **swamp pine,** from the tree *P. caribaea,* **Caribbean pine,** which grows along the southern coasts of the United States and the Caribbean countries, is a yellow pine. It is one of the most rapidly growing forest trees in the United States and produces one of the heaviest, hardest, and strongest of all the conifers or softwoods. In Central America it is called **ocoté.** Slash pine is next to longleaf pine as a source of turpentine and rosin. As heartwood does not develop until the tree is 20 or more years old, slash pine forms a valuable source of paper pulp. The term **Arkansas pine** in the lumber trade includes mixtures of shortleaf, longleaf, slash, loblolly, and pond pines. **Lodgepole pine,** *P. contorta,* is from a small slow-growing tree of western United States, Canada, and Alaska. It is also called **knotty pine,** scrub pine, and jack pine. The wood is moderatly lightweight, yellow to brown in color, with a narrow white sapwood. It is straight-grained with resin ducts, and has large shrinkage. It is used for poles, ties, mine timbers, and rough construction. Also known as **jack pine** is the medium-size *P. banksiana* of central Canada, which is used for creosoted telephone poles. **Spruce pine,** or **cedar pine,** is a large tree, *P. glabra,* growing in a narrow area from southern Louisiana to Florida. **Virginia pine,** *P. virginiana,* also called **Jersey pine** and **scrub pine,** is a plentiful tree of the Atlantic states. The wood is soft, very knotty, and not durable. It is used for firewood, but much is used in low-cost houses.

Ponderosa pine, also called **western yellow pine, western white pine,** and **Oregon white pine,** is from the tree *P. ponderosa.* The tree grows to a height of 175 ft (53 m) and a diameter of 6 ft (2 m). It grows throughout the mountain states from Mexico to Canada, and is also a source of turpentine and rosin. A similar western pine, **Jeffrey pine,** *P. jeffreyi,* contains heptane instead of turpentine in the oleoresin, and is a more economic source of this material than petroleum. The lumber is usually mixed with ponderosa pine and sold as such. It is a moderately soft and lightweight wood with the heartwood light reddish brown, and quite similar to yellow pine. **Loblolly pine,** *P. taeda,* is called North Carolina pine, **Oldfield pine,** and **sap pine.** It grows from Virginia to northern Florida and to Texas; it is adapted to extensive areas and is easily propagated, receiving the name of **field pine.** It is a type of shortleaf pine distinguished by three leaves or straws in each cluster, rough bark, and small prickly burrs. It grows to a diameter of 12 in (30 cm) in 12 years. **Pitch pine** is the **pond pine,** *P. rigida,* of the southern states, but all yellow pines are called pitch pine in the export trade. **Norway pine,** of the north central states, is *P. resinosa.* The yellow pines are harder and more difficult to work than white pine. They are resinous and more durable. They also take a better polish and show a more figured grain. They are valued for flooring and for general construction. White pine has a specific gravity, kiln-dried, of 0.38, and compressive strength perpendicular to the grain of 780 lb/in^2 (5 MPa); **western white pine,** *P. monticola,* has a specific gravity of 0.42 and a compressive strength of 750 lb/in^2 (5 MPa). **Deal** is a European name for the wood of the tree *P. sylvestris,* also known as **Danzig pine, Baltic pine, Scotch fir, Scotch pine,** and **northern pine.** It is popularly called **Scots pine** in England, and is one of the most plentiful of the European conifers, especially in Norway, Sweden, and Finland. It gives a straight pole, up to 70 ft (21 m) long, valued for telegraph poles.

Paraná pine is a soft, yellowish-white wood with rose veins from the tree *Araucaria brasiliensis* of southern Brazil. In Argentina it competes with American softwoods and is called **Brazilian pine** and **araucarian pine.** The specific gravity is 0.865. The tree is very tall, with branches only at the top, and a notable feature of the wood is the absence of knots. In the United States it is used for telephone-pole crossbars, and to replace birch for such articles as paintbrush handles. Other species of araucarian pine, or **Antarctic pine,** grow in southern Chile and Argentina. **Araucaria oil** is distilled from the wood. It is a viscous reddish oil of roselike odor containing a high percentage of eudeomol and some geraniol. It has a more durable scent than guaiac wood oil for soaps.

The **araucarian pines** of New Guinea, *A. khinkii* and *A. cunninghamia,* are called **hoop pine. New Caledonia pine** is from the tree *A. cooki,* growing to a height of 200 ft (61 m) with no lower branches. White pine of New Zealand is from the very large tree *Podocarpus dacrydioides,* called also **kahikatea.** The sapwood is white, and the small heartwood is yellow. The wood is straight-grained, inodorous, easily worked, but not durable. It weighs 29 lb/ft^3 (465 kg/m^3). It is used for boxes, crates, and packing. Another species, from the tree *P. ferrugineus,* called **miro,** is brownish in color, fine-grained, easily worked, and has high strength. The trees average 65 ft (20 m) in height and 20 in (51 cm) in diameter. The New Zealand species known as **black pine,** or **matai,** is from the *P. spicatus.* The wood is yellowish brown, straight-grained, and weighs 38 lb/ft^3 (609 kg/m^3). **Red pine,** or **rimu,** is the chief timber of New Zealand, used for furniture, millwork, and kraft pulp. The tree *Dacrydium cupressinum* averages 100 ft (30 m) in height and 30 in (76 cm) in diameter. The wood is reddish brown with streaks, straight-grained, easily worked, and weighs 37 lb/ft^3 (593 kg/m^3). **Silver pine, pink pine,** and **yellow silver pine** of New Zealand are from several other species of *Dacrydium* obtained only in limited amounts. The name silver is applied to the shiny white woods, and the darker and mottled woods are called pink. They are very durable cypresslike woods. **Mercus pine** is the wood of the tropical pine tree, *P. merkusii,* of the East Indies, India, and the Philippines. It is called **Tinyu pine** in India and **Mindoro**

pine in the Philippines. The wood is used in general construction. The tree yields a superior turpentine.

PINE OIL. An oil obtained from the wood of the *Pinus palustris,* or longleaf pine, in the steam extraction of wood turpentine. It is used as a cold solvent for varnish gums and for nitrocellulose lacquers, and as a frothing agent in the flotation of ores. In paints and varnishes it aids dispersion of metallic pigments and improves the flow. It is also used in metal polishes and in liquid and powder scrubbing soaps, as the oil is a powerful solvent of dirt and grease. When free from water, pine oil has a yellowish color, but it is water-white when it contains dissolved water. It has an aromatic characteristic odor, and is distinct from the pine oils distilled from pine leaves and needles and used in medicine. The distillate of the gum of the Jeffrey and Digger pines of California, called **abietine** in medicine, contains 96% heptane and is used as a cleaning agent and insecticide, and as a constituent of standard gasolines for measuring detonation of engines. Pine oil is obtained mainly from old trunks and branches, and is a product formed by hydrolysis. Pine-oil disinfectants are made with steam-distilled pine oil. **Yarmor** is a refined pine oil of Hercules Inc. which is used to increase the detergency of soaps, for dyes, and as a solvent for oils and greases. **Hercosol** is a solvent made from pine oil by the same company. Synthetic pine oil made from gum turpentine by this company has a mild pine odor, a specific gravity of 0.9186, and a flash point of 154°F (68°C). It is technically the same as the natural and has the same uses. **Pine-root oil** was produced in Japan on a large scale for the manufacture of fuel oils. The terpenes of the pine oil are converted to aromatic and hydroaromatic compounds by catalytic reaction. The edible **pine kernels** of Europe are the seeds from the large cones of the *P. pinea* of southern Europe and Cyprus. **Pine-needle oil** is distilled from the **Siberian fir** tree, *Abies sibirica,* of northeastern Russia. It is also known as **Siberian pine oil.** It contains a high percentage of **bornyl acetate** and is used in soaps and perfumes.

PLANE WOOD. The wood of the plane tree, *Platanus orientalis,* native to Europe, and *P. occidentalis,* of North America. The latter species is also called **buttonwood** and **buttonball.** It is a yellowish, compact wood with a fine, open grain. The weight is about 40 lb/ft^3 (641 kg/m^3). It resembles maple and gives a beautiful grain when quartered. It is employed in cabinetwork.

PLASTIC ALLOYS. A mechanical blend of two or more resins, usually thermoplastics. The resins most widely used in alloys are polyvinyl chloride (PVC), ABS, and polycarbonate. These three can be combined with each other or with other resins. **ABS-polycarbonate alloys** extend the exceptionally high impact strength of carbonate plastics to section thicknesses over ⅙6 in (0.16 cm). **ABS-polyurethane alloys** combine the excellent abrasion resistance and toughness of the urethanes with the lower cost and rigidity of ABS.

ABS-PVC alloys are available commercially in several grades. One of the established grades provides self-extinguishing properties, thus eliminating the need for intumescent (nonburning) coatings in ABS applications, such as power tool housings, where self-extinguishing materials are required. A second grade possesses an impact strength about 30% higher than general-purpose ABS. ABS-PVC alloys also can be produced in sheet form. The sheet materials have improved hot strength, which allows deeper draws than are possible with standard rubber-modified PVC base sheet. They also are nonfogging when exposed to the heat of sunlight. Some properties of ABS-PVC alloys are lower than those of the base resins. Rigidity, in general, is somewhat lower, and tensile strength is more or less dependent on the type and amount of ABS in the alloy.

Another sheet material, an alloy of about 80% PVC and the rest acrylic plastic, combines the nonburning properties, chemical resistance, and toughness of vinyl plastics with the rigidity and deep drawing merits of the acrylics. The **PVC-acrylic alloy** approaches some metals in its ability to withstand repeated blows. Because of its unusually high rigidity, sheets ranging in thickness from 0.60 to 0.187 in (1.5 to 0.5 cm) can be formed into thin-walled, deeply drawn parts.

PVC is also alloyed with chlorinated polyethylene (CPE) by end users to gain materials with improved outdoor weathering or to obtain better low-temperature flexibility. The **PVC-CPE alloy** applications include wire and cable jacketing, extruded and molded shapes, and film sheeting. **Acrylic-base alloys** with a polybutadiene additive have also been developed, chiefly for blow-molded products. The acrylic content can range from 50 to 95% , depending on the application. Besides blow-molded bottles, the alloys are suitable for thermoformed products such as tubs, trays, and blister pods. The material is rigid and tough and has good heat-distortion resistance up to 180°F (82°C). Another group of plastics, polyphenylene oxide (PPO), can be blended with polystyrene to produce a **PPO-polystyrene alloy** with improved processing traits and lower cost than nonalloyed PPO. The addition of polystyrene reduces tensile strength and heat deflection temperature somewhat and increases thermal expansion.

PLASTIC BRONZE. A name applied by makers of bearing bronzes to copper alloys that are sufficiently plastic to assume the shape of the shaft and make a good bearing by running in. These bronzes have a variety of composition, but the plasticity is always obtained by the addition of lead, which in turn weakens the bearing. In some cases the lead content is so high, and the tin content so low, that the alloy is not a bronze. These copper-lead alloys are referred to as **red metals.** The plastic bronze ingot marketed by one large foundry for journal bearings contains 65 to 75% copper, 5 to 7 tin, and the balance lead. **Semiplastic bronze** usually contains above 75% copper and not more than 15 lead. **ASTM alloy No. 7** has about 10% lead, 10 tin, 1 zinc, 1 antimony, and 78 copper. The compressive strength is 12,500 lb/in^2 (85 MPa).

PLASTIC LAMINATES. Resin-impregnated paper or fabric, produced under heat and high pressure. Also referred to as **high-pressure plastic laminates.** Two major categories are decorative thermosetting laminates and industrial thermosetting laminates. Most of the **decorative thermosetting laminates** are a paper base, and are known generically as **papreg.** Decorative laiminates are usually composed of a combination of phenolic- and melamine-impregnated sheets of paper. The final properties of the laminate are related directly to the properties of the paper from which the laminate is made.

Early laminates were designated by trade names, such as **Bakelite** of Union Carbide, **Textolite** of General Electric, **Micarta** of Westinghouse, **Phenolite** of National Vulcanized Fiber Co., **Condensite** and **Dilecto** of Continental Diamond Fiber Co., **Haveg** of Haveg Corp., **Spauldite** of Spaulding Fiber Co., and **Synthane** of Synthane Corp. **Formica,** of Formica Corp., designated various types of laminates with a decorative facing layer for such uses as tabletops. Trade names now usually include a number or symbol to describe the type and grade. Textolite, for example, embraces more than 70 categories of laminates subdivided into use-specification grades, all produced in many sizes and thicknesses. **Textolite 11711** is an electronic laminate for such uses as multilayer circuit boards. It is made with polyphenolene oxide resin, and may have a copper or aluminum cladding. The tensile strength is up to 10,000 lb/in^2 (68 MPa), and the dielectric strength is 400 volts per mil (16×10^6 volts per meter). **Phenolyte Y240** is a paper-base laminate bonded with a polyester-modified melamine which gives high dielectric strength and arc resistance together with good punching or blanking characteristics in thicknesses up to ⅛ in (0.32 cm). **Doryl H17511,** of Westinghouse, has glass fabric laminations bonded with a modified phenolic resin based on diphenyl oxide and polyphenyl ether. This laminate contains a flexural strength of 27,400 lb/n^2 (184 MPa) at 480°F (249°C). **Luxwood,** of Formica Corp., for furniture, is a ⅟16-in (0.16-cm) laminate with photographic reproductions of wood grains on the face, while **Beautywood** is this material in thicker sizes for wall panels.

Industrial thermosetting laminates are available in the form of sheet, rod, and rolled or molded tubing. Impregnating resins commonly used are phenolic, polyester, melamine, epoxy, and silicone. The base material, or reinforcement, is usually one of the following: paper, woven cotton or linen, asbestos, glass cloth, or glass mat. NEMA (National Electrical Manufacturing Association) has published standards covering over 25 standard grades of these laminates. Each manufacturer, in addition to these, usually produces a range of special grades.

Laminating resins may be marketed under one trade name by the resin producer and other names by the molders of the laminate. **Paraplex P resins,** of Rohm & Haas, for example, comprise a series of polyester solutions in monomeric styrene which can be blended with other resins to give varied qualities. But **Panelyte,** of St. Regis Paper Co., refers to the laminates which are made with phenolic, melamine, silicone, or other resin, for a variety of applications. **Rockite,** of F. A. Hughes & Co., Ltd., and **Tufnol,** of Ellison Insulations, Ltd., are British phenol resins.

PLASTIC POWDER COATINGS. Although many different plastic powders can be applied as coatings, vinyl, epoxy, and nylon are most often used. Vinyl and epoxy provide good corrosion and weather resistance as well as good electrical insulation. Nylon is used chiefly for its outstanding wear and abrasion resistance. Other plastics frequently used in powder coating include chlorinated polyethers, polycarbonates, acetals, cellulosics, acrylics, and fluorocarbons.

Several different methods have been developed to apply these coatings. In the most popular process, **fluidized bed,** parts are preheated and then immersed in a tank of finely divided plastic powders, which are held in a suspended state by a rising current of air. When the powder particles contact the heated part, they fuse and adhere to the surface, forming a continuous, uniform coating. Another process, **electrostatic spraying,** works on the principle that oppositely charged materials attract each other. Powder is fed through a gun, which applies an electrostatic charge opposite to that applied to the part to be coated. When the charged particles leave the gun, they are attracted to the part where they cling until fused together as a plastic coating. Other powder-application methods include flock and flow coating, flame and plasma spraying, and a cloud-chamber technique.

PLASTICS. A major group of materials that are primarily noncrystalline hydrocarbon substances composed of large molecular chains whose major element is carbon. The three terms—**plastics, polymers,** and **synthetic resins (or resins)**—are sometimes used interchangeably to identify these materials. However, the term **plastics** has now come to be the commonly used designation.

The first commercial plastic, **Celluloid,** was developed in 1868 to replace ivory for billiard balls. **Phenolic plastics,** developed by Baekeland

and named Bakelite after him, were introduced around the turn of the century. A plastic material, as defined by the Society of the Plastics Industry, is "Any one of a large and varied group of materials consisting wholly or in part of combinations of carbon with oxygen, hydrogen, nitrogen, and other organic and inorganic elements which, while solid in the finished state, at some stage in its manufacture is made liquid, and thus capable of being formed into various shapes, most usually through the application, either singly or together, of heat and pressure."

There are two basic types of plastics based on intermolecular bonding. **Thermoplastics,** because of little or no cross-bonding between molecules, soften when heated and harden when cooled, no matter how often the process is repeated. **Thermosets,** on the other hand, have strong intermolecular bonding. Therefore, once the plastic is set into permanent shape under heat and pressure, reheating will not soften it.

Within these major classes, plastics are commonly classified on the basis of base **monomers.** There are over two dozen such monomer families or groups. Plastics are also sometimes classified roughly into three stiffness categories: rigid, flexible, and elastic. Another method of classification is by the "level" of performance or the general area of application, using such categories as engineering, general-purpose, and specialty plastics, or the two broad categories of engineering and commodity plastics.

Some of the major characteristics of plastics that distinguish them from other materials, particularly metals, are: (1) they are essentially noncrystalline in structure; (2) they are nonconductors of electricity and are relatively low in heat conductance; (3) they are, with some important exceptions, resistant to chemical and corrosive environments; (4) they have relatively low softening temperatures; (5) they are readily formed into complex shapes; and (6) they exhibit viscoelastic behavior—that is, after an applied load is removed, plastics tend to continue to exhibit strain or deformation with time.

Polymers can be built of one, two, or even three different monomers, and are termed **homopolymers, copolymers,** and **terpolymers,** respectively. Their geometrical form can be linear or branched. Linear or unbranched polymers are composed of monomers linked end-to-end to form a molecular chain that is like a simple string of beads or a piece of spaghetti. Branched polymers have side chains of molecules attached to the main linear polymer. These branches can be composed either of the basic linear monomer or of a different one. If the side molecules are arranged randomly, the polymer is **atactic;** if they branch out on one side of the linear chain in the same plane, the polymer is **isotactic;** and if they alternate from one side to the other, the polymer is **syndiotactic.**

Few plastics in use are totally composed of polymer resins. Nearly all contain one or more additive materials to modify or control properties, or to reduce costs. **Fillers** are probably the most common of the additives. They are usually used to either provide bulk or modify certain properties. Generally, they are inert and thus do not react chemically with the resin during processing. The fillers are often cheap and serve to reduce costs by increasing bulk. For example, **wood flour,** a common low-cost filler, sometimes makes up 50% of a plastic compound. Other typical fillers are chopped fabrics, asbestos, talc, gypsum, and milled glass. Besides lowering costs, fillers can improve properties. For example, asbestos increases heat resistance, and cotton fibers improve toughness.

Plasticizers are added to plastics compounds either to improve flow during processing by reducing the glass transition temperature or to improve properties such as flexibility. Plasticizers are usually liquids that have high boiling points, such as certain phthalates. Substances which are themselve polymers of low molecular weight, such as polyesters, are also used as plasticizers. **Stabilizers** are added to plastics to help prevent breakdown or deterioration during molding or when the polymer is exposed to sunlight, heat, oxygen, ozone, or combinations of these. Thus there is a wide range of compounds, each designated for a specific function. Stabilizers can be metal compounds, based on tin, lead, cadmium, barium, and others. And phenols and amines are added antioxidants that protect the plastic by diverting the oxidation reactions to themselves.

Catalysts, by controlling the rate and extent of the polymerization process in the resin, allow the curing cycle to be tailored to the processing requirements of the application. Catalysts also affect the shelf life of the plastics. Both metallic and organic chemical compounds are used as catalysts. **Colorants,** added to plastics for decorative purposes, come in a wide variety of pigments and dyestuffs. The traditional colorants are metal-base pigments such as cadmium, lead, and selenium. More recently, liquid colorants, composed of dispersions of pigments in a liquid, have been developed. **Fire retardants** are added to plastic products that must meet fire-retardant requirements, because polymer resins are generally flammable, except for such notable exceptions as polyvinyl chloride. In general, the function of fire retardants is limited to the spread of fire. They do not normally increase heat resistance or prevent the plastic from charring or melting. Some fire-retardant additives include compounds containing chlorine or bromine, phosphate-ester compounds, antimony thrioxide, alumina trihydrate, and zinc borate.

Reinforcements obtained with plastics are not normally considered additives. They are used in plastics primarily to improve mechanical properties, particularly strength. Although asbestos and some other materials are used, glass fibers are the predominant reinforcement for plastics.

Plastics are produced in a variety of different forms. Most common are **plastic moldings,** which range in size from less than one inch to several feet (two centimeters to several meters). Thermoplastics, such as polyvinylchloride and polyethylene, are widely used in the form of **plastic film** and **plastic sheeting.** The term *film,* is used for thicknesses up to and including 10 mils (0.25 cm), while sheeting refers to thicknesses over that.

Both thermosetting and thermoplastic materials are used as **plastic coatings** on metal, wood, paper, fabric, leather, glass, concrete, ceramics, or other plastics. There are many coating processes, including knife or spread coating, spraying, roller coating, dipping, brushing, calendering, and the fluidized-bed process. Thermosetting plastics are used in **high-pressure laminates** to hold together the reinforcing materials that comprise the body of the finished product. The reinforcing materials may be cloth, paper, wood, or glass fibers. The end product may be plain flat sheets, or decorative sheets as in counter tops, rods, tubes, or formed shapes.

PLATINUM. A whitish-gray metal, symbol Pt. It is more ductile than silver, gold, or copper, and is heavier than gold. The melting point is 3190°F (1754°C), and the specific gravity is 21.45. The hardness of the annealed metal is 45 Brinell, with a tensile strength of 17,000 lb/in^2 (117 MPa); when hard-rolled the Brinell hardness is 97 and tensile strength 34,000 lb/in^2 (234 MPa). The electric conductivity is about 16% that of copper. The metal has a face-centered cubic lattice structure, and it is very ductile and malleable. It is resistant to acids and alkalies, but dissolves in aqua regia. Platinum is widely used in jewelry, but because of its heat resistance and chemical resistance it is also valued for electric contacts and resistance wire, thermocouples, standard weights, and laboratory dishes. It is generally too soft for use alone, and is almost always alloyed with harder metals of the same group, such as osmium, rhodium, and iridium. An important use of the metal is in the form of gauze as a catalyst. **Platinum gauze** is of high purity in standard meshes of 45 to 80 per inch (18 to 31 per centimeter), with wire from 0.0078 to 0.003 in (0.020 to 0.008 cm) in diameter. **Dental foil** is of 99.99% purity of maximum softness. **Platinum foil** for other uses is made in thicknesses down to 0.0002 in (0.0005 cm). **Platinum powder** comes in fine submesh particle size. It is made by chemical reduction and has a minimum purity of 99.9% with amorphous particles 0.3 to 3.5 μm in diameter. Atomized powder has spherical particles of 50 to 200 mesh, and is 99.9% pure and free-flowing. **Platinum flake** has the powder particles in tiny laminar platelets which overlap in the coating film. The particles in **Platinum flake No. 22,** of the Metz Refining Corp., have an average diameter of 3 μm and thickness of 0.1 μm.

Because of the high resistance of the metal to atmospheric corrosion even in sulfur environments, **platinum coatings** are used on springs and other functioning parts of instruments and electronic devices where precise operation is essential. Electroplating may be done with an electrolyte bath of **platinum dichloride,** $PtCl_2$, or **platinum tetrachloride,** $PtCl_4$. Hard plates may be produced with an alkaline bath of **platinum diamine nitrite,** $Pt(NH_3)_2(NO_2)_2$. Coatings are also produced by vapor deposition of platinum compounds; thin coatings, 0.0002 in (0.0005 cm) or less, are made by painting the surface with a solution of platinum powder in an organic vehicle and then firing to drive off the organic material to leave an adherent coating of platinum metal.

Platinum occurs native in small flat grains or in pebbles usually in alluvial sands, and the native metal generally contains other metals of the platinum group. The largest nugget ever found came from South America and weighed 2 lb (0.9 kg). The chief sources of the metal are Russia and Colombia, with smaller amounts from Alaska, Canada, and South Africa. Some platinum is obtained from the copper-nickel ores of Canada and South Africa. There are no commercial ores of the metal, but the rare mineral **sperrylite** is found in Wyoming and in Ontario. It is a **platinum arsenide,** $PtAs_2$, found in small grains of a tin-white metallic luster. The only other known natural compound is the rare mineral **cupperite,** which is a **platinum sulfide,** PtS. The Russian platinum is 99.8 to 99.9% pure, with 0.05 to 1.10% iridium. Platinum is sold by the troy ounce (0.03 kg), 1 in^3 (16 cm^3) of the metal weighing 11.28 troy oz (0.34 kg).

PLATINUM ALLOYS. **Platinum-iridium alloys** are employed for instruments, magneto contacts, and jewelry. The alloys are hard, tough, and noncorrosive. An alloy of 5% iridium and 95 platinum, when hard-worked, has a Brinell hardness of 170; an alloy with 30% iridium has a hardness of 400. The 5 and 10% alloys are used for jewelry manufacture; the 25 and 30% alloys are employed for making surgical instruments. An alloy of 80% platinum and 20 iridium is used for magneto contact points, and the 90–10 alloy is widely used for electric contacts in industrial control devices. The addition of iridium does not alter the color of the platinum. The 5% alloy dissolves readily in aqua regia; the 30% alloy dissolves slowly.

Platinum-rhodium alloys are used for thermocouples for temperatures above 1100°C. The standard thermocouple used is platinum versus platinum–10% rhodium. Other thermocouples for higher operating temperatures use platinum-rhodium alloys in both elements. The alloys of platinum-rhodium are widely used in the glass industry, particularly as glass fiber extrusion bushings. Rhodium increases the high temperature strength of platinum without reducing its resistance to oxidation. **Platinum-rhodium gauze** for use as a catalyst in producing nitric acid from ammonium contains 90% platinum and 10 rhodium. **Platinum-rhenium alloys** are efficient catalysts for reforming operations on aromatic compounds. The platinum alloys have lower electric conductivity than pure platinum, but are generally harder and more wear-resistant, and have high melting points. A platinum-rhenium alloy with 10% rhenium has an electric conductivity of only 5.5% that of copper compared with 16% for pure platinum. Its melting point is 1850°C, and the Rockwell T hardness of the cold-rolled metal is 91 compared with 78 for cold-rolled platinum. **Platinum-ruthenium alloy,** with 10% ruthenium, has a melting point of 1800°C, and an electric conductivity 4% that of copper.

PLYWOOD. A laminated wood made up of thin sheets of wood glued together with the grains of successive layers set at right angles to give strength in both directions. Plywood is an outgrowth of the laminated wood known as veneer, which consists of an outside sheet of hardwood glued to a base of lower-cost wood. The term **veneer** actually refers only to the facing layer of selected wood used for artistic effect, or for economy in the use of expensive woods. Veneers are generally marketed in strip form in thicknesses of less than ⅛ in (0.32 cm) in mahogany, oak, cedar, and other woods. The usual purpose of plywood now is not aesthetic but to obtain high strength with low weight. The term laminated wood generally means heavier laminates for special purposes, and such laminates usually contain a heavy impregnation of bonding resin which gives them more of the characteristics of the resin than of the wood.

Plywood usually comes in 4- by 8-ft (1.2- by 2.4-m) panels, and is always built up with an odd number of layers. The cross-ply construction gives strength in both directions, and also gives symmetrical shrinkage stresses. A three-ply softwood panel of equal ply thicknesses may shrink about 0.080 in (0.203 cm) in width and 0.100 in (0.254 cm) in length, but increasing the thickness of the core ply can equalize the shrinkage, or equalization may be obtained by increasing the number of plies. The odd number of plies gives a symmetry of construction about the core ply. Low-cost plywoods may be bonded with starch pastes, animal glues, or casein, and are not water-resistant, but are useful for boxes and for interior work. Waterproof plywood for paneling and general construction is now bonded with synthetic resins, but when the plies are heavily impregnated with the resin and the whole cured into a solid sheet, the material is known as a hardboard or as a laminated plastic rather than a plywood.

For construction purposes, where plywood is employed because of its unit strength and nonwarping characteristics, the plies may be of a single type of wood and without a hardwood face. The Douglas Fir Plywood Association sets up four classes of construction plywood under general trade names. **Plywall** is plywood in wallboard grade; **Plypanel** is plywood in three standard grades for general uses; **Plyscord** is unsanded plywood with defects plugged and patched on one side; **Plyform** is plywood in a grade for use in concrete forms.

The bulk of commercial plywood comes within these classes, the variations being in the type of wood used, the type of bonding adhesive, or the finish. **Etch wood,** for example, is a paneling plywood with the face wire brushed to remove the soft fibers and leave the hard grain for two-tone finish. **Paneling plywoods** with faces of mahogany, walnut, or other expensive wood have cores of lower-cost woods, but the woods of good physical qualities are usually chosen. The tensile strength of a white ash three-ply plywood parallel to the grain of the faces is about 6,200 lb/ in^2 (42 MPa), that of a mahogany plywood is 6,400 lb/in^2 (44 MPa), and that of a walnut plywood is 8,200 lb/in^2 (56 MPa). **Stabilite,** of the Georgia-Pacific Corp., is a wood laminate with the veneers impregnated with phenolic resin and bonded together with the grain of the plies parallel. It has the density of hardwood and the workability of cherry wood.

The **K-Veneer** used during the Second World War as a substitute for plywood was a $\frac{3}{16}$-in (0.478-cm) fir or hemlock sheet bonded to a heavy kraft paper. The **Welchboard,** of the West Coast Plywood Co., is $\frac{3}{8}$-in (0.953-cm) plywood with a smooth grainless surface produced by curing on one side under heat and pressure a mixture of wood pulp and synthetic resin. A great variety of trade-named plywoods are marketed for paneling, but they do not always have the typical characteristics of plywood, and are often **paneling boards** rather than plywoods. Some have metal faces, or they may have special-purpose cores or backings. **Fybr-Tech,** developed by Technical Plywoods for aircraft paneling, has a $\frac{1}{64}$-in (0.038-cm) walnut veneer on a $\frac{1}{4}$-in (0.635-cm) balsa wood core, with a 0.005-in (0.013-cm) vulcanized fiber back. The weight is 0.2 lb/ft^2 (0.98 kg/m^2). One of the earliest aluminum-faced plywoods was the English **Plymax** of Venesta, Ltd. **Plymetl,** of the Haskelite Mfg. Corp., has a core of plywood with facings of aluminum or steels. **Metal-faced plywoods** are strong and can be riveted. **Met-L-Wood,** of the Met-L-Wood Corp., has two layers of light wood separated by sheet metal, designed for truck sides. **Metalite,** developed by the United Aircraft Corp., has thin sheets of strong aluminum alloy bonded to both sides of a balsa wood core, the grain of the wood being perpendicular to the metal faces. **Siding-panel 15,** of Weyerhaeuser Co., has a Douglas fir plywood core and a facing of 0.01 in (0.025 cm) embossed aluminum with a vinyl resin finish. **Flexwood,** of U.S. Plywood Corp., consists of very thin sheets of veneer glued under heat and pressure to cotton sheeting, used as an ornamental covering for walls. **Algonite** is Masonite faced with fancy veneers. **Protekwood,** of this company, designed for protection against vermin attack, has a sheet of hardwood between two sheets impregnated with asphalt and resin, and bonded with a urea-formaldehyde resin to a total thickness of $\frac{5}{32}$ in (0.396 cm). **Parkwood,** of the Parkwood Corp., is a flexible **woven veneer** made with thin strips of mahogany or other fancy wood pressed between sheets of transparent cellulose acetate or other plastic. **Novoply,** of this company, for panels, furniture, and structural parts, has a core of resin-impregnated wood chips bonded between hardwood veneers. Sheets are up to ¾ in (1.91 cm) thick, and have high strength. **Randalite,** of the same company, is birch veneer bonded to a kraft liner.

POISON GASES. Substances employed in chemical warfare for disabling people, and in some cases used industrially as fumigants. They are all popularly called gases, but many are liquids or solids. Normally, information on military gases is kept secret, but the tear gases used by police are also poisonous, often causing serious damage to eyes, throat, and lungs. **Anesthetic gases** have so far not been used in chemical warfare, but are used in medicine. One of the simplest of these, **nitrous oxide,** N_2O, called **laughing gas,** produces a deep sleep. **Fluorthane,** or **ethyl fluoride,** is a volatile liquid like ether, but is nonexplosive, and is used to replace ether in surgery. **Cyclopropane** is a potent anesthetic. It is not less than 99% by volume of C_3H_6.

Poison gases are classified according to their main effect on the human system, but one gas may have several effects. They are grouped as follows: **lethal gases,** intended to kill, such as phosgene; **lachrymators,** or **tear gases,** which have a powerful irritating effect on the eyes causing temporary blindness and swelling of the eyes with a copious flow of tears; **vesicants,** or skin blisterers, such as lewisite and mustard gas; **sternutatory gases,** which induce sneezing; and **camouflage gases,** which are harmless, but cause soldiers to suffer the inconvenience of wearing gas masks and thus reduce their morale. Some of the gases have a sour, irritating odor and are also classed as **harassing agents.** Gases are also sometimes designated as **casualty agents** and further subdivided into persistent and nonpersistent. A **systemic gas** is one that interferes with one phase of the system, such as carbon monoxide, which paralyzes the respiratory function of the blood. A **laryrinthic gas** is one that affects an organ of the body, such as **dichlormethyl ether,** which affects the ears.

Effects of persistent gases, such as mustard, remain over the ground for as long as 7 days, but phosgene is quickly decomposed by dampness. **Obscuring agents,** such as white phosphorus, and the **toxic smokes,** such as diphenylaminochloroarsine, are also classed as war gases. Dusts of materials having catalytic properties, but not poisonous themselves, may be used to penetrate gas masks and create poisons, such as carbon monoxide, within the mask. Carbon and oil smokes may be used to choke the filters of gas masks and cause their removal. Absorbents used in gas masks are usually activated charcoal and soda lime. These will absorb or disassociate most of the toxic gases, but will not stop carbon monoxide. A mixture of powdered oxides of copper, manganese, silver, and cobalt, called **Hopcalite,** is used as a catalyst to oxidize carbon monoxide.

Lethal gases are divided into four classes: actual poisons, such as hydrocyanic acid, which kill with little pain; **asphyxiating gases,** such as phosgene, diphosgene, and chloropicrin, which affect the membranes of the lungs, destroying them and allowing blood to fill the air sacs; poisons which destroy the lining of the air passages and block the passages to the lung tissues, such as mustard gas and ethyldichloroarsine; and poisons which affect the nose and throat, causing great pain, headache, and vomiting, such as diphenylchloroarsine. **Mustard gas,** $(CH_2ClCH_2)_2S$, also known as **blister gas, yperite,** and **yellow cross,** is an oily liquid which boils at 210°C and vaporizes easily in the air. It destroys the cornea of the eyes, blisters the skin, affects the lungs, and causes discharge from the nose and vomiting. One part in 14 million parts air is toxic and cannot be detected in dilutions by smell. Another powerful vesicant is **Bromlost,** which is **dibromethyl sulfide,** $(CH_2BrCH_2)_2S$. It is a solid melting at 21°C. **Sulvanite** is **ethylsulfuryl chloride,** $ClSO_3C_2H_5$. It is a colorless liquid boiling at 135°C.

Lewisite, $CHCL{:}CH \cdot AsCL_2$, is a liquid boiling at 190°C. It is a powerful vesicant causing painful blisters on the skin, pain in the eyes, nose irritation, permanent impairment of eyesight, and arsenic poisoning. It forms a heavy mist and has been called **dew of death. Chloropicrin,** called **aquinite, klop,** and **nitrochloroform,** is **nitrotrichloromethane,** CCl_3NO_2. It is a persistent lachrymatory and lethal poison. It is a colorless liquid boiling at 112°C, with a specific gravity of 1.692. It is used as a soil fumigant to control insects and fungi. **Tonite** is **chloroacetone,** CH_3COCH_2Cl, a clear liquid vaporizing at 119°C. It is a powerful lachrymator and skin blisterer. As it is very reactive, it is also used in the synthesis of pharmaceuticals, dyes, and organic chemicals.

Bromoacetone is a colorless liquid of the composition $CH_2BrCOCH_3$, with a specific gravity of 1.631 and boiling point 126°C. It is thrown in bombs or shells and disseminated as a mist which attacks the eyes. **Bromobenzyl cyanide** is a solid of the composition $BrC_6H_4CH_2CN$, with a melting point of 25°C; when impure it is a liquid, but it is not purified, as it is easily decomposed. It is very persistent and because of its odor is classed as a harassing agent. It was called **camite** by the French. Another tear gas, **chloroacetophenone,** $C_6H_5COCH_2Cl$, is a white, crystalline solid, specific gravity 1.321, and melting point 59°C, which, when thrown as a vapor, has a sweet, locustlike odor but produces pains in the eyes and temporary blindness. The tear gas called **Mace,** used for riot control, contains this substance. In contact with the skin it causes severe dermatitis and often permanent damage to the eyes. The French gas **fraissite** is **benzyl iodide,** $C_6H_5CH_2I$, a liquid boiling at 226°C. The gas **papite** is acrolein with stannic chloride. **Caderite** is **benzyl bromide** with stannic chloride. **Xylyl bromide,** $CH_3C_6H_4CH_2Br$, is a colorless liquid boiling at 216°C. When disseminated as a mist from explosive bombs, it causes a copious flow of tears. The tear gas known as **CS gas** is **chlorobenzol malononitrile**

and is destructive to nasal and lung tissues. Benzyl chloride also has been used as a tear gas.

Martonite, a powerful lachrymator, is a mixture of 20% chloroacetone and 80 bromoacetone. **Bretonite** is **iodoacetone,** CH_3COCH_2I, a brownish liquid boiling at 102°C, mixed with stannic chloride as a lachrymator. **Manguinite** is **cyanogen chloride,** CNCl, which boils at 13°C, and is a lachrymator. Mixed with arsenic trichloride to make it more toxic, it was used under the name of **vitrite. Campillit** is **cyanogen bromide,** CNBr, a white solid melting at 52°C and vaporizing at 61.3°C. The fumes are highly toxic, paralyzing the nerve centers. **Diphenylchloroarsine,** or **blue cross** $(C_6H_5)_2AsCl$, is a **sneezing gas** which penetrates gas masks, forcing their removal. It affects chiefly the nose and throat, but is used with other more violent gases. **Adamsite** is a greenish granular solid of the composition $(C_6H_4)_2NHAsCl$, which has a pleasant odor but burns the nose and throat. Many of the lachrymators have important industrial uses. **Phenyl isocyanate,** C_6H_5NCO, is a water-white liquid of specific gravity 1.101 and boiling point 162°C, used for the production of alkyd resins, ureas, urethanes, and other chemicals. **Decontaminants,** used for combating the effects of poison gases, are neutralizing chemicals. The decontaminant known as STB, or **supertropical bleach,** is a mixture of chlorinated lime and ground quicklime.

POLONIUM. A rare metallic element, symbol Po, belonging to the group of radioactive metals, but emitting only alpha rays. The melting point of the metal is about 1100°C. It is used in meteorological stations for measuring the electrical potential of the air. **Polonium-plated metal** in strip and rod forms has been employed as a static dissipator in textile coating machines. The alpha rays ionize the air near the strip, making it a conductor and drawing off static electric charges. **Polonium 210** is obtained by irradiating bismuth, 100 lb (45 kg) yielding 1 g (0.04 oz) of polonium 210. It is used as a heat source for emergency auxiliary power such as in spacecraft. The metal is expensive, but can be produced in quantity from bismuth.

POLYCARBONATE RESINS. Made by reacting bisphenol and phosgene, or by reacting a polyphenol with methylene chloride and phosgene. The monomer may be $OC_6H_4C(CH_3)_2C_6H_4OC{:}O$. The molecular structure is in double-linked zigzag chains that give high rigidity. The resin is thermoplastic. It is crystalline with rhombic crystals.

Polycarbonate is a linear, low-crystalline, transparent, high-molecular-weight plastic. It is generally considered to be the toughest of all plastics. In thin sections, up to about 3/16 in (0.478 cm), its impact strength is as high as 16 ft·lb (24 kg·m). In addition, polycarbonate is one of the hardest plastics. It also has good strength and rigidity, and, because of its high modulus of elasticity, is resistant to creep. These properties, along with its excellent electrical resistivity, are maintained over a temperature range of about −275 to 250°F (−170 to 121°C). It has negligible moisture absorption, but it also has poor solvent resistance, and, in a stressed condition, will craze or crack when exposed to some chemicals. It is generally unaffected by greases, oils, and acids. Polycarbonate plastics are easily processed by extrusion, by injection, blow, and rotational molding, and by vacuum forming. They have very low and uniform mold shrinkage. With a white light transmission of almost 90% and high impact resistance, they are good glazing materials. They have more than 30 times the impact resistance of safety glass. Other typical applications are safety shields and lenses. Besides glazing, polycarbonate's high impact strength makes it useful for air-conditioner housings, filter bowls, portable tool housings, marine propellers, and housings for small appliances and food-dispensing machines.

Lexan resin, of General Electric, with a molecular weight of 18,000, has a tensile strength of 10,500 lb/in^2 (71 MPa), with elongation of 60%, dielectric strength to 2,500 volts per mil (98 × 10^6 volts per meter), and Rockwell hardness M70. The deformation temperature under load is 290°F (143°C), and the specific gravity is 1.20. The material is transparent, and a ⅛-in (0.318-cm) thickness transmits 85% of the light. This material is called **Makrolon** in Germany. The **Merlon,** of the Mobray Chemical Co., is a polycarbonate resin, and the **Plestar film,** of Ansco, is polycarbonate. **Lexan 145,** of General Electric, is polycarbonate powder for use in emulsion coatings which have high strength and toughness and are nontoxic. When this type of resin is compounded with 40% of glass fiber, its tensile strength is doubled and its coefficient of thermal expansion is greatly reduced. **Lexan FL 1800,** a structural foam resin with high resistance to combustion, meets both service performance requirements and flammability standards for materials used in enclosure for large data process systems. **Carbonate resins** are also used in mixtures with an acrylonitrile-butadiene-styrene copolymer without copolymerization. **Cycolac KM,** of Borg-Warner Corp., is this material. It retains the general characteristics of the carbonate with added flexural strength.

POLYESTER RESINS. A large group of synthetic resins produced by condensation of acids such as maleic, phthalic, or itaconic with an alcohol or glycol such as allyl alcohol or ethylene glycol to form an unsaturated polyester which, when polymerized, will give a cross-linked, three-dimensional molecular structure, which in turn will copolymerize with an unsaturated hydrocarbon such as styrene or cyclopentadiene to form a copolymer of complex structure of several monomers linked and cross-linked. At least one of the acids or alcohols of the first reaction must be unsaturated. The polyesters made with saturated acids and saturated hydroxy compounds are called **alkyd resins,** and these are largely limited to the production of protective coatings and are not copolymerized.

The resins undergo polymerization during cure without liberation of water, and do not require high pressure for curing. Through the secondary stage of modification with hydrocarbons a very wide range of characteristics can be obtained. The most important use of the polyesters is as laminating and molding materials, especially for glass-fiber-reinforced plastic products. The resins have high strength, good chemical resistance, high adhesion, and capacity to take bright colors. They are also used, without fillers, as casting resins, for filling and strengthening porous materials such as ceramics and plaster of Paris articles, and for sealing the pores in metal castings. Some of the resins have great toughness, and are used to produce textile fibers and thin plastic sheet and film. Others of the resins are used with fillers to produce molding powders that cure at low pressure of 500 to 900 lb/in^2 (3 to 6 MPa) with fast operating cycles.

Polyester laminates are usually made with a high proportion of glass-fiber mat or glass fabric, and high-strength reinforced moldings may also contain a high proportion of filler. A resin slurry may contain as high as 70% calcium carbonate or calcium sulfate, with only about 11% of glass fiber added, giving an impact strength of 24,000 lb/in^2 (165 MPa) in the cured material. Bars and structural shapes of glass-fiber reinforced polyester resins of high tensile and flexural strengths are made by having the glass fibers parallel in the direction of the extrusion. The **Glastrusions** of the Hugh C. Marshall Co., in the form of rods and tubes, are made by having the glass-fiber rovings carded under tension, then passing through an impregnating tank, an extruding die, and a heat-curing die. The rods contain 65% glass fiber and 35 resin. They have a flexural strength of 64,000 lb/in^2 (441 MPa), and a Rockwell M hardness of 65.

Physical properties of polyester moldings vary with the type of raw materials used and the type of reinforcing agents. A standard glass-fiber-filled molding may have a specific gravity from 1.7 to 2.0, a tensile strength of 4,000 to 10,000 lb/in^2 (27 to 68 MPa) with elongation of 16 to 20%, a flexural strength to 30,000 lb/in^2 (206 MPa), a dielectric strength to about 400 volts per mil (16×10^6 volts per meter), and a heat distortion temperature of 350 to 400°F (177 to 204°C). The moldings have good acid and alkali resistance. But, since an almost unlimited number of fatty-type acids are available from natural fatty oils or by synthesis from petroleum, and the possibilities of variation by combination with alchohols, glycols, and other materials are also unlimited, the polyesters form an ever-expanding group of plastics.

The **Mylar film** of Du Pont is a polyester made by the condensation of terephthalic acid and ethylene glycol. The extremely thin film, 0.00025 to 0.0005 in (0.00063 to 0.0013 cm), used for capacitors and for insulation of motors and transformers, has a high dielectric strength, up to 6,000 volts per mil (236×10^6 volts per meter). It has a tensile strength of 20,000 lb/in^2 (137 MPa) with elongation of 70%. It is highly resistant to chemicals, and has low water absorption. The material is thermoplastic, with a melting point at about 490°F (254°C). **Polyester fibers** are widely used in clothing fabrics. The textile fiber produced from dimethyl terephthalate is known as **Dacron.** The English textile fibers called **Teron** and **Terylene** are similar materials, and **Melinex** is the English name for the film. **Mylar 50T,** used for magnetic sound-recording tape, has the molecules oriented by stretching to give high strength. The 0.005-in (0.013-cm) tape has a breaking strength of 120 oz (3.4 kg) per ¼ in (0.64 cm) of width. **Urylon,** a Japanese fiber, has a low specific gravity, 1.07, and a high melting point, 250°C. It is produced from azelaic acid. The polyester film called **Terefilm,** of the Acme Backing Corp., used for insulation and for magnetic tape, is a cyclohexylene dimethylene terephthalate. The dielectric strength of the 0.0005-in (0.0013-cm) film is 8,000 volts per mil (315×10^6 volts per meter), with tensile strength of 20,000 lb/in^2 (137 MPa), and heat distortion temperature 340°F (171°C). **Electronic tape** may also have a magnetic-powder coating on the polyester. But where high temperatures may be encountered, as in spacecraft, the magnetic coating is applied to metal tapes. **Densimag,** of the Whittaker Corp., has the magnetic coating on a nonmagnetic stainless-steel tape, the two having a thickness of 0.00105 in (0.00267 cm). It will operate at temperatures to 600°F (316°C).

The **Koplac resins** of the Koppers Co., Inc., are polyesters based on styrene. **Vibrin 135,** of the Naugatuck Chemical Co., is a polyester resin made with triallyl cyanurate and modified with maleic anhydride. Moldings reinforced with glass fiber have a tensile strength of 38,000 lb/in^2 (261 MPa), and retain a strength of 23,000 lb/in^2 (158 MPa) at 500°F (260°C). **Vibrin 136A** has higher strength and very high radar transparency. It is used for radomes and nose cones.

Transparent thermoplastic polyester resins are made by copolymerizing esters of itaconic acid with vinyl chloride, methacrylate, or acrylonitrile. **Itaconic acid,** $CH_2{:}CCH_2(COOH)_2$, is made from anhydrous glucose. **Pimelic acid,** $HO_2C(CH_2)CO_2H$, made from petroleum as a white crystalline solid, is also used for making polyester and polyamide resins. Another of the many acids used for these resins is **glutaric acid,** $HO_2C(CH_2)_3CO_2H$, produced from acrylein. **Glutaric anhydride,** $O{:}HC(CH_2)_3HC{:}O$, is also used, and its cross-linking ability is employed for insolubilizing starches and proteins to give water resistance to paints and paper coatings.

Het acid, of Hooker Chemical, is a complex chlorinated phthalic acid produced by hydrolyzing the product of the condensation of maleic anhydride with hexachlorocyclopentadiene made from pentane. This acid is reacted with glycols and maleic anhydride to give a hard polyester resin which is then cross-linked with styrene to give the liquid **Hetron resin,** which will cure with heat and a catalyst to an insoluble solid. The resin contains 30% chlorine. It is used for making laminated or reinforced plastics. Another chlorinated polyester resin is **FR resin** of the Interchemical Co. It is flame-resistant, cures at normal temperatures, and is used for such lay-up lamination work as boat building and tank construction. Some of the polyester-type resins have rubberlike properties, with higher tensile strengths than the rubbers and superior resistance to oxidation. **Vulcollan,** of the Goodyear Tire & Rubber Co., is such a resin with higher wear resistance and chemical resistance than GRS rubber. It is made by reacting adipic acid with ethylene glycol and propylene glycol and then adding diisocyanate to control the solidifying action. It can be processed like rubber, but solidifies more rapidly. **Chemigum SL,** of the same company, is a **polyester rubber.** The polyesters also offer a great variety of possibilities in textile fibers. **Kodel,** of Eastman Chemical Products, Inc., is a white polyester fiber that is easily dyed. It is resistant to pilling, which is the tendency of surface fibers to form balls, and it has high dimensional stability and heat resistance. A flame-resistant polyester fiber, **Trevira 271,** is produced by Hoechst Fibers Industries. It is made from **polyethylene terephthalate** polymers that have been modified by the inclusion of a flame retardant in the polymer structure itself. If subjected directly to flame, the manufacturer says that fabrics made of this fiber melt and shrink away. When flame is removed, the fabrics self-extinguish.

Thermoplastic polyester is a crystalline plastic molding compound. One type is made by polycondensation of 1,4-butanediol and dimethyl-terephthalate (DMT) to produce **polytetramethylene terephthalate (PTMT).** They are most commonly referred to as **polybutylene terephthalate (PBT).** The many different grades that are produced by a number of different companies can be divided into four major types: unmodified, flame retardant, glass-fiber reinforced, and mineral filled. Unmodified types have the greatest elongation and usually the shortest molding cycles. Flame-retardant grades can maintain their flame retardance in very thin sections—at $\frac{1}{32}$ in (0.8 mm), for example. Thermoplastic polyesters reinforced with glass fibers are among the toughest plastics materials. Along with good heat resistance, they provide excellent impact strength and mechanical properties. Glass loadings range from 10 to 50%. To counter warpage associated with glass-fiber-reinforced grades, mineral fillers can be added. Although some strength is sacrificed, these grades have shrinkage of 0.5% or less.

A notable disadvantage of thermoplastic polyesters is notch sensitivity and unsuitability for long-time hot-water immersion. However, water resistance at room temperature is excellent. Some producers of PBT materials are Celanese Plastics **(Celanex)**, Eastman Chemical **(Tenite polyterephthalate)**, and General Electric **(Valox)**,.

A more recently developed thermoplastic polyester is a modified **polyethylene terephthalate (PET)** with the trade name **Rynite.** Produced by Du Pont, the material, reinforced with glass fibers, provides unusually high temperature resistance, high tensile strength, good stiffness, and high impact strength. Its areas of application are automobile parts, hardware, and consumer goods.

An extensive group of polyesters and alkyd resins with good heat stability can be made from **pyromellitic dianhydride,** $C_{10}H_2O_6$, a benzene tetracarboxylic dianhydride. It is marketed by Princeton Chemical Research, Inc., as **PMDA,** a white powder with melting point of 287°C. It reacts with alcohols, benzene, and other hydrocarbons. It is produced from **mellitic acid**, $C_6(COOH)_6$, which has a melting point of 288°C. It occurs in brown coal, peat, and the mineral **mellite,** or **honeystone,** which is hydrous aluminum mellate. **Tetrahydrophthalic anhydride** is easier to combine with styrene than phthalic or maleic anhydride, and gives coating resins that are flexible and have quicker cure with high gloss.

POLYETHYLENES. A group of polyolefin polymers derived from ethylene by polymerization by heat and pressure. **Polyethylene plastics** are one of the lowest-cost and most widely used plastics. As a group, they are noted for toughness, excellent dielectric strength, and chemical resistance. Another outstanding characteristic is their low water absorption and permeability, which is the reason for their wide use in sheet form as moisture barriers. They are white in thick sections, but otherwise the range varies from translucent to opaque. They feel waxy. The many available types, ranging from flexible to rigid materials, are classified by density (specific gravity) into three major groups: low density, 0.910 to 0.925; medium density, 0.926 to 0.940; and high density, 0.941 to 0.959. The variations in properties among these three groups are directly related to density. As density increases, polymer cross-bonding or branching and crystallinity increase. Thus stiffness, tensile strength, hardness, and heat and chemical resistance increase with density in polyethylenes. Low-density polyethylenes are flexible, tough, and less translucent than high-density grades. High-density grades, often called **linear polyethylene** grades, are stronger, more rigid, and have high creep resistance under load, but they have lower impact resistance. Typical uses of low-density polyethylenes include blow-molded bottles and containers, gaskets, paintbrush handles, and flexible-film packaging. High-density grades are used for wire insulation, beverage cases, dishpans, toys, and the film used for boil-in-bag packaging. In general, polyethylenes are not used in load-bearing applications because of their tendency to creep. However, a special type, high-molecular-weight polyethylene, is used for machine parts, bearings, bushing, and gears.

Polyethylenes can be blended or combined with other monomers—propylene, ethyl acrylate, and vinyl acetate—to produce copolymers to improve such properties as stress-crack resistance and clarity and to increase flexibility. They can also be modified by exposure to high-energy radiation, which produces cross-linking and thereby increases heat resistance and stiffness. **Zetafax resin,** of Dow Chemical Corp., is an **ethylene-acrylic resin.** It has good adhesion to metals and high chemical resistance. **Zetabon** is a Zetafax-coated metal tape for cable shielding. The **Ultrathene UE630-81A,** of U.S. Industrial Chemicals Co., is an **ethylene vinyl acetate** copolymer for rubberlike packaging film. The **Aircoflex resins,** of Air Reduction Co., are similar copolymers for use as pigment binders, paper and textile coatings, and adhesives. They have higher molecular weight and better stability than polyethylene.

Polythene and **Alathon** are names for polyethylene of Du Pont, in the forms of molding powder, rod, sheet, tubes, foil for packaging, and paper coatings. **Agilene,** of the American Agile Corp., is polyethylene. The plastic can be cross-linked by irradiation, and irradiated polyethylene parts become thermoset and have increased strength, toughness, and higher heat resistance. **Irrathane,** of the General Electric Co., is irradiated polyethylene. The plastic can also be cross-linked chemically by heating with carbon black and a diperoxide. For piping, this method increases strength, improves weather resistance, and eliminates stress cracking. **Irrathene tape SPT,** of this company, is a flexible irradiated polyethylene self-sealing insulating tape for corona-resistant electric cables. It withstands temperatures to 260°F (127°C).

Fortiflex A, of the Celanese Corp., used for rigid chemical piping, is polyethylene of specific gravity 0.96, having a linear crystalline structure. It has a tensile strength of 4,500 lb/in^2 (30 MPa), flexural strength of 5,500 lb/in^2 (37 MPa), Rockwell hardness R40, and will withstand operating temperatures above 400°F (204°C). For piping and wire covering, polyethylene is also compounded with small amounts of carbon black to give high resistance to weathering. **Fortiflex F-087** has a specific gravity of 0.938 and transverse elongation of 600%. It is used for packaging film and takes printing ink well. **Microthene ML708,** of U.S. Industrial Chemicals Co., is a high-density powder of 350 mesh for rotational molding of thin-walled parts. **Dylan,** of the Koppers Co., is a low-pressure linear polyethylene used for **polyethylene fibers,** with a density of 0.95, a tensile strength of 3,500 lb/in^2 (23 MPa), elongation of 225%, and softening point at 255°F (123°C). The fibers and fabrics are marketed under trade names. **Reevon,** of Reeves Bros., Inc., is an upholstery fabric woven of polyethylene monofilament. **Polyethylene foam** is light in weight, has negligible water absorption, and is used in sheet and film for thermal insulation, and for wire insulation. **Orthofoam** and **Metafoam,** of Ludlow Papers, are polyethylene foams in sheets from 0.016 to 0.035 in (0.041 to 0.089 cm) thick. The low-density film has a tensile strength to 1,500 lb/in^2 (9 MPa), and the high-density material has a tensile strength to 12,000 lb/in^2 (82 MPa) with elongation of 2.5 to 5%. **Polyethylene DGDA-2580,** of the Union Carbide Corp., for extruded insulation on electric cables, gives uniformly dispersed closed cells so that the material has about 30% gas by volume. Extruded coatings have a smooth surface, a tensile strength of 2,800 lb/in^2 (18 MPa), and a dielectric constant of 1.5 to 1.7.

Polyethylene film has high resistance to oils, greases, and fatty acids; it also has good tear strength and fold endurance, and the light weight gives a large area per pound. It is thus widely used for packaging. **Tenite 161M,** of Eastman Chemical Products, Inc., is an extruding grade giving a film with a density of 0.923, tensile strength of 2,200 lb/in^2 (14 MPa), and elongation of 600%. When polyethylene film is irradiated and stretched biaxially it can be shrunk as much as 20% in all directions by applying a blast of hot air or dipping in water at a temperature of 180°F (82°C), and such films are used for packaging meats and poultry where a tight, close fit is desired. **Cryovac L,** of W. R. Grace & Co., is a film of this kind. High-density polyethylene has a high concentration of hydrogen atoms which are capable of slowing down or stopping fast neutrons, and sheets made with a small amount of boron to stop also the low-energy neutrons are used for atomic shielding where light weight is necessary. **Panelyte sheet,** of the St. Regis Paper Co., for this purpose, is made of **Petrothene 100,** of the U.S. Industrial Chemicals Co., a polyethylene containing 2% boron.

Polyethylene rubbers are rubberlike materials made by cross-linking with chlorine and sulfur, or they are ethylene copolymers. **Chlorosulfonated polyethylene** is white spongy material. It has chlorine atoms and sulfonyl chloride groups spaced along the molecule. It is used to blend with rubber to add stiffness, abrasion resistance, and resistance to ozone, and also for wire covering. **Hypalon S-2,** of Du Pont, is this material. **Plaskon CPF200,** of Allied Chemical Corp., is a chlorinated polyethylene containing about 73% chlorine. It is used for coatings and has high resistance to acids and alkalies. **Ethylene-propylene rubber,** produced by various companies, is a chemically resistant rubber of high tear strength. The **ethylene butadiene resin** of Phillips Petroleum Co. can be vulcanized with sulfur to give high hardness and wide temperature range. For greater elongation a terpolymer with butene can be made.

Polyethylene of low molecular weight is used for extending and modifying waxes, and also in coating compounds especially to add toughness, gloss, and heat-sealing properties. **Epolene N-11,** of the Eastman Chemical Products, Inc., for blending with waxes, has a molecular weight of 1,500, a density of 0.925, and a softening point at 103°C. **Epolene N,** used in paste polishes, has a molecular weight of 2,500 to 3,000. **Epolene LVE,** used in paper and textile coatings, is a low-density polyethylene with a molecular weight of 1,500, but **Epolene HDE,** used for self-polishing floor waxes to add hardness to the film, has the same molecular weight but a high density, 0.956. Such materials are called **polyethylene wax,** but they are not chemical waxes. They can be made emulsifiable by oxidation, and they can be given additional properties by copolymerization with other plastics. **Elvax,** of Du Pont, is such a copolymer of ethylene with vinyl acetate. It is compatible with vegetable and paraffin waxes, and when added to these waxes it increases adhesiveness, gloss, toughness, and heat sealing. Wax polyethylene compounds for paper coatings may be sold under trade names. **Ladcote,** of the L. A. Dreyfus Co., is such a compound. **Chemetron 100,** of the Chemetron Corp., is a modified **ethylene wax.** It is an **ethylene stearamide,** and comes as powder or beads. It improves luster, pigment dispersion, and mar resistance in lacquers. The **polymethylene waxes** are microcrystalline and have sharper melting points than the ethylene waxes. They are more costly, but have high luster and durability. **Polybutylene plastics** are rubberlike polyolefins with superior resistance to creep and stress cracking. Films of this resin have high tear resistance, toughness, and flexibility, and are used widely for industrial refuse bags. Chemical and electrical properties are similar to those of polyethylene and polypropylene plastics. **Polymethyl pentene** is a moderately crystalline polyolefin plastic resin that is transparent even in thick sections. Almost optically clear, it has a light transmission value of 90%. Parts molded of this plastic are hard and shiny with good impact strength down to −20°F (−29°C). Specific gravity (0.83) is the lowest of any commercial solid plastic. A major use is for molded food containers for quick frozen foods that are later heated by the consumer.

POLYPHENYLENE OXIDE. A plastic that is notable for its high strength and broad temperature resistance. There are two major types: **phenylene oxide (PPO)** and **modified phenylene oxide (Noryl).** These materials have a deflection temperature ranging from 212 to 345°F (100 to 174°C) at 264 lb/in^2 (2 MPa). Their coefficients of linear thermal expansion are among the lowest for engineering thermoplatics. Room-temperature strength and modulus of elasticity are high and creep is low. In addition, they have good electrical resistivity. Their ability to withstand steam sterilization and their hydrolytic stability make them suitable for medical instruments, electric dishwashers, and food dispensers. They are also used in the electrical and electronic fields and for business-machine housings.

Tensile strength and modulus of phenylene oxides rank high among engineering thermoplastics. They are processed by injection-molding, extrusion, and thermoforming techniques. The foam grades, with their high rigidity, are suitable for large structural parts. Because of good dimensional stability at high temperatures and under moisture conditions, these plastics are readily plated without blistering.

POLYSTYRENE. Often referred to as **styrene resin,** it is used for molding, in lacquers, and for coatings, formed by the polymerization of monomeric styrene, which is a colorless liquid of the composition $C_6H_5CH:CH_2$, specific gravity 0.906, and boiling point 145°C. It is made from ethylene, and is ethylene with one of the hydrogen atoms replaced by a phenyl group. It is also called **phenyl ethylene** and **vinyl benzene.** As it can be made by heating **cinnamic acid,** $C_6H_5CH:CHCO_2H$, an acid found in natural balsams and resins, it is also called **cinnamene.** In the form of **vinyl toluene,** which consists of mixed isomers of methyl styrene, the material is reacted with drying oils to form alkyd resins for paints and coatings.

The polymerized resin is a transparent solid very light in weight with a specific gravity of 1.054 to 1.070. The tensile strength is 4,000 to 10,000 lb/in^2 (27 to 68 MPa), compressive strength 12,000 to 17,000 lb/in^2 (82 to 117 MPa), and dielectric strength 450 to 600 volts per mil (18 to 24×10^6 volts per meter). Polystyrene is notable for water resistance and high dimensional stability. It is also tougher and stronger at low temperatures than most other plastics. It is valued as an electrical insulating material, and the films are used for cable wrapping.

When produced from **methyl stryene,** parts have a hardness to Rockwell M83, with tensile strengths to 8,900 lb/in^2 (61 MPa), and have a stiffness that makes them suitable for such products as cabinets and housings. Dielectric strength is also high, above 800 volts per mil (32×10^6 volts per meter), and the resin is thus used for electronic parts. The heat distortion temperature is 215°F (101°C).

Styrenes are subject to creep. Therefore the long-term bearing strength (over 2 weeks) is only about one-third the short-time tensile strength. Since their maximum useful service temperature is about 160°F (71°C), their use is restricted chiefly to room-temperature applications. Because of their low cost and ease of processing, polystyrenes are widely used for consumer products. The impact grades and glass-filled types are used quite widely for engineering parts and semistructural applications. Also, **polystyrene foams** are highest in volume use of all the foam plastics. Because of good processing characteristics, polystyrenes are produced in a wide range of forms. They can be extruded, injection-, compression-, and blow-molded, and thermoformed. They are also available as film sheet and foam.

Polystyrenes can be divided into the following major types: general-purpose grades, the lowest in cost, are characterized by clarity, colorability, and rigidity. They are applicable where appearance and rigidity, but not toughness, are required. Common uses are wall tiles, compact cases, knobs, brush backs, and container lids. Impact grades of polystyrenes are produced by physically blending styrene and rubber. Grades are generally specified as medium, high, and extra-high. As impact strength increases, rigidity decreases. Medium-impact grades are used where a combination of moderate toughness and translucency is desired, for example, in such products as containers, closures, and small radio cabinets. High-impact polystyrenes have improved heat resistance and surface gloss. They are used for refrigerator door liners and crisper trays, containers, toys, and heater ducts in automobiles. The extra-high-impact grades are quite low in stiffness, and their use is limited to parts subject to high-speed loading.

Styrene can be polymerized with butadiene, acrylonitrile, and other resins. The terpolymer, **acrylonitrile-butadiene-styrene** (abbreviated **ABS),** is one of the common combinations. **Styrene-acrylonitrile (SAN)** has excellent resistance to acids, bases, salts, and some solvents. It also is among the stiffest of the thermoplastics, with a tensile modulus of 400,000 to 550,000 lb/in^2 (2,757 to 3,791 MPa). Styrene resins for molding are now marketed under a wide variety of trade names, with or without fillers and reinforcing agents. Many of these are copylymer resins, or are modified with plasticizers or cross-linking agents. **Victron,** of the U.S. Rubber Co., is a clear transparent polystyrene. **Lustron,** of Monsanto, is polystyrene in various grades, and **Stymer** is a polystyrene resin for sizing textiles. **Piccotex,** of the Pennsylvania Industrial Chemical Corp., is a styrene copolymer in solid form soluble in mineral spirits for use in paints, coatings, and adhesives. **Styron** and **Styraloy,** of the Dow Chemical Co., are polystyrene molding resins, and **Tyril,** of this company, originally called **Styrex,** is a styrene-acrylonitrile copolymer.

Loalin, of the Catalin Corp., is a polystyrene with a specific gravity 1.05 to 1.07. It is crystal clear, and will take light pastel colors. In the clear form it transmits 90% light. It is water-resistant and has a dielectric strength of 500 to 700 volts per mil (1,970 to 2,758 volts per meter). It is not affected by alcohol, acids, or alkalies, but is soluble in aromatic hydrocarbons. It is preferably injection-molded. **Exon 860,** of the Firestone Plastics Co., is a soft grade of polystyrene that molds easily into products of high flexibility. The molded material has a tensile strength of 6,000 lb/in^2 (41 MPa) with elongation of 50%, Rockwell hardness of R100, and dielectric strength of 510 volts per mil (20×10^6 volts per meter). **Fibertuff,** of the Koppers Co., marketed in pellets for injection molding, is 60% polystyrene and 40% glass fiber. Molded parts have a specific gravity of 1.33, a tensile strength of 11,000 lb/in^2 (75 MPa), heat distortion point of 220°F (104°C), and high impact resistance.

Styrene-butylene resins are copolymers that mold easily and produce thermoplastic products of low water absorption and good electrical properties. They have strength equal to the vinyls with greater elongation. **Foamed polystyrene** is available in blocks and heavy sheets for thermal insulation. It weighs about 1 lb/ft^3 (16 kg/m^3) and is rigid. Flexible **styrene foam** is also made into very thin sheets for wrapping frozen foods. It is grease-resistant and a good insulator, and is low in cost. Styrene is now best known for its use in synthetic rubbers, but the difference between resins and rubbers is chiefly in flexibility.

PONTIANAK. A gum from the trees *Dyera costulata* and *D. laxifolia* of Borneo and Malaya. The commercial pontianak is a grayish-white mass like burned lime, and contains 60% water, with only 10 to 25% rubberlike materials. It is a rubber, but has a high content of resin similar to balata and gutta percha, and is classed with the lower guttas. It is used in the friction compounds employed for coating transmission belting, in insulations and varnishes, for mixing with gutta percha, and also to adulterate or replace chicle for chewing gums. It is also called **jelutong. Pontianak copal** is from varieties of *Agathis* trees of Borneo. Its peculiar turpentine-like qualities come from the method of tapping. It is valued for varnishes.

POPLAR. The wood of several species of the tree *Populus*. The **black poplar,** or **English poplar,** *P. Nigra,* of Europe, is a large tree with blackish bark. The wood is yellowish white with a fine, open grain. It is soft and easy to work. The weight is about 25 lb/ft^3 (400 kg/m^3). It is used for paneling, inlaying, packing cases, carpentry, and paper pulp. **Lombardy poplar** is a hybrid variety of this species. It is a tall, columnar tree that is male only and can be propagated only from rootstocks. It is grown in the United States for shelter belts, but in some countries is grown in fruit districts as a wood for packing boxes. **White poplar,** *P. alba,* is a larger tree native to the United States. The wood is similar to that of the black poplar. **Cottonwood** is another species of poplar. **Gray poplar** is from the tree *P. canescens,* of Europe. The color of the wood is light yellow. It has a tough, close texture somewhat resembling that of maple. It is used for carpentry and flooring. The wood of the canary whitewood is called **Virginia poplar,** or simply poplar, but belongs to a different family of trees. Aspen is also called poplar.

PORCELAIN. Porcelains and **stoneware** are highly vitrified ceramics that are widely used in chemical and electrical products. **Electrical porcelains,** which are basically classical clay-type ceramics, are conventionally divided into low- and high-voltage types. The high-voltage grades are suitable for voltages of 500 and higher, and are capable of withstanding extremes of climatic conditions. **Chemical porcelains** and stoneware are produced from blends of clay, quartz, fledspar, kaolin, and certain other materials. Porcelain is more vitrified than stoneware and is white in color. A hard glaze is generally applied. Stonewares can be classified into two types: a dense, vitrified body for use with corrosive liquids, and a less dense body for use in contact with corrosive fumes. **Chemical stoneware** may range from 30 to 70% clay, 5 to 25 fledspar, and 30 to 60 silica. The vitrified and glazed product will have a tensile strength up to 2,500 lb/in^2 (16 MPa) and a compressive strength up to 80,000 lb/ in^2 (551 MPa). **Industrial stoneware** is made from specially selected or blended clays to give desired properties.

Both chemical porcelains and stoneware resist all acids except hydrofluoric. Strong, hot, caustic alkalies mildly attack the surface. These ceramics generally show low thermal-shock resistance and tensile strength. Their universal chemical resistance explains their wide use in the chemical and processing industries for tanks, reactor chambers, condensers, pipes, cooling coils, fittings, pumps, ducts, blenders, filters, and so on.

Ceratherm, of the U.S. Stoneware Co., is an acid-resistant and heat-shock-resistant ceramic having a base of high-alumina clay. It is strong and nonporous, and is used for pump and chemical-equipment linings. The chemical stoneware of the General Ceramics Corp., made with a mullite-zircon body, is white, dense, strong, and thermal-shock-resistant. The ceramic marketed under the name of **Prestite** by the Westinghouse Electric Corp. for insulators and molded electrical parts is blended of flint, feldspar, kaolin, and ball clay. It is nonporous and moistureproof without a glaze, has high dielectric strength, a tensile strength of 5,000 lb/in^2 (34 MPa), and a compressive strength of 48,000 lb/in^2 (330 MPa). The ceramic produced by the Rostone Corp. under the name of **Rosite,** for molded electrical parts and panels, is a calcium-aluminum-silicate mixed with asbestos. It will withstand temperatures to 900°F (482°C), is resistant to alkalies, and has a compressive strength up to 15,000 lb/in^2 (103 MPa).

POROUS METALS. Metals with uniformly distributed controlled pore sizes, in the form of sheets, tubes, and shapes, used for filtering liquids and gases at elevated temperatures. They are made by powder metallurgy, and the pore size and density are controlled by the particle size and the pressure used. Stainless steel, nickel, bronze, silver, or other metal powders are used, depending on the corrosion requirements of the filter. Pore sizes offered by Purolator Products, Inc., can be as small as 0.2 μm, but the most generally used filters have pores of 4, 8, 12, and 25 μm. Pore sizes have a uniformity within 10%. The density range is from 40 to 50% of the theoretical density of the metal. Standard **filter sheet** is 0.30 to 0.60 in (0.76 to 1.52 cm), but thinner sheets are available. Union Carbide produces sheet as thin as 0.004 in (0.010 cm), and with void fractions as high as 90%, for fuel cells and catalytic reactors. **Porous steel,** of the Micro Metallic Corp., is made from 18–8 stainless steel, with pore openings from 20 μm to 65 μm. The fine-pore sheet has a minimum tensile strength of 10,000 lb/in^2 (68 MPa), and the coarse has a strength of 7,000 lb/in^2 (48 MPa). **Felted metal,** developed by the Armour Research Foundation, is porous sheet made by felting metal fibers, pressing, and sintering. It gives a high strength-to-porosity ratio, and the porosity can be controlled over a wide range. In this type of porous metal the pores may be from 0.001 to 0.015 in (0.003 to 0.038 cm) in diameter, and of any metal to suit the filtering conditions. A felted fiber filter of Type 430 stainless steel with 25% porosity has a tensile strength of 25,000 lb/in^2 (172 MPa).

POTASH. Also called **pearl ash.** A white alkaline granular powder, which is a **potassium carbonate,** K_2CO_3 or K_2CO_3:H_2O, used in soft soaps, for wool washing, and in glass manufacture. It is produced from natural deposits in Russia and Germany and also produced from wood and plant ashes. The American production is largely from potash salts of New Mexico, from the brines of Searles Lake, Calif., and from solar evaporation in Utah. The material as produced by the Hooker Chemical Co. is a free-flowing white powder of 91 to 94% K_2CO_3, or is the hydrate at 84%, or calcined at 99% purity. The specific gravity of potash is 2.33 and melting point 909°C. **Hartsalz,** mined in the Carpathian Mountains and used for producing potash, is a mixture of sodium chloride, potassium chloride, and magnesium sulfate. It is also a source of magnesium. The extensive potassium mineral deposits at Strassfurt and Mülhausen contain **sylvite,** KCl; **carnallite,** $KCl \cdot MgCl_2 \cdot 6H_2O$; **kainite,** $K_2SO_4 \cdot MgSO_4 \cdot MgCl_2 \cdot 6H_2O$; and **leonite,** $K_2SO_4 \cdot MgSO_4 \cdot 4H_2O$. There are at least a billion tons of the potash mineral **Wyomingite** in the deposits near Green River, Wyo. It is a complex mineral containing leucite, phlogopite, diopside, kataphorite, and apatite. It has 11.4% K_2O, with sodium oxide, magnesium oxide, phosphorus pentoxide, and other oxides. The sylvite ore mined at Carlsbad, N.M., contains KCl and NaCl. It is electrically refined to 99.95% KCl, and used to produce caustic potash. Electrolysis of the chloride solution yields caustic potash.

POTASSIUM. An elementary metal, symbol K, and atomic weight 39.1, also known as **kalium.** It is silvery white in color, but oxidizes rapidly in the air and must be kept submerged in ether or kerosene. It has a low melting point, 63.7°C, and a boiling point at 760°C. The specific gravity is 0.819. It is soluble in alcohol and in acids. It decomposes water with great violence. Potassium is obtained by the electrolysis of potassium chloride. Potassium metal is used in combination with sodium as a heat-exchange fluid in atomic reactors and high-temperature processing equipment. A

potassium-sodium alloy of the Mine Safety Appliance Co. contains 78% potassium and 22 sodium. It has a melting point of −11°C and a boiling point of 784°C, and is a silvery mobile liquid. The thermal conductivity at 200°C is 0.06. The **cesium-potassium-sodium alloys** of this company are called **BZ Alloys. Potassium hydride** is used for the photosensitive deposit on the cathode of some photoelectric cells. It is extremely sensitive and will emit electrons under a flash so weak and so rapid as to be imperceptible to the eye. **Potassium diphosphate,** KH_2PO_4, a colorless, crystalline, or white powder soluble in water, is used as a lubricant for wool fibers to replace olive oil in spinning wool. It has the advantages that it does not become rancid like oil and can be removed without scouring. Potassium, like sodium, has a broad range of use in its compounds, giving strong bonds. Metallurgically it is listed as having a body-centered cubical molecular structure, but the atoms arrange themselves in pairs in the metal as K_2, and the structure is cryptocrystalline.

POTASSIUM CHLORATE. Also known as **chlorate of potash** and **potassium oxymuriate.** A white crystalline powder, or lustrous crystalline substance, of the composition $KClO_3$, employed in explosives, chiefly as a source of oxygen. It is also used as an oxidizing agent in the chemical industry, as a cardiac stimulant in medicine, and in toothpaste. It melts at 357°C and decomposes at 400°C with the rapid evolution of oxygen. It is odorless but has a slightly bitter saline taste. The specific gravity is 2.337. It is not hygroscopic, but is soluble in water. It imparts a violet color to the flame in pryotechnic compositions.

Potassium chloride is a colorless or white crystalline compound of the composition KCl, used for molten salt baths for the heat treatment of steels. It is also used in fertilizers and in explosives. The specific gravity is 1.987. A bath composed of three parts potassium chloride and two barium chloride is used for hardening carbon-steel drills and other tools. Steel tools heated in this bath and quenched in a 3% sulfuric acid solution have a very bright surface. A common bath is made up of potassium chloride and common salt and can be used for temperatures up to 900°C.

POTASSIUM CYANIDE. A white amorphous or crystalline solid of the composition KCN, employed for carbonizing steel for cases hardening and for electroplating. The specific gravity is 1.52, and it melts at about 1550°F (843°C). It is soluble in water and is extremely poisonous, giving off the deadly hydrocyanic acid gas. For cyaniding steel the latter is immersed in a bath of molten cyanide and then quenched in water, or the cyanide is rubbed on the red-hot steel. For this use, however, sodium cyanide is usually preferred, because of its lower cost and the higher content of CN in the latter. Commercial potassium cyanide is likely to contain a proportion of sodium cyanide. **Potassium ferrocyanide,** or **yellow prussiate of potash,** can also be used for case-hardening steel. It has the composition $K_4Fe(CN)_6$ and comes in yellow crystals or powder. The nitrogen as well as the carbon enters the steel to form the hard case. **Potassium ferricyanide,** or **red prussiate of potash,** is a bright-red granular powder of the composition $K_3Fe(CN)_6$, used in photographic reducing solutions, in etching solutions, in blueprint paper, and in silvering mirrors. **Redsol crystals,** of the American Cyanamid Co., is the name of this chemical for use as a reducer and mild oxidizing agent, or toner, for photography. **Potassium cyanate,** KCNO, is a white crystalline solid used for the production of organic chemicals and drugs. It melts at 310°C. The **potassium silver cyanide** used for silver plating comes in white, water-soluble crystals of the composition $KAg(CN)_2$. **Sel-Rex,** of the Bart-Messing Corp., is this material.

POTASSIUM NITRATE. Also called **niter** and **saltpeter,** although these usually refer to the native mineral. A substance of the composition KNO_3, it is used in explosives, for bluing steel, and in fertilizers. A mixture of potassium nitrate and sodium nitrate is used for steel-tempering baths. The mixture melts at 250°C. Potassium nitrate is made by the action of potassium chloride on sodium nitrate, or **Chile saltpeter.** It occurs in colorless prismatic crystals, or as a crystalline white powder. It has a sharp saline taste and is soluble in water. The specific gravity is 2.1 and the melting point is 337°C. It is found in nature in limited quantities in the alkali region of the western United States. Potassium nitrate contains a large percentage of oxygen which is readily given up and is well adapted for pyrotechnic compounds. It gives a beautiful violet flame in burning. It is used in flares and in signal rockets. **Potassium nitrite** is a solid of the composition KNO_2 used as a rust inhibitor, for the regeneration of heat-transfer salts, and for the manufacture of dyes.

POTATO. The bulbous tubers of the roots of the annual plant *Solanum tuberosum,* native to Peru but now grown in many parts of the world. It is used chiefly as a direct food, but is also employed for making starch and alcohol. The potato was brought to Europe in 1580, and received the name of **Irish potato** when brought to New England in 1719 by Irish immigrants. The plant is hardy and has a short growing cycle, making it adaptable to many climates. The tuber contains about 78% water, 18 starch with some sugar, 2 proteins, 1 potash, and only about 0.1 fats. The average water loss in storage is about 11%. There are more than 500 varieties of potato cultivated. **Dehydrated potato,** produced as powder, flake, and porous granules, is widely used in restaurants and institutions and is marketed under various trade names for home use.

The **sweet potato** is the root bulb of the trailing perennial vine *Ipomoea batata,* native to tropical America, but now grown extensively in the southern United States and in warm climates. In South America it is known by the Carib name **batata.** Like the white potato, the sweet potato has a high water content, but is rich in sugars. There are many varieties and two general types: one with a dry mealy yellow flesh and the other with a soft gelatinous flesh higher in sugars. The latter type is called **yam** in the United States, but the true yam is a larger tuber from the climbing plant *Diascorea alata,* grown widely in the West Indies. The sweet potato is used as a direct food, but large amounts are also employed for making preserves, starch, and flour for confectionery. **Alamalt** is cooked and toasted sweet potato ground to a powder for use in confectionery. It adds flavor as well as sugar to the confectionery. **Sweet-potato flake,** used in foodstuffs, is cooked and dehydrated sweet potato in orange-colored flakes with the flavor of candied sweet potato. It is reconstituted with milk or water.

The **taro** is the root tuber of the large leafy plant *Colocasia esculenta,* which constitutes one of the chief foods of southeast Asia and Polynesia. There are more than 300 varieties grown. The tuber is high in starch, has more proteins than the potato, but has an acrid taste until cooked. The pasty starch food known as **poi** in the Pacific Islands is made from taro. In Micronesia the taro is called **jaua,** and the **mwang plant,** *Cyrtosperma chamissonis,* is called taro. This plant is larger, and the rootstock weighs as much as 50 lb (23 kg), while the taro does not exceed 5 lb (2 kg). Taro matures in 6 months, while the mwang requires 2 years. The **dasheen** is a variety of taro grown in the southern United States. The **yautia,** grown in the West Indies, resembles the taro, but is from the large plant *Xanthosma sagittifolium.* It is high in starch and has more food value than the potato.

PRECIOUS METALS. A general term for the expensive metals that are used for coinage, jewelry, and ornaments. The name is limited to gold, silver, and platinum. Expense or rarity alone is not the determining factor; rather, a value is set by law, with the coin having an intrinsic metal value as distinct from a copper coin, which is merely a token with little metal value. The term noble metal is not synonymous, although a metal may be both precious and noble, as platinum. Although platinum was once used in Russia for coinage, only gold and silver fulfill the three requisites for coinage metals. Platinum does not have the necessary wide distribution of source. The **noble metals** are gold, platinum, iridium, rhodium, osmium, and ruthenium. Unalloyed, they are highly resistant to acids and corrosion. Radium and certain other metals are more expensive than platinum but are not classed as precious metals. Because of the expense of the platinum noble metals, they may be alloyed with gold for use in chemical crucibles. **Platino** is an alloy of 89% gold and 11 platinum. **Palau** is the name of an alloy of gold and palladium, and **rhotanium** is a rhodium-gold alloy.

PREFINISHED METALS. Sheet metals that are precoated or treated at the mill so as to eliminate or minimize final finishing by the user. The metals are made in a ready-to-use form with a decorative and/or functional finish already applied. A large number and variety of prefinished metals have been developed. **Prepainted metals** are produced using almost every organic coating on most common ferrous and nonferrous metals. Extra durability or special decorative effects are provided by **plastic-metal laminates.** Polyvinyl chloride, polyvinyl fluoride, and polyester are the plastics commonly used.

Black-coated steel is used to give a high thermal emittance in electronic equipment. The base metal is aluminum-deoxidized steel containing 0.13% carbon. 0.45 manganese, 0.04 max phosphorus, and 0.05 max sulfur. The steel is coated with a 5% by weight layer of nickel oxide, which is reduced in a hydrogen furnace to form a spongy layer of nickel. This sponge is impregnated with a carbon slurry to form a black carbonized surface.

Preplated metals have a polished surface of another metal placed on one side, or have the finish metal bonded and rolled on the base metal. In general, they do not differ from clad metals except that the prime purpose is to obtain a finished stamped or drawn article directly without the operations of polishing and plating. Thus, the base metal is usually a softer and more workable metal than the cladding, and the applied plate must have a ductility that will permit drawing and bending without fracture. One of the earliest groups of metals of this class included the **Brassoid, Nickeloid,** and **Chromaloid** of the American Nickeloid Co., which were brass, nickel, and chromium bonded to zinc sheet. Prefinished metals are now available with almost any metal plated or bonded to almost any other metal, or single metals may be had prefinished or in colors and patterns. They come in bright or matte finishes, and usually have a thin paper coating on the polished side which is easily stripped off before or after forming. The metals are sold under a variety of trade names and are used for decorative articles, appliances, advertising displays, panels, and mechanical parts.

PRESERVATIVES. Chemicals used to prevent oxidation, fermentation, or other deterioration of foodstuffs. The antioxidants, inhibitors, and stabilizers used to retard deterioration of industrial chemicals are not usually called preservatives. The most usual function of a preservative is to kill bacteria, and this may be accomplished by an acid, an alcohol, an aldehyde, or a salt. A legal requirement under the Food and Drug Act is that a preservative must be nontoxic in the quantities permitted. **Sugar** is the most commonly used preservative for fruit products. **Sodium chloride** is used for protein foods. **Sodium nitrate** is reduced to sodium nitrite in curing meats, and the nitrite has an inhibitory action on bacterial growth, the effect being greatest in acid flesh. **Potassium sorbate,** $KOCOCH{:}CH{:}CHCH_3$, a white water-soluble powder, inhibits the growth of many molds, yeasts, and bacteria which cause food deterioration, and is used in cheese, syrups, pickles, and other prepared foods.

The inhibitory effect of organic acids is due chiefly to the undissociated molecule. **Acetic acid** is normally more toxic to bacteria than lactic acid, but when sugar is present the reverse is true, and citric acid then has little toxicity. The inhibitory action of inorganic acids is due mainly to the pH change which they produce. The antimicrobe effect of the **vanillic acid** esters generally increases with increasing molecular weight. Only small quantities of chemicals are usually needed for preservation. **Isobutyl vanillate,** an ester of vanillic acid, is effective as a preserving agent in milk and some other foods when only 0.10 to 0.15% is used. Preservatives are also maketed for external application to foodstuffs in storage, though these are more properly classed as fumigants.

PRIMER. A surfacing material employed in painting or finishing to provide an anchorage or adhesion of the finishing material. A primer may be colorless, or it may be with color. In the latter case it is sometimes called an **undercoat.** A primer is distinct from the filler coat used on woods to fill the pores and thus economize on the more expensive finish. Primers for industrial or production finishing are of two types: air dry and baking. The air-dry types have drying-oil vehicle bases, and are usually called **paints.** They may be modified with resins. They are not used as extensively as the baking-type primers, which have resin or varnish vehicle bases. These dry chiefly by polymerization. Some primers, known as **flash primers,** are applied by spraying, and dry by solvent evaporation within 10 min. In practically all primers, the pigments impart most of the anticorrosion properties to the primer, and, along with the vehicle, determine its compatability and adherence with the base metal.

Primer coats of **red lead paint** were formerly much used on construction steel to give corrosion resistance, but chromate or phosphate primers are now more common. **Barium potassium chromate** gives a pale-yellow coating with good anticorrosion properties for steel, aluminum, and magnesium. **Zinc yellow paints** may also be used as primer coats on metal. **Zinc chromate** is used as a primer on steel. It has a tendency to dissolve when moisture penetrates the paint, and this dissolved chromate retards corrosion of the steel. **Zinc phosphate primers** applied to iron and steel give corrosion resistance and improve paint adhesion. **Manganese phosphate** forms a dense crystalline coating on steel, which acts as a corrosion-resistant base for paint. A mixture of 95% zinc powder and 5% epoxy resin binder in a solvent gives a gray metallic finish, and the zinc blocks corrosion by galvanic action. In addition to the pigment, various corrosion inhibitors may be used in primer paints. **Ammonium ferrous phosphate,** $NH_4FePO_4 \cdot H_2O$, has a platelike structure which gives impermeability to the film as well as adding corrosion resistance. It is greenish in color. A primer is especially required in the finishing of sheet-metal objects that are likely to receive dents or severe service, but they are not usually necessary for castings or roughened surfaces. For sheet-metal work, baked enamels were formerly much used for the primers for lacquer finishes, but synthetic resin primers give good adhesion and are less expensive.

PROPYLENE PLASTICS. An important group of synthetic plastics employed for molding resins, film, and texture fibers. Developed in 1957 in Italy and Germany, they are produced as **polypropylene** by catalytic polymerization of propylene, or may be copolymers with ethylene or other material. **Propylene** is a **methyl ethylene,** $CH_3CH:CH_2$, produced in the cracking of petroleum, and also used for making isopropyl alcohol and other chemicals. The boiling point is −48.2°C. It belongs to the class of unsaturated hydrocarbons known as **olefins,** which are designated by the word ending *-ene.* Thus propylene is known as **propene** as distinct from propane, the corresponding saturated compound of the group of **alkanes** from petroleum and natural gas. These unsaturated hydrocarbons tend to polymerize and form gums, and are thus not used in fuels although they have antiknock properties.

In polypropylene plastics the carbon atoms linked in the molecular chain between the CH_2 units have each a CH_3 and an H attached as side links, with the bulky side groups spiraled regularly around the closely packed chain. The resulting plastic has a crystalline structure with increased hardness and toughness and a higher melting point. This type of stereosymmetric plastic has been called **isotactic plastic.** It can also be produced with butylene or styrene, and the general term for the plastics is **polyolefins.** Copolymers of propylene are termed **polyallomers.**

Polypropylene is low in weight. The molded plastic has a density of 0.910, a tensile strength of 5,000 lb/in^2 (34 MPa), with elongation of 150% and hardness of Rockwell R95. The dielectric strength is 1,500 volts per mil (59×10^6 volts per meter), dielectric constant 2.3, and softening point 150°C. Blown bottles of polypropylene have good clarity and are nontoxic. The melt flow is superior to that of ethylene. A unique property is their ability in thin sections to withstand prolonged flexing. This characteristic has made polypropylenes popular for "living hinge" applications. In tests, they have been flexed over 70 million times without failure.

The many different grades of polypropylenes fall into three basic groups: homopolymers, copolymers, and reinforced and polymer blends. Properties of the homopolymers vary with molecular-weight distribution and the degree of crystallinity. Commonly, copolymers are produced by adding other types of olefin monomers to the propylene monomers to improve properties such as low-temperature toughness. Copolymers are also made by radiation grafting. Polypropylenes are frequently reinforced with glass or asbestos fibers to improve mechanical properties and increase resistance to deformation at elevated temperatures.

Tenite polypropylene, of Eastman Chemical Products, Inc., is used for molded parts, film, fibers, pipe, and wire covering. The **polypropylene film** of the Avisun Corp., called **Olefane,** used for packaging, has a specific gravity of 0.89. It is resistant to moisture, oils, and solvents, is crystal clear, and is flexible. It withstands temperatures to 250°F (121°C). The 0.001-in (0.003-cm) film has 31,000 ft/lb (20,830 m/kg). **Dynafilm 200,** of U.S. Industrial Chemicals Co., is polypropylene laminated with polyethylene to give easy heat sealing for packaging. The 0.001-in (0.003-cm) film has a strength of 3,000 lb/in^2 (20 MPa) and 400% elongation. **Dynafilm 300** has the appearance, feel, and machine-handling properties of a Cellophane film. It consists of oriented polypropylene coated on both sides with vinyl acetate. The film comes in thicknesses from 0.0008 to 0.0012 in (0.0020 to 0.0030 cm) and heat-seals at 100°F (38°C).

Polypropylene fiber was originally produced in Italy under the name of **Merkalon.** Unless modified, it is more brittle at low temperatures and has less light stability than polyethylene, but it has about twice the strength of high-density linear polyethylene. Monofilament fibers are used for filter fabrics, and have high abrasion resistance and a melting point at 310°F (154°C). Multifilament yarns are used for textiles and rope. **Polypropylene rope** is used for marine hawsers, will float on water, and does not absorb water like Manila rope. It has a permanent elongation, or set, of 20%, compared with 19% for nylon and 11% for Manila rope, but the working elasticity is 16%, compared with 25% for nylon and 8% for Manila. The tensile strength of the rope is 59,000 lb/ in^2 (406 MPa). Fine-denier multifilament **polypropylene yarn** for weaving and knitting dyes easily and comes in many colors. **Chlorinated polypropylene** is used in coatings, paper sizing, and adhesives. It has good heat and light stability, high abrasion resistance, and high chemical resistance.

PROTEIN. A nitrogen organic compound of high molecular weight, from 3,000 to many millions. Proteins are made up of complex combinations of simple amino acids, and they occur in all animal and vegetable matter, but are also made synthetically. They form a necessary constituent of foods and feeds, and are also used for many commercial products, but some proteins are highly poisonous. The poison of the cobra and that of the jellyfish are proteins.

Different types of plant and animal life have different types of proteins. At least 10 different proteins are known to be essential to human body growth and maintenance, but many others may have subsidiary functions since the amino acids are selective chelating agents, separating copper, iron, and other elements from the common sodium, calcium, and potassium compounds entering the system. The simple proteins are made up entirely of the acids, but the complex or conjugated proteins also contain carbohydrates and special groups, while the cystine of hair and wool also contains sulfur. The constituent amino acids of the protein molecule are linked together with a peptide bond, and the linkage forms the backbone of the molecule, but the arrangement is not similar to the high polymers usually associated with plastics with one type of polymer, or group, repeating itself. The linkage is formed by the loss of carbon dioxide rather than by the loss of water as in plastics.

The simplest proteins are the **protamines** with molecular weights down to about 3,000. They are strongly basic and water-soluble, and contain no sulfur. **Clupeine** in herring and **salmine** in salmon and trout are examples. The **histones** which occur in white blood corpuscles contain sulfur and are more complex. The albumins of eggs and milk are soluble in water and coagulate with heat. They also occur in plants, as in the **leucosin** of wheat. **Prolamines** are **vegetable proteins,** as the zein of corn and the **gliadin** of wheat. They are not an adequate human protein food without **animal proteins.**

Biologically, the edible proteins are classified as first-class and second-class, the first being from animal and the second from vegetable origin. Meat and fish proteins are both complete, or first-class proteins, but the digestibility of fish protein is slightly higher than that of beef protein, while oyster protein is high in growth-promoting value. The synthesis of globulin and antibody formation for resistance to disease depends upon the utilization of various amino acids most readily obtained from first-class proteins. The term **protein isolates** used in the food industry refers to proteins from soybean or other sources and containing 85% or more of protein.

Lysine, essential for human nutrition, is found in soy proteins, but it is also made synthetically as a water-soluble white powder, and is added to bakery products to raise the protein value to nearly that of animal protein. **Isoleucine,** a bitter amino acid, occurs in casein. It is an **aminomethyl valeric acid** which is fermented by yeast to amyl alcohol. **Sclero proteins,** or **albuminoids,** contain much sulfur, are insoluble in water, resist hydrolysis, and are the most complex of the proteins. They occur in skin, ligaments, horn, wool, and silk. The complex indigestible protein of poultry feathers, used for making brush fibers, is also broken down to produce digestible proteins used in feeds. **Glycine,** or **glycocoll,** is an **aminoacetic acid,** H_2NCH_2COOH, formed by the hydrolysis of complex proteins and also made synthetically. The **methylated glycine,** called **betaine,** occurs in plants and is obtained from sugar-beet molasses.

Hydrolized proteins are used in flavoring foodstuffs. The Japanese condiment **adjinimoto,** made from wheat gluten, is largely **sodium glutamate,** a salt of **glutamic acid,** $C_5H_9O_4$, which also occurs in seeds and beets. **Monosodium glutamate** is sold under trade names such as **Zest** of A. E. Staley Mfg. Co. and **Ac'cent** of Ac'cent International. It is a white crystalline powder derived from soybeans and sugar beets. In small amounts it has no flavor, but intensifies the taste and flavor of foodstuffs. This effect may be greatly increased by adding a chemical compound such as disodium inosinate or disodium guanylate to replace part of the glutamate. **Mertaste,** of Merck and Co., is such a compound which enhances the flavor and acts as a synergist. In large amounts the glutamates have an offensive odor and flavor and are considered toxic. **Insulin,** used in medicine, contains **glutamine,** the half amide of glutamic acid, and also **cystine,** the disulfide of **cysteic acid,** an aminopropionic acid essential for nutrition. **Cystein** is aminomercaptopropionic acid. It is used with whey in bakery dough to react with yeast enzymes and give faster rising at lower temperatures. **Royal jelly,** used in face creams, is a protein complex high in vitamin B. It is a secretion of bees to nourish the egg of the queen bee, but has no apparent therapeutic value.

Wheat gluten, made from flour as a spray-dried powder, contains about 82% protein. It is used as an additive to improve the texture and the shelf life of baked goods. **Soybean protein** is marketed as a highly refined, odorless, and tasteless powder for use in confectionery and other foods to retain freshness and add food value. **Proset-Flake,** of the J. A. Jenks Co., is a soybean protein in flake form to give firmer texture. **Mushroom powder** is used to add proteins to foodstuffs where the flavor is also desirable, as in sauces. It is made by sheeting the pulped mushroom, *Agaricus campestris,* on a double-drum drier and then grinding the sheet to a fine powder. **Animal protein factor,** used in animal feeds, and marketed commercially in fish solubles, is an amino-acid combination containing several vitamins. It is also used in feeds for single-stomach animals, such as the hog, which cannot synthesize within themselves all of the amino acids necessary for health. It is also used in poultry feeds, as it contains the hatchability factor not adequately supplied by grains. **DL-methionine,** of the Dow Chemical Co., is a synthetic amino acid of the composition $CH_3SCH_2CH_2CH(NH_2)CO_2H$, used in feeds for fattening poultry. **MPF granules** of General Mills, Inc., are granules of concentrated proteins, vitamins, and minerals for adding to bakery and meat products. The whey from milk in powder form is added to bread to increase the protein content, and it also refines the texture of bakery products. In porous granule form it is used as a **flow agent** to add butter, margarine, or oils to bakery products as well as protein. **Kraflow,** of Kraft Foods, consists of **whey granules** which will absorb eight times their volume of oils.

Protein plastics are produced by the isolation of precipitation of proteins from animal or vegetable products and hardening or condensing into stable compounds that can be molded into sheet or fiber. The oldest of the protein plastics is **casein plastic** which was used to replace bone for toothbrush handles and buttons. It is still an important molding plastic as it is tough and can be made in pastel shades, but it is more costly than many other molding plastics. The proteins from soybean meal or other vegetable products are condensed with aldehydes or with various mineral salts or acids to form plastics. These plastics are distinct from those made from the fatty acids of soybean or other oils. **Peanut fiber,** under the name of **Ardil,** is made in England by precipitating the protein from peanut meal with an acid at low temperatures so as not to denature the meal. It is then dissolved in caustic soda and formed through spinnerets into a hardening bath. The fiber is soft and resilient, moisture-absorbent like wool, moth-resistant, and it will dye easily. It is mixed with wool in weaving fabrics. **Vicara,** of the Virginia-Carolina Chemical Corp., is a protein fiber produced from the zein of corn. The fiber is light yellow in color, soft, tough, and strong. In fabrics it has the warmth of wool, is resistant to mildew, and will withstand temperatures to 310°F (154°C). It can be blended with cotton, wool, or rayons. Proteins obtained by alkaline extraction from cottonseed are also used to produce woollike fibers. **Azlon** is a general name for protein fibers. **Chromated protein** finishes that provide a corrosion-resistant undercoat on iron or steel are produced by coating the metal with casein or albumin and then impregnating with a chromate solution which hardens the film.

PUMICE. A porous, frothlike volcanic glass which did not crystallize due to rapid cooling, and frothed with the sudden release of dissolved gases. Powdered or ground pumice is used as an abrasive for fine polishing, in metal polishes, in scouring compounds and soaps, and in plaster and lightweight concrete and pozzuolanic cement. In very fine powder it is called **pounce** when used for preparing parchment and tracing cloth. **Pouncing paper** is paper coated with pumice used for pouncing, or polishing, felt hats. Pumice is grayish white in color, and the fine powder will float on the surface of water. The natural lump pumice contains 65 to 75% silica, 12 to 15 alumina, and 4 to 5 each of soda and potash. It is produced chiefly in California and New Mexico. **Pumicite** is a **volcanic ash** similar in composition to pumice, found in large beds in Nebraska, Kansas, and Colorado. Its chief distinction is that it is fine-grained and has sharp edges suitable for abrasive purposes. The natural material is dried, pulverized, and screened so that 98.8% will pass a 325 mesh screen. **Seismotite** is a trade name of the Cudahy Co. for pumice used as an abrasive in scouring compounds. **Slag pumice,** or **artificial pumice,** is made in Germany by treating molten slag with less water than is required for granulation. It is used as an aggregate in lightweight concrete and as a heat insulator. **Obsidian** will change into pumice when melted, but obsidian is a general name for **volcanic glass** and varies in composition. It is an extrusive igneous rock that gets its glassy nature from its method of cooling, and some obsidian has a composition similar to that of granite. It is colored black from magnetite, or brown to red from hematite. Obsidian was used by the ancients for instruments and by the American Indians for arrowheads and knives. Semitransparent smoky-colored obsidian nodules of Arizona ore are called **marekanite,** and are cut for Indian silver jewelry.

Hawaiian obsidian, or **tachylite,** also known as **basalt glass,** is a volcanic glass from Oahu, Hawaii. It is jet black, takes a fine polish, and is used for making ornamental articles. The type of obsidian found in Oregon and California, known as **perlite,** is flash-roasted to form a bubblelike expanded material about 15 times the original size which is crushed to a white fluffy powder. Perlite contains about 75% silica. In California it has been called **calite.** The powder weighs only 4 to 12 lb/ft^3 (64 to 192 kg/m^3), and is used in lightweight wallboard, acoustical tile, insulation, and as a lightweight aggregate in concrete. **Perlalex,** of the Alexander Film Co., is the finely ground powder used for removing smears from drawing paper. **Rhyolite perlite** of California is expanded to ovaloid particles of 590- to 840-μm size containing complete vacuums and weighing only 0.78 lb/ft^3 (12.4 kg/m^3). This material has about 70% silica, 15 alumina, 2 Fe_2O_3, 1.5 CaO, 2.75 Na_2O, 1 MgO, 4 K_2O, and 4 water. **Grellex,** of the Great Lakes Carbon Co., is expanded California perlite in fine particle size for use in plastics, adhesives, and insulation.

Other materials besides perlite may be expanded by heat. **Expanded clay** is made from common brick clays by grinding and screening to 48 to 80 mesh and feeding through a gas burner at a temperature of about 2700°F (1482°C). The ferric oxide is changed to ferrous oxide, liberating oxygen and CO_2. Strong, light bubbles about 0.020 in (0.051 cm) in diameter are formed. **Kanamite,** of the Kanium Corp., is this material. It weighs 17 to 25 lb/ft^3 (272 to 401 kg/m^3), and is used as an aggregate for lightweight concrete.

PURPLEHEART. The wood of several species of trees, notably *Peltogyne paniculata* of the Guianas. The color of the wood is brown, the heartwood turning purple on exposure. The grain is open and fine. The wood weighs about 53 lb/ft^3 (849 kg/m^3), is very hard, strong, and durable. It is used for machine and implement parts, inlays, furniture, and turnery.

PUTTY. A mixture of calcium carbonate with linseed oil, with sometimes white lead added. It is used for cementing window glass in place and also as a filler for patterns. Litharge is often added to putty for steel sash.

Another putty for steel contains red lead, calcium carbonate, and linseed oil. The dry pigment for putty, **whiting putty,** according to ASTM specifications, contains 95% calcium carbonate and 5 tinting pigment. **White lead putty** contains 10% or more white lead mixed with the calcium carbonate. Federal specifications call for a minimum of 11% boiled or processed linseed oil with a maximum of 89% pigment as a white lead–whiting putty. **Putty powder** is a mixture of lead and tin oxides, or a mixture of tin oxide and oxalic acid, or it may be merely an impure form of tin oxide. It is used in enameling and for polishing stone and glass, and as a mild abrasive for dental polishes. **Calking putty,** used for setting window and door frames, is made of asbestos fibers, pigments, and drying oils, or with rubber or resins. The older calking compounds used in the building industry for sealing between window and door frames and masonry were made with drying oils and inert fillers, but they had poor adhesion and weathering qualities. **Calking compounds** are now composed of synthetics with usually a polysulfide rubber and a lead peroxide curing agent. They are heavy pastes of 75 to 95% solids. **Koplac 1251-5,** of the Koppers Co., Inc., for automobile body patching and general utility putties, is a low-viscosity polyester resin with 32% styrene. In putty formulations it is used with 50% talc and a catalyst. It hardens rapidly at room temperature.

PYRARGYRITE. An ore of silver, known also as dark **ruby silver.** It is a sulfantimonite of silver, Ag_3SbS_3, containing 22.3% antimony and 59.8 silver. It is found in various parts of Europe, and in Mexico, Colorado, Nevada, and New Mexico. It occurs in crystals or massive, and also in grains. Its hardness is 2.5 and specific gravity 5.85. The color is dark red to black, showing ruby red in thin splinters. **Proustite** is another ore of silver occurring in silver veins associated with other metals. It is found in the mines of Peru, Mexico, Chile, and in Nevada and Colorado. It is also called light ruby silver and is a sulfarsenite of silver of the composition Ag_3AsS_3, containing theoretically 65.4% silver. It occurs massive, compact, in disseminated grains. The hardness is 2 to 2.5, specific gravity 5.55, and the color is ruby red with an adamantine luster.

PYRETHRUM. The dried flowers of several species of chrysanthemum, of which the *Chrysanthemum cinerariaefolium* of Yugoslavia and Japan is the best known. It is a slender perennial, about 15 in (38 cm) high, with daisylike flowers. The powder is used as an insecticide chiefly in sprays. The crude pyrethrum from Kenya contains 1.3% **pyrethrin** as compared with only 0.9% in the Japanese. **Persian powder** is pyrethrum from the species *C. coccineum* of southwestern Asia. **Lethane,** of the Rohm & Haas Co., is a synthetic aliphatic diacyanate used as a substitute for pyrethrum. It is 30 to 40% more powerful than pyrethrum in insect sprays. Pyrethrin contains

pyrethronic acid and **cyclopentane,** $(CH_2)_5$. The active principles of natural pyrethrum flowers have been designated as pyrethrin and **cinerin,** and the synthetic material marketed by the U.S. Industrial Chemicals, Inc., is a homolog of cinerin. In high concentration, it is more effective than natural pyrethrum. A substitute for pyrethrum for the control of corn worms is **styrene bromide,** or **bromostyrene,** an oily liquid in the composition $C_6H_5CBr{:}CH_2$. **Ryanodine,** of the composition $C_{25}H_{35}NO_9$, is three times more toxic than pyrethrin. It is extracted from the stem wood of the **ryania,** a shrub of Trinidad.

PYROPHORIC ALLOYS. Metals which produce sparks when struck by steel, used chiefly for gas and cigarette lighters. The original pyrophoric alloy, or **sparking metal,** was known as **Auer metal.** It was patented by Auer von Welsbach in 1903, and contained 35% iron and 65 misch metal. The French **kunheim metal** contained 10% magnesium and 1 aluminum instead of iron. A very durable alloy for cigarette lighters is **zirconium-lead alloy** containing 50% of each metal. Titanium can replace part of the zirconium, and tin can replace part of the lead, but alloys with less than 25% zirconium are not pyrophoric. The 50–50 alloy has a crystalline structure. Some liquids are pyrophoric. **Trimethyl aluminum,** $Al(CH_3)_3$, a colorless liquid made by sodium reduction of methyl aluminum chloride, is used as a pyrophoric fuel.

PYROPHYLLITE. An aluminum silicate mineral found in North Carolina, used as a substitute for talc. It is similar to talc in structure and appearance, but its composition, $Al_2Si_4O_{10}(OH)_2$, is more nearly like kaolin. It is white, gray, or brown, with a pearly or greasy luster, specific gravity 2.8, and hardness 1 to 2. Compact varieties of the mineral are made into slate pencils and crayons. A fine-grained compact rock mined in South Africa, composed of about 90% pyrophyllite, with rutile and other minerals, is called **wonderstone,** and is used for tabletops and switchboard panels. It is resistant to weathering, acids, and heat, and it can be planed, sawed, or turned in the lathe. It then becomes harder on exposure. Wonderstone is an ancient indurated clay resembling fireclay in which the colloidal matter has been destroyed by heat, pressure, and age. Unfired refractory bricks are made of dry-pressed pyrophyllite. They have high spalling resistance and do not shrink. **Pyrax,** of the R. T. Vanderbilt Co., is pyrophyllite in fine white powder of 100 mesh, with specific gravity 2.6, used as a filler in rubber. **Silical,** of Herron Bros. & Meyer, is pyrophyllite from Newfoundland in fine powder form as a dusting talc for rubber. The Japanese employ great quantities of pyrophyllite in the making of firebrick, fireclay, and crucibles. The mineral used averages 86.7% pyrophyllite, 12.8 kaolin, and 0.5 diaspore.

QUARTZ. The most common variety of silica. It occurs mostly in grains or in masses of a white or gray color, but often colored by impurities. Pure crystalline quartz is colorless and is called **rock crystal.** Quartz usually crystallizes in hexagonal prisms or pyramids. Many crystals are obtained from nodules, called **geodes,** which are rounded hollow rocks with the crystals grown on the inside surface of the cavity. These rocks range in size from very small to hundreds of pounds. The crystals are not always quartz, but may be grown from minor constituents of the rocks. A geode in limestone usually has a shell of silica and the interior crystals are of quartz or calcite, but some geodes contain crystals of gem quality containing metal coloring constituents. The hard, rigid **beta quartz crystals** have a latticelike molecular structure in which each silicon atom is linked to four separate oxygen atoms, each oxygen atom being linked to two different silicon atoms. The formula of quartz crystal, therefore, is $(Si_2O_7)_x$, which is a pattern of the lattice, while the unit crystal of silica is SiO_2 and the silica grains are cryptocrystalline. The grains in sand are often less than 0.04 in (0.10 cm), but crystals up to 20 in (51 cm) have been found. The specific gravity is 2.65. Pure crystals have a dielectric strength of 1,500 volts per mil (59×10^6 volts per meter) and a dielectric constant of 3.8, with good corona resistance.

Quartz crystals have the property of generating an electric force when placed under pressure, and conversely, of changing dimensions when an electric field is applied. This property is termed piezoelectric. A **piezoelectric crystal** is made up of molecules that lack both centers and planes of symmetry. Many materials other than quartz have this property, such as rochelle salt and ammonium dihydrogen phosphate, but most of them are water-soluble or lack hardness. The best quartz crystals are hexagonal prisms with three large and three small cap faces. For electric use the crystals must have no bubbles, cracks, or flaws, and they should be free from twinning, or change in the atomic plane. **Brazilian quartz** crystals are cut into plates of different sizes to initiate and receive various frequencies on multiple-message telephone wires, and to obtain selectivity in radio apparatus. Owing to its peculiar refractive powers, quartz crystal is also employed for the plates in polarization instruments and in lenses. Quartz crystals for radio-frequency control are marketed in three forms: rough-sawed blanks, cut to specified angles; semifinished blanks, machine-lapped to approximate size; and electrically finished blanks, finished by hand and electrically tested. **Synthetic quartz crystals** of large size and high purity and uniformity are grown from seed crystals suspended in an alkaline silica solution at high temperature and pressure. The synthetic crystals are purified by imposing an electric current across the crystal at 500°C, which sweeps out the sodium and lithium impurities by electrolysis. The addition of lithium nitride to the sodium-hydroxide solution used in the hydrothermal growing process increases the Q value of the crystal to the range of natural quartz crystals. These synthetic crystals are used in precision oscillators and highly selective wave filters.

Barium titanate crystals are used to replace quartz for electronic use. **Ethylenediamine tartrate** crystals may be used to replace quartz for telephone and sonar work.

Quartz is harder than most minerals, being 7 Mohs, and the crushed material is much used for abrasive purposes. Finely ground quartz is also used as a filler, and powdered quartz is employed as a flux in melting metals. When quartz is fused, it loses its crystalline structure and becomes a **silica glass** with a specific gravity of 2.2, compressive strength 210,000 lb/in^2 (1,447 MPa), tensile strength 4,000 lb/in^2 (27 MPa), hardness 5, and dielectric strength 410 volts per mil (16×10^6 volts per meter). The chemical formula of this material is sometimes given as SiO_3, but is really SiO_2 repeated in a lattice structure but different from that of quartz crystal. **Fused quartz,** or **quartz glass,** is used for bulbs, optical glass, crucibles, and for tubes and rods in furnaces. Its softening and working temperature is about 3040°F (1671°C), and it fuses at 3193°F (1755°C). The translucent material, made from sand, has a specific gravity of 2.7, with much lower strength. It will withstand rapid changes of temperature without breaking. Fused quartz made from rock crystal is transparent to visible light, while fused silica is normally translucent or opaque. **Vitreosil,** of the Thermal Syndicate, Ltd., is fused quartz, containing 99.8% silica. It comes opaque, translucent, and transparent. It transmits ultraviolet and short wavelengths, has high electrical resistance, and a coefficient of expansion about one-seventeenth that of ordinary glass. **Quartz tubing** for electronic use comes in round, square, hexagon, and other shapes. The softening point is 1667°C. Tubing as small as 0.003 in (0.008 cm), produced by the Monsanto Co., is flexible and as strong as steel.

Quartz fiber originally was made by extruding the molten quartz through a stream of high-pressure hot air which produced a fluffy mass of fine fibers of random lengths. Quartz fibers are now made with many differing compositions and methods of manufacture. Fibers used for wool or mat have a diameter of 1 to 15 μm. Those used for continuous filament may be as small as 0.0035 in (0.009 cm). **Astroquartz,** of J. P. Stevens & Co., may be corded as a thread with outside diameters from 0.014 to 0.020 in (0.036 to 0.051 cm). The thread is used for sewing insulation blankets and separation curtains. **Quartz yarn** made from these filaments is used for weaving into tape and fabric. **Quartz paper,** or **ceramic paper,** developed by the Naval Ordnance Laboratory and used to replace mica for electrical insulation, is made from quartz fiber by mixing with bentonite and sheeting on a papermaking machine. It has high dielectric strength, and will withstand temperatures to 3000°F (1649°C). **Micro quartz,** of the L.

O. F. Glass Fibers Co., is felted fine quartz fibers for insulation. The felted material has a density of 3 lb/ft^3 (48 kg/m^3), and is for service temperatures to 2000°F (1093°c).

Since quartz crystallizes more slowly than many other minerals, the natural crystals may include other minerals which were crystallized previously. **Sagenite** is a form of crystalline quartz containing hairlike crystals crossing in a netlike manner. A variety of fibered quartz with a pale amethyst color which shows deep red by transmitted light, found in Russia and Colorado, is called **onegite. Rutilated quartz** is clear quartz penetrated by rutile crystals. A smoky, dark quartz of this type is the **Venushair stone. Aventurine** is a form of quartz crystal containing the inclusion in the form of flakes or spangles. It comes from the Ural Mountains and from India and is prized for gems. For costume jewelry it is made synthetically in great quantities under the name of **goldstone** by melting the inclusions into quartz glass. Amethyst, **topaz,** and many other gemstones are quartz. The golden-yellow topaz of Mexico and Brazil is a type of quartz called **citrine.** The yellow variety called **imperial topaz** in Brazil is rare, but yellow-brown stones are common. **Pink topaz** is also rare, but can be made by heating yellow-brown stones with a risk of breaking. Inferior-colored amethysts may also be made into yellow or orange-colored citrine by heating. The **rose quartz** of South Dakota is prized in the beautiful rose color, but in the large deposits the shades may run from milky white through pale pink to deep rose-red. The best stones are used for gems, as are also the translucent pink crystals from Maine. Other grades are cut into vases, ornaments, and architectural facings. **Chalcedony** is a cryptocrystalline quartz with a waxy luster deposited in rock veins from colloidal solution, or in concentric rings on rocks. Its fibers are biaxial instead of the uniaxial of quartz. The chalcedony of South Dakota known as **beckite** fluoresces under ultraviolet light. Chalcedony was an ancient gemstone, and was used for intaglios and seals and for figurines and vases. Some chalcedony from New Mexico and Arizona is stained and cut for costume jewelry. **Chrysoprase** is a translucent, apple-green variety of cryptocrystalline quartz colored with hydrated nickel silicate found in Silesia. It is highly valued for mural decorations and as a gemstone.

The so-called **massive topaz** used as a refractory material instead of kyanite is not true topaz or quartz. The massive topaz mined in North Carolina contains about 50% Al_2O_3 and 40 SiO_2, with iron oxide. When calcined for refractory use, it has the same composition as kyanite. The topaz from the wolframite mines of São Paulo, Brazil, used for refractories, has a high alumina content and a high fluorine content. The purer crystals have a melting point of 1880°C. The quartz known as **cristobalite,** used as a refractory, differs from ordinary quartz only in the crystal structure. It has a melting point of 3140°F (1725°C). **Jasper** is a variety of quartz colored

red with iron oxide. It is cut and polished as an ornamental building stone. **Egyptian jasper** is brown in color with dark zones. In ancient times many of the gemstones were silica stones, and the **Athiaenon stone** from Cyprus was jasper of bright colors. The **jasper iron ore** of Michigan has an iron content of about 33% with less silica than taconite, making it easier to crush, but concentration must be done by flotation, which is more expensive than the magnetic separation of taconite.

Quartzite is a rock composed of quartz grains cemented together by silica. It is firm and compact and breaks with uneven, splintery fractures. Most of the quartzites used are made up of angular grains of quartz and are white or light in color with a glistening appearance. It often resembles marble, but is harder and does not effervesce in acid. Quartzite is employed for making silica brick, abrasives, and siliceous linings for tube mills. It is also rather widely used as a structural stone and as a broken stone for roads. It is found as a widely distributed common rock. **Medina quartzite,** from Pennsylvania, contains 97.8% silica. The melting point is about 1700°C.

QUASSIA. Also known as **bitterwood.** The wood of the Jamaica quassia tree, *Picroena excelsa,* and of the **Surinam quassia,** *Quassia amara,* of the West Indies and northern South America. The **Jamaica quassia** is a large tree, sometimes called **bitter ash** because the leaves resemble those of the common ash. The wood is yellow, light, dense, and tough. It is odorless, but has an intensely bitter taste. The wood is imported mostly as chips for the production of the extract which is used in medicine as a bitter tonic, and also in insecticides. It is also used as an ingredient in stock-feed tonics for cattle. In tropical countries the wood is valued for furniture, as it is resistant to insects. The wood of the Surinam quassia is darker in color, heavier, and harder, but has similar properties. **Quassin,** extracted from quassia, is used to denature alcohol.

QUEBRACHO. The wood of the **quebracho colorado,** or **red quebracho** tree, *Aspidospera quebracho,* found only along the west bank of the Parana and Paraguay Rivers in Argentina and Paraguay. It contains about 24% tannin. The wood is exceedingly hard and has a brownish-red color often spotted and stained almost black. Quebracho is valued as a firewood in Argentina, and is also used for crossties and posts, but is too brittle for structural work. It takes a fine polish and is very durable, carvings of this wood being in perfect condition after 300 years. The weight is 78 lb/ft^3 (1,250 kg/m^3). **Quebracho extract,** from the wood, is a hard, resinous, brownish-black, and extremely bitter solid containing 62% soluble tannins. One metric ton of wood yields about 250 kg of solid extract. The liquid extract contains 25 to 35% tannin. It is employed in tanning leather and is rapid-acting, but is seldom used alone as it makes a dark leather. It is mixed with alum and salt, or with chestnut extract. Some extract is used in boiler compounds, but one of the larger uses has been for the treatment of oil-well-drilling muds. **Aerosol Q,** of the American Cyanamid Co., is powdered quebracho and an organic colloid for oil-well muds. **White quebracho,** *Schinopois lorentsii,* is a smaller tree than the red quebracho, growing over a wider area of Argentina, Brazil, and Paraguay. It produces a similar tannin. Some **urunday extract** is produced in Argentina for export instead of quebracho. The **urunday wood** is red in color and very hard, but not as brittle as quebracho, and is valued locally for cabinetwork. The tannin from the wood is similar to quebracho extract.

RADIOACTIVE METALS. Metallic elements which emit radiations that are capable of penetrating matter opaque to ordinary light. They give out light and appear luminous, also having an effect on the photographic plate. The metal radium is the most radioactive of all the natural elements, and was much used for luminous paints for the hands of watches and instrument pointers, but, because of the emission of dangerous gamma rays, is now largely replaced for this purpose by radioactive isotopes of other metals. These isotopes, such as **cobalt-60,** used as a source of gamma rays, and **crypton-85,** for beta rays, are marketed selectively. Radioactive metals are used in medicine, for luminous paints, for ionization, for the breaking of particle bonds for powdering minerals, for polymerization and other chemical reactions, and for various electronic applications.

The metals which are naturally radioactive, such as uranium and thorium, all have high atomic weights. The radiating power is atomic and is unaffected in combinations. **Radium** and other radioactive metals are changing substances. Radium gives out three types of rays; some of the other elements give out only one or two. The measure of the rate of radioactivity is the **curie,** which is the equivalent of the radioactivity of one gram of radium.

Each radioactive metal has a definite breakdown period, measured in half-life. **Actinium,** which is **element 89,** has a half-life of 21.7 years. It emits alpha particles to decay to **actinium K,** which is the radioactive isotope of **francium,** and then emits beta particles. Radioactive metals break down successively into other elements. By comparison of changing atomic weights, it has been deduced that the metal lead is the ultimate product, and uranium the parent metal under present existing stability conditions. But heavier metals, now no longer stable under present conditions, have been produced synthetically, notably plutonium. The heavy **element 103** was first produced in 1961 and named **lawrencium** in honor of the inventor of the cyclotron. Not all radiation produces radioactive materials, and by controlled radiation useful elements may be introduced into alloys in a manner not possible by metallurgy. The crystal lattice of an alloy can be expanded, or atoms displaced in the lattice, thus altering the properties of the alloy. In like manner the molecules of plastics may be cross-linked or otherwise modified by the application of radiation. For example, ethylene bottles may be irradiated after blowing to give higher strength and stiffness. Radioactive isotopes are also used widely in chemistry and in medicine, and as sources of electric power.

RADIUM. The best-known radioactive metal, symbol Ra, scattered in minute quantities throughout almost all classes of rocks, but commercially obtainable only from the uranium ores monazite, carnotite, and uraninite. It is a breakdown product, and it disintegrates with a half-life of 1,590 years. The metal is white, but it tarnishes rapidly in the air. The melting point is about 700°C. It was discovered in 1898 by Curie, and the original source was from the pitchblende of the Sudetenland area of Austria after extraction of thorium oxide, but most of the present supply comes from the carnotite of Zaire and from the pitchblende of western Canada. One gram of radium and 7,800 lb (3,538 kg) of uranium are obtained from 370 tons (336 metric tons) of pitchblende. The ratio of radium to uranium in any uranium ore is about 1:3,000,000. Radium is marketed in the form of bromides or sulfate in tubes and is extremely radioactive in these forms.

In a given interval of time a definite proportion of the atoms breaks up with the explusion of α, β, and γ rays. When an alpha particle is emitted from radium, the atom from which it is emitted becomes a new substance, the inert gas **radon,** or **element 86,** with a half-life of 3.82 days. During its short life it is a definite elemental gas, but it deposits as three isotopes in solid particles, decaying through polonium to lead. Radium is most widely known for its use in therapeutic medicine. It is also used for making inspections of metal castings in place of X-rays. **Radium-beryllium powder** is marketed for use as a neutron source.

RAMIE. A fiber used for cordage and for various kinds of coarse fabrics, obtained from the plant *Urtica nivea,* of temperate climates, and *U. tenacissima,* of tropical climates. The former plant has leaves white on the underside, and the latter has leaves all green. The name **rhea** is used in India to designate the latter species. It is also grown in China, Egypt, Brazil, and Florida. The plants grow in tall slender stalks like hemp and belong to the nettle family. The bast fibers underneath the bark are used, but are more difficult to separate than hemp fiber owing to the insolubility of the adhesive gums. The fibers are eight times stronger than cotton, four times stronger than flax, and nearly three times stronger than hemp. They are fine and white, and are as silky as jute. They are not very flexible and are not in general suitable for weaving, but their high wet strength, absorbent qualities, and resistance to mildew make the fibers suitable for warp yarns in wool and rayon fabrics. The yarn is used also for strong, wear-resistant canvas for such products as fire hose. The fiber is also valued for marine gland packings and for twine. The composition is almost pure cellulose, and the tow and waste are used for making cigarette paper. **China grass** is the hand-cleaned but not degummed fiber. It is stiff and greenish yellow in color. **Grass cloth** is woven fabric made in China from ramie. **Swatow grass cloth,** imported into the United States, is made of ramie fibers in parallel strands, not twisted into yarns.

RAPE OIL. An oil obtained from seeds of the mustard family, *Cruciferae.* The genus *Brassica,* a form of turnip, species of which are referred to as *B. campestris, B. rapa, B. napus,* and *B. hirta,* is grown in India, Pakistan, Europe, and Canada. **Rapeseed** is one of the principal oil seeds of the world. It is widely used as an edible oil, for making factice, and for mixing with lubricating and cutting oils and for quenching oils. The seeds are very small, an ounce having as many as 40,000 seeds. The seeds contain 40% oil. The edible oil is cold-pressed and refined with caustic soda. The burning and lubricating oils are refined with sulfuric acid. The iodine value is about 100, the specific gravity 0.915, and the flash point 455°F (235°C). The oil contains palmitic, oleic, linoleic, and stearic acids and 43 to 50% of the typical acid, **erucic acid,** also called **brassidic acid,** $C_{21}H_{41}COOH$. It has a melting point of 34°C. It occurs also in grape seed oil. **Colza oil** is a rape oil extracted from French seed, used to mix with mineral oils to make cutting oils. The name colza now refers to any refined rape oil. **Chinese colza oil,** from the *B. campestris chinoleifera,* contains the mustard volatile oil. The specific gravity is 0.91, saponification value 174, and iodine number 100.3. From 15 to 20% of blown rapeseed oil is mixed with mineral oil for lubricating marine engines. **Crambe seed oil,** from *Crambe abyssinica,* an Asiatic mustard, contains 55 to 60% erucic acid. The erucic can be broken down to perlargonic acid used as a substitute for dibasic acids such as azelaic and brassylic acids. **Cameline oil,** called also **dodder oil** and **German sesame oil,** has the same uses as rape oil. It is from the plant *Camelina sativa* grown in central Europe. The seeds contain 35% oil which contains oleic and palmitic acids and also erucic acid. The seed itself is high in mineral and protein content and is used in birdseed mixtures.

RARE-EARTH METALS. A group of trivalent metallic elements that occur together, also called **rare earths,** because of the difficulty of extracting them rather than because of their rarity. They include the elements 57 through 71, from lanthanum to luterium, and also **yttrium, element 39,** and **thorium, element 90,** since these are also found together in monazite, the chief ore. The **cerium metals** are a group of rare-earth metals consisting of elements with atomic numbers 57 through 63, and including the metal **cerium.** This group is also referred to as the **light rare earths.** The metal **ytterbium** (atomic number 70) also may be included in this group because of its light weight.

Thorium is separated by a relatively easy process, and the others remain grouped as the cerium metals, to be extracted as metals or compounds for special purposes justifying high costs. The separate metals are regularly marketed in pellets and in 325 mesh powder of 99.9% purity for pyrophoric and electronic uses, and as oxides of 99.9% purity. Cerium metal has an iron-gray color, is only slightly harder than lead, and is malleable. It has a specific gravity of 6.77, and a melting point at 1480°F (804°C).

After extraction of the thorium oxide from monazite, the chief rare-earth ore, the residual matter is reduced by converting the oxides to chlorides and then removing the metals by electrolysis. The product obtained is an alloy containing about 50% cerium together with lanthanum, didymium, and the other rare metals. It is usually called **misch metal,** the German name for mixed metal, and its original use was for making pyrophoric alloys. **Cerium standard alloy** of the Cerium Metals Corp. is a misch metal containing 50 to 55% cerium, 22 to 25 lanthanum, 15 to 17 neodymium, and the balance a mixture of yttrium, terbium, illinium, praseodymium, and samarium, with 0.5 to 0.8% iron.

Misch metal is used in making aluminum alloys, and in some steels and irons. In cast iron it opposed graphitization, and produces a malleableized iron. It removes the sulfur and the oxides, and completely degasifies steel. In stainless steel it is used as a precipitation hardening agent. An important use of misch metal is in magnesium alloys for aircraft castings. From 3 to 4% of misch metal is used with 0.2 to 0.6% zirconium, both of which refine the grain and give sound castings of complex shapes. The cerium metals also add heat resistance, and castings have service temperatures to 500°F (260°C). **Magnesium alloy EK30A** is an alloy of this type with 3% cerium metals, 0.55 zirconium, and the balance magnesium. The tensile strength is 23,000 lb/in^2 (158 MPa) with elongation of 3% and specific gravity 1.76. **Magnesium alloy EZ33A** is a modification containing also 3% zinc.

Ceria, cerium oxide, or **ceric oxide,** CeO_2, is a pale-yellow heavy powder of specific gravity 7.65, used in coloring ceramics and glass for producing distortion-free optical glass. It is used also for decolorizing crystal glass, but when the glass contains titania it produces a canary-yellow color. **Cerious oxide,** Ce_2O_3, is a greenish powder of specific gravity 7.0, and refractive index 2.19. About 3% of the oxide in glass makes the glass completely absorbent to ultraviolet rays. It is also an excellent opacifier for ceramics. **Cerium fluoride,** CeF_3, is used in arc carbons to increase brilliance. **Cerious nitrate,** $Ce(No_3)_3 \cdot 6H_2O$, is a red crystalline powder used in gas-mantle manufacture. Cerium salts are used for coloring glass. **Ceric**

titanate, $Ce(TiO_3)_2$, gives a golden-yellow color, and **ceric molybdate** gives a blue color.

Neodymium has a specific gravity of 7.01 and a melting point of 1875°F (1024°C). It is used in magnesium alloys to increase strength at elevated temperatures, and is used in some glasses to reduce glare. **Neodymium glass,** containing small amounts of neodymium oxide, is used for color television filter plates since it transmits 90% of the blue, green, and red light rays and no more than 10% of the yellow. It thus produces truer colors and sharper contrasts in the pictures and decreases the tendency toward gray tones. **Praseodymium** has a specific gravity of 6.77, and a melting point at 1715°F (935°C). **Lanthanum** is a white metal, malleable and ductile, with a specific gravity of 6.16, melting at 1688°F (919°C). Like the other cerium metals, it oxidizes easily in the air and is easily soluble in acids. **Lanthanum oxide,** La_2O_3, is a white powder used for absorbing gases in vacuum tubes. **Lanthanum boride,** LaB_6, is a crystalline powder used as an electron emitter for maintaining a constant, active cathode surface. It has high electric conductivity.

Didymium is not an element, but is a mixture of rare earths without cerium. It averages 45% La_2O_3, 38% **neodymium oxide,** Nd_2O_3; 11% **praseodymium oxide,** Pr_6O_{11}, 4% **samarium oxide,** Sm_2O_3; and other oxides. It is really the basic material from which the rare metals are produced. In glass it gives a neutral gray color, and it is used in glass for welders' goggles as it absorbs yellow light and reduces glare and eye fatigue. It is available as **didymium carbonate,** a pink powder soluble in acids, as **didymium oxide,** a brown acid-soluble powder, and as **didymium chloride** in pink lumps soluble in water and in acids.

Dysprosium has a specific gravity of 8.56 and a melting point of 2700°F (1482°C). Its corrosion resistance is higher than that of other cerium metals. It also has good neutron-absorption ability, with a neutron cross section of 1,100 barns. It is used in nuclear-reactor control rods, in magnetic alloys, and in ferrites for microwave use. The metal is paramagnetic. It is also used in mercury vapor lamps. With argon gas in the arc area it balances the color spectrum and gives a higher light output. **Samarium** has a higher neutron cross section, 5,500 barns, and is used for neutron absorption in reactors. Samarium has a specific gravity of 7.54, and a melting point at 1925°F (1052°C).

Ytterbium metal is produced in lumps and ingots. It has a specific gravity of 6.96 and a melting point at 1515°F (824°C). **Yttrium** is more abundant in nature than lead, but is difficult to extract. It is found associated with the elements 57 to 71, although its atomic number is 39. It has a silvery luster, a specific gravity of 4.47, and a melting point at 1550°F (843°C). It is the lightest of the cerium metals except scandium. The metal is corrosion-resistant to 400°C. It has a hexagonal close-packed crystal structure. **Ytterbium oxide,** Yb_2O_3, and **yttrium oxide,** Y_2O_3, are the usual commercial forms of these metals. The two metals occur in the mineral **gadolinite,** or **ytterbite,** $4BeO \cdot FeO \cdot Y_2O_3 \cdot 2SiO_2$, which also contains **gadolinium, erbium, europium, holmium** and **rhenium.** The mineral is found in Greenland, Sweden, Norway, and Colorado. Erbium has been obtained only in small quantities as a dark-gray powder. The metal has a specific gravity of 9.06 and a melting point at about 2700°F (1482°C). It forms the rose-red **erbium oxide,** or **erbia,** Er_2O_3, and other highly colored reddish salts. At high temperatures erbia glows with a greenish light.

Yttrium also occurs in the scarce mineral **nuevite** found in California. The mineral also carries titanium, tantalum, iron, and quartz, and is similar to the **keilhauite** found in Norway. **Fergusonite,** a brown mineral with a vitreous fracture, found sparsely in the Appalachian hills from New England to South Carolina, is a columbate and tantalate of yttrium with cerium, erbium, and uranium. In Europe it is known as **bragite** and **tyrite.** The mineral known as **bastnasite** in California is a fluorocarbonate of cerium and lanthanum, and the deposit at Mountain Pass, Calif., is sufficient to supply commercial needs of all the cerium metals. About 0.01% of rare-earth metals remain in the waste after apatite ores are processed in the making of phosphoric acid fertilizer. These metals are extracted by solvent or ion-exchange methods. Yttrium oxide is available as a fine white powder. **Yttralox,** of General Electric Co., is a transparent ceramic made from yttrium oxide and has a melting point above 4000°F (2204°C). It is used for special high-temperature lenses, infrared windows, lasers, and high-intensity lamps. For the brilliant reds for television phosphors small amounts of europium are added to **yttrium vanadate.**

Gadolinium oxide, or **gadolinia,** G_2O_3, has high neutron absorption, and is used for shields in atomic-power plants. As a molded ceramic it has a density of 7.0, low thermal conductivity, and a melting point of 2350°C. **Lutetium** is a heavy, refractory metal with a specific gravity of 9.85 and a melting point at about 3100°F (1704°C). **Lutetium oxide,** Lu_2O_3, of 99% purity, is produced for atomic uses. **Holmium** has a specific gravity of 8.8 and a melting point at about 2700°F (1482°C). It is used in glass to transmit radiant energy for wavelength-calibration instruments. **Thulium** is a heavy metal with a specific gravity of 9.32 and a melting point at about 2800°F (1538°C). It is used for radiographic applications. **Thulium 170** is a soft-gamma-ray emitter and is used as a radiation source. **Thulium oxide,** Tm_2O_3, is radioactive and is used as a power source for small thermoelectric devises. **Scandium** is a silvery-white metal found in the mineral **thortveitite,** $(ScYt)_2Si_2O_7$, of Norway, which contains 42% **scandium oxide,** Sc_2O_3. It also occurs in small amounts in the fergusonite of Montana, and in lepidolite, muscovite, beryl, and the amphiboles. The metal has a melting point of 1400°C, and a specific gravity slightly higher than aluminum.

RARE GASES. A general name applied to the five elements helium, neon, argon, krypton, and xenon. They are rare in that they are highly rarified gases at ordinary temperatures and are found dissipated in minute quantities in the atmosphere and in some substances. All have zero valence and normally make no chemical combinations, but by special catalyzations, except in the case of **helium,** the outer proton bonding may be broken and compounds produced. For torch shielding in the metallurgical industry these elements are referred to as **inert gases.**

Neon is procured from the air by liquefaction. When it is energized, it emits light and is used for signs and in glow lamps. The specific gravity, compared with air, is 0.674. It liquefies at −248°C. It is colorless, but gives a reddish-orange glow in lamps to which an electric current is applied. Neon is also used in voltage-regulating tubes for radio apparatus, and will respond to low voltages. In television the neon lamp will give fluctuations from full brilliancy to total darkness as many as 100,000 times a second. Colored electric advertising signs are often referred to as neon signs, but the colors other than orange are produced by different gases. **Argon** gives a purple light when an electric current is passed through it. It occurs free in the atmosphere to the extent of 0.935%. Its liquefying point is about −187°C. It is obtained by passing atmospheric nitrogen over red-hot magnesium, forming magnesium nitride and free argon. It is also obtained by separation from industrial gases. Argon is employed in incandescent lamps to give increased light and to prevent vaporization of the filament, and is used instead of helium for shielding electrodes in arc welding and as an inert blanket for nuclear fuels.

Krypton, which occurs in the air to the extent of 1 part in 1 million, gives a pale-violet light. It is a heavy gas, with a specific gravity of 2.896. It is used as a filler for fluorescent lamps to decrease filament evaporation and heat loss and to permit higher temperatures in the lamp. The 3-billion-candlepower aircraft-approach lights first used on the Berlin airlift contained krypton. **Krypton 85,** obtained from atomic reactions, is a beta-ray emitter with a half-life of 9.4 years. It is used in luminous paints for activating phosphors and also as a source of radiation. As marketed by the Tracerlab Co., it comes combined in solid form with a hydroquinone to give higher concentration of energy and more convenience in use. American Atomics Corp. markets the isotope in disks of 0.23 and 0.5 in (0.58 to 1.27 cm) in diameter, encased in acrylic plastic for use with phosphors as luminous sources. The light is yellow-green. **Xenon,** another gas occurring in the air to the extent of 1 part in 11 million, gives a sky-blue to green light. It is the heaviest of the rare gases, with a specific gravity of 4.561 compared with 1 for air. Its liquefying point is about −108°C. When atomic reactors are operated at high power, xenon tends to build up as a reaction product, poisoning the fuel and reducing the reactivity. **Xenon lamps** for military use give a clear white light known as sunlight plus northsky light. This color does not change with the voltage, and thus the lamps require no voltage regulators. An 800-watt xenon lamp delivers 2,000 lumens, 4 times as much as a 1,000-watt incandescent lamp. The xenon lamp reflects each half cycle, so that shutterless projectors are possible. Krypton and xenon have lower thermal conductivity and lower electrical resistance than argon. A helical arc of xenon is used to activate ruby optical masers. Xenon is a mild anesthetic, the accumulation from air helping to induce natural sleep, but it cannot be used in surgery since the quantity needed produces asphyxiation.

RARE METALS. A term given to metals that are rare in the sense that they are difficult to extract and are rare and expensive commercially. They include the elements **astatine, technetium,** and **francium.** The silvery metal technetium, element 43, has been produced by bombardment of molybdenum with neutrons. It is available in the form of **technetium carbonyl,** $Tc_2(CO)_{10}$, which is stable in air and soluble in most organic solvents but reacts with halogens. Although **radium** is a widely distributed metal it is classed as a rare metal. All of the ultraheavy metallic elements, such as plutonium, which are produced synthetically, are classed as rare metals. They are called **transuranic metals** because they are above the heavy metal uranium weight. They are all radioactive.

Element 99, called **einsteinium,** was originally named **ekaholmium** because it appears to have chemical properties similar to holmium. It is produced by bombarding uranium 238 with stripped nitrogen atoms. It decays rapidly to form the lighter **berkelium,** or element 97. **Neptunium,** element 93, **californium,** element 98, and **illinium,** element 61, are also made atomically. The latter also has the names **florentium** and **promethium.**

Plutonium is made from uranium 238 by absorption of neutrons from recycled fuel. The metal, 99.8% pure, is obtained by reduction of **plutonium fluoride,** PuF_4, or **plutonium chloride,** $PuCl_3$. It has a melting point of 64°C. The surface reacts in the air to form the nonadherent **plutonium oxide,** PuO_2, which becomes airborne and is pyrophoric and poisonous. **Plutonium 238,** and also **239** and **240,** emit chiefly alpha rays. Plutonium 238 has a low radiation level and is used as a heat source for small water-circulating heat exchanges for naval underseas diving suits.

Plutonium 241 emits beta and gamma rays. Since all the allotropic forms are radioactive, it is a pure nuclear fuel while uranium is only 0.7% directly useful for fission. It is thus necessary to dilute plutonium for control. For fuel elements it may be dispersed in stainless steel and pressed into pellets at about 1600°F (871°C), or pellets may be made of **plutonium carbide. Plutonium-iron alloy,** with 9.5% iron, melts at 410°C. It is encased in a tantalum tube for use as a reactor fuel. **Plutonium-aluminum** alloy is also used. These alloys have hard compounds of PuFe and PuAl in the matrix, and there is no solubility of the plutonium. While plutonium 241 has a half-life of only 14 years, the beta emitters plutonium 239 and 240 have half-lives of 24,300 and 16,600 years, respectively. Element 102, called **nobelium,** has a half-life of only 12 min. Other transuranic metals produced synthetically are **americium,** element 95, and **curium,** element 96. Curium is used as a heat source in remote applications. **Curium 244** is obtained as **curium nitrate** in the reprocessing of spent reactor fuel. It is converted to **curium oxide.** The by-product americium is used as a component in neutron sources. Other transuranic metals that have been produced by nuclear reactions and synthesis include **fermium** (element 100), **mendelevium** (element 101), **lawrencium** (element 103), **rutherfordium** or **kurchatovium** (104), and **hahnium** or **nielsbohrium** (105).

RATANY. Also known by the original Inca name of **payta,** and in medicine as **krameria.** The root of the shrub *Krameria triandria,* which grows in Peru and is used for tanning leather and in medicine as an astringent. The root comes in diameters up to 1 in (2.5 cm), and in pieces up to 3 ft (0.9 m) in length. It contains about 40% tannin which is extracted by hot water. It gives the leather a deep-brown color, and is usually mixed with other tannins.

RATTAN. The long slender stem of the palm *Calamus rotang* and other species, of Sri Lanka, Malaya, and Laos. There are more than 40 varieties. The Malay word is rottan, meaning cane. It is tough, flexible, strong, and durable, and is used for canes, umbrella handles, and furniture. When split it is used for car seats, baskets, baby-carriage bodies, furniture, whips, and heavy cordage. Commercial rattans are in pieces 5 to 20 ft (1.5 to 6.1 m) long. A substitute for rattan is **jacitará,** from the plant *Desmoncus macroacanthus,* of Brazil. It is used for seating. Vinylidene chloride plastic is now widely used as a substitute for rattan for seating.

RAYON. A general name for artificial-silk textile fibers or yarns made from **cellulose nitrate, cellulose acetate,** or cellulose derivatives. The general name was adopted after the Federal Trade Commission had ruled against the use of such terms as **art silk, fiber silk, chardonnet silk,** and **artificial silk.** Some of the products are referred to preferably by the manufacturer's trade name, such as Celanese for cellulose-acetate yarns and fabrics. In general, the name rayon is limited to the viscose, cuprammonium, and acetate fibers, or to fibers having a cellulose base. Other synthetic-fiber groups have their own group names, such as azlon for the protein fibers and nylon for the polymeric amine fibers, in addition to individual trade names. High-tenacity rayons designed for tire-cord use have the general name of **Tyrex.**

Viscose rayon is made by treating the cellulose with caustic soda and then with carbon disulfide to form cellulose xanthate, which is dissolved in a weak caustic solution to form the viscose. With the cuprammonium process the cellulose is digested in an ammonia solution of copper sulfate, and the solution forced through the spinnerets into dilute acid for hardening. Rayons manufactured by the different processes vary both chemically and physically. They are resistant to caustic solutions which would destroy natural silk. They are also mildewproof, durable, and easily cleaned. But they do not have the permeability and soft feel of silk. The **acetate rayons** are more resistant than the viscose or cuprammonium. The lack of permeability of the fibers is partly overcome by having superfine fibers so that the yarns are permeable. The one-denier viscose staple produced by the FMC Corp. by stretching the fiber after it leaves the spinnerets is finer than Egyptian cotton and can thus be made into yarns that are permeable between the fibers. Fabrics made from the superfine yarns have the appearance of sheer silk. **Multicell rayon** of this company, for making nonwoven fabrics and lining and filter papers on regular papermaking machines without the addition of a binder, is a multicellular short-staple fiber cut to uniform ¼-in (0.64-cm) length. The fibers lock themselves firmly in place with contacting fibers, and the 1.5-denier fiber makes a soft, opaque sheet of paper. The density of the sheets and the strength per unit weight decrease with increase in the fiber denier. **Fiber 40,** of this company, is a type of rayon called **Avril,** made by a special pulping process which decreases the tendency to shrink or felt when the fiber is wet. **Avlin** is a multicellular rayon fiber that gives a tight, firm bulk to fabrics. **Fabray,** of Stearnes & Foster, is a thin lightweight porous nonwoven fabric of rayon fiber used for throw-away garments.

The objectionable high gloss of synthetic fibers is reduced by pigmentation. **Glos** was an early name for rayon because of this gloss, but the name has now been abandoned. Mixtures of rayon and other fibers have some use for dress fabrics as well as underwear. The material is also used for automotive tire fabrics because of its strength. **Stable fiber** is fiber cut to length for the spinning system to be used. The **Lanese fiber** of the Celanese Corp. of America is 3-, 5.5-, 8-, and 12-denier acetate rayon cut to lengths of 1 ⅛ to 2 in (2.9 to 5.1 cm). **Seraceta,** of the FMC Corp., is a continuous-filament acetate rayon. **Matesa** is the name of continuous-filament cuprammonium rayon of the American Bemberg Corp.

High-tenacity rayon is produced by stretching the fibers so that the molecular chains run parallel to the filament axis, and a number of small crystalline regions act as anchors for the cellulose chains. **Tire cord** stretched in this way has the tensile strength increased from 27 to 37 lb (12 to 17 kg). **Rayocord,** of Rayonier, Inc., is a high-tenacity rayon for tire cords and reinforcing fabrics for plastics. It is made from sulfite cellulose, is stretched to produce oriented fiber, and has twice the strength of ordinary rayon. **Super Cordura,** of Du Pont, used for tire cords, is high-tensile rayon made from a blend of wood cellulose and cotton linters. The **Nytex tire cord** of the Sieberling Rubber Co. is a combination nylon-rayon cord. The nylon adds strength, and the rayon eliminates the flat-spotting tendency of the nylon.

RED BRASS. The standard wrought metal known as red brass, or **rich low brass,** of the brass mills contains 85% copper and 15 zinc. It is one of the most ductile and malleable of the brasses, and the working stresses can be relieved without softening the metal by heating for a half hour at 275°C. It is also valued for its fine reddish color, ability to take a high polish, and corrosion resistance. It is essentially a drawing and stamping metal and, unless it contains a little lead, will not machine well. The annealed metal has a tensile strength of 38,000 lb/in^2 (261 MPa), with elongation 45%, while the hard-rolled metal has a strength of 70,000 lb/ in^2 (482 MPa), with elongation 5%, and Rockwell B hardness 75. The melting point is 1875°F (1024°C), weight 0.316 lb/in^3 (8,747 kg/m^3), coefficient of expansion 0.0000104 in/in(°F) [0.0000189 cm/cm(°C)], and electric conductivity 34.7% that of copper. It is produced regularly in the form of sheets and tubes, and is used for jewelry, nameplates, dials, drawn and stamped hardware and mechanical parts, and corrosion-resistant hot-water pipes. **Plumrite 85,** of the Bridgeport Brass Co., is this alloy in tubes for piping. **Revere alloy 130,** of the Revere Copper & Brass, Inc., is this metal in sheets and tubes. **Redalloy,** of the Chase Brass & Copper Co., contains 85% copper, 14 zinc, and 1 tin. The tensile strength is 42,000 lb/in^2 (290 MPa) and elongation 48%.

The high-copper alloy known in the mills as **commercial bronze** contains 90% copper and 10 zinc. It is widely used for making costume jewelry, weather stripping, stamped hardware, forgings, and screws. It has a bronze color, a weight of 0.318 lb/in^3 (8,802 kg/m^3), and electric conductivity 43.6% that of copper. The tensile strength, soft, is 38,000 lb/in^2 (261 MPa), with elongation 40%. The tensile strength of the hard-rolled metal is 60,000 lb/in^2 (413 MPa) and elongation 5%. **Revere alloy 120** is this metal. Federal specifications for gilding metal call for this alloy, and it is also known as government gilding metal. But standard **gilding metal** employed for making cheap jewelry and small-arms ammunition contain 95 to 97% copper. It has a golden-red color, is stronger and harder than copper, but has only about half the electric conductivity of copper. The English **cap copper,** used for cartrige caps, has 97% copper and 3 zinc. **BES gilding metal** is in three grades, 80, 85, and 90% copper, but the 80–20 alloy has a definite golden-yellow color and is not in the class of red brasses. The 80–20 brass is one of the standard alloys of the brass mills under the name of **low brass.** It was early used for jewelry under the name of **Dutch metal.** It is very ductile and is easily drawn, takes a high polish, and has high strength. It is still much used for cheap jewelry. **Chain bronze,** used for flat-link jack chains, and also for costume jewelry, contains 87 to 89% copper, 0.60 to 1.25 tin, and the balance zinc. The tin increases the hardness, and it has good strength and corrosion resistance.

Many old names used for designating the golden or reddish high-copper brasses, especially for cheap jewelry making, are still in occasional use. **Pinchbeck metal,** originally made by C. Pinchbeck, an English jeweler, contained 88% copper and 12 zinc. This is the same as the **Guinea gold** used for traders' jewelry. **Manila gold,** or **traders' gold,** was about the same with some lead to heighten the color. **Ormolu gold,** a name still used by brass platers, was any composition that would give a golden color. **Rich gold metal** was the 90–10 alloy, and this alloy is still being used for decorative purposes under the name of **copper-rich brass. Manheim gold,** containing 83% copper, 10 zinc, and 7 tin, was a German alloy for cheap jewelry. It is a considerably harder alloy than ordinary red brasses. **Tournay metal** was a French alloy widely used for buttons when brass buttons were in vogue. It was essentially the same as the original **tombac metal** used by the Chinese for buttons, and was the 85–15 alloy containing considerable arsenic to give a brilliant grayish tone to the metal. The alloy is quite brittle. The name tombac is the Malay word tombaga used to designate gold-colored jewelry alloys. **Chrysochalk** is another old name for gilding metal containing enough lead to give a dull gold tone. Japanese low-priced jewelry alloys often contained silver or gold to balance the color and make them resistant to tarnishing. **Shadke** was a high-copper alloy with some gold, and **Shaku-do** contained about 4% gold and 1 silver. The gold color of the red brasses is enhanced by pickling in nitric acid.

RED LEAD. A common lead pigment, also erroneously called **minium.** It is a **lead tetroxide,** Pb_3O_4, forming a bright-red or orange-red powder of specific gravity 0.096, insoluble in water. As a pigment it has great covering power and brilliancy, but red lead which has not been completely oxidized and contains litharge must be applied immediately after mixing to avoid combination with the oil. It is used as a heavy protective paint for iron and steel. Red lead is also used in storage-battery plates and in lead glass. With linseed oil it is used as a lute in pipe fitting. **Orange mineral** is a pure form of red lead made from white lead and has an orange color. Chemically it is the same as red lead, but it has a different structure, giving it a more brilliant color. Red lead is made from lead metal by drossing and then heating in a furnace. **Fume red lead** is a fine grade made from **fume litharge,** which is made by oxidizing molten lead and passed off as a yellow smoke or fume. Fume red lead is notable for the extreme fineness of its particles, and bulks greater than ordinary red lead. Fume red lead is marketed for pigment as **superfine red lead.**

REDWOOD. Also called **sequoia, California redwood,** and **Humboldt redwood.** The wood of the tree *Sequoia sempervirens,* native to the West Coast of the United States. The wood is light, soft, and spongy but has comparatively high strength and is resistant to decay and insect attack. The trees grow in a narrow coastal strip in California, and are of an immense size, reaching a diameter of 20 ft (6.1 m) and a height of 350 ft (107 m) in 2,000 years. The so-called Big tree, however, is the *S. gigantea* or *S. washingtoniana.* It grows in the mountains at elevations of 5,000 to 8,500 ft (1,524 to 2,591 m), but is not cut for lumber. Planks of redwood can be readily obtained 6 ft (1.8 m) in width. The specific gravity is 0.374 to 0.387. It has a tensile strength of from 7,000 to 11,000 lb/in^2 (48 to 75 MPa). The heartwood varies from light-cherry to dark-mahogany in color, and the narrow sapwood is almost white. The wood is used in all kinds of common construction. **Redwood bark fiber** is the shredded fiber of the bark of the redwood. It has excellent felting properties, is water-resistant and fire-resistant, and is an excellent insulator for house walls and refrigerators. It is also used in wool mixtures for blankets and overcoatings. The fiber is short, but has a natural twist that facilitates spinning. The fiber contains a high content of lignin, and makes a good filler for heat-curing plastics. **Palco wool,** of the Pacific Lumber Co., is redwood bark fiber, and **Palco board** is a lightweight insulation board for cold-storage chambers made from the shredded bark fibers. Palcotan, of the same company, is a sodium salt of **palcotannic acid,** a weak sulfonated tannic acid, used as a binder in ceramic clays, a stabilizer for asphalt emulsions, and in latex and paste adhesives. **Redwood tannin,** produced by soaking redwood chips in hot water and dissolving out the tannin with ethyl acetate, is used as an oxidation inhibitor in hydrocarbons.

REFRACTORIES. Materials, usually ceramics, employed where resistance to very high temperature is required, as for furnance linings and metal-melting pots. Materials with a melting point above Seger cone 26, or 1580°C, are called refractory, and those with melting points above cone 36, or 1790°C, are called highly refractory. But, in addition to the ability to resist softening and deformation at the operating temperatures, other factors are considered in the choice of a refractory, especially load-bearing capacity and resistance to slag attack and spalling. Heat transfer and electrical resistivity are sometimes also important. Many of the refractories are derived directly from natural minerals, but synthetic materials are much used.

Clay is the oldest and most common of the refractories. The natural refractories are kaolin, chromite, bauxite, zirconia, and magnesite, often marketed under trade names. Refractories may be acid, such as silica, or basic, such as magnesite or bauxite, for use in acid- or basic-process steel furnaces. Graphite and chromite are neutral refractories. Magnesia is insoluble in the slag of open-hearth furnaces and is used for linings. Magnesia fuses at 3929°F (2165°C), chromite at 3722°F (2049°C), and alumina at 3670°F (2048°C). The fusing point of the refractory, however, is usually dependent on the binder, as all binders or impurities lower the melting point. **Arco refractory brick,** of the General Refractories Co., with 60 to 80% alumina, will withstand temperatures from 3290 to 3335°F (1810 to 1835°C). The melting point of a 99.5% pure alumina is given as 3725°F (2041°C), and the decomposing point of a 98% pure silicon carbide as 4175°F (2301°C). **Chrome-magnesite bricks** are made usually with 75% chrome ore and 25 dead-burned magnesite.

The chief artificial refractories are silicon carbide and aluminum oxide. **Refrax,** of the Carborundum Co., is silicon carbide held together by crystallization without a binder. It withstands temperatures to 2240°C, at which point it decomposes. The crushing strength of the brick is 12,500 lb/in^2 (85 MPa). This type of material can be made only in simple shapes, but is also made into rolls for roller-type furnaces. **Refrax FS** is fused silica bonded with silicon nitride. It is used for formed parts up to 23 in (58 cm) in diameter for such applications as brazing fixtures, and has dimensional stability and thermal shock resistance to 2250°F (1232°C). **Silfrax** is in grades with 40 to 78% SiC, and porosities from 9 to 18%. **Monofrax,** of the same company, is a refractory block for lining glass furnaces. It is composed of 98% alpha and beta alumina crystals interlocked in a dense structure weighing 200 lb/ft^3 (3,204 kg/m^3), with only a small amount of bond and impurities. A grade with more open structure weighs 175 lb/ft^3 (2,804 kg/m^3), and the **Monofrax K** contains 80% alumina and the balance chromite crystals saturated with alumina. These materials are resistant to abrasion to temperatures above 3000°F (1649°C). They have porosities from 20 to 29%. **Aluminite** is furnished by the Philip Carey Co. in blocks for temperatures to 2000°F (1093°C). **Korundal XD brick,** of the Harbison-Walker Refractories Co., is corundum bonded with mullite. For open-hearth and electric-furnace roofs it will withstand a 25-lb/in^2 (0.17-MPa) load at a temperature of 3000°F (1649°C) without spalling. The **Lo-Sil brick** of Kaiser Aluminum, for aluminum-melting furnaces, is made with a high alumina content with very low silica to avoid pickup of silicon in the molten aluminum.

The **silica brick** used in coke ovens has a high coefficient of expansion below about 155°F (68°C), so that ovens must be heated gradually over a period of 6 to 8 weeks to prevent cracking of the brick. **Fire sand** is a sand composed of 98% silica and is very refractory. The natural silica refractories used to replace fireclay for high temperatures should contain at least 97% silica and not yield too fine a powder on crushing. In order of merit the materials used are chalcedony, old quartzites, and vein quartz. The refractory known in Europe as **klebsand,** used for steel furnace linings, has 87% SiO_2, 8.6 Al_2O_3, 0.3 TiO_2, and some ferric oxide. **Ganister** is a natural refractory mineral used for furnace linings. In compact form it was used for furnace hearths. It contains about 95% silica, about 1.5 Al_2O_3, and a small amount of lime as a binder. The chief deposits are in the quartzites of Wisconsin, Alabama, and Colorado. An artifical ganister is made with silica and clay. **Ganisand,** of the Quigley Furnace Specialties Co., Inc., is a ganister having a fusing point at 3250°F (1788°C). **Dinas silica** is an English ganister with about 97% silica, having a melting point of 1680°C. Silica brick made from quartzite containing 98% silica has lime added as bond, and the resulting brick usually contains about 96% silica, 2 lime, and 2 alumina and ferric oxide. The bricks are rigid under load, and are resistant to attack by acid slags and to spalling under rapid temperature changes. **Vega silica brick,** of Harbison-Walker Refractories Co., has no more than 0.4% of oxides other than lime, and will withstand temperatures of 3090°F (1699°C) under load. **Foamed silica blocks** for lining tanks and for refractory insulation are made from pure fused silica. They have a density of 10 to 15 lb/ft^3 (160 to 240 kg/m^3) with an impermeable closed-cell structure, have a compressive strength of 130 to 210 lb/in^2 (0.9 to 1.4 MPa), and will withstand temperatures to 2200°F (1204°C). **Foamsil,** of the Pittsburgh Corning Corp., is this material.

Pinite, from Nevada, is a secondary mineral derived from the alteration of feldspar and other rocks, and is used for kiln linings in cement plants. It is a hydrous silicate of alumina and potash, and the massive material resembles steatite. It will bond alone like clay and has low shrinkage. At 1125°C, the mineral inverts to mullite. **Agalmatolite** is a massive pinite and can be used in the same way. **Bull-dog** is an old name for a refractory which is a mixture of ferric oxide and silica made by roasting tap cinder with free access of air. **Tap cinder** is a basic silicate of iron.

REFRACTORY CEMENT. A large proportion of the commercial refractory cements used for furnace and oven linings and for fillers are fireclay-silica-ganister mixtures with a refractory range of 2600 to 2800°F (1427 to 1538°C). Cheaper varieties may be mixtures of fireclay and crushed brick, fireclay and sodium silicate, or fireclay and silica sand. An important class of refractory cements is made of silicon carbide grains or silicon carbide–fire sand with clay bonds or synthetic mineral bonds. The temperature range of these cements is 2700 to 3400°F (1482 to 1871°C). **Silicon carbide cements** are acid-resistant and have high thermal and electric conductivity. For crucible furnaces the silicon carbide cements are widely used except for molten iron. Alumina and **alumina-silica cements** are very refractory and have high thermal conductivity. Calcined kaolin, diaspore clay, mullite, sillimanite, and combinations of these make cements that are neutral to most slags and to metal attacks. They are electrical insulators.

Chrome-ore cements are difficult to bond unless mixed with magnesite. **Plastic 695** of the Basic Refractories, Inc., is a **chrome-magnesite cement** made of treated magnesite and high-grade chrome ore. It sets quickly and forms a hard, dense structure. The melting point is above 3600°F (1982°C). It is used particularly for hot repairs in open-hearth furnaces. **Magnefer** is the name of a dead-burned dolomite refractory of the same company, while **Basifrit** is a magnesia refractory for resurfacing. **Zircon-magnesite cement** is made with 25% refined zircon sand, 10 milled zircon, 15 fused magnesia, and 50 low-iron dead-burned magnesite bonded with sodium silicate. A wide range of refractory cements of varying compositions and characteristics is sold under trade names, and these are usually selected by their rated temperature resistance. For example, **Hadesite,** of the Eagle-Picher Lead Co., is composed of refractory clay and aggregates, mineral wool, and a binder, and is recommended for temperatures up to 1900°F (1038°C). **Carbofrax cement,** of the Carborundum Co., is silicon carbide with a small amount of binder in various grades for temperatures from 1600 to 3200°F (871 to 1760°C), depending on the fineness and the bond. **Firefrax cement** of the same company is an aluminum silicate, sometimes used in mixtures with ganister for lining furnaces. It is for temperatures to 3000°F (1649°C). **Alfrax cement** is fused silica also in various grades for temperatures from 1650 to 3300°F (899 to 1816°C). **Mullfrax cement** has a base of electric furnace mullite, and is for temperatures from 2200 to 3200°F (1204 to 1760°C), while **Mullite S cement** is of converted kyanite for ferrous and nonferrous melting furnaces for temperatures to 3150°F (1732°C). **Ankorite,** of the Harbison-Walker Refractories Co., is a high-alumina hot-setting refractory mortar for laying superduty firebrick. **Thermolith,** of this company, is chrome-base cold-setting mortar for laying magnesite, chrome, and forsterite brick.

REFRACTORY HARD METALS. True chemical compounds of two or more metals in the form of crystals of very high melting point and high hardness. Because of their ceramic-like nature, they are often classified as ceramics. These materials were originally called **Hartstoffe** in Germany, and they do not include the hard metallic carbides, some of which, with metal binders, have similar uses, nor do they include the hard cermets. The refractory hard metals may be single large crystals, or crystalline powder bonded to itself by recrystallization under heat and pressure. In general, parts made from them do not have binders, or contain only a small percentage of stabilizing binder. The **intermetallic compounds,** or **intermetals,** are marketed regularly as powders of particle size from 150 to 325 mesh for pressing into mechanical parts or for plasma-arc deposition as **refractory coatings,** and the powders are referred to chemically, such as **borides, ber-**

yllides, and **silicides.** The oxides and carbides of the metals are also used for sintering and for coatings, and the oxides are called cermets.

Zirconium boride is a microcrystalline gray powder of the composition ZrB_2. When compressed and sintered to a density of about 5.3, it has a Rockwell A hardness of 90, a melting point of 2980°C, and a tensile strength of 35,000 to 40,000 lb/in_2 (241 to 275 MPa). It is resistant to nitric and hydrochloric acids, to molten aluminum and silicon, and to oxidation. At 2200°F (1204°C) it has a transverse rupture strength of 55,000 lb/in^2 (379 MPa). It is used for crucibles and for rocket nozzles.

Chromium boride occurs as very hard crystalline powders in several phases, the CrB orthorhombic crystal melting at 1500°C, the hexagonal crystal Cr_2B melting at 1850°C, and the tetragonal crystal Cr_3B_2 melting at 1960°C. Chromium boride parts produced by powder metallurgy have a specific gravity of 6.20 to 7.31, with a Rockwell A hardness of 77 to 88. They have good resistance to oxidation at high temperatures, are stable to strong acids, and have high heat-shock resistance up to 2400°F (1316°C). The transverse rupture strength is from 80,000 to 135,000 lb/in^2 (551 to 930 MPa). **Colmonoy,** of the Wall Colmonoy Corp., is chromium boride, CrB, used for oil-well drilling. A sintered material, used for gas turbine blades, contains 85% CrB with 15% nickel binder. It has a Rockwell A hardness of 87 and a transverse rupture strength of 123,000 lb/in^2 (848 MPa).

Molybdenum boride, Mo_2B, has a specific gravity of 9.3, a Knoop hardness of 1,660, and a melting point of about 1660°C. **Tungsten boride,** W_2B, has a specific gravity of 16.7, and a melting point of 2770°C. **Titanium boride,** TiB_2, is light in weight with a specific gravity of 4.5. It has a melting point at about 4700°F (2593°C). Molded parts made from the powder have a Knoop hardness of 3,300 and a flexural strength of 35,000 lb/in_2 (241 MPa), and they are resistant to oxidation to 1800°F (982°C) with a very low oxidation rate above that point to about 2500°F (1371°C). They are inert to molten aluminum. Intermetal powders of **beryllium-tantalum, beryllium-zirconium,** and **beryllium-columbium** are also marketed, and they are light in weight and have high strength. Sintered parts resist oxidation to 3000°F (1649°C).

Molybdenum disilicide, $MoSi_2$, has a crystalline structure in tetragonal prisms, and has a Knoop hardness of 1,240. The decomposing point is above 1870°C. It can be produced by sintering molybdenum and silicon powders, or by growing single crystals from an arc melt. The specific gravity of the single crystal is 6.24. The tensile strength of the sintered parts is 40,000 lb/in^2 (275 MPa) and compressive strength 333,000 lb/in^2 (2,295 MPa). The resistivity is 29 $\mu\Omega \cdot cm$. It is used in rod form for heating elements in furnaces. The material is brittle, but can be bent to shape at temperatures above 2000°F (1093°C). **Super Hot Rod,** of the Norton Co., is molybdenum disilicide rod. **Kanthal Super,** of Kanthal Corp., is a similar material. In an inert atmosphere the operating temperature is 1600°C. Furnace gases containing active oxygen raise the operating temperature to about 1700°C, while gases containing active hydrogen lower it to about 1350°C.

Tungsten disilicide, WSi_2, is not as hard and not as resistant to oxidation at high temperatures, but has a higher melting point, 2050°C. **Titanium nitride,** TiN, is a light-brown powder with a cubic lattice crystal structure. Sintered bars are extremely hard and brittle, with a hardness above Mohs 9 and a melting point of 2950°C. It is not attacked by nitric, sulfuric, or hydrochloric acids, and is resistant to oxidation at high temperatures. **Aluminum nitride,** AlN, when molded into shapes and sintered, forms a dense, nonporous structure with a hardness of Mohs 6. It resists the action of molten iron or silicon to 3100°F (1704°C), and molten aluminum to 2600°F (1427°C), but is attacked by oxygen or carbon dioxide at 1400° (760°C).

Nickel aluminide is a chemical compound of the two metals, and, when molded and sintered into shapes, has good oxidation resistance and heat shock at high temperatures. **Borolite 1505,** of the Borolite Corp., is a nickel aluminide sintered material with a specific gravity of 5.9, and a transverse rupture strength of 150,000 lb/in^2 (1,034 MPa) at 2000°F (1093°C), which is twice that of cobalt-bonded titanium carbide at the same temperature. The melting point is 3000°F (1649°C). It resists oxidation at 2000°F (1093°C). It is used for highly stressed parts in high-temperature equipment. **Metco 405,** of Metco, Inc., is this compound in wire form for welding, flame coating, and hard surfacing. **Metco 404** is aluminum powder coated with nickel. This powder may be mixed with zirconia or alumina which will increase the hardness and heat resistance of the nickel aluminide coating. **Columbium aluminide** is used as a refractory coating as it is highly resistant for long periods at 2600°F (1427°C). **Tin aluminide** is oxidation-resistant to 2000°F (1093°C), but a liquid phase is formed at about this point.

Tribaloy intermetallic materials, developed by Du Pont, are composed of various combinations of nickel, cobalt, molybdenum, chromium, and silicon. For example, one composition is 50% nickel, 32 molybdenum, 15 chromium, 3 silicon. Another is 52% cobalt, 28 molybdenum, 17 chromium, and 3 silicon. The structure consists of a hard intermetallic Laves phase in a softer matrix. Supplied as alloy metal powder, welding rod, or casting stock, they can be cast, deposited as a hard-facing surface, plasma-sprayed, or consolidated by powder metallurgy. The materials have exceptional wear and corrosion resistance properties in corrosive media and in air up to 2000°F (1093°C). Typical applications are in pumps, valves, bearings, seals, and other parts for chemical process equipment. Also the materials are suited for marine and saltwater applications and for parts subject to wear in atomic energy plants.

REFRIGERANTS. Gases, or very-low-boiling-point liquids, used for the heat-absorbent cycle in refrigerating machines. The ideal refrigerant, besides having a low boiling point, should be noncorrosive, nonflammable, and nontoxic. It must be free of water, since as little as 40 parts per million of water may cause freezing in the system. Ammonia is a common refrigerant. Ethyl chloride, methyl ether, carbon dioxide, and various chlorinated and fluorinated hydrocarbons marketed under trade names such as the **Freons** of Du Pont and the **Genetrons** of the Allied Chemical Corp. are also used. **Dichlorodifluoromethane,** a nonflammable, colorless, odorless gas of the composition CF_2Cl_2, is one of the Freons. It liquefies at −21.7°F (−30°C). **Freon E3** is a fluorated hydrocarbon which boils at 306°F (152°C), and is stable to 570°F (299°C). It is pourable at 175°F (79°C). It is used for dielectric insulating in electronic equipment. **Trichloromonofluoromethane,** CCl_3F, is used as a refrigerant in industrial systems employing centrifugal compressors, and in indirect expansion-type air-conditioning systems. The boiling point is 74.7°F (24°C), and freezing point −168°F (−111°C). The condensing pressure at 86°F (30°C) is 18.3 lb/in^2 (0.13 MPa), and the net refrigerating effect of the liquid is 67.5 Btu/lb (371 kcal/kg). **Genetron 11** is this material. **Genetron 101** has the composition CH_3:$CClF_2$, with boiling point at −9.2°C and freezing point at −130.8°C.

RESINS. Historically, resins is the term applied to an important group of substances obtained as gums from trees or manufactured synthetically. It is also frequently used interchangeably with the term plastics. The common resin of the pine tree is called **rosin.** The natural resins are soluble in most organic solvents, and are used in varnishes, adhesives, and various compounds. **Oleoresins** are natural resins containing essential oils of the plants. **Gum resins** are natural mixtures of true gums and resins and are not as soluble in alcohol. They include rubber, gutta percha, gamboge, myrrh, and olibanum. Some of the more common natural resins are rosin, dammar, mastic, sandarac, lac, and animi. **Fossil resins,** such as amber and copal, are natural resins from ancient trees, which have been chemically altered by long exposure. The synthetic resins differ chemically from natural resins, and few of the natural resins have physical properties that make them suitable for mechanical parts.

Galbanum, used in medicine, is a gum resin from the perennial herb *Ferula galbaniflua* of western Asia. It comes in yellowish to brownish tears. **Myrrh** is a yellow reddish aromatic gum resin from the *Commiphora malmol* and other species of small trees of India, Arabia, and northeast Africa. There are more than 80 species. The Hebrew word mur means bitter, and the gum has a bitter taste. It consists of a mixture of complex acids and alcohols but mainly the mucilaginous **arabin,** with from 3 to 8% of a volatile oil of the formula $C_{10}H_{14}O$. It is used in incense and perfumes, and in medicine as a tonic. The gum is called **mulmul** and **ogo** in eastern Africa and **herabol** in India. **Sweet myrrh,** or **bisabol myrrh,** is a very ancient perfume and incense. It comes from the tree *C. erythraea* of Arabia. **Herabol myrrh** is from the small tree *C. myrrha* of India and is a brown to black solid. It is used in medicine as a stimulant and antiseptic and in perfumes and incense. **Asafetida** is a gum resin with a foul odor and acrid taste obtained from the roots of the perennial herb *F. assafoetida* and other species. It is used in Asia for flavoring foods, but in the United States is employed in medicines and perfumes.

Creosote bush resin is an amber-colored, soft, sticky resin from the leaves and small twigs of the **greasewood bush**, *Larrae tridentata,* or **creosote bush,** *L. divaricata,* of the desert regions of Mexico and the southwestern United States. It is used in adhesives, insecticides, core binders, insulating compounds, and pharmaceuticals. When distilled, the resin yields **nordihydroguaiaretic acid,** $(C_6H_3 \cdot Cl \cdot COOH)_2CH_2$, called **NDGA,** formerly used as a preservative but now prohibited by the FDA for use in foods. It is a white crystalline solid melting at 184°C, soluble in fats and slightly soluble in hot water. **Okra gum** is from the pods of the plant *Hibiscus esculentus,* native to Africa but now grown in many countries. It is edible, and is used as a thickening agent in foodstuffs and pharmaceuticals. It has antioxidant properties, and also acts as a stabilizer. As a gelling agent it forms a network molecular structure with branched chains, giving slipperiness valued for foodstuff spreads. It is a tan-colored, water-soluble powder. Okra gum is also used in plating baths as a brightener.

Ammoniac is a gum resin from the stems of the *Dorema ammoniacum,* a desert perennial plant of Persia and India. It forms in hard, brittle, brownish-yellow tears, and has a peculiar fetid odor and an acrid taste. It is used in adhesives, in perfumery, and as a stimulant in medicine. In pharmacy it is called **ammoniacum. Oil ammoniac** is a yellow liquid distilled from the gum. The specific gravity is 0.890, boiling point 275°C, and it is soluble in alcohol and benzol.

RESISTANCE WIRE. The standard alloy for electrical resistance wire for heaters and electrical appliances is **nickel-chromium,** but **nickel-manganese** and other alloys are used. For consumer products made in large quantities, cost and the relative scarcity of supply of the alloying elements are important considerations. For high-temperature furnaces, tungsten, molybdenum, and alloys of the more expensive high-melting metals are employed. The much-used alloy with 80% nickel and 20 chromium resists scaling and oxidation to 2150°F (1177°C), but it is subject to an intergranular corrosion known as green rot, which may occur in chromium above 1500°F (816°C) unless modified with other elements. The 80–20 alloy has a resistivity of 108 × 10^{-8} Ω·m. The tensile strength of the annealed wire is 100,000 lb/in^2 (689 MPa), with elongation of 35%, and the hardness is Rockwell B80. The specific gravity is 8.412. In many appliances high elongation is undesirable because of sag in the wire.

In times of nickel stringency, or for cost reduction, various **nickel-chromium-iron** alloys are used. An alloy of 60% nickel, 16 chromium, and 24 iron has a resistivity of 675 Ω with oxidation resistance to 1950°F (1066°C). **Tophet C,** of the Wilber B. Driver Co., is this alloy. The alloy with 30% nickel, 20 chromium, and 50 iron is resistant to 1560°F (849°C). The resistivity of the low-nickel, chromium-iron alloys is high, and the heat resistance is ample for some types of appliances, but the strength is lower, with a tendency for the hot wire to sag. **Cromel AA,** of the Hoskins Mfg. Corp., is an 80–20 nickel-chromium alloy for continuous service to 2150°F (1127°C). It is modified with small amounts of cobalt, manganese, columbium, silicon, and iron, the columbium stabilizing the chromium to prevent green rot. It is also resistant to carbon pickup which tends to make the chromium-iron alloys brittle. The resistance of the wire is 116 × 10^{-8} Ω·m.

The **chromium-aluminum-iron** alloys have high resistivity and high oxidation resistance, but have a tendency to become brittle. **Hoskins alloy 870** contains 22.5% chromium, 5.5 aluminum, 0.5 silicon, 0.10 carbon, and the balance iron. The resistivity is 142 × 10^{-8} Ω·m. It is used as wire or ribbon in furnaces to 2350°F (1288°C). The **Kanthal alloys** marketed as wire and ribbon by the Kanthal Corp. have 20 to 25% chromium with some cobalt and aluminum, and the balance iron. **Kanthal A,** with 5% aluminum, will withstand temperatures to 2370°F (1299°C), has a resistivity of 139 × 10^{-8} Ω·m, and is resistant to sulfuric acid. The tensile strength is 118,000 lb/in^2 (813 MPa) with elongation of 12 to 16%. **Kanthal A-1,** for furnaces, has a resistance of 872 Ω and an operating temperature to 2505°F (1373°C). The **Nikrothal alloys** of this company are nickel-base modifications of Kanthal. They have higher tensile strength, up to 200,000 lb/in$_2$ (1,378 MPa), to permit rapid winding of tape without breakage. **Nikrothal 6** has 60% nickel, 15 chromium, and 25 iron. **Heating tape,** of the Rogers Corp., designed for heating rocket batteries, is also used for panel heating and is operable at continuous temperatures up to 250°F (121°C). It has three flat wires of copper-nickel alloy encapsulated in Mylar tape 0.008 in (0.020 cm) thick and 0.375 in (0.953 cm) wide in lengths to 250 ft (76 m). The rating is 2 W/ft (6.6 W/m), and the dielectric strength is 2,400 V.

A series of alloys of the Westinghouse Electric Corp., called **Hirox alloys,** contain 6 to 10% aluminum, 3 to 9 chromium, up to 4% manganese, with the balance iron except for small additions of boron and zirconium to reduce the size of the aluminum-iron grains and refine the structure. The alloy with 9% aluminum and 9 chromium has a resistivity of 850 Ω and a tensile strength of 118,000 lb/in^2 (813 MPa). At 1300°F (704°C) the tensile strength is 15,000 lb/in^2 (103 MPa) with elongation of 94%. Wire will give continuous service in air at 2350°F (1288°C) without failure.

Resistance alloys are marketed under a wide variety of trade names, and they are generally specified for specific uses rather than by composition. Controlled resistivity over a temperature range instead of high heat resistance is desired for instrument use, while a definite coefficient of expansion is required for spark-plug wire and other uses where the wire is embedded. In some cases, good heat resistance with selected low resistivity is desired. **Oxalloy 28,** of Sylvania Electric Products, Inc., is copper wire clad with 28% by weight of chromium-iron alloy. It withstands continuous service at 1300°F (704°C). The resistivity at 1100°F (593°C) is 8.6×10^{-8} Ω·m. **Neyoro G,** of the J. M. Ney Co., used for fine resistance wire in potentiometers and electronic applications where high cost is not a factor, has a high gold content with platinum, silver, and copper. The drawn wire has a tensile strength of 185,000 lb/in^2 (1,275 MPa) and high corrosion resistance.

Copper-manganese alloys have high resistivity, an alloy with 96 to 98% manganese having a resistance of more than 1,000 μΩ/cm^3. But when the manganese content is high, they are brittle and difficult to make into wire. Addition of nickel makes them ductile, but lowers the resistivity. A typical alloy contains 35% manganese, 35 nickel, and 30 copper. A resistance alloy developed by the National Bureau of Standards, called **Therlo,** contains 85% copper, 9.5 manganese, and 5.5 aluminum. **Fecraloy,** of the Wilber B. Driver Co., has 15% chromium, 5 aluminum, and the balance iron. It is for temperatures to 1400°F. **Sparkaloy,** of the same company, is a **spark-plug wire,** and is a manganese-nickel alloy. The spark-plug wire of the Hoskins Mfg. Co., called **Hoskins alloy 667,** contains 4% manganese, 1 silicon, and the balance nickel. The resistance is 25 × 10$^-$;8 Ω·m, specific gravity 8.4, and coefficient of expansion 0.0000151 in (0.00004 cm) per degrees Celsius to 500°C. **Manganin,** of the Wilber B. Driver Co., contains 80 to 85% copper, 2 to 5 nickel, and 12 to 15 manganese. It has a tensile strength of 70,000 lb/in^2 (482 MPa) and a resistance of 48 × ;10^{-8} Ω·m. It is used for coils and shunt wires in electric instruments, and also in sheet form for instrument springs. **Tophet A,** of the same company, is a standard 80–20 nickel-chromium alloy. The tensile strength is 120,000 lb/in^2 (827 MPa) and resistance 108×10^{-8} Ω·m. **Incoloy,** developed by the International Nickel Co. to replace high-nickel alloys for moderate-temperature resistance wire, contains 34% nickel, 21.5 chromium, and the balance chiefly iron. The physical properties at normal temperature are about the same as those for Inconel.

Calorite, of the General Electric Co., has 65% nickel, 8 manganese, 12 chromium, and 15 iron. **Excello metal,** of H. Boker & Co., Inc., contains 85% nickel, 14 chromium, and 0.5 each manganese and iron. It is used in electric heaters for temperatures up to 2000°F (1093°C). **Alumel,** of the Hoskins Mfg. Co., intended for temperatures up to 1250°C, has 94% nickel, 2.5 manganese, 0.5 iron, and small amounts of other elements. **Calido,** of the Driver-Harris Co., contains 59% nickel, 16 chromium, and 25 iron. **Nichrome V,** of the same company, is the 80–20 alloy. **Nichrome S** contains 25% nickel, 17 chromium, and 2.5 silicon. It is marketed in sheet form for temperatures up to 1800°F (982°C). **Comet metal,** of the same company, used for rheostats, contains 30% nickel, 5 chromium, and the balance iron. It has high strength, up to 160,000 lb/in^2 (1,103 MPa), and a resistivity of 570 Ω. The Driver-Harris resistance alloy known as **Karma** contains 20% chromium, 3 iron, 3 aluminum, 0.30 silicon, 0.15 manganese, 0.06 carbon, and the balance nickel. Its melting point is 2552°F (1400°C) and its resistivity 800 Ω, and the annealed wire has a tensile strength of 130,000 lb/in^2 (896 MPa) with elongation 25%. **Hytemco,** of the same company, is an iron-nickel alloy used for low-temperature wire. The resistance is 2.0×10^{-8} Ω·m. **Magno** is a 95% nickel, 5 manganese alloy of the same company; **Climax metal** has 74% iron, 25 nickel, and 1 manganese.

RESORCINOL. A colorless crystalline material with the composition $C_6H_4(OH)_2$ and melting point 110°C. It is very soluble in water and in alcohol. It is used in the production of plastics; in the manufacture of fluorescein; in the production of xanthane and azo dyes, particularly the fast **Alsace green;** in medicine as an antiseptic; and in making the explosive lead styphnate. Resorcinol polymerizes with formaldehyde to form the **resorcinol-formaldehyde plastics** that will cure at room temperature and with only slight pressure. They are used in strong adhesives for plywood and wood products, and do not deteriorate from acid action as some other plastics do. **Resorcinol adhesives** remain water-soluble during the working period for 2 to 4 h and are then insoluble and chemical-resistant. **Pencolite G-1215,** of the Pennsylvania Coal Products Co., is a resorcinol-formaldehyde adhesive that cures at 75°F (24°C) and gives a bonding strength of 3,000 lb/in^2 (20 MPa). The resorcinol adhesives may be modified with phenol to reduce cost and still retain the low-temperature setting. **Phenac 703,** of the American Cyanamid Co., is a phenol-modified resorcinol resin used as an adhesive for plywoods. Derivatives of resorcinol are used in medicine as specific bactericides, such as **diverinol,** which is propyl resorcinol, and **olivertol,** which is amyl resorcinol.

RHENIUM. An elementary metal, symbol Re, present in small quantities in many minerals. Rhenium has a specific gravity of 21.4, being almost twice as heavy as lead. The melting point is 5756°F (3180°C). It is a hard, silvery-white metal which takes a high polish. As a plating metal the white color is darker than that of rhodium. The 99.99% pure metal has a tensile strength of 80,000 lb/in^2 (551 MPa). It has good chemical resistance, but is soluble in nitric acid. The crystal structure is closely packed hexagonal, making it more difficult to work than the cubic-structured tungsten, but the crystal grains are tiny, and small amounts of rhenium added to tungsten give better ductility and improved high-temperature strength to the tungsten for lamp filaments and electronic wire.

Rhenium is obtained from molybdenite, which contains 0.0001 to 0.05% of the metal. But the usual source is from the flue dusts and from the sublimed **rhenium oxide,** Re_2O_7, of stack gases in the smelting of copper and other ores. It is precipitated from the flue dust of molybdenum-bearing copper ores in the form of **potassium perrhenate,** $KReO_4$, or of **ammonium perrhenate,** NH_4ReO_4. The Russians also obtain some rhenium from the Ural platinum ores. The stable **rhenia,** used in alloying is **rhenium trioxide,** ReO_3, which comes as a red powder.

The metal is obtained as a dense silvery powder which can be compacted, sintered, and cold-rolled with frequent annealing. It is marketed in the form of rod, strip, foil, and wire. **Rhenium-tungsten alloys,** with up to about 35% rhenium, are ductile and are used for high heat resistance. The melting point of the alloy, 5650°F (4023°C), is lower than that of pure tungsten but has higher recrystallization temperature and retains its strength at all temperatures. Rhenium has a higher electrical resistivity than tungsten, has high arc resistance, and does not become brittle after prolonged heating as tungsten does. The alloys are used for electronic filaments and small structural parts. Rhenium metal is used as an undercoat on graphite nozzles under the coating of tungsten to prevent the formation of tungsten carbide and thus give the full heat resistance of the tungsten. Rhenium or rhenium-tungsten electric contact give very long service life. Rhenium-tungsten versus tungsten thermocouples are good for service to 2800°C. The **tungsten-rhenium alloy** of Englehard Industries, Inc., for thermocouple use, contains 26% rhenium.

RHODIUM. A rare metal, symbol Rh, found in platinum ores such as the nickel-copper ores of Canada and **pyroxinite** of South Africa. It is very hard and is one of the most infusible of the metals. The melting point is 3542°F (1948°C). It is insoluble in most acids, including aqua regia, but is attacked by chlorine at elevated temperatures and by hot fuming sulfuric acid. Liquid rhodium dissolves oxygen, and ingots are made by argon-arc melting. At temperatures above 1200°C rhodium reacts with oxygen to form a **rhodium oxide,** Rh_2O_3. The specific gravity is 12.44. Rhodium is used to make the nibs of writing pens, to make resistance windings in high-temperature furnaces, for high-temperature thermocouples, as a catalyst, and for laboratory dishes. It is the hardest of the platinum-group metals, the annealed metal having a Brinell hardness of 135, and the rolled metal a hardness of 390. Rhodium is also valued for electroplating jewelry, electric contacts, appliances, surgical instruments, and reflectors. For electrodepositing rhodium, an electrolyte of **rhodium sulfate,** $Rh_2(SO_4)_3$, is used. The coatings are wear-resistant and tarnish-resistant. The plated metal has a pinkish-white luster of high corrosion resistance and a light reflectivity of 80%. Decorative finishes are seldom more than 0.0002 in (0.0005 cm) thick, but plates for electric contacts may be up to 0.005 in (0.013 cm). The electrical resistivity is 4.69, and the Vickers hardness is about 600. **Rhodium carbonyl** has the general formula Rh(CO), but with several variations in number of CO groups. They are used for depositing **rhodium coatings.**

The most important alloys of rhodium are **rhodium-platinum.** They form solid solutions in any proportion, but alloys of more than 40% rhodium are rare. Rhodium is not a potent hardener of platinum but increases its high-temperature strength. It is easily workable and does not tarnish or oxidize at high temperatures. These alloys are used for thermocouples and in the glass industry. A **rhodium-iron** alloy with equiatomic proportions of the metal has an ordered crystal structure and changes from antiferromagnetic to ferromagnetic in an electric field. Rhodium is sold by the troy ounce (31 g), 1 in^3 (16 cm^3) weighing 6.56 troy oz (204 g).

RICE. The white seed of the large annual grass *Oryza sativa,* growing to a height of 2 to 4 ft (0.6 to 1.2 m). It is a tropical plant native to Asia, but grows in hot, moist regions well into the temperate zones and is cultivated in many parts of the world. Rice fields are flooded after planting to control weeds. The rice seed grows in an inflorescence composed of a number of fine branches, each terminating in a single grain enclosed in a brown husk. Rice forms the staple food of more than half the population of the world. **Wild rice** was used by the Indians of North America before the first Asiatic rice was brought to South Carolina in 1694. Rice is high in starch and low in proteins. It is used as a direct food, also as flour, cereal, in puddings, and for the manufacture of starch and for alcoholic beverages. **Rice hulls** are used as stock feed, and **rice straw** is used for packing, hats, and other articles. Rice in the husk before hulling is known by the Hindu name of **paddy. Brown rice** is rice that has been cleaned but not polished. Broken grains are sold in the India trade as **coodie** or **khood,** and about 20% of the rice produced from paddy is broken. **Patna rice** does not refer to the Patna district of India, but to a variety of rice with bold and hard grains especially suited for soups as it holds its shape in boiling. **Malekized rice,** developed by the General American Transportation Co., is produced by steaming unpolished rice to force the soluble part of the bran and the vitamins into the core of the grain, and then sealing the rice kernel by gelatinization, after which it is polished. The treated rice holds it shape and does not become gummy when cooked, and the nutritional value is improved. The beverage known in Japan as **saké** is **rice wine** containing 14% alcohol, made by fermenting rice with the mold **tané koji. Rice bran oil,** used as a salad and cooking oil and in lubricants, is produced from rice bran. By wet-milling whole rice in a rice oil solution a higher yield of oil and wax is obtained together with a yield of proteins. It is clear, odorless, and neutral, with a pleasant flavor, and is resistant to oxidation and rancidity. **Rice wax** is produced from rice bran by hot hexane extraction after cold extraction of the oil. It is a hard, brown, lustrous wax with a melting point of 79°C, used in polishes. **Synthetic rice,** used in Japan as a rice extender, is made from wheat flour, potato starch, and powdered rice.

ROSEWOOD. The wood of several species of *Dalbergia* of northern South America, but chiefly from the **jacaranda tree,** *D. nigra.* It is used for fine cabinetwork, pianos, novelties, and expensive furniture. It should not be confused with the wood of the tree *Physocalymma floridum,* which also comes from Brazil and is there called **pao rosa** or rosewood. The color of rosewood is dark brown to purple, and it takes a beautiful polish. It has a characteristic fragrance. Very hard with a coarse, even grain, it weighs 54 lb/ft^3 (865 kg/m^3). The tree grows to a height of 125 ft (38 m). Another Brazilian wood, **caroba,** from the large tree *Jacaranda copaia,* is also called **jacaranda** and is sometimes confused with rosewood. The tree has purple flowers while the true rosewood has white flowers. Caroba wood is chocolate-colored, and is used for fine furniture and knife handles. **Indian rosewood** is from the tree *D. sissoo* of India. It is also called **sissoo,** and is a beautiful, brown hardwood employed for carvings. In Europe it is also used for parquet floors. **Borneo rosewood,** also known as **ringas,** is the wood of several species of trees of the genus *Melanorrhoea* of Borneo. The wood has a deep-red color with light and dark streaks. It has a close texture suitable for carving. **Satinee** is a type of rosewood from the tree *Ferolia guianensis* of the order *Rosaceae,* native to tropical America, particularly the Guianas. The wood is reddish brown, weighs 54 lb/ft^3 (865 kg/m^3), is fairly hard, has a fine grain, and takes a lustrous polish. It is used for cabinetwork. **Bois de rose oil,** or **rosewood oil,** is not from rosewood, but is extracted from the heartwood of the tree *Aniba panurensis* of Brazil and the Guianas, though the wood of this tree is also used as a cabinetwood. The oil is also called **Cayenne linaloe,** or **Cayenne oil.** It has a delicate rose odor with a suggestion of orange and mignonette valued in perfumes. It contains a high percentage of **linalol,** a colorless alcohol with a soft, sweet odor, also found in the rose, lilac, lily, lavender, petit-grain, and other plants. It also contains geraniol. **Linaloe oil,** or **Mexican linaloe,** is distilled from the heartwood of the trees *Bursera delpechianum* and *B. aloeoxylon.* It contains less linalol and also terpineol and geraniol. Linalol is closely related to geraniol and nerol. Bois de rose is also made synthetically from geraniol. **Oriental linaloe** is distilled from selected highly perfumed parts of the wood of the large tree *Aquilaria agollocha* of eastern India, Burma, and Java. The odor of the oil is like rose, ambergris, and sandalwood. Like the linaloes of the American continent the oil is a pathological product and comes only from old trees. It is also called **aloe wood oil** and **agar attar,** and is a very ancient perfume. The beautifully figured and fragrant reddish wood of this tree, called **aloes wood, eagle wood,** and **paradise wood,** is used for ornamental articles. True original rosewood oil known as **rhodium oil** was distilled from the wood of the plant *Convolvulus scoparius* of the Canary Islands.

ROSIN. The common resin of several varieties of the pine tree, found widely distributed in North America and Europe. It is obtained by cutting a longitudinal slice in the tree and allowing the exudation to drip into containers. The liquid resin is then distilled to remove the turpentine, and the residue forms what is known as **gum rosin,** or **pine gum. Wood rosin** is obtained by distillation of old pine stumps. It is darker than gum rosin, and is inferior for general use.

Rosin contains seven acids with very similar characteristics, but consists chiefly of **abietic acid,** $C_{19}H_{29}COOH$. Normally, when gum rosin is heated, the natural **pimaric acid** isomerizes to form abietic acid, but in the production of turpentine and rosin from pine sap, the turpentine is removed by steam distillation, and various acids are then extracted. Pimaric acid is closely related to abietic acid. It reacts with maleic and anhydride, and the **maleopimaric acid** is used in printing inks and coatings. Rosin has a specific gravity of about 1.08, a melting point of about 82°C, and it is soluble in alcohol, turpentine, and alkalies. It is used in varnishes, paint driers, soluble oils, paper sizing, and belt dressings; for compounding with rubber and other resins; and for producing many chemicals.

Rosin is generally graded commercially by letters according to color. The darkest grade is B, and the lightest is W. Extra grades are A, nearly black, and WW, water white. Thirteen color grades are designated under the Naval Stores Act. The dark grades of wood rosin are considered inferior. They have high melting point and low acid number and are used for making rosin oil, for battery wax, thermoplastics, and dark varnish, and for linoleum manufacture. The ruby-red wood rosin, obtained by extraction from fat pine wood, has high acid number, 155, and low melting point, 175°F (79°C). It is used for printing inks, paper size, and adhesives. Rosin is usually marketed in barrels of 280 lb (127 kg). **Naval stores** is an old name for rosin and turpentine. **Pelletized rosin** consists of free-flowing dustless pellets produced by coating droplets of molten rosin with inert powder. **Colophony** is an old pharmacy name for rosin before distillation of rosin oil. Rosin was referred to by early writers as **Greek pitch,** but the ancient incendiary known as **Greek fire** was tow or pine sawdust impregnated with rosin, pitch, and sulfur. **Burgundy pitch** was originally the resin of the Norway spruce, *Picea abies,* used in medicine, but the name was later applied to a rosin, rubber, and mineral oil compound used for friction tape.

Hardened rosin is a weak resinate made by adding 6 to 8% high-calcium lime to melted rosin. It is used in some varnishes. **Fosfo rosin,** of Newport Industries, Inc., is a **lime-hardened rosin.** It is an FF rosin treated with 4.75% calcium hydrate, which raises the melting point, decreases the free rosin acids, and decreases the tendency to crystallize. It is used in paints, varnishes, and molded products. **Soda-treated rosins,** with about 1% Na_2O, but no free alkali, are used for soap, paper size, and disinfectants. **Rosin size** is alkali-treated rosin in dry powder or emulsion forms for sizing paper. **Dresinite** is such a sodium or potassium salt of rosin. **Cyfor,** of the American Cyanamid Co., is a rosin size fortified with a synthetic resin to give increased water and acid resistance to paper. **Rosin ester,** or **ester gum,** is prepared by heating rosin with glycerin. It is lighter in color than rosin, has a higher softening point, and has a much lower acid number, usually 7 to 9. It is used with tung oil in enamels and varnishes and in adhesives. **Resin V,** of the Advance Solvents & Chemical Corp., is a rosin glycerin ester gum. Rosin esterified with glycerin has lower molecular weight, and is not as stable as rosin esterified with pentaerythritol or other tetrahydric alcohol, but modified rosin ester gums develop hardness quickly in nitrocellulose and are used for such purposes as furniture lacquers. **Cellolyn 102,** of the Hercules Co., is a modified ester gum of this type, and **Lewisol 28** is a maleic alkyd modified rosin ester used for hard, glossy furniture lacquers. **Hydroabietyl alcohol,** used as a plasticizer and tackifier for rubber and for sizing textiles, is a colorless, tacky liquid made by reduction of rosin. The 85% alcohol has a specific gravity of 1.008 and a flash point of 187°C. It is also used for making rosin esters. **Abitol,** of Hercules, is this material. **Abalyn** is a methyl ester of abietic acid, **methyl abietate,** made by treating rosin with methyl alcohol. It is a liquid rosin used as a plasticizer.

Hydrogenated rosin has greater resistance to oxidation than common rosin, has less odor and taste, and has a pale color that is more stable to light. It is used in protective coatings, in paper size, in adhesives, in soaps, and as a tackifier and plasticizer in rubber. Because of its saturated nature it cannot be used for rosin-modified plastics. The average acid number is 162, saponification value 167, and softening point 157°F (69°C). **Stabelite resin,** of Hercules, may be glycerol ester or ethylene glycol ester of hydrogenated rosin. **Vinsol resin** is a hard, high-melting, dark resin produced from the distillation of wood, or is the black residue left after rosin is extracted with petroleum solvents. It is soluble in alcohols, has a melting point of 115°C, and is used for insulating varnishes where light color is not essential, and for compounding in thermoplastics. **Hercolyn** is Abalyn hydrogenated to saturate the double bonds with hydrogen. **Flexalyn,** of the same company, is a pale-colored, very tacky, semisolid resin produced by the esterification of rosin acids with diethylene glycol, and has a complex chemical structure. It is used in adhesives to give added tack and strength. The **Pentalyn resins** of Hercules are pentaerythritol esters. The **Pentalyn M** is a phenol-formaldehyde-modified pentaerythritol ester; it has a melting point of 165°C, and when used in linseed oil varnishes gives a tough coating.

Rosin is hardened by polymerization to form a dipolymer of abietic acid. The product is then pale in color, and has a lower acid number and a higher melting point than rosin. **Poly-pale resin,** of Hercules, is a polymerized rosin with melting point of 208 to 217°F (98 to 103°C), acid number 152 to 156, and saponification value 157 to 160. It can be substituted for natural copals in paints, and in gloss oil it gives water resistance and high viscosity. In the making of metallic resinates it gives higher melting points, higher viscosity, and better solubility than natural rosin. Another modified rosin of this company is **Dymerex resin.** It consists chiefly of dimeric rosin acids, is highly soluble, is resistant to oxidation, and has a high softening point at 282°F. It is used in synthetic resins and protective coatings. **Rosin amine D,** of the same company, is a primary amine made from rosin. It is a yellow viscous liquid that wets glass and siliceous materials. It is soluble in most organic solvents, and emulsifies in water. It is used in cutback asphalts, in road asphalts, in asphalt cements, in ceramic inks, in foundry core binders, and in paper pulp to improve adhesion of resins. **Nuroz** is a polymerized rosin of Newport Industries, Inc., with a melting point of 76°C and acid value 161. It has high resistance to oxidation, and is used in varnishes and soaps.

Rosin oil is an oil produced by the dry distillation of rosin at a temperature of 200 to 360°C. There are two qualities of the oil: a light spirit, **pinolin,** which forms from 1 to 5% of the rosin, and a bluish, heavy oil, which forms 80 to 84%. It contains abietic acid and has an acid value of about 28. The commercial oil has a specific gravity of 1.020 with a flash point 160 to 170°C. The refined oil is a yellow liquid with a pleasant odor and is used for blending with turpentine. It is also employed as a plasticizer in rubber, as a tack producer in rubber cements, and in synthetic molding resins. When treated with lime, it may be used to mix with lubricating oils. The light distillate is used sometimes in pharmacy under the name of **oil of amber.** Blended rosin oil is a mixture with mineral oils.

ROUGE. A hydrated iron oxide used for polishing metals and in break-in lubricants for aluminum bronze bearings. It has a hardness of 5.5 to 6.5, and is made by calcining ferrous sulfate and driving off the sulfur. The color is varying shades of red; the darker the color, the harder the rouge. The grains are rounded, unlike the grains of crocus. The pale-red rouge is used for finishing operations; the other grades are used for various polishing of metal surfaces. **Stick rouge** is made of finely crushed powder. Although the word rouge means red, materials of other colors are used for buffing and are called rouge. **Black rouge,** also called **Glassite,** is magnetic iron oxide made by precipitating ferrous sulfate with caustic soda. It is used for buffing but is not popular because it stains the skin. **Green chrome rouge** is **chromium oxide,** CrO, made by the strong heating of chromic hydroxide. It is used for buffing stainless steels. When used as a paint pigment, it is called **Guignet's green. Satin rouge** is a name applied to lampblack when used as a polishing medium in the form of brick for polishing silverware. **Crocus** is a name applied to mineral powders of a deep-yellow, brown, or red color made into cakes with grease for polishing. **Polishing crocus** is usually red ferric oxide. **Crocus cloth** is a fabric coated with red iron oxide marketed in sheets and used for polishing metals.

RUBBER. A gum resin exudation of a wide variety of trees and plants, but especially of the tree *Hevea brasiliensis* and several other species of *Hevea* growing in all tropical countries and cultivated on plantations in southern Asia, Indonesia, Sri Lanka, Zaire, and Liberia, from which **natural rubber elastomers** are made.

The gum resin was formerly referred to as **India rubber,** and the name given it by Charles Goodyear was **gum elastic.** The first highly compounded rubber for insulation, developed in 1867, was called **Kerite.** Brazilian rubber is sometimes called **Pará rubber. Caoutchouc** was an early name for the crude rubber then cured over a fire into a dark, solid mass for shipment. **Castilla rubber,** or **castilloa,** is from the large tree *Castilla elastica,* and was the original rubber of the Carib and Mayan Indians, but was cultivated only in Mexico and in Panama where it was called **Panama rubber.** The latex and rubber are identical with **hevea rubber** after purification . **Euphorbia rubber** is from vines of the genus *Euphorbia,* of which there are 120 species in tropical Africa. Much **mangabeira rubber** was formerly produced in the Amazon Valley. It is the latex of the mangabeira tree, which comprises various species of the genus *Hancoria* and yields the edible fruit **mangaba.** The latex is coagulated with alum or sodium chloride, but the native Indians coagulated it with the latex of the **caxiguba** tree, *Ficus anthelmintica,* giving a better rubber. The rubber is softer than hevea rubber, but ages better. The low-grade **Assam rubber** is from a species of **ficus tree,** *F. elastica,* of India and Malaya. **Ceara rubber** comes from the small, rapid-growing tree, *Manihot glaziovii,* native to the semidesert regions of Brazil but now grown in India and Sri Lanka. The rubber is of good grade.

Rubber latex is a colloidal emulsion of the gathered sap, containing about 35% of rubber solids, blended from various sources to give average uniformity. The latex is coagulated with acid and milled into ribbed sheets called **crepe rubber,** or into sheets exposed to wood smoke to kill bacteria and called **smoked sheet rubber.** These sheets constitute the commercial **crude rubber,** although much rubber latex is used directly, especially for dipped goods such as gloves, toys, and ballons, for coatings, and for making foam rubber. Rubber has the property of being vulcanized with sulfur and heat, removing the tackiness and making it harder and more elastic in the low-sulfur compounds. All natural rubber except adhesive rubbers is thus vulcanized rubber. Ordinary **soft rubber** contains only 3 to 6% sulfur, but usually also contains softeners, fillers, antioxidants, or other compounding agents, giving varying degrees of elasticity, strength, and other qualities. When as much as 30% sulfur is added, the product is called **hard rubber. Vapor-cured rubber** is rubber vulcanized by sulfur chloride fumes and neutralized with magnesium carbonate. It is used for thin goods only. **Acid-cured rubber** is rubber cured in a bath of sulfur chloride in a solvent.

The tensile strength of rubber of low vulcanization is 800 to 1,200 lb/in^2 (6 to 8 MPa) of the original cross section. A good soft rubber can be stretched as much as 1,000% without rupture, and will return to close to the original length with little permanent set. The specific gravity is about 1.05, but with fillers may be as high as 1.30. When the term **vulcanized rubber** is now used, it generally refers to **hard rubber** vulcanized to a rigid but resilient solid, used for electrical parts and tool handles. **Ace hard rubber,** of the American Hard Rubber Co., has a specific gravity of 1.27, a tensile strength of 8,700 lb/in^2 (60 MPa), dielectric strength of 485 volts per mil (19×10^6 volts per meter), distortion temperature 172°F (78°C), and water absorption 0.04%. **Vulcanite** and **Ebonite** are old names for hard rubber. **Reclaimed rubber** is produced largely from old tires and factory scrap. It is usually lower in cost than new rubber, but it is easier to process and is employed in large quantities even when the price is higher. It is sold in sheets, slabs, pellets, and powder, but much of the **rubber**

powder, or **granulated rubber,** used for adhesives and molding is not reclaimed rubber but is made by spray-drying latex. **Vultex** is a natural rubber powder in paste form for coatings and adhesives. **Mealorub** is a rubber powder developed by the Indonesian Rubber Research Institute for mixing with asphalt for road surfacing.

Several types of modified natural rubber are used in the production of coatings, protective films, and adhesives. These types are **chlorinated rubber, rubber hydrochloride,** and **cyclized rubber** or **isomerized rubber.** Chlorinated rubber, for example, modified with any one of a number of plastic resins, provides maximum protection against a wide range of chemicals, and the coatings are widely used in chemical plants, in gas works, and as tank-car linings.

Red rubber is now simply rubber colored red, but was originally rubber vulcanized with antimony pentasulfide which broke down with the heat of vulcanization, yielding sulfur to the rubber and coloring it red with the residual antimony trisulfide. Many trade-name accelerators, fillers, and stiffeners are marketed for rubber compounding. **Crumb rubber** is any rubber in the form of porous particles that can be dissolved easily without milling, cutting, or pelletizing. It is used in adhesives and plastics. **Magnetic rubbers,** produced in sheets and strips of various magnetic strengths, are made of synthetic rubbers compounded with magnetic metal powders.

RUBIDIUM. A rare metallic element, symbol Rb, atomic weight 85.45, belonging to the group of alkali metals. The chief occurrence of rubidium is in the mineral lepidolite. There is no real rubidium ore, but the element is widely disseminated over the earth in tiny quantities. It is a necessary element in plant and animal life, and is found in tea, coffee, tobacco, and other plants. It is a silvery-white metal, with a specific gravity of 1.53, melting point of 38.5°C, and boiling at 696°C. It takes fire easily in the air, and decomposes water. Of all the alkali metals it is next to cesium in highest chemical activity. It can be obtained by electrolysis, but has few industrial applications owing to its rarity. Its chief use is in electronics. For photoelectric cells it is preferred to cesium, and a very thin film is effective. Like potassium, it has a weak radioactivity by the emission of beta particles, the beta emission being only about one-thousandth that of an equal weight of uranium.

RUBY. A red variety of the mineral **corundum** which ranks with the best grades of precious stones as a gemstone, while the off-color stones are used for watch and instrument bearings. Most of the best rubies come from Upper Burma, Thailand, and Cambodia, but the center of natural-ruby cutting is near Bombay. Some deep-red rubies are found in East Africa, and they also occur in western North Carolina. The carmine-red, or pigeon's-blood, stones are the most highly prized. Before the advent of the synthetic ruby, the larger stones were more valuable than the diamond. The pink-to-deep-red colors of the ruby are due to varying percentages of chromic oxide. **Star rubies** contain also a small amount of titania which precipitates along crystallographic planes of the hexogonal crystal and shows as a movable six-ray star when the gem is cut with the axis normal to the base of the stone. **Spinel ruby** is not corundum but magnesium aluminate and is a spinel, often occurring in the same deposits.

Synthetic rubies are equal in all technical qualitities to the natural, and synthetic star rubies surpass the natural stones in perfection and quality. Most of the ruby used for instrument bearings is synthetic corundum colored with chromic oxide, since the instrument makers prefer the red color, but the name ruby is often applied regardless of color. The American practice in bearing manufacture is to start with a cylindrical rod of the diameter of the desired bearing and to slice to the required thickness. The single crystal rods are flame-polished. For industrial uses the name ruby is often applied to the synthetic material even when it is not red in color. Synthetic rubies with 0.05% chromium are also used for **lasers,** or light amplifiers, to produce high-intensity light pulses in a narrow beam for communications. Lasers have a wide range of uses, and various other materials are used for specific purposes. Crystals of **potassium dihydrogen phosphate** are used to control the direction of narrow beams of light. The crystal is mounted on the face of the cathode-ray tube. Platelets of zinc oxide with a phosphor, when activated at 15,000 to 20,000 volts by a pulsated electron stream, give a beam of pure ultraviolet light at 375 nm for use in chemical synthesis.

Rubies are also used in **masers** to detect radio signals for space rockets at great distances. The word maser means microwave application by stimulated emission of radiation. Ruby has the same physical and chemical properties as the sapphire and corundum. But the color inclusions do affect the electronic properties. The **ruby-sapphire crystals** produced by Linde for **optical masers** are grown with a core of ruby and an overlay sheath of sapphire. The ruby core containing 0.05% chromic oxide emits a beam of extremely high frequency, 4.2×10^{14} cycles, for sending messages. The sapphire has a high refractive index, 1.76, and when the sheath is surrounded by a helical xenon arc, the light from the arc excites the chromium atoms in the ruby to emit a concentrated beam of parallel rays from one end.

The cutting of rubies, sapphires, and other hard crystals into tiny shaped bearings was formerly a specialized hand industry, and large stocks of cut stones were kept in the National Stockpile for wartime emergencies. But the slicing and shaping of hard crystals are now widely dispersed on a production basis to meet the needs of the electronic industries, and the equipment is regularly manufactured. For example, the **Accu-Cut wheels** of Aremco Products, for slicing and notching crystals and hard electronic ceramics, are metal discs 0.010 to 0.030 in (0.025 to 0.076 cm) thick and 1 to 4 in (2.54 to 10.2 cm) in diameter, with diamond grit metallurgically bonded to the periphery.

RUTHENIUM. A hard, silvery-white metal, having a specific gravity of 12.2, a melting point of 2450°C, and a Brinell hardness of 220 in the annealed state. The metal is obtained from the residue of platinum ores by heat reduction of **ruthenium oxide,** RuO_2, in hydrogen. Ruthenium is the most chemically resistant of the platinum metals, and is not dissolved by aqua regia. It is used as a catalyst to combine nitrogen in chemicals. As **ruthenium tetroxide,** RuO_4, it is a powerful catalyst for organic synthesis, oxidizing alcohols to acids, ethers to esters, and amides to imides. Ruthenium has a hexagonal lattice structure. It has a hardening effect on platinum, 5% addition of ruthenium raising the Brinell hardness from 30 to 130, and the electrical resistivity to double that of pure platinum. **Ruthenium-platinum alloys** are used for electric contacts, electronic wires, chemical equipment, and jewelry. The alloy with 5% ruthenium has a tensile strength, annealed, of 60,000 lb/in^2 (413 MPa) with elongation of 34% and Brinell hardness of 130. The hard metal has a hardness of Brinell 210. The alloy with 10% ruthenium has a tensile strength of 85,000 lb/in^2 (586 MPa), and a hardness of Brinell 190 in the soft condition and Brinell 280 when hard-drawn.

RYE. The seed of the plant *Secale cereale* used as a food grain, but in the United States and Great Britain it is valued chiefly for the production of whiskey and alcohol and for feeding animals. Only 4% of the world production is in the United States. The grain looks like wheat, and the stalks of the plant are slender and tough, growing to a height of up to 6 ft (1.8 m). But flour made from the grain produces bread that is dark in color, bitter, and soggy. When used for flour in the United States, it is mixed with wheat and other flours. The plant has the advantage that it will grow on poor soil, in arid regions, at high altitudes, and in regions of severe winter. It is thus a grain of poor agricultural countries, and has been called the **grain of poverty. Rye straw** is the dried and sun-bleached stalks of the plant. It is very tough and resilient, and is the most valued of all the commercial **straw** derived from grains. It is used for packing, bedding, and for the manufacture of strawboard.

SAGO FLOUR. A starch extracted from the pith of the sago palm, *Metroxylon sagu,* of Indonesia and Malaya, and also from the aren palm. Sago is valued industrially for sizing and filling textiles because, like tapioca, it holds mineral fillers better than other starches. It gives a tougher and more flexible feel than tapioca, but its tan color limits its use. From 600 to 800 lb (272 to 363 kg) of crude sago is obtained from a tree, which is destroyed in the process. **Pearl sago,** used for food, is the same material made into dough and forced through a sieve. The aren palm, or **sugar palm,** is the species which yields arenga fiber. It contains only 20% as much sago as the sago palm, but a juice, called **taewak,** is produced from the cut flower stems and is used to make **palm wine,** or **arak.** The juice is also boiled down to produce a brown **palm sugar** used for sweetening.

SALMON OIL. A pale-yellow oil obtained as a by-product in the salmon-canning industry, and employed as a drying oil for finishes and also used in soaps. There are different classes of the oil, depending upon the type of salmon. The oil contains an average of 23.5% **arachidonic acid,** $C_{19}H_{31}COOH$, 16.2% clupanodonic, 11.5 linoleic, 17.1 oleic, 15 palmitic, 10.6 **palmitoleic acid,** $C_{15}H_{29}COOH$, 4 myristic, and 2 stearic. The specific gravity is 0.926. It has a high iodine number, up to 160, but does not form an elastic skin on drying, and is not a good varnish oil untreated. It is, however, a valuable source of fatty acids for paint-oil blends and for plastics. The salmon is a valuable food fish and is extensively canned. There are five commercial species of North Pacific salmon of the genus *Oncorhynchus;* the **steelhead trout,** or **salmon trout,** *Salmo gairdneri,* is of the Atlantic. The **Atlantic salmon,** caught off Newfoundland, is *S. salar*. The **red salmon,** or **sockeye salmon,** is *O. nerka;* the **pink salmon** is *O. gorbuscha;* and the **Chinook,** or **king salmon,** is *O. tschawytscha.* The catch is in the rivers on both sides of the Alaska peninsula and in the Columbia River where the fish enter the rivers to spawn. **Australian salmon,** which is the chief fish canned in Australia, is of a different genus. *Arripis trutta.*

SALT. The common name for **sodium chloride,** known in mineralogy as **halite** but chemically a salt; it is any compound derived from an acid by replacing hydrogen atoms of the acid with the atoms of a metal. **Common salt,** or sodium chloride, is widely used as a preservative, for flavoring food, in freezing mixtures, for salt-brine quenching baths, and for the manufacture of soda ash and many chemicals. Common salt has such a variety of uses that its curve of consumption practically parallels the curve of industrial expansion. It is a stable compound of the composition NaCl, containing theoretically 60.6% chlorine, but it usually contains impurities such as calcium sulfate and calcium and magnesium chlorides. The hardness of salt is 2.5, specific gravity 2.1 to 2.6, and melting point 1472°F (800°C). It is colorless to white, but when impure may have shades of yellow, red, or blue. It occurs in crystalline granular masses with cubical cleavage, known as **rock salt,** or **mineral salt.** Vast deposits of salt are found underground in Louisiana, and a large area of Kansas is underlain with a salt deposit reaching 800 ft (244 m) in thickness. The rock salt occurring in immense quantities on the island of Hormoz in the Persian Gulf contains 97.4% NaCl, 1.83 $CaCO_3$, and only very small amounts of magnesium chloride, iron oxide, and silica. **Bay salt** is an old name for salt extracted from seawater, now known as **solar salt.** Seawater also contains more than 20% of magnesium chloride and magnesium, calcium, and potassium sulfates, which are extracted to give a purity of at least 99% sodium chloride. A short ton (907 kg) of **seawater** contains about 55 lb (25 kg) of common salt. But some **sea salt** containing all the original elements is marketed for corrosion tests. The **Sea-Rite salt** of the Lake Products Co. is a synthetic sea salt containing all the elements of natural sea salt except those of less than 0.0004%. From the salt wells of Michigan, magnesium, bromine, and other elements are extracted, and the salt brine is an important source of these elements.

Commercial salt is marketed in many grades, depending chiefly on the size of the grain. The term **industrial salt** refers rather to the method of packing and shipping than to a grade distinct from **domestic salt,** but most of the industrial salt is rock salt; the bulk of the domestic salt is evaporated salt. Producers of salt for the food-processing industries usually guarantee a quality 99.95% pure since small amounts of calcium, magnesium, copper, and iron in the salt may give a bitter taste, discolor some foods, or cause oxidation rancidity in foods containing fats. **Crystal Flake salt,** of Diamond Salt, is 99.5% pure with less than 1.5 parts per million of copper or iron. **Micronized salt,** of the same company, is 99.9% pure with thin flake particles of super fineness which have a solubility five times greater than granulated salt. **Dendritic salt,** of Morton Salt, is highly purified salt evaporated by a process that produces tiny dendritic crystals instead of the regular cubic form. It has a faster dissolving rate. This company produces **Calacid salt,** a formulated salt in tablet form for regulating acidity in canned foods. It contains 60.9% sodium chloride with citric acid, calcium sulfate dihydrate, and sodium carbonate for effervescence. **Sec 100,** of Diamond Salt, is a noncaking, free-flowing salt for use in automatic food-processing equipment. It contains 5 ppm of prussiate of soda which blocks the interchemical action of the salt crystals without decreasing solubility, and 0.04% glycerin to increase the free-flowing properties.

Domestic consumption of salt for direct human consumption is large in all countries. It is required in the bloodstream up to about 3.5% and is rapidly exhausted in hot weather. **Salt tablets,** used in hot weather, or by workers in steel mills, are made with about 70% salt and 30 dextrose. To make domestic salt free-flowing in humid weather, 2% of calcium sulfate may be added. Salt obtained by simple evaporation of seawater contains salt-resistant bacteria which are capable of developing in salted hides or fish and injuring the material. Mineral salt is thus preferred for these purposes. The glasslike salt crystals, called **struvite,** that form in canned fish and some other products in storage, are not common salt, but are crystals of magnesium ammonium phosphite hexahydrate, $Mg(NH_4)PO_4 \cdot 6H_2O$. They are harmless, but objectionable in appearance, and common salt is added to inhibit their growth. In making **salt brines** for steel treating and other industrial purposes, 100 parts of water at ordinary temperature will dissolve 36 parts of common salt, but no more than 15% is ordinarily used because of the corrosive effect.

Sodium hypochlorite, or **sodium oxychlorite,** NaOCl, is a stable, noncorrosive salt used in tanneries. **Merclor D** is a trade name of Monsanto for this material in water solution. **Javel water** is a name given in the laundry industry to a water solution of sodium hypochlorite used as a bleach. Sodium hypochlorite when used as a bleach in the textile industry is called **chemic. Sodium chlorite,** $NaClO_2$, is a white to yellow crystalline water-soluble powder used as a bleaching agent for textiles and paper pulp. It is stable up to 150°C. It yields ClO_2 in the solution, is an oxidizing agent, and attacks the coloring matter without injuring the fibers. It is also used for waterworks purification. **Textone** is a trade name for sodium chlorite as a bleach for textiles. **Sodium chlorate,** $NaClO_3$, is used in large quantities as a weed killer and for cotton defoliation, and is also used for paper pulp and textile bleaching. It comes in water-soluble colorless crystals melting at 250°C. It is used as the electrolyte in the chemical machining of metals.

SAND. An accumulation of grains of mineral matter derived from the disintegration of rocks. It is distinguished from gravel only by the size of the grains or particles, but is distinct from clays which contain organic materials. Sands that have been sorted out and separated from the organic material by the action of currents of water or by winds across arid lands are generally quite uniform in size of grains. Usually commercial sand is obtained from river beds, or from sand dunes originally formed by the action of winds. Much of the earth's surface is sandy, and these sands are usually quartz and other siliceous materials. The most useful commercially are **silica sands,** often above 98% pure. Silica sands for making glass must be free from iron. The sand mined near Hot Springs, Ark., called **amosil,** is 99.5% pure silica, and comes in transparent rounded grains of 3 μm average size. **Beach sands** usually have smooth spherical to ovaloid particles from the abrasive action of waves and tides and are free of organic matter. The white beach sands are largely silica but may also be of zircon, monazite, garnet, and other minerals, and are used for extracting various elements. **Monazite sand** is the chief source of thorium. The **black sands** of Oregon contain chromate, and those of Japan contain magnetite. Kyanite is found in the Florida sands.

Sand is used for making mortar and concrete and for polishing and sandblasting. Sands containing a little clay are used for making molds in foundries. Clear sands are employed for filtering water. Sand is sold by the cubic yard or ton but is always shipped by weight. The weight varies from 2,600 to 3,100 lb/yd3 (1,538 to 1,842 kg/m^3), depending on the composition and size of grain. Construction sand is not shipped great distances, and the quality of sands used for this purpose varies according to local supply. **Standard sand** is a silica sand used in making concrete and cement tests. The grains are free of organic matter and will pass through a 20-mesh sieve, but will be retained on a 30 mesh. **Engine sand,** or **traction sand,** is a high-silica sand of 20 to 80 mesh washed free of soft bond and fine particles, used to prevent the driving wheels of locomotives or cars from slipping on wet rails.

Molding sand, or **foundry sand,** is any sand employed for making molds for casting metals, but especially refers to sands that are refractory and also have binding qualities. Pure silica is ideal for heat resistance, but must contain enough alumina to make it bind together. Molding sands may contain from 80 to 92% silica, up to about 15% alumina, about 2% iron oxide, and not more than a trace of lime. Some molding sand contains enough clay or loam to bond it when tamped into place. The amount of bond in **Grant sand** and in **Tuscarawa sand** is 17 to 18%. About 33% of these natural sands pass through a No. 100 screen, and 20 through a 150 screen. The finer the grain, the smoother the casting, but fine-grained sand is not suitable for heavy work because of its impermeability to the gases. Sands without natural bond are more refractory and are used for steel molding. Sands for steel casting must be silica sands containing 90% silica, or preferably 98%, and are mixed with 2 to 10% fireclay. For precision casting, finely ground aluminum silicate is used in the silica sand mixes, and it requires less bonding agent. **Calamo,** of the Harbison-Walker Refractories Co., is aluminum silicate for this purpose.

Zircon sand has high heat resistance, and is used for alloy steel casting. **Zircon flour** is finely milled zirconite sand used as a mold wash. **Zirconite sand** for molding is 100 to 200 mesh in its natural state. It is 70% heavier than silica sand, and has a higher heat conductivity that gives more rapid chilling of castings. The zirconite sands have melting points from 3650 to 3850°F (2010 to 2121°C). Common molding sands may contain from 5 to 18% of clay materials, and may be mixtures of sand, silt, and clay, but they must have the qualities of refractoriness, cohesiveness, fineness of grain, and permeability. To have refractory quality they must be free of calcium carbonate, iron oxide, and hydrocarbons. **Core sands** also have these qualities, but they are of coarser grain, and always require a bond that will bake solidly but will break down easily at the temperature of pouring. About 25% of a medium molding sand will be retained on a 150-mesh sieve, and about 10% on a 200-mesh sieve. Sand with rounded grains is preferred, and the grains must be very uniform in size to prevent filling. When a molding sand is burned out, it is made suitable for reuse by adding bond, but when fireclay is used as a bond, it adheres to the sand grains and makes it unsuitable for reuse. **Parting sand** is a round-grained sand without bond used on the joints of molds. **Foundry parting** is usually tripoli or bentonite. Cores are made with sand mixed with core oils. **Greensand cores** are unbaked cores made with molding sand.

Sandblast sand is sand employed in a blast of air for cleaning castings, removing paint, cleaning metal articles, giving a dull rough finish to glass or metal goods, or renovating the walls of stone or brick buildings. Sandblast sand is not closely graded, and the grades vary with different producers. The U.S. Bureau of Mines gives the following usual range: No. 1 sand should pass through a 20-mesh and be retained on a 48-mesh screen; No. 2 should pass through a 10-mesh and be retained on a 28-mesh screen; No. 3 through a 6-mesh and be retained on a 14-mesh screen; No. 4 through a 4-mesh and be retained on an 8-mesh screen. No. 1 sand is used for light work where a smooth finish is desired; No. 4 sand is employed for rough cast-iron and cast-steel work. Sharp grains cut faster, but rounded grains produce smoother surfaces. The sand is usually employed over and over, screening out the dust. The dust and fine used sand may be blasted wet. This is known as mud blasting and produces a dull finish.

SANDALWOOD. The heartwood of the evergreen tree *Santalum album* and other species of southern Asia. The heartwood is usually equivalent to about one-third of the log. It is sweet-scented and is used for chests, boxes, and small carved work. The chips and sawdust are used for incense and for oil production. **Sandalwood oil** is a yellowish essential oil of specific gravity 0.953 to 0.985, distilled from the wood, which yields 5 to 7% of the oil. It is used in medicine, perfumery, and soaps. West Indian sandalwood oil is called **anyris oil. Australian sandalwood** oil is from the tree *S. spicatum.* It has a very strong and lasting sandalwood odor. **Sandela,** of Givaudan-Delawanna, Inc., is a synthetic sandalwood oil. It is a polycyclic alcohol product with the odor and properties of the natural oil.

SANDARAC. Known also as **white gum,** or **Australian pine gum.** A white, brittle resin obtained as an exudation from various species of the coniferous tree *Callitris,* known as **Cyprus pine.** The North African sandarac is from the tree *C. quadrivalis* of the Atlas Mountains, and resembles the resin from the Australian tree *C. arenosa.* The trees in Morocco are tapped from May to June, and two months later the small tears of gum are gathered. Sandarac is used in varnishes and is soluble in turpentine and alcohol. It melts at 135 to 140°C. It gives a hard, white spirit varnish used for coating labels and for paper and leather finishes. Ground sandarac, under the name of **pounce,** was formerly used as a pouncing powder and for smoothing parchment and tracing cloth, but is now replaced by pumice.

SANDPAPER. Originally a heavy paper coated with sand grains on one side, used as an abrasive, especially for finishing wood. Sharp grains obtained by crushing quartz later replaced sand, and the product was called flint paper. But most abrasive papers are now made with aluminum oxide or silicon carbide, although the term sandpapering is still employed in wood polishing. Quartz grains, however, are still much used on papers for the wood industries. For this purpose the quartz grains are in grades from the 20 mesh, known as No. 3½, through No. 3, 2½, 2, 1½, 0, 00, and 000. All of the No. 000 grains pass through a 150-mesh sieve, with 25% retained on a 200-mesh sieve and 80% on a 325-mesh sieve. Good **sandpaper quartz** will contain at least 98.9% silica. The paper used is heavy, tough, and flexible, usually 70- or 80-lb paper, and the grains are bonded with a strong glue. A process is also employed to deposit the grains on end by electrostatic attraction so that the sharp edges of the grains are presented to the work.

SANDSTONE. A consolidated sand rock, consisting of sand grains united with a natural cementing material. The size of the particles and the strength of the cement vary greatly in different natural sandstones. The most common sand in sandstone is quartz, with considerable feldspar, lime, mica, and clayey matter. The cementing material is often fine chalcedony. **Silica sandstones** are hard and durable but difficult to work. **Calcareous sandstone,** in which the grains are cemented by calcium carbonate, is called **freestone** and is easily worked, but it disintegrates by weathering. Freestone is homogeneous and splits almost equally well in both directions. **Chert,** formerly used as an abrasive and, when employed in building and paving, known under local names as **hearthstone, firestone,** and **malmstone,** is a siliceous stone of sedimentary origin. It has a radiating structure and splintery fracture and is closely allied to flint. In color it is light gray to black or banded. The colors of sandstones are due to impurities, pure siliceous and calcareous stones being white or cream-colored. The yellow to red colors usually come from iron oxides, black from manganese dioxide, and green from glauconite. **Crab Orchard stone** of Tennessee is high in silica with practically no CaO, and is often beautifully variegated with red and brown streaks. It splits in uniform slabs and is used for facing. The compressive strength is high, up to 24,000 lb/in^2 (165 MPa), and the weight is 165 lb/ft^3 (2,643 kg/m^3). The water absorption is less than 2%.

About half of the commercial sandstone block in the United States comes from Ohio. It weighs 140 lb/ft^3 (2,243 kg/m^3), and has a compressive strength of 10,000 lb/in^2 (68 MPa), but the average of much other sandstone is 135 lb/ft^3 (2,163 kg/m^3) with a compressive strength of 12,000 lb/in^2 (82 MPa). Sandstones for building purposes are produced under innumerable names, usually referring to the locality. The **bluestone** of New York state is noted for its even grain and high crushing strength, up to 19,000 lb/in^2 (131 MPa). It contains about 70% silica sand with clay as the binder. **Amherst sandstone** from Ohio contains up to 95% silica with 1% aluminum oxide, and is colored gray and buff with iron oxides. **Flexible sandstone,** which can be bent, comes from North Carolina. It is **itacolumite,** and has symmetrically arranged quartz grains which interlock and rotate against one another in a binder of mica and talc.

Holystone is a block of close-grained sandstone, formerly used for rubbing down the decks of ships and still used for rubbing down furniture and concrete work. **Briar Hill stone** and **Macstone** are trade names for building blocks consisting of lightweight concrete faced with a slab of sandstone. **Kemrock,** of the Kewaunee Mfg. Co., is a sandstone impregnated with a black furfural resin and baked to a hard finish. It is used for tabletops and chemical equipment to resist acids and alkalies. The term **reservoir rock** refers to friable porous sandstone that contains oil or gas deposits. The porosity of such sandstone or compacted sand of Pennsylvania is from 15 to 20%, while that of California and the Gulf Coast is 25 to 40%. A sandstone of 20% porosity may contain as much as 75,000 bbl of oil per acre-foot (1,234 m^3).

SANDWICH MATERIALS. A type of **laminar composite** composed of a relatively thick, low-density core between faces of comparatively higher density. **Structural sandwiches** can be compared to I beams. The facings correspond to the flanges, the object being to place a high-density, high-strength material as far from the neutral axis as possible, thus increasing the section modulus. The bulk of a sandwich is the core. Therefore it is usually lightweight for high strength-to-weight and stiffness-to-weight ratios. However, it must also be strong enough to withstand normal shear and compressive loadings, and it must be rigid enough to resist bending or flexure.

Core materials can be divided into three broad groups: cellular, solid, and foam. Paper, reinforced plastics, impregnated cotton fabrics, and metals are used in cellular form. Balsa wood, plywood, fiberboard, gypsum, cement-asbestos board, and calcium silicate are used as solid cores. **Plastic foam cores**—especially polystyrene, urethane, cellulose acetate, phenolic, epoxy, and silicone—are used for thermal-insulating and architectural applications. **Foamed inorganics** such as glass, ceramics, and concrete also find some use. **Foam cores** are particularly useful where the special properties of foams are desired, such as insulation. And the ability to foam in place is an added advantage in some applications, particular in hard-to-get-at areas.

Of all the core types, however, the best for structural applications are the **rigid cellular cores.** The primary advantages of the cellular core are that (1) it provides the highest possible strength-to-weight ratio, and (2) nearly any material can be used, thereby satisfying virtually any service condition.

There are, essentially, three types of cellular cores: honeycomb, corrugated, and waffle. Other variations include small tubes or cones and mushroom shapes. All these configurations have certain advantages and limitations. **Honeycomb sandwich materials,** for example, can be isotropic, and they have a high strength-to-weight ratio, good thermal and acoustical properties, and excellent fatigue resistance. **Corrugated-core sandwich** is anisotropic and does not have as wide a range of application as honeycomb, but it is often more practical than honeycomb for high production and fabrication into panels.

Theoretically, any metal that can be made into a foil and then welded, brazed, or adhesive bonded can be made into a cellular core. A number of materials are used, including aluminum, glass-reinforced plastics, and paper. In addition, stainless steel, titanium, ceramic, and some superalloy cores have been developed for special environments.

One of the advantages of sandwich construction is the wide choice of facings, as well as the opportunity to use thin sheet materials. The facings carry the major applied loads and therefore determine the stiffness, stability, and, to a large extent, the strength of the sandwich. Theoretically, any thin, bondable material with a high tensile- or compressive-strength–weight ratio is a potential facing material. The materials most commonly used are aluminum, stainless steel, glass-reinforced plastics, wood, paper, and vinyl and acrylic plastics, although magnesium, titanium, beryllium, molybdenum, and ceramics have also been used.

SAPELE. The figured woods of various species of trees of tropical Africa which are mixed with khaya and exported from West Africa as **African mahogany.** Sapele woods are harder and heavier then red khaya, but shrink and swell more than khaya with changes in moisture. They are also darker in color with a purplish tinge. **Sapele mahogany,** also called **scented mahogany** and **West African cedar,** is from the *Entandrophragma cylindricum,* a very large tree growing on the Ivory Coast and in Ghana and Nigeria. On the Ivory Coast it is called **aboundikro.** Another species, *E. angolense,* is called **Tiama mahogany** on the Ivory Coast, and in Nigeria is known as **brown mahogany** and **gedunohor.** A less heavy wood, from the tree *E. utile,* is known on the Ivory Coast as **Sipo mahogany** and in the Cameroons as **Assie mahogany.** It is one of the chief woods exported as mahogany from the Cameroons. The wood known on the Ivory Coast as **heavy mahogany** and **omu** in Nigeria is from the tree *E. candollei,* and is much heavier than other sapeles. **Nigerian pearwood,** from species of *Guarea,* notably *G. cedrata* and *G. thompsonii,* is also exported as African mahogany. The woods are more properly called **guarea.** The color is pale pink to reddish. The weight is about the same as sapele. The wood is of a finer texture than khaya, but is not figured like sapele of khaya. Another wood marketed as African mahogany is **lingue,** from the tree *Afzelia africana* of the west coast of Africa from Senegal to Nigeria. The wood is light brown, turning dark when seasoned, and is beautifully figured.

SAPONIN. Glucosides of the empirical formala $C_{32}H_{54}O_{18}$ which have the property of frothing with water. They are found in soap bark, soap nut, licorice, and other plants; when separated out, saponin is a white amorphous powder of a disagreeable odor. Before the advent of the synthetic detergents saponin was important for replacing soaps in washing compounds where high sudsing was undersirable, and it was used in industrial scouring compounds, soapless shampoos, and tooth powders. It is still used in some detergents, in fire extinguishers, as an emulsifying agent, and for synthesizing other complex chemicals. Saponin is not a single compound, but is a great group of **alicyclic compunds,** or five-membered or more highly complex ring compounds having **aliphatic,** or fatty acid, properties. The saponins occur directly in plants where they have a triterpene structure and may be either converted to or derived from a great variety of acids, vitamins, and other products by photosynthesis or catalyzation. They are closely related to the styrols of animal life, and in both plant and animal life slight catalytic rearrangements with nitrogen produce the natural **venoms** and poisonous compounds. The saponins thus form one of the most useful of the basic chemical groups for biological and pharmaceutical work. Chemically, they are called **polymethylene compounds,** and can be synthesized from petroleum. In the drug industry they are called **sapogenines.**

The saponins can be obtained from many plants. **Soap nut** is the fruit of the trees *Sapidus mukorossi* and *S. laurifolia* of northern India known locally as **ritha.** The soap nut has been used as a detergent in washing fabrics since ancient times. The nut has 56% of pericarp and 44 of seed, and the saponin is found in the pericarp. It is extracted with solvents from the dried powdered fruit. Saponin is soluble in water but insoluble in petroleum spirits. **Soapbark,** also called **morillo bark,** is the dried inner bark of the tree *Quillaja saponaria* of the west coast of South America. It was used by the Incas, and the botanical name comes from the Inca word quillean, meaning to wash. The bark produces suds in water, but the powdered bark is highly sternutatory owing to fine crystals of calcium oxalate. It is marketed in brownish-white pieces, and is used as a source of saponin. It has been used in beverages to produce froth, but is highly toxic, affecting the heart and respiration. In medicine it is called **quillaja** and is used as an irritant and expectorant. **Soapwort** consists of the leaves of the plant *Saponaria officinalis,* growing in North America. The leaves contain saponin which dissolves out in water to produce a lather useful for cleaning silk and fine woolens. **Soapberry** is the fruit of the tropical tree *Sapindus saponaria,* used in hair and toilet preparations. The soapberry of the American Southwest consists of the fleshy berries of the small tree *S. drummondii.* **Soaproot** is the bulb root of the plant *Chlorogalum pomeridianum* of California. **Mexican soaproot** is the thick rootstock of the *Yucca baccata* and of the **wild date,** *Y. glauca,* growing in the dry regions of Mexico and the southwestern United States. Both plants are called **yucca,** and in Mexico they are called **amole.** The Indians used the roots, which were called **vegetable soap** by the settlers, for washing. In the processing of yucca leaves to obtain fiber about 20% of a powder is obtained which contains 3% saponin. **Yucca powder** is used in scouring compounds or for the extraction of saponin. The long stout stems of the **soap plant,** *Chenopodium californicum,* also yield saponin.

SAPPHIRE. A transparent variety of the mineral corundum. When it has the beautiful blue hue for which it is noted, it ranks with the diamond, ruby, and emerald among precious gemstones. The off-color stones are cut for pointers and wearing points of instruments. The specific gravity of sapphire is 3.98, and the hardness is 9 Mohs. The blue color is from iron and titanium oxides and is rarely uniform throughout the stone in the natural material. The green is produced with cobalt, and the yellow comes from nickel and magnesium. The **pink sapphire** contains a tiny proportion of chromic oxide, and larger amounts produce the dark-red ruby. The best gem sapphires come from India. A valuable **black sapphire** comes from Thailand. Industrial stones are found in Montana. Most natural sapphires are small, but a large one known as the **Star of Artaban** weighs 300 carats.

Synthetic sapphire is produced by flame-fusing a pure alumina powder made from calcined ammonium aluminum sulfate. The fused material forms a boule as a single crystal. The average boule is 200 carats, but sometimes they are as large as 400 carats, or about ¾ in (1.9 cm) in diameter and 2 in (5.1 cm) long. The rods are single crystals up to 0.23 in (0.58 cm) in diameter and 18 in (45.7 cm) long. **Sapphire balls,** for bearings and valves, are produced by Linde to great accuracy in diameters from 1⁄16 to ⅝ in (0.16 to 1.59 cm). The balls are single crystals with a hardness from 1,525 to 2,000 Knoop, a coefficient of friction of 0.140, and a compressive strength of 300,000 lb/in^2 (2,068 MPa). The material has a melting point of 2030°C, and is resistant to acids and alkalies. It also has a very low coefficient of expansion, and is used for ring and plug gages, and for such wear parts as the thread guides on textile machines. Stones free from strains, and as large as ¾ in (1.9 cm) square, for use as lenses, prisms, and optical windows, are made by the Bell Laboratories by recrystallization at high temperature and pressure. They transmit light better than quartz into the infrared and ultraviolet areas, and sapphire is used as an **infrared detector** in antiaircraft missiles. The dielectric constant of sapphire is also high, about 10.6.

Sapphire whiskers are **alumina fibers** 1 μm in diameter and ⅛ in (0.3 cm) long. They may be matted without bond into **ceramic papers** for electrical insulation and filters. As a reinforcement for light metals, **sapphire fibers** increase tensile and fatigue strengths. **Whiskerloy AA20,** of Melpar, Inc., is an aluminum alloy with 20% sapphire fiber. It has a tensile strength of 48,000 lb/in^2 (330 MPa) and a modulus of 18 million lb/in^2 (0.12 million MPa).

SCHEELITE. An ore of the metal tungsten, occurring usually with quartz in crystalline rocks associated with wolframite, fluorite, cassiterite, and some other minerals. It is found in various parts of the United States, Brazil, Asia, and Europe. Scheelite is **calcium tungstate,** $CaWO_4$, containing theoretically 80.6% tungsten trioxide and 19.4 lime. It is called **powellite** when it contains some molybdenum to replace a part of the tungsten. It occurs massive granular or in crystals. The color is white, yellow, brown, or green, with a vitreous luster. Chinese scheelite from Kiangsi averages 65% WO_3 and can be used directly for adding tungsten to steel. **Tungstic acid** is a yellow powder of the composition H_2WO_4 made from the ore by treating with hydrochloric acid. It is not soluble in water, but is soluble in alkalies and in hydrofluoric acid, and is used as a mordant in dyeing, in plastics, and for making tungsten wire by reducing. Tungstic acid is also obtained as a by-product in the manufacture of alkalies from the brine of Owens Lake, and is a source of tungsten, 1,000 lb (454 kg) of acid yielding about 800 lb (363 kg) of metallic tungsten. **Phosphotungstic acid,** $H_3[P(W_3O_{10})_4] \cdot 5H_2O$, called **heavy acid,** is used as a catalyst in difficult synthesis operations on complex ring-compound chemicals. The molecular weight is 2,879, of which three-fourths is tungsten. The three hydrogens produce the strong acid activity. The acid is soluble in water and in organic solvents.

Pure crystals of scheelite suitable for scintillation-counter phosphors for gamma-ray detection are found, but the natural crystal is rare. Calcium tungstate is grown synthetically as a clear, water-white crystal of tetragonal structure in rods and boules with the axis oriented perpendicular to the growth axis of the rod. It has a specific gravity of 6.12, a Mohs hardness of 4.5 to 5, a melting point of 1535°C, and a refractive index of 1.9368. It has a blue luminescence under ultraviolet light. The crystals can be made in shiny crystalline scales, and the material is also used in fluorescent pigments. **Cadmium tungstate,** $CdWO_4$, is similarly grown in clear, yellowish-green monoclinic crystals with a refractive index of 2.25, and is superior to calcium tungstate for scintillation counters. The crystals can also be grown with a cleavage much like mica, and it is used in fluorescent pigments.

SCOURING ABRASIVE. Natural sand grains or pulverized quartz employed in scouring compounds and soaps, buffing compounds, and metal polishes. Federal specifications require that the abrasive grains used in grit cake soap and scouring compounds shall all pass a No. 100 screen; the grains for scouring compounds for marble floors must all pass a No. 100, and 95% pass a No. 200 screen. For ceramic floors 90% must pass a No. 80, and 95% must pass a No. 60 screen. Very fine air-floated quartz is employed in metal polishes, and all grains pass a 325-mesh screen, but the extremely fine powders of metal oxides for polishes and fine finishes are generally called soft abrasives and are not classed as scouring materials.

SCREW STOCK. A common term for soft steel with free-cutting qualities used for screws and small turned parts made in the screw machine. It usually contains a larger percentage of sulfur than ordinary soft steel. Sulfur makes steel hot-short, or brittle, at red heat, and reduces the tensile strength, but it aids machinability and is called for in steel for simple parts where strength is not important. **SAE screw stock** specifications call for 0.08 to 0.155% sulfur, 0.09 to 0.13 phosphorus, 0.060 to 0.90 manganese, and up to 0.25 carbon. SAE 1112 and SAE 1120 steels are produced regularly as free-cutting screw stock. High-manganese screw stock is an open-hearth, high-sulfur steel in the medium-manganese class having good cutting and case-hardening properties and higher strength. A grade of managanese screw stock under the name of **Ryco steel** is marketed by Joseph T. Ryerson & Son, Inc., having a tensile strength of 90,000 lb/in^2 (620 MPa) and elongation of 20%. **Silcut steel** is a free-cutting steel produced by W. T. Flather, Ltd. **Super-cut steel** and **Freecut steel** of the Union Drawn Steel Div., Republic Steel Corp., are high-sulfur, bessemer cold-drawn steels with tensile strengths up to 100,000 lb/in^2 (689 MPa), elongation 18%, and hardness 196 Brinell. **USS steel MX,** of the U.S. Steel Corp., is an SAE 1113 bessemer steel made to close control limits to give uniform machining. **Free-cutting steels** are produced by almost all steel mills, containing sulfur, selenium, or lead. **La-Led X steel,** of the La Salle Steel Co., is a leaded carbon steel containing also tellurium. It can be turned at cutting speeds more than twice those used for regular sulfur steels. Many of the regular grades of stainless and alloy steels are also obtainable with free-cutting qualities without perceptible impairment of the physical characteristics. The 4100 Series steels containing small amounts of tellurium are marketed as free-cutting **tellurium steel.** The machinability is increased without altering other physical properties or losing tensile strength. The name screw stock also refers to free-cutting brass rod containing a small percentage of lead.

SEAL OIL. An oil resembling sperm oil obtained from the blubber of the **oil seal,** *Phoca vitulina,* a sea mammal native to the Atlantic Ocean. The oil has a saponification value as high as 195, and an iodine value up to 150, and was once valued for lubricating and cutting oils but is now scarce. In the nineteenth century as many as 400 ships at a time operated from Newfoundland in seal catching, but the unrestricted catch resulted in the destruction of the herds, and North Atlantic sealing was reduced to three ships by the middle of the twentieth century. The industry now centers around South Georgia in the South Atlantic as an adjunct to the whale industry, but considerable oil and **seal meal** come as by-products of the Alaskan fur-seal industry.

Some seal oil is obtained from Steller's sea lion, a large-eared seal occurring from southern California to the Bering Sea. The adult male weighs up to 2,200 lb (998 kg). The blubber is about 75% oil, with an iodine value of 143 and saponification value of 190. From 40 to 50% of the carcass is a dense, dark-red, edible meat, but in the United States **seal meat** is used only in animal foods. **Seal leather,** from the skin, is used for fancy specialty articles, but it has too many defects for general use. The product known as **sealskin** is a valuable fur skin from the fur seal, about 80% of which are caught off the Pribilof Islands where they return in June to breed. No killing is now permitted at sea. Each bull seal has as many as 50 females, and the killing is usually restricted to the surplus males. About 30% of the skins are black fur which brings the highest price. Next in value is the **Matara fur,** or dark-brown, which is 60% of the catch. The **Safari fur** is light brown.

SEAWEED. A plant growing in the sea, belonging to the extensive plant division known as **algae.** About 17,000 varieties of seaweed are listed, but only a few are exploited commercially. Algae are non-seed-bearing plants containing photosynthetic pigments. They have no vascular or food-conveying system, and must remain submerged in the medium from which they acquire their food. They occur in both fresh and salt waters.

The **brown seaweeds,** which are the true kelps, grow in temperate and polar waters. They produce **algin, fucoidin,** and **laminarin.** The red seaweeds are the **carrageens,** which produce **carrageenin,** and the **agarophytes,** which yield **agar** and **agaroid.** They grow in warm waters. But color is an indefinite classification; the chlorophyll in the green Irish moss is often so masked by other pigments that the weed may be purplish black. All of the seaweed colloids, or phycocolloids, are polysaccharides, having galactose units linked in long chains of molecular weights from 100,000 to 500,000, varying in their chemical structure. They are anionic polyelectrolytes, with negative radicals on each repeating polymer unit. **Irish moss,** also called **chondrus** (pronounced chone-droosh), **pearl moss,** and carrageen, is a dwarf variety of brown seaweed, *Chondrus crispus,* and *Gigantina mamillosa,* found off the west coast of Ireland and in New England. The weed used mostly for alginic acid is the brown kelp *Laminaria saccharium, L. ditata,* and other species, found off the Hebrides. It is a cold-water plant.

The seaweed *G. stellata,* of the North Atlantic, is also used to produce agar and algin. It is bleached and treated to produce gelatin used in foodstuffs, as a clarifying agent, and as a sizing for textiles. It is a better suspending and gelatinizing medium than agar for foodstuffs and cosmetic emulsions. At least 25 mineral salts are known to be present in seaweed as well as several vitamins. In the utilization of the seaweed as gelatin or alginate these are left in the **kelp meal** which is marketed as poultry and stock feed. In Asia the whole plant is cooked and eaten. **Seaweed flour,** made in Germany from **Iceland seaweed,** *Phaeophyceen,* is the ground dry seaweed containing all the minerals and vitamins. It is mixed with wheat and rye flours to make **algenbrot,** a bread with higher food value and better keeping qualities than ordinary wheat bread. But more than 8% gives a peculiar flavor to the bread. The Irish name **dulse** is applied to the dried or cooked seaweed, *Rhodymenia palmata,* used in the Canadian Maritime Provinces for food. It is purple in color and rich in iodine and mineral salts. Other species, known as **laver** and **murlins,** are also used in Iceland, Ireland, and Scotland for food. When used for producing iodine in Scotland, the seaweed goes under the general name of **tangle.** Much of the 4,500 miles (7,607 km) of coastline of Scotland contains brown kelp. The kelp found along the Chilean and Peruvian coasts, when dried for making algin products, is called by the Quechua name **cochayuyo. Kombu,** used by the Japanese for food, is a brown seaweed from the coast of Hokkaido.

Dry seaweed contains up to 30% **alginic acid;** the water-soluble salts of this acid are called algin. It belongs to the group of complex open-chain **uronic acids** which occur widely in plant and animal tissues and are related to the proteins and pectins. All the algins are edible, but they pass unchanged through the alimentary tract and add no food value. Carrageenin is much used as a stabilizer for chocolate in milk. Laminarin is used as **laminarin sulfate** as a blood-clotting agent. **Sodium alginate** is used as a stabilizer and ice-crystal retarder in ice cream, as a emulsifier in medicines, and to replace gum arabic. It is a colorless water-soluble gum made by dissolving algin in sodium carbonate solution and neutralizing with hydrochloric acid. **Protan jelly,** used for coating fish for freezing, is algin in a dilute edible acid. When frozen, the jelly is impervious to air and prevents oxidation. It can be washed off with water.

Kelgin, of the Kelco Co., is sodium alginate used as a foodstuff stabilizer, and **Keltrex** is the material in granular form for textile coating. **Kelset,** of the same company, is a algin used as a suspending agent for foodstuffs. It does not alter flavor or texture and is stable up to 200°F (93°C). **Dariloid** is sodium alginate to replace gelatin in ice cream; **Kelcosol** is an alginate to replace starch in foodstuffs. One part of algin can replace six parts of starch, and it does not smother flavor as starch does. **Protakyp K,** of Croda, Inc., used as a thickener for textile printing inks, is an alginate compatible with gums. **Viscobond,** of Stein, Hall & Co., is a modified sodium alginate for finishing cellulosic textiles. **Kim-Ko gel,** of the Krim-Ko Corp., is a light-buff scaly powder easily soluble in water, made from Irish moss, used as a collodial gelling agent. It has a pH of 6.4 to 7.2. **Carrigar** is a purified alginate for pharmaceuticals and foodstuffs. It contains the natural mineral salts and has food value. It has high capacity for water absorption, making rigid sugar-free jellies with less than 2% in solution. **Algaloid** and **Agagel,** of T. H. Duche & Sons, Inc., are algins of this type.

Alginic fibers are silklike fibers made by forcing a sodium alginate solution through spinnerets into a calcium chloride bath and insolubilizing with beryllium acetate, but the fiber is soluble in sodium soaps and the fabrics must be dry-cleaned. Soluble alginic yarns are used for making fancy fabrics where uneven spacing of threads is desired without change in the loom. The alginate yarn is washed out of the fabric after weaving, leaving the desired spacing.

SELENIUM. An elementary metal, symbol Se, found native in cavities in Vesuvian lavas and in some shales. The volcanic tuff of Wyoming contains 150 parts per million of selenium, and the black shale of Idaho has up to 1 lb (0.45 kg) of selenium per ton (0.9 metric ton). It also occurs in many minerals, chiefly in **cucairite,** $(AgCu)_2Se$, **naumannite,** Ag_2Se, **zorgite,** $(ZnCu)_2Se$, and in crooksite and clausthalite. Production in the United States and Canada is largely as a by-product of copper refining, the blister copper anodes containing 0.03 to 0.04%, and the refinery slimes having a content of 8 to 9%. The commercial recovery is 0.66 lb/ ton (0.30 kg/0.9 metric ton) of copper. In England it is recovered from the residues of roasting iron sulfide ores in sulfuric acid production.

Like sulfur, selenium exists in various forms. Six allotropic forms are recognized, but four well-defined forms are usually listed. **Amorphous**

selenium, produced by reducing selenous acid, is a finely divided, brick-red powder with a specific gravity of 4.26. It yields the vitreous form on heating. **Vitreous selenium** is a brownish-black, brittle, glassy mass with a specific gravity of 4.28. It is a dielectric and is electrified by friction. The monoclinic **crystalline selenium** is produced by crystallization from carbon disulfide, and is a deep-red glass material with a specific gravity of 4.46, and a melting point of 175°C. The hexagonal crystalline selenium is produced by heating the monoclinic. It is a stable metal, and is a good conductor of electricity. It has a specific gravity of 4.79, and melts at 217°C. All of the forms become gaseous at 688°C. Selenium is marketed as a blackish powder, the high grade being 99.99% pure, and the commercial grade 99.5% pure.

Selenium metal is odorless and tasteless, but the vapor has a putrid odor. The material is highly poisonous, and is used in insecticides and in ship-hull paints. Foods grown on soils containing selenium may have toxic effects, and some weeds growing in the western states have high concentrations of selenium and are poisonous to animals eating them. Selenium burns in air with a bright flame to form **selenium dioxide,** SeO_2, which is in white, four-sided, crystalline needles. The oxide dissolves in water to form **selenous acid,** H_2SeO_3, resembling sulfurous acid but very weak. Oxidation of this acid forms **selenic acid,** H_2SeO_4, a strong acid resembling sulfuric acid. By burning loco weed and converting to the acids, selenium has been extracted from the weeds.

The photoelectric properties of selenium make it useful for light-measuring instruments and for electric eyes. Amorphous or vitreous selenium is a poor conductor of electricity, but when heated, it takes the crystalline form and its electrical resistance is reduced, and it changes electrical resistance when exposed to light. The change of electric conductivity is instantaneous, even the light of small lamps have a marked effect since the resistance varies directly as the square of the illumination. The pure amorphous powder is also used for coating nickel-plated steel or aluminum plates in rectifiers for changing alternating current to pulsating direct current. The coated plates are subjected to heat and pressure to change the selenium to the metallic form, and the selenium coating is covered with a layer of cadmium-bismuth alloy. Selenium rectifiers are smaller and more efficient than copper oxide rectifiers, but they require more space than silicon rectifiers and are limited to an ambient temperature of 85°C.

Selenium is also used in steels to make them free-machining, up to 0.35% being used. **Selenium steels** are not as susceptible to corrosion as those with sulfur and are stronger. Up to 0.05% of selenium may also be used in forging steels to control the gas and produce a more homogenous metal. From 0.6 to 0.85 oz (0.017 to 0.024 kg) of selenium per ton (0.9 metric ton) of glass may be used in glass to neutralize the green tint of iron compounds. Large amounts produce pink and ruby glass. Selenium gives the only pure red color for signal lenses. Pigment for glass may be in the form of the black powder, **barium selenite,** $BaSeO_3$, or as **sodium selenite,** Na_2SeO_3, and may be used with cadmium sulfide. Selenium is also used as an accelerator in rubber and to increase abrasion resistance. **Vandex,** of R. T. Vanderbilt Co., is a selenium powder used as a rubber vulcanizer. **Novac,** of Herron Bros. & Meyer, used for curing synthetic rubbers, is selenium dibutyl dithiocarbonate in the form of a liquid easily dispersed in the rubber. **Selsun,** of the Abbott Laboratories, is **selenium sulfide** suspended in a detergent, used to control dandruff in hair. In copper alloys selenium improves machinability without hot-shortness. **Selenium copper** is a free-cutting copper containing about 0.50% selenium. It machines easily, and the electric conductivity is nearly equal to that of pure copper. The tensile strength of annealed selenium copper is about 30,000 lb/in^2 (206 MPa). Small amounts of selenium salts are added to lubricating oils to prevent oxidation and gumming.

SEMICONDUCTORS. Materials which are capable of being partly conductors of electricity and partly insulators, and are used in rectifiers for changing alternating current to pulsating direct current, and in transistors for amplifying currents. They can also be used for the conversion of heat energy to electric energy, as in the solar battery. In an electric conductor the outer rings of electrons of the atoms are free to move and provide a means of conduction. In a semiconductor the outer electrons, or **valence electrons,** are normally stable, but, when a **doping element** that serves to raise or lower energy is incorporated, the application of a weak electric current will cause displacement of valence electrons in the material. Silicon and germanium, each with a single stable valence of four outer electrons, are the most commonly used semiconductors. Elements such as boron, with a lower energy level but with electrons available for bonding and thus accepting electrons into the valence ring, are called **hypoelectronic elements.** Elements such as **arsenic,** which have more valence electrons than are needed for bonding and may give up an electron, are called **hyperelectronic elements.** Another class of elements, like **cobalt,** can either accept or donate an electron, and these are called **buffer atoms.** All of these types of elements constitute the doping elements for semiconductors.

In a nonconducting material, used as an **electrical insulator,** the energy required to break the valence bond is very high, but there is always a limit at which an insulator will break the bond and become a conductor with high current energy. The resistivity of a conductor rises with increased temperature, but in a semiconductor the resistivity decreases with temperature rise, and the semiconductor becomes useless beyond its temperature limit. **Germanium** can be used as a semiconductor to about 200°F (93°C), **silicon** can be used to about 400°F (204°C), and **silicon carbide** can be used to about 650°F (343°C).

Some materials have a complete lack of measurable electrical resistance at very low temperatures, and are called **superconductors.** A very low power source will start a current in a superconductor, and the current will continue to flow indefinitely after the power source is removed. Most metals except columbium lose their conductive properties in magnetic fields, and the superconductors usually contain a high percentage of columbium. The superconductors are used for magnets and memory elements in high-speed computers where space saving is important. A **superconductive alloy** of North American Aviation, Inc., contains 75% columbium and 25 zirconium. **Niostan,** of National Research Corp., is columbium ribbon coated with a thin layer of **columbium stannide.** Its efficiency permits smaller units in computing equipment used in subzero temperatures. In operation, the superconductors are kept at subzero temperatures.

Metals for use as semiconductors must be of great purity, since even minute quantities of impurities would cause erratic action. The highly purified material is called an **intrinsic metal,** and the desired electron movement must come only from the doping element, or **extrinsic conductor,** that is introduced. The semiconductors are usually made in single crystals, and the positive and negative elements need be applied only to the surfaces of the crystal, but methods are also used to incorporate the doping element uniformly throughout the crystal.

The process of electron movement, although varying for different uses and in different intrinsic materials, can be stated in general terms. In the silicon semiconductor the atoms of silicon with four outer valence electrons bind themselves together in pairs surrounded by eight electrons. When a doping element with three outer electrons, such as boron or indium, is added to the crystal it tends to take an electron from one of the pairs, leaving a hole and setting up an unbalance. This forms the **p-type semiconductor.** When an element with five outer electrons, such as antimony or bismuth, is added to the crystal it gives off electrons, setting up a **conductive band** which is the **n-type semiconductor.** Fusing together the two types forms a *p-n* junction, and a negative voltage applied to the *p* side attracts the electrons of the three-valence atoms away from the junction so that the crystal resists electronic flow. If the voltage is applied to the *n* side it pushes electrons across the junction and the electrons flow. This is a **diode,** or **rectifier,** for rectifying alternating current into pulsating direct current. When the crystal wafers are assembled in three layers, *p-n-p* or *n-p-n,* a weak voltage applied to the middle wafer increases the flow of electrons across the whole unit. This is a **transistor.** Germanium and silicon are bipolar, but silicon carbide is unipolar and does not need a third voltage to accelerate the electrons.

Semiconductors can be used for rectifying or amplifying, or they can be used to modulate or limit the current. By the application of heat to ionize the atoms and cause movement they can also be used to generate electric current; or in reverse, by the application of a current they can be used to generate heat or remove heat for heating or cooling purposes in air conditioning, heating, and refrigeration. But for uses other than rectifying or altering electric current the materials are usually designated by other names and not called semiconductors. **Varistors** are materials, such as silicon carbide, whose resistance is a function of the applied voltage. They are used for such applications as frequency multiplication and voltage stabilization. **Thermistors** are thermally sensitive materials. Their resistance decreases as the temperature increases, which can be measured as close as 0.001°C, and they are used for controlling temperature or to control liquid level, flow, and other functions affected by rate of heat transfer. They are also used for the production or the removal of heat in air conditioning, and may then be called **thermoelectric metals.**

Indium antimonide, InSb, has a cubic crystal structure, with three valence electrons for each indium atom and five for each antimony atom. Between each atom and its four nearest neighbors there are four electron-pair bonds, and there is an average of four electrons per atom in the compound. It is used for **infrared detectors** and for amplifiers in galvanomagnetic devices. **Indium arsenide,** InAs, also has a very high electron mobility, and is used in thermistors for heat-current conversion since the number of electrons free to constitute the electric current increases about 3% with each 1°F rise in temperature. It can be used to 1500°F (816°C). Some materials can be used only for relatively low temperatures. Copper oxide and pure selenium have been much used in current rectifiers, but they are useful only at moderate temperatures, and they have the disadvantage of requiring much space. **Indium phosphide,** InP, has a mobility higher than that of germanium, and can be used in transistors above 600°F (316°C). **Aluminum antimonide,** AlSb, can be used at temperatures to 1000°F (538°C). In **lead selenide,** PbSe, the mobility of the charge-carrying electrons decreases with rise in temperature, increasing the resistivity, and it is used in thermistors.

The thermoelectric generation, set up when the junction of two dissimilar thermoelectric metals is heated and which is used in thermostats for temperatures measuring and control, is essentially the same as the energy conversion and heat pumping with *p*-type and *n*-type materials. The difference is in mechanical applications. When a semiconductor is operated thermoelectrically as a heat pump, the electric charge passing through the heat-absorbing junction is carried by electrons in the *n*-type material and holes in the *p*-type material, and the charge carriers both move away from the junction and carry away heat, thus reducing temperature at the junction. By reversing the current, heat is produced. Each material has a definite temperature difference, or gradient, and the efficiency is proportional to the temperature difference across the material, while the power rating is proportional to the square of the temperature difference. Thus, a material with low efficiency may have a high power rating if it can be operated at a high enough temperature, but some materials do not maintain chemical stability at high temperatures. Also, for many uses it is undesirable to operate at advanced temperatures.

Bismuth telluride, Bi_2Te_3, maintains its operating properties between −50 and 400°F (−46 and 204°C), which is the most useful range for both heating and refrigeration. When doped as a *p*-type conductor it has a temperature difference of 1115°F (601°C) and an efficiency of 5.8%. When doped as an *n*-type conductor the temperature difference is lower, 450°F (232°C), but the efficiency within this range is more than doubled. **Lead telluride,** PbTe, has a higher efficiency, 13.5%, and a temperature difference of 1080°F (582°C), but it is not usable below 350°F (177°C), and is employed for conversion of the waste heat from atomic reactors at about 700°F (371°C).

Gallium arsenide has high electron mobility, and can be used as a semiconductor. When polycrystalline semiconductors are used in thin films against a metal barrier the minimum grain size of the deposited film must equal the thickness of the film so that the carrier is not intercepted by a grain boundary. **Cadmium sulfide,** CdS, is thus deposited as a semiconductor film for **photovoltaic cells,** or **solar batteries,** with film thickness of about 2 μm. When radioactive isotopes, instead of solar rays, are added to provide the activating agent the unit is called an **atomic battery,** and the large area of transparent backing for the semiconductor is not needed.

Manganese telluride, MnTe, with a temperature difference of 1800°F (982°C), has also been used as a semiconductor. Many other materials can be used, and semiconductors with temperature differences at different gradients can be joined in series electrically to obtain a wider gradient, but the materials must have no diffusion at the junction. If intermetal compounds are of such a nature as to have a *p-n* balance, no doping is needed, but usually they are not in perfect balance, causing scattering, and a balancing is necessary. Materials for thermoelectric use are usually doped higher than for semiconductors, but increased doping reduces resistivity, and for high emf and low power only small amounts are used.

Cesium sulfide, CeS, has good stability and thermoelectric properties at temperatures to 2000°F (1093°C), and has a high temperature difference, 2030°F (1110°C). It can thus be used as a high-stage unit in conversion devices. High conversion efficiency is necessary for **transducers,** while a high dielectric constant is desirable for capacitors. A low thermal conductivity makes it easier to maintain the temperature gradient, but for some uses a high thermal conductivity is desirable. **Silver-antimony-tel-**

luride, $AgSbTe_2$, has a high energy-conversion efficiency for converting heat to electric current, and it has a very low thermal conductivity, about 1% that of germanium.

Mechanical stress as well as heat stress produces an electric charge in balanced semiconductors, and they can be used for controlling pressure. The semiconductor-type intermetals are also used in magnetic devices, since the ferroelectric phenomenon of heat conversion is the electrical analog of ferromagnetism. **Chromium-manganese-antimonide** is nonmagnetic below about 250°C and magnetic above the temperature. Various compounds have different critical temperatures. Below the critical temperature the distance between the atoms is less than that which determines the line-up of magnetic forces, but with increased temperature the atomic distance becomes greater and the forces swing into a magnetic pattern.

Organic semiconductors fall into two major classes: well-defined substances, such as molecular crystals and crystalline complexes, isotatic and syndiotactic polymers; and disordered materials, such as atatic polymers and pyrolitic materials. Few of these materials have yet found commercial application. **Anthracene,** an example of the molecular crystal type, has perhaps been given the most study. Its transport properties are not unlike those of silicon and germanium. **Amorphous silicon** containing hydrogen is promising for use in solar cells because of its low cost and suitable electrical and optical properties.

SENNA. The dried small leaves and the pods of the bushy plant *Cassia acutifolia,* the **Alexandrian senna,** and *C. angustifolia,* the **Tinnevelly senna,** of India, Arabia, and North Africa. The plants are cultivated in India, but the Sudan material comes mostly from wild plants. The sundried leaves and pods are shipped in bales. They are used directly as a laxative by steeping in water, or the extract is used in pharmaceuticals. It contains the yellowish noncrystalline **cathartine,** a powerful purgative. Another species of the plant, *C. auricula,* yields **avarem bark,** which is an important tanning material in India. It is similar to algarobilla in action.

SERPENTINE. A mineral of the theoretical formula $3MgO \cdot 2SiO_2 \cdot 2H_2O$, containing 43% magnesium oxide. It is used for building trim and for making ornaments and novelties. The chips are employed in terrazzo and for roofing granules. Actually, the stone rarely approaches the theoretical formula, and usually contains 2 to 8% iron oxide with much silica and aluminum. It has an asbestoslike structure. The attractively colored and veined serpentine of Vermont is marketed under the name of **verde antique marble.** The massive verde antique of Pennsylvania is used with dolomite in refractories. **Antigorite** is a form of serpentine found in California which has a platy rather than a fibrous structure. The serpentine of Columbus County, Ga., contains 36 to 38% MgO and 2 to 5 chrome ore. It is used as a source of magnesia.

SESAME OIL. A pale-yellow, odorless, bland oil obtained from the seeds of the tropical plant *Sesamum orientale* and other species, grown in India, China, Africa, and Latin America, and used for soaps, foodstuffs, and for blending industrial oils. It is distinct from German sesame oil. The seeds from different species and localities vary greatly in size and color, from yellowish white to reddish brown to black. The oil from Nigeria is called **benne oil,** and the seed **benniseed.** In India it is known as **til oil,** and the seed as **til seed.** In Madras it is called **gingelli.** In Mexico it is called **ajonjoli.** The seeds contain 50% of oil of a specific gravity of 0.920 to 0.925, with saponification value 188 to 193.

Sesame is one of the most ancient of food grains, but has been little grown in the United States because of the difficulty of harvesting and collecting the seed. **Sesame seed** is used as a flavor garnish for breads and bakery products. **Sesame protein,** extracted from the seed and marketed as a powder, contains the amino acids methionine and cystine which are low in other vegetable foods. It is low in lycine. It is valued for blending in foodstuffs. The oil contains the natural antioxidants **sesamol** and **sesamolin,** making it very stable, and it is also highly unsaturated. The protein is marketed as a flour containing 60% protein and less than 1% fat, and in granules containing 55% protein and up to 18% fat. **Sesa-Lac 86,** of John Kraft Sesame Corp., is a spray-dried blend of sesame protein and whey for use in beverages, bakery products, and prepared foods, giving a more complete amino acid combination than milk powder.

SHALE. A rock formed by deposition of colloidal particles of clay and mud, and consolidated by pressure. It is fine-grained and has a laminated structure, usually containing much sand colored by metal oxides. Unlike sandstones, shales are not usually porous, most shales being hard, slatelike rocks. Slate is a form of shale that has been subjected to intense pressure. Some shales are calcareous or dolomitic and are used with limestone in making portland cement. These are called **marlstone. Oil shale** is a hard shale with veins of greasy solid known as **kerogen,** which is oil mixed with organic matter. Crude **shale oil** is a black, viscous liquid containing up to 2% nitrogen and a high sulfur content. But when oil shale is heated above 750°F (399°C), the kerogen is cracked into gases condensable to oils, gases, and coke. Some shales also yield resins and waxes, and the **Kvarmtorp shale** of Sweden contains small amounts of uranium, vanadium, and molybdenum. The regular commercial by-products of **Swedish shale oil** recovery are sulfur, fuel gas, ammonium sulfate, tar, and lime. Oil shales are widely distributed in many parts of the world and are regularly distilled in most of the countries of Europe, the yield varying from 15 to 100 gal (57 to 379 L) per short ton (907 kg). Scottish shales give an average yield of 24.5 gal (93 L) crude oil and 35.7 lb (16 kg) ammonium sulfate per short ton (907 kg). Shale occurs in strata and is mined like coal. The oil shales of Colorado contain **dawsonite,** a hydrated carbonate of aluminum and sodium and also **nacolite,** a sodium bicarbonate mineral. Alumina, soda ash, and other by-products can be produced from the shales. **Bituminous shale** was originally called **boghead coal** in England, and **torbane mineral** in Scotland. In the Green River Basin of northwestern Colorado about 1,000 mi^2 (2,590 km^2) is underlain by oil shale 500 ft (152 m) thick averaging 15 gal/ton (0.06 L/kg). The lower portion, about 100 ft (30 m) thick, averages 30 gal of oil per ton (0.13 L/kg). Recovery in a continuous retort extracts 94% of the oil, which is then cracked by heat and treated with hydrogen to remove impurities and improve the quality before it is sent to the refinery. It is estimated that the shale of the Mahogany Ledge in Colorado, extending into Utah and Wyoming, has a content of 1.2 trillion barrals of extractable oil. No oil is visible in the shale, but it is present as the solid kerogen which yields oil when heated. Other deposits occur in Nevada, Tennessee, Indiana, Ohio, and Kentucky. **Oil sands** of Alberta, Utah, and California are free-flowing sands impregnated with bituminous oil. A deposit on Athabasca River covers 1,800 mi^2 (4,662 km^2) and is 165 ft (50 m) thick. Vast quantities of oil are available from these sands.

SHARK LEATHER. A durable nonscuffing leather used for bookbindings, handbags, and fancy shoes, made from the skin of sharks. The shark is the largest of the true fishes, but has a skin unlike fishskin. When tanned, the surface is hard, the epidermis thicker than cowhide, and the long fibers lie in a cross weave. The shark is split on the back instead of the belly as in cowhides, and the skins measure form 3 to 20 ft^2 (0.3 to 2 m^2), averaging 10 ft^2 (1 m^2). The hard denticle, called the shagreen, is usually removed, after which the leather is pliable but firm, the exposed grain not pulling out. **Shagreen leather** is a hard, strong leather with the grain side covered with globular granules made to imitate the sharkskin. Eastern shark leather has a deep grain with beautiful markings. The eastern shark includes about a dozen species of shark caught off the Florida and Cuban coasts except the **nurse shark** and the **sawfish** which are graded separately. The **whale shark** attains a length of 50 ft (15 m) and a weight of several tons. It is an offshore species, feeding on small organisms, and is harmless to people. The **basking shark** and the **white shark** grow to 40 ft (12 m). The nurse shark measures 6 to 10 ft (1.8 to 3 m). **Olcotrop leather,** from a species of shark, has a smooth, fine grain with regular markings. **Galuchat leather,** or **pearl sharkskin,** is from the Japanese ray. It is used for trim on pocketbooks. **Boroso sharkskin, rousette leather,** or **Morocco leather,** is from a small shark of the Mediterranean, but the name is also applied to a vegetable-tanned Spanish goatskin on which a pebbly grain is worked up by hand boarding. It is now made from ordinary goatskin by embossing.

Most of the sharkskin is now a by-product of the catch for oil, which is used for medicinal purposes. The shark liver is about one-fourth the total weight of the animal, and **shark-liver oil** is 30 times higher in vitamin A than cod-liver oil. The oil is also used for soap, lubricant, and heat-treating oil, though normally it is too expensive for these purposes. The Mexican shark-oil industry centers at Mazatlán, and about 25 species are caught off the West Coast. **Vitamin oil** from South Africa is from the liver of the **stockfish,** *Merluccius capensis,* which is an important food fish. The liver contains 30% oil.

SHEEPSKIN. The skin of numerous varieties of sheep, employed for fine leather for many uses. The best sheepskins come from the sheep yielding the poorest wool. When the hair is short, coarse, and sparse, the nourishment goes into the skin. The merino types having fine wool have the poorest pelts. Wild sheep and the low-wool crossbreds of India, Brazil, and South Africa have close-fibered, firm pelts comparable in strength with some kidskin, and retain the softness of sheepskin. This type of sheepskin from the hair sheep is termed **cabretta,** and is used almost entirely for making gloves and for shoe uppers. None is produced in the United States. The lambs grown in the mountains of Wales, Scotland, and the western United States also furnish good skins. The commercial difference between sheepskin and **lambskin** is one of weight only. Sheepskins usually run 3 to 3½ lb (1.4 to 1.6 kg) per skin without wool, and lambskins are those below 3 lb (1.4 kg). Sheepskins are tanned with alum, chrome, or sumac. The large, heavy skins from Argentina and Australia are often split, and the grain side tanned in sumac for bookbinding and other goods; the flesh side is tanned in oil or formaldehyde and marketed as chamois. The fine-grained sheepskins from Egypt, when skived and specially treated, are known as **mocha leather. Uda skins** and **white fulani skins,** from Nigeria sheep, are used for good-quality grain and suede glove leather. **Sheepskin shearlings** are skins taken from heavy-wooled sheep a few weeks after shearing. The wool is about 1 in (2.54 cm) in length. They are tanned with the wool on, and the leather is used for aviation flying suits and for coats.

SHELLAC. A product of the *Tachardia lacca,* an insect that lives on various trees of southern Asia. The larvae of the lac insect settle on the branches, pierce the bark, and feed on the sap. The lac secretion produced by the insects forms a coating over their bodies and makes a thick incrustation over the twig. Eggs developed in the females are deposited in a space formed in the cell, and the hatched larvae emerge. This swarming continues for 3 weeks and is repeated twice a year. The incrustation formed on the twigs is scraped off, dried in the shade, and is the commercial **stick lac.** It contains woody matter, lac resin, lac dye, and bodies of insects. **Seed lac** is obtained by screening, grinding, and washing stick lac. The washing removes the lac dye. **Lac dye** was once an important dyestuff, giving about the same colors as cochineal but not as strong. It gives a fast bright-red tint to silk and to wool, but is now replaced by synthetics. **Ari lac** is stick lac collected before the young insects have swarmed, and contains living insects. Lac harvested after the swarming is called **phunki lac** and contains dead bodies of the insects. Average yield of stick lac from kusum trees is 12 lb (5.4 kg), from the ber tree 3 lb (1.4 kg), and from the palas tree 2 lb (0.9 kg). About 80% of lac production is in the state of Bihar in India, but it is also obtained from Bengal, the Central Provinces, and Assam. The stick lac from Burma and Thailand is brought to India for making shellac.

Shellac is prepared from seed lac by melting or by extraction with solvents. The molten material is spread over a hot cylinder, stretched, and the cooled sheet broken into flakes of shellac. **Button lac** is made by dropping molten lac on a flat surface which spreads it into button-shaped cakes 3 to 4 in (7.6 to 10.2 cm) in diameter. **Kiri** is the refuse from the filtering bags. It is marketed in pressed cakes and contains 50 to 60% lac with resin and dirt. The yield of shellac from stick lac is about 57%. When pure, shellac varies from pale orange to lemon yellow in color, but the color of commerical shellac may be from a high content of common rosin. **White shellac** is made by bleaching with alkalies. **Garnet lac** is the material with lac dye left in. Color may also be balanced with pigments. **Orange shellac** contains up to 1% powdered orpiment, and the yellow may have smaller quantities. Shellac is composed of polyhydric acids which condense with loss of water to form long-chain esters, thus giving polyester resins in the final coating. It also contains resin and wax. **Aleuritic acid,** $OH \cdot CH_3(CH_2)_5(CH \cdot OH)_2(CH_2)_7COOH$, extracted from shellac, reacts with alcohols to produce odoriferous esters used in perfumes. It is a yellowish solid melting at 101°C. It can be made synthetically, and is also used in cellulose lacquers.

Hard lac has the soft constituents removed by solvent extraction. For electrical use the wax content should be below 3½%. By solvent extraction of the seed lac the wax may be reduced to 1%. Shellac is graded by color and by its freedom from dirt. The first grade contains no rosin, but other grades may contain up to 12%. Most Indian exports of seed lac to the United States are of the special grade which has a high bleach index. **Cut shellac** is shellac dissolved in alcohol, but usually mixed with a high percentage of rosin. Shellac has good adhesive properties and high dielectric strength, and is used in adhesives, varnishes, floor waxes, insulating compounds, and in some molding plastics. Hard-face wax polishes contain a high percentage of shellac, up to 80%, to conserve carnauba and other waxes. **Shellac plastics,** usually for electrical purposes, are sold under trade names such as **Electrose,** of the Insulation Mfg. Co., and **Harvite,** of the Siemon Co. **Shellac substitutes** are made by blending natural resins with nitrile rubber, or with modified synthetic resins. **Beckasite P-720,** of Reichhold Chemicals, Inc., is a rosin-modified maleic resin. It is soluble in alcohol and in water-ammonia solutions, and is used in wax emulsions. **Waterez 1550,** of the same company, is a shellac substitute made by reacting phthalic anhydride with a polyol. It is used in self-polishing floor waxes. It can be removed easily with alkali cleansers. Sucrose esters are also used to replace shellac.

SHOCK-RESISTANT STEELS. A general name for steels used for tools and parts that are required to withstand much pounding. There are two general types. One type contains chromium, vanadium, and a small amount of molybdenum, with usually fairly high manganese; the other type contains up to 2% silicon, with usually some molybdenum. The silicon steels are used for pneumatic tools and for such purposes as coining dies. A **chromium-silicon steel** marketed by the Temkin Steel & Tube Co. under the name of **Sicromo steel** is a combination of both types, and has high strength and resistance. It contains 2.25 to 2.75% chromium, 0.5 to 1 silicon, 0.4 to 0.6 molybdenum, and 0.15 carbon. The tensile strength, annealed, is 60,000 lb/in^2 (413 MPa), elongation 30%, and hardness 170 Brinell. The low-alloy chromium-silicon steels with about 1% chromium, 1 silicon, 0.60 to 0.70 manganese, and 0.40 to 0.45 carbon have tensile strengths above 200,000 lb/in^2 (1,378 MPa) when heat-treated. They have good fatigue resistance and are also used for springs. A steel of the Allegheny Ludlum Steel Co. called **shoe die steel,** for cutting block dies, or **clicker dies,** for cutting leather and paper, contains 0.70% chromium, 0.35 molybdenum, 0.60 manganese, 0.20 silicon, and 0.53 carbon. Shoe die steel is marketed in beveled shapes. **MSM steel** of A. Milne & Co., for punches, chisels, and shear blades, has 0.50% carbon, 2.0 silicon, 0.70 manganese, and 0.25 molybdenum. **Omega steel,** of the Bethlehem Steel Co., for chisels, shear blades, and swaging tools, contains 0.60% carbon, 1.85 silicon, 0.70 manganese, 0.20 vanadium, and 0.45 molybdenum. When hardened in oil, it has a high resistance to impact. **Bearcat steel** of this company is an air-hardening, shock-resistant, nondeforming steel for rivet sets, chisels, punches, gripper dies, and die-casting dies. It contains 0.50% carbon, 0.70 manganese, 0.25 silicon, 3.25 chromium, and 1.40 molybdenum. **Hy-Tuf steel,** of the Crucible Steel Co. of America, for shock-resistant tools, gears, and parts, is an oil-hardening steel containing 1.30% manganese, 1.50 silicon, 1.80 nickel, 0.40 molybdenum, and 0.25 carbon. When heat-treated it has a tensile strength of 230,000 lb/in^2

(1,585 MPa), elongation of 13%, and Rockwell C hardness of 46. **Super Hy-Tuf** contains 1.30 manganese, 2.30 silicon, 1.40 chromium, 0.20 vanadium. 0.35 molybdenum, and 0.40 carbon. When hardened it has a tensile strength of 294,000 lb/in^2 (2,027 MPa), elongation of 10%, and Rockwell C hardness of 54. This steel is used for such applications as aircraft landing-gear forgings. **Halvan steel,** of this company, for chisels, punches, and rivet sets, contains 0.50% carbon, 0.80 manganese, 1 chromium, and 0.20 vanadium. A grade of Crucible's **La Belle steel,** for coining dies, jewelry dies, and cold-header dies, has 0.95% carbon, 0.35 manganese, and 0.45 silicon. It is in reality a straight carbon steel with deep-hardening characteristics capable of withstanding high impact pressures per unit area, and has been called **cold-striking steel. La Belle HT steel,** for rivet sets, shear blades, and cold-header dies, has 2.25% silicon, 1.35 manganese, 1.35 chromium, 0.40 molybdenum, 0.30 vanadium, and 0.43 carbon. It gives a Rockwell hardness of C58, and has high wear resistance and toughness. Rigid melting and processing procedures in the production of these steels to ensure uniformity are more important than exact composition. **Ludlum 602 steel,** for pneumatic tools, has 0.45% carbon, 0.70 manganese, 1.70 silicon, 0.40 molybdenum, and 0.12 vanadium, while **Seminole hard steel** of the same company has 1.3% chromium, 2 tungsten, 0.25 vanadium, and 0.48 carbon. **Atsil steel,** of the Atlantic Steel Co., for shear blades, punches, and pneumatic tools, contains 0.50% carbon, 0.60 manganese, 0.50 tungsten, 0.30 molybdenum, and 1.30 silicon.

SILICA. A mineral of the general composition SiO_2, **silicon dioxide,** which is the most common of all materials, and in the combined an uncombined states is estimated to form 60% of the earth's crust. Many sands, clays, and rocks are largely composed of small silica crystals. When pure, silica is colorless to white. The unit crystal, or molecule, of ordinary silica has the formula SiO_2, and the single crystal grains are thus molecularly cryptocrystalline with no electron bonded lattice. But the chemical formula of fused silica and quartz is given as Si_2O_7, which is the pattern of a continuous lattice in which each silicon atom is surrounded by four oxygen atoms and each oxygen atom is surrounded by four silicon atoms. The varieties of natural silica are **crystalline silica,** such as quartz and **tridymite; cryptocrystalline silica** (minute crystals), such as flint, chert, chalcedony, and agate; and **Amorphous silica,** such as opal. Silica is insoluble in water when anhydrous and is also insoluble in most acids except hydrofluoric. Crystallized silica in the form of quartz has a hardness of 7 Mohs and a specific gravity of 2.65. Amorphous **silica glass** has a density of 2.21. It is a transparent fused silica. **Vitreous silica** is a silica glass of high transparency. When impurities are no more than 1 ppm, it is the most transparent of the glasses, and has high transmission of ultraviolet rays.

Pure **fused silica** has a melting point of 1750°C, but softens slightly at

1400°C. In chemical and heat ware it is used up to 1100°C. The coefficient of expansion is very low, 0.00000054 per degree Celsius, and the dielectric strength is 500 volts per mil (20×10^6 volts per meter). The infrared transmission of a 96% pure silica glass extends to 4 μm while retaining the lower reach of the band to 0.4 μm. Fused silica is used for chemical parts as it withstands severe thermal shock and is resistant to acid except hydrofluoric and hot phosphoric. **Amersil,** of the Amersil Co., in the form of pipes and shapes, will withstand continuous temperatures of 2700°F (1482°C). Fused silica parts may be made by pressing and sintering silica powder or by casting. Large cast parts for crucibles, molds, and furnace hearths, of the Glassrock Co., are made by remelting a powder produced by melting 99.9% pure silica sand and then crushing and grinding the glass. Cast parts have a tensile strength of 1,500 lb/in^2 (10 MPa), compressive strength of 20,000 lb/in^2 (137 MPa), and withstand repeated heating and cooling from 2000°F (1093°C). The material is white in color.

Fibrous silica, used for high-temperature insulation in jet aircraft, is produced from silica minerals in the same manner as rock wool and then extracting the nonsilica content of the fiber. **Refrasil,** of the H. I. Thompson Co., is this material. The fibers have a diameter of 0.00023 in (0.00058 cm), fuse at 3100°F (1704°C), and will withstand continuous temperatures to 2000°F (1093°C). It is produced as fibers, batts, cloth, and cordage. **Silica fiber** in diameters as small as 0.00003 in (0.000076 cm) comes in random matted form or in rovings. **Irish Refrasil** is 98% silica and has a green color. It is used for ablative protective coatings. It resists temperatures to 2800°F (1588°C). **Silica flour,** made by grinding sand, is used in paints, as a facing for sand molds, and for making flooring blocks. **Silver bond silica,** of the Tamms Silica Co., is a water-floated silica flour of 98.5% SiO_2, ground to 325 mesh. In zinc and lead paints it gives a hard surface. **Pulverized silica,** made from crushed quartz, is used to replace tripoli as an abrasive. **Ultrafine silica,** a white powder having spherical particles of 4 to 25 mm, is made by burning silicon tetrachloride. It is used in rubber compounding, as a grease thickener, and as a flatting agent in paints. **Aerosil,** of Cabot Corp., is this material. **Silica powder,** of Linde, is a white amorphous powder with maximum particle size of 50 nm. The natural amorphous silica of the Illinois Minerals Co. comes in an average particle size of 1.5 μm with no particles large than 10 μm. **Quso,** of the Philadelphia Quartz Co., is a soft white powder with small particles, 10 to 20 nm. It is used in cosmetics and paper coatings and as an anticaking agent in pharmaceuticals. As a filler in plastics it gives a plasticizing action that aids extrusion. It comes in grades with a pH of 5.9 and 8.2. These fine silicas are also marketed as dust-free agglomerate particles which disperse easily in solution to the discrete hydrophyllic particle. **Arc silica,** of Pittsburgh Plate Glass, used as a flatting agent in clear lacquers, is produced directly from silica sand in an arc furnace at 3000°C. It has crystals of 0.015 μm agglomerated into translucent grains, 2 to 3 μm. **Valron,** of Du Pont, originally called **Estersil,** is ester-coated silica powder of 8 to 10 mm particle size, for use as a filler in silicone rubbers, printing inks, and plastics. **Ludex,** of the same company, is another **colloidal silica** with the fine particles negatively charged by the incorporation of a small amount of alkali. It forms a **sol,** or high-concentration solution, without gelling. **Min-U-Sil,** of the Pennsylvania Glass Sand Corp., for making molded ceramics, has tiny crystalline particles. **Syton,** of Monsanto, is a water dispersion of colloidal silica for treating textiles. Translucent silica particles deposited on the fibers increase the coefficient of friction, giving uniformly high strength yarns.

A **polymer-impregnated silica, Polysil,** produced by Westinghouse, has twice the dielectric strength of porcelain as well as better strength. It is also cheaper to make, and its composition can be tailored to meet specific environmental and operating conditions.

Silica aerogel is a fine, white, semitransparent silica powder, the grains of which have a honeycomb structure, giving extreme lightness. It weighs 2.5 lb/ft^3 (40 kg/m^3) and is used as an insulating material in the walls of refrigerators, as a filler in molding plastics, as a flatting agent in paints, as a bodying agent in printing inks, and as a reinforcement for rubber. It is produced by treating sand with caustic soda to form sodium silicate, and then treating with sulfuric acid to form a jellylike material called **silica gel** which is washed and ground to a fine dry powder. It is also called **synthetic silica. Syloid,** of the Davison Chemical Div., is this material. It is a fluffy white powder with a pH of 7.2. **Silica hydrogel,** of the same company, is a colorless, translucent, semisolid **hydrated silica** of the composition $SiO_2 \cdot xH_2O$, bulking about 44 lb/ft^3 (705 kg/m^3). It contains 28% solids and 72% water. It becomes fluid by mixing with water, and regels on standing. It is used for paper and textile coatings, ointments, and for water suspensions of silica. **Hi-Sil,** of Pittsburgh Plate Glass, and **Santocel,** of Monsanto, are silica gels. **Mertone WB-2,** of the same company, is silica gel used as a coating material for blueprint papers to deepen the blue and increase legibility. When silica gel is used as a pigment, the vehicle surrounds the irregular particle formation, producing greater rigidity and hardness of paint surface than when a smooth pigment is used. For insulation use, the thermal conductivity of silica gel powder is given as 0.1 $Btu/(h)(ft^2)(°F)$ $[(0.000057W/(cm^2)(°C)]$ at −115°F (−81°C), and 0.30 Btu (0.00017W) at 500°F (260°C).

Silicon monoxide, SiO, does not occur naturally but is made by reducing silica with carbon in the electric furnace and condensing the vapor out of contact with the air. It is lighter than silica, having a specific gravity of 2.24, and is less soluble in acid. It is brown powder valued as a pigment for oil painting as it takes up a higher percentage of oil than ochres or red lead. It combines chemically with the oil. **Monox** is a trade name for silicon monoxide. **Fumed silica** is a fine translucent powder of the simple amorphous silica formula made by calcining ethyl silicate. It is used instead of carbon black in rubber compounding to make light-colored products, and to coagulate oil slicks on water so that they can be burned off. It is often called **white carbon,** but the "white carbon black" of Cabot Corp. called **Cab-O-Sil,** used for rubber, is a silica powder made from silicon tetrachloride. **Cab-O-Sil EH5,** a fumed colloidal form, is used as a thickener in resin coatings. The thermal expansion of amorphous fused silica is only about one-eighth that of alumina. Refractory ceramic parts made from it can be heated to 2000°F (1093°C) and cooled rapidly to subzero temperatures without fracture.

SILICON. A metallic element used chiefly in its combined forms; pure silicon metal is used in transistors, rectifiers, and electronic devices. It is a semiconductor, and is superior to germanium for transistors as it will withstand temperatures to 300°F (149°C) and will carry more power. Rectifiers made with silicon instead of selenium can be smaller, and will withstand higher temperatures. Its melting point when pure is about 2615°F (1434°C), but it readily dissolves in molten metals. It is never found free in nature, but combined with oxygen it forms silica, SiO_2, one of the most common substances in the earth. Silicon can be obtained in three modifications. **Amorphous silicon** is a brown-colored powder with a specific gravity of 2.35. It is fusible and dissolves in molten metals. When heated in the air, it burns to form silica. **Graphitoidal silicon** consists of black glistening spangles, and is not easily oxidized and not attacked by the common acids, but is soluble in alkalies. **Crystalline silicon** is obtained in dark, steel-gray globules or crystals of six-sided pyramids of specific gravity 2.4. It is less reactive than the amorphous form, but is attacked by boiling water. All these forms are obtainable by chemical reduction. Silicon is an important constituent of commercial metals. Molding sands are largely silica, and silicon carbides are used as abrasives. Commercial silicon is sold in the graphitoidal flake form, or as ferrosilicon, and silicon-copper. The latter forms are employed for adding silicon to iron and steel alloys. Commercial refined silicon contains 97% pure silicon and less than 1 iron. It is used for adding silicon to aluminum alloys and for fluxing copper alloys. High-purity silicon metal, 99.95% pure, made in an arc furnace, is too expensive for common uses, but is employed for electronic devices and in making silicones. For electronic use, silicon must have extremely high purity, and the pure metal is a nonconductor with a resistivity of 300,000 $\Omega \cdot cm$. For semiconductor use it is "doped" with other atoms yielding electron activity for conducting current. **Epitaxial silicon** is higher purified silicon doped with exact amounts of impurities added to the crystal to give desired electronic properties. Thus, silicon doped with boron has resistivities in grades from 1,000 to 10,000 Ω·cm. **Silicon ribbon** of Westinghouse, for semiconductors, consists of dendritic silicon crystals grown into thin continuous sheets ½ in (1.3 cm) wide, thus eliminating the need to saw slices from ingots. Pure single-crystal silicon ribbon of Dow Chemical is as thin as 1.25 μm and is made as a membrane formed by surface tension between two growing dendritic crystals.

Silicon does not possess a metallic-type lattice structure and, like antimony, is a semimetal and lacks plasticity, but is more akin to the diamond in structure. Because of its feeble electronegative nature, it has a greater tendency to form compounds with nonmetals than with metals. Silicon forms **silicon hybrides** of the general formula Si_xH_{2x+2}, similar to the paraffin hydrocarbons, but they are very unstable and ignite in the air. But a mixture of ferrosilicon and sodium hydroxide, called **hydrogenite,** which yields hydrogen gas when water is added, is used for filling balloons. Silicon, like carbon, has a valence of 4 and links readily to carbon in SiC chain formations. The SiC bond acts like the C-C bond of organic chemistry, but silicon does not enter into animal or plant structures. **Silicon nitride,** used for heat-resistant ceramic parts, is made by furnace nitriding preformed shapes of silicon metal powder. The metal in the part is completely nitrided to the composition Si_3N_4, with a density of 3.44, Rockwell hardness A99, and a compressive strength of 90,000 lb/in² (620 MPa). It sublimes at 3450°F (1899°C), but the parts resist oxidation to 3000°F (1649°C) and have high thermal shock resistance to 2200°F (1204°C). The thermal expansion is low. The material is also used as a coating for corrosion resistance at high temperature. **Roydazide,** of Materials Research Corp., is silicon nitride for coatings and molded parts. **Noralide,** of Norton Co., is a silicon nitride that is hot-pressed at temperatures over 3100°F (1704°C) to produce ball and roller bearings that have good fatigue resistance and excellent friction and wear properties.

SILICON BRONZE. A name applied to two classes of **copper-silicon alloys.** One of these is a copper, or nearly pure copper, fluxed with silicon, in which little or no silicon remains in the final metal. This material is used for strong electric wires and has high strength and resistance to corrosion. A standard alloy of this class contains 98.55% copper, 1.40 tin, and 0.05 silicon. The tensile strength, hard-drawn, is 92,000 lb/in² (634 MPa). The second class of alloy contains a considerable amount of silicon and may have nickel, tin, and other elements. This type of alloy usually depends for hardness on the formation of silicides of nickel or iron. These alloys can be heat-treated and age-hardened.

But the term silicon bronze is used frequently for many copper-base alloys, some of which may be brasses or aluminum bronzes. **Duronze 708,** of Bridgeport Brass, contains 91.5% copper, 6.75 aluminum, and 1.75 silicon. It is for high-strength forgings. The annealed alloy has a tensile strength of 85,000 lb/in^2 (586 MPa), with elongation of 30%. It retains its strength to 600°F (316°C), as well as at subzero temperatures for valves for liquid oxygen. **Silnic bronze,** of the Chase Brass & Copper Co., contains 0.45 to 0.75% silicon with enough nickel, 1.6 to 2.2%, to put the nickel into solution as a nickel silicide and prevent segregation of the hard silicon crystals. It is age-hardening, by heating to 850°F (454°C), and then has a tensile strength to 100,000 lb/in^2 (689 MPa). It comes in rod and wire, and is very corrosion-resistant.

Silicon forms solid alpha solutions in low brass up to about 2.8% silicon, above which point a beta phase is formed, and this mixed-phase structure is present in many high-silicon bronzes used for bearings. **Tombasil alloy,** developed in Germany for bearings and marketed by the Ajax Metals Co., is tombac metal with the addition of silicon. The tensile strength is 65,000 lb/in^2 (448 MPa), with elongation of 15% and Brinell hardness of 135. **White Tombasil,** of H. Kramer & Co., is a modification of this alloy for marine hardware which has a lustrous silver-gray finish without plating. **P.M.G. metal,** of Vickers-Armstrong, has 2% iron, 3 to 4 silicon, 2 zinc, and the balance copper. The forged metal has a tensile strength of 94,000 lb/in^2 (648 MPa), elongation 17%, and Brinell hardness 153. Cast P.M.G. metal has a tensile strength of about 50,000 lb/in^2 (344 MPa) with elongation 15 to 23%, Brinell hardness 90 to 125, and specific gravity 8.44. Castings will withstand high liquid pressures and are corrosion-resistant.

A grade of **Olympic bronze,** of the Chase Brass & Copper Co., contains 3% silicon and 1 zinc. The tensile strength is from 56,000 to 110,000 lb/in^2 (386 to 758 MPa) and elongation 5 to 65%. **Olympic bronze G** contains 22% zinc and 1 silicon, with the balance copper. It is a strong **silicon brass,** and the annealed sheet has a strength of 65,000 lb/in^2 (448 MPa) with elongation 50%. **Doler brass,** of the Doehler Die Casting Co., is a silicon brass for die castings. A group of copper-silicon-manganese alloys is produced by the American Brass Co. under the name of **Everdur metal,** originally patented by Charles Jacobs and called **Jacobs' alloy. Everdur 1000** for castings contains about 94.9% copper, 4 silicon, and 1.1 manganese. The tensile strength, cast, is 45,000 lb/in^2 (310 MPa) and elongation 15%. The alloys combine high strength and toughness with acid and corrosion resistance. The wrought alloys have some physical characteristics similar to those of mild steel, but they have the disadvantage as compared with tin bronze that they have no proportionality limit and suffer permanent deformation at a low load and low impact. The hard-drawn rods have a minimum tensile strength of 70,000 lb/in^2 (482 MPa). The electric conductivity of the cast metal is 7% that of copper, and of hard-wrought metal 12% that of copper. A European alloy, under the name of **Kuprodur,** containing 0.5% silicon, 0.75 nickel, and the balance copper, has good strength at elevated temperatures and was used for locomotive firebox plates.

Manganese up to about 1% may be added to improve the working qualities, and the ratio of silicon to manganese is kept at 3:1. When hard-drawn these alloys can have tensile strengths up to 150,000 lb/in^2 (1,034 MPa). **Webert alloy,** of the American Brass Co., contains small amounts of silicon and manganese, and is used for pressure die castings. Tensile strength is 85,000 lb/in^2 (586 MPa). **Arcoloy,** of the American Radiator Co., is a casting alloy containing 97.25% copper, 2.63 silicon, 0.12 iron, and 0.01 phosphorus. **Cusiloy A,** of the Scovill Mfg. Co., has 95.5% copper, 3 silicon, 1 iron, and 0.5 tin. It is marketed in various forms, and the annealed wire has a tensile strength of 60,000 lb/in^2 (413 MPa) and elongation 50%. **Herculoy 418,** of Revere Copper & Brass, Inc., contains 96.5% copper, 3 silicon, and 0.5 tin. The tensile strength of the annealed sheets is 60,000 lb/in^2 (413 MPa) with elongation 50%, and of the cold-rolled sheets 85,000 lb/in^2 (586 MPa), with elongation 10%. The weight is 0.308 lb/in^3 (8,525 kg/m^3), melting point 1820°F (993°C), but the electric conductivity is only 8.1% that of copper. It is resistant to many chemicals, and is used as a structural metal in the chemical industries.

Alloys of this type but with lower silicon contents are softer and more ductile and are suited to cold heading, spinning, and drawing. **Herculoy 419** has 2% silicon and 0.25 tin. **Herculoy 421** has 1.5% silicon and 0.25 manganese. This alloy in annealed sheet form has a tensile strength of 38,000 lb/in^2 (261 MPa) and elongation 35%, while the cold-rolled sheet has a tensile strength of 60,000 lb/in^2 (413 MPa) and elongation 10%. The copper-silicon alloys are marketed in sheets, rods, bars, shapes, tubes, forgings, and ingots for castings. Both the silicon-tin and the silicon-manganese types are single-phase, and the castings are dense and fine-grained, suitable for such uses as valves.

SILICON CARBIDE. A bluish-black, crystalline, artificial mineral of the composition SiC having a hardness of 2,500 on the Knoop scale. It is used as an abrasive as loose powder, coated abrasive cloth and paper, wheels, and hones. It will withstand temperatures to its decomposing point of 4175°F (2301°C) and is valued as a refractory. It retains its strength at high temperatures and has low thermal expansion, and its heat conductivity is 10 times that of fireclay. Silicon carbide is made by fusing sand and coke at a temperature above 4000°F (2204°C).

Unlike aluminum oxide, the crystals of silicon carbide are large, and they are crushed to make the small grains used as abrasives. They are harder than aluminum oxide, and as they fracture less easily, they are more suited for grinding hard cast irons and ceramics. The standard grain sizes are usually from 100 to 1,000 mesh. The crystalline powder in grain sizes from 60 to 240 mesh is also used in lightning arrestors. **Carborundum,** of the Carborundum Co., **Crystolon,** of the Norton Co., and **Carbolon,** of the Exolon Co., are trade names for silicon carbide. Many other trade names are used, such as **Carborite, Carbolox, Carbolite, Carbobrant, Storalon, Sterbon,** and **Natalon. Ferrocarbo** is a silicon carbide of the Carborundum Co. in briquettes for adding to the iron cupola charge. It breaks down in the cupola above 2000°F (1093°C) to form nascent carbon and silicon for adding to the iron and also for deoxidizing. It produces more uniform iron castings. **Alsimag 539,** of the American Lava Corp., is a fine-grained silicon carbide in the form of molded parts for brazing fixtures and furniture for kilns for high-temperature sintering. The **siliconized graphites** produced by Pure Carbon Co., named **Purebide,** are graphite materials with surfaces chemically converted to silicon carbide. They have the wear resistance of silicon carbide, but retain some of the lubricity of graphite. Cost savings are achieved by machining graphite into intricate shapes before conversion, and subsequently impregnating parts to control leakage or modify strength/wear properties.

When used as a refractory in the form of blocks or shapes, silicon carbide may be ceramic bonded or self-bonded by recrystallization. A standard silicon carbide brick has about 90% SiC, with up to 8% silica. The specific gravity is about 3.2. It has very high resistance to spalling. The thermal conductivity of 109 Btu (12×10^4 J) is about the same as that of mullite, and the coefficient of expansion is about 4.7×10^{-6} per degree Celsius. **Carbex** is a **silicon carbide firebrick** of the General Refractories Co. **Refrax silicon carbide** of the Carborundum Co. is bonded with silicon nitride. It is used for hot-spray nozzles, heat-resistant parts, and for lining electrolytic cells for smelting aluminum. **Silicon carbide KT,** of the same company, is molded without a binder. It has 96.5% SiC with about 2.5% silica. The specific gravity is about 3.1, and it is impermeable to gases. Parts made by pressing or extruding and then sintering have a flexural strength of 24,000 lb/in^2 (165 MPa) and compressive strength of 150,000 lb/in^2 (1,034 MPa). The Knoop hardness is 2,740. It is made in rods, tubes, and molded shapes, and the rough crystal surface can be diamond-ground to a smooth close tolerance. The operating temperature in inert atmospheres is to 4000°F (2204°C), and in oxidizing atmospheres to 3000°F (1649°C). For reactor parts, it has a low neutron-capture cross section and high radiation stability. The thermal conductivity is 2.5 times that of stainless steel. The **Crystolon R** of the Norton Co. is a stabilized silicon carbide bonded by recrystallization. It has a specific gravity of 2.5, a tensile strength of 5,500 lb/in^2 (37 MPa), compressive strength of 25,000 lb/in^2 (172 MPa), and Knoop hardness of 2,500. The porosity is 21%. It is for parts subject to temperatures to 4200°F (2316°C), and it withstands high thermal shock. **Crystolon C** is a self-bonding silicon carbide for coating molded graphite parts to give high wear and erosion resistance. The coatings, 0.003 to 0.020 in (0.008 to 0.051 cm) thick, produced by high-temperature chemical reaction, form an integral part of the grapite surface.

Silicon carbide foam is a lightweight material made of self-bonded silicon carbide foamed into shapes. The low-density foam weighs 17 lb/ ft^3 (272 kg/m^3), has a porosity of 90%, and has a tensile strength of 30 lb/in^2 (0.2 MPa), with compressive strength of 30 lb/in^2 (0.2 MPa). The high-density foam of 33 lb/ft^3 (529 kg/m^3) has a tensile strength of 85 lb/in^2 (0.6 MPa) and compressive strength of 750 lb/in^2 (5 MPa). Its porosity is 80%. It is inert to hot chemicals. It can be machined.

Silicon carbide crystals are used for semiconductors at temperatures above 650°F (343°C). As the cathode of electronic tubes instead of a hot-wire cathode, the crystals take less power and need no warm-up. In the silicon carbide crystal both the silicon and the crystalline carbon have the covalent bond in which each atom has four near neighbors and is bonded to each of these with two electrons symmetrically placed between the atoms, but since there is an electronegative difference between silicon and carbon, there is some ionic bonding which results in a lesser mobility for lattice scattering. The silicon carbide semiconductor crystals of Westinghouse have less than 1 part of impurities to 10 million, and the junction is made by diffusing aluminum atoms into the crystal at a temperature of 3900°F (2149°C), making a *p*-type junction.

Silicon carbide fiber is used as a reinforcement in light metals and in plastics to increase strength and stiffness. The fiber is as small as 7 μm in diameter and usually in short-length particles called whiskers. **Whiskerloy AS15,** of Melpar, Inc., is aluminum alloy 7075 with a content of 15% fiber. It has a tensile strength of 91,000 lb/in^2 (627 MPa), and a modulus of 18 million lb/in^2 (0.12 million MPa). **Whiskerloy MS20** is magnesium alloy AZ31B with 20% carbide fiber. It has a modulus of 13 million lb/in^2 (0.09 million MPa).

SILICON CAST IRON. An **acid-resistant cast iron** containing a high percentage of silicon. When the amount of silicon in cast iron is above 10%, there is a notable increase in corrosion and acid resistance. The acid resistance is obtained from the compound Fe_3Si, which contains 14.5% silicon. The usual amount of silicon in acid-resistant castings is from 12 to 15%. The alloy casts well but is hard and cannot be machined. These castings usually contain 0.75 to 0.85% carbon; amounts in excess of this decrease the acid resistance. Too much carbon also separates out as graphite in silicon irons, causing faulty castings. Increasing the content of silicon in iron reduces the melting point progressively from 1530°C for pure iron to 1250°C for iron containing 23% silicon. A 14 to 14.5% **silicon iron** has a silvery-white structure, a compressive strength of about 70,000 lb/in^2 (482 MPa), Brinell hardness 299 to 350, and is resistant to hot sulfuric acid, nitric acid, and organic acids. Silicon irons are also very wear-resistant, and are valued for pump parts and for parts for chemical machinery. They are marketed under many trade names. **Duriron,** of the Duriron Co., contains 14.5% silicon and 1 carbon and manganese. The tensile strength is 16,000 lb/in^2 (110 MPa) and weight 0.253 lb/in^3 (7,003 kg/m^3).

SILICON-COPPER. An alloy of silicon and copper used for adding silicon to copper, brass, or bronze, and also employed as a deoxidizer of copper and for making hard copper. Silicon alloys in almost any proportion with copper, and is the best commercial hardener of copper. A 50– 50 alloy of silicon and copper is hard and extremely brittle and black in color. A 10% silicon, 90 copper alloy is as brittle as glass; in this proportion silicon copper is used for making the addition to molten copper to produce hard, sound copper castings of high strength. The resulting copper alloy is easy to run in the foundry and does not dross. Silicon-copper grades in 5, 10, 15, and 20% silicon are also marketed, being usually sold in slabs notched for breaking into small sections for adding to the melt. A 10% silicon-copper melts at 1500°F (816°C); a 20% alloy melts at 1152°F (623°C).

SILICON-MANGANESE. An alloy employed for adding manganese to steel, and also as a deoxidizer and scavenger of steel. It usually contains 65 to 70% manganese and 12 to 25 silicon. It is graded according to the amount of carbon, generally 1, 2, and 2.5%. For making steels low in carbon and high in manganese, silicomanganese is more suitable than ferromanganese. A reverse alloy, called **manganese-silicon,** contains 73 to 78% silicon and 20 to 25 manganese, with 1.5 max iron and 0.25 max carbon. It is used for adding manganese and silicon to metals without the addition of iron. Still another alloy is called **ferromanganese-silicon,** containing 20 to 25% manganeses, about 50 silicon, and 25 to 30 iron, with only about 0.50 or less carbon. This alloy has a low melting point, giving ready solubility in the metal.

Silicon-spiegel is an alloy of silicon and manganese with iron employed for making additions of silicon and manganese to open-hearth steels, and also for adding manganese to cast iron in the cupola. A typical analysis gives 25 to 30% manganese, 7 to 8 silicon, and 2 to 3 carbon. Both the silicon and manganese act as strong deoxidizers, forming a thin fusible slag, making clean steel.

SILICON STEEL. All grades of steel contain some silicon and most of them contain from 0.10 to 0.35% as a residual of the silicon used as a deoxidizer. But from 3 to 5% silicon is sometimes added to increase the magnetic permeability, and larger amounts are added to obtain wear-resisting or acid-resisting properties. Silicon deoxidizes steel, and up to 1.75% the elastic limit and impact resistance are increased without loss of ductility. Silicon steels within this range are used for structural purposes and for springs, giving a tensile strength of about 75,000 lb/in^2 (517 MPa) and elongation 25%. A common low-silicon structural steel contains up to 0.35% silicon and 0.20 to 0.40 carbon, but the **structural silicon steels** are ordinarily **silicon-manganese steel,** with the manganese above 0.50%. Low-carbon steels used as structural steels are made by careful control of carbon, manganese, and silicon and with special mill heat treatment. **Lukens LT-75,** of Lukens Steel, contains 0.2% carbon, up to 1.35 manganese, and 0.3 silicon. The tensile strength is 90,000 lb/ in^2 (620 MPa), with elongation of 24%. European silicon structural steels contain 0.80% or more silicon, with manganese above 0.50%, and very low carbon. The silicon alone is a graphitizer, and to be most effective needs the assistance of manganese or other carbide-forming elements. It is useful in high-strength low-alloy steels, and has a wide range of utility when used with expert technique in alloy steels. Considerable addition of silicon above 1.75% increases the hardness and the corrosion resistance, but reduces the ductility and makes the steel brittle. The lower grades can be rolled, however, and silicon-steel sheet is used for electric transformer laminations. Silicon forms a chemical combination with the metal, forming an iron silicide.

The value of silicon steel as a **transformer steel** was discovered by Hadfield in 1883. Silicon increases the electrical resistivity and also decreases the hysteresis loss, making silicon steel valuable for magnetic circuits where alternating current is used. **Electrical steel,** or **electric sheet,** is sheet steel for armatures and transformers, in various grades from 1 to 4.5% silicon. **Hipersil** is a high-permeability silicon steel of Westinghouse. **Cubex,** of the same company, is a silicon steel containing 3% silicon which has been processed so that each cubical crystal of the steel structure is oriented with the faces symmetrical, giving alignment in four directions instead of the normal two. The steel is easily magnetized across as well as along the sheet. In transformers it lowers energy losses, and also gives greater flexibility in designing shapes. Westinghouse also produces a silicon iron which is double oriented, with the cubical crystals of the iron in exact alignment in all directions with the sides of the cubes parallel to the sides and ends of the sheets. It gives high permeability with low induction loss. The **relay steel,** of the Allegheny Ludlum Steel Co., used for relays and mangnets, contains 0.5 to 2.75% silicon. **Orthosil,** of the Thomas & Skinner Steel Products Co., is silicon steel sheet, 0.004 in (0.010 cm) thick, for electrical laminations.

A tough and strong tool steel for forming tools, pneumatic tools, and long punches is made with the addition of silicon and some molybdenum. **Solar steel,** of Carpenter steel, has 1% silicon, 0.50 molybdenum, 0.40 manganese, and 0.50 carbon. It is water-hardening, and has a breaking strength of 323,000 lb/in^2 (2,225 MPa), with elongation 4.5%. **Silman steel,** of the Vanadium Alloys Steel Co., has 2.1% silicon, 0.25 chromium, 0.30 vanadium, 0.85 manganese, and 0.55 carbon. The silicon gives it wear-resistant properties, making it suitable for shear blades and punches; the other alloying elements give toughness and resistance to fatigue. These steels are often referred to as **shock-resistant steel. Black giant steel,** of the Bethlehem Steel Co., for bending, drawing, and stamping dies, has 1.4% carbon, 2.2 silicon. 0.90 manganese, 0.45 chromium, and 0.5 tungsten. The high silicon precipitates some of the carbon in graphitic form, giving the steel a low coefficient of friction and high resistance to wear. **Dargraph steel,** of Darwin & Milner, Inc., for wear plates, gages, and cutting dies, has 1.15% silicon, 0.80 manganese, 0.25 molybdenum, 0.20 chromium, and 1.45 carbon. It has some carbon uniformly distributed as free graphite. Silicon steels with balanced amounts of other elements to give high strength without the normal brittleness of plain silicon steels are used for forgings for aircraft. But the silicon-manganese steels with manganese up to 1% are tough, strong, and wear-resistant. **SAE steel 9255,** with 2% silicon and 0.75 manganese, is such a steel. The **Super Hy-Tuf steel,** of the Crucible Steel Co., has 2.3% silicon, 1.4 chromium, 1.3 manganese, 0.35 molybdenum, 0.20 vanadium, and 0.47 carbon. When heat-treated, it has a tensile strength of 300,000 lb/in^2 (2,068 MPa) with elongation 13%.

SILICONES. A group of resinlike materials in which silicon takes the place of the carbon of the organic synthetic resins. Silicon is quadrivalent like carbon. But, while carbon also has a valence of 2, silicon has only one valence of 4, and the angles of molecular formation are different. The two elements also differ in electronegativity, and silicon is an **amphoteric element,** having both acid and basic properties. The molecular formation of the silicones varies from that of the common plastics, and they are designated as **inorganic plastics** as distinct from the **organic plastics** made with carbon.

In the long-chain organic synthetic resins the carbon atoms repeat themselves, attaching on two sides to other carbon atoms, while in the silicones the silicon atom alternates with an oxygen atom so that the silicon atoms are not tied to each other. The simple **silane** formed by silicon and hydrogen corresponding to methane, CH_4, is also a gas, as is methane, and has the formula SiH_4. But, in general, the silicones do not have the SiH radicals, but contain CH radicals as in the organic plastics. Basically, silicon is treated with methyl chloride and a catalyst to produce a gas mixture of silanes, $(CH_3)_x(SiCl)_{4-x}$. After condensing, three silanes are fractioned, methyl chlorosilane, dimethyl dichlorosilane, and trimethyl trichlorosilane. These are the common building blocks of the **siloxane** chains, and by hydrolyzing them cyclic linear polymers can be produced with acid or alkali catalysts to give fluids, resins, and rubbers. **Silicone resins** have, in general, more heat resistance than organic resins, have higher dielectric strength, and are highly water-resistant. Like organic plastics, they can be compounded with plasticizers, fillers, and pigments. They are usually cured by heat. Because of the quartzlike structure, molded parts have exceptional thermal stability. Their maximum continuous-use service temperature is about 500°F (260°C). Special grades exceed this and go as high as 700 to 900°F (371 to 482°C). Their heat-deflection temperature for 265 lb/in^2 (1.8 MPa) is 900°F (482°C). Their moisture absorption is low, and resistance to petroleum products and acids is good. Nonreinforced silicones have only moderate tensile and impact strength, but fillers and reinforcements provide substantial improvement. Because silicones are high in cost, they are premium plastics and are generally limited to critical or high-performance products such as high-temperature components in the aircraft, aerospace, and electronics fields.

A great variety of molecular combinations are available in the silicone polymers, giving resins of varying characteristics, and those having CH radicals with silicon bonds are termed **organosilicon polymers. Silicon tetramethyl,** $Si(CH_3)_4$, is a liquid boiling at 26°C. **Trichlorosilane,** $HSiCl_3$, is also called **silicochloroform,** and corresponds in formation to chloroform. By replacing the hydrogen atom of this compound with an alkyl group, the **alkylchlorosilanes** are made which have high adhesion to metals and are used in enamels. **Methyl chlorosilane,** $(CH_3)_2SiCl_2$, is a liquid used for waterproofing ceramic electrical insulators. The material reacts with the moisture in the ceramic, forming a water-repellent coating of methyl silicone resin and leaving a residue of hydrochloric acid which is washed off. **Dry-Film 9977,** of the General Electric Co., is a liquid mixture of dimethyl dichlorosilane and methyl trichlorosilane for this same purpose. **Silicone SC-50,** of the same company, used in concrete and in gypsum plaster and in water paints to impart water repellency, is a **sodium methyl siliconate. Velvasil,** of this company, is a **dimethyl siloxane** used in cosmetics as a water repellent, and in lipsticks for smear resistance. The dimethyl polysiloxanes are very stable fluids used widely in cosmetics, but valued also for use in shock absorbers, transformers, and in automatic control systems. The **Sun Filter polish,** of the Westley Industries, is this material. It is resistant to sunlight and to alkaline chemicals and withstands temperatures to 300°F (149°C). The **silicone cement** of Charles Englehard, Inc., to give strong, heat-resistant bonds to metals, glass, and ceramics, is a polysiloxane with mineral fillers. The strong cements developed by the

Naval Ordnance Laboratory for making chemical bonds between the glass fibers and the resin in plastic laminates are made by reacting allyl trichlorosilane with phenol resorcinol, or xylenol.

Silicone insulating varnishes will withstand continuous operating temperatures at 350°F (177°C) or higher. **Silicone enamels** and paints are more resistant to chemicals than most organic plastics, and when pigmented with mineral pigments will withstand temperatures up to 1000°F (538°C). For lubricants the liquid silicones are compounded with graphite or metallic soaps and will operate between −50 and 500°F (−46 and 260°C). **Thermocone,** of the Joseph Dixon Crucible Co., is a black, chemical-resistant paint which withstands temperatures to 1000°F (538°C). It is a liquid silicone containing graphite flake. The silicone liquids are stable at their boiling points, between 750 and 800°F (399 and 427°C), and have low vapor pressures, so that they are also used for hydraulic fluids and heat-transfer media. **Silicone oils,** used for lubrication and as insulating and hydraulic fluids, are methyl silicone polymers. They retain a stable viscosity at both high and low temperatures. As hydraulic fluids they permit smaller systems to operate at higher temperatures. In general, silicone oils are poor lubricants compared with petroleum oils, but they are used for high temperatures, 150 to 200°C, at low speeds and low loads.

Silicone resins are blended with alkyd resins for use in outside paints, usually modified with a drying oil. **Resin XR-807,** of Dow Corning, is such a silicone-alkyd containing 25% silicone resin, used to produce paints of high weather and sunlight resistance. A catalyst is added to give air drying. **Silicone-alkyd resins** are also used for baked finishes, combining the adhesiveness and flexibility of the alkyd with the heat resistance of the silicone. A **phenyl ethyl silicone** is used for impregnating glass-fiber cloth for electrical insulation and it has about double the insulating value of ordinary varnished cloth.

Silicone rubber is usually a long-chain **dimethyl silicone** which will flow under heat and pressure, but can be vulcanized by cross-linking the linear chains. Basically, it consists of alternate silicon and oxygen atoms with two methyl groups attached to each silicon atom. The tensile strength is 300 lb/in^2 (2 MPa), but with fillers it is raised to 600 lb/in^2 (4 MPa). It is usually compounded with silica and pigments. It is odorless and tasteless, is resistant to most chemicals but not to strong acids and alkalies, will resist heat to 500°F (260°C), and will remain flexible to − 70°F (−57°C). The dielectric strength is 500 volts per mil (20×10^6 volts per meter). **Silastic,** of Dow Corning, is a silicone rubber. It is a white rubbery material with an elongation of 70 to 150%, and has good adhesion to various materials. It is used for electric insulation, for coating fabrics, and for gaskets and other parts. **Silastic RTV 601,** used for encapsulation of electrical units, has a specific gravity of 1.29, tensile strength of 300 lb/in^2 (2 MPa), elongation of 150%, and dielectric strength of 550 volts per mil (22 × 10^6 volts per meter). The hardness is Shore A40.

Ordinary silicone rubber has the molecular group ($H \cdot CH_2 \cdot Si \cdot CH_2 \cdot H$) in a repeating chain connected with oxygen linkages, but in the **nitrile-silcone rubber** of General Electric one of the end hydrogens of every fourth group in the repeating chain is replaced by a C:N radical. These polar nitrile groups give a low affinity for oils, and the rubber does not swell with oils and solvents. It retains strength and flexibility at temperatures from −100°F (−73°C) to above 500°F (260°C), and is used for such products as gaskets and chemical hose. This material, called **N.S. Fluids** in the form of water-white to yellow liquids having 3 to 23% nitrile content, is used for solvent-resistant lubricants and as antistatic plasticizers. As lubricants, they retain a nearly constant viscosity at varying temperatures. **Fluorosilicones** have fluoroalkyd groups substituted for some of the methyl groups attached to the siloxane polymer of dimcthyl silicone. They are fluids, greases, and rubbers, incompatible with petroleum oils and insoluble in most solvents. The greases are the fluids thickened with lithium soap, or with a mineral filler.

SILK. The fibrous material in which the silkworm, or larva of the moth **Bombyx mori,** envelops itself before passing into the chrysalis state. Silk is closely allied to cellulose and resembles wool in structure, but unlike wool it contains no sulfur. The natural silk is covered with a wax or silk glue which is removed by scouring in manufacture, leaving the glossy **fibroin,** or raw-silk fiber. The fibroin consists largely of the amino acid **alanine,** $CH_3CH(NH_2)CO_2H$, which can be synthesized from pyruvic acid. **Silk fabrics** are used mostly for fine garments, but are also valued for military powder bags because they burn without a sooty residue.

The fiber is unwound from the cocoon and spun into threads. Each cocoon has from 2,000 to 3,000 yd (1,829 to 2,743 m) of thread. The chief silk-producing countries are China, Japan, India, Italy, and France. **Floss silk** is a soft silk yarn practically without twist, or is the loose waste silk produced by the worm when beginning to spin its cocoon. **Hard silk** is thrown silk from which the gum has not been discharged. **Soft silk** is thrown silk yarn, degummed, dyed or undyed. **Souple silk** is dyed skein silk from which little gum has been discharged. It is firmer but is less lustrous. **Organizine silk** is from the best grade of cocoons. **Marabout silk,** used for making imitation feathers, is a white silk, twisted and dyed without discharging the gum. **Silk waste** is silk other than that reeled from the cocoon. It includes cocoons not fit for reeling, partly unwound cocoons, broken filaments, mill waste, and discarded noils. It is used in the spun-silk yarn industry. **Noils** consist of the short, staple knotty combings.

In China the cultivation of the silkworm is claimed to date back to 2640

B.C. Silk was first woven in Rome about 50 B.C. The eggs of the silkworm were smuggled into Europe in the year 552. **Sericulture,** or silkworm culture, is a highly developed industry. The larvae, which have voracious appetites, are fed on mulberry leaves for 24 days, after which they complete their cocoons in 3 to 4 days. In from 7 to 70 days these are heated to kill the chrysalis to prevent bursting of the shell. The reeling is done by hand and by machine. **Wild silk** is from a night peacock moth which does not feed on the mulberry. It is coarser and stronger, but darker in color and less lustrous. **Tussah silk** is a variety of wild silk from South China and India. **Charka silk** is raw silk produced in Bengal on native hand-reeling machines. **Byssus silk** is a long fiber from a mussel of Sardinia and Corsica which spins the thread to attach itself to rocks. The fiber is golden brown, soft, lustrous, and elastic, and not dissolved by acids or alkalies. It was formerly used for fine garments but is no longer obtained commercially. **Canton silk** is soft and fluffy, but is greenish in color and lacks firmness. It is from the *B. textor,* and is used for weft yarns and in crepes. The silk grown in India and known as **Indian silk** is the finest of all silks with fibers 0.0004 in (0.0016 cm) compared with 0.001 in (0.003 cm) for Japanese silk. Before the war Japan produced most of the silk of the world from a cultivated moth of the tussah variety, *Antheria yama mai.* **Shantung silk** is from a tussah moth, *A. pernyi,* which feeds on oak leaves.

The fabric called **shantung** is a rough-textured plain-woven silk with irregular fillings. It is heavier and more bumpy than pongee. **Grosgrain** is a heavy close-woven corded fabric of silk. It is used for tapestry, and in narrow widths for ribbons. **China silk,** or **habutai,** is an unweighted all-silk fabric of close, firm, but uneven texture woven of low-quality, unthrown **raw silk** in the gum, but it is also imitated with textiles with a silk warp and a rayon filling. The light-weight grades of 3, 3½, and 4 momme (76, 65, and 56 m^2/kg) are classed as sheer fabrics, and are used for impregnated fabrics for umbrellas, raincoats, and hospital sheetings. Unimpregnated habutai is used for curtains, lampshades, handkerchiefs, and caps. Heavy-weight habutai of 12 momme (18m2/kg) is used for parachutes. **Pongee** is a rough-textured plain-woven silk fabric with irregular filling yarns. It is made in natural color or dyed, and like China silk has a gummy feel. **Bolting cloth,** for screening flour, is a fine, strong, silk fabric. The yarn is a fine-thread, hard-twist tram thrown in the gum from high-quality raw silk. The fabric has a lino weave with two warp threads swiveled around the weft. It comes in various meshes, the finest having 166 to 200 threads per linear inch. It is produced on handlooms in Switzerland and France. **Cartridge cloth** is a thin strong fabric for powder bags for large-caliber guns. It is made of silk waste and noils. The silk is consumed in the explosion without leaving residues that would cause premature explosion of the subsequent charge. It also does not deteriorate in storage in contact with the powder.

The **kente cloth** of Ghana is a silk fabric of fine weave in delicate colors, hand-woven in long narrow strips which are sewn together to make a pattern. **Satin** is a heavy silk fabric with a close twill weave in which the fine warp threads appear on the surface and the weft threads are covered up by the peculiar twill. Common satin is of eight-leaf twill, the weft intersecting and binding down the warp at every eighth pick, but 16 to 20 twills are also made. In the best satins a fine quality of silk is used. It was originally called **zayton,** derived from the Arab name of the Chinese trading post where the fabric was produced. Varieties of imitation satin are made with a cotton weft. Satins are dyed to many colors, and much used for linings and trimings.

Qiana, of Du Pont, originally called Fiber Y, produces **synthetic resin fabrics** with the feel and drape of silk. They are resilient and take dyes readily. The fiber is a polyamide based on an alicyclic diamine. **A-Tell** is a Japanese textile fiber of great silkiness. It is a polyethylene oxybenzoate, the molecule having both ester and ether linkages. Another Japanese fiber is 50% polyvinyl chloride and 50% polyvinyl alcohol. Called **Cordelan,** it produces fabrics with the feel of wool.

SILVER. A white metal, symbol Ag, very malleable and ductile, and classed with the precious metals. It occurs in the native state, and also combined with sulfur and chlorine. Copper, lead, and zinc ores frequently contain silver; about 70% of the production of silver is a by-product of the refining of these metals. Mexico and the United States produce more than half of the silver of the world. Canada, Peru, and Bolivia are also important producers. Although nearly 90% of the silver produced in Arizona comes from copper ores, most of that produced in California is a by-product of gold quartz mining. Silver is the whitest of all the metals and takes a high polish, but easily tarnishes in the air because of the formation of a silver sulfide. It does not corrode. It has the highest electric and heat conductivity. The specific gravity is 10.7, and the melting point is 1762°F (964°C). When heated above the boiling point, it passes off as a green vapor. It is soluble in nitric acid and in hot sulfuric acid. The tensile strength of cast silver is 41,000 lb/in^2 (282 MPa) with Brinell hardness 59. The metal is marketed on a troy-ounce value.

Since silver is a very soft metal, it is not normally used industrially in its pure state, but is alloyed with a harderner, usually copper. **Sterling silver** is the name given to a standard high-grade alloy containing a minimum of 925 parts in 1,000 of silver. It is used for the best tableware, jewelry, and electric contacts. This alloy of 7.5% copper work-hardens, and requires annealing between rollings. Silver can also be hardened by alloying with other elements. The old alloy **silanca** contained small amounts of zinc and antimony, but the name sterling silver is applied only to the specific **silver-copper alloy.**

The standard types of commercial silver are fine silver, sterling silver, and coin silver. **Fine silver** is at least 99.9% pure, and is used for plating, making chemicals, and for parts produced by powder metallurgy. **Coin silver** is usually an alloy of 90% silver and 10% copper, but when actually used for coins the composition and weight of the coin are designated by law. Silver and gold are the only two metals which fulfill all the requirements for coinage. The so-called **coins** made from other metals are really official tokens, corresponding to paper money, and are not true coins. Coin silver has a Vickers hardness of 148 compared with the hardness of of 76 for hard-rolled pure silver. It is also used for silverware, ornaments, plating, for alloying with gold, and for electric contacts. When about 2.5% of the copper in coin silver is replaced by aluminum the alloys can be age-hardened to 190 Vickers.

Silver is not an industrial metal in the ordinary sense. It derives its coinage value from its intrinsic aesthetic value for jewelry and plate, and in all civilized countries silver is a controlled metal.

Silver powder, 99.9% purity, for use in coatings, integrated circuits, and other electrical and electronic applications, is produced in several forms. Amorphous powder is made by chemical reduction and comes in particle size from 0.9 to 15 μm. Powder made electrolytically is in dendritic crystals with particle sizes from 10 to 200 μm. Atomized powder has spherical particles and may be as fine as 400 mesh. **Silver-clad powder** for electric contacts is a copper powder coated with silver to economize on silver. **Silver flake** is in the form of laminar platelets and is particularly useful for conductive and reflective coatings and circuitry. The tiny flat plates are deposited in overlapping layers permitting a metal weight saving as much as 30% without reduction in electrical properties. **Silver flake No. 3,** of the Metz Refining Co., has an average particle diameter size of 6 μm with thickness of 0.2 μm. **Nickel-coated silver powder,** for contacts and other parts made by powder metallurgy, comes in grades with ¼, ½, 1, and 2% nickel by weight.

The **porous silver** of the Pall Corp. comes in sheets in standard porosity grades from 2 to 55 μm. It is used for chemical filtering. **Doré metal** used for jewelry is silver containing some gold, but the material known as doré metal, obtained as a by-product in the production of selenium from copper slimes, is a mixture of silver, gold, and platinum. Silver plating is sometimes done with a **silver-tin alloy** containing 20 to 40 parts silver and the remainder tin. It gives a plate having the appearance of silver but with better wear resistance. Silver plates have good reflectivity in the high wavelengths, but the reflectivity falls off at about 350 nm, and is zero at 3,000, so that it is not used for heat reflectors.

Silver-clad sheet, made of a cheaper nonferrous sheet with a coating of silver rolled on, is used for food-processing equipment. It is resistant to organic acids but not to products containing sulfur. **Silver-clad steel,** used for machinery bearings, shims, and reflectors, is made with pure silver bonded to the billet of steel and then rolled. For bearings, the silver is 0.010 to 0.35 in (0.025 to 0.889 cm) thick, but for reflectors the silver is only 0.001 to 0.003 in (0.003 to 0.008 cm) thick. The **silver-clad stainless steel** of the American Cladmetals Co. is stainless-steel sheet with a thin layer of silver rolled on one side to give an electrically conductive surface.

Silver iodide is a pale-yellow powder of the composition AgI, best known for its use as a nucleating agent and for seeding rain clouds. **Silver nitrate,** formerly known as **lunar caustic,** is a colorless, crystalline, poisonous, and corrosive material of the composition $AgNO_3$. It is used for silvering mirrors, for silver plating, in indelible inks, in medicine, and for making other silver chemicals. The high-purity material is made by dissolving silver in nitric acid, evaporating the solution and crystallizing the nitrate, and then redissolving the crystals in distilled water and recrystallizing. It is an active oxidizing agent. **Silver chloride,** AgCl, is a white granular powder used in silver-plating solutions. This salt of silver and other halogen compounds of silver, especially **silver bromide,** AgBr, are used for photographic plates and films. The image cast on the plate by the lens breaks down the atomic structure of the compound in proportion to the intensity of the light waves received and the time of exposure. Electrons gather on the positive lower side of the bromide grains, causing the formation of black threads of silver when the film is placed in a developing solution of **ferrous oxalate,** FeC_2O_4, or other reducing chemical. The comparative values, or tones, in the picture come from the different color wavelengths in the white light and the different intensities of the incoming waves. Measured in second-units, the action of violet light, the shortest wavelength, on the compound is more than 40 times greater than the action of the long wavelength of red light. To prevent further action by light, the film is transferred to a fixing bath of sodium thiosulfate which dissolves out the unreduced silver bromide.

Silver chloride crystals in sizes up to 10 lb (4.5 kg) are grown synthetically. The crystals are cubical, and can be heated and pressed into sheets instead of being sawed. The specific gravity is 5.56, index of refraction 2.071, and melting point 455°C. They are slightly soluble in water and soluble in alkalies. The crystals transmit more than 80% of the wavelengths from 50 to 200 μm. **Cerargyrite,** sometimes called **horn silver,** an ore of the metal silver, found in the upper zone of silver veins in Nevada, Colorado, Idaho, Peru, Chile, and Mexico, is a silver chloride containing theoretically 75.3% silver, with sometimes some mercury. The hardness is 2.3 and specific gravity 5.8. It is massive, resembling wax, with a pearl-gray color.

Silver sulfide, Ag_2S, is a gray-black, heavy powder used for inlaying in metal work. It changes its crystal structure at about 355°F (179°C), with a drop in electrical resistivity, and is used for self-resetting circuit breakers. **Silver potassium cyanide,** $KAg(CN)_2$, is a white, crystalline, poisonous solid used for silver-plating solutions. **Silver tungstate,** Ag_2WO_4, **silver manganate,** $AgMnO_4$, and other silver compounds are produced by Alfa Inorganics in high-purity grades for electronic and chemical uses.

SILVER SOLDER. High-melting-point solder employed for soldering joints where more than ordinary strength and, sometimes, electric conductivity are required. The economy of silver solder is in the lower cost of labor in producing a tight, strong joint. Most silver solders are **copper-zinc brazing alloys** with the addition of silver, and the soldering is done with a blowtorch. They may contain from 9 to 80% silver, and the color varies from brass yellow to silver white. Cadmium may also be added to lower the melting point. Silver solders do not necessarily contain zinc, and may be alloys of silver and copper in proportions arranged to obtain the desired melting point and strength. A silver solder with a relatively low melting point contains 65% silver, 20 copper, and 15 zinc. It melts at 1280°F (693°C), has a tensile strength of 64,800 lb/in^2 (447 MPa), and elongation 34%. The electric conductivity is 21% that of pure copper. A solder melting at 1400°F (760°C) contains 20% silver, 45 copper, and 35 zinc. **ASTM silver solder No. 3** is this solder with 5% cadmium replacing an equal amount of the zinc. It is a general-purpose solder. **ASTM silver solder No. 5** contains 50% silver, 34 copper, and 16 zinc. It melts at 1280°F (693°C), and is used for electrical work and refrigeration equipment.

Any tin present in silver solders makes them brittle; lead and iron make the solders difficult to work. Silver solders are malleable and ductile and have high strength. They are also corrosion-resistant and are especially valuable on food machinery and apparatus where lead is objectionable. Small additions of lithium to silver solders increase the fluidity and wetting properties, especially for brazing stainless steels or titanium. **Sil-Fos,** of Handy & Harmon, is a phosphor-silver brazing solder with a melting point of 1300°F (704°C). It contains 15% silver, 80 copper, and 5 phosphorus. Lap joints brazed with Sil-Fos have a tensile strength of 30,000 lb/in^2 (206 MPa). The phosphorus in the alloy acts as a deoxidizer, and the solder requires little or no flux. It is used for brazing brass, bronze, and nickel alloys. The grade made by this company under the name of **Easy solder** contains 65% silver, melts at 1325°F (718°C), and is a color match for sterling silver. **TL silver solder** of the same company has only 9% silver and melts at 1600°F (871°C). It is brass yellow in color, and is used for brazing nonferrous metals. **Sterling silver solder,** for brazing sterling silver, contains 92.8% silver, 7 copper, and 0.2 lithium. The flow temperature is 1650°F (899°C).

A **lead-silver solder** recommended by the Indium Corp. of America to replace tin solder contains 96% lead, 3 silver, and 1 indium. It melts at 310°C, spreads better than ordinary lead-silver solders, and gives a strength of 4,970 lb/in^2 (34 MPa) in the joint. **Silver-palladium alloys** for high-temperature brazing contain from 5 to 30% palladium. With 30% the melting point is about 1232°C. These alloys have exceptional melting and flow qualities and are used in electronic and spacecraft applications.

SISAL. The hard, strong, light-yellow to reddish fibers from the large leaves of the sisal plant, *Agave sisalana,* and the henequen plant, *A. fourcroydes,* employed for making rope, cordage, and sacking. About 80% of all binder twine is normally made from sisal, but sisal ropes have only 75% of the strength of Manila rope and are not as resistant to moisture. Sisal is a tropical plant, and grows best in semiarid regions. The agave plant is native to Mexico, but most of the sisal comes from Haiti, East Africa, and Indonesia. The retting, separation, and washing of the fiber is done by machine, and less than 5% of the weight of the leaf results in good fiber. **Mexican sisal** is classed in seven grades from the Superior white fiber 105 cm in length to the Grade C-1, short-spotted fiber 60 cm in length. **Yucatan sisal,** or **henequen,** is from the henequen plant and is reddish in color, stiffer and coarser, and is used for binder twine. The Indian word henequen means knife, from the knifelike leaves. The plant is more drought-resistant than sisal. Henequen also comes from Indonesia as the spotted or reddish grades of sisal. **Maguey,** or **cantala,** is from the leaves of the *A. cantala* of India, the Philippines, and Indonesia. It is used principally for binder twine. The fibers are white, brilliant, stiff, and light in weight. The fibers are not as strong as sisal, but have a better appearance and greater suppleness. **Zapupe fiber,** of Mexico, is from the *A. zapupe.* The fiber is similar to sisal, finer and softer than henequen. **Salvador sisal,** of El Salvador, is from the *A. letonae.* The leaves are more slender than Mexican sisal, and the fiber is softer and finer. It is used for cordage and fabrics.

The fibers of sisal are not as long or as strong as those of Manila hemp, and they swell when wet, but they are soft and are preferred for binder twine either alone or mixed with Manila hemp. **Sisal fiber** is also used instead of hair in cement plasters for walls and in laminated plastics. **Corolite** is a molded plastic of the Columbia Rope Co. made with a mat of sisal fibers so as to give equal strength in all directions. **Agave fibers** from other varieties of the plant are used for various purposes, notably **tampico,** from the *A. rigida,* which yields a stiff, hard, but pliant fiber employed for circular power brushes, and **istle,** a similar stiff brush fiber from several plants. Tampico is valued for polishing wheels, as the fibers hold the grease buffing compositions, and it is not brittle but abrades with flexibility. **Jaumave istle** is from the *A. funkiana* of Mexico. It yields long, uniform fibers finer than tampico. **Lechuguilla** is a type of istle from the *A. lechuguilla..*

There are at least 50 species of agave in Mexico and the southwestern

United States which yield valuable by-products in addition to fiber. From some varieties saponin is obtained as a by-product. From a number of thick-leaved species the buds are cut off, leaving a cavity from which juice exudes. This juice is fermented to produce **pulque,** a liquor with a ciderlike taste containing about 7% alcohol. The juice contains a sugar, **agavose,** $C_{12}H_{22}O_{11}$, which is used in medicine as a laxative and diuretic. **Agava,** of Agava Products, Inc., is a dark-brown viscous liquid extracted from the leaves of agave plants, used as a water conditioner for boiler-water treatment. It is a complex mixture of sapogenines, enzymes, chlorophyllin, and polysaccharides.

A fine strong fiber is obtained from the long leaves of the **pineapple,** *ananus comosus,* native to tropical America. The plant is grown chiefly for its fruit, known in South America under its Carib name **ananá** and marketed widely as canned fruit and juice, preserves, and confections. **Pineapple concentrate** is also sold as a flavor enhancer, as much as 10% being added to apricot, cherry, or other fruit juices without altering the original flavor. For fiber production the plants are spaced widely for leaf development and are harvested before the leaves are fully mature. The retted fibers are long, white, and of fine texture and may be woven into water-resistant fabrics. The very delicate and expensive **piña cloth** of the Philippines is made from **pineapple fiber.** The fabrics of Taiwan are usually coarser and harder.

SLAG. The molten material that is drawn from the surface of iron in the blast furnace. Slag is formed from the earthy materials in the ore and from the flux. Slags are produced in the melting of other metals, but iron blast-furnace slag is usually meant by the term. Slag is used in cements and concrete, for roofing, and as a ballast for roads and railways. Finely crushed slag is used in agriculture for neutralizing acid soils. **Blast-furnace slag** is one of the lightest concrete aggregates available. It has a porous structure and, when crushed, is angular. It is also crushed and used for making pozzuolana and other cements. Slag contains about 32% silica, 14 alumina, 47 lime, 2 magnesia, and small amounts of other elements. It is crushed, screened, and graded for marketing. **Crushed slag** weighs 1,900 to 2,100 lb/yd^3 (1,127 to 1,245 kg/m^3), or about 30% lighter than gravel. **Honeycomb slag** weighs only about 30 lb/ft^3 (481 kg/m^3). The finest grade of commercial slag is from $\frac{3}{16}$ in (0.48 cm) to dust; the run-of-crusher slag is from 4 in (10 cm) to dust. **Basic phosphate slag,** a by-product in the manufacture of steel from phosphatic ores, is finely ground and sold for fertilizer. It contains not less than 12% phosphoric oxide, P_2O_5, and is known in Europe as **Thomas slag. Foamed slag** is a name used in England for honeycomb slag used for making lightweight, heat-insulating blocks. A superphosphate cement is made in Belgium from a mixture of basic slag, slaked lime, and gypsum.

SLATE. A shale having a straight cleavage. Most shales are of sedimentary origin, and their cleavage was the result of heavy or long-continued pressure. In some cases slates have been formed by the consolidation of volcanic ashes. The salty cleavage does not usually coincide with the original stratification. Slate is of various colors: black, gray, green, and reddish. It is used for electric panels, blackboards, slate pencils, tabletops, roofing shingles, floor tiles, and treads. The terms **flagstone** and **cleftstone** are given to large flat sections of slate used for paving, but the names are also applied to blue sandstones cut for this purpose. Slate is quarried in large blocks, and then slabbed and split. The chief slate-producing states are Pennsylvania, Vermont, Virginia, New York, and Maine. **Roofing slates** vary in size from 12 by 6 in (30 by 15 cm) to 24 by 14 in (61 by 36 cm), and from ⅛ to ¾ in (0.32 to 1.91 cm) in thickness, and are usually of the harder varieties. The roofing slate from coal beds is black, fine-grained, and breaks into brittle thin sheets. It does not have the hardness or weather resistance of true slate. As late as 1915 more than 85% of all slate mined was used for roofing, but the tonnage now used for this purpose is small. **Ribbon slate,** with streaks of hard material, is inferior for all purposes. Lime impurities can be detected by the application of dilute hydrochloric acid to the edges and noting if rapid effervescence occurs. Iron is a detriment to slates for electric purposes. The average compressive strength of slate is 15,000 lb/in^2 (103 MPa) and the weight 175 lb/ft^3 (2,804 kg/m^3). **Slate granules** are small graded chips used for surfacing prepared roofing. **Slate flour** is ground slate, largely a by-product of granule production. It is used in linoleum, calking compounds, and in asphalt surfacing mixtures. **Slate lime** is an intimate mixture of finely divided calcined slate and lime, about 60% by weight lime to 40 slate. It is employed for making porous concrete for insulating partition walls. The process consists in adding a mixture of slate lime and powdered aluminum, zinc, or magnesium to the cement. The gas generated on the addition of water makes the cement porous.

SMOKE AGENTS. Chemicals used in warfare to produce an obscuring cloud or fog to hide movements. Smokes may be harmless and are then called **screening smokes,** or **smoke screens,** or they may be toxic and called **blanketing clouds.** There are two types of smokes: those forming solid or liquid particles, and those forming fogs or mists by chemical reaction. The first naval smoke screens were made by limiting the admission of air to the fuel in the boilers, and the first Army smoke pots contained mixtures of pitch, tallow, saltpeter, and gunpowder. The British **smoke candles** contained 40% potassium nitrate, 29 pitch, 14 sulfur, 8 borax, and 9 coal dust. They gave a brown smoke, but one that lifted too easily.

Fog or military screening may be made by spraying an oil mixture into the air at high velocity. The microscopic droplets produce an impenetrable fog which remains for a long period. **White phosphorous** gives a dense white smoke by burning to the pentoxide and changing to phosphoric acid in the moisture of the air. Its vapor is toxic. **Sulfuric trioxide,** SO_3, is an effective smoke producer in humid air. It is a mobile, colorless liquid vaporizing at 45°C to form dense white clouds with an irritating effect. The French **opacite** is **tin tetrachloride,** or **stannic chloride,** $SnCl_4$, a liquid that fumes in the air. When hydrated, it becomes the crystalline pentahydrate, $SnCl_4 \cdot 5H_2O$. The smoke is not dense, but it is corrosive and it penetrates gas masks. **Sulfuryl chloride,** SO_2Cl_2, is a liquid that decomposes on contact with the air into sulfuric and hydrochloric acids. **F.S. smoke** is made with a mixture of chlorosulfonic acid and sulfur trioxide. **Silicon tetrachloride,** $SiCl_4$, is a colorless liquid that boils at 60°C, and fumes in the air, forming a dense cloud. Mixed with ammonia vapor it resembles a natural fog. The heavy mineral known as **amang,** separated from Malayan tin ore, containing ilmenite and zircon, is used in smoke screens. **Titanium tetrachloride,** $TiCl_4$, is a colorless to reddish liquid boiling at 136°C. It is used for smoke screens and for skywriting from airplanes. In most air it forms dense, white fumes of **titanic acid,** H_2TiO_3, and hydrogen chloride. The commercial liquid contains about 25% titanium by weight.

A common smoke for airplanes is **oleum.** It is a mixture of sulfur trioxide in sulfuric acid, which forms fuming sulfuric acid, or **pyrosulfuric acid,** $H_2S_2O_7$. The dense liquid is squirted in the exhaust manifold. **Zinc smoke** is made with mixtures of zinc dust or zinc oxide with various chemicals to form clouds. **H.C. smoke** is zinc chloride with an oxidizing agent to burn up residual carbon so that the smoke will be gray and not black. **Signal smoke** is colored smoke used for ship distress signals, and for aviation marking signals. They are mixtures of a fuel, an oxidizing agent, a dye, and sometimes a cooling agent to regulate the rate of burning and to prevent decomposition of the dye. Unmistakable colors are used so that the signals may be distinguished from fires, and the dyes are mainly anthraquinone derivatives.

SNAKESKINS. The snakeskins employed for fancy leathers are in general the skins of large tropical snakes which are notable for the beauty or oddity of their markings. Snakeskins for shoe-upper leathers, belts, and handbags are glazed like kid and calfskin after tanning. Small cuttings are used for inlaying on novelties. The leather is very thin, but is remarkably durable and is vegetable-tanned and finished in natural colors, or is dyed. **Python skins** are used for ladies' shoes. **Regal python skins** from Borneo, the Philippines, and the Malay Peninsula sometimes measure 30 ft (9 m) in length and have characteristic checked markings. Diamond-backed rattlesnakes are raised on snake farms in the United States. The meat is canned as food, and the skins are tanned into leather. Only the back is used for leather, as the belly is colorless.

SOAP. A cleansing compound produced by saponifying oils, fats, or greases with an alkali. When caustic soda is added to fat, glycerin separates out, leaving **sodium oleate,** $Na(C_{17}H_{33}O_2)$, which is soap. But since oils and fats are mixtures of various acid glycerides, the soaps made directly from vegetable and animal oils may be mixtures of oleates, palmitates, linoleates, and laurates. **Soap oils** in general, however, are those oils which have greater proportions of nearly saturated fatty acids, since the unsaturated fractions tend to oxidize to form aldehydes, ketones, or other acids, and turn rancid. If an excess of alkali is used, the soap will contain free alkali, and the greater the proportion of the free alkali the coarser is the action of the soap. ASTM standards for milled toilet soap permit only 0.17% free alkali. **Sodium soaps** are always harder than **potassium soaps** with the same fat or oil. Hard sodium soaps are used for chips, powders, and toilet soaps. Soft caustic potash soaps are the liquid, soft, and semisoft pastes. Mixtures of the two are also used. Soaps are made by either the boiled process or the cold process. **Chip soap** is made by pouring the hot soap onto a cooled revolving cylinder from which the soap is scraped in the form of chips or ribbons which are then dried to reduce the moisture content from 30 to 10%. **Soap flakes** are made by passing chips through milling rollers to make thin, polished, easily soluble flakes.

Powdered soap is made from chips by further reducing the moisture and grinding. **Milled soaps** are made from chips by adding color and perfumes to the dried chips and then passing through milling rollers and finally pressing in molds. **Toilet soaps** are made in this way. Soap is used widely in industrial processing, and much of the production has consisted of chips, flakes, powdered, granulated, and scouring powders. Soaps have definite limitations of use. They are unstable in acid solutions and may form insoluble salts. In hard waters they may form insoluble soaps of calcium or magnesium unless a phosphate is added. Many industrial cleansers, therefore, may be balanced combinations of soaps, synthetic detergents, phosphates, or alkalies, designed for particular purposes.

About half of all soap is made with tallow, 25% with coconut oil, and the remainder with palm oil, greases, fish oils, olive oil, soybean oil, or mixtures. A typical soap contains 80% mixed oils and 20 coconut oil, with not over 0.2 free alkali. Auxiliary ingredients are used in soap to improve the color, for perfuming, as an astringent, or for abrasive or harsh cleaning purposes. Phenol or chloronated compounds are used in **antiseptic soap.** The soft soaps and liquid soaps of USP grade have a therapeutic value and may be sold under trade names.

Solvents are added to industrial soaps for scouring textiles or when used in soluble oils in the metal industry. Zinc oxide, benzoic acid, and other materials are used in facial soaps with the idea of aiding complexion. Excessive alkalinity in soaps dries and irritates the skin, but **hand grit soap** usually has 2 to 5% alkaline salts such as borax or soda ash and 10 to 25%

abrasive minerals. Softer **hand soap** may contain marble flour. Silicate of soda, used as a filler, also irritates the skin. **Face soap,** or **toilet soaps,** contain coloring agents, stabilizers, and perfuming agents. For special purposes, **cosmetic soaps** contain medications. **Deodorant soaps** contain antibacterial chemicals, such as **triclosan,** which inhibit the production of bacteria on the skin. Experts disagree on whether antibacterial ingredients are harmful to the skin. Some, such as **Dove,** are a blend of detergents and soap. **Castile soap** is a semitransparent soap made with olive oil. **Marseilles soap** and **Venetian soap** are names for castile soap with olive oil and soda. Ordinary soft soaps used as bases for toilet soap are made with mixtures of linseed oil and olive oil. Linseed oil, however, gives a disagreeable odor. Soybean oil, corn oil, and peanut oil are also used, although peanut oil, unless the arachidic acid is removed, makes a hard soap. **Tall oil soaps** are sodium soaps made from the fatty acids of tall oil. They are inferior to sodium oleate in detergency, but superior to sodium rosinate.

Saddle soap is any soap used for cleaning leather goods which has the property of filling and smoothing the leather as well as cleaning. The original saddle soaps were made of palm oil, rosin, and lye, with glycerin and beeswax added. Oils for the best soaps are of the nondrying type. High-grade **soft soap** for industrial use is made with coconut or palm kernel oil with caustic potash. But soft soap in paste form is generally made of low titer oils with caustic soda, usually linseed, soybean, or corn oil. The lauric acid of coconut oil gives the coconut-oil soaps their characteristic of profuse lathering, but lauric acid affects some skins by causing itching, and soaps with high coconut-oil content and low titer are also likely to break down in hot water and wash ineffectively. Palm-kernel oil develops free acids, and upon aging the soap acquires the odor of the oil. Palm oil produces a crumbly soap. It does not lather freely, but is mild to the skin. Olive oil is slow-lathering, but has good cleansing powers. It is often used in textile soaps. Cottonseed oil is used in some laundry soaps, but develops yellow spots in the soap. Corn oil with potash makes a mild soft soap. Soybean oil also makes a soft soap. Rosin is used to make yellow laundry soaps. ASTM standards for **bar soap** permit up to 25% rosin. Sulfonated oils do not give as good cleansing action as straight oils, but are used in shampoos where it is desirable to have some oil or greasiness. Blending of various oils is necessary to obtain a balance of desired characteristics in a soap. Hand soaps may be made with trisodium phosphate or with disodium phosphate, or **sodium perborate,** $NaBO_3 \cdot H_2O$, known as **perborin,** all of which are crystalline substances which are dissolved in water solution. **Soap powder** is **granular soap** made in a vacuum chamber or by other special processes. It usually contains 15 to 20% soap and the balance sodium carbonate. **Scouring powder** is an intimate mixture of soap powder and an insoluble abrasive such as pumice. **Floating soaps** are made light by blowing air through them while in the vats. **Soapless shampoos** and tooth powders contain saponin or chemical detergents.

SOAPSTONE. A massive variety of impure talc employed for electric panels, gas-jet trips, stove linings, tank linings, and as an abrasive. It can be cut easily and becomes very hard when heated because of the loss of its combined water. The waste product from the cutting of soapstone is ground and used for the same purposes as talc powder. **Steatite** is a massive stone rich in talc that can be cut readily, while soapstone may be low in talc. When free of iron oxide and other impurities, block steatite is used for making spacer insulators for electronic tubes and for special electrical insulators. **Block steatite** suitable for electrical insulation is mined in Montana, India, and Sardinia. Steatite is also ground and molded into insulators. It can be purified of iron and other metallic impurities by electrolytic osmosis. When fluxed with alkaline earths instead of feldspar, the molded steatite ceramics have a low loss factor at high frequencies, and have good electrical properties at high temperatures. The white-burning refractory steatite of the Red Sea coast of Egypt averages 60% silica and 30.5 magnesia, with 1% iron oxide and 1.5 CaO.

Alberene stone, quarried in Virginia, is blue-gray in color. The medium-hard varieties are used for building trim and for chemical laboratory tables and sinks, and the hard varieties are employed for stair treads and flooring. Alberene stone marketed by the Alberene Stone Corp. as a basic refactory substitute for chrome or magnesite for medium temperatures has a fusion point of 2400°F (1316°C). **Virginia greenstone** is a gray-green soapstone resistant to weathering, used as a building stone. **Talc crayons** for marking steel are sticks of soapstone.

SODA ASH. The common name for anhydrous **sodium carbonate,** Na_2CO_3, which is the most important industrial alkali. It is a grayish-white lumpy material which loses any water of crystallization when heated. For household use in hydrous crystallized form, $Na_2CO_3 \cdot 10H_2O$, it is called **washing soda, soda crystals,** or **sal soda,** as distinct from **baking soda,** which is **sodium hydrogen carbonate,** or **sodium bicarbonate,** $NaHCO_3$. Sal soda contains more than 60% water. Another grade, with one molecule of water, $Na_2CO_3 \cdot H_2O$, is the standard product for scouring solutions. Federal specifications call for this product to have a total alkalinity not less than 49.7% Na_2O. Commercial high-quality soda ash contains 99% min Na_2CO_3, or 58 min Na_2O. It varies in size of particle and in bulk density, being marketed as extra-light, light, and dense. The extra-light has a density of 23 lb/ft^3 (368 kg/m^3) and the dense is 63 lb/ft^3 (1,009 kg/m^3). **Laundry soda** is soda ash mixed with sodium bicarbonate, with 39 to 43% Na_2O. **Modified sodas,** used for cleansing where a mild detergent is required, are mixtures of sodium carbonate and sodium bicarbonate. They are used in both industrial and household cleaners. **Tanners' alkali,** used in processing fine leathers, and **textile soda,** used in fine wool and cotton textiles, are modified sodas. **Flour bland,** used by the milling industry in making free-flowing, self-raising food flours, is a mixture of sodium bicarbonate and tricalcium phosphate.

Soda ash is made by the Solvay process, which consists in treating a solution of common salt with ammonia and with carbon dioxide and calcining the resulting filter cake of sodium bicarbonate to make **light soda ash. Dense soda ash** is then made by adding water and recalcining. Soda ash is less expensive than caustic soda and is used for cleansing, for softening water, in glass as a flux and to prevent fogging, in the wood-pulp industry, for refining oils, in soapmaking, and for the treating of ores. **Caustic ash,** a strong cleaner for metal scouring and for paint removal, is a mixture of about 70% caustic soda and 30 soda ash. **Flake alkali,** of the Columbia Chemical Div., Pittsburgh Plate Glass Co., contains 71% caustic soda and 29 soda ash. Soda ash is also used as a flux in melting iron to increase the fluxing action of the limestone, as it will carry off 11% sulfur in the slag. **Soda briquettes,** used for desulfurizing iron, are made of soda ash formed into pellets with a hydrocarbon bond. **Hennig purifier** is soda ash combined with other steel-purifying agents made into pellets.

The natural hydrous sodium carbonate of Egypt and Libya is called **nitron.** Natural soda ash is obtained in Wyoming from beds 5 to 10 ft (1.5 to 3.0 m) thick located 1,200 ft (366 m) undergound, which contain 47% Na_2CO_3 and 36 $NaHCO_3$, designated as **trona,** $Na_2CO_3 \cdot NaHCO_3 \cdot 2H_2O$. By calcination the excess CO_2 is driven off, yielding soda ash. The salt brine of Owens Lake, Calif., is an important source of soda ash. The brine, which contains 10.5% Na_2CO_3 and 2.5 **sodium borate decahydrate,** is concentrated and treated to precipitate the trona. The Salt Lake area of Utah is a source of trona. Soda ash and sodium carbonate may be sold under trade names. **Purite** is sodium carbonate of the Mathieson Alkali Works.

SODIUM. A metallic element, symbol Na and atomic weight 23, occurring naturally only in the form of its salts. The most important mineral containing sodium is the chloride, NaCl, which is common salt. It also occurs as the nitrate, Chile saltpeter, as a borate in borax, and as a fluoride and a sulfate. When pure, sodium is silvery white and ductile, and it melts at 97.8°C and boils at 883°C. The specific gravity is 0.928. It can be obtained in metallic form by the electrolysis of salt. When exposed to the air, it oxidizes rapidly, and it must therefore be kept in airtight containers. It has a high affinity for oxygen, and it decomposes water violently. It also combines directly with the halogens, and is a good reducing agent for the metal chlorides. Sodium is one of the best conductors of electricity and heat. The element has five isotopes, and **sodium 24,** made by neutron irradiation of ordinary sodium, is radioactive. It has a half-life of 15 h and decays to stable **magnesium 24** with the emission of one beta particle and two gamma rays per atom.

The metal is a powerful desulfurizer of iron and steel even in combination. For this purpose it may be used in the form of **soda-ash pellets** or in alloys. **Desulfurizing alloys** for brasses and bronzes are **sodium-tin,** with 95% tin and 5 sodium, or **sodium-copper.** The **sodium-lead** of Humphrey-Wilkinson, Inc., used for adding sodium to alloys, contains 10% sodium, and is marketed as small spheroidal shot. The same company also markets **sodium marbles,** which are spheres of pure sodium up to 1 in (2.54 cm) in diameter coated with oil to reduce handling hazard. The **sodium bricks** of Gray Chemical contain 50% sodium metal powder dispersed in a paraffin binder. They can be handled in the air, and are a source of active sodium. Sodium in combination with potassium is used as a heat-exchange fluid in reactors and high-temperature processing equipment. A **sodium-potassium alloy** of the Mine Safety Appliances Co., containing 56% sodium and 44 potassium, has a melting point at 19°C and a boiling point at 825°C. It is a silvery mobile liquid with a heat conductivity at 200°C of 0.063. **High-surface sodium,** of U.S. Chemicals Co., is sodium metal absorbed on common salt, alumina, or activated carbon to give a large surface area for use in the reduction of metals or in hydrocarbon refining. Common salt will adsorb up to 10% of its weight of sodium in a thin film on its surface, and this sodium is 100% available for chemical reaction. It is used in reducing titanium tetrachloride to titanium metal. **Sodium vapor** is used in electric lamps. When the vapor is used with a fused alumina tube it gives a golden-white color. A 400-watt lamp produces 42,000 lumens and retains 85% of its efficiency after 6,000 h.

Sodium compounds are widely used in industry, particularly sodium chloride, sodium hydroxide, and soda ash. **Sodium bichromate,** $Na_2Cr_2O_7 \cdot 2H_2O$, a red crystalline powder, is used in leather tanning, textile dyeing, wood preservation, and in pigments. When heated, it changes to the anhydrous form which melts at 356°C and decomposes at about 400°C. In a hot water solution with sulfuric acid, sodium bichromate gives a golden-brown brasslike finish to zinc parts. The sodium bichromate liquor from alkali production is used for making pigments. When combined with lead compounds the bichromate precipitate is yellow. The addition of **iron blue,** or ferric cyanide, develops greens. **Sodium metavandate,** $NaVO_3$, is used as a corrosion inhibitor to protect some chemical-processing piping. It dissolves in hot water, and a small amount in the water forms a tough impervious coating of magnetic iron oxide on the walls of the pipe. **Sodium iodide** crystals are used as scintillation probes for the detection and analysis of nuclear energies. **Sodium oxalate** is used as an antienzyme to retard tooth decay. In the drug industry sodium is used to compound with pharmaceuticals to make them water-soluble salts. Sodium is a plentiful element, easily available, and is one of the most widely used.

SODIUM CYANIDE. A salt of hydrocyanic acid of the composition NaCN, used for carbonizing steel for case hardening, for heat-treating baths, for electroplating, and for the extraction of gold and silver from their ores. For carburizing steel it is preferred to potassium cyanide because of its lower cost and its higher content of available carbon. It contains 53% CN, as compared with 40% in potassium cyanide. The nitrogen also aids in forming the hard case on the steel. The 30% grade of sodium cyanide, melting at 1156°F (679°C), is used for heat-treating baths instead of lead, but it forms a slight case on the steel. Sodium cyanide is very unstable, and on exposure to moist air liberates the highly poisonous **hydrocyanic acid** gas, HCN. For gold and silver extraction it easily combines with the metals, forming soluble double salts, $NaAu(CN)_2$. Sodium cyanide is made by passing a stream of nitrogen gas over a hot mixture of sodium carbonate and carbon in the presence of a catalyst. It is a white crystalline powder, soluble in water. The white **copper cyanide** used in electroplating has the composition $Cu_2(CN)_2$, containing 70% copper. It melts at 474.5°C and is insoluble in water, but is soluble in sodium cyanide solution. **Sodium ferrocyanide,** or yellow **prussiate of soda,** is a lemon-yellow crystalline solid of the composition $Na_4Fe(CN)_6 \cdot 10H_2O$, used for carbonizing steel for case hardening. It is also employed in paints, in printing inks, and for the purification of organic acids; in minute quantities, it is used in salt to make it free-flowing. It is soluble in water. **Calcium cyanide** in powder or granulated forms is used as an insecticide. It liberates 25% of hydrocyanic acid gas. **Cyanogas,** of the American Cyanamid Co., is gaseous HCN from calcium cyanide.

SODIUM HYDROXIDE Known commonly as **caustic soda,** and also as **sodium hydrate. Lye** is an old name used in some industries and in household uses. It is a white, massive crystalline solid of the composition NaOH used for scouring and cleaning baths, for etching aluminum, in quenching baths for heat-treating steel, in cutting and soluble oils, in making soaps, and in a wide variety of other applications. It is usually a by-product in the production of chlorine from salt. The specific gravity is 2.13 and melting point 318°C. It is soluble in water, alcohol, and in glycerin. Sodium hydroxide is sold in liquid and in solid or powder forms on the basis of its Na_2O content. A high-grade commercial caustic soda contains 98% min NaOH equivalent to 76 min Na_2O. The liquid contains 50% min NaOH. **Pels,** of PPG Industries, is a caustic soda in bead form. It is less irritating to the skin when used in detergents. **Phosflake,** of Pittsburgh Plate Glass, used in washing machines, is a mixture of caustic soda and trisodium phosphate. **Caustic potash** is **potassium hydroxide,** KOH, which has the same uses but is more expensive. Caustic potash is a white, lumpy solid. It is soluble in water and makes a powerful cleansing bath for scouring metals. It is marketed as solid, flake, granular, or broken, and also in 40 to 50% liquid solutions. It is also used in soaps and for bleaching textiles. When used in steel-quenching baths, it gives a higher quenching rate than water alone and does not corrode the steel as a salt solution does.

SODIUM NITRATE. Also called **soda niter** and **Chile saltpeter.** A mineral found in large quantities in the arid regions of Chile, Argentina, and Bolivia, where the crude nitrate with iodine and other impurities is called **caliche.** It is used for making nitric and sulfuric acids, for explosives, as a flux in welding, and as a fertilizer. The composition is $NaNO_3$. It is usually of massive granular crystalline structure with a hardness of 1.5 to 2 and specific gravity of 2.29. It is colorless to white, but sometimes colored by impurities. It is readily soluble in water. In other parts of the world it occurs in beds with common salt, borax, and gypsum. Sodium nitrate is also made by nitrogen fixation and is marketed granulated, in crystals, or in sticks. It is colorless and odorless, and it has a specific gravity of 2.267 and a melting point of 316°C. It has a bitter, saline taste. **Sodan,** of the Allied Chemical & Dye Corp., used for spraying on soils, is a clear liquid solution of sodium nitrate and ammonium nitrate containing 20% nitrogen. **Norway saltpeter,** used in fertilizers and explosives, is **calcium nitrate,** $Ca(NO_3)_2$, in colorless crystals soluble in water. Calcium nitrate of fine crystal size is used as a coagulant for rubber latex.

SODIUM SILICATE. A water-soluble salt commonly known as **water glass** or **soluble glass.** Chemically, it is **sodium metasilicate** of the composition Na_2SiO_3 or $NaSiO_3 \cdot 9H_2O$. Two other forms of the silicate are also available, sodium **sequisilicate,** $3Na_2O \cdot 2SiO_2$, and **sodium orthosilicate,** $2Na_2O \cdot SiO_2$. All of these are noted for their powerful detergent and emulsifying properties, and for their suspending power. The material has good adhesion, and large quantities are used in water solutions for industrial adhesives. When solid, sodium silicate is glassy in appearance and dissolves in hot water. It melts at 1018°C. It is obtained by melting sodium carbonate with silica, or by melting sand, charcoal, and soda. The fused product is ground and dissolved in water by long boiling. **Potassium silicate** is made in the same way, or a complex soluble glass is made by using both sodium and potassium carbonates. Potassium silicate is more soluble than sodium silicate. **Kasil,** of the Philadelphia Quartz Co., is a potassium silicate in fine powder containing 71% SiO_2 and 28.4 K_2O. It is used in ceramic coatings and refractory cements. The **Corlok** of the Pennsylvania Chemicals Corp. is potassium silicate free of fluorides and sodium compounds. It is resistant to strong oxidizing acids, has good bond strength, and is used as a cement for acid tanks. **Ammonium silicate** has an ammonium group instead of the sodium. **Quram 220,** of the Philadelphia Quartz Co., is this material in the form of white powder or in opalescent solution. The intermediate silica grades act like sodium silicates and are used as binders for refractory ceramics.

Sodium silicate is marketed as a viscous liquid or in powder form. It is used as a detergent, as a protection for wood and porous stone, as a fixing agent for pigments, for cementing stoneware, for lute cements for such uses as sealing light bulbs, for waterproofing walls, greaseproofing paper containers, for coating welding rods, as a filler for soaps, and as a catalyst for high-octane gasoline. It increases the cleansing power of soaps but irritates the skin. However, it is used in cleansing compounds because it is a powerful detergent. **Brite Sil,** of Allegheny Industrial Chemicals, is a spray-dried sodium silicate powder which dissolves more easily and more uniformly.

Sodium silicte is also used for insulating electric wire. It is applied in solution and the coated wire is then heated, leaving a flexible coating. Mixed with whiting, it is used as a strong cement for grinding wheels. Sodium metasilicate marketed by the Philadelphia Quartz Co. as a cleaner of metals is a crystalline powder. Hot solutions of this salt in water are caustic and will clean grease readily from metals. **Drymet,** of the Cowles Detergent Co., is the anhydrous sodium metasilicate. It is a fine white powder with total alkalinity of 51% Na_2O. It is easily soluble in water, and is used as a detergent in soap powders to give free-flowing, noncaking properties. The anhydrous material for a given detergent strength weighs little more than half the weight of the hydrous powder. **Dryorth** is the anhydrous material of 60% alkalinity. It is powerful detergent and grease remover. **Crystamet,** of the same company, is the material with 42% water of crystallization. It is a free-flowing white powder. **Penchlor,** of the Pennsylvania Salt Mfg. Co., is an **acidproof cement** made by mixing cement powder with a sodium silicate solution. It is used for lining chemical tanks and drains. **Aquagel,** of the Silica Products Co., is a hydrous silicate of alumina, used in the same manner for waterproofing concrete.

SOLDER. An alloy of two or more metals used for joining other metals together by surface adhesion without melting the base metals as in welding alloys and without requiring a high-temperature flame as for the high-melting brazing alloys. However, with skill in application, there is often no definite temperature line between the **soldering alloys** and the brazing alloys, and a torch flame may be used for soldering. A requirement for a true solder is that it have a lower melting point than the metals being joined, and also have an affinity for, or be capable of uniting with, the metals to be joined.

The most common solder is called **half-and-half,** or **plumbers' solder,** and is composed of equal parts of lead and tin. It melts at 360°F (182°C). The weight of this solder is 0.318 lb/in^3 (8,802 kg/m3), the tensile strength 5,500 lb/in^2 (39 MPa), and the electric conductivity 11% that of copper. **SAE Solder No. 1** has 49.5 to 50.0% tin, 50 lead, 0.12 max antimony, and 0.08 max copper. It melts at 357.8°F (181°C). Much commercial half-and-half, however, usually contains larger proportions of lead and some antimony, with less tin. These mixtures have higher melting points, and solders with less than 50% tin have a wide melting range and do not freeze quickly. Sometimes a wide melting range is desired, in which case a **wiping solder** with 38 to 45% of tin is used. **Solder No. 4663,** of Westinghouse, is a narrow-melting-range solder, melting from 183 to 185°C, while **Solder No. 1580** is a wiping solder with a melting range from 183 to 231°C. The first contains 60% tin and 40 lead, and the second has 42% tin and 58 lead. **Slicker solder** is the best quality of plumbers' solder, containing 63 to 66% tin and the balance lead. The earliest solders were the Roman solder called **argentarium,** containing equal parts of tin and lead, and **tertiarium,** containing 1 part tin and 2 lead. Both alloys are still in use, and throughout early industrial times tertiarium was known as **tinman's solder.**

Good-quality solders for electrical joints should have at least 40% tin, as the electric conductivity of lead is only about half that of tin, but conductivity is frequently sacrificed for better wiping ability, and the wiping solders are usually employed for electrical work. **Soft solders** should not contain zinc because of poor adhesion from the formation of oxides. Various melting points to suit the work are obtained with solders by varying the proportions of the metals. The **low-melting solders** are those that have a melting point at 230°C or lower, and the **high-melting solders** melt at higher temperatures. The flow point, at which the solder is entirely liquid, is often considerably above the melting point. Tin added to lead lowers the melting point of the lead until at 356°F (180°C), or 68% tin, the melting point rises as the tin content increases until the melting point of pure tin is reached. A standard solder with 48% tin and 52 lead melts at 360°F (182°C). A 45–55 solder melts at 440°F (227°C). Cheap solders may contain much less tin, but they have lower adhesion. **SAE solder No. 4** contains 22.5 to 23.5% tin, 75 lead, and 2 max antimony. It melts at 370.4°F (188°C).

Solders with low melting points are obtained from mixtures of lead, tin, cadmium, and bismuth. **Bismuth solder** is also more fluid, as the bismuth lowers the surface tension. Bismuth, however, hardens the alloy, although to a lesser extent than antimony. A bismuth solder containing equal parts of lead, tin, and bismuth melts at 284°F (140°C). The **Cerrolow alloys** of the Cerro de Pasco Corp. are bismuth solders containing sufficient indium to be designated as **indium solders. Cerrolow 147,** which melts at 142°F (61°C), contains 48% bismuth, 25.6 lead, 12.8 tin, 9.6 cadmium, and 4 indium. **Cerrolow 105,** melting at 100°F (38°C), contains 42.9% bismuth, 21.7 lead, 8 tin, 5 cadmium, 18.3 indium, and 4 mercury. Cadmium solders have low melting points, are hard, and are usually cheaper than tin solders, but they have the disadvantage of blackening and corroding, and the fumes are toxic. **Cadmium-zinc solders** were used in wartime because of the scarcity of tin. A solder containing 80% lead, 10 tin, and 10 cadmium has about the same strength as a 50–50 tin-lead solder and has greater ductility, but is darker in color. **Cadmium-tin solder,** with high cadmium, is used to solder magnesium alloys. Soft solders for soldering brass and copper, especially for electric connections, may be of tin hardened with antimony. **Solder wire,** marketed by the American Brass Co. for this purpose, contains 95% tin and 5 antimony. Thallium may be used in high-lead solders to increase strength and adhesion.

Hard solder may be any solder with a melting point above that of the tin-lead solders; more specifically, hard solders are the brazing solders, silver solders, or aluminum solders applied with a brazing torch. **Aluminum solders** may contain up to 15% aluminum. A solder prepared by the National Bureau of Standards contains 87% tin, 8 zinc, and 5 aluminum. It has good strength and ductility. **Alcoa solder 805,** for joining aluminum to steel or other metals, has 95% zinc and 5 aluminum. The melting range is 715 to 725°F (379 to 385°C). For soldering aluminum to aluminum, an alloy of 91% tin and 9 zinc is used.

The solder known as **Richard's solder** is a yellow brass with 3% aluminum and 3 phosphor tin. Solders with nickel content are used for soldering nickel silver, and silver and gold solders are used for jewelry work. Silver solder in varying proportions is also used as a high-melting-point solder for general work, and small amounts of silver are sometimes used in lead-tin solders to conserve tin, but the melting point is high. **Lead-silver solders** with more than 90% lead and some silver, in use during the war emergency, had high melting points and poor spreading qualities. Indium improves these solders, and a solder with 96% lead, 3 silver, and 1 indium has a melting point of 310°C and a tensile strength of 4,970 lb/in^2 (33 MPa). **Cerroseal 35,** of the Cerro de Pasco Corp., contains 50% tin and 50 indium. It melts at 240°F (116°C), has low vapor pressure, and will adhere to ceramics. **Alkali-resistant solders** are **indium-lead alloys.** A solder with 50% lead and 50 indium melts at 360°F (182°C), and is very resistant to alkalies, but lead-tin solders with as low as 25% indium are resistant to alkaline solutions, have better wetting characteristics, and are strong. Indium solders are expensive. Adding 0.85% silver to a 40% tin soft solder gives equivalent wetting on copper alloys to a 63% tin solder, but the addition is not effective on low-tin solders. A **gold-copper solder** used for making high-vacuum seals and for brazing difficult metals such as iron-cobalt alloys contains 37.5% gold and 62.5 copper. The **silver-palladium solders** have high melting points, 1232°C for a 30% palladium alloy, good flow, and corrosion resistance. A **palladium-nickel alloy** with 40% nickel has a melting point about 1237°C. The brazing alloys containing palladium are useful for a wide range of metals, or metal to ceramic joints.

Cold solder, used for filling cracks in metals, may be a mixture of a metal powder in a pyroxylin cement with or without a mineral filler, but the strong cold solders are made with synthetic resins, usually epoxies, cured with catalysts, and with no solvents to cause shrinkage. The metal content may be as high as 80%. **Devcon F,** of the Devcon Corp., for repairing holes in castings, has 80% aluminum powder and 20 epoxy resin. It is cured by heat at 150°F (66°C), giving high adhesion. **Epoxyn solder,** of Co-Polymer Chemicals, Inc., is aluminum powder in an epoxy resin in the form of a putty for filling cracks or holes in sheet metal. It cures with a catalyst. The metal-epoxy mixtures give a shrinkage of less than 0.2%, and they can be machined and polished smooth.

SOLVENT. A material, usually a liquid, having the power of dissolving another material and forming a homogeneous mixture called a **solution.** The mixture is physical, and no chemical action takes place. A solid solution is such a mixture of two metals, but the actual mixing occurs during the liquid or gaseous state. Some materials are soluble in certain other materials in all proportions, while others are soluble only up to a definite percentage and the residue is precipitated out of solution. Homogeneous mixtures of gases may technically be called solutions, but are generally referred to only as mixtures.

The usual industrial applications of solvents are for putting solid materials into liquid solution for more convenient chemical processing, for thinning paints and coatings, and for dissolving away foreign matter as in dry-cleaning textiles. But they may have other uses, such as absorbing dust on roadways and killing weeds. They have an important use in separating materials, for example, in the extraction of oils from seeds. In such use, a **clathrate** is a solid compound added to the solution containing a difficult-to-extract material, but which is trapped selectively by the clathrate. The solid clathrate is then filtered out and processed by heat or chemicals to separate the desired compound. **Antifoamers** are chemicals, such as the silicones, added to solvents to reduce foam so that processing equipment can be used to capacity without spillover. **Antifoam 71,** of General Electric, is a silicone emulsion that can be used in foodstuffs in proportions up to 100 ppm. **Solvent-solvents** are solvents used for second-stage extraction of difficult-to-extract metals such as gold, uranium, and thorium. **Tributyl phosphine oxide,** $(C_4H_9)_3PO$, a white crystalline powder, is such a material used in benzene or kerosene solution for extracting metals from the acids employed in ore extraction.

The usual commercial solvents for organic substances are the alcohols, ether, benzene, and turpentine, the latter two being common solvents for paints and varnishes containing gums and resins. The so-called **coal-tar solvents** are light oils from coal tar, distilling off between 145 and 180°C, with specific gravities ranging from 0.850 to 0.890. **Solvent oils,** from coal tar, are amber to dark liquids with distillation ranges from about 150 to 340°C, with specific gravities from 0.910 to 0.980. They are used as solvents for asphalt varnishes and bituminous paints. **Shingle stains** are amber to dark grades of solvent oils of specific gravities from 0.910 to 0.930.

A valuable solvent for rubbers and many other products is **carbon bisulfide,** CS_2, also called **carbon disulfide,** made by heating together carbon and sulfur. It is flammable and toxic. When pure it is nearly odorless. The specific gravity is 1.2927 and boiling point 46.5°C. **Ethyl acetate,** $CH_3COOC_2H_5$, made from ethyl alcohol and acetic acid, is an important solvent for nitrocellulose and lacquers. It is liquid boiling at 77°C. One of the best solvents for cellulose is **cuprammonium hydroxide.** Amyl and other alcohols, amyl acetate, and other volatile liquids are used for quick-drying lacquers, but many synthetic chemicals are available for such use. **Dioxan,** a water-white liquid of specific gravity of 1.035 and composition $CH_2CH_2OCH_2CH_2O$, is a good solvent for cellulose compounds, resins, and varnishes, and is used also in **paint removers,** which owe their action to their solvent power. **Ethyl lactate,** used as a solvent for cellulose nitrate, is a liquid with boiling point of 150°C and specific gravity of 1.03. **Octyl alcohol,** a liquid of the composition $CH_3(CH_2)_6CH_2OH$, specific gravity 1.429, and boiling point of 195°C, has a high solvent power for nitrocellulose and resins. **Diafoam,** of the Resinous Products & Chemical Co., is a secondary octyl alcohol used as a defoaming agent in plastics and lacquers. **Methyl hexyl ketone,** $CH_3(CH_2)_5COCH_3$, is a powerful high-boiling solvent which also acts as a dispersing agent in inks, dyestuffs, and perfumes. It is a water-white liquid boiling at 173°C.

The chlorinated hydrocarbons have powerful solvent action on fats, waxes, and oils and are used in degreasing. Water is a solvent for most acids and alkalies and for many organic and inorganic materials. Acids or alkalies that decompose the material are not solvents for the material. Solvents are used to produce a solution that can be applied, as in the case of paints, and the evaporation of the solvent then leaves the material chemically unchanged. They may also be employed to separate one substance from another, by the selection of a solvent that dissolves one substance but not the other. **Dichlorethyl ether,** a yellowish liquid with a chloroformlike odor, of the composition $ClCH_2CH_2OCH_2CH_2Cl$, is a good solvent for fats and greases and is used in scouring solutions and in soaps. **Dichlorethylene** is a liquid of the composition $C_2H_2Cl_2$, specific gravity 1.278, and boiling point about 52°C. It is used as a solvent for the extraction of fats and for rubber.

Dichloromethane, known also as **methylene chloride** and **carrene,** is a colorless, nonflammable liquid of the composition CH_2Cl_2, boiling at 39.8°C. It is soluble in alcohol, and is used in paint removers, as a dewaxing solvent for oils, for degreasing textiles, and as a refrigerant. A low-boiling solvent for oils and waxes is **butyl chloride,** $CH_3CH_2CH_2CH_2Cl$. It is a water-white liquid of specific gravity 0.8875, boiling at 78.6°C. **Isocrotyl chloride** is a liquid of the composition $CH_3{:}C(CH_2)_2 \cdot CHCl$, with specific gravity of 0.919 and boiling point 68°C, used for cleaning and degreasing. **Cyclohexane,** $(CH_2)_6$, made by the hydrogenation of benzene, is a good solvent for rubbers, resins, fats, and waxes. It is a water-white, highly flammable liquid of specific gravity 0.777, boiling point 80.8°C, and flash point 10°F (−12°C). This solvent is marketed in England as **Sextone.** The **Nadene** of the Allied Chemical Co. is **cyclohexanone,** $CH_2(CH_2)_4C \cdot O$. It is a powerful general solvent, and is used as a coupling agent for immiscible compounds. The **Sulfolanes** of the Shell Oil Co. are selective solvents for separating mixtures having different degrees of saturation, and can be removed easily by water wash. **Dimethyl sulfolane** is produced from pentadiene by reacting with SO_2 and hydrogenation. **Cyclohexanol,** also called **hexalin** and **hexahydrophenol,** $C_6H_{11}OH$, is a solvent for oils, gums, waxes, rubber, and resins. It is made by the hydrogenation of phenol, and is a liquid boiling at 158°C.

Dichlorethyl formula, $CH_2(OCH_2 \cdot CH_2 \cdot Cl)_2$, is a water-insoluble high-boiling solvent for cellulose, fats, oils, and resins. The boiling point is 218.1°C, and specific gravity 1.234. The **nitroparaffins** constitute a group of powerful solvents for oil, fats, waxes, gums, and resins. Blended with alcohols they are solvents for cellulose acetate, producing good flow and hardening properties for nonblushing lacquers. **Nitromethane,** CH_3NO_2, is a water-white liquid, with specific gravity 1.139, boiling point 101°C, and freezing point −29°C. It is also used as a rocket fuel. At 500°F (260°C) it explodes into a hot mixture of nitrogen, hydrogen, carbon monoxide, carbon dioxide, and water vapor, but with a catalyst the disintegration can be controlled into a smooth continuous explosion. **Nitroethane,** $CH_3CH_2NO_2$, has a specific gravity of 1.052, boiling point 114°C, and freezing point −90°C.

A **plasticizer** is a liquid or solid that dissolves in or is compatible with a resin, gum, or other material and renders it plastic, flexible, or easy to work. A sufficient quantity of platicizer will result in a viscous mixture which consists of a suspension of solid grains of the resin or gum in the liquid plasticizer. The plasticizer is in that sense a solvent, but unlike an ordinary solvent the platicizer remains with the cured resin to give added properties to the materials, such as flexibility. **Dibutyl phthalate,** a water-white oily liquid of specific gravity 1.048, boiling point 340°C, and composition $C_{16}H_{22}O_4$, is a plasticizer for Buna N rubber and polyvinyl chloride plastics. **Monoplex DOA,** of the Rohm & Haas Co., used to give flexibility to vinyl resins at low temperatures, is **diisooctyl adipate.** It has a flash point at 400°F (204°C), and freezing point at 55°C. An **aprotic solvent** is a solvent that contains no hydrogen, such as **selenium oxychloride,** a liquid of the composition $SeOCl_2$. Such solvents are used in electronic applications where the energy deflection of free protons would be undesirable. **Phosphorus oxychloride,** $POCl_3$, is an aprotic solvent used with neodymium in **liquid lasers** to give high light-beam efficiency.

SORBITOL. A **hexahydric alcohol,** $(CH_2OH)_5CHOH$, which occurs naturally in many fruits, but is now made on a large scale by the direct hydrogenation of corn sugar, or dextroglucose. It is a white, odorless, crystalline powder of faint sweet taste. It melts at 97.7°C, and is easily dissolved in water. It is used as a humectant, softener, and blending agent; for the production of synthetic resins, plasticizers, and drying oils; and as an emulsifier in cosmetics and pharmaceuticals. It is digestible and nutritive and is used in confectionery to improve texture and storage life by inhibiting crystal growth of the sugar, and also in dietary foods as a substitute for sugar. **Sorbo** and **Arlex,** of the Atlas Chemical Industries, are water solutions of sorbitol. **Mannitol** is an isomer form of the alcohol and is produced in granular form for pharmaceuticals and foodstuffs as a binder. In the form of a free-flowing powder it is used as an **anticaking agent** in pharmaceuticals and foodstuffs where a silica or other mineral-based agent is undesirable. The **polysorbates** are esters of sorbitol. **Polysorbate 80,** of Hodag Chemical Corp, is such a material used as an emulsifier in prepared mixed food for improving texture and stability. **Hex,** a metal-cleaning and protective agent of Fields Paint Mfg. Co., is a phosphoric acid ester of sorbitol.

SOUND INSULATORS. Materials employed, chiefly in walls, for reducing the transmission of noise. Insulators are used to impede the passage of sound waves, as distinct from isolators used under machines to absorb the vibrations that cause the sound. For factory use the walls, partitions, and ceilings offer the only mediums for the installation of sound insulators. All material substances offer resistance to the passage of sound waves, and even glass windows may be considered as insulators. But the term refers to the special materials placed in the walls for this specific purpose. Insulators may consist of mineral wool, hair felt, foamed plastics, fiber sheathing boards, or simple sheathing papers. Sound insulators are marketed under a variety of trade names, such as **Celotex,** made from bagasse, and **Fibrofelt,** made from flax or rye fiber. Wheat straw is also used for making insulating board. Sound insulators are often also heat insulators. **Linofelt,** of Union Fiber, is a sound- and heat-insulating material used for walls. It consists of a quilt of flax fiber between tough waterproof paper. It comes in sheets ⅝₁₆ to ¾ in (0.80 to 1.91 cm) thick. **Torofoleum** is a German insulating material made from peat moss treated with a waterproofing agent. It will withstand temperatures up to 230°F (110°C), is porous, and weighs less than 1 lb/bd ft (0.005 kg/m^3).

Vibration insulators, or **isolators,** to reduce vibrations that produce noises, are usually felt or fiberboards placed between the machine base and the foundation, but for heavy pressures they may be metal wire helically wound or specially woven, deriving their effectiveness from the form rather than the material. **Keldur,** of International Products, is a fibrous insulating material made up in sheets ¾ in (1.91 cm) thick, with a resilient binder. **Korfund isolator,** of the Korfund Co., is a resilient mat of cork treated with oil and bound in a steel frame. It will take loadings up to 4,000 lb/ft^2 (19,528 kg/m^2). **Vibro-Insulator,** of B. F. Goodrich, is an isolator of Ameripol synthetic rubber. Plastic foams in sheet or flexible tape form are also used as isolators for instruments.

SOYBEAN OIL. Also known as **soya bean oil.** A pale-yellow oil obtained by expression from the seeds of the plant *Glycine soya,* native to Manchuria but grown in the United States. It is primarily a food oil but has an undesirable off-flavor unless highly purified. It is also used as a drying oil for linoleum, paints, and varnishes, or for mixing with linseed oil, although the untreated oil has only half the drying power of linseed oil. It is also used in core oils and in soaps. The bean contains up to 20% oil. The average yield factor is 15%, but by trichlorethylene extraction a bushel of beans will yield 11 lb (5 kg) of oil and 46 lb (21 kg) of high-protein meal containing less than 1% oil. The oil content decreases in warm climates. Southern-grown soybeans contain 2 to 5% less oil than those grown in Illinois. The usual conversion factor is 8.5 lb (4 kg) of oil and 48 lb (22 kg) of meal per bushel of beans. The oil is easy to bleach, has good consistency as a food oil, and does not become rancid easily, but has less flavor stability than many other oils. There are 280 varieties of the bean grown in the United States and 2,500 varieties listed. The pods contain two or three beans which range in color from light straw through gray and brown to nearly black. Most varieties are straw-colored or greenish yellow. The stalks and leaves of the plant contain much nitrogen, and about half of the crop is usually plowed under for fertilizer.

The specific gravity of the oil is about 0.925, iodine value 134, and it should have a maximum of not more than 1.5% free fatty acids and not more than 0.3% moisture and volatile matter. The fractionated oil yields 15% cut soybean oil of an iodine value of 70 to 90, used for soaps, lubricants, and rubber compounding; 72% selected-acid oil of an iodine value of 145 to 155, used for varnish and paint oils alone or in blends with other oils, or for glycerin making; 13% bottoms, used for soaps, lubricants, and giving a by-product pitch used in insulation and mastic flooring. **Snowflake oil,** of the Archer-Daniels-Midland Co., is a heavy-bodied oxidized soybean oil for paints. It has a specific gravity of 0.986 to 0.989 and iodine number from 64 to 95. **Soyalene,** of the same company, is an alkali-refined soybean oil for varnishes. The specific gravity is 0.924, and the iodine number is 130. **Epoxidized soybean oil** is used in vinyl and alkyd resins as a plasticizer and to increase heat resistance. A very large use of soybean oil is in the making of margarine.

Soybean meal is the product obtained by grinding the soybean chips from the expeller process, or the **soybean oil cake** from the hydraulic process. The meal is marketed as stock feed or fertilizer. It is chiefly used as a protein feed for dairy cattle, but it is inferior to fish meal for poultry, as it lacks the mineral salts and vitamins of fish meal. Soybean meal hardened with formaldehyde is used as a filler with wood flour in plastics to give better flow in molding. **Gelsoy** is a protein gel extracted from soybean meal. It is used in foodstuffs as a thickening agent, and is also used as a strong adhesive.

Soybean flour for bakery food products for the American market is made from meal that has been treated by acidulated washing to remove the soluble enzymes and sugars that carry the taste. Meal produced by heat processing averages 40% protein and 20% fats, while meal from solvent extraction has 42 to 50% protein and a maximum of 2.5% fats. Further processing of the meal to remove sugars and other materials varies the final protein content of the flour, and meals from different types of beans vary in content. A soybean flour of Power Protein, Inc. has 40% protein and 20% fats with the sugars that give easy solubility for blending. It is used as a partial replacement for milk powder and wheat flour in baked goods. The **Promax** and **Isopro** soybean flours, of Griffith Laboratories, for high-protein additions to foodstuffs, contain 70% protein with all flavor removed, and are high in lysine. They have a pH of 5.5 and 7.0, respectively. **Soy proteins,** of General Mills, used in canned soups and meat products, is toasted to eliminate all enzyme activity. It contains 50% protein with 2% lecithin and 3% lysine. **Promine,** of the Central Soya Co., is a 93% concentrate of soybean proteins, used for thickening and enriching soup mixes. **Supro 610,** of Ralston Purina Co., is a spray-dried powder, 95% protein, with a light cream color and no bitter flavor.

SPECULUM METAL. An alloy formerly used for mirrors, and also in optical instruments. It contains 65 to 67% copper and the remainder tin. It takes a beautiful polish and is hard and tough. Speculum metal should have a maximum of crystals of Cu_4Sn, containing 66.6% copper. An old Roman mirror contained about 64% copper, 19 tin, and 17 lead, and an Egyptian mirror contained 85% copper, 14 tin, and 1 iron. The old Greek mirrors were carefully worked out with 32% tin and 68 copper. They had 70% of the reflecting power of silver, with a slight red excess of reflection that gave a warm glow, without the blue of nickel or antimony. This alloy is now plated on metals for reflectors. A modern telescope mirror contains 70% copper and 30 tin. **Chinese speculum** contains about 8% antimony and 10 tin. **Speculum plate** advocated by the Tin Research Institute for electroplating, to give a hard, white, corrosion-resistant surface for food-processing equipment and optical reflectors, has 55% copper and 45 tin. It is harder than nickel, and retains its reflectivity better than silver.

SPERM OIL. The waxy oil extracted from the head cavity of the sperm whale, *Physeter breviceps* and *P. catadon,* and the **Bottlenose whale,** *P. macrocephalus.* Sperm whales have teeth and feed in deep water on squid and large animal life. The male sperm whale attains a length of 60 ft (18 m) and the female about 38 ft (12 m). The spermaceti is first separated out, leaving a clear yellow oil. It is purified by being pressed at a low temperature. It is graded according to the temperature of pressing. A good grade of sperm oil has a specific gravity of 0.875 to 0.885, and a flash point above 440°F (227°C). Inferior grades of sperm oil may be from sperm-whale blubber. Commercial sperm oil is likely to be one-third head oil and two-thirds body oil. Sperm oil differs from fish oil and whale oil in consisting chiefly of liquid waxes of the higher fatty alcohol esters and not fats. Sperm oil absorbs very little oxygen from the atmosphere and resists decomposition even at temperatures above 400°F (204°C), and it will pour below its cloud point of 38 to 45°F (3 to 7°C). It wets metal surfaces easily. It is thus a valuable lubricating oil. It was formerly used as a lamp oil, burning with a white shining flame. It is also an excellent soap oil. **Sperm 42,** of Werner G. Smith, Inc., is a sperm oil with carbon chains of C_{10} to C_{22}, and it is emulsifiable in cold or warm water. **Sulfonated sperm oil** is used as a wetting agent for textiles, and it is also valued for cutting oils, crankcase oil, and high-pressure lubricants. **Smithol 25,** of the same company, is a synthetic fatty-acid oil resembling sperm oil and having the same uses. It is a light-colored odorless oil with high viscosity, a low pour point at −16°F (−27°C), and an iodine value of 105.

Spermaceti is the white crystalline flakes of fatty substance, or wax, that separate out from sperm oil on cooling after boiling. It is **cetyl palmitate,** a true wax, and does not yield glycerin when saponified. It is purified by pressing, and the triple-refined is snow white. It is also separated out from **dolphin-head oil.** Spermaceti is odorless and tasteless, has a melting point of 43°C, and is insoluble in water, but soluble in hot alcohol. It burns with a bright flame. It was formerly used for candles but now is employed chiefly as a fine wax for ointments and compounds. Sperm oil and spermaceti are inedible and indigestible. **Cetyl alcohol,** $C_{16}H_{33}OH$, originally obtained from spermaceti, is now made synthetically from **ethyl palmitate.**

SPICE. An aromatic vegetable substance, generally a solid used in powdered form, employed for flavoring foods. There is no sharp dividing line between **flavors** and spices, but in general a spice is a material that is used to stimulate the appetite and increase the flow of gastric juices. Spices are not classed as foods in themselves, having little food value, but as food accessories. Pepper is distinctly a spice, though not grouped with the spices. Some spices are also used widely as flavors and in perfumes, and also in medicine either for antiseptic or other values or to disguise the unpleasant taste of drugs. A **condiment** is a strong spice, or a spice of sharp taste, although the word is often erroneously applied to any spice. A **savory** is a fragrant herb or seed used for flavor in cooking. Spices are obtained from the stalks, bark, fruits, flowers, seeds, or roots of plants. **Microground spices,** used to give uniform distribution in the quantity manufacture of foodstuffs, are spices ground to microscopic fineness in a roller mill. The most popular spices in the United States, in the order of quantity used, are: cinnamon, nutmeg, ginger, cloves, allspice, poppy seed, and caraway seed. Since ground spices lose flavor rapidly by the loss of volatile oils, the particles are sometimes coated with dextrose or a water-soluble gum. The **Spisoseals,** of Dodge & Olcott, Inc., are ground spices with coated particles.

Allspice, also known as **pimento** and **Jamaica pepper,** is the dried unripe fruit of the small evergreen tree *Pimenta officinalis* of the myrtle family growing in the West Indies and tropical America. The fruit is a small berry which when dried is wrinkled and reddish brown. It has a flavor much like a combination of clove, nutmeg, and cinnamon. **Pimento oil** is a fragrant essential oil distilled from the berries, which contain 4%. It contains eugenol and cineol and is used in flavors, in bay rum, and in carnation perfumes. **Coriander** is the dried fruit of the perennial plant *Coriandrum sativum* grown in the Mediterranean countries and India. It is one of the oldest of spices and has a pleasant aromatic taste. **Oil of coriander,** extracted from the dried seed, is used in medicine, beverages, and flavoring extracts. It has a higher aromatic flavor than the fruit. **Savory** is a fragrant herb of the mint family, *Satureia hortensis,* used in cooking, and in medicine as a carminative. It contains **carvacrol,** a complex phenol also occurring in caraway and camphor. The word savory also designates other herbs used directly in foods as flavors.

Celery seed, used as a savory, is from the plant *Apium graveolens.* The best-quality leafstalks, known as **celery,** are bleached white and eaten raw or cooked. The plant is widely grown for seed in France and Spain. **Celery-seed oil** is a pale-yellow oil extracted from the seeds and used as a flavor and in perfumery. **Fennel** is the dried oval seed of the perennial plants *Foeniculum vulgare* and *F. dulce.* The stalks of the latter are blanched and eaten as a vegetable in Europe. Fennel is used as a flavoring in confectionery and liqueurs, and as a carminative in medicine. **Fennel oil** is a pale yellowish essential oil with specific gravity of 0.975, distilled from the seed. It has an aromatic odor and a camphorlike taste with a secondary sweetish, spicy taste. It contains **fenchone,** $C_{10}H_{16}O$, an isomer of camphor, with also pinene, camphene, and **anethole,** or **anise camphor,** $C_3H_5C_6H_4OCH_3$. The latter is used in dentifrices and pharmaceuticals. **Fenugreek** is the seed from the long pods of the annual legume *Trigonella foenum-graecum,* native to southern Europe. It is used in curries, in medicine, and for making artificial maple flavor. **Oregano,** used as an ingredient in chili powder and as a spice in a variety of dishes, is the pungent herb *Coleus amboinica.*

Dill seed, from the herb *Anethum graveolens,* of the parsley family, is used as a condiment for pickles. **Dill leaves** are used as seasoning for soups, sauces, and pickles. **Dill oil,** extracted from the whole herb, is used as a flavor in the food industry. It resembles caraway oil and has a finer flavor than **dill-seed oil,** which is more plentiful, but **dill flavor** prepared from the whole seed is stronger. Dill is grown in the central United States and in Central Europe. **Cardamon** is the highly aromatic and delicately flavored seeds of the large perennial herb *Elettaria cardamomum,* of India, Ceylon, and Central America. The seeds are used in pickles, curries, and cakes, and the oil is employed as a flavor. **Garlic** is the root bulb of the lily *Allium sativum,* used as a condiment, and also used in medicine as an expectorant under the name of **allium.** It contains **allyl sulfide,** a liquid of the composition $(CH_2{:}CHCH_2)_2S$, which gives it a pungent odor and taste. **Allicine,** extracted from garlic, is used in medicine as an antibacterial. It is an oily liquid with a sharp garlic odor.

Cumin is the seed of the *Cuminum cyminum,* the true cumin, and *Nigella sativa,* the **black cumin,** both of India. The seed is used in confectionery and in curries. A kind of black cumin known as **shiah zira,** from the plant *Carum indicum* of India, is superior in taste and fragrance to ordinary cumin. **Caraway** is the spicy seed of the biennial herb *C. carvi* of Europe and North Africa. The seeds are used on cookies. **Caraway oil,** distilled from the seeds, contains carvone and limonene, and in combination with cassia gives a pleasant odor. It is used in soap, perfumes, and mouthwashes. **Sage** is the grayish-green hairy leaves of the shrublike plant *Salvia officinalis* used as a spice. It is cultivated extensively in the Mediterranean region. **Oil of sage** is used in perfumery. **Clary sage oil** is distilled from the flowers of the *S. sclarea* of France, Italy, and North Africa. It has the odor of a mixture of ambergris, neroli, and lavender, and is used in flavoring vermouth liquor and muscatel wines, and also in eau de cologne. **Sassafras** is sometimes classed as a spice but is a flavor. It is the aromatic spicy bark of the root of the tree *Sassafras albidum* which grows wild in the eastern United States. It is used mostly for making root beer, but also for flavoring tobacco, and in patent medicines. **Sassafras oil** is an oil extracted from the whole roots, which contain 2% of the yellow oil, and is used in medicine, perfumery, and soaps. It produces artificial heliotrope. The oil contains **safrol,** $C_{10}H_{10}O_2$, also produced from brown camphor oil. **Brazilian sassafras oil,** or **ocotea oil,** is distilled from the root of the tree *Ocotea cymbarum,* also of the laurel family. The root yields about 1% of an oil which contains 90% safrol, and has the odor and flavor of American sassafras oil. **Sarsaparilla** is an oil obtained from the long brown roots of the climbing vine *Smilax regellii* of Honduras, *S. aristolochiaefolia* of Mexico, and other species, all growing in tropical jungles. The roots are used in medicine. The oil is used as a flavor. It is odorless, but has an acrid sweet taste. It contains saponins.

Wintergreen oil is from the leaves of the small evergreen plant *Gaultheria procumbens* of the Middle Atlantic states. The oil does not exist in the plant but is formed by the reaction between a glucoside and an enzyme when the chopped leaves are steeped in water. It is largely **methyl salicylate,** $C_8H_8O_3$. It is used in flavoring candies and soft drinks and in medicine. **Hop oil,** used to give hop flavor to cereal beverages, and also in perfumes, is obtained from **lupulin,** a glandular powder found in the female inflorescence of the hop plant, *Humulus lupulus.* **Hops** are used directly in making beer, and the oil is produced from the discard hops which contain 0.75% oil. **Anise seed** is from the annual plant *Pimpinella anisum* grown in the Mediterranean countries and in India. The best grades come from Spain. The seed is used in flavoring in the baking industry. The distilled oil, **anise oil,** is used in perfumes and in soaps, and in the liqueur known as **anisette.** The oil contains choline, and is used in medicine as a carminative and expectorant.

SPINEL. A **magnesium aluminate,** $MgO \cdot Al_2O_3$, occurring as octahedral crystals of varying colors due to impurities of iron, manganese, or chromium. The best transparent stones are used as gems. Spinel is found as crystals or rolled pebbles in gem gravels with corundum stones; the **ruby spinel** often occurs with the true ruby. It has a deep-red color, but the variety **almandine** is violet.

Synthetic spinel was originally made in Germany to replace ruby and sapphire for instrument bearings because it is easier to cut and thus conserves diamond abrasive. Spinel is produced by Linde in the forms of drawing dies, gages, wearing parts, orifices, and balls. The composition is $MgO \cdot$

$3\frac{1}{2}Al_2O_3$, and the crystal structure is cubic. The specific gravity is 3.61, the melting point is about 2040°C, and the Knoop hardness is 1,175 to 1,380, designated as Mohs 8. Like corundum, it is not attacked by common acids or by sodium hydroxide. The spinel powder from which the crystals are flame-grown is made by calcining a mixture of pure ammonium sulfate and ammonium magnesium sulfate. Much synthetic spinel is used for synthetic gems, the colors being obtained with metal oxides. Small amounts of chromic oxide give the tinted crystals of sapphire, while up to 6% is used for the dark ruby colors. Blue is obtained with oxides of iron and titania, and green is from cobalt oxide. **Golden topaz** is colored with nickel and magnesium oxides. The **aquamarine spinel** is tinted with a complex mixture of nickel, cobalt, vanadium, and titanium oxides.

SPODUMENE. A mineral of the composition $Li_2O \cdot Al_2O_3 \cdot 4SiO_2$, with some potassium and sodium oxides.It is the chief ore of the metal lithium, but it requires a higher temperature for sintering than lepidolite, and the sinter is more difficult to leach. It is found in South Dakota and the Carolinas, and has an average content of 4% Li_2O, ranging from 2.9 to 7.6%. Crystals of spodumene in South Dakota are 8 to 10 ft (2 to 3 m) long and 1 ft (0.3 m) in thickness, appearing like logs of wood, with as high as 6.5% lithium. The specific gravity is 3.13 to 3.20, and the melting point is 1395 to 1425°C. Spodumene is three times more active than feldspar as a flux in ceramics, giving fluidity, increasing surface tension, and eliminating pinholes. A mixture of 25% spodumene with 75 feldspar is an active vitrifying agent in ceramics. The melting point of the mixture is 1110°C, which is below the usual minimum temperature used for chinaware; it thus forms a glaze. **Lithospar** is a name for feldspar and spodumene from the pegmatities of King's Mountain, North Carolina. In Germany lithium is obtained from the lithium mica **zinnwaldite,** which is a mixture of potassium-aluminum orthosilicate and lithium orthosilicate with some iron, and contains less than 3% Li_2O. **Kryolithionite,** a mineral found in Greenland, has the composition $Na_3Li_3(AlF_6)_2$ and contains up to 11.5% Li_2O. It has a crystal structure resembling garnet. A transparent emerald-green spodumene in small crystals, known as **hiddenite,** is found in North Carolina, and is cut into gemstones.

SPONGE. The cellular skeleton of a marine animal of the genus *Spongia,* of which there are about 3,000 known species, only 13 of which are of commercial importance. It is employed chiefly for wiping and cleaning, as it will hold a great quantity of water in proportion to its weight, but it also has many industrial uses such as applying glaze to pottery. Sponges grow like plants, attached to rocks on the sea bottom. They are prepared for use by crushing to kill them, scraping off the rubbery skin, macerating in water to remove the gelatinous matter, and bleaching in the sun. Tarpon Springs, Fla., is the center of American sponge fishing, but most of the best sponges have come from the Mediterranean and Red Seas.

The prepared sponge is an elastic fibrous structure chemically allied to silk. It has sievelike membranes with small pores leading into pear-shaped chambers. The best sponges are spheroidal, regular, and soft. Commercial sponges for the American market must have a diameter of 4.5 in (11.4 cm) or more. Most of the Florida sponges are the **sheepswool sponge,** *Euspongia lachne,* used for cleaning and industrial sponging. The **Rock Island sponge,** from Florida, and the **Key wool sponge** are superior in texture and durability to the **Bahama wool sponge,** which is coarser, more open, and less absorbent. The **Key yellow sponge** is the finest grade. The **grass sponge,** *E. graminea,* of the Caribbean, is inferior in shape and texture. The fine **honeycomb sponge,** *Hippiospongia equina,* of the Mediterranean Sea, is of superior grade and has been preferred as a bath sponge. About 80% of the North African catch consists of honeycomb sponges, with the remainder **Turkey cup sponge,** *E. officinalis,* and **zimocca sponge,** *E. zimocca.* The Turkey cup is rated as the finest, softest, and most elastic of the sponges, but the larger of the zimocca sponges are too hard for surgical use, and are employed for industrial cleaning. Sponges for industrial and household uses have now been largely replaced by foamed rubbers and plastics.

SPONGE IRON. Iron made from ferrous sand and pressed into briquettes, which can be charged directly into steel furnaces instead of pig iron. It was originally made on a large scale in Japan where only low-grade sandy ores were available. Sponge iron is made by charging the sand continuously into a rotary furnace to drive off the light volatile products and reduce the iron oxide to metallic iron, which is passed through magnetic separators, and the finely divided iron is briquetted. Unbriquetted sponge iron, with a specific gravity of 2, is difficult to melt because of the oxidation, but the briquetted material, with a specific gravity of 6, can be melted in electric furnaces. Sponge iron to replace scrap in steelmaking is also made from low-grade ores by reducing the ore with coke-oven gas or natural gas. It is not melted, but the oxygen is driven off, leaving a spongy granular product. As it is very low in carbon, it is also valuable for making high-grade alloy steels.

A form of sponge iron employed as a substitute for lead for coupling packings was made in Germany under the name of **sinterit.** The reduction is carried out in a reducing atmosphere at a temperature of 1200 to 1350°C, instead of heating the iron oxide with carbon. Since the porous iron corrodes easily, it is coated with asphalt for packing use. **Iron sponge,** employed as a purifier for removing sulfur and carbonic acid from illuminating gas, is a sesquioxide of iron obtained by heating together iron ore and carbon. It has a spongy texture filled with small cells.

SPRENGLE EXPLOSIVES. Chlorate compounds that have been rendered reasonably safe from violent explosion by separating the chlorate from the combustible matter. The potassium chlorate, made into porous cartridges and dipped, just before use, in a liquid combustible such as nitrobenzene or dead oil, was called **rack-a-rock.** Sprengle explosives were formerly used as military explosives, are very sensitive to friction and heat, and are now valued only for mining or when it is desired to economize on nitrates. **Cheddite** is a French explosive consisting of a chlorate with an oily material, such as castor oil, thickened by a nitrated hydrocarbon dissolved in it. A typical cheddite has 80% potassium chlorate, 8 castor oil, and 12 mononitronaphthalene. With sodium chlorate it is less sensitive to detonation and more powerful but is hygroscopic. Potassium chlorate cheddite is a soft, yellowish, fine-grained material, and is a slow, mild explosive which will split rocks rather than shatter them. **Minelite** is a chlorate with paraffin wax. **Steelite** is a chlorate explosive with rosin. **Prométhée** is another French chlorate explosive. In this explosive the oxygen carrier consists of 95% potassium chlorate and 5 manganese dioxide, and the combustible contains 50% nitrobenzene, with turpentine and naphtha. It is extremely sensitive and will explode by friction. **Silesia** is a German high explosive used for blasting. It is potassium chlorate with rosin, with some sodium chlorate to make it less sensitive.

SPRING STEEL. A term applied to any steel used for making springs. The majority of springs are made of steel, but brass, bronze, nickel silver, and phosphor bronze are used where corrosion resistance or electric conductivity is desired. Carbon steels, with from 0.50 to 1.0% carbon, are much used, but vanadium and chrome-vanadium steels are also employed, especially for heavy car and locomotive springs. Special requirements for springs are that the steel be low in sulfur and phosphorus, and that the analysis be kept uniform. For flat or spiral springs that are not heat-treated after manufacture, hard-drawn or rolled steels are used. These may be tempered in the mill shape. Music wire is widely employed for making small spiral springs. A much-used straight-carbon spring steel has 1% carbon and 0.30 to 0.40 manganese, but becomes brittle when overstressed. **ASTM carbon steel** for flat springs has 0.70 to 0.80% carbon and 0.50 to 0.8 manganese, with 0.04 max each of sulfur and phosphorus. Motor springs are made of this steel rolled hard to a tensile strength of 250,000 lb/in^2 (1,723 MPa). **Watch spring steel,** for mainsprings, has high carbon, 1.15%, and low manganese, 0.15 to 0.25, rolled hard and giving an elastic limit above 300,000 lb/in^2 (2,068 MPa).

Silicon steels are used for springs. They have high strength and impact resistance. These steels average about 0.40% carbon, 0.75 silicon, and 0.95 manganese, with or without copper, but the silicon may be as high as 2%. **Flexo steel,** of Carpenter Technology, used for automobile leaf springs and for recoil springs, contains 2% silicon, 0.75 manganese, and 0.60 carbon. The elastic limit is 100,000 to 300,000 lb/in^2 (689 to 2,068 MPa), depending on the drawing temperature, with hardness 250 to 600 Brinell.

Manganese steels for automotive springs contain about 1.25% manganese and 0.40 carbon, or about 2% manganese and 0.45 carbon. When heat-treated, the latter has a tensile strength of 200,000 lb/in^2 (1,378 MPa) and elongation 10%. Part of the manganese may be replaced by silicon and the **silicon-manganese steels** have tensile strengths as high as 270,000 lb/in^2 (1,861 MPa). The addition of chromium or other elements increases the elongation and improves the physical properties. **Uma spring steel,** of the Republic Steel Corp., is a chromium-manganese steel with 1 to 1.2% chromium, 0.80 to 1 manganese, and about 0.50 carbon. In the rolled condition it has an ultimate strength of 135,000 lb/in^2 (930 MPa) and Brinell hardness up to 332. Manganese steels are deep-hardening but are sensitive to overheating. The addition of chromium, vanadium, or molybdenum widens the hardening range.

Wire for coil springs ranges in carbon from 0.50 to 1.20%, and in sulfur from 0.028 to 0.029. Bessemer wire contains too much sulfur for spring use. Cold working is the method for hardening the wire and for raising the tensile strength. A 0.85% carbon rod, with an ultimate strength of 140,000 lb/in^2 (965 MPa), when drawn with four or five passes through dies will have a strength of 235,000 lb/in^2 (1,620 MPa). Wire drawn down to a diameter of 0.015 in (0.038 cm) may have an ultimate strength of 400,000 lb/in^2 (2,757 MPa). The highest grades of wire are referred to as music wire. The second grade is called **hard-drawn spring wire.** The latter is a less expensive basic open-hearth steel with manganese content of 0.80 to 1.10%, and an ultimate strength up to 300,000 lb/in^2 (2,068 MPa). Specially treated carbon steels for springs are sold under trade names such as **Enduria,** of Bethlehem Steel. **Resilla** is a silicon-manganese spring steel of this company.

For springs for jet engines and other applications where resistance to high temperatures is required, stainless steel and high-alloy steels are used. But, while these may have the names and approximate compositions of standard stainless steels, for spring-wire use their manufacture is usually closely controlled. For example, when the carbon content is raised in high-chromium steels to obtain the needed spring qualities, the carbide tends to collect in the grain boundaries and cause intergranular corrosion unless small quantities of titanium, columbium, or another element is added to immobilize the carbon. **Blue Label stainless,** of Carpenter Steel, is Type

302 steel of highly controlled analysis marketed in 0.0025 to 0.312 in (0.006 to 0.792 cm) round diameters and in square and rectangular wire for coil springs. **Alloy NS-355,** of National-Standard, is a stainless steel having a typical analysis of 15.64% chromium, 4.38 nickel, 2.68 molybdenum, 1 manganese, 0.32 silicon, 0.12 copper, with the carbon at 0.14. The modulus of elasticity is 29.3 million lb/in^2 (205,100 MPa) at 80°F (27°C) and 24 million lb/in^2 (168,000 MPa) at 800°F (427°C). **Armco 17-7 PH steel,** of Armco Steel, has 17% chromium, 7 nickel, 1 aluminum, and 0.07 carbon. The wire has a tensile strength up to 345,000 lb/in^2 (2,378 MPa). **Spring wire** for high-temperature coil springs may contain little or no iron. **Alloy NS-25,** of National-Standard, for springs operating at 1400°F (760°C), contains about 50% cobalt, 20 chromium, 15 tungsten, and 10 nickel, with not more than 0.15 carbon. The annealed wire drawn to a 30% reduction has a tensile strength of 240,000 lb/in^2 (1,654 MPa) with elongation of 8%. **Matreloy,** of Materials Research Corp., for high-temperature springs, contains 39% chromium, 4 molybdenum, 2 titanium, 1 aluminum, and the balance nickel. The rolled metal has a yield strength of 275,000 lb/in^2 (1,896 MPa), and at 1400°F (760°C) it retains a strength of 120,000 lb/in^2 (827 MPa).

SPRUCE. The wood of various coniferous trees of northern Europe and North America. Spruce is a leading commercial wood of northern Europe and is exported from the Baltic region as **white fir** and **white deal.** It is also called **Norway spruce** and **spruce fir.** The wood is white, and has a straight, even grain. It is tough and elastic, and is more difficult to work than pine. The weight is 36 lb/ft^3 (577 kg/m^3). Norway spruce is *Picea abies,* and this tree yields the **Jura turpentine** of Europe. Spruce is used for making paper pulp, for packing boxes, and as a general-utility lumber. White spruce is from the tree *P. canadensis,* of the United States and Canada. It has quite similar characteristics. **Red spruce,** *P. rubra,* is the chief lumber spruce in the eastern United States. It is also called **yellow spruce, West Virginia spruce,** and **Canadian spruce. Black spruce,** *P. mariana,* of New England, eastern Canada, and Newfoundland, is used for making paper pulp. It is also called **blue spruce, bog spruce,** and **spruce pine. White spruce,** or **shingle spruce,** is from *P. glauca.* It is also called **skunk spruce** because of the peculiar odor of the foliage. All of these three species are called **eastern spruce,** and they grow from Nova Scotia to Tennessee and westward to Wisconsin except that red spruce does not grow in the lake states. All are mountain trees and are slow-growing. **Silver spruce, yellow spruce, Sitka spruce,** or **western spruce,** is from the enormous tree, *P. sitchensis,* of the West Coast of the United States and Canada. It is soft and light in weight, but strong, close-grained, and very free from knots. The wide sapwood is creamy white, and the heartwood pinkish to brownish.

The weight is less than eastern spruce, but it has high strength in proportion to weight. The trees reach a height of 280 ft (85 m) and a diameter of 10 ft (3 m) in 600 years, but the growth is rapid in early life. The wood is used for boxes, crates, millwork, and paper pulp. It is particularly adapted for groundwood pulp, giving higher strength in paper than most groundwood. The various species of commercial spruce have an average specific gravity, when kiln-dried, of 0.40, a compressive strength of 840 lb/in^2 (6 MPa) perpendicular to the grain, and a shearing strength of 750 lb/in^2 (5 MPa) parallel to the grain. It combines stiffness and strength per unit weight, and has a uniform texture free from pitch. **Japanese spruce** is from *Abies mariesii,* and **Himalayan spruce** is from *P. morinda.* The latter resembles Norway spruce.

Spruce gum is the gum exudation of the *Picea rubra, P. mariana,* and *P. canadensis* of the northeastern United States and Canada. It exudes from cuts in the trees as a transparent viscous liquid which hardens when it loses the volatile oil. It occurs on all parts of the tree, and the nodules of gum are sometimes as large as an egg. **Spruce oil** is extracted from the needles. In Colonial days the young twigs were boiled, and the liquid, mixed with molasses, was used as a beverage. The gum is brown or reddish black in color, and has a turpentinelike odor and a bitter pungent taste. It is used in cough medicines and chewing gum.

SQUILL. Also known as **red squill** and **sea onion.** A reddish powder used chiefly for the control of rats in warehouses and docks. It is obtained from the onionlike bulb of the perennial plant *Urginea maritima,* which grows on the beaches of Italy and other Mediterranean countries. The bulb is pear shaped, from 1 to 6 lb (0.5 to 15 kg) in weight and 6 to 12 in (15 to 30 cm) in diameter. The outer scales are dry, brittle, and reddish brown, and the inner scales are cream color to deep purple. **Red squill powder** is a powerful emetic to man or animals other than rats or mice. As rats and mice do not vomit, they are poisoned by it, while it is harmless to poultry and domestic animals. It is also used in medicine. It contains calcium oxalate and in contact with the skin it gives a sensation like nettle poisoning. **White squill** is another variety used in medicine as an emetic, heart tonic, and expectorant. The substitute for red squill known as **Antu,** of Du Pont, is **naphthyl thiourea,** a gray powder of little odor or taste about 100 times more poisonous to rats than squill and not normally injurious to domestic animals. Another poison more toxic to rats than squill is **Ratbane 1080,** which is **sodium fluoracetic acid** made synthetically. The poison occurs as natural fluoracetic acid in the **gifblaar** plant, *Dichapetalum cymosum,* of South Africa, which has been used locally for killing rodents. The poison, however, also kills domestic animals, and can thus be employed only in restricted places. The rodent poison known as **Warfarin** is a complex dicoumarol made under license of the Wisconsin Alumni Research Foundation. **Pival,** of the Atlantic Research Corp., is 2-pivalyl-1,3-indandione. It kills rats but is not toxic to other animals. **Pivalyn** is the same material in a water-soluble sodium salt form.

STAINLESS STEEL. A large and widely used family of **iron-chromium alloys** known for their corrosion resistance—notably their "nonrusting" quality. This ability to resist corrosion is attributable to a surface chromium oxide film that forms in the presence of oxygen. The film is essentially insoluble, self-healing, and nonporous. A minimum chromium content of 12% is required for the film's formation, and 18% is sufficient to resist the most severe atmospheric-corrosive conditions. However, for other reasons, the chromium content of stainless steels goes as high as 30% . Other elements, such as nickel, aluminum, silicon, and molybdenum, may also be present. There are nearly 50 compositions that are recognized by the American Iron and Steel Institute (AISI) as standard stainless grades. They are divided into three categories, according to the major microstructure present: austenitic, ferritic, and martensitic. In addition to the classic grades, a number of specialty stainless steels have been developed over the years. Principal among these are precipitation-hardening compositions and grades that are hardenable by cold work.

Stainless steels were first made in America in 1914 under English and German patents. The original composition was 13.5% chromium and 0.35 carbon. The original Krupp austenitic, or **KA steel,** or simple **austenitic steel,** had 20% chromium and 7 nickel, which was later balanced at 18-8. The **eighteen-eight steels,** or **chromium-nickel steels,** were called **super stainless steels** in England to distinguish them from the plain chromium steels.

The **martensitic steels,** or hardenable stainless steels, are the **AISI Type 400 steels.** They are the chromium steels in various grades with chromium up to 27%, and carbon up to 1.25%. The general utility **chromium stainless steel** is AISI 410 with 12% chromium and 0.15 max carbon. These steels are ferromagnetic and air-hardening. Unless modified, they have a blue-gray color, as distinct from the whiter color of the nickel-bearing stainless steels, and they are thus not preferred for food-processing equipment. They are highly corrosion-resistant, but they stain and assume a deep-blue tinge after heating, and are thus not normally used for cooking utensils. The **semiaustenitic steels** are the chromium steels with the addition of about 4.5% nickel. When solution-annealed they contain a proportion of unhardenable delta ferrite, and are soft and easily worked. When heat-aged, carbon precipitates from the austenite to the austenite-ferrite boundaries, and they can be hardened to high tensile strengths. Even small amounts of nickel give a precipitation effect.

The **precipitation-hardened steels** are produced by a controlled method of quenching with a pressure water spray uniformly distributed between the pinch rolls. It gives a deep uniform hardening that increases tensile and flexural strengths.

Free-cutting stainless steels are usually regular grades with the addition of a small amount of sulfur or selenium. Sulfur does not impair the resistance to corrosion, but it decreases ductility.

Ferritic stainless steels, the **AISI Type 400** series, contain approximately 12 to 28% chromium, but no nickel. The basic composition is **AISI Type 430** with 14% chromium, widely used for automotive trim. Since ferritic grades are low in carbon, not easily hardened by heat treatment, and only moderately hardened by cold working, they are lower in strength than the other two major stainless grades. They are always magnetic and retain their basic microstructure up to the melting point. Because of high chromium content, the ferritic grades have good corrosion resistance and excellent resistance to oxidation at high temperatures. However, they are susceptible to embrittlement, particularly the higher chromium compositions, at temperatures in the 750 to 1000°F (399 to 538°C) range. **Stabilized ferritic stainless steels,** sometimes referred to as **superferritics,** combine high resistance to chlorine-induced stress-corrosion cracking with good weldability and weld toughness. They are used chiefly for chemical and oil processing equipment. **AISI Type 420 steel,** with 13% chromium, 1% each of silicon and manganese, and more than 0.15 carbon is much used for cutlery.

Many modified grades of these steels are marketed under trade names. **Colonial stainless FMS,** of the Colonial Steel Co., has 0.12% max carbon and up to 0.35 sulfur. It is easily machined, and is used for valve parts, screws, and gun barrels. **U.S.S. steel 12A1,** of the U.S. Steel Corp., is a 13.5% chromium steel with 0.10 to 0.30 aluminum and 0.08 max carbon. It is used for pipe and tubes in refineries. The aluminum acts as a ferrite stabilizer, and it combines high strength and ductility. It requires heat treatment to obtain greatest corrosion resistance, but the steel does not harden. **U.S.S. steel 27,** for severe oxidation and reducing conditions up to 2100°F (1149°C), contains 23 to 30% chromium, 0.10 to 0.25 nitrogen, and 0.20 max carbon. The nitrogen eliminates embrittlement. The steel is ferritic, and has higher physical qualities than a plain chromium steel. **Radianite steel,** of the Labtrobe Steel Corp., has up to 18% chromium, 0.75 carbon, with 0.50 each of silicon and manganese. It keeps a fine cutting edge, and retains a high polish. **Lusterite** is this steel with higher carbon, and is a **keen-edge steel** for surgical instruments and knives. **Hy-Glo steel** of the same company has 17% chromium and 0.60 carbon. It has a tensile strength up to 250,000 lb/in^2 (1,723 MPa), and hardness to 600 Brinell. In general, the stainless steels do not hold a fine, keen-cutting edge as do the vanadium cutlery steels.

The **AISI Type 300 steels** are the chromium-nickel **austenitic stainless steels** in grades containing up to 30% chromium and up to 20% nickel. The standard 18–8 grade is **AISI Type 302 steel** with no more than 0.15% carbon and up to 2% manganese. **AISI Type 303 steel** is this grade with sulfur or selenium added for free machining. These steels do not harden by heat treatment, but can be rolled hard. The annealed Type 302 steel has a tensile strength of 80,000 lb/in^2 (551 MPa) with elongation of 50% and Brinell hardness of 180, but when hard-rolled it has a tensile strength to 200,000 lb/in^2 (1,378 MPa). The steels are tough, ductile, and nonmagnetic. Precipitation of chromium carbide at the grain boundaries, making them subject to intergranular corrosion, can be eliminated by keeping the carbon very low or by adding small amounts of columbium or titanium. Small amounts of nitrogen inhibit grain growth, and the nickel may be lowered in nitrogen-bearing steels. Copper improves the resistance to sulfuric acid, and small amounts of molybdenum increase the resistance to hot chemicals.

The austenitic stainless steels are favored for structural uses in aircraft and transport equipment because the high strength and corrosion resistance permit smaller sections to be used for weight saving. **AISI Type 304** is most used as it is generally available in all wrought forms and is easily welded. It contains 18% chromium, 8 nickel, 2 manganese, and 1 silicon with carbon kept below 0.08%. Where corrosive conditions exist **AISI Type 316** is used. It has 16% chromium, 10 nickel, 2 manganese, 1 silicon, and 2 molybdenum. Standard specifications for these steels call for a minimum tensile strength, annealed, of 85,000 lb/in^2 (586 MPa), and yield strength of 30,000 lb/in^2 (206 MPa). The cold-rolled steel tempered to ½-hard has a tensile strength of 150,000 lb/in^2 (1,034 MPa), yield strength of 110,000 lb/in^2 (758 MPa), elongation of 16%, and Rockwell hardness C32.

The austenitic steels are normally hot-short, but small additions of the cerium metals may be used to improve hot-working. Varying the alloying elements in these steels makes the number of possible types and grades almost unlimited. In addition to the regular AISI grades, most producers market the steels under trade names to denote guaranteed uniformity or special characteristics. **Sharonsteel 302,** of the Sharon Steel Corp., is an 18–8 steel that can be cold-worked to a tensile strength of 275,000 lb/in^2 (1,896 MPa). **Stainless U steel,** of the Colonial Steel Co., is an 18–8 steel with the addition of 1.5% molybdenum and 1.25 copper. As rolled, it has a tensile strength of 100,000 lb/in^2 (689 MPa) and elongation of 40%. **Cooper alloy 17SM,** of the Cooper Alloy Foundry Co., is an 18–8 steel with 1.5% silicon, 3 molybdenum, and 0.08 max carbon. **Vickers FV 520 steel,** of Firth-Vickers Stainless Steels, Ltd., for aircraft construction, contains 16% chromium, 6 nickel, 1.5 molybdenum, 1.5 copper, 0.3 titanium, and 0.07 carbon. It is precipitation-hardened, and has a tensile strength from 124,000 to 184,000 lb/in^2 (854 to 1,268 MPa), with elongation from 23 to 45%. The French steel, **Uranus B6,** for high resistance to hot concentrated sulfuric acid, has 20% chromium, 25 nickel, 1.5 copper, and 4.5 molybdenum. **Maxilvry steel,** of Edgar Allen & Co., Ltd., a chromium-nickel steel with copper for deep drawing, has a tensile strength of 90,000 lb/in^2 (620 MPa) with elongation of 68%. **Bethalon steel,** of the Bethlehem Steel Co., and **Allegheny metal,** of the Allegheny-Ludlum Steel Co., are 18–8 type steels in various grades. **Sta-Gloss, Hi-Gloss,** and **Duro Gloss** are stainless steels of the Jessop Steel Co. **Stainless W steel,** of the U.S. Steel Corp., has 17% chromium, 7 nickel, 0.70 titanium, 0.20 aluminum, 0.50 silicon, 0.50 manganese, and 0.70 carbon. When hardened to Rockwell C47 the tensile strength is 225,000 lb/in^2 (1,551 MPa) and elongation 10%. **Armco 18–9 steel,** which can be swaged and cold-headed with little work-hardening, has 18% chromium, 9 nickel, 3.5 copper, 2 manganese, 1 silicon, and not over 0.10 carbon. With a 50% cold reduction it has a tensile strength of 152,000 lb/in^2 (1,047 MPa). **Nitronic steels** are nitrogen-strengthened chromium-manganese-nickel austenitic stainless steels of Armco Steel. They have roughly twice the annealed strength of standard grades, very low magnetic permeability even after cold work, and good to excellent corrosion resistance.

All of the 18–8 type steels have a brilliant luster when polished. The heat conductivity is low, about 5% that of copper, so that where conductance is required, as for cooking utensils, they are usually laminated with copper. They are ductile and malleable, but harden rapidly with cold work. The cast steels are varied in composition to suit particular requirements. **Circle L22 steel,** of the Lebanon Steel Foundry, for cast parts for chemical equipment, contains 19.5% chromium, 9 nickel, 1.25 silicon, 0.75 manganese, with low carbon, 0.07 max, to prevent formation of carbides in welding. Cast parts have a tensile strength of 75,000 lb/in^2 (517 MPa), elongation of 50%, and Brinell hardness of 135. **Circle L22M** and **Circle L22XM** are variations of this steel with 3% of molybdenum to increase hot chemical resistance, and with a small amount of selenium to improve machining. **Durco KA2S steel,** of the Duriron Co., is a similar steel with similar variations. Small amounts of cobalt are sometimes added to this type of steel to increase the resistance to intergranular corrosion. For structural parts for atomic equipment, boron may be added as a neutron absorber. Boron increases the yield strength of the steel, but decreases the corrosion resistance and drastically decreases the ductility. Thus, only the isotope boron 10 is added, since it is 10 times more effective against neutrons than regular boron. The chromium-nickel high alloys used for severe chemical resistance at high temperatures are often called stainless steels, but they are not steels. **Ni-O-Nel,** of the International Nickel Co., Inc., used for chemical equipment to resist strong mixed chemicals and hot gases, contains 42% nickel, 21.5 chromium, 3 molybdenum, 2.25 copper, small amounts of silicon, manganese, titanium, and carbon, and the balance iron. The alloy is resistant to hot sulfuric, nitric, and phosphoric acids, and to ammonium hydroxide, but such metals are expensive and are not classed with the commercial stainless steels.

Chromium-manganese steels, with manganese used to replace the nickel of the 18–8 steels, were early used in Germany to replace the chromium-nickel stainless steels, but they were difficult to work and not as corrosion-resistant as the 18–8 steels. Modified with nickel, they now constitute the **AISI Type 200 steels,** and in general they have higher hardness and yield strength than the 300 series. **AISI Type 201 steel** has 17% chromium, 4.5 nickel, 6.5 manganese, and 0.15 max carbon. It has a tensile strength of 115,000 lb/in^2 (792 MPa), yield strength 55,000 lb/in^2 (379 MPa), and elongation 55%. **AISI Type 202 steel** has 18% chromium, 5 nickel, 8.5 manganese, and 0.15 max carbon. Its tensile strength is 105,000 lb/in^2 (723 MPa) and elongation 55%. These steels are lower in cost than the 18–8 steels, and have comparable working properties and corrosion resistance. **TRC steel,** of the Allegheny-Ludlum Steel Co., contains 15% chromium 16.5 manganese, 1 nickel, and 0.10 carbon. It also contains 0.15 nitrogen to stabilize the austenite. The annealed steel has a tensile strength of 102,000 lb/in^2 (703 MPa) with elongation of 57%, and when cold-rolled the tensile strength is 150,000 lb/in^2 (1,034 MPa) with elongation of 12%. It has a white color, is highly corrosion resistant, and can be worked about the same as 18–8 steel. It is used for railway cars and truck trailers. **Allegheny 216** contains a 19.75% chromium, 8.25 manganese, and 6 nickel. The annealed yield strength is 108,000 lb/ in^2 (744 MPa) with elongation of 51%. The tensile and fatigue strengths at elevated temperatures are superior to those of the 18–8 steels, and cutting speeds for machining can be half those of regular stainless steels. **Tenelon,** of U.S. Steel, is an austenitic **manganese stainless steel** with corrosion resistance about equal to that of the 18–8 steels and higher physical properties at elevated temperatures. It contains 17% chromium, 14.5 manganese, 1 silicon, 0.10 carbon, and 0.40 nitrogen. It has a tensile strength of 125,000 lb/in^2 (861 MPa) with elongation of 45%, and at 1000°F (538°C) the tensile strength is 81,000 lb/in^2 (558 MPa).

Diffused stainless steel is a sheet steel with a low-carbon ductile steel core and a diffused chromium-iron alloy surface. It is produced from low-carbon steel by heating the sheets in a retort containing a chromium compound which diffuses into the metal at a temperature of about 2000°F (1093°C). The chromium alloys with the steel, the alloy on the surface containing as much as 40% chromium, which tapers off to leave a ductile nonchromium core in the sheet. **Black stainless steel,** for electronic applications, is produced by immersing sheet steel in a bath of molten potassium dichromate and sodium dichromate. The steel has a shiny black finish.

Metal fibers are used for weaving into fabrics for arctic heating clothing, heated draperies, chemical-resistant fabrics, and reinforcement in plastics and metals. **Stainless-steel yarn** made from the fibers is woven into **stainless-steel fabric** that has good crease resistance and retains its physical properties to 800°F (427°C). The fiber may be blended with cotton or wool for static control, particularly for carpeting.

STARCH. A large group of natural carbohydrate compounds of the empirical formula $(C_6H_{10}O_5)_x$, occurring in grains, tubers, and fruits. The common cereal grains contain from 55 to 75% starch, and potatoes contain about 18%. Starches have a wide usage for foodstuffs, adhesives, textile and paper sizing, gelling agents, and fillers; in making explosives and many chemicals; and for making biodegradable detergents such as sodium tripolyphosphate. Starch is a basic need of all peoples and all industries. Much of it is employed in its natural form, but it is also easily converted to other forms and more than 1,000 different varieties of starch are usually on the American market at any one time.

Most of the commercial starch comes from corn, potatoes, and mandioca. Starches from different plants have similar chemical reactions, but all have different granular structure, and the differences in size and shape of the grain have much to do with the physical properties. **Cornstarch** has a polygonal grain of simple structure. It is the chief food starch in the Western world, although sweet-potato starch is used where high gelatinization is desired, and tapioca starch is used to give quick tack and high adhesion in glues. **Tapioca starch** has rounded grains truncated on one side and is of lamellar structure. It produces gels of clarity and flexibility, and, since it has no cereal flavor, it can be used directly for thickening foodstuffs. **Rice starch** is polygonal and lamellar, and has very small particles. It makes an opaque stiff gel and is also valued as a **dusting starch** for bakery products, although it is expensive for this purpose. **White-potato starch** has conchoidal or ellipsoidal grains of lamellar structure. When cooked, it forms clear solutions easily controlled in viscosity, and gives tough resilient films for coating paper and fabrics. Prolonged grinding of grain starches reduces the molecular chain, and the lower weight then gives greater solubility in cold water. Green fruits, especially bananas, often contain much starch, but the ripening process changes the starch to sugars.

In general, starch is a white, amorphous powder having a specific gravity from 0.499 to 0.513. It is insoluble in cold water but can be converted to **soluble starch** by treating with a dilute acid. When cooked in water, starch produces an adhesive paste. Starch is easily distinguished from dextrins as it gives a blue color with iodine while dextrins give violet and red. The starch molecule is often described as a chain of glucose units, with the adhesive **waxy starches** as those with coiled chains. But starch is a complex member of the great group of natural plant compounds consisting of starches, sugars, and cellulose, and originally named carbohydrates because the molecular formula could be written as $C_n(H_2O)_x$, but not all now-known carbohydrates can be classified in this form, and many now-known acids and aldehydes can be indicated by this formula.

Starch can be fractionated into two polymers of high molecular weight. **Amylose** is a straight-chain fraction having high adhesive properties for coatings and sizings, and **amylopectin** is a branched-chain fraction best known as a suspending agent for foodstuffs. Amylose is chemically identical with cellulose, but the chain units of the molecule have an alpha linkage and are coiled, while the cellulose molecule is rigid. It has a molecular weight of 150,000, while amylopectin has a molecular weight above 1 million. The 1–4 alpha linkage of amylopectin with random branches at the 6-carbon position makes the material easily dispersible in cold water but resistant to gelling. Amylopectin is thus best suited for thickening, but, since it can be combined and cross-linked with synthetic resins and is highly resistant to deterioration, it is used with resins for water-resistant coatings for paper and textiles.

Tapioca is the starch from the root of the large tuber *Manihot utilissima,* now grown in most tropical countries. It is called **cassava** in southern Asia, **manioc** in Brazil, **mandioca** in Paraguay, and **yuca** in Cuba. This perennial vegetatively propagated shrub was cultivated as far back as 2,500 years ago, and there is some indirect evidence that it has been grown for 4,000 years in the Americas. Its fresh roots contain 30 to 40% dry matter and have a starch content of approximately 85% of the dry matter. It is used in enormous quantities for food in some countries, and in some areas much is used for the production of alcohol. In the United States it is valued for adhesives and coatings, and only a small proportion in globules and flakes, known as **pearl tapioca,** is used in foodstuffs. **Gaplek,** used for cattle feed in Asia, is not the starch, but is dried and sliced cassava root. Tapioca starch may be sold under trade names. **Kreamgel,** of Morningstar-Paisley, Inc., used as a thickener for canned soups, sauces, and pastries, is refined tapioca that gives clear solutions without imparting odor or flavor.

Potato starch, produced from the common **white potato,** *Solanum tuberosum,* has been the most important starch in Europe, but in the United States it is usually more expensive than cornstarch. It forms heavier hot pastes than tapioca. It is also free of flavor, and is used as a thickener in foods. It does not crystallize easily. **Arogum,** of Morningstar-Paisley, Inc., is potato starch used to give tough resilient coatings on paper and textiles, and **Arojel P** is pregelatinized potato starch used as a beater additive to improve the strength and scuff resistance of kraft paper. **Sweet-potato starch** is from the tuber *Opomoea batata.* An average of 10 lb (4.5 kg) of starch is produced per bushel. The root has poor shipping qualities, and the starch is expensive, but it has excellent colloidal qualities and gelatinizes completely at 165°F (74°C). It is used in some foodstuffs. It has a pleasant sweetish flavor, and in Latin countries great quantities are marketed in the form of a stiff gel as a dessert sweet known as **dulce de batata.**

Arrowroot starch is from the tubers of the *Maranta arundinacea* of the West Indies. It is easily digested, and is used in cookies and other food products, especially baby foods. **Florida arrowroot** is from the *Zamia floridana.* **East Indian arrowroot** is from the plant *Curcuma angustifolia,* which belongs to the ginger family. Arrowroot from St. Vincent, used in instant-pudding mixes and icings, is marketed as a precooked powder of about 200 mesh. It swells in cold water, and does not add flavor.

The starches do not crystallize like sugar, and they may be added to some confections to minimize crystallization. They are also used as binders in candies and in tablet sugar, but any considerable quantity in such products is considered as an adulterant. Metabolism of starch in the human system requires conversion to sugars, and the taking in of excessive quantities of uncooked starch is undesirable. **Modified starches** are starches with the molecule altered by chemical treatment to give characteristics suitable for particular industrial rquirements. The modified starches and especially prepared starches are usually sold under trade names. **Superlose,** of Stein-Hall & Co., is amylose from cornstarch, and **Auperlose** is amylose from potato starch. **Ramalin,** of the same company, is amylopectin. **Amylon,** of the National Starch & Chemicals Corp., is cornstarch containing 57% amylose, and **Kosul** is cornstarch high in amylopectin. **Textaid,** of the same company, is a modified starch which reacts with water to form a grainy structure. It is used in comminuted meat products to give a firm texture. The **ColFlo** thickening agents, stable and soluble in frozen foods, are modified waxy cornstarches, high in amylopectin. **Pregelatinized starches** are pre-heat-treated starches that require no cooking for use in dry food mixes or adhesives. **Snow Flake starch,** of Corn Products Sales Co., is a cornstarch of this type.

Wheat starch is a fine white starch made by separating out the gluten of wheat flour by wash flotation. It is used in prepared mixes for foam-type cakes and pie crusts to improve texture, add volume, and reduce the amount of shortening needed. It replaces up to 30% of the wheat-flour content of the mix. **Starbake starch,** of Hercules, is wheat starch. **Paygel,** of General Mills, is also wheat starch, but **alant starch,** or **inulin,** $(C_6H_{10}O_5)_6 \cdot H_2O)$, is not a starch in the ordinary sense, but is an insoluble sugar which occurs as the reserve polysaccharide in many plants. It is obtained from the roots of the **artichoke,** *Helianthus tuberosis,* native to America but now grown widely in Europe. Unlike starch, the molecule has fructose units held in glucoside linkage, and hydrolysis converts it to fructose.

Starch acetate, or **acetylated starch,** is used for textile sizing, in adhesives, and for greaseproofing paper. The insertion of acetate radicals reduces the tendency of the molecular chains to cling together. The acetylated starches are gums which gelatinize at lower temperatures than starch, and produce stable, nonlumping pastes which give strong, flexible films. **Miralloid** and **Mira-Film,** of A. E. Staley Mfg. Co., are acetylated cornstarches. **Morgum,** of Morningstar-Paisley, Inc., is a hydroxyethyl **etherized starch** which gives high film strength in coatings. The **Kofilms** of General Mills are acetylated cornstarches which give greaseproof, craze-resistant coatings on paper and textiles.

Laundry starches are usually ordinary starches, but silicone resin emulsions may be added to starches to permit higher ironing temperatures, improve slipperiness, and improve the hand of the starched fabric. The so-called **permanent starches,** for household use, that are not removed by washing, are not starch, but are emulsions of polyvinyl acetate. **Oxidized starch,** a resistant starch for coatings, is made by the chloro-oxidation of a starch solution. **Sumstar 190,** of the Miles Chemical Co., is a **diallyl starch** made by acid oxidation of cornstarch. Small amounts of the powder added to kraft, tissue, or toweling pulp increase the wet and dry strengths and the folding endurance of the papers. An **ammoniated starch** called **Q-Tac starch,** of the Corn Products Sales Co., is cornstarch reacted with quaternary ammonium groups. A less than 1% solution improves paper strength. **Sulfonated starches** are used as dirt-suspending agents with detergents for cleaning textiles. **Nu-Film** is a starch of this type. **Clear Flo starch,** of the National Starch Products Co., is a modified starch containing a carboxyl group and a sulfonic acid group in the molecule. It has high hydrating capacity, and gelatinizes sharply at low temperatures. It is used in adhesives and water paints. **Cato starch,** of the same company, is a carboxymethyl starch used in paper sizing to add strength. **Dry Flo starch,** of the same company, is modified to contain a **hydrophobic radical,** such as $\cdot CH_2$, which makes the material insoluble in water but soluble in oils. It is used in paints.

Many enzymes hydrolize starch to maltose, but some enzymes convert the starch to the hard, tough glucosides known as **mannans,** such as the **mannose** of the ivory nut. **Phospho mannan,** produced by the fermentation of starch, is such a material used in adhesives. **Granular starch,** used in enzyme-conversion processing, is in dense granular particles produced by flash drying. **Easy-Enz starch,** of the Corn Products Co., is such a starch. **Cationic starch** is a starch with the molecules of stable negative polarity to give higher adhesion on the cellulose fibers of paper or textiles. **Molding starch,** for adding to sugar candies to give sharp molding chracteristics, is starch containing an edible oil.

The **phosphate starch** of the American Maize Products Co. is an orthophosphate ester of cornstarch, marketed in sodium salt form as a light-tan dry powder. It has high thickening power and makes a clearer paste than cornstarch. It has superior water-binding properties at low temperatures. Frozen foods made with it do not curdle or separate when thawed, and canned foods thickened with the starch can be stored for long periods without clouding. It is also used as a briquetting binder for charcoal.

Starch sponge is an edible starch in the form of a coarse-textured, porous, crispy, spongelike material, used for confections by impregnating with chocolate or sweets. In crushed form it is added to candy or cookies. It is produced by freezing, thawing, and pressing starch paste. The freezing insolubilizes the starch so that no soluble starch goes off when the water is pressed out. **Nitrostarch,** or **starch nitrate,** $C_{12}H_{12}O_{10}(NO_2)_3$, is a fine white powder made by treating starch with mixed acid. It is highly explosive and is used for blasting, as a military explosive, and in signal lights. **Grenite** is nitrostarch mixed with an oil binder for use in grenades. **Trojan explosive** is a mixture of 40% nitrostarch with ammonium and sodium nitrates and some inert material to reduce the sensitiveness.

STATUARY BRONZE. Copper alloys used for casting statues, plaques, and ornamental objects that require fine detail and a smooth, reddish surface. Most of the famous large bronze statues of Europe contain from 87 to 90% copper, with varying amounts of tin, zinc, and lead. Early Greek statues contained from 9 to 11% tin with as much as 5% lead added apparently to give greater fluidity for sharper casting. A general average bronze will contain 90% copper, 6 tin, 3 zinc, and 1 lead. Statuary bronze for cast plaques used in building construction contains 86% copper, 2 tin, 2 lead, 8 zinc, and 2 nickel. The nickel improves fluidity and hardens and strengthens the alloy, and the lead promotes an oxidized finish on exposure. The statuary bronze used for hardware has 83.5% copper, 4 lead, 2 tin, and 10 zinc.

STEARIC ACID. A hard, white, waxlike solid of the composition $CH_3(CH_2)_{16}COOH$, obtained from animal and vegetable fats and oils by splitting and distilling. The hard cattle fats are high in stearic acid, but other fats and oils contain varying amounts. It is also called **octodecanoic acid,** and can be made by hydrogenation of oleic acid. Stearic acid has a specific gravity of 0.922 to 0.935, and a melting point at about 130°F (54°C), and it is soluble in alcohol but insoluble in water. It is marketed in cakes, powder, and flakes. **Emory 3101-D** is **isostearic acid** which has the solubility and physical properties of oleic acid while retaining the heat and oxidation stability of stearic acid. **Pearl stearic acid** is the material in free-flowing bead powder. The acid is used for making soaps, candles, paint driers, lubricating greases, and buffing compositions, and for compounding in rubbers, cosmetics, and coatings.

Successive pressings remove liquid oils, thus raising the melting point and giving a whiter, harder product of lower iodine value. **Oleo oil** is a yellow oil obtained by cold-pressing the first-run cattle tallow. **Tallow oil** is the oil following the first two grades of oleo oil. **Oleostearin,** used for treating leather, is the stearin remaining after extraction of the oils.

Stearin is the glyceride of stearic acid. **Acetostearins** are the monoglycerides acetylated with acetic anhydride. They are closely related to fats, but are nongreasy and are plastic even at low temperatures. The highly acetylated stearins melt below body temperature and are edible. Acetostearins are used as plasticizers for waxes and synthetic resins to improve low-temperature characteristics. **Stearite** is a trade name for **synthetic stearic acid** made by the hydrogenation of unsaturated animal and fish oils. It is used in rubber compounding, as it is more uniform than ordinary stearic acid. **Hystrene,** of the Trendex Co., is purified and hardened stearic acid in grades of 70, 80, and 97% stearic acid, with the remainder palmitic acid, used for candles, cosmetics, and stearates. **Intarvin** is a synthetic edible fat made from stearic acid by converting it to **margaric acid,** or **daturic acid,** $C_{16}H_{33}COOH$, and then esterifying with glycerin. It is used as a fat for diabetics as it does not undergo the beta oxidation to lose two carbon atoms at a time and produce acetoacetic acid in the system as do the even-carbon food acids.

Wilmar 272, of Wilson & Co., is refined stearic acid in flake form for use in candles and coatings. **Flexchem B,** of Swift & Co., is a **sodium stearate,** $NaC_{18}H_{35}O_2$, in the form of a water-soluble white powder which is insoluble in oils. It is used as a bodying agent in cosmetic creams. **Myvacet,** of the Distillation Products Industries, is an acetostearin used as an edible plastic coating for poultry, cheese, and frozen fish and meats to prevent loss of the natural color and flavor. It is a white waxy solid with melting points from 99 to 109°F (37 to 43°C), but it also comes as an oil with congealing point at 45°F (7°C) for use as a release agent on bakery equipment.

Stearin pitch is a brown-to-black by-product residue obtained in the splitting and distillation of fats and oils in the manufacture of soaps, candles, and fatty acids. While the word stearin implies that it contains only stearic acid, it usually comes from a variety of oils and has mixed acids, and it may take the name of the oil, such as **linseed pitch** or **palm pitch.** It is used in varnishes and cold-molding compositions.

STEEL. Iron with carbon chemically dissolved in it. Steels with less than 0.15% carbon are technically classed as **irons,** although the name steel is used even when the carbon is very low, especially when the material is heavily alloyed. Alloys with high proportions of other elements, and a relatively small amount of iron, are still called steel if the iron and carbon are important influencing elements.

Steel is graded according to the percentage of carbon in it, and is roughly divided into four groups: **low-carbon steel,** with 0.15 to 0.30% carbon; **medium-carbon steel,** with 0.30 to 0.60% carbon; **high-carbon steel,** with 0.60 to 0.90% carbon; and very high carbon steel, with 0.90 to 1.5% carbon. Steel with more than 1.25% carbon is very brittle unless there are some other balancing elements. The low-carbon steels machine easily and can be forged readily. Those containing above 0.90% carbon are difficult to forge and machine. In general, the low-carbon steels are used for construction parts and the high-carbon steels for tools and high-strength parts. All steels contain some manganese, usually at least 0.30%, left after deoxidizing and desulfurizing with ferromanganese. The latter forms MnS and MnO, which pass off in the slag, but sometimes an excess of manganese is used for hard steels as the manganese is a carbide-forming element. Manganese is added to the steels used for electromagnetic devices to retard particle growth, thus reducing magnetic aging. Sulfur in steel forms the weak and soft sulfide FeS, which weakens the steel and promotes hot-shortness or brittleness at red heat. But sometimes a very small amount of sulfur is left in the steel to aid machinability. Oxygen in the steel forms the embrittling compound FeO and is undesirable in any quantity. Normally, phosphorus is considered detrimental in steels, but small amounts, up to 0.20%, in low-carbon steels increase the hardness, strength, and also increase corrosion resistance. **Dirty steel** is steel with inclusions of iron oxide from scrap used in the converter or of aluminum oxide from the aluminum used in deoxidizing. The melting point of steel varies with the carbon, but is always higher than that of cast iron with the same amount of combined carbon. The average weight of carbon steel is 0.283 lb/in^3 (7,833 kg/m^3). One of the simplest of the plain carbon steels is **AISI steel C1015,** with 0.15% carbon, and this steel has a tensile strength in the soft hot-rolled condition of 50,000 lb/in^2 (344 MPa), elongation 28%, and Brinell hardness 100. It is taken as a standard for comparing the strength and other physical properties of other steels and alloys.

Open-hearth steel is made by fusing cast iron with steel scrap or wrought iron in a regenerative furnace. Cementation consists in heating bars of wrought iron in contact with carbon. The product is known as **blister steel,** as the surface of the bars is covered with blisters. Blister steel has a crystalline fracture decreasing toward the center of the bar where there is less carbon. **Shear steel** is produced from this blister steel by cutting and piling together, heating to a high temperature, and rolling or hammering into bars. Shear steel was originally the only form of commercial steel.

Bessemer steel is made by decarbonizing cast iron by forcing a powerful blast of air through the molten iron. **Crucible steel** is made by melting wrought iron in a crucible with charcoal and ferromanganese, or special steel mixes in a crucible furnace. The slag separates by gravity from the molten metal in the crucibles, and the oxides combine with carbon and manganese or aluminum, and boil off. The metal is poured into ingots which are reheated and rolled into bars or other forms. The crucible process permits a high degree of control and reduces sulfur and phosphorus. **Electric steel** is made in either the induction or the arc-type furnace, and is of a uniform quality and of a higher strength and ductility than open-hearth steel of the same carbon content.

Many variations are now used in the production of steels. Oxygen may be used in the blast instead of air to give greater efficiency and better control. Calcium carbide may be used as a fuel in basic-oxygen furnaces. It permits use of more scrap and improves the efficiency of melting but shortens the life of the refractory lining. Degassing may be done in a vacuum chamber with a jet of inert gas to remove the hydrogen. Or, steel may be produced without blast-furnace reduction by compressing iron oxide in rollers at about 2200°F (1204°C) and then hot-strip rolling and cold-rolling. Ordinary steel has only carbon as the influencing element, except that residual manganese, sulfur, and some other elements are present in very small amounts.

Ordinary steel used for **wrought-steel pipe** has a tensile strength of 52,000 lb/in^2 (358 MPa), yield point 30,000 lb/in^2 (206 MPa), and elongation 20%. The tensile strength of a typical open-hearth steel varies from about 50,000 lb/in^2 (344 MPa) for a 0.12% carbon steel to 110,000 lb/ in^2 (758 MPa) for a 0.55% carbon steel, with elongations of 29 and 12%, respectively. Ordinary forgings are made from steels of 0.15 to 0.30% carbon, and strong forgings from steel with 0.35 to 0.40% carbon, but most strong forgings are now made with alloy steels. Raising the carbon content from 0.40 to 0.45% will increase the possible hardness of the steel about 10%, but greater variations in strength and hardness of a steel may come from even small additions of alloying elements plus heat treatment, and very high strengths are obtainable in these steels. Very high carbon can be used to give high strength and abrasion resistance when the steel contains enough carbide-forming elements to absorb the carbon. **Aircraft-quality steel** refers to the quality of steel, and may be of any type of steel. It is also called **magnaflux steel,** and each bar is tested with a magnetic powder to expose any surface or subsurface irregularities.

SAE steels is the designation for the standard grades of steel approved by the Society of Automotive Engineers. These steels are made regularly by the various mills and are known by their designating numbers. The first number indicates the class of steel, as follows: carbon steel, 1; nickel-carbon, 2; nickel-chromium, 3; molybdenum, 4; chromium, 5; chromium-vanadium, 6; tungsten, 7; nickel-chromium molybdenum, 8; and silicon-manganese steel, 9. The second figure indicates the average percentage of the predominating alloying element. The last two or three figures indicate the approximate carbon content in hundredths of 1%. Thus **SAE 2350**

steel is a nickel steel containing 3% nickel (2.75 to 3.25) and 0.50 carbon (0.45 to 0.55). The managanese steels, with manganese from 1.60 to 1.90%, are designated by the letter T before the initial 1. Thus **SAE T1350 steel** contains 0.45 to 0.55% carbon and 1.60 to 1.90 manganese. The SAE steels are made to close specifications of manganese, sulfur, and phosphorus content, and since they are very uniform in quality and usually carried in stock, they have been widely adopted for use in all kinds of products.

Steel-making processes used to produce steel mill products, such as plates, sheets, and bars, have an important effect on a steel's properties. Steels are often identified as to the degree of deoxidation resulting during steel production. **Killed steels,** because they are strongly deoxidized, are characterized by high composition and property uniformity. They are used for forging, carburizing, and heat-treating applications. **Semikilled steels** have variable degrees of uniformity, intermediate between those of killed and rimmed steels. They are used for plates, structural sections, and galvanized sheets and strips. **Rimmed steels** are only slightly deoxidized during solidification. Carbon concentration is highest at the center of the ingot. Because the ingot's outer layer is relatively ductile, these steels are ideal for rolling. Sheets and strips made from rimmed steels have excellent surface-quality and cold-forming characteristics. **Capped steels** have a thin low-carbon rim which gives them surface qualities similar to those of rimmed steels. Their cross-sectional uniformity approaches that of semikilled steels.

Steels are also classified as air melted, vacuum melted, or vacuum degassed: **Air-melted steels** are produced by conventional melting methods, such as open hearth, basic oxygen, and electric furnace. **Vacuum-melted steels** are produced by induction vacuum melting and consumable electrode vacuum melting. **Vacuum-degassed steels** are air-melted steels that are vacuum processed before solidification. This produces steels with lower gas content, fewer nonmetallic inclusions, and less center porosity and segregation. They are more costly, but have better mechanical properties, such as ductility and impact and fatigue strengths.

Steel-mill products are produced from various primary forms such as heated blooms, billets, and slabs. These primary forms are first reduced to finished or semifinished shape by hot-working operations. If the final shape is produced by hot-working processes, the steel is known as **hot rolled steel.** If it is finally shaped cold, the steel is known as **cold finished steel,** or, more specifically as **cold rolled steel,** or **cold drawn steel.** Hot-rolled mill products are usually limited to low and medium, nonheat-treated, carbon-steel grades. They are the most economical steels, have good formability and weldability, and are used widely for large structural shapes. Cold-finished shapes, compared to hot-rolled products, have higher strength and hardness, better surface finish, and lower ductility.

STEEL POWDER. Finely divided steel powder used for molding under hydraulic pressure into various parts which are then sintered at a temperature of 1075°C. The tensile strength of the molded and sintered products is about the same as that of cast steel of the same composition. It has the advantage that there is no metal loss in machining, and accurate gears, cams, and other parts can be made without machining. **Sinterloy** is the name of a steel powder marketed by Charles Hardy, Inc., in compositions of 0.05, 0.40, and 0.80% carbon. Some grades also contain nickel and chromium. **Durex iron,** of the Moraine Products Div., General Motors Corp., is iron or steel made by compressing iron powder containing a small percentage of carbon and sintering in special furnaces with controlled atmospheres. It can be controlled by pressure to produce parts with densities of 70 to 95% that of pure iron or with a porosity range from 5 to 30% by volume when it is desired to impregnate with oil. It is used for mechanical parts or the porous variety for bearings. **Porex** is a trade name of this company for porous metals made from powders and used as filters. The production of products by the compression and sintering of powders is called **powder metallurgy. P.M.P. iron,** of the Powdered Metals Corp., used for bearings, contains 90% iron and 10 copper. With a density of 5.5 lb/in^3 (152,240 kg/m^3) it has a tensile strength of 30,000 lb/in^2 (206 MPa) and will absorb 30% of oil by volume. When compressed to a higher density, it has greater strength, and is used for cams and machine parts.

Stainless-steel powder has a typical analysis of 17.36% chromium, 8.89 nickel, 0.11 carbon, 0.47 silicon, 0.022 phosphorus, and 0.008 sulfur. It is prealloyed, and the powder has the original grain boundaries broken down. **Prealloyed steel powders** are made in a wide variety of compositions of carbon and alloy steels. The particles are spherical and are in meshes 50 to 100. Physical properties of parts made by powder metallurgy can be controlled by choice of powder size and by the pressure. The stainless-steel powder of the American Electro Metal Corp. has spongelike particles, and it comes in mixed screen sizes from 100 to 325 mesh to give high green strength in the pressed part. The **Steelmet powders** of P. R. Mallory & Co., Inc., are prealloyed powders with controlled particle sizes to yield pressed parts with 95% of theoretical density. **Steelmet 100** has 1% manganese, 1 nickel, and the balance iron. The sintered parts have a tensile strength of 55,000 lb/in^2 (379 MPa), elongation 25%. **Steelmet 600** has 2% copper and 0.25 nickel. The tensile strength of the parts is 95,000 lb/in^2 (654 MPa) with elongation of 16%. Practically any type of alloy powder can now be had in any particle size down to air-float powders below 15 nm.

STEEL WOOL. Long, fine fibers of steel used for abrasive purposes, chiefly for cleaning utensils and for polishing. It is made from low-carbon bessemer wire of high tensile strength, usually having 0.10 to 0.20% carbon and 0.50 to 1 manganese. The wire is drawn over a track and shaved by a stationary knife bearing down on it, and may be made in a continuous piece as long as 100,000 ft (30,480 m). Steel wool usually has three edges but may have four or five, and strands of various types are mixed. There are nine standard grades of steel wool, the finest of which has no fibers greater than 0.0005 in (0.0027 cm) thick, the most commonly used grade having fibers that vary between 0.002 and 0.004 in (0.006 and 0.010 cm). Steel wool comes in batts, or in flat ribbon form on spools usually 4 in (10 cm) wide. **Stainless-steel wool** is also made, and **copper wool** is marketed for some cleaning operations.

STILLINGIA OIL. A drying oil obtained from the kernels of the seeds of the tree *Stillingia sebifera,* cultivated in China. The seeds contain about 23% of a light-yellow oil resembling linseed oil but of somewhat inferior drying power. The oil has a specific gravity of 0.943 to 0.946 and iodine value of 160. It has the peculiar property of expanding with great force at the congealing point. Stillingia oil is edible, but deteriorates rapidly, becoming bitter in taste and disagreeable in odor. **Stillingia tallow,** also known as **Chinese vegetable tallow,** is obtained by pressing from the coating, or mesocarp, of the seeds. Sometimes the whole seed is crushed, producing a softer fat than the true tallow. The tallow contains palmitic and oleic acids and is used in soaps and for mixing with other waxes. Some stillingia trees are grown in Texas.

STRIPPABLE COATINGS. Coatings that are applied for temporary protection and can be readily removed. They are composed of such resins as cellulosics, vinyl, acrylic, and polyethylene; they can be water-base, solvent-base, or hot-melt. The choice of base depends on the surface to be protected. Water-base grades are neutral to plastic and painted surfaces, whereas solvent-base types affect those surfaces. Clear **vinyl strippable coatings,** perhaps the most widely used, are usually applied by spraying in thicknesses of 30 to 40 mils. **Acrylic strippable coatings** impart a clear, high-gloss, high-strength temporary film to metal parts. **Polyethylene strippable coatings** are relatively low-cost and can be used on almost all surfaces except glass. **Cellulosic strippable coatings** are designed for hot-dip application. Film thicknesses range widely and can go as high as 200 mils. The mineral oil often present in these coatings exudes and coats the metal surface to protect it from corrosion over long periods.

STRONTIUM. A metallic element of the alkaline group. It occurs in the minerals **strontianite,** $SrCO_3$, and **celestite,** $SrSO_4$, and resembles barium in its properties and combinations, but is slightly harder and less reactive and is not as white in color. It has a specific gravity of 2.54 and a melting point at about 770°C, and it decomposes water. The metal is obtained by electrolysis of the fused chloride, and small amounts are used for doping semiconductors. Its compounds have been used for deoxidizing nonferrous alloys, and were used in Germany for desulfurizing steel. But the chief uses have been in signal flares to give a red light, and in hard, heat-resistant greases. **Strontium 90,** produced atomically, is used in ship-deck signs as it emits no dangerous gamma rays. It gives a bright sign, and the color can be varied with the content of zinc, but it is short-lived. Strontium is very reactive and used only in compounds.

Strontium nitrate is a yellowish-white crystalline powder, $Sr(NO_3)_2$, produced by roasting and leaching celestite and treating with nitric acid. The specific gravity is 2.96, the melting point is 645°C, and it is soluble in water. It gives a bright crimson flame, and is used in railway-signal lights and in military flares. It is also used as a source of oxygen. The **strontium sulfate** used as a brightening agent in paints is powdered celestite. The ore of Nova Scotia contains 75% strontium sulfate. **Strontium sulfide,** SrS, used in luminous paint, gives a blue-green glow, but it deteriorates rapidly unless sealed in. **Strontium carbonate,** $SrCO_3$, is used in pyrotechnics, ceramics, and ceramic permanent magnets for small motors. **Strontium hydrate,** $Sr(OH)_2 \cdot 8H_2O$, loses its water of crystallization at 100°C and melts at 375°C. It is used in making lubricating greases and as a stabilizer in plastics. **Strontium fluoride** is produced in single crystals by Semi Elements, Inc., for use as a laser material. When doped with samarium it gives an output wavelength around 650 nm.

STYRAX. A grayish-brown, viscous, sticky, aromatic balsam obtained from the small tree *Liquidambar orientalis* of Asia Minor. It is also called **Levant styrax.** It is used in cough medicines and for skin diseases, as a fixative for heavy perfumes, and for flavoring tobacco and soaps. American styrax is obtained by tapping the sweet gum, *L. styraciflua,* of Alabama, a tree producing 8 oz (0.2 kg) of gum per year. It is a brownish semisolid and has the same uses as Levant styrax. It is shipped from Central America under the name **liquidambar,** and in the southern United States is called **sweet gum** and **storax.** The gum is not present in large amounts in the wood, but its formation is induced by cuts. **Benzoin** is another balsam obtained from several species of *Styrax* trees. It is a highly aromatic solid with an odor like vanilla, and is used in medicine and in perfumes and incense. **Sumatra benzoin** is from the tree *S. benzoin* and comes in reddish-brown lumps or tears. In medicine it was originally called **gum Benjamin. Siam benzoin,** from southern Asia, is from the trees *S. tonkinense* and *S. benzoides.* It is in yellowish or brownish tears. The Sumatra benzoin contains cinnamic acid, while the Asiatic gum contains benzoic acid. **Benzoic**

acid, or **phenylformic acid,** C_6H_5COOH, formerly produced from benzoin, is now made synthetically from benzol and called **carboxybenzene.** It is a white crystalline solid melting at 122°C, soluble in water and in alcohol. It is used as a food preservative, as an antiseptic, for flavoring tobacco, as a weak acid mordant in printing textiles, and in the manufacture of dyestuffs, pharmaceuticals, and cosmetics. Because it is poisonous, not more than 0.1% is used in food preserving in the form of its salt, **Benzoate of soda,** or **sodium benzoate,** C_6H_5COONa, which is a white crystalline powder. **Sorbic acid,** $CH_3CH:(CH)_2:CHCOOH$, a solid melting at 134°C, occurs in unripe apples, but is made synthetically. As a preservative and antimold agent it is more effective than benzoic acid, is nontoxic, and is readily absorbed in the human system. It is used in cheese and other foods. For food preservation it is used in the form of the water-soluble salt **potassium sorbate.** In a concentration of 0.2% it does not affect taste or aroma. **Anisic acid,** $CH_3OC_6H_4COOH$, used for pharmaceuticals, is the methyl ether of hydroxybenzoic acid. It is produced synthetically from carbon tetrachloride and phenol, and is a solid melting at 184°C. It is also called **methoxybenzoic acid, umbellic acid,** and **dragonic acids.**

SUEDE. Also called **napped leather.** A soft-finished, chrome-tanned leather made from calf, kid, or cowhide splits, or from sheepskin. It is worked on a staking machine until it is soft and supple, and then buffed or polished on an abrasive wheel. It has a soft nap on the polished side and may be dyed in any color. Suede is used for shoe uppers, coats, hats, and pocketbooks, but is now largely imitated with synthetic fabrics. **Artificial suede,** or **Izarine,** of the Atlas Powder Co., has a base of rubberized fabric. Fine cotton fibers dyed in colors are cemented to one side, and the underside of the sheet is beaten to make the fibers stand out until the cement hardens. The fabric looks and feels like fine suede. Some suede is also made by chemical treatment of sheepskins without staking. It has a delicate softness, but is not as wear-resistant as calfskin suede.

SUGAR. A colorless to white or brownish crystalline sweet material produced by evaporating and crystallizing the extracted juice of the sugarcane or the sugar beet. Refined sugar is practically pure **sucrose,** $C_{12}H_{22}O_{11}$, and in addition to being a sweetening agent for many foods it is a valuable carbohydrate food, and also a food preservative. When used with cooked fruits to make jams and jellies, it is both a preservative and an added food. Lack of sugar in the diet develops ketosis, the disease of diabetics, and results in the wasting away of muscles, using up of reserve fats, and the production of poisonous ketones. When the blood-sugar level is low a feeling of hunger is induced which may not be satisfied even by overeating. A small amount of sugar curbs the appetite and obviates surplus eating of proteins and fats that create obesity. Natural **brown sugar** contains about 2% of the minerals found in the plant, calcium, iron, phosphorus, magnesium, and potassium, and, although these are valuable as foods, they are lost in the refining process.

Sugar is at present most valued as a food and for the production of by-product alcohol from the residue molasses, but the sucrose molecule is a convenient starting point for the production of many chemicals. However, the production and distribution of sugar have been hemmed in by restrictive laws based on its use for food. The sucrose molecule has two complex rings, a glucosido and a fructose. It can be regarded as a type of **fructo-sido-glucose,** but the fructose in sucrose has a different structure, a **furanose,** or five-ring form, instead of the **pyranose,** or six-ring structure, of ordinary fructose. Hydrolysis of sucrose with acid gives dextrorotatory glucose and fructose, and the mixture is called **invert sugar. Numoline,** of SuCrest Corp., is a noncrystallizing invert sugar made by hydrolyzing sucrose to split the molecule into levulose and dextrose. It is used in confectionery and bakery products. Oxidation of sucrose produces oxalic acid and also **saccharic acid,** $(HCOH)_4(COOH)_2$, which can be reduced to adipic acid. Glycerin can be made from sugar by hydrogenation to sorbitol and then splitting. Thus, because of the great versatility of the sucrose molecule, and the ease with which sugar can be grown, sugar is one of the most valuable chemical raw materials. **Sucrose benzoate** is a benzoic acid derivative of sucrose used as a plasticizer and modifier for synthetic resins for lacquers and inks.

Sugarcane, *Saccharum officinarum,* is a tropical plant, originating in Asia and first brought to the Canary Islands in 1503 and thence to the West Indies. The plant will not withstand frost, but can be grown in a few favored regions outside of the tropics such as Louisiana. It is now grown on plantations in Cuba, Hawaii, Brazil, the Philippines, Indonesia, Puerto Rico, Peru, and many other countries. The cane or stalks of the plant are crushed to extract the juice, which is then concentrated by boiling, crystallized, and clarified with activated carbon or other material. The yield of sugar in Hawaii is about 14 tons (12,698 kg) of raw sugar per acre (4,047 m^2). Analysis of sugarcane gives an average of 13.4% sucrose by weight of cane. The average yield by milling is 91% of the contained sucrose, but yields as high as 98.8% are obtained by diffusion extraction of the cut-cane chips.

The **sugar beet** is a white-rooted variety of the common beet, *Beta vulgaris,* and grows in temperate climates. It is cut up and boiled to extract the juice, and the production and refining of the sugar are essentially the same as for cane sugar. There is no difference in the final product, although **raffinose,** or **melitriose,** $C_{18}H_{32}O_{16}$, a tasteless trisaccharide, occurs in the sugar beet, and may not be completely changed to sucrose by hydrolysis, so that a greater quantity may sometimes be needed to obtain equal sweetening effect. The pectins and starches of the sugar beet are not extracted by the use of the slicing and diffusion method.

Refined commercial sugar contains 99.98% sucrose, and is graded by screening to crystal size. The best qualities are the larger crystals from the first and second runs. The soft sugars are from further crystallizing, until the noncrystallizing brown sugars are reached. Raw sugar testing 96° by the polariscope is the grade used as a basis for raw-sugar quotations. Commercial sugar may have starch added. The ultrafine 6X **confectioners sugar** usually contains 4% cornstarch as a noncaking agent, and block sugars may contain starch as a binding agent, but starch reduces the sweetening power.

Cane sugar is the high-grade syrup or liquid sugar, while **molasses** is the heavy residual syrup left after the crystallization. **Edible molasses** is the yellow to brownish, light purified residue syrup. **Blackstrap molasses** is the final, inedible, unpurified residue heavy syrup, used for the production of ethyl alcohol. It contains 50 to 60% sugar by weight, mostly sucrose but some glucose. A purified grade which retains the minerals is marketed as an edible blackstrap molasses.

Molasses powder, used for bakery products, is made by spray drying. It is a free-flowing, noncaking powder. **Liquid sugar,** much used in food manufacturing because it saves handling costs, comes in various liquid densities and in various degrees of invert. The liquid sugars are usually not pure sucrose, and are called **multisugars.** For food manufacturing the calcium and other minerals may be left in, and they then have a yellow color. Multisugars with 90% sucrose and 10% levulose and dextrose crystallize in hard aggregate clusters, desirable in some confections. **Flo-Sweet,** of Refined Syrups & Sugars, Inc., is liquid sugar. **Sucrodex,** of the same company, is liquid sugar containing one-third dextrose and two-thirds sucrose, with a solubility of 72% compared with only 45% for dextrose and 67% for sucrose. **Inverdex,** for canning and for fountain syrups, is about 85% invert sugar and 15 dextrose. **Amberdex,** used for cakes and cookies, is an amber-colored 50–50 mixture of sucrose and dextrose with the edible minerals left in. **Caramel,** used for flavoring and coloring foodstuffs and liquors, has a deep-brown color and a characteristic taste. It is burnt sugar marketed as a liquid or powder.

The **papelon** of South America is solidified edible molasses. **Gur** is unrefined brown sugar of India, and the **pilancillo** of Mexico is unrefined brown sugar. **Treacle** is an English name for edible molasses. The refuse from sugarcane, called bagasse, is used as fuel and for making paper and insulating board. **Beet pulp,** after extraction of the juice, is marketed as cattle feed. Despite restrictive controls over the world supply of sugar, much sucrose is not being used in the production of chemical products.

Nonionic detergents, which are odorless, biodegradable powders with low toxicity, are made by reacting sucrose with fatty acid esters of volatile alcohols. Allyl sucrose is used as a shellac substitute. **Sucrose acetate butyrate,** of Eastman Chemical Products, Inc., is a clear viscous liquid boiling at 550°F (288°C), used as a plasticizer in synthetic resins to improve extrusion and to give flexibility and adhesiveness in coatings. As much as 70% is used in nitrocellulose to give tough, flexible melt coatings. **Nitto ester,** of the Dai Nippon Sugar Mfg. Co., is **sucrose ester** made with sugar and stearic acid. It is used as a food additive.

A type of edible sugar syrup is also obtained from the juice of a variety of **sorghum grass,** *Sorghum vulgare,* native to South Africa, but now grown in the southern United States. The juice or syrup, called **sorghum syrup,** or **sorgo syrup,** is light in color, has a characteristic delicate flavor, and contains gums and starch, which prevent crystallization. It also contains other sugars besides sucrose, and considerable mineral salts of value as foods. The total sugar in the juice is from 9 to 17%, varying with the age of the plant. It is used in some sections to replace sugar, and is employed in some confectionery to give a distinctive flavor.

Apple syrup, or **apple honey,** used as a sweetening agent in the food industry, for curing hams, and as a substitute moistening agent for tobacco, is made from cull apples. The reduced syrup is treated to remove the bitter calcium malate. It contains 75% solids of which 65% consists of the sugars levulose, dextrose, and sucrose. **Palm sugar,** or **jaggary,** is the evaporated sap of several varieties of palm, including the coconut and the palms from which kittool, gomuti, and palmyra fibers are obtained. The sap contains about 14% sugar. It is much used in India and the Pacific Islands. The **palm wine** known as **arrack** is made by fermenting the juice, called **taewak,** of the flower stems of the **aren palm** of Java. A liter of taewak yields about 0.09 kg of brown palm sugar. **Wood molasses** is made by concentrating and neutralizing the dilute sugar solution produced by pressure hydrolysis of wood chips using dilute sulfuric acid at high temperature. The molasses has a slightly bitter taste, but is used for stock feed and for industrial purposes. **Wood sugars** contain **xylose,** $CHO(HCOH)_3CH_2OH$, which belongs to the great group of **pentosans** occurring in plant life. They have the same general formula with different numbers of the (HCOH) group. Oxidation converts them to the respective acid, as **xylonic acid** from xylose, or arabinic acid from the arabinose of gum arabic. They can also be converted to the lactones, and are related to the furanes, so that the wood sugars have a wide utility for the production of chemicals.

Other plants yield **sweetening agents,** but few are of commercial importance. The leaves of the **caá heé,** a small plant of Paraguay, are used locally for sweetening Paraguayan tea. The name, pronounced kah-áh aye-áye, means sweet herb, and it has a more intense sweetening effect than sugar. **Miracle Fruit powder,** of the International Minerals and Chemical Corp., is a complex protein-based chemical derived from the fruit pulp of the **Agbayun** shrub, *Synsepalum dulcificum,* of West Africa. It has a strong sweetening effect and a pleasant natural flavor. The 6-carbon sugar derivative known as **glucoronic lactone,** used as an antiarthritic drug, is derived from dextrose. **Amino sugar,** or **glucosamine,** has an NH_2 group in the molecule in place of the alpha hydroxyl group of glucose. This sugar occurs in marine animals.

Synthetic sweetening agents of no food value are used in diabetic foods and in dietetic foods for the treatment of obesity. Many of these synthetic sweeteners are toxic in excess and are cumulative in the human system. Thus, **dietary foods** that depend on the substitution of chemicals in place of sugar should be taken only with caution and under medical direction. **Saccharin,** produced from coal tar, is **benzoic sulfinide,** $C_6H_4SO_2NHCO$. It is 450 times sweeter than sugar, and has no food value, but it has a disagreeable aftertaste. It is a water-insoluble white powder, but its salts, **sodium saccharin** and **calcium saccharin,** are soluble in water, and are 300 times sweeter than sugar. Saccharin is also used as a pH indicator, and as a brightener in nickel-plating baths.

The **cyclamates** were widely used in beverages and diet foods, but are now recognized as toxic drugs and are restricted. **Sodium cyclohexylsulfamate,** or **sodium cyclamate,** $Na(C_6H_{12}NO_3S)_2 \cdot 2H_20$, is used in dietetic foods and in some soft drinks as it has no food value. It is 30 times sweeter than sugar, but at the 25% sweetening level of sugar it has an undesirable aftertaste, and at the sugar-sweetness level the off-taste predominates. For both sugar-free and salt-free diets, the calcium salt, **calcium cyclamate,** is used. **Sucaryl,** of the Abbott Laboratories, is sodium cyclamate, and **Cyclan,** of Du Pont, is calcium cycalmate. **Hexamic acid,** a white crystalline powder which is cycle hexysulfamic acid, is used as a supplement sweetener and intensifier with the cyclamates and saccharin. **Aspartame,** also known as **Nutrasweet,** is a low-calorie sweetener used alone or in combination with saccharin in some breakfast cereals, diet soft drinks, and other ready-mixed beverages. **Peryllartine** is the sweetest known substance, being 2,000 times sweeter than sucrose. It is a complex aldehyde derived from **terpenylic acid,** which occurs in combined forms in turpentine and many essential oils.

SUGAR PINE. The common name of the wood of the *Pinus lambertiana,* a coniferous tree growing in California and Oregon. The tree grows ordinarily to a height from 150 to 175 ft (46 to 53 m) with a diameter of 4 to 5 ft (1.2 to 1.5 m). Occasional trees are more than 200 ft (61 m) in height and 12 ft (3.7 m) in diameter, and are often free of limbs up to 75 ft (23

m) from the ground. It is the largest of the pines. Sugar pine is durable, has moderate strength and fairly even grain, and is not subject to excessive shrinkage or warping. Because of the latter quality it has come into use to replace the scarcer eastern pines for patterns. It does not darken on exposure as western pine does. It is widely employed for construction work and for **factory lumber** for doors, frames, boxes, and wooden articles. Sugar pine is classified into three standard classes of grades according to freedom from knots and faults as select, commons, and factory, or shop. The selects are designated as No. 1 and 2 clear, C select, and D select. The commons are graded as Nos. 1, 2, 3, and 4; the factory as No. 3 clear, No. 1 shop, No. 2 shop, and No. 3 shop. The shops are judged with the idea that they will be cut up into small pieces, and are consequently classified by the area of clear cuttings that can be obtained.

SULFAMIC ACID. A white crystalline odorless solid of the composition $HSO_3 \cdot NH_2$, very soluble in water, but only slightly soluble in alcohol. The melting point is 178°C. The acid is stronger than other solid acids, approaching the strength of hydrochloric. It is used in bating and tanning leather, giving a silky, tight grain in the leather. An important use is for cleaning boiler and heat-exchanger tubes. It converts the calcium carbonate scale to the water-soluble calcium sulfamate, which can then be flushed off and combined with sodium chloride; it also converts the rust to ferric chloride and then to the water-soluble iron sulfamate. **Ammonium sulfamate** is the ammonia salt of the acid, used as a cleanser and anodizer of metals, as a weed killer, and for flameproofing paper and textiles. **Lead ammonium sulfamate,** $Pb(SO_3NH_2)_2$, used in lead plating, is very soluble in water and has high throwing power. **Aminoethylsulfamic acid,** $NH_2CH_2CH_2OSO_3H$, is used for treating paper and textile fibers to increase wet strength and water repellency. **Tobias acid,** used in making azo dyes, is **naphthylamine sulfonic acid,** $NH_2C_{10}H_6SO_3H$, in white needles decomposing at 230°C.

SULFONATED OIL. A fatty oil that has been treated with sulfuric acid, the excess acid being washed out, and only the chemically combined acid remaining. The oil is then neutralized with an alkali. Sulfonated oils are water-soluble and are used in cutting oils and in fat liquors for leather finishing. **Sulfonated castor oil** is called **Turkey red oil. Leatherlubric** is the trade name of E. F. Houghton & Co. for sulfonated sperm oil used for leather. **Solcod** is the sulfonated cod oil of the same company. **Sulfonated stearin** and **sulfonated tallow** are also used in leather dressing. They are cream-colored pastes readily soluble in hot water. **Mahogany soap** is a name for oil-soluble petroleum sulfonates used as dispersing and wetting agents, corrosion inhibitors, emulsifiers, and to increase the oil absorption of mineral pigments in paints. **Petronate,** of L. Sonneborn Sons, Inc., is a petroleum sulfonate containing 62% sulfonates, 35 mineral oil, and 3 water. **Phosphorated oils,** or their sulfonates, may be used instead of the sulfonates as emulsifying agents or in treating textiles and leathers. They are more stable to alkalies. **Phosoils,** of the Beacon Co., are phosphorated vegetable oils. **Aquasol,** of the American Cyanamid Co., is a sulfonated castor oil used as an emulsifying agent. **Cream softener** is a name used in the textile industry for sulfonated tallow.

SULFUR. One of the most industrially useful of the elements. Its occurrence in nature is little more than 1% that of aluminum, but it is easy to extract and is relatively plentiful. In economics, it belongs to the group of "S" materials—salt, sulfur, steel, sugars, starches—whose consumption is a measure of the industrialization and the rate of industrial growth of a nation. Sulfur is obtained from volcanic deposits in Sicily, Mexico, Chile, and Argentina, and along the Gulf Coast in Louisiana, Texas, and Mexico it is obtained from great underground deposits in the cap rock above salt domes. Offshore deposits worked in the Gulf of Mexico are 2,000 ft (610 m) under the bottom. It is also obtained by the distillation of iron pyrites, as a by-product of copper and other metal smelting, and from natural gas. The National Sulphur Co. produces sulfur from gypsum. The **sterri** exported from Sicily for making sulfuric acid is broken rock rich in sulfur. **Brimstone** is a very ancient name still in popular use for solid sulfur, but the District Court of Texas has ruled that sulfur obtained from gas is not subject to tax as brimstone.

Sulfur forms a crystalline mass of a pale-yellow color, with a hardness of 1.5 to 2.5 Mohs, a specific gravity of 2.05 to 2.09, and melting point at 232°F (111°C). It forms a ruby vapor at about 780°F (416°C). When melted and cast, it forms **amorphous sulfur** with a specific gravity of 1.955. The tensile strength is 160 lb/in^2 (1 MPa), and compressive strength is 3,300 lb/in^2 (22 MPa). Since ancient times it has been used as a lute for setting metals into stone. Sulfur also condenses into light flakes known as **flowers of sulfur,** and the **hydrogen sulfide gas,** H_2S, separated from sour natural gas, yields a sulfur powder. **Flotation sulfur** is a fine free-flowing sulfur dust with particle sizes less than 4 μm, recovered in gas production from coal. Commercial crude Sicilian sulfur contains from 2 to 11% of impurities and is sold in three grades. Refined sulfur is marketed in crystals, roll, or various grades of powder, and the Sicilian superior grade is 99.5% pure. This is the grade used in rubber manufacture. The sulfur of the Texas Gulf Sulphur Co. is 99.5% pure and is free of arsenic, selenium, and tellurium. **Crystex,** of the Stauffer Chemical Co., is a sulfur, 85% insoluble in carbon bisulfide, used in rubber compounding. The **sulfur powder** of Electronic Space Products, Inc., used for semiconductors, is 99.9999% pure.

Sulfur has twice the atomic weight of oxygen but has many similar properties and has great affinity for most metals. It has 6 valence electrons, but also has valences of 2 and 4. The **crystalline sulfur** is orthorhombic, which converts to monoclinic crystals if cooled slowly from 120°C. This form remains stable below 120°C. When molten sulfur is cooled suddenly it forms the amorphous sulfur which has a ring molecular structure and is plastic, but converts gradually to the rhombic form. Sulfur has a wide variety of uses in all industries. It is used for making gunpowder and for vulcanizing rubber, but for most uses it is employed in compounds, especially as sulfuric acid or sulfur dioxide. A vast number of so-called thio compounds have been produced. The **thio alcohols,** or **mercaptans,** have an SH group instead of the OH of true alcohols, and they do not react like alcohols, but the **thio esters** are made directly from the mercaptans. **Thionyl chloride,** $SOCl_2$, a yellow liquid, is a typical compound used as a source of sulfur in synthesis. Most of the thio compounds have an offensive odor. **Vegetable sulfur** does not contain sulfur, but is **lycopodium,** a fine yellow powder from the spores of the **club moss,** a fernlike plant, *Lycopodium clavatum,* which grows in North America and Europe. It belongs to the group of **lipochromes,** or coloring matter of plants related to lycopene and carotene.

Sulfur dioxide, or **sulfurous acid anhydride,** is a colorless gas of the composition SO_2, used as a refrigerant, as a preservative, in bleaching, and for making other chemicals. It liquefies at about −10°C. As a refrigerant it has a condensing pressure of 51.7 lb (23.5 kg) at 86°F (30°C). The gas is toxic and has a pungent, suffocating odor, so that leaks are detected easily. It is corrosive to organic materials but does not attack copper or brass. The gas is soluble in water, forming **sulfurous acid,** H_2SO_3, a colorless liquid with suffocating fumes. The acid form is the usual method of use of the gas for bleaching.

SULFURIC ACID. An oily, highly corrosive liquid of the composition H_2SO_4, having a specific gravity of 1.834, and boiling at 338°C. It is miscible in water in all proportions, and the color is yellowish to brown according to the purity. It may be made by burning sulfur to the dioxide, oxidizing to the trioxide, and reacting with steam to form the acid. It is a strong acid, oxidizing organic materials and most metals. Sulfuric acid is used for pickling and cleaning metals, in electric batteries and plating baths, for making explosives and fertilizers, and for many other purposes. In the metal industries it is called **dipping acid,** and in the automotive trade it is called **battery acid. Fuming sulfuric acid,** of 100% purity, was called **Nordhauser acid.** The grade of sulfuric acid known as **oil of vitriol,** or **vitriol,** is 66°Be, or 93.2% acid. **Sulfur trioxide,** or **sulfuric anhydride,** SO_3, is the acid minus water. It is a colorless liquid boiling at 46°C, and forms sulfuric acid when mixed with water. It is used for sulfonation. **Sulfan,** of the Allied Chemical Corp., is sulfuric anhydride. **Chlorosulfonic acid,** $HClSO_3$, has equal amounts of sulfur trioxide and hydrochloric acid, and is a vigorous dehydrating agent, used also in chlorosulfonating organic compounds. It has a specific gravity of 1.752, and boils at 311°F (155°C). Mixed with sulfur trioxide it has been called **FS smoke** for military smoke screens.

Niter cake, which is **sodium acid sulfate,** $NaHSO_4$, or **sodium bisulfate,** contains 30 to 35% available sulfuric acid and is used in hot solutions for pickling and cleaning metals. It comes in colorless crystals or white lumps, with a specific gravity of 2.435 and melting point 300°C. **Sodium sulfate,** or **Glauber's salt,** is a white crystalline material of the composition $Na_2SO_4 \cdot 10H_2O$, used in making kraft paper, rayon, and glass. It was first produced from Hungarian spring water by Johann Glauber, and when obtained from mineral springs is called **crazy water crystals.** The **burkeite,** sodium sulfate–sodium carbonate double salt, which separates out of Searles Lake brine, is used to produce sodium sulfate and other chemicals as by-products. **Salt cake,** Na_2SO_4, is impure sodium sulfate used in the cooking liquor in making paper pulp from wood. It is also used in freezing mixtures. **Synthetic salt cake,** used for making kraft pulp, is produced by sintering soda ash and sulfur. **Chrome cake** is a greenish by-product salt cake which contains some chromium as an impurity. It is used in papermaking. **Kaiseroda** is a German name for salt cake of high purity obtained as a by-product from the production of magnesium chloride from potash minerals. **Sodium sulfite,** Na_2SO_3 or $Na_2SO_3 \cdot 7H_2O$, is a white to tan crystalline powder very soluble in water but nonhygroscopic. **Santosite,** of Monsanto, is a grade of sodium sulfite containing 93% sodium sulfite with the balance chiefly sodium sulfate.

Sodium sulfide, Na_2S, is a pink flaky solid, used in tanneries for dehairing, and in the manufacture of dyes and pigments. The commercial product contains 60 to 62% Na_2S, 3.5 NaCl, and other salts, and the balance water of crystallization. **Sodium sulfhydrate,** NaSH, is in lemon-yellow flakes. It has much less alkalinity than sodium sulfide, and is used in tanneries in unhairing solutions, and for making thiourea and other chemicals. It contains 62.6% by weight of sulfur, and is an economical material for sulfonating. **Sodium dithionate,** $Na_2S_2O_6 \cdot 2H_2O$, is used in leather tanning, as an assist in textile dyeing and printing, and for making other chemicals. It comes in transparent prismatic crystals of bitter taste. **Sodium thiosulfate,** $Na_2S_2O_3 \cdot 5H_2O$, known as **hypo,** is a white crystalline compound having a density of 1.73 and a melting point of 45°C. It is used in photography to fix films, plates, and papers. **White vitriol** is zinc sulfate in colorless crystals soluble in water and melting at 39°C. It is used for making zinc salts, as a mordant, for zinc plating, and as a preservative in adhesives.

SUMAC. The dried ground leaves of the bush *Rhus coriaria* of Sicily, or *R. typhina* of the eastern United States, used for tanning leather. The Sicilian leaves contain up to 30% tannin, and the American leaves up to 38%. It contains gallotannin and ellagitannin, and gives a rapid tan. Sumac produces a light strong leather of fine soft grain and has a bleaching action which can produce a white leather. It is used for book and hatband leathers. Sumac grows profusely in the eastern states, but the gathering of the leaves is not organized commercially.

SUNFLOWER OIL. A pale-yellow drying oil with a pleasant odor and taste obtained from the large seeds of the common sunflower plant, *Helianthus annuus,* of which there are many varieties. The plant is native to Peru but is now grown in many parts of the world, particularly in California, Canada, Argentina, Chile, Uruguay, and Russia. It requires boron in the soil. The specific gravity of the oil is 0.925. Sunflower oil is used in varnish and soap manufacture or as a food oil. The by-product cake is used chiefly for cattle feed, but **sunflower meal** is also blended with wheat flour or cornmeal in foods. It is higher in vitamin B than soybean flour. **Sunflower seeds** are also used as poultry feed. **Madia-seed oil** is quite similar to sunflower oil and has the same uses. It is obtained from the seeds of the plant *Madia sativa,* native to California. The seeds contain 35% oil, and the cold-pressed oil has a pleasant taste. **Watermelon-seed oil,** produced in Senegal as **bereff oil,** is an edible oil similar to sunflower. It contains about 43% linoleic acid, 27 oleic, 19.5 stearic, and 5 palmitic acid.

The leaves of selected varieties of some species of sunflower contain from 1 to 6% **sunflower rubber** and up to 8 resin. The *H. occidentalis, H. giganteus, H. maximiliani,* and *H. strumosus* are cultivated in Russia both for the oil seed and for the rubber in the leaves. These perennials yield leaves up to 10 years. Another similar rubber-bearing plant of southern Russia is the *Asclepias cornuti,* known as **vatochnik.** It is a perennial, producing leaves for 10 to 15 years. The leaves yield 1 to 6.5% rubber and large percentages of resin.

SUN HEMP. The bast fiber of the plant *Crotalaria juncea.* It is used for cordage and rope in place of jute, but is lighter in color and is more flexible, stronger, and more durable than jute. It resembles true hemp, but is not as strong. It is more properly called **sann hemp** from the Hindu word sann. It is also known as **sunn fiber, Indian hemp,** and **Bombay hemp.** The plant, which is a shrub, is cultivated extensively in India. It grows to a height of about 8 ft (2.4 m), with slender branches yielding the fiber. The method of extraction is the same as for true hemp. The best fibers are retained locally for making into cloth. It is also used in the United States for making cigarette paper, and for oakum. **Madras hemp** is from another species of the same plant.

SUPERALLOYS. Also designated as **high temperature–high strength alloys.** The name applies broadly to iron-base, nickel-base, and cobalt-base alloys that combine exceptional high-temperature mechanical properties with excellent oxidation resistance. The operational temperature range for superalloys is roughly between 1000 and 2000°F (538 and 1093°C). In general, iron-base alloys are the least costly and the most easily worked of the three major groups. They are used at the lower end of the range. While room-temperature strength may exceed that of other superalloys, it falls off rapidly above 1200°F (649°C). Thus most uses are in the 1000 to 1500°F (538 to 816°C) range. While many of the **iron-base superalloys** now have standard designations in the **AISI 600 steel** series, they are still better known by trade names such as **A-286 alloy, Discaloy, N-155 alloy,** and **alloy 16-25-6.** These and the other alloys are **chromium-nickel-iron steels.** They are of two major types. One group consists of non-heat-treated chromium-nickel-iron steels commonly used in the as-worked or in the hot- or cold-worked condition. They are primarily low in carbon with one or more carbide-forming elements such as molybdenum. They have higher creep stress than regular austenitic stainless steels. They also have excellent surface stability provided proper precautions are taken to prevent carbide precipitation. Alloys in the other group are generally heat-treated. Optimum high-temperature properties are obtained by precipitation-hardening treatments. Their high-temperature strengths are somewhat higher than those of the non-heat-treated group. Major use of the iron-base superalloys is in jet engines for such parts as ducts, bolts, exhaust covers, turbine blades, and tail cones.

Most **nickel-base superalloys** contain over 50% nickel, from 10 to 25% chromium, and lesser amounts of other elements, such as cobalt, molybdenum, aluminum, titanium, boron, zirconium, and iron. The **Inconels** are in this group. Also in this group are a number of **nickel-chromium-cobalt alloys** that are rich in cobalt and chromium. Most of them are age-hardened for optimum strength. **Cobalt-base superalloys** contain around 60% cobalt or higher, about 10 to 30% nickel, around 25% chromium, and up to around 15% tungsten. They were developed chiefly for jet engine parts.

Some typical trade-named superalloys follow. **Nimonic alloy,** of the Mond Nickel Co., Ltd., for jet turbine blades, has 75% nickel, 21 chromium, 2.5 titanium, 0.7 iron, 0.6 aluminum, and 0.5 carbon. It has a tensile strength of 17,200 lb/in^2 (119 MPa) at 1500°F (816°C). **Alloy 545,** of Westinghouse Electric Corp., contains 26% nickel, 13.5 chromium, 3 titanium, 1.75 molybdenum, 1.5 manganese, and 0.20 boron, with small amounts of aluminum, copper, and silicon. It was designed for jet-engine turbine disks to resist high centrifugal forces above 1200°F (649°C). This alloy has a tensile strength of 162,000 lb/in^2 (1,116 MPa), and at 1200°F

(649°C) it is 120,000 lb/in^2 (827 MPa) with a Brinell hardness of 300. **Alloy K42B** is a nickel-chromium-cobalt alloy with high strength and creep resistance at 1350°F (732°C). **Discaloy,** for turbine blades, is an austenitic alloy containing 26% nickel, 13.5 chromium, 3.2 molybdenum, 1.6 titanium, 0.8 silicon, 0.7 manganese, 0.03 carbon, and the balance iron. At a temperature of 1200°F (649°C) the tensile strength is 100,000 lb/in^2 (689 MPa). **Refract-alloy 80** has 20% each of nickel, chromium, and cobalt, 10 molybdenum, 5 tungsten, and 15 iron. **Alloy S-816,** for turbine buckets, has 20% nickel, 20 chromium, 43 cobalt, 4 molybdenum, 4 tungsten, 4 columbium, and 5 iron. **Waspaloy** has 55% nickel, 20 chromium, 13 cobalt, 4 molybdenum, 3 iron, and some titanium and aluminum. **Kromarc 55** is an austenitic steel that is easily welded and will retain high strength to 1200°F. It has 20% nickel and 16 chromium, with manganese, molybdenum, and a small amount of silicon. **Kromarc 58,** for high-strength welded structures, has a similar composition but with additions of vanadium, zirconium, and boron. It will retain a tensile strength of 80,000 lb/in^2 (551 MPa) at 1200°F (649°C). **Crucible CG-27,** of the Crucible Steel Co., contains 38% nickel, 13 chromium, 5.75 molybdenum, 2.5 titanium, 1.6 aluminum, 0.7 columbium, 0.01 boron, 0.05 carbon, and the balance iron. The yield strength is 135,000 lb/in^2 (930 MPa) with elongation of 20%, and at 1200°F (649°C) it retains a strength of 80,000 lb/in^2 (551 MPa).

The **Udimet 500** of Special Metals is a vacuum-cast nickel-base alloy containing 18% chromium, 18 cobalt, 4 molybdenum, 3 aluminum, 3 titanium, 2 iron, 0.2 silicon, 0.2 manganese, 0.08 boron, and the balance nickel. The tensile strength is 110,000 lb/in^2 (758 MPa), and at 1700°F (927°C) the strength is 25,000 lb/in^2 (172 MPa). **Alloy RA303** of Rolled Alloys, Inc., is resistant to oxidation at temperatures to 2100°F (1149°C), though the normal tensile strength of 105,000 lb/in^2 (723 MPa) drops to 10,000 lb/in^2 (68 MPa) at 1950°F (1066°C). It contains 45% nickel, 26 chromium, 3 tungsten, 3 cobalt, 3 molybdenum, 2 manganese, 1.25 silicon, 0.50 copper, and 0.08 max carbon. **Haynes alloy 25,** of the Union Carbide Corp., has exceptional rolling and deep-drawing qualities and retains high strength to 1800°F (982°C). It has 50% cobalt with the balance chromium, nickel, and tungsten. Metals & Controls Corp. produces this alloy in thin gage down to a thickness of 0.001 in (0.003 cm), and the cold-rolled sheet has a tensile strength of 320,000 lb/in^2 (2,206 MPa), elongation 2%, and Rockwell hardness C55. **Haynes alloy 56,** for service temperatures to 1900°F (1038°C), has 13% nickel, 21 chromium, 11.5 cobalt, 4.5 molybdenum, 1.5 tungsten, 1.5 manganese, 1 silicon, 0.75 columbium, 0.30 carbon, and 0.10 nitrogen, with the balance iron. The tensile strength is 129,500 lb/in^2 (889 MPa). At a temperature of 1600°F (871°C) the strength is 43,700 lb/in^2 (300 MPa) with elongation of 42%, and at 2000°F (1093°C) it still retains a strength of 15,200 lb/in^2 (105 MPa). **Haynes alloy**

150 is a cobalt-base alloy with no tungsten but 20% each of chromium and iron. It has a melting point of 2600°F (1427°C) and is used for cast and wrought parts. **Hastelloy N,** of the same company, was designed for resistance to molten fluoride salts, and it also has good oxidation resistance to 1800°F (982°C). It contains 16.5% molybdenum, 7 chromium, 5 iron, 0.8 manganese, 0.5 silicon, 0.01 boron, 0.5 aluminum and titanium, 0.06 carbon, and the balance nickel. The tensile strength of the wrought metal is 115,000 lb/ in^2 (792 MPa) with elongation of 50%, and that of the castings is 87,000 lb/in^2 (599 MPa) with elongation of 22%. At 1700°F (927°C) the strength is 34,000 lb/in^2 (234 MPa).

The **Supertherm alloy** of the Electro-Alloys Div., Abex Corp., designed for furnace conveyor castings to withstand temperatures above 1800°F (982°C) for long periods, has 35% nickel and 26 chromium, with cobalt, tungsten, and iron. **Allegheny alloy D-979,** of the Allegheny-Ludlum Steel Corp., for wrought parts and forgings to give good strength and creep resistance at service temperatures to 1600°F (871°C), contains 43.5% nickel, 14.3 chromium, 3.7 tungsten, 3.8 molybdenum, 2.9 titanium, 1 aluminum, 0.46 silicon, 0.25 manganese, 0.01 boron, 0.07 carbon, and the balance iron. Another structural alloy of this company, **Allegheny alloy AF-71,** for service temperatures to 1500°F (816°C), contains none of the scarcer metals nickel and cobalt. It has 12.6% chromium, 18.4 manganese, 3 molybdenum, 0.24 silicon, 0.18 boron, 0.80 vanadium, 0.19 carbon, and the balance iron. **Pandex alloy,** of the Latrobe Steel Co., for jet turbine blades and fuel nozzles at operating temperatures to 1300°F (704°C), contains 26% nickel, 15 chromium, 2 titanium, 1.5 manganese, 1.25 molybdenum, 0.30 vanadium, 0.75 silicon, 0.20 aluminum, and 0.05 carbon. For the temperature range around 1200°F (649°C) for jet-engine and gas turbine parts, the chromium-type, nonnickel steels are modified to give better forging, welding, and machining qualities. **Carpenter steel H-46,** of Carpenter Technology, has 12% chromium, 0.65 manganese, 0.40 silicon, 0.45 nickel, 0.65 molybdenum, 0.30 vanadium, 0.40 columbium, 0.08 nitrogen, and 0.18 carbon. At 1000°F (538°C) it retains a strength of 100,000 lb/in^2 (689 MPa). **Linco steel** and **Hychrom 5616,** of the Latrobe Steel Co., are intended to give higher strength and corrosion resistance at about 1200°F (649°C) than the Type 400 stainless steels. The first has 11.5% chromium, 2.75 molybdenum, 1.1 manganese, 0.35 nickel, 0.35 silicon, 0.25 vanadium, and 0.30 carbon, while the second, for higher strength, has 13% chromium, 3 tungsten, 2 nickel, 0.35 manganese, 0.35 silicon, and 0.18 carbon.

SUPERBRONZE. A name applied to brasses containing both aluminum and manganese. They are ordinarily high brasses modified with 2 to 3% manganese and 1 to 6 aluminum, with sometimes also some iron. They have greatly increased strength and hardness over the original brasses, but the ductility is reduced and they are difficult to work and machine. The early superbronze was known as **Heusler alloy.** Muntz metal is also frequently modified with manganese, iron, and aluminum. The alloys are used where high strength and corrosion resistance are required, and they are often marketed under trade names. **Tensilite,** of the American Manganese Bronze Co., is a bronze of this type. A grade containing about 64% copper, 30 zinc, 2.5 manganese, 3 aluminum, and 1 iron has a tensile strength, cast, up to 100,000 lb/in^2 (689 MPa) and elongation 16%. **Mallory metal** is a superbronze of the P. R. Mallory Co., but **Mallory 100** metal is a high-conductivity, heat-treatable, strong alloy containing 2.6% cobalt, 0.40 beryllium, and the balance copper. It has a tensile strength of 90,000 lb/in^2 (620 MPa) and an elongation of 8 to 15%. The name superbronze is a shop term rather than a technical classification, and thus the name is often applied to any hard, high-strength, heat-treatable copper-base alloy.

SUPERPOLYMERS. Several plastics developed in recent years that maintain mechanical and chemical integrity above 400°F for extended periods. They are **polyimide, polysulfone, polyphenylene sulfide, polyarylsulfone, novaloc epoxy,** and **aromatic polyester.** In addition to high-temperature resistance, they have in common high strength and modulus of elasticity, and excellent resistance to solvents, oils, and corrosive environments. They are also high in cost. Their major disadvantage is processing difficulty. Molding temperatures and pressures are extremely high compared to conventional plastics. Some of them, including polyimide and aromatic polyester, are not molded conventionally. Because they do not melt, the molding process is more of a sintering operation. Their high price largely limits them to specialized uses in the aerospace and nuclear-energy fields. One indication of the high-temperature resistance of the superpolymers is their glass transition temperature of well over 500°F (260°C), as compared to less than 350°F (177°C) for most conventional plastics. In the case of polyimides, the glass transition temperature is greater than 800°F (427°C) and the material decomposes rather than softens when heated excessively. Polysulfone has the highest service temperature of any melt-processable thermoplastic. Its flexural modulus stays above 300,000 lb/in^2 (2,068 MPa) at up to 320°F (160°C). Even at such temperatures it does not discolor or degrade. Aromatic polyester, a homopolymer also known as **polyoxybenzoate,** does not melt, but at 800°F (427°C) can be made to flow in a nonviscous manner similar to metals. Thus, filled and unfilled forms and parts can be made by hot sintering, high-velocity forging, and plasma spraying. Notable properties are high thermal stability, good strength at 600°F (316°C), high thermal conductivity, good wear resistance, and extra-high compressive strength. Aromatic polyesters have also been developed for injection and compression molding. They have long-term thermal stability and a strength of 3,000 lb/in^2 (20 MPa) at 550°F (288°C). At room temperature polyimide is the stiffest of the group with a top modulus of elasticity of 7.5 million lb/in^2 (51,675 MPa), followed by polyphenylene sulfide with a modulus of 4.8 million lb/in^2 (33,072 MPa). Polyarysulfone has the best impact resistance of the superpolymers with an impact strength of 5 ft·lb/in (0.27 kg·m/cm) (notch).

Polyetherimide, with the trade name **Ultem,** is an amorphous thermoplastic that can be processed with conventional thermoplastic processing equipment. Its continuous-use temperature is 340°F (170°C) and its deflection temperature is 400°F (200°C) at 264 lb/in^2 (2 MPa). The polymer also has inherent flame resistance without use of additives. This feature, along with its resistance to food stains and cleaning agents, make it suitable for aircraft panels and seat component parts. Tensile strength ranges from 15,000 to 24,000 lb/in^2 (103 to 165 MPa). Flexural modulus at room temperature is 480,000 lb/ in^2 (3,300 MPa).

Polyimide foam is a spongy, lightweight, flame-resistant material that resists ignition up to 800°F (427°C) and then only chars and decomposes. Some formulations result in harder materials that can be used as lightweight wallboard or floor panels while retaining fire resistance.

SYCAMORE. The wood of the tree *Acer pseudo-platanus,* which is also classed as a kind of maple, especially in England. The species cut as sycamore in the United States is largely *Platanus occidentalis.* The wood has a close, firm, tough texture and is yellowish in color, with a reddish-brown heartwood. The light-colored sapwood is up to 3 in (8 cm) thick in commercial trees. When quartered, the wood resembles quartered oak. The weight is about 38 lb/ft^3 (609 kg/m^3). The surface is lustrous and takes a fine polish. It is used for veneers, flooring, furniture, cooperage, and for handles and rollers. Two other species grown in the Southwest are **California sycamore,** *P. racemosa,* and **Arizona sycamore,** *P. wrightii.*

TALC. A soft friable mineral of fine colloidal particles with a soapy feel. It is a **hydrated magnesium silicate,** $4SiO_2 \cdot 3MgO \cdot H_2O$, with a specific gravity of 2.8, and a hardness of 1 Mohs. It is white when pure, but may be colored gray, green, brown, or red with impurities. The pure white talc of Italy has been valued since ancient times for cosmetics. Talc is now used for cosmetics, for paper coatings, as a filler for paints and plastics, and for molding into electrical insulators, heater parts, and chemical ware. The massive block material, called **steatite talc,** is cut into electrical insulators. It is also called **lava talc. Talc dust** is a superfine, 400-mesh powder from the milling of steatite talc. It has an oily feel and is used in cosmetics. The more impure block talcs are used for firebox linings and will withstand temperatures to 1700°F (927°C). Gritty varieties contain carbonate minerals and are in the class of soapstones. Varieties containing lime are used for making porcelain.

Talc of specified purity and particle size is marketed under trade names. **Asbestine,** of C. K. Williams & Co., is a talc powder of 325 mesh for use as a filler. **Ceramitalc,** of the International Talc Co., Inc., is a talc powder used as a source of magnesia and to prevent crazing in ceramics. The **Sierra Fibrene** of Innis, Speiden & Co. is a California talc milled to 400 mesh. It is white, has a platy structure, and as an extender in paints it wets easily and promotes pigment dispersion. **French chalk** is a high-grade talc in massive block form used for marking. The mineral occurs in the United States in the Appalachian region from Vermont to Georgia. The **Georgia talc** for making **crayons** is mined in blocks. **Attasorb** and **Permagel,** of the Minerals & Chemicals Corp., is finely powdered, cream-colored, hydrated **magnesium-aluminum silicate** from the mineral **attapulgite,** used for emulsifying and as a flatting agent and extender in paints. The material is also used in starch adhesives to improve shear strength. **Attacote** is the material in superfine particle size for use as an anticaking agent for hygroscopic chemicals. **Veegum F,** of the R. T. Vanderbilt Co., is a fine white colloidal magnesium-aluminum silicate used as a suspending agent for oils and waxes.

Cordierite is a talclike mineral with a high percentage of magnesia used for refractory electronic parts. It is found sparsely in Norway, Finland, and Connecticut, usually in granite and gneiss, or in vitrified sandstones. When heated to 1440°C it is converted to sillimanite and glass. Synthetic cordierite is made by the Muscle Shoals Electrochemical Corp. by mixing pure silica, magnesia, and alumina in various porportions and stabilizing with calcia. It is marketed as a powder for producing refractory insulating parts.

Agalite is a mineral having the same composition as talc but with a less soapy feel. It is used as a filler in writing papers, but is more wearing on the paper rolls than talc. The talc of northern New York, known as **rensselaerite,** does not have the usual talc slip, and has a fibrous nature. The hydrous aluminum silicate **pyrophyllite,** found in California, is similar to talc but with the magnesium replaced by aluminum. In mixtures with talc for wall tile it eliminates crazing. It is also substituted for talc as a filler for paints and paper. **Thix,** of the National Lead Co., used as a thickening agent in emulsion paints, in cosmetics, and in textile finishes, is a refined, hydrous magnesium silicate marketed as a 200-mesh powder. It contains 56% silica, 26 magnesia, 2.8 calcia, 2.5 NaO_2, and 1.1 lithia.

Magnesium silicate, used as a filler in rubber and plastics, and also as an alkaline bleaching agent for oils, waxes, and solvents, is a white, water-insoluble powder of the composition $MgSiO_4$, having a pH of 7.5 to 8.5.

In the cosmetic trade it is known as **talcum powder. Magnesol,** of the Westvaco Chemical Co., is finely ground magnesium silicate. **Brite-Sorb 30,** of Allegheny Industrial Chemical Co., is a synthetic magnesium silicate with high adsorption and filtering power. The **magnesium trisilicate** used in pharmaceuticals as an antacid is of extreme fineness, the superbulking grade having 65% of the particles less than 5 μm in size. The material known as **killas** from the tin mines of Cornwall is a slaty schist. It is finely ground and used like talc.

TALL OIL. An oily resinous liquid obtained as a by-product of the sulfite paper-pulp mills. The alkali saponifies the acids, and the resulting soap is skimmed off and treated with sulfuric acid to produce tall oil. The name comes from the Swedish **talloel,** meaning pine oil. The crude oil is brown, but the refined oil is reddish yellow and nearly odorless. It has a specific gravity of 0.98, flash point of 360°F (182°C), and acid number about 165. The oil from Florida paper mills contains 41 to 45% rosin, 10 to 15 pitch, and the balance chiefly fatty acids. The fatty acids can be obtained separately by fractionating the crude whole oil. The oil also contains up to 10% of the phytosterol **sitosterol,** *used in making the drug cortisone.*

Tall oil is used in scouring soaps, asphalt emulsions, cutting oils, insecticides, animal dips, in making factice, and in plastics and paint oils. It is marketed in processed and concentrated form. The **Flextal** of the Farac Oil & Chemical Co. is processed tall oil containing 60% rosin acids. Detergents are made by reacting tall oil with ethylene oxide. Saturated alcohols are produced by high-pressure hydrogenation of tall oil. The high linoleic acid content makes tall oil suitable for making drying oils. **Lumitol** is a German vinyl plastic produced by reacting tall oil with acetylene. It is used for coatings. **Smithco RT,** of the Archer-Daniels-Midland Co., used for varnishes and paints, is refined tall oil esterified with glycerin. **Smithco PE** is tall oil esterified with pentaerythritol. **Ardex PE,** of the same company, is a varnish oil that dries quickly to a hard film, made by esterifying tall oil with pentaerythritol. **Sulfonated tall oil** is used to replace sulfonated castor oil in coating mixes for paper to increase folding strength. **Opoil** is a crude tall oil of the National Southern Products Co., and **Facoil** is the refined oil with 60% fatty acid content and low rosin acid content. **Acolin, Acosix,** and **Aconon** are grades of refined tall oil of Newport Industries, Inc. **Pamac,** of Hercules, consists of tall oil monobasic fatty acids, used in resin coatings.

TALLOW. A general name for the heavy fats obtained from all parts of the bodies of sheep and cattle. The best grades of internal fats, or **suet,** are used for edible purpose, but the external fats are employed for lubricants, for mixing with waxes and vegetable fats, for soaps and candles, and for producing chemicals. The tallows have the same general composition as lard, but are higher in the harder saturated acids, with about 51% of palmitic and stearic acids, and lower in oleic acid. The edible grades known as **premier jus,** prime, and edible, are white to pale yellow, almost tasteless, and free from disagreeable odor, but the nonedible or industrial tallows are yellow to brown unless bleached. The best grade of industrial tallow is Packers No. 1. **White grease, yellow grease,** and **brown grease** may be hog fat or they may be tallows with a titer below 40°C, the titer being the only commercial distinction between tallow and fat. Tallow is thus all animal fat above 40°C titer. **Beef tallow** is used to produce stearic acid, for leather dressing, lubricating greases, and for making soap. **Mutton tallow** contains less liquid fat and is harder, but it becomes rancid more easily. Tallow for industrial use is generally highly purified and chemically treated, and marketed under trade names. **Adogen 442,** of the Archer-Daniels-Midland Co., used as a softener for textiles, is a dimethyl **hydrogenated tallow.** It comes as a nearly white, odorless paste in isopropanol and water, and is dispersible in water or in organic solvents.

TANNING AGENTS. Materials, known as **tannins,** used for the treatment of skins and hides to preserve the hide substance and make it resistant to decay. The tanned leather is then treated with fats or greases to make it soft and pliable. Tannins may be natural or artificial. The natural tannins are chiefly vegetable, but some mineral tanning agents are used. The vegetable tannins are divided into two color classes: the **catechol** and the **pyrogallol.** The catechol tannins are cutch, quebracho, hemlock, larch, gambier, oak, and willow. The pyrogallol tannins are gallnuts, sumac, myrobalans, chestnut, valonia, divi-divi, and algarobilla. Catechol tannin is distinguished by giving a greenish-black precipitate with ferric salts; the pyrogallol tannins give a bluish-black precipitate. The catechol tannins, in general, produce leathers that are more resistant to heat and decay than the pyrogallols. Some tannins contain considerable coloring or dye matter, but the color that a tannin imparts to leather may be lightened or darkened by raising or lowering the acidity of the tannin bath. In the ink industry the catechol tannins are known as iron-greening, and the pyrogallol tannins as iron-bluing, and the latter are used for making writing inks. Catechol is also produced synthetically from coal tar. It is a water-soluble **dihydric phenol** in white crystalline granules known as ortho-**dihydroxybenzene,** $C_6H_4(OH)_2$. It is used in some inks, and for making dyestuffs, medicinals, and antioxidants.

Alum tanning is an ancient process but was introduced into Europe only about the year 1100, and the alum- and salt-tanned leather was called **Hungary leather. Formaldehyde** is also used as a tanning agent. Formaldehyde was patented as a tanning agent in 1898. A later patent covered a rapid process of tanning sheepskins with alcohol and formalin and then neutralizing in a solution of soda ash. Unlike vegetable agents, formaldehyde does not add weight to the skin. It is often used as a pretanning agent to lessen the astringency of the vegetable tannin and increase its rate of diffusion. **Melamine** resins are used for tanning to give a leather that is white throughout and does not yellow with age. Leather may also be tanned with chromic acid or chrome salts, which make the fibers insoluble and produce a soft, strong leather. Chrome alum, sodium or potassium dichromates, or products in which chromic acid has been used as an oxidizing agent may be used. **Chrome tanning** is rapid and is used chiefly for light leathers. **Tanolin** is a name for basic **chromium chloride** marketed in crystal form for use in the chrome tanning of leather. **Santotan KR** is a trade name of Monsanto for basic **chromium sulfate,** $Cr_2(SO_4)_2(OH)_2$, used as a one-bath chrome-tanning agent. This material is also used for treating magnesium-alloy parts to give a gray to black surface color. **Panchrome,** an English tanning agent, is a sulfur dioxide dichromate. **Chromalin** is a glycerin-reduced dichromate. Chrome-tanned leather is more resistant to heat than vegetable-tanned leathers, withstanding temperatures to 200°F (93°C). Chrome tanning is used for shoe-upper leathers and for gloves, beltings, and packings. **Iron-tanned leather** is produced by pretanning with formaldehyde, then tanning with ferric salts and trisodium phosphate, and neutralizing with a solution of phthalic anhydride and sodium carbonate. The leather is soft, will absorb much oil and grease, and is suitable for use where a pliable leather is desired. Glutar aldehyde gives a soft bulky leather suitable for garments. It may be blended with chrome or vegetable tanning agents.

In tanning processes various supplementary materials may be used to give special properties to the leathers. Glucose or starch may be used to make the leather more plump. Hydrochloric acid is used in two-bath chrome tanning to enhance the feel and appearance of the leather. Synthetic tannins, or **syntans,** are largely condensation products made by condensing sulfonated phenols with formaldehyde. **Neradol D** is such a syntan. **Tansyn** is the trade name of an English syntan of this kind. **Permanol,** of the Monsanto Company, is a sulfonic acid condensation syntan in liquid form used to produce light-fast white leathers. The free sulfuric acid is completely neutralized. Syntans do not add weight to leather and are seldom used alone. They are marketed under trade names. **Leukanol,** of the Rohm & Haas Co., has a bleaching action and is used in combination with vegetable tannins to speed up the tanning and to give a light-colored leather. **Orotan,** of the same company, is a sulfonated phenol formaldehyde which makes a good shoe leather when used alone. **Tanigan,** a German tannin, is a complex condensation product produced from water pulp–mill liquor and formaldehyde or diphenyl methane.

TANTALUM. A white, lustrous metal resembling platinum. It is one of the most acid-resistant of the metals and is classed as a noble metal. The specific gravity is 16.6, or about twice that of steel, and the melting point is very high, 2850°C. It is very ductile and can be rolled down from 0.300 to 0.0015 in (0.762 to 0.004 cm) without annealing, or drawn into extremely fine wire. Because of its high melting point the metal is not melted, but the powder obtained by chemical extraction or electrolysis is pressed into billets, sintered in a vacuum, and then rolled. The metal is used especially for chemical equipment. The tensile strength of the sheet metal is 50,000 lb/in^2 (344 MPa), and of the drawn wire 130,000 lb/in^2 (896 MPa). The annealed metal has a hardness of 75 Brinell. It is resistant to all acids except hydrofluoric, and is not dissolved by aqua regia. It will dissolve, however, in a mixture of nitric and hydrofluoric acids, and also reacts with chlorine above 175°C. When heated in the air to about 400°C, it becomes blue; at a higher temperature, it becomes black. At very high temperatures it absorbs oxygen, hydrogen, or nitrogen, and becomes very brittle. It will absorb 740 times its own volume of hydrogen, producing a coarse, brittle substance. Tantalum can be tempered or hardened to about 600 Brinell by heating in the air to absorb gases, and will hold a fine cutting edge on tools. It can also be hardened by the addition of silicon to a hardness close to that of the diamond, but any alloying is difficult because of the high melting point.

Tantalum is used as a filament in electric light bulbs. The metal becomes incandescent at 1700°C, or 400° below that of tungsten, so that a tantalum lamp is cooler. It is also used in radio tubes operating at high temperatures, and in vacuum tubes to absorb gases. It is lower in cost than platinum, and is used for surgical instruments and gauze, pens, instruments, and acid-resistant chemical equipment. It is also used in alloy special steels to give increased resistance to scaling at high temperatures, but since it has a large neutron cross section it must be kept very low in steels used for atomic reactors. **Tantalum-tungsten alloys,** of Stauffer Chemical Co., used for rocket motor parts, have melting points to 6150°F (3399°C), and retain tensile strengths to 15,000 lb/in^2 (103 MPa) at 4500°F (2482°C). **Tantalum alloy T-111,** used in the fine tubing by Superior Tube Co., has 8% tungsten and 2% hafnium. The tubes, in diameters down to 0.012 in (0.030 cm), have high corrosion resistance and a tensile strength of 90,000 lb/in^2 (620 MPa). **Tantalum alloy T-222,** of Westinghouse Electric Corp., contains 9.6% tungsten, 2.4 hafnium, and 0.01 carbon. It has a tensile strength of 110,000 lb/in^2 (758 MPa), and at 3500°F (1927°C) will retain a strength of 14,000 lb/in^2 (96 MPa). It comes in plate or rod. Tantalum coils are used to heat acid baths. A tube with a wall thickness of 0.020 in (0.051 cm) will withstand operating steam pressure up to 150 lb/in^2 (1 MPa). The metal has the property of passing an alternating electric current in one direction only, and is thus used for current rectifiers. As an anode, tantalum reacts instantly with oxygen in acid solutions, forming a stable oxide film which is current-blocking; this property is used in rectifiers and electrolytic condensers.

Tantalum carbide is an extremely hard, heavy, brownish, crystalline material of high melting point, 3875°C, used for the same purposes as tungsten carbide, as an abrasive and cutting material. It has the composition TaC_2, and in hardness it ranks close to the diamond. For use in cutting tools the carbide is ground to 325 mesh, mixed with a binder of powdered cobalt, iron, or nickel, molded to shape, and sintered at high heat. **Ramet** is the trade name of the Fansteel Metallurgical Corp. for tantalum carbide cutting materials, and **Tantaloy** is a name for a sintered alloy in bar form for flowing on faces of tools with a welding torch. Tantalum carbide filaments for incandescent lamps are used at temperatures to 6020°F (3323°C) while tungsten filaments have a maximum-use temperature to 5660°F (3126°C). They give a 25% increase in light brilliance, and have longer life. But tantalum carbide decomposes in the nitrogen-argon atmosphere used for tungsten, and a hydrogen-halogen acid atmosphere is used. **Tantalum-hafnium carbide,** $4TaC \cdot HfC$, for refractory parts, has an extremely high melting point, 3942°C.

TANTALUM ORES. The most important ore of the metal tantalum is **tantalite.** When pure, its composition is $FeO \cdot Ta_2O_5$, but the American ore may contain only from 10 to 40% **tantalic oxide,** Ta_2O_5, and the Australian ore may contain as high as 70%. The ore is marketed on the basis of 60% tantalic oxide content. Tantalite occurs usually as a black crystalline mineral with a specific gravity up to 7.3. It often contains manganese, tin, titanium, and sometimes tungsten; the tantalum may be replaced by columbium, which is similar to it. When the columbium content in the ore predominates, the mineral is called **columbite.** Tantalite also contains small amounts of germanium. The tantalite of the Congo usually contains tin. The ore from the Lukushi Basin contains 58% Ta_2O_5, 16.5 Cb_2O_5, 12.5 MnO, 4.5 Fe_2O_3, and 1.6 SnO_2, with some zirconium and titanium oxides. **Thoreaulite** of that region contains 72 to 74% Ta_2O_5 and 20 to 22% SnO_2. Tantalum metal is produced from tantalite by dissolving in acid and separating the tantalum salts from the columbium by precipitation. The tantalum salts are reduced to powdered metal, which is then compressed into rods and sintered and rolled. The tantalite ore of Manitoba is embedded in pegmatite and the crude ore contains about 0.25% Ta_2O_5.

A tantalum ore that is abundant at Wodgina, Western Australia, is **mangano tantalite** which contains about 69% tantalic oxide, 15 **columbium pentoxide,** Cb_2O_5, and 14 manganese protoxide, with a little tin oxide. The specific gravity of the ore is 6.34. **Microlite,** an ore found at Wodgina and in the McPhee Range of Western Australia, contains 76% Ta_2O_5 and

4 to 7 Cb_2O_5. **Tanteuxenite,** another Western Australian ore, contains 24 to 47% Ta_2O_5 and 4 to 14 Cb_2O_5. **Tapiolite,** of Australia, contains 82% Ta_2O_5 and 2 Cb_2O_5. **Euxenite,** of Idaho, contains about 28% columbium-tantalum oxide. The mineral **pyrochlore,** of Canada, is composed of complex oxides of tantalum, columbium, sodium, and calcium, and the metal oxides are obtained by acid extraction.

TAR. A black solid mass obtained in the destructive distillation of coal, peat, wood, petroleum, or other organic material. When coal is heated to redness in an enclosed oven, it yields volatile products and the residue coke. Upon cooling the volatile matter, tar and water are deposited, leaving the coal gases free. Various types of coal yield tars of different qualities and quantities. Anthracite gives little tar, and cannel coal yields large quantities of low-gravity tar. In the manufacture of gas the tar produced from bituminous coal is a viscous black liquid containing 20 to 30% free carbon, and is rich in benzene, toluene, naphthalene, and other aromatic compounds. In the dry state this tar has a specific gravity of about 1.20. Tar is also produced as a by-product from coke ovens.

Coal tars are usually distilled to remove the light aromatics which are used for making chemicals, and the residue tar, known as **treated tar,** or **pitch,** is employed for roofing, road making, and for bituminous paints and waterproofing compounds. **Coal-tar pitch** is the most stable bituminous material for covering underground pipes. **Tarvia** is the trade name of a refined coal tar, marketed by the Barrett Co. in various grades. **Tarmac** is practically the same material marketed by the Koopers Co. **Bituplastic,** of the same company, used for coating pipes and structures, is a refined coal-tar pitch that is odorless and quick-drying. **Bituvia** is a road tar, produced in various grades by the Reilly Tar & Chemical Corp. **Coal-tar carbon** amounts to about 32% of the original tar. It is marketed in lump form for chemical use. The fixed carbon content is 92.5 to 95.6%, sulfur about 0.30, and volatile matter 3 to 6. **Calcined carbon,** from coal tar, contains less than 0.5% sulfur and 0.5 volatile matter.

The lightest distillate of coal tar, benzol, is used as an automotive fuel. **Coal-tar oils** are used as solvents and plasticizers. They consist of various distillates or fractions up to semisolids. **Tar oil** from brown coal tars was used for diesel fuel oil by extracting the phenols with methyl alcohol. **Bardol B,** of Allied Chemical, is a clear yellow coal-tar fraction of specific gravity 1.0 to 1.04, used as a plasticizer for synthetic rubber, while **Carbonex** is a solid black tar hydrocarbon in flake form used as a rubber plasticizer. The softening point is between 205 and 220°F (96 to 104°C). **Xylol** is a water-white liquid of specific gravity 0.860 to 0.870, distilling between 135 and 185°C. It is a mixture of **xylenes,** which are dimethyl benzene, $C_6H_4Me_2$.

Naphthalene and anthracene are among the distillates. **Anthracene** is a colorless crystalline product of the composition C_6H_4:(C_2H_2):C_6H_4 and melting point 217°C used for the production of dyes, resins, plasticizers, tanning agents, and inhibitors. Crystals of anthracene are used for scintillation counters for gamma-ray detection. **Quinoline,** called also **benzazine** and **chinoline,** is a liquid with a tar odor. It has a double-ring molecular structure of the empirical formula C_9H_7N, and boils at 237°C. It is used for making antiseptics, pharmaceuticals, insecticides, and rubber accelerators.

Pine tar is a by-product in the distillation of pinewood. It is a viscous black mass and is much used for roofing. It is also sometimes called pitch, but pitch is the tar with the pine-tar oil removed, known as **pine pitch. Tarene,** of National Rosin Oil Products, Inc., is a dry, free-flowing powder made by absorbing pine tar into a synthetic hydrous calcium silicate which absorbs four times its own weight of liquid tar. It is used for formulating with rubbers. **Navy pitch** and **ship pitch** are names that refer to specification pine pitch for marine use. It is medium hard to solid, of a specific gravity of 1.08 to 1.10, of a melting point not less than 148°F (64°C), completely soluble in benzol, and of uniform black color, or red brown in thin layers. **Wood tar** from the destructive distillation of other woods is a dark-brown viscous liquid used as a preservative, deriving this property from its content of creosote. **Stockholm tar,** a name now out of commercial use, was a term employed in shipbuilding for the tar obtained from the crude distillation of pine stumps and roots.

TEA. The dried leaves of the shrubs *Camellia sinensis* and *Thea sinensis,* grown chiefly in southern Asia, Japan, Sri Lanka, Russian Caucasia, and Indonesia but also in Peru and in Tanzania. The plant requires a warm subtropical humid climate. Tea leaves are valued for making the beverage tea which contains the alkaloid caffeine and is stimulating. The leaves contain more caffeine than coffee berries but the flavor is different. Like coffee, also, it contains tannin, which dissolves out when the tea leaves are steeped too long, and is an astringent. In well-prepared tea the tannins have been oxidized to the brown and red tannin which is not easily soluble and does not enter the properly steeped beverage to any great extent although it gives the beverage color. The flavor and aroma of tea depend largely upon the age of the leaves when picked and the method of drying. **Green tea** is made by drying the fresh leaves in the sun or artificially, while **black tea** is made by first fermenting the leaves and then drying. Rolling is done to break the leaves and release the juices. The **oolong tea** of Taiwan is partly fermented and is intermediate between green and black. **Pouchong tea** is graded by mixing oolong with aromatic flowers such as jasmine. Tea is also graded by the size and age of the leaf. Flowery orange pekoe is the smallest leaf, **orange pekoe tea** the second, then pekoe, pekoe souchong, and souchong. Tea also varies with varieties grown in different climates so that Japan tea, China tea, and Ceylon have different flavors.

Commercial tea is usually a blend of different varieties to give uniformity under one trade name. The blending of tea is considered an art. **Brick tea,** made in China, is produced from coarse leaves and twigs which have been fermented. They are mixed with tea dust, treated with rice water, and pressed into bricks. **Cake tea,** or **puerh tea,** is produced in Yunnan. The leaves are panned, sun-dried, and steamed, and then pressed into circular cakes. **Tablet tea** is selected tea dust pressed into small tablets. **Tea waste** is the final dust from the tea siftings, and is used for the production of caffeine. **Teaseed oil,** or **sasanqua oil,** is from the seeds of another species of the tea plant, *Thea sasanqua,* of Asia. The seeds contain 58% of a pale-yellow oil with a specific gravity of 0.916 used for lubrication, hair oil, soap, and pharmaceutical preparations. **Paraguayan tea,** or **yerba maté,** used in immense quantities as a beverage in Argentina, Paraguay, Brazil, and some other South American countries, consists of the dried smoked leaves of the small evergreen tree *Ilex paraguayensis,* native to Paraguay and southern Brazil. The growing region is about the Upper Paraná River. It was an ancient beverage of the Indians, and cultivation began on a large scale under the early missionaries. It contains a higher percentage of caffeine than tea or coffee, 3.88%, but less tannin. The flavor of the steeped beverage is different from that of tea. **Cassine** is a tealike beverage obtained from the twigs and leaves of two species of holly, *Ilex cassine* and *I. vomitoria,* found in the southern United States from Virginia to Texas. It was called **Yaupon** by the Indians and used medicinally and in religious rites. During the Civil War it was used in the South as a tea substitute. The beverage has an odor similar to tea but has a dark color with a sharp, bitter taste. It contains caffeine, tannin, and essential oils.

TEAK. The wood of the tree *Tectona grandis,* of southern Asia. It resembles oak in appearance, is strong and firm, and in England is called **Indian oak.** It contains an oil that gives it a pleasant odor and makes it immune to the attacks of insects. It is used for boxes, chests, home furnishings, and for woodwork on ships. The color is golden yellow, the grain is coarse and open, and the surface is greasy to the touch. It is one of the most durable of woods, and also has small shrinkage. The weight is 40 lb/ft^3 (641 kg/m^3). In Burma large plantations grow teak for export. Trees grow to a height of 100 ft (30 m) with a diameter of 3 ft (0.9 m). The growth is slow, a 2-ft (0.6-m) tree averaging 150 years of age. The wood marketed as **African teak,** known also as **iroko,** is from the tree *Chlorophora excelsa,* of West Africa, and is unlike true teak. It is a firm, strong wood with a brownish color and a coarse, open grain. The weight is somewhat less than teak, and it is harder to work, but it is resistant to decay and to termite attack, and is used in ship construction. **Surinam teak** is the wood of the tree *Hymenea courbaril* of the Guianas and the West Indies. It is also called **West Indian locust.** The wood is dark brown in color, hard, heavy, and difficult to work. It is not very similar to teak and not as durable. **Seacoast teak,** or **bua bua,** is a hard, yellow, durable wood from species of the tree *Guettarda* of Malaya. **Australian teak,** from New South Wales, is from the tree *Flindersia australis.* It is yellowish red in color, close-grained, and hard with an oily feel resembling teak, but more difficult to work. **In wood,** of Burma, also called **eng teak,** is from the tree *Dipterocarpus tuberculatus,* from which gurjun balsam is obtained. The wood is reddish-brown, and it is not as durable as teak. Two woods of Brazil are used for the same purposes as teak: the **itaúba,** *Silvia itauba,* a tree growing to a height of about 75 ft (23 m) in the upland forests of the lower Amazon, and **itaúba preta,** *Oreodaphne bookeriana,* a larger tree growing over a wider area. The first is a greenish-yellow wood with compact texture and rough fiber, formerly prized for shipbuilding. The second resembles teak more closely, and is used for cabinetwork.

TELLURIUM. An elementary metal, symbol Te, obtained as a steel-gray powder of 99% purity by the reduction of **tellurium oxide,** or **tellurite,** TeO_2, recovered from the residues of lead and copper refineries. It is also marketed in slabs and sticks, and is sometimes known as **sylvanium.** It occurs also with gold in Washington and Colorado as **gold telluride,** $AuTe_2$. The specific gravity is about 6.2 and melting point 450°C. The chief uses are in lead to harden and toughen the metal, and in rubber as an accelerator and toughener. Less than 0.1% tellurium in lead makes the metal more resistant to corrosion and acids, and gives a finer grain structure and higher endurance limit. Tellurium-lead pipe, with less than 0.1% tellurium, has a 75% greater resistance to hydraulic pressure than plain lead. A **tellurium lead,** patented in England, contains 0.05% tellurium and 6 antimony. **Tellurium copper** is a free-machining copper containing about 1.0% tellurium. It machines 25% more easily than free-cutting brass. The tensile strength, annealed, is 30,000 lb/in^2 (206 MPa), and the electric conductivity is 98% that of copper. A **tellurium bronze** containing 1% tellurium and 1.5 tin has a tensile strength, annealed, of 40,000 lb/in^2 (275 MPa), and is free-machining. Tellurium is used in small amounts in some steels to make them free-machining without making the steel hot-short as do increased amounts of sulfur. But telurium is objectionable for this purpose because inhalation of dust or fumes by workers causes garlic breath for days after exposure, although the material is not toxic. As a secondary vulcanizing agent with sulfur in rubber, tellurium in very small proportions, 0.5 to 1%, increases the tensile strength and aging qualities of the rubber. It is not as strong an accelerator as selenium, but gives greater heat resistance to the rubber. **Telloy** is the trade name of the R. T. Vanderbilt Co. for **tellurium powder** ground very fine for rubber compounding.

TERNEPLATE. Bessemer or open-hearth steel plate having on each side a thin coating of an alloy of 20% tin and 80 lead, although other proportions may be used. Terne is an old name meaning dull and refers to the color as compared with bright tinplate. Terneplate is made by the dip process and is used for roofing, construction work, and to replace the more expensive tinplate for uses not in contact with foodstuffs. The coating is measured by pounds per double base box containing approximately 436 ft^2 (41 m^2) or 112 sheets, 20 by 28 in (0.5 by 0.7 m). **Long ternes** are those with coatings of 8, 12, and 15 lb (4, 5, and 7 kg), not heavier than No. 14 gage or lighter than No. 30 gage. **Short ternes** are those with coatings of 8 lb (4 kg) or lighter, or in very heavy coatings from 15 to 40 lb (7 to 18 kg), for roofing. The name long terne is also used to designate flat sheets of larger size, up to 48 by 120 in (1.2 by 3.0 m) used for manufacturing purposes. The usual roofing material is 40 lb (18 kg), and the coating is 25% tin and 75 lead. Industrial terneplate usually comes in base boxes of 112 sheets, 14 by 20 in (0.4 by 0.5 m), furnished as standard, deep-drawing, and extra-deep-drawing. When copper steel is specified, it has at least 0.18% copper.

Lead-coated steel is now much used instead of terneplate for building and for stamped and formed parts. Lead alone does not adhere well to steel, but the lead dip used contains small amounts of other elements. **Weiralead,** of the Weirton Steel Co., is a lead-coated steel in gages 16 to 28. The lead coating contains small amounts of tin, silver, and antimony, and it will not peel when bent or deep-drawn. **Amaloy,** of the American Machine & Foundry Co., is a lead alloy with 1% tin, used for hot-dipping iron and steel parts. **Plate-Loy,** of the American Smelting & Refining Co., is a hot-dip plate, and the coating contains 2 to 4% tin, 1 antimony, 1.5 to 2 zinc, and the balance lead. It adheres well to the steel, and the plate can be seamed and soldered.

TERRA COTTA. A general English term applied to fired, unglazed, yellow, and red clay wares; in the United States it refers particularly to the red and brown square and hexagonal tiles made from common brick clay, always containing iron. Some special terra cottas are nearly white, while for special architectural work other shades are obtained. The clays are washed, and only very fine sands are mixed with them in order to secure a fine open texture and smooth surface. Terra cotta is used for roofing and for tile floors, for hollow building blocks, and in decorative construction work. Good, well-burned terra cotta is less than 1½ in (3.8 cm) thick. Terra cotta is very light, 120 lb/ft^3 (1,922 kg/m^3), and will withstand fire and frost.

TETRACHLORETHANE. A colorless liquid of the chemical formula $CHCl_2 \cdot CHCl_2$ employed as a solvent for organic compounds such as oils, resins, and tarry substances. It is an excellent solvent for sulfur, phosphorus, iodine, and various other elements. It is used as a paint remover and bleacher, as an insecticide, and in the production of other chlorine compounds. It is also called **acetylene tetrachloride,** and is made by the combination of chlorine with acetylene. Tetrachlorethane boils at 144°C, freezes at −36°C, is nonflammable, and has a specific gravity of 1.601. It is narcotic and toxic, and the breathing of the vapors is injurious. Mixed with dilute alkalies, it forms explosive compounds. In the presence of moisture it is very corrosive to metals. Mixed with zinc dust and sawdust, it is employed as a smoke screen.

THALLIUM. A soft bluish-white metal resembling lead but not as malleable. The specific gravity is 11.85, and melting point 578°F (302°C). At about 600°F (316°C) it ignites and burns with a green light. The electric conductivity is low. It tarnishes in the air, forming an oxide coating. It is attacked by nitric acid and by sulfuric acid. The metal has a tensile strength of 1,300 lb/in^2 (9 MPa) and a Brinell hardness of 2. **Thallium-mercury alloy,** with 8.5% thallium, is liquid with a lower freezing point than mercury alone, −60°C, and is used in low-temperature switches. **Thallium-lead alloys** are corrosion-resistant, and are used for plates on some chemical-equipment parts.

The metal occurs in copper pyrites and zinc ores, and the chief source is the flue dust of smelters from sphalerite ores. Four rare minerals are ores of thallium: **vrbaite,** $Tl_2S \cdot 3(AsSb)_2S_3$, is found in Macedonia; **lorandite,** $Tl_2S \cdot 2As_2S_3$, is found in Macedonia and Wyoming; **hutchinsonite,** $PbS \cdot (TlAg)_2S \cdot 2As_2S_3$, occurs in Switzerland and Sweden; and **crooksite,** $(CuTlAg)_2Se$, is found in Sweden. The salts of thallium are highly poisonous, the sulfide being used as a rat poison. **Thallium oxysulfide** is used in light-sensitive cells. It is also sensitive to infrared rays, and is used for dark signaling. **Thallium sulfate,** $Tl_2(SO_4)_3$, is a crystalline powder used as an insecticide. It is more toxic than lead compounds. Thallium also gives high refraction to optical glass. Thallium bromide iodide crystals, grown synthetically, are used for infrared spectrometers.

The so-called **alkali-halide crystals** used in the **discriminator circuits** of **scintillators** for gamma spectrometry contain thallium. They separate the slow-decaying pulses of protons produced as fast neutrons from the electron pulses produced by gamma absorption. A French crystal, called **Scintibloc,** is **sodium iodide thallide,** NaI(Tl). The **cesium iodide thallide** crystal, CsI(Tl), gives a very blue light under electron excitation.

THERMOPLASTIC ELASTOMERS. A group of polymeric materials having some characteristics of both plastics and elastomers. Also called **elastoplastics.** Requiring no vulcanization or curing, they can be processed on standard plastics processing equipment. They are lightweight, resilient materials that perform well over a wide temperature range. There are a half-dozen different types of elastoplastics. The **olefinics,** or **TPOs,** are produced in durometer hardnesses from 54A to 96A. Specialty flame-retardant and semiconductive grades are also available. The TPOs are used in autos for paintable body filler panels and air deflectors, and as sound-deadening materials in diesel-powered vehicles. Producers of olefinics include Uniroyal (**TPR**), DuPont (**Somel**), Exxon Chemical (**TPV** and **Vistaflex**), Hercules (**Pro-fax**), and B. F. Goodrich Chemical (**Telcar**). The **styrenics** are block copolymers, composed of polystyrene segments in a matrix of polybutadiene or polyisoprene. Lowest in cost of the elastoplastics, they are available in crumb grades and molding grades, and are produced in durometer hardnesses from 35A to 95A. Manufactured by Shell Chemical (**Kraton**) and Philips Petroleum (**Solprene**), they are used for shoe soles, sealants, tubing, and sheeting.

Thermoplastic urethanes, high-priced specialty materials, are of three types: **polyester-urethane, polyether-urethane,** and **caproester-urethane.** All three are linear polymeric materials, and therefore do not have the heat resistance and compression set of the cross-linked urethanes. They are produced chiefly in three durometer hardness grades— 55A, 80A, and 90A. The soft 80A grade is used where high flexibility is required, and the hard grade, 70D, is used for low-deflection load-bearing applications. Producers of these elastoplastics are B. F. Goodrich Chemical (**Estane**), American Cyanamid (**Cyanaprene**), Mobay Chemical (**Texin**), Uniroyal (**Pellethane**), K. J. Quinn (**Q-thene**), Upjohn, and Hooker Chemical.

Copolyether-ester thermoplastic elastomers of DuPont are high-performance, high-cost materials, trade-named **Hytrel.** The five basic grades by durometer hardness are 40D, 47D, 55D, 63D, and 72D. Their applications include tubing and hose, V belts, couplings, oil-field parts, and jacketing for wire and cable. Their chief characteristic is toughness and impact resistance over a broad temperature range. A thermoplastic elastomer of polyvinyl chloride and nitrile, produced by Uniroyal and known as **TPR 3700,** provides excellent oil resistance. **Trans-Pip** is a **trans-1, 4 polyisoprene** thermoplastic elastomer of Polystar Co., developed as a replacement for refined balata rubber in golf-ball covers.

THERMOSTAT METALS. The metals used for indicating very high temperatures, called also **thermocouple metals** and **thermoelectric metals,** consist of two different metals or alloys joined at one end. Application of heat to the coupled end will set up an electric current in the circuit. The voltage generated is very small, usually less than 50 mV, but the electromotive force is proportional to the heat at the junction, and when connected by wires to a sensitive galvanometer will indicate the temperature on a graduated dial. Accuracy within a range of 2°F (− 17°C) can be obtained for high-temperature readings. Copper-constantan or nickel-constantan may be used for temperatures to about 1650°F (899°C). A thermocouple wire of the Englehard Industries, Inc., for temperature indication from 1200 to 2800°C, consists of tungsten versus **tungsten-rhenium alloy** of 74% tungsten and 26 rhenium. The millivolt output at 2800°C is 43.25; at 2000°C it is 34.13; and at 1200°C it drops to 18.25. A thermocouple of this company that gives a higher electromotive force at lower temperatures, 48.02 mV at 1200°C, is called **Platinel,** and is used in turbojet engines for measuring temperatures from 350 to 1260°C. It consists of a gold-palladium-platinum alloy wire versus a gold-palladium alloy wire. Besides the capability of generating an electromotive force by the difference between the two metals, the metals must be capable of withstanding the high temperatures without deterioration and have a stability within an accepted drift in voltage indication of not more than 0.75%.

Various intermetallic crystals are used as thermoelectrics for transforming electric current into heat energy or, in reverse, as **heat pumps,** for refrigeration. The efficiency is proportional to the temperature difference induced across the crystal, and different crystals have their maximum efficiencies at definite temperature operating limits. Polycrystalline **bismuth telluride,** Bi_2Te_3, has a wide temperature difference, 1115°F (601°C). Some materials with low efficiency at low temperatures have high power rating at high temperatures, and others operate efficiently only at lower temperatures. Thus, the materials may be set up in parallel for heat pumps. However, the thermoelectric metals for thermostat use need to generate only slight electromotive force.

The **thermometals** used as temperature controls in electrical appliances, for temperatures from about −40 to 1000°F (−40 to 538°C), are bimetals consisting of two metals or alloys with different rates of thermal expansion welded together so that a change in temperature bends or deflects the bimetal. In heat-control or indicating devices the deflection is measured to indicate the temperature, or the deflection is utilized for mechanical or electrical action. A wide variety of metals is used for thermometals. The requirements are corrosion resistance, heat resistance, and uniform pull proportional to the temperature change. Thermometals welded at the contact surfaces are sold under trade names. **Highflex,** of the H. A. Wilson Co., is an all-steel bimetal with a temperature range of maximum sensitivity from 50 to 300°F (10 to 149°C). **Saflex,** of the same company, is relatively inactive up to 400°F (204°C), but has high deflection between 500 and 800°F (260 and 427°C). **Muflex,** used where high permeability is required, has the high expanding side of pure iron and the low expanding side of Invar.

THORIUM. A soft, ductile, silvery-white metal occurring in nature to about the same extent as lead but so widely disseminated in minute quantities difficult to extract that it is considered as a rare metal. It was once valued for use in incandescent gas mantles in the form of **thorium nitrate,** $Th(NO_3)_4$, but is now used chiefly for nuclear and electronic applications. **Thorium powder** is produced by calcium reduction of thorium oxide. The impure powder burns in the air with great brilliance. Pure thorium metal in sheet form has a specific gravity of 11.7, a melting point of 3090°F (1699°C), and a tensile strength of about 35,000 lb/in^2 (241 MPa). Even small amounts of impurities affect the physical properties greatly, and cold working increases the strength. The metal is dissolved by aqua regia or by hydrochloric acid.

Natural thorium consists largely of the alpha-emitting isotope **thorium 232,** and is a powerful emitter of alpha rays. Thorium produces fissile material, uranium 233, only when triggered by another fission material. Under neutron bombardment it forms **protactinium** which is nonfissile but decays slowly into fissile uranium 233. But, in rapid burning, the buildup of protactinium may be converted into the nonfissile uranium 234.

Thorium 230 is found in minerals that contain uranium and radium and was originally considered as a separate metal under the name of **ionium.** It is radioactive, emitting alpha rays. It has a half-life of 76,000 years, slowly converting to radium. The original production was from the fractionation of uranium ores. It was used as an additive in spark-plug wire, but is too expensive for this purpose.

The chief **thorium ore** is the mineral **monazite,** occurring as sand or in granular masses, usually as sea sand. It is the chief source of thorium oxide and of the rare-earth metals. Most of the monazite comes from Brazil, India, and the East Indies. The monazite sands of Brazil contain 8% **thorium oxide,** or **thoria,** ThO_2. The ore of India may have as high as 10%, but is marketed on the basis of 8% oxide and 60 rare-earth metals. Thoria has a high melting point, 3050°C, but its use as a refractory ceramic is limited because of its high cost and radioactivity. Monazite contains about 3.5 g of **mesothorium** per 1,000 tons (907 metric tons) and usually has 30 to 35% of the oxides of lanthanum, yttrium, neodymium, and praseodymium, and a small amount of europium. Mesothorium was originally considered a separate element, but is an isotope of thorium, with an atomic weight of 228 and a half-life of 6.7 years. The radiations from mesothorium are the same as from radium, alpha, beta, and gamma rays. As it decomposes, it forms **radiothorium,** which is identical in chemical properties with thorium but emits a powerful alpha radiation. It is used in luminous paints and is a safer activator for this purpose than radium, but is scarcer and more expensive, and has a shorter life.

The type of monazite called **uranothorite,** from the Bancroft area of Canada, contains from 0.04 to 0.27% thorium oxide. The thorium is recovered from the waste liquors of the uranium-treatment plant. The rare mineral **thorite,** found in Norway, is a **thorium silicate,** $ThSiO_2$. It occurs in crystals or massive, orange to black in color, and has a resinous luster, and a specific gravity of about 5. **Thoria-urania ceramics** are used for reactor-fuel elements. They are reinforced with columbium or zirconium fibers to increase thermal conductivity and shock resistance. **Thorium-tungsten alloys** have been used for very-high-voltage electronic filaments. The **incandescent mantle,** invented by Welsbach in 1893, and widely used during the period of gas lighting, consisted of a mixture of 98 to 99% thorium nitrate and 1 to 2 cerium oxide. The nitrate is converted to thorium oxide on ignition, with an increase of 10 times its original volume, and glows in the gas flame with an intense white light.

THUYA. The wood of the tree *Thuya plicata,* also known as **western red cedar, giant arbor vitae, shinglewood,** and **Pacific red cedar.** The tree grows in cool, humid coast regions from Alaska to northern California, and the wood is widely used for shingles, poles, and tanks. It is lightweight, soft, and weak, with a straight coarse grain, but is durable. The sapwood is white and the heartwood reddish. The tree grows to great size, reaching to 200 ft (61 m) in height and 16 ft (5 m) in diameter at the age of 1,000 years. **Northern white cedar** is the wood of the tree **T. occidentalis,** of the northeastern United States. It is also called **white cedar, arbor vitae, swamp cedar,** or simply **cedar.** The wood is soft, knotty, brittle, and weak, but very durable. It is used for shingles, poles, posts, and lumber for small boats. The sapwood is white and the heartwood light brown. The trees have a diameter of 1 to 3 ft (0.3 to 0.9 m), and a height of 25 to 75 ft (8 to 23 m). **Thuya leaf oil,** used as a fixative in perfumery, is a colorless oil with a bornyl acetate odor, distilled from the leaves.

TIN. A silvery-white lustrous metal with a bluish tinge. It is soft and malleable, and can be rolled into foil as thin as 0.0002 in (0.0051 cm). Tin melts at 232°C. Its specific gravity is 7.298, close to that of steel. Its tensile strength is 4,000 lb/in^2 (27 MPa). Its hardness is slightly greater than that of lead, and its electric conductivity is about one-seventh that of silver. It is resistant to atmospheric corrosion, but is dissolved in mineral acids. The cast metal has a crystalline structure, and the surface shows dendritic crystals when cast in a steel mold. **Tin pest** is the breaking up of the metal into a gray powder which occurs below 19°C, and the metal is not used for applications at very low temperatures.

Tin is used in brasses, bronzes, and babbitts, and in soft solders. Tin with 0.4% copper is used as foil and for collapsible tubes. One of the most important uses is for the making of tinplate, and as an electroplating material. **Electroplated tin** has a fine white color, gives a durable protective finish, and also has a lubricating effect as a bearing surface. **Standard tin** of the London Metal Exchange must contain over 99.75% tin. The common grade is known as **Grade A tin. Straits tin** is 99.895% pure. Federal specifications for **pig tin** are 99.80% min. **Block tin** is virgin tin cast in stone molds. Even small traces of impurities have an influence on the physical properties of tin. Lead softens the metal; arsenic and zinc harden it. An addition of 0.3% nickel doubles the tensile strength; 2% copper increases the strength 150%. Pure tin melts sharply, but small amounts of impurities broaden the melting point. **Tin powder,** used for making sintered alloys, is 99.8% pure, in powder from 100 to 300 mesh. The **tin crystals** used in the chemical industry are **tin chloride,** or **stannous chloride,** $SnCl_2 \cdot 2H_2O$, coming as large colorless crystals or white water-soluble flakes, melting at 246°C. They are also used for immersion tinning of metals, and for sensitizing glass and plastics before metallizing. The chief source of tin is the mineral cassiterite, but Nigerian columbite may contain up to 6% tin oxide. The principal tin-producing countries are Indonesia, Malaya, Bolivia, China, and Nigeria, but tin mines have been worked in Cornwall since ancient times, and tin is also found in Canada and in irregular quantities in some other areas.

Tin oxide, or **stannous oxide,** is a fine black crystalline powder of the composition SnO, made by oxidizing tin powder. It is used as an opacifier in ceramic enamels, as a ceramic color, as an abrasive, and as a coating for conductive glass. As a color in ceramics it is light-stable and acid-resistant. With magnesium and cobalt oxides it gives a sky-blue color called **cerulean blue.** It is also used with copper oxide to produce ruby glass.

Stannic oxide, SnO_2, is a white powder used in ceramic glazes as an opacifier and for color. As little as 1 to 2% gives fluidity and high luster to glass. With chromates and lime it gives pinks and maroons in enamels, and with vanadium compounds it gives yellows. With gold chloride it gives brilliant-red jewelry enamels. **Protectatin** is the name of the Tin Research Institute for a thin, invisible film of oxide on tinplate to protect against sulfur staining and to give a base for paint. It is produced by dipping the tinplate in a solution of trisodium phosphate, sodium dichromate, and sodium hydroxide. **Potassium stannate,** $K_2SnO_3 \cdot 3H_2O$, or **sodium stan-**

nate, $Na_2SnO_3 \cdot 3H_2O$, may be used for immersion tinning of aluminum. Both come as white water-soluble crystals. The term **organotin** usually refers to butyl compounds of the metal used as catalysts, or heat and light stabilizers in vinyl polymers. **Stan-Guard 100,** of Chas. Pfizer & Co., is a liquid **butyl-tin** compound containing sulfur and used as a stabilizer in rigid PVC sheet. A **butyl-tin maleate** powder is effective as a light stabilizer. **Hollicide LT-125,** of R. M. Hollingshead Corp., is a water-soluble organo**trialkyl-tin** used as a bacteriocide in paper and textile processing. It can be used over a wide pH range.

TINPLATE. Soft-steel plate containing a thin coating of pure tin on both sides. A large proportion of the tinplate used goes into the manufacture of food containers because of its resistance to the action of vegetable acids and its nonpoisonous character. It solders easily, and also is easier to work in dies than terneplate, so that it also is preferred over terneplate for making toys and other cheap articles in spite of a higher cost. Commercial tinplate comes in boxes of 112 sheets, 14 by 20 in (0.36 by 0.51 m) and is designated by the net weight per box when below 100 lb (45 kg). Heavy tinplate above 100 lb (45 kg) goes by number, as steel does, or by letter symbols. The weight of tin may be as high as 1.7% of the total weight of the sheet. **Coke plates** carry as little tin as is necessary to protect and brighten the plate for temporary use. The tin of the coat forms compounds of $FeSn_2$, Fe_2Sn, and FeSn with the iron of the plate, and on a coke plate this compound is 0.00006 to 0.00015 in (0.00015 to 0.00038 cm) thick. Best cokes carry more tin than do the standard cokes. **Charcoal plates** have heavier coats of tin designated by the letter A. The AAAAAA, or 6A, has the heaviest coating. Tinplate is made by the hot-dip process using palm oil as a flux, or by a continuous electroplating process. A base box contains 31,360 in^2 (20 m^2) of tinplate, and standard-dip tinplate has 1½ lb (0.7 kg) of tin per base box, while electrolytic plate has only ¼ lb (0.1 kg) of tin per base box and much **electrolytic tinplate** for container use has only 0.10 lb (0.05 kg) of tin per base box. Electrotinning gives intimately adherent coatings of any desired thickness, and the plate may have a serviceable coat as thin as 0.00003 in (0.00008 cm), or about one-third that of the thinnest possible dipped plate. A slight cold rolling of electrolytic tinplate gives a bright and smooth finish.

Taggers was originally a name for tinplate that is undersized, or below the gage of the plate in the package, but the name **taggers tin** is also applied to light-gage plate. These sizes are No. 38 gage, 55 lb (25 kg); No. 37, 60 lb (27 kg); and No. 36, 65 lb (29 kg). **Ductilite,** of the Wheeling Steel Corp., is a tinplate that is not made by hot rolling in packs, but is cold-rolled from single hot-rolled strip steel. It is of uniform gage and does not have the thin edges of pack-rolled plate. It also has a uniform grain structure. **Weirite,** of the Weirton Steel Co., is cold-reduced coke tinplate. **Black plate,** used for cans in place of tinplate where the tin protection is not necessary, is not black, but is any sheet steel other than tinplate or terneplate in tinplate sizes. It may be chemically treated to resist rust or corrosion.

TITANATES. Compounds made by heating a mixture of an oxide or carbonate of a metal and titanium dioxide. High dielectric constants, high refractive indices, and ferroelectric properties contribute primarily to their commercial importance. **Barium titanate** crystals, $BaTiO_3$, are made by die-pressing titanium dioxide and barium carbonate and sintering at high temperature. This crystal belongs to the class of **perovskite** in which the closely packed lattice of barium and oxygen ions has a barium ion in each corner and an oxygen ion in the center of each face of a cube with the titanium ion in the center of the oxygen octahedron. For piezoelectric use the crystals are subjected to a high current, and they give a quick response to changes in pressure or electric current. They also store electric charges, and are used for capacitors. **Glennite 103,** of the Gulton Mfg. Co., is a **piezoelectric ceramic** molded from barium titanate modified with temperature stabilizers. **Bismuth stannate,** $Bi_2(SnO_3) \cdot 5H_2O$, a crystalline powder that dehydrates at about 140°C, may be used with barium titanate in capacitors to increase stability at high temperatures. **Ceramelex,** of the Erie Resistor Corp., is molded polycrystalline barium titanate. **Lead zirconate–lead titanate** is a piezoelectric ceramic that can be used at higher temperatures than barium titanate. **Lead titanate,** $PbTiO_3$, is used as a less costly substitute for titanium oxide. It is yellowish in color and has only 60% of the hiding power, but is very durable and protects steel from rust. **Butyl titanate,** of Henley & Co., Inc., is a yellow viscous liquid used in anticorrosion varnishes and for flameproofing fabrics. It is a condensation product of the tetrabutyl ester of ortho-titanic acid, and contains about 36% titanium dioxide. **Calcium titanate,** $CaTiO_3$, occurs in nature as the mineral perovskite. As a ceramic it has a room-temperature dielectric constant of about 160. It is frequently used as an addition to barium titanate or by itself as a temperature-compensating capacitor. **Magnesium titanate,** $MgTiO_3$, crystallizes as an ilmenite rather than a perovskite structure. It is not ferroelectric, and is used with titanium dioxide to form temperature-compensating capacitors. It has also been used as an addition agent to barium titanate. **Strontium titanate,** $SrTiO_3$, has a cubic perovskite structure at room temperature. It has a dielectric constant of about 230 as a ceramic, and is commonly used as an additive to barium titanate to decrease the Curie temperature. By itself it is used as temperature-compensating material because of its negative temperature characteristics. The **strontium titanate** of the Engle Mfg. Co., used as a brilliant diamondlike gemstone, is a strontium mesotrititanate. Stones are made up to 4 carats. The refractive index is 2.412. It has a cubic crystal similar to the diamond but the crystal is opaque in the X-ray spectrum. **Titanate fibers** can be used as reinforcement in thermoplastic moldings. The fibers, called **Fybex,** produced by LNP Corp., can also be used in plated plastics to reduce thermal expansion, warpage, and shrinkage. Titanate fibers in plastics also provide opacity.

TITANIUM. A metallic element, symbol Ti, occurring in a great variety of minerals. It was first discovered as an element in 1791 in a black magnetic sand at Manachin, Cornwall, England, and called **menachite,** from the name of the sand, **menachinite.** Its chief commercial ores are rutile and ilmenite. In rutile it occurs as an oxide. It is an abundant element, but is difficult to reduce from the oxide. High-purity titanium metal has a yield strength of 35,000 lb/in^2 (241 MPa) with elongation of 55%. It can be produced only in protected atmospheres. Commercially pure titanium contains up to 0.3% oxygen, 0.1 nitrogen, and up to 0.2% each of carbon and iron. It is 99% min pure. The minimum yield strength is 70,000 lb/in^2 (482 MPa), with minimum elongation of 15%. The tensile strength may be up to 105,000 lb/in^2 (723 MPa). The specific gravity is 4.54, and melting point 1660°C. It is paramagnetic and has low electric conductivity.

The commercial metal is produced from **sponge titanium,** which is made by converting the oxide to titanium tetrachloride and then reducing with molten magnesium. The metal can also be produced in dendritic crystals of 99.6% purity by electrolytic deposition from titanium carbide. Despite its high melting point, titanium reacts readily in copper and in other metals, and is much used for alloying and for deoxidizing. It is a more powerful deoxidizer of steel than silicon or manganese. An early German deoxidizing alloy known as **Badin metal** contained about 9% aluminum, 19 silicon, 5 titanium, and the balance iron. **Titanium copper,** used for deoxidizing nonferrous metals, is made by adding titanium to molten copper, and the congealed alloy is broken into lumps.

One of the chief uses of the metal has been in the form of titanium oxide as a white pigment. It is also valued as titanium carbide for hard facings and for cutting tools. Small percentages of titanium are added to steels and alloys to increase hardness and strength by the formation of carbides or oxides or, when nickel is present, by the formation of **nickel titanide.** The first **titanium alloys** in the United States were produced in 1945 by the Bureau of Mines.

Titanium is one of the few allotropic metals (steel is another); that is, it can exist in two different crystallographic forms. At room temperature, it has a close-packed hexagonal structure, designated as the alpha phase. At around 1625°F (884°C), the alpha phase transforms to a body-centered cubic structure, known as the beta phase, which is stable up to titanium's melting point of about 3050°F (1677°C). Alloying elements promote formation of one or the other of the two phases. Aluminum, for example, stabilizes the alpha phase; that is, it raises the alpha to the beta transformation temperature. Other alpha stabilizers are carbon, oxygen, and nitrogen. Beta stabilizers such as copper, chromium, iron, molybdenum, and vanadium lower the transformation temperature, therefore allowing the beta phase to remain stable at lower temperatures, and even at room temperature. Titanium's mechanical properties are closely related to these allotropic phases. For example, the beta phase is much stronger, but more brittle than the alpha phase. Titanium alloys therefore can be usefully classified into three groups on the basis of allotropic phases: the alpha, beta, and alpha-beta alloys.

Titanium and its alloys have attractive engineering properties. They are about 40% lighter than steel and 60% heavier than aluminum. The combination of moderate weight and high strengths, up to 200,000 lb/in^2 (1,378 MPa), gives titanium alloys the highest strength-to-weight ratio of any structural metal—roughly 30% greater than aluminum or steel. Furthermore, this exceptional strength-to-weight ratio is maintained from −420°F (−216°C) up to 1000°F (538°C). A second outstanding property of titanium materials is corrosion resistance. The presence of a thin, tough oxide surface film provides excellent resistance to atmospheric and sea environments as well as a wide range of chemicals, including chlorine and organics containing chlorides. Being near the cathodic end of the galvanic series, titanium performs the function of a noble metal. Other notable properties are a higher melting point than iron, low thermal conductivity, low coefficient of expansion, and high electrical resistivity.

Fabrication is relatively difficult because of titanium's susceptibility to hydrogen, oxygen, and nitrogen impurities, which cause embrittlement. Therefore elevated-temperature processing, including welding, must be performed under special conditions that avoid diffusion of gases into the metal. Thanks to extensive research and development supported by the federal government for a number of years after World War II, effective processing methods have been developed that minimize the fabricating problems.

Commercially pure titanium and many of the titanium alloys are now available in most common wrought mill forms, such as plate, sheet, tubing, wire, extrusion, and forging. Castings can also be produced in titanium and some of the alloys for surgical implants, marine hardware, and chemical equipment such as compressors and valve bodies. Two casting processes—investment casting and graphite-mold (rammed graphite) casting—are used. Because of titanium's highly reactive nature in the presence of such gases as oxygen, the casting must be done in a vacuum furnace.

Although this limits the size of parts, titanium castings up to at least 60 in (1.5 m) in diameter have been produced.

There are about a half-dozen grades of **commercially pure titanium,** which have titanium contents from 98.9 to 99.5%. Because the small amounts of impurities significantly affect mechanical properties, they can be considered as alloys of the alpha type. Although not nearly as strong as the more highly alloyed types, commercially pure titaniums have a broad range of strengths—from about 40,000 lb/in^2 (275 MPa) to nearly 100,000 lb/in^2 (689 MPa). A special corrosion-resistant **titanium-palladium alloy,** with 0.15 to 0.20% palladium, has improved resistance to mildly reducing media such as dilute hydrochloric and sulfuric acids. The **titanium alpha alloys** contain such alloying elements as aluminum, tin, columbium, zirconium, vanadium, and molybdenum in amounts varying from about 1 to 10%. They are nonheatable, having good stability up to 1000°F (538°C) and as low as −420°F (−251°C). They have a good combination of weldability, strength, and toughness. The **titanium alloy 5 Al–2.5 Sn,** perhaps the most widely used alpha alloy, has been employed in numerous space and aircraft applications. It has a strength at room temperature of 120,000 lb/in^2 (827 MPa), acceptable ductility, and is useful at temperatures up to 800 and 1000°F (427 and 538°C). In addition, it has good oxidation resistance, and good weldability and formability. Under the alpha classification are five **near-alpha titanium alloys.** Of these **Ti–8 Al–1 Mo–1 V alloy** is most common. Special annealing cycles improve creep strength and fracture toughness while maintaining good strength levels.

The **titanium alpha-beta alloys** are the largest and most widely used titanium alloys. Because these alloys are a two-phase combination of alpha and beta alloys, their behavior falls in a range between the two single-phase alloys. They are heat-treatable, useful up to 800°F (427°C), and more formable than alpha alloys, but less tough and more difficult to weld. The most popular alloy in this group is **titanium alloy 6 Al–4 V.** Its volume of use equals that of all other titanium materials combined. It can be heat-treated up to 170,000 lb/in^2 (1,172 MPa), has good impact and fatigue strength, and unlike other alpha-beta alloys, is weldable. The 6% aluminum, 6% vanadium, and 2% tin alloy is heat-treatable to 190,000 lb/in^2 (1,309 MPa). The **titanium beta alloys** have exceptionally high strengths—over 200,000 lb/in^2 (1,378 MPa). However, their lack of toughness and low fatigue strength limits their use. They retain an unusually high percentage of strength up to 600°F (316°C), but cannot be used at much higher temperatures, and they become brittle at temperatures below −100°F (−73°C).

The **TiTech alloys,** of Titanium Technology Corp., are various grades of titanium casting alloys produced by controlled methods to replace forgings.

The first attempt at chemical synthesis of titanium alloys was made at the Naval Ordnance Laboratory, but the project was limited to the production of binar alloys by melt-furnace methods. The **Ti-Ni alloys** produced are intermetals, or chemical combinations of titanium and nickel. The molecule is reversible, and the alloys can be hardened or annealed by heat treatment like steel. They are suitable for cutting tools or for parts requiring high hardness and strength.

TITANIUM CARBIDE. A hard crystalline powder of the composition TiC made by reacting titanium dioxide and carbon black at temperatures above 1800°C. It is compacted with cobalt or nickel for use in cutting tools and for heat-resistant parts. It is lighter in weight and less costly than tungsten carbide, but in cutting tools it is more brittle. When combined with tungsten carbide in sintered carbide tool materials, however, it reduces the tendency to cratering in the tool. A general-purpose cutting tool of this type contains about 82% tungsten carbide, 8 titanium carbide, and 10% cobalt binder. **Kentanium,** of Kennametal, Inc., is titanium carbide in various grades with up to 40% of either cobalt or nickel as the binder, used for high-temperature, erosion-resistant parts. For highest oxidation resistance only about 5% cobalt binder is used. **Kentanium 138,** with 20% cobalt, is used for parts where higher strength and shock resistance are needed, and where temperatures are below about 1800°F (982°C). This material has a tensile strength of 45,000 lb/in^2 (310 MPa), compressive strength of 550,000 lb/in^2 (3,789 MPa) and Rockwell hardness A90. **Kentanium 151A,** for resistance to molten glass or aluminum, has a binder of 20% nickel. **Titanium-carbide alloy,** of the Ford Motor Co., for tool bits, has 80% titanium carbide dispersed in a binder of 10% nickel and 10% molybdenum. The material has a hardness of Rockwell A93, and a dense, fine-grained structure. **Ferro-Tic,** of the Chromalloy Corp., has the titanium carbide bonded with stainless steel. It has a hardness of Rockwell C55. **Machinable carbide** is titanium carbide in a matrix of Ferro-Tic C tool steel. **Titanium carbide tubing** is produced in round or rectangular form 0.10 to 3 in (0.25 to 7.6 cm) in diameter, by TEEG Research, Inc. It is made by vapor deposition of the carbide without a binder. The tubing has a hardness above 2,000 Knoop and a melting point of 5880°F (3249°C).

Grown single crystals of titanium carbide of the Linde Co. have the composition $TiC_{0.94}$, with 19% carbon. The melting point is 5882°F (3250°C), density 4.93, and Vickers hardness 3230.

TITANIUM ORES. The most common titanium ores are ilmenite and rutile. Ilmenite is an iron-black mineral having a specific gravity of about 4.5, and containing about 52% **titanic oxide,** or titania, TiO_2. The ore of India is sold on the basis of titanium dioxide content, and the high-grade ore averages about 60% TiO_2, 22.5 iron, and 0.4 silica. **Ilmenite** is a **ferrotitanate,** $FeO \cdot TiO_2$, but much of the material called ilmenite is **arizonite,** $Fe_2O_3 \cdot 3TiO_2$. Titanium ores are widely distributed and plentiful. Ilmenite is found in northern New York, Florida, North Carolina, and in Arkansas, but the most extensive, accessible resources are found in Canada. The Quebec ilmenite contains 30% iron. The concentrated ore has about 36% TiO_2, and 41% iron, and is smelted to produce pig iron and a slag containing 70% TiO_2 which is used to produce titanium oxide. The beach sands of Senegal are mixed ores, the ilmenite containing 55 to 58% TiO_2, and the **zirconiferous quartz** containing 70 to 90% zirconia. The beach sands of Brazil are washed to yield a product averaging 71.6% ilmenite, 13 zircon, and 6 monazite. The Indian ilmenite also comes from beach sands. The ore of New York state averages 19% TiO_2.

Rutile is a **titanium dioxide,** TiO_2, containing theoretically 60% titanium. Its usual occurrence is crystalline or compact massive, with a specific gravity of 4.18 and 4.25 and a hardness 6 to 6.5. The color is red to brown, occasionally black. Rutile is found in granite, gneiss, limestone, or dolomite. It is obtained from beach sand of northern Florida, and Espirito Santo, Brazil, and is also produced in Virginia, and in Australia and India. Rutile, and also **brookite** and **Octahedrite,** or **anatase,** are produced in Arkansas and Massachusetts. The best Virginia concentrates are 92.5 to 98% TiO_2, but some are 42% from rock originally showing 18.5% TiO_2 in a body of feldspar. Rutile is marketed in the form of concentrates on the basis of 79 to 98.5% titanium oxide. It is used as an opacifier in ceramic glazes and to produce tan-colored glass. It is also employed for welding rod coatings. On welding rods it aids stabilization of the arc and frees the metal of slag. **Tanarc,** used on welding rods as a replacement for rutile, is made from slag from Canadian titaniferous hematite, and contains 70% TiO_2.

TITANIUM OXIDE. The white **titanium dioxide,** or **titania,** of the composition TiO_2, which is an important paint pigment. The best quality is produced from ilmenite, and is higher in price than many white pigments but has great hiding power and durability. Off-color pigments, with a light buff tone, are made by grinding rutile ore. The pigments have fine physical qualities and may be used wherever the color is not important. Titania is also substituted for zinc oxide and lithopone in the manufacture of white rubber goods, and for paper filler. The specific gravity is about 4. Mixed with blanc fixe it is also marketed under the name of **Titanox. Zopaque,** of the Chemical & Pigment Co., is a pure titanium oxide for rubber compounding. The **Ti-Pure** of Du Pont is commercially pure titanium dioxide for pigment use. **Duolith,** of this company, is titanated lithopone pigment containing 15% titanium dioxide, 25 zinc sulfide, and 60 barium sulfate.

Titania crystals are produced by Linde in the form of pale-yellow, single-crystal boules for making optical prisms and lenses for applications where the high refractive index is needed. The crystals are also used as electric semiconductors, and for gemstones. They have a higher refractive index than the diamond, and the cut stones are more brilliant but are much softer. The hardness is about 925 Knoop, and the melting point is 1825°C. The refractive index of the rutile form is 2.7 and that of the anatase is 2.5, while the synthetic crystals have a refractive index of 2.616 vertically and 2.903 horizontally.

Titanium oxide is a good refractory and electrical insulator. The finely ground material gives good plasticity without binders, and is molded to make resistors for electronic use. The **micro sheet** of the Glenco Corp. is titanium oxide in sheets as thin as 0.003 in (0.008 cm) for use as a substitute for mica for electrical insulation where brittleness is not important. **Titania-magnesia ceramics** were made in Germany in the form of extruded rods and plates and pressed parts.

TOBACCO. The leaf of an unbranched annual plant of the genus *Nicotiana,* of which there are about 50 species and many varieties. It is used for smoking, chewing, snuff, insecticides, and for the production of the alkaloid nicotine. Commercial crops are grown in about 60 countries, but about a third of world production is in the United States. Only two species are grown commercially, *N. tabacum,* a tropical plant native to the West Indies and South America, and *N. rustica,* grown by the Indians of Mexico and North America before 1492. About 85% of world production is now from *N. tabacum,* and there are more than 100 varieties of this plant.

Tobacco was not known in Europe until it was brought from the West Indies by Columbus. Plants for cultivation were brought to Spain in 1558, and by 1586 smoking had become a general practice in western Europe. The first commercial shipments were made from Virginia in 1618, the growing of cultured varieties having begun in 1612. Smoking of tobacco was practiced by the Indians from Canada to Patagonia, and the natives of Haiti used powdered tobacco leaf as **snuff** under the name of **cohoba.** Like Indian corn, the tobacco plant had been domesticated for centuries and the original wild ancestor of the plant is not known. Some Indian tribes, such as the Tobacco nation of southwest Ontario, specialized in the growing of tobacco types.

The quality of the tobacco leaf varies greatly with the soil and climate, the care of the plant, and the curing of the leaf; the nicotine content develops in the curing process. The narcotic effects are due to the alkaloid **nicotine,** $C_{10}H_{14}N_2$, a complex pyrrolidine, which is a heavy, water-white oil. The nicotine is absorbed through the mucous membranes of the nose and throat. The aroma and flavor come from the essential oils in the leaf developed during fermentation and curing. The more harmful effects to the eyes and respiratory system come from the **pyridine** C_5H_5N, a toxic aromatic compound that also occurs in coal tar, and from other elements of the smoke and not from the alkaloid. The burning of the tars may also produce **carcinogen** compounds which are complex condensed benzene-ring nuclei injurious to tissues.

Although *N. tabacum* is a less hardy plant than *N. rustica,* it adapts itself to a wide variety of climates and soils, and the types generated in given areas do not normally reproduce the same type in another area. The variety developed in the Near East, known as **Turkish tobacco** and valued as an aromatic blend for cigarettes, is a small plant with numerous leaves only about 3 in (7.6 cm) long, while the American tobaccos grown from the same species have leaves up to 3 ft (0.9 m) long. The nicotine content of Turkish tobacco is from 1 to 2%, while that of flue-cured Virginia tobacco is 2.5 to 3%, and that of burley and fire-cured American types is up to 4.5%. **Perique,** a strong black tobacco much used in French and British pipe mixtures, is cultivated only in a small area of southern Louisiana. Other tobaccos brought into the area become perique in the second year, but when transplanted back they do not thrive. *N. rustica* was the first tobacco grown in Virginia, but the tobacco now grown in the area and known as **Virginia tobacco** is *N. tabacum* brought from the West Indies, but now differing in type from West Indian tobacco. **Makhorka tobacco,** a black air-cured type grown in Russia and Poland and very high in nicotine, is from *N. rustica.* Strong, black, highly fermented tobaccos high in nicotine, and considered as inferior in the United States, are preferred in France and some other countries.

Types of tobacco are based on color, flavor, strength, and methods of curing and fermentation, while grades are based on size, aroma, and texture, but the geographical growing area also determines characteristics. Commercial purchasing is done by the area and the Department of Agriculture type classification: fire-cured, dark air-cured, flue-cured, cigar wrapper, cigar binder, cigar filler, burley, Maryland, and perique, all of which are from *N. tabacum.* Grading is done by specialists, and a single-area crop may produce more than 50 grades. In the manufacture of cigarettes, blending is done to attain uniformity, and some of the flavor and aroma may be from added ingredients. **Air-cured tobaccos** are alkaline, while **flue-cured tobaccos** are acid and the nicotine is less readily given off. *N. rustica* may contain as high as 10% nicotine, and is thus more desirable for insecticide use or for the extraction of nicotine, but some strains of *N. tabacum* have been developed for smoking with as little as 0.3% nicotine.

Tobacco seed oil has an iodine value of 140 to 146, and is a valuable drying oil, but the production is low because the seed heads are topped in cultivation and seeds are developed only on the sucker growths. **Tobacco sauce,** used for flavoring chewing and smoking tobaccos, contains up to 10% nicotine, but since the nicotine is not desired in the flavoring it is usually extracted for industrial use. Nicotine can be oxidized easily to **nicotinic acid** and to **nicotinonitrile,** both of which are important as antipellagra vitamins. Most of the nicotine used for insecticide is marketed as **nicotine sulfate** in water solution containing 40% nicotine. It is used as a sheep dip and as a contact insecticide. **Tobacco dust** is used for the control of plant lice. **Anabasine,** obtained in Russia from the Asiatic shrub *Anabasis aphylla,* has the same chemical composition as nicotine and is an isomer of nicotine. It is marketed in the form of a solution of the sulfate as an insecticide. It can also be obtained from *N. glauca,* a wild tree tobacco native to Mexico and the southeastern United States, or is made synthetically under the name of **neonicotine.**

TOLU BALSAM. A yellowish-brown semisolid gum with a pleasant aromatic odor and taste, obtained from the tree *Myroxylon balsamum,* or *Toluifera balsamum,* of Venezuela, Colombia, and Peru. It is used in medicine, chiefly in cough syrups, and also as a fixative in perfumes. **Balsam of Peru,** or **black balsam,** is a reddish-brown viscous aromatic liquid from bark of the tall tree *M. pereirae* of El Salvador. It is used in cough medicines and skin ointments, as an extender for vanilla, and as a fixative in perfumes. Some white-colored balsam is also obtained from the fruit of the tree. **Peru balsam** contains benzyl benzoate, benzyl cinnamate, and some vanillin.

TOLUOL. Also called **toluene, methyl benzene,** and **methyl benzol.** A liquid of the composition $C_6H_5CH_3$, resembling benzene but with a distinctive odor. It is obtained as a by-product from coke ovens and from coal tar. It occurs also in petroleum, with from 0.20 to 0.70% in Texas crude oil, which is not sufficient to extract. But toluol may be produced by dehydrogenation of petroleum fractions. It is used as a solvent, and for making explosives, dyestuffs, and many chemicals, and in aviation gasoline to improve the octane rating. Industrially pure toluol from coal tar distills off between 108.6 and 112.6°C, and is a water-white liquid with a specific gravity of 0.864 to 0.874, flash point 35 to 40°F (2 to 4°C), and freezing point about −95°C. The fumes are poisonous. **Monochlorotoluene,** used as a solvent for rubber and synthetic resins, is a colorless liquid of the composition $CH_3C_6H_4Cl$, boiling at about 160°C and freezing at −45°C. **T oil** is a sulfur toluene condensation product made under a British patent and used as a plasticizer for chlorinated rubber. **Notol No. 1,** of the Neville Co., is a coal-tar hydrocarbon high in aromatics used as a substitute for toluol as a lacquer solvent. The specific gravity is 0.825 and the boiling point between 177 and 280°F (81 and 138°C). **Tollac,** of the same company, is another hydrocarbon substitute for toluol. **Methyl cyclohexane,** $C_6H_{11}CH_3$, is a water-white liquid with a distilling range of 100 to 103°C, produced by hydrogenating toluol. It is used as a solvent for oils, fats, waxes, and rubbers. **Methyl cyclohexanol,** $C_6H_{10}CH_3OH$, another toluol derivative, is used as a cellulose ester solvent and as an antioxidant in lubricants. It is a straw-colored viscous liquid distilling between 155 and 180°C. **Polyvinyl toluene** is a methyl form of styrene. It is polymerized with terphenyl stilbene to form plastic scintillators to count radiation isotopes.

TONKA BEAN. Called in northeastern Brazil **cumarú bean.** The kernel of the pit of the fruit of the **sarrapia tree,** *Dipteryx odorata* or *Coumarouna odorata,* of northern South America, used for the production of **coumarin** for flavoring and scenting. It has an aroma resembling vanilla. The trees often reach a height of 100 ft (30 m), and begin to bear in 3 years. The fruit is like a mahogany-colored plum, but with a fibrous pulp. The pits, or nuts, contain a single shiny black seed 1 in (2.5 cm) or longer. The chief production is in Venezuela, Brazil, Colombia, Trinidad, and the Guianas. The tonka bean from the tree *D. oleifera* of Central America has an unpleasant odor. Before shipping, the beans are soaked in rum or alcohol to crystallize the coumarin. The ground beans are again soaked in rum, and the aromatic liquid is used to spray on cigarette tobacco. The coumarin extract is also used as a perfume or flavor in soaps, liqueurs, and confectionery. The essential oil produced from the seed is called **cumarú oil.** A substitute for tonka bean is **deer's-tongue leaf,** which is the long leaf of the herb *Trilisa odoratissima* growing wild on the edges of swamps from Carolina to Florida. The leaf has a strong odor of coumarin when dry, and contains coumarin. It is used in cigarette manufacture, in flavoring, and to produce synthetic vanilla.

TOOL STEEL. A high-carbon steel used for making tools. It has the property of becoming extremely hard by quenching from a temperature of 1400 to 1800°F (760 to 982°C). It can then be drawn to any degree of hardness by heating at lower temperatures. The hardening properties depend upon the formation of hard reversible carbides. The early tool steels were chance combinations in which the carbon of the fuel combined with the iron, molybdenum, or other contained elements. One of the earliest recorded tool steels was the ancient **Chalybeate steel,** originally referring to steel from the Chalybes in Pontus. The unqualified term tool steel does not usually include special alloy steels containing nickel, manganese, and other metals, nor high-speed steels. However, tool steel for special purposes may contain many other elements besides carbon. The possibilities of percentage combinations of vanadium, nickel, manganese, chromium, silicon, tungsten, and other elements in alloy tool steels are infinite; and as there are hundreds of trade-named steels on the market, the name

carbon tool steel is used to designate tool steel containing only carbon, and with other elements below perceptible amounts. Tool steels may be divided into two general classes, water-hardening and oil-hardening. The **water-hardened steels,** which cool quickly in the quenching, have a hard exterior with a tough shock-resistant core. This hard case wears off with repeated grinding. The **oil-hardened steels** cool more slowly, thus hardening evenly throughout the metal. In general, the oil-hardening steels contain significant amounts of manganese, chromium, tungsten, vanadium, or molybdenum. **Desegatized steels** are tool steels that have been given special treatment to produce an even dispersion of free carbides, eliminating danger of a brittle central mass.

Tool steel may contain from 0.65 to 1.50% carbon, the lower-carbon grades, up to 0.90 carbon, being used for punches, hammers, chisels, and other tools requiring some degree of elasticity, and the high-carbon grades are used for dies, drills, and edge tools. Files, saws, and engraving tools may contain up to 1.60% carbon. The manganese content is 0.20 to 0.30. **Razor steel** was steel with 1.5% carbon, but razors are now usually made of alloy steels. **Orthopedic steel** contains 0.95% carbon, and the annealed steel has a tensile strength of 145,000 lb/in^2 (999 MPa) with elongation of 12%. Beyond 1% carbon, there is an excess of carbon and the steels become very brittle when hardened. Theoretically, the maximum point of solution of the Fe_3C in a plain carbon steel is at 0.85% carbon. When other elements are present, other carbides are formed, giving greater hardness and strength above this point. The ideal maximums of phosphorus and sulfur in a tool steel are 0.025%, with silicon at 0.20 and manganese at 0.25, but in special steels the silicon and manganese are increased. Some water-hardening carbon tool steels have higher content of silicon to give wear resistance for dies, liners, and bushings. The silicon also adds fatigue resistance. **Graph-Sil steel,** of the Timken Steel & Tube Co., has 1.5% carbon, 0.85 to 0.95 silicon, and 0.40 max manganese.

The lower temperature ranges are used for hardening high-carbon steels and thin pieces. Some steelmakers grade carbon steels by divisions as low as 5 points of carbon. **Pompton tool steel,** of the Allegheny-Ludlum Steel Co., has 19 grades from 0.50 to 1.45% carbon. A free-machining steel of the same company, **Oilgraph EZ,** contains 1.15% silicon, 0.80 manganese, 0.20 chromium, 0.25 molybdenum, 0.10 sulfur, with 1.45 carbon and 0.30 graphite. The graphite, evenly dispersed in the steel, makes easy machining, gives resistance to galling, and improves the finish. Modern carbon tool steels for ordinary water hardening, with or without residual vanadium, develop remarkable physical properties. Ryerson **VD die steel,** as quenched, has a hardness of 725 Brinell, or 96 Scleroscope. **Granada steel,** of the Crucible Steel Co., is a general-purpose water-hardening tool steel containing 1% carbon, 0.30 manganese, and 0.25 silicon.

Oil-hardening tool steels usually contain about 1% manganese, but mild-alloy tool steels may contain less, with other elements. **CM** and **CMM tap steels,** of the Colonial Steel Co., are oil-hardening, keen-edge steels; CM contains 0.50% chromium, 0.60 manganese, and 1.20 carbon; CMM contains 0.50% chromium, 0.85 manganese, 0.60 molybdenum, and 1.20 carbon. These steels have high torsional strength. Additions of small amounts of molybdenum with higher silicon give oil-hardening steels of high strength and toughness. **Halcomb SS steel,** of the Crucible Steel Co., is such a steel for drills, taps, and broaches. It contains 1.20% chromium, 0.30 molybdenum, 0.35 manganese, and 1 carbon. **Versatool,** of the same company, is an intermediate steel that has high hardness and wear resistance and maintains its properties at temperatures close to those of high-speed steel. It contains 4.25% chromium, 2.5 molybdenum, 1.15 vanadium, 0.3 tungsten, 2 silicon, 0.3 manganese, and 1 carbon. It is used for shears and cutting dies. **Ketos steel** is a general-purpose steel containing 0.90% carbon, 1.35 manganese, 0.35 silicon, 0.50 chromium, and 0.50 tungsten. **BTR steel,** of Bethlehem Steel, is a general-purpose oil-hardening die and tool steel that is nondeforming and wear-resistant. It contains 1.20% manganese, 0.50 tungsten, 0.50 chromium, 0.20 vanadium, and 0.90 carbon.

Tool steel comes regularly in round, square, and octagon bars, and in flats, but drawn shapes are also available. Tool steels require more care in forging than low-carbon machinery steels, and they are more difficult to machine. **A.S.V. steel,** of Firth-Sterling Steel, is made by a patented winged ingot form of casting in order to eliminate any porosity in the center. **Maraged steel** is a name given by International Nickel to high-alloy steels with high strength and ductility developed by a martensitic aging treatment. Such a steel with 18% nickel, 7 cobalt, 5 molybdenum, 0.5 titanium, and 0.05 max carbon is resistant to corrosion cracking under stress and has a tensile strength of 400,000 lb/in^2 (2,757 MPa) and yield strength of 250,000 lb/in^2 (1,723 MPa). It is easily cold-formed and welded. This composition can be varied to give different grades with differing properties.

Ceramicast steel, of the Lebanon Steel Foundry, is accurately cast steel requiring no machining. The ceramic mold is made by pouring a liquid mixture of hydrolyzed ethyl silicate and sillimanite over the pattern, removing the pattern after the mixture has set, and firing to give a rigid ceramic mold. **Carbon-vanadium tool steels** are produced in all carbon contents with about 0.20% vanadium. They have a uniform fine grain, and constitute a class of "super" carbon tool steels. Some steels for special purposes contain more vanadium. **Colhead steel,** of the Vanadium Alloys Steel Co., for cold-heading dies, has 0.45% vanadium and 1 carbon. **Vatool,** of Henry Disston & Sons, Inc., is a vanadium steel for taps. **Shim steel** may be either carbon or stainless steel in gages as thin as 0.0015 in (0.0038 cm), with extra close tolerances. **Micro-Shim steel,** of the American Silver Co., is low-carbon steel or stainless steel in thicknesses as low as 0.0005 in (0.0013 cm). Tool steels are now usually refined alloy steels, and producers such as Carpenter Technology market them on a performance basis rather than by composition.

TRAGACANTH GUM. An exudation of the shrub *Astragalus gummifer* of Asia Minor and Iran, used in adhesives or for mucilage, for leather dressing, for textile printing, and as an emulsifying agent. To obtain the gum a small incision is made at the base of the shrub, from which the juice exudes and solidifies into an alteration product, not merely the dried juice. The gum derived from the first day's incision, known as **fiori,** is the best quality, and is in clear fine ribbons or white flakes. The second incision produces a yellow gum known as **biondo.** The third incision produces the poorest quality, a dark gum known as **sari.** Rainy weather during the incision period may cause a still inferior product. Tragacanth is insoluble in alcohol but is soluble in alkalies and swells in water. **Karaya gum** from southern Asia is from various species of *Sterculia* trees, especially *S. urens,* of India. It is also known as **Indian gum, Indian hog gum,** and **hog tragacanth.** The sticky gum is dried, and the chunks are broken and the pieces sorted by color. A single chunk may have colors varying from clear white to dark amber and black. The color is caused by tannin or other impurities. The No. 3 grade, the lowest, has up to 3% insoluble impurities. The gum is marketed in flakes and as a white, odorless, 150-mesh powder. The chief constituent is galactan. In general, the gum is more acid than tragancanth and is likely to form lumpy gels unless finely ground. It is widely used as a thickening and suspending agent for foodstuffs, drugs, cosmetics, adhesives, and for textile finishes.

The granules of **water-soluble gums,** such as karaya, tragacanth, and acacia, are swelled by water and dispersed in the water in microscopic particles to form cells or filamentlike structures which hold the water like a sponge and will not settle out. This type of colloidal dispersion is called a **hydrasol,** and when thick and viscous is called a **gel.** From 2 to 3% of karaya or other gum will form a gel in water. These gums will gel in cold water, while gelatin requires hot water for dissolving. In a gel there is continuous structure with molecules forming a network, while in a **sol** the particles are in separate suspension and a sol is merely a dispersion. Some dispersions, such as albumen, cross-link with heat; some, like guar gum, cross-link with alkalies; some, like pectin, link with sugar and an acid. Gums with weak surface forces form weak gels which are **pastes** or mucilage, and a high concentration is needed to produce a solid. Karaya has great swelling power, and is used in medicine as a bulk laxative. **Ghatti gum,** from the abundant tree *Anogeissus latifolia* of India, is entirely soluble in water to form a viscous mucilage. It is twice as effective as gum arabic as an emulsifier, but is less adhesive. It comes in colorless to pale-yellow tears of vitreous fracture, called also Indian gum, and is used in India for textile finishing. **Aqualized gum,** of Glyco products Co., Inc., is tragacanth or karaya chemically treated to give more rapid solubility. Water-soluble gums are also produced synthetically. **Polyox gum,** of the Union Carbide Chemicals Co., is a polymer of polyethylene oxide containing carboxylic groups giving water solubility when the pH is above 4.0. In paper coating with ammonia the ammonia evaporates to leave a water-insoluble, grease-resistant film that is heat-sealing. It is also used in latex paints and in cosmetics.

Another water-soluble gum which forms a true gel with a continuous branched-chain molecular network is **okra gum,** produced by Morning star-Paisley, Inc., as a 200-mesh tan powder. It is edible, and is used for thickening and stabilizing foods and pharmaceuticals. It is also used in plating baths for brightening nickel, silver, and cadmium plates. It is extracted from the pods of the **okra,** *Hibiscus esculentus,* a plant of the cotton family. In the southern states the pods, called **gumbo,** are used in soups. The refined gum, after extraction of the oils and sugars, contains 40.4% carbon, 6.1 hydrogen, and 2.1 nitrogen, with the balance insoluble cellulose.

TRIPOLI. A name given to finely granulated, white, porous, siliceous rock, used as an abrasive and as a filler. True tripoli is an infusorial diatomaceous earth known as **tripolite,** and is a variety of opal, or **opaline silica.** In the abrasive industry it is called **soft silica.** It is quarried in Missouri, Illinois, eastern Tennessee, and Georgia. Pennsylvania rottenstone is not tripoli, although it is often classed with it. The material marketed for oil-well drilling mud by the Corona Products, Inc., under the name of **Opalite,** is an amorphous silica. The Missouri tripoli ranges in color from white to reddish, and the crude rock has a porosity of 45%, and contains 30% or more of moisture. It is air-dried and then crushed and furnace-dried. Tripoli is used in massive form for the manufacture of filter stones for filtering small supplies of water. Missouri tripoli is also used for the manufacture of foundry parting. Finely ground tripoli, free from iron oxide, is used as a paint filler and in rubber. The grade of tripoli known as O.G. (once ground) is used for buffing composition, D.G. (double ground) for foundry partings, and the air-float product for metal polishes. Tripoli grains are soft, porous, and free from sharp cutting faces, and give a fine polishing effect. It is the most commonly used polishing agent. The word **silex,** which is an old name for silica and is also used to designate the pulverized flint from Belgium, is sometimes applied to finely ground white tripoli employed as an inert filler for paints. Much Illinois fine-grained tripoli is used for paint, and for this purpose should be free from iron oxide. **Rottenstone** is a soft, friable, earthy stone of light-gray to olive color, used as an abrasive for metal and wood finishing. It resembles Missouri tripoli and is derived from the weathering of siliceous-argillaceous limestone, with generally from 80 to 85% alumina, 4 to 15 silica, and 5 to 10 iron oxides. Rottenstone was largely imported from England, but a variety is found in Pennsylvania. It is finely ground and is marketed either as a powder or molded into bricks. The latter form is used with oil on ragwheel polishing. A 250-mesh powder is used as a filler in molding compounds.

TRISODIUM PHOSPHATE. A white crystalline substance of the composition $Na_3PO_4 \cdot 12H_2O$, also known as **phosphate cleaner,** used in soaps, cleaning compounds, plating, textile processing, and boiler compounds. The commercial grade is not less than 97% pure, with total alkalinity of 16 to 19% calculated as Na_2O. The anhydrous trisodium phosphate is 2.3 times as effective as the crystalline form, but requires a longer time to dissolve. **Disodium phosphate** is a white crystalline product of the composition $Na_2HPO_4 \cdot 12H_2O$ used for weighting silk, boiler treatment, cheese making, and in cattle feeds. The medicinal, or USP, grade has only seven molecules of water and has a different crystal structure. The commercial grade is 99.4% pure, and is readily soluble in water. **Sodium tetraphosphate,** $Na_6P_4O_{13}$, contains 39.6% Na_2O and 60.4 P_2O_5. It is the sodium salt of **tetraphosphoric acid,** and is marketed in beads that are mildly alkaline and highly soluble in water. The specific gravity is 2.55, and melting point 600°C. It is used in the textile industry as a water softener and to accelerate cleansing operations. It removes lime precipitation and sludge and saves soap. **Quandrafos,** of the American Cyanamid Co., used to replace quebracho for reducing the viscosity of oil-well drilling mud, is sodium tetraphosphate, containing 63.5% of P_2O_5. It makes the calcium and magnesium compounds inactive, and 0.06% of the material controls 16.1% of water in reducing viscosity. It also gives smooth flow with minimum water in paper coating and textile printing. **Metafos,** of the same company, has a higher percentage of P_2O_5, 67%, and a lower pH, for use in textile printing where low alkalinity is needed. **Sodium pyrophosphate,** $Na_4P_2O_7$, is added to soap powders to increase the detergent effect and the lathering. It is also used in oil-drilling mud. The crystalline form, $Na_4P_2O_7 \cdot 10H_2O$, is very soluble in water and is noncaking, and is used in household cleaning compounds. **Sodium tripolyphosphate,** $Na_5P_3O_{10}$, is a water-soluble, white powder used as a detergent, a water softener, and a deflocculating agent in portland cement to govern the viscosity of the shale slurry without excessive use of water. Large quantities of these phosphates are used in the processing of chemicals, textiles, and paper; and since they are toxic containments of ground and surface waters, mill wastes must be deactivated before they are discharged.

TULIPWOOD. Also called **yellow poplar, whitewood,** and **canary whitewood.** The wood of the tree *Liriodendron tulipifera* of Canada and the eastern United States. The tree grows to a height of 250 ft (76 m) and to diameters of more than 10 ft (3 m). It is used for furniture, veneer, millwork, toys, woodenware, boxes, crates, and pulpwood. Owing to its close texture and even coefficient of expansion it has been used for expansion blocks in humidity regulators. It is yellowish, soft, and durable. It weighs about 30 lb/ft^3 (481 kg/m^3). The lumber may be mixed with **cucumber magnolia,** *Magnolia acuminata,* and **evergreen magnolia,** *M. grandifolia,* but **magnolia woods** are lighter in color.

TUNG OIL. A drying oil which has almost double the rapidity of linseed oil. It is used for enamels and varnishes, in brake linings, plastic compounds, and linoleum, and for making pigment for India ink. Tung oil is pressed from the seeds of the *Aleurites montana* and *A. fordii.* The names **wood oil** or **China wood oil** are loosely and erroneously used to designate tung oils, but true wood oil is an oleoresin from the **Keruing tree** of Malaya used for waterproofing and calking boats, while tung oil is never from the wood. The oil has a powerful purgative action, and the Chinese word tung means stomach. The **Chinese tung oil** is from the nuts of the tree *A. montana,* the China wood oil tree, and the *A. fordii.* The latter tree is more hardy than the *A. montana,* which requires a hot climate. The American tung oil is from the nuts of the tree *A. fordii* of the Gulf states which gives an annual production of about 30 lb (14 kg) of oil per tree. The tree grows to a height of 25 ft (8 m), and bears for 5 years. The seeds, or nuts, contain 50 to 55% oil. This tree is also grown in South Africa and in Argentina.

The color of tung oil varies from golden yellow to dark brown according to the degree of heat used in extraction. It has a pungent odor resembling that of bacon fat. A good grade of raw tung oil should have a specific gravity between 0.943 and 0.940, a saponification value of 190, and an iodine value of 163. The oil contains about 72% **eleostearic acid,** which has a very high iodine value, 274, and gives to the oil a greater drying power than is indicated by the iodine value of the oil itself. The oil has the property of drying throughout at a uniform rate instead of forming a skin as linseed oil does, but it dries flat instead of glossy like linseed oil and is inclined to produce a wrinkled surface. It is mixed with rosin, since rosin has great affinity for it, and the two together are suitable for gloss varnishes. The oil from *A. montana,* or **mu oil,** has a higher percentage of eleostearic acid than that from *A. fordii.* The **Japanese tung oil** is from the nuts of the larger tree *A. cordata.* The oil is superior to Chinese tung oil and is seldom exported. It does not gelatinize like Chinese tung oil when heated. It is used in Japan for varnishes, waterproofing paper, and soaps. The saponification value is 193 to 195, iodine No. 149 to 159, and specific gravity 0.934 to 0.940. The kernels of the nuts yield about 40% oil. The tree is grown also in Brazil and thrives in hot climates. **Candlenut oil** is from the seed nuts of the *A. moluccana* of Oceania and southern Asia. It received its name from the fact that the Polynesians used the nuts as candles to light their houses. The oil is variously known as **kukui, kekune,** and **lumbang,** and as an artist's paint oil is called **walnut oil** or **artist's oil.** The nut resembles the walnut but has a thicker shell. The oil has a specific gravity of 0.923, iodine value 165, and is between linseed and soybean oil in properties. It is high in linoleic and linolenic acids. The variety known as soft lumbang oil, or **bagilumbang oil,** from the tree *A. trisperma* of the Philippines, resembles tung oil and is high in eleostearic acid. The chief production of lumbang oil is in the Fiji Islands.

The **safflower,** *Carthamus tinctorius,* is grown in California, France, and India, and in the latter country it is grown on a large scale for seeds which yield up to 35% of the clear, yellowish **safflower oil** used in paints, leather dressings, and for foods. The oil has a high content, 73%, of linoleic acid, the highest of essential polyunsaturated acids of any vegetable food oil. It is odorless, with a bland taste. **Safflower 22,** of the Pacific Vegetable Oil Corp., is a conjugated paint oil made by isomerizing safflower oil. It has a rapid drying rate, color retention, and an ability to produce wrinkled finishes by adjustment of the amount of drier. It can thus replace tung oil. It takes up maleic anhydride readily, and is used for making modified alkyd finishes. **Wecoline SF,** of Drew Chemical, is a concentrate of safflower fatty acids with 67.3% linoleic and only 0.2 linolenic acid, for compounding in coatings. **Saff,** of Abbott Laboratories, is an emulsion of safflower oil used as a drug to lower blood cholesterol. The heads of the plant are dried and used as food colors, for dyeing textiles, and for cosmetic rouge.

TUNGSTEN. A heavy white metal with a specific gravity of 19.6, weighing 0.697 lb/in^3 (19,290 kg/m^3), and having the highest melting point, 6170°F (3410°C), of all the metals. **Wolframite** is the chief ore of the metal tungsten. Its composition is $(FeMn)WO_3$. When the manganese tungstate is low, the ore is called **ferberite;** when the iron tungstate is low, it is called **hübnerite.** The ore is concentrated by gravity methods to a concentrate containing 60 to 65% **tungstic oxide,** WO_3. To extract pure WO_3 from the concentrate it is fused with sodium carbonate, Na_2CO_3, to form **sodium tungstate,** Na_2WO_3, which is dissolved in water. When an acid is added to the solution, the WO_3 precipitates out as a yellow powder. The metallic tungsten is obtained by reducing, and is then pressed into bars and sintered. Wolframite occurs usually bladed or columnar in form. It has a specific gravity of 7.2 to 7.5, a hardness of 5, a black color, and a submetallic luster. It is found in the mountain states, Alaska, China, and Argentina, but is also widely distributed in various parts of the world in small quantities. Chinese **wolfram concentrates** contain 65% tungstic oxide; the Arizona concentrates contain an average of 67%. California and Nevada concentrates are scheelite containing from 60 to 67% tungstic oxide. The **sanmartinite** of Argentina is a variety containing zinc.

Tungsten has a wide usage for alloy steels, magnets, heavy metals, electric contacts, rocket nozzles, and electronic applications. Tungsten resists oxidation at very high temperatures, and it is not attacked by nitric, hydrofluoric, or sulfuric acid solutions. Flame-sprayed coatings are used for nozzles and other parts subject to heat erosion.

Tungsten is usually added to iron and steel in the form of **ferrotungsten,** made by electric-furnace reduction of the oxide with iron or by reducing tungsten ores with carbon and silicon. Standard grades with 75 to 85% tungsten have melting points from 3200 to 3450°F (1760 to 1899°C). **Tungsten powder** is usually in sizes from 200 to 325 mesh, and may be had in a purity of 99.9%. Parts, rods, and sheet are made by powder metallurgy, and rolling and forging are done at high temperature. The rolled metal may have a tensile strength as high as 500,000 lb/in^2 (3,445 MPa) and hardness of Brinell 290, whereas drawn wire may have a tensile strength to 590,000 lb/in^2 (4,065 MPa). The tungsten powder of Kennametal, Inc., for use in spray coatings for radiation shielding, has a particle size less than 40 μm and density of 165 grains/in^3 (0.65 g/cm^3). The needle-shaped powders of General Electric Co. for powder metallurgy have particle diameters of 4.8 to 5.8 μm. The **tungsten wire** of Universal-Cyclops Steel Corp. for spark plug and electronic use is made by powder metallurgy swaging. It has uniform density and comes in diameters from 0.02 to 0.50 in (0.05 to 1.27 cm). Tungsten wire as fine as 0.00018 in (0.00046 cm) is used in electronic hardware. **Tungsten whiskers,** which are extremely fine fibers, are used in copper alloys to add strength. Copper wire which normally has a tensile strength of 30,000 lb/in^2 (206 MPa) will have a strength of 120,000 lb/in^2 (827 MPa) when 35% of tungsten whiskers are added. **Tungsten yarns** of Union Carbide Corp. are as small in diameter as 0.0005 in (0.0013 cm) and are made up of fine fibers of the metal. The yarns are flexible and can be woven into fabrics. Continuous **tungsten filaments,** usually 10 to 15 μm in diameter, are used for reinforcement in metal, ceramic, and plastic structural composites. Finer filaments of tungsten are used as cores, or substrates, in **boron filaments.**

The metal is now also produced as arc-fused grown crystals, usually no larger than ⅜ in (0.952 cm) diameter and 10 in (25.4 cm) long, and worked into rod, sheet, strip, and wire. **Tungsten crystals,** 99.9975% pure, are produced by Linde and Westinghouse. Tungsten crystals are ductile even at very low temperatures, and wire as fine as 0.003 in (0.008 cm) and strip as thin as 0.005 in (0.013 cm) can be cold-drawn and cold-rolled from the crystal. The crystal metal has nearly zero porosity and its electric and heat conductivity are higher than ordinary tungsten. The normal electric conductivity is about 33% that of copper, but that of the crystal tungsten is 15% higher. In microscopy the molecules of tungsten appear as body-centered cubes, but in the pure metal the atoms normally bond uniformly in six directions forming a double lattice so that each grain forms a true single crystal. At elevated temperatures tungsten forms many compounds in chemicals and alloys. The **tungsten-aluminum alloy** of Du Pont is a chemical compound and is made by reducing tungsten hexachloride with molten aluminum.

Tungsten retains a tensile strength of about 50,000 lb/in^2 (344 MPa) at 2500°F (1371°C), but because of its great weight is normally used in aircraft or missile parts only as coatings, usually sprayed on. It is also used for X-ray and gamma-ray shielding. Electroplates of tungsten or tungsten alloys give surface hardnesses to Vickers 700 or above. **Cobalt-tungsten alloy,** with 50% tungsten, gives a plate that retains a high hardness at red heat. **Tungsten RhC,** of Cleveland Refractory Metals, is a **tungsten-rhenium carbide alloy** containing 4% rhenium carbide. It is used for parts requiring high strength and hardness at high temperatures. The alloy retains a tensile strength of 75,000 lb/in^2 (517 MPa) at 3500°F (1927°C). **Ammonium metatungstate,** used for electroplating, is a white powder of the composition $(NH_4)_6H_2W_{12}O_{40}$. It is readily soluble in water and gives solutions of 50% tungsten content. **Tungsten hexafluoride** is used for producing tungsten coatings by vapor deposition. At a temperature of 900°F (482°C) the gas mixed with hydrogen deposits a tungsten plate. **Tungsten hexachloride,** WCl_6, is also used for depositing **tungsten coatings** at a temperature of 900°C in a hydrogen atmosphere. Smooth, dense **tungsten platings** can be deposited from **tungsten carbonyl,** $W(CO)_6$, at a temperature of 150°C. The carbonyl is made by reacting the hexachloride with carbon monoxide.

TUNGSTEN CARBIDE. An iron-gray powder of minute cubical crystals with a Mohs hardness above 9.5 and a melting point of about 5400°F (2982°C). It is produced by reacting a hydrocarbon vapor with tungsten at high temperature. The composition is WC, but at high heat it may decompose into W_2C and carbon, and the carbide may be a mixture of the two forms. Other forms may also be produced, W_3C and W_3C_4. Tungsten carbide is used chiefly for cutting tool bits and for heat- and erosion-resistant parts and coatings.

Briquetting of tungsten carbide into usable form was first patented in Germany and produced by the Krupp Works under the name of **Widia metal.** It is made by diffusing powdered cobalt through the finely divided carbide under hydraulic pressure, and then sintering in an inert atmosphere at about 1500°C. The briquetted material is then ground to shape, and the pieces are brazed to tools. They will withstand cutting speeds from 3 to 10 times those of high-speed steel, and will turn manganese steel with a hardness of 550 Brinell, but are not shock-resistant. Pressed and sintered parts usually contain 3 to 20% cobalt binder, but nickel may also be used as a binder. The compressive strengths may be as high as 700,000 lb/in^2 (4,823 MPa) with rupture strengths to 200,000 lb/in^2 (1,378 MPa) or higher.

One of the earliest of the American bonded tungsten carbides was **Carboloy,** of the General Electric Co., used for cutting tools, gages, drawing dies, and wear parts. The sintered materials are now sold under many trade names such as **Dimondite, Firthite,** and **Firthaloy,** of the Firth-Sterling Steel Co.; **Armide,** of the Armstrong Bros. Tool Co.; **Wilcoloy,** of the H. A. Wilson Co.; and **Borium** and **Borod,** of the Stoody Co. But the carbides are now often mixed carbides. **Carboloy 608** contains 83% chromium carbide, 2% tungsten carbide, and 15% nickel binder. It is lighter in weight than tungsten carbide, is nonmagnetic, and has a hardness to Rockwell A93. It is used for wear-resistant parts, and resists oxidation to 2000°F (1092°C). Titanium carbide is more fragile, but may be mixed with tungsten carbide to add hardness for dies. **Cutanit,** of Firth-Sterling, is such a mixture. **Kennametal K601,** of Kennametal, Inc., for seal rings and wear parts, is a mixture of tantalum and tungsten carbides without a binder. It has a compressive strength of 675,000 lb/in^2 (4,650 MPa), rupture strength of 100,000 lb/in^2 (689 MPa), and Rockwell hardness A94. **Kennametal K501** is tungsten carbide with a platinum binder for parts subject to severe heat erosion. **Strauss metal,** of the Allegheny Ludlum Steel Co., is tungsten carbide. **Tungsten carbide LW-1,** of the Linde Co., is tungsten carbide with about 6% cobalt binder used for flame-coating metal parts to give high-temperature wear resistance. Deposited coatings have a Vickers hardness to 1450, and resist oxidation at 1000°F (538°C). **Tungsten carbide LW-1N,** with 15% cobalt binder, has a much higher rupture strength, but the hardness is reduced to 1150. **Metco 35C** is a fine powder of tungsten carbide and cobalt for flame spraying to produce a hard wear-resistant coating of carbide in a matrix of cobalt.

TUNGSTEN STEEL. Any steel containing tungsten as the alloying element imparting the chief characteristics to the steel. It is one of the oldest of the alloying elements in steel, the celebrated ancient Eastern sword steels having had tungsten in them. Tungsten increases the hardness of steel, and gives it the property of red hardness, stabilizing the hard carbides at high temperatures. It also widens the hardening range of steel, and gives deep hardening. Very small quantities serve to produce a fine grain and raise the yield point. The tungsten forms a very hard carbide and an iron tungstite, and the strength of the steel is also increased, but it is brittle when the tungsten content is high. When large percentages of tungsten are used in steel, they must be supplemented by other carbide-forming elements. Tungsten steels, except the low-tungsten chromium-tungsten steels, are not suitable for construction, but they are widely used for cutting tools, because the tungsten forms hard abrasion-resistant particles in high-carbon steels. Tungsten also increases the acid resistance and corrosion resistance of steels. The alloys are difficult to forge, and cannot be readily welded when the tungsten exceeds 2%. Standard **SAE tungsten steels,** SAE 71360 and SAE 71660, contain 12 to 15 and 15 to 18% tungsten, respectively, with 3 to 4 chromium and 0.50 to 0.70 carbon. **SAE steel 7260** contains 1.5 to 2% tungsten and 0.50 to 1 chromium.

When the tungsten content is high, particularly when the steel also contains manganese the steel can be hardened by air cooling. These alloy steels have a close, uniform texture, and tools made from them will keep an edge even when hot. In annealing tungsten steels the hard stable carbide WC may be formed; to prevent this, chromium is used as an inhibitor. A low tungsten steel for taps and cutters contains 1.5 to 2% tungsten, 1 to 1.3 carbon, with a small amount of chromium and vanadium. **Maxtack steel,** of A. Milne & Co., for tack and nail dies, has 10% tungsten, 2 chromium, 2.5 manganese, 1 silicon, and 2.25 carbon. It is self-hardening. Steels with 2 to 3% tungsten have high wear resistance and impact resistance. They are used for cold-drawing dies, air hammers, and general tools. **Graph-Tung,** of the Timken Steel & Tube Co., has 2.6 to 3% tungsten, 0.50 molybdenum, 0.70 silicon, and 1.5 carbon. **O.K. steel** of William Jessop & Sons, used for chisels and punches, has 2% tungsten with only 0.40 carbon. It is hard but very shock-resisting. A chisel steel, **J-S steel,** of Firth Sterling, Inc., has 2.25% tungsten, 1.4 chromium, 0.50 carbon, and a trace of vanadium. It is also used for cutters, punches, and shear blades for hot and cold work. It will retain a good cutting edge. **Wizard steel,** of the Ziv Steel & Wire Co., used for pneumatic tools, riveter dies, and swaging dies, has 1% tungsten, 1 chromium, 0.35 carbon, and a small amount of molybdenum. At a hardness of 55 Rockwell it has great toughness and resistance to shock. **High-Wear 64 steel,** of Carpenter Steel, which gives abrasion resistance and long life in blanking and drawing dies and in molds for compacting metals, contains 4% tungsten, 1 molybdenum, 0.90 chromium, 0.25 silicon, and 1.5 carbon. It is air-hardened at 1550°F (843°C) and is not intended for hot-work tools. **LT steel,** of Firth Sterling, Inc., for hot-forging dies, contains 9.5% tungsten, 3.5 chromium, 0.5 vanadium, and 0.33 carbon. It is hardened in air or oil from 2150°F (1177°C) and tempered at 1100°F (593°C) to give a hardness of Rockwell C52. **XDL steel** of the same company, for severe hot work, has 14% tungsten, 3.4 chromium, 0.5 vanadium, and 0.38 carbon. It is tempered to just above the maximum service heat, 1000°F (538°C) giving a hardness of Rockwell C54, and 1300°F (704°C) giving a hardness of C42.

TURPENTINE. Also called in the paint industry **oil of turpentine.** An oil obtained by steam distillation of the oleoresin which exudes when various conifer trees are cut. Longleaf pine and slash pine are the main sources. It also includes oils obtained by distillation and solvent extraction from stumpwood and waste wood. Longleaf sapwood contains about 2% oleoresin, heartwood 7 to 10%, and stumpwood 25%. Most oleoresin is obtained from the sapwood of living trees, but it is not the sap of the tree. Heartwood resin is obtained only when the cut wood is treated with solvents. The oleoresin yields about 20% oil of turpentine and 80 rosin; both together are known as **naval stores.**

Wood turpentine, called in the paint industry **spirits of turpentine,** is obtained from waste wood, chips, or sawdust by steam extraction or by destructive distillation. Wood turpentine forms more than 10% of all American commercial turpentines. Wood turpentine has a peculiar characteristic sawmill odor, and the residue of distillation has a camphorlike odor different from gum turpentine. It differs very little in composition, however, from the true turpentine. Steam-distilled wood turpentine contains about 90% terpenes, of which 80% is alpha pinene and 10% is a mixture of beta pinene and camphene. Some wood turpentine is produced as a by-product in the manufacture of cellulose. **Sulfate turpentine** is a by-product in the making of wood pulp. It varies in composition as the less stable beta pinene is affected by the pulping process, and it is used largely in chemical manufacture. By hydrogenation it produces cymene from which **dimethyl styrene** is made. This material can be copolymerized to produce vinyl resins.

Turpentine varies in composition according to the species of pine from which it is obtained. It is produced chiefly in the United State, France, and Spain. The turpentine of India comes from the **chir pine,** *Pinus longifolia,* of the southern slopes of the Himalayas, also valued for lumber, and the **khasia pine,** *P. khasya.* The gum of the chir pine is different from American gum, and the turpentine, unless carefully distilled, is slower-drying and greasy. French and Spanish turpentine, or **Bordeaux turpentine,** is from the **maritime pine,** *P. pinaster,* which is the chief source, and from **Aleppo pine,** *P. halepensis,* and **Corsican pine,** *P. lavicia.* In Portugal, the **stone pine,** *P. pinea,* is the source. The French maritime pine is also grown on plantations in Australia. Aleppo pine of Greece was the source of the naval stores of the ancients. **Venetian turpentine,** or **Venice turpentine,** is from the Corsican pine or European larch. It produces a harder film than American turpentine. Artificial Venice turpentine is made by mixing rosin with turpentine. European pines do not give as high a yield as American longleaf and slash pines.

American turpentine oil boils at 154°C, and the specific gravity is 0.860. It is a valuable drying oil for paints and varnishes, owing to its property of rapidly absorbing oxygen from the atmosphere and transferring it to the linseed or other drying oil, which leaves a tough and durable film of paint. Turpentine is also used in the manufacture of artificial camphor and rubber, and in linoleum, soap, and ink. **Gum thus,** used in artists' oil paints, is thickened turpentine, although gum thus was originally made from olibanum. Turpentine is often adulterated with other oils of the pine or with petroleum products, and the various states have laws regulating its adulteration for paint use.

The **pinene** in European turpentine is levorotatory while that in the American is dextro. **Pinonic acid** is acetyl dimethyl cyclobutane acetic acid. It is produced by oxidation of the pinene, and is a white powder used as a cross-linking agent for making heat-stable plastics.

Terpene alcohol, or **methylol pinene,** $C_{11}H_{17}OH$, is produced by condensing the beta pinene of gum turpentine with formaldehyde. **Nopol,** of the Glidden Co., is terpene alcohol. It has the chemical reactions of both a primary alcohol and pinene, and is used in making many chemicals. It is a water-insoluble liquid of specific gravity 0.963, boiling at 235°C. **Terpineol** is a name for refined terpene alcohols used largely for producing essential oils and perfumes. **Piccolyte resin,** of the Pennsylvania Industrial Chemical Corp., is a terpene thermoplastic varnish resin made from turpentine. The grades have melting points from 10 to 125°C. **Myrcene** is a polyolefin with three double bonds, which can be used as a substitute for butadiene in the manufacture of synthetic rubbers, or can be reacted with maleic anhydride or dibasic acids to form synthetic resins. It is made by isomerizing the beta pinene of gum turpentine. **Camphene** is produced by isomerizing the alpha pinene of turpentine. Camphor is then produced by oxidation of camphene in acid. Camphene was also the name of a lamp oil of the early nineteenth century made from distilled turpentine and alcohol. It gave a bright white light, but was explosive. The insecticide known as **Toxaphene,** of Hercules Inc., is made by chlorinating camphene to 68% chlorine, or to the empirical formula $C_{10}H_{20}Cl_8$. It is a yellow waxy powder with a piney odor, melting at 65 to 90°C. It is soluble in petroleum solvents.

TURQUOISE. An opaque blue gemstone with a waxy luster. It is a hydrous phosphate of aluminum and copper oxides. It is found in the western United States in streaks in volcanic rocks, but most of the turquoise has come from the Kuh-i-Firouzeh, or turquoise mountain, of Iran, which is a vast deposit of brecciated porphyry, or feldspar igneous rock. The valuable stones are the deep blue. The pale blue and green stones were called **Mecca stones** because they were sent to Mecca for sale to pilgrims. **Bone turquoise,** or **odontolite,** used for jewelry, is fossil bone or tooth, colored by a phosphate of iron.

TYPE METAL. Any metal used for making printing type, but the name generally refers to **lead-antimony-tin** alloys. The antimony has the property of expanding on cooling, and thus fills the mold and produces sharp, accurate type. The properties required in a type metal are ability to make sharp, uniform castings, strength and hardness, fairly low melting point, narrow freezing range to facilitate rapid manufacture in type-making machines, and resistance to drossing. A common type metal is composed of 9 parts lead to 1 antimony, but many varieties of other mixtures are also used. The antimony content may be as high as 30%, 15 to 20% being frequent. A common **monotype metal** has 72% lead, 18 antimony, and 10 tin. Larger and softer types are made of other alloys, sometimes containing bismuth; the hardest small type contains 3 parts lead to 1 antimony. A low-melting-point, soft-type metal contains 22% bismuth, 50 lead, and 28 antimony. It will melt at about 310°F (154°C). Copper, up to 2%, is sometimes added to type metal to increase the hardness, but is not ordinarily used in metals employed in rapid-acting type machines. Some monotype metal has about 18% antimony, 8 tin, and 0.1 copper, but standard **linotype metal** for pressure casting has 79% lead, 16 antimony, and 5 tin. **Stereotype metal,** for sharp casting and hard-wearing qualities, is given as 80.0% lead, 13.5 antimony, 6 tin, and 0.5 copper. **Intertype metal** has 11 to 14% antimony and 3 to 5 tin. A typical formula for **electrotype metal** is 94% lead, 3 tin, and 3 antimony. The Brinell hardness of machine-molded type ranges from 17 to 23, and that of stereotype metal is up to 30. As constant remelting causes the separation of the tin and lead, and the loss of tin, or impoverishment of the metal, new metal must be constantly added to prevent deterioration of a standard metal into an inferior alloy. For many years lead-antimony-tin alloys have been used as a weld seam filler in auto and truck bodies. In this application they are commonly referred to as **body solder.** Because of advances in printing technology and auto manufacturing, use of these lead alloys is steadily declining.

ULTRAHIGH-STRENGTH STEELS. The highest strength steels available. Arbitrarily, steels with tensile strengths of around 200,000 lb/in^2 (1,378 MPa) or higher are included in this category, and more than 100 alloy steels can be thus classified. They differ rather widely among themselves in composition and/or the way in which the ultrahigh strengths are achieved.

Medium-carbon low-alloy steels were the initial ultrahigh-strength steels, and within this group, a **chromium-molybdenum steel** (4130) grade and a **chromium-nickel-molybdenum steel** (4340) grade were the first developed. These steels have yield strengths as high as 240,000 lb/in^2 (1,654 MPa) and tensile strengths approaching 300,000 lb/in^2 (2,068 MPa). They are particularly useful for thick sections because they are moderately priced and have high hardenability. Several types of **stainless steels** are capable of strengths above 200,000 lb/in^2 (1,378 MPa), including a number of martensitic, cold-rolled austenitic, and semiaustenitic grades. The typical martensitic grades are types 410, 420, and 431, as well as certain **age-hardenable alloys.** The **cold-rolled austenitic stainless steels** work-harden rapidly and can achieve 180,000 lb/ in^2 (1,241 MPa) yield and 200,000 lb/in^2 (1,378 MPa) ultimate strength. The strength can also be increased by cold working at cryogenic temperatures. **Semiaustenitic stainless steels** can be heat-treated for use at yield strengths as high as 220,000 lb/in^2 (1,516 MPa) and ultimate strengths of 235,000 lb/in^2 (1,620 MPa). **Maraging steels,** a relatively new family, contain 18 to 25% nickel plus substantial amounts of cobalt and molybdenum. Some newer grades contain somewhat less than 10% nickel and between 10 and 14% chromium. Because of the low carbon (0.03% max) and nickel content, maraging steels are martensitic in the annealed condition, but are still readily formed, machined, and welded. By a simple aging treatment at about 900°F (482°C), yield strengths of as high as 300,000 and 350,000 lb/in^2 (2,068 and 2,413 MPa) are attainable, depending on specific composition. In this condition, although ductility is fairly low, the material is still far from being brittle.

There are two types of ultrahigh-strength, low-carbon, hardenable steels. One, a **chromium-nickel-molybdenum steel,** named **Astralloy,** with 0.24% carbon is air-hardened to a yield strength of 180,000 lb/in^2 (1,241 MPa) in heavy sections when it is normalized and tempered at 500°F (260°C). The other type is an **iron-chromium-molybdenum-cobalt steel** and is strengthened by a precipitation hardening and aging process to levels of up to 245,000 lb/in^2 (1,654 MPa) in yield strength. High-alloy **quenched-and-tempered steels** are another group that have extra high strengths. They contain 9% nickel, 4% cobalt, and from 0.20 to 0.30% carbon, and develop yield strengths close to 300,000 lb/in^2 (2,068 MPa) and ultimate strengths of 350,000 lb/in^2 (2,413 MPa). Another group in this high-alloy category resembles **high-speed tool steels,** but are modified to eliminate excess carbide, thus considerably improving ductility. These so-called **matrix steels** contain tungsten, molybdenum, chromium, vanadium, cobalt, and about 0.5% carbon. They can be heat-treated to ultimate strengths of over 400,000 lb/in^2 (2,757 MPa)—the highest strength presently attainable in steels, except for heavily **cold-worked high-carbon steel** strips used for razor blades and drawn wire for musical instruments, both of which have tensile strengths as high as 600,000 lb/in^2 (4,136 MPA).

URANIUM. An elementary metal, symbol U. It never occurs free in nature but is found chiefly as an oxide in the minerals pitchblende and carnotite where it is associated with radium. The metal has a specific gravity of 18.68 and atomic weight 238.2. The melting point is about 1133°C. It is hard but malleable, resembling nickel in color, but related to chromium, tungsten, and molybdenum. It is soluble in mineral acids.

Uranium has three forms. The alpha phase, or orthorhombic crystal, is stable to 660°C; the beta, or tetragonal, exists from 660 to 760°C; and the gamma, or body-centered cubic, is from 760°C to the melting point. The cast metal has a hardness of 80 to 100 Rockwell B, work-hardening easily. The metal is alloyed with iron to make **ferrouranium,** used to impart special properties to steel. It increases the elastic limit and the tensile strength of steels, and is also a more powerful deoxidizer than vanadium. It will denitrogenize steel and has also carbide-forming qualities. It has been used in high-speed steels in amounts of 0.05 to 5% to increase the strength and toughness, but because of its importance for atomic applications its use in steel is now limited to the by-product nonradioactive isotope uranium 238. The **green salt** used in atomic work is **uranium tetrafluoride,** UF_4. **Uranium hexafluoride,** UF_6, is a gas used to separate uranium isotopes.

Metallic uranium is used as a cathode in photoelectric tubes responsive to ultraviolet radiation. Uranium compounds, especially the uranium oxides, were used for making glazes in the ceramic industry and also for paint pigments. It produces a yellowish-green fluorescent glass, and a beautiful red with yellowish tinge is produced on pottery glazes. **Uranium dioxide,** UO_2, is used in sintered forms as fuel for power reactors. It is chemically stable, and has a high melting point at about 2760°C, but a low thermal conductivity. For fuel use the particles may be coated with about 0.001 in (0.003 cm) of aluminum oxide. This coating is impervious to xenon and other radioactive isotopes so that only the useful power-providing rays can escape. These are not dangerous at a distance of about 6 in (15 cm), and thus less shielding is needed. For temperatures above 2300°F (1260°C) a coating of pyrolitic graphite is used.

Uranium has isotopes from 234 to 239, and **uranium-235,** with 92 protons and 143 neutrons, is the one valued for atomic work. The purified natural metal contains only about one part U^{235} to about 140 parts of U^{238}, and about 100,000 lb (45,360 kg) of uranium fluoride, UF_6, must be processed to obtain one pound (0.45 kg) of $U^{235}F_6$. **Uranium-238,** after the loss of three alpha particles of total mass 12, changes into **radium-226.** The lead of old uranium minerals came from Ra^{226} by the loss of five alpha particles, and is **lead-206,** while the lead in thorium metals is **lead-208. Lead-207** comes from the decay of actinum, and exists only in small quantities.

Natural uranium does not normally undergo fission because of the high probability of the neutron being captured by the U^{238} which then merely ejects a gamma ray and becomes U^{239}. When natural uranium is not in concentrated form, but is embodied in a matrix of graphite or heavy water, it will sustain a slow chain reaction sufficient to produce heat. In the fission of U^{235}, neutrons are created which maintain the chain reaction and convert U^{238} to plutonium. About 40 elements of the central portion of the periodic table are also produced by the fission, and eventually these products build up to a point where the reaction is no longer self-sustaining. The slow, nonexplosive disintegration of the plutonium yields neptunium. **Uranium-233** is made by neutron bombardment of thorium. This isotope is fissionable and is used in thermonuclear reactors.

Uranium yellow, also called **yellow oxide,** is a **sodium diuranate** of the composition $Na_2U_2O_7 \cdot 6H_2O$, obtained by reduction and treatment of the mineral pitchblende. It is used for yellow and greenish glazing enamels and for imparting an opalescent yellow to glass, which is green in reflected light. **Uranium oxide** is an olive-green powder of the composition U_3O_8, used as a pigment. **Uranium trioxide,** UO_3, is an orange-yellow powder also used for ceramics and pigments. It is also called **uranic oxide.** As a pigment in glass it produces a beautiful greenish-yellow **uranium glass. Uranium pentoxide,** U_2O_5, is a black powder, and **uranous oxide,** UO_2, is used in glass to give a fine black color. **Sodium uranate,** Na_2UO_4, is a yellow to orange powder used to produce ivory to yellow shades in pottery glazes. The uranium oxide colors give luster and iridescence, but because of the application of the metal to atom work the uses in pigments and ceramics are now limited.

URANIUM ORES. The chief source of radium and uranium is **uraninite,** or **pitchblende,** a black massive or granular mineral with pitchlike luster. The mineral is a combination of the oxides of uranium, UO_2, UO_3, and U_3O_8, together with small amounts of lead, thorium, yttrium, cerium, helium, argon, and radium. The process of separation of radium is chemically complicated. Uraninite is found with the ores of silver and lead in central Europe. In the United States it occurs in pegmatite veins, in the mica mines of North Carolina, and in the carnotite of Utah and Colorado. The richest ores come from the Congo and from near Great Bear Lake, Canada. About 370 tons (336 metric tons) of Great Bear Lake ore produce 1 g of radium and 7,800 lb (3,538 kg) of uranium, and also small amounts of polonium, ionium, silver, and radioactive lead. Numerous minor uranium ores occur in many areas. A low-grade ore of 0.1% U_3O_8 can be upgraded to as high as 5% by ion exchange. Black mud from the fjords of Norway contains up to 2 oz (0.06 kg) of uranium per long ton (1 metric ton). **Tyuyamunite,** found in Turkman, averages 1.3% U_3O_8, with radium, vanadium, and copper. **Autunite,** or **uranite,** is a secondary mineral from the decomposition of pitchblende. The composition is approximately $P_2O_5 \cdot 2UO_3 \cdot CaO \cdot 8H_2O$. It is produced in Utah, Portugal, and South Australia. **Torbernite,** or **copper uranite,** $Cu(UO_2)_2P_20_8 \cdot 12H_2O$, is a radioactive mineral of specific gravity 3.22 to 3.6 and hardness 2 to 2.5. **Sengierite** is a copper-uranium mineral found in the Congo. It occurs in small green crystals. **Casolite** is a yellow earthy **lead uranium silicate,** $3(PbO \cdot UO_3 \cdot SiO_2)4H_2O$. **Pilbarite** is a **thorium lead uranate. Umohoite,** found in Utah, contains 48% uranium, with molybdenum, hydrogen, and oxygen. The name of the ore is a combination of the symbols of the contained elements. Uranium is also recovered chemically from phosphate rock. The phosphate waste rock of Florida contains from 0.1 to 0.4% of U_3O_8. Most uranium ores contain less than 0.3% U_3O_8. Solvent methods of extraction are used.

UREA. Also called **carbamide.** A colorless to white crystalline powder, $NH_2 \cdot CO \cdot NH_2$, best known for its use in plastics and fertilizers. The chemistry of urea and the carbamates is very complex, and a very great variety of related products are produced. Urea is produced by combining ammonia and carbon dioxide, or from **cyanamide,** $NH_2 \cdot C \cdot N$. It is a normal waste product of animal protein metabolism, and is the chief nitrogen constituent of urine. It was the first organic chemical ever synthesized commercially. It has a specific gravity of 1.323, and a melting point at 132°C.

The formula for urea may be considered as $O \cdot C(NH_2)_2$, and thus as an amide substitution in **carbonic acid,** $O \cdot C(OH)_2$, an acid which really exists only in its compounds. The urea-type plastics are called **amino resins.** The **carbamates** can also be considered as deriving from **carbamic acid,** NH_2COOH, an **aminoformic acid** that likewise appears only in its compounds. The carbamates have the same structural formula as the bicarbonates, so that **sodium carbamate** has an NH_2 group substituted for each OH group of the sodium bicarbonate. The **urethanes,** used for plastics and rubber, are **alkyl carbamates** made by reacting urea with an alcohol, or by reacting isocyanates with alcohols or carboxy compounds. They are white powders of the composition $NH_2COOC_2H_5$, melting at 50°C.

Isocyanates are esters of **isocyanic acid,** $H \cdot N \cdot C \cdot O$, which does not appear independently. The **dibasic diisocyanate** of General Mills, Inc., is made from a 36-carbon fatty acid. It reacts with compounds containing active hydrogen. With modified polyamines it forms **polyurea resins,** and with other diisocyanates it forms a wide range of urethanes. **Tosyl isocyanate,** of the Upjohn Co., for producing urethane resins without a catalyst is toluene sulphonyl isocyanate. The sulphonyl group increases the reactivity.

Methyl isocyanate, CH_3NCO, known as **MIC**, is a colorless liquid with a specific gravity of 0.9599. It reacts with water. With a flash point of less than 20°F (−6.6°C), it is flammable and a fire risk. It is a strong irritant and highly toxic. One of its principal uses is as an intermediate in the production of pesticides.

Urea is used with acid phosphates in fertilizers. It contains about 45% nitrogen and is one of the most efficient sources of nitrogen. Urea reacted with malonic esters produces **malonyl urea** which is the **barbituric acid** that forms the basis for the many soporific compounds such as **luminal, phenobarbital,** and **amytal.** The malonic esters are made from acetic acid, and **malonic acid** derived from the esters is a solid of the composition $CH_2(COOH)_2$ which decomposes at about 160°C to yield acetic acid and carbon dioxide.

For plastics manufacture, substitution on the sulfur atom in thiourea is easier than on the oxygen in urea. **Thiourea,** $NH_2 \cdot CS \cdot NH_2$, also called **thiocarbamide, sulfourea,** and **sulfocarbamide,** is a white, crystalline, water-soluble material of bitter taste, with a specific gravity of 1.405. It is used for making plastics and chemicals. On prolonged heating below its melting point, 182°C, it changes to **ammonium thiocyanate,** or **ammonium sulfocyanide,** a white, crystalline, water-soluble powder of the composition NH_4SCN, melting at 150°C. This material is also used in making plastics, as a mordant in dyeing, to produce black nickel coatings, and as a weedkiller. **Permafresh,** of Warwick Chemical Co., used to control shrinkage and give wash-and-wear properties to fabrics, is **dimethylol urea,** $CO(NHCH_2OH)_2$, which gives clear solutions in warm water.

Urea-formaldehyde resins are made by condensing urea or thiourea with formaldehyde. They belong to the group known as **aminoaldehyde resins** made by the interaction of an amine and an aldehyde. An initial condensation product is obtained which is soluble in water, and is used in coatings and adhesives. The final condensation product is insoluble in water and is highly chemical-resistant. Molding is done with heat and pressure. The urea resins are noted for their transparency and ability to take translucent colors. Molded parts with cellulose filler have a specific gravity of about 1.50, tensile strength from 6,000 to 13,000 lb/in² (41 to 89 MPa), elongation 15%, compressive strength to 45,000 lb/in² (310 MPa), dielectric strength to 400 volts per mil (16×10^6 volts per meter), and heat distortion temperature to 280°F (138°C). Rockwell hardness is about M118. **Urea resins** are marketed under a wide variety of trade names. The **Uformite resins** of Rohm & Haas are water-soluble thermosetting resins for adhesives and sizing. The **Urac resins,** of American Cyanamid, and the **Casco resins** and **Cascamite,** of the Borden Co., are urea-formaldehyde.

Weldwood, of the U.S. Plywood Corp., is a urea-formaldehyde adhesive. Other urea-formaldehyde resins are **Polybond, Arodure, Plaskon,** and **Synvarite.**

URETHANES. Also termed **polyurethanes.** A group of plastic materials based on polyether or polyester resin. The chemistry involved is the reaction of a diisocyanate with a hydroxyl-terminated polyester or polyether to form a higher molecular weight prepolymer which in turn is chain-extended by adding difunctional compounds containing active hydrogens, such as water, glycols, diamines, or amino alcohols. The urethanes are block-polymers capable of being formed by a literally indeterminate number of combinations of these components. The urethanes have excellent tensile strength and elongation, good ozone resistance, and good abrasion resistance. Combinations of hardness and elasticity unobtainable with other systems are possible in urethanes, ranging from Shore hardnesses of 15 to 30 on the "A" scale (printing rolls, potting compounds) through the 60 to 90 "A" scale for most industrial or mechanical goods applications, to the 70 to 85 Shore "D" scale. Urethanes are fairly resistant to many chemicals such as aliphatic solvents, alcohols, ether, certain fuels, and oils. They are attacked by hot water, polar solvents, and concentrated acids and bases.

Urethane foams are made by adding a compound that produces carbon dioxide or by reaction of a diisocyanate with a compound containing active hydrogen. Foams can be classified somewhat according to modulus as flexible, semiflexible or semirigid, and rigid. No sharp lines of demarcation have been set on these different classes as the gradation from the flexibles to the rigids is continuous. Densities of flexible foams range from about 1.0 lb/ft^3 (16 kg/m^3) at the lightest to 4 to 5 lb/ft^3 (64 to 80 kg/m^3) depending on the end use. Applications of flexible foams range from comfort cushioning of all types, e.g., mattresses, pillows, sofa seats, backs and arms, automobile topper pads, and rug underlay, to clothing interliners for warmth at light weight. Densities of rigid urethane foams range from about 1.5 to 50 lb/ft_3 (24 to 800kg/m^3).

Urethane elastomers are made with various isocyanates, the principal ones being **TDI (tolylene diisocyanate)** and **MDI (4,4′-diphenylmethane diisocyanate)** reacting with linear polyols of the polyester and polyether families. Various chain extenders, such as glycols, water, diamines, or aminoalcohols, are used in either a prepolymer or a one-shot type of system to form the long-chain polymer.

Textile fibers of urethane were first made in Germany under the name of **Igamide.** The **Fiber K,** or **Lycra,** of Du Pont, and the **Vyrene,** of the U.S. Rubber Co., are flexible **urethane fibers** used for flexible garments. They are more durable than ordinary rubber fibers or filaments, and are

30% lighter in weight. They are resistant to oils and to washing chemicals, and also have the advantage that they are white in color. **Spandex fibers** are stretchable fibers produced from a fiber-forming substance in which a long chain of synthetic molecules are composed of a segmented polyurethane. Stretch before break of these fibers is from 520 to 610%, compared to 760% for rubber. Recovery is not as good as in rubber. Spandex is white and dyeable. Resistance to chemicals is good but it is degraded by hypochlorides.

There are six basic types of **polyurethane coatings,** or **urethane coatings,** as defined by the American Society for Testing and Materials Specification D16. Types 1, 2, 3, and 6 have long storage life and are formulated to cure by oxidation, by reaction with atmospheric moisture, or by heat. Types 4 and 5 are catalyst-cured and are used as coatings on leather and rubber and as fast-curing industrial product finishes. Urethane coatings have good weathering characteristics as well as high resistance to stains, water, and abrasion.

VALVE STEEL. Used for automobile intake valves, this steel contains about 2% chromium, 3.75 silicon, 0.3 manganese, and 0.40 carbon, and the steel for exhaust valves has up to 24% chromium, 5 nickel, 3 molybdenum, 1 each of manganese and silicon, and 0.40 carbon. These steels are hot-sprayed with aluminum, and heated to 1450°F (788°C) to diffuse the aluminum into the surface to give an iron-aluminum compound about 0.001 in (0.0025 cm) thick which is resistant to hot corrosive gases.

VANADIUM. An elementary metal, symbol V, widely distributed, but found in commercial quantities in only a few places, chiefly Peru, Zimbabwe, Southwest Africa, and the United States. The common ores of vanadium are carnotite, patronite, roscoelite, and vanadinite. Much of the commercial vanadium comes from Peruvian patronite and shales. Some Russian vanadium comes from the mineral **tyuyamunite,** the calcium analog of carnotite. This analog also occurs in American carnotite as a greenish-yellow powder. Titaniferous ores of South Africa also furnish vanadium. But more than 60% of the known resources are in the United States. Carnotite occurs in Utah and Colorado, and the Arizona ore is vanadinite. The most important ore in the United States is **roscoelite.** It is a muscovite mica in which part of the aluminum has been replaced by vanadium. It occurs in micalike scales varying in color from green to brown. It has a specific gravity of 2.9. The ore mined in Colorado contains about 1.5% vanadium oxide, V_2O_5, and this oxide is extracted and marketed for making ferrovanadium and vanadium compounds. The slag from Idaho phosphorus workings contains up to 5% vanadium, which is concentrated to 13% and extracted as vanadium pentoxide. It is now also being recovered from petroleum. Venezuelan crude oil, containing 130 ppm vanadium, yields 2,000 lb (907 kg) of vanadium pentoxide per 1 million gal (3,785 m^3) of oil.

Vanadium is a pale-gray metal with a silvery luster. Its specific gravity is 6.02, and it melts at 3236°F (1780°C). It does not oxidize in the air and is not attacked by hydrochloric or dilute sulfuric acid. It dissolves with a blue color in solutions of nitric acid. It is marketed by the Vanadium Corp., 99.5% pure, in cast ingots, machined ingots, and buttons. The as-cast metal has a tensile strength of 54,000 lb/in^2 (372 MPa), yield strength of 45,000 lb/in^2 (310 MPa), and elongation 12%. Annealed sheet has a tensile strength of 78,000 lb/in^2 (537 MPa), yield strength 66,000 lb/in^2 (455 MPa), and elongation 20%, while the cold-rolled sheet has a tensile strength of 120,000 lb/in^2 (827 MPa) with elongation of 2%. Vanadium metal is expensive, but is used for special purposes such as for springs of high flexural strength and corrosion resistance. The greatest use of vanadium is for alloying. **Ferrovanadium,** for use in adding to steels, usually contains 30 to 40% vanadium, 3 to 6 carbon, and 8 to 15 silicon, with the balance iron, but may also be had with very low carbon and silicon. **Vanadium-boron,** for alloying steels, is marketed by the Vanadium Corp. as a master alloy containing 40 to 45% vanadium, 8 boron, 5 titanium, 2.5 aluminum, and the balance iron, but the alloy may also be had with no titanium. **Van-Ad alloy,** of the Chicago Development Corp., for adding vanadium to titanium alloys, contains 75% vanadium and the balance titanium. It comes as fine crystals. The **vanadium-columbium alloys** developed by Union Carbide, containing 20 to 50% columbium, have a tensile strength above 100,000 lb/in^2 (689 MPa) at 700°C, 70,000 lb/in^2 (482 MPa) at 1000°C, and 40,000 lb/in^2 (275 MPa) at 1200°C.

Vanadium salts are used to color pottery and glass and as mordants in dyeing. **Red cake,** or crystalline **vanadium oxide,** is a reddish-brown material, containing about 85% **vanadium pentoxide,** V_2O_5, and 9% Na_2O, used as a catalyst and for making vanadium compounds. Vanadium oxide is also used to produce yellow glass; the pigment known as **vanadium-tin yellow** is a mixture of vanadium pentoxide and tin oxide.

VANADIUM STEEL. Vanadium was originally used in steel as a cleanser, but is now employed in small amounts, 0.15 to 0.25%, especially with a small quantity of chromium, as an alloying element to make strong, tough, and hard low-alloy steels. It increases the tensile strength without lowering the ductility, reduces grain growth, and increases the fatigue-resisting qualities of steels. Larger amounts are used in high-speed steels and in special steels. Vanadium is a powerful deoxidizer in steels, but is too expensive for this purpose alone. Steels with 0.45 to 0.55% carbon and small amounts of vanadium are used for forgings, and cast steels for aircraft parts usually contain vanadium. In tool steels vanadium widens the hardening range, and by the formation of double carbides with chromium makes hard and keen-edge die and cutter steels. **Vasco vanadium steel,** of the Vanadium-Alloys Steel Co., contains 0.20% vanadium with 0.80 chromium in the various carbon grades from 0.50 to 1%. The higher carbon steels, for gages and rollers, have somewhat more chromium. All of these steels are classed as **chromium-vanadium steel.** The carbon-vanadium steels for forgings and castings, without chromium, have slightly higher manganese. Plain **carbon-vanadium steel** is regularly marketed in all standard carbon grades. It is finer graded, tougher, and keener edged than plain carbon steel. **Colonial No. 7, Red star steel,** and **Elvandi,** of the Vanadium-Alloys Steel Co., are steels of this type. **Python steel,** a shock-resistant steel of the Allegheny Ludlum Steel Co., has 0.25% vanadium, 0.90 carbon, 0.30 manganese, and 0.25 silicon. In case-hardening steels the vanadium forms an integrated bonding between the case and the core which gives shock resistance to the steel.

Vanadium steels require higher quenching temperatures than ordinary steels or nickel steels. **SAE 6145 steel,** with 0.18% vanadium and 1 chromium, has a fine grain structure and is used for gears. It has a tensile strength of 116,000 to 292,000 lb/in^2 (799 to 2,013 MPa) when heat-treated, with a Brinell hardness 248 to 566, depending on the temperature of drawing, and an elongation 7 to 26%. In cast vanadium steels it is usual to have from 0.18 to 0.25% vanadium with 0.35 to 0.45 carbon. Such castings have a tensile strength of about 80,000 lb/in^2 (551 MPa) and an elongation 22%. A nickel-vanadium cast steel has much higher strength, but high-alloy steels with only small amounts of vanadium are not usually classed as vanadium steels.

VANILLA BEANS. The seed pods of a climbing plant of the orchid family of which there are more than 50 known species. It is native to Mexico, but now also grown commercially in Malagasy, Réunion, Mauritius, and tropical America. It is used for the production of the flavor **vanilla.** The species grown for commercial vanilla is *Vanilla planifolia,* a tall climbing herb with yellow flowers. It grows in humid tropic climates. The flowers are pollinated by hand to produce 30 to 40 beans per plant. The green beans are cured immediately in ovens to prevent spoilage after a sweating process. During the curing the glucoside is changed by enzyme action into **vanillin,** which crystallizes on the surface and possesses the characteristic odor and flavor. The dark-brown cured pods are put up in small packs in tin containers. **Vanilla extract** is made by percolating the chopped bean pods in ethyl alcohol, and then concentrating the mixture by evaporating the alcohol at a low temperature to avoid impairing the flavor.

The species *V. pompana* is more widely distributed, but is not as fragrant.

The vanilla grown in Tahiti has an odor of heliotrope which must be removed. At least 15 species of vanilla grow in the Amazon and Orinoco valleys. Vanilla was used by the Aztecs for flavoring chocolate. It is now used for the same purpose, and as a flavor for ice cream, puddings, cakes, and other foodstuffs.

Vanillin is also produced synthetically from **eugenol** derived from clove oil. It is also made from **coniferin,** $C_{16}H_{22}O_8 \cdot 2H_2O$, a white crystalline material of melting point 185°C obtained from the sapwood of the northern pine. It is produced in Wisconsin from pulp-mill waste liquors by hydrating into sugars and oxidizing to vanillin. But the synthetic vanillin does not give the full, true flavor of vanilla, as a blend of other flavors is present in the natural product. The demand for vanilla as a flavor is always greater than the supply, so that even the grades rated as pure vanilla extract may be so adulterated or diluted as to lose the full, rich flavor.

Ethyl vanillate, $C_6H_3(OH)(OCH)_3(COOC_2H_5)$, is made from Wisconsin sulfite liquor. It is used in cheese to prevent mold, and as a preservative in tomato and apple juice. **Lioxin,** of the Ontario Paper Co., is an impure 97% vanillin made from sulfite lignin. It is not suitable for use as a flavor, but is used as an odor-masking agent, as a brightener in zinc-plating baths, as an antifoam agent in lubricating oils, and for making syntans. **Vanitrope,** of Shulton, Inc., is a synthetic aromatic with a flavor 15 times more powerful than vanillin but with a resinous note resembling that of coumarin. It differs from vanillin chemically by having no aldehyde group, and is a **propenyl guaethol** related to eugenol. It is used as a vanilla extender. A blend of Vanitrope and vanilla, called **Nuvan,** is used as a low-cost vanilla flavor. **Vanatone** and **Vanarine,** of Fritzsche Bros., Inc., are blends of vanillin with aldehydes and esters to increase the flavor tone.

VAPOR-DEPOSITED COATINGS. Thin single or multilayer coatings applied to base surfaces by deposition of the coating metal from its vapor phase. Most metals and even some nonmetals, such as siliconoxide, can be vapor-deposited. **Vacuum-evaporated films,** or **vacuum metallized films,** of aluminum are most common. They are applied by vaporizing aluminum in a high vacuum and then allowing it to condense on the object to be coated. Vacuum-metallized films are extremely thin, ranging from 0.002 to 0.1 mil (0.00005 to 0.003 mm). In addition to vacuum evaporation, vapor-deposited films can be produced by **ion sputtering, chemical-vapor plating,** and a **glow-discharge** process. In ion sputtering, a high voltage applied to a target of the coating material in an ionized gas media causes target atoms (ions) to be dislodged and then to condense as a **sputtered** coating on the base material. In chemical-vapor plating, a film is deposited when a metal-bearing gas thermally decomposes on contact with the heated surface of the base material. And in the glow-discharge process, applicable only to polymer films, a gas discharge deposits and polymerizes the plastic film on the base material.

VARNISH. A solution of a resin in drying oil, which when spread out in a thin film dries and hardens by evaporation of the volatile solvent, or by the oxidation of the oil, or both. A smooth, glossy coating is left on the surface. Varnishes do not contain pigments; when mixed with pigments, they become enamels. The most commonly used resin is ordinary rosin, and the most common drying oils are linseed and tung oils. **Spirit varnishes** are those in which a volatile liquid, such as alcohol or ether, is used as a solvent for the resin or oil. They dry by the evaporation of the solvent. **Oleoresinous varnishes** are those in which the resin is compounded with an oxidizable oil, such as linseed oil. The gums used in varnish, such as copal, dammar, and kauri, produce hardness and gloss to the film, and the **fossil resins,** such as kauri, give greater hardness and luster to varnishes than do the natural resins. The oils, such as tung and linseed, make it elastic and durable.

Other important ingredients of varnishes are driers, such as manganese oxide, to hasten the action of the drying oil, and thinning agents, or reducers, such as turpentine, naphtha, and benzol. Hydrated lime is added to varnishes to neutralize the acid in the resin, and to clarify and harden the varnish to prevent it from becoming sticky in warm weather. **Spar varnishes** are those made to withstand weather conditions. **Gloss oil** is a solution of hardened rosin in benzine or in turpentine with sometimes a small amount of tung oil to give a tougher film. It gives a high gloss but is not durable. **Long varnishes** are those containing 20 to 100 gal (76 to 379 L) of oil to 100 lb (45 kg) of resin; a **short varnish** is one with less oil. The short varnishes are hard and more glossy, but not as flexible or durable. Ordinarily, quick-drying varnish made with a natural resin is less durable than slow-drying; hardness and gloss are not guarantees of good varnish.

Varnish was originally only a colorless or nearly colorless coating material for furniture and fancy wood products to give a smooth, glossy surface for protection and to bring out the texture of the wood, and **marine varnish** was a high grade of spar varnish. Any color used was merely to accent the original color of the wood or to imitate the color of another wood of similar grain. **Insulating varnishes** were colorless varnishes for protecting drawings, paintings, and other products from moisture, or for electrical insulating. But the term varnish has come to mean any light-bodied quick-drying glossy finish as distinct from heavily pigmented glossy enamels. **Synthetic varnishes** may now contain synthetic resins in oils, or they may be made entirely with synthetic resins in solvents. **Electrical varnishes** are likely to be silicone, epoxy, or polyester resins that give good dielectric strength and adhesion.

VEGETABLE FATS. When specifically used, the term refers particularly to semisolid vegetable oils that are used chiefly for food. Vegetable oils and fats usually contain only small quantities of the fat-soluble vitamins A, D, and E, and after refining, they are usually devoid of vitamins. Thus, they are a better food in the producing countries. Climate in which the plant is grown has an effect on the nature of the oils. Warm climates favor the development of oleic acid while colder climates favor the less palatable linolenic acid. The low-melting-point oils are more easily assimilated in the body, but when these are hydrogenated to a melting point above 45°C they become difficult to assimilate. Most of the more edible vegetable fats, as distinct from the more liquid food oils, are tropical products. **Suari fat** is a hard white fat with a pleasant taste obtained from the kernels of the seeds of the *Caryocar brasiliense* and other species of tropical America. The kernels yield 60 to 70% fat of a specific gravity of 0.989, melting point 30 to 37°C, and iodine value 41 to 50. **Ucuhuba tallow,** used in soaps and for candles, is a fat from the seeds of the trees *Virola surinamensis* and *V. sebifera* of Brazil. **Mahuba fat** is a hard edible fat from the fruit of the tree *Acrodicilidium mahuba* of Brazil. **Gamboge butter,** known locally as **gurgi** and **murga,** is from the seeds of the fruit of the trees *Garcinia morella, G. hanburii,* and other species of Sri Lanka and India. The melting point is 34 to 37°C, specific gravity 0.90 to 0.913, and saponification value 196. It is used as a soap and food oil, and locally as an illuminant and as an ointment. From these trees also comes the gum resin **gamboge** used in medicine as a cathartic, and also used as a brown dyestuff. It is alcohol-soluble.

Sierra Leone butter, also called **kanga** or **lamy,** is a pale-yellow fat from the seeds of the fruit of the tree *Pentadeama butyracea* of West Africa. The melting point is about 40°C, specific gravity 0.857 to 0.860, and saponificatin value 186 to 198. It is a soap oil. **Mafura tallow** is a bitter-tasting heavy fat from the nuts of the tree *Trichilia emitica* of East Africa. It is used for soap, candles, and ointments. The specific gravity is 0.902, melting point about 40°C, and saponification value 201. **Shea nut oil,** also known as **shea butter, Bambuk butter, Galam butter,** and by various local names as **karité, kade,** and **kedempó,** is a fat obtained from the kernels of the fruit of the large tree *Bassis butyrospermum* of tropical West Africa. The kernels contain 45 to 55% fat, which when refined is white, stiffer than lard, with little odor or taste. The melting point is 33 to 42°C. It contains oleic and stearic acids and 3 to 4% lauric acid. It is used in Europe in butter substitutes, and as a substitute for cocoa butter, and also in candles. **Malabar tallow,** also called **dhupa fat** and **pinay tallow,** used in Europe for soap and candles, and in India for food, is from the kernels of the seed of the evergreen **pinne tree,** *Vateria indica* of south India. The tree also yields white dammar or Indian copal. The seeds give about 25% of a greenish-yellow, odorless, and tasteless fat of specific gravity 0.890, melting point

40°C, and saponification value 190. The fat is extracted by grinding the roasted seed, boiling in water, and skimming off. It is bleached by exposure.

Vegetable tallow, also called **bayberry tallow** and **myrtle wax,** used extensively in Europe for soapmaking and in the United States for blending in candles and with waxes, is a waxy fat obtained from the outside coating of the berries of species of *Myrica* bushes of America, Europe, and Africa by boiling the berries in water. The berries yield 15 to 25% tallow. The species *M. cerifera* and *M. carolinensis* grow in the eastern coastal states, and the *M. mexicana* grows in Central America. The melting point of the tallow is 40 to 46°C, specific gravity 0.995, and saponification value 205 to 212. The Central American product contains about 58% myristic acid, 36 palmitic, and 1.3 oleic acid. **Ocuba wax** is a waxy fat, but not chemically a wax, obtained from the seeds of the fruit of the shrub *Myristica ocuba* of Brazil. The seeds yield about 20% fat with a specific gravity of 0.920 and melting point of 40°C, which is used in candles. The fruit nut is surrounded by a thick skin which yields a water-soluble pink dye known as **ocuba red.**

Mahubarana fat is a pale-yellow solid oil of melting point 40 to 44°C obtained from the kernels of the fruit of trees of the genus *Boldoa.* The kernels contain 65% oil, which is used for soaps and candles. **Mocaya butter** is a fat from the kernels of the nuts of the **Paraguayan palm,** *Acrocomia sclerocarpa,* of tropical South America. The tree resembles a coconut palm, but the nuts grow in bunches. The pulp of the fruit contains 60% of a yellow oil similar to palm oil. The kernels yield 53 to 65% of the mocaya fat, which is softer than palm kernel oil. The specific gravity is 0.865, and saponification value 240. It has the same uses as palm kernel oil.

VEGETABLE OILS. An important class of oils obtained from plants, used industrially as drying oils, for lubricants, in cutting oils, for dressing leather, and for many other purposes. Many of the oils find wide usage in food products. Large tracts of land are under cultivation in all parts of the world for the production of the seeds and fruits from which the oils are obtained. Linseed, cottonseed, palm, olive, and castor beans are examples of these, and the oils are obtained by crushing. In some cases the oilbearing material, copra or soybean, may be dehydrated before crushing, making it simpler to extract the oil, and giving a better residue meal for animal feed. The chief distinction between vegetable oils and fats is a physical one, oils being fluid at ordinary temperatures. Vegetable oils can be thickened for various uses by oxidation, by blowing air through them, or they can be solidified by hydrogenation.

In the making of plastics and chemicals from fatty acids derived from vegetable oils the cost of the oil may be as much as 50% of the final cost of the product, and price economics plays an important part in the choice of the raw material. Oils produced in countries subject to political and economic disturbances may have sudden and great price fluctuations. Thus, domestic soybean oil may be substituted for castor oil in making nylon even though more chemical operations are needed, or the acids may be synthesized from petroleum hydrocarbons.

Food oils are chosen by their content of essential fatty acids, but taste is an important factor. Linseed oil is not used for food in the Unitd States, although it has high food value and contains both linoleic and linolenic acids. Safflower oil, high in linoleic acid, ranks high as a food oil, only 1.35 g of oil being required to provide 1 g of essential fatty acids. Olive oil, high in oleic acid with only one double bond, requires 14.2 g of oil for 1 g of essential acid. But olive oil requires less linoleic acid to counteract its effect than an equivalent amount of a saturated acid with no double bond. Butter requires the consumption of 20 g to obtain 1 g of essential acids, while soybean, corn, and cottonseed oils, used in margarine, rank high as food oils.

Considerable oil is extracted from the kernels of the stones or pits of cherries, apricots, and other fruits as a by-product of the canning and drying of fruits. **Cherry kernel oil** is from cherry pits which contain 30 to 38% oil. The cold-pressed oil is yellow and has a pleasant flavor. It is used in salad oils and in cosmetics. The hot-pressed oil is used in soaps. The oil contains 47% oleic acid, 40 linoleic, 4 palmitic, 3 stearic, and some arachidic and myristic acids. **Grapeseed oil** is obtained by pressing the by-product grape seeds from the wine industry. The seeds contain 10 to 15% oil, valued in Europe as an edible oil, but used in the United States mostly for paints and soaps. The oil contains about 52% linoleic acid, 32% oleic acid, and palmitic, stearic, and arachidic acids. The hot-pressed oil is dark green and not sweet, but the cold-pressed refined oil is colorless and has a nutlike taste. **Tomato seed oil** is from the seeds of the **tomato,** *Lycopersicon esculentum,* the seeds being by-products of the manufacture of **tomato juice** and **tomato puree,** vast quantities of which are produced in the United States from the pulp. The tomato plant is a perennial native to Central and South America, and was grown by the Aztecs under the name of **tomatl.** There are many varieties, and the fruits are true berries. The common red varieties are 2 to 3 in (5 to 8 cm) in diameter and contain a large number of seeds in the pulp. The seeds yield 17% oil by cold pressing, or 33% by solvent extraction. The cold-pressed oil is a clear liquid of 0.920 specific gravity, with an agreeable odor and bland taste. The iodine number is 113, and saponification value 192. It is used in salad oils, margarine, soaps, and as a semidrying oil for paints.

VELVET. A closely woven silk fabric with a short pile on one side formed by carrying the warp threads over wires and then cutting open the loops.

Velvet is made in a great variety of qualities and weights, and may have a cotton back in the cheaper grades, or be made in wool. True velvet is all silk, but because of the number of imitations in other materials this variety is usually designated as **silk velvet.** Velvet is dyed in various colors, the depth of color shown by the pile, giving it an air of richness. Its largest use is in dress goods and hangings, but it is used industrially for upholstery, fancy linings, and trim.

Plush is a name for fabrics woven of cotton, silk, linen, or wool, having a pile deeper than that of velvet. It is used for upholstery. Originally the pile of plush consisted of mohair or worsted yarns, but there is now no distinction except in the length of the pile. **Upholstery plush** is made in brocade designs by burning the pile with rollers to form a lower background. Plush is also dyed and curled to imitate furs.

Velveteen is an imitation velvet, woven of cotton. In the best grades the pile is of mercerized yarns. Velveteen is woven into two systems of filling yarns and one system of warp yarns, the pile being made with the cut filling yarns instead of the warp yarns as in velvet. It belongs to the class of **fustians** which also includes **moleskin** and **corduroy.** The latter is a sturdy pile fabric with heavy warp rib, dyed in the piece. It is also made in wool. The ribs run lengthwise, while **whipcord,** a hard-woven worsted fabric, has fine ribs running diagonally on the face. Velveteen is used for apparel, linings for jewelry and silverware boxes, shoe uppers, artificial flowers, and covering material. It is made either plain back or twill back, the plain back having a tendency to loosen and drop the pile.

VERMICULITE. A foliated mineral employed in making plasters and board for heat, cold, and sound insulation, as a filler in calking compounds, and for plastic mortars and refractory concrete. The mineral is an alteration product of **biotite** and other micas, and is found in Colorado, Wyoming, Montana, and the Transvaal. It occurs in crystalline plates, specific gravity 2.3, and hardness 1.5, measuring sometimes as much as 9 in (23 cm) across and 6 in (15 cm) in thickness. The color is yellowish to brown. Upon calcination at 1750°F (954°C), vermiculite expands at right angles to the cleavage into threads with a vermicular motion like a mass of small worms; hence its name. The volume increases as much as 16 times, and the color changes to a silvery or golden hue. It is ground into pellet form. Plaster made with 60% vermiculite, 30 plaster of Paris, and 10 asbestos will withstand red heat without disintegrating. **Therm-O-Flake brick,** of the Illinois Clay Products Co. is made of granules of **exfoliated vermiculite** bonded with a chemical. It is lightweight, tough, and will withstand temperatures up to 2000°F (1093°C). The corklike pellets of vermiculite used for insulating fill in house walls are called **mica pellets. Zonolite,** of W. R. Grace & Co., is an exfoliated vermiculite. **Zonolite Verxite,** of the same company, is a spongy granular powder form of **Verxite,** a thermally expanded vermiculite. It is used as a blending agent in animal feeds. A sound-absorbing building tile marketed by Johns-Manville under the name of **Rockoustile** is made of exfoliated mica. An expanded vermiculite of extremely fine mesh, under the name of **Mikolite,** of the Mikolite Co., is used as an extender in aluminum paint and in lubricating oils. **Exfoliated mica** is a name for expanded vermiculite. **Terra-Lite,** of the Zonolite Co., is fluffy powdered vermiculite for conditioning soils. It holds water and prevents soil crusting, and helps to maintain soil temperature below the critical 80°F (27°C).

VINYL RESINS. A group of products varying from liquids to hard solids, made by the polymerization of ethylene derivatives, employed for finishes, coatings, and molding resins, or it can be made directly by reacting acetic acid with ethylene and oxygen. In general, the term vinyl designates plastics made by polymerizing vinyl chloride; vinyl acetate, or vinylidene chloride, but may include plastics made from styrene and other chemicals. The term is generic for compounds of the basic formula RCH:CR′CR″. The simplest are the polyesters of vinyl alcohol, such as vinyl acetate. This resin is lightweight, with a specific gravity of 1.18, and is transparent, but it has poor molding qualities and its strength is no more than 5,000 lb/in^2 (34 MPa). The vinyl resins are brittle at low temperatures. **Elvacet,** of Du Pont, and **Lemac,** of the American Monomer Corp., are vinyl acetate molding resins. But the vinyl halides, CH_2:CHX also polymerize readily to form **vinylite resins,** which mold well, have tensile strengths to 9,000 lb/in^2 (62 MPa), high dielectric strength, and high chemical resistance, and a widely useful range of resins is produced by copolymers of vinyl acetate and vinyl chloride. Various grades of **Bakelite, Geon, Tygon,** and other resins are these chloride acetate coopolymers.

Vinyl alcohol, CH_2:CHOH, is a liquid boiling at 35.5°C. **Polyvinyl alcohol** is a white, odorless, tasteless powder which on drying from solutions forms a colorless and tough film. The material is used as a thickener for latex, in chewing gum, and for sizes and adhesives. It can be compounded with plasticizers and molded or extruded into tough and elastic products. **Hydrolized polyvinyl alcohol** has greater water resistance, higher adhesion, and its lower residual acetate gives lower foaming. **Soluble film,** for packaging detergents and other water-dispersible materials to eliminate the need of opening the package, is a clear polyvinyl alcohol film. Textile fibers are also made from polyvinyl alcohol, either water-soluble or insolubilized with formaldehyde or another agent. **Vinal,** of the Air Reduction Chemical Co., is a polyvinyl alcohol textile fiber which is hot-drawn by a semimelt process and insolubilized after drawing. The fiber has a high degree of orientation and crystallinity, giving good strength and hot-water resistance. Polyvinyl alcohol fibers are called **Vinylon** in Europe and **Kuravilon** in Japan.

Vinyl alcohol reacted with an aldehyde and an acid catalyst produces a group of polymers known as **vinyl acetal resins,** and separately designated by type names, as polyvinyl butyral and polyvinyl formal. The polyvinyl alcohols are called **Solvars,** and the polyvinyl acetates are called **Gelvas.** The vinyl ethers range from **vinyl methyl ether,** CH_2:$CHOCH_3$, to **vinyl ethylhexyl ether,** from soft compounds to hard resins. **Vinyl ether** is a liquid which polymerizes, or can be reacted with hydroxyl groups to form acetal resins. **Gantrez M,** of General Aniline & Film Corp., is a water-soluble polyvinyl methyl ether for use in paints, inks, and adhesives. **Alkyd vinyl ethers** are made by reacting acetylene with an alcohol under pressure, producing **methyl vinyl ether, ethyl vinyl ether,** or **butyl vinyl ether.** They have reactive double bonds which can be used to copolymerize with other vinyls to give a variety of physical properties. The **polyvinyl formals, Formvars,** are used in molding compounds, wire coatings, and impregnating compounds. It is one of the toughest of the thermoplastics. The Formvar of the Shawinigan Products Corp. has a specific gravity of about 1.3, a tensile strength up to 12,000 lb/in^2 (82 MPa), elongation from 5 to 20%, Rockwell hardness M85, and dielectric strength of 450 volts per mil (17 $\times$ 10^6 volts per meter). This type of plastic is resistant to alkalies but is attacked by acids.

A **plastisol** is a vinyl resin dissolved in a plasticizer to make a pourable liquid without a volatile solvent for casting. The poured liquid is solidified by heating. **Plastigels** are plastisols to which a gelling agent has been added to increase viscosity. The **polyvinyl acetals, Alvars,** are used in lacquers, adhesives, and phonograph records. The transparent **polyvinyl butyrals, Butvars,** are used as interlayers in laminated glass. They are made by reacting polyvinyl alcohol with **butyraldehyde,** C_3H_7CHO. **Vinal** is a general name for vinyl butyral resin used for laminated glass.

Vinyl acetate is a water-white mobile liquid with boiling point 70°C, usually shipped with a copper salt to prevent polymerization in transit. The composition is CH_3:COO:CH:CH_2. It may be polymerized in benzene and marketed in solution, or in water solution for use as an extender for rubber, and for adhesives and coatings. The higher the polymerization of the resin, the higher the softening point of the resin. The formula for **polyvinyl acetate resin** is given as $(CH_2\text{:}CHOOCCH_3)_x$. It is a colorless, odorless thermoplastic with density of 1.189, unaffected by water, gasoline, or oils, but soluble in the lower alcohols, benzene, and chlorinated hydrocarbons. Polyvinyl acetate resins are stable to light, transparent to untraviolet light, and are valued for lacquers and coatings because of their high adhesion, durability, and ease of compounding with gums and resins. Resins of low molecular weight are used for coatings, and those of high molecular weight for molding. **Darex** and **Everflex,** of W. R. Grace & Co., are paint and coating resins, and **Vinylite** and **Vinyloid,** of Union Carbide, are vinyl acetate resins for molding. **Wood Glu,** of Paisley Products, Inc., is a milky water dispersion of polyvinyl acetate used as an adhesive for wood and paper. Vinyl acetate will copolymerize with maleic acrylonitrile, or acrylic esters. With ethylene it produces a copolymer latex of superior toughness and abrasion resistance for coatings. **Aircoflex 500,** of Air Reduction Co., Inc., is a vinyl acetate ethylene copolymer latex.

Vinyl chloride, CH_2CHCl, also called **ethenyl chloride** and **chloroethylene,** produced by reacting ethylene with oxygen from the air and ethylene dichloride, is the basic material for the polyvinyl chloride resins. It is a gas. The plastic was produced originally in Germany under the name of **Igelite** for cable insulation and as **Vinnol** for tire tubes. The tensile strength of the plastic may vary from the flexible resins with about 3,000 lb/in^2 (20 MPa) to the rigid resin with a tensile strength to 9,000 lb/in^2 (62 MPa) and Shore hardness of 90. The dielectric strength is high, up to 1,300 volts per mil (52×10^6 volts per meter). It is resistant to acids and alkalies. **Polyvinyl chloride** usually comes as a white powder for molding or extruding, but **PVC pearls,** of the Escambia Chemical Corp., is the material made by a water-suspension process in the form of white porous particles capable of absorbing easily a high proportion of plasticizer. **Polvin** and **Opalon,** of Monsanto, are polyvinyl chloride resins, and **Ultron** is polyvinyl chloride film. Unplasticized polyvinyl chloride is used for rigid chemical-resistant pipe. **Kraloy D-500,** of the Kraloy Plastic Pipe Co., and **Carlon V,** of the Carlon Products Corp. are polylvinyl chloride in rigid pipe. **Vyflex sheet,** of Kaykor Industries, Inc., is rigid unplasticized polyvinyl chloride in sheets for making acid tanks, ducts, and flumes. **Polyvinyl chloride sheet,** unmodified, may have a tensile strength of 8,200 lb/in^2 (57 MPa), flexural strength of 12,600 lb/ in^2 (86 MPa), and a light transmission of 78%. But in bends or corners of moldings, or in folds of sheet, the resin shows opaque whitening. This can be eliminated by modifying the resin with an acrylic. The resins have improved processing qualities and flexibility, especially for blown bottles and extruded parts, but the light transmission is reduced. **Acryloid KM-607,** of Rohm & Haas, is a white acrylic powder. When used in polyvinyl sheet to a content of 20% it gives a tensile strength of 6,700 lb/in^2 (46 MPa) and a light transmission of 72%.

Vinylidene chloride plastics are derived from ethylene and chlorine polymerized to produce a thermoplastic with softening point of 240 to 280°F (116 to 138°C). The resins are noted for their toughness and resistance to water and chemicals. The molded resins have a specific gravity of 1.68 to 1.75, tensile strength 4,000 to 7,000 lb/in^2 (27 to 48 MPa), and flexural strength 15,000 to 17,000 lb/in^2 (103 to 117 MPa). **Saran** is the name of a vinylidene chloride plastic of the Dow Chemical Co., extruded in the form of tubes for handling chemicals, brines, and solvents to temperatures as high as 275°F (135°C). It is also extruded into strands and woven into a box-weave material as a substitute for rattan for seating. **Velon,** of the Firestone Tire & Rubber Co., is this material for screens and fabrics. **Saran latex,** a water dispersion of the plastic, is used for coating and impregnating fabrics. For coating food-packaging papers, it is waterproof and greaseproof, is odorless and tasteless, and gives the papers a high gloss. Saran is also produced as a strong transparent film for packaging. **Saranex,** of Dow Chemical Co., is **Saran film** laminated with polyethylene sheet to give heat-sealing qualities. **Saran bristles** for brushes are made in diameters from 0.010 to 0.020 in (0.025 to 0.051 cm). **Zetek,** a textile fiber of the B. F. Goodrich Co., is a polyvinylidene cyanide.

Vinyl benzoate is an oily liquid of the composition $CH_2{:}CHOOCC_6H_5$, which can be polymerized to form resins with higher softening points than those of polyvinyl acetate, but are more brittle at low temperatures. These resins, copolymerized with vinyl acetate, are used for water-repellent coatings. **Vinyl crotonate,** $CH_2{:}CHOOCCH{:}CHCH_3$, is a liquid of specific gravity of 0.9434. Its copolymers are brittle resins, but it is used as a cross-linking agent for other resins to raise the softening point and to increase abrasion resistance. **Vinyl formate,** $CH_2{:}CHOOCH$, is a colorless liquid which polymerizes to form clear **polyvinyl formate resins** that are harder and more resistant to solvents than polyvinyl acetate. The monomer is also copolymerized with ethylene monomers to form resins for mixing in specialty rubbers. **Methyl vinyl pyridine,** $(CH_3)(CHCH_2)C_5H_3N$, is produced by the Phillips Chemical Co. for use in making resins, fibers, and oil-resistant rubbers. It is a colorless liquid boiling at 64.4°C. The active methyl groups give condensation reactions, and it will copolymerize with butadiene, styrene, or acrylonitrile. **Polyvinyl carbazole,** under the name of **Luvican,** was used in Germany as a mica substitute for high-frequency insulation. It is a brown resin, softening at 150°C.

An adhesive to replace rubber cement was made in Germany by combining **Oppanol C,** a high-molecular-weight polyvinyl isobutylene ether, with **Igovin,**, a low-molecular-weight polyvinyl isobutyl ether, and compounding with zinc oxide and wool grease. **Marvinol VR-10,** developed by the Glenn L. Martin Co., for coatings and impregnations, was produced by the reaction of acetylene and hydrogen chloride with a catalyst. It is also used for casting into films. The **Marvinol resins** of the U.S. Rubber Co. are polyvinyl chloride. The vinyl plastics are much used for wall tile and sheet wall coverings. They are adaptable to bright colors, are nonstaining, and easily cleaned. **Kalitex** is a vinyl sheet wall covering of the U.S. Plywood Corp., having an embossed burlap-weave pattern with colors on the reverse side. **Vinyon,** of Union Carbide, is a vinyl chloride-acetate fiber in various grades. Since it is resistant to strong acids and alkalies, **Vinyon fiber** is made into filter cloth for temperatures not above 160°C. It is also used in wool mixtures. It is produced by the copolymerization of vinyl chloride and vinyl acetate. **Vinyon N** is a vinyl chloride-acrylonitrile copolymer marketed as a fine, silklike textile fiber. The fiber has high strength, an elongation of 30%, and is nonflammable. It has a light yellow color and is easily dyed. **Carilan** is a Japanese vinyl acetate fiber.

The possibility of variation in the vinyl resins by change of the monomer, copolymerization, and difference in compounding is so great that the term vinyl resin is almost meaningless when used alone. The resins are marketed under a continuously increasing number of trade names. In general, each resin is designed for specific uses, but not limited to those uses. **Advagum,** of the Advance Solvents & Chemical Corp., is a highly plasticized vinyl copolymer used as an extender for rubbers. **Pliovac,** of the Goodyear Tire & Rubber Co., is a high-molecular-weight vinyl chloride for coatings, tiling, and extrusions. **Victron,** of the Naugatuck Chemical Co., and **Saflex,** of the Monsanto Company, are vinyl acetal resins. **Butacite,** of Du Pont, is a clear polyvinyl butyral for laminated glass. **Formex,** of the General Electric Co., is a vinyl formal resin for insulating wire. **Boltaron 6200,** of Bolta Products, Inc., is rigid unplasticized polyvinyl chloride in sheets, rods, and pipes, resisting thermal distortion below 175°F (80°C). **Ingerin** is a German polyvinyl ether, and **Cosal** is made from vinyl isobutyl ether. **Mipolam** is a German vinyl polymer for floor coverings.

Kynar, of Pennwalt, is a polymer of **vinylidene fluoride,** $CH_2:CF_2$, with a high molecular weight, about 500,000. It is a hard, white thermoplastic resin with a slippery surface, and has a high resistance to chemicals. It resists temperatures to 650°F (343°C), and does not become brittle at low temperatures. It extrudes easily, and has been used for wire insulation, gaskets, seals, molded parts, and piping.

VITAMINS. Organic chemical compounds which are vital building units, coenzymes, or catalyzing agents in the growth and maintenance of animal bodies. They are produced by extraction from vegetable or animal products or made synthetically, and marketed in solid or extract form for use in foodstuffs and pharmaceuticals. **Vitamin A,** called **carotene** because of its abundance in carrots, is an orange-yellow needle-shaped crystalline substance with a complex molecular structure having the empirical formula $C_{40}H_{56}$. It is soluble in fats, but poorly soluble in water. Yellow and leafy green vegetables are rich sources of carotene-bearing pigments, and carotene accompanies the green chlorophyll coloring of all plants. The more intense the green or yellow coloring, the greater the carotene content. **Lycopene,** the red coloring agent of tomatoes, has the same empirical formula as carotene, and both contain eight isoprene units, but it has a different structure. The color is due to large numbers of conjugated double bonds, and different colors are from different arrangements. **Cryptoxanthin,** one of the four yellow carotene-carrying pigments, occurs in yellow corn, egg yolk, and green grasses. Animals convert carotene of green plants into vitamin A which is then obtained commercially from the tissues, especially from the liver. Deficiency of vitamin A in the human body causes night blindness, muscular weakness, and defective tooth structure, but an excess can cause body deformities and stillbirth.

Vitamin B is a complex of several vitamins, including **vitamin B_1** and **vitamin G.** The former cannot be formed in the normal processes of the human body and must be supplied in the diet. Plants manufacture and store it in the seed. Lack of vitamin B_1 causes beriberi, fatigue, stiffness, headache, nervousness, and loss of appetite and, when chronic, causes enlargement of the heart, polyneuritis, and loss of coordination of the muscular movements. The crystalline vitamin B_1 is called **thiamine chloride,** and in Europe is called **aneurin.** It is water-soluble, insoluble in most fats, and is destroyed by heat in the presence of moisture. In alkaline solutions the destruction is rapid. It is essential to the health of every living cell. Greater amounts are needed when much starch or sugar foods are eaten in order to prevent the formation of pyruvic acid, which produces noxious breath. **Pyruvic acid,** CH_3COCO_2H, a liquid boiling at 165°C, also called **glucic acid, pyroacemic acid,** and **propanone acid,** is **onion flavor. Onions** are root plants of the genus *Allium,* of which there are more than 200 species. They constitute a valuable food product but contain varying amounts of pyruvic acid, from 5.3 μmol/mil in the yellow **Spanish onion** to 18.6 in the strong **Ebenezer onion.**

Riboflavin is the accepted name for **vitamin B_2,** or **vitamin G.** The orange-yellow needle-shaped crystals have a green fluorescence. Riboflavin, $C_{27}H_{20}N_4O_6$, is water-soluble. It is gradually destroyed by exposure to light, and is destroyed by many chemicals, or by high temperatures in the presence of alkalies. It is present in meats, eggs, barley malt, yeast, milk, green leafy vegetables, and grasses. Deficiency of riboflavin results in ill health, loss of hair, and dermatosis. **Nicotinic acid,** or **niacin,** is the pellagra-preventing member of the vitamin B complex. It can be made from the nicotine of the variety of tobacco *Nicotiana rustica.* Coffee contains some niacin, and meat extracts are rich in both niacin and riboflavin. **Biotin,** originally named **vitamin H,** is also a member of the B group, and has an enzyme action on starches and sugars. It occurs widely in nature as a **phytohormone** for the growth of organisms and plants. It is extracted from yeast, egg yolk, and liver by adsorbing on carbon. **Vitamin B_6,** $C_8H_{11}NO_3HCl$, called **pyridoxine,** is required to enable the human system to assimilate proteins. A deficiency causes nausea, muscular weakness, and anemia. **Vitamin B_{12},** or **cobalamin,** is a high-molecular-weight complex containing five-membered nitrogen nuclei. It can replace protein as a growth factor.

Vitamin C, $C_6H_8O_6$, known also as **ascorbic acid,** or **ceritamic acid,** is unstable and is easily oxidized, especially in the presence of alkalies or in iron or copper vessels, so that in foods that have been long exposed to the air or overcooked it loses its value. It is thus probably the only vitamin likely to be deficient in the American diet, but the need is easily satisfied with fresh fruits, tomatoes, and green vegetables, and it is now added to frozen and canned foods as it also preserves the natural color of the products. **Isoascorbic acid,** or **erythorbic acid,** has the same composition as ascorbic acid but with the OH and H reversed on one carbon. It has the same antioxidant value, and is a lower-cost chemical, but has no vitamin C activity. It is used in meats to preserve the red color, and in canned foods to prevent discoloring. **Mercate 5,** of Merck & Co., Inc., is isoascorbic acid for these purposes, and **Mercate 20** is **sodium isoascorbate. Cebicure,** for curing meats, is ascorbic acid.

The synthetic ascorbic acid is not claimed to be a complete cure for scurvy. The natural vitamin from lemons and limes also contains bioflavins which counteract the skin hemorrhage of scurvy. The juice from the **acerola plant** of Puerto Rico, used for scurvy, is 80 times richer in vitamin C than orange juice.

Vitamin D regulates the metabolism of calcium and phosphorus in the human body. Without it the body is subject to rickets, soft bones, or ill-formed bones and teeth. It is also used to counteract the germ of tuberculosis. It is found in fish and fish-liver oils and in some fruits. The vitamin D concentrate of General Mills, Inc., is made by the activation of crystalline ergosterol with low-velocity electrons, in vegetable oil. **Calciferol,** or **vitamin D_2,** is a synthetic antiarchitic marketed in crystalline form or in solution in corn oil. Its melting point is 116°C. Vitamin D_2 is formed in the body from cholesterol by the action of sunlight on the skin. **Vitamin E** is so widely distributed in foods that the effect is not well known. It is also called **tocopherol** as it is a tocopherol acetate. **Tofanin,** of Winthrop-Stearns, Inc., is vitamin E. **Vitamin K_5** is very stable. It is used in the foodstuffs industry instead of sulfur dioxide to control fermentation without affecting flavor, and in medicine to coagulate blood.

Vitamin P is found in capsicum and in lemon peel, and is used as a preventive of rheumatic fever. Although proper quantities of vitamins are necessary in the human body, overdoses are often toxic and poisonous. An excess of vitamin C, for example, causes irritability, vertigo, and vomiting. An excess of vitamin D causes metastatic calcification, or deposition of calcium in the arteries and kidneys, and concentrated vitamin D is classed as a toxic drug. Since metabolism may vary with each person and is also affected by physical condition, vitamins should not be taken as supplementary drugs without medical advice.

VULCANIZED FIBER. A wood, paper, or other cellulose fiberboard impregnated with a gelatinizing medium. It is not vulcanized in the same sense that rubber is vulcanized. It is made by various processes, and the medium may be sulfuric acid, zinc chloride solution, or cuproammonium solution. It may also be made by impregnating the cellulose fiber with a phenol-furfural resin dissolved in alcohol or other solvent. After dipping in the solution, the fiber is washed to remove excess alcohol, and then dipped in a zinc chloride solution which hydrolizes it, and it is then washed free of the chloride, dried, and rolled. The original vulcanized fiber, patented in 1899 and called **Cellulith,** was sulfite wood pulp molded into sheets or formed parts. The modern fiber in the hard grades is a tough, resilient, hornlike material in standard gray, red, and black colors. Soft flexible grades are made for washers and gaskets. The four major NEMA grades are: electrical insulation, commercial, bone (high density), and trunk. The commercial grade is in thicknesses from 0.005 to 1 in (0.013 to 2.54 cm), with lengthwise tensile strength of 7,500 lb/in^2 (51 MPa), flexural strength of 14,000 lb/in^2 (96 MPa), compressive strength 20,000 lb/in^2 (137 MPa), and dielectric strength of 250 volts per mil (10×10^6 volts per meter). Unless impregnated with a synthetic resin it is not resistant to alkalies. The bone quality is a dense material with a specific gravity of 1.4, capable of being machined. The **hard vulcanized fiber** of the Spaulding Fibre Co. was made from cotton rags gelatinized in a zinc chloride solution and built up in layers. The shear strength is to 15,000 lb/in^2 (103 MPa), and compressive strength 30,000 lb/in^2 (206 MPa). **Codite,** of the Continental-Diamond Fibre Co., was a hard, tough fiber in the form of tubing. **Bone fiber** is characterized by being dense and hard, while trunk fiber is tough and abrasion-resistant. Because of the moderate cost vulcanized fiber still has many uses, but practically all the material for electrical use is now of the insoluble type made with synthetic resin impregnation and having higher dielectric strength. **Fish paper,** for electrical use, was originally vulcanized fiber in thicknesses down to 0.004 in (0.010 cm), but is now likely to be a resin impregnate. **Shoe fiber** is vulcanized fiber in leather color used for reinforcement in shoes. It is very resilient, and can be die-cut and nailed. Much of the fiber generally called vulcanized fiber is now impregnated with synthetic resins to meet conditions of chemical resistance, strength, and electrical properties. Vulcanized fiber is produced in the form of sheets, coils, tubes, and rods. Sheets are made in thicknesses of 0.0025 to 2 in (0.0064 cm to 5 cm) and approximately 48 by 80 in (122 by 203 cm) in size. Outside diameters of tubes range from $\frac{3}{16}$ to $4\frac{3}{8}$ in (0.478 to 11 cm). Rods are produced $\frac{3}{32}$ to 2 in (0.239 to 5 cm) in diameter.

VULCANIZED OILS. Vegetable oils vulcanized with sulfur and used for compounding with rubber for rubber goods, or as a rubber substitute. Castor oil, corn oil, rapeseed oil, and soybean oil are used. Vulcanized oil is a white to brown, spongy, odorless cake, or a sticky plastic, with specific gravity of 1.04. The material was invented in France in 1847, and was known as **factice. Factice cake** is solidified vulcanized oils, cut in slab form. It is an oil modifier of rubber to add softness and plasticity. It also has some elasticity. **Brown factice** is made by treating the oil with sulfur or sulfur chloride at 160 to 200°C. The softer grades are made with blown oils and low sulfur. The harder grades contain up to 20% sulfur. **White factice** is made from rapeseed oil, which is high in a characteristic acid, erucic acid, by slow addition of sulfur chloride up to 25% sulfur content. **Erasing rubbers** are rubber compounded with white factice or the factice alone. **Black factice** has mineral bitumen added to brown factice. **Neophax** is the trade name of the Stamford Rubber Supply Co. for brown factice, and **Amberex** is the name for light-tan-colored factice. **Factex** of the same company is partly vulcanized oil dispersed in water. It produces a nontacky elastic film. When mixed with rubber latex to the extent of 30%, it gives a velvety feel to the vulcanized product and does not decrease the strength greatly. **Factice sheet** is specially processed factice made by treating warm oil with sulfur and then with sulfur chloride. The strength and elasticity are higher. **Mineral rubber** was a name applied to vulcanized oils mixed with bitumens, especially gilsonite.

WALNUT. A hardwood from the tree *Juglans regia,* native to Europe and Asia Minor, but now growing in many other places. The wood is firm, with a fine to coarse open grain, and a lustrous surface. The weight is about 45 lb/ft^3 (721 kg/m^3). The color is dark brown to black, and it takes a beautiful polish. Walnut has great strength toughness, and elasticity. It also has great uniformity of texture and does not split easily. It is particularly adapted for carving. Walnut is valued as a cabinet wood, for fine furniture, and gunstocks. The wood from *J regia* is called **English walnut,** and the beautifully figured wood from Iran is known as **Circassian walnut. Black walnut,** or **American walnut,** is from the tree *J. nigra,* of the eastern United States. The color is darker, and it has a more uniform color than European walnut. It is handsomely grained and has the same general characteristics and uses as European walnut. It has a specific gravity, kiln-dried, of 0.56, a shearing strength parallel to the grain of 1,000 lb/in^2 (6 MPa), and a compressive strength perpendicular to the grain of 1,730 lb/in^2 (10 MPa). **Butternut,** from the tree *J. cinerea,* resembles closely the wood of the black walnut except for its color, which is yellowish gray. The supply of this wood is limited.

East India walnut is the wood of the tree *Albizzia lebbek* of tropical Asia and Africa, used for furniture, paneling, and interior decorative work. It is hard, dense, close-grained, and has a weight of about 50 lb/ ft^3 (801 kg/ m^3). The color is dark brown with gray streaks. The logs come as large as

30 in (76 cm) square and 20 ft (6 cm) long. The shipments may be mixed with the wood of the *A. procera,* the **white siris wood** of India. This wood has a brown walnut color, is lustrous, and resembles true walnut more than does the East India walnut. **Mahoe,** also called **blue mahoe** and **majagua,** is the wood of the tree *Hibiscus elatus,* of tropical America. It has been used to replace true walnut for gunstocks and in cabinetwork. The wood has a gray-blue color, an aromatic odor, and is hard with a coarse, open grain. **Brazilian walnut,** or **frejó,** from the tree *Cordia goeldiana,* is a strong, tough, straight-grained wood used for cooperage and cabinetwork.

African walnut, or **amonilla,** is from the tree *Lovoa klaineana* of Nigeria. The wood is brown, and has a fine texture and an interlocking grain that shows a striped figure when quartersawed. It weighs about 40 lb/ft^3 (641 kg/m^3). It is used for flooring, paneling, veneers, and cabinetwork. The Brazilian wood known as **imbuia,** from the tree *Nectandra villosa,* is a close match to true walnut, and is valued for cabinetwork, flooring, and furniture. The heartwood has an olive to brown color and takes a high polish. There are as many as 50 species of *Nectandra* trees in Brazil, varying widely in characteristics. The **canela preta,** from the tree *N. mollis,* is a wood with large satiny stains on a dark-yellow background. It has a silvery luster when polished, has a spicy scent, and is very durable. It resembles bleached walnut.

Walnut oil is yellowish oil obtained by pressing the nut kernels of the common walnut. It is a good drying oil and is used especially for artists' paints. The specific gravity is 0.919 to 0.929 and iodine value 148. It is soluble in alcohol. The oil from the candlenut is also called walnut oil. **Walnut-shell flour,** made from the refuse shells of the walnut industry of California, is used as a filler in molded plastics and in synthetic adhesives to increase bonding strength. It contains cellulose with about 28% lignin, 5 furfural, 9 pentosans, 6 methyl hydroxylamine, and 2.5 sugars and starch. In colonial times walnut bark was used as a cathartic. It contains a **juglone,** or **nucin,** a complex **naphthoquinone,** $C_{10}H_6O_2$, a reddish crystalline compound also called **lapachol** as it occurs also in **lapacho,** an important hardwood of Argentina and Paraguay.

WALRUS HIDE. The skin of the walrus, a marine mammal, *Odontobaenus rosmarus* and *O. abesus,* native to the North Atlantic and Pacific Oceans. The animals sometimes have a length of 16 ft (5 m), and a weight up to 2,000 lb (907 kg), and the hide is obtainable in large pieces. They congregate in herds on the icebergs of the North. The skin is tanned and makes a leather with a beautiful natural grain. It is also very tough and was formerly much used for coach traces. It is now employed where a tough and ornamental leather is needed, but the animals are now scarce and the killing of them is controlled by law. Imitation **walrus leather** is made from specially tanned and heavy, embossed sheepskins, and is used mostly for bags and ornamental articles.

WATER. A nearly transparent liquid of the composition H_2O. The specific gravity of pure water is taken as 1.0 at 4°C, and water is used as the standard for measuring the specific gravity of other liquid and solid materials. The boiling point is 100°C, and the freezing point is 0°C. The first essential use of water is for drinking and for the watering of plants to sustain life, but the largest consumptive use is in industrial processing, and a large supply of water is essential for manufacturing.

The per capita intake of water for human drinking is taken as 1 gal (3.8 L) per day, but, because of waste, the amount is larger. The consumption of water from the municipal systems of large American cities exceeds 150 gal (568 L) per capita per day, and that figure includes some industrial use. The employment of water for hydroelectric power is not considered as a consumptive use. About 80% of the supply in the United States is from surface water, and about 20% from groundwater.

Quality of water is important in many industrial operations. Factories may obtain water from municipal systems, from groundwater pumped from wells, or from surface water from streams. As industries concentrate, it becomes more important to protect water supply by dams and watersheds, and by preventing the pollution of streams by the return of unclean water. Typical municipal waters contain from 30 to 1,000 ppm of dissolved minerals, chiefly silica, iron, calcia, magnesia, potassium, sulfates, chlorides, and nitrites. Organic matter is also present in the water in varying amounts. So-called pure lake water averages above 150 ppm. Thus, pure water for chemical processing may require ion-exchange purification after filtering. Water for atomic reactors is thus purified to not more than 0.08 ppm.

The water molecule is one of the simplest of the chemical combinations of the natural elements, but it embraces such a vast complexity that it can be taken as an example to illustrate the basic principles of the combining habits of all the elements. The unit molecule, $H \cdot O \cdot H$, appears to be of ovaloid shape over a wide range of temperatures. The bonding is very strong. With active metallic inclusions in water, the disintegration of the molecule may begin at an energy equivalent of about 1000°C, but complete dissociation into its elements, hydrogen and oxygen, requires a temperature above 2000°C. Water is a very stable **oxygen hydride.**

The molecule is too minute for measurement of hardness by any known methods, but it is deduced to be extremely hard. When a stream of water is projected through a tiny orifice at high pressure it will cut through a tree like a power saw. At the energy levels of the liquid stage the molecules are very mobile and roll on each other, but when water drips from a faucet or falls as rain it forms into rounded droplets. The cause is usually given as **surface tension,** but may be the electromagnetic attraction of the exposed proton pairs on the oxygen atoms. When the molecules are lubricated with a chemical the water flows more easily. This property is used in fire-fighting hose. **Polyox FRA,** of Union Carbide, is a polymer derivative of ethylene oxide used for this purpose. When a 1% solution of the resin is injected into the line the stream of water is projected more than twice the normal distance. Water can also be coated to form **dry water** of semisolid consistency in the form of droplets which are dry to the touch. This property is utilized in water-based cosmetic creams by adding silicic acid. When the cosmetic is applied to the face silicic acid dehydrolizes to leave an adherent coating of extremely fine amorphous silica on the skin, and the thin film of released water evaporates quickly to give a cool fresh feel.

Lumping and caking of flours and other powders are usually caused by absorption of water moisture from the air, but the addition of as little as 0.5% of an **anticaking agent** such as silicic acid will coat and dry the absorbed moisture and prevent caking. The material known as **anomalous water** has properties similar to those of pure water treated with silicic acid. It forms in dry globules of low mobility, and has a wider temperature range in the liquid state. It is produced by condensing distilled water in vacuum in a small quartz (latticed silica) tube. The accumulation is extremely slow, requiring 18 h to fill a tube 0.010 in (0.025 cm) in diameter. It is also called **polywater,** though the water molecule does not polymerize in the ordinary meaning of the term. The water molecule can also be absorbed within the molecules of many chemicals, both solid and liquid, and such materials are called **hygroscopic.** For example, phosphophenyl methyl phosphinic acid has a unit molecular group in the shape of an irregular toroid. With only enough water to fill the spaces in the toroid, the unit is rounded and mobile, and the material becomes a viscous liquid which is dissimilar to a water solution of a material. Large proportions of water may be encased in the cells and lattices of proteins or other organic materials, making the mass into a solid or semisolid in which as much as 97% of the volume is water. This process is used in the making of puddings such as **Jello** in which a small amount of edible gelatin encases the water. **Water-filled plastics** are usually thermoplastic resins in which 80% or more of water is used as a filler. The water is stirred into the resin and the casting sets to a hard, firm solid which can be machined, nailed, or screwed. The castings are used for models, ornaments, and such products as lamp stands.

There is no room within the water molecule for the containment of any other atom or molecule, but the liquid mass may be likened to the arrangement of a great pile of uniformly sized, ovaloid Danish pebbles, and the mass has comparatively large spaces between and among the units. All natural waters contain much foreign matter including oxygen and other gases from the air, organic material from plant and animal life, minerals picked up from contact with the earth, and often high proportions of large inorganic molecules such as **astrakanite,** $Na_2SO_4 \cdot MgSO_4 \cdot 4H_2O$, and carnallite. **Groundwater,** pumped from deep wells for industrial and municipal uses, was once considered the purest of waters, containing chiefly mineral salts, but in populated areas underground waters are now contaminated with seepage from fertilizers, insecticides, industrial chemicals, and sewage which includes froth-forming detergents. Water in streams, in addition to the natural minerals and plant-decay matter, now usually contains high contents of industrial and agricultural chemicals and sewage, plus biological compounds from the decayed proteins and albumins of human and animal wastes. Many of these can only be removed by costly special ion-exchange methods. Bacteria may be inactivated by the addition of chlorine, but the skatolelike odors may remain. The content of air in natural water, given in terms of O_2, is a minimum of 9 ppm. This contained air is a biological necessity in water for human consumption and for the maintenance of fish and other marine life. The chemical process of decay of sewage in water depletes this oxygen and thus tends to destroy marine life. **Seawater,** and water in inland lakes such as the Dead Sea where evaporation greatly exceeds runoff, contains high amounts of mineral salts. The variety and percentage content vary in different areas and at different temperatures. The density of seawater at 68°F (20°C) may be 69 lb/ft^3 (1105 kg/m^3) compared with about 62 lb/ft^3 (993 kg/m^3) for natural fresh water. About 80 elements have been found in seawater, and it is probable that all of the natural elements occur to some extent. About 30% of all commercial halite, or common salt, is now produced from seawater. By ordinary solar evaporation at about 80°F (27°C), the proportion of common salt, NaCl, precipitating after the calcite and gypsum, may be about 12% of the total salts. As much as 65% of commercial magnesium metal is produced from seawater by precipitation and reduction of contained magnesium compounds such as epsomite, kainite, kieserite, **bischofite,** $MgCl_2 \cdot 6H_2O$, and the magnesium sulfate known as **hexahydrite,** $MgSO_4 \cdot 6H_2O$. At the energy level of solidification the water molecules arrange themselves in precise order close together, and the frozen water, or **ice,** can be split in straight cleavage from a line scratched on the surface. In the solid assembly of molecules there is no available space, and the contained impurities of water are thrown out in freezing, except in the dendritic snow molecule.

WATER REPELLENTS. Chemicals used for treating textiles, leather, and paper such as washable wallpaper, to make them resistant to wetting by water. They are different from waterproofing materials in that they are used where it is not desirable to make the material completely waterproof, but to permit the leather or fabric to "breathe." Water repellents must not form acids that would destroy the material, and they must set the dyes rather than cause them to bleed on washing. There are two basic types, a durable type that resists cleaning, and a renewable type that must be replaced after the fabric is dry-cleaned. **Zelan,** a pyridinium-resin compound of Du Pont, is representative of the first type. **Quilon,** of the same company, is used for paper, textiles, and glass fabric, and forms a strong chemical bond to the surface of the material by an attachment of the chromium end of the molecule through the covalent bond to the negatively charged surface. It is a stearatochromic chloride. The second type is usually an emulsion of a mineral salt over which a wax emulsion is placed; the treatment may be a one-bath process, or it may be by two separate treatments. **Aluminum acetate** is one of the most common materials for this purpose. Basic aluminum acetate is a white amorphous powder of the composition $Al(OH)(OOC \cdot CH_3)$. It is only slightly soluble in water but is soluble in mineral acids. **Niaproof,** of Union Carbide, is a concentrated aluminum acetate for waterproofing textiles, and **Ramasit,** of the General Dyestuff Corp., and **Migasol,** of the Ciba Co., are similar materials. **Zirconium acetate,** a white crystalline material of the composition $ZrOH(C_2H_3O_2)_3$, and also its sodium salt, are used as water repellents. **Zirconyl acetate,** $ZrO(C_2H_3O_2)_2$, a light-yellow solution containing 13% ZRo_2, is used for both water repellency and flame resistance of textile fibers. **Intumescent agents** are repellent coatings that swell and snuff out fire when they become hot. **Latex 744B,** of the Dow Chemical Co., is a repellent of this type. It is a vinyl water emulsion compounded with pentaerythritol, dicyandiamide, and monosodium phosphate, and is used on textiles, wallboard, and fiber tile.

WATER SOFTENERS. Chemical compounds used for converting soluble scale-forming solids in water into insoluble forms. In the latter condition they are then removed by setting or filtration. The hardness of water is due chiefly to the presence of carbonates, bicarbonates, and sulfates of calcium and magnesium, but many natural waters also contain other metal complexes which need special treatment for removal. Temporary **hard waters** are those that can be softened by boiling; permanent hard waters are those that require chemicals to change their condition. Sodium hydroxide is used to precipitate magnesium sulfate. Caustic lime is employed to precipitate bicarbonate of magnesium, and sodium aluminate is used as an accelerator. Barium carbonate may also be used. Prepared water softeners may consist of mixtures of lime, soda ash, and sodium aluminate, the three acting together. **Sodium aluminate,** $Na_2AL_2O_4$, is a water-soluble white powder melting at 1650°C, which is also used as a textile mordant, for sizing paper, and in making milky glass. Reynolds Metals

Co. produces this material in flake form with iron content below 0.0056% for paints, water softeners, and paper coatings. Alum is used in settling tanks to precipitate mud, and zeolite is used extensively for filtering water. The liquids added to the washing water to produce fluffier textiles are **fabric softeners** and not water softeners. They are usually basic quaternary ammonium compounds such as distearyl dimethyl ammonium chloride with 16 and 18 carbon atoms, which are cationic, or positively charged. A thin coating is deposited on the negatively charged fabric, giving a lubricated cloth with a fluffy feel.

Water is also softened and purified with **ion-exchange agents,** which may be specially prepared synthetic resins. **Cation-exchange agents** substitute sodium for calcium and magnesium ions and produce soft waters. When the water is treated with a hydrogen derivative of a resin, the metal cations form acids from the salts. The carbonates are converted to carbonic acid which goes off in the air. When it is treated again with a basic resin derivative, or **anion-exchange agent,** the acids are removed. Water receiving this double treatment is equal to distilled water. Salt-cycle anion-exchange substitutes chloride ions for other anions in the water, and when combined with cation exchange it produces sodium chloride in the water in place of other ions.

In electrolytic ion exchangers for converting seawater to fresh water, the basic cell is divided into three compartments by two membranes, one permeable only to cations and the other only to anions. The sodium ions migrate toward the cathode and the chlorine ions go toward the anode leaving fresh water in the center compartment. **Ion-exchange membranes** for electrodialysis (salt splitting or separation), and also used in fuel cells, are theoretically the same as powdered exchange resins but with an inorganic binder. Such a membrane resin of the Armour Research Foundation is made by the reaction of zirconyl chloride and phosphoric acid, giving a chain molecule of zirconium-oxygen with side chains of dihydrogen phosphate. **Zeo-Karb,** a sulfonated coal, and **Zeo-Rex,** a sulfonated phenol-formaldehyde resin, are cation exchangers of the Permutit Co., while **De-Acidite** and **Permutit A** of the same company are anion exchangers. **Amberlite IRA-400,** of Rohm & Haas, is a strongly basic alkyl amine which will split neutral salts in the water and also remove silica. The German **Wofatit P** exchanger is a sodium salt of a phenol-formaldehyde resin. Ion-exchange agents are also used for refining sugar, glycerin, and other products, and for the purification of acids and the separation of metals. An **eluting agent** is a solvent used to elutriate the resin beds in the separation of metals, that is, to separate the heavier from the lighter particles, causing a metal ion on the resin to change place with hydrogen or with an ammonium group in the elutriant.

WAX. A general name for a variety of substances of animal and vegetable origin, which are fatty acids in combination with higher alcohols instead of with glycerin as in fats and oils. They are usually harder then fats, less greasy, and more brittle, but when used alone do not mold as well. Chemically, the waxes differ from fats and oils in being composed of high-molecular-weight fatty acids with high-molecular-weight alcohols. The most familiar wax is beeswax from the honeybee, but commercial beeswax is usually greatly mixed or adulterated. Another **animal wax** is spermaceti from the sperm whale. **Vegetable waxes** include Japan wax, jojoba oil, candelilla, and carnauba wax. **Mineral waxes** include paraffin wax from petroleum, ozokerite, ceresin, montan wax. The mineral waxes differ from the true waxes and are mixtures of saturated hydrocarbons.

The animal and vegetable waxes are not plentiful materials, and are often blended with or replaced by hydrocarbon waxes or waxy synthetic resins. But waxes can be made from common oils and fats by splitting off the glycerin and reesterifying selected mixtures of the fatty acids with higher alcohols. **Hywax 122,** of the Werner G. Smith Co., is a self-emulsifiable wax composed of cetyl, myristyl, and stearyl esters derived from animal and vegetable oils. **Opalwax,** of Du Pont, is a **synthetic wax** produced by the hydrogenation of castor oil. It has about the same hardness as carnauba, specific gravity of 0.98, and melting point of 86 to 88°C, but it lacks the luster of carnauba. It is odorless, and of a pearl-white color. It is very resistant to most solvents, and is used for insulation, coatings, candles, and carbon paper. **Acrawax,** of Glyco Products Co., Inc., is a somewhat similar substitute for carnauba with higher melting point. **Stroba wax,** of the same company, is a synthetic wax with a base of stearic acid and lime. The melting point is 103 to 106°C. It is used in polishes, insulation, and as a flatting agent. Synthetic wax under the name of **Pentawax 286,** of the Heyden Newport Chemical Corp., is a true wax in that it is a combination of fatty acids with an alcohol. It is made from the long-chain acids of vegetable oils with pentaerythritol. It has a higher melting point than carnauba, 110°F (43°C), but does not form a self-polishing liquid wax like carnauba. **Sheerwax,** of Ru-Jac, Inc., is made by catalytic hydrogenation of vegetable oils. It has the hardness and high-melting point of carnauba wax and can be had in white color. Waxes are employed in polishes, coatings, leather dressings, sizings, waterproofing for paper, candles, and varnishes. They are softer and have lower melting points than resins, are soluble in mineral spirits and in alcohol, and insoluble in water.

Some plastics have wax characteristics, and may be used in polishes and coatings or for blending with waxes. Polyethylene waxes are light-colored, odorless solids of low molecular weight, up to about 6,000. Mixed in solid waxes to the extent of 50%, and in liquid waxes up to 20%, they add gloss and durability and increase toughness. In emulsions they add stability. **Alcowax,** of the Allied Chemical Corp., is a polyethylene of low molecular weight, about 2,000. It is a translucent solid melting at about 100°C. **Marlex 20,** of the Phillips Petroleum Co., is a methylene polymer used to blend with vegetable or paraffin waxes to increase the melting point, strength, and hardness. **Santowax R,** of Monsanto, is a mixture of terphenols. It is a light-buff, waxy solid, highly soluble in benzene, and with good resistance to heat, acids, and alkalies. It is used to blend with natural waxes in candles, coatings, and insulation. **Epolene wax,** of Eastman Chemical Products, Inc., is a polyethylene. Waxes are not digestible, and the so-called **edible waxes** used as water-resistant coatings for cheese, meats, and dried fruits are not waxes, but are modified glycerides. **Monocet,** of the Glyco Products Co., Inc., is such a material. It is a white, odorless, tasteless waxy solid melting at 40°C, and is an acetylated monoglyceride of fatty acids.

WEAR-RESISTANT STEEL. Many types of steel have wear-resistant properties, but the term usually refers to **high-carbon, high-alloy steels** used for press dies subject to abrasion and for wear-resistant castings. They are generally cast and ground to shape. They are mostly sold under trade names for specific purposes. The excess carbon of the steels is in spheroidal form rather than as graphite. One of the earlier materials of this kind for drawing and forming dies, **Adamite,** of the Mackintosh-Hemphill Co., was a **chromium-nickel-iron alloy** with up to 1.5% chromium, nickel equal to half that of the chromium, and from 1.5 to 3.5% carbon with silicon from 0.5 to 2%. The Brinell hardness ranges from 185 to 475 as cast, with tensile strengths to 125,000 lb/in^2 (861MPa). The softer grades can be machined and then hardened, but the hard grades are finished by grinding. **Kinite,** of the Kinite Corp., has about 13% chromium, 1.5 carbon, 1.1 molybdenum, 0.70 cobalt, 0.55 silicon, 0.50 manganese, and 0.40 nickel. It is used for blanking dies, forming dies, and cams. **Martin steel,** of the Detroit Alloy Steel Co., has 13% chromium, about 1 molybdenum, 0.80 cobalt, 0.35 vanadium, and 1.5 carbon. **Circle T15 steel,** of Firth Sterling, Inc., for extreme abrasion resistance in cutting tools, is classed as a super-high-speed steel. It has 13.5% tungsten, 4.5 chromium, 5 cobalt, 4.75 vanadium, 0.50 molybdenum, and 1.5 carbon. Its great hardness comes from the hard vanadium carbide and the complex tungsten-chromium carbides, and it has full red-hardness. The property of abrasion or wear resistance in steels generally comes from the hard carbides, and is thus inherent with proper heat treatment in many types of steel.

WELDING ALLOYS. Usually in the form of rod, wire, or powder used for either electric or gas welding, for building up surfaces, or for hard-facing surfaces. In the small sizes in continuous lengths welding alloys are called

welding wire. Nonferrous rods used for welding bronzes are usually referred to as **brazing rods,** as the metal to be welded is not fused when using them. Welding rods may be standard metals or special alloys, coated with a fluxing material or uncoated, and are normally in diameters from 3/32 to 1/4 in (0.239 to 0.635 cm). Compositions of standard welding rods follow the specifications of the American Welding Society. Molded carbon, in sizes from 1/8 to 1 in (0.318 to 2.54 cm) in diameter, is also used for arc welding. **Intensarc** is the trade name of the National Carbon Co., Inc., for carbon rods. Low-carbon steel rods for welding cast iron and steel contain less than 0.18% carbon. High-carbon rods produce a hard deposit that requires annealing, but these are also used for producing a hard filler. High-carbon rods, with 0.85 to 1.10% carbon, will give deposits with an initial hardness of 575 Brinell, whereas high-manganese rod deposits will be below 200 Brinell but will work-harden to above 500 Brinell. For high-production automatic welding operations, carbon-steel wire may have a thin coating of copper for easy operation and nonspattering. Stainless-steel rods are marketed in various compositions. The **Flexarc welding rods** of the Westinghouse Electric Corp. comprise a range of stainless steels with either titania-lime or straight-lime coatings. **Stainless C,** of the Lincoln Electric Co., is an 18–8 type of stainless steel having also 3.5% molybdenum. **Aluminum-weld,** of the same company, is a 5% silicon aluminum rod for welding silicon-aluminum alloys, and the **Tungweld rods,** for hard surfacing, are steel tubes containing fine particles of tungsten carbide. **Kennametal KT-200,** of Kennametal, Inc., has a core of tungsten carbide and a sheathing of steel. It gives coatings with a hardness of Rockwell C63. **Chromang,** of the Arcos Co., for welding high-alloy steels, is an 18–8 stainless steel modified with 2.5 to 4% manganese.

The **Amsco welding rods** of the American Manganese Steel Co. are grades of high-manganese steel giving hardnesses from 500 to 700 Brinell, and **Toolface** is a high-speed steel rod of the same company for facing worn cutting tools. **Superloy,** of the Resisto-Loy Co., used for facing surfaces where extreme hardness is required, has the alloy granules in a soft steel tube. The welded deposit has a composition of 30% chromium, 8 cobalt, 8 molybdenum, 5 tungsten, 0.05 boron, and 0.20 carbon. **Tung-Alloy, Resisto-Loy,** and **Isorod** are other **hard-facing rods** of the same company. Resisto-Loy has a nonferrous content. The **Croloy welding rods** of the Babcock & Wilcox Co., for welding alloy steels without high preheating, are low-alloy chrome-molybdenum steels. **Stoodite** is a high-manganese steel in the form of rods marketed by the Stoody Co. for hard facing. **Chromend 9M,** of the Arcos Corp., is a rod for arc-welding hard deposits. It contains 8 to 10% chromium and 1.5 molybdenum, giving a weld with a hardness of 400 Brinell. **Elkonite,** of the P. R. Mallory Co., is the name of a group of welding alloys made especially for welding machines. They are, in general, sintered tungsten or molybdenum carbides, combined with copper or silver, and are electrodes for spot welding rather than welding rods. **Tungsten electrodes** may be pure tungsten, thoriated tungsten, or zirconium tungsten, the latter two used for direct current. **Thoriated tungsten** gives high arc stability, and the thoria also increases the machinability of the tungsten. **Zirconium tungsten** provides adhesion between the solid electrode and the molten metal to give uniformity in the weld.

Thermit is a mixture of aluminum powder and iron oxide used for welding large sections of iron or steel or for filling large cavities. The process, originally developed by the Goldschmidt Thermit Co., consists of the burning of the aluminum to react with the oxide, which sets free the iron in molten form. To ignite the aluminum and start the reaction a temperature of about 2800°F (1538°C) is required, which is reached with the aid of a gas torch or ignition powder, and the exothermic temperature is about 4600°F (2538°C). **Cast-iron thermit,** used for welding cast iron, is thermit with the addition of about 3% ferrosilicon and 20 steel punchings. **Red thermit** is made with red oxide, and **black thermit** with black oxide. **Railroad thermit** is thermit with additions of nickel, manganese, and steel.

The **Stellite hard-facing rods,** of Union Carbide, are cobalt-based alloys that retain hardness at red heat and are very corrosion-resistant. The grades have tensile strengths to 105,000 lb/in^2 (723 Mpa) and hardnesses to Rockwell C52. **Inco-Weld A,** of the International Nickel Co., is welding wire for stainless steels and for overlays. It contains 70% nickel, 16 chromium, 8 iron, 2 manganese, 3 titanium, and not more than 0.07 carbon. The annealed weld has a tensile strength of 80,000 lb/in^2 (551 MPa) with elongation of 12%. **Nickel welding rod** is much used for cast iron, and the operation is brazing, with the base metal not melted. The **Nickel silver 828** of the American Brass Co., for brazing cast iron, contains 46.5% copper, 43.38 zinc, 10 nickel, 0.10 silicon, and 0.02 phosphorus. The deposit matches the color of the iron. **Colmonoy 23A,** of Wall Colmonoy Corp., is a nickel alloy **welding powder** for welding cast iron and filling blow holes in iron castings by torch application. It has a composition of 2.3% silicon, 1.25 boron, 0.10 carbon, not over 1.5 iron, and the balance nickel, with a melting point of 1950°F (1066°C). For welding on large structures where no heat treatment of the weldment is possible, the welding rods must have balanced compositions with no elements that form brittle compounds. **Rockide rods,** of Norton, are metal oxides for hard surfacing.

WETTING AGENTS. Chemicals used in making solutions, emulsions, or compounded mixtures, such as paints, inks, cosmetics, starch pastes, oil emulsions, dentifrices, and detergents, to reduce the surface tension and give greater ease of mixing and stability to the solution. In the food industries chemical wetting agents are added to the solutions for washing fruits and vegetables to produce a cleaner and bacteria-free product. Wetting agents are described in general as chemicals having a large hydrophilic group associated with a smaller hydrophilic group. Some liquids naturally wet pigments, oils, or waxes, but others require a proportion of a wetting agent to give mordant or wetting properties. Pine oil is a common wetting agent, but many are complex chemicals. They should be powerful enough not to be precipitated out of solutions in the form of salts, and they should be free of odor or any characteristic that would affect the solution. **Aerosol wetting agents,** of the American Cyanamid Co., are in the form of liquids, waxy pellets, or free-flowing powders. **Aerosol OS** is a sodium salt of an alkyl naphthalene sulfonic acid. It is a yellowish-brown powder soluble in most organic solvents. This salt was called **Nekal** in Germany. The **Dresinols** of Hercules, Inc., are sodium or ammonium dispersions of modified rosin, with 90% of the particles below 1 μm in size. **Polyfon,** of the West Virginia Pulp & Paper Co., is a sodium lignosulfonate produced from lignin waste liquor. It is used for dye and pigment dispersion, oil-well drilling mud, ore flotation, and boiler feed-water treatment.

WHALE OIL. An oil extracted by boiling and steaming the blubber of several species of whale that are found chiefly in the cold waters of the extreme north and south. Whales are mammals, and are predaceous, living on animal food. The blubber blanket of fat protects the body, and the tissues and organs also contain deposits of fat. Most whale oil is true fat, namely, the glycerides of fatty acids, but the head contains a waxy fat. In the larger animals the meat and bones yield more fat than the blubber. Both the whalebone whales and the toothed whales produce whale oil. The **bluehead whales** of the south, *Silbaldus musculus,* are the largest and also yield most oil per weight. The whaling industry is under international control, and allocations are made on the basis of blue whale units averaging 20 tons (18 metric tons) of oil each. The blue whale is about 25 ft (8 m) long at birth and reaches 70 ft (21 m) in 2 years. This species often reaches 100 ft (30 m) with a weight of about 150 tons (136 metric tons) and will yield about 27 tons (24 metric tons) of oil. The **gray whale,** or **California whale,** of the northern Pacific, is a small 50-ft (15 m) species. The **Greenland whale** of the north, *Balaena mysticetus,* and the **finback whale** of the south, *Balaenoptera physalus,* produce much oil. The **beluga,** or **white whale,** *Delphinapterus leucas,* and the **narwhal,** *Monodon monoceros,* of the North Polar seas, produce **porpoise oil.** Both species of porpoise measure up to 20 ft (6 m) in length.

Whale oil is sold according to grade, which depends upon its color and keeping qualities. The latter in turn depends largely upon proper cooking at extraction. No. 0 and No. 1 grades are fine pale-yellow oils, No. 2 is amber, No. 3 is pale brown, and No. 4 is the darkest oil. Grade 1 has less than 1% free fatty acids, while grade 4 has from 15 to 60% with a strong fishy odor. The specific gravity is 0.920 to 0.927, saponification value 180 to 197, and iodine value 105 to 135. Whale oil contains oleic, stearic, palmitic, and other acids in varying amounts. But whales are now so scarce that the former uses of the oils and meat are restricted, particularly in the United States.

Whale oils of the lower grades were used for quenching baths for heat-treating steels, and also in lubricating oils. The best oils are used in soaps and candles, or for preparing textile fibers for spinning, or for treating leather. In Europe whale oil is favored for making margarine because it requires less hydrogen than other oils for hardening, and the grouping of 16 to 22 carbon atom acids gives the hardened product greater plasticity over a wider temperature range. **Sod oil** is oil recovered from the treatment of leather in which whale or other marine mammal oil was used. It contains some of the tannins and nitrogenous matter which make it more emulsifiable and more penetrant than the original oil.

Whale meat was used for food in Japan, and in dog food in the United States. When cured in the air, the outside is hard and black, but the inside is soft. In young animals the flesh is pale; in older animals it is dark red. It has a slight fishy flavor, but when cooked with vegetables is almost indistinguishable from beef. It contains 15 to 18% proteins. **Whale-meat extract** is used in bouillon cubes and dehydrated soups. It is 25% weaker than beef extract. **Whale liver oil** is used in medicine for its high vitamin A content. It also contains **kitol,** which has properties similar to vitamin A but is not absorbed in all animal metabolism. **Whalebones** are the elastic, hornlike strips in the upper jaw of the Greenland whale and some other species. The strips are generally from 8 to 10 ft (2 to 3 m) long and number up to 600. Those from the **bowhead whale** of the Arctic Ocean are the longest slabs, measuring up to 13 ft (4 m) in length to 10 to 12 in (25 to 30 cm) wide at the bottom. **Finback whalebone** is less than 4 ft (1 m) in length. The **humpback whale,** *Megaptera longimana,* of the northern Pacific, is a **baleen whale** with no teeth and with plates of baleen in the mouth to act as a sieve. It grows to a length of 50 ft (15 m). Whalebone is lightweight, very flexible, elastic, tough, and durable. It consists of a conglomeration of hairy fibers covered with an enamel-like fibrous tissue. It is easily split and when softened in hot water is easily carved. Whalebone has a variety of uses in making whips, helmet frames, ribs, and brush fibers. **Baleen** is a trade name for strips of whalebone used for whips, and for products where great flexibility and elasticity are required.

WHEAT. The edible seed grains of an annual grass of the genus *Triticum,* of which there are many species and thousands of varieties. Wheat was the basic food grain of the early civilizations of the Near East, and has remained the chief grain of the white races except in cold climates where rye grows better. The plains of the United States, Canada, Argentina, Australia, southern Russia, the Danube Valley, and northern India are the great wheat areas.

The types grown commercially are chiefly common wheat and durum wheat. **Common wheat,** *T. vulgare,* is the chief source of **wheat flour.** It has a stout head from which the grains can be separated easily. The hundreds of varieties are divided roughly into hard wheats and soft wheats, and red wheats and white wheats. The hard wheats usually have smaller grains, but are richer in proteins.

Spring wheat is wheat that is sown in the spring and harvested in the late summer. **Winter wheat** is sown in the fall to develop a root system before winter, and is then harvested in the early summer. It is more resistant and gives a higher yield. **Durum wheat,** *T. durum,* has a thick head with long beards, and large hard grains rich in gluten. The plant is hardy and drought-resistant, but the flour is too glutenous for American bread, and is much used for macaroni and in mixtures. Seven classes of wheat are designated in the official grain standards of the U.S. Department of Agriculture: hard red spring wheat; durum wheat; red durum wheat; hard red winter wheat; soft red winter wheat; white winter wheat; and mixed wheat. Each class permits mixtures of varieties. The minimum test weight of wheat is required to be 60 lb/bu (778 kg/m^3).

Most of the wheat production is ground for edible flour. Since wheat varies with the variety, the climate, and the soil, uniformity in the flour could formerly be obtained only by blending wheats from different areas to obtain an average, but uniformity is now obtained by an air-spinning process which separates the milled flour into fractions according to protein-starch ratios, and then combining for the flour of uniform ratio. These are called **turbo-flours.** Wheat flour is not normally a uniform product even from one area, as it is made up of starch granules, fractured endosperm cells, and protein fragments. **Pregelatinized flour** is used for canned goods to reduce the time needed for dextrinizing. **Wheat-flour paste,** for textile coatings, is **hydroxyethylated flour** made by treating wheat flour with ethylene oxide. It requires little cooking to form a starchy product.

Wheat is also used for making beer, and at times is employed for producing starch and alcohol. Some wheat is used for stock feed, but most of the wheat for this purpose is of lower and condemned grades. **Buck-wheat** consists of the seed grains of the *Fagopyrum esculentum,* a plant of the same family as the rhubarb and dock. It is native to Asia and is one of the chief foods in Russia, but is used only in mixed flours in the United States. The flour is more starchy and has less protein than wheat. It is also darker in color and has a different flavor.

WHETSTONE. Stones of regular fine grains composed largely of chalcedony silica, often with minute garnet and rutile crystals. They are used as fine abrasive stones for the final sharpening of edge tools. Whetstones are also sometimes selected, fine sandstones from the grindstone quarries. The **chocolate whetstone** from New Hampshire is mica schist. The finest whetstones are called **oilstones.** A fine-grained **honestone,** known as **coticule,** comes from Belgium, and is used for sharpening fine-edged tools. It is compact, yellow in color, and contains minute crystals of yellow manganese garnet, with also potash mica and tourmaline. Coticule is often cut double with blue-gray **phyllite** rock adhering to and supporting it. **Scythestones** are made from Ohio and Indiana sandstones, and from the schist of Vermont. **Rubbing stones** are fine-grained Indiana sandstones.

WHISKERS. Very fine single crystal fibers that range from 3 to 10 μm in diameter and have length-to-diameter ratios of from 50 to 10,000. Since they are single crystals, their strengths approach the calculated theoretical strengths of the materials. **Alumina whiskers,** which have received the most attention, have tensile strengths up to 3 million lb/in^2 (0.02 million MPa) and a modulus of elasticity of 62 million lb/in^2 (0.5 million MPa). Other potential whisker materials are **silicon carbide, silicon nitride, boron carbide,** and **beryllia.**

WHITE BRASS. A bearing metal which is actually outside of the range of the brasses, bronzes, or babbitt metals. It is used in various grades, the specification adopted by SAE being tin, 65%; zinc, 28 to 30; and copper, 3 to 6. It is used for automobile bearings, and is close-grained, hard, and tough. It also casts well. A different alloy is known under the name of white brass in the cheap jewelry and novelty trade. It has no tin, small proportions of copper, and the remainder zinc. It is a high-zinc brass, and varies in color from silvery white to yellow, depending upon the copper content. An old alloy formerly used for casting buttons, known as **Birmingham platina,** or **platina,** contained 75% zinc and 25 copper. It had a white color but was very brittle. A yellowish metal known as **bath metal,** once widely used for casting buttons, candlesticks, and other articles, was a brass containing 55% copper and 45 zinc. **White nickel brass** is a grade of nickel silver. The white brass used for castings where a white color is desired may contain up to 30% nickel. The 60:20:20 alloy is used for white plaque castings for buildings. The high-nickel brasses do not cast well unless they also contain lead. Those with 15 to 20% nickel and 2 lead are used for casting hardware and valves. **White nickel alloy** is a cupronickel containing some aluminum. **White copper** is a name sometimes used for cupronickel or nickel brass. The nickel brasses known as **German silver** are copper-nickel-zinc white alloys used as a base metal for plated silverware, for springs and contacts in electrical equipment, and for corrosion-resistant parts. The alloys are graded according to the nickel content. Extra-white metal, the highest grade, contains 50% copper, 30 nickel, and 20 zinc. The lower grade, called fifths, for plated goods, has a yellowish color. It contains 57% copper, 7 nickel, and 36 zinc. All of the early German silvers contained up to 2% iron which increased the strength, hardness, and whiteness but is not desirable in the alloys used for electrical work. Some of the early English alloys also contained up to 2% tin, but tin makes the alloys brittle. The Federal Trade Commission prohibits the use of the term German silver in the marketing of silver-plated ware, but the name still persists in other industries.

WHITE CAST IRON. A cast iron with virtually no graphite. The carbon is present in a matrix of fine pearlite as large particles of iron carbide, thus providing a material that is high in compressive strength, very hard, and abrasion-resistant, but low in tensile strength and impact resistance. Hardness of unalloyed white iron ranges between 300 and 575 Bhn, but tensile strength is only 30,000 lb/in^2 (206 MPa). For somewhat higher hardness, higher tensile and impact strength, and other special service properties, white irons are alloyed with nickel, chromium, and molybdenum. For example, nickel and chromium in a white cast iron commonly known as **Ni-Hard** provide a martensitic matrix. Hardness is increased to around 600 Bhn and tensile and impact strength are more than doubled. **Flint-metal,** of the Taylor-Wharton Iron & Steel Co., contains 4 to 5% nickel, 1.25 to 1.75 chromium, and 3 to 3.5 carbon. It has a Brinell hardness of 600. **High-chromium white irons** and **chromium-molybdenum white irons** combine excellent wear resistance with oxidation and corrosion resistance at elevated as well as normal temperatures.

By controlling composition and cooling rates, castings can be made with a white-iron surface layer and a core of gray iron, thus providing a duplex structure that combines excellent abrasion resistance with relatively good toughness. This metal is known as **chilled iron.** Castings of white and chilled iron are mainly used in applications that involve resistance to wear and abrasion. Typical uses include parts for crushers, grinders, slurry pumps, railroad car wheels and brake shoes, rolling mill rolls, and machinery handling abrasive materials.

WHITE GOLD. The name of a class of jewelers' white alloys used as substitutes for platinum. The name gives no idea of the relative value of the different grades, which vary widely. Gold and platinum may be alloyed together to make a white gold, but the usual alloys consist of from 20 to 50% nickel, with the balance gold. Nickel and zinc with gold may also be used for white golds. The best commercial grades of white gold are made by melting the gold with a white alloy prepared for the purpose. This alloy contains nickel, silver, palladium, and zinc. The 14-karat white gold contains 14 parts pure gold and 10 white alloy. A superior class of white gold is made of 90% gold and 10 palladium. High-strength white gold contains copper, nickel, and zinc with the gold. Such an alloy, containing 37.5% gold, 28 copper, 17.5 nickel and 17 zinc, when aged by heat treatment, has a tensile strength of about 100,000 lb/in^2 (689 MPa) and an elongation of 35%. It is used for making jewelry, has a fine, white color, and is easily worked into intricate shapes. **White-gold solder** is made in many grades containing up to 12% nickel, up to 15 zinc, with usually also copper and silver, and from 30 to 80 gold. The melting points of eight grades of Handy & Harman are from 695 to 845°C.

WHITE METALS. Although a great variety of combinations can be made with numerous metals to produce white or silvery alloys, the name usually refers to the lead-antimony-tin alloys employed for machine bearings, packings, and linings, to the low-melting-point alloys used for toys, ornaments, and fusible metals, and to the type metals. **Slush castings,** for ornamental articles and hollow parts, are made in a wide variety of soft white alloys, usually varying proportions of lead, tin, zinc, and antimony, depending on cost and the accuracy and finish desired. These castings are made by pouring the molten metal into a metal mold without a core, and immediately pouring the metal out, so that a thin shell of the alloy solidifies against the metal of the mold and forms a hollow product. A number of white metals are specified by the ASTM for bearing use. These vary in a wide range from 2 to 91% tin, 4.5 to 15 antimony, up to 90 lead, and up to 8 copper. The alloy containing 75% tin, 12 antimony, 10 lead, and 3 copper melts at 184°C, is poured at about 375°C, and has an ultimate compressive strength of 16,150 lb/in^2 (111 MPa) and a Brinell hardness of 24. The alloy containing 10% tin, 15 antimony, and 75 lead melts at 240°C, and has a compressive strength of 15,650 lb/in^2 (108 MPa) and a Brinell hardness of 22. The first of these two alloys contains copper-tin crystals; the second contains tin-antimony crystals. A white bearing metal produced by the American Smelting & Refining Co. under the name of **Asarcoloy** is composed of cadmium with 1.3% nickel. It contains $NiCd_7$ crystals, and is harder and has higher compressive strength than babbitt, and a low coefficient of friction. It has a melting point of 604°F (317°C). **SAE alloy 18** is such a **cadmium-nickel alloy** with also small amounts of silver, copper, tin, and zinc. A **bismuth-lead alloy** containing 58% bismuth and 42 lead melts at 123.5°C. It casts to exact size without shrinkage or expansion, and is used for master patterns and for sealing.

Various high-tin or reverse bronzes have been used as corrosion-resistant metals, especially before the advent of the chromium, nickel, and aluminum alloys for this purpose. **Trabuk** was a corrosion-resistant, high-tin bronze with about 5% nickel. **Fahry's alloy** was a **reverse bronze** containing 90% tin and 10 copper, used as a bearing metal, and the **Jacoby metal** used for machine parts had 85% tin, 10 antimony, and 5 copper. The scarcity and high cost of tin have made these alloys obsolete. The bearing alloy known in England as **motor bronze** is a babbitt with about double the amount of copper of a standard babbitt. One analysis gives tin, 84%; antimony, 7.5; copper, 7.5; and bismuth 1. An old alloy, used in India for utensils and known as **bidery metal,** contained 31 parts zinc, 1 lead, and 2 copper, fluxed with resins. It was finished with a velvety-black color by treating with a solution of copper sulfate. A **white metal sheet** now much used for making stamped and formed parts for costume jewelry and electronic parts, is zinc with up to 1.5% copper and up to 0.5% titanium. The titanium with the copper prevents coarse-grain formation, raising the recrystallization temperature. The alloy weighs 22% less than copper, and it plates and solders easily. **Zilloy-20,** of the New Jersey Zinc Co., is pure zinc with no more than 1% of other elements. In rolled strip it has a tensile strength up to 27,000 lb/in^2 (186 MPa) and elongation of 35%.

WILLOW. The wood of the trees *Salix coerulea* and *S. alba,* native to Europe, but grown in many other places. It is best known as a material for cricket bats made in England. The American willows are known as **black willow,** from the tree *S. nigra,* and **western black willow,** from the tree *S. lasiandra.* The wood is also employed for making artificial limbs, and for articles where toughness and nonshrinking qualities are valued. The wood is brownish yellow in color, has a fine, open grain, and weighs about 30 lb/ft^3 (481 kg/m^3). It is of the approximate hardness of cherry and birch. **Japanese willow** is from the tree *S. urbaniana.* It has a closer and finer texture, and a browner color. Black willow has a maximum crushing strength parallel to the grain of about 1,500 lb/in^2 (9 MPa). **Salicin,** also called **salicoside** and **saligenin,** is a glucoside extracted from several species of willow bark of England and also from the American aspen. It is a colorless crystalline material of the composition $(OH)_4C_6H_7 \cdot OO \cdot C_6H_4CH_2OH$, decomposing at 201°C, and soluble in water and in alcohol. It is used in medicine as an antipyretic and tonic, and as a reagent for nitric acid. It hydrolyzes to glucose and salicyl alcohol, and the latter is oxidized to **salicylic acid,** $C_6H_4(OH)COOH$. **Aspirin, acetyl salicylic acid,** is used as an antipyretic and analgesic.

WIRE CLOTH. Stiff fabrics made of fine wire woven with plain loose weave, used for screens to protect windows, for guards, and for sieves and filters. Steel and iron wire may be used, either plain, painted, galvanized, or rustproofed, or various nonferrous metal wires are employed. It is usually put up in rolls in widths from 18 to 48 in (46 to 122 cm). **Screen cloth** is usually 12, 14, 16, and 18 mesh, but wire cloth in copper, brass, or Monel metal is made regularly in meshes from 4 to 100. The size of wire is usually from 0.009 to 0.065 in (0.023 to 0.165 cm) in diameter. Wire cloth for fine filtering is made in very fine meshes. **Mesh** indicates the number of openings per inch, and has no reference to the diameter of wire. A 200-mesh cloth has 200 openings each way on a square inch, or 40,000 openings per square inch (6.4 cm^2). Wire cloth as fine as 400 mesh, or 160,000 openings per square inch (6.4 cm^2), is made by the Newark Wire Cloth Co., by wedge-shaped weaving, although 250 wires of the size of 0.004 in (0.010 cm) when placed parallel and in contact will fill the space of 1 in (2.5 cm). Very fine mesh wire cloth must be woven at an angle since the globular nature of most liquids will not permit passage of the liquid through microscopic square openings. The wire screen cloth of the Michigan Wire Cloth Co., for filtering and screening, has elongated openings. One way the 0.0055-in (0.0140-cm) wire count is 200 per inch (2.5 cm), while the other way the 0.007-in (0.018-cm) warp wire is 40 per inch (2.5 cm).

Wire fabrics for reentry parachutes are made of heat-resistant nickel-chromium alloys, and the wire is not larger than 0.005 in (0.013 cm) in diameter to give flexibility to the cloth. Wire fabrics for ion engines to operate in cesium vapor at temperatures to 2400°F (1316°C) are made with tantalum, molybdenum, or tungsten wire, 0.003 to 0.006 in (0.008 to 0.015 cm) in diameter, with a twill weave. Meshes to a fineness of 350 by 2,300 can be obtained. Porosity uniformity is controlled by pressure calendering of the woven cloth, but for extremely fine meshes in wire cloth it is difficult to obtain the uniformity that can be obtained with porous sintered metals.

Where accuracy of sizing is not important, as in gravel or ore screening, wire fabric is made with oblong or rectangular openings instead of squares to give faster screening. High-manganese steel wire is used for rock screens. For window screening in tropical climates or in corrosive atmospheres plastic filaments are sometimes substituted for the standard copper or steel wire. **Lumite screen cloth,** of the Chicopee Mfg. Co., is woven of vinylidene chloride monofilament 0.015 in (0.038 cm) in diameter in 18 and 20 mesh. The impact strength of the plastic cloth is higher than that of metal wire cloth, but it cannot be used for screening very hot materials. **Lektromesh,** of the C. O. Jelliff Corp., is copper or nickel screen cloth of 40 to 200 mesh made in one piece by electrodeposition. It can be drawn or formed more readily than wire screen, and circular or other shapes can be made with an integral selvage edge.

WIRE GLASS. A sheet glass used in building construction for windows, doors, floors, and skylights, having woven wire mesh embedded in the center of the plate. It does not splinter or fly apart like common glass when subjected to fire or shock, and has higher strength than common glass. It is made in standard thicknesses from ⅛ to ⅜ in (0.318 to 0.953 cm) and in plates 60 by 110 in (1.5 by 2.8 m) and 61 by 140 in (1.5 by 3.6 m). Underwriters' specifications call for a minimum thickness of ¼ in (0.635 cm). Wire glass is made with plain, rough, or polished surfaces, or with ribbed or cobweb surface on one side for diffusing the light and for decorative purposes. It is also obtainable in corrugated sheets, usually 27¾ in (70.5 cm) wide. Wire glass ¼ in (0.635 cm) thick weighs 2.25 lb/ft^2 (11 kg/m^2). Plastic-coated wire mesh may be used to replace wire glass for hothouses or skylights where less weight and fuller penetration of light rays are desired. **Cel-O-Glass,** of Du Pont, is a plastic-coated wire mesh in sheet form.

WOLLASTON WIRE. Any wire made by the Wollaston process of fine-wire drawing. It consists of inserting a length of bare drawn wire into a close-fitting tube of another metal, the tube and core then being treated as a single rod and drawn through dies down to the required size. The outside jacket of metal is then dissolved away by an acid that does not affect the core metal. **Platinum wire** as fine as 0.00005 in (0.00013 cm) in diameter is made commercially by this method, and gold wire as fine as 0.00001 in (0.00002 cm) in diameter is also drawn. Wires of this fineness are employed only in instruments. They are marketed as composite wires, the user dissolving off the jacket. **Taylor process wire** is a very fine wire made by the process of drawing in a glass tube. The process is used chiefly for obtaining fine wire from a material lacking ductility, such as antimony, or extremely fine wire from a ductile metal. The procedure is to melt the metal or alloy into a glass or quartz tube, and then draw down this tube with its contained material. Wire as fine as 0.00004 in (0.00012 cm) in diameter is made, but only in short lengths.

WOOD. A general name applied to the cut material derived from trees. A **tree,** as distinguished from a **bush,** is designated by the U.S. Forest Service as a woody plant with a single erect stem 3 in (7.6 cm) or more in diameter at 4½ ft (1.4 m) above the ground, and at least 12 ft (3.7 m) high. But this definition is merely empirical since in the cold climate of northern Canada perfect, full-grown trees 10 to 15 years old may be only 6 in (15 cm) high. **Timber,** in general, refers to standing trees, while **lumber** is the sawed wood used for construction purposes. In construction work the word timber is often applied to large pieces of lumber used as beams.

Wood is an organic chemical compound composed of approximately 49% carbon, 44 oxygen, 6 hydrogen and 1 ash. It is largely cellulose and lignin. The wood of white pine is about 50% cellulose, 25 lignin, and the remainder sugars, resin, acetic acid, and other materials. Wood is produced in most trees by a progressive growth from the outside. In the spring, when sap flows rapidly, a rapid formation of large cells takes place, followed by a slower growth of hard and close cells in the summer. In some woods, such as oak, there is a considerable difference in quality and appearance between the spring and summer woods. In some long-lived trees, such as Douglas fir, there is a decreasing in strength between the outside wood with narrow rings and the wide-ringed wood of the interior. **Heartwood** is the dark center of the tree which has become set, and through which the sap has ceased to flow. **Sapwood** is the outer, live wood of the tree; unless treated, it has low decay resistance. The grain of sawed lumber results from sawing across the annual growth rings, varied to produce different grains.

Wood is seasoned either by exposing it to the air to dry, or by kiln drying. The former method is considered to give superior quality, but it requires more time, is expensive, and is indefinite. Numerous tests made at the U.S. Forest Products Laboratory did not reveal any superiority in air-dried wood when kiln drying was well done. Solvent seasoning is a rapid process consisting in circulating a hot solvent through the wood in a closed chamber. California redwood, when seasoned with acetone at 130°F (54°C), yields tannin and some other chemicals as by-products. Seasoned wood, when dry, is always stronger than the unseasoned wood. **Tank woods** are selected for resistance to the liquids to be contained. Tanks for vinegar and foodstuffs containing vinegar, such as pickles, are of white oak, cypress, or western red cedar. Beer tanks are of white oak or cypress. Tanks for aging wine are of redwood, oak, or fir.

The term **log** designates the tree trunk with the branches removed. **Balk** is a roughly squared log; **plank** is a piece cut to rectangular section 11 in (28 cm) wide; **deal** is a piece 9 in (23 cm) wide; and **batten** is a piece 7 in (18 cm) wide. **Board** is a thin piece of any width less than 2 in (5 cm) thick. **Flitch** is half a balk cut in two lengthwise. **Scantling** is a piece sawed on all sides. **Shakes** are longitudinal splits or cracks in the wood due to shrinkage or decay.

All woods are divided into two major classes on the basis of the type of tree from which they are cut. **Hardwoods** are from broad-leaved, deciduous trees. **Softwoods** are from conifers, which have needle- or scalelike leaves, which are, with few exceptions, evergreens. These terms do not refer to the relative hardnesses of the woods in these two classes. Hardwood lumber is available in three basic categories: **factory lumber, dimension lumber,** or **dimension parts,** and **finished market products.** The important difference between factory lumber and dimension parts is that factory lumber grades reflect the proportion of the pieces that can be cut into useful smaller pieces while the dimension grades are based on use of the entire piece. Finished market products are graded for their end use with little or no remanufacturing. Examples of finished market products are flooring, siding, ties, timbers, trim, molding, stair treads, and risers. The rules adopted by the National Hardwood Lumber Association are considered standard in grading factory lumber. The grades from the highest to the lowest quality are as follows: firsts, the top quality, and seconds, both of which are usually marketed as one grade called firsts and seconds (FAS); selects; and common grades No. 1, No. 2, No. 3A, and No. 3B. Sometimes a grade is further specified, such as FAS one face, which means that only one face is of the FAS quality. Another designation, WHND, sometimes used, means that wormholes are not considered defects in determining the grade. Dimension lumber, generally graded under the rules of the Hardware Dimension Manufacturers Association, are of three classes: solid dimension flat stock, kiln-dried dimension flat stock, and solid dimension squares. Each class may be rough, semifabricated, or fabricated. Rough dimension blanks are usually kiln-dried and are supplied sawn and ripped to size. Surfaced or semifabricated stock has been further processed by gluing, surfacing, etc. Fabricated stock has been completely processed for the end use. Solid dimension flat stock has five grades: clear—two faces, clear—one face, paint, core, and sound. Squares have three grades if rough (clear, select, sound) and four if surfaced (clear, select, paint, second).

There are two major categories of softwood lumber—construction and remanufacture. **Construction lumber** is of three general types: stress-graded; non-stress-graded, also referred to as **yard lumber;** and **appearance lumber. Stress-graded lumber** is structural lumber never less than 2 in (5 cm) thick, intended for use where definite strength requirements are specified. The allowable stresses specified for stress-graded lumber depend on the size, number, and placement of defects. Because the location of defects is important, the piece must be used in its entirety for the specified strength to be realized. Stress-graded products include timbers, posts, stringers, beams, decking, and some boards. Typical **non-stress-graded lumber** items include boards, lath, battens, crossarms, planks, and foundation stock. **Boards,** sometimes referred to as **commons,** are one of the more important non-stress-graded products. They are separated into three to five different grades, depending upon the species and lumber manufacturing association involved. Grades may be described by number (No. 1, No. 2) or by descriptive terms (construction, standard). First-grade boards are usually graded primarily for serviceability, but appearance is also considered. Second- and third-grade boards are often used together for such purposes as subfloors and sheathing. Fourth-grade boards are not selected for appearance but for adequate strength. The appearance category of construction lumber includes trim, siding, flooring, ceiling, paneling, casing, and finish boards. Most appearance lumber grades are designated by letters and combinations of letters, and are also often known as select grades. Typical grades of **lumber remanufacture** are the factory grades and industrial clears. Factory Select and Select Shop are typical high grades of factory lumber, followed by No. 1, No. 2, and No. 3 shop. Industrial clears are used for cabinet stock, door stock, and other products where excellent appearance, mechanical and physical properties, and finishing characteristics are important. The principal grades are B&BTR, C, and D.

Metallized wood is wood treated with molten metal so that the cells of the wood are filled with the metal. Fusible alloys, with melting points below the scorching point of the wood, are used. The wood is immersed in molten metal in a closed container under pressure. The hardness, compressive strength, and flexural strength of the wood are increased, and the wood becomes an electric conductor lengthwise of the grain. Woods are also metallized with a surface coating of metal by vacuum deposition.

Sugar pine is one of the most widely used **pattern woods** for foundry patterns. It replaces eastern white pine, which is scarcer and now usually more costly. Poplar is used for patterns where a firmer wood is desired; cherry or maple is employed where the pattern is to be used frequently or will be subject to severe treatment. Densified wood is also used for patterns required to be very wear-resistant. Mahogany is used for small and intricate patterns where a firm texture and freedom from warpage are needed. However, for small castings made in quantities on gates, aluminum or brass is more frequently used.

Excelsior is an old trade name, still used, for continuous, curly, fine wood shavings employed as a packing material for breakable articles. It is light and elastic, and it is also used as a cushioning and stuffing material. It is usually made from poplar, aspen, basswood, or cottonwood. A cord of wood produces about 1,500 lb (680 kg), but it may be made as a by-product from other woodworking. It is also called **wood fiber** and **wood wool,** but these terms more properly refer to fibers of controlled size and length used with a resin binder for molding into handles, knobs, and other **imitation wood** parts.

Some wood for special purposes comes from roots or from bushes. The **briar** used for tobacco pipes is from the root of the **white heath,** *Erica arborea,* of North Africa. Substitutes for briar are the burls of the laurel and rhododendron. **Yareta,** used for fuel in the copper region of Chile, is a mosslike woody plant which grows on the sunny northern mountain slopes at altitudes above 12,000 ft (3,658 m) and requires several hundred years to reach a useful size.

WOOD FLOUR. Finely ground dried wood employed as a filler and as reinforcing material in molding plastics and in linoleum, and as an absorbent for nitroglycerin. It is made largely from light-colored softwoods, chiefly pine and spruce, but maple and ash flours are preferred where no resin content is desired. Woods containing essential oils, such as cedar, are not suitable. Wood flour is produced from sawdust and shavings by grinding in burr mills. It has the appearance of wheat flour. The sizes commonly used are 40, 60, and 80 mesh; the finest is 140 mesh. Grade 1, used as a filler in rubber and plastics, has a particle size of 60 mesh and a specific gravity of 1.25, but 80 and 100 mesh are also used for plastic filler. Since wood flour absorbs the resin or gums when mixed in molding plastics and sets hard, it is sometimes mixed with mineral powders to vary the hardness and toughness of the molded product. The **Hygeia wood flour,** of the Penn-Rillton Co., for use as a filler in plastics, is a dust-free 100-mesh powder made from oak. The char point is 410°F (210°C) and ignition point is 600°F (316°C).

Vast quantities of **sawdust** are obtained in the sawmill areas. Besides being used as a fuel, it is also employed for packing, for finishing metal parts in tumbling machines, for making particleboard, and for distilling to obtain resins, alcohols, sugars, and other chemicals. Some sawdust is pulped, and as much as 20% of such pulp can be used in kraft paper without loss of strength. Hickory, walnut, and oak sawdusts are used for meat smoking, or for the making of **liquid smoke,** which is produced by burning the sawdust and absorbing the smoke into water. For the rapid production of bacon and other meats, immersion in liquid smoke imitates the flavor of smoked meat. **Charsol,** of Arrow Products Corp., is maple smoke absorbed in water and refined to remove the creosote. Some sawdust is used for agricultural mulch and fertilizer by chemical treatment to accelerate decay. **Bark fuel,** of the Southern Extract Co., is shredded bark, flash-dried and pelletized with powdered coal. **Kube-Kut,** of Michael Wood Products, Inc., is maple scrap wood cut into fine cubical particles for use in tumble polishing. **Particleboard,** made by compressing sawdust or wood particles with a resin binder into sheets, has uniform strength in all directions, and a smooth grainless surface. When used as a core for veneer panels it requires no cross-laminating. **Versacore,** of the Weyerhaeuser Co., is a veneer building board with an oak particleboard core. Mechanical pulp for newsprint can be made from sawdust, but the quantity available is usually not sufficient. The material known as **ground wood,** of fine-mesh fibers, is made from cord wood, about one ton of fibers being produced from one cord of pulpwood. **Plastic wood,** usually marketed as a paste in tubes for filling cavities or seams in wood products, is wood flour or wood cellulose compounded with a synthetic resin of high molecular weight that will give good adhesion but not penetrate the wood particles to destroy their nature. The solvent is kept low to reduce shrinkage. When cured in place the material can be machined, polished, and painted. **Plastiplate,** of Duorite Plastics, is wood powder with a phenolic resin binder and a curing accelerator. The cured material has a specific gravity of 1.05 and a flexural strength of 3,000 lb/in^2 (20 MPa).

WOOD PRESERVATIVES. These fall into two general classes: oils, such as **creosote** and petroleum solutions of **pentachlorophenol;** and **waterborne salts** that are applied as water solutions. **Coal tar creosote,** a black or brownish oil made by distilling coal tar, is the oldest and still one of the more important and useful wood preservatives. Its advantages are high toxicity to wood-destroying organisms; relative insolubility in water and low volatility, which impart to it a great degree of permanence under the most varied use conditions; ease of application; ease with which its depth of penetration can be determined; general availability and relatively low cost; and long record of satisfactory use.

Creosotes distilled from tars other than coal tar are used to some extent for wood preservation. For many years, either **cold tar** or **petroleum oil** has been mixed with cold tar creosote in various proportions to lower preservative costs.

Water-repellent solutions containing chlorinated phenols, principally pentachlorophenol, in solvents of the mineral spirit type have been used in commercial treatment of wood by the millwork industry since about 1931. Pentachlorophenol solutions for wood preservation generally contain 5% (by weight) of this chemical although solutions with volatile solvents may contain lower or higher concentrations. Preservative systems containing water-repellent components are sold under various trade names, principally for the dip or equivalent treatment of window sash and other millwork. According to federal specifications the preservative chemicals may not contain less than 5% of pentachlorophenol.

Standard wood preservatives used in water solution include **acid copper chromate, ammoniacal copper arsenite, chromated copper arsenate, chromated zinc chloride** and **fluor chrome arsenate phenol**. These preservatives are often employed when cleanliness and paintability of the treated wood are required. The chromated zinc chloride and fluor chrome arsenate phenol formulations resist leaching less than preservative oils, and are seldom used where a high degree of protection is required for wood in ground contact or for other wet installations. Several formulations involving combinations of copper, chromium, and arsenic have shown high resistance to leaching and very good performance in service. The ammoniacal copper arsenite and chromated copper arsenate are included in specifications for such items as building foundations, building poles, utility poles, marine piling, and piling for land and fresh water use.

WOOL. The fine, soft, curly hair or fleece of the sheep, alpaca, vicuña, certain goats, and a few other animals. The specific designation "wool" always means the wool of sheep. **Sheep's wool** is one of the most important commercial fibers because of its good physical qualities and its insulating value, especially for clothing, but it now constitutes only about 10% of the textile fiber market. It is best known for its use in clothing fabrics, called **woolens.** These are designated under a variety of very old general trade names such as a loosely woven fabric called **flannel,** or the fine, smooth fabric known as **broadcloth. Cheviot** is a close-napped, twill-woven fabric, and **tweed** is a woolen fabric with a coarse surface, usually with a herring-bone-twill weave. **Serge** is a twill-woven worsted fabric. **Worsteds** are wool fabrics made from combed-wool yarn, usually from long, smooth wool. Wool is also employed for packings and for insulation, either loose or felted, and for making felts. The average amount of wool shorn from sheep in the United States is 8.1 lb (3.7 kg) per animal.

Wool differs from hair in fineness and in its felting and spinning properties. The latter are due to the fine scales of the wool fibers. The finest short-staple wool has as many as 4,000 scales to the inch (2.5 cm), and the average long-staple wool has about 2,000 scales per inch (2.5 cm). These scales give wool its cohesive qualities. Some animals have both wool and hair, while others have wool only when young. There is no sharp dividing line between wool and hair.

Wool quality is by fineness, softness, length, and scaliness. Fiber diameters vary from 0.0025 to 0.005 in (0.0064 to 0.013 cm). Long wools are generally heavy. Fibers below 3 in (7.6 cm) in length are known as **clothing wool,** and those from 3 to 7 in (7.6 to 17.8 cm) are called **combing wools.** Long wools are fibers longer than 7 in (17.8 cm). The term **apparel wool** generally means clothing wool of fine weaving quality from known sources. **Fleece wool** is the unscoured fiber. It may contain as high as 65% grease and dirt, but this is the form in which wool is normally shipped because it then has the protection of the wool fat until it is manufactured. Wool is very absorbent to moisture and will take up about 33% of its weight of water, and in some areas moisture and dirty grease are added to fleece wool to increase weight. **Carpet wools** are usually long nonresilient fibers from sheep bred in severe climates, such as the **Mongolian wool.** The only breed of sheep developed for wool alone is the Merino. In Australia the **corriedale** and the **polworth sheep** are dual-purpose animals for wool and meat.

The finest of sheep wools comes from the **merino sheep,** but these vary according to the age of breeding of the animal. The **Lincoln sheep** produces the longest fiber. It is lustrous but very coarse. Luster of wool depends upon the size and smoothness of the scales, but the chemical composition is important. The molecular chains are linked with sulfur, and when sulfur is fed to the sheep in some deficient areas the quality of the wool is improved. Crimpiness in wool is due to the open formation of the scales. A fine Merino will have 24 crimps per inch (2.5 cm), whereas a coarse crossbred will have only 6 per inch (2.5 cm). Strength of wool fibers often depends upon the health of the animal and the feeding.

One quarter of the world production of wool is in Australia. Argentina ranks second in production, with the United States third. But the United States is a lamb-eating nation, and a large proportion of the animals are slaughtered when 4 to 8 months old, and most of the others are kept only one season for one crop of wool. New Zealand, Uruguay, Russia, and England are also important producers. England is the center of wool-sheep breeding, with more varieties than any other country. In general, warm climates produce fine wools, and hot climates produce thin, wiry wools, but the fundamental differences come from the type of animal and the feeding. The **reused wool** from old cloth was originally called **shoddy,** but the name has an opprobrious signification in the United States, and is not used by manufacturers to designate the fabrics made from reclaimed wool. Shoddy is used in mixtures with new wool for clothing and other fabrics. **Extract wool** is shoddy that is recovered by dissolving out the cotton fibers of the old cloth with sulfuric acid. Short fibers of shoddy, less than ½ in (1.27 cm) are known as **mungo fibers.** They are used in woolen blends to obtain a napped effect. **Reprocessed wool** is fiber obtained from waste fabric which has not been used. **Noils** are short fibers produced in the combing of wool tops for making worsteds. They are used for woolen goods and felt. **Zeset,** of Du Pont, a shrink-proofing agent for wools, is a variant of **Surlyn T,** a terpolymer of 70% ethylene, 6 methacryloyl chloride, and 24 vinyl acetate. It prevents shrinkage and pilling under ordinary laundry methods, does not affect color, and increases the tensile strength of the fiber. But all resinous additives tend to harden the fiber and lessen the drape and feel. Conversely, each dry cleaning of wool fabric decreases the natural oil content and hardens the fiber.

WOOL GREASE. A brownish waxy fat of a faint, disagreeable odor, obtained as a by-product in the scouring of wool. The purified grease was formerly known as **degras** and was used for leather dressing, in lubricating and slushing oils, and in soaps and ointments, but it is now largely employed for the production of lanolin and its derivatives, chiefly for cosmetics. Wool grease contains **lanoceric acid,** $C_{30}H_0O_4$; **lanopalmic acid,** $C_{15}H_{30}O_3$; and **lanosterol,** a high alcohol related to cholesterol. All of these can be broken down into derivatives.

Lanolin is a purified and hydrated grease, also known as **lanain,** and in pharmacy as **lanum** and **adeps lanae.** It has a melting point at about 40°C, and is soluble in alcohol. Lanolin is basically a wax consisting of esters of sterol alcohols combined with straight-chain fatty acids, and with only a small proportion of free alcohols. It contains about 95% of fatty acid esters, but its direct use as an emollient depends on the 5% of free alcohols and acids. However, more than 30 derivatives are obtained from lanolin, and these are used in blends to give specific properties to cosmetics. They are often marketed under trade names, and some of the ingredients may be synthesized from raw materials other than wool grease, or chemically altered from wool-grease derivatives.

A variety of products used in cosmetics and pharmaceuticals are made by fractionation or chemical alteration of lanolin. They are also useful in compounding plastics and industrial coatings, but are generally too scarce and expensive for these purposes. **Ethoxylated lanolin** and ethoxylated **lanolin alcohols** are used in water-soluble emulsions and conditioners. **Solulan,** of Chemectants, Inc., is a general trade name for these materials. **Lanolin oil** and **lanolin wax** are made by solvent fractionation of lanolin. **Viscolan** and **Waxolan,** of the same company, are these products. **Isopropyl lanolates,** trade-named **Amerlate** by American Cholesterol Products, Inc., are soft hydrophylic solids which liquefy easily and are used in cosmetics as emollients, emulsifiers, and pigment dispersants. **Amerlate LFA** is derived from lanolin hydroxy acids containing iso-acids. The high hydroxyl content produces the emollient and emulsifying qualities. **Barium lanolate,** made by saponification, is used as an anticorrosion agent. It is antiphobic and is also used as an anticaking agent. In a 25% barium concentration it is used for hard lubricating grease.

Ethoxylan, of N. I. Malmstrom & Co., is an ethylene oxide derivative of lanolin, soluble in water and in alcohol, and used in shampoos. **Ceralan,** of the Robison-Wagner Co., is a waxy solid melting at 55°C to an amber-colored viscous liquid. It is a mixture of monohydroxyl alcohols, obtained by splitting lanolin, and contains 30% sterol, and free cholesterol. It forms water-in-oil emulsions, and is used in cosmetics as a dispersing and stiffening agent and as an emollient. **Acetylated lanolin,** usually sold under trade names, is made by reacting lanolin with polyoxyethylenes. They are clear, nongreasy liquids soluble in water, oils, and in alcohol. The acetylated lanolin is hydrophobic and oil-soluble, and is used as an odorless, nontacky emollient in cosmetics.

Veriderm, of the Upjohn Co., is a substitute for lanolin as an emollient.

It contains about the same percentage of triglycerol esters of fatty acids, free cholesterol, and saturated and unsaturated hydrocarbons, as occurs in the natural human skin oils. Cholesterol is one of the most important of the complex **sterols,** or **zoosterols,** from animal sources. It is produced from lanolin, but also from other sources, and used in drugs and cosmetics. **Amerchol L-101,** of the American Cholesterol Products, Inc., is a liquid nonionic cholesterol containing other sterols. Wool grease from the scouring of wool was originally called **Yorkshire grease. Moellon degras** is not wool grease, but is a by-product of chamois leather making. The sheepskins are impregnated with fish oil and, when the tanning is complete, they are soaked in warm water and the excess oil is pressed out to form the moellon degras.

WROUGHT IRON. Commercially pure iron made by melting white cast iron and passing an oxidizing flame over it, leaving the iron in a porous condition which is then rolled to unite it into one mass. As thus made it has a fibrous structure, with fibers of slag through the iron in the direction of rolling. It is also made by the Aston process of shotting bessemer iron into a ladle of molten slag. Modern wrought iron has a fine dispersion of silicate inclusions which interrupt the granular pattern and give it a fibrous nature.

The value of wrought iron is in its corrosion resistance and ductility. It is used chiefly for rivets, staybolts, water pipes, tank plates, and forged work. Minimum specifications for **ASTM wrought iron** call for a tensile strength of 40,000 lb/in^2 (275 MPa), yield strength of 24,000 lb/in^2 (165 MPa), and elongation of 12%, with carbon not over 0.08%, but the physical properties are usually higher. **Wrought iron 4D,** of the A. M. Byers Co., has only 0.02% carbon with 0.12 phosphorus, and the fine fibers are of a controlled composition of silicon, manganese, and phosphorus. This iron has a tensile strength of 48,000 lb/in^2 (330 MPa), elongation 14%, and Brinell hardness 105. **Mn wrought iron** has 1% manganese for higher impact strength.

Ordinary wrought iron with slag may contain frequent slag cracks, and the quality grades are now made by controlled additions of silicate, and with controlled working to obtain uniformity. But for tanks and plate work, ingot iron is now usually substituted. **Merchant bar iron** is an old name for wrought-iron bars and rods made by faggoting and forging. **Iron-fibered steel,** of the Edgar Allen Steel Co., Inc., is soft steel with fine iron wire worked into it. **Staybolt iron** may be wrought iron, but was originally puddled charcoal iron. **Lewis iron,** of Joseph T. Reyerson & Son, Inc., for staybolts, is highly refined, puddled iron with a tensile strength of 52,000 lb/in^2 (358 MPa) and elongation of 30%.

The **Norway iron** formerly much used for bolts and rivets was a **Swed-**

ish charcoal iron brought to America in Norwegian ships. This iron, with as low as 0.02% carbon, and extremely low silicon, sulfur, and phosphorus, was valued for its great ductility and toughness and also for its permeability qualities for transformer cores. Commercial wrought iron is now usually ingot iron or fibered low-carbon steel.

YARNS. Assemblages or bundles of fibers twisted or laid together to form continuous strands. They are produced with either filaments or staple fibers. Single strands of yarns can be twisted together to form ply or plied yarns, and ply yarns in turn can be twisted together to form cabled yarn or cord. Important yarn characteristics related to behavior are fineness (diameter or linear density) and number of twists per unit length. The measure of fineness is commonly referred to as yarn number. Yarn numbering systems are somewhat complex, and they are different for different types of fibers. Essentially, they provide a measure of fineness in terms of weight per unit or length per unit weight.

Cotton yarns are designated by numbers, or **counts.** The standard count of cotton is 840 yd/lb. Number 10 yarn is therefore 8,400 yd/lb. A No. 80 sewing cotton is 80 × 840, or 67,200 yd/lb.

Linen yarns are designated by the **lea** of 300 yd. A 10-count linen yarn is 10 × 300, or 3,000 yd/lb.

The size or count of spun **rayon yarns** is on the same basis as cotton yarn. The size or count of rayon filament yarn is on the basis of the **denier,** the rayon denier being 450 m weighing 5 cg. If 450 m (492.12 yd) of yarn weigh 5 cg, it has a count of 1 denier. If it weighs 10 cg, it is No. 2 denier. Rayon yarns run from 15 denier, the finest, to 1,200 denier, the coarsest.

Reeled **silk yarn** counts are designated in deniers. The **international denier** for reeled silk is 500 m of yarn weighing 0.05 g. If 500 m weigh 1 g, the denier is No. 20. Spun silk count under the English system is the same as the cotton count. Under the French system the count is designated by the number of **skeins** weighing 1 kg. The skein of silk is 1,000 m.

A **ply yarn** is one that has two or more yarns twisted together. A two-ply yarn has two separate yarns twisted together. The separate yarns may be of different materials, such as cotton and rayon. A six-ply yarn has six separate yarns. A ply yarn may have the different plies of different twists to give different effects. Ply yarns are stronger than single yarns of the same diameter. Tightly twisted yarns make strong, hard fabrics. Linen yarns are not twisted as tightly as cotton because the flux fiber is longer, stronger, and not as fuzzy as the cotton. **Filament rayon yarn** is yarn made from long, continuous rayon fibers, and it requires only slight twist. Fabrics made from filament yarn are called **twalle. Monofilament** is fiber heavy enough to be used alone as yarn, usually more than 15 denier. **Tow** consists of multifilament reject strands suitable for cutting into staple lengths for spinning. Spun rayon yarn is yarn made from **staple fiber,** which is rayon filament cut into standard short lengths.

YUCCA FIBER. The fiber obtained from the leaves of a number of desert plants of the genus *Yucca* of the lily family native to the southwestern United States and northern Mexico. The fiber is similar to fibers from agave plants and is often confused with them and with istle. The heavier fibers are used for brushes, and the lighter fibers are employed for cordage and burlap fabrics. In Mexico the word **palma** designates yucca fibers and grades of istle as well as palm-leaf fibers. **Palma samandoca** is fiber from the plant *Samuela carnerosana,* the **date yucca.** It is also called **palma istle. Palmilla fiber** is from the *Y. elata.* **Palma pita** is a fiber from the *Y. treculeana.* **Pita fiber** used for coffee bags in Colombia and Central America is from a different plant. Other yucca fibers come from the plants *Y. glauca, Y. baccata,* and *Y. gloriosa.* Some varieties of the *Y. baccata* also yield edible fruits. The roots of species of yucca yield saponin which is also obtained as a by-product in extracting the yucca fiber.

ZINC. A bluish-white crystalline metal with a specific gravity of 7.13, melting at 419.5°C and boiling at 907°C. The commercially pure metal has a tensile strength, cast, of about 9,000 lb/in^2 (62 MPa) with elongation of 1%, and the rolled metal has a strength of 24,000 lb/in^2 (165 MPa) with elongation of 35%. But small amounts of alloying elements harden and strengthen the metal, and it is seldom used alone. Zinc is used for galvanizing and plating; for making brass, bronze, and nickel silver; for electric batteries; for die castings; and in alloyed sheets for flashings, gutters, and stamped and formed parts. The metal is harder than tin, and an electrodeposited plate has a Vickers hardness of about 45. Zinc is also used for many chemicals.

The old name **spelter,** often applied to slab zinc, came from the name spailter used by Dutch traders for the zinc brought from China. The first zinc produced in the United States in 1838 came from New Jersey ore. **Sterling spelter** was 99.5% pure. Special high-grade zinc is distilled, with a purity of 99.99%, containing no more than 0.006% lead and 0.004 cadmium. High-grade zinc, used in alloys for die casting, is 99.9% pure, with 0.07 max lead. **Brass special zinc** is 99.10% pure, with 0.6 max lead and 0.5 max cadmium. **Prime western zinc,** used for galvanizing, contains 1.60 max lead and 0.08 max iron. **Zinc crystals** produced by Semi-Elements, Inc., for electronic uses, are 99.999% pure metal.

On exposure to the air, zinc becomes coated with a film of carbonate and is then very corrosion-resistant. **Zinc foil** comes in thicknesses from 0.001 to 0.006 in (0.003 to 0.015 cm). It is produced by electrodeposition on an aluminum drum cathode and stripping off on a collecting reel. But most of the **zinc sheet** contains a small amount of alloying elements to increase the physical properties. Slight amounts of copper and titanium reduce grain size in the sheet zinc. In cast zinc the hexagonal columnar grain extends from the mold face to the surface or to other grains growing from another mold face, and even very slight additions of iron can control this grain growth. Aluminum is also much used in alloying zinc. In zinc used for galvanizing, a small addition of aluminum prevents formation of the brittle alloy layer, increases ductility of the coating, and gives a smoother surface. Small additions of tin give bright spangled coatings.

Zinc has 12 isotopes, but the natural material consists of 5 stable isotopes, of which nearly half is **zinc 64.** The stable isotope **zinc 67,** occurring to the extent of about 4% in natural zinc, is sensitive to tiny variations in transmitted energy, giving off electromagnetic radiations which permit high accuracy in measuring instruments. It measures gamma-ray vibrations with great sensitivity, and is used in the nuclear clock.

Zinc powder, or **zinc dust,** is a fine gray powder of 97% min purity usually in 325-mesh particle size. It is used in pyrotechnics, in paints, as a reducing agent and catalyst, in rubbers as a secondary dispersing agent and to increase flexing, and to produce **Sherardized steel.** Sherardizing consists in hot-tumbling the steel parts in a closed drum with the zinc powder. It is a form of galvanizing, and controlled zinc coatings of 0.1 to 0.4 oz/ft^2 (0.4 to 1.8 g/cm^2) of surface give good corrosion protection. In paints, zinc powder is easily wetted by oils. It keeps the zinc oxide in suspension, and also hardens the film. **Mossy zinc,** used to obtain color effects on face brick, is a spangly zinc powder made by pouring the molten metal into water. **Feathered zinc** is a fine grade of mossy zinc. **Photoengraving zinc** for printing plates is made from pure zinc with only a small amount of iron to reduce grain size and alloyed with not more than 0.2% each of cadmium, manganese, and magnesium. **Cathodic zinc,** used in the form of small bars or plates fastened to the hulls of ships or to underground pipelines to reduce electrolytic corrosion, is zinc of 99.99% purity and with the iron kept below 0.0014% to prevent polarization. **Merrillite,** of the Pacific Smelting Co., is high-purity zinc dust.

ZINC-BASE ALLOYS. Alloys of zinc are mostly used for die castings for decorative parts and for functional parts where the load-bearing and shock requirements are relatively low. Since the zinc alloys can be cast easily in high-speed machines, producing parts that weigh less than brass and have high accuracy and smooth surface that require minimum machining and finishing, they are widely used for such parts as handles, and for gears, levers, pawls, and other parts for small assemblies. Zinc alloys for sheet contain only small amounts of alloying elements, with 92 to 98% zinc, and the sheet is generally referred to simply as zinc or by a trade name. The

Modified zinc sheet of the New Jersey Zinc Co., used for stamped, drawn, or spun parts for costume jewelry and electronics, contains up to 1.5% copper and 0.5 titanium. The titanium raises the recrystallization temperature, permitting heat treatment without coarse grain formation.

Hartzink had 5% iron and 2 to 3 lead, but iron forms various chemical compounds with zinc and the alloy is hard and brittle. Copper reduces the brittleness. **Germania bearing bronze** contained 1% iron, 10 tin, about 5 each of copper and lead, and the balance zinc. **Fenton's alloy** had 14 tin, 6 copper, and 80 zinc, and **Ehrhard's bearing metal** contained 2.5% aluminum, 10 copper, 1 lead, and a small amount of tin to form copper-tin crystals. **Binding metal,** for wire-rope slings, has about 2.8% tin, 3.7 antimony, and the balance zinc. **Pattern metal,** for casting gates of small patterns, was almost any brass with more zinc and some lead added, but is now standard die-casting metal.

Zinc-base alloys for die castings are now quite narrowly standardized, with about 4% aluminum, with or without much copper, and with small amounts of controlling elements. **Zinc die castings** have a wide range of use because of their low cost, good strength, ease of production of complicated shapes to accurate dimensions, and smooth surface with ease of finishing. They are valued for gears and parts of miniature motors and instruments. In automobiles they are used for such parts as carburetor bodies, handles, and latches. The two principal die-casting alloys are ASTM **zinc alloy A640A** (SAE903) and **zinc alloy AC41A** (SAE925). Both contain a fraction of a percent of magnesium and 3.5 to 4.3% aluminum. AC41A also contains 0.75 to 1.25% copper. The properties of both alloys are about the same. Tensile strength ranges between 40,000 and 47,000 lb/in^2 (275 and 324 MPa). Impact strength exceeds that of most cast irons and aluminum castings. The alloys are close to the weight of steel and are relatively soft. Their excellent castability allows production of smooth-surface castings in sections as thin as 0.015 in (0.038 cm). Machinability is good, but zinc alloys are difficult to weld and solder. Applications of zinc die castings range in size from watch hairspring wedges to automobile panels 6 ft (2 m) long and 40 lb (18 kg) in weight.

Zinc alloys for permanent-mold and sand castings are similar to those used for die castings. Zinc-alloy sand castings are widely used for forming, drawing, stretching, blanking, and trimming large shapes made from light-gage sheet aluminum, magnesium, steels, and other metals. **Zinc-aluminum casting alloys,** initially developed for sand and permanent-mold castings, are also used for die castings. In general they provide somewhat higher performance properties (strength, hardness, wear resistance, and bearing properties).

SAE alloy 903 has a similar composition but with copper limited to 0.10 maximum. Its physical properties are somewhat lower, but these properties are more stable, and it has better dimensional stability. It is, therefore, more generally used.

Manganese-zinc alloys, with up to 25% manganese, for high-strength extrusions and forgings, are really 60–40 brass with part of the copper replaced by an equal amount of manganese, and are classed with manganese bronze. They have a bright white color and are corrosion-resistant. **Zam metal,** of Hanson-Van Winkle-Munning Co., for zinc-plating anodes, is zinc with small percentages of aluminum and mercury to stabilize against acid attack. **Zinc solders** are used for joining aluminum. The **tin-zinc solders** have 70 to 80% tin, about 1.5 aluminum, and the balance zinc. The working range is 500 to 590°F (260 to 310°C). **Zinc-cadmium solder** has about 60% zinc and 40 cadmium. The pasty range is between 510 and 599°F (266 and 315°C).

A group of wrought alloys, called **superplastic zinc alloys,** have elongations of up to 2500% in the annealed condition. These alloys contain about 22% aluminum. One grade can be annealed and air-cooled to a strength of 71,000 lb/in^2 (489 MPa). Parts of these alloys have been produced by vacuum forming and by a compression molding technique similar to forging but requiring lower pressures.

ZINC CHEMICALS. With the exception of the oxide, the quantities of zinc compounds consumed are not large compared with many other metals, but zinc chemicals have a very wide range of use, being essential in almost all industries and for the maintenance of animal and vegetable life. Zinc is a complex element and can provide some unusual conditions in alloys and chemicals.

Zinc oxide, ZnO, is a white, water-insoluble, refractory powder melting at about 1975°C, having a specific gravity of 5.66. It is much used as a pigment and accelerator in paints and rubbers. Its high refractive index, about 2.01, absorption of untraviolet light, and fine particle size give high hiding power in paints, and make it also useful in such products as cosmetic creams to protect against sunburn. Commercial zinc oxide is always white, and in the paint industry is also called **zinc white** and **Chinese white.** But with a small excess of zinc atoms in the crystals, obtained by heat treatment, the color is brown to red.

In paints, zinc oxide is not as whitening as lithopone, but it resists the action of ultraviolet rays and is not affected by sulfur atmospheres, and is thus valued in outside paints. **Leaded zinc oxide,** consisting of zinc oxide and basic lead sulfate, is used in paints, but for use in rubber the oxide must be free of lead. In insulating compounds zinc oxide improves electrical resistance. In paper coatings it gives opacity and improves the finish. **Zinc-white paste** for paint mixing usually has 90% oxide and 10 oil. **Zinc oxide stabilizers,** composed of zinc oxides and other chemicals, can be added to plastic molding compounds to reduce the deteriorating effects of sunlight and other types of degrading atmospheres.

Zinc oxide crystals are used for transducers and other piezoelectric devices. The crystals are hexagonal and are effective at elevated temperatures, as the crystal has no phase change up to its disassociation point. The resistivity range is 0.5 to 10 $\Omega \cdot$cm. Zinc oxide has luminescent and light-sensitive properties which are utilized in phosphors and ferrites. But the oxygen-dominated zinc phosphors used for radar and television are modifications of zinc sulfide phosphors. The **zinc sulfide phosphors** which produce luminescence by exposure to light are made with zinc sulfide mixed with about 2% sodium chloride and 0.005% copper, manganese, or other activator, and fired in a nonoxidizing atmosphere. The cubic crystal structure of zinc sulfide changes to a stable hexagonal structure at 1020°C, but both forms have the phosphor properties. **Zinc sulfide** is a white powder of the composition $ZnS \cdot H_2O$, and is also used as a paint pigment, for whitening rubber, and for paper coating. **Cryptone,** of the New Jersey Zinc Co., is zinc sulfide for pigment use in various grades, some grades containing barium sulfate, calcium sulfide, or titanium dioxide.

Zinc is an **amphoteric element,** having both acid and basic properties, and it combines with fatty acids to form metallic soaps, or with the alkali metals or with ammonia to form **zincates. Sodium zincate** is used for waterproofing asbestos-cement shingles. **Zinc stearate,** $Zn(C_{18}H_{35}O_2)_2$, is a **zinc soap** in the form of a fine white powder used in paints and in rubber. A USP grade of 325 mesh is used in cosmetics. **Aquazinc,** of the Beacon Co., and **Liquizinc,** of Rubba, Inc., are zinc stearate dispersions in water used as an antitack agent in milling rubber. **Zinc acetate,** $Zn(C_2H_3O_2)_2$, is a white solid partly soluble in water, used as a mordant, as a wood preservative, in porcelain glazes, and as a mild antiseptic in pharmaceuticals.

Zinc sulfate, $ZnSO_4 \cdot 7H_2O$, is the chief material for supplying zinc in fertilizers, agricultural sprays, and animal feeds. For these purposes it is used in the form of white vitriol containing 22% zinc, or as the monohydrate, $ZnSO_4 \cdot H_2O$, containing 37% zinc. **Zinc chloride,** a white, crystalline, water-soluble powder, $ZNCl_2$, was formerly an important preservative for wood, and railway crossties treated with the material were called **Burnettized wood.** But it is highly soluble, and leaches out of the wood, and is now chromated and copperized with sodium bichromate and cupric chloride. **Copperized CZC,** of the Koppers Co., Inc., for treating wood against rot and termites, is copperized chromated zinc chloride. Zinc chloride is also used for vulcanizing fiber, as a mordant, in mercerizing cotton, in dry batteries, in disinfecting, and in making many chemicals. **Spirits of salts** and **butter of zinc** are old names for the material.

Zinc chromate, used chiefly as a pigment and called **zinc yellow** and **buttercup yellow,** is stable to light and in sulfur atmospheres, but has a lower tinting strength than chrome yellow, although it is less subject to staining and discoloration. It is a crystalline powder of specific gravity 3.40. It is only slightly soluble in water, but will absorb 24 lb (11 kg) of linseed oil per 100 lb (45.8 kg). Zinc chromates are made by reacting zinc oxide with chromate solutions, and they may vary, but the usual composition is $4ZnO \cdot 4CrO_3 \cdot K_2O \cdot 3H_2O$. The **zinc bichromate,** $ZnCr_2O_7$, is an orange-yellow pigment. The **zinc peroxide** used in dental pastes and cosmetics as a mild antiseptic is a white powder, ZnO_2, containing 8.5% active oxygen.

ZINC ORES. The metal zinc is obtained from a large number of ores, but the average zinc content of the ores in the United States is only about 3% , so that they are concentrated to contain 35 to 65% before treatment. The sulfide ores are marketed on the basis of 60% zinc content, and the oxide ores on the basis of 40% zinc content. **Sphalerite,** or **zinc blende,** is the most important ore and is found in quantities in Missouri and surrounding states and in Europe. Sphalerite is a zinc sulfide, ZnS, containing theoretically 67% zinc. It has a massive crystalline or granular structure and a hardness of about 4. When pure, its color is white; it colors yellow, brown, green, to black with impurities. The ores from New York state are ground and concentrated by flotation to an average of 58% zinc and 32 sulfur, which is then concentrated by roasting to 68% zinc and 1 sulfur. It is then sintered to remove lead and cadmium and finally smelted with coke and the zinc vapor condensed. The Silesian zinc blende, known as **wurtzite,** contains 15% zinc, 2 lead, and some cadmium.

Calamine is found in New Jersey, Pennsylvania, Missouri, and Europe. It is the ore that was formerly mixed directly with copper for making brass. The ore usually contains only about 3% zinc, and is concentrated to 35 to 45%, and then roasted and distilled. Calamine is **zinc silicate,** $2ZnO \cdot SiO_2 \cdot H_2O$. It is a mineral occurring in crystal groups of a vitreous luster, and may be white, greenish, yellow, or brown. The specific gravity is 3.4 and hardness 4.5 to 5. It occurs in Arkansas with **smithsonite,** a **zinc carbonate** ore, $ZnCo_3$. **Franklinite** is an ore of both the metals zinc and manganese. Its approximate composition is $(FeZnMo)O \cdot (FeMn)_2O_3$, but it shows wide variation in the proportions of the different elements. It is found in the zinc deposits of New Jersey. The zinc is converted into zinc white, and the residue is smelted to form spiegeleisen. The mineral franklinite occurs in massive granular structure with a metallic luster and an iron-black color.

The ore **zincite** is used chiefly for the production of the zinc oxide known as zinc white employed as a pigment. Zincite has the composition ZnO, containing theoretically 80.3% zinc. The mineral has usually a massive granular structure with a deep-red to orange streaked color. It may be translucent or almost opaque. Deep-red specimens from the workings at

Franklin, N.J., are cut into gemstones for costume jewelry. **Willemite** is an anhydrous silicate, Zn_2SiO_4, containing theoretically 58.5% zinc. When manganese replaces part of the zinc the ore is called **troostite.** It is in hexagonal prisms of white, yellow, green, or blue colors; manganese makes it apple green, brown, or red. The specific gravity is about 4, and the hardness 5.5. The crushed ore is used in making fluorescent glass. The ore is widely dispersed in the United States.

ZIRCONIA. A white crystalline powder which is **zirconium oxide,** ZrO_2, with a specific gravity of 5.7, hardness 6.5, and refractive index 2.2. When pure, its melting point is about 5000°F (2760°C), and it is one of the most refractory of the ceramics. It is produced by reacting zircon sand and dolomite at 2500°F (1371°C) and leaching out the silicates. The material is used as fused or sintered ceramics and for crucibles and furnace bricks. From 4.5 to 6% of CaO or other oxide is added to convert the unstable monoclinic crystal to the stable cubic form with a lowered melting point.

Fused zirconia, used as a refractory ceramic, has a melting point of 4620°F (2549°C) and a usable temperature to 4450°F (2454°C). The **Zinnorite** fused zirconia of the Norton Co. is a powder that contains less than 0.8% silica and has a melting point at 4900°F (2704°C). A **sintered zirconia** with a density of 5.4 has a tensile strength of 12,000 lb/in^2 (82 MPa), compressive strength of 200,000 lb/in^2 (1,378 MPa), and Knoop hardness of 1,100. **Zircoa B,** of the Zirconium Corp., is stabilized cubic zirconia used for making ceramics. **Zircoa A** is the pure monoclinic zirconia used as a pigment, as a catalyst, in glass, and as an opacifier in ceramic coatings.

Zirconia brick for lining electric furnaces has no more than 94% zirconia, with up to 5% calcium oxide as a stabilizer, and some silica. It melts at about 4300°F (2371°C), but softens at about 3600°F (1982°C). The **IBC 4200** brick of Ipsen Industries, Inc., is zirconia with calcium and hafnium oxides for stabilizing. It withstands temperatures to 4200°F (2316°C) in oxidizing atmospheres and to 3000°F (1849°C) in reducing atmospheres. **Zirconia foam** of the National Beryllia Corp. is marketed in bricks and shapes for thermal insulation. With a porosity of 75% it has a flexural strength above 500 lb/in^2 (3 MPa) and a compressive strength above 100 lb/in^2 (0.7 MPa). For use in crucibles, zirconia is insoluble in most metals except the alkali metals and titanium. It is resistant to most oxides, but with silica it forms $ZrSiO_4$, and with titania it forms $ZrTiO_4$. Since structural disintegration of zirconia refractories comes from crystal alteration the phase changes are important considerations. The monoclinic material, with a specific gravity of 5.7, is stable to 1850°F (1010°C) and then inverts to the tetragonal crystal with a specific gravity of 6.1 and volume change of 7%. It reverts when the temperature again drops below 1850°F (1010°C). The cubic material, with a specific gravity of 5.55, is stable at all temperatures to the melting point which is not above 4800°F (2649°C) because of the contained stabilizers. A lime-stabilized zirconia refractory with a tensile strength of 20,000 lb/ in^2 (138 MPa) has a tensile strength of 10,000 lb/in^2 (68 MPa) at 2370°F (1299°C). **Stabilized zirconia** has a very low coefficient of expansion, and white-hot parts can be plunged into cold water without breaking. The thermal conductivity is only about one-third that of magnesia. It is also resistant to acids and alkalies, and is a good electrical insulator.

Zirconia is produced from the zirconium ores known as **zircon** and **baddeleyite.** The latter is a natural zirconium oxide, but is obtainable commerically only from Minas Gerais, Brazil. It is also called **zirkite** and **Brazilite.** Zircon is **zirconium silicate,** $ZrO_2 \cdot SiO_2$, and comes chiefly from beach sands. The commercial sands are produced in Florida, Brazil, India, Sri Lanka, Australia, and western Africa. The sands are also called **zirkelite and zirconite,** or merely **zircon sand.** The white zircon sand from India has a zorconia content of 62%, and contains less than 1% iron. Beach sands of New South Wales are naturally concentrated to an average of 74% zircon, but Australian zircon is shipped on a basis of 65% zirconia. Zircon sand may be used directly for making firebricks, as an opacifier in ceramics, and for mold facings. Clear **zircon crystals** are valued as gemstones since the high refractive index gives great brilliance. The colorless natural crystals are called **Matura diamonds,** and the yellow-red are known as **jacinth.**

Zirconia fiber, used for high-temperature textiles, is produced from zirconia with about 5% lime for stabilization. The fiber is polycrystalline, has a melting point of 4700°F (2593°C), and will withstand continuous temperatures above 3000°F (1649°C). These fibers are produced by Union Carbide as small as 3 to 10 μm and are made into fabrics for filter and fuel cell use. **Zirconia fabrics** are woven, knitted, or felted of short-length fibers and are flexible. **Ultratemp adhesive,** of Aremco Products, for high-heat applications, is **zirconia powder** in solution. At 1100°F (593°C) it adheres strongly to metals and will withstand temperatures to 4400°F (2427°C). **Zircar,** of Union Carbide, is zirconia fiber compressed into sheets to a density of 20 lb/ft^3 (320 kg/m^3). It will withstand temperatures up to 4500°F (2482°C) and has low thermal conductivity. It is used for insulation and for high-temperature filtering.

ZIRCONIUM. A silvery-white metal having a specific gravity of 6.5 and melting at about 1850°C. It is more abundant than nickel, but is difficult to reduce to metallic form as it combines easily with oxygen, nitrogen, carbon, and silicon. The metal is obtained from zircon sand by reacting with carbon and then converting to the tetrachloride, which is reduced to a sponge metal for the further production of shapes. The ordinary **sponge**

zirconium contains about 2.5% hafnium, which is closely related and difficult to separate. The commercial metal usually contains hafnium, but reactor-grade zirconium, for use in atomic work, is hafnium-free.

Commercially pure zirconium is not a high-strength metal, having a tensile strength of about 32,000 lb/in^2 (220 MPa), elongation 40%, and Brinell hardness 30, or about the same physical properties as pure iron. But it is valued for atomic-construction purposes because of its low neutron-capture cross section, thermal stability, and corrosion resistance. It is employed mostly in the form of alloys but may be had from Atomergic Chemetals Co. in 99.99% pure single-crystal rods, sheets, foil, and wire for superconductors, surgical implants, and vacuum-tube parts. The neutron cross section of zirconium is 0.18 barn, compared with 2.4 for iron and 4.5 for nickel. The cold-worked metal, with 50% reduction, has a tensile strength of about 82,000 lb/in^2 (545 MPa), with elongation of 18% and hardness of Brinell 95. The unalloyed metal is difficult to roll, and is usually worked at temperatures to 900°F (482°C).

The metal has a close-packed hexagonal crystal structure, which changes at 862°C to a body-centered cubic structure which is stable to the melting point. At 300 to 400°C the metal absorbs hydrogen rapidly, and above 200°C it picks up oxygen. At about 400°C it picks up nitrogen, and at 800°C the absorption is rapid, increasing the volume and embrittling the metal. The metal is not attacked by nitric, sulfuric, or hydrochloric acids, but is dissolved by hydrofluoric acid. **Zirconium powder** is very reactive, and for making sintered metals it is usually marketed as **zirconium hydride,** ZrH_2, containing about 2% hydrogen which is driven off when the powder is heated to 300°C. For making sintered parts, alloyed powders are also used. **Zirconium copper,** containing 35% zirconium, **zirconium nickel,** with 35 to 50% zirconium, and **zirconium cobalt,** with 50% zirconium, are marketed by Metal Hydrides, Inc., as powders of 200 to 300 mesh.

Small amounts of zirconium are used in many steels. It is a powerful deoxidizer, removes the nitrogen, and combines with the sulfur, reducing hot-shortness and giving ductility. **Zirconium steels** with small amounts of residual zirconium have a fine grain, and are shock-resistant and fatigue-resistant. In amounts above 0.15% the zirconium forms zirconium sulfide and improves the cutting quality of the steel. **Zirconium alloys** generally have only small amounts of alloying elements to add strength and resist the pickup of hydrogen. **Zircoloy 2,** of the Westinghouse Electric Corp., for reactor structural parts, has 1.5% tin, 0.12 iron, 0.10 chromium, 0.05 nickel and the balance zirconium. The tensile strength is 68,000 lb/in^2 (468 MPa), elongation 37%, and hardness Rockwell B89; at 600°F (316°C) it retains a strength of 30,000 lb/in^2 (206 MPa).

Small amounts of zirconium in copper give age-hardening and increase the tensile strength. Copper alloys containing even small amounts of zirconium are called **zirconium bronze.** They pour more easily than bronzes with titanium and they have good electric conductivity. Zirconium-copper master alloy for adding zirconium to brasses and bronzes is marketed in grades with 12.5 and 35% zirconium. A **nickel-zirconium** master alloy, of the Electro Metallurgical Co., has 40 to 50% nickel, 25 to 30 zirconium, 10 aluminum, and up to 10 silicon and 5 iron. **Zirconium-ferrosilicon,** for alloying with steel, contains 9 to 12% zirconium, 40 to 47 silicon, 40 to 45 iron, and 0.20 max carbon, but other compositions are available for special uses. **SMZ alloy,** of the same company, for making high-strength cast irons without leaving residual zirconium in the iron, has about 75% silicon, 7 manganese, 7 zirconium, and the balance iron. A typical zirconium copper for electrical use is **Amzirc** of the American Metal Climax, Inc. It is oxygen-free copper with only 0.15% of zirconium added. At 400°C it has a conductivity of 37% IACS, tensile strength of 52,000 lb/in^2 (358 MPa), and elongation of 9%. The softening temperature is 580°C.

Zirconium carbide, ZrC_2, is produced by heating zirconia with carbon at about 2000°C. The cubic crystalline powder has a hardness of Knoop 2,090, and melting point of 3540°C. The powder is used as an abrasive and for hot-pressing into heat-resistant and abrasion-resistant parts. **Zirconium oxychloride,** $ZrOCl_2 \cdot 8H_2O$, is a cream-colored powder soluble in water that is used as a catalyst, in the manufacture of color lakes, and in textile coatings. **Zirconium fused salt,** used to refine aluminum and magnesium, is **zirconium tetrachloride,** a hygroscopic solid with 86% $ZrCl_4$. **Zirconium sulfate,** $Zr(SO_4)_2 \cdot 4H_2O$, comes in fine, white, water-soluble crystals. It is used in high-temperature lubricants, as a protein precipitant, and for tanning to produce white leathers. **Soluble zirconium** is **sodium zirconium sulfate,** used for the precipitation of proteins, as a stabilizer for pigments, and as an opacifier in paper. **Zirconium carbonate** is used in ointments for poison ivy, as the zirconium combines with the hydroxy groups of the urushiol poison and neutralizes it. **Zirconium hydride** has been used as a neutron moderator, although the energy moderation may be chiefly from the hydrogen.

Zirconium alloys with high zirconium content have few uses except for atomic applications. **Zircoloy tubing** is used to contain the uranium oxide fuel pellets in reactors since the zirconium does not have grain growth and deterioration from radiation. **Zirconium ceramics** are valued for electrical and high-temperature parts and refractory coatings. The **zirconium oxide powder** of the Norton Co., for flame-sprayed coatings, comes in either hexagonal or cubic crystal forms. **Zirconium silicate,** $ZrSi_2$, comes as a tetragonal crystal powder. Its melting point is about 3000°F (1649°C) and hardness about 1,000 Knoop.

2

Nature and Properties of Materials

THE NATURE OF MATTER AND MATERIALS

Elements, or **atoms,** are the basic building blocks of all tangible materials in the universe. There are 92 elements, or material atoms, almost all of which are stable, from hydrogen, atomic number 1, or element 1, to uranium, or element 92. Elements of higher atomic weight than uranium are made, but they are unstable, their time decay being measured progressively as half-life.

The atom gets its name from the Greek word atomos, meaning indivisible, and it is not divisible by ordinary chemical means. The elements are used either alone or in combination for the making of useful products. They combine either as mechanical mixtures or as **chemical compounds.** In a **mixture** each element retains its original nature and energy, and the constituents of the mixture can be separated by mechanical means. In chemical compounds of two or more elements the original elements lose their separate identities; the new substance formed has entirely different properties, and the atomic energy stored within the compound is not equal to the sum of the elemental energies. The atoms in chemical compounds are bonded by electrons. An **alloy** is usually a combination of chemical compounds and mixtures, the metal mixtures in the matrix being gaged by their maximum fused or liquid solubility, known as the **eutectic point.** With the elements the number of different compounds, or useful substances, that can be made by varying the combinations of elements and the proportions is infinite.

The known atoms are arranged progressively in a **Periodic Table** by **atomic number,** based on the **atomic weight** of the element with hydrogen as the unit of mass, though oxygen may be taken as the point of calculation. The atom is not a solid, but a region of energy particles in motion. At various energy levels the geometrical shape of the electron orbit changes, and the apparent ring, or **electron shell** structure, is the energy-level extension of the orbital pattern. The distances and space covered are so vast in relation to the size of the particle, and the speeds are so great, that the interior of the atom might be considered mostly as empty space. As a single atom is a billion or more times the size of an electron, it is estimated that if the space within the atom could be removed, a thimbleful of atoms would weigh millions of tons. If the copper atom were magnified 10 billion times, the electrons that the chemist employs to connect it with another atom of a molecule would still be too tiny to be seen. Thus, a solid metal used for construction is a region of relatively vast space populated by energy particles in perpetual motion.

The term **space chemistry** was first used at the beginning of the twentieth century by the Dutch physicist van't Hoff, the founder of modern physical chemistry, but the subject was not new. It may be said that modern atomic science, equipped with advanced experimental methods and testing instruments, has taken up where the Greeks, working only with geometry and the theoretical deductions of metaphysics, left off at their School of Numbers about 450 B.C. The Greeks reasoned that all matter came from one source, made from a qualitatively indeterminable primordial unit, the **monad,** now known as **energy.** It was stated to be incorporeal, composed of No Thing, but vital and always in motion. This idea of a nonmaterial basis of tangible materials, now necessary for modern scientific analysis of materials, is intrinsic in human logic. It came to the Greeks from the Ionians, survivors of the Cretan civilization antedating 3000 B.C., and appears in the Hebraic Genesis, in the Sanskrit Vedas, and in the Taoism of ancient China. Energy is in harmonic motion, in waves or rays, and may be said to become a particle of **mass** when the frequency is 1, that is, a closed unit cycle. All materials give off light when activated, and **light rays** have the fastest known speed, 186,000 mi/s (299,274 m/s).

More than 70 new elements, to element 168, have been projected, though not all have been synthesized. These are higher elements made by additions to natural elements. Atoms may also be broken down by the application of high energy. The process, known as **fission,** is usually by electric energy built up to extremely high voltage by resonant pulsation in a magnetic field in a manner akin to that of the generation of lightning in the clouds. More than 30 subatomic particles have been isolated. **Fissionable elements** are normally considered to be only those of high atomic weight and radioactivity, and relative unstability, but all elements are fissionable.

A subatomic unit may be considered as both a wave and a **particle.** The **nucleus** of the atom is a relative term. The **proton** is identical with the nucleus of the hydrogen atom, and is one unit of positive electricity. The nuclei of all other elements consist of combinations of protons and neutrons. The electrons of the various atoms appear to orbit around the nucleus, but the electron, though considered a negatively charged particle, is also a **beta ray,** and the axis of its vortex motion is in calculable relativity to the respective positron. A spheron may contain one or more neutrons, and atoms having different numbers of neutrons are called **isotopes** and are of different atomic weights and have different physical properties.

The **helium atom** of mass 4 and positive charge 2 has two protons and two neutrons with the protons apparently in opposite polarity, and it has zero valence. This combination is called an **alpha particle.** Alpha particles are emitted at high velocity from radioactive elements, expelling the detached electrons, and when captured are deposited as helium. These usually come from outer-ring spherons and not necessarily from the inner nucleus. The expelled electrons are beta rays. When these collide with a nucleus, high-frequency **X-rays** break off. **Gamma rays** are emitted from some radioactive elements. The difference between X-rays and gamma rays is their origin and wavelength. Gamma rays come from the nucleus, while X-rays come from electrons striking matter. Few of the high-energy X-rays coming from the sun penetrate the atmosphere.

A detached positron has only a momentary existence. In conjunction with an electron, it forms an atomlike structure known as positronium.

Gamma rays from the sun come only in infrequent bursts, and the **cosmic rays** from space are also entirely protons, or stripped ions of hydrogen. Cosmic rays appear to travel at about the speed of light. **Mesons** from cosmic rays appear to carry unit charges as beta rays do, but they have more energy and greater range. While beta rays are stopped in the human skin, mesons can cause damage throughout the body. High-energy cosmic rays are stopped by the atmosphere, and only a small proportion penetrate to the earth's surface.

The **neutron** is a particle of neutral charge with a mass approximately that of a proton. A neutron has a mass 1,838 times that of an electron, while a proton has a mass 1,836 times that of an electron. High-energy bombardment of nuclei or an individual nucleus yields electron positrons, mesons, and neutrinos. In recent work, these seemingly fundamental particles have been subdivided into quarks and gluons.

In the technology of producing and processing materials the atom is not subdivided, although in some operations of electrochemistry and electronics the electron is detached, and particles and rays are also employed, especially for activation. In respect to the combining of the elements, **metallurgy** is high-energy chemistry. In a solid metal, as in other materials, the atom does not appear alone, and the physical properties of a metal or alloy derive chiefly from the molecular structure.

Elements having one, two, or three outside valence electrons are **metals.** In chemical reactions they can release these electrons and form positive metal ions. The elements having five, six, or seven outer electrons are **nonmetals.** An element with four outer electrons is a **semimetal** and can react as either a metal or a nonmetal. An element with eight outer electrons is said to have **zero valence** and is normally inactive, but by special energy application, or catalyzation, the linkage of the spherons can be broken and the electrons freed for chemical reactions.

The elements that make up all the planets and the stellar systems of the universe appear to be the same as those of the earth. There are many theories for the original formation of the material elements, but the subject pertains to astronomy rather than to materials technology, and involves the mathematics of progressive assembly of energy waves into monoquantic vortices which constitute mass. While elements do not have life in the same sense as the term is used for animals and plants, they do have intrinsic habits that can be controlled and altered by changing the environmental conditions. Elements are gregarious, and atoms separate only when activated by extremes of energy as with high heat, and they tend to congregate even when dissipated in water or in air.

Elements have orderly calculable habits of combining into **molecules,** or geometrically shaped units bonded to their own kind or to atoms of other elements. Compounding the elements into useful materials is done by the addition or subtraction of energy with consideration of time and space. Even the automatic reactions of two elements in proximity, known as **chemical affinity,** and the seeming holding action of **stabilizing agents,** depend upon a transfer of energy. **Energy, time,** and **space** are nonmaterial, and their limits and bounds are not presently comprehensible to the human mind. Thus, most of the knowledge of the internal workings of the extremely minute atoms is analogical, or reasoned by inference from test comparisons of effects. The tools of the technologist in working with the atoms, therefore, are the theories. A **theory** is considered the best available knowledge at any given time, based on the analytical coordination of the results of many tests. A theory, in the light of further test experiments, may prove inadequate for more advanced operations, but old theories are often not discarded because they may be correct within certain bounds and thus may continue to be useful as the easiest way to accomplish limited objectives.

The term **crystal** is usually applied only to molecular structures which at normal temperatures are hard solids that form into pronounced geometric shapes or are capable of being split on precise planes, and solids without apparent planes are termed **amorphous.** But the crystal shapes tabulated for metals usually represent merely the typical position pattern of the atoms. **Single crystals,** used for electronic purposes, may be cut from natural crystals, grown by flame melting, or grown chemically by application of heat and pressure. **Seed crystals** used to initiate growth are grains or particles made up of many molecules, while a **unit crystal** is the unit molecule or, in some cases, the unit pattern of the lattice, and these determine the shape and nature of the structure. In microscopy the structures of aluminum and silver appear optically as similar cubes, but the unit crystal of aluminum in the solid state forms both a cube and a lattice, while the unit crystal of silver forms no cube and does not lattice, and the metal grains are cryptocrystalline. Usually, the smaller the grain size, the nearer the approach to the physical properties of the single crystal, so that large single crystals are sometimes made by compacting extremely fine powders.

All elements convert progressively from solid to gaseous form by the application of energy, usually by heat application, and vice versa by the extraction of heat. The terms **solid, liquid,** and **gas** are **phase changes** depending on the mobility of the molecule caused by changes in its three-dimensional shape. A **gaseous element** is one that is a gas at ordinary temperatures and pressures, such as hydrogen. At extremely low temperatures a **hydrogen crystal** should be a hard white metal of cryptocrystalline structure with straight planes of cleavage. **Liquid hydrogen** for rocket fuel normally has a molecule of conical shape in spin. When catalyzed by hot platinum it changes to an ovaloid shape which can pass through a smaller molecular sieve, and it also requires 20% less storage space per unit of fuel. These forms are called **ortho hydrogen** and **meta hydrogen,** but are both H_2.

Phase changes often occur within the solid stage, and the change in dimensions of the material, called **creep,** is the effect from change in volume of the molecules. With some materials the liquid stage is so short as to be undetectable, appearing to pass directly from solid to vapor, and this transition is called **sublimation.** All molecules have energy transition points at which they break down to free the original elements or to interact and combine with other available elements to form new compounds. For example, iron molecules, having free electrons, disintegrate easily in the presence of air or moisture to form iron oxides. This process is called corrosion; in organic materials it is called decay. The molecule of gold has no free electrons and, because of its high energy, is not broken down easily by the influence of other elements. Thus it is said to be noncorrosive. In the case of aluminum, oxygen from the air cross-links the free electrons on the surface of the grains and protects the metal from further corrosion.

In metallurgy and the metalworking industries the elements are normally not used alone in a pure state, and as solids and liquids only in molecular forms. In the casting of metals and alloys from a melt the time of solidification is short and, without the application of high energy as in the form of high pressure, there is no growth into large single crystals. Growth is usually into particles, or grains, which may be single crystals or irregular conglomerates of unit crystals. In the contraction of cooling, however, grain boundaries may be so close as to be undetectable even at a magnification of 2 million to 1, and thus the impurities are likely to be in the unmatched open spaces among the crystals, and not interstitial. But with some latticing molecules, such as copper, there is room within the lattice for smaller atoms or molecules, such as those of beryllium, without interference with the paths of bonding electrons, while in the aluminum lattice there appears to be no such room.

Organic and other chemicals are usually produced from the elements by **synthesis,** that is, built up by progressive steps logically deduced from known data and theories concerning the natural habits and characteristics of the atoms and their elementary groups. A compound may thus be written as a **chemical formula** which expresses graphically the specific number and locations of the atomic elements in the compound. In some degree this system is also used in the production of **ceramics,** i.e., compounds or compound mixtures based on metallic oxides, where the resultant material is expressed in percentage proportions of the crystal formulas. Alloys are usually made by batch-mixing the elements, and the resultant material is expressed in weight percentages of the contained elements, not in terms of the molecular structure on which the physical properties of the alloy depend.

The Natural Elements

Name	Atomic number	Symbol	Atomic weight O = 16.000	Melting point, °C
Actinium	89	Ac	......	1800
Aluminum	13	Al	26.97	660.0
Antimony	51	Sb	121.76	630.5
Argon	18	A	39.944	−189.3
Arsenic	33	As	74.91	814
Astatine	85	At	......	470
Barium	56	Ba	137.36	704
Beryllium	4	Be	9.02	1280
Bismuth	83	Bi	209.00	271.3
Boron	5	B	10.82	2300
Bromine	35	Br	79.916	−7.2
Cadmium	48	Cd	112.41	320.9
Calcium	20	Ca	40.08	850
Carbon	6	C	12.00	3700
Cerium	58	Ce	140.13	600
Cesium	55	Cs	132.91	28
Chlorine	17	Cl	35.457	−101
Chromium	24	Cr	52.01	1800
Cobalt	27	Co	58.94	1490
Columbium (niobium)	41	Cb	92.91	2000
Copper	29	Cu	63.57	1083.0
Dysprosium	66	Dy	162.46	
Erbium	68	Er	167.64	
Europium	63	Eu	152.0	
Fluorine	9	Fl	19.00	−223
Francium	87	Vi		
Gadolinium	64	Gd	157.3	
Gallium	31	Ga	69.72	29.78
Germanium	32	Ge	72.60	958
Gold	79	Au	197.2	1063.0
Hafnium	72	Hf	178.6	1700
Helium	2	He	4.002	−271.4
Holmium	67	Ho	163.5	
Hydrogen	1	H	1.0078	−259.2

The Natural Elements (continued)

Name	Atomic number	Symbol	Atomic weight O = 16.000	Melting point, °C
Illinium	61	Il	140.0	
Indium	49	In	114.76	156.4
Iodine	53	I	126.92	114
Iridium	77	Ir	193.1	2454
Iron	26	Fe	55.84	1535
Krypton	36	Kr	83.7	−157
Lanthanum	57	La	138.92	826
Lead	82	Pb	207.22	327.4
Lithium	3	Li	6.940	186
Lutecium	71	Lu	175.0	
Magnesium	12	Mg	24.32	650
Manganese	25	Mn	54.93	1260
Mercury	80	Hg	200.61	−38.87
Molybdenum	42	Mo	96.0	2625
Neodymium	60	Nd	144.27	840
Neon	10	Ne	20.183	−248.6
Nickel	28	Ni	58.69	1455
Nitrogen	7	N	14.008	−210.0
Osmium	76	Os	191.5	2700
Oxygen	8	O	16.0000	−218.8
Palladium	46	Pd	106.7	1554
Phosphorus	15	P	31.02	44.1
Platinum	78	Pt	195.23	1773.5
Polonium	84	Po	......	1800
Potassium	19	K	39.096	63
Praseodymium	59	Pr	140.92	940
Protoactinium	91	Pa	231	
Radium	88	Ra	226.05	700
Radon	86	Rn	222	−71
Rhenium	75	Re	186.31	3000
Rhodium	45	Rh	102.91	1966
Rubidium	37	Rb	84.44	39
Ruthenium	44	Ru	101.7	2450
Samarium	62	Sm	105.43	1300
Scandium	21	Sc	45.10	1200
Selenium	34	Se	78.96	220
Silicon	14	Si	28.06	1430
Silver	47	Ag	107.880	960.5
Sodium	11	Na	22.997	97.7
Strontium	38	Sr	87.63	770
Sulfur	16	S	32.06	119.2
Tantalum	73	Ta	180.88	3000
Technetium	43	Ma	97.8	2300
Tellurium	52	Te	127.61	450
Terbium	65	Tb	159.2	327
Thallium	81	Tl	204.39	300

The Natural Elements (continued)

Name	Atomic number	Symbol	Atomic weight O = 16.000	Melting point, °C
Thorium	90	Th	232.12	1700
Thulium	69	Tm	169.4	
Tin	50	Sn	118.70	231.9
Titanium	22	Ti	47.90	1820
Tungsten	74	W	184.0	3410
Uranium	92	U	238.14	1850
Vanadium	23	V	50.95	1735
Xenon	54	Xe	131.3	−112
Ytterbium	70	Yb	173.04	1500
Yttrium	39	Y	88.92	1490
Zinc	30	Zn	65.38	419.5
Zirconium	40	Zr	91.22	1700

WAVES AND COLORS AS MATERIAL ELEMENTS

Electromagnetic Radiations

Tangible materials and **radiations** have a common energy origin, and thus bear a cosmic relation, but radiation is not matter in the ordinary sense of the term. Radiation is caused by vibrations, and is characterized by wavelengths rather than mass as is ordinary matter. Waves of high frequency and short wavelength result from the vibration of extremely small particles, such as electrons of the material atom, while those of low frequency and long wavelength arise from slow vibrations, such as those from a coil in a magnetic field.

Radiations are produced when materials are broken down or changed to another form, and there is then an actual loss of mass equal to the amount of energy emitted. In reverse, matter is produced when energy in the form of radiation is directed upon matter, and an actual increase in the mass of the matter results. All materials in nature are being constantly bombarded with various radiations, but it requires such an extremely large amount of energy to produce the most minute quantities of matter that the continuous changes in most materials are not noticeable in any historic period of time.

The spectrum of electromagnetic radiations extends from wavelengths of many hundred-millionths of a centimeter, or infinitely small, to wavelengths of many kilometers, or infinitely large. The velocity of these waves is the same for all lengths of wave, 186,000 mi/s. In the spectrum, the light waves which make objects visible to the human eye form only a small part.

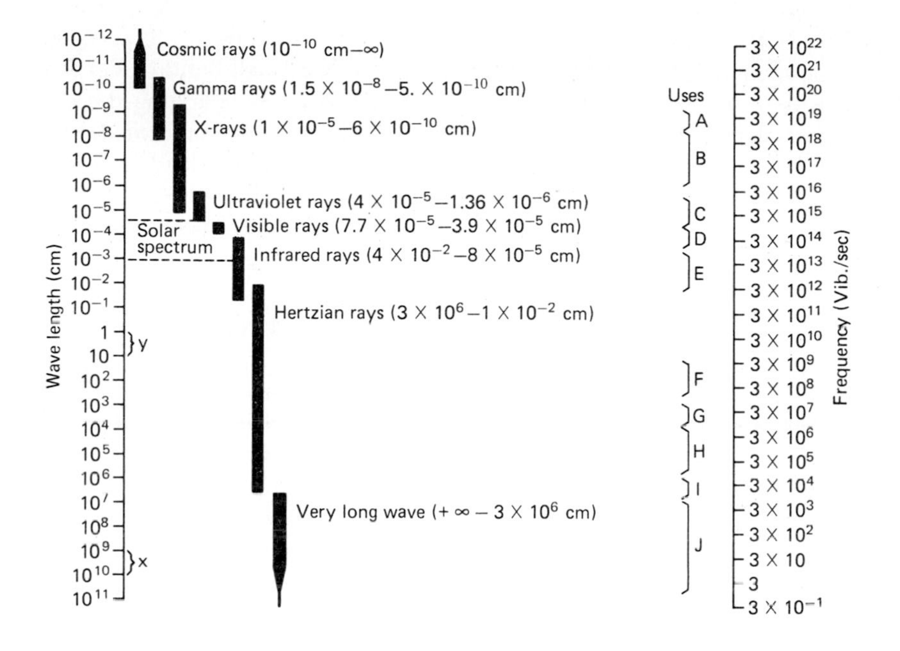

Spectrum of Electromagnetic Radiations

Uses:

A-Radiology, radiotherapy, x-ray diagnosis

B-Physical and chemical analysis of matter

C-Therapeutic applications

D-Light

E-Heating, industrial baking

F-FM, radar, television, sonar

G-Short wave radio

H-Wireless

I-Long wave radio

J-Electric power, alternating current

Explanation:

The chart is logarithmic, the distance x from 10^9 to 10^{10} being a billion times the distance y from 1 to 10

$10^1 = 10$	$10^3 = 1000$	$10^{-1} = 0.1$	$10^{-3} = 0.001$
$10^2 = 100$	$10^4 = 10{,}000$	$10^{-2} = 0.01$	$10^{-4} = 0.0001$

Wavelengths and frequencies. (*From J. F. Moulton, Jr.*)

The human eye can see through only such materials as these light waves will penetrate. But electrical eyes can be made to operate in other wavelengths and record vision not seen by the human eye. Not all animals see with the same wavelengths, and some animals do not have normal eyes but receive vibrations through special receiving parts of the body. Different materials transmit, absorb, or reflect radiations differently. Quartz and glass, normally called transparent, transmit only a small band of light and heat waves, but will not pass very short radiations. By changing the composition of the glass the heat waves can be blocked, or some of the very short waves can be passed through. Some materials, like lead, will block the very short waves, and can be used for X-ray shields. Other materials, like beryllium, will pass only very short waves, and can be used for selective windows.

Silver will reflect 90% of visible light, while tin reflects only 70%, but silver loses reflectivity in sulfur atmospheres. Gold reflects only 61% of visible light, but has high reflectivity of infrared rays, useful for electronic purposes. All materials are sensitive to particular light waves and emit electrons when struck by those waves. Zinc is sensitive to very short ultraviolet light; cesium is sensitive to green light; potassium is sensitive to blue light. This property is the basis of electronic color selectors. It is also the basis for the operation of photoelectric cells, in which the liberated electrons constitute an electric current. Such cells are widely used as automatic switches and for electronic conversion of light intensities to sound waves.

Element Colors at Incandescence

Flame colorations caused by heating materials to incandescence indicate the presence of certain elements, as the light from each element in burning has a predominance of rays or wavelengths that are characteristic of that particular element. Some elements, such as sodium, show a distinct bright color because of a predominance of wavelengths within that color range in the visible spectrum, while others show pale or intermediate colors difficult to distinguish, usually because the rays have no predominating wavelength within the visible spectrum but are mixtures of many wavelengths. Other elements, such as iron, have a predominance of rays that are not in the visible band, with wavelengths shorter or longer than those visible to the eye. Flame coloration is used in metallurgical laboratories to determine the content of alloys by burning small pieces and studying the light with a refractive prism. This property of the elements is also utilized in making carbon electrodes for electric-arc lights to give the full white light of sunshine, or short waves for therapy or industrial use, or long wavelengths for heat. For example, carbon alone gives a predominance of short wavelengths with the visible rays predominantly on the red side of the spectrum. When cerium metals are blended with the carbon, the visible light is balanced with the blue-violet to give a more even white light. When the carbon is blended with iron, nickel, and aluminum, which are all on the low-wave side of the spectrum, lower-zone ultraviolet rays are obtained.

Predominant Flame Colors of Materials

Element	Color	Element	Color
Lithium	Deep red	Antimony	Blue-green
Strontium	Crimson	Copper	Green-blue
Calcium	Yellow-red	Arsenic	Light blue
Sodium	Bright yellow	Lead	Light blue
Barium	Yellow-green	Selenium	Blue
Molybdenum	Green-yellow	Indium	Deep blue
Zinc	Light green	Potassium	Purple-red
Boron	Green	Rubidium	Violet
Tellurium	Deep green	Cesium	Bluish purple
Thallium	Greenish blue		

Reflecting Powers of Various Metal Surfaces

	White light directly reflected, percent	Color Silver = 0	
Silver	90	0	
Chromium	61	Blue-green	12 units
Nickel	50	Red	16 units
Stainless steel	49	Blue-green	3 units
White bronze speculum	70	Red	1 unit

Reflecting Power of Various Colors in Paints

Color	Light reflection, percent	Color	Light reflection, percent
Flat white	85–89	Sky blue	58
Bone white	69–70	Light orchid	57
Canary yellow	68–72	Buff	47
Light ivory	70	Pea green	40
Aluminum	70	Tan	34
Cream	65–69	Peacock blue	34
Light green	66	Steel gray	30
Ivory	61–63	Brown	9
Peach	58–59		

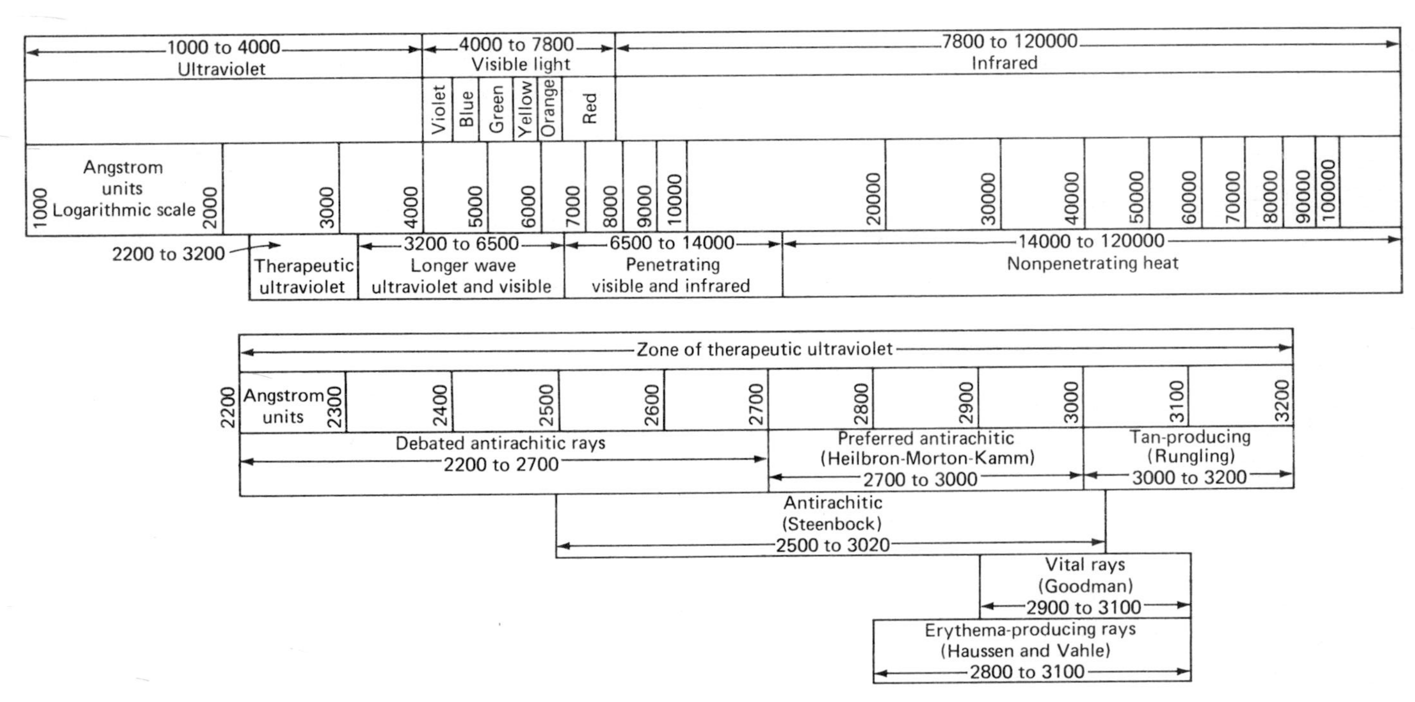

Heat and light rays from the carbon arc. Controlled rays are obtained with arc carbons by varying the core content of the carbon. With cerium metals in the core, a light approximating sunlight is obtained. Iron in the core gives only one-quarter the visible light of the plain carbon with the same current and voltage, but it gives strong ultraviolet rays. A carbon containing iron, nickel, and aluminum gives powerful ultraviolet rays between 250 and 302 nm. A carbon with strontium in the core gives penetrating infrared heat rays above 650 nm. *(Chart from National Carbon Co.)*

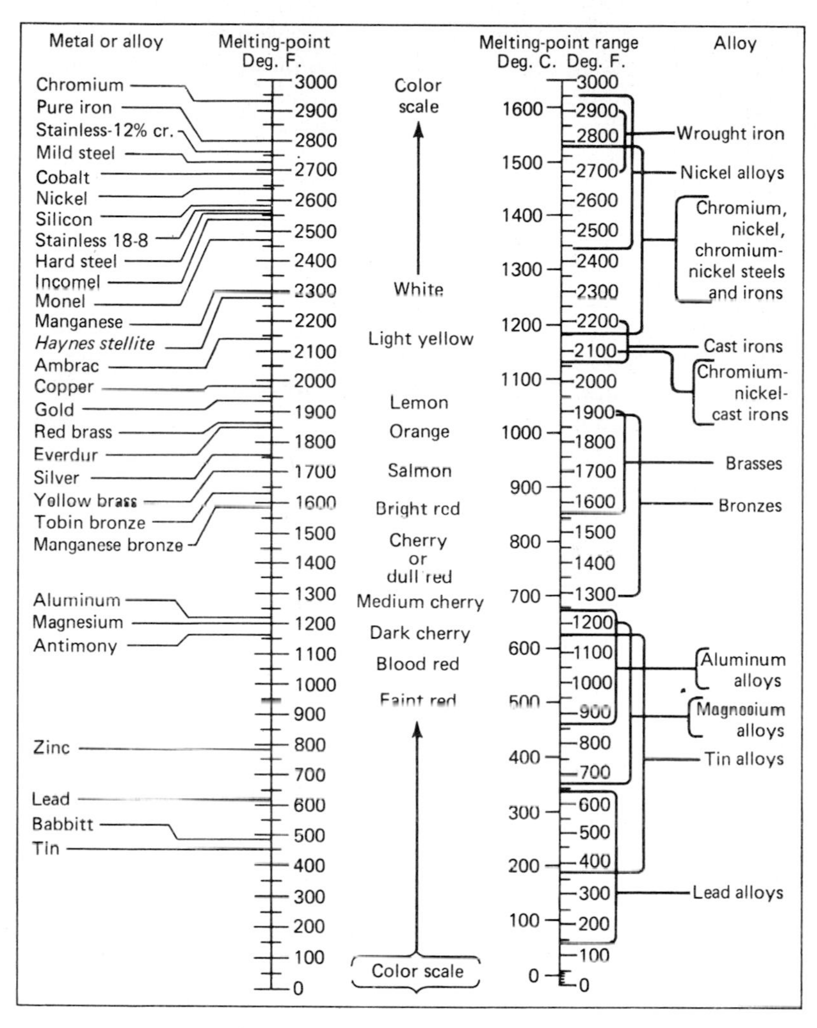

Metal melting range and color scale. *(Chart from the Linde Air Products Co.)*

Terms Used in Material Color Designation

Hue is the predominant light wavelength reflected by the coloring material, and determines the **color** designation.

Brightness, or **value,** is the percentage of light reflected. A brilliant white approaches 100%, and a jet black approaches 0%. Black is the absence of light waves; white is a combination of all the various wavelengths. White light is broken down by refraction into separate wave bands, or hues, as in the natural rainbow. **Chroma** refers to the **intensity** of a color. **Tint** refers to color modified toward white, **shade** to one toward black.

The color circle is composed of 12 colors spaced at equal intervals: yellow, orange, red, violet, blue, green, etc., with intermediates between each. **Pigment colors** are obtained usually by **subtractive** mixing; for example, when blue and yellow are mixed, the blue absorbs the red, orange, and yellow rays, and the yellow absorbs the blue and violet rays, and the resulting color is green.

Under proper illumination it is possible to detect with the eye exceedingly slight color differences, the number of distinguishable colors being estimated, by the U.S. Bureau of Standards, at 10,000,000.

Colors or hues vary slightly with different batches of paints, dyes, etc. For this reason products that must be matched exactly in hue are usually finished from the same batch or lot. Color matching of metals is also often important. For example, for installation of kitchens or other building equipment the stainless steel should preferably be from one lot since the color shades vary with the proportions of chromium, nickel, or manganese. These are "white" metals, but chromium has a blue tone, nickel has a yellow tone, and manganese has a purple tone. Welding alloys and solders are also matched to the color of the base metal by varying the proportions of metals with different tints.

Visibility at a distance varies with different colors. Red can be seen and recognized at long distances while blue can be seen at only short distances. The order of visibility of colors at a distance is red, green, white, yellow, blue. The legibility of a color, however, also varies with the background. Black on yellow is more legible than black on white, whereas green, red, or blue on white is more legible than black on white. Visibility and legibility are important in signs, packages, or products that must be distinguished easily.

Harmony of color or **tone design** is a complicated art. It comprises the color relationship to convey pleasing emotional reaction, and includes various terms. A **rich color** is a hue at its fullest intensity. A **warm color** is one in which the red-orange predominates. A **cool color** is one in which the blue-green predominates. In general, warm colors are pleasing or exciting, while cool colors are not so pleasing or are restful to the senses. A **receding color** is one giving the illusion of withdrawing into distance by a gradation toward another tone or hue. Color, from the standpoint of harmony and design, is a sensation effect. It is not inherent in the pigments, dyes, or other materials, but is the sensation effect from those light rays reflected to the eye by the material.

PROPERTY OF FLAVOR IN MATERIALS

The quality of many materials is judged by the flavor. Flavor is the resultant of three senses: taste, smell, and feeling. Some materials, such as salt and quinine, may be detected by taste alone. Some, such as coffee and butter, depend largely upon smell. Flower perfumes are detected by smell alone. The flavor of fruits depends upon both taste and smell, and without smell would be only sour or sweet. Pepper has little or no taste, but is detected by the aroma and by the sense of feeling.

The four standard components of taste are: sweet, sour, salt, and bitter. Taste buds are located in the tongue. The tip, back, and edges of the tongue can detect all four sensations, but the center of the tongue can detect only a sour taste, and the surrounding area can detect only salty and sour. **Sour taste** is caused by hydrogen ions, and **salty taste** is due to cations from the alkali metals, accented when anions from the halogens are present. **Sweetness and bitterness** may or may not be from ions, and the stimuli are more complex. But all taste is electrochemical, translated to the nerves as sensations. Some materials when injected into the blood can be tasted when the blood reaches the tongue.

Odor detection is electrochemical but not entirely so, since molecules which have the same shape may have the same odor though unrelated chemically. The sense of smell is due to oscillations of the valence electrons in the molecules of the substance. The molecules of substances inhaled stimulate the tiny olfactory hairs high in the nasal cavity, and the effect is translated to the nerves as impressions or sensations. The sense of taste usually requires considerable material to register the sensation, but only the most minute molecular qualities are required to register smell. A normal person can detect a vast variety of odors, but for convenience the four fundamental odors have been designated as fragrant, acid, burnt, and caprylic. A material is then designated with a four-digit number to indicate the degree of each odor, each odor thus having ten degrees or variations, thus giving 9,999 variants. Various materials are taken as standards for the numbers, or 40 standards. The sense of smell is so discriminating that it can detect separate odors in highly complicated mixtures. Most odors are mixtures, and the art involved in the perfumery industry is to form harmonies that give a resultant pleasant sensation.

Flavors that affect the touch sensation are described as pungent, sharp, acrid, and cool. These are caused by actual pain as the biting of an acid or the cooling effect of deoxidation. **Greasiness** and **oiliness** are sensations of feel that affect taste but are not a part of it. **Texture** also is a feeling sense and not a part of flavor. The sensation of puckery of the tannin of some fruits is a definite constriction of membrane and is not taste. All of these have an effect upon the desirability of the material as a food, but in a manner apart from flavor. Too much sweetness, sourness, or saltiness will clog the taste buds, and they must be then rested before a true flavor can be detected, but the recovery is rapid. Temperature also has an effect, and true flavors are detected only at about the temperature of the body.

The judgment and grading of coffee, tea, butter, etc., are done solely by the senses of experts in comparison with standards.

FUNDAMENTALS OF BIOTIC MATERIALS

The **biotics** constitute an extensive group of organic materials that are actual living microorganisms and cannot be expressed as chemical formulas. All of these minute bits of living matter are of plant and animal origin and may be considered as chemical factories. It is the chemicals that certain species secrete under certain conditions that make them industrially and medicinally useful. The chemical secretions are enzymatic and catalytic in character and thus enable various chemicals to react on contact. In industry they are used as chemical activating agents, as ferments, as leavening agents, and in various processing. In medicine they are known as **antibiotics** and are used to destroy the biotics, or bacteria, of diseases which locate themselves in the human body.

Biotics are found everywhere in myriad quantities. A cubic centimeter of raw earth may contain as many as 50,000 **fungi,** or **microphytes,** 500 million bacteria, and 250 million **actinomycetes,** the latter being living organisms that may be ascribed, with reservations, to either the plant or the animal kingdom, and are distinguished by their mass of long silky filaments. All these organisms lead a junglelike existence, attacking everything, decomposing plant debris to make humus, liberating nitrogen from proteins, liberating oxygen from rocks, liberating carbon dioxide and water from organic acids in plants and soil, and also preying on one another as jungle animals do. Without these organisms the life cycles of all living things could not be maintained. This same teeming population also furnishes the individual types of organisms that aid man in medicine and industry.

The number of species of these microorganisms is innumerable. Each species has at least one enemy species that it destroys on contact or by which it is itself destroyed. Each apparently has a certain definite range of activity, beyond the bounds of which it is useless. Where one biotic is used in the manufacture of a certain product, a similar but different species used under the same conditions produces an entirely different product, or may affect the quantitative yield of the product desired. The formation of ethyl alcohol by the fermentation of starch or sugar is caused by a biotic which is then itself killed when the alcohol produced has arrived at a certain concentration. Other biotics may be killed by the heat that their own work produces, as at the "crisis" point of certain fevers. In general, biotics can withstand excessive cold but are usually killed by relatively low heats.

Besides the use of biotics for the medicinal and industrial applications that are now known, there are also believed to be enormous possibilities for their use in large-scale chemical processing in the future. But the isolation of microorganisms is tedious laboratory work involving the extraction of a pure strain from cultures containing many species, and once separated, the proper conditions to promote rapid multiplication must be discovered. Even at this stage, a biotic found to be useful in the manufacture of a certain product may simultaneously manufacture another unwanted product, difficult and costly to separate from the desired one. But from each biotic some kind of chemical is secreted, and that chemical is the ultimate end of biotic research.

UNITS OF MEASURE

Useful Conversion Factors

1 acre = 43,560 square feet = 0.40469 hectare
1 nanometer = 0.001 micrometer = 0.03937 millionths of an inch
1 ardeb (Egypt) = 5.44 bushels
1 arshin (Russia) = 28 inches
1 barrel (U.S.A.), cement = 376 pounds
1 barrel, oils = 42 gallons
1 berkovets (Russia) = 361.13 pounds
1 board foot = 144 cubic inches
1 bolt, cloth = 10 yards = 36.576 meters
1 buncal (Indonesia) = 1.49 troy ounces
1 bushel = 2,150.4 cubic inches
1 bushel, imperial (British) = 1.0315 U.S. bushels
1 candy (India) = 784 pounds
1 carat, metric = 0.200 gram
1 carat (U.S.A.) = 0.2056 gram
1 chittak (India) = 900 grains
1 circular mil = 0.0000007845 square inch
1 cuarteron (Spain), oil = 0.133 liquid quart
1 cuartillo (Mexico), liquid = 0.482 liquid quart
1 cubic foot = 1,728 cubic inches
1 cuffisco (Sicily), oil = 5.6 gallons
1 dram = 1.7718 grams
1 dinero (Spain) = 18.5 grains
1 drachma (Turkey) = 49.5 grains
1 ell = 48 inches
1 feddan (Egypt) = 1.038 acres
1 firkin = 9 U.S. gallons = 34.068 liters
1 flask, mercury = 75 pounds = 34.02 kilograms
1 foot = 12 inches = 0.3048 meter
1 gallon = 231 cubic inches

1 gallon, imperial (British) = 1.20094 U.S. gallons
1 gallon, proof (British) = 1.37 U.S. proof gallons
1 gill = 0.25 pint = 0.118292 liter
1 grain = 0.06480 gram
1 gram = 15.43 grains = 0.03527 avoirdupois ounce
1 gram (Libya) = 165.3 pounds
1 hamlah (Egypt) = 165.1 pounds
1 hectare = 2.471044 acres
1 hogshead = 63 U.S. gallons
1 hundred weight (British) = 112 pounds
1 inch = 0.0833 foot = 2.54005 centimeters
1 iron, leather thickness measure = 1/48 inch
1 kantar (Egypt) = 99.034 pounds
1 keel (British), coal = 21.2 long tons
1 kilogram = 2.205 pounds
1 koku (Japan) = 47.65 gallons = 5.119 bushels
1 kun (Korea) = 1.323 pounds
1 kwan (Japan) = 1,000 momme = 8.267 pounds
1 ligne, metal button measure = 1/40 inch
1 liter = 1.057 liquid quarts
1 lug (Bahamas) = 30 pounds avoirdupois
1 meter, square = 1.196 square yards
1 mil, square = 0.000001 square inch
1 micrometer = 0.001 millimeter = 0.00003937 inch
1 mil = 0.001 inch = 0.0254 millimeter
1 mile = 5,280 feet = 1.69035 kilometers
1 mile, square = 640 acres
1 millimeter = 0.03937 inch
1 ounce, avoirdupois = 28.35 grams = 0.0625 pound
1 ounce, troy = 31.1 grams
1 peck = 0.25 bushel = 8.8096 liters
1 perch = 1 square rod = 30.25 square yards
1 picul (China) = 100 catties = 133⅓ pounds
1 picul (Indonesia) = 136.2 pounds
1 picul (Japan) = 132.3 pounds
1 pint, dry measure = 33.6 cubic inches
1 pint, liquid measure = 28.875 cubic inches
1 pood (Russia) = 36.11 pounds
1 pound, avoirdupois = 16 ounces = 7,000 grains = 0.4536 kilogram
1 pound, troy = 12 ounces = 0.37324 kilogram
1 pound, Venetian = 1.058 avoirdupois pounds
1 quart, dry measure = 2 pints, dry = 1.1012 liters

1 quart, liquid = 57.749 cubic inches = 0.9463 liter
1 quintal (British) = 112 pounds
1 quintal, metric = 110 kilograms = 220.5 pounds
1 ream, paper measure = 500 sheets = 20 quires
1 rod = 5.5 yards
1 scruple, apothecary weight = 20 grains = 1.296 grams
1 shih tan (China) = 50 kilograms = 110.231 pounds
1 standard (British), timber = 1,980 board feet
1 standard (U.S.A.), timber = 16⅔ cubic feet
1 tank (India), gemstones and pearls = 24 rati = 0.145 ounce
1 tierce, thin-staved cask = 42 gallons = 310 to 370 pounds
1 ton, long = 2,240 pounds = 1016.047 kilograms
1 ton, metric = 1,000 kilograms = 0.9842 long ton = 1.102 short tons
1 ton, short = 2,000 pounds = 0.8929 long ton
1 vedro (Russia) = 3,249 gallons
1 yard = 3 feet = 0.9144 meter

Metric Length Measurements

Unit	Inches	Feet	Millimeters	Centimeters	Meters
One inch	1	0.0833	25.4	2.54	0.0254
One foot	12	1	304.8	30.48	0.3048
One millimeter	0.03937	0.00328	1	0.1	0.001
One centimeter	0.3937	0.0328	10	1	0.01
One meter	39.37	3.2809	1000	100	1
One yard	36	3	914.4	91.44	0.9144

Standard Paper Sizes

Folio note	5½ by 8½ in (14.0 by 21.5 cm)
Pocket note	6 by 9½ in (15.2 by 24.1 cm)
U.S. government writing	8 by 10½ in (20.3 by 26.7 cm)
Commercial writing	8½ by 11 in (21.6 by 27.9 cm)
Legal cap	8½ by 14 in (21.6 by 35.6 cm)
Foolscap	12 by 16 in (30.5 by 40.6 cm)
Denny	16 by 21 in (40.6 by 53.3 cm)
Folio	17 by 22 in (43.2 by 55.9 cm)
Royal	19 by 24 in (48.3 by 70.0 cm)
Super royal	20 by 28 in (50.8 by 71.2 cm)
Elephant	23 by 28 in (58.4 by 71.2 cm)
Imperial	23 by 31 in (58.4 by 78.7 cm)

Temperature Conversion Scale

To change a temperature in degrees centigrade to degrees Fahrenheit, multiply by $\frac{9}{5}$ and add 32, thus, $F = \frac{9}{5} C + 32$. To change degrees Fahrenheit to degrees centigrade, subtract 32 and multiply by $\frac{5}{9}$, thus $C = \frac{5}{9} (F - 32)$.

C	F	C	F	C	F	C	F	C	F
0	32	230	446	460	860	690	1274	920	1688
5	41	235	455	465	869	695	1283	925	1697
10	50	240	464	470	878	700	1292	930	1706
15	59	245	473	475	887	705	1301	935	1715
20	68	250	482	480	896	710	1310	940	1724
25	77	255	491	485	905	715	1319	945	1733
30	86	260	500	490	914	720	1328	950	1742
35	95	265	509	495	923	725	1337	955	1751
40	104	270	518	500	932	730	1346	960	1760
45	113	275	527	505	941	735	1355	965	1769
50	122	280	536	510	950	740	1364	970	1778
55	131	285	545	515	959	745	1373	975	1787
60	140	290	554	520	968	750	1382	980	1796
65	149	295	563	525	977	755	1391	985	1805
70	158	300	572	530	986	760	1400	990	1814
75	167	305	581	535	995	765	1409	995	1823
80	176	310	590	540	1004	770	1418	1000	1832
85	185	315	599	545	1013	775	1427	1005	1841
90	194	320	608	550	1022	780	1436	1010	1850
95	203	325	617	555	1031	785	1445	1015	1859
100	212	330	626	560	1040	790	1454	1020	1868
105	221	335	635	565	1049	795	1463	1025	1877
110	230	340	644	570	1058	800	1472	1030	1886
115	239	345	653	575	1067	805	1481	1035	1895
120	248	350	662	580	1076	810	1490	1040	1904
125	257	355	671	585	1085	815	1499	1045	1913
130	266	360	680	590	1094	820	1508	1050	1922
135	275	365	689	595	1103	825	1517	1055	1931
140	284	370	698	600	1112	830	1526	1060	1940
145	293	375	707	605	1121	835	1535	1065	1949
150	302	380	716	610	1130	840	1544	1070	1958
155	311	385	725	615	1139	845	1553	1075	1967
160	320	390	734	620	1148	850	1562	1080	1976
165	329	395	743	625	1157	855	1571	1085	1985
170	338	400	752	630	1166	860	1580	1090	1994
175	347	405	761	635	1175	865	1589	1095	2003
180	356	410	770	640	1184	870	1598	1100	2012
185	365	415	779	645	1193	875	1607	1105	2021
190	374	420	788	650	1202	880	1616	1110	2030
195	383	425	797	655	1211	885	1625	1115	2039
200	392	430	806	660	1220	890	1634	1120	2048
205	401	435	815	665	1229	895	1643	1125	2057
210	410	440	824	670	1238	900	1652	1130	2066
215	419	445	833	675	1247	905	1661	1135	2075
220	428	450	842	680	1256	910	1670	1140	2084
225	437	455	851	685	1265	915	1679	1145	2093

Hardness Numbers

The Brinell method of determining hardness is by the indentation effect of a hard ball pressed into the surface of the metal to be tested. Tables of hardness numbers corresponding to the various indentation measurements are furnished by the makers.

The Scleroscope, or "Shore," method measures hardness by a comparison of the effect of the drop and rebound of a diamond-tipped hammer dropped from a fixed height. The resulting rebound is then read on a graduated scale.

The Rockwell hardness tester measures hardness by determining the depth of penetration under load of a steel ball or diamond cone in the material being tested. Rockwell hardness is expressed as a number, which is read on a graduated gage.

The Mohs hardness scale for abrasives and minerals is measured by scratch comparison, the mineral talc being taken as 1 and the diamond as 10 on the scale. This method is only an approximation for mineral comparison, and the Knoop indentor is used for measuring comparative hardness of hard materials.

The Vickers method is similar to the Brinell and Rockwell methods except that a diamond in the form of a pyramid is used as the penetrator. It is thus suitable for measuring metals of high hardness.

The Bierbaum microcharacter, or Bierbaum number, is used to determine the hardness by scratch. The width of a scratch made by drawing the point of a cube-shaped diamond across the surface under a 3 g load is measured with a microscope and determines the degree of hardness.

Index of Refraction

Index of refraction indicates the relative amount of light transmitted by a material. As the index of refraction increases, the transmitted light decreases. The amount of light reflected back may be considered as in inverse proportion to the amount transmitted, though much of the light may be dissipated. A vacuum transmits 100% of the light and reflects 0%, and has a refractive index of 1.00. A polished diamond with parallel sides will transmit only 83% of the light, and the reflected light makes the diamond shine. The sparkle of angle-cut diamonds and highly refractive cut glass is caused by the dissipated or deflected light emerging from the angles. (See the table on page 931.)

Acidity and Alkalinity Scale

The degree of acidity or alkalinity of solutions is expressed by the pH value. Water is considered neutral and is given a pH value of 7. Values

Approximate Relationship of Vickers, Shore (Scleroscope), Rockwell, and Brinell Hardness Numbers

Vickers or Firth	Shore or Scleroscope	Rockwell		Brinell	Approximate tensile strength of steels, lb/in^2	Hardness class of steel
		C	B			
1,220	96	68	...	780	329,000–380,000 (2,268–2,619 MPa)	Hard to file
1,114	94	67	...	745		
1,021	92	65	...	712		
940	89	63	...	682		
867	86	62	...	653		
803	84	60	...	627	165,000–317,000 (1,137–2,185 MPa)	Machining operations difficult
746	81	58	...	601		
694	78	56	...	578		
649	75	55	...	555		
608	73	53	...	534		
587	71	51	...	514		
551	68	50	...	495		
534	66	48	...	477		
502	64	47	...	461		
474	62	46	...	444		
460	60	44	...	429		
435	58	43	...	415		
423	56	42	...	401		
401	54	41	...	388		
390	52	39	...	375		
380	51	38	...	363		
361	49	37	...	352		
344	48	36	...	341	78,000–159,000 (537–1,096 MPa)	Commercial machine range
335	46	35	...	331		
320	45	34	...	321		
312	43	32	...	311		
305	42	31	...	302		
291	41	30	...	293		
285	40	29	...	285		
278	38	28	...	277		
272	37	27	...	269		
261	36	26	...	262		
255	35	25	...	255		
250	34	24	100	248		
240	33	23	99	241		
235	32	22	99	235		
226	32	21	98	229		
221	31	20	97	223		
217	30	18	96	217		
213	30	17	95	212		
209	29	16	95	207		
197	28	14	93	197		
186	27	12	91	187		
177	25	10	89	179		
171	24	8	87	170		
154	23	4	83	156		
144	21	0	79	143		

Index of Refraction

Material	Index	Material	Index
Diamond	2.42	Nylon	1.53
Ruby	1.80	Polyethylene plastic	1.52
Sapphire (synthetic)	1.77	Pyrex (borosilicate glass)	1.52
Iceland spar	1.66	Window glass (soda-lime)	1.52
Flint glass (dense leaded)	1.66	Acrylic plastics	1.50
Flint glass (dense barium)	1.62	Cellulose acetate	1.48
Vinyledene chloride	1.61	Cellulose acetate butyrate	1.47
Polystyrene plastic	1.59	Ethyl cellulose	1.47
Flint glass (light leaded)	1.58	Fused quartz	1.46
Flint glass (light barium)	1.57	Water	1.33
Amber	1.55	Ice	1.30
Quartz crystal	1.54	Air	1.0003
Urea-formaldehyde	1.54	Vacuum	1.00

below 7 are acid, each declining value being 10 times more acid than the previous value. A pH of 6 is 10 times more acid than a pH of 7, and a pH of 3 is 10,000 times more acid than a pH of 7. Solutions having values from 7 to 14 are alkaline by the same multiples of 10. A pH of 14 is 10 million times more alkaline than a pH of 7. Chemical indicators used to indicate the acidity or alkalinity of solutions are shown in the acidity-alkalinity table.

Acidity-Alkalinity Indicators

	pH range	
Meta cresol purple	1.2–2.8	Red to yellow
Thymol blue	1.2–2.8	Red to yellow
Bromophenol blue	3.0–4.6	Yellow to blue
Bromocresol green	4.0–5.6	Yellow to blue
Chlorophenol red	5.2–6.8	Yellow to red
Bromothymol blue	6.0–7.6	Yellow to blue
Phenol red	6.8–8.4	Yellow to red
Cresol red	7.2–8.8	Yellow to red
Thymol blue	8.0–9.6	Red to blue

Viscosity of Liquids

The **viscosity** of a liquid is its resistance to change in its form, or flow caused by the internal friction of its particle components. Thus, the higher the viscosity, the less fluid it is. When a liquid is hot, there is less internal friction owing to the greater mobility and distance between the molecules, and a liquid will flow more readily than when it is cold. Thus, all comparisons of viscosity should be at the same temperature. **Kinematic viscosity** is the ratio of viscosity to density. **Specific viscosity** is the ratio of the viscosity of any liquid to that of water at the same temperature. The reciprocal of viscosity is called **fluidity.** Viscosity is usually expressed in poises or centipoises, a poise being equal to 1 g/(cm)(s).

Liquid	Viscosity, centipoises	Liquid	Viscosity, centipoises
Benzene (0°C)	0.906	Linseed oil (30°C)	33.1
Carbon tetrachloride (0°C)	1.35	Soybean oil (30°C)	40.6
Mercury (0°C)	1.68	Sperm oil (15°C)	42.0
Ethyl alcohol (0°C)	1.77	Sulfuric acid (0°C)	48.4
Water (0°C)	1.79	Castor oil (10°C)	2,420
Phenol (18°C)	12.7	Rape oil (0°C)	2,530

The **specific gravity** of a liquid is the relative weight per unit volume of the liquid compared with the weight per unit volume of pure water. Water is arbitrarily assigned the value of 1.000 g/cm^3. All liquids heavier than water thus have specific gravities greater than 1.000; liquids lighter than water have values less than 1.000. Usually, the specific gravity of a liquid is measured at 15°C or at room temperature. In practice, measurements are taken with a series of weighted and graduated glass cylinders called **hydrometers.** These float vertically, and the markings are usually in degrees Baumé.

Color Determination of Lubricating Oils

Color determination of lubricating oils and petrolatum is made by comparison with standard colored disks. Light is dispersed through a 4-oz (0.1-kg) sample bottle of the oil to be tested, and the color is compared visually with the gelatin colors on the glass. The colors on the standard glass disks of the National Petroleum Association are as follows:

1	Lily white	4	Orange pale
1½	Cream white	4½	Pale
2	Extra pale	5	Light red
2½	Extra lemon pale	6	Dark red
3	Lemon pale	7	Claret red
3½	Extra orange pale		

Gasoline and Fuel Oil Rating

The **cetane number** of a diesel fuel is numerically equal to the percentage by volume of cetane in a mixture of cetane and *a*-methylnaphthalene which will match the fuel in ignition quality. Cetane has the composition $CH_3(CH_2)_{14}CH_3$, and *a*-methylnaphthalene $CH_3 \cdot C_{10}H_7$. The cetane number of a fuel is given as the nearest whole number. Thus, if it required 49.8% of cetane in the mixture to match, the number would be 50.

The octane number of a fuel is the whole number nearest to the percentage by volume of isooctane. $(CH_3)_3C{:}CH_2CH(CH_3)_2$, in a blend of isooctane and normal heptane, $CH_3(CH_2)_5CH_3$, that the fuel matches in knock characteristics.

Phenol Coefficient

Phenol is used as the standard for measuring the bacteria-killing power of all other disinfectants, and the relative bacteria-killing power is expressed as the **phenol coefficient.**

The phenol coefficient is the ratio of the dilution required to kill the Hopkins strain of typhoid bacillus in a specified time compared with the dilution of phenol required for the same organism in the same time. Usually, 2.5- and 15-minute time limits are used, and the coefficient is calculated from the average of the two. For example, if 1:80 and 1:110 dilutions of phenol kill in 2.5 and 15 minutes, respectively, as the necessary dilutions of the disinfectant under test are 1:375 and 1:650, then the phenol coefficient of the disinfectant is 5.3.

PHYSICAL AND MECHANICAL PROPERTIES

Definitions of Physical and Chemical Properties

Acid number. The weight in milligrams of potassium hydroxide required to neutralize the fatty acid in 1 g of fat or fatty oil.

Aliphatic. Having a straight, chainlike molecular structure.

Anhydrous. Having no water of crystallization in the molecule. A hydrated compound contains water of crystallization which can be driven off by heating.

Aromatic. Having a ringlike molecular structure.

Brittleness. The property of breaking without perceptible warning or without visible deformation.

Bursting strength. The measure of the ability of a material, usually in sheet form, to withstand hydrostatic pressure without rupture.

Compressibility. The extent to which a material, such as for gaskets, is compressed by a specified load. **Permanent set** is the unit amount, in percentage of the compressibility, that the material fails to return to the original thickness when the load is removed. **Recovery** is the amount in percentage of return to the original thickness in a given time, and is usually less under a prolonged load.

Conductivity. The relative rate at which a material conducts heat or electricity at normal temperature [60°F (16°C)]. Silver is the standard of reference, as it is the best of the known conductors of both heat and electricity.

Creep rate. The rate (in/in/h, or % h) at which strain, or deformation, occurs in a material under stress or load. **Creep strength** is the maximum tensile or compressive strength that can be sustained by a material for a specified time at a specified temperature without rupturing. **Creep recovery** is a measure, in percentage, of the decrease in strain, or deformation, when the load is removed.

Ductility. The property of being permanently deformed by tension without rupture, that is, the ability to be drawn from a large to a small size.

Elasticity. The ability of a material to resume its original form after the removal of the force which has produced a change in form. A substance is highly elastic if it is easily deformed and quickly recovers.

Elastic limit. The greatest unit stress that a material is capable of withstanding without permanent deformation.

Elongation. The increase in length of a bar or section under test expressed as a percentage difference between the original length and the length at the moment of rupture.

Factor of safety. The ratio of the ultimate strength of a material to its working stress.

Fatigue strength. The measure in pounds per square inch (megapascals) of the load-carrying ability without failure of a material subjected to a loading repeated a definite number of times. Fatigue strength is usually higher than the prolonged service tensile strength. **Fatigue life** is a measure of the useful life, or the number of cycles of loading, of a specified magnitude that can be withstood by a material without failure.

Flash point. The minimum temperature at which a material or its vapor will ignite or explode.

Flow, or creep. The gradual continuous distortion of a material under continued load, usually at high temperatures.

Fusibility. The ease with which a material is melted.

Hardness. A property applied to solids and very viscous liquids to indicate solidity and firmness in substance or outline. A hard substance does not readily receive an indentation.

Hygroscopic. Readily absorbing and retaining moisture.

Impact strength. The force in foot-pounds (Joules) required to break a material when applied with a sudden blow.

Iodine value. The number of grams of iodine absorbed by 100 g of fat or fatty oil. It gives a measure of the chemical unsaturation of an oil or fat. High iodine value, 117 to 206, in vegetable oils, indicates suitability of the oil for use in paints. Low iodine value, not subject to oxidation, indicates nondrying suitable for soaps.

Malleability. The property of being permanently deformed by compression without rupture, that is, the ability to be rolled or hammered into thin sheets.

Modulus of elasticity. The ratio of the unit stress to unit strain within the elastic limit without fracture.

Modulus of rigidity. When an elastic material is subjected to a shearing stress, a displacement takes place; the ratio of the unit shearing stress to the displacement per unit length is the modulus of rigidity.

Plasticity. The property in a material of being deformed under the action of a force and not returning to its original shape upon the removal of the force.

Porosity. The ratio of the volume of the interstices of a material to the volume of its mass.

Reduction of area. The percentage difference between the area of a bar before being subjected to stress and the area of the bar after rupture.

Resilience. The energy of elasticity—the energy stored in a material under strain within its elastic limit which will cause it to resume its original shape when the stress is removed. The modulus of resilience is the capacity of a unit volume to store energy up to the elastic limit.

Saponification value. The number of milligrams of potassium hydroxide required to saponify 1 gram of fatty oil or grease.

Shrinkage. The diminution in dimensions and mass of a material.

Softening point. The Vicat softening point for thermoplastic materials is the temperature at which a flat-ended needle of one square millimeter area will penetrate a specimen to a depth of 1 mm under a load of 1,000 g when the temperature of the specimen is raised at a constant rate of 50°C/h.

Solubility. Capacity for being dissolved in a liquid so that it will not separate out on standing, except the excess over the percentage which the liquid (solvent) will dissolve. A **suspension** is a physical dispersion of particles sufficiently large that physical forces control their dissolution in the liquid. A **colloidal solution** is a dispersion of particles so finely divided that surface phenomena and kinetic energy control their behavior in the liquid. A colloidal solution is close to a molecular combination.

Specific gravity. The ratio of the weight of a given volume of a material to the weight of an equal volume of pure water at 4°C.

Specific heat. The number of calories required to raise 1 g of a material 1°C in temperature.

Stiffness. The material property which is measured by the rate at which the stress in a material increases with the strain.

Strain. The distortion set up in a material by the action of an external force.

Strength. The ability to resist physical forces imposed upon a material.

Stress. Internal forces set up in a material by the action of an external force.

Tensile strength. The maximum tensile load per square unit of original cross section that a material is able to withstand. Tensile strength is the most common measure of the strength and ductility of metals.

Thermal conductivity. The number of calories transmitted per second between opposite faces of a cube, 1 cm by 1 cm by 1 cm, when the difference between the opposite faces of the cube is 1°C.

Thermal expansion. The coefficient of linear thermal expansion is the increase in unit length with each change of 1° in temperature.

Thermoplastic. Capable of being molded and remolded without rupture by heat and pressure at temperatures slightly above normal. When a material sets under heat and pressure into a hard solid not capable of being remolded, it is called **thermosetting.**

Toughness. The relative degree of resistance to impact without fracture; the property of a material which enables it to absorb energy while being stressed above its elastic limit but without being fractured.

Ultimate strength. The stress, calculated on the maximum value of the force and the original area of cross section, which causes fracture of the material.

Yield point. The minimum tensile stress required to produce continuous deformation in a solid material.

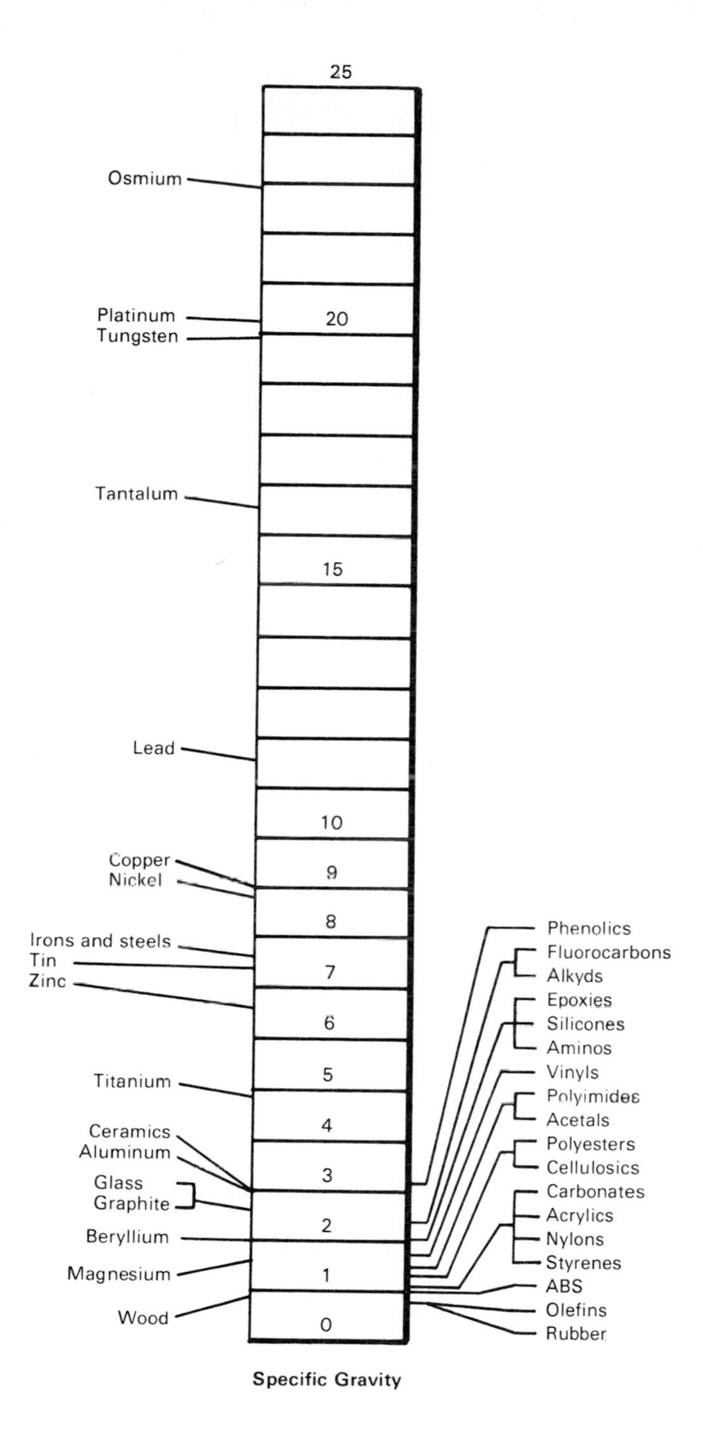

Specific Gravity

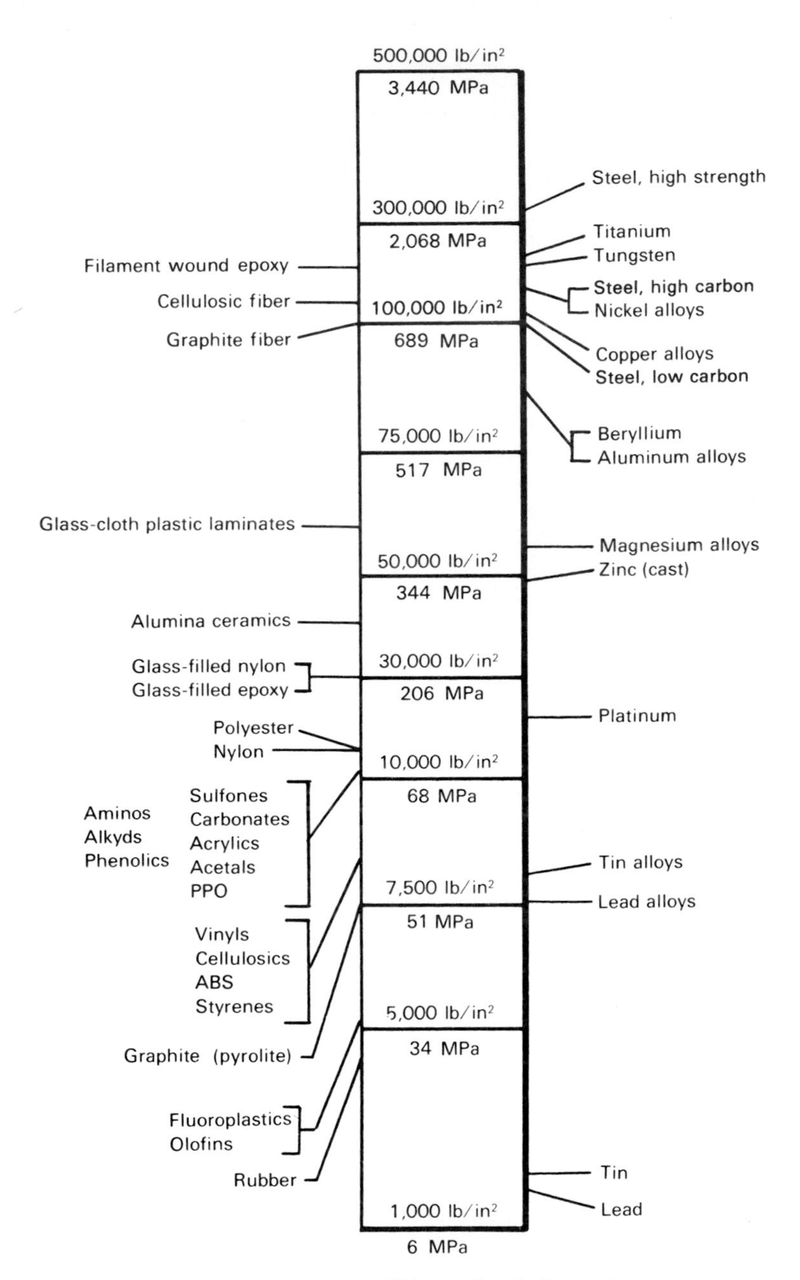

Maximum Ultimate Tensile Strength

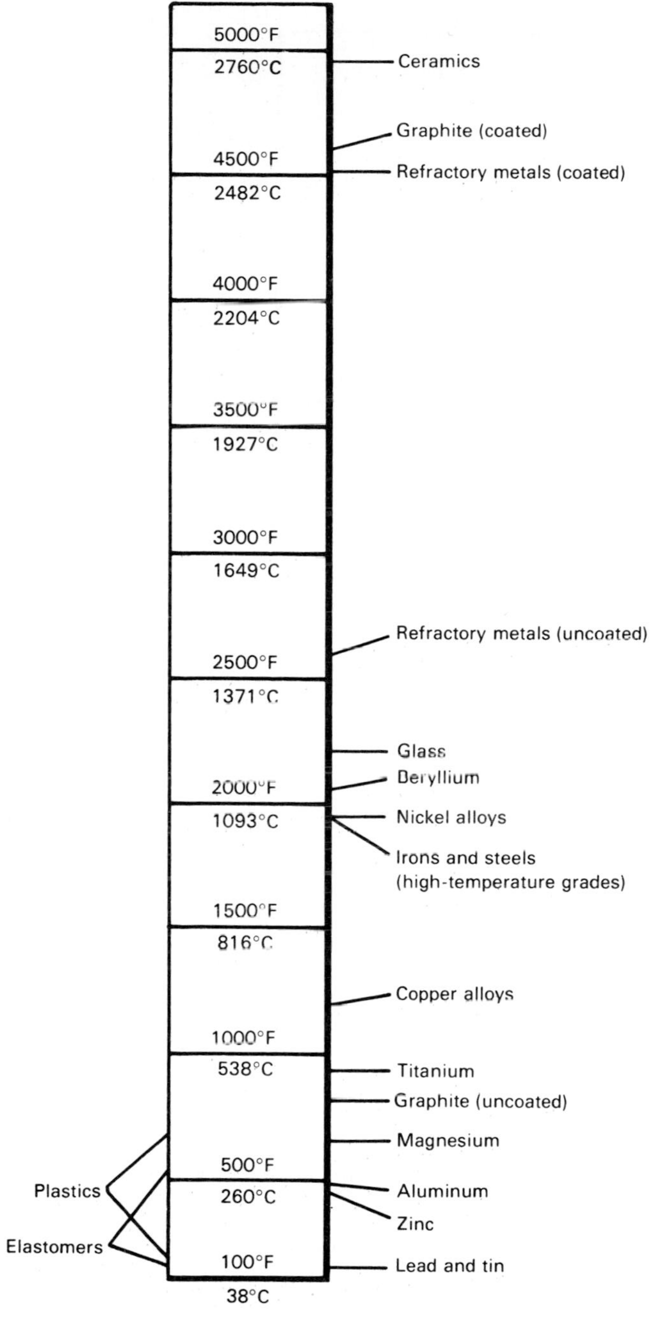

Highest Useful Temperature

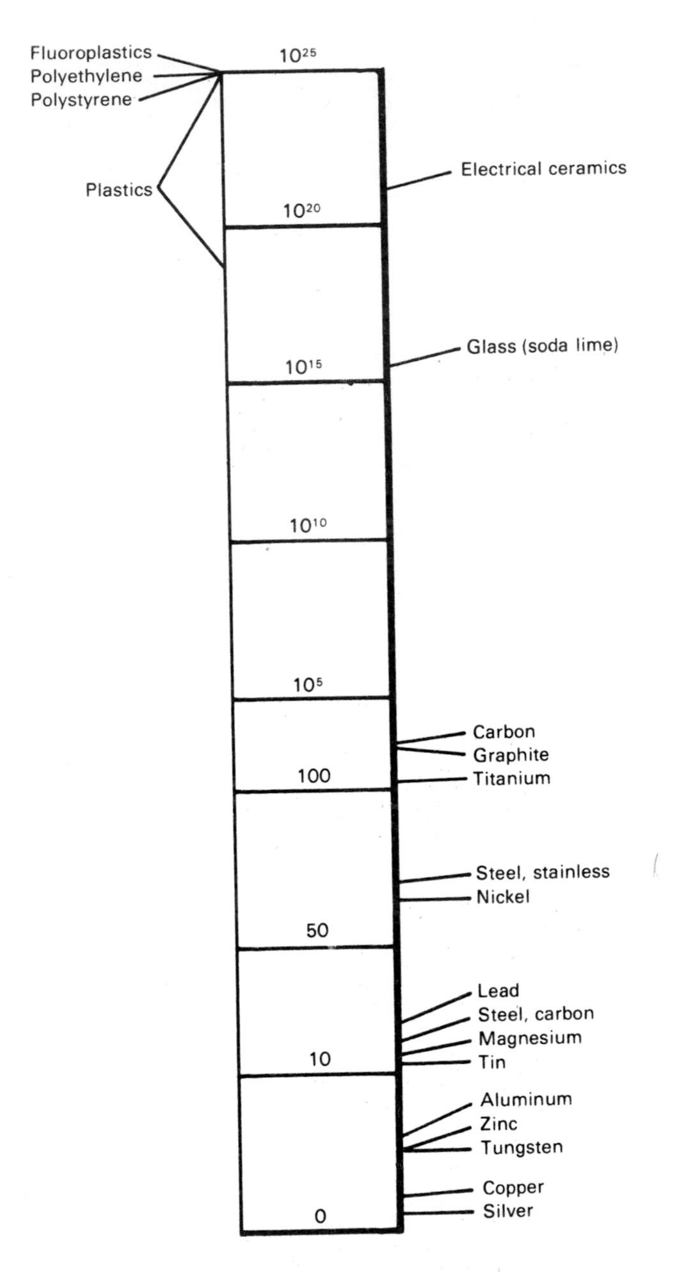

Electrical Resistivity ($\mu\Omega \cdot$cm)

Modulus of Elasticity in Tension of Typical Materials

	lb/in^2	MPa
Lead (cast)	700,000	4,826
Lead (hard-drawn)	1,000,000	6,895
Phenolic plastic (fabric laminated)	1,000,000	6,895
Pine (static bending)	1,200,000	8,280
Ash (static bending)	1,300,000	8,957
Phenolic plastic (paper base)	2,100,000	14,470
Tin (cast)	4,000,000	27,560
Tin (rolled)	5,700,000	39,275
Glass	8,000,000	55,120
Brass	9,000,000	62,010
Aluminum (cast)	10,000,000	68,950
Copper (cast)	11,000,000	75,790
Zinc (cast)	11,000,000	75,790
Zinc (rolled)	12,000,000	82,680
Cast iron (soft gray iron)	12,000,000	82,680
Brass (cast)	13,000,000	89,570
Bronze (average)	13,000,000	89,570
Phosphor bronze	13,000,000	89,570
Manganese bronze (cast)	14,000,000	96,460
Slate	14,000,000	96,460
Copper (soft, wrought)	15,000,000	103,350
Cast iron (average, with steel scrap)	16,000,000	110,240
Clock brass	16,600,000	114,370
Copper (hard-drawn)	18,000,000	124,400
Cast iron (hard, white iron)	20,000,000	137,780
Malleable iron	23,000,000	158,470
Wrought iron	27,000,000	186,030
Carbon steel	30,000,000	206,700
Alloy steel (nickel-chromium)	30,000,000	206,700
Nickel	30,000,000	206,700
Tungsten	60,000,000	413,400

Order of Ductility of Metals

1. Gold	6. Aluminum
2. Platinum	7. Nickel
3. Silver	8. Zinc
4. Iron	9. Tin
5. Copper	10. Lead

Temperatures Available for Melting and Welding

Direct electric arc	4000°C
Oxygen-acetylene torch	3500°C
Electric furnace	3000°C
Aluminum-iron oxide powder	2800°C
Combustion furnace	1700°C
Oxygen-hydrogen torch	1450°C
Plasma arc welding	33,000°C
Electron beam welding	>10,000°C
Laser beam welding	>10,000°C

Mohs Hardness Scale for Minerals

Original Mohs scale		Modified Mohs scale		
Hardness number	Mineral	Hardness number	Material	Bierbaum number
1	Talc	1	Talc	1
2	Gypsum	2	Gypsum	
3	Calcite	3	Calcite	15
4	Fluorite	4	Fluorite	
5	Apatite	5	Apatite	
6	Orthoclase	6	Orthoclase	
		7	Vitreous silica	
7	Quartz	8	Quartz or stellite	
8	Topaz	9	Topaz	
		10	Garnet	
		11	Fused zirconia	
9	Corundum	12	Fused alumina	
		13	Silicon carbide	
		14	Boron carbide	
10	Diamond	15	Diamond	10,000 (1 micrometer scratch)

Hardness Grades in Woods

1. Exceedingly hard	Lignum-vitae, ebony
2. Extremely hard	Boxwood, lilac, jarrah, karri
3. Very hard	Whitethorn, blackthorn, persimmon
4. Hard	Hornbeam, elder, yew
5. Rather hard	Ash, holly, plum, elm
6. Firm	Teak, chestnut, beech, walnut, apple, oak
7. Soft	Willow, deal, alder, Australian red cedar, birch, hazel
8. Very soft	White pine, poplar, redwood

Knoop Indentor Hardness of Hard Materials

Diamond	6,000–6,500
Boron carbide	2,250–2,260
Silicon carbide	2,130–2,140
Sapphire	1,600–2,100
Aluminum oxide (corundum)	1,635–1,680
Tungsten carbide (cobalt binder)	1,000–1,500
Spinel	1,200–1,400
Topaz	1,250
Quartz	710 700
Hardened steel	400–800
Glass	300–600

Comparative Hardness of Hard Abrasives

(Scale: Diamond 10, corundum 9)

South American brown bort	10.00
South American Ballas	9.99
Congo yellow (cubic crystals)	9.96
Congo clear white (cubic crystals)	9.95
Congo gray opaque (cubic crystals)	9.89
South American carbonadoes	9.82
Boron carbide	9.32
Black silicon carbide	9.15
Green silicon carbide	9.13
Tungsten carbide (13% cobalt)	9.09
Fused alumina (3.14% TiO_2)	9.06
Fused alumina	9.03
African crystal corundum	9.00
Rock-crystal quartz	8.94

Thermal Conductivity of Materials[1]

Conductivity measured in British thermal units transmitted per hour per square foot of material 1 in thick, per degree Fahrenheit difference in temperature of the two faces.

Silver	2,920.0	Diatomite block	0.58
Copper	2,588.0	Magnesia, 85 percent	0.51
Steel, 1.0 carbon	328.0	Wood pulp board	0.39
Building stone	12.50	Bagasse board	0.35
Slate, shingles	10.37	Cork, ground	0.31
Concrete, 1:2:4	6.10	Flax fiber	0.31
Glass, plate	5.53	Diatomite powder	0.308
Brickwork, mortar bond	4.00	Mineral wool	0.296
Gypsum plaster	2.32	Asbestos sheet	0.29
Brick, dry	1.21	Vermiculite	0.263
Air space, 3½ in.	1.10	Wool	0.261
Pine wood	0.958	Hair felt	0.26
Clay tile	0.60	Cotton, compressed	0.206

[1]From Paul M. Tyler, U.S. Bureau of Mines.

Linear Expansion of Metals

Unit length increase per degree centigrade rise in temperature

Cast iron	0.000010
Steel	0.000011
Cobalt	0.000012
Bismuth	0.000013
Gold	0.000014
Nickel	0.000014
Copper	0.000017
Brass	0.000019
Silver	0.000019
Tobin bronze	0.000021
Aluminum	0.000024
Zinc	0.000026
Tin	0.000027
Lead	0.000028
Cadmium	0.000029
Magnesium	0.000029

Melting Points of Materials Commonly Used for Heat-treating Baths

Material	Melting points	
	°F	°C
35 percent lead, 65 percent tin	358	181
50 percent sodium nitrate, 50 percent potassium nitrate	424	218
Tin	450	232
Sodium nitrate	586	308
Lead	620	327
Potassium nitrate	642	339
45 percent sodium chloride, 55 percent sodium sulfate	1154	623
Sodium chloride (common salt)	1474	801
Sodium sulfate	1618	881
Barium chloride	1760	960

Forging Temperatures of Steels

	Maximum forging temperatures		Burning temperatures	
	°F	°C	°F	°C
1.5 percent carbon steel	1920	1049	2080	1138
1.1 percent carbon steel	1980	1082	2140	1171
0.9 percent carbon steel	2050	1121	2230	1221
0.7 percent carbon steel	2140	1171	2340	1282
0.5 percent carbon steel	2280	1249	2460	1349
0.2 percent carbon steel	2410	1321	2680	1471
0.1 percent carbon steel	2460	1349	2710	1488
Silico-manganese spring steel	2280	1249	2460	1349
3 percent nickel steel	2280	1249	2500	1371
3 percent nickel-chromium steel	2280	1249	2500	1371
Air-hardening Ni-Cr steel	2280	1249	2500	1371
5 percent nickel (case-hardening) steel	2320	1271	2640	1449
Chromium-vanadium steel	2280	1249	2460	1349
High-speed steel	2370	1299	2520	1382
Stainless steel	2340	1282	2520	1382
Austenitic chromium-nickel steel	2370	1299	2590	1421

Average Fatty Acid Composition and Constants of Fats and Oils

	Chemical formula	Coconut	Palm kernel	Tallow	Lard	Palm	Olive	Cottonseed	Corn	Peanut
Saturated acids										
Caproic	$C_6H_{12}O_2$	0.2	Trace							
Caprylic	$C_8H_{16}O_2$	8.0	3.0							
Capric	$C_{10}H_{20}O_2$	7.0	6.0							
Lauric	$C_{12}H_{24}O_2$	48.0	50.0							
Myristic	$C_{14}H_{28}O_2$	17.5	15.0	2.0	1.0	1.0	Trace	0.5		
Palmitic	$C_{16}H_{32}O_2$	8.8	7.5	30.0	26.0	42.5	9.0	21.0	7.5	7.0
Stearic	$C_{18}H_{36}O_2$	2.0	1.5	21.0	11.5	4.0	2.3	2.0	3.5	5.0
Arachidic	$C_{20}H_{40}O_2$						0.2	Trace	0.5	4.0
Behenic	$C_{22}H_{44}O_2$									
Lignoceric	$C_{24}H_{48}O_2$					Trace			0.2	3.0
Unsaturated acids										
Myristoleic	$C_{14}H_{26}O_2$									
Palmitoleic	$C_{16}H_{30}O_2$									
Oleic	$C_{18}H_{34}O_2$	6.0	16.0	45.0	58.0	43.0	82.5	33 .0	46.3	60.0
Linoleic	$C_{18}H_{32}O_2$	2.5	1.0	2.0	3.5	9.5	6.0	43.5	42.0	21.0
Linolenic	$C_{18}H_{30}O_2$									
Elaeostearic	$C_{18}H_{30}O_2$									
Ricinoleic	$C_{18}H_{34}O_3$									
C_{20} unsaturated	$C_{20}H_{2(20-x)}O_2$									
C_{22} unsaturated	$C_{20}H_{2(22-x)}O_2$									
Constants										
Saponification value		251–264	240–250	196–200	195–200	196–206	185–200	192–200	188–193	185–192
Iodine number		8–10	16–23	35–44	50–69	48–58	74–94	100–115	116–130	83–95
Titer—°C		20–23	20–23	37–46	36–43	38–47	18–25	32–38	18–20	28–32

Average Fatty Acid Composition and Constants of Fats and Oils (continued)

	Soybean	Sunflower	Linseed	Perilla	Castor	Tung	Whale	Menhaden	Sardine	Herring
					Saturated acids					
Caproic										
Caprylic										
Capric										
Lauric										
Myristic							8.0	7.0	5.0	7.0
Palmitic	6.5	3.5	5.0	7.5		4.0	11.0	16.0	14.0	8.0
Stearic	4.5	3.0	3.5	Trace	2.0	1.5	2.5	1.0	3.0	Trace
Arachidic	0.7	0.6	Trace							
Behenic										
Lignoceric	Trace	0.4								
					Unsaturated Acids					
Myristoleic							1.5	Trace	Trace	Trace
Palmitoleic							17.0	17.0	12.0	18.0
Oleic	33.5	34.0	5.0	8.0	8.6	15.0	34.0	27.0	10.0	9.0
Linoleic	52.5	58.5	61.5	38.0	3.5		9.0	Trace	15.0	13.0
Linolenic	2.3		25.0	46.5			Trace			
Elaeostearic						79.5				
Ricinoleic					85.9					
C_{20} unsaturated							5.0	20.0	22.0	20.0
C_{22} unsaturated							12.0	12.0	19.0	25.0
					Constants					
Saponification value	189–194	189–194	189–196	188–197	175–183	189–195	185–195	189–193	189–193	179–194
Iodine number	124–148	120–136	179–204	185–206	82–86	160–170	110–136	148–185	170–190	130–144
Titer—°C	20–21	17–20	19–21	12–17		37–38	22–24	31–33	28–34	23–27

Relative Values of Electrical Insulating Materials

The usual comparisons of insulating values of materials are made on the basis of their dielectric strengths. The **dielectric strength** of a material is the voltage that a material of a given thickness will resist. It is usually given in volts per mil (1 mil equals 0.001 in) or volts per meter. In any higher voltage the dielectric strength will permit a spark to pass through the material. The quoted dielectric strengths, however, are generally the minimum for the materials.

Material	Dielectric strength		Material	Dielectric strength	
	$\times 10^6$ Volts per meter	Volts per mil		$\times 10^6$ Volts per meter	Volts per mil
Mica, muscovite	39.4	1,000	Buna rubbers	20.3	515
Glass	35.5	900	Vinylidene chloride	19.7	500
Mica, phlogopite	31.5	800	Fish paper	19.7	500
Electrical porcelain	31.5	800	Methyl methacrylate	18.9	480
Steatite	29.6	750	Cellulose acetate	15.8	400
Hard rubber	27.6	700	Casein plastic	15.8	400
Silicone rubber	23.6	600	Shellac	15.8	400
Polystyrene	23.6	600	Varnished cambric	15.8	400
Pyroxylin	23.6	600			

Relative Electric Conductivity of Elements

Element	Conductivity	Element	Conductivity
Silver	100.00	Iron	14.57
Copper	97.61	Platinum	14.43
Gold	76.61	Tin	14.39
Aluminum	63.00	Tungsten	14.00
Tantalum	54.63	Osmium	13.98
Magnesium	39.44	Titanium	13.73
Sodium	31.98	Iridium	13.52
Beryllium	31.13	Ruthenium	13.22
Barium	30.61	Nickel	12.89
Zinc	29.57	Rhodium	12.60
Indium	26.98	Palladium	12.00
Cadmium	24.38	Steel	12.00
Calcium	21.77	Thallium	9.13
Rubidium	20.46	Lead	8.42
Cesium	20.00	Columbium	5.13

Relative Electric Conductivity of Elements (continued)

Lithium	18.68	Vanadium	4.95
Molybdenum	17.60	Arsenic	4.90
Cobalt	16.93	Antimony	3.59
Uranium	16.47	Mercury	1.75
Chromium	16.00	Bismuth	1.40
Manganese	15.75	Tellurium	0.001

The Electrochemical Series of Elements

In this table, the elements are electropositive to the ones which follow them, and will displace them from solutions of their salts.

1. Cesium	23. Nickel	45. Silicon
2. Rubidium	24. Cobalt	46. Titanium
3. Potassium	25. Thallium	47. Columbium
4. Sodium	26. Cadmium	48. Tantalum
5. Lithium	27. Lead	49. Tellurium
6. Barium	28. Germanium	50. Antimony
7. Strontium	29. Indium	51. Carbon
8. Calcium	30. Gallium	52. Boron
9. Magnesium	31. Bismuth	53. Tungsten
10. Beryllium	32. Uranium	54. Molybdenum
11. Ytterbium	33. Copper	55. Vanadium
12. Erbium	34. Silver	56. Chromium
13. Scandium	35. Mercury	57. Arsenic
14. Aluminum	36. Palladium	58. Phosphorus
15. Zirconium	37. Ruthenium	59. Selenium
16. Thorium	38. Rhodium	60. Iodine
17. Cerium	39. Platinum	61. Bromine
18. Didymium	40. Iridium	62. Chlorine
19. Lanthanum	41. Osmium	63. Fluorine
20. Manganese	42. Gold	64. Nitrogen
21. Zinc	43. Hydrogen	65. Sulfur
22. Iron	44. Tin	66. Oxygen

Specific Gravities and Densities

For rough estimating of common construction materials

	Specific gravity	Density, lb/ft^3	Density, kg/m^3
Aluminum	2.7	165	2,643
Bronze	8.0	509	8,154
Cast iron	7.2	450	7,209
Copper	8.9	556	8,907
Glass	2.5	160	2,563
Lead	11.38	710	11,374
Magnesium	1.74	109	1,746
Nickel	8.9	556	8,907
Steel	7.8	490	7,850
Zinc	7.5	440	7,049
Ash, dry	0.63	40	641
Cedar, dry	0.36	22	352
Fir, dry	0.56	32	513
Maple, dry	0.65	43	689
Redwood, dry	0.42	26	417
White pine, dry	0.41	26	417
Granite	2.6	165	2,643
Limestone	2.5	165	2,643
Sandstone	1.8	110	1,762
Pressed brick	2.2	140	2,243
Common brick	1.9	120	1,922
Terra cotta	1.9	120	1,922
Concrete	2.3	144	2,307
Portland cement	3.0	183	2,932
Mortar	1.7	103	1,650
Earth, dry, loose		76	1,218
Earth, dry, packed		95	1,522
Sand and gravel		60	961
Asbestos		153	2,451
Marble	2.7	170	2,723
Shale		92	1,474
Tar	1.2	75	1,202
Bluestone	2.5	159	2,547

Index

ABOUT THE AUTHORS

George S. Brady (deceased), a Materials Consultant, was the originator and longtime prime mover behind the *Materials Handbook.* His engineering career comprised service as an engineer on the construction of the Panama Canal, service in World War II as Chief Materials Consultant to the Office of Civilian Supply and to the Board of Economic Warfare, and subsequent service as Materials Advisor to the Federal Bureau of Supply and Consultant to the Office of Defense Mobilization.

After publication of the tenth edition of the *Handbook*, Colonel Brady transferred preparation of the book to **Henry R. Clauser**. Mr. Clauser is a Materials Consultant, presently to government and industry. A Fellow of the American Association for the Advancement of Science, he has served on the National Materials Advisory Board and on national committees of the American Institute of Chemical Engineers, the American Society of Mechanical Engineers, and the American Society of Testing & Materials. He has also published articles on materials, processes, and design engineering in such journals as *Scientific American, Mechanical Engineering, Iron Age*, the *Harvard Business Review, Progressive Architecture*, and *Materials Engineering.* He is the author of *Industrial and Engineering Materials* (McGraw-Hill).